Student Study Guide/ Solutions Manual

to accompany

Organic Chemistry with Biological Topics

Sixth Edition

Prepared by

Janice Gorzynski Smith
University of Hawai'i at Mānoa

Erin R. Smith

STUDENT STUDY GUIDE/SOLUTIONS MANUAL TO ACCOMPANY
ORGANIC CHEMISTRY WITH BIOLOGICAL TOPICS, SIXTH EDITION

Published by McGraw-Hill Education, 2 Penn Plaza, New York, NY 10121.

Some ancillaries, including electronic and print components, may not be available to customers outside the United States.

This book is printed on acid-free paper.

1 2 3 4 5 6 7 8 9 LHN 23 22 21 20

ISBN 978-1-260-51646-3
MHID 1-260-51646-6

mheducation.com/highered

Contents

Chapter 1: Structure and Bonding

Chapter Review

Important facts

- **The general rule of bonding:** Atoms strive to attain a complete outer shell of valence electrons (Section 1.2). H "wants" 2 electrons. Second-row elements "want" 8 electrons.

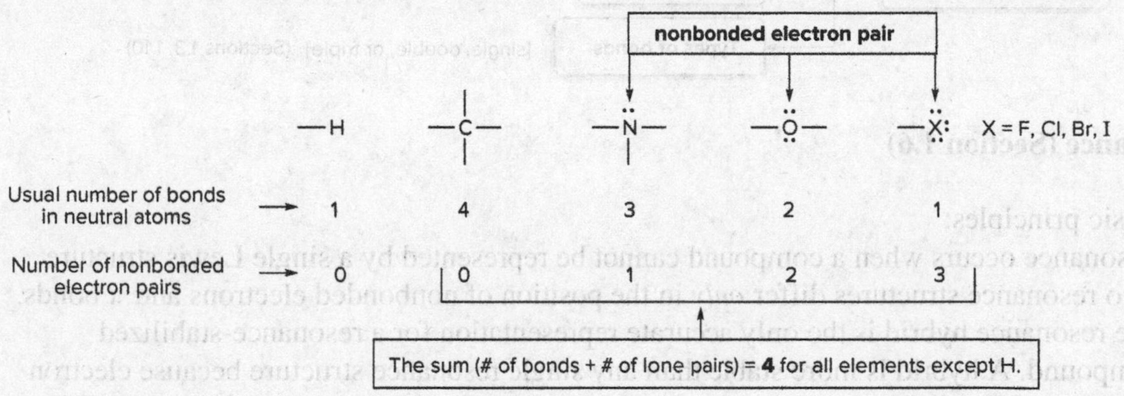

- **Formal charge** (FC) is the difference between the number of valence electrons on an atom and the number of electrons it "owns" (Section 1.3C). See Sample Problem 1.3 for a stepwise example.

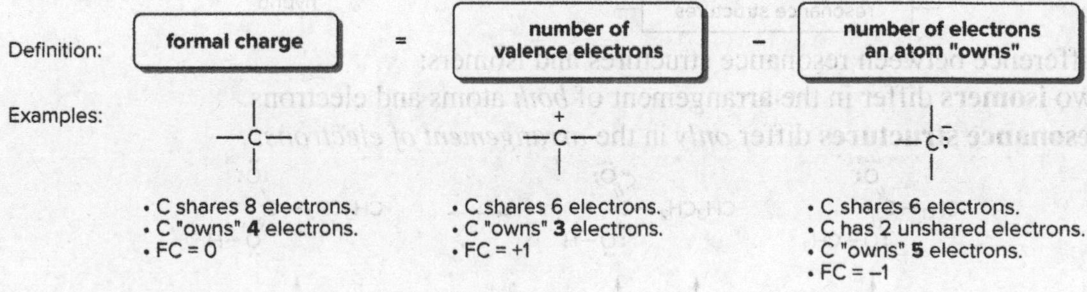

- **Curved arrow notation** shows the movement of an electron pair. The tail of the arrow always begins at an electron pair, either in a bond or a lone pair. The head points to where the electron pair "moves" (Section 1.6).

Use an electron pair to form a double bond.

Move an electron pair to O.

- **Electrostatic potential plots** are color-coded maps of electron density, indicating electron rich (red) and electron deficient (blue) regions (Section 1.12).

The importance of Lewis structures (Sections 1.3–1.5)

A properly drawn Lewis structure shows the number of bonds and lone pairs present around each atom in a molecule. In a valid Lewis structure, each H has two electrons, and each second-row element has no more than eight. This is the first step needed to determine many properties of a molecule.

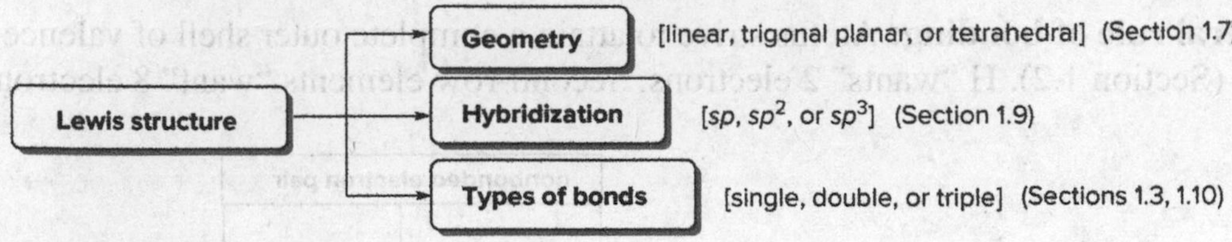

Resonance (Section 1.6)

The basic principles:

- Resonance occurs when a compound cannot be represented by a single Lewis structure.
- Two resonance structures differ *only* in the position of nonbonded electrons and π bonds.
- The resonance hybrid is the only accurate representation for a resonance-stabilized compound. A hybrid is more stable than any single resonance structure because electron density is delocalized.

$CH_3CH_2-C(=O)O^-$ ⟷ $CH_3CH_2-C(O^-)=O$; hybrid with δ– on each O

resonance structures

delocalized charges

delocalized π bonds

hybrid

The difference between resonance structures and isomers:

- Two **isomers** differ in the arrangement of *both* atoms and electrons.
- **Resonance structures** differ *only* in the *arrangement of electrons*.

$CH_3-C(=O)-O-CH_3$; $CH_3CH_2-C(=O)-O-H$ ⟷ $CH_3CH_2-C(O^-)=O^+-H$

isomers

resonance structures

Geometry and hybridization

The number of groups around an atom determines both its geometry (Section 1.7) and hybridization (Section 1.9).

Number of groups	Geometry	Bond angle (°)	Hybridization	Examples
2	linear	180	*sp*	HC≡CH
3	trigonal planar	120	sp^2	$CH_2=CH_2$
4	tetrahedral	109.5	sp^3	CH_4, NH_3, H_2O

Drawing organic molecules (Section 1.8)

- Shorthand methods are used to abbreviate the structure of organic molecules.

skeletal structure = isooctane ($CH_3-C(CH_3)(H)-C(H)(H)-C(CH_3)(CH_3)-CH_3$) = condensed structure $(CH_3)_2CHCH_2C(CH_3)_3$

- A carbon bonded to four atoms is tetrahedral in shape. The best way to represent a tetrahedron is to draw two bonds in the plane, one in front, and one behind.

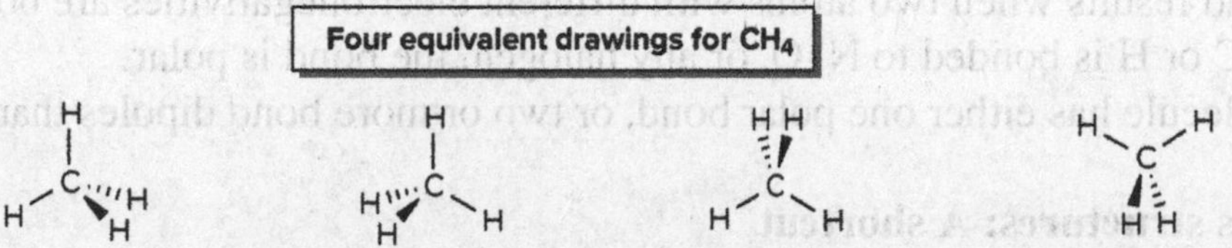

Each drawing has two solid lines, one wedge, and one dashed wedge.

Bond length

- Bond length decreases as you go from left to right across a row and increases down a column of the periodic table (Section 1.7A).

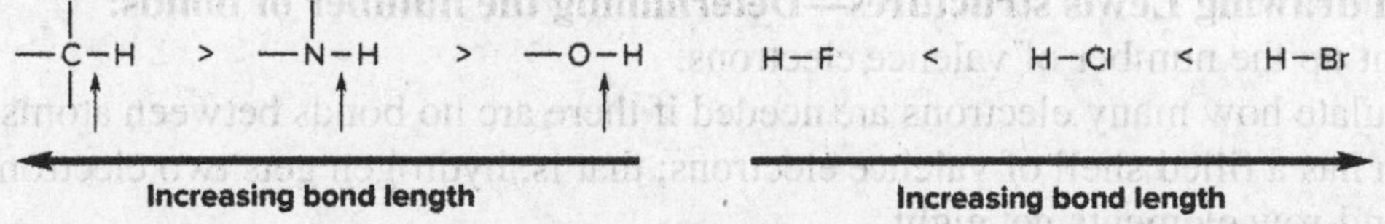

- Bond length decreases as the number of electrons between two nuclei increases (Section 1.11A).

CH_3-CH_3 < $CH_2=CH_2$ < $H-C\equiv C-H$

Increasing bond length

- Bond length increases as the percent *s*-character decreases (Section 1.11B).

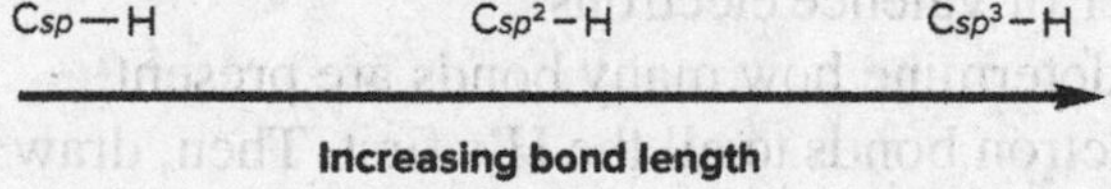

- Bond length and bond strength are inversely related. Shorter bonds are stronger bonds (Section 1.11).

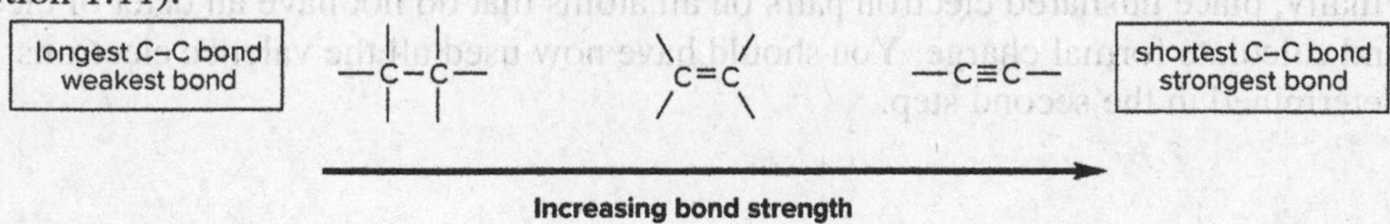

Increasing bond strength

- Sigma (σ) bonds are generally stronger than π bonds (Section 1.10).

$$\text{—C—C—} \qquad \text{C=C} \qquad \text{—C}\equiv\text{C—}$$

1 strong σ bond

1 stronger σ bond
1 weaker π bond

1 stronger σ bond
2 weaker π bonds

Electronegativity and polarity (Sections 1.12, 1.13)

- Electronegativity increases from left to right across a row and decreases down a column of the periodic table.
- A polar bond results when two atoms with different electronegativities are bonded together. Whenever C or H is bonded to N, O, or any halogen, the bond is polar.
- A polar molecule has either one polar bond, or two or more bond dipoles that reinforce.

Drawing Lewis structures: A shortcut

Chapter 1 devotes a great deal of time to drawing valid Lewis structures. For molecules with many bonds, it may take quite a while to find acceptable Lewis structures by using trial-and-error to place electrons. Fortunately, a shortcut can be used to figure out how many bonds are present in a molecule.

Shortcut on drawing Lewis structures—Determining the number of bonds:

[1] Count up the number of valence electrons.

[2] Calculate how many electrons are needed if there are no bonds between atoms and every atom has a filled shell of valence electrons; that is, hydrogen gets two electrons, and second-row elements get eight.

[3] Subtract the number obtained in Step [1] from the sum obtained in Step [2]. **This difference tells how many electrons must be shared** to give every H two electrons and every second-row element eight. Because there are two electrons per bond, dividing this difference by two tells how many bonds are needed.

To draw the Lewis structure:

[1] Arrange the atoms as usual.

[2] Count up the number of valence electrons.

[3] Use the shortcut to determine how many bonds are present.

[4] Draw in the two-electron bonds to all the H's first. Then, draw the remaining bonds between other atoms, making sure that no second-row element gets more than eight electrons and that you use the total number of bonds determined previously.

[5] Finally, place unshared electron pairs on all atoms that do not have an octet of electrons, and calculate formal charge. You should have now used all the valence electrons determined in the second step.

Example: Draw all valid Lewis structures for CH_3NCO using the shortcut procedure.

[1] Arrange the atoms.

```
   H
H  C  N  C  O
   H
```

- In this case the arrangement of atoms is implied by the way the structure is drawn.

[2] Count up the number of valence electrons.

3 H's	x	1 electron per H	=	3 electrons
2 C's	x	4 electrons per C	=	8 electrons
1 N	x	5 electrons per N	=	5 electrons
1 O	x	6 electrons per O	=	+ 6 electrons
				22 electrons total

[3] Use the shortcut to figure out how many bonds are needed.

- Number of electrons needed if there were no bonds:

3 H's	x	2 electrons per H	=	6 electrons
4 second-row elements	x	8 electrons per element	=	+ 32 electrons
				38 electrons needed if there were no bonds

- Number of electrons that must be shared:

38 electrons
− 22 electrons
16 electrons must be shared

- Every bond requires two electrons, so 16/2 = **8 bonds are needed.**

[4] Draw all possible Lewis structures.

- Draw the bonds to the H's first (three bonds). Then add five more bonds. Arrange them between the C's, N, and O, making sure that no atom gets more than eight electrons. There are three possible arrangements of bonds; that is, there are three resonance structures.
- Add additional electron pairs to give each atom an octet and check that all 22 electrons are used.

```
                     H                       H
                     |                       |
                   H-C-N≡C-O   ------->    H-C-N≡C-Ö:
                     |                       |
                     H                       H
   H
   |
 H-C  N  C  O  --->  H                       H
   |                 |                       |
   H               H-C-N=C=O   ------->    H-C-N̈=C=Ö
                     |                       |
                     H                       H

                     H                       H
                     |                       |
                   H-C-N-C≡O   ------->    H-C-N̈-C≡O:
                     |                       |
                     H                       H
```

Bonds to H's added.

All bonds drawn in.
Three arrangements possible.

Electron pairs drawn in.
Every atom has an octet.

- **Calculate the formal charge on each atom.**

$H_3C-N\equiv C-\ddot{O}:$ (N: +1, O: −1) ⟷ $H_3C-N=C=O$ ⟷ $H_3C-\ddot{N}-C\equiv O:$ (N: −1, O: +1)

- **You can evaluate the Lewis structures you have drawn. The middle structure is the best resonance structure, because it has no charged atoms.**

Note: This method works for compounds that contain second-row elements in which every element gets an octet of electrons. It does NOT necessarily work for compounds with an atom that does not have an octet (such as BF_3), or compounds that have elements located in the third row and later in the periodic table.

Practice Test on Chapter Review

1.a. Which compound(s) contain a labeled atom with a +1 formal charge? All lone pairs of electrons have been drawn in.

1. =N(CH₃)₂ 2. O–CH₃ 3. CH₃—C̈—CH₃

4. Both (1) and (2) have labeled atoms with a +1 charge.
5. Compounds (1), (2), and (3) all contain labeled atoms with a +1 formal charge.

b. Which of the following compounds is a valid resonance structure for **A**?

A: ÖCH₃ 1. ÖCH₃ 2. ÖCH₃ 3. ÖCH₃

4. Both (1) and (2) are valid resonance structures for **A**.
5. Cations (1), (2), and (3) are all valid resonance structures for **A**.

c. Which species contains a labeled carbon atom that is sp^2 hybridized?

1. + 2. C=CH₂ 3. −

4. Both (1) and (2) contain labeled sp^2 hybridized atoms.
5. Species (1), (2), and (3) all contain labeled sp^2 hybridized carbon atoms.

d. Which of the following compounds has a net dipole?

1. $CH_3CH_2NHCH_2CH_3$
2. $CH_3CH_2CH_2OH$
3. $FCH_2CH_2CH_2F$
4. Compounds (1) and (2) both have net dipoles.
5. Compounds (1), (2), and (3) all have net dipoles.

2. Rank the labeled bonds in order of increasing bond length. Label the shortest bond as **1**, the longest bond as **4**, and the bonds of intermediate length as **2** and **3**.

B C A D HC≡C

3. Answer the following questions about compounds **A–D**.

$\ddot{N}CH_2CH_3$ [1] $C=\ddot{O}$ [2]

A B C D

a. What is the hybridization of the labeled atom in **A**?
b. What is the molecular shape around the labeled atom in **B**?
c. In what type of orbital does the lone pair in **C** reside?
d. What orbitals are used to form bond [1] in **D**?
e. Which orbitals are used to form the carbon–oxygen double bond [2] in **D**?

4. Draw an acceptable Lewis structure for CH_3NO_3. Assume that the atoms are arranged as drawn.

```
  H     O
H C O N O
  H
```

5. Follow the curved arrows and draw the product with all the needed charges and lone pairs.

Answers to Practice Test

1. a. 4
   b. 1
   c. 1
   d. 5

2. **A** – 1
   B – 4
   C – 2
   D – 3

3. a. sp^3
   b. trigonal planar
   c. sp^2
   d. Csp^3–Csp^2
   e. Csp–Osp^2, Cp–Op

4. One possibility:

5.

Answers to Problems

1.1 Two isotopes differ in the number of neutrons. Two isotopes have the same number of protons and electrons, group number, and number of valence electrons. The **mass number** is the number of protons and neutrons. The **atomic number** is the number of protons and is the same for all isotopes.

	Nitrogen-14	Nitrogen-13
a. number of protons = atomic number for N = 7	7	7
b. number of neutrons = mass number – atomic number	7	6
c. number of electrons = number of protons	7	7
d. group number	5A	5A
e. number of valence electrons	5	5

1.2 **Ionic bonds** form when an element on the far left side of the periodic table transfers an electron to an element on the far right side of the periodic table. **Covalent bonds** result when two atoms *share* electrons.

a. F–F (covalent)

b. $Li^+ Br^-$ (ionic)

c. H–C–C–H with H's above and below each C: All C–H and C–C bonds are covalent.

d. Na^+ $:\ddot{N}$–H with H below N: ionic; Both N–H bonds are covalent.

e. Na^+ $^-:\ddot{O}$–C–H with H's above and below C: ionic; All other bonds are covalent.

1.3 Atoms with one, two, three, or four valence electrons form one, two, three, or four bonds, respectively. Atoms with five or more valence electrons form [8 – (number of valence electrons)] bonds.

a. O 8 – 6 valence e^- = 2 bonds

b. Al 3 valence e^- = 3 bonds

c. Br 8 – 7 valence e^- = 1 bond

d. Si 4 valence e^- = 4 bonds

1.4 [1] Arrange the atoms with the H's on the periphery.
[2] Count the valence electrons.
[3] Arrange the electrons around the atoms. Give the H's 2 electrons first, and then fill the octets of the other atoms.
[4] Assign formal charges (Section 1.3C).

a. [1] H C C H, with H H above and H H below the carbons

[2] Count valence e^-.
2 C x 4 e^- = 8
6 H x 1 e^- = 6
total e^- = 14

[3] H–C–C–H with H's above and below each C. All 14 e^- used. All second-row elements have an octet.

b. [1] H C N H, with H above C and H H below

[2] Count valence e^-.
1 C x 4 e^- = 4
5 H x 1 e^- = 5
1 N x 5 e^- = 5
total e^- = 14

[3] H–C–N–H with H above C and H below C and N. 12 e^- used. N needs 2 more electrons for an octet. → H–C–$\ddot{N}$–H with H above C and H below C and N.

c. [1] H C C Br (with H H above and H H below)

[2] Count valence e⁻.
2 C x 4 e⁻ = 8
5 H x 1 e⁻ = 5
1 Br x 7 e⁻ = 7
total e⁻ = 20

[3] H–C–C–Br (with H atoms) → H–C–C–Br: (with lone pairs)

14 e⁻ used.
Br needs 6 more electrons for an octet.

Complete octet.

1.5 Follow the directions from Answer 1.4.

a. H_2CO — H C O, H

Count valence e⁻.
1 C x 4 e⁻ = 4
2 H x 1 e⁻ = 2
1 O x 6 e⁻ = 6
total e⁻ = 12

H–C–O (H below) → H–C=O: (H below)

6 e⁻ used.

Complete O and C octets.

b. $HOCH_2CO_2H$ — H O C C O H (H, O above; H below)

Count valence e⁻.
2 C x 4 e⁻ = 8
4 H x 1 e⁻ = 4
3 O x 6 e⁻ = 18
total e⁻ = 30

H–O–C–C–O–H → H–Ö–C–C–Ö–H (with :O: double bonded)

16 e⁻ used.

Complete octets.

1.6 Follow the directions from Answer 1.4.

a. HCN — H C N

Count valence e⁻.
1 C x 4 e⁻ = 4
1 H x 1 e⁻ = 1
1 N x 5 e⁻ = 5
total e⁻ = 10

H–C–N → H–C≡N:

4 e⁻ used.

Complete N and C octets.

b. C_3H_4 — H C C C H (H above, H below)

Count valence e⁻.
3 C x 4 e⁻ = 12
4 H x 1 e⁻ = 4
total e⁻ = 16

H–C–C–C–H → H–C–C≡C–H

12 e⁻ used.

Complete octets.

1.7 **Formal charge** (FC) = number of valence electrons – [number of unshared electrons + (½)(number of shared electrons)]

a. $[NH_4]^+$: 5 – [0 + (1/2)(8)] = +1

b. $CH_3–N≡C:$
5 – [0 + (1/2)(8)] = +1 (N)
4 – [0 + (1/2)(8)] = 0 (C of CH_3)
4 – [2 + (1/2)(6)] = –1 (terminal C)

c. :Ö=Ö–Ö:
6 – [2 + (1/2)(6)] = +1 (central O)
6 – [4 + (1/2)(4)] = 0 (left O)
6 – [6 + (1/2)(2)] = –1 (right O)

d. CH_3–N=Ö with :Ö: on N
5 – [0 + (1/2)(8)] = +1 (N)
6 – [4 + (1/2)(4)] = 0 (double-bonded O)
6 – [6 + (1/2)(2)] = –1 (single-bonded O)
4 – [0 + (1/2)(8)] = 0 (C)

1.8

a. CH_3O^-

[1] 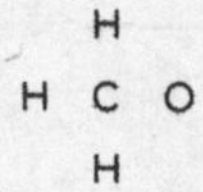

[2] Count valence e⁻.

$1\,C \times 4\,e^- =$	4
$3\,H \times 1\,e^- =$	3
$1\,O \times 6\,e^- =$	6
total e⁻ =	13

Add 1 for (–) charge = 14

[3] 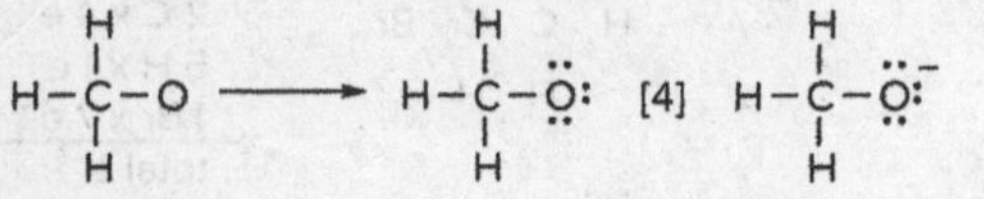

8 e⁻ used. Assign charge.

b. HC_2^-

[1] H C C

[2] Count valence e⁻.

$2\,C \times 4\,e^- =$	8
$1\,H \times 1\,e^- =$	1
total e⁻ =	9

Add 1 for (–) charge = 10

[3] 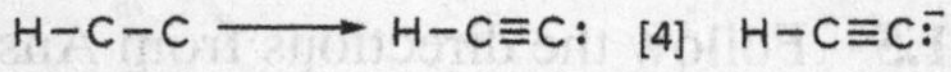

4 e⁻ used. Assign charge.

c. $(CH_3NH_3)^+$

[1] 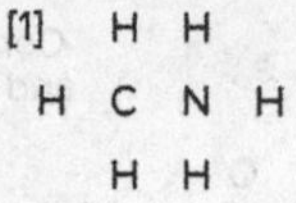

[2] Count valence e⁻.

$1\,C \times 4\,e^- =$	4
$6\,H \times 1\,e^- =$	6
$1\,N \times 5\,e^- =$	5
total e⁻ =	15

Subtract 1 for (+) charge = 14

[3] 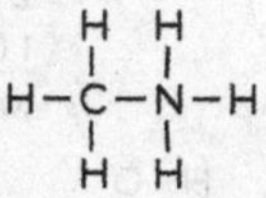

14 e⁻ used.

[4] 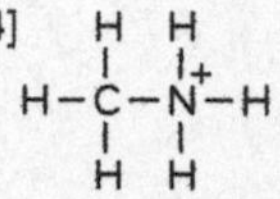

Assign charge.

d. $(CH_3NH)^-$

[1] 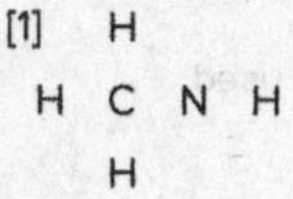

[2] Count valence e⁻.

$1\,C \times 4\,e^- =$	4
$4\,H \times 1\,e^- =$	4
$1\,N \times 5\,e^- =$	5
total e⁻ =	13

Add 1 for (–) charge = 14

[3] 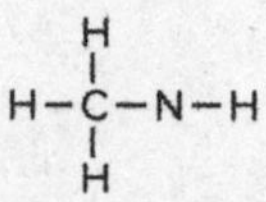

10 e⁻ used.

[4] 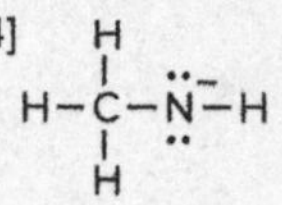

Complete octet and assign charge.

1.9

a. 

6 – [2 + (1/2)6] = +1

b. 

6 – [2 + (1/2)6] = +1

c. 

6 – [4 + (1/2)4] = 0

1.10

a. $C_2H_4Cl_2$ (two isomers)

Count valence e⁻.

$2\,C \times 4\,e^- =$	8
$4\,H \times 1\,e^- =$	4
$2\,Cl \times 7\,e^- =$	14
total e⁻ =	26

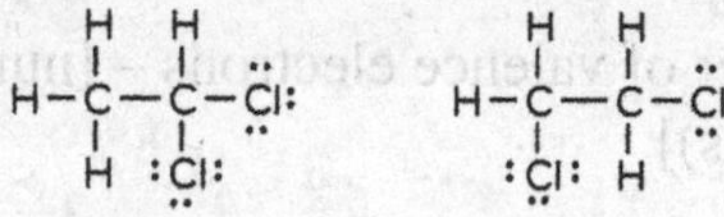

b. C_3H_8O (three isomers)

Count valence e⁻.

$3\,C \times 4\,e^- =$	12
$8\,H \times 1\,e^- =$	8
$1\,O \times 6\,e^- =$	6
total e⁻ =	26

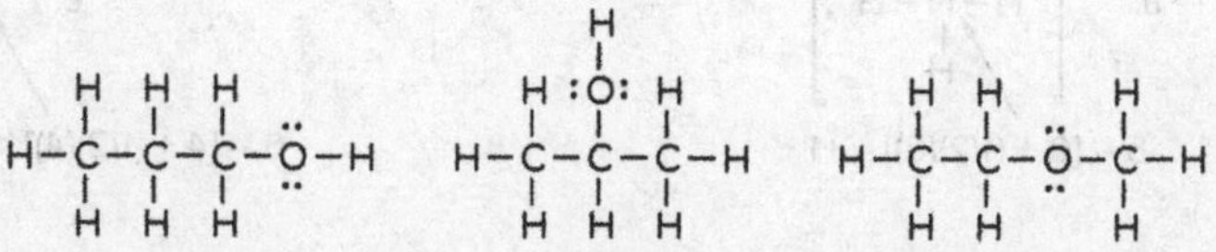

c. C_3H_6 (two isomers)

Count valence e⁻.

$3\,C \times 4\,e^- =$	12
$6\,H \times 1\,e^- =$	6
total e⁻ =	18

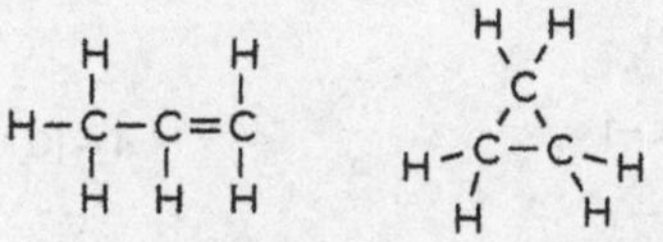

1.11 Two different definitions:

- **Isomers** have the same molecular formula and a *different* arrangement of atoms.
- **Resonance structures** have the *same* arrangement of atoms and a *different* arrangement of electrons.

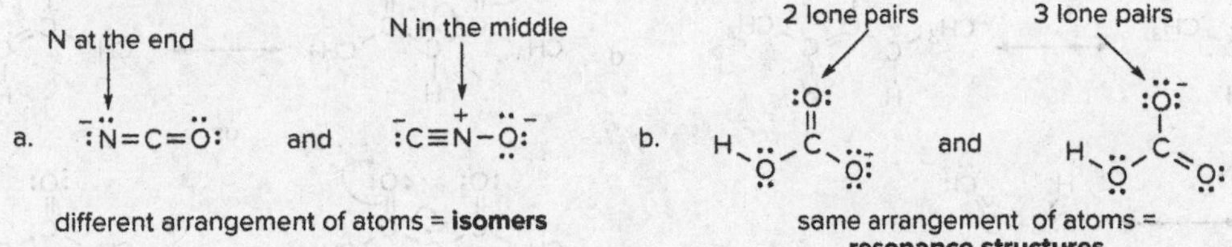

1.12 **Isomers** have the same molecular formula and a *different* arrangement of atoms. **Resonance structures** have the same molecular formula and the *same* arrangement of atoms.

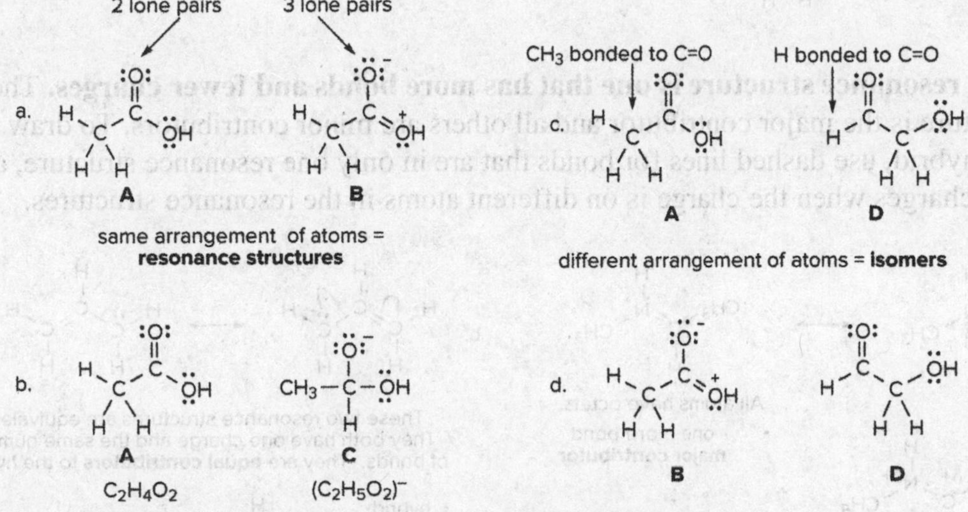

1.13 Curved arrow notation shows the movement of an electron pair. The tail begins at an electron pair (a bond or a lone pair) and the head points to where the electron pair moves.

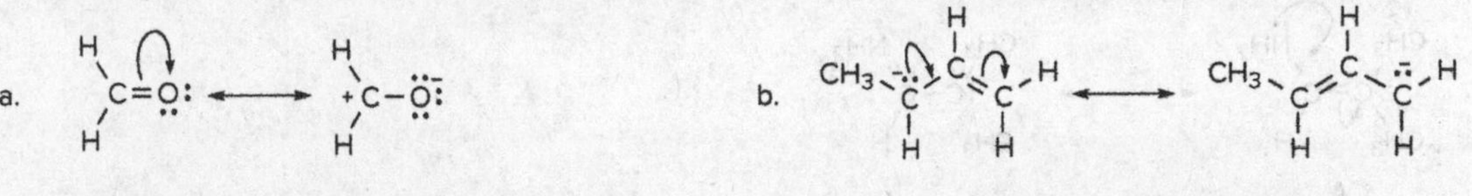

1.14 Compare the resonance structures to see which electrons have "moved." **Use one curved arrow to show the movement of each electron pair.**

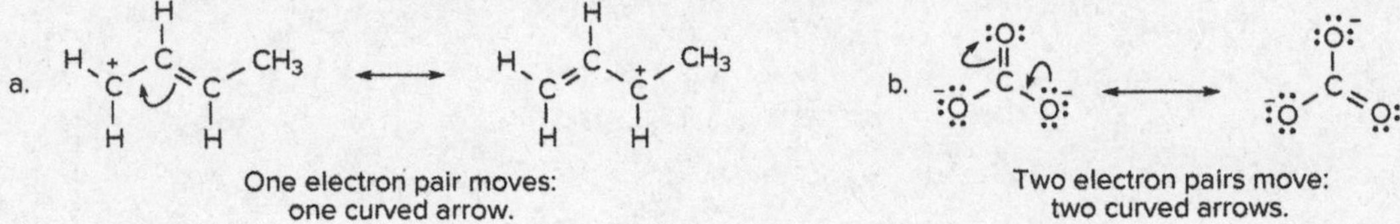

1.15 To draw another resonance structure, **move electrons only in multiple bonds and lone pairs** and keep the number of unpaired electrons constant.

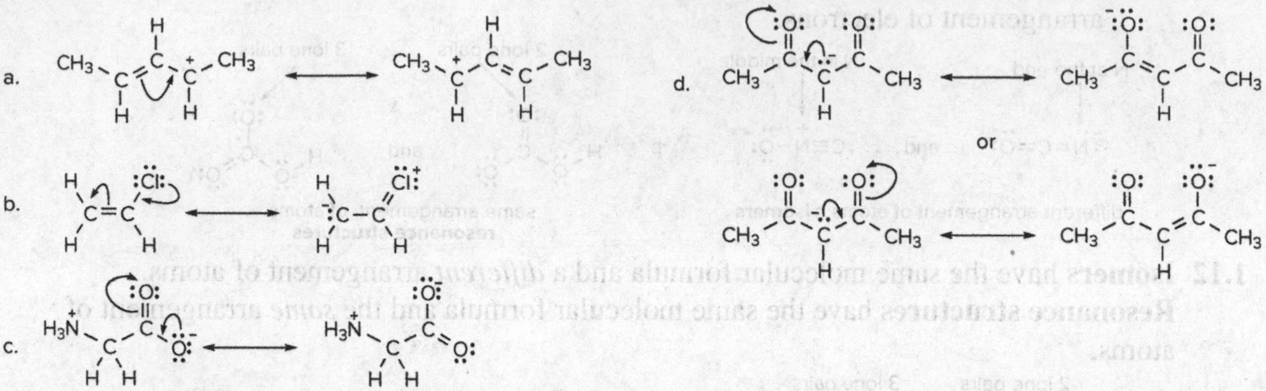

a. b. c. d.

or

1.16 A "better" resonance structure is one that has more bonds and fewer charges. The better structure is the major contributor and all others are minor contributors. To draw the resonance hybrid, use dashed lines for bonds that are in only one resonance structure, and use partial charges when the charge is on different atoms in the resonance structures.

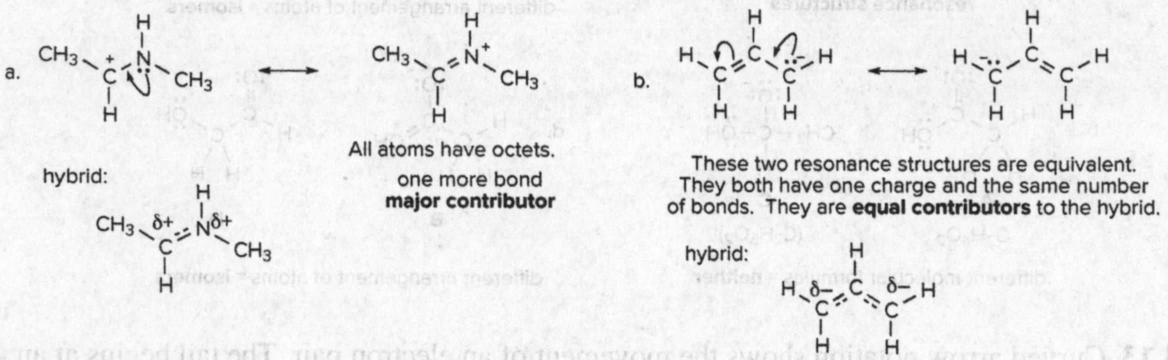

a.

All atoms have octets.

one more bond
major contributor

hybrid:

b.

These two resonance structures are equivalent. They both have one charge and the same number of bonds. They are **equal contributors** to the hybrid.

hybrid:

1.17

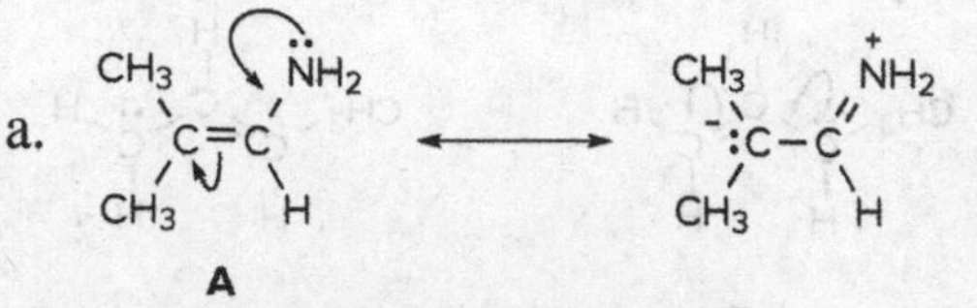

a.

A

b. The N atom in **B** has four atoms and no lone pairs, so there is no way to move the electrons to give the N another bond.

1.18 To predict the geometry around an atom, **count the number of groups (atoms + lone pairs),** making sure to draw in any needed lone pairs or hydrogens: 2 groups = linear, 3 groups = trigonal planar, 4 groups = tetrahedral.

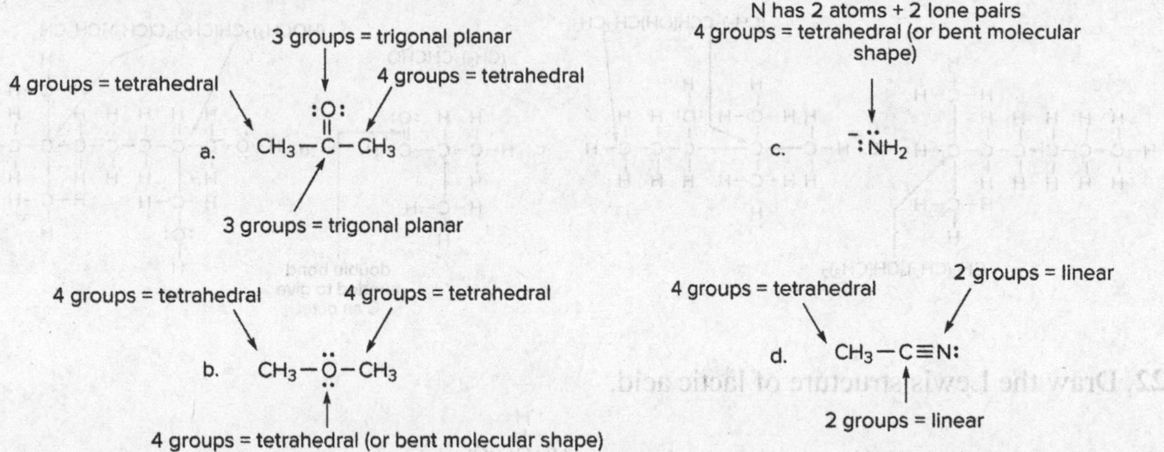

1.19 To predict the bond angle around an atom, **count the number of groups (atoms + lone pairs),** making sure to draw in any needed lone pairs or hydrogens: 2 groups = 180°, 3 groups = 120°, 4 groups = 109.5°.

2 groups = 180°

a. CH_3–C≡C–Cl:

2 groups = 180°

This C has 3 groups, so both angles are 120°.

b. CH_2=CH–Cl:

c. CH_3–CH_2–Cl:

This C has 4 groups, so both angles are 109.5°.

1.20 To predict the geometry around an atom, use the rules in Answer 1.18.

4 groups tetrahedral (or bent molecular shape)

3 groups trigonal planar

4 groups tetrahedral

2 groups linear

4 groups tetrahedral (or bent molecular shape)

1.21 Reading from left-to-right, draw the molecule as a Lewis structure. Always check that carbon has four bonds and all heteroatoms have an octet by adding any needed lone pairs.

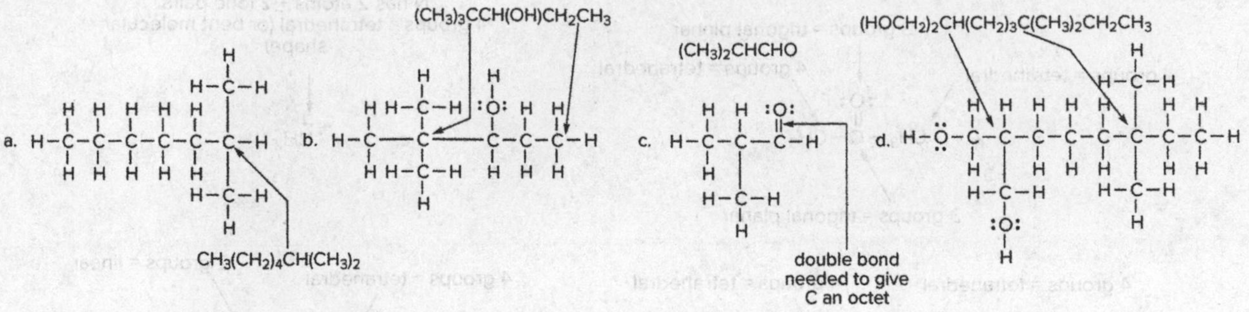

a. $CH_3(CH_2)_4CH(CH_3)_2$

b. $(CH_3)_3CCH(OH)CH_2CH_3$

c. $(CH_3)_2CHCHO$

double bond needed to give C an octet

d. $(HOCH_2)_2CH(CH_2)_3C(CH_3)_2CH_2CH_3$

1.22 Draw the Lewis structure of lactic acid.

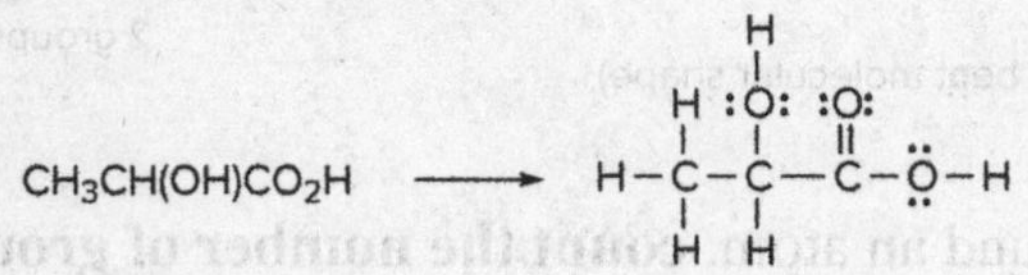

$CH_3CH(OH)CO_2H$ ⟶

1.23 In shorthand or skeletal drawings, **all line junctions or ends of lines represent carbon atoms.** The carbons are all tetravalent.

a. 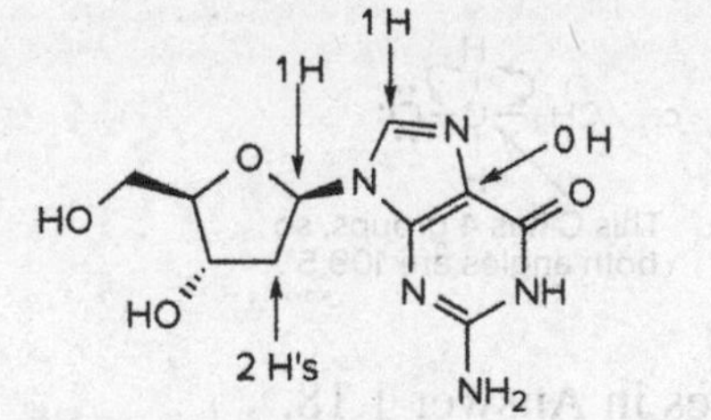 1 H, 1 H, 2 H's, $C_{10}H_{13}N_5O_4$

b. 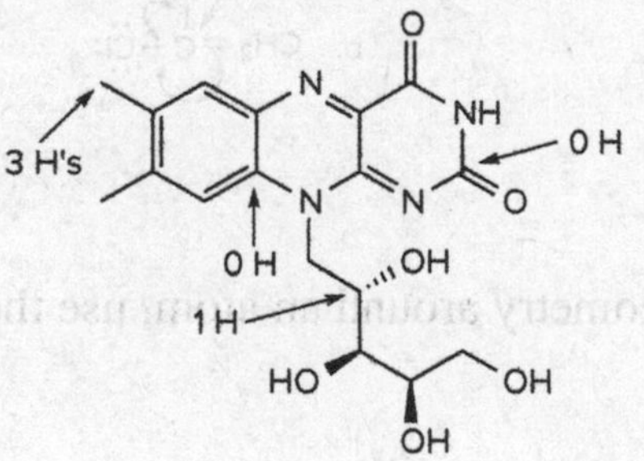 3 H's, O H, O H, 1 H, $C_{17}H_{20}N_4O_6$

1.24 In shorthand or skeletal drawings, **all line junctions or ends of lines represent carbon atoms.** Convert by writing in all carbons, and then adding hydrogen atoms to make the carbons tetravalent.

a.

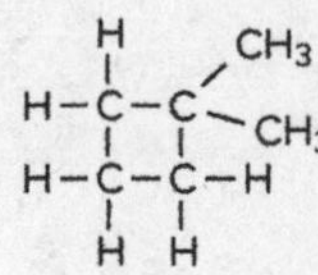

b.

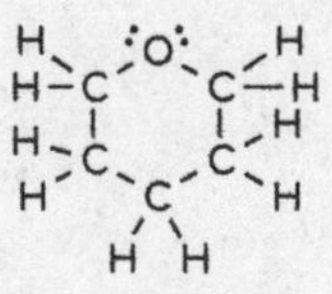

c.

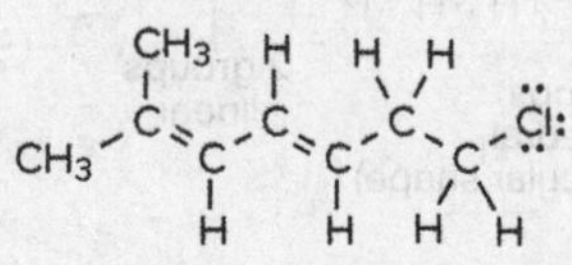

d.

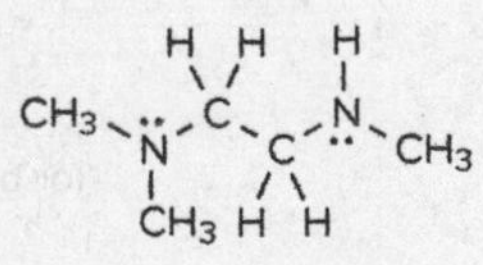

1.25

a. H N N Cl N

$C_{18}H_{26}ClN_3$

b. HO HO O OH NH_2

$C_9H_{11}NO_4$

1.26 A charge on a carbon atom takes the place of one hydrogen atom. **A negatively charged C has one lone pair, whereas a positively charged C has none.**

a. positive charge no lone pairs no H's needed

b. H negative charge one lone pair one H needed

c. H positive charge no lone pairs one H needed

d. O H negative charge one lone pair one H needed

e. negative charge one lone pair no H's needed

1.27

a. N+ N

b. N

c. O N+ O

d. O O NH_3

1.28 Draw each indicated structure. Recall that in the skeletal drawings, a carbon atom is located at the intersection of any two lines and at the end of any line.

a. $CH_3O(CH_2)_2COCH{=}C(CH_3)_2$ =

c. = $(CH_3)_2CH(CH_2)_2CONHCH_3$

b. CH_3–C–C–CH_2CH_2Cl / H_2N–C–C–H = Cl H_2N

d. HO O O = $HO(CH_2)_2CH{=}CHCO_2CH(CH_3)_2$

1.29 To determine the orbitals used in bonding, **count the number of groups** (atoms + lone pairs): 4 groups = sp^3, 3 groups = sp^2, 2 groups = sp, H atom = 1s (no hybridization). All covalent single bonds are σ bonds, and all double bonds contain one σ and one π bond.

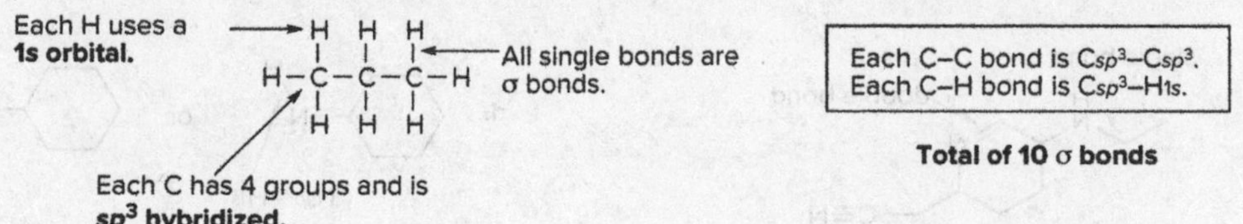
Each H uses a
1s orbital.
All single bonds are
σ bonds.
Each C has 4 groups and is
sp³ hybridized.
Each C–C bond is Csp³–Csp³.
Each C–H bond is Csp³–H1s.
Total of 10 σ bonds

1.30 Draw a valid Lewis structure and count groups around each atom: 4 groups = sp^3, 3 groups = sp^2, 2 groups = sp, H atom = $1s$ (no hybridization).

(a) (d) (c) (b)

a. C_{sp^3}–O_{sp^3}
b. O_{sp^3}–H_{1s}
c. N_{sp^3}–H_{1s}
d. C_{sp^3}–N_{sp^3}

All lone pairs are in sp^3 hybrid orbitals.

1.31 To determine the hybridization, **count the number of groups** around each atom: 4 groups = sp^3, 3 groups = sp^2, 2 groups = sp, H atom = $1s$ (no hybridization).

a. $CH_3-C\equiv CH$ — 4 groups sp^3; 2 groups sp

b. CH_3, =N — 3 groups sp^2; 3 groups sp^2

c. $CH_2=C=CH_2$ — 3 groups sp^2; 2 groups sp

1.32

sp^2 hybridized O; HO, O, HO, OH, OH; C_{sp^2}–C_{sp^3}; sp^3 hybridized O

a. Isocitric acid has three sp^2 hybridized C's, labeled with open circles.
b. One O atom is sp^2 hybridized and one O atom is sp^3 hybridized, as shown.
c. The highlighted C–C bond is formed from one sp^2 hybridized C and one sp^3 hybridized C.
d. There are 19 σ bonds in isocitric acid, including the σ bonds that are part of the three double bonds.
e. Isocitric acid has three π bonds that are part of the double bonds.

1.33 Single bonds are weaker and longer than double bonds, which are weaker and longer than triple bonds. Increasing percent *s*-character increases bond strength and decreases bond length.

a. double bond; triple bond
The triple bond is shorter than the double bond.

b. single bond; double bond; C≡N
The C=N bond is shorter than the C–N bond.

c. C_{sp^2}–H_{1s}, 33% *s*-character, shorter bond or C_{sp^3}–H_{1s}, 25% *s*-character

d. N_{sp^2}–H_{1s}, 33% *s*-character, shorter bond or N_{sp^3}–H_{1s}, 25% *s*-character

1.34 Increasing percent *s*-character decreases bond length.

[1] Csp^3–Csp^2 [2] Csp^2–Csp [3] Csp–Csp^3 [4] Csp^3–Csp^3

[2] < [3] < [1] < [4]

1.35 Electronegativity increases from left-to-right across a row of the periodic table and decreases down a column. Look at the relative position of the atoms to determine their relative electronegativity.

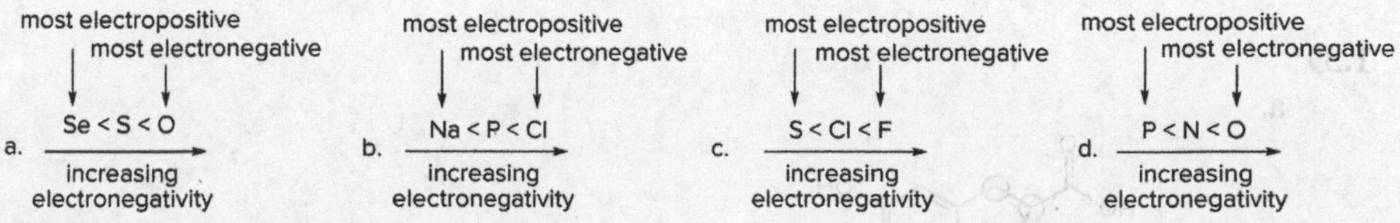

1.36 Dipoles result from unequal sharing of electrons in covalent bonds. More electronegative atoms "pull" electron density toward them, making a dipole. **Dipole arrows point toward the atom of higher electron density.**

a. δ+ H–F δ− b. δ+ B–C δ− c. δ− C–Li δ+ d. δ+ C–Cl δ−

1.37 Polar molecules result from a net dipole. To determine polarity, draw the molecule in three dimensions around any polar bonds, draw in the dipoles, and look to see whether the dipoles cancel or reinforce.

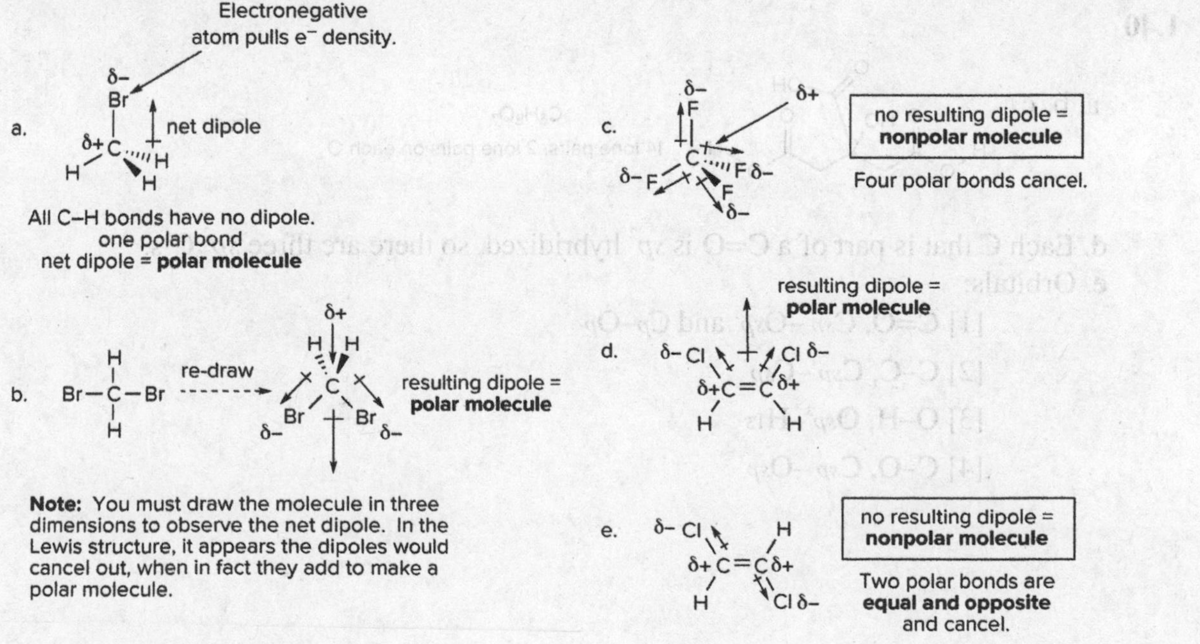

1.38 a, b. Convert the skeletal structure to a Lewis structure. Each O atom needs two lone pairs and each N atom needs one for an octet. Count groups around an atom to determine hybridization and shape.

*sp*² hybridized and trigonal planar

*sp*³ hybridized and tetrahedral

*sp*³ hybridized and trigonal pyramidal

c. All C–O, O–H, N–H, and C–N bonds are polar because the electronegativity difference between the atoms is large.

1.39

a.

skeletal structure

b. Circled carbons are sp^3 hybridized. All other carbons are sp^2 hybridized.

c. Each N is surrounded by three atoms and a lone pair, making it sp^3 hybridized and trigonal pyramidal in molecular shape.

d.

11 polar bonds shown in bold

1.40

a, b, c.

$C_6H_8O_7$

14 lone pairs, 2 lone pairs on each O

d. Each C that is part of a C=O is sp^2 hybridized, so there are three sp^2 C's.

e. Orbitals:

[1] C=O, Csp^2–Osp^2 and Cp–Op

[2] C–C, Csp^3–Csp^2

[3] O–H, Osp^3–$H1s$

[4] C–O, Csp^3–Osp^3

1.41

a. The molecular formula for 2'-deoxyadenosine is $C_{10}H_{13}N_5O_3$.
b. Each N atom has one lone pair and each O has two, so there are 11 lone pairs.
c. Skeletal structure:

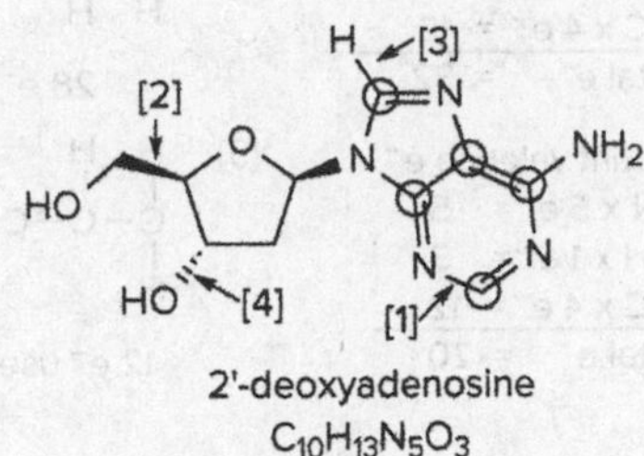

2'-deoxyadenosine
$C_{10}H_{13}N_5O_3$

d. There are five sp^2 hybridized C's, labeled with open circles.
e. Orbitals used to form indicated bonds:

[1] $N_{sp^2}–C_{sp^2}$ [2] $C_{sp^3}–C_{sp^3}$ [3] $C_{sp^2}–H_{1s}$ [4] $C_{sp^3}–O_{sp^3}$

1.42 Formal charge (FC) = number of valence electrons – [number of unshared electrons + (½)(number of shared electrons)]. C is in group 4A.

a. 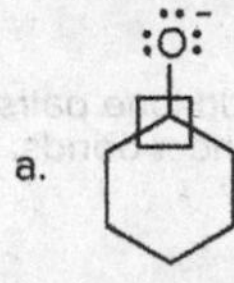

4 – [0 + (1/2)(6)] = +1

b. 4 – [1 + (1/2)(6)] = 0

c. 4 – [2 + (1/2)(4)] = 0

d. :ÖCH₃ 4 – [0 + (1/2)(6)] = +1

e. :ÖCH₃ 4 – [0 + (1/2)(8)] = 0

f. 4 – [2 + (1/2)(6)] = –1

1.43 Formal charge (FC) = number of valence electrons – [number of unshared electrons + (½)(number of shared electrons)]. N is in group 5A and O is in group 6A.

a. 5 – [4 + (1/2)(4)] = –1

b. :N=N=N: 5 – [0 + (1/2)(8)] = +1; 5 – [4 + (1/2)(4)] = –1; 5 – [4 + (1/2)(4)] = –1

c. 6 [2 + (1/2)(6)] = +1

d. –N=O: 5 – [2 + (1/2)(6)] = 0; 6 – [4 + (1/2)(4)] = 0

1.44 Follow the steps in Answer 1.4 to draw Lewis structures.

a. CH_2N_2

valence e⁻	
1 C x 4 e⁻	= 4
2 H x 1 e⁻	= 2
2 N x 5 e⁻	= 10
total e⁻	= 16

H–C=N=N: or H–C–N≡N:

b. CH_3NO_2

valence e⁻	
1 C x 4 e⁻	= 4
3 H x 1 e⁻	= 3
1 N x 5 e⁻	= 5
2 O x 6 e⁻	= 12
total e⁻	= 24

H–C–N–O: or H–C–N=O:

c. CH_3CNO

valence e⁻	
2 C x 4 e⁻	= 8
3 H x 1 e⁻	= 3
1 N x 5 e⁻	= 5
1 O x 6 e⁻	= 6
total e⁻	= 22

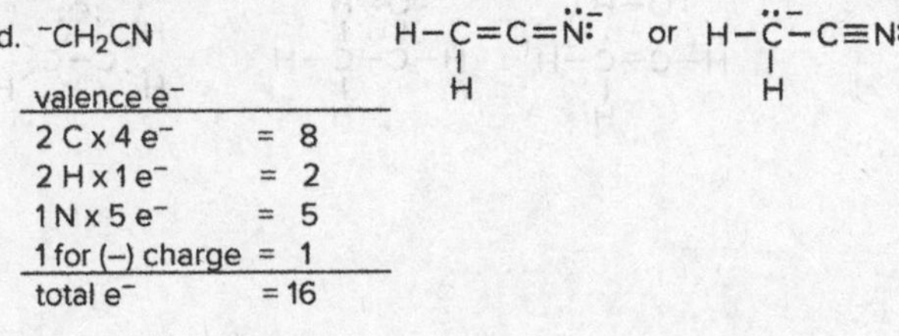

d. $^-CH_2CN$

valence e⁻	
2 C x 4 e⁻	= 8
2 H x 1 e⁻	= 2
1 N x 5 e⁻	= 5
1 for (–) charge	= 1
total e⁻	= 16

1.45 Follow the steps in Answer 1.4 to draw Lewis structures.

a. $(CH_3CH_2)_2O$ [1] H C C O C C H skeleton [2] Count valence e⁻.
1 O x 6 e⁻ = 6
10 H x 1 e⁻ = 10
4 C x 4 e⁻ = 16
total e⁻ = 32
[3] H–C–C–O–C–C–H, 28 e⁻ used. [4] Add lone pairs.

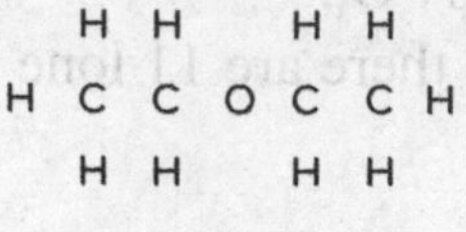

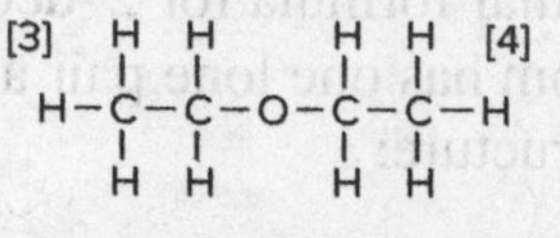

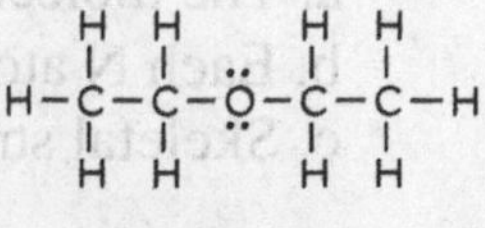

b. CH_2CHCN [1] C C C N skeleton [2] Count valence e⁻.
1 N x 5 e⁻ = 5
3 H x 1 e⁻ = 3
3 C x 4 e⁻ = 12
total e⁻ = 20
[3] C–C–C–N, 12 e⁻ used. [4] C=C–C≡N: Add lone pairs and π bonds.

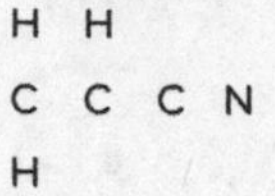

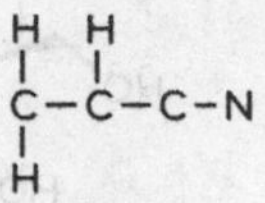

c. $(HOCH_2)_2CO$ [1] H O C C C O H skeleton [2] Count valence e⁻.
3 O x 6 e⁻ = 18
6 H x 1 e⁻ = 6
3 C x 4 e⁻ = 12
total e⁻ = 36
[3] H–O–C–C–C–O–H, 22 e⁻ used. [4] Add lone pairs and π bonds.

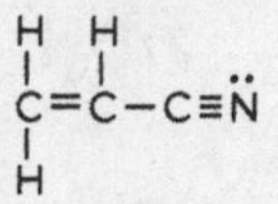

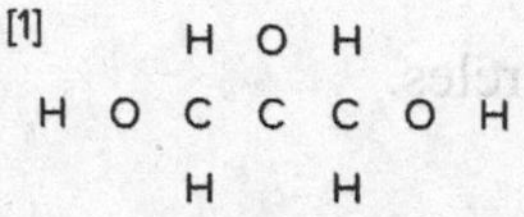

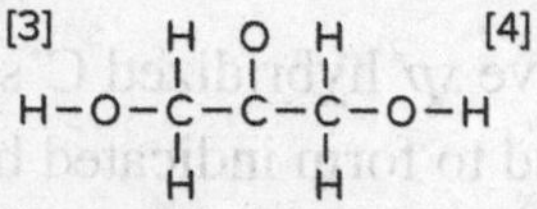

d. $(CH_3CO)_2O$ [1] H C C O C C H skeleton [2] Count valence e⁻.
3 O x 6 e⁻ = 18
6 H x 1 e⁻ = 6
4 C x 4 e⁻ = 16
total e⁻ = 40
[3] H–C–C–O–C–C–H, 24 e⁻ used. [4] Add lone pairs and π bonds.

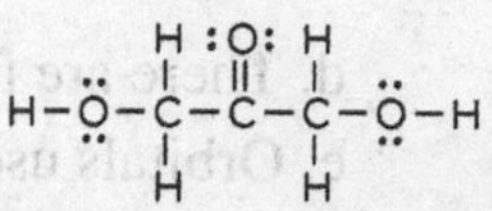

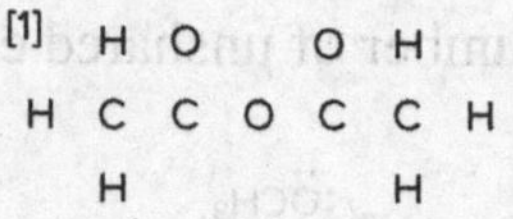

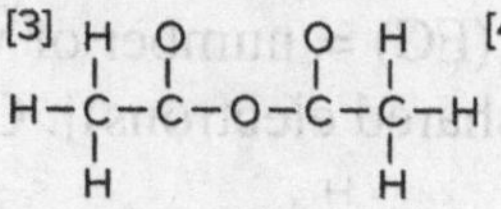

1.46

a. [structure with $\overset{+}{N}H_2$, H_2N, N, :O:, and $\ddot{O}$:⁻]

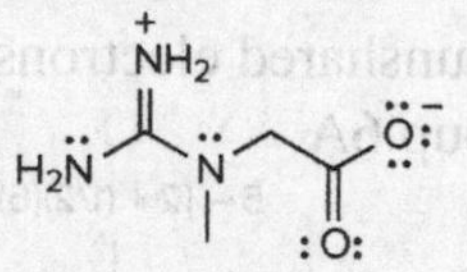

b. Two of the possible resonance structures:

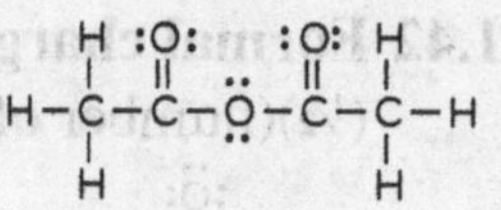

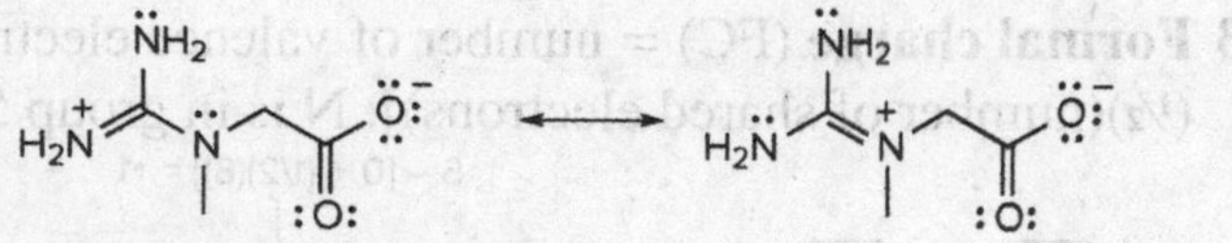

[H_2N–C(=NH$_2$)–N(CH$_3$)–CH$_2$–C(=O)–O⁻ ⟷ $H_2\overset{+}{N}$=C(–NH$_2$)–N(CH$_3$)–CH$_2$–C(=O)–O⁻]

1.47 Isomers must have a different arrangement of atoms. Compounds are drawn as Lewis structures with no implied geometry.

a. Two isomers of molecular formula C_3H_7Cl

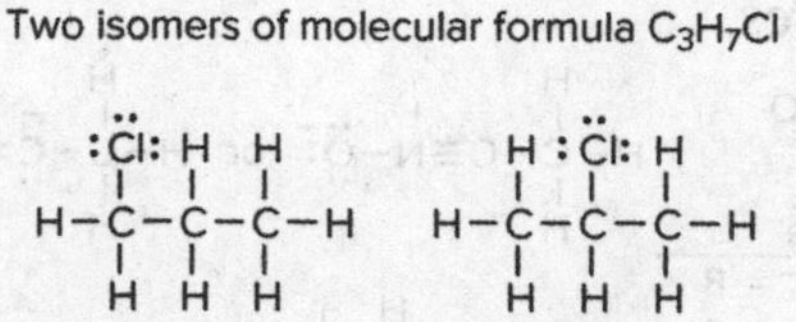

H–C(Cl)H–CH$_2$–CH$_3$ (Cl on end carbon); H–CH$_2$–CH(Cl)–CH$_3$ (Cl on middle carbon)

b. Three isomers of molecular formula C_2H_4O

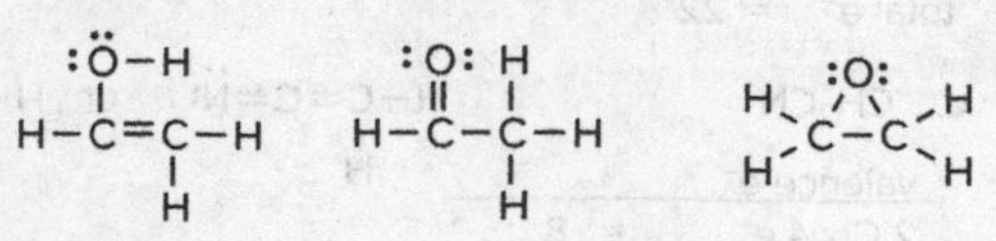

:Ö–H on H–C=C–H; H–C(=O:)–C–H; three-membered ring H$_2$C–O–CH$_2$

c. Four isomers of molecular formula C_3H_9N

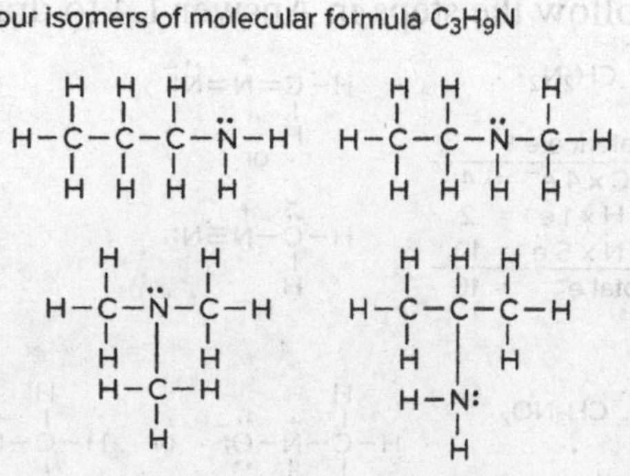

H–C–C–C–N̈–H; H–C–C–N̈–C–H; H–C–N̈(–CH$_3$)–C–H; H–C–C(–N̈H$_2$)–C–H

1.48 Nine isomers of C_3H_6O:

1.49 Use the definition of isomers and resonance structures in Answer 1.11.

one fewer H
one more H
one more H
one fewer H

A

B
isomer
Two labeled C's differ.

C
resonance structure

D
isomer

E
resonance structure

1.50 Use the definitions of isomers and resonance structures in Answer 1.11.

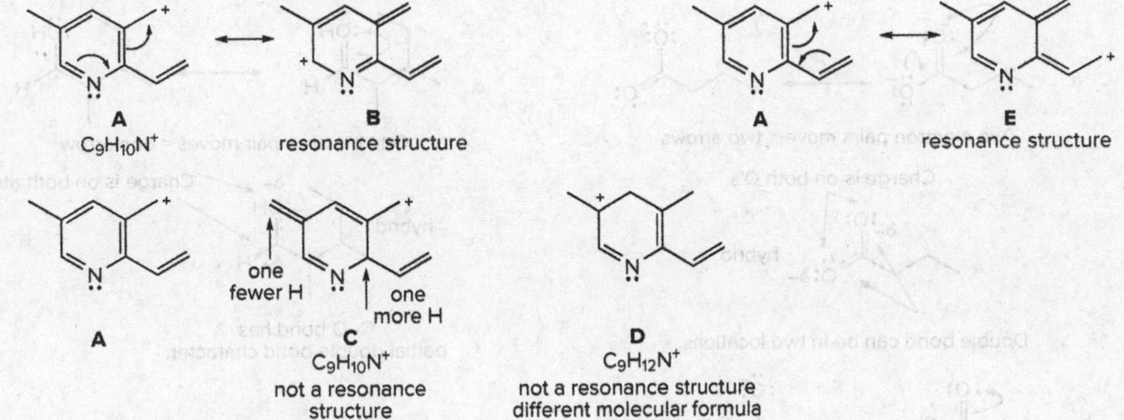

1.51 Use the definitions of isomers and resonance structures in Answer 1.11.

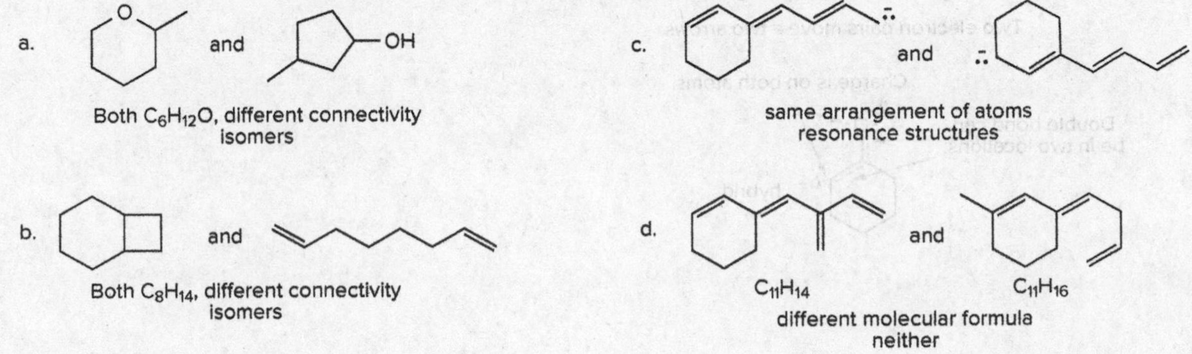

1.52 Compare the resonance structures to see what electrons have "moved." **Use one curved arrow to show the movement of each electron pair.**

a. H, N, CH_3 ⟷ H, N, CH_3

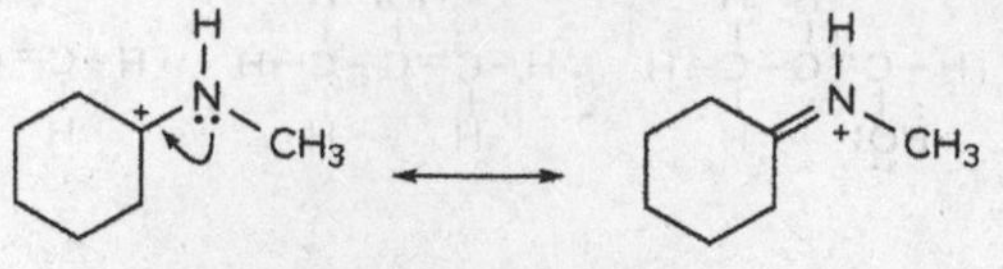

One electron pair moves = one arrow

b. Four electron pairs move = four arrows

1.53 Curved arrow notation shows the movement of an electron pair. The tail begins at an electron pair (a bond or a lone pair) and the head points to where the electron pair moves.

a. –N≡N: ⟷ –N=N:

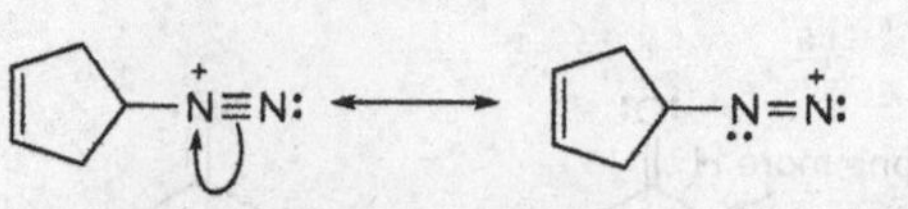

c. NH_2 ⟷ NH_2

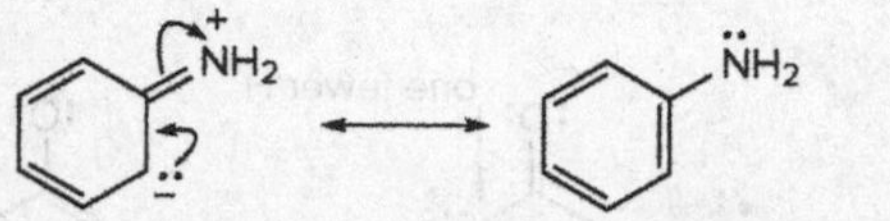

b. :O: ⟷ :O:

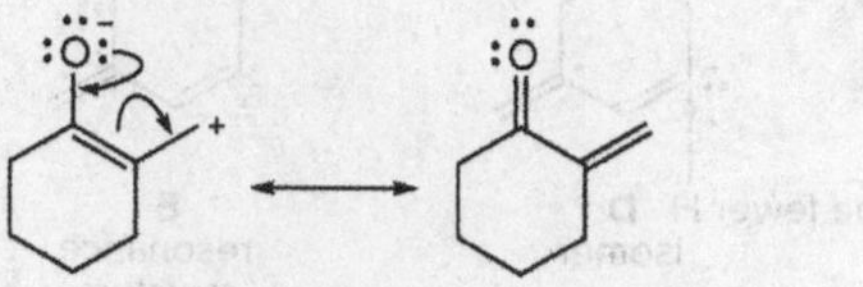

1.54 Use the rules in Answer 1.15.

a. :O: O: ⟷ :O: O:

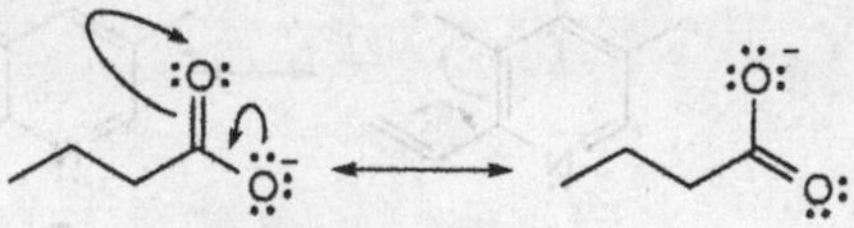

Two electron pairs move = two arrows

Charge is on both O's.

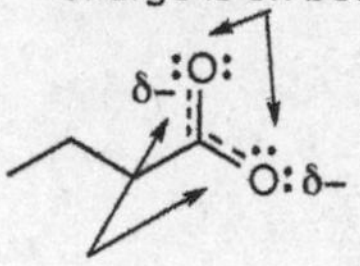

δ– :O: O: δ– hybrid

Double bond can be in two locations.

c. :OH H ⟷ :OH H

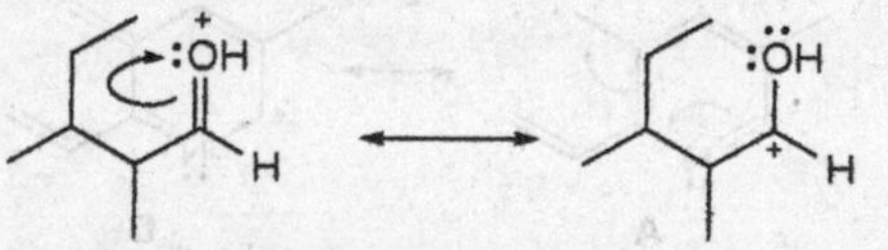

One electron pair moves = one arrow

δ+ :OH δ+ H hybrid

Charge is on both atoms.

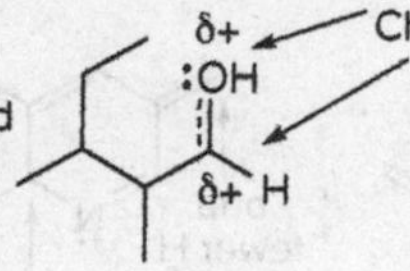

C–O bond has partial double bond character.

b. :O: ⟷ :O:

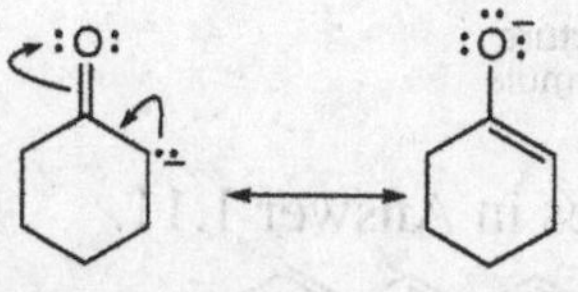

Two electron pairs move = two arrows

Charge is on both atoms.

Double bond can be in two locations.

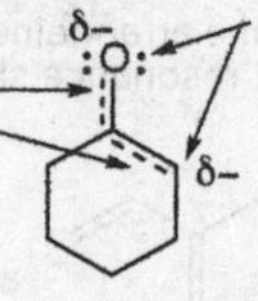

δ– :O: δ– hybrid

1.55 For the compounds where the arrangement of atoms is not given, first draw a Lewis structure. Then use the rules in Answer 1.15.

a. O_3

Count valence e⁻.
3 O x 6 e⁻ = 18
total e⁻ = 18

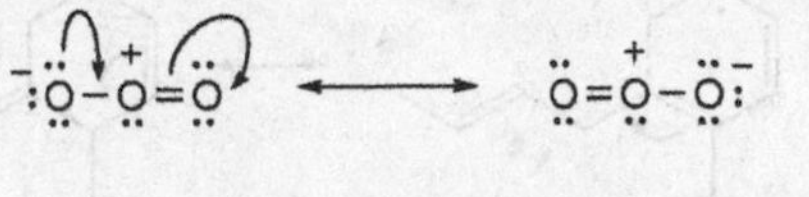

b. NO_3^- (a central N atom)

Count valence e⁻.
1 N x 5 e⁻ = 5
3 O x 6 e⁻ = 18
(–) charge = 1
total e⁻ = 24

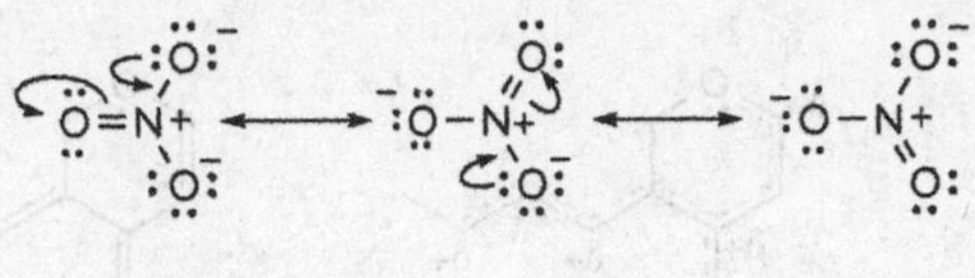

c. N_3^-

Count valence e⁻.
3 N x 5 e⁻ = 15
(–) charge = 1
total e⁻ = 16

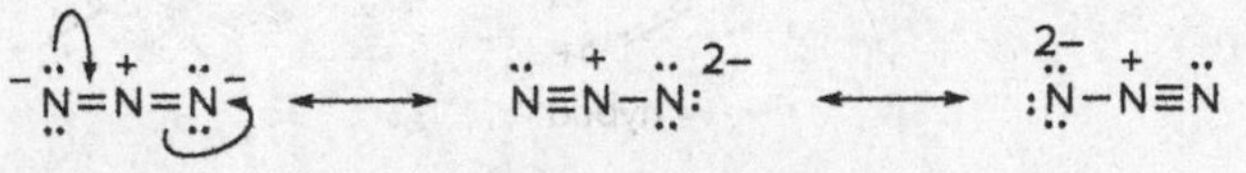

d.

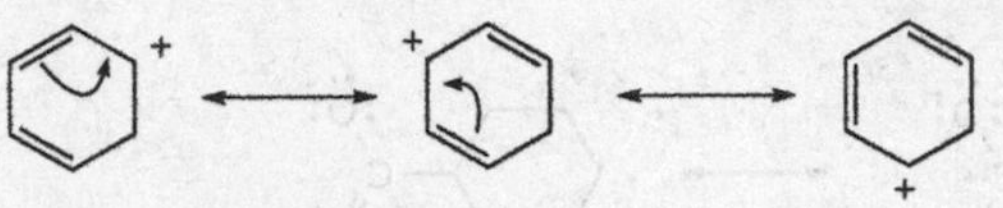

e.

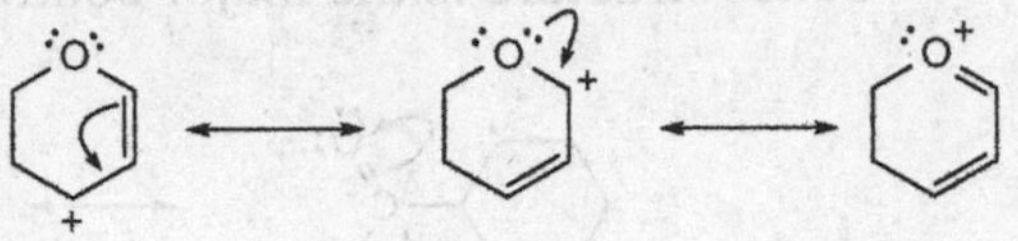

1.56 a. No additional Lewis structures can be drawn for **B**.

b. Additional Lewis structures can be drawn for **A, C,** and **D**, which are all examples of X=Y–Z: resonance.

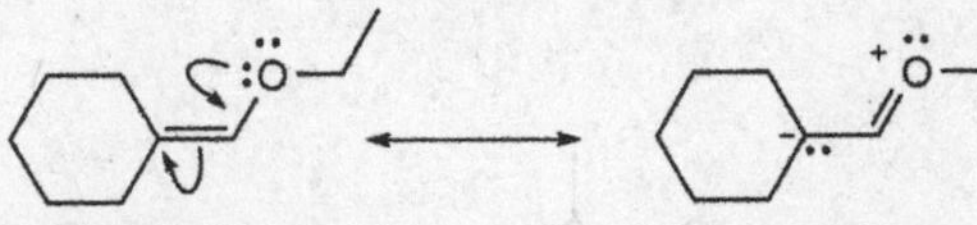

A

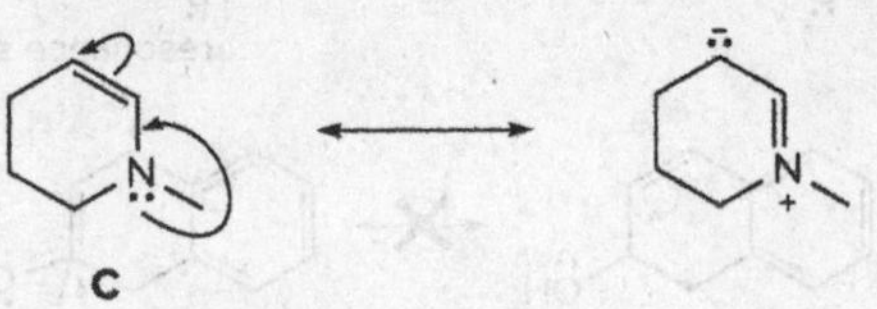

C

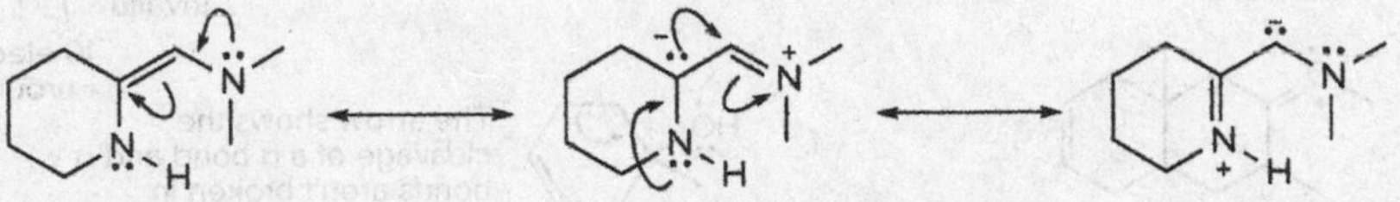

D

1.57 To draw the **resonance hybrid,** use the rules in Answer 1.16.

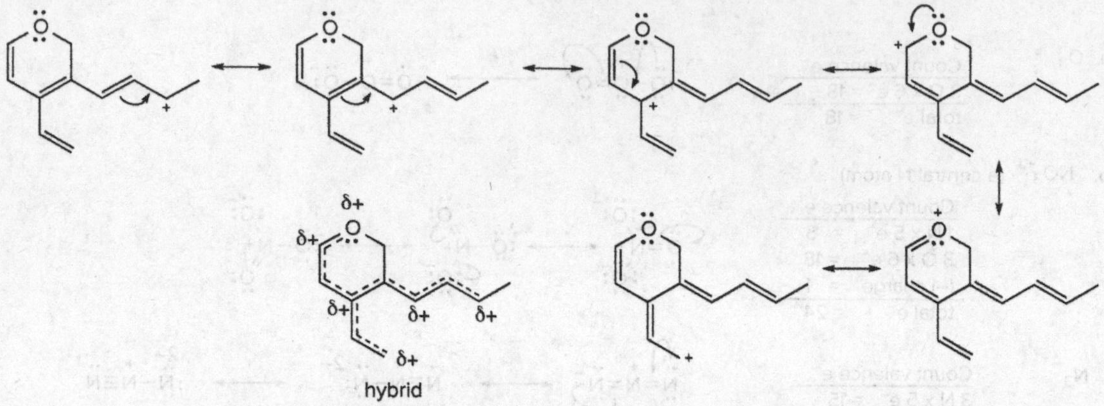

δ+ δ+ δ+ δ+ δ+ δ+

hybrid

1.58 A "better" resonance structure is one that has more bonds and fewer charges. The better structure is the major contributor and all others are minor contributors.

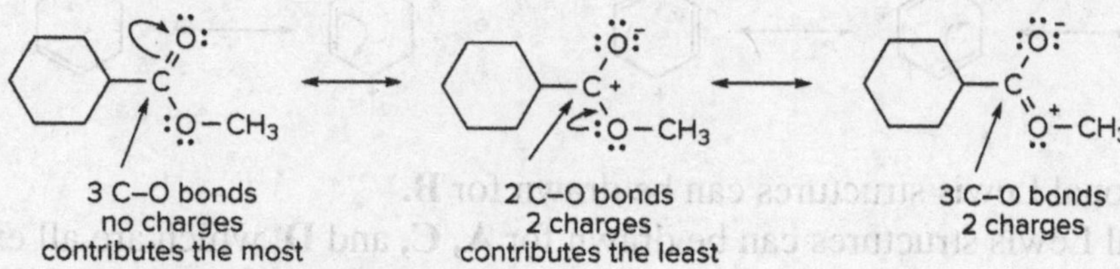

:O–CH3 — 3 C–O bonds, no charges, contributes the most

:O–CH3 — 2 C–O bonds, 2 charges, contributes the least

O–CH3 — 3 C–O bonds, 2 charges

1.59

a. H H N — invalid — 10 electrons around N

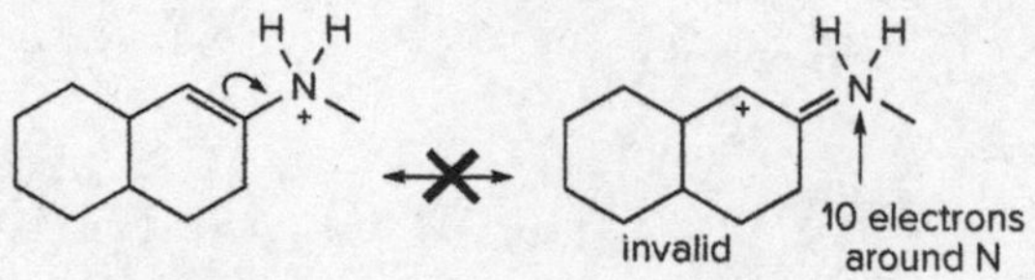

b. N: — resonance structure

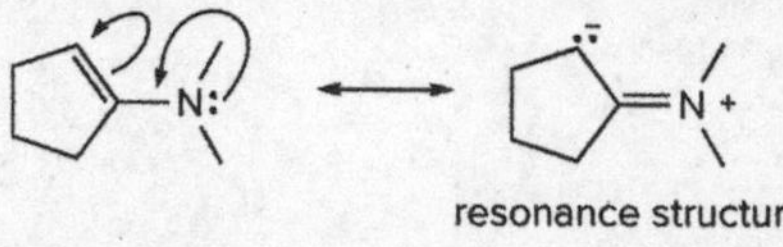

c. resonance structure

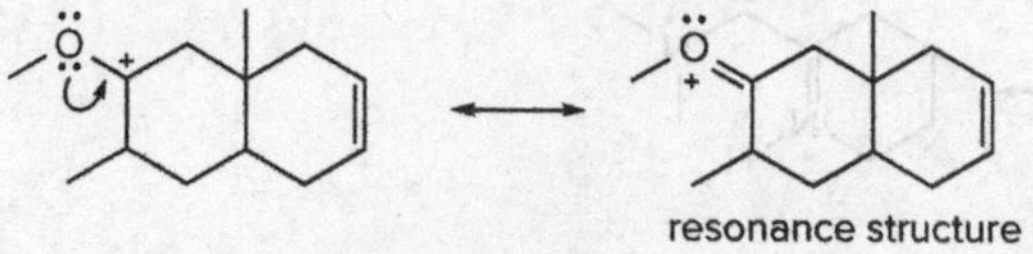

d. :O: — resonance structure

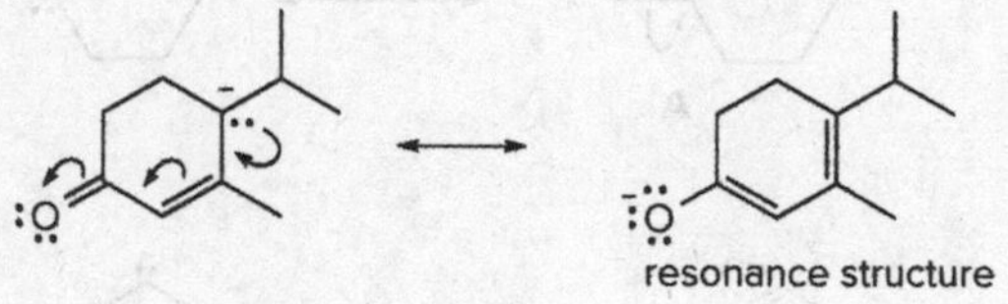

e. OH — invalid — H — 10 electrons around C

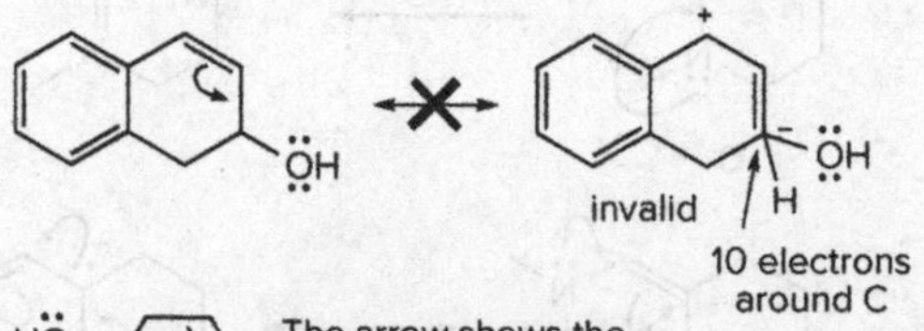

f. HO — The arrow shows the cleavage of a σ bond and σ bonds aren't broken in drawing resonance structures.

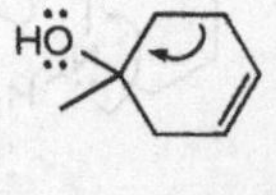

1.60 Use the rules in Answer 1.19.

a. CH_3Cl

4 groups = 109.5°

c. 3 groups = 120°

120°

4 groups = 109.5°

3 groups = 120°

e. All C atoms have 3 groups = 120°.

120°

b. 4 groups = ~109.5°

4 groups = ~109.5°

d. 4 groups = 109.5°

both C's surrounded by 2 groups = 180°

4 groups = ~109.5°

1.61 To predict the geometry around an atom, use the rules in Answer 1.18.

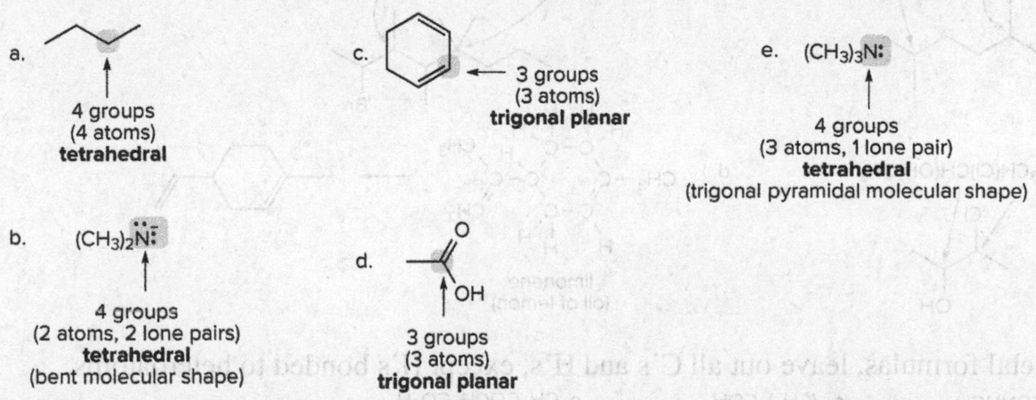

1.62 In shorthand or skeletal drawings, **all line junctions or ends of lines represent carbon atoms.** The C's are all tetravalent. All H's bonded to C's are drawn in the following structures. C's labeled with (*) have no H's bonded to them.

a.

b.

1.63 In shorthand or skeletal drawings, **all line junctions or ends of lines represent carbon atoms.** Convert by writing in all C's, and then adding H's to make the C's tetravalent.

a. **menthol**
(isolated from peppermint oil)

b. **myrcene**
(isolated from bayberry)

c. **ethambutol**
(drug used to treat tuberculosis)

d. **estradiol**
(a female sex hormone)

1.64 In skeletal formulas, leave out all C's and H's, except H's bonded to heteroatoms.

a. $(CH_3)_2CHCH_2CH_2CH(CH_3)_2$

b. $CH_3CH(Cl)CH(OH)CH_3$

c. $CH_3(CH_2)_2C(CH_3)_2CH(CH_3)CH(CH_3)CH(Br)CH_3$

d. limonene
(oil of lemon)

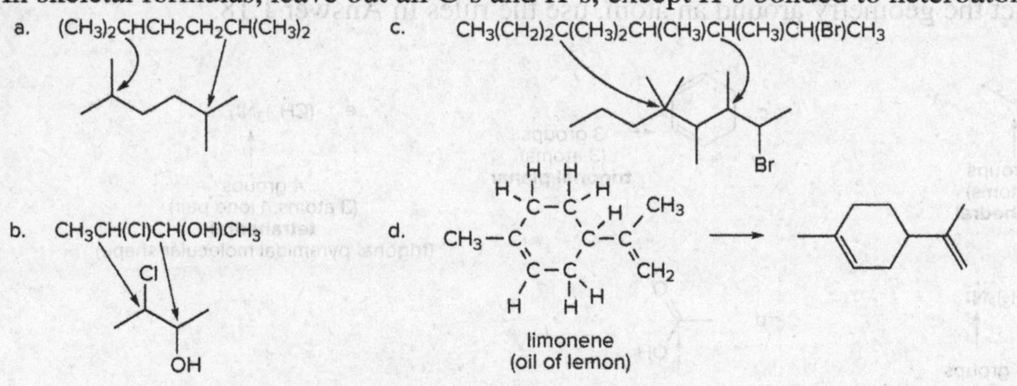

1.65 In skeletal formulas, leave out all C's and H's, except H's bonded to heteroatoms.

a. $CH_3CONHCH_3$

b. CH_3COCH_2Br

c. $(CH_3)_3COH$

d. CH_3COCl

e. $CH_3COCH_2CO_2H$

f. $HO_2CCH(OH)CO_2H$

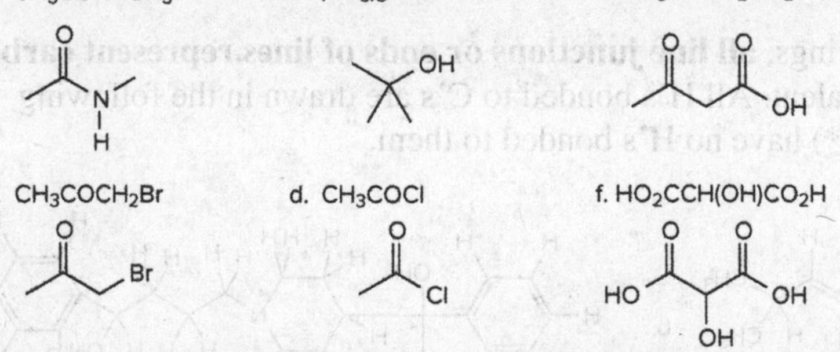

1.66 A charge on a C atom takes the place of one H atom. A negatively charged C has one lone pair, and a positively charged C has none.

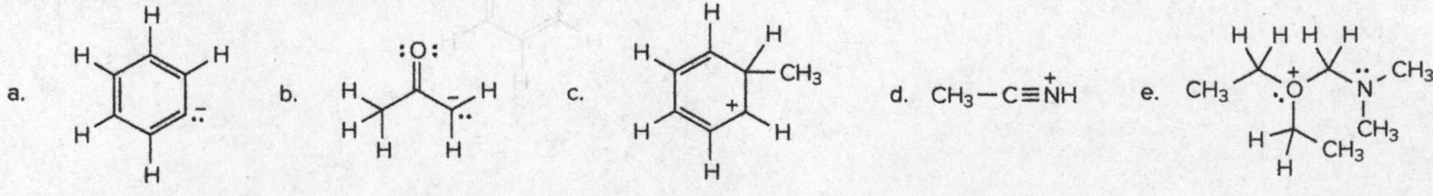

1.67 To determine the hybridization around the labeled atoms, use the procedure in Answer 1.31.

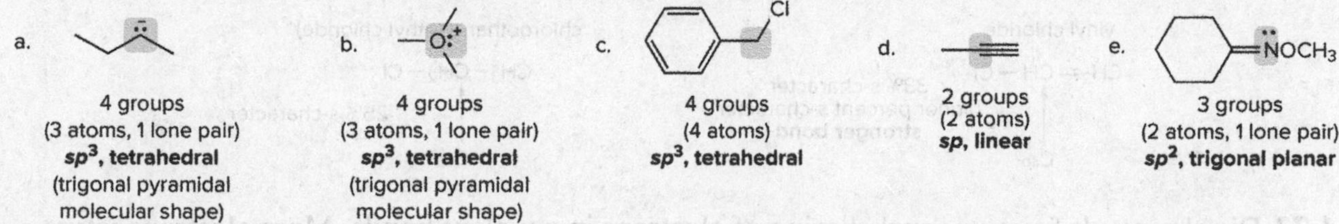

1.68 To determine which orbitals are involved in bonding, use the procedure in Answer 1.29.

1.69

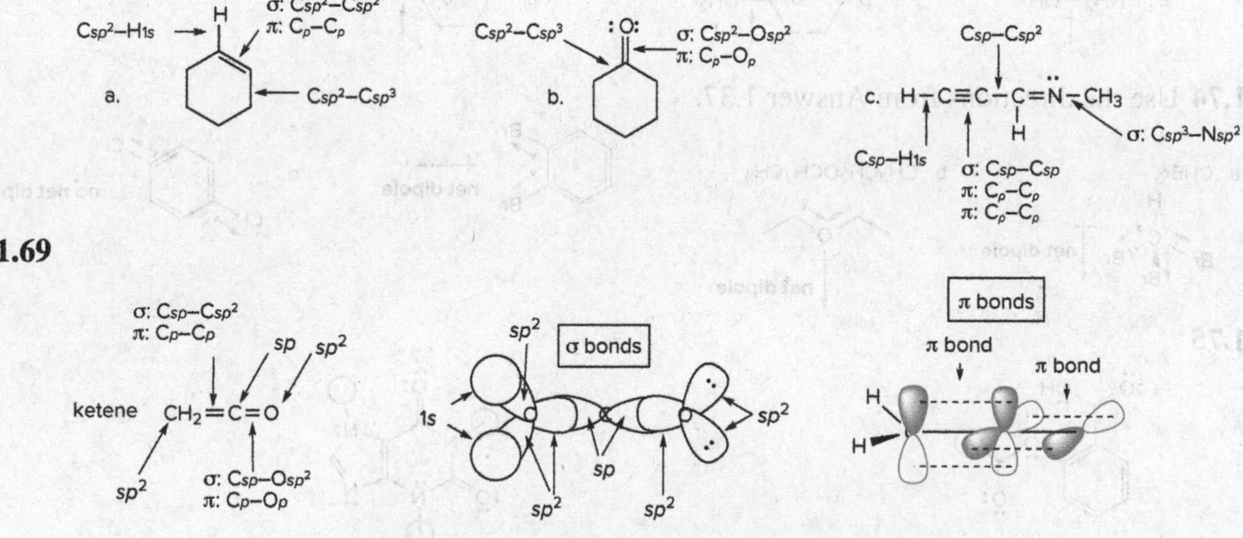

1.70 To determine relative bond length, use the rules in Answer 1.33. All C's and H's are drawn in for emphasis.

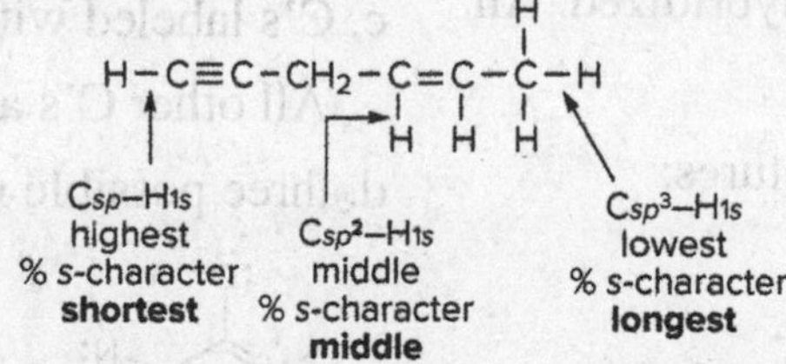

1.71

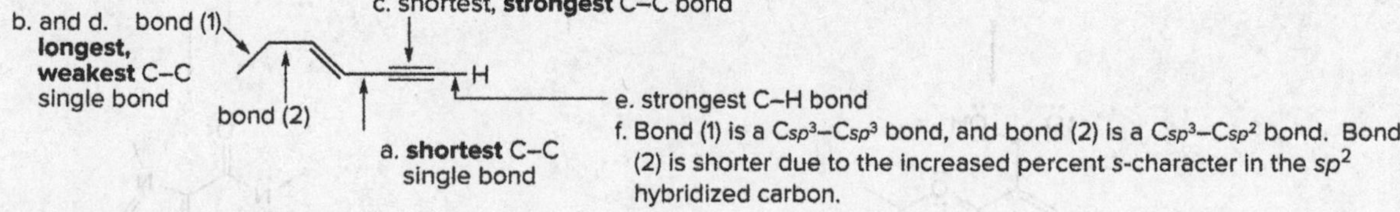

1.72 Percent *s*-character determines the strength of a bond. **The higher percent *s*-character of an orbital used to form a bond, the stronger the bond.**

vinyl chloride

$CH_2{=}CH{-}Cl$

Csp^2

33% *s*-character
higher percent *s*-character
stronger bond

chloroethane (ethyl chloride)

$CH_3{-}CH_2{-}Cl$

Csp^3

25% *s*-character

1.73 Dipoles result from unequal sharing of electrons in covalent bonds. More electronegative atoms "pull" electron density toward them, making a dipole.

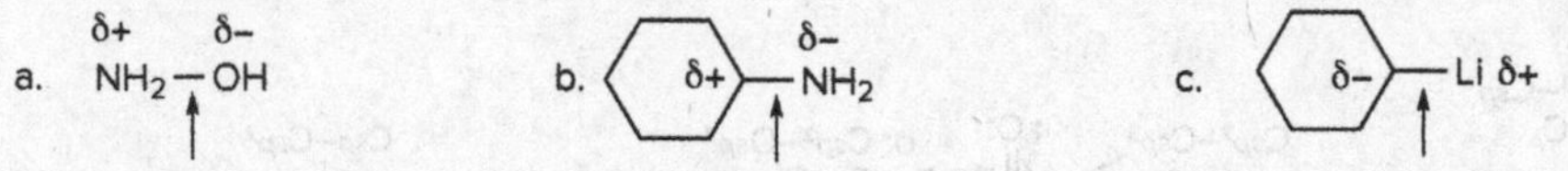

a. $\overset{\delta+}{NH_2}{-}\overset{\delta-}{OH}$

b. (cyclohexyl δ+)–NH_2 δ−

c. (cyclohexyl δ−)–Li δ+

1.74 Use the directions from Answer 1.37.

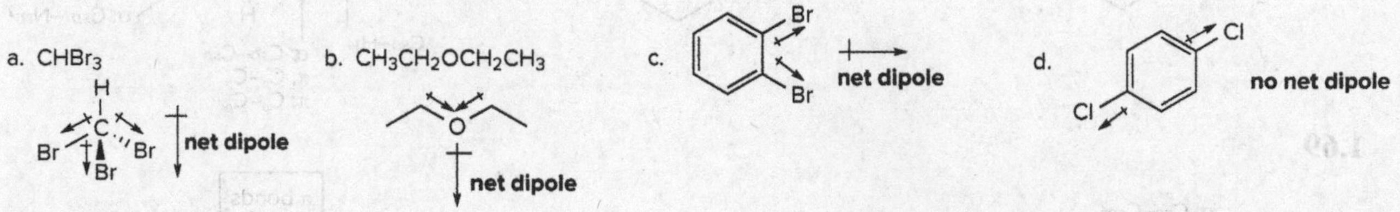

a. $CHBr_3$ — **net dipole**

b. $CH_3CH_2OCH_2CH_3$ — **net dipole**

c. (1,2-dibromobenzene) — **net dipole**

d. (1,4-dichlorobenzene) — **no net dipole**

1.75

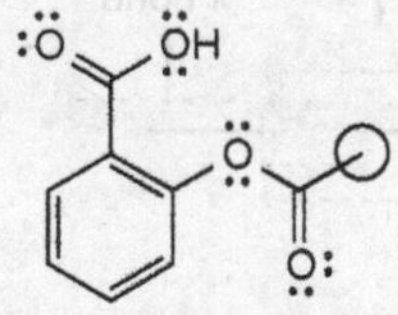

aspirin

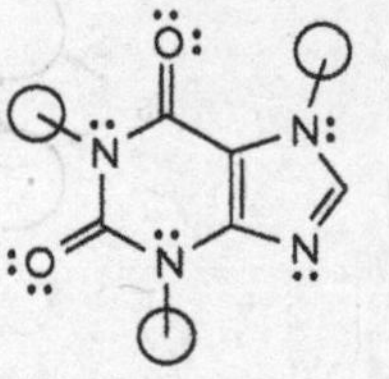

caffeine

a. molecular formula $C_9H_8O_4$

b. eight lone pairs total

c. C labeled with a circle is sp^3 hybridized. All other C's are sp^2 hybridized.

d. three possible resonance structures:

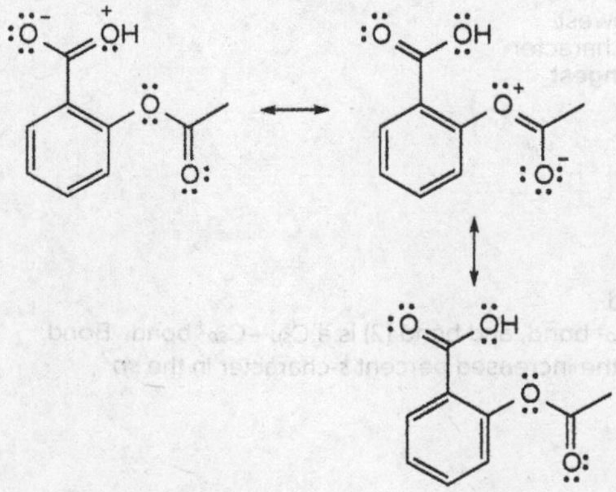

a. molecular formula $C_8H_{10}N_4O_2$

b. eight lone pairs total

c. C's labeled with a circle are sp^3 hybridized. All other C's are sp^2 hybridized.

d. three possible resonance structures:

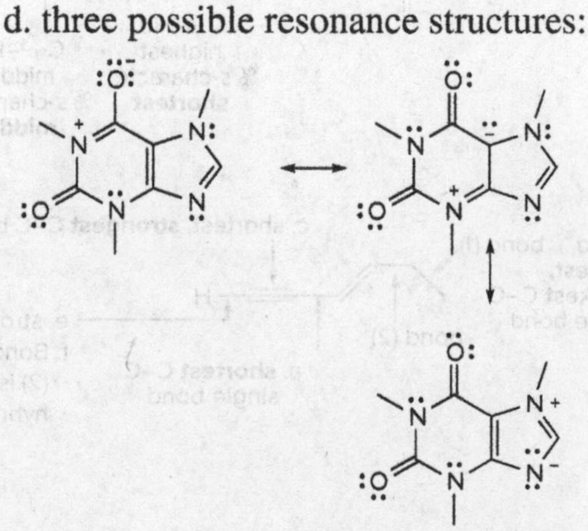

1.76

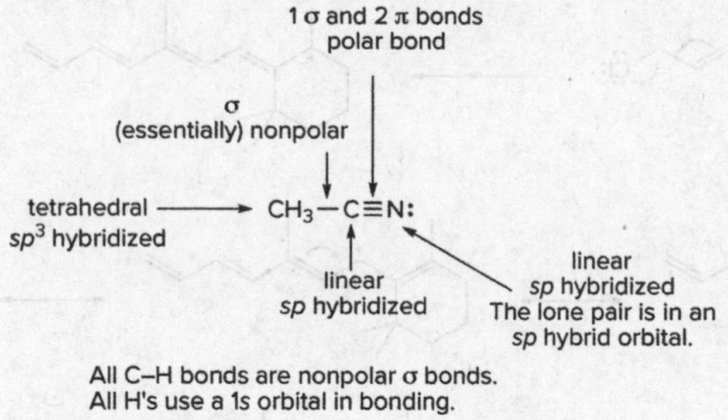

1.77

a. Each N has one lone pair; the S atom and uncharged O's have two lone pairs; negatively charged O's have three lone pairs. Total = 47 lone pairs.

b. H's have only two electrons around them and the P atoms each have 10 electrons, so these atoms do not follow the octet rule.

c, d. The hybridization of the highlighted atoms is shown, as well as the type of orbital that the indicated lone pairs occupy.

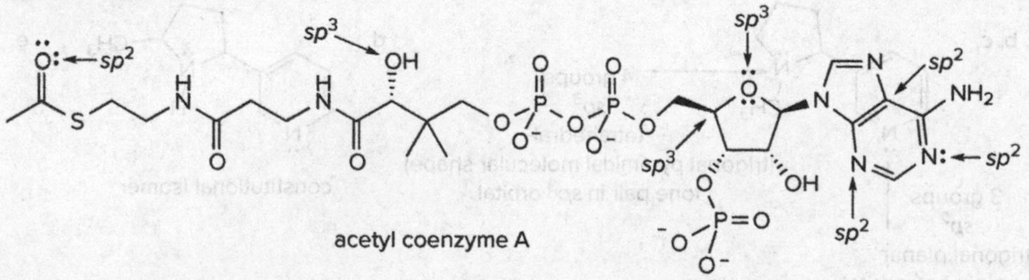

e. Three additional resonance structures:

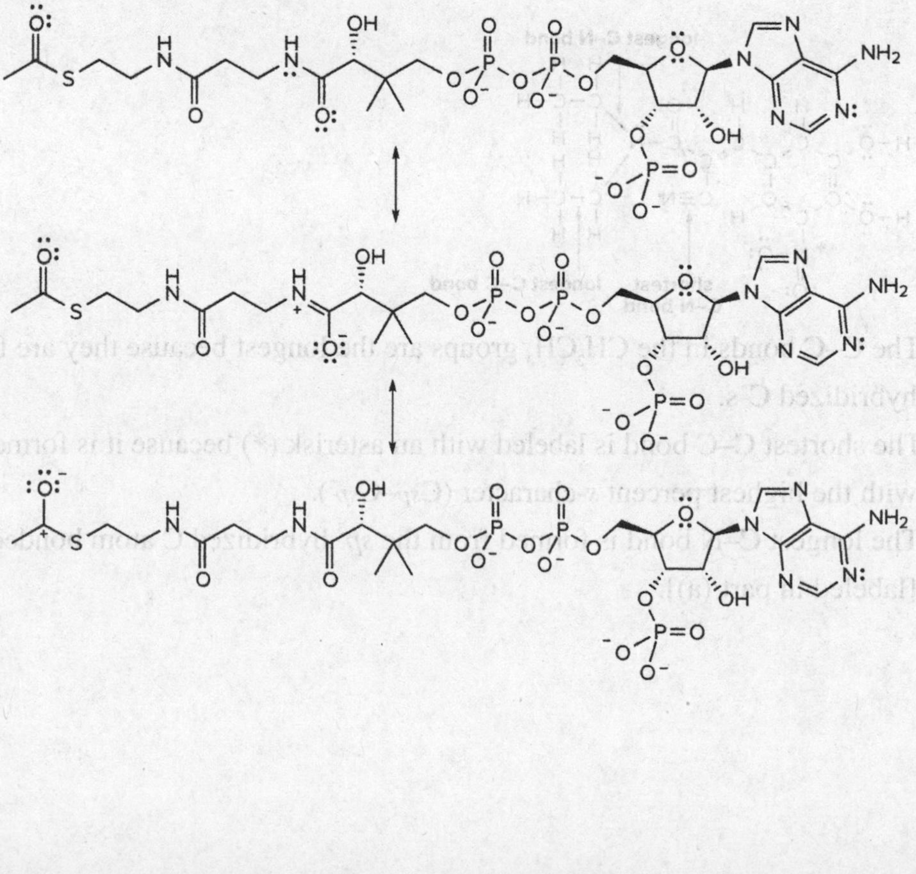

1.78

a. A → B + :Cl:⁻

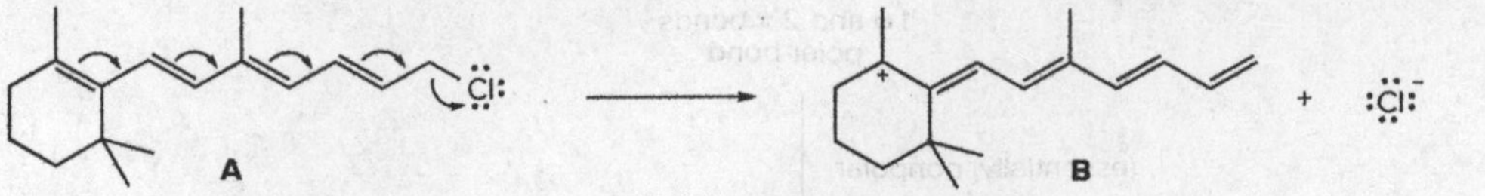

b. B (resonance structures)

hybrid: δ+ δ+ δ+ δ+ δ+

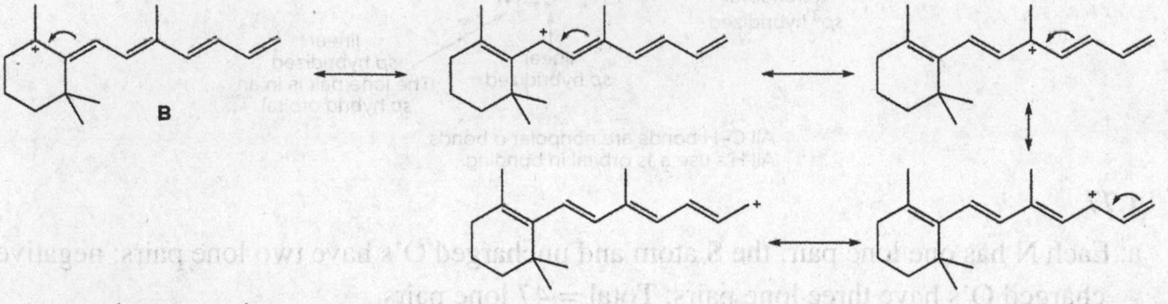

c.

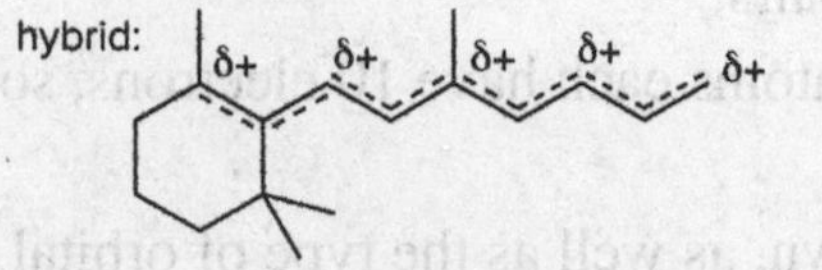

1.79

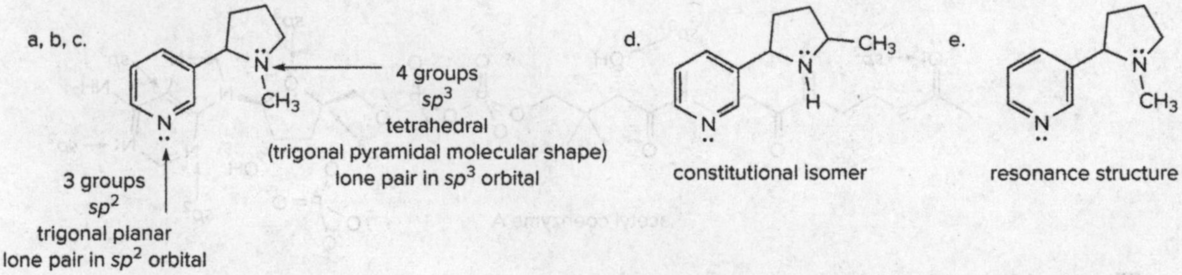

a, b, c.

4 groups
sp^3
tetrahedral
(trigonal pyramidal molecular shape)
lone pair in sp^3 orbital

3 groups
sp^2
trigonal planar
lone pair in sp^2 orbital

d. constitutional isomer

e. resonance structure

1.80 a.

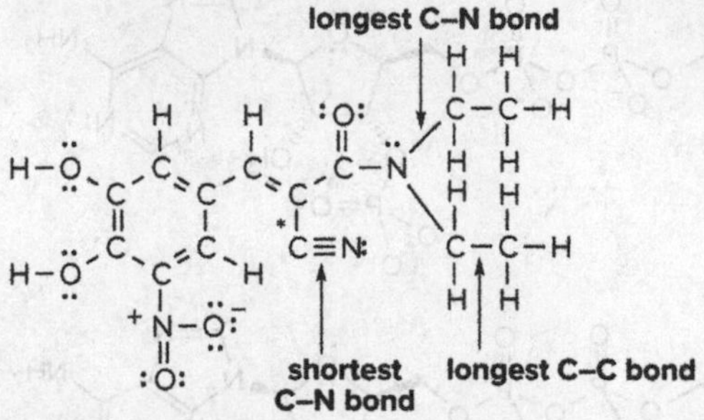

longest C–N bond

shortest C–N bond

longest C–C bond

b. The C–C bonds in the CH_2CH_3 groups are the longest because they are formed from sp^3 hybridized C's.

c. The shortest C–C bond is labeled with an asterisk (*) because it is formed from orbitals with the highest percent *s*-character (Csp–Csp^2).

d. The longest C–N bond is formed from the sp^3 hybridized C atom bonded to a N atom [labeled in part (a)].

e. The shortest C–N bond is the triple bond (C≡N); increasing the number of electrons between atoms decreases bond length.

f.

g.

1.81

$\overset{+}{C}H_3$ 3 groups
sp^2 trigonal planar
plot **A**

The blue region is evidence of the electron-poor cation.

$\bar{:}CH_3$ 4 groups
sp^3 tetrahedral
(The molecular shape is trigonal pyramidal.)
plot **B**

The red region is evidence of the electron-rich anion.

1.82 If the N atom is sp^2 hybridized, the lone pair occupies a *p* orbital, which can overlap with the π bond of the adjacent C=O. This allows electron density to delocalize, which is a stabilizing feature.

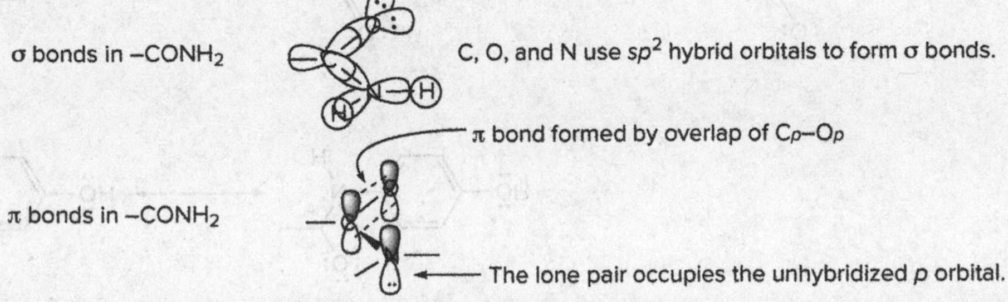
σ bonds in –CONH$_2$
C, O, and N use sp^2 hybrid orbitals to form σ bonds.
π bond formed by overlap of Cp–Op
π bonds in –CONH$_2$
The lone pair occupies the unhybridized *p* orbital.

1.83

a.

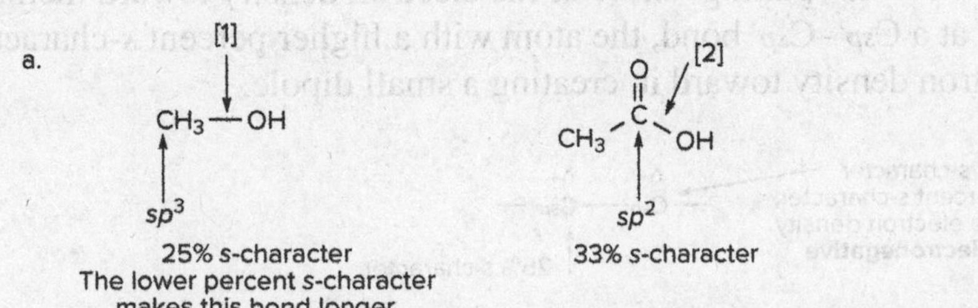
[1]
CH$_3$—OH
sp^3
25% *s*-character
The lower percent *s*-character makes this bond longer.
[2]
CH$_3$ C OH
sp^2
33% *s*-character

b.

[3]

[4]

two resonance structures

hybrid

Bonds [3] and [4] are both equivalent in length, because the anion is resonance stabilized, and the C–O bond of the hybrid is a composite of one single bond and one double bond. Both resonance structures contribute equally to the hybrid. Because each C–O bond in the hybrid has partial double bond character, it is shorter than the C–O bond labeled [2].

1.84 Ten additional resonance structures are drawn. (There are more possibilities.)

1.85 Polar bonds result from unequal sharing of electrons in covalent bonds. Normally we think of more electronegative atoms "pulling" more of the electron density toward them, making a dipole. In looking at a Csp^2–Csp^3 bond, the atom with a higher percent *s*-character will "pull" more of the electron density toward it, creating a small dipole.

33% *s*-character
higher percent *s*-character
pulls more electron density
more electronegative

δ– δ+
Csp^2 — Csp^3

25% *s*-character

1.86

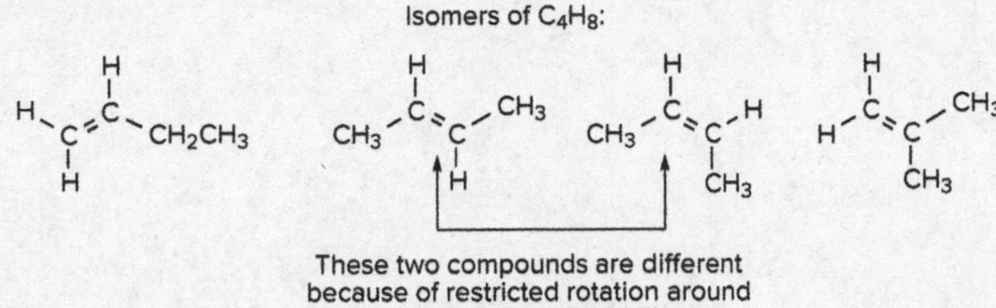

Isomers of C_4H_8:
These two compounds are different because of restricted rotation around the C=C (Section 8.2B).

1.87 Carbocation **A** is more stable than carbocation **B** because resonance distributes the positive charge over two carbons. Delocalizing electron density is stabilizing. **B** has no possibility of resonance delocalization.

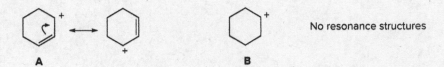

A

B

No resonance structures

1.88

a. [1]

b. [2] [3]

X

phenol

Chapter 2: Acids and Bases

Chapter Review

A comparison of Brønsted–Lowry and Lewis acids and bases

Type	Definition	Structural feature	Examples
Brønsted–Lowry acid (2.1)	proton donor	a proton	HCl, H_2SO_4, H_2O, CH_3CO_2H, TsOH
Brønsted–Lowry base (2.1)	proton acceptor	a lone pair or a π bond	^-OH, $^-OCH_3$, H^-, $^-NH_2$, $CH_2{=}CH_2$
Lewis acid (2.8)	electron pair acceptor	a proton, or an unfilled valence shell, or a partial (+) charge	BF_3, $AlCl_3$, HCl, CH_3CO_2H, H_2O
Lewis base (2.8)	electron pair donor	a lone pair or a π bond	^-OH, $^-OCH_3$, H^-, $^-NH_2$, $CH_2{=}CH_2$

Acid–base reactions

[1] A Brønsted–Lowry acid donates a proton to a Brønsted–Lowry base (2.2).

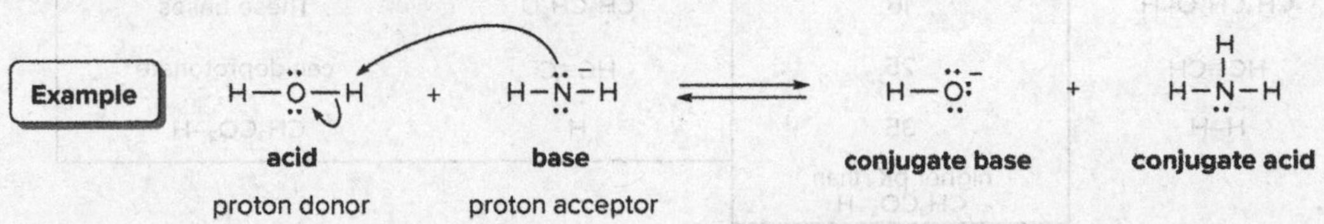

[2] A Lewis base donates an electron pair to a Lewis acid (2.8).

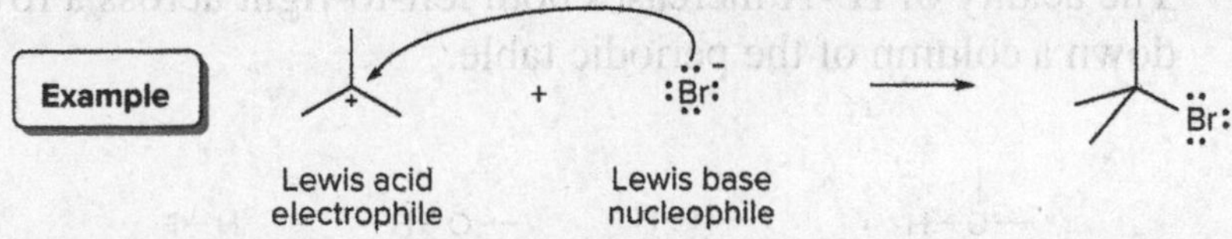

- Electron-rich species react with electron-poor ones.
- Nucleophiles react with electrophiles.

Important facts

- Definition: **pK_a = –log K_a.** The **lower the pK_a**, the **stronger** the acid (2.3).

NH_3	versus	H_2O
pK_a = 38		pK_a = 15.7
		lower pK_a = stronger acid

- The stronger the acid, the weaker the conjugate base (2.3).

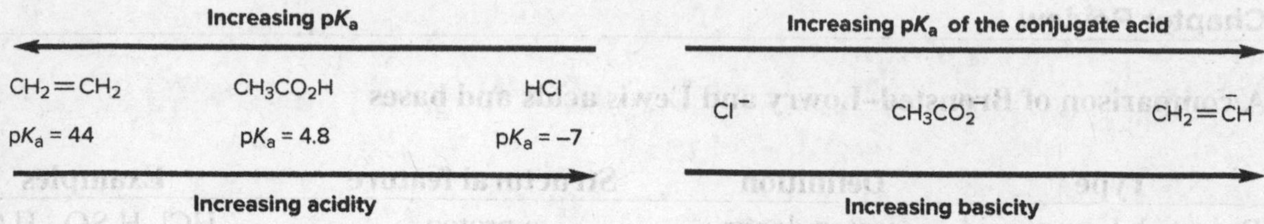

- In proton transfer reactions, equilibrium favors the weaker acid and the weaker base (2.4).

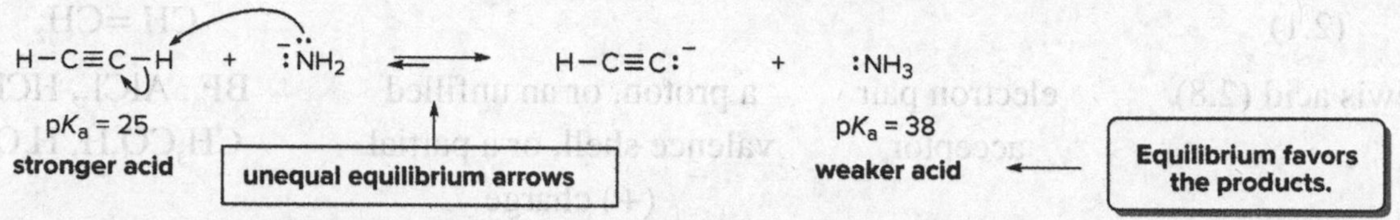

- An acid can be deprotonated by the conjugate base of any acid having a **higher pK_a** (2.4).

Acid	pK_a	Conjugate base	
CH_3CO_2–H	4.8	$CH_3CO_2^-$	
CH_3CH_2O–H	16	$CH_3CH_2O^-$	These bases
HC≡CH	25	HC≡C⁻	can deprotonate
H–H	35	H^-	CH_3CO_2–H.
	higher pK_a than CH_3CO_2–H		

Factors that determine acidity (2.5)

[1] **Element effects** (2.5A) The acidity of H–A increases both left-to-right across a row and down a column of the periodic table.

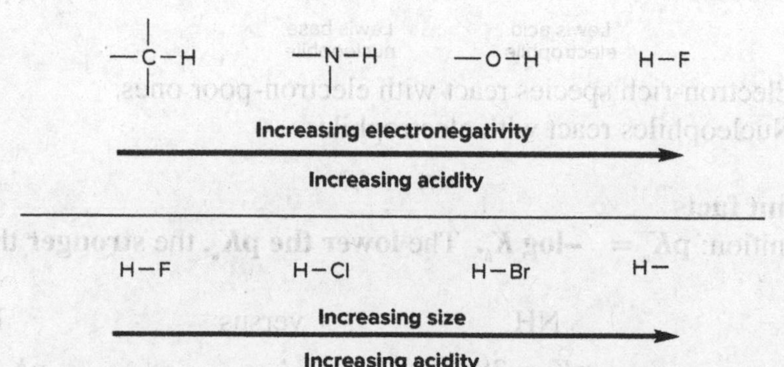

[2] **Inductive effects** (2.5B)	The acidity of H–A increases with the presence of electron-withdrawing groups in A.

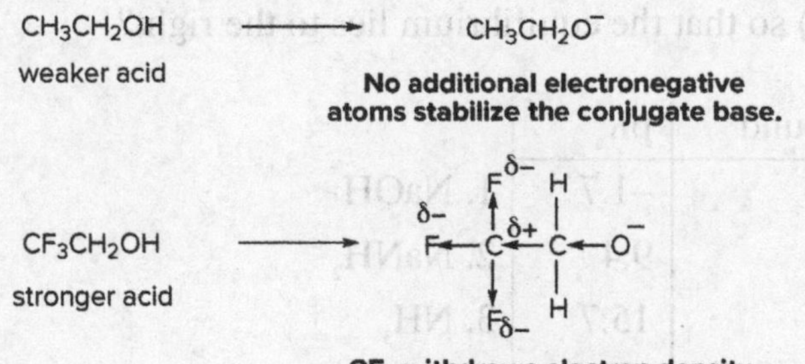

[3] **Resonance effects** (2.5C)	The acidity of H–A increases when the conjugate base A:⁻ is resonance stabilized.

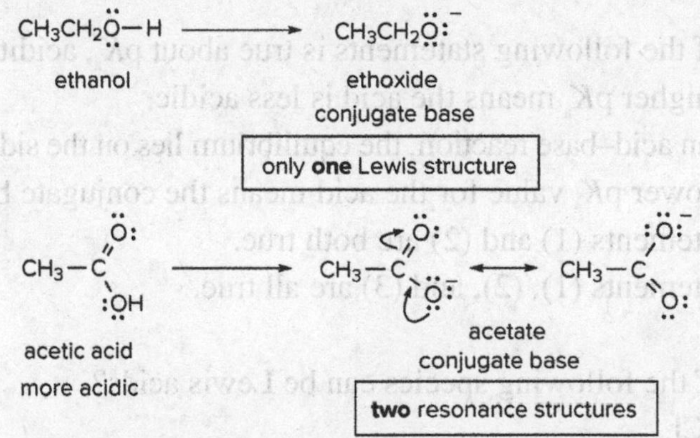

[4] **Hybridization effects** (2.5D)	The acidity of H–A increases as the percent *s*-character of the A:⁻ increases.

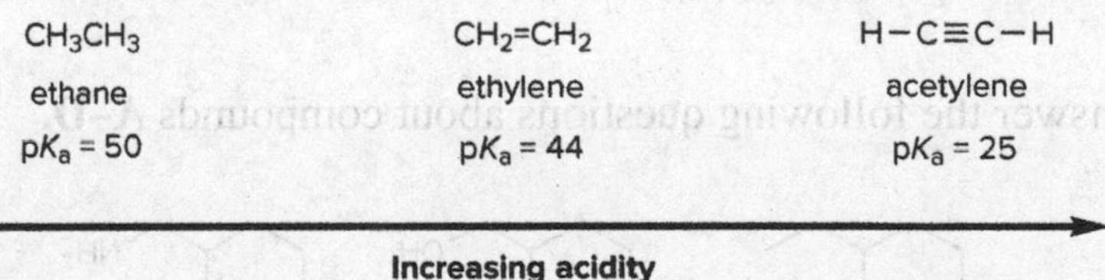

Practice Test on Chapter Review

1.a. Given the pK_a data, which of the following bases is strong enough to deprotonate C_6H_5OH (pK_a = 10) so that the equilibrium lies to the right?

Compound	pK_a
H_3O^+	–1.7
NH_4^+	9.4
H_2O	15.7
NH_3	38

1. NaOH
2. $NaNH_2$
3. NH_3
4. Compounds (1) and (2) are strong enough to deprotonate C_6H_5OH.
5. Compounds (1), (2), and (3) are all strong enough to deprotonate C_6H_5OH.

b. Which of the following statements is true about pK_a, acidity, and basicity?

1. A higher pK_a means the acid is less acidic.
2. In an acid–base reaction, the equilibrium lies on the side of the acid with the higher pK_a.
3. A lower pK_a value for the acid means the conjugate base is more basic.
4. Statements (1) and (2) are both true.
5. Statements (1), (2), and (3) are all true.

c. Which of the following species can be Lewis acids?

1. BCl_3
2. CH_3OH
3. $(CH_3)_3C^+$
4. Both (1) and (2) can be Lewis acids.
5. Species (1), (2), and (3) can all be Lewis acids.

2. Answer the following questions about compounds **A–D**.

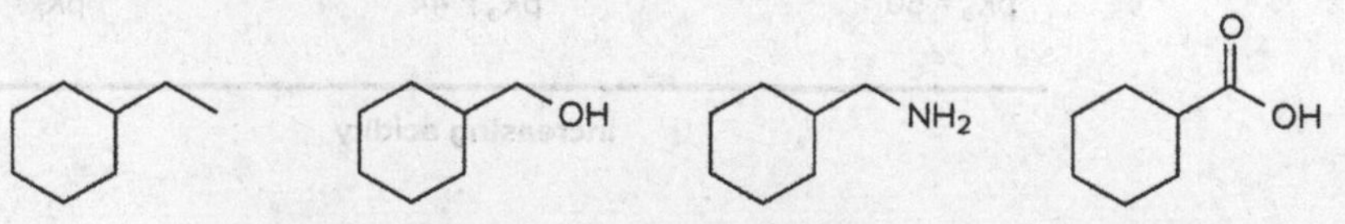

OH NH2 OH O

A B C D

a. Which compound is the strongest acid?
b. Which compound forms the strongest conjugate base?
c. The conjugate base of **C** is strong enough to remove a proton on which compound(s), so that the equilibrium favors the products?

3. (a) Which compound is the strongest Brønsted–Lowry acid? (b) Which compound is the weakest Brønsted–Lowry acid?

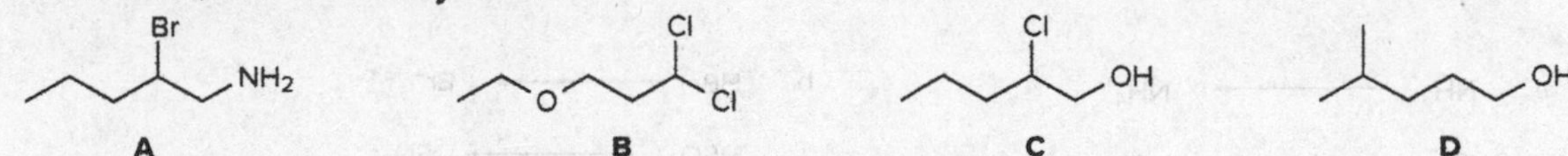

4. Draw all the products formed in the following reactions.

a. (benzene ring)–CH₂–CH(CH₃)–NHCH₃ + HBr ⟶ b. HO–(benzene ring)(Cl)(NO₂) + $Na^{+}\ {}^{-}NH_2$ ⟶

5. Draw the product(s) formed in the following Lewis acid–base reaction.

$(CH_3)_2\overset{+}{C}H$ + CH_3OH ⟶

Answers to Practice Test

1. a. 4 b. 4 c. 5

2. a. **D** b. **A** c. **B, D**

3. a. **C** b. **B**

4. a. (benzene ring)–CH₂–CH(CH₃)–$\overset{+}{N}H_2CH_3$ + Br^-

b. $Na^{+}\ {}^{-}O$–(benzene ring)(Cl)(NO_2) + NH_3

5. $(CH_3)_2CH$–$\overset{+}{\ddot{O}}$(H)–CH_3

Answers to Problems

2.1 Brønsted–Lowry acids are **proton donors** and must contain a hydrogen atom. **Brønsted–Lowry bases** are **proton acceptors** and must have an available electron pair (either a lone pair or a π bond).

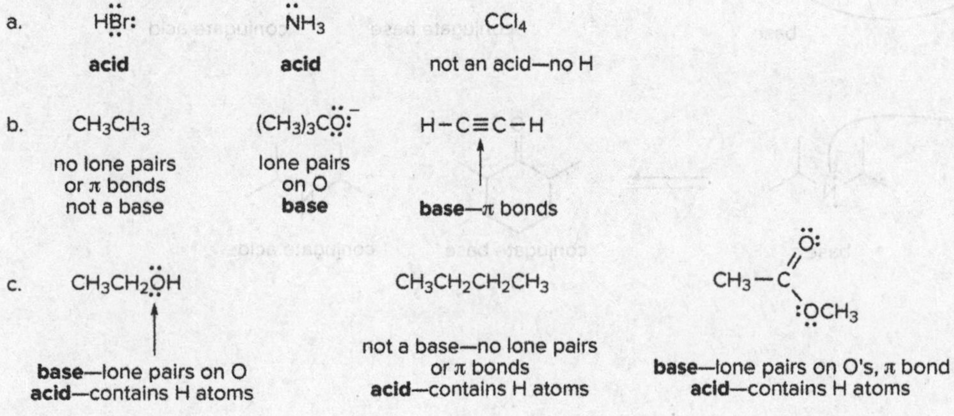

2.2 A Brønsted–Lowry base accepts a proton to form the conjugate acid. A Brønsted–Lowry acid loses a proton to form the conjugate base.

a. $NH_3 \longrightarrow NH_4^+$

$Cl^- \longrightarrow HCl$

$(CH_3)_2C{=}O \longrightarrow (CH_3)_2C{=}\overset{+}{O}H$

b. $HBr \longrightarrow Br^-$

$HSO_4^- \longrightarrow SO_4^{2-}$

$CH_3OH \longrightarrow CH_3O^-$

2.3 a. True.

$$\underset{\text{base}}{CH_2{=}CH_2} + H^+ \longrightarrow \underset{\text{conjugate acid}}{CH_3CH_2^+}$$

b. False. $CH_3CH_2^-$ cannot be the conjugate base of $CH_3CH_2^+$ because they both have the same number of H's and a conjugate base must have one fewer H.

c. False. $CH_2{=}CH_2$ and $CH_3CH_2^-$ differ by the presence of H^-, not H^+.

d. True.

$$\underset{\text{acid}}{CH_2{=}CH_2} \xrightarrow{\text{Remove } H^+} \underset{\text{conjugate base}}{CH_2{=}\overset{-}{C}H}$$

e. True.

$$\underset{\text{base}}{CH_3CH_2^-} + H^+ \longrightarrow \underset{\text{conjugate acid}}{CH_3CH_3}$$

2.4 The Brønsted–Lowry base accepts a proton to form the conjugate acid. The Brønsted–Lowry acid loses a proton to form the conjugate base. Use curved arrows to show the movement of electrons (***NOT protons***). Re-draw the starting materials if necessary to clarify the electron movement.

a. H–Cl: (acid) + diethyl ether (base) ⇌ :Cl:⁻ (conjugate base) + protonated diethyl ether (conjugate acid)

b. ketone with α-H (acid) + ethoxide (base) ⇌ enolate (conjugate base) + ethanol (conjugate acid)

c. 2-methylcyclohexanone α-H (acid) + diisopropylamide (base) ⇌ enolate (conjugate base) + diisopropylamine (conjugate acid)

2.5 To draw the products:
[1] Find the acid and base.
[2] Transfer a proton from the acid to the base.
[3] Check that the charges on each side of the arrows are balanced.

a. $Cl_3C\text{–}COOH$ (acid) + $^{-}OCH_3$ (base) ⇌ $Cl_3C\text{–}COO^{-}$ + $HO\text{–}CH_3$ (–)1 charge on each side

b. $Cl_3C\text{–}C{\equiv}C\text{–}H$ (acid) + H^{-} (base) ⇌ $Cl_3C\text{–}C{\equiv}C^{-}$ + H_2 (–)1 charge on each side

c. cyclopentyl–NH_2 (base) + $H\text{–}Cl$ (acid) ⇌ cyclopentyl–NH_3^{+} + Cl^{-} net neutral on each side

d. $CH_3CH_2CH_2OH$ (base) + $H\text{–}OSO_3H$ (acid) ⇌ $CH_3CH_2CH_2OH_2^{+}$ + $^{-}OSO_3H$ net neutral on each side

2.6 Draw the products in each reaction as in Answer 2.4.

a. alcohol (OH) + HCl → protonated alcohol ($\overset{+}{O}H_2$) + Cl^-

b. $CH_3CH_2OCH_2CH_3$ + HCl → $CH_3CH_2\overset{+}{O}(H)CH_2CH_3$ + Cl^-

c. amine (N) + HCl → protonated amine ($\overset{+}{N}$–H) + Cl^-

d. pyrrolidine (NH) + HCl → pyrrolidinium ($\overset{+}{N}H_2$) + Cl^-

2.7 The smaller the pK_a, the stronger the acid. The larger the K_a, the stronger the acid.

a. $CH_3CH_2CH_3$ or CH_3CH_2OH

$CH_3CH_2CH_3$: pK_a = 50

CH_3CH_2OH: pK_a = 16, smaller pK_a, **stronger acid**

b. phenol (OH) or toluene (CH_3)

phenol: $K_a = 10^{-10}$, larger K_a, **stronger acid**

toluene: $K_a = 10^{-41}$

2.8 To convert from K_a to pK_a, take (–) the log of the K_a; **pK_a = –log K_a**.
To convert pK_a to K_a, take the antilog of (–) the pK_a.

a. $K_a = 10^{-10}$ → pK_a = 10; $K_a = 10^{-21}$ → pK_a = 21; $K_a = 5.2 \times 10^{-5}$ → pK_a = 4.3

b. pK_a = 7 → $K_a = 10^{-7}$; pK_a = 11 → $K_a = 10^{-11}$; pK_a = 3.2 → $K_a = 6.3 \times 10^{-4}$

2.9 Because **strong acids form weak conjugate bases**, the basicity of conjugate bases increases with increasing pK_a of their acids. Find the pK_a of each acid from Table 2.1 and then rank the acids in order of increasing pK_a. This will also be the order of increasing basicity of their conjugate bases.

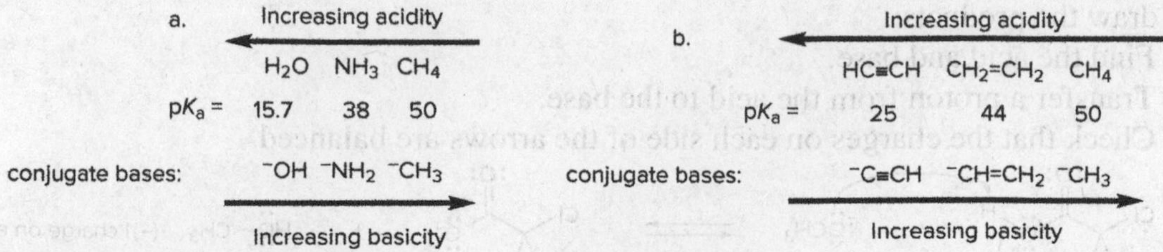

2.10 Use the definitions in Answer 2.9 to compare the acids. The smaller the pK_a, the larger the K_a and the stronger the acid. When a stronger acid dissolves in water, the equilibrium lies further to the right.

HCO_2H	$(CH_3)_3CCO_2H$
formic acid	**pivalic acid**
pK_a = 3.8	pK_a = 5.0

a. smaller pK_a = larger K_a
b. smaller pK_a = stronger acid
c. weaker acid = stronger conjugate base
d. stronger acid = equilibrium further to the right

2.11 To estimate the pK_a of the indicated bond, find a similar bond in the pK_a table (H bonded to the same atom with the same hybridization).

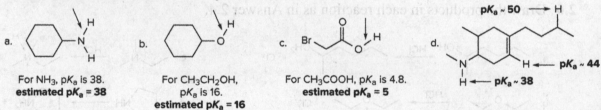

a. For NH_3, pK_a is 38.
estimated pK_a = 38

b. For CH_3CH_2OH, pK_a is 16.
estimated pK_a = 16

c. For CH_3COOH, pK_a is 4.8.
estimated pK_a = 5

d. **pK_a ~ 50**; **pK_a ~ 44**; **pK_a ~ 38**

2.12 Label the acid and the base and then transfer a proton from the acid to the base. To determine if the reaction will proceed as written, compare the pK_a of the acid on the left with the conjugate acid on the right. **The equilibrium always favors the formation of the weaker acid and the weaker base.**

a. $CH_2=CH_2$ (acid, pK_a = 44, **weaker acid**) + $H:^-$ (base) ⇌ $CH_2=\ddot{C}H^-$ (conjugate base) + H_2 (conjugate acid, pK_a = 35)
Equilibrium favors the **starting materials.**

b. CH_4 (acid, pK_a = 50, **weaker acid**) + $:\ddot{O}H^-$ (base) ⇌ $:CH_3^-$ (conjugate base) + $H_2\ddot{O}:$ (conjugate acid, pK_a = 15.7)
Equilibrium favors the **starting materials.**

c. CH_3CO_2H (acid, pK_a = 4.8) + $CH_3CH_2\ddot{O}:^-$ (base) ⇌ $CH_3CO_2^-$ (conjugate base) + $CH_3CH_2\ddot{O}H$ (conjugate acid, pK_a = 16, **weaker acid**)
Equilibrium favors the **products.**

d. $:\ddot{Cl}:^-$ (base) + $CH_3CH_2\ddot{O}H$ (acid, pK_a = 16, **weaker acid**) ⇌ $H\ddot{Cl}:$ (conjugate acid, pK_a = −7) + $CH_3CH_2\ddot{O}:^-$ (conjugate base)
Equilibrium favors the **starting materials.**

2.13 An acid can be deprotonated by the conjugate base of any acid with a higher pK_a.

CH_3CN
pK_a = 25
Any base having a conjugate acid with a pK_a higher than 25 can deprotonate this acid.

Base	Conjugate acid	pK_a
NaH	H_2	35
Na_2CO_3	HCO_3^-	10.2
NaOH	H_2O	15.7
$NaNH_2$	NH_3	38
$NaHCO_3$	H_2CO_3	6.4

Only NaH and $NaNH_2$ are strong enough to deprotonate acetonitrile.

2.14 The acidity of H–Z **increases left-to-right across a row and down a column** of the periodic table.

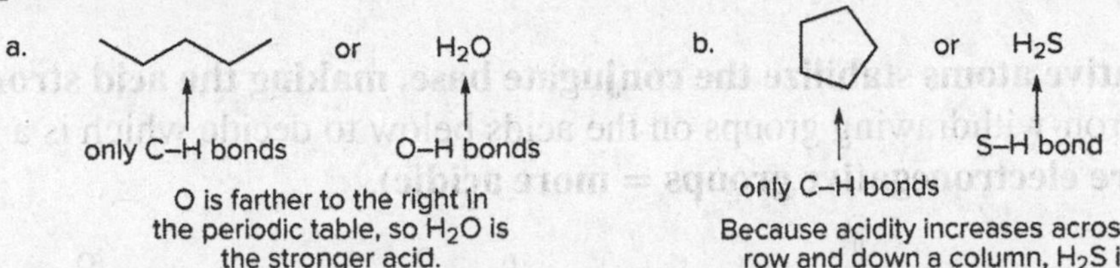

2.15

Because acidity increases across a row and down a column, the order of acidity is N–H < O–H < S–H.

2.16 Look at the element bonded to the acidic H and decide its acidity based on the periodic trends. **Farther to the right and down the periodic table is more acidic.**

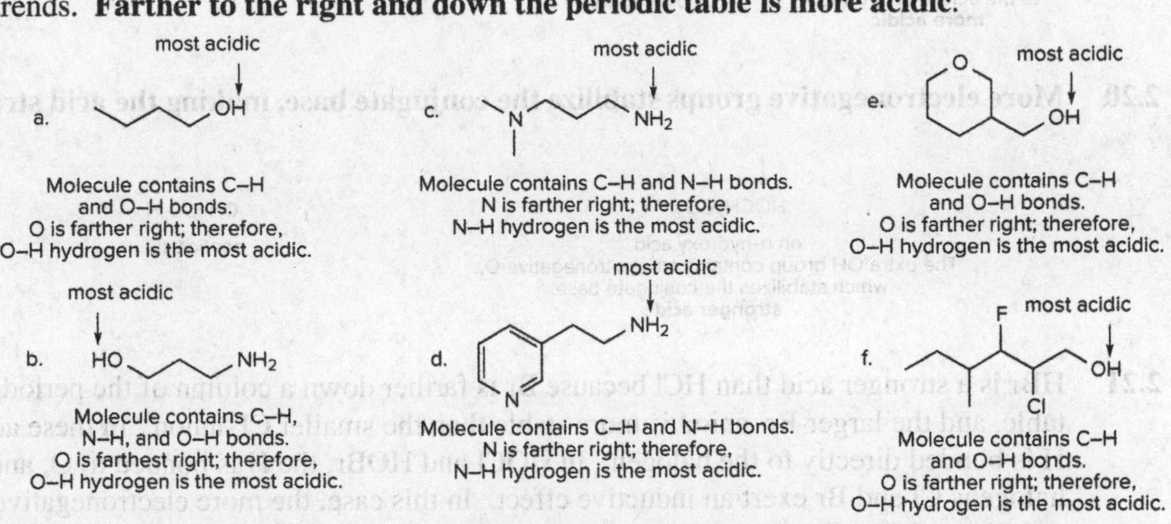

2.17 The acidity of HA increases left-to-right across the periodic table. Pseudoephedrine contains C–H, N–H, and O–H bonds. The O–H bond is most acidic.

pseudoephedrine skeletal structure

2.18 Compare the most acidic protons in each molecule to rank the compounds.

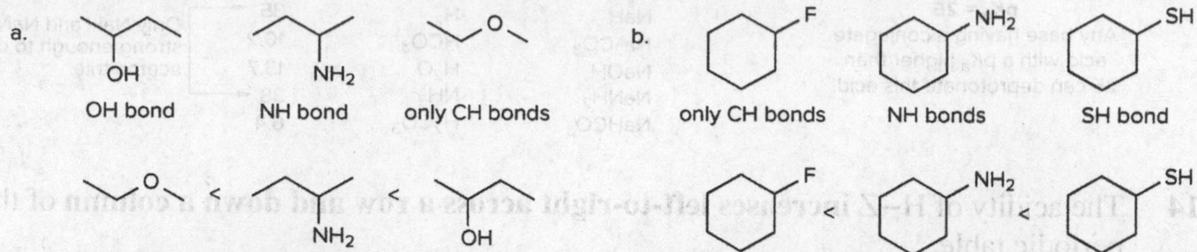

2.19 **More electronegative atoms stabilize the conjugate base, making the acid stronger.** Compare the electron-withdrawing groups on the acids below to decide which is a stronger acid **(more electronegative groups = more acidic).**

a. (Cl-substituted cyclopentanecarboxylic acid) or (F-substituted cyclopentanecarboxylic acid) — more acidic

F is more electronegative than Cl, making the O–H bond in the acid on the right **more acidic.**

b. (Cl on carbon closer to OH) or (Cl farther from OH)

Cl is closer to the acidic O–H bond. **more acidic**

Cl is farther from the O–H bond.

c. (3-methylbutanoic acid) or (NO_2-substituted acid) — more acidic

NO_2 is electron withdrawing, making the O–H bond in the acid on the right **more acidic.**

2.20 **More electronegative groups stabilize the conjugate base, making the acid stronger.**

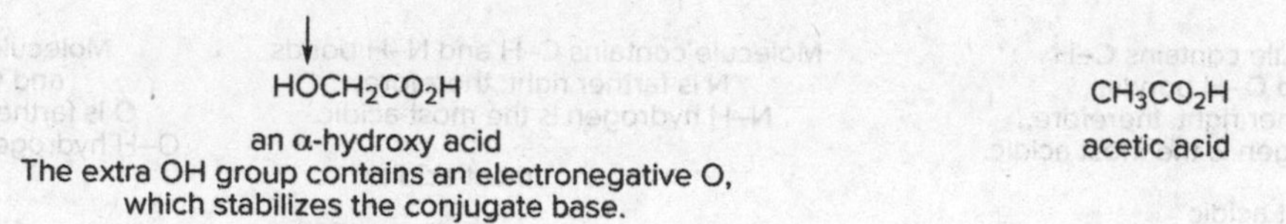

2.21 HBr is a stronger acid than HCl because Br is farther down a column of the periodic table, and the larger Br^- anion is more stable than the smaller Cl^- anion. In these acids the H is bonded directly to the halogen. In HOCl and HOBr, the H is bonded to O, and the halogens Cl and Br exert an inductive effect. In this case, the more electronegative Cl stabilizes ^-OCl more than the less electronegative Br stabilizes ^-OBr. Thus, HOCl forms the more stable conjugate base, making it the stronger acid.

2.22 The acidity of an acid increases when the conjugate base is resonance stabilized. Compare the conjugate bases of acetone and propane to explain why acetone is more acidic.

acetone
pK_a = 19.2

base

One resonance structure places the (–) charge on the more electronegative O atom. This is especially good.

2 resonance structures
more stable conjugate base
Acetone is more acidic.

$CH_3CH_2CH_3$ —base→ $CH_3CH_2\ddot{C}H_2$

propane
pK_a = 50

only one Lewis structure
less stable conjugate base

(Any C–H bond in the starting material can be removed.)

2.23 The acidity of HX bonds increases across a row of the periodic table, so C–H < N–H < O–H. Acidity also increases with resonance stabilization of a conjugate base.

H_aO OH_c H_b

loss of H_a

least acidic C–H bond
highest pK_a

loss of H_c

negative charge on O
not resonance stabilized

negative charge on O
resonance stabilized
most acidic
lowest pK_a

In order of increasing pK_a (that is, decreasing acidity):
$H_c < H_a < H_b$

2.24 **Increasing percent *s*-character makes an acid more acidic.** Compare the percent *s*-character of the carbon atoms in each of the C–H bonds in question. A stronger acid has a weaker conjugate base.

a.

H on *sp* hybridized C
50% *s*-character
more acidic

or

H on sp^3 hybridized C
25% *s*-character

base

base

or

stronger conjugate base

b.

H on sp^2 hybridized C
33% *s*-character
more acidic

or

H on sp^3 hybridized C
25% *s*-character

base

base

or

stronger conjugate base

2.25 To compare the acids, first **look for element effects.** Then identify electron-withdrawing groups, resonance, or hybridization differences.

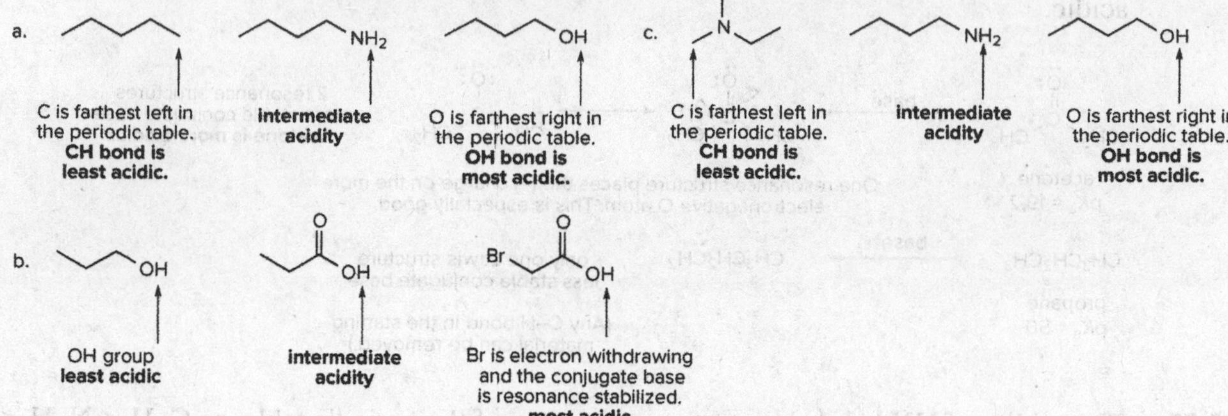
a.
NH2
OH
C is farthest left in the periodic table. **CH bond is least acidic.**
intermediate acidity
O is farthest right in the periodic table. **OH bond is most acidic.**
c.
N
NH2
OH
C is farthest left in the periodic table. **CH bond is least acidic.**
intermediate acidity
O is farthest right in the periodic table. **OH bond is most acidic.**
b.
OH
O
OH
Br
O
OH
OH group **least acidic**
intermediate acidity
Br is electron withdrawing and the conjugate base is resonance stabilized. **most acidic**

2.26

:O:
:Ö:⁻

A is resonance stabilized so A is the weaker base.

A

:O:
H
:Ö:⁻

The negative charge is localized on O in B, so **B** is the stronger base.

B

2.27 Look at the element bonded to the acidic H and decide its acidity based on the periodic trends. **Farther to the right and down the periodic table is more acidic.**

a.

most acidic

OH

O

THC
tetrahydrocannabinol

The molecule contains C–H and O–H bonds. O is farther right; therefore, O–H hydrogen is the most acidic.

b.

O

most acidic

COOH

ketoprofen

The molecule contains C–H and O–H bonds. O is farther right; therefore, O–H hydrogen is the most acidic.

2.28 Draw the products of proton transfer from the acid to the base.

a. $(CH_3)_2CH\ddot{O}{-}H$ (acid) + $Na^+\ H{:}^-$ (base) $\rightleftharpoons$ $(CH_3)_2CH\ddot{O}{:}^-\ Na^+$ (conjugate base) + H_2 (conjugate acid)

b. $(CH_3)_2CH\ddot{O}{-}H$ (base) + $H{-}OSO_3H$ (acid) $\rightleftharpoons$ $(CH_3)_2CH\overset{+}{\ddot{O}}H_2$ (conjugate acid) + HSO_4^- (conjugate base)

c. $(CH_3)_2CH\ddot{O}{-}H$ (acid) + $Li^+\ {}^-\ddot{N}[CH(CH_3)_2]_2$ (base) $\rightleftharpoons$ $(CH_3)_2CH\ddot{O}{:}^-\ Li^+$ (conjugate base) + $H\ddot{N}[CH(CH_3)_2]_2$ (conjugate acid)

d. $(CH_3)_2CH\ddot{O}-H$ + $H-OCOCH_3$ ⇌ $(CH_3)_2CH\overset{+}{\ddot{O}}H_2$ + $^{-}OCOCH_3$

base — acid — conjugate acid — conjugate base

2.29 To cross a cell membrane, amphetamine must be in its neutral (not ionic) form.

amphetamine ($:NH_2$) → protonation by HCl in the stomach → ($^+NH_3$) → deprotonation in the intestines → ($:NH_2$) absorption here in the neutral form

2.30 **Lewis bases are electron pair donors**: they contain a lone pair or a π bond.

a. $\ddot{N}H_3$ — yes—has lone pair

b. $CH_3CH_2CH_3$ — no—no lone pair or π bond

c. $H:^-$ — yes—has lone pair

d. $H-C\equiv C-H$ — yes—has 2 π bonds

2.31 **Lewis acids are electron pair acceptors.** Most Lewis acids contain a proton or an unfilled valence shell of electrons.

a. BBr_3 — yes — unfilled valence shell on B

b. CH_3CH_2OH — yes — contains a proton

c. $(CH_3)_3C^+$ — yes — unfilled valence shell on C

d. Br^- — no — no proton — no unfilled valence shell

2.32 Label the Lewis acid and Lewis base and then draw the curved arrows.

a. BF_3 + $CH_3\ddot{O}CH_3$ → $F_3\bar{B}-\overset{+}{O}(CH_3)_2$ (new bond)

Lewis acid — unfilled valence shell on B

Lewis base — lone pairs on O

b. carbocation + $:\ddot{O}H^-$ → alcohol

Lewis acid — unfilled valence shell on C

Lewis base — lone pairs on O

2.33 A Lewis acid is also called an **electrophile.** When a Lewis base reacts with an electrophile other than a proton, it is called a **nucleophile.** Label the electrophile and nucleophile in the starting materials and then draw the products.

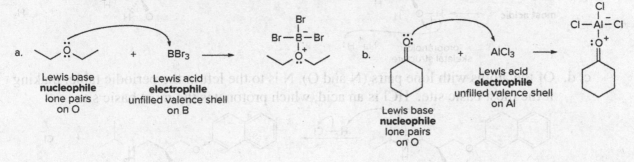

a. diethyl ether + BBr_3 → $Br_3\bar{B}-\overset{+}{O}Et_2$

Lewis base — **nucleophile** — lone pairs on O

Lewis acid — **electrophile** — unfilled valence shell on B

b. cyclohexanone + $AlCl_3$ → $Cl_3\bar{Al}-\overset{+}{O}$=cyclohexane

Lewis acid — **electrophile** — unfilled valence shell on Al

Lewis base — **nucleophile** — lone pairs on O

2.34 Draw the product of each reaction by using an electron pair of the Lewis base to form a new bond to the Lewis acid.

a. $CH_3CH_2-\ddot{N}(CH_2CH_3)-CH_2CH_3$ + $B(CH_3)_3$ ⟶ $CH_3CH_2-\overset{+}{N}(CH_2CH_3)(CH_2CH_3)-\overset{-}{B}(CH_3)_3$

Lewis base **nucleophile** lone pair on N

Lewis acid **electrophile** unfilled valence shell on B

b. $CH_3CH_2-\ddot{N}(CH_2CH_3)-CH_2CH_3$ + $^+C(CH_3)_3$ ⟶ $CH_3CH_2-\overset{+}{N}(CH_2CH_3)(CH_2CH_3)-C(CH_3)_3$

Lewis base **nucleophile** lone pair on N

Lewis acid **electrophile** unfilled valence shell on C

c. $CH_3CH_2-\ddot{N}(CH_2CH_3)-CH_2CH_3$ + $AlCl_3$ ⟶ $CH_3CH_2-\overset{+}{N}(CH_2CH_3)(CH_2CH_3)-\overset{-}{Al}Cl_3$

Lewis base **nucleophile** lone pair on N

Lewis acid **electrophile** unfilled valence shell on Al

2.35 A curved arrow begins at the Lewis base and points toward the Lewis acid.

:O: (ketone) + H_3O^+ ⟶ :OH (protonated ketone) + $H_2\ddot{O}$

Lewis base contains a lone pair

Lewis acid contains a proton

2.36 a, b. Because acidity increases from left-to-right across a row of the periodic table and propranolol has C–H, N–H, and O–H bonds, the O–H bond is most acidic. NaH is a base that removes the most acidic OH proton.

most acidic ⟶ H–O

propranolol skeletal structure + $Na^+ H:^-$ ⟶ ^-O (alkoxide) + Na^+ + H_2

c, d. Of the atoms with lone pairs (N and O), N is to the left in the periodic table, making it the most basic site. HCl is an acid, which protonates the most basic site.

propranolol (HO, N–H) + H–Cl ⟶ $\overset{+}{N}H_2$ (protonated propranolol) + Cl^-

2.37 a, b. Using periodic trends, the N–H bond of amphetamine is most acidic. NaH is a base that removes an acidic proton on N, forming H_2 in the process.

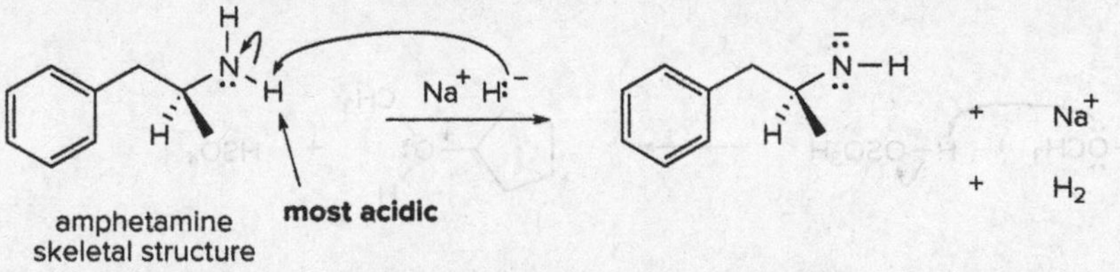

amphetamine skeletal structure

most acidic

Na^+ $H:^-$

+ Na^+

+ H_2

c. HCl protonates the lone pair on N (the most basic site).

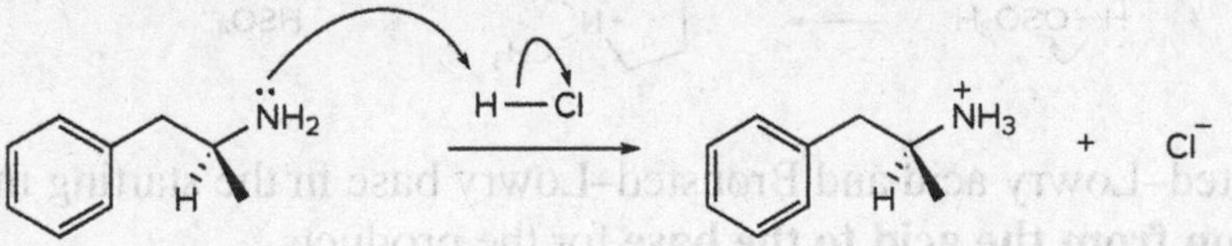

$\ddot{N}H_2$ H—Cl → $\overset{+}{N}H_3$ + Cl^-

2.38 To draw the conjugate acid of a Brønsted–Lowry base, **add a proton to the base.**

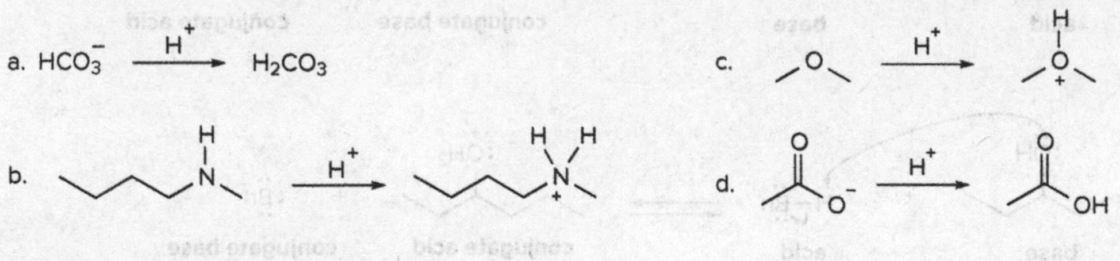

a. HCO_3^- $\xrightarrow{H^+}$ H_2CO_3

b. $\xrightarrow{H^+}$

c. $\xrightarrow{H^+}$

d. $\xrightarrow{H^+}$ OH

2.39 To draw the conjugate base of a Brønsted–Lowry acid, **remove a proton from the acid.**

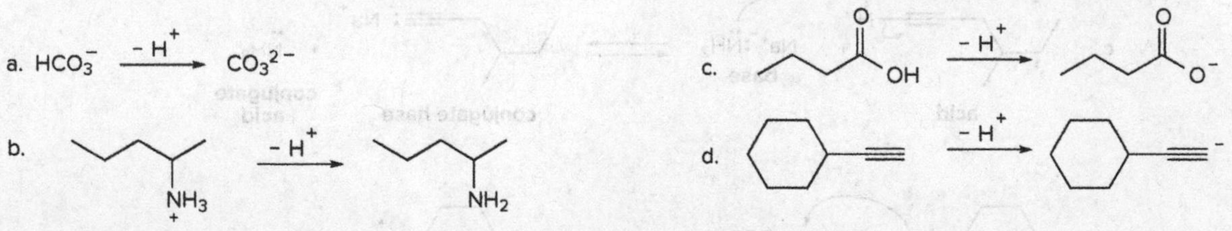

a. HCO_3^- $\xrightarrow{-H^+}$ CO_3^{2-}

b. NH_3^+ $\xrightarrow{-H^+}$ NH_2

c. OH $\xrightarrow{-H^+}$ O^-

d. $\xrightarrow{-H^+}$

2.40

a. A $\xrightarrow{[1]}$ B $\xrightarrow{[2]}$ C $\xrightarrow{[3]}$ D + H_3O^+

$H-\overset{+}{O}H_2$; + $H_2\ddot{O}$:; + $H_2\ddot{O}$:

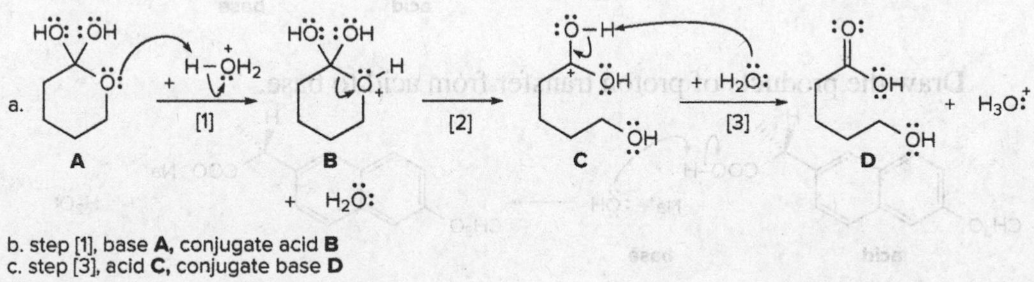

b. step [1], base **A**, conjugate acid **B**
c. step [3], acid **C**, conjugate base **D**

2.41 To draw the products of an acid–base reaction, transfer a proton from the acid (H_2SO_4 in this case) to the base.

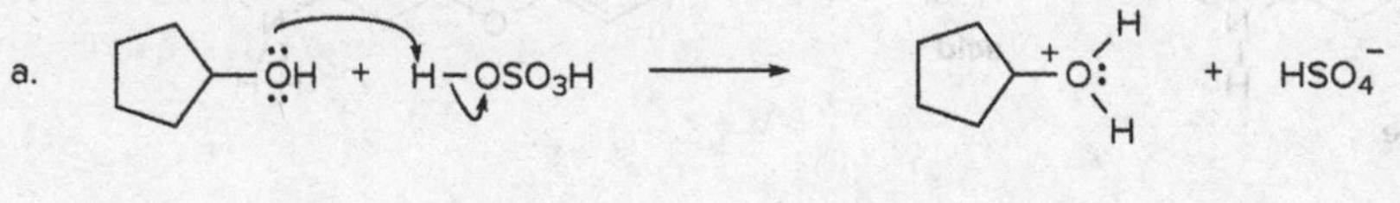

a. $\ddot{O}H$ + $H-OSO_3H$ ⟶ $\overset{+}{O}H_2$ + HSO_4^-

b. Cyclopentyl–$\ddot{N}H_2$ + $H-OSO_3H$ ⟶ Cyclopentyl–$\overset{+}{N}H_3$ + HSO_4^-

c. Cyclopentyl–$\ddot{O}CH_3$ + $H-OSO_3H$ ⟶ Cyclopentyl–$\overset{+}{O}(CH_3)H$ + HSO_4^-

d. N-methylpyrrolidine ($\ddot{N}-CH_3$) + $H-OSO_3H$ ⟶ $\overset{+}{N}(H)CH_3$ + HSO_4^-

2.42 Label the Brønsted–Lowry acid and Brønsted–Lowry base in the starting materials and **transfer a proton from the acid to the base** for the products.

a. carboxylic acid (O–H) + $CH_3\ddot{O}:^-$ ⇌ carboxylate ($\ddot{O}:^-$) + $CH_3\ddot{O}H$

acid, base, conjugate base, conjugate acid

b. $:\ddot{O}H$ (alcohol) + $H-\ddot{B}r:$ ⇌ $:\overset{+}{O}H_2$ + $:\ddot{B}r:^-$

base, acid, conjugate acid, conjugate base

c. terminal alkyne (≡–H) + $Na^+\ :\ddot{N}H_2^-$ ⇌ alkynide (≡: Na^+) + $\ddot{N}H_3$

acid, base, conjugate base, conjugate acid

d. cyclohexanone (=$\ddot{O}$:) + $H-OSO_3H$ ⇌ (=$\overset{+}{\ddot{O}}H$) + HSO_4^-

base, acid, conjugate acid, conjugate base

2.43 Draw the products of proton transfer from acid to base.

a. CH_3O–naphthalene–CH(H)–COO–H + $Na^+\ :\ddot{O}H^-$ ⟶ CH_3O–naphthalene–CH(H)–$COO^-\ Na^+$ + $H_2\ddot{O}:$

acid, base

b. CF_3–C₆H₄–O–CH(Ph)–CH₂CH₂–$\ddot{N}(H)$–CH_3 + $H-Cl$ ⟶ CF_3–C₆H₄–O–CH(Ph)–CH₂CH₂–$\overset{+}{N}(H)(H)$–CH_3 + Cl^-

base, acid

2.44 NaH removes the most acidic H.

a. [thiophene compound with OH] — OH: most acidic H; NaH → product with $:\ddot{O}:^-$

b. [N–H amide compound] — N–H: most acidic H; NaH → product with $:\ddot{N}^-$

c. [H_2N thiophene ester] — H_2N: most acidic H; NaH → product with $H\ddot{N}^-$

2.45 To convert pK_a to K_a, take the antilog of (–) the pK_a.

a. H_2S $pK_a = 7.0$ $K_a = 10^{-7}$

b. $ClCH_2COOH$ $pK_a = 2.8$ $K_a = 1.6 \times 10^{-3}$

c. HCN $pK_a = 9.1$ $K_a = 7.9 \times 10^{-10}$

2.46 To convert from K_a to pK_a, take (–) the log of the K_a; $\mathbf{pK_a = -log\ K_a}$.

a. $C_6H_5CH_2\overset{+}{N}H_3$ $K_a = 4.7 \times 10^{-10}$ $pK_a = 9.3$

b. $C_6H_5\overset{+}{N}H_3$ $K_a = 2.3 \times 10^{-5}$ $pK_a = 4.6$

c. CF_3COOH $K_a = 5.9 \times 10^{-1}$ $pK_a = 0.23$

2.47 An acid can be deprotonated by the conjugate base of any acid with a higher pK_a.

C_6H_5OH
pK_a = 10

Base	Conjugate acid	pK_a	Deprotonate?
H_2O	H_3O^+	–1.7	no
NaOH	H_2O	15.7	yes
$NaNH_2$	NH_3	38	yes
CH_3NH_2	$CH_3NH_3^+$	10.7	yes (low concentration)
$NaHCO_3$	H_2CO_3	6.4	no
NaSH	H_2S	7.0	no
NaH	H_2	35	yes

2.48 Draw the products and then compare the pK_a of the acid on the left and the conjugate acid on the right. **The equilibrium lies toward the side having the acid with a higher pK_a (weaker acid).**

a. $CH_3\ddot{N}H_2 + H\text{–}OSO_3H \rightleftharpoons CH_3\overset{+}{N}H_3 + HSO_4^-$ **products favored**

$pK_a = -9$ $pK_a = 10.7$

b. $CH_3CH_2COOH + Na^+ :\ddot{Cl}:^- \rightleftharpoons CH_3CH_2COO^- Na^+ + H\ddot{Cl}:$ **starting material favored**

$pK_a \sim 5$ $pK_a = -7$

c. C_6H_5OH (pK_a = 10) + $Na^+ HCO_3^-$ ⇌ $C_6H_5O^- Na^+$ + H_2CO_3 (pK_a = 6.4) **starting material favored**

d. $H-C\equiv C-H$ (pK_a = 25) + $CH_3CH_2^- Li^+$ ⇌ $H-C\equiv C:^- Li^+$ + CH_3CH_3 (pK_a = 50) **products favored**

2.49 Use Appendix C to determine or estimate the pK_a's of the compounds.

C_6H_5COOH: pK_a = 4.2; $C_6H_{11}NH_3^+$: pK_a = 10.7; $C_6H_{11}NH_2$: pK_a ~ 40 (estimated)

a. $C_6H_{11}NH_2$ (pK_a ~ 40, weaker acid, reactants favored) + $C_6H_5COO^-$ ⇌ $C_6H_{11}NH^-$ + C_6H_5COOH (pK_a = 4.2)

b. $C_6H_{11}NH^-$ + C_6H_5COOH (pK_a = 4.2) ⇌ $C_6H_{11}NH_2$ (pK_a ~ 40, weaker acid, products favored) + $C_6H_5COO^-$

c. $C_6H_{11}NH_3^+$ (pK_a = 10.7, weaker acid, reactants favored) + $C_6H_5COO^-$ ⇌ $C_6H_{11}NH_2$ + C_6H_5COOH (pK_a = 4.2)

d. $C_6H_{11}NH_2$ + C_6H_5COOH (pK_a = 4.2) ⇌ $C_6H_{11}NH_3^+$ (pK_a = 10.7, weaker acid, products favored) + $C_6H_5COO^-$

2.50

a. Using the pK_a values for the three H's in H_3PO_4 (2.1, 6.9, and 12.4), the two OH groups bonded to P have approximate pK_a values of 2.1 and 6.9.

b. Because H_2O, the conjugate acid of ^-OH, has a pK_a of 15.7, ^-OH is a strong enough base to remove two H's from **A,** forming dianion **B.**

A —(2 NaOH)→ B + 2 H_2O (pK_a = 15.7)

c. Because $C_5H_6N^+$, the conjugate acid of pyridine, has a pK_a of 5.3, pyridine is a strong enough base to remove only one H from **A**, forming anion **C.**

A + pyridine → C + pyridinium (pK_a = 5.3)

2.51 Compare element effects first and then resonance, hybridization, and electron-withdrawing groups to determine the relative strengths of the acids.

a. propane, ethylamine (NH_2), ethanol (OH) — acidity increases across a row; **weakest acid** (propane), **strongest acid** (ethanol)

b. 2-chloropropane (only C–H bonds, **weakest acid**), 2-propanol (O–H bond), 2-butanethiol (S–H bond, **strongest acid**)

c. propane (only C–H bonds, **weakest acid**), propanol (O–H bond), Cl-substituted propanol (O–H bond and electron-withdrawing Cl, **strongest acid**)

d. butane (all sp^3 C–H, **weakest acid**), 2-butene (sp^2 C–H), 1-butyne (sp C–H, **strongest acid**)

2.52 Take into account the factors that determine acidity.

- H_c is more acidic than H_a because H_c is bonded to an sp^2 hybridized C atom. whereas H_a is bonded to an sp^3 hybridized C atom. H_a and H_c are the least acidic protons with the highest pK_a's.
- H_e is more acidic than H_d because H_e is bonded to an O atom and OH bonds are more acidic than NH bonds. Loss of both H_d and H_e forms resonance-stabilized conjugate bases.
- H_b is bonded to a C, but removal of H_b forms a resonance-stabilized anion. H_b is intermediate in acidity.

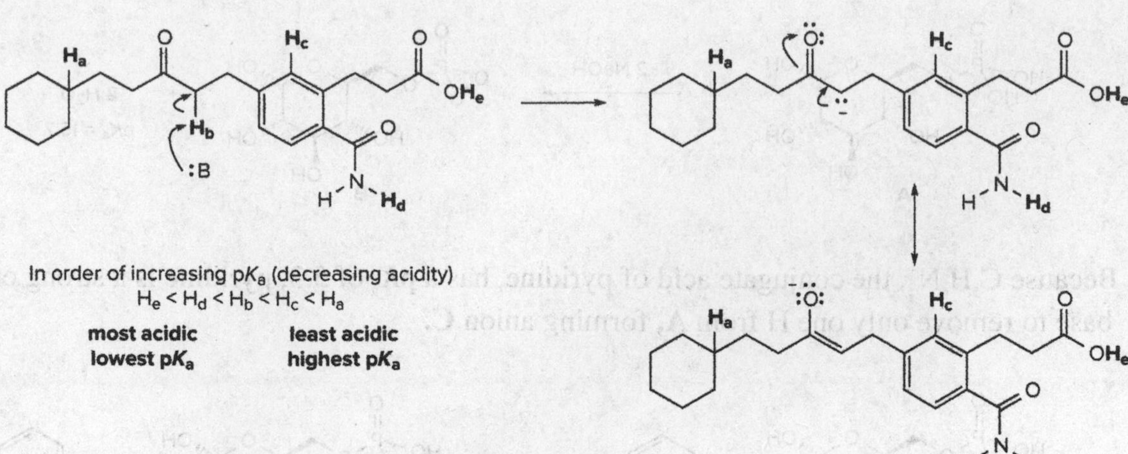

In order of increasing pK_a (decreasing acidity)

$H_e < H_d < H_b < H_c < H_a$

most acidic
lowest pK_a

least acidic
highest pK_a

2.53

Br OH NH O F OH

A	B	C	D
OH bond and electronegative Br	only CH bonds	NH bond	OH bond

In order of increasing acidity:

B < C < D < A

The weakest acid (**B**) forms the strongest conjugate base.

2.54 The strongest acid has the weakest conjugate base.

a. Draw the conjugate acid.

Increasing acidity of conjugate acids:

$CH_3CH_3 < CH_3NH_2 < CH_3OH$

increasing basicity: $CH_3O^- < CH_3NH^- < CH_3CH_2^-$

b. Draw the conjugate acid.

Increasing acidity of conjugate acids:

$CH_4 < H_2O < HBr$

increasing basicity: $Br^- < HO^- < CH_3^-$

c. Draw the conjugate acid.

Increasing acidity of conjugate acids:

OH < OH < Cl OH

increasing basicity:

Cl O⁻ < O⁻ < O⁻

d. Draw the conjugate acid.

Increasing acidity of conjugate acids:

< <

increasing basicity:

$C \equiv C:^-$ < <

2.55

a. The most acidic proton in uracil is the H atom on the N between the two carbonyls because its conjugate base is the most resonance stabilized.

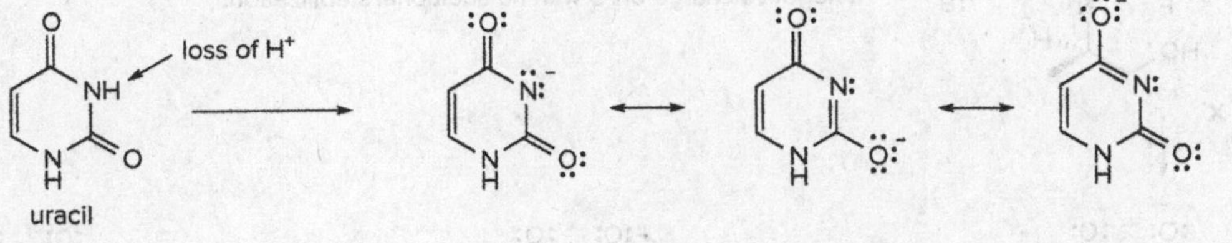

loss of H^+

uracil

b. If uracil is treated with two equivalents of base, both NH protons would be removed to form a dianion.

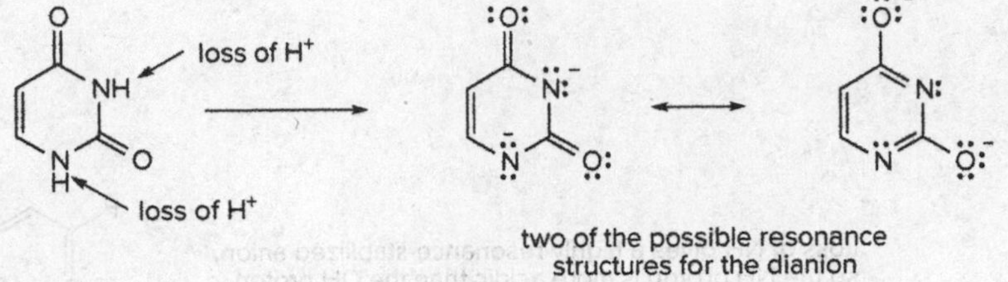

loss of H^+

loss of H^+

two of the possible resonance structures for the dianion

2.56

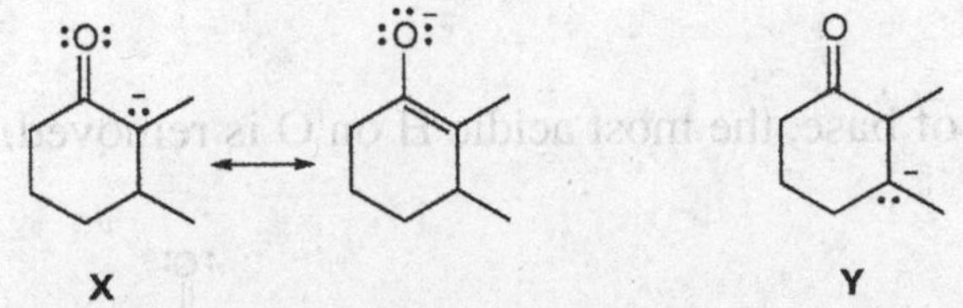

X

Y

The negative charge in **X** is delocalized by resonance, making **X** more stable than **Y**. This makes **Y** the stronger base.

2.57 To draw the conjugate acid, look for the most basic site and protonate it. To draw the conjugate base, look for the most acidic site and remove a proton.

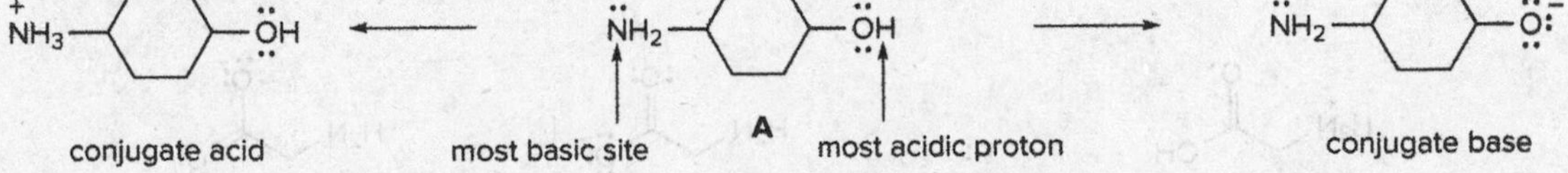

conjugate acid | most basic site | **A** | most acidic proton | conjugate base

2.58

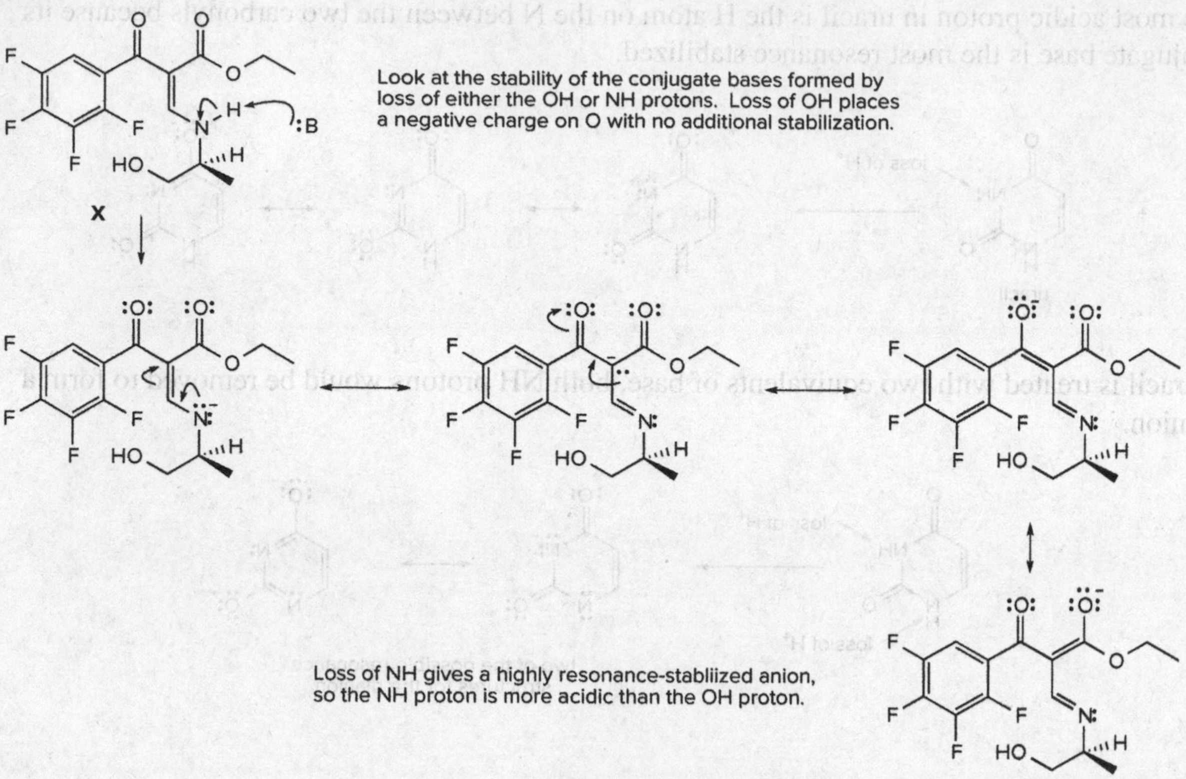

Look at the stability of the conjugate bases formed by loss of either the OH or NH protons. Loss of OH places a negative charge on O with no additional stabilization.

Loss of NH gives a highly resonance-stabliized anion, so the NH proton is more acidic than the OH proton.

2.59

a. With one equivalent of base, the most acidic H on O is removed.

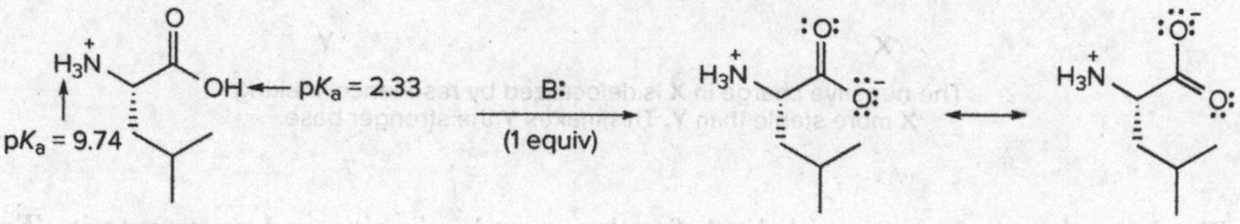

$pK_a = 9.74$ $pK_a = 2.33$ B: (1 equiv)   leucine

b. With two equivalents of base, both the OH and NH protons are removed.

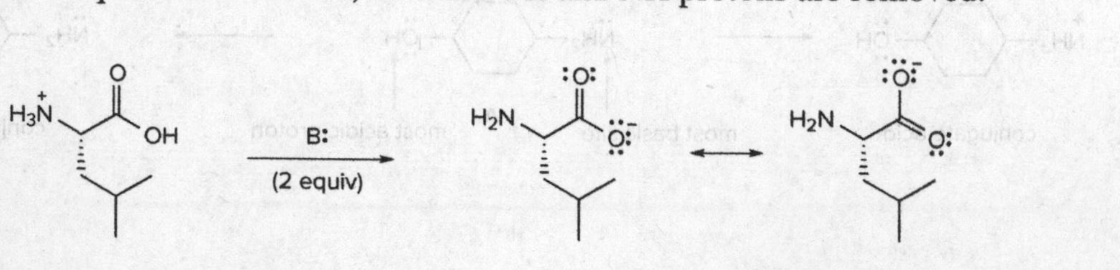

B: (2 equiv)

c. Because H_2O, the conjugate acid of ^-OH, has a pK_a of 15.7, ^-OH is a strong enough base to remove both protons from leucine.

2.60 Remove the most acidic proton to form the conjugate base. Protonate the most basic electron pair to form the conjugate acid.

a.

only O–H bond
most acidic proton

COOH

ibuprofen

conjugate base:

COO$^-$

b.

most basic electron pair → :N

Increasing basicity:

—Ö—   —N̈—

cocaine

conjugate acid:

$H-N^+$

2.61 Compare the isomers.

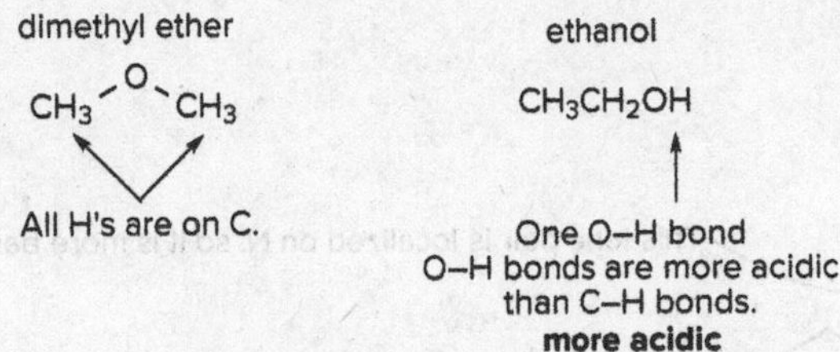

2.62

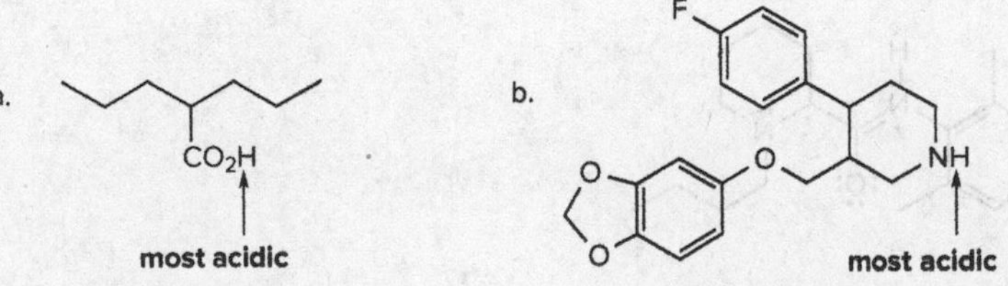

c.

most acidic
OH
CH_3O

a. The molecule contains C–H and O–H bonds. O is farther right in the periodic table; therefore, the O–H hydrogen is the most acidic.

b. The molecule contains C–H and N–H bonds. N is farther right in the periodic table; therefore, the N–H hydrogen is the most acidic.

c. The molecule contains C–H, N–H, and O–H bonds. O is farthest right in the periodic table; therefore, the O–H hydrogen is the most acidic.

2.63 Use element effects, inductive effects, and resonance to determine which protons are the most acidic. The H's of the CH_3 group are least acidic, because they are bonded to an sp^3 hybridized C and the conjugate base formed by their removal is not resonance stabilized.

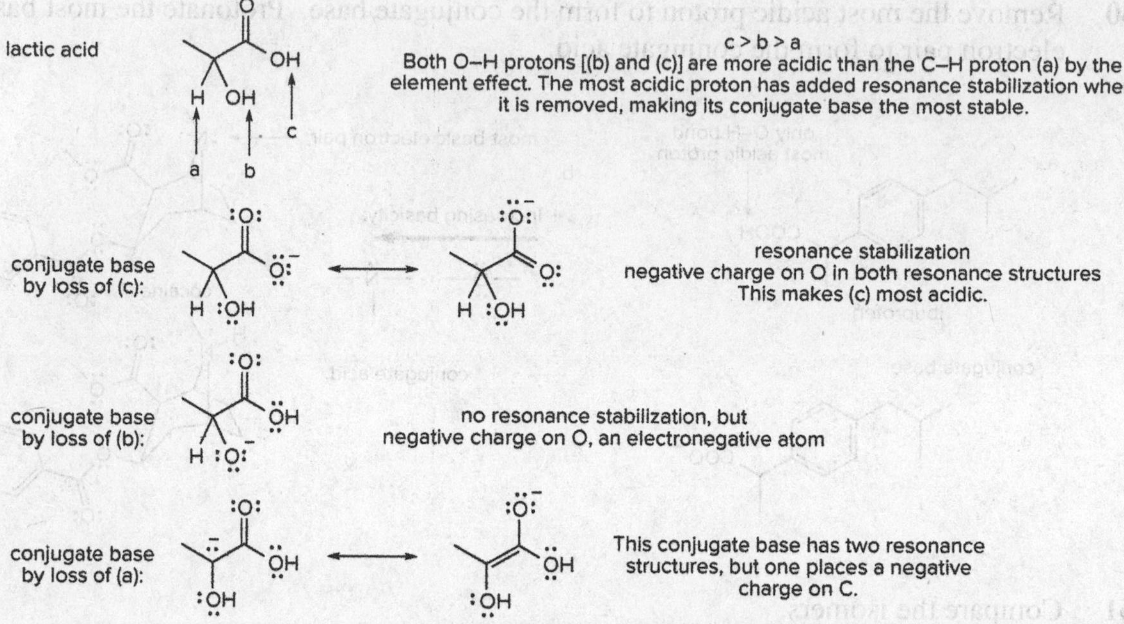
lactic acid
c > b > a
Both O–H protons [(b) and (c)] are more acidic than the C–H proton (a) by the element effect. The most acidic proton has added resonance stabilization when it is removed, making its conjugate base the most stable.
a
b
c
conjugate base by loss of (c):
resonance stabilization
negative charge on O in both resonance structures
This makes (c) most acidic.
conjugate base by loss of (b):
no resonance stabilization, but negative charge on O, an electronegative atom
conjugate base by loss of (a):
This conjugate base has two resonance structures, but one places a negative charge on C.

2.64

This lone pair is localized on N, so it is more basic.

bupivacaine

This lone pair is delocalized by resonance, so it is less available to donate to an acid, making it less basic.

2.65 ***Lewis bases* are electron pair donors:** they contain a lone pair or a π bond. ***Brønsted–Lowry bases* are proton acceptors:** to accept a proton they need a lone pair or a π bond. This means Lewis bases are also Brønsted–Lowry bases.

a. lone pairs on O **both**

b. $CH_3CH_2{-}Cl$: lone pairs on Cl **both**

c. **neither** = no lone pairs or π bonds

d. π bonds **both**

2.66 A ***Lewis acid* is an electron pair acceptor** and usually contains a proton or an unfilled valence shell of electrons. A ***Brønsted–Lowry acid* is a proton donor** and must contain a hydrogen atom. All Brønsted–Lowry acids are Lewis acids, though the reverse may not be true.

a. H_3O^+	b. Cl_3C^+	c. BCl_3	d. BF_4^-
both contains a H	**Lewis acid** unfilled valence shell on C	**Lewis acid** unfilled valence shell on B	**neither** no H or unfilled valence shell

2.67 Label the Lewis acid and Lewis base and then draw the products.

a. $:\ddot{Cl}:^-$ + BCl_3 → Cl_3B^-–Cl (new bond)
Lewis base Lewis acid

b. acid chloride (Lewis acid) + $:\ddot{O}H^-$ (Lewis base) → tetrahedral product (new bond)

2.68 A Lewis acid is also called an **electrophile.** When a Lewis base reacts with an electrophile other than a proton, it is called a **nucleophile.** Label the electrophile and nucleophile in the starting materials and then draw the products.

a. diethyl sulfide + $AlCl_3$ → $^-AlCl_3$–S^+(Et)₂
nucleophile electrophile

b. acetone + BF_3 → $C{=}O^+{-}BF_3^-$
nucleophile electrophile

c. cyclohexyl cation + $H_2\ddot{O}$ → cyclohexyl–$\overset{+}{O}H_2$
electrophile nucleophile

2.69

A B C D

a. A conjugate acid–base pair differ in the presence of a proton. **A** and **B** (or **B** and **D**) represent a conjugate acid–base pair.

b. **A** and **D** are resonance structures because the atom placement is the same, but the electron pairs are placed differently.

c. **A** and **C** (or **C** and **D**) are constitutional isomers because they have the same molecular formula (C_5H_9NO), but the bonds are different. **A** and **D** have an N–H bond, whereas **C** has an O–H bond.

2.70

a. :ÖH + H–Br → (proton transfer) → $\overset{+}{O}H_2$ + :Br:⁻ → Br + $H_2\ddot{O}$:

nucleophile

proton transfer

electrophile

b. + Br⁺ → Br + :Br:⁻ → (proton transfer) → Br + HBr

nucleophile electrophile

proton transfer

2.71 Draw the products of each reaction. In part (a), ⁻OH pulls off a proton and thus acts as a Brønsted–Lowry base. In part (b), ⁻OH attacks a carbon and thus acts as a Lewis base.

a. :ÖH → $H_2\ddot{O}$: +

b. ⁻:ÖH + → ÖH

2.72 Answer each question about esmolol.

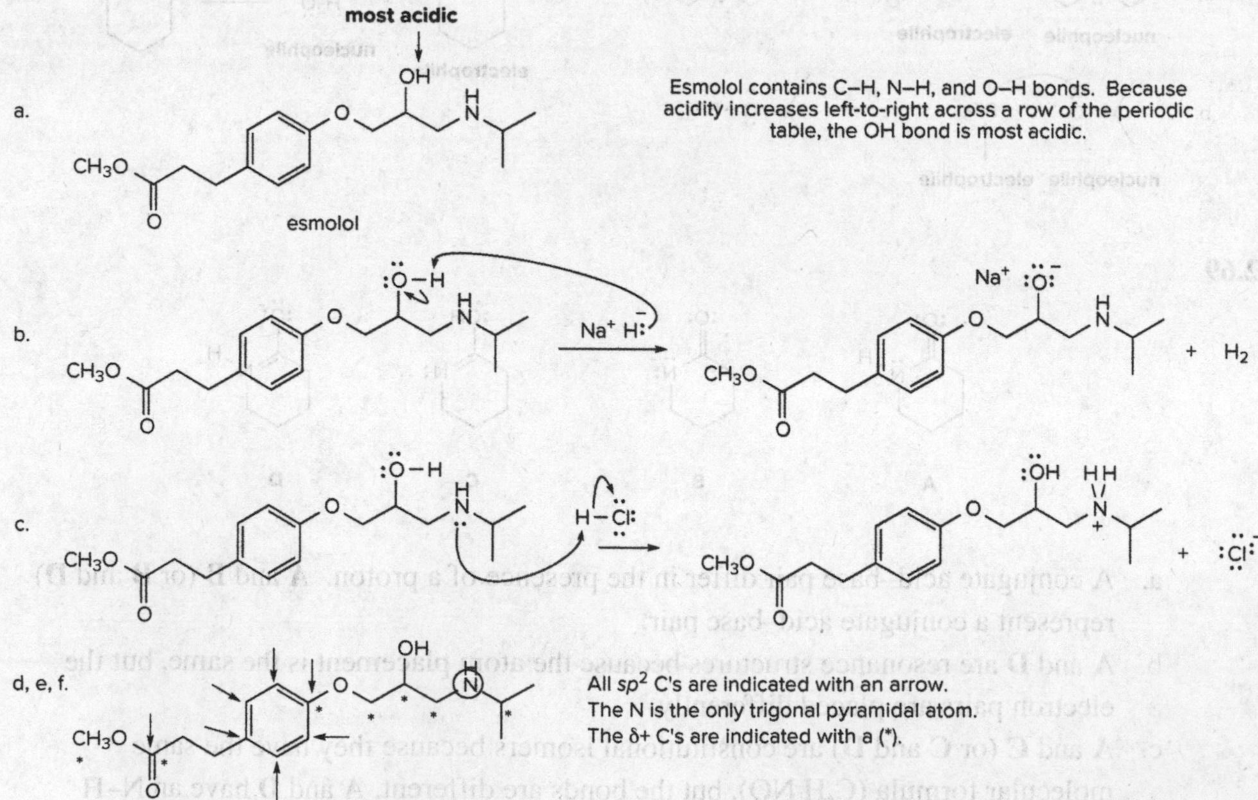
most acidic
a.
OH
H
N
O
CH_3O
O
esmolol
Esmolol contains C–H, N–H, and O–H bonds. Because acidity increases left-to-right across a row of the periodic table, the OH bond is most acidic.
b.
Na⁺ H:⁻
Na⁺ :Ö:⁻
+ H_2
c.
H–Cl:
:ÖH
+ :Cl:⁻
d, e, f.
All sp^2 C's are indicated with an arrow.
The N is the only trigonal pyramidal atom.
The δ+ C's are indicated with a (*).

2.73 Draw resonance structures to determine which conjugate base is more stable.

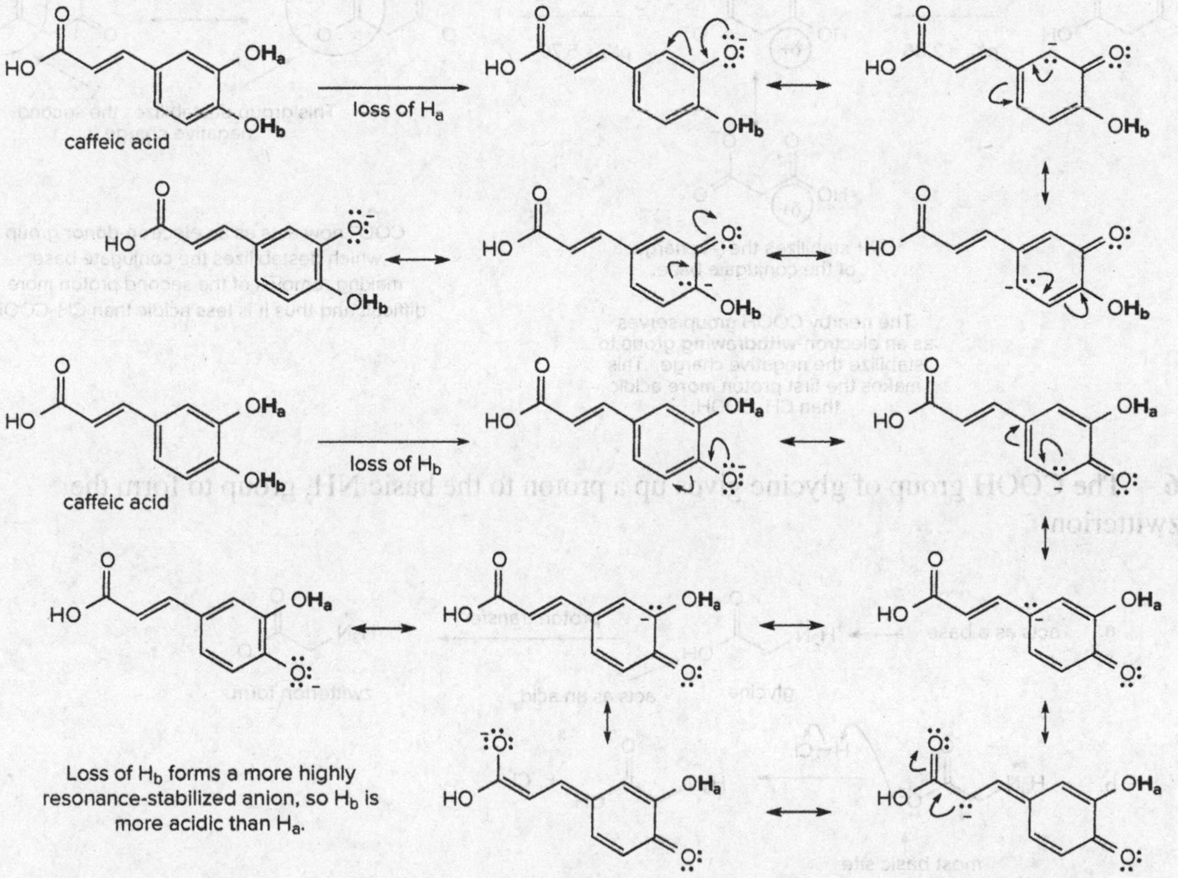

caffeic acid

loss of H_a

caffeic acid

loss of H_b

Loss of H_b forms a more highly resonance-stabilized anion, so H_b is more acidic than H_a.

2.74 Draw the product of protonation of either O or N and compare the conjugate acids. When acetamide reacts with an acid, the O atom is protonated because it results in a resonance-stabilized conjugate acid.

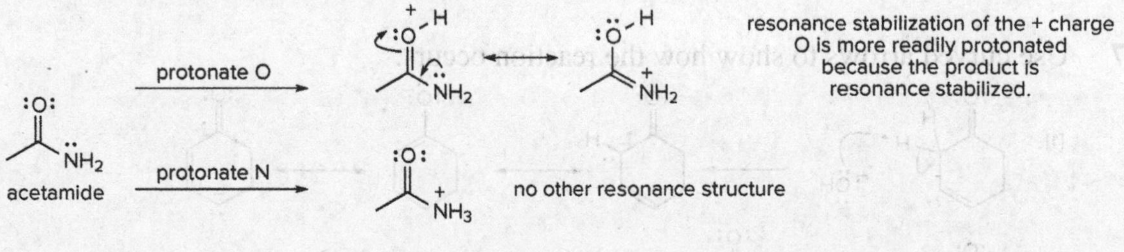

acetamide

protonate O

protonate N

resonance stabilization of the + charge O is more readily protonated because the product is resonance stabilized.

no other resonance structure

2.75

HO–C(=O)–CH$_2$–C(=O)–OH → (pK_a = 2.86) HO–C(=O)–CH$_2$–C(=O)–O$^-$ (δ+) → (pK_a = 5.70) $^-$O–C(=O)–CH$_2$–C(=O)–O$^-$

δ+ stabilizes the (–) charge of the conjugate base.

The nearby COOH group serves as an electron-**withdrawing** group to stabilize the negative charge. This makes the first proton **more** acidic than CH_3COOH.

This group destabilizes the second negative charge.

COO^- now acts as an electron-**donor** group which destabilizes the conjugate base, making removal of the second proton more difficult and thus it is **less** acidic than CH_3COOH.

2.76 The COOH group of glycine gives up a proton to the basic NH_2 group to form the zwitterion.

a. acts as a base → H_2N–CH$_2$–COOH (glycine) ← acts as an acid —proton transfer→ H_3N^+–CH$_2$–COO$^-$ (zwitterion form)

b. H_3N^+–CH$_2$–COO$^-$ + H–Cl → H_3N^+–CH$_2$–COOH + Cl^-

most basic site

c. H_3N^+–CH$_2$–COO$^-$ + Na^+ $^-$:OH → H_2N–CH$_2$–COO$^-$ + Na^+ + H_2O:

most acidic site

2.77 Use curved arrows to show how the reaction occurs.

[1] cyclohexenone + $^-$:OH → [resonance structures of enolate anion]

[2] enolate + H–Ö–H → product + $^-$:OH

Protonate the negative charge on this carbon to form the product.

2.78 Compare the OH bonds in vitamin C and decide which one is the most acidic.

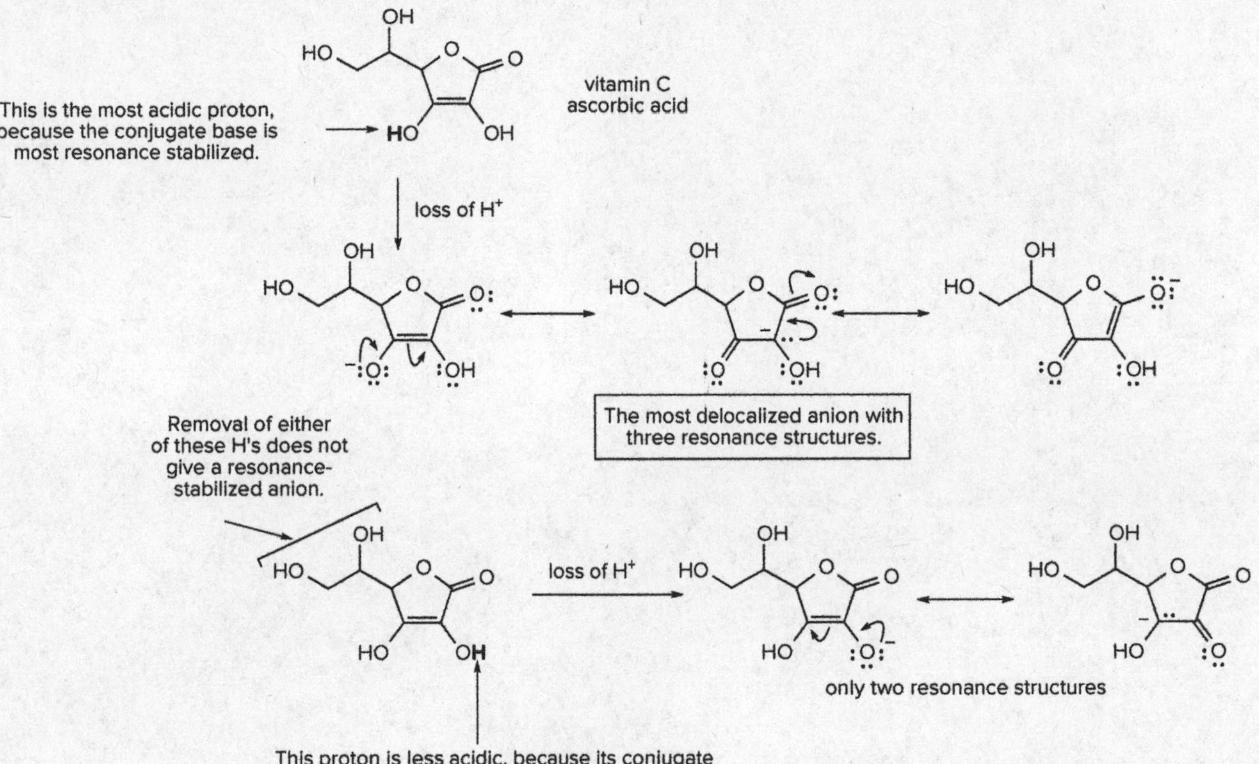
vitamin C
ascorbic acid
This is the most acidic proton, because the conjugate base is most resonance stabilized.
loss of H+
The most delocalized anion with three resonance structures.
Removal of either of these H's does not give a resonance-stabilized anion.
loss of H+
only two resonance structures
This proton is less acidic, because its conjugate base is less resonance stabilized.

Chapter 3: Introduction to Organic Molecules and Functional Groups

Chapter Review

Classifying carbon atoms, hydrogen atoms, alcohols, alkyl halides, amines, and amides (3.2)

- Carbon atoms are classified by the number of carbons bonded to them; a 1° carbon is bonded to one other carbon, and so forth.
- Hydrogen atoms are classified by the type of carbon atom to which they are bonded; a 1° hydrogen is bonded to a 1° carbon, and so forth.
- Alkyl halides and alcohols are classified by the type of carbon to which the OH or X group is bonded; a 1° alcohol has an OH group bonded to a 1° carbon, and so forth.
- Amines and amides are classified by the number of carbons bonded to the nitrogen atom; a 1° amine has one carbon–nitrogen bond, and so forth.

Types of intermolecular forces (3.3)

Type of force	Cause	Examples
van der Waals (VDW)	Due to the interaction of temporary dipoles • Larger surface area, stronger forces • Larger, more polarizable atoms, stronger forces	All organic compounds
dipole–dipole (DD)	Due to the interaction of permanent dipoles	$(CH_3)_2C{=}O$, H_2O
hydrogen bonding (HB or H-bonding)	Due to the electrostatic interaction of a H atom in an O–H, N–H, or H–F bond with another N, O, or F atom.	H_2O
ion–ion	Due to the interaction of two ions	NaCl, LiF

Increasing strength ↓

Physical properties

Property	Observation
Boiling point (3.4A)	• For compounds of comparable molecular weight, the stronger the forces the higher the bp.

VDW, MW = 72, bp = 36 °C

:O: H — VDW, DD, MW = 72, bp = 76 °C

ÖH — VDW, DD, HB, MW = 74, bp = 118 °C

Increasing strength of intermolecular forces
Increasing boiling point

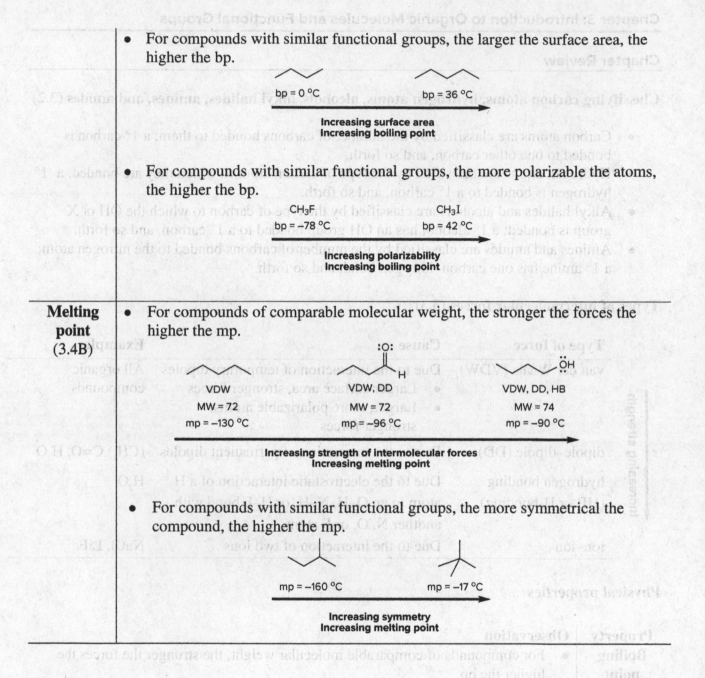

	• For compounds with similar functional groups, the larger the surface area, the higher the bp. bp = 0 °C → bp = 36 °C Increasing surface area Increasing boiling point • For compounds with similar functional groups, the more polarizable the atoms, the higher the bp. CH_3F bp = −78 °C → CH_3I bp = 42 °C Increasing polarizability Increasing boiling point
Melting point (3.4B)	• For compounds of comparable molecular weight, the stronger the forces the higher the mp. VDW, MW = 72, mp = −130 °C → VDW, DD, MW = 72, mp = −96 °C → VDW, DD, HB, MW = 74, mp = −90 °C Increasing strength of intermolecular forces Increasing melting point • For compounds with similar functional groups, the more symmetrical the compound, the higher the mp. mp = −160 °C → mp = −17 °C Increasing symmetry Increasing melting point

Solubility (3.4C)	Types of water-soluble compounds: • Ionic compounds • Organic compounds having ≤ 5 C's and an O or N atom for hydrogen bonding (for a compound with one functional group) Types of compounds soluble in organic solvents: • Organic compounds regardless of size or functional group • Examples: 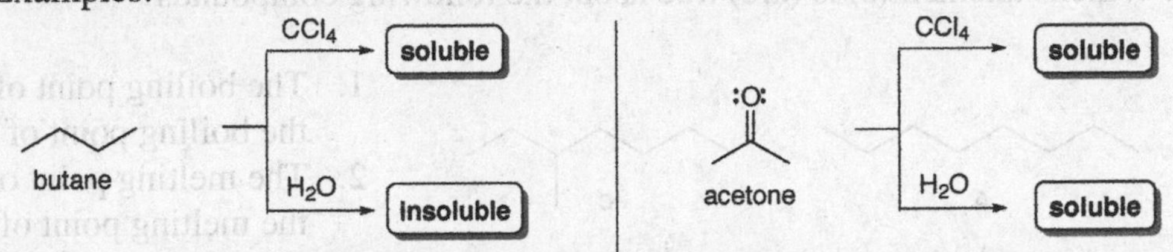 Key: VDW = van der Waals, DD = dipole–dipole, HB = hydrogen bonding MW = molecular weight

Reactivity (3.8)

- **Nucleophiles react with electrophiles.**
- Electronegative heteroatoms create electrophilic carbon atoms that react with nucleophiles.
- Lone pairs and π bonds are nucleophilic sites that react with electrophiles.

Practice Test on Chapter Review

1.a. Which of the following compounds exhibits dipole–dipole interactions?

1. CH_2Cl_2
2. $(CH_3)_2O$
3. $CH_3CH_2CH_2—OH$

4. Compounds (1) and (2) both exhibit dipole–dipole interactions.

5. Compounds (1), (2), and (3) all exhibit dipole–dipole interactions.

b. Which of the following compounds can hydrogen bond both to another molecule like itself and to water?

1. (cyclic structure with O, O and OCH_3)
2. (cyclic structure with NH)
3. $CH_3C{\equiv}N$

4. Compounds (1) and (2) can each hydrogen bond to another molecule like itself and water.

5. Each of the three compounds, (1), (2), and (3), can hydrogen bond to another molecule like itself and to water.

c. Which of the following compounds has the highest boiling point?

1. $CH_3CH_2CH_2OCH_2CH_3$
2. $CH_3(CH_2)_4CH_3$
3. $CH_3(CH_2)_4NH_2$
4. $(CH_3)_2CHCH(CH_3)NH_2$
5. $(CH_3)_2CHCH(CH_3)_2$

d. Which statement(s) is (are) true about the following compounds?

A C B D

1. The boiling point of **A** is higher than the boiling point of **B**.
2. The melting point of **B** is higher than the melting point of **C**.
3. The boiling point of **A** is higher than the boiling point of **D**.
4. Statements (1) and (2) are both true.
5. Statements (1), (2), and (3) are all true.

2. Consider the anti-hypertensive agent atenolol drawn below.

E H_2N C H_a A N F B O D O H_c H_b

a. What is the hybridization of the N atom labeled with **A?**
b. What is the shape around the C atom labeled with **F?**
c. What orbitals are used to form the bond labeled with **E?**
d. Which bond (**B, C,** or **D**) is the longest bond?
e. Would you predict atenolol to be soluble in water?
f. Which of the labeled H atoms (H_a, H_b, or H_c) is most acidic?

Answers to Practice Test

1. a. 5
   b. 2
   c. 3
   d. 5

2. a. *sp*^{3}
   b. trigonal planar
   c. C_{sp^2}–C_{sp^3}
   d. **D**
   e. yes
   f. H_b

Answers to Problems

3.1

$CH_3CH_2-\ddot{O}H \xrightarrow{H-OSO_3H} CH_3CH_2-\overset{+}{O}H_2 + HSO_4^-$ $\qquad CH_3CH_3 \xrightarrow{H_2SO_4}$ no reaction

$CH_3CH_2-\ddot{O}-H \xrightarrow{Na^+\ H:^-} CH_3CH_2-\ddot{O}:^-\ Na^+ + H_2$ $\qquad CH_3CH_3 \xrightarrow{NaH}$ no reaction

3.2 To classify a carbon atom as 1°, 2°, 3°, or 4°, **determine how many carbon atoms it is bonded to (1° C = bonded to one other C, 2° C = bonded to two other C's, 3° C = bonded to three other C's, 4° C = bonded to four other C's).** Re-draw if necessary to see each carbon clearly.

To classify a hydrogen atom as 1°, 2°, or 3°, **determine if it is bonded to a 1°, 2°, or 3° C (a 1° H is bonded to a 1° C; a 2° H is bonded to a 2° C; a 3° H is bonded to a 3° C).** Re-draw if necessary.

a. [1] 1° C, 1° C, 2° C's — re-draw
[2] 1° C, 3° C, 1° C, 1° C — re-draw
[3] 4° C's. All other C's are 1° C's. — re-draw
[4] 1° C's, 4° C, 1° C, 3° C. All other C's are 2° C's. — re-draw

b. [1] $CH_3CH_2CH_2CH_3$ — 1° H's, 1° H's, 2° H's
[2] $CH_3-\underset{CH_3}{\overset{CH_3}{C}}-H$ — 1° H's, 1° H's, 1° H's, 3° H
[3] $CH_3-\underset{CH_3}{\overset{CH_3}{C}}-\underset{CH_3}{\overset{CH_3}{C}}-CH_3$ — all 1° H's
[4] 1° H's, 3° H. All others are 2° H's.

3.3 Use the definition of 1°, 2°, 3°, or 4° carbon atoms from Answer 3.2.

a. sucrose — 1°, 1°, 1°, 1°
All other C's are 2°.

b. amoxicillin — 2°, 1°, 3°, 1°, 1°, 2°, 2°

3.54 Compare the intermolecular forces of crack and cocaine hydrochloride. Stronger intermolecular forces increase both the boiling point and the water solubility.

cocaine (crack)
neutral organic molecule

cocaine hydrochloride
a salt

The molecules are identical except for the ionic bond in cocaine hydrochloride. Ionic forces are extremely strong forces, and therefore the cocaine hydrochloride salt has a much **higher boiling point and is more water soluble.** Because the salt is highly water soluble, it can be injected directly into the bloodstream, where it dissolves. Crack is smoked because it can dissolve in the organic tissues of the nasal passages and lungs.

3.55

a.

b.

five functional groups that have many opportunities for H-bonding
water soluble

ionic salt
more water soluble

c. Because the hydrochloride salt is ionic and therefore more water soluble, it is more readily transported in the bloodstream.

3.56

a.

acetyl coenzyme A

choline

b. The thioester is the electrophile. The OH group of choline is the nucleophile.

3.57 Use the rules from Answer 3.27.

a. nucleophilic; δ+; δ–; electrophilic

b. nucleophilic; δ+; δ+; electrophilic

c. All the C=C's are nucleophilic.

d. CH_3; δ+; nucleophilic; electrophilic (All lone pairs on O and Cl are nucleophilic.)

3.58

a. + Br^- ⟶ NO; nucleophilic; nucleophilic

b. + ^-CN ⟶ YES; electrophilic; nucleophilic

c. + ^-OH ⟶ NO; nucleophilic; nucleophilic

d. + H_3O^+ ⟶ YES; nucleophilic; electrophilic

3.59 More rigid cell membranes have phospholipids with *fewer* C=C's. Each C=C introduces a bend in the molecule, making the phospholipids pack less tightly. Phospholipids without C=C's can pack very tightly, making the membrane less fluid and more rigid.

The double bonds introduce kinks in the chain, making packing of the hydrocarbon chains less efficient. This makes the cell membrane formed from them more fluid.

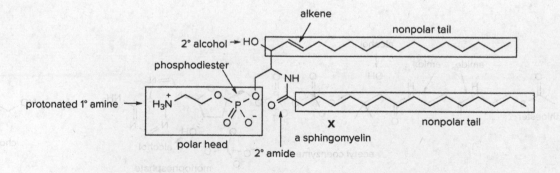

3.60

alkene

nonpolar tail

2° alcohol ⟶ HO

phosphodiester

NH

protonated 1° amine ⟶ $H_3\overset{+}{N}$

polar head

X

a sphingomyelin

nonpolar tail

2° amide

3.61 **B** is ionic, so it does not "dissolve" in the nonpolar interior of a cell membrane and it cannot cross the blood–brain barrier. **A** is a neutral organic molecule, so it can enter the nonpolar interior of the cell membrane and cross the blood–brain barrier to act as an anesthetic.

3.62

a, b.

3° amine
3° amide
3° amide
3° amide
2° amide
2° amide
synthadotin

c. Synthadotin can hydrogen bond to itself using the two N–H bond (boxed in).

d. All N and O atoms are nucleophilic sites.

e. All carbonyl C's are electrophilic sites. Any C bonded to an electronegative N atom is also an electrophilic site.

f. When synthadotin is treated with HCl, the most basic atom, the amine N, is protonated.

H–Cl

Cl^-

3.63

a, b.

2° amine
ester
carboxylic acid
3° amide
aromatic ring
aromatic ring
quinapril

quinapril

c. Quinapril can hydrogen bond to water at all N and O atoms (boxed in).

d. Quinapril can hydrogen bond to acetone at its O–H and N–H bonds (shown with arrows).

e. The most acidic H is the lone H bonded to O, which is part of a carboxylic acid.

f. The most basic site is the N atom of the amine (circled).

3.64

a.

amide

amine

aromatic ring

aromatic ring

e. An isomer that can hydrogen bond should have a higher boiling point.

hydrogen bonding possible

b, c, f, g.

most basic atom

The N atom of the amide is less basic because the lone pair is part of resonance with the C=O.

Atoms that can hydrogen bond to H_2O are boxed in.
Electrophilic carbons are labeled with (*).

most acidic H, $pK_a \sim 25$
All H's are bonded to C's. Removal of the H on the C adjacent to the C=O results in a resonance-stabilized anion.

d. Fentanyl exhibits van der Waals and dipole–dipole interactions but no hydrogen bonding, because there is no H bonded to O or N.

3.65

A

B

The OH and CHO groups are close enough that they can intramolecularly H-bond to each other. Because the two polar functional groups are involved in intramolecular H-bonding, they are less available for H-bonding to H_2O. This makes **A** less H_2O soluble than **B,** whose two functional groups are both available for H-bonding to the H_2O solvent.

The OH and the CHO are too far apart to intramolecularly H-bond to each other, leaving more opportunity to H-bond with solvent.

3.66

a. melting point

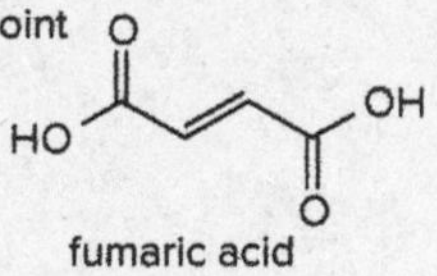

HO OH O O

fumaric acid

Fumaric acid has its two larger COOH groups on opposite ends of the molecule, and in this way it can pack better in a lattice than maleic acid, giving it a **higher mp.**

b. solubility

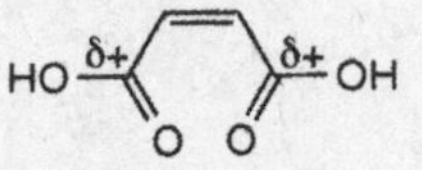

HO δ+ δ+ OH O O

maleic acid

Maleic acid is more polar, giving it greater **H_2O solubility.** The bond dipoles in fumaric acid cancel.

c. removal of the first proton (pK_{a1})

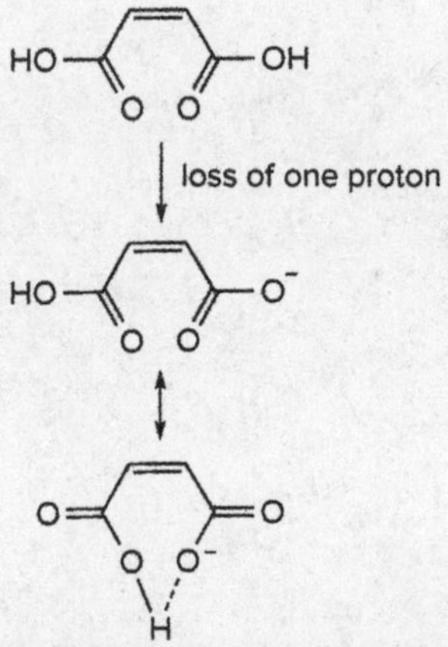

loss of one proton

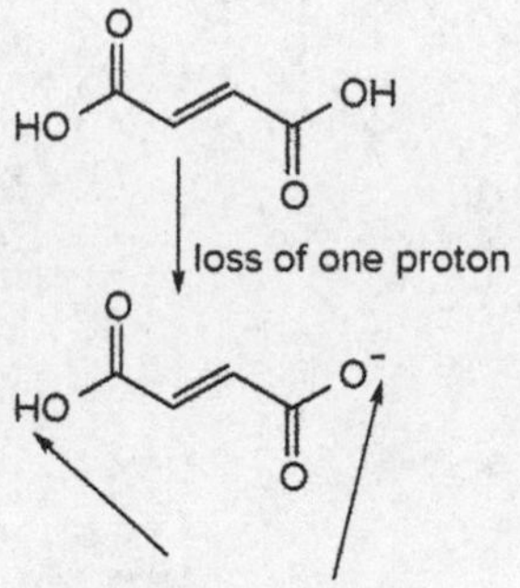

loss of one proton

Intramolecular H-bonding is not possible here.

In maleic acid, intramolecular H-bonding stabilizes the conjugate base after one H is removed, making maleic acid more acidic than fumaric acid.

d. removal of the second proton (pK_{a2})

Now the dianion is held in close proximity in maleic acid, and this destabilizes the conjugate base. Thus, removing the second H in maleic acid is harder, making it a weaker acid than fumaric acid for removal of the second proton.

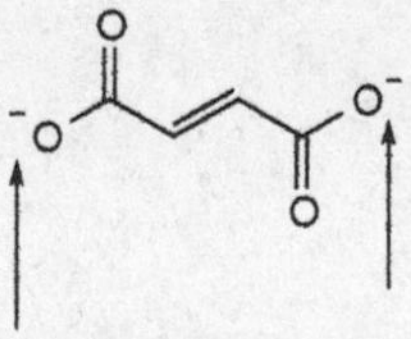

The two negative charges are much farther apart. This makes the dianion from fumaric acid more stable and thus pK_{a2} is lower for fumaric acid than maleic acid.

Chapter 4: Alkanes

Chapter Review

General facts about alkanes (4.1–4.3)

- Alkanes are composed of **tetrahedral, *sp*^{3}** hybridized C's.
- There are two types of alkanes: acyclic alkanes having molecular formula $\mathbf{C_nH_{2n+2}}$, and cycloalkanes having molecular formula $\mathbf{C_nH_{2n}}$.
- Alkanes have only **nonpolar C–C and C–H bonds** and no functional group, so they undergo few reactions.
- Alkanes are named with the suffix ***-ane.***

Names of alkyl groups (4.4A)

CH_3— = methyl

CH_3CH_2— = ethyl

$CH_3CH_2CH_2$— = propyl

$(CH_3)_2CH$— = isopropyl

$CH_3CH_2CH_2CH_2$— = butyl

$CH_3CH_2CHCH_3$ = *sec*-butyl

$(CH_3)_2CHCH_2$— = isobutyl

$(CH_3)_3C$— = *tert*-butyl

Conformations in acyclic alkanes (4.9, 4.10)

- Alkane conformations can be classified as **staggered, eclipsed, anti,** or **gauche** depending on the relative orientation of the groups on adjacent carbons.

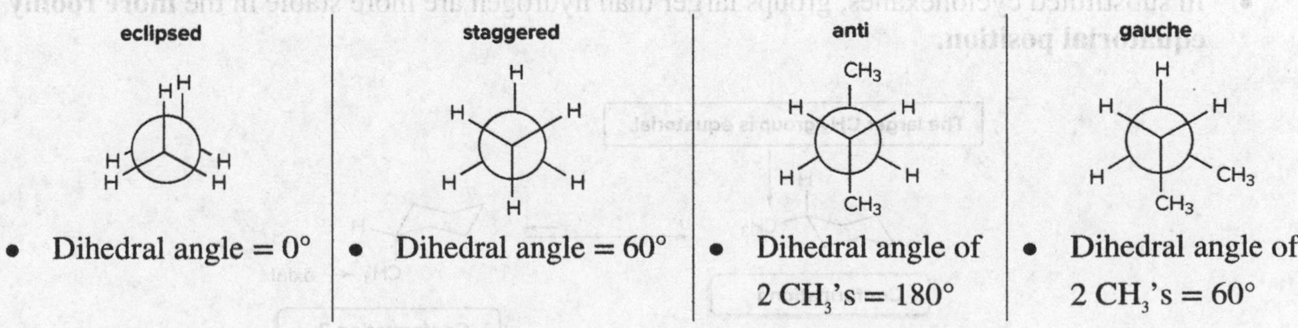

- A staggered conformation is **lower in energy** than an eclipsed conformation.
- An anti conformation is **lower in energy** than a gauche conformation.

Types of strain

- **Torsional strain**—an increase in energy due to eclipsing interactions (4.9).
- **Steric strain**—an increase in energy when atoms are forced too close to each other (4.10).
- **Angle strain**—an increase in energy when tetrahedral bond angles deviate from 109.5° (4.11).

Two types of isomers

[1] **Constitutional isomers**—isomers that differ in the way the atoms are connected to each other (4.1A).
[2] **Stereoisomers**—isomers that differ only in the way atoms are oriented in space (4.13B).

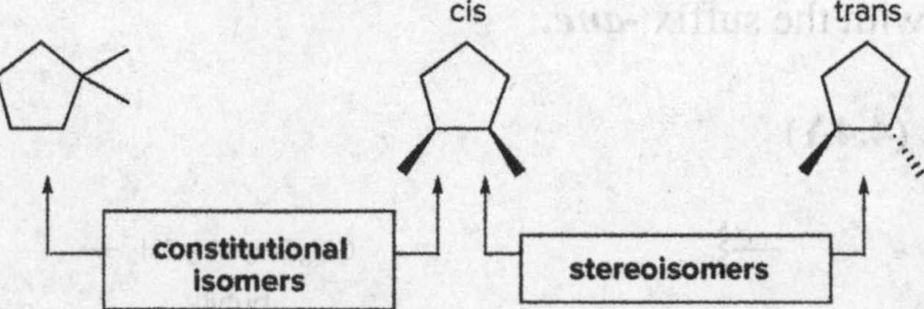

Conformations in cyclohexane (4.12, 4.13)

- Cyclohexane exists as **two chair conformations** in rapid equilibrium at room temperature.
- Each carbon atom on a cyclohexane ring has **one axial** and **one equatorial hydrogen.** Ring-flipping converts axial H's to equatorial H's, and vice versa.

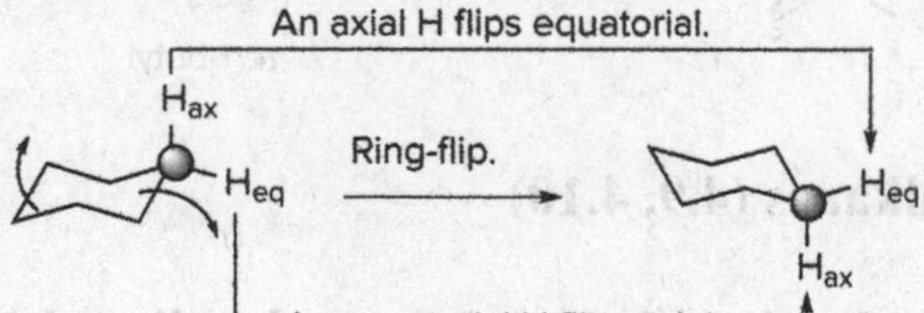

- In substituted cyclohexanes, groups larger than hydrogen are more stable in the **more roomy equatorial position.**

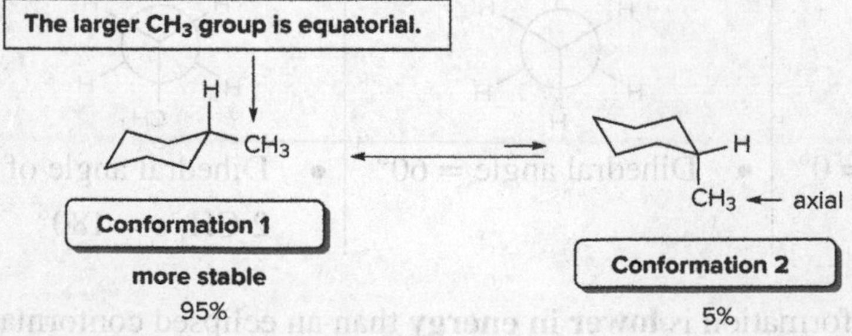

- Disubstituted cyclohexanes with substituents on different atoms exist as two possible stereoisomers.
  - The **cis** isomer has two groups on the **same side** of the ring, either both up or both down.
  - The **trans** isomer has two groups on **opposite sides** of the ring, one up and one down.

trans isomer cis isomer

Oxidation–reduction reactions (4.14)

- **Oxidation** results in an **increase in the number of C–Z bonds** or a **decrease in the number of C–H bonds.**

CH_3CH_2–OH (ethanol) → CH_3COOH (acetic acid)

Increase in C–O bonds = **oxidation**

- **Reduction** results in a **decrease in the number of C–Z bonds** or an **increase in the number of C–H bonds.**

$H_2C{=}CH_2$ (ethylene) → $H_3C{-}CH_3$ (ethane)

Increase in C–H bonds = **reduction**

Practice Test on Chapter Review

1.a. Which statement is true about compounds **A**–**D** below?

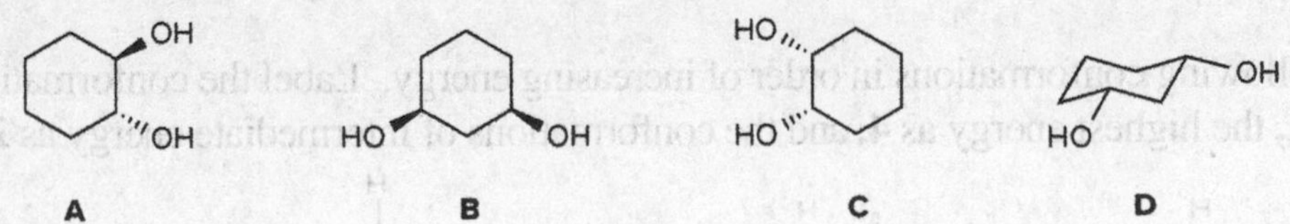

1. **A** and **C** are stereoisomers.
2. **B** and **D** are identical.
3. **A** and **B** are stereoisomers.
4. Statements (1) and (2) are both true.
5. Statements (1), (2), and (3) are all true.

b. Which of the following statements is true about *cis*-1-isopropyl-2-methylcyclohexane?

1. The more stable conformation has the isopropyl group in the equatorial position and the methyl group in the axial position.
2. The more stable conformation has both the methyl and isopropyl groups in the equatorial position.
3. *cis*-1-Isopropyl-2-methylcyclohexane is a stereoisomer of *trans*-1-isopropyl-3-methyl-cyclohexane.
4. Statements (1) and (2) are true.
5. Statements (1), (2), and (3) are all true.

c. Rank the following conformations in order of ***increasing energy.***

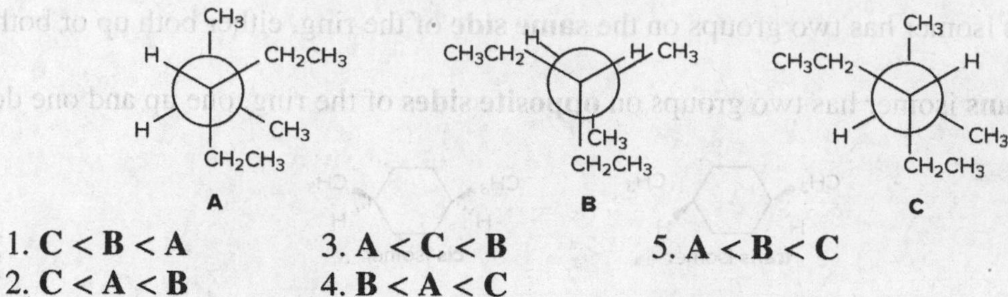

1. C < B < A **3. A < C < B** **5. A < B < C**
2. C < A < B **4. B < A < C**

2. Give the IUPAC name for each of the following compounds.

a.

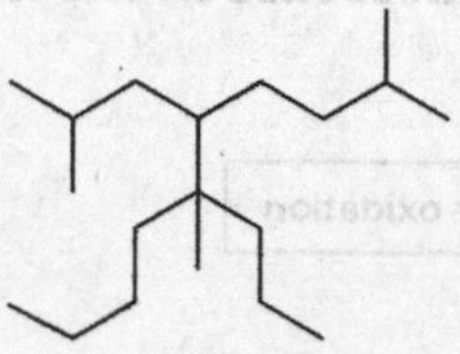

b.

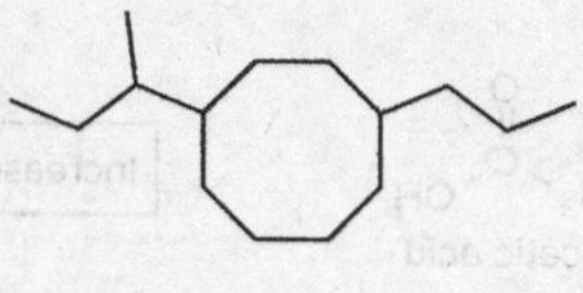

3. How are the molecules in each pair related? Are they constitutional isomers, stereoisomers, identical, or not isomers?

a. and

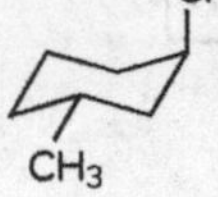

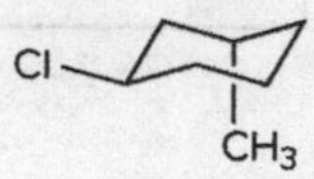

c. 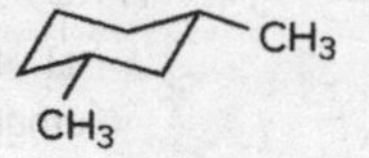 and

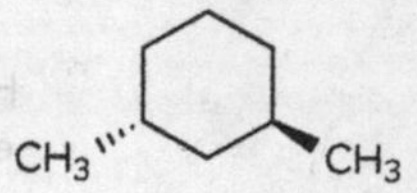

b. 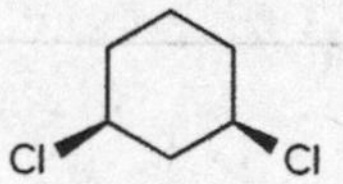 and

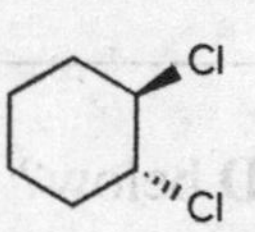

4. Rank the following conformations in order of increasing energy. Label the conformation of lowest energy as **1,** the highest energy as **4,** and the conformations of intermediate energy as **2** and **3.**

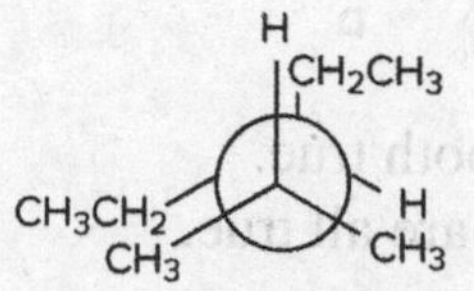

A 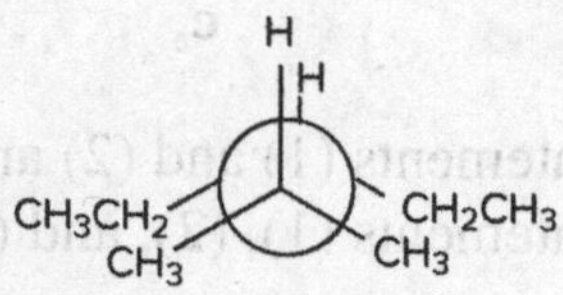 B 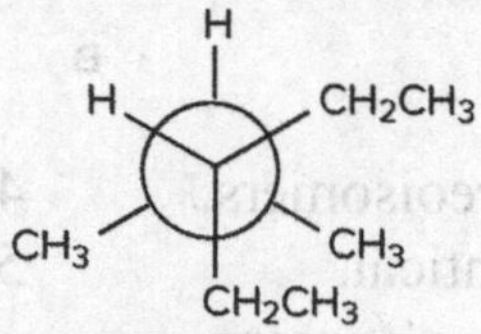 C 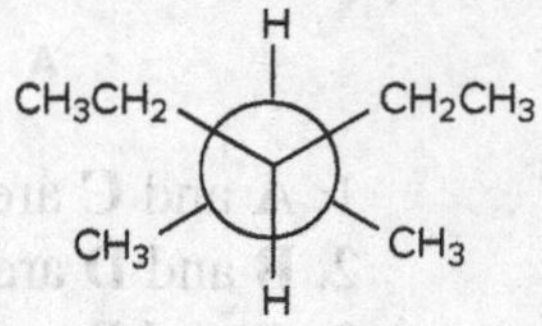 D

5. Consider the following disubstituted cyclohexane drawn below:

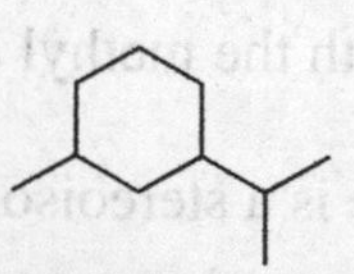

a. Draw the more stable chair conformation for the cis isomer.
b. Draw the more stable chair conformation for the trans isomer.

Answers to Practice Test

1. a. 4	2. a. 5-isobutyl-2,6-dimethyl-6-propyldecane	3. a. identical	4. **A–3** **B–4** **C–2** **D–1**	5. a.
b. 1	b. 1-*sec*-butyl-4-propylcyclooctane	b. constitutional isomers		
c. 2		c. stereoisomers		b. 

Answers to Problems

4.1 The general molecular formula for an acyclic alkane is $\mathbf{C_nH_{2n+2}}$.

Number of C atoms = *n*	**2*n* + 2**	Number of H atoms
27	2(27) + 2 =	56
31	2(31) + 2 =	64

4.2 2-Methylbutane has 4 C's in a row with a 1 C branch.

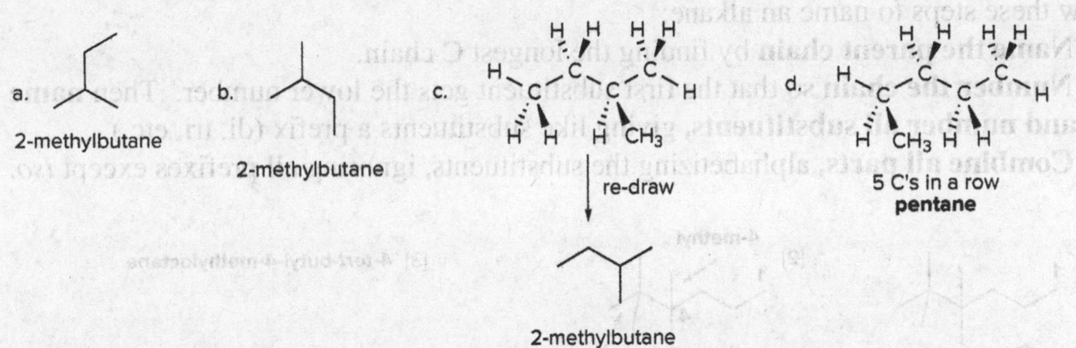

4.3 **Constitutional isomers differ in the way the atoms are connected to each other.** To draw all the constitutional isomers:

[1] Draw all of the C's in a long chain.

[2] Take off one C and use it as a substituent. (Don't add it to the end carbon: this re-makes the long chain.)

[3] Take off two C's and use these as substituents, etc.

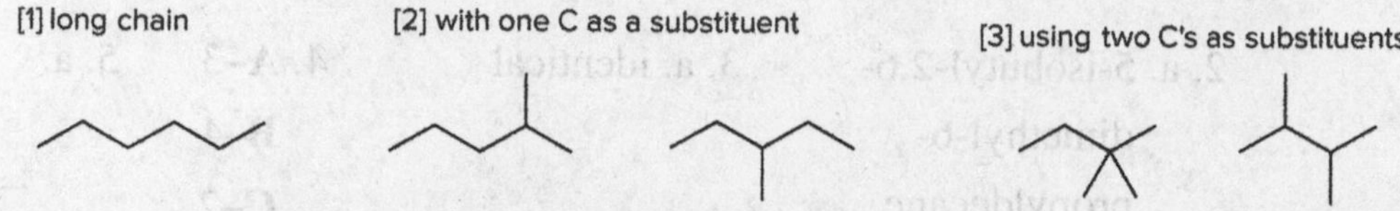

4.4 Draw each alkane to satisfy the requirements.

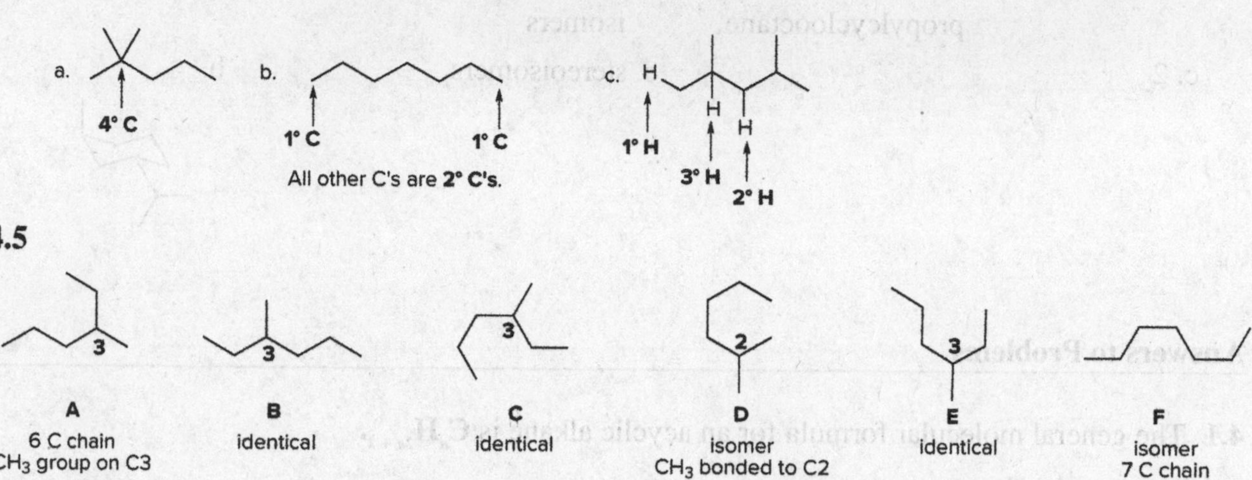

4.6 Use the steps from Answer 4.3 to draw the constitutional isomers.

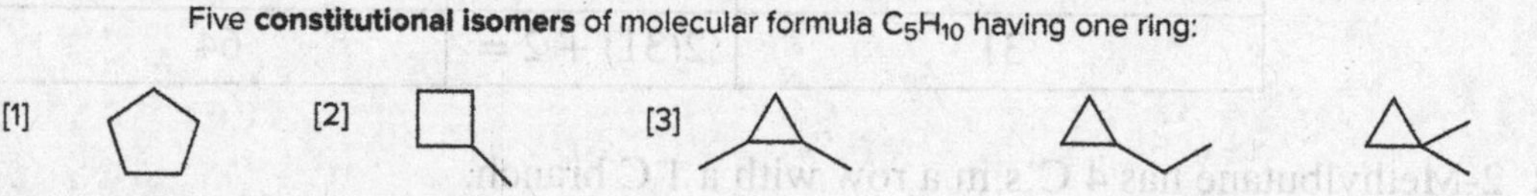

4.7 Follow these steps to name an alkane:

[1] **Name the parent chain** by finding the longest C chain.

[2] **Number the chain** so that the first substituent gets the lower number. Then **name and number all substituents,** giving like substituents a prefix (di, tri, etc.).

[3] **Combine all parts,** alphabetizing the substituents, ignoring all prefixes except *iso*.

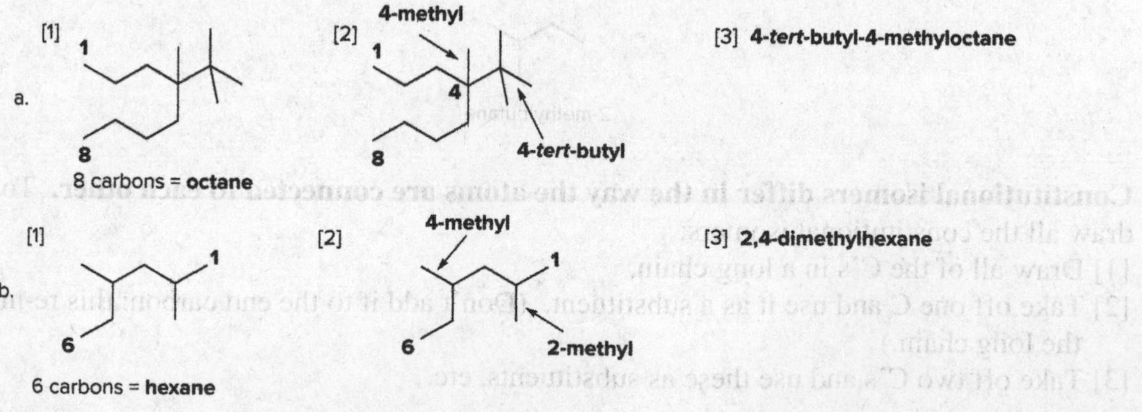

c. [1] 9 carbons = **nonane** [2] **6-isopropyl**, **3-methyl** [3] **6-isopropyl-3-methylnonane**

d. [1] 7 carbons = **heptane** [2] **4-methyl**, **2-methyl** [3] **2,4-dimethylheptane**

4.8 Use the steps in Answer 4.7 to name each alkane.

a. $(CH_3)_3CCH_2CH(CH_2CH_3)_2$

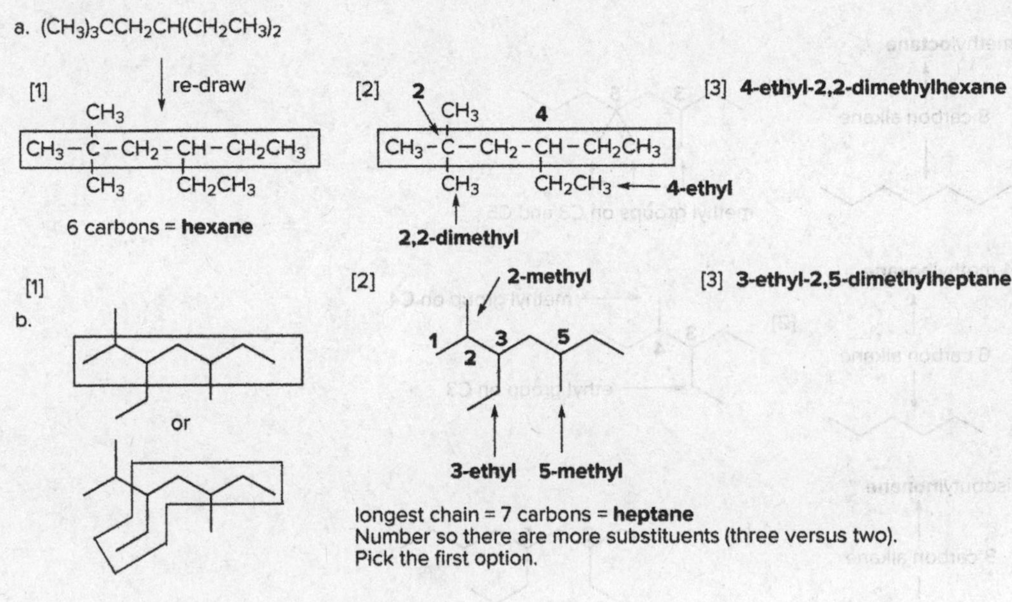

longest chain = 7 carbons = **heptane**
Number so there are more substituents (three versus two).
Pick the first option.

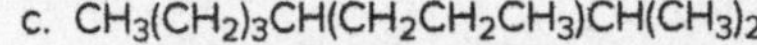

c. $CH_3(CH_2)_3CH(CH_2CH_2CH_3)CH(CH_3)_2$

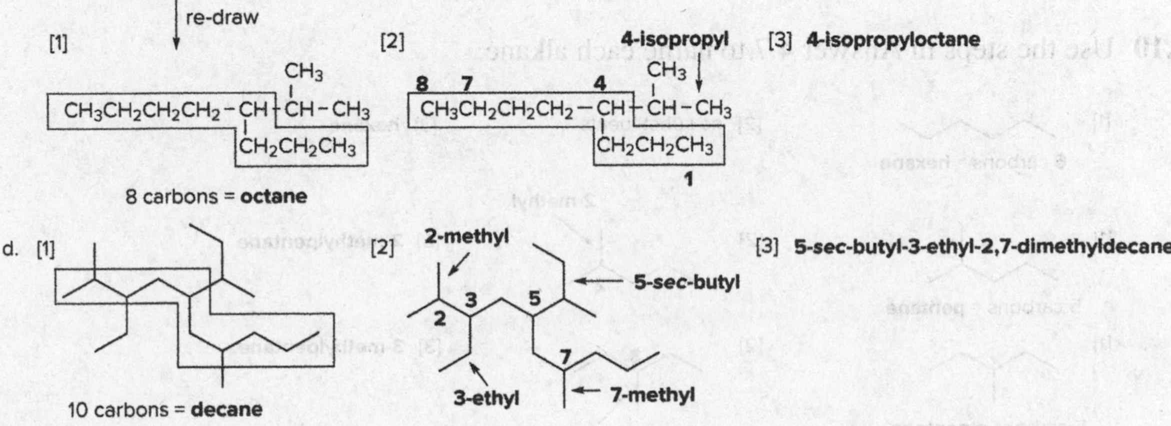

4.9 To work backwards from a name to a structure:

[1] Find the parent name and draw that number of C's. Use the suffix to identify the functional group (*-ane* = **alkane**).

[2] Arbitrarily number the C's in the chain or ring. Add the substituents to the appropriate C's.

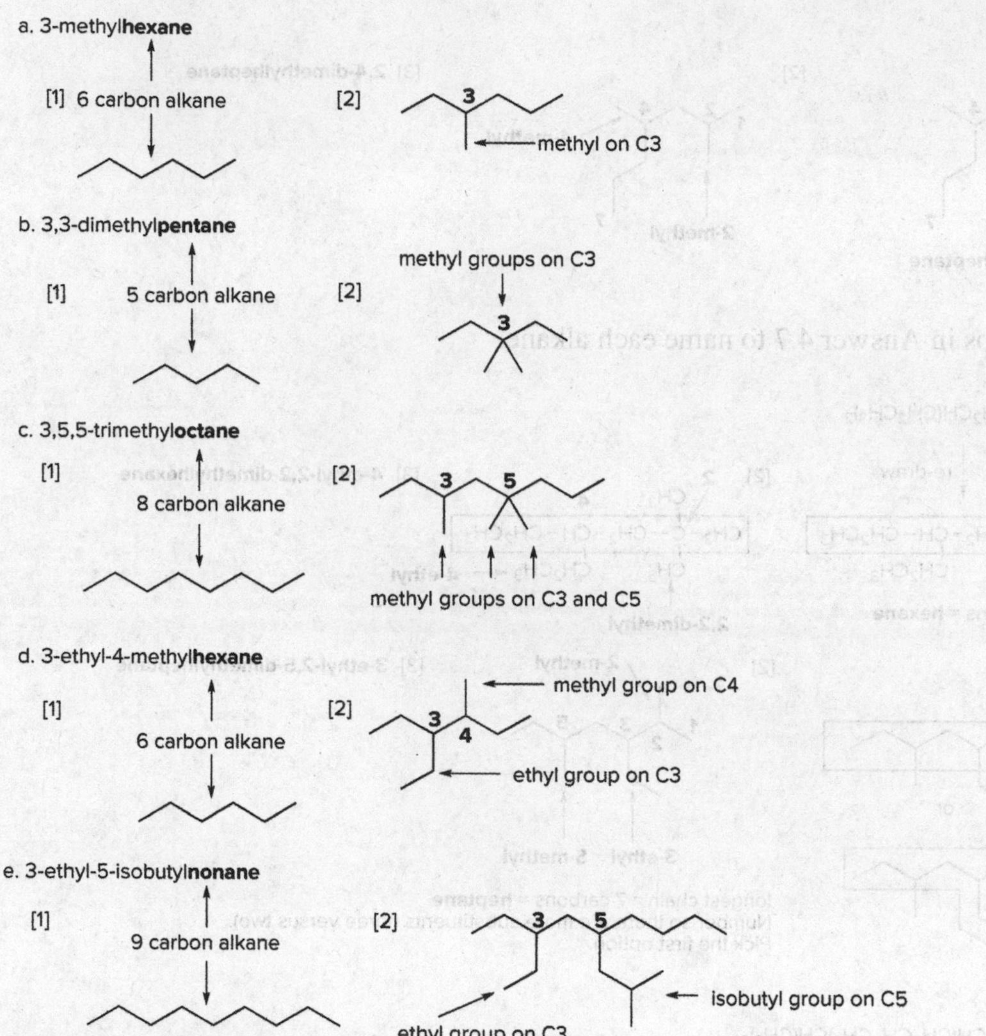

4.10 Use the steps in Answer 4.7 to name each alkane.

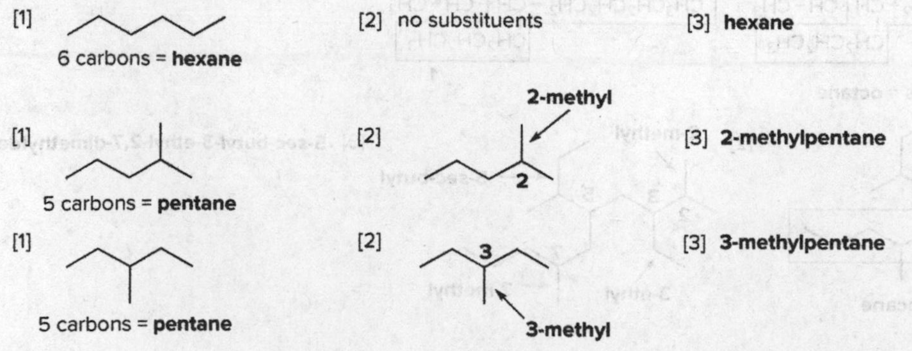

[1] 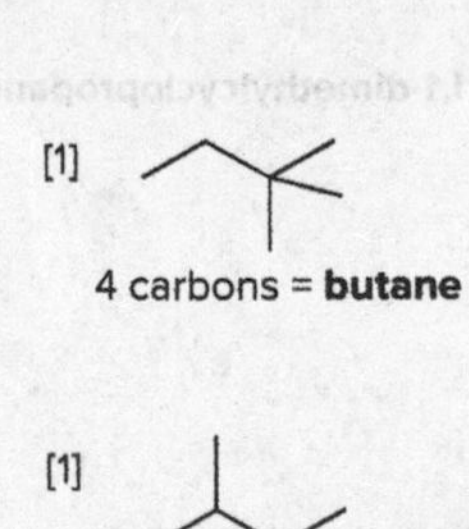

[2] 2,2-dimethyl; 2

[3] **2,2-dimethylbutane**

[1] 4 carbons = **butane**

[2] 2 3; 2,3-dimethyl

[3] **2,3-dimethylbutane**

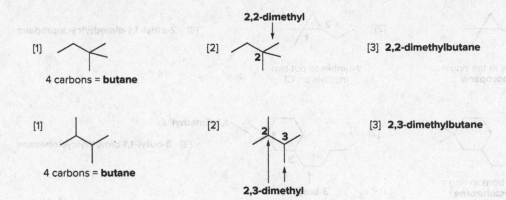

4.11 Follow these steps to name a cycloalkane:

[1] **Name the parent cycloalkane** by counting the C's in the ring and adding *cyclo-*.

[2] **Numbering:**

a. **Number around the ring** beginning at a substituent and giving the second substituent the lower number.

b. **Number to assign the lower number to the substituents alphabetically.**

c. **Name and number all substituents,** giving like substituents a prefix (di, tri, etc.).

[3] **Combine all parts,** alphabetizing the substituents, ignoring all prefixes except *iso*.

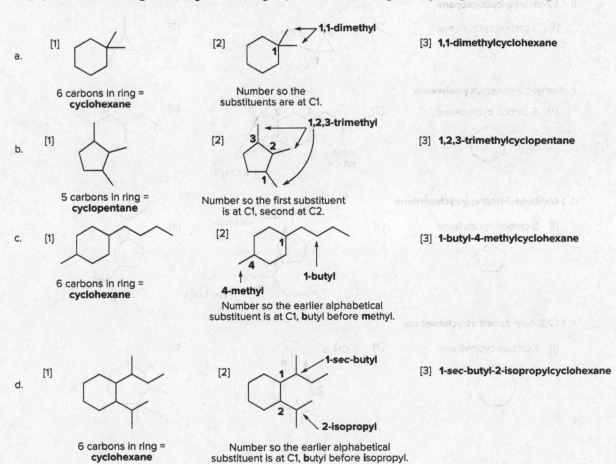

a. [1] 6 carbons in ring = **cyclohexane**

[2] 1,1-dimethyl; 1. Number so the substituents are at C1.

[3] **1,1-dimethylcyclohexane**

b. [1] 5 carbons in ring = **cyclopentane**

[2] 1,2,3-trimethyl; 3 2 1. Number so the first substituent is at C1, second at C2.

[3] **1,2,3-trimethylcyclopentane**

c. [1] 6 carbons in ring = **cyclohexane**

[2] 1; 4; 1-butyl; 4-methyl. Number so the earlier alphabetical substituent is at C1, **b**utyl before **m**ethyl.

[3] **1-butyl-4-methylcyclohexane**

d. [1] 6 carbons in ring = **cyclohexane**

[2] 1-*sec*-butyl; 1; 2; 2-isopropyl. Number so the earlier alphabetical substituent is at C1, **b**utyl before **i**sopropyl.

[3] **1-*sec*-butyl-2-isopropylcyclohexane**

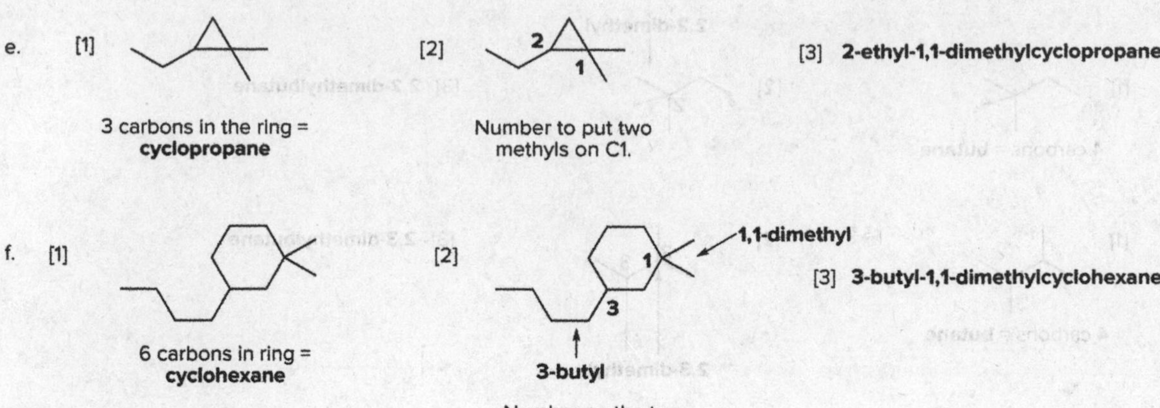
e. [1]
[2]
2
1
[3] 2-ethyl-1,1-dimethylcyclopropane
3 carbons in the ring = cyclopropane
Number to put two methyls on C1.
f. [1]
[2]
1,1-dimethyl
1
3
[3] 3-butyl-1,1-dimethylcyclohexane
6 carbons in ring = cyclohexane
3-butyl
Number so the two methyls are at C1.

4.12 To draw the structures, use the steps in Answer 4.9.

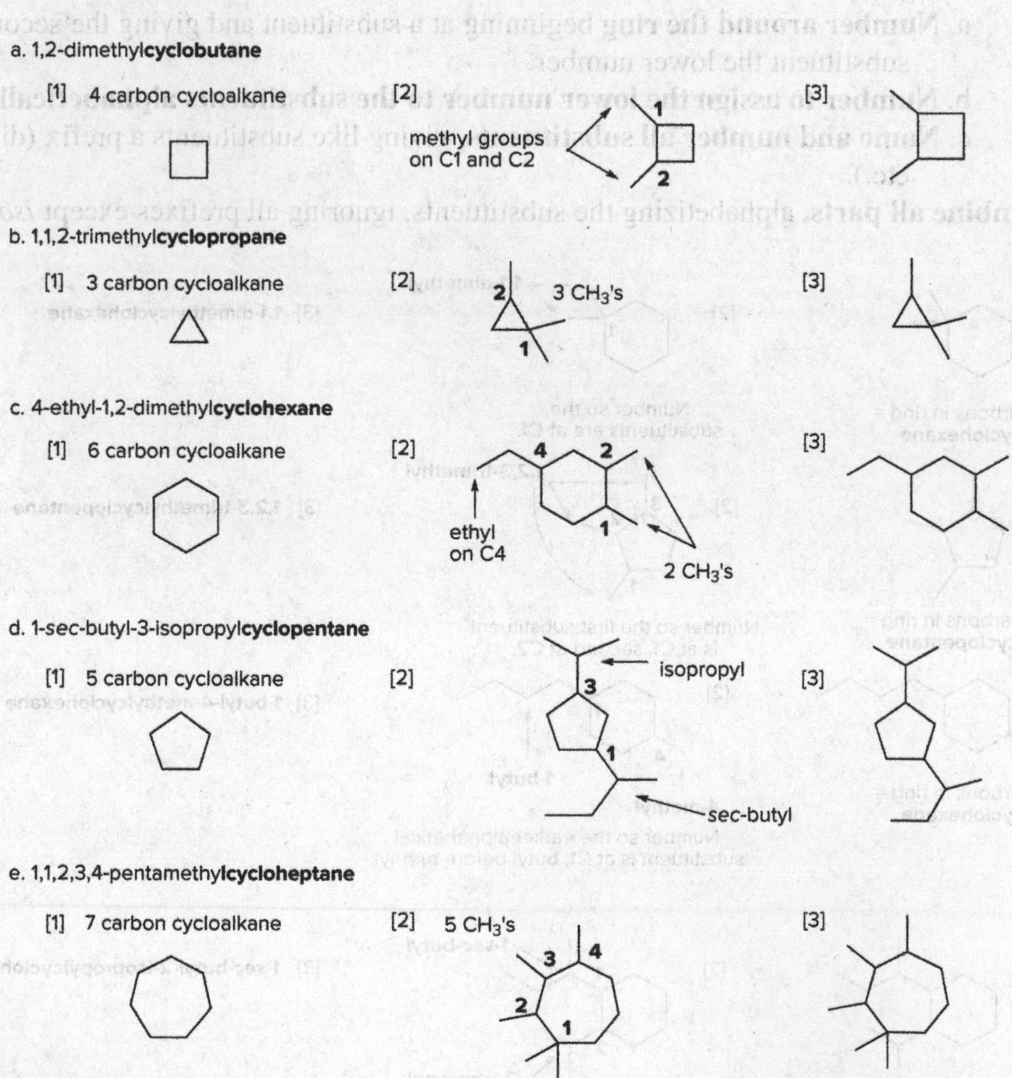
a. 1,2-dimethylcyclobutane
[1] 4 carbon cycloalkane
[2]
methyl groups on C1 and C2
1
2
[3]
b. 1,1,2-trimethylcyclopropane
[1] 3 carbon cycloalkane
[2]
2
3 CH3's
1
[3]
c. 4-ethyl-1,2-dimethylcyclohexane
[1] 6 carbon cycloalkane
[2]
4
2
1
ethyl on C4
2 CH3's
[3]
d. 1-sec-butyl-3-isopropylcyclopentane
[1] 5 carbon cycloalkane
[2]
isopropyl
3
1
sec-butyl
[3]
e. 1,1,2,3,4-pentamethylcycloheptane
[1] 7 carbon cycloalkane
[2] 5 CH3's
3
4
2
1
[3]

4.13 Compare the number of C's and surface area to determine relative boiling points.
Rules:
[1] Increasing number of C's = increasing boiling point.
[2] Increasing surface area = increasing boiling point (branching decreases surface area).

8 C's linear, largest number of C's, no branching, **highest bp** — 7 C's linear — 7 C's one branch — 7 C's three branches
increasing branching, decreasing surface area, **decreasing bp**

Increasing boiling point: $(CH_3)_3CCH(CH_3)_2$ < $CH_3CH_2CH_2CH_2CH(CH_3)_2$ < $CH_3(CH_2)_5CH_3$ < $CH_3(CH_2)_6CH_3$

4.14 To draw a Newman projection, visualize the carbons as one in front and one in back of each other. The C–C bond is not drawn.

a. C in front
b. C in front
c. C in front
d. C in front

4.15 Re-draw each Newman projection as a skeletal structure, remembering that a Newman projection shows the substituents bonded to two C's but not the C's themselves.

2-methylpentane

A: CH_3–C(H)(CH_3)–$CH_2CH_2CH_3$ (C in back) — 2-methylpentane

B: CH_3–C(H)(CH_3)–C(H)(H)–CH_2CH_3 (C in back) — 2-methylpentane

C: CH_3CH_2–C(H)(CH_3)–C(H)(H)–CH_3 (C in back) — 3-methylpentane

4.16

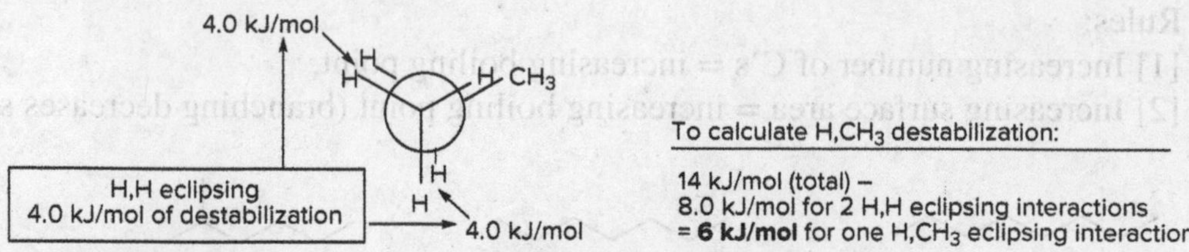

4.17 To determine the energy of conformations, keep two things in mind:
[1] Staggered conformations are more stable than eclipsed conformations.
[2] Minimize steric interactions: keep large groups away from each other.
The highest energy conformation is the eclipsed conformation in which the two largest groups are eclipsed. The lowest energy conformation is the staggered conformation in which the two largest groups are anti.

rotation here

60° → 60° → 60° → 60° → 60°

1 staggered **most stable**

2 eclipsed

3 staggered **most stable**

4 eclipsed **least stable**

5 staggered

6 eclipsed **least stable**

4.18 Use the criteria in Answer 4.17 to determine the relative energy.

A B C D

Staggered conformations **A** and **D** are lower in energy than eclipsed conformations **B** and **C**. **D** is lower in energy than **A** because it has two gauche interactions and **A** has three. **C** is higher in energy than **B** because all the larger groups (CH_3 and CH_2CH_3) are eclipsed.
Ranking: **D < A < B < C**

4.19 To determine the most and least stable conformations, use the rules from Answer 4.17.

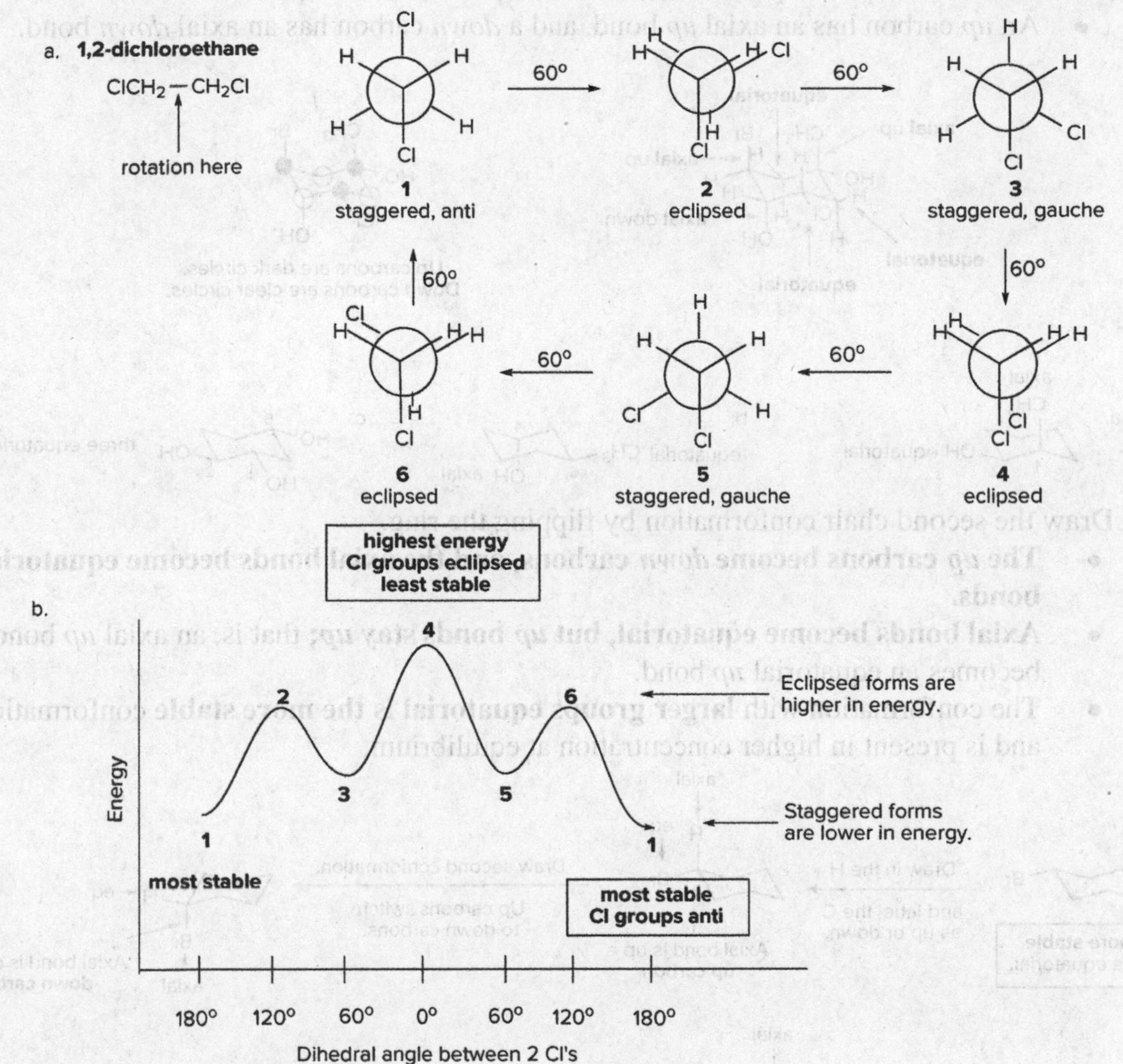

4.20 Add the energy increase for each eclipsing interaction to determine the destabilization.

a.

1 H,H interaction = 4.0 kJ/mol
2 H,CH_3 interactions (2 x 6.0 kJ/mol) = 12.0 kJ/mol

Total destabilization = 16 kJ/mol

b.

3 H,CH_3 interactions (3 x 6.0 kJ/mol) = **18 kJ/mol**

Total destabilization

4.21 Two points:

- **Axial bonds point up or down, whereas equatorial bonds point out.**
- **An *up* carbon has an axial *up* bond, and a *down* carbon has an axial *down* bond.**

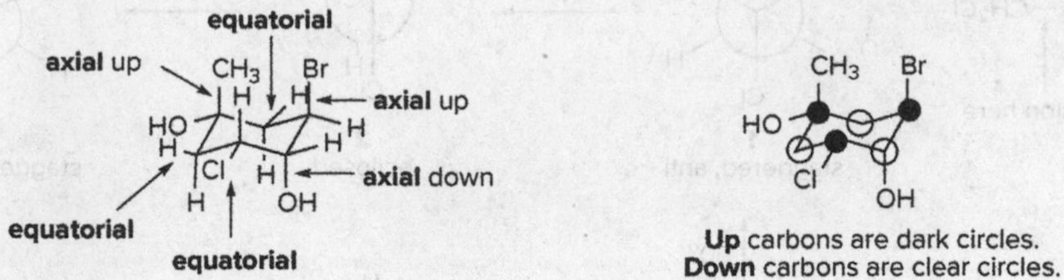

4.22

a. axial CH_3; 1, 2; OH equatorial

b. equatorial CH_3; 4; 6; OH axial

c. HO; 5; OH; HO; 1; 2; three equatorial OH's

4.23 Draw the second chair conformation by flipping the ring.

- **The *up* carbons become *down* carbons, and the axial bonds become equatorial bonds.**
- **Axial bonds become equatorial, but *up* bonds stay *up***; that is, an axial *up* bond becomes an equatorial *up* bond.
- The conformation with **larger groups equatorial is the more stable** conformation and is present in higher concentration at equilibrium.

a. Br → Draw in the H and label the C as up or down. → axial H, eq Br; Axial bond is up = **up carbon** → Draw second conformation. Up carbons switch to down carbons. → H eq, Br axial; Axial bond is down = **down carbon**

more stable
Br is equatorial.

b. Cl → Draw in the H and label the C as up or down. → axial Cl, H eq; Axial bond is up = **up carbon** → Draw second conformation. Up carbons switch to down carbons. → eq Cl, axial H; Axial bond is down = **down carbon**

more stable
Cl is equatorial.

c. CH_2CH_3 → Draw in the H and label the C as up or down. → H, CH_2CH_3 eq; Axial bond is up = **up carbon** → Draw second conformation. Up carbons switch to down carbons. → eq H, CH_2CH_3; Axial bond is down = **down carbon**

more stable
CH_2CH_3 is equatorial.

4.24

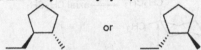

CH_3 ax
CH_2CH_3
eq
more stable
The larger ethyl group is in the roomier equatorial position.
CH_3 eq
CH_2CH_3
ax

4.25 Wedges represent "up" groups in front of the page, whereas dashed wedges are "down" groups in back of the page. Cis groups are on the same side of the ring, whereas trans groups are on opposite sides of the ring.

a. ***cis*-1,2-dimethylcyclopropane**

or

cis = same side of the ring
both groups on wedges or
both on dashed wedges

b. ***trans*-1-ethyl-2-methylcyclopentane**

or

trans = opposite sides of the ring
one group on a wedge,
one group on a dashed wedge

4.26 Cis and trans isomers are stereoisomers.

***cis*-1,3-diethylcyclobutane**

cis = same side of the ring
both groups on wedges or
both on dashed wedges

a. ***trans*-1,3-diethylcyclobutane**

trans = opposite sides of the ring
one group on a wedge,
one group on a dashed wedge

b. ***cis*-1,2-diethylcyclobutane**

constitutional isomer
different arrangement of atoms

4.27

a.

groups on same side
cis isomer

groups on opposite sides
trans isomer
(one possibility)

b. cis:

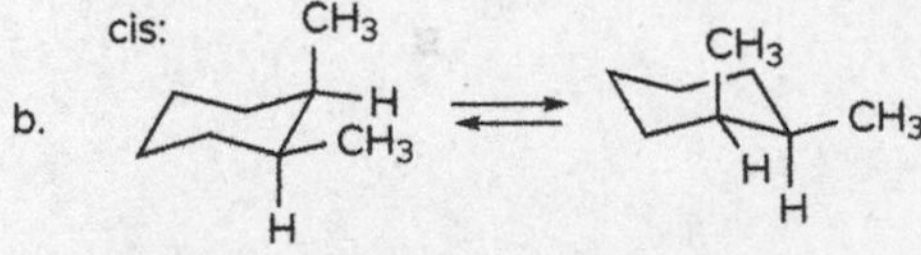

two chair conformations for the **cis isomer**

Same stability because they both have one equatorial, one axial CH_3 group.

c. trans:

CH_3
H
H
CH_3
H
CH_3
CH_3
H

both groups equatorial
more stable

two chair conformations for the **trans isomer**

d. The **trans isomer is more stable** because it can have both methyl groups in the more roomy **equatorial** position.

4.28 To classify a compound as a cis or trans isomer, **classify each non-hydrogen group as up or down. Groups on the same side = cis isomer; groups on opposite sides = trans isomer.**

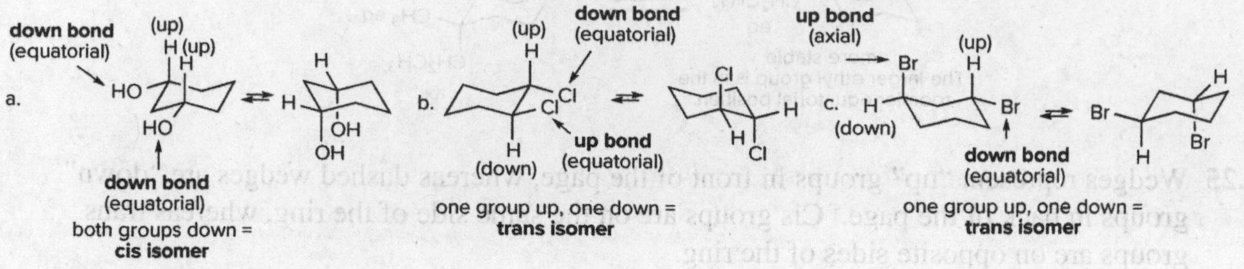

a. down bond (equatorial) (up) H (up) HO H ⇌ H HO OH OH
down bond (equatorial)
both groups down = **cis isomer**

b. (up) H down bond (equatorial) Cl ⇌ Cl H Cl H (down) up bond (equatorial)
one group up, one down = **trans isomer**

c. up bond (axial) Br (up) H H (down) Br ⇌ Br H Br
down bond (equatorial)
one group up, one down = **trans isomer**

4.29

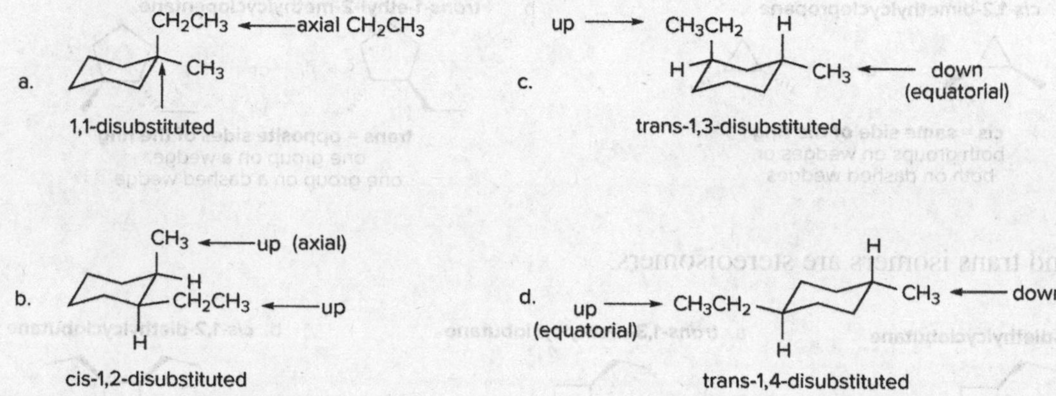

a. CH_2CH_3 ← axial CH_2CH_3, CH_3
1,1-disubstituted

b. CH_3 ← up (axial), H, CH_2CH_3 ← up, H
cis-1,2-disubstituted

c. up → CH_3CH_2 H, H, CH_3 ← down (equatorial)
trans-1,3-disubstituted

d. H, CH_3 ← down, up → CH_3CH_2 (equatorial), H
trans-1,4-disubstituted

4.30

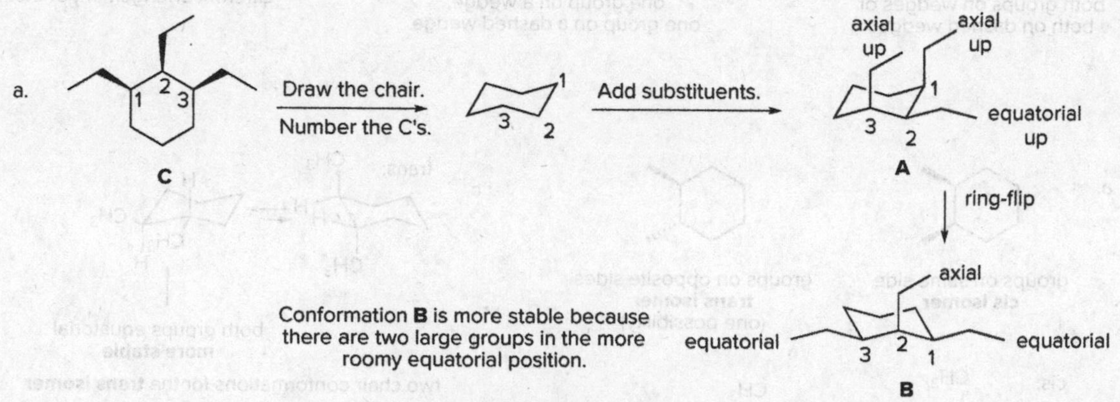

a. C → Draw the chair. Number the C's. → Add substituents. → A (axial up, axial up, equatorial up) ⇌ ring-flip → B (axial, equatorial, equatorial)

Conformation **B** is more stable because there are two large groups in the more roomy equatorial position.

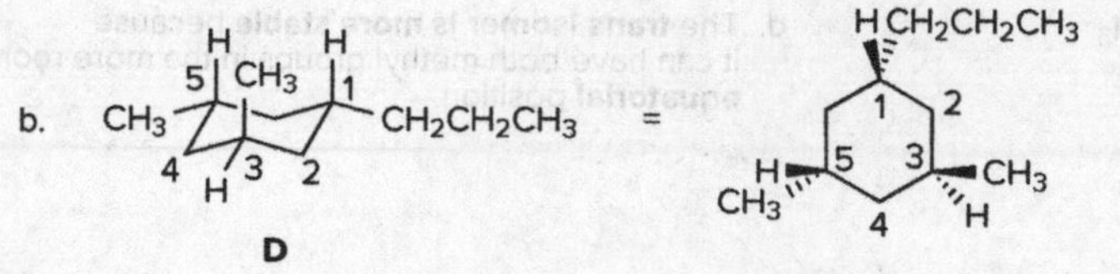

b. D: CH_3, H, H, CH_3, $CH_2CH_2CH_3$, H = H $CH_2CH_2CH_3$, CH_3, H, CH_3, H

4.31 *Oxidation* results in an *increase* in the number of C–Z bonds, or a *decrease* in the number of C–H bonds.
Reduction results in a *decrease* in the number of C–Z bonds, or an *increase* in the number of C–H bonds.

a. (benzaldehyde → benzoic acid)

Decrease in the number of C–H bonds.
Increase in the number of C–O bonds.
Oxidation

c. (acetone → 2,2-propanediol, OH/OH)

No change in the number of C–O or C–H bonds. **Neither**

b. (bicyclic ketone → bicyclic alkane)

Decrease in the number of C–O bonds.
Increase in the number of C–H bonds.
Reduction

d. (cyclohexanone → cyclohexanol, OH)

Decrease in the number of C–O bonds.
Increase in the number of C–H bonds.
Reduction

4.32 The products of a combustion reaction of a hydrocarbon are always the same: $CO_2 + H_2O$.

a. $CH_3CH_2CH_3 + 5\ O_2 \xrightarrow{\text{flame}} 3\ CO_2 + 4\ H_2O$ + heat

b. (cyclohexane) $+ 9\ O_2 \xrightarrow{\text{flame}} 6\ CO_2 + 6\ H_2O$ + heat

4.33 Re-draw the ball-and-stick model as a chair form.

a. (chair: H, H, H, 1, 2, 4, Br, CH_3, $CH(CH_3)_2$; $CH(CH_3)_2$ equatorial; Br equatorial; CH_3 axial)

b. CH_3 on C1 and Br on C2 are both down, making them cis.

c. Br on C2 and $CH(CH_3)_2$ on C4 are both down, making them cis.

d. Second chair form:

(chair: H, 1, CH_3, H, 2, Br, 4, H, $CH(CH_3)_2$)

This chair form is less stable because two groups [Br and $CH(CH_3)_2$] are in the more crowded axial position.

4.34

a. b.

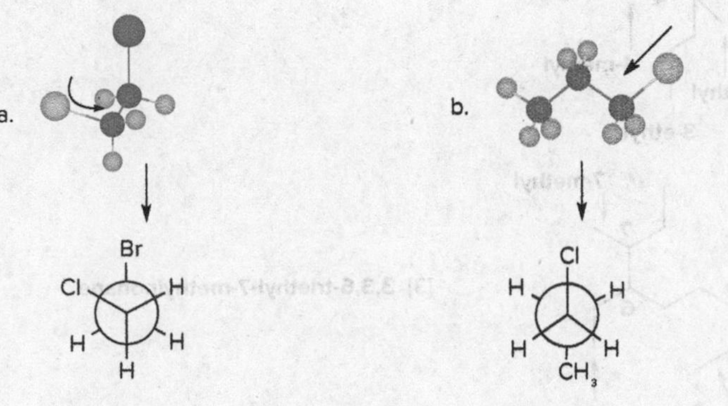

c.

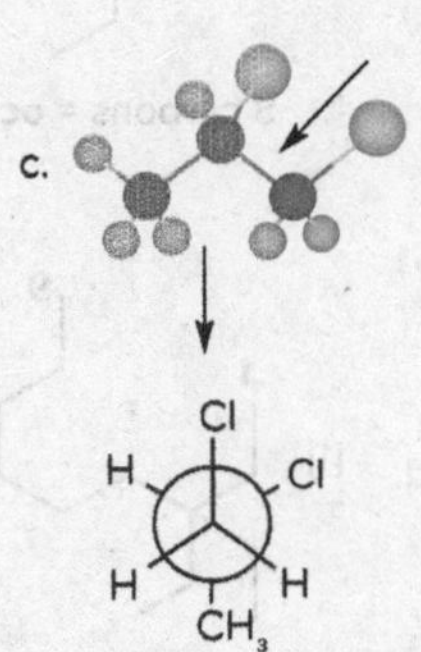

4.35 a. Five constitutional isomers of molecular formula C_4H_8:

b. Nine constitutional isomers of molecular formula C_7H_{16}:

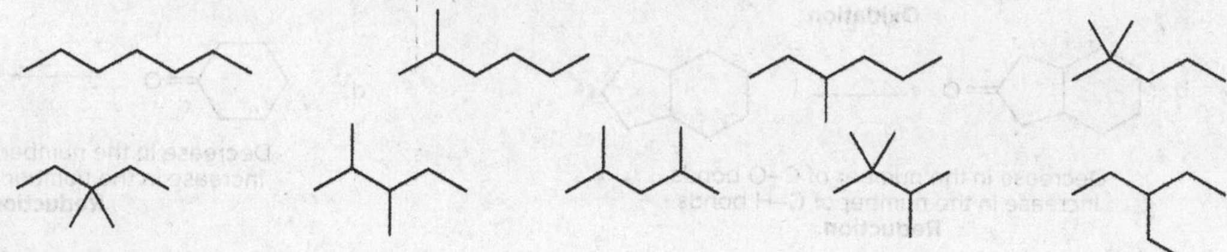

c. Twelve constitutional isomers of molecular formula C_6H_{12} containing one ring:

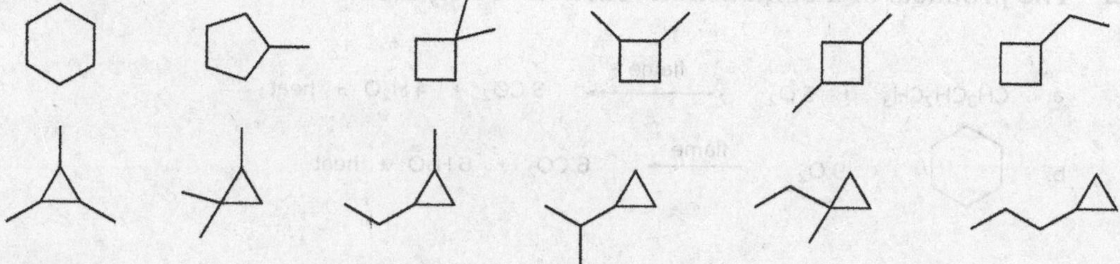

4.36 Use the steps in Answers 4.7 and 4.11 to name the alkanes.

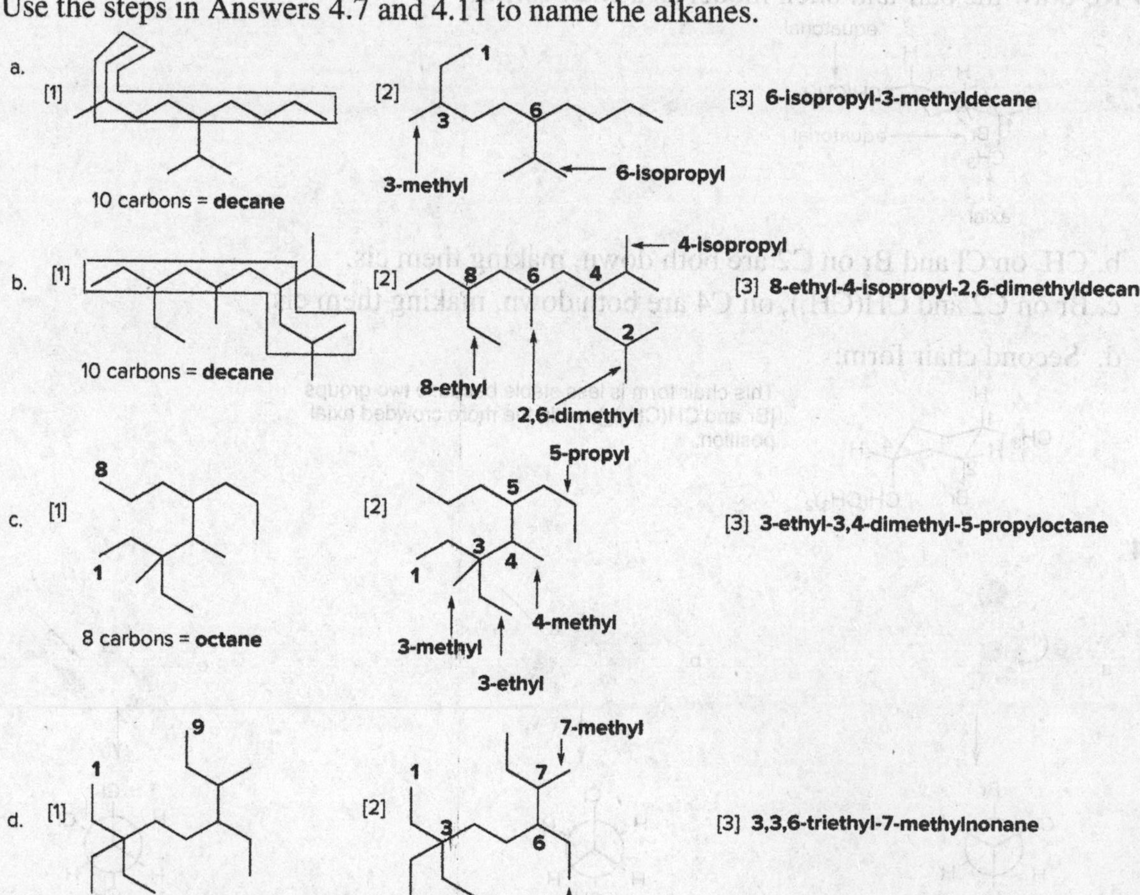

e. 3-cyclobutyl **3-cyclobutylpentane**

f. 1-*sec*-butyl 2-isopropyl **1-*sec*-butyl-2-isopropylcyclopentane**

g. 1-isobutyl 3-isopropyl **1-isobutyl-3-isopropylcyclohexane**

h. 4-isopropyl 2-methyl 6-propyl 5-isobutyl **5-isobutyl-4-isopropyl-2-methyl-6-propyldecane**

i. 2-*sec*-butyl 1,1-dimethyl 5-ethyl **2-*sec*-butyl-5-ethyl-1,1-dimethylcyclohexane**

j.

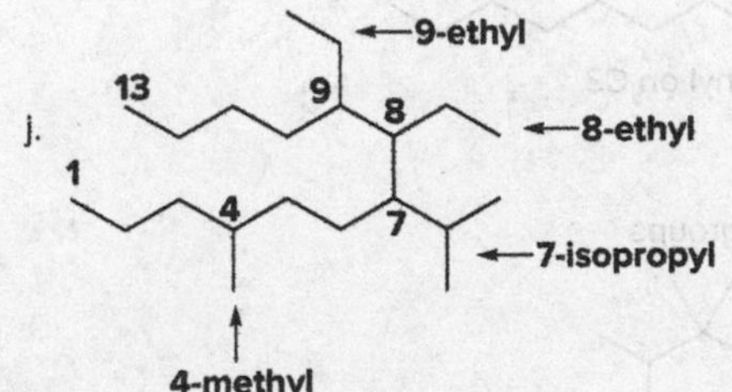

8,9-diethyl-7-isopropyl-4-methyltridecane

4.37 Use the steps in Answer 4.9 to draw the structures.

a. 3-ethyl-2-methyl**hexane**

[1] 6 C chain

[2] methyl on C2 ← ethyl on C3

b. *sec*-butyl**cyclopentane**

[1] 5 C ring

[2]

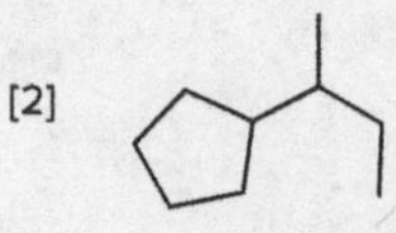

c. 4-isopropyl-2,4,5-trimethyl**undecane**

[1] 11 C chain

[2]

isopropyl on C4

methyls on C2, C4, and C5

d. cyclobutylcycloheptane

[1] 7 C cycloalkane

[2]

e. 3-ethyl-1,1-dimethyl**cyclohexane**

[1] 6 C cycloalkane

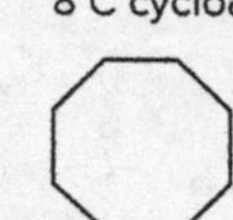

[2]

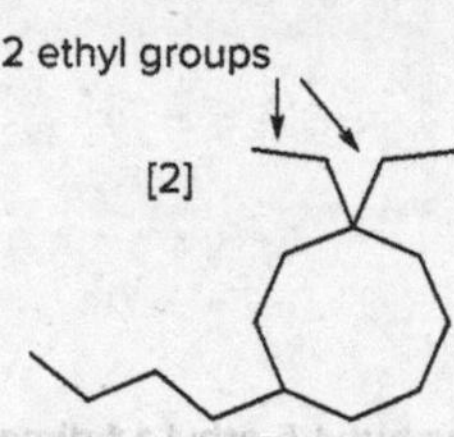

f. 4-butyl-1,1-diethyl**cyclooctane**

[1] 8 C cycloalkane

[2]

2 ethyl groups

g. 6-isopropyl-2,3-dimethyl**dodecane**

[1] 12 C alkane

[2]

methyl on C2

isopropyl on C6

methyl on C3

h. 2,2,6,6,7-pentamethyl**octane**

[1] 8 C alkane

[2]

5 methyl groups

i. *cis*-1-ethyl-3-methyl**cyclopentane**

[1] 5 C ring

[2]

ethyl on C1

methyl on C3

or

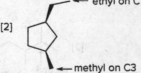

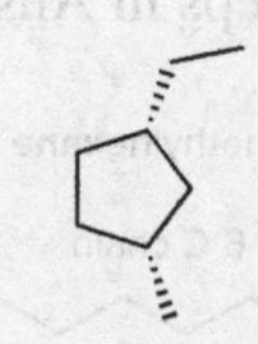

j. *trans*-1-*tert*-butyl-4-ethyl**cyclohexane**

[1] 6 C ring [2]

4.38 Correct each name.

a. Incorrect name: 7-ethyl-3,6-dimethylnonane

Proper numbering: 9 8 7 6 5 4 3 2 1

Number to give the second substituent in the chain the lower number:
3-ethyl-4,7-dimethylnonane

c. Incorrect name: 3-ethyl-1,4-dimethylcycloheptane

Ring was numbered incorrectly.
Proper numbering:

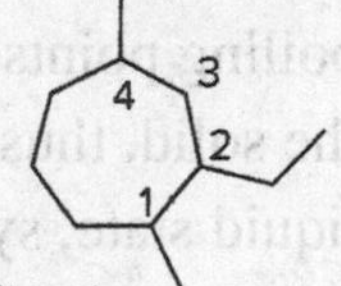

Number to give the second substituent on the ring the lower number:
2-ethyl-1,4-dimethylcycloheptane

b. Incorrect name: 4-ethyl-3-isopropylheptane

Proper numbering: 3 4 5 6 7 2 1

Choose the 7 C chain with more substituents:
3,4-diethyl-2-methylheptane

d. Incorrect name: 1-ethyl-3-methyl-5-isopropylcyclohexane

Begin at ethyl and number the ring to give the lower number to the substituent that comes earlier in the alphabet. Alphabetize the "i" of isopropyl before the "m" of methyl:
1-ethyl-3-isopropyl-5-methylcyclohexane

3 1 5

4.39

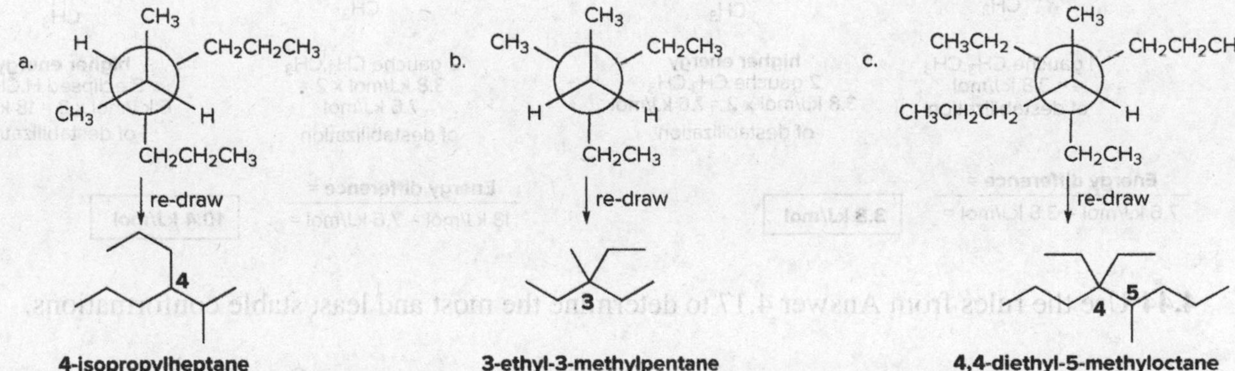

4.40 Use the rules from Answer 4.13.

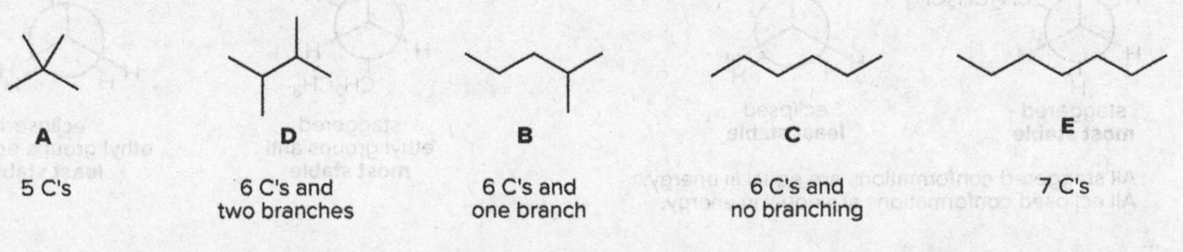

4.41 a.

$CH_3(CH_2)_6CH_3$
no branching = higher surface area
higher boiling point

$(CH_3)_3CC(CH_3)_3$
branching = lower surface area
lower boiling point
more spherical, better packing =
higher melting point

b. There is a 159°C difference in the melting points, but only a 20°C difference in the boiling points because the symmetry in $(CH_3)_3CC(CH_3)_3$ allows it to pack more tightly in the solid, thus requiring more energy to melt. In contrast, once the compounds are in the liquid state, symmetry is no longer a factor, the compounds are isomeric alkanes, and the boiling points are closer together.

4.42 The mineral oil can prevent the body's absorption of important fat-soluble vitamins. The vitamins dissolve in the mineral oil, and are thus not absorbed. Instead, they are expelled with the mineral oil.

4.43

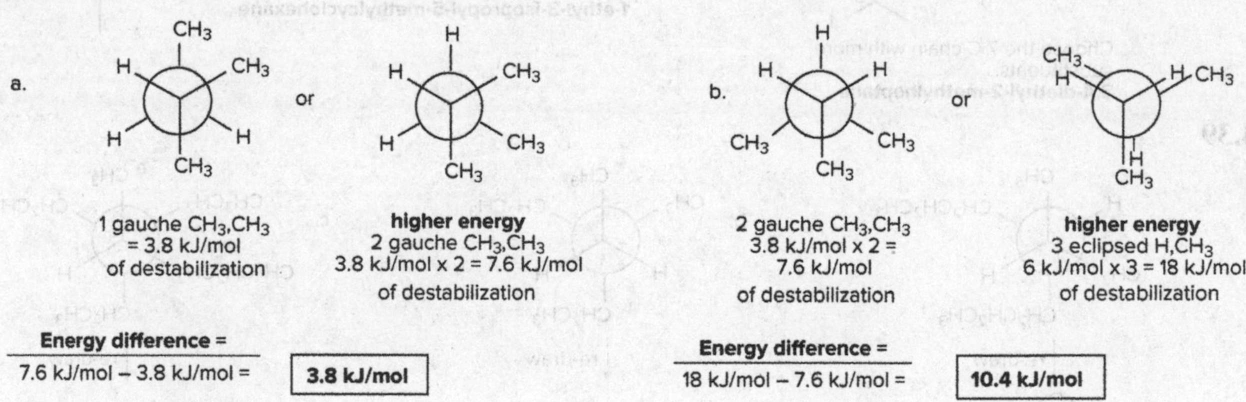

4.44 Use the rules from Answer 4.17 to determine the most and least stable conformations.

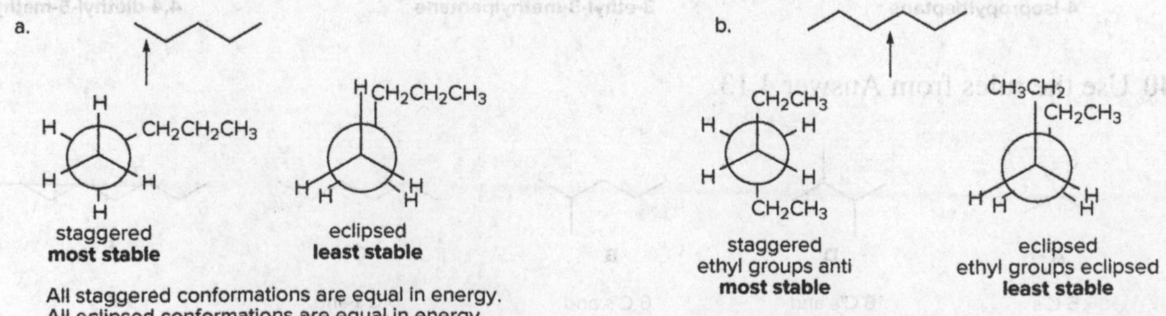

4.45 Use the rules from Answer 4.17 to rank the conformations.

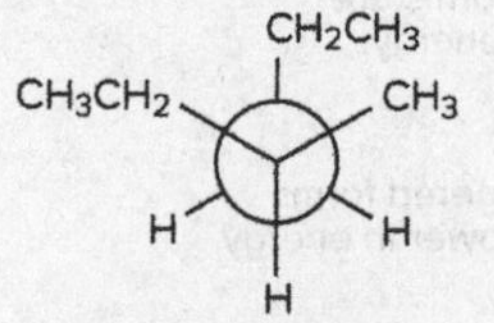

A
staggered
two gauche interactions
highest energy
staggered conformation

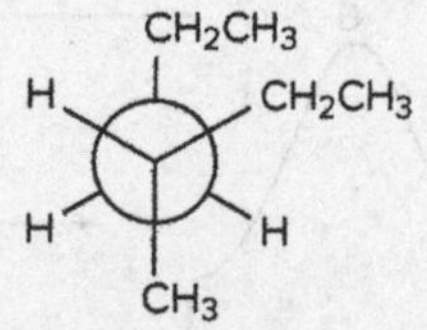

B
staggered
two ethyl groups gauche

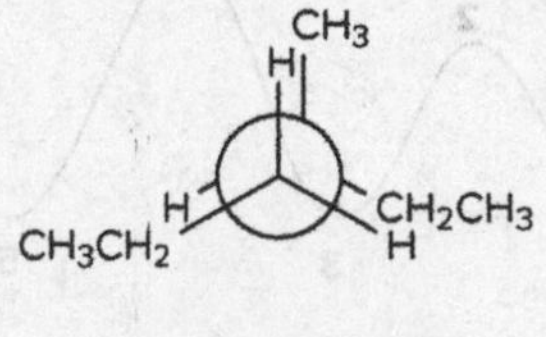

C
eclipsed conformation
highest in energy

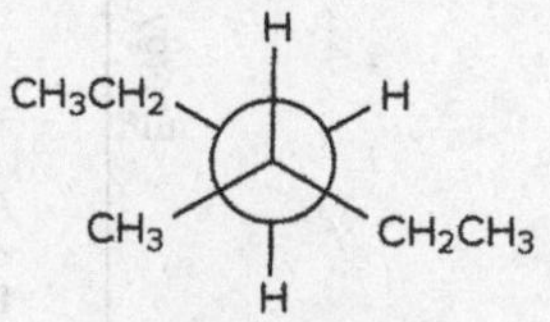

D
staggered
ethyl groups anti
lowest in energy

order: **D** < **B** < **A** < **C**
lowest energy — highest energy

4.46

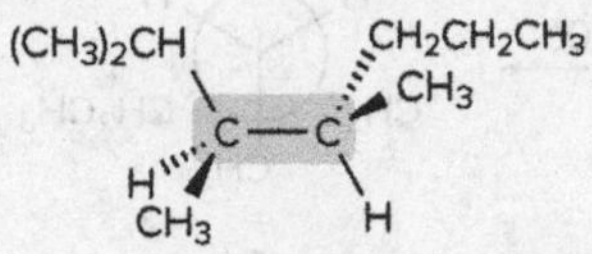

A
staggered
second most stable
Isopropyl and propyl are gauche.

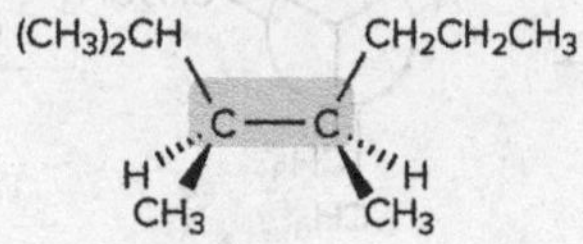

B
eclipsed
highest energy
largest groups eclipsed
least stable

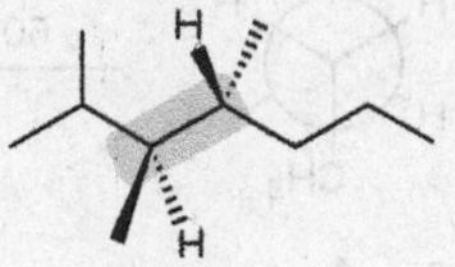

C
staggered
most stable
largest groups anti
two CH_3's anti

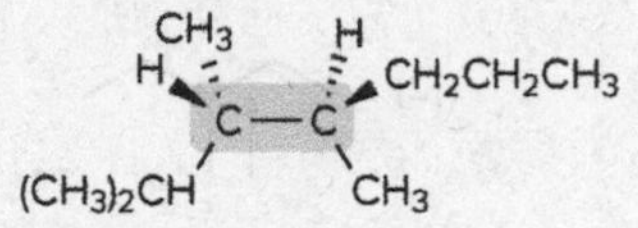

D
eclipsed
second highest in energy

Increasing stability: **B** < **D** < **A** < **C**

4.47

[1]

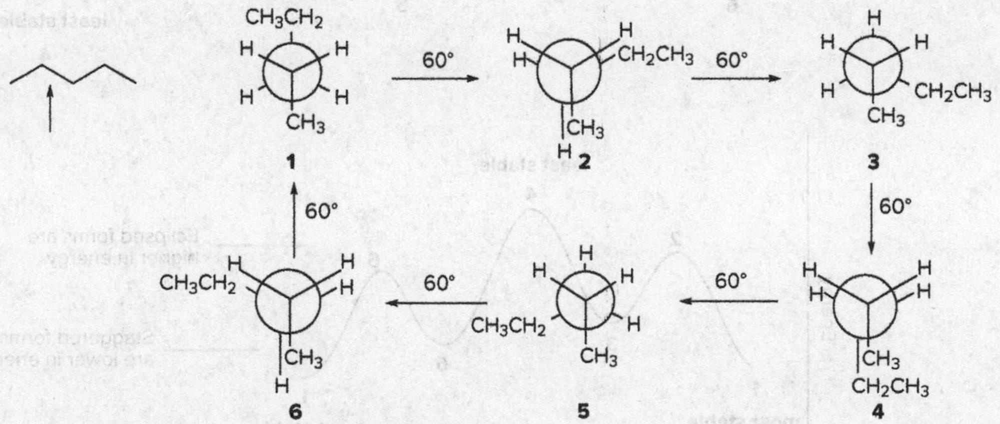

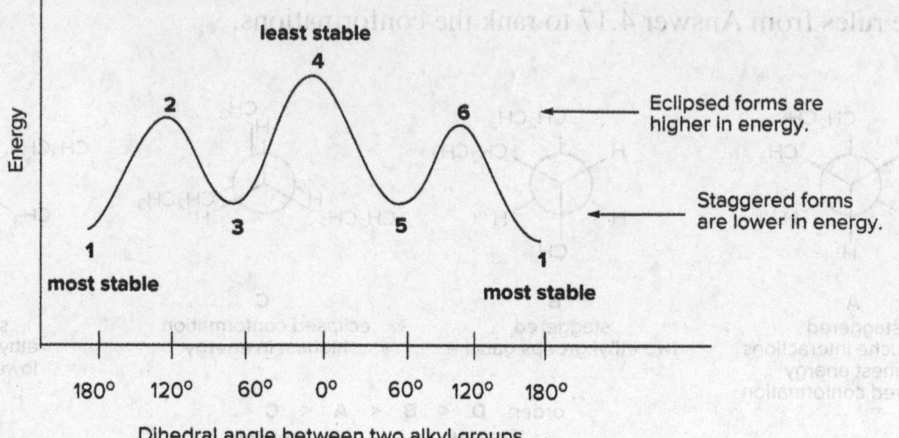

[2]

most stable

1: CH_3CH_2, H, H, H, CH_3, CH_3

60°

2: H, H, H, CH_2CH_3, CH_3, CH_3

60°

3: H, H, CH_3, CH_2CH_3, CH_3

60°

4: H, CH_3, H, H, CH_3, CH_2CH_3 — least stable

60°

5: CH_3, H, H, CH_3CH_2, H, CH_3

60°

6: CH_3CH_2, H, H, CH_3, CH_3, H

60°

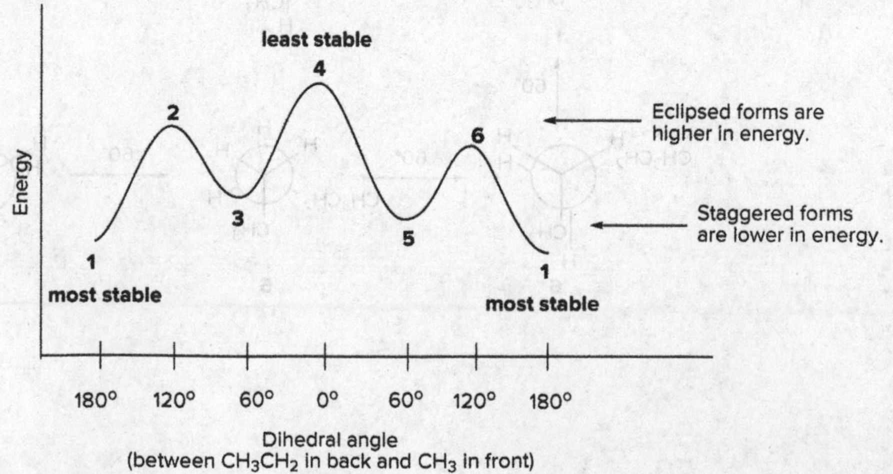

4.48 Two types of strain:

- *Torsional strain* is due to eclipsed groups on adjacent carbon atoms.
- *Steric strain* is due to overlapping electron clouds of large groups (e.g. gauche interactions).

a. two sites
three bulky methyl groups close = **steric strain**

b. eclipsed conformation = **torsional strain**

c. two bulky ethyl groups close = **steric strain**
eclipsed conformation = **torsional strain**

4.49 The barrier to rotation is equal to the difference in energy between the highest energy eclipsed and lowest energy staggered conformations of the molecule.

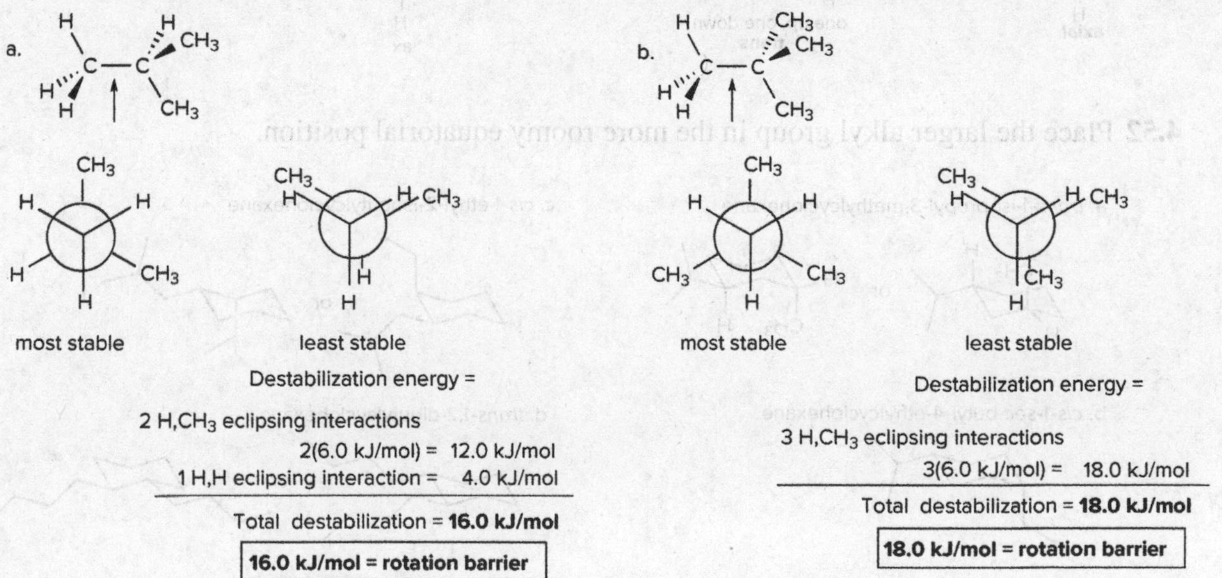

a.
most stable
least stable
Destabilization energy =
2 H,CH3 eclipsing interactions
2(6.0 kJ/mol) = 12.0 kJ/mol
1 H,H eclipsing interaction = 4.0 kJ/mol
Total destabilization = 16.0 kJ/mol
16.0 kJ/mol = rotation barrier
b.
most stable
least stable
Destabilization energy =
3 H,CH3 eclipsing interactions
3(6.0 kJ/mol) = 18.0 kJ/mol
Total destabilization = 18.0 kJ/mol
18.0 kJ/mol = rotation barrier

4.50

most stable

least stable

2 H,H eclipsing interactions = 2(4.0 kJ/mol) = 8.0 kJ/mol

Because the barrier to rotation is 15 kJ/mol, the difference between this value and the destabilization due to H,H eclipsing is the destabilization due to H,Cl eclipsing.

15.0 kJ/mol – 8.0 kJ/mol = 7.0 kJ/mol

destabilization due to H,Cl eclipsing

4.51

[1]

a. axial H, OH axial, HO eq, H eq

b. H up, OH, down HO, H — one up, one down = **trans**

c. HO, OH

d. ax H ax, OH, eq HO, H eq ⇌ H eq, eq HO, OH ax, H ax

[2]

a. axial Br, H eq, CH_3 eq, H axial

b. up Br, H, CH_3 up, H — both up = **cis**

c. Br, CH_3

d. ax Br, H eq, CH_3 eq, H ax ⇌ ax CH_3, eq Br, H eq, H ax

[3]

a. axial H, OH eq, HO eq, H axial

b. H, HO up, OH down, H — one up, one down = **trans**

c. OH, HO

d. ax H, eq HO, OH eq, H ax ⇌ ax OH, eq H, H eq, OH ax

4.52 Place the larger alkyl group in the more roomy equatorial position.

a. *trans*-1-isopropyl-3-methylcyclohexane

H, CH_3, H or H, CH_3, H

b. *cis*-1-*sec*-butyl-4-ethylcyclohexane

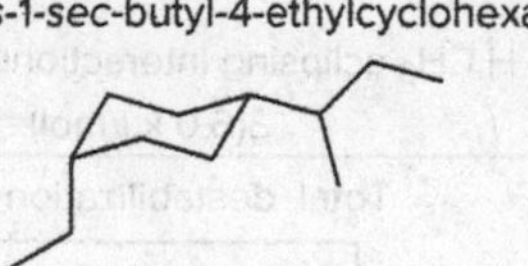

c. *cis*-1-ethyl-2-isobutylcyclohexane

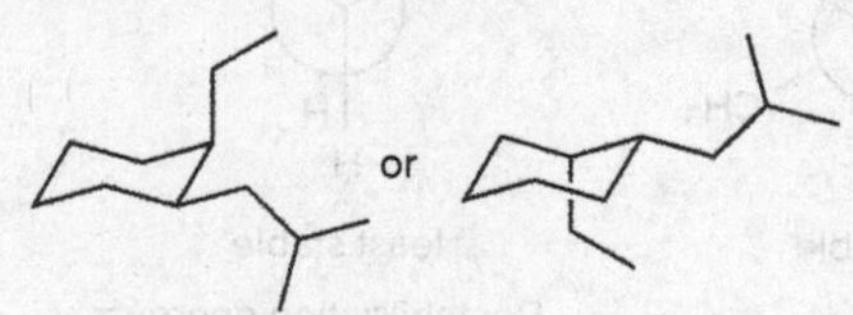

d. *trans*-1,2-dibutylcyclohexane

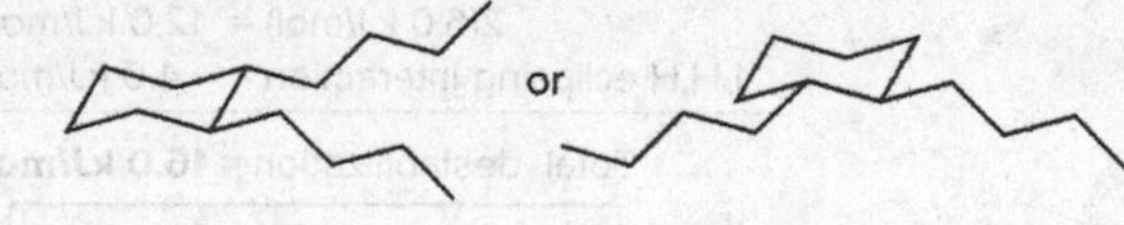

4.53 A **cis isomer** has two groups on the **same side** of the ring. The two groups can be drawn both up or both down. Only one possibility is drawn. A **trans isomer** has one group on one side of the ring and one group on the other side. Either group can be drawn on either side. Only one possibility is drawn.

[1] [2] [3]

a. cis trans a. cis trans a. cis trans

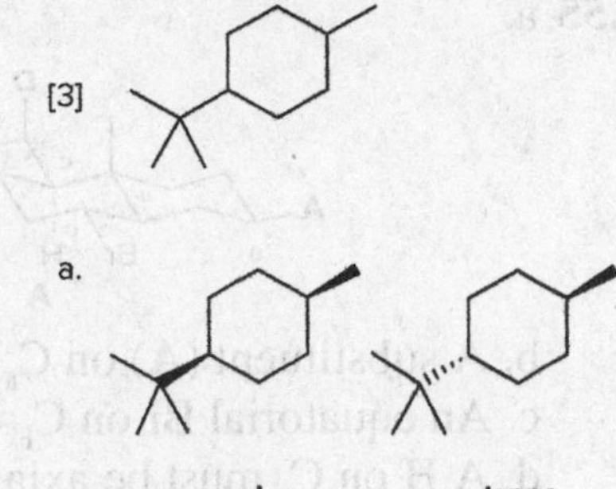

b. cis isomer

both groups equatorial **more stable**

larger group equatorial **more stable**

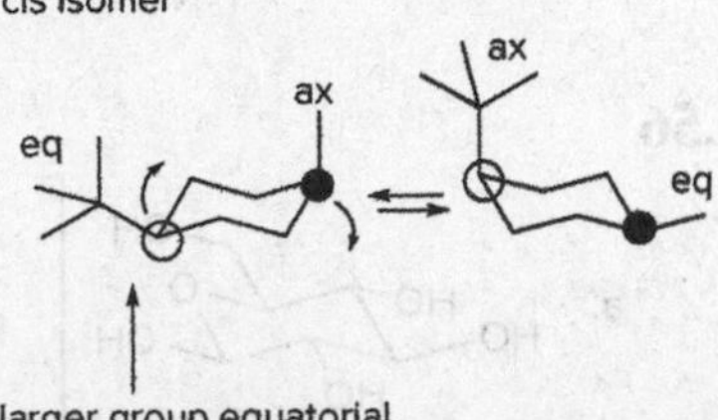

c. trans isomer

larger group equatorial **more stable**

both groups equatorial **more stable**

both groups equatorial **more stable**

d. The cis isomer is more stable than the trans because one conformation has both groups equatorial.

d. The trans isomer is more stable than the cis because one conformation has both groups equatorial.

d. The trans isomer is more stable than the cis because one conformation has both groups equatorial.

4.54

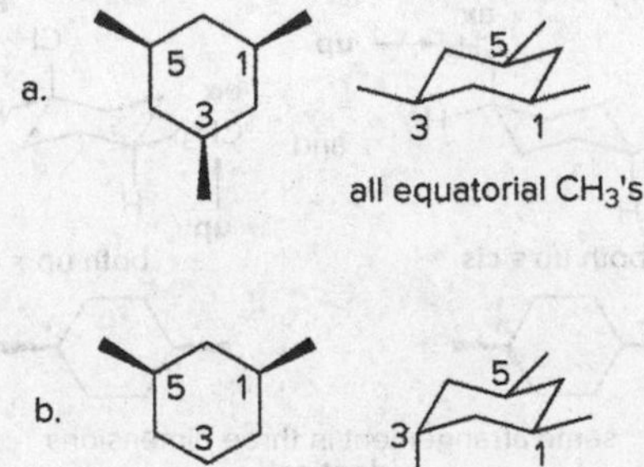

all equatorial CH_3's

c. 2 equatorial CH_3's 1 axial CH_3

b. 2 equatorial CH_3's 1 axial CH_3

d. 3 equatorial CH_3's

4.55 a.

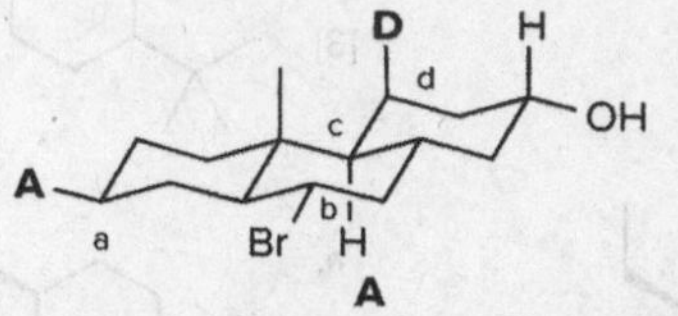

D, H, d, OH, c, A, b, a, Br, H, A

CH$_3$ is axial and up. OH is equatorial and down, so the two groups are trans.

b. A substituent (**A**) on C$_a$ that is cis to the existing CH$_3$ must be up, so it is equatorial.
c. An equatorial Br on C$_b$ is down, so it is cis to the OH group.
d. A H on C$_c$ must be axial and down, making it cis to the OH group.
e. A substituent (**D**) on C$_d$ that is trans to the OH must be up and axial.

4.56

a. HO, HO, HO, OH, O, OH — **most stable** All groups are equatorial. [or HO, HO, HO, OH, O, OH]

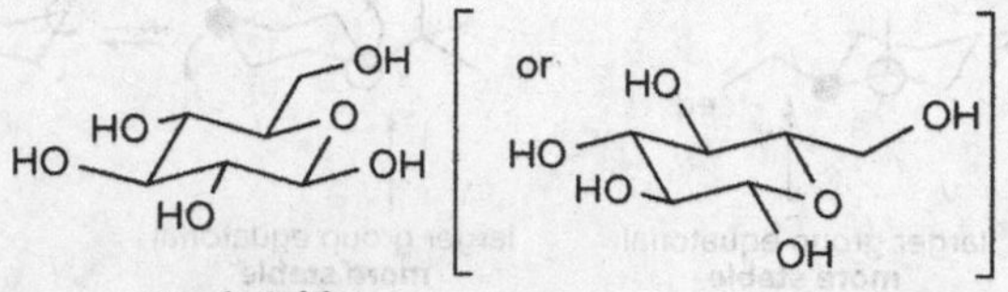

c. OH, HO, O, OH, OH, HO — **constitutional isomer**

b. HO, HO, O, OH, HO, OH [or HO, HO, O, OH, HO, OH]

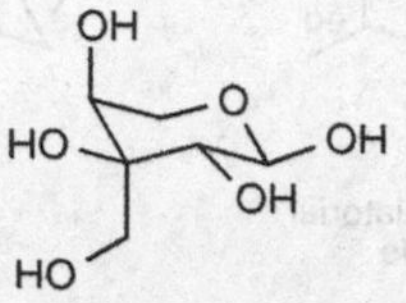

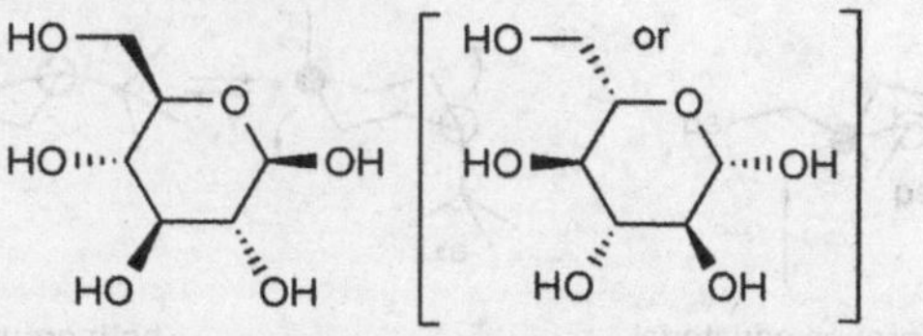

d. OH OH, HO, O, OH, OH

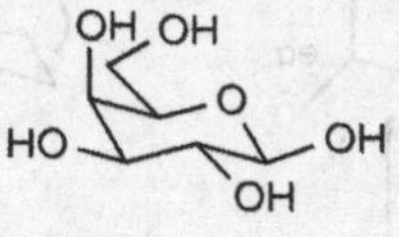

4.57

a. and H, CH$_3$, CH$_3$, H =

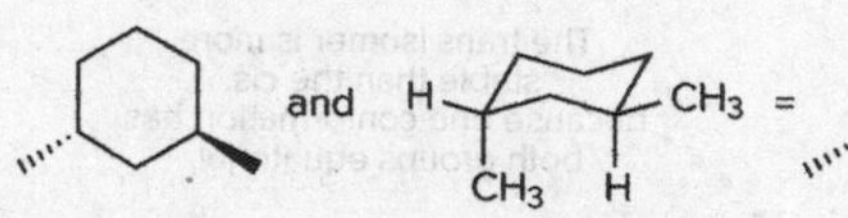

1 down, 1 up = **trans**

1 down, 1 up = **trans**

same arrangement in three dimensions **identical**

c. and CH$_2$CH$_3$, CH$_3$CH$_2$

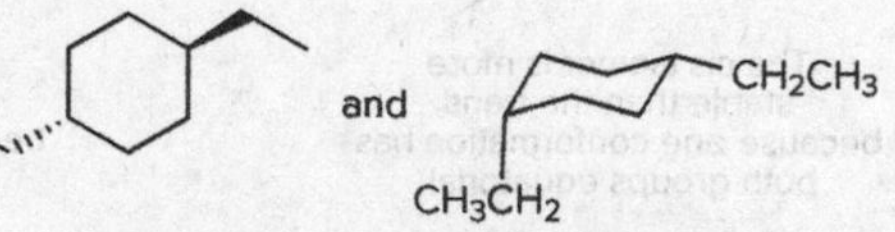

1 down, 1 up = **trans**

both down = **cis**

different arrangement in three dimensions **stereoisomers**

b. and

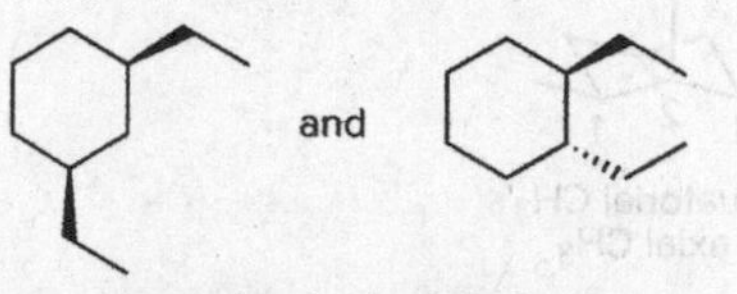

same molecular formula C$_{10}$H$_{20}$ different connectivity **constitutional isomers**

d. ax CH$_3$ ← up, eq CH$_3$CH$_2$, H, H, ↑ up — both up = **cis**

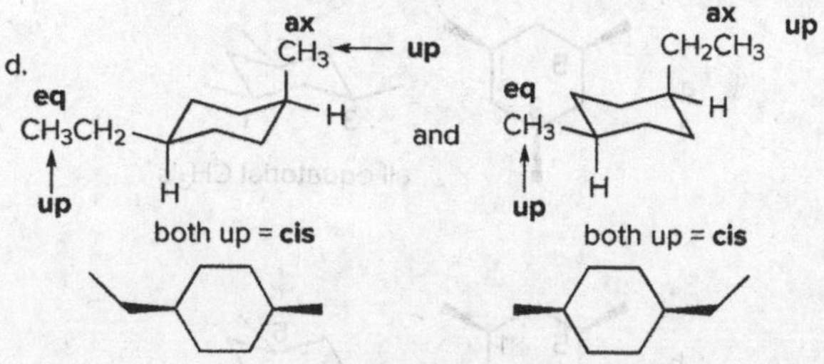

and

ax CH$_2$CH$_3$ up, eq CH$_3$, H, H, ↑ up — both up = **cis**

same arrangement in three dimensions **identical**

4.58

5 4 3 2 1

A
3-ethyl-2,4-dimethylpentane

B
identical

$(CH_3)_2CHC(CH_3)_2CH(CH_3)_2$

C
isomer

H
$(CH_3)_2CH$ CH_2CH_3
CH_3 CH_3
H

D
identical

4.59

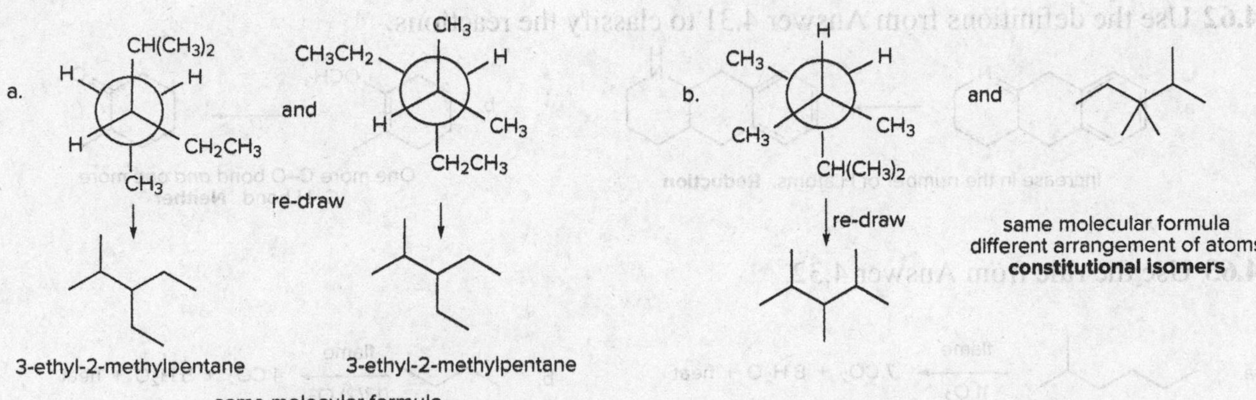

4.60

a. **A** and **B**: constitutional isomers
A and **C**: identical
B and **D**: stereoisomers

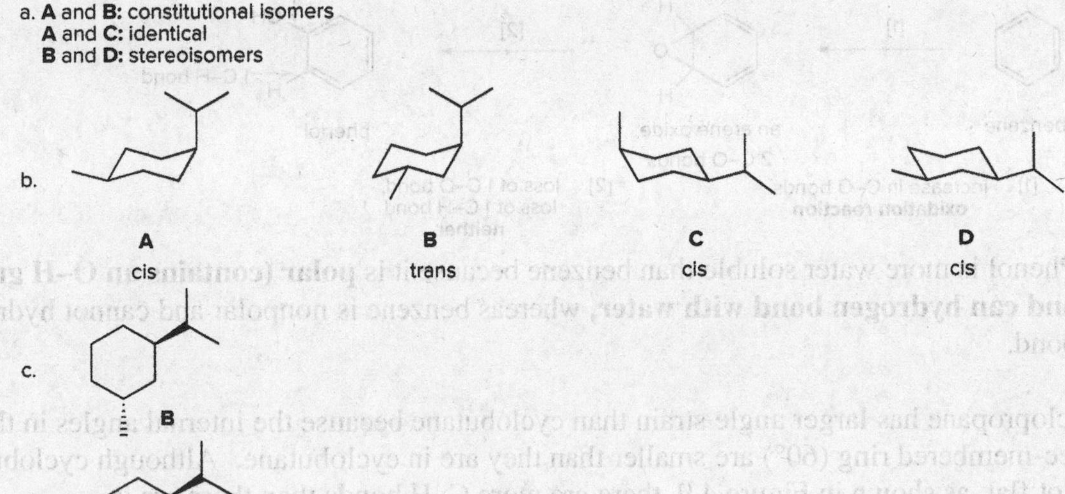

4.61

Three constitutional isomers of C_7H_{14}:

1,1-dimethylcyclopentane

1,2-dimethylcyclopentane

1,3-dimethylcyclopentane

trans or cis (1,2-dimethylcyclopentane: trans, cis)

(1,3-dimethylcyclopentane: trans, cis)

4.62 Use the definitions from Answer 4.31 to classify the reactions.

a. Increase in the number of H atoms. **Reduction**

b. One more C–O bond *and* one more C–H bond. **Neither**

4.63 Use the rule from Answer 4.32.

a. $\xrightarrow[11\ O_2]{\text{flame}}$ $7\ CO_2 + 8\ H_2O$ + heat

b. $\xrightarrow[(13/2)\ O_2]{\text{flame}}$ $4\ CO_2 + 5\ H_2O$ + heat

4.64

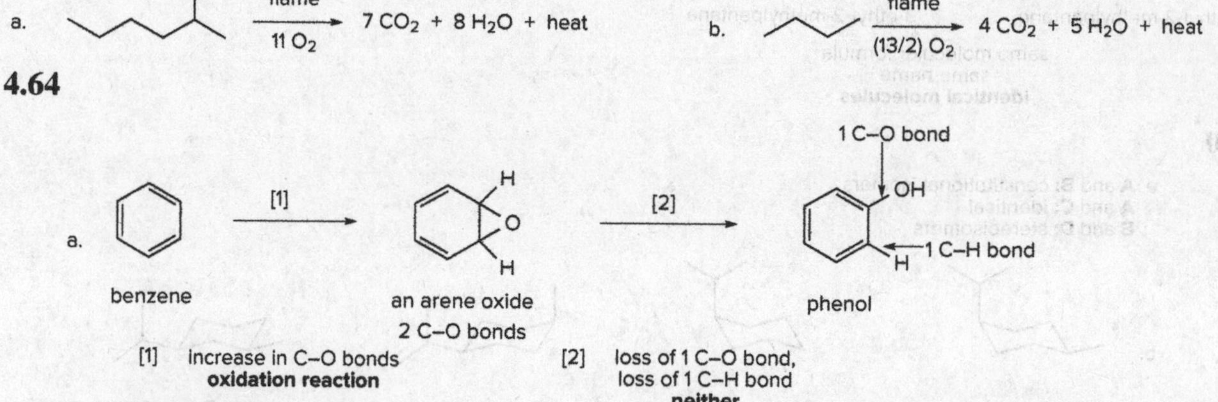

b. Phenol is more water soluble than benzene because it is **polar (contains an O–H group) and can hydrogen bond with water,** whereas benzene is nonpolar and cannot hydrogen bond.

4.65 Cyclopropane has larger angle strain than cyclobutane because the internal angles in the three-membered ring (60°) are smaller than they are in cyclobutane. Although cyclobutane is not flat, as shown in Figure 4.9, there are more C–H bonds than there are in cyclopropane, so there are more sites of torsional strain. Thus cyclopropane has more angle strain but less torsional strain. The result is that both cyclopropane and cyclobutane have roughly similar strain energies.

4.66 The amide in the four-membered ring has 90° bond angles giving it angle strain, which makes it more reactive.

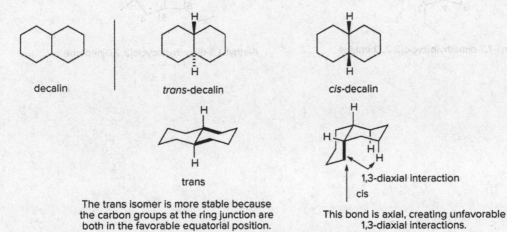

amide

penicillin G

strained amide
more reactive

4.67

Example:

Although I is a much bigger atom than Cl, the C–I bond is also much longer than the C–Cl bond. As a result, the eclipsing interaction of the H and I atoms is not very much different in magnitude from the H,Cl eclipsing interaction.

longer bond

4.68

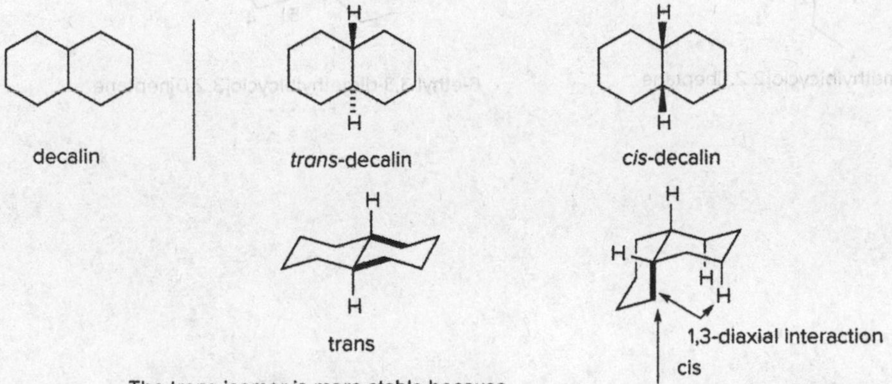

The trans isomer is more stable because the carbon groups at the ring junction are both in the favorable equatorial position.

This bond is axial, creating unfavorable 1,3-diaxial interactions.

4.69

a, b.

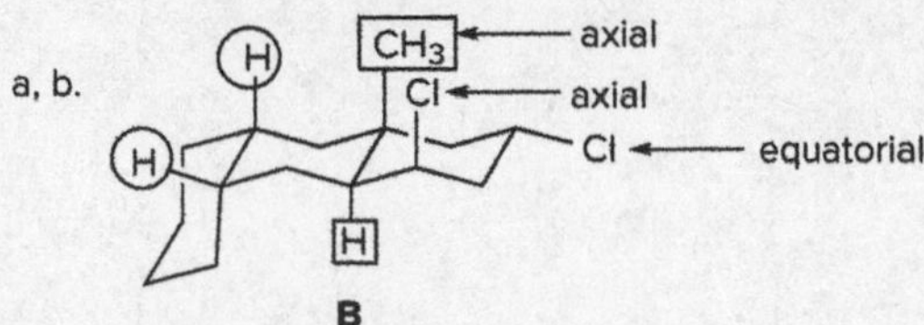

c. The circled H's at one ring fusion are cis. The boxed in CH_3 and H at the second ring fusion are trans.

4.70

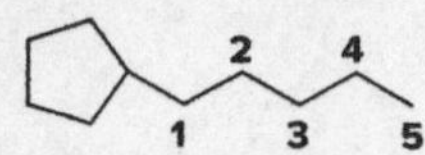

pentylcyclopentane

(1,1-dimethylpropyl)cyclopentane

(2-methylbutyl)cyclopentane

(2,2-dimethylpropyl)cyclopentane

(1-methylbutyl)cyclopentane

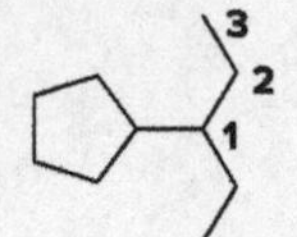

(1-ethylpropyl)cyclopentane

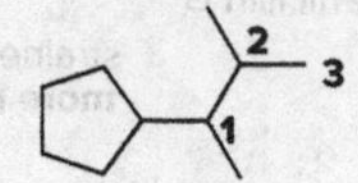

(1,2-dimethylpropyl)cyclopentane

(3-methylbutyl)cyclopentane

4.71

a.

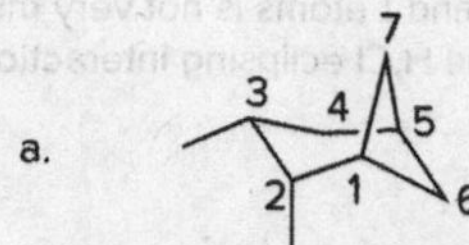

2,3-dimethylbicyclo[3.1.1]heptane

c.

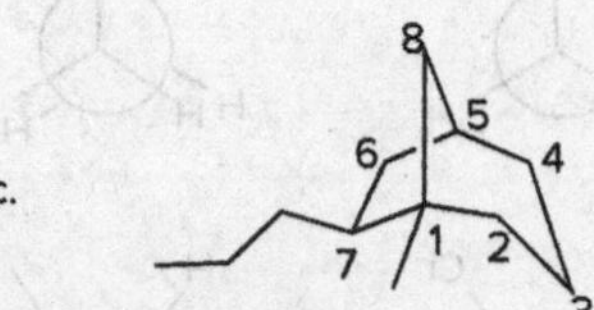

1-methyl-7-propylbicyclo[3.2.1]octane

b.

2-ethyl-7,7-dimethylbicyclo[2.2.1]heptane

d.

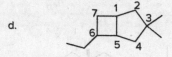

6-ethyl-3,3-dimethylbicyclo[3.2.0]heptane

Chapter 5: Stereochemistry

Chapter Review

Isomers are different compounds with the same molecular formula (5.2, 5.11).

[1] **Constitutional isomers**—isomers that differ in the way the atoms are connected to each other. They have:

- different IUPAC names
- the same or different functional groups
- different physical and chemical properties.

[2] **Stereoisomers**—isomers that differ only in the way atoms are oriented in space. They have the same functional group and the same IUPAC name except for prefixes such as cis, trans, *R*, and *S*.

- **Enantiomers**—stereoisomers that are nonsuperimposable mirror images of each other (5.4).
- **Diastereomers**—stereoisomers that are not mirror images of each other (5.7).

A, B: enantiomers

C, D: enantiomers

A and **B** are diastereomers of **C** and **D**.

Assigning priority (5.6)

- Assign priorities (1, 2, 3, or 4) to the atoms bonded directly to the stereogenic center in order of decreasing atomic number. The atom of *highest* atomic number gets the *highest* priority (1).
- If two atoms on a stereogenic center are the *same,* assign priority based on the atomic number of the atoms bonded to these atoms. *One* atom of higher atomic number determines a higher priority.
- If two isotopes are bonded to the stereogenic center, assign priorities in order of decreasing *mass* number.
- To assign a priority to an atom that is part of a multiple bond, treat a multiply bonded atom as an equivalent number of singly bonded atoms.

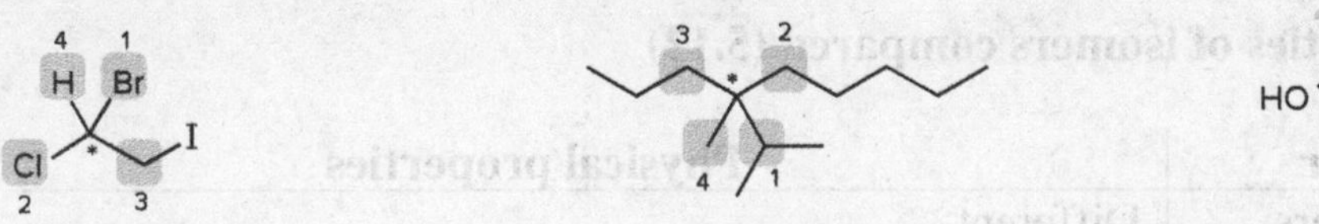

- The stereogenic center is bonded to Br, Cl, C, and H.
- The stereogenic center is *not* bonded directly to I.

- $CH(CH_3)_2$ gets the highest priority because the C is bonded to two other C's.

- OH gets the highest priority because O has the highest atomic number.
- CO_2H (three bonds to O) gets higher priority than CH_2OH (one bond to O).

Some basic principles

- When a compound and its mirror image are **superimposable,** they are **identical achiral compounds.** A plane of symmetry in one conformation makes a compound achiral (5.3).
- When a compound and its mirror image are **not superimposable,** they are **different chiral compounds** called **enantiomers.** A chiral compound has no plane of symmetry in any conformation (5.3).
- A **tetrahedral stereogenic center** is a carbon atom bonded to four different groups (5.4, 5.5).
- For ***n*** **stereogenic centers,** the maximum number of stereoisomers is 2^n (5.7).

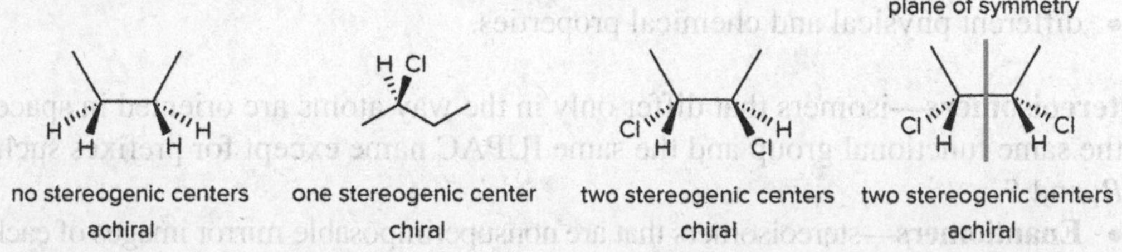

Optical activity is the ability of a compound to rotate plane-polarized light (5.12).

- An optically active solution contains a chiral compound.
- An optically inactive solution contains one of the following:
  - an achiral compound with no stereogenic centers.
  - a meso compound—an achiral compound with two or more stereogenic centers.
  - a racemic mixture—an equal amount of two enantiomers.

The prefixes *R* and *S* compared with *d* and *l*

The prefixes *R* and *S* are labels used in nomenclature. Rules on assigning *R,S* are found in Section 5.6.

- An enantiomer has every stereogenic center opposite in configuration.
- A diastereomer of a compound with two stereogenic centers has one stereogenic center with the same configuration and one that is opposite.

The prefixes *d* (or +) and *l* (or –) tell the direction a compound rotates plane-polarized light (5.12).

- *d* (or +) stands for dextrorotatory, rotating polarized light clockwise.
- *l* (or –) stands for levorotatory, rotating polarized light counterclockwise.

The physical properties of isomers compared (5.12)

Type of isomer	Physical properties
Constitutional isomers	Different
Enantiomers	Identical except the direction of rotation of polarized light
Diastereomers	Different
Racemic mixture	Possibly different from either enantiomer

Equations

- Specific rotation (5.12C):

$$\text{specific rotation} = [\alpha] = \frac{\alpha}{l \times c}$$

α = observed rotation (°)
l = length of sample tube (dm)
c = concentration (g/mL)

[dm = decimeter; 1 dm = 10 cm]

- Enantiomeric excess (5.12D):

$$ee = \text{\% of one enantiomer} - \text{\% of other enantiomer}$$
$$= \frac{[\alpha]\ \text{mixture}}{[\alpha]\ \text{pure enantiomer}} \times 100\%$$

Practice Test on Chapter Review

1.a. Which of the following statements is true for compounds **A–D** below?

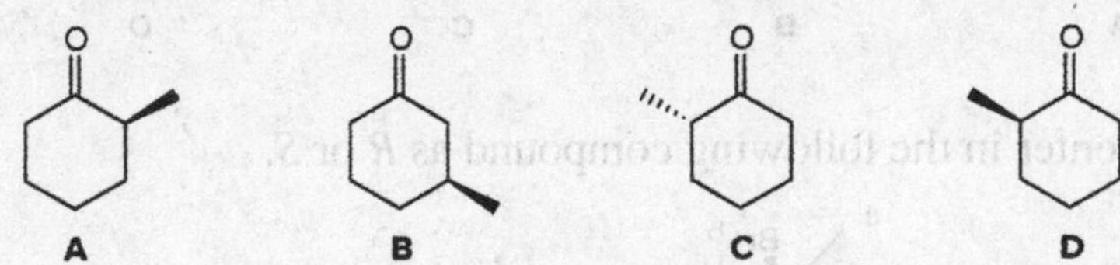

1. **A** and **B** are separable by physical methods such as distillation.
2. **A** and **C** are separable by physical methods such as distillation.
3. **A** and **D** are separable by physical methods such as distillation.
4. Statements (1) and (2) are both true.
5. Statements (1), (2), and (3) are all true.

b. Which of the following statements is true about compounds **A–C** below?

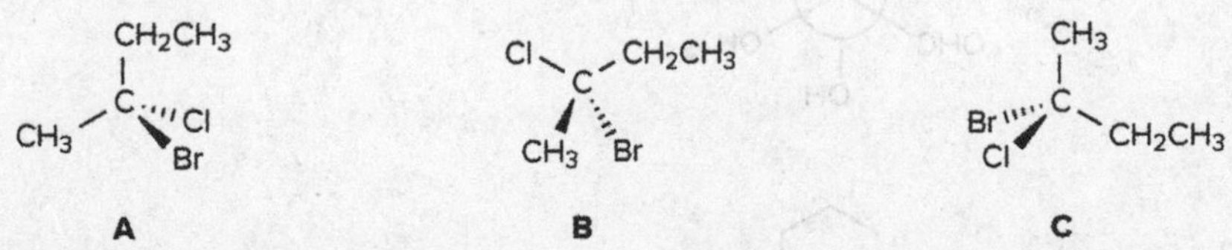

1. **A** and **B** are enantiomers.
2. **A** and **C** are enantiomers.
3. An equal mixture of **B** and **C** is optically active.
4. Statements (1) and (2) are true.
5. Statements (1), (2), and (3) are all true.

c. Which compound is a diastereomer of **A**?

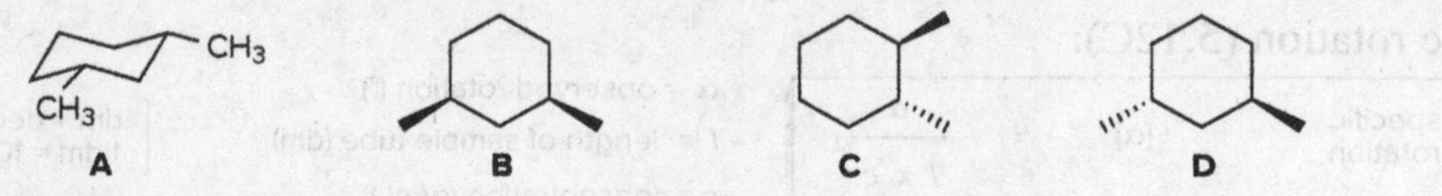

1. **B** only
2. **C** only
3. **D** only
4. Both **B** and **C**
5. Compounds **B, C,** and **D**

2. Rank the following four groups around a stereogenic center in order of decreasing priority. Rank the highest priority group as **1**, the lowest priority group as **4**, and the two groups of intermediate priority as **2** and **3**.

$—CH_2Cl$	$—CH_2CH_2Br$	$—COOH$	$—CH_2OH$
A	**B**	**C**	**D**

3. Label each stereogenic center in the following compound as *R* or *S*.

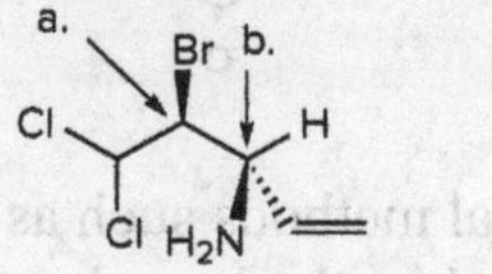

a. b.

4. State how the compounds in each pair are related to each other. Choose from constitutional isomers, enantiomers, diastereomers, or identical compounds.

a. and

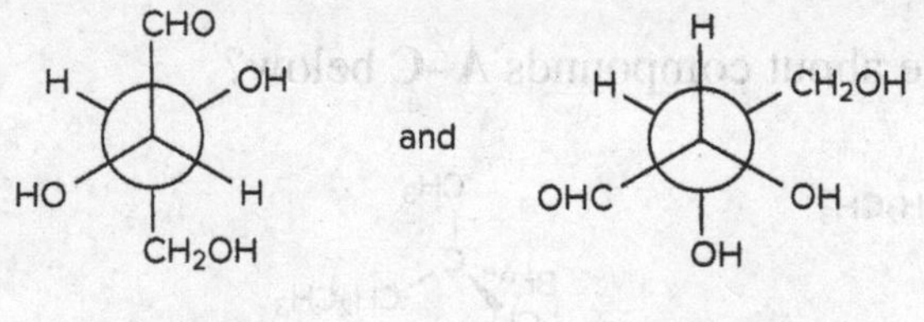

b. and

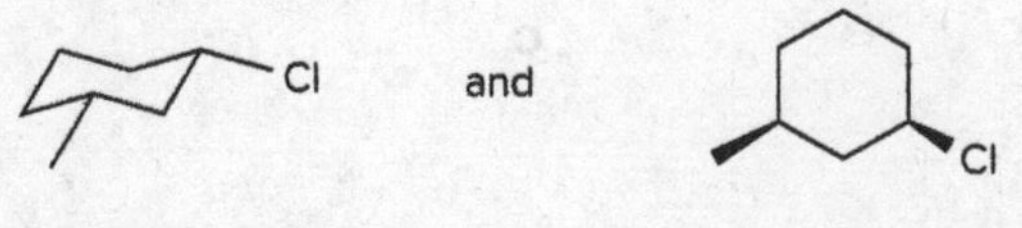

c. and

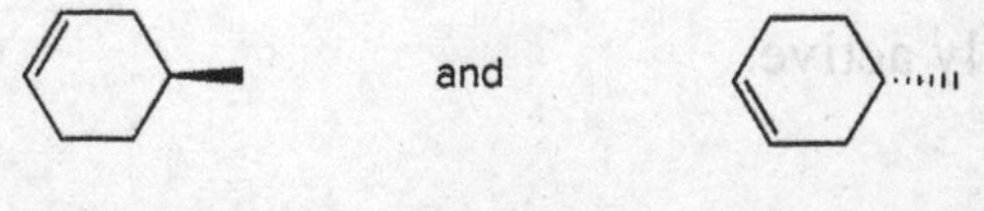

5. The enantiomeric excess of a mixture of **A** and **B** is 62% with **A** in excess. How much of **A** and **B** are present in the mixture?

Answers to Practice Test

1. a. 1	2. **A**–1	3. a. *S*	4. a. diastereomers	5. 81% **A**
b. 2	**B**–4	b. *S*	b. enantiomers	19% **B**
c. 3	**C**–2		c. identical	
	D–3			

Answers to Problems

5.1 Constitutional isomers have atoms bonded to different atoms.
Stereoisomers differ only in the three-dimensional arrangement of atoms.

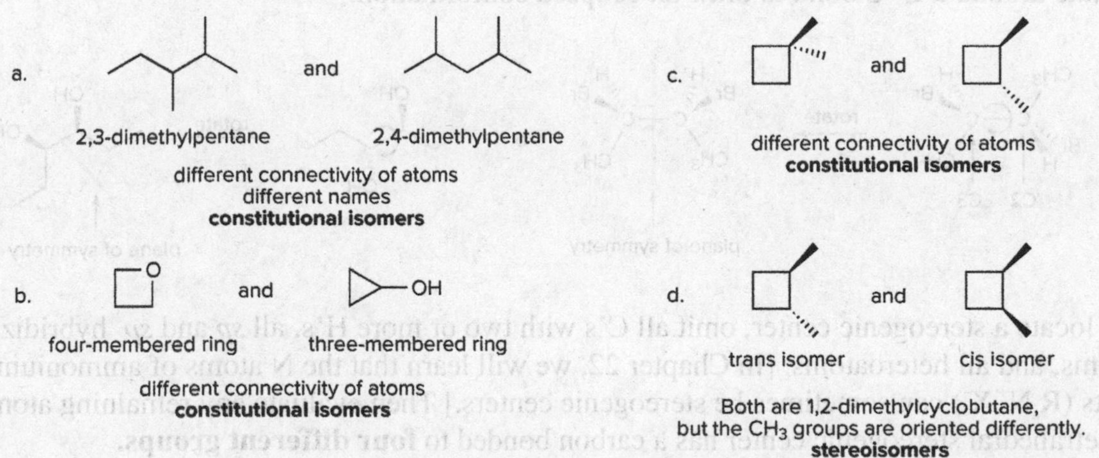

5.2 Draw the mirror image of each molecule by drawing a mirror plane and then drawing the molecule's reflection. **A chiral molecule is one that is not superimposable on its mirror image.** A molecule with one stereogenic center is always chiral. A molecule with zero stereogenic centers is not chiral (in general).

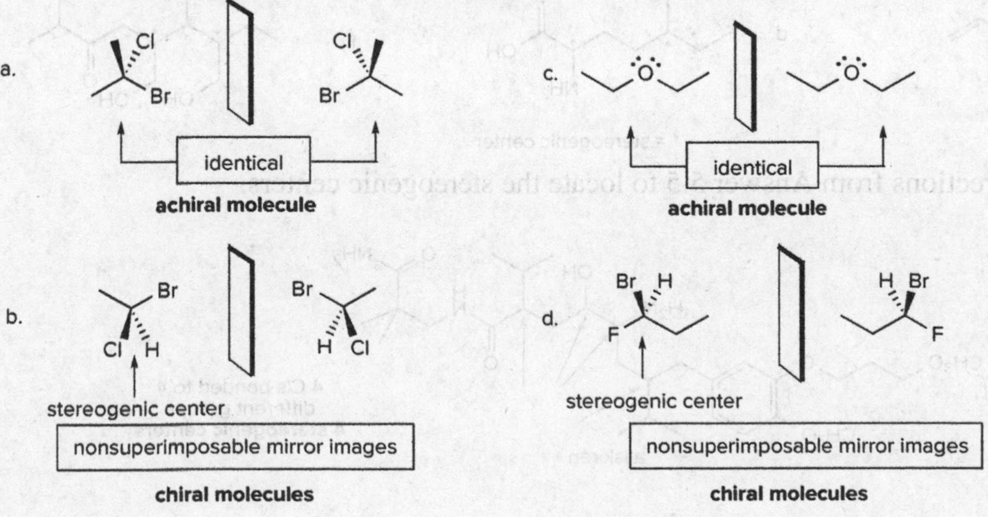

5.3 A plane of symmetry cuts the molecule into **two identical halves.**

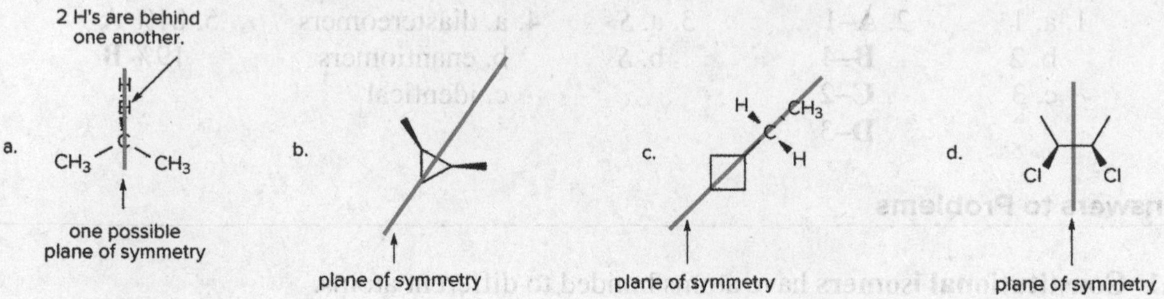

a. 2 H's are behind one another. CH_3 C CH_3 — one possible plane of symmetry

b. plane of symmetry

c. H CH_3 C H — plane of symmetry

d. Cl Cl — plane of symmetry

5.4 Rotate around a C–C bond to draw an eclipsed conformation.

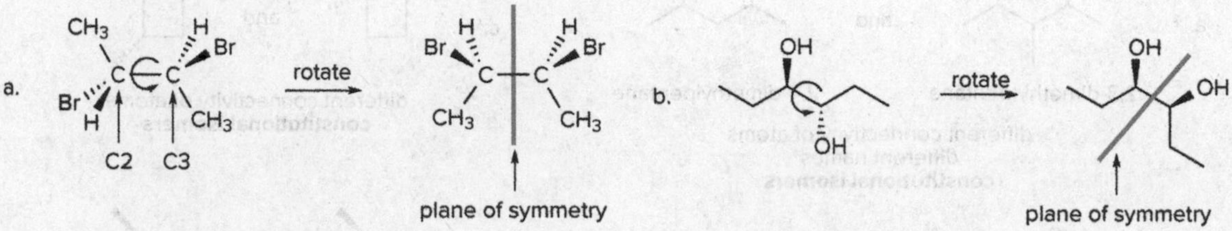

a. CH_3 H Br C C Br H CH_3 C2 C3 — rotate → H H Br C C Br CH_3 CH_3 — plane of symmetry

b. OH OH — rotate → OH OH — plane of symmetry

5.5 To locate a stereogenic center, omit all C's with two or more H's, all *sp* and *sp*^{2} hybridized atoms, and all heteroatoms. [In Chapter 22, we will learn that the N atoms of ammonium salts ($R_4N^+X^-$) can sometimes be stereogenic centers.] Then evaluate any remaining atoms. A tetrahedral stereogenic center has a carbon bonded to **four different groups.**

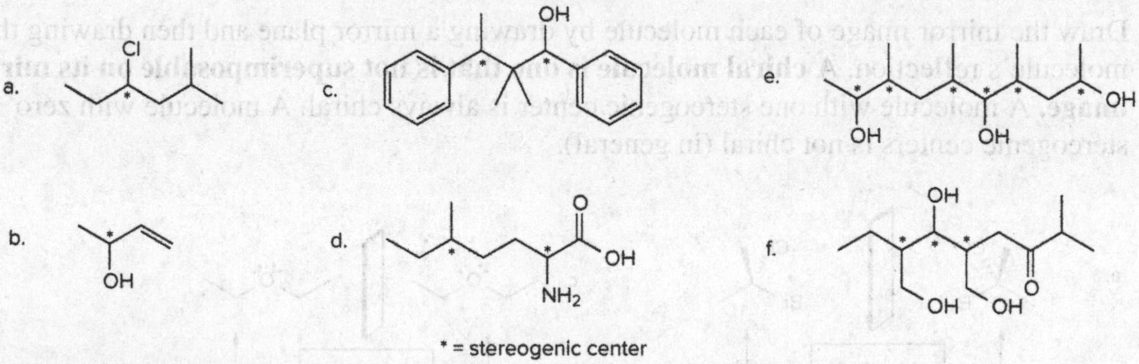

a. Cl

b. OH

c. OH

d. O OH NH_2

e. OH OH OH

f. OH O OH OH

* = stereogenic center

5.6 Use the directions from Answer 5.5 to locate the stereogenic centers.

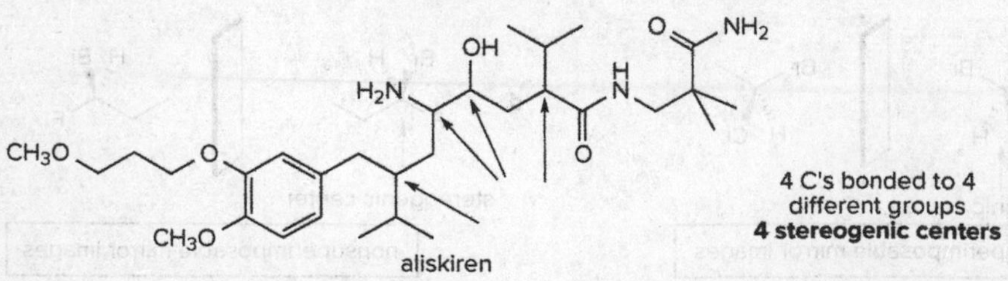

CH_3O O CH_3O H_2N OH H N O O NH_2

aliskiren

4 C's bonded to 4 different groups

4 stereogenic centers

5.7 To draw the enantiomer, change the position of groups on the wedges and dashed wedges.

a. HO, HO, O, OH, H, NH_2

b. N, N, O, HO, O, Cl

c. H_2N, O, OH

5.8 Find the C bonded to four different groups in each molecule. At the stereogenic center, draw two bonds in the plane of the page, one in front (on a wedge), and one behind (on a dashed wedge). Then draw the mirror image (enantiomer).

a. **stereogenic center** Cl H | H Cl — mirror images *nonsuperimposable* **enantiomers**

b. **stereogenic center** OH, HO H | HO, H OH — mirror images *nonsuperimposable* **enantiomers**

c. **stereogenic center** S, O, OH, H_2N H | HO, O, S, H NH_2 — mirror images *nonsuperimposable* **enantiomers**

5.9 Use the directions from Answer 5.5 to locate the stereogenic centers.

a. C bonded to H and 3 different C's: **1 stereogenic center**

b. Cl, Cl — Each labeled C is bonded to: H, Cl, CH_2, CHCl: **2 stereogenic centers**

c. O — 4 C's bonded to 4 different groups: **4 stereogenic centers**

d. O, O, O, O, N, H, CO_2H — C bonded to H, 2 different O's and 1 C: **1 stereogenic center**

e. 3 C's bonded to 4 different groups: **3 stereogenic centers**

f. O, O, OCH_3 — 3 C's bonded to 4 different groups: **3 stereogenic centers**

5.10

a. O, O, OH, O, O — simvastatin

All stereogenic C's are circled. Each C is sp^3 hybridized and bonded to 4 different groups.

b. O, H, H, H | O, H, H, H — enantiomer

5.11 Assign priority based on atomic number: atoms with a higher atomic number get a higher priority. If two atoms are the same, look at what they are bonded to and assign priority based on the atomic number of these atoms.

a. $-CH_3$, $-CH_2CH_3$ (higher priority: $-CH_2CH_3$)

c. –H, –D (higher mass, higher priority: –D)

e. $-CH_2CH_2Cl$, $-CH_2CH(CH_3)_2$ (higher priority: $-CH_2CH_2Cl$)

b. –, –Br (higher priority: –Br)

d. $-CH_2Br$, $-CH_2CH_2Br$ (higher priority: $-CH_2Br$)

f. $-CH_2OH$, $-CHO$ = $-\overset{H}{C}=O$ = $-\overset{H}{\underset{O}{C}}-\underset{C}{O}$

2 H's, 1 O; **2 O's**, 1 H; 2 C–O bonds

C bonded to 2 O's has **higher priority.**

5.12 Rank by decreasing priority. Lower atomic number = lower priority.

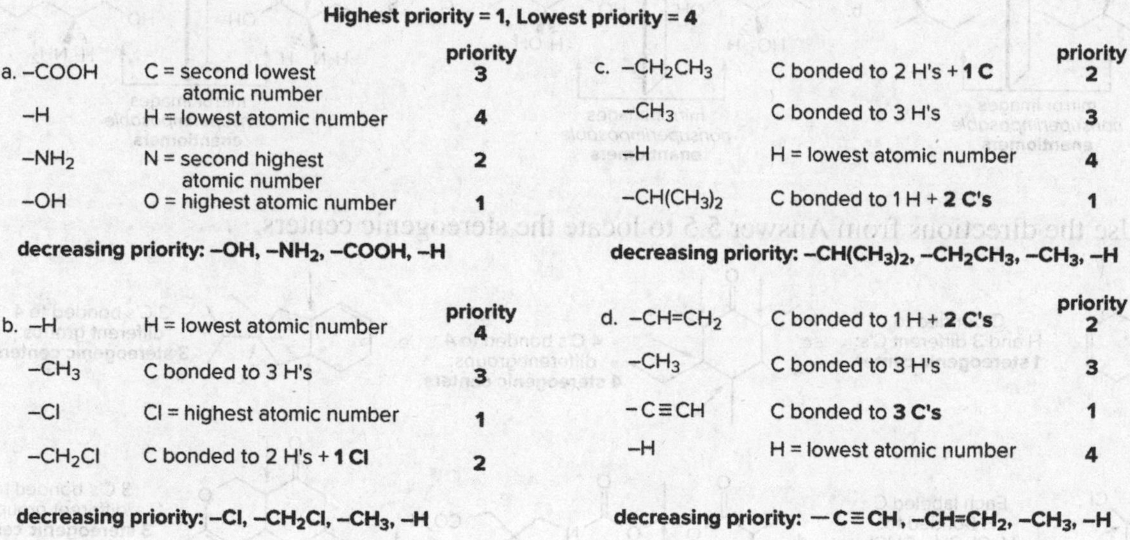

Highest priority = 1, Lowest priority = 4

a.		priority
–COOH	C = second lowest atomic number	3
–H	H = lowest atomic number	4
$-NH_2$	N = second highest atomic number	2
–OH	O = highest atomic number	1

decreasing priority: –OH, $-NH_2$, –COOH, –H

c.		priority
$-CH_2CH_3$	C bonded to 2 H's + **1 C**	2
$-CH_3$	C bonded to 3 H's	3
–H	H = lowest atomic number	4
$-CH(CH_3)_2$	C bonded to 1 H + **2 C's**	1

decreasing priority: $-CH(CH_3)_2$, $-CH_2CH_3$, $-CH_3$, –H

b.		priority
–H	H = lowest atomic number	4
$-CH_3$	C bonded to 3 H's	3
–Cl	Cl = highest atomic number	1
$-CH_2Cl$	C bonded to 2 H's + **1 Cl**	2

decreasing priority: –Cl, $-CH_2Cl$, $-CH_3$, –H

d.		priority
$-CH=CH_2$	C bonded to 1 H + **2 C's**	2
$-CH_3$	C bonded to 3 H's	3
$-C\equiv CH$	C bonded to **3 C's**	1
–H	H = lowest atomic number	4

decreasing priority: $-C\equiv CH$, $-CH=CH_2$, $-CH_3$, –H

5.13

a. HO H; 1, 2, 3; *S*

b. F, Br; *R*, *S*

c. HO H; O; H; *R*, *R*

5.14 To assign *R* or *S* to the molecule, first rank the groups. The lowest priority group must be oriented behind the page. If tracing a circle from (1) → (2) → (3) proceeds in the clockwise direction, then the stereogenic center is labeled *R;* if the circle is counterclockwise, then it is labeled *S*.

a. Cl 2, Br 1, 3, **S**

c. 3, 2, ***R***, 1

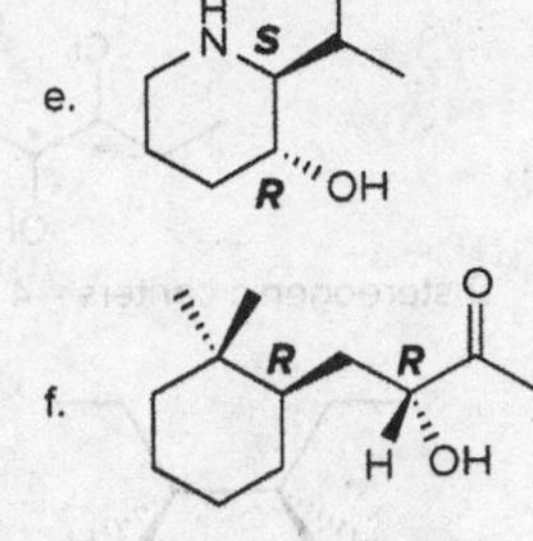

b. HO 2, 3, H OH 1

lowest priority forward
clockwise
It looks like *R*, but reverse answer.
R*→*S

d. Br 3, Cl 2, OH 1

lowest priority forward
clockwise
It looks like *R*, but reverse answer.
R*→*S

5.15

FADH$_2$

5.16 a, b. Re-draw lisinopril as a skeletal structure, locate the stereogenic centers, and assign *R*,*S*.

three stereogenic centers
All have the *S* configuration.

5.17 The maximum number of stereoisomers = 2^n, where n = the number of stereogenic centers.

a. 3 stereogenic centers
2^3 = 8 stereoisomers

b. 8 stereogenic centers
2^8 = 256 stereoisomers

5.18

a.

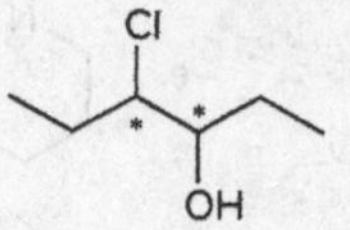

2 stereogenic centers = 4 possible stereoisomers

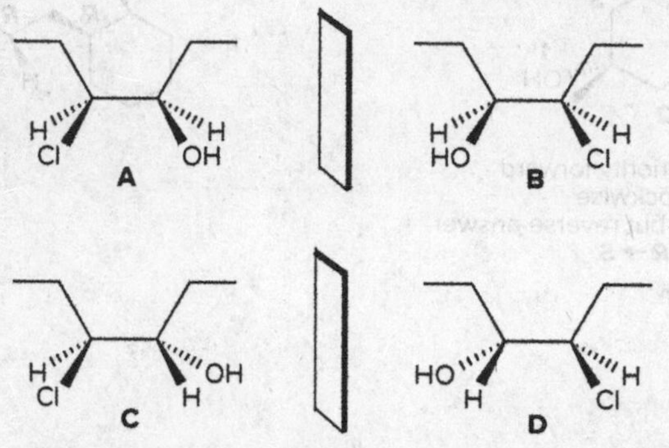

A B

C D

b.

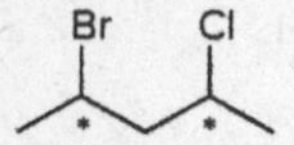

2 stereogenic centers = 4 possible stereoisomers

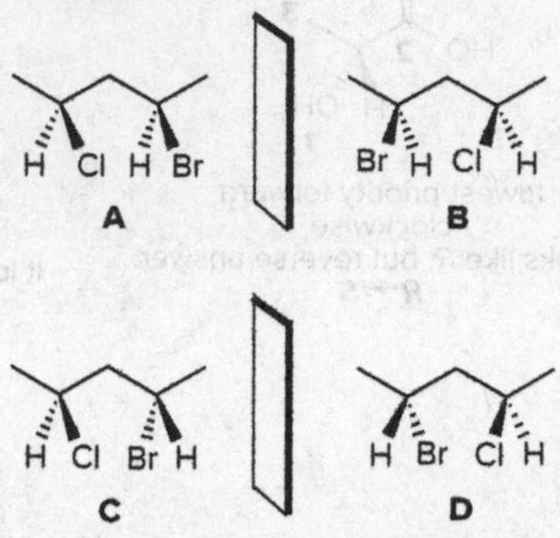

A B

C D

5.19

E rotate same as C

F rotate same as A

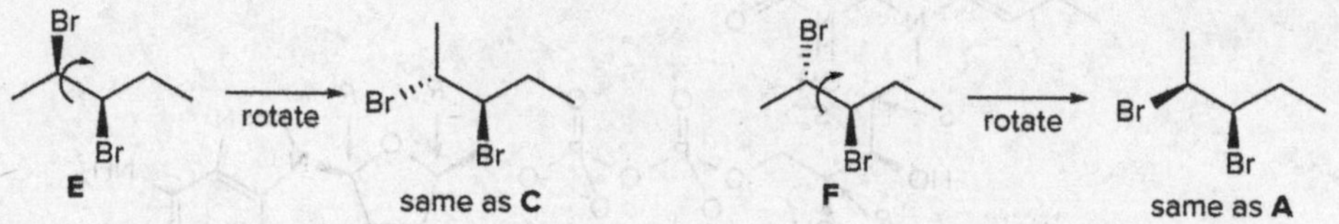

5.20

a.

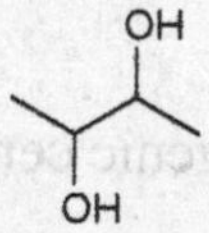

2 stereogenic centers = 4 possible stereoisomers

A B

C C

identical

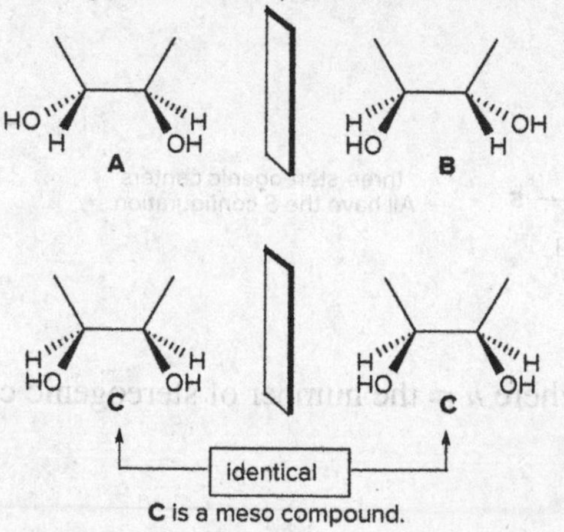

C is a meso compound.

A and **B** are enantiomers.
Pairs of diastereomers: **A** and **C**, **B** and **C**.

b.

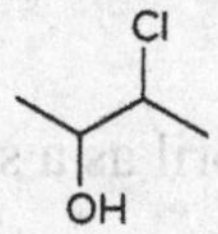

2 stereogenic centers = 4 possible stereoisomers

A B

C D

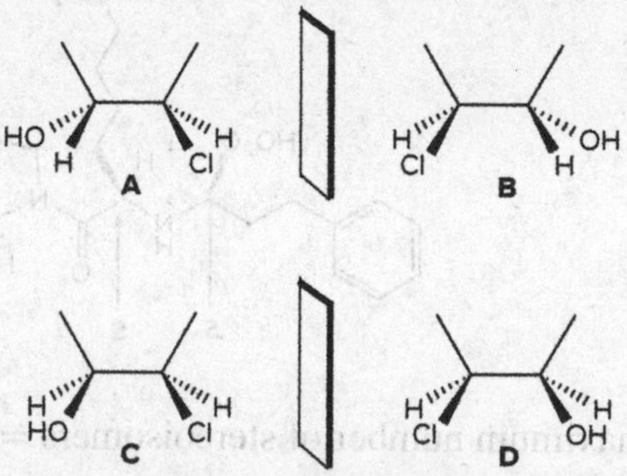

Pairs of enantiomers: **A** and **B**, **C** and **D**.
Pairs of diastereomers: **A** and **C**, **A** and **D**, **B** and **C**, **B** and **D**.

5.21 **A meso compound must have at least two stereogenic centers. Usually a meso compound has a plane of symmetry.** You may have to rotate around a C–C bond to see the plane of symmetry clearly.

a. HO H H OH

2 stereogenic centers
plane of symmetry
meso compound

b. OH OH

2 stereogenic centers
no plane of symmetry
not a meso compound

c. H Br Br H → rotate → H Br Br H

2 stereogenic centers
plane of symmetry
meso compound

5.22 The enantiomer has the exact opposite *R*,*S* designations. For diastereomers, at least one of the *R*,*S* designations is the same, but not all of them.

a. (2*R*,3*S*)-hexane-2,3-diol and (2*R*,3*R*)-hexane-2,3-diol
One changes; one remains the same:
diastereomers

b. (2*R*,3*R*)-hexane-2,3-diol and (2*S*,3*S*)-hexane-2,3-diol
Both *R*'s change to *S*'s:
enantiomers

c. (2*R*,3*S*,4*R*)-hexane-2,3,4-triol and (2*S*,3*R*,4*R*)-hexane-2,3,4-triol
Two change; one remains the same:
diastereomers

5.23 The enantiomer must have the exact opposite *R*,*S* designations. For diastereomers, at least one of the *R*,*S* designations is the same, but not all of them.

a. sorbitol (*R*, *R*, *R*, *S*)

b. **A** (*R*, *R*, *S*, *S*)
One changes; three remain the same.
diastereomer

c. **B** (*S*, *S*, *S*, *R*)
All stereogenic centers change.
enantiomer

5.24 Meso compounds generally have a plane of symmetry. They cannot have just one stereogenic center.

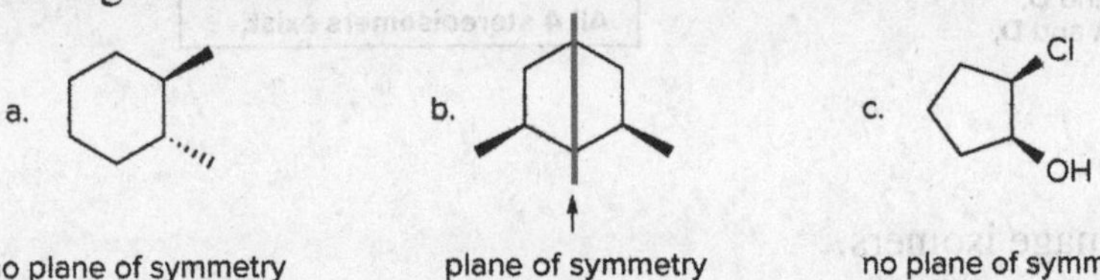

a. no plane of symmetry
not a meso compound

b. plane of symmetry
meso compound

c. Cl OH
no plane of symmetry
not a meso compound

5.25

a. 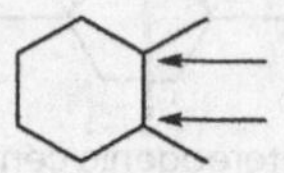 2 stereogenic centers = 4 stereoisomers maximum

Draw the cis and trans isomers:

cis

A

identical

trans

B

C

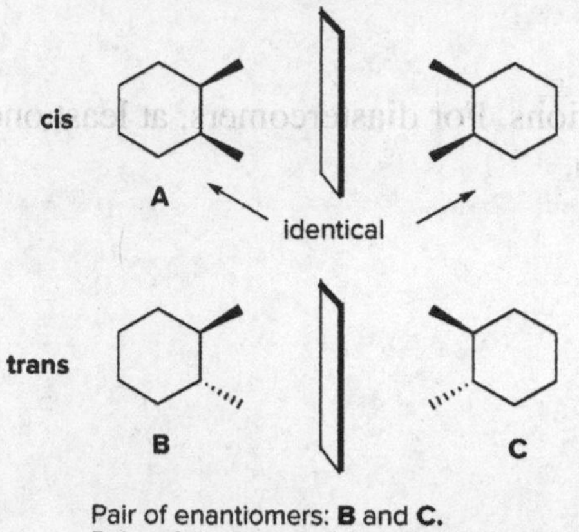

Pair of enantiomers: **B** and **C.**
Pairs of diastereomers: **A** and **B, A** and **C.**

Only 3 stereoisomers exist.

c.

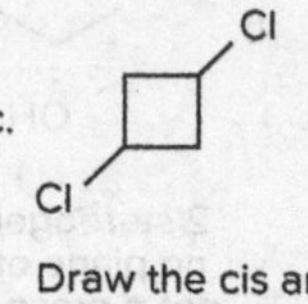

Cl

Cl

Draw the cis and trans isomers:

Cl

Cl

A

identical

Cl

Cl

Cl

Cl

B

identical

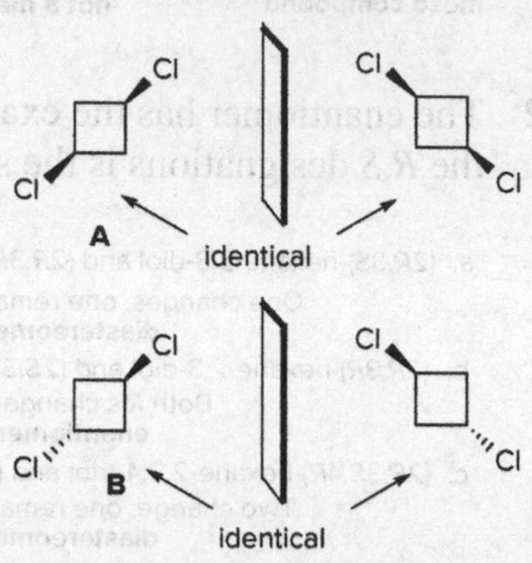

Pair of diastereomers: **A** and **B.**

Only 2 stereoisomers exist.

b. 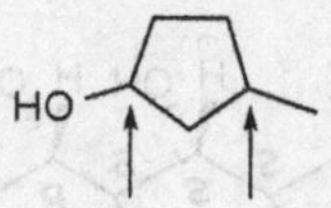 HO

2 stereogenic centers = 4 stereoisomers maximum

Draw the cis and trans isomers:

cis

OH

HO

A

B

trans

OH

HO

C

D

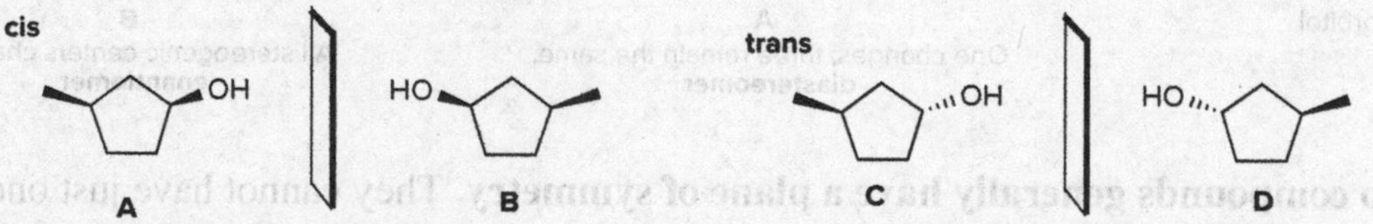

Pairs of enantiomers: **A** and **B, C** and **D.**
Pairs of diastereomers: **A** and **C, A** and **D,**
B and **C, B** and **D.**

All 4 stereoisomers exist.

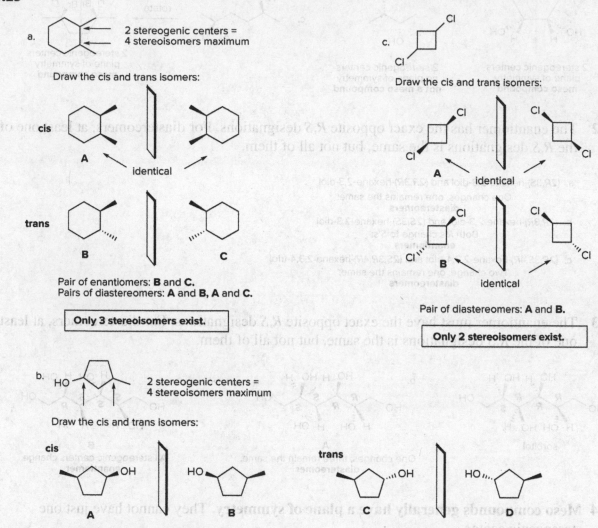

5.26 Four facts:

- **Enantiomers** are mirror image isomers.
- **Diastereomers** are stereoisomers that are not mirror images.
- **Constitutional isomers** have the same molecular formula but the atoms are bonded to different atoms.
- **Cis and trans isomers** are always diastereomers.

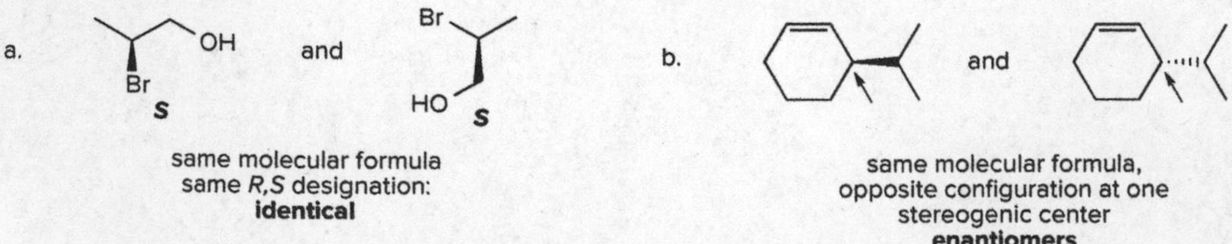

c. HO–(cyclohexane)–OH and (cyclohexane)–OH with HO

1,4- isomer 1,3-isomer

constitutional isomers

d. (cyclohexane)–OH with HO (**trans**) and (cyclohexane)···OH with HO (**cis**)

Both 1,3 isomers, cis and trans: **diastereomers**

5.27

(*S*)-alanine

$[\alpha] = +8.5$

mp = 297 °C

a. Mp = same as the *S* isomer.

b. The mp of a racemic mixture is often different from the melting point of the enantiomers.

c. –8.5, same as *S* but opposite sign

d. Zero. A racemic mixture is optically inactive.

e. Solution of pure (*S*)-alanine: **optically active**
Equal mixture of (*R*)- and (*S*)-alanine: **optically inactive**
75% (*S*) – and 25% (*R*)-alanine: **optically active**

5.28

$$[\alpha] = \frac{\alpha}{l \times c}$$

α = observed rotation
l = length of tube (dm)
c = concentration (g/mL)

$$[\alpha] = \frac{10°}{1 \text{ dm} \times (1 \text{ g}/10 \text{ mL})} = +100 = \textbf{specific rotation}$$

5.29 Enantiomeric excess = *ee* = % of one enantiomer – % of other enantiomer.

a. 95% – 5% = **90% *ee*** b. 85% – 15% = **70% *ee***

5.30

a. 90% *ee* means 90% excess of **A** and 10% racemic mixture of **A** and **B** (5% each); therefore, **95% A and 5% B.**

b. 99% *ee* means 99% excess of **A** and 1% racemic mixture of **A** and **B** (0.5% each); therefore, **99.5% A and 0.5% B.**

c. 60% *ee* means 60% excess of **A** and 40% racemic mixture of **A** and **B** (20% each); therefore, **80% A and 20% B.**

5.31

$$ee = \frac{[\alpha]\text{ mixture}}{[\alpha]\text{ pure enantiomer}} \times 100\%$$

a. $\frac{+10}{+24} \times 100\% = 42\%\ ee$

b. $\frac{[\alpha]\text{ solution}}{+24} \times 100\% = 80\%\ ee$

$[\alpha]$ solution = +19.2

5.32

a. $\frac{[\alpha]\text{ mixture}}{+3.8} \times 100\% = 60\%\ ee$

$[\alpha]$ mixture = +2.3

b. % one enantiomer – % other enantiomer = *ee*
80% – 20% = 60% *ee*

80% dextrorotatory (+) enantiomer
20% levorotatory (–) enantiomer

5.33 • **Enantiomers have the same physical properties** (mp, bp, solubility), and rotate the plane of polarized light to an equal extent, but in opposite directions.
• **Diastereomers have different physical properties.**
• **A racemic mixture is optically inactive.**

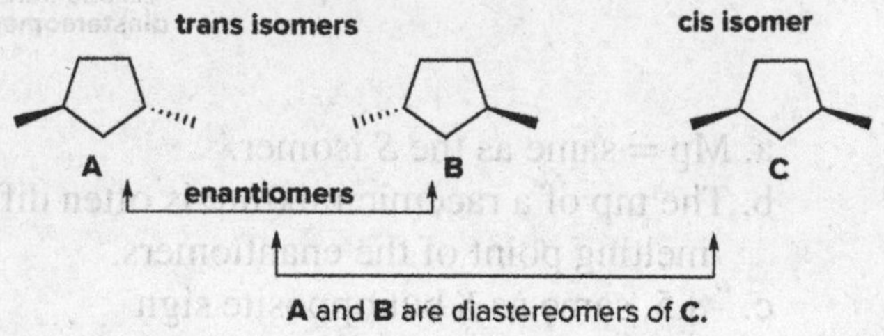

trans isomers

cis isomer

A B C

enantiomers

A and B are diastereomers of C.

a. The bp's of **A** and **B** are the same. The bp's of **A** and **C** are different.
b. Pure **A**: optically active
Pure **B**: optically active
Pure **C**: optically inactive
Equal mixture of **A** and **B**: optically inactive
Equal mixture of **A** and **C**: optically active
c. There would be two fractions: one containing **A** and **B** (optically inactive), and one containing **C** (optically inactive).

5.34

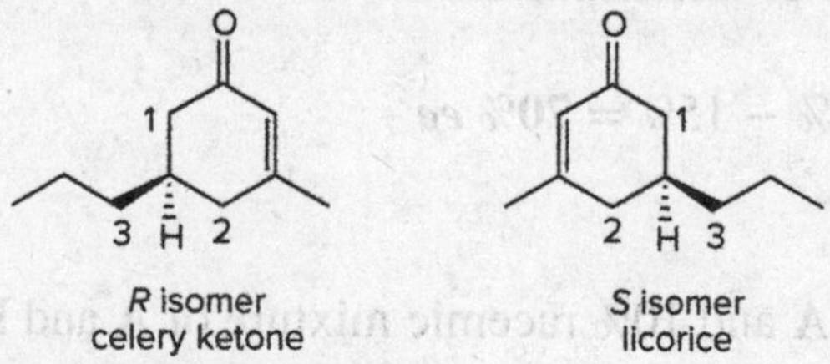

R isomer
celery ketone

S isomer
licorice

5.35

a, b.

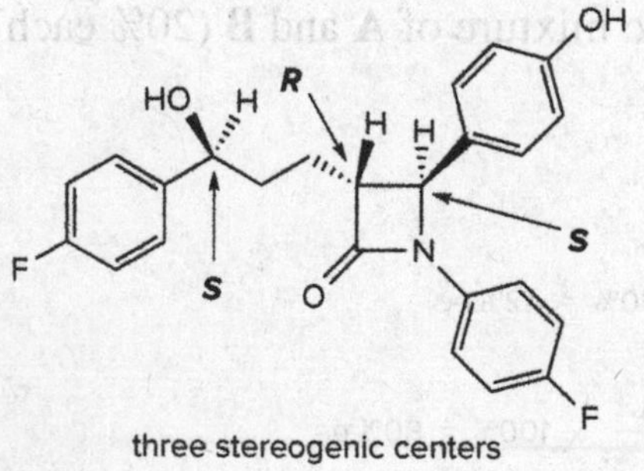

three stereogenic centers

5.36

A identical B

C D

a. **A and B, same *R,S* assignment, identical**
b. **A and C, opposite *R,S* assignment, enantiomers**
c. **A and D, one stereogenic center different, diastereomers**
d. **C and D, one stereogenic center different, diastereomers**

5.37 Use the definitions from Answer 5.1.

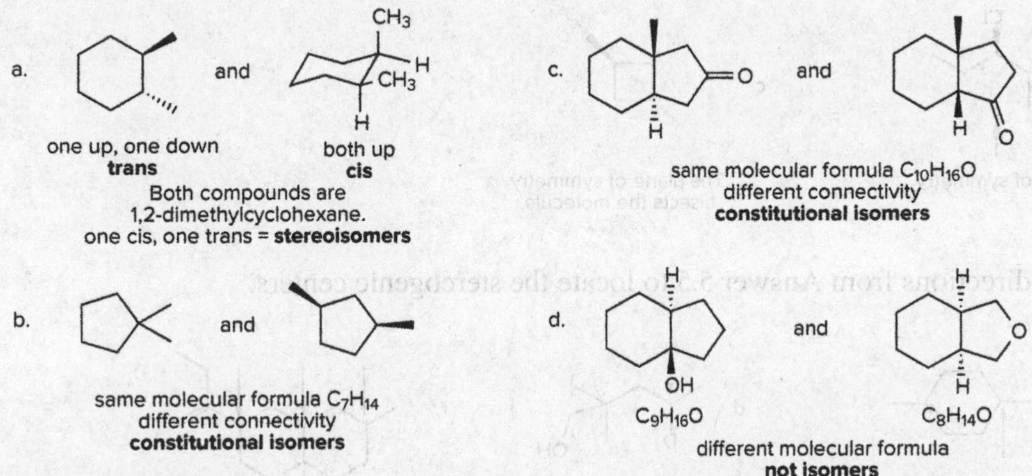

5.38 Use the definitions from Answer 5.2.

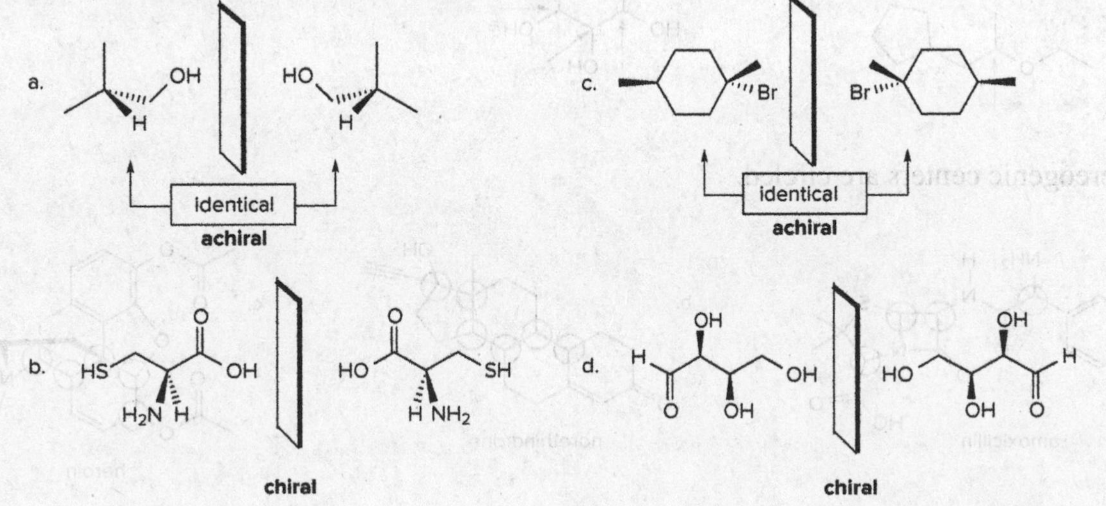

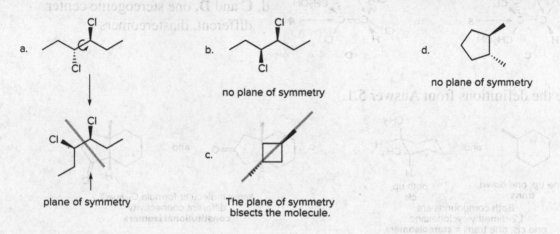

5.39

a. b. c.

R isomer

S enantiomer

R identical

S enantiomer

5.40 A plane of symmetry cuts the molecule into **two identical halves.**

a. b. d.

no plane of symmetry

no plane of symmetry

c.

plane of symmetry

The plane of symmetry bisects the molecule.

5.41 Use the directions from Answer 5.5 to locate the stereogenic centers.

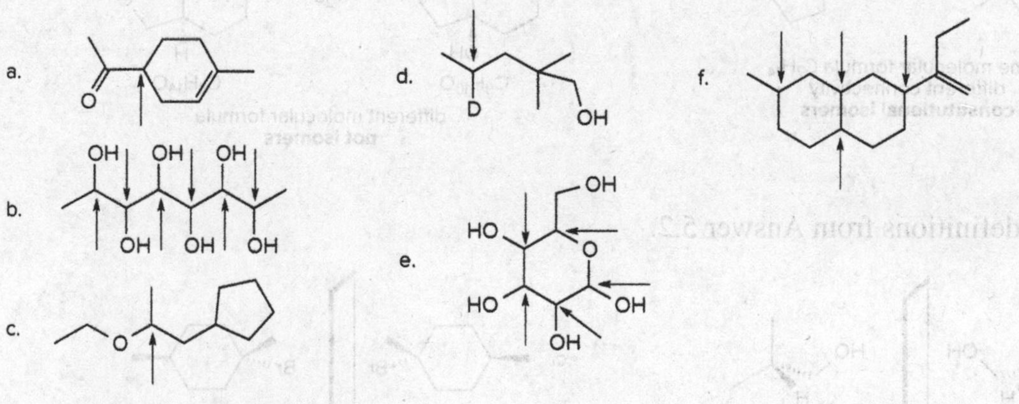

5.42 Stereogenic centers are circled.

a. b. c.

NH_2 H N S O HO

amoxicillin

OH

norethindrone

N O

heroin

5.43

a. amphetamine

H_2N H | H NH_2

b. ketoprofen

OH | OH | OH

5.44 Assign priority based on the rules in Answer 5.11.

a. $-CD_3$, $-CH_3$
↑
D higher mass than H
higher priority

b. $-CH(CH_3)_2$, $-CH_2OH$
↑
C bonded to O
higher priority

c. $-CH_2Cl$, $-CH_2CH_2CH_2Br$
↑
C bonded to Cl
higher priority

d. $-CH_2NH_2$, $-NHCH_3$
↑
higher atomic number
higher priority

5.45 Assign priority based on the rules in Answer 5.11.

a. $-F > -OH > -NH_2 > -CH_3$

b. $-(CH_2)_3CH_3 > -CH_2CH_2CH_3 > -CH_2CH_3 > -CH_3$

c. $-NH_2 > -CH_2NHCH_3 > -CH_2NH_2 > -CH_3$

d. $-COOH > -CHO > -CH_2OH > -H$

e. $-Cl > -SH > -OH > -CH_3$

f. $-C{\equiv}CH > -CH{=}CH_2 > -CH(CH_3)_2 > -CH_2CH_3$

5.46 Use the rules in Answer 5.14 to assign *R* or *S* to each stereogenic center.

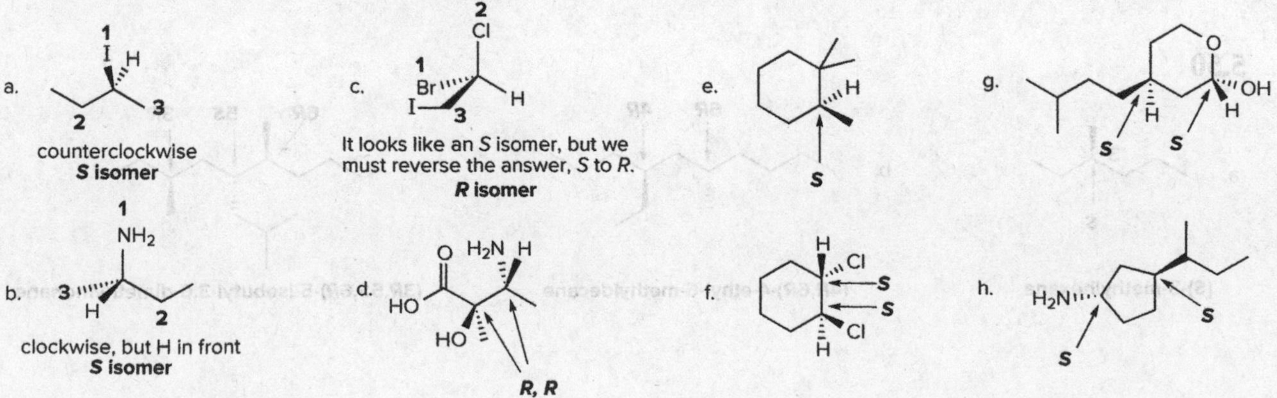

5.47

a. CH3, H, Br, HO, CH=CH2, CH2CH3

re-draw

H Br R
R HO

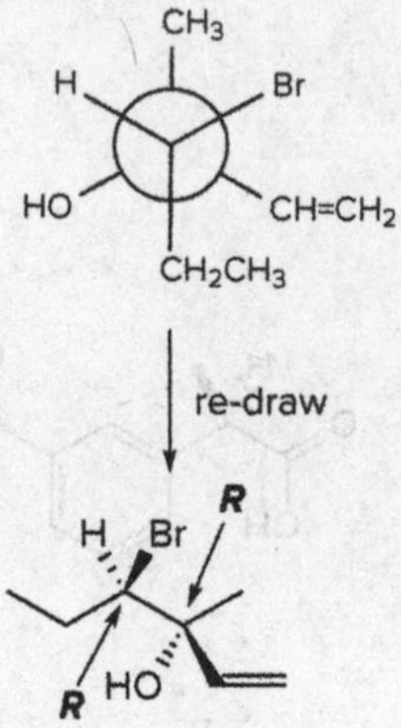

b. CHO, HO, H, H, OH, CH2CH2Br

re-draw

H OH CHO
H
OH
Br S S

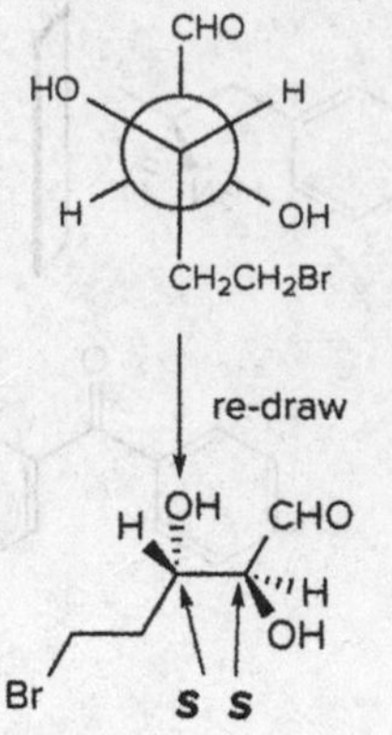

5.48

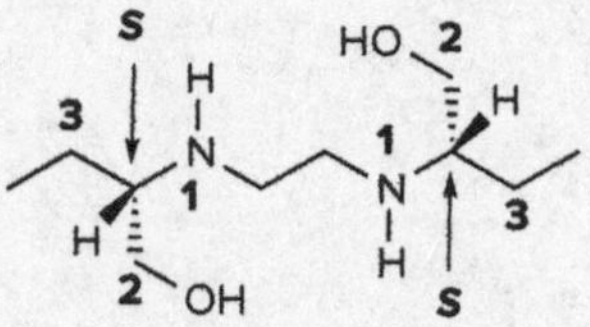

S, H, 3, N, 1, H, 2, OH, HO 2, H, 1, N, H, 3, S

5.49

a. (*R*)-3-methylhexane

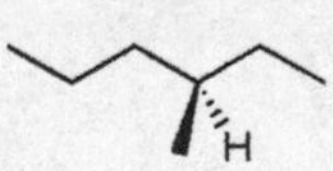

H

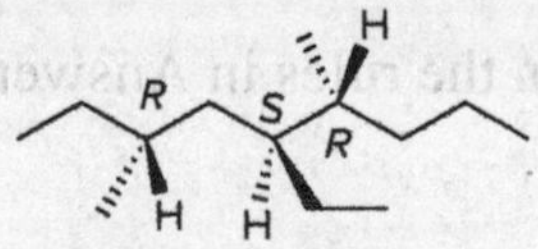

b. (4*R*,5*S*)-4,5-diethyloctane

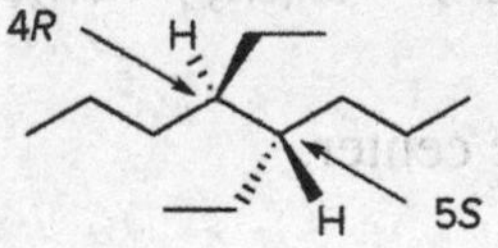

4*R* H
H 5*S*

c. (3*R*,5*S*,6*R*)-5-ethyl-3,6-dimethylnonane

H
R S R
H H

d. (3*S*,6*S*)-6-isopropyl-3-methyldecane

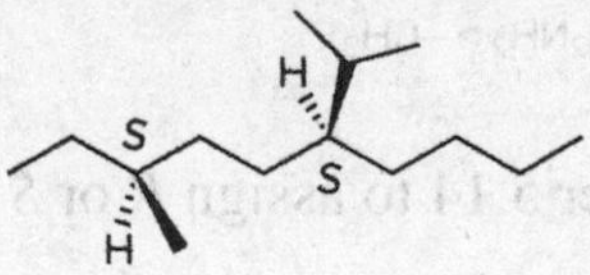

H
S S
H

5.50

a. 3
S

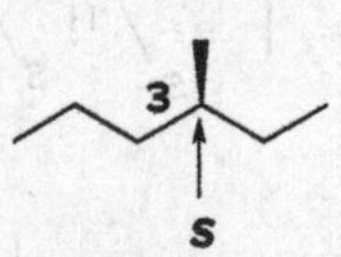

(*S*)-3-methylhexane

b. 6*R* 4*R*

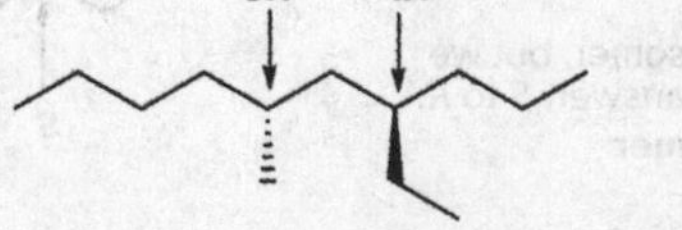

(4*R*,6*R*)-4-ethyl-6-methyldecane

c. 6*R* 5*S* 3*R*

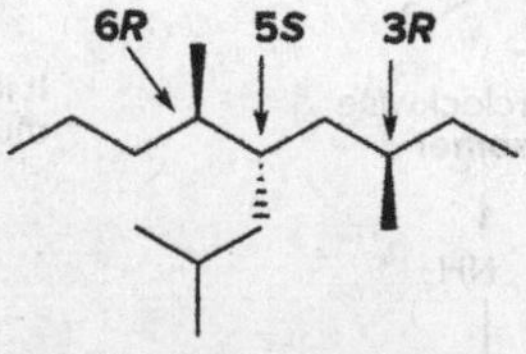

(3*R*,5*S*,6*R*)-5-isobutyl-3,6-dimethylnonane

5.51

a. 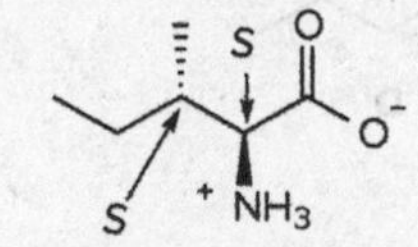 S, S, $^{+}NH_3$, O^-

b. 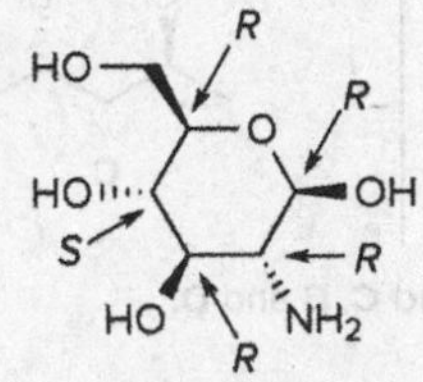 HO, R, R, OH, HO, S, R, NH_2, HO, R

c. 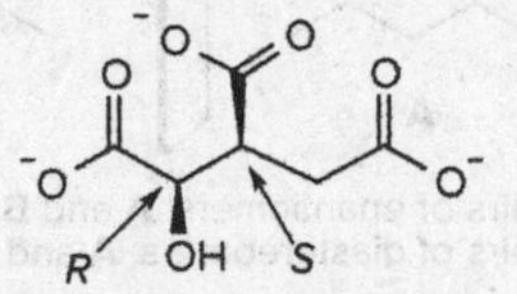 O^-, O, O, O, O^-, O^-, R, OH, S

5.52

a. 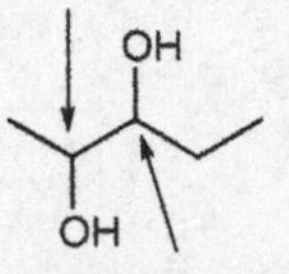 OH, OH

2 stereogenic centers
2^2 = 4 possible stereoisomers

b. 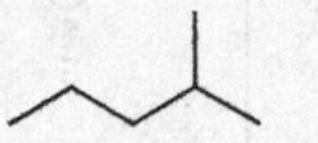 0 stereogenic centers

c. 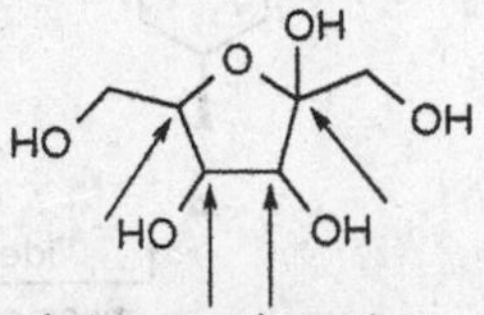 HO, OH, O, OH, HO, OH

4 stereogenic centers
2^4 = 16 possible stereoisomers

5.53

a. 1*R*,2*S*

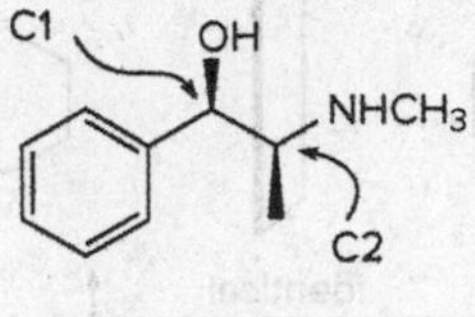

C1, OH, $NHCH_3$, C2

ephedrine

b. 1*S*,2*S*

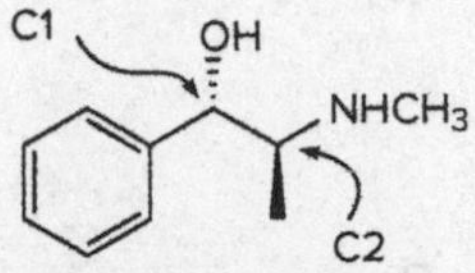

C1, OH, $NHCH_3$, C2

pseudoephedrine

c. Ephedrine and pseudoephedrine are diastereomers. One stereogenic center is the same; one is different.

d. OH, *R*, $NHCH_3$, *S* ← e. enantiomer of (–)-ephedrine

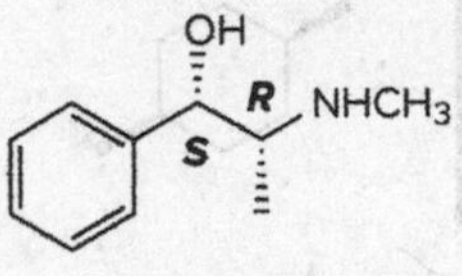

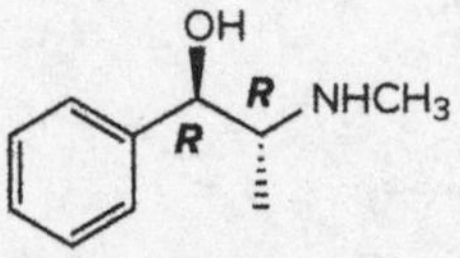

OH, *R*, $NHCH_3$, *R* ← diastereomer of (–)-ephedrine

5.54

a.

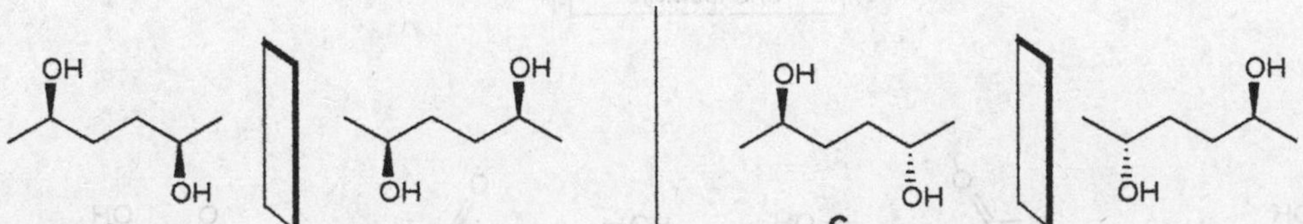

OH, OH — **A**; OH, OH — **B**; OH, OH — **C**; OH, OH

identical
meso compound

Pair of enantiomers: **A** and **B**.
Pairs of diastereomers: **A** and **C**, **B** and **C**.

b.

A B C D

Pairs of enantiomers: **A** and **B**, **C** and **D**.
Pairs of diastereomers: **A** and **C**, **A** and **D**, **B** and **C**, **B** and **D**.

c.

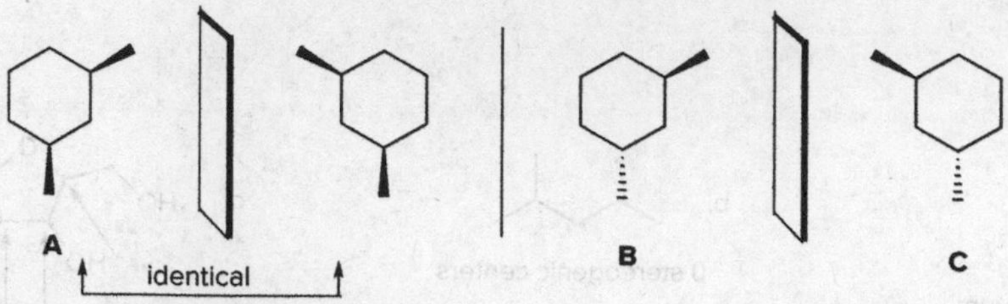

A B C

identical

meso compound

Pair of enantiomers: **B** and **C**.
Pairs of diastereomers: **A** and **B**, **A** and **C**.

d.

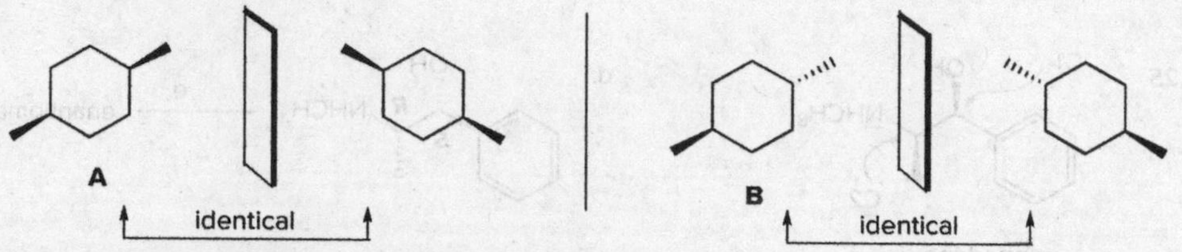

A B

identical identical

Pair of diastereomers: **A** and **B**.
Meso compounds: **A** and **B**.

5.55

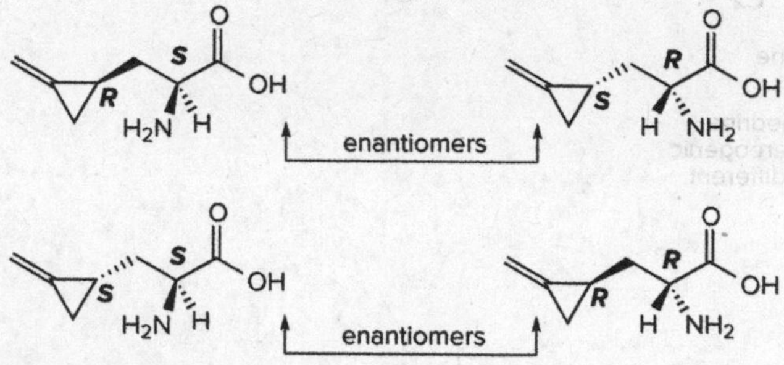

enantiomers

enantiomers

5.56

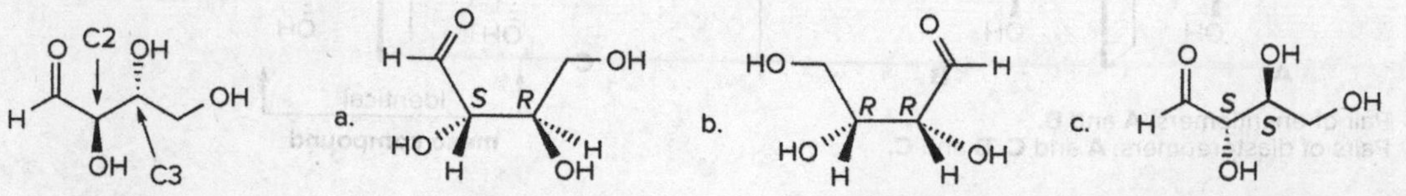

D-erythrose
2*R*,3*R*

a. 2*S*,3*R*
diastereomer

b. 2*R*,3*R*
identical

c. 2*S*,3*S*
enantiomer

5.57 Re-draw each Newman projection and determine the *R,S* configuration. Then determine how the molecules are related.

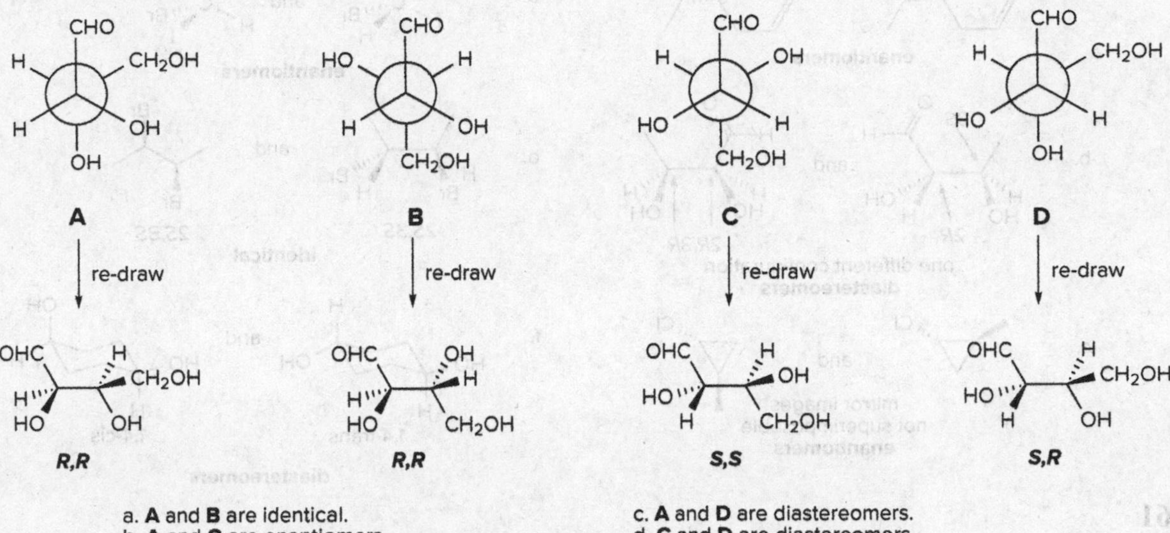

a. **A** and **B** are identical.
b. **A** and **C** are enantiomers.
c. **A** and **D** are diastereomers.
d. **C** and **D** are diastereomers.

5.58

trans R NH_2 **A** — cis R NH_2 **B** — trans R NH_2 **C** — trans S NH_2 **D** — NH_2 **E**

E: NH_2 group is in a different location.

A (trans, *R*) and **B** (cis, *R*) are diastereomers.
A (trans, *R*) and **C** (trans, *R*) are identical molecules.
A (trans, *R*) and **D** (trans, *S*) are enantiomers.
A (trans, *R*) and **E** are constitutional isomers.

5.59

OH S **A** — S OH **B** identical — OH **C** bonded to different C constitutional isomers — OH **D** enantiomers

5.60

a. and

enantiomers

b. 3S 2R and 2R,3R

one different configuration

diastereomers

c. and

mirror images
not superimposable
enantiomers

d. and

enantiomers

e. and

2S,3S 2S,3S

identical

f. and

1,4-trans 1,4-cis

diastereomers

5.61

A B C D

a. A chiral compound is optically active. **A, B,** and **C** are optically active.

b. Constitutional isomers and diastereomers have different physical properties. **A** and **C** are enantiomers, so they have the same boiling point. There would be three fractions: **A** and **C, B,** and **D.**

c. Only the fraction with **B** is optically active.

5.62

a. **A** and **B** are constitutional isomers.
A and **C** are constitutional isomers.
B and **C** are diastereomers (cis and trans).
C and **D** are enantiomers.

b.

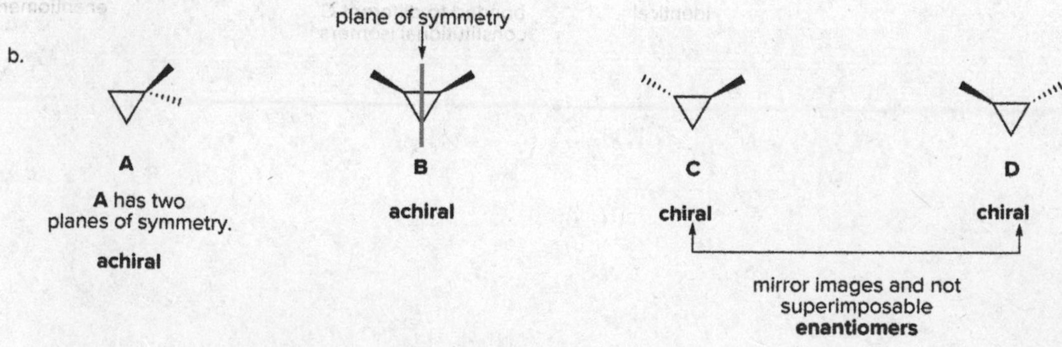

c. Alone, **C** and **D** would be optically active.
d. **A** and **B** have a plane of symmetry.
e. **A** and **B** have different boiling points.
B and **C** have different boiling points.
C and **D** have the same boiling point.
f. **B** is a meso compound.
g. An equal mixture of **C** and **D** is optically inactive because it is a racemic mixture.
An equal mixture of **B** and **C** would be optically active.

5.63

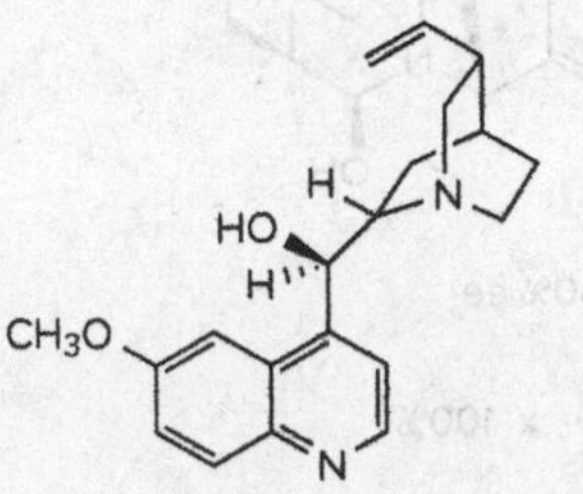

quinine

$$ee = \frac{[\alpha]\text{ mixture}}{[\alpha]\text{ pure enantiomer}} \times 100\%$$

quinine = **A**
quinine's enantiomer = **B**

a.

$\frac{-50}{-165} \times 100\% = 30\%\ ee$

$\frac{-83}{-165} \times 100\% = 50\%\ ee$

$\frac{-120}{-165} \times 100\% = 73\%\ ee$

b. 30% *ee* = 30% excess one compound (**A**)
remaining 70% = mixture of 2 compounds (35% each **A** and **B**)
Amount of **A** = 30 + 35 = **65%**
Amount of **B** = **35%**

50% *ee* = 50% excess one compound (**A**)
remaining 50% = mixture of 2 compounds (25% each **A** and **B**)
Amount of **A** = 50 + 25 = **75%**
Amount of **B** = **25%**

73% *ee* = 73% excess of one compound (**A**)
remaining 27% = mixture of 2 compounds (13.5% each **A** and **B**)
Amount of **A** = 73 + 13.5 = **86.5%**
Amount of **B** = **13.5%**

c. $[\alpha] = +165$
d. 80% – 20% = 60% ee

e. $60\% = \frac{[\alpha]\text{ mixture}}{-165} \times 100\%$

$[\alpha]$ mixture = –99

5.64

a.

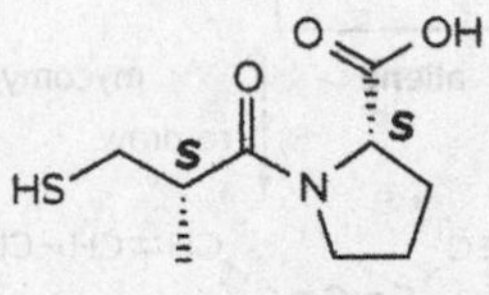

b.

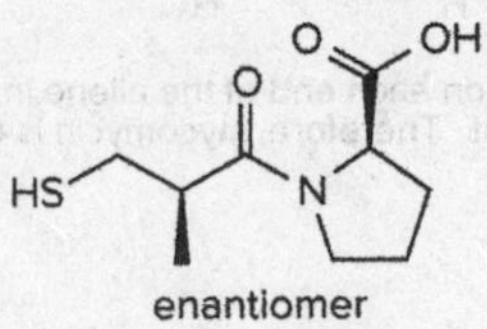

enantiomer

c. One equivalent of base removes the most acidic proton to form:

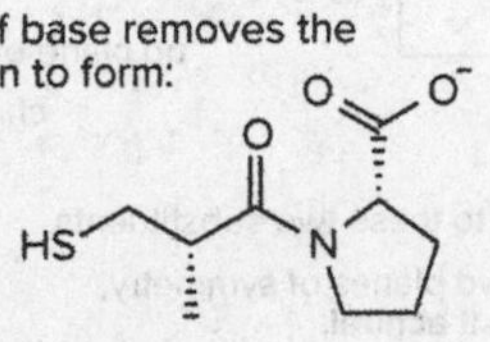

d. With two equivalents of base, the two most acidic protons are removed to form:

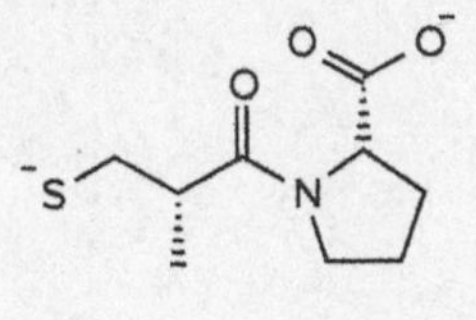

5.65

a.

HO NH O O O S H HO N N O O OH

7 stereogenic centers circled

b. maximum number of stereoisomers = 2^7 = 128 isomers

c. To draw the enantiomer, change all wedges to dashed wedges and vice versa:

d. To draw a diastereomer, change the orientation of groups at only one stereogenic center :

e. ee = 75% – 25% = 50% ee

$$ee = \frac{[\alpha] \text{ solution}}{[\alpha] \text{ pure}} \times 100\%$$

$$50\% = \frac{[\alpha]}{+41.5} \times 100\%$$

$$[\alpha] = +20.8$$

f. $$ee = \frac{[\alpha] \text{ solution}}{[\alpha] \text{ pure}} \times 100\% = \frac{+10.5}{+41.5} \times 100\% = 25.3\%\ ee$$

5.66 Allenes contain an *sp* hybridized carbon atom doubly bonded to two other carbons. This makes the double bonds of an allene perpendicular to each other. When each end of the allene has two like substituents, the allene contains two planes of symmetry and it is achiral. When each end of the allene has two different groups, the allene has no plane of symmetry and it becomes chiral.

CH_3 C=C=C H H CH_3

A

These two substituents are at 90° to these two substituents.

Allene **A** contains two planes of symmetry, making it **achiral.**

CH_3 C=C=C CH_3 H H

B

no plane of symmetry

chiral

HC≡C–C≡C–CH=C=CH–CH=CH–CH=CHCH$_2$CO$_2$H

allene mycomycin

re-draw

HC≡C–C≡C, C=C=C, CH=CH–CH=CHCH$_2$CO$_2$H H H

The substituents on each end of the allene in mycomycin are different. Therefore, mycomycin is **chiral.**

5.67

discodermolide

a. The 13 tetrahedral stereogenic centers are circled.

b. Because there is restricted rotation around a C–C double bond, groups on the end of the double bond cannot interconvert. Whenever the substituents on each end of the double bond are different from each other, the double bond is a stereogenic site. Thus, the following two double bonds are isomers:

These compounds are isomers.

There are three stereogenic sites due to the double bonds in discodermolide, labeled with arrows.

c. The maximum number of stereoisomers for discodermolide must include the 13 tetrahedral stereogenic centers and the three double bonds.
Maximum number of stereoisomers = 2^{16} = 65,536.

5.68

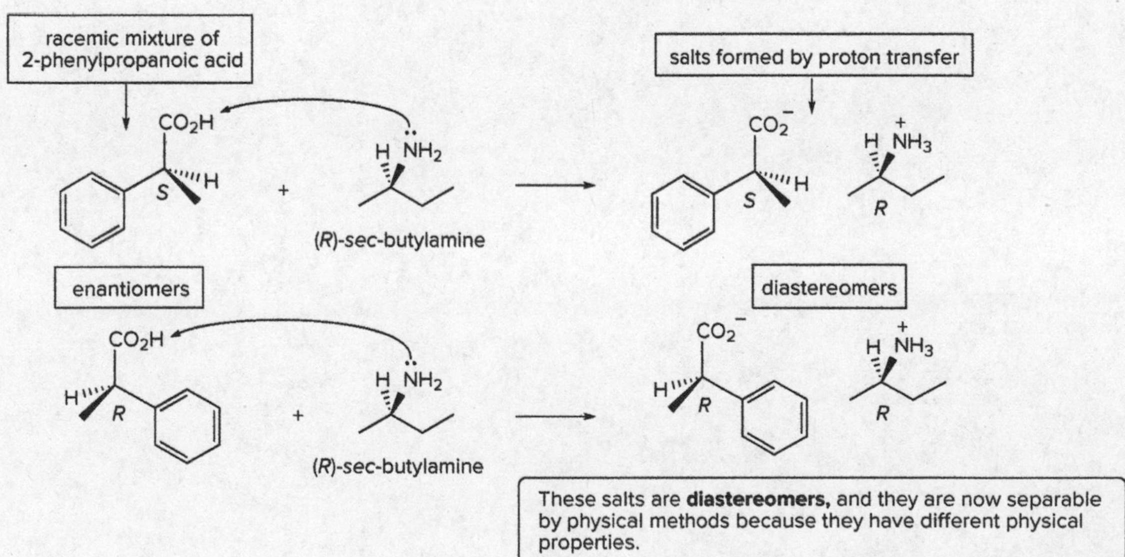

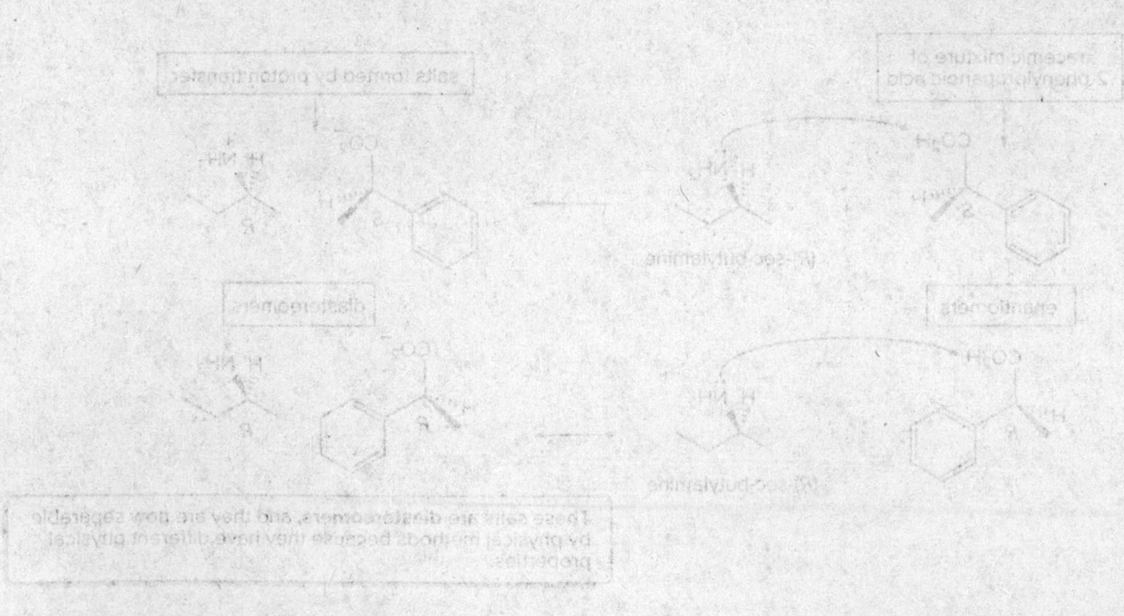

Chapter 6: Understanding Organic Reactions

Chapter Review

Writing organic reactions (6.1)

- Use curved arrows to show the movement of electrons. Half-headed arrows are used for single electrons and full-headed arrows are used for electron pairs.

half-headed arrows | full-headed arrow

- Reagents can be drawn either on the left side of an equation or over an arrow. Catalysts are drawn over or under an arrow.

Types of reactions (6.2)

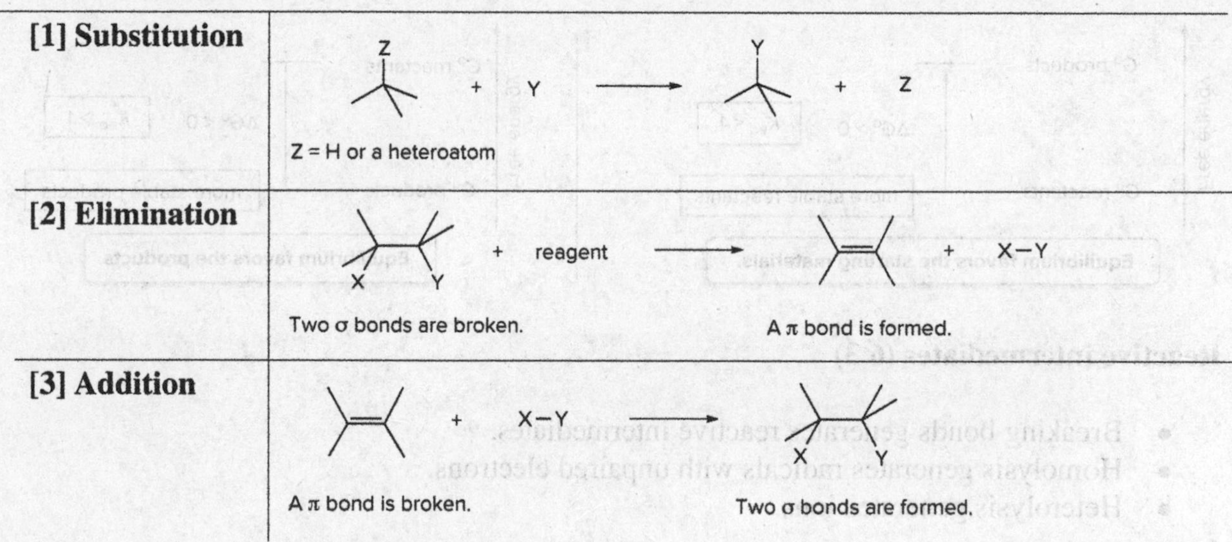

Important trends

Values compared	Trend
Bond dissociation energy and **bond strength**	The ***higher*** the bond dissociation energy, the ***stronger*** the bond (6.4). **Increasing size of the halogen** → CH_3-F CH_3-Cl CH_3-Br CH_3-I $\Delta H°$ = 456 kJ/mol 351 kJ/mol 293 kJ/mol 234 kJ/mol ← **Increasing bond strength**

E_a **and reaction rate**	The ***larger*** the energy of activation, the ***slower*** the reaction (6.9A).
E_a **and rate constant**	The ***higher*** the energy of activation, the ***smaller*** the rate constant (6.9B).

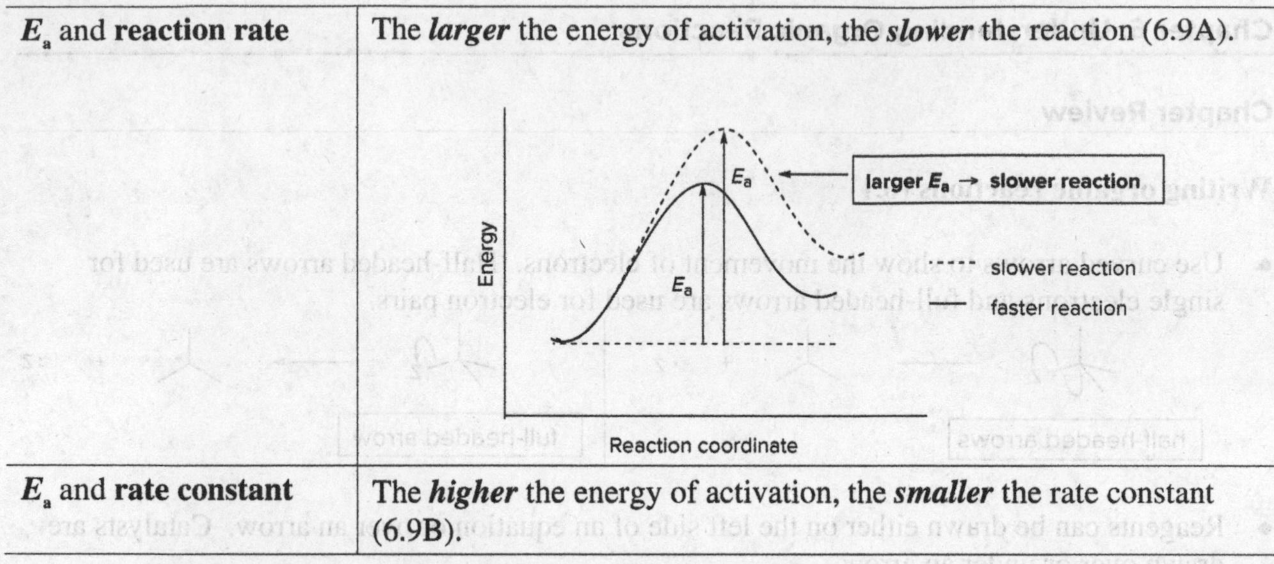

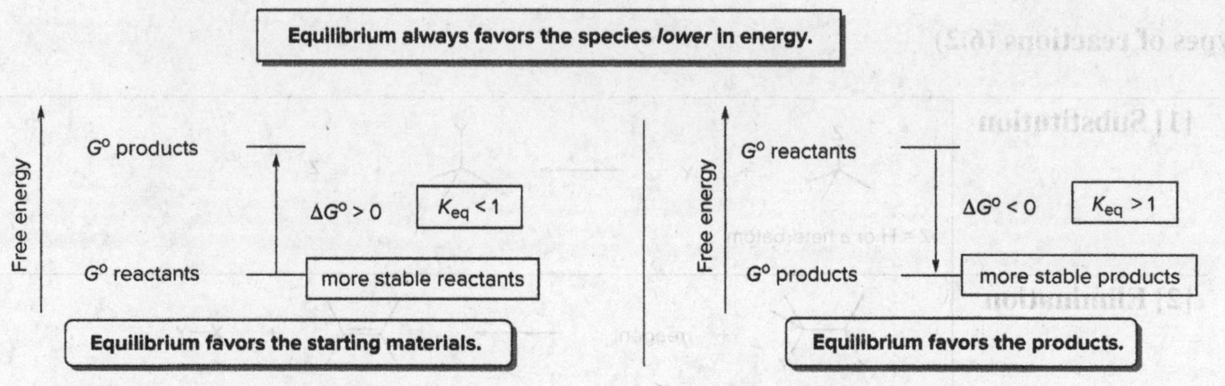

Reactive intermediates (6.3)

- Breaking bonds generates reactive intermediates.
- Homolysis generates radicals with unpaired electrons.
- Heterolysis generates ions.

Reactive intermediate	General structure	Reactive feature	Reactivity
radical		unpaired electron	electrophilic
carbocation		positive charge; only six electrons around C	electrophilic
carbanion		net negative charge; lone electron pair on C	nucleophilic

6.40 Reactions resulting in an increase in entropy are favored. When a single molecule forms two molecules, there is an increase in entropy.

a. [alkane] ⟶ [alkene] + [alkane] — increased number of molecules; $\Delta S°$ is positive. products favored

b. $CH_3CO_2CH_3$ + H_2O ⟶ CH_3CO_2H + CH_3OH — no change in the number of molecules; neither favored

6.41

a. Write the net (coupled) reaction by summing the substances in both the phosphorylation and hydrolysis equations.

hydrolysis: phosphoenolpyruvate + ~~H_2O~~ ⟶ pyruvate + ~~HPO_4^{2-}~~ $\Delta G°' = -61.9$ kJ/mol

phosphorylation: ADP + ~~HPO_4^{2-}~~ ⟶ ATP + ~~H_2O~~ $\Delta G°' = 30.5$ kJ/mol

coupled reaction: phosphoenolpyruvate + ADP ⟶ pyruvate + ATP $\Delta G°' = -31.4$ kJ/mol

b. As shown in part (a), the sum of the energies for phosphorylation and hydrolysis gives the energy of the coupled reaction; $\Delta G°' = -31.4$ kJ/mol.

c.

phosphoenolpyruvate —(ADP → ATP; Mg^{2+}; pyruvate kinase)→ pyruvate

6.42 Use the directions in Answer 6.17 to draw the transition state. Nonbonded electron pairs are drawn in at reacting sites.

a. cyclohexanol + $:\ddot{N}H_2^-$ ⟶ cyclohexoxide + NH_3

transition state: [cyclohexyl–$\ddot{O}^{\delta-}$- - H - - $NH_2^{\delta-}$]‡

b. carbocation + $H_2\ddot{O}:$ ⟶ alkene + $H_3\ddot{O}^+$

transition state: [$C^{\delta+}$ ··· H ··· $OH_2^{\delta+}$]‡

6.43

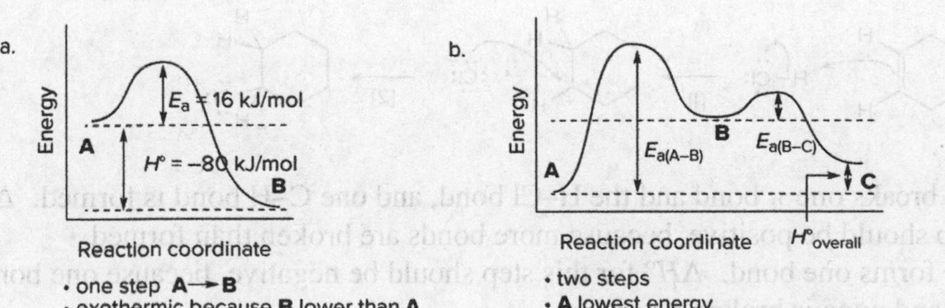

a.
- one step A → B
- exothermic because **B** lower than **A**

b.
- two steps
- **A** lowest energy
- **B** highest energy
- **$E_{a(A-B)}$ is rate-determining,** because the transition state for Step [1] is higher in energy.

6.44

a. $CH_3{-}H$ + $\cdot\ddot{Cl}$: ⟶ $\cdot CH_3$ + $H\ddot{Cl}$:

b. $\cdot Cl + CH_4$ ⟶ $\cdot CH_3$ + HCl

[1] Bonds broken		**[2] Bonds formed**		**[3] Overall $\Delta H°$ =**
	$\Delta H°$ (kJ/mol)		$\Delta H°$ (kJ/mol)	+ 435 kJ/mol
$CH_3{-}H$	+ 435 kJ/mol	H−Cl	− 431 kJ/mol	− 431 kJ/mol
				ANSWER: + 4 kJ/mol

c.

Energy

E_a = 16 kJ/mol

$\Delta H°$ = 4 kJ/mol

Reaction coordinate

d. E_a for the reverse reaction is the difference in energy between the products and the transition state, 12 kJ/mol.

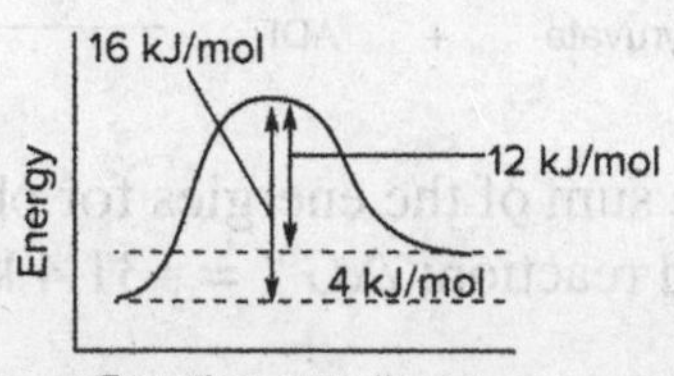

6.45

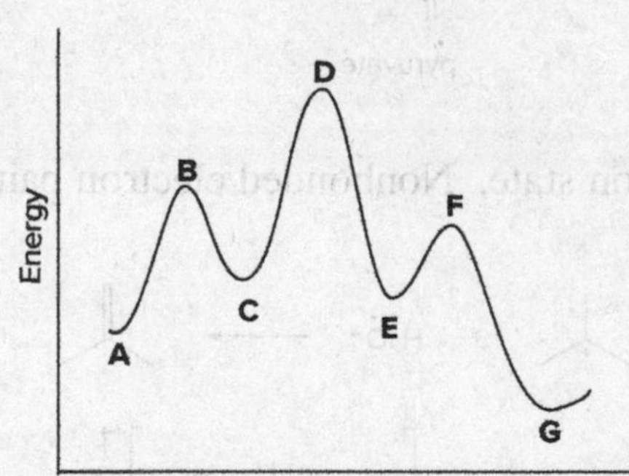

a. **B, D,** and **F** are transition states.
b. **C** and **E** are reactive intermediates.
c. The overall reaction has **three steps.**
d. **A–C** is endothermic.
C–E is exothermic.
E–G is exothermic.
e. The overall reaction is exothermic.

6.46

cyclohexene + $H{-}\ddot{Cl}$: —[1]⟶ carbocation (+) + :$\ddot{Cl}$:$^-$ —[2]⟶ chlorocyclohexane

a. Step [1] breaks one π bond and the H–Cl bond, and one C–H bond is formed. $\Delta H°$ for this step should be positive, because more bonds are broken than formed.

b. Step [2] forms one bond. $\Delta H°$ for this step should be negative, because one bond is formed and none is broken.

c. Step [1] is rate-determining, because it is more difficult.

d. Transition state for Step [1]: Transition state for Step [2]:

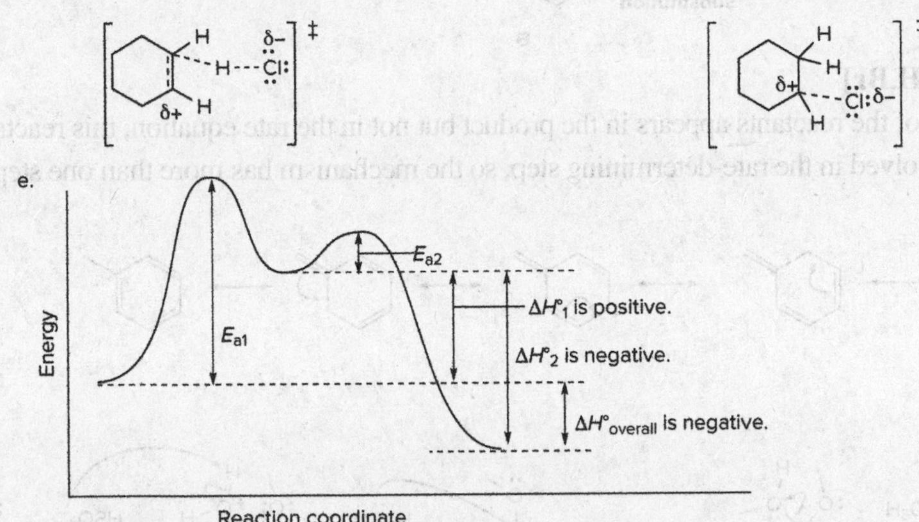

6.47 E_a, concentration, catalysts, rate constant, and temperature affect reaction rate so (c), (d), (e), (g), and (h) affect rate.

6.48

a. **rate = $k[CH_3Br][NaCN]$**
b. Double $[CH_3Br]$ = **rate doubles.**
c. Halve [NaCN] = **rate halved.**
d. Increase both $[CH_3Br]$ and [NaCN] by factor of 5 = [5][5] = **rate increases by a factor of 25.**

6.49

a. acetyl chloride + CH_3O^- —[1] slow→ intermediate —[2] fast→ methyl acetate + Cl^-

b. Only the slow step is included in the rate equation: **Rate = $k[CH_3O^-][CH_3COCl]$**
c. CH_3O^- is in the rate equation. Increasing its concentration by 10 times would increase the rate by **10 times.**
d. When both reactant concentrations are increased by 10 times, the rate increases by **100 times (10 × 10 = 100).**
e. This is a **substitution reaction** (OCH_3 substitutes for Cl).

6.50

a. A + CH_3OH → (substitution) B + HBr

b. **Rate = *k*[C_7H_7Br]**

Because one of the reactants appears in the product but not in the rate equation, this reactant cannot be involved in the rate-determining step, so the mechanism has more than one step.

c.

6.51

a. A (acetal) ⇌ [1] B + HSO_4^- ⇌ [2] C + CH_3OH ⇌ [3] D ⇌ [4] E (hemiacetal) + H_2SO_4

b. C

c. The electrophile is **C**; H_2O is the nucleophile.

d. Steps [1] and [4] are Brønsted–Lowry acid–base reactions.

6.52

a. A + B → C

b. Two π bonds in **A** are broken and one π bond in **B** is broken. Two new σ bonds in **C** are formed (in bold), as well as a new π bond.

c. The reaction should be exothermic because more energy is released in forming two new C–C σ bonds than is required to break two C–C π bonds.

d. Entropy favors the reactants for two reasons. There are two molecules of reactant and only one product. The reactants are both acyclic and the product has a ring with fewer degrees of freedom.

e. The Diels–Alder reaction is an addition reaction because π bonds are broken and new σ bonds are formed.

6.53

a. The first mechanism has one step: **Rate =** $k[(CH_3)_3CI][^-OH]$

b. The second mechanism has two steps, but only the first step would be in the rate equation, because it is slow and therefore rate-determining: **Rate =** $k[(CH_3)_3CI]$

c. Possibility [1] is second order; possibility [2] is first order.

d. These rate equations can be used to show which mechanism is plausible by changing the concentration of ^-OH. If this affects the rate, then possibility [1] is reasonable. If it does not affect the rate, then possibility [2] is reasonable.

e.

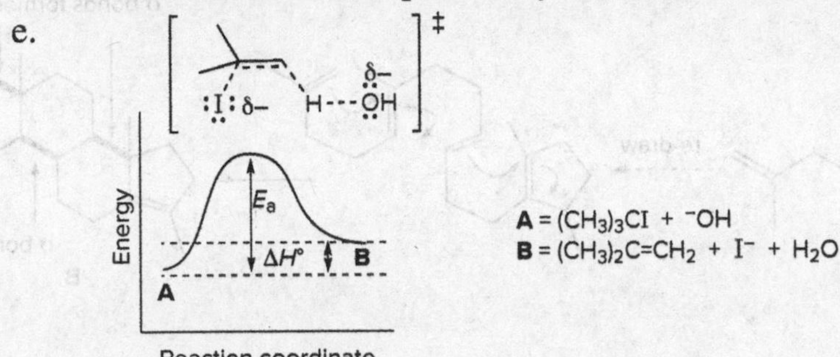

f.

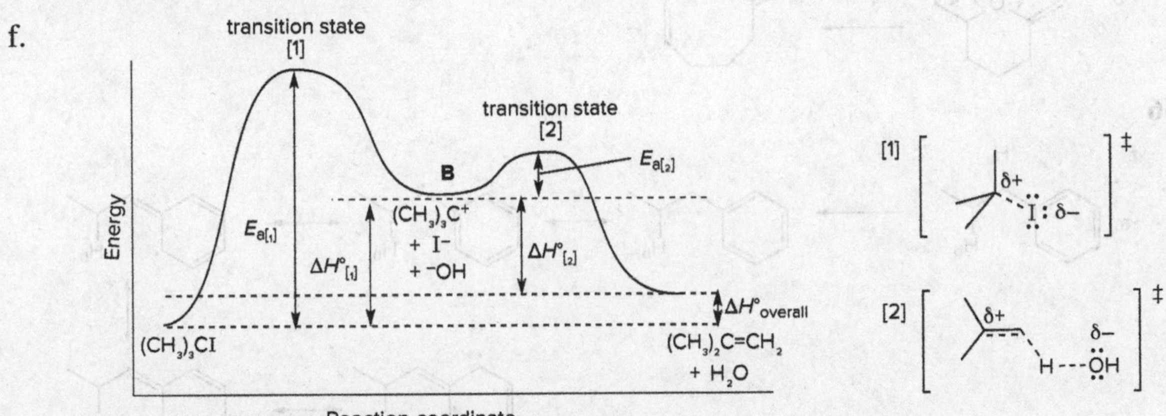

6.54 The difference in both the acidity and the bond dissociation energy of CH_3CH_3 versus HC≡CH is due to the same factor: percent *s*-character. The difference results because one process is based on homolysis and one is based on heterolysis.

Bond dissociation energy:

CH_3CH_2–H: *sp*[3] hybridized, 25% *s*-character

HC≡C–H: *sp* hybridized, 50% *s*-character. Higher percent *s*-character makes this bond shorter and stronger.

Acidity: To compare acidity, we must compare the stability of the conjugate bases:

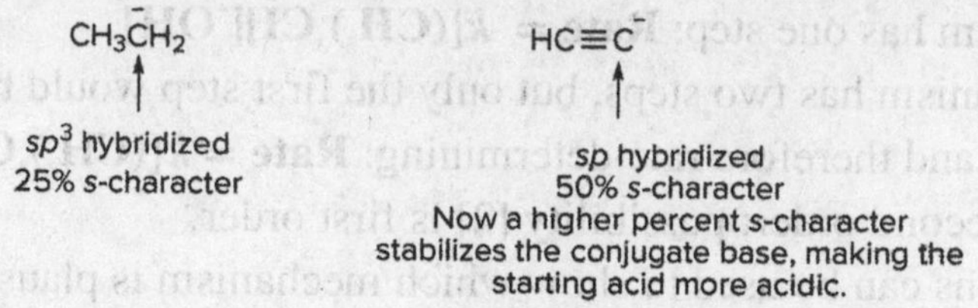

CH_3CH_2
sp^3 hybridized
25% *s*-character

HC≡C
sp hybridized
50% *s*-character
Now a higher percent *s*-character stabilizes the conjugate base, making the starting acid more acidic.

6.55 a. Re-draw **A** to see more clearly how cyclization occurs.

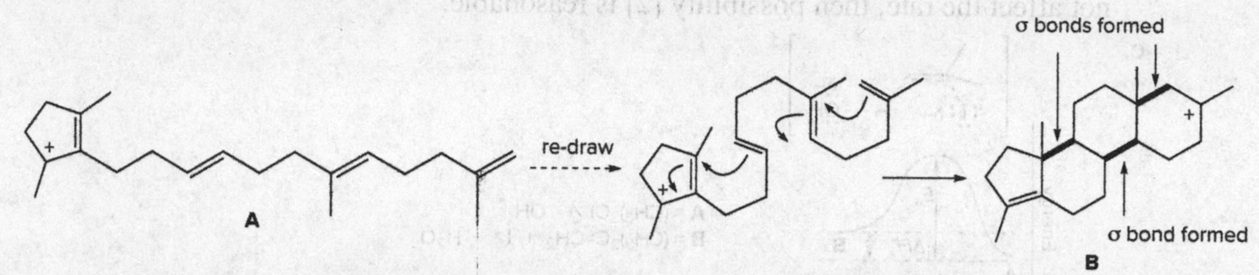

A
re-draw
σ bonds formed
σ bond formed
B

b.

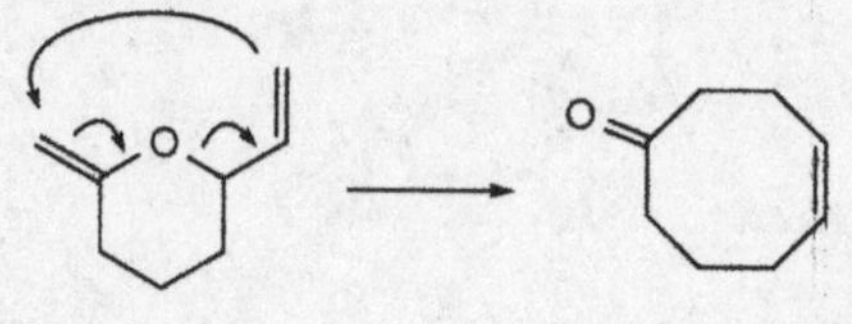

6.56

a. H_a H_b

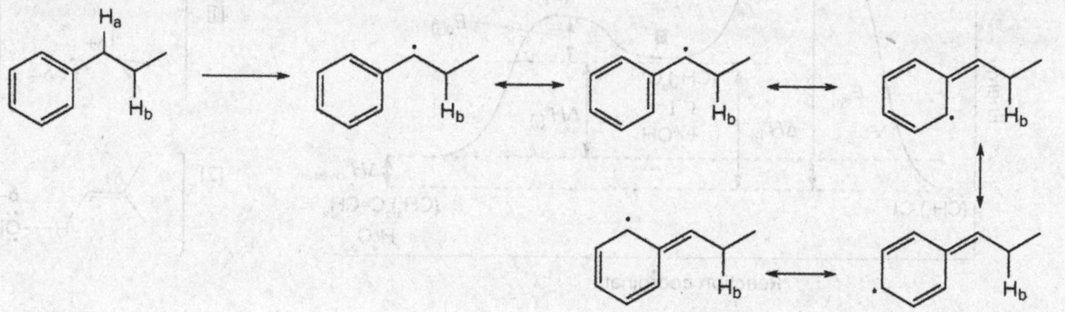

b. H_a H_b

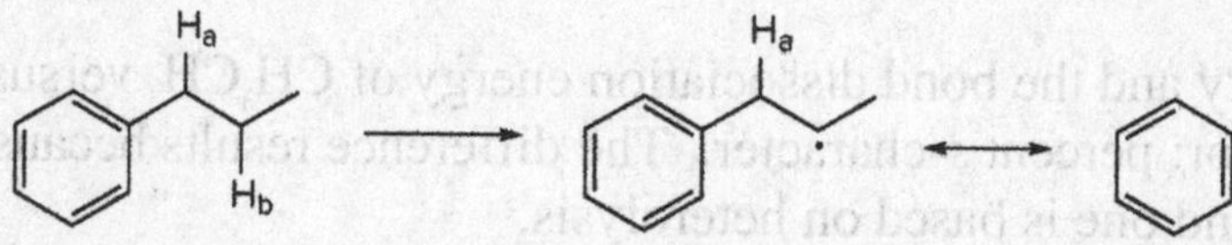

c. C–H_a is weaker than C–H_b because the carbon radical formed when the C–H_a bond is broken is highly resonance stabilized. This means the bond dissociation energy for C–H_a is lower.

6.57

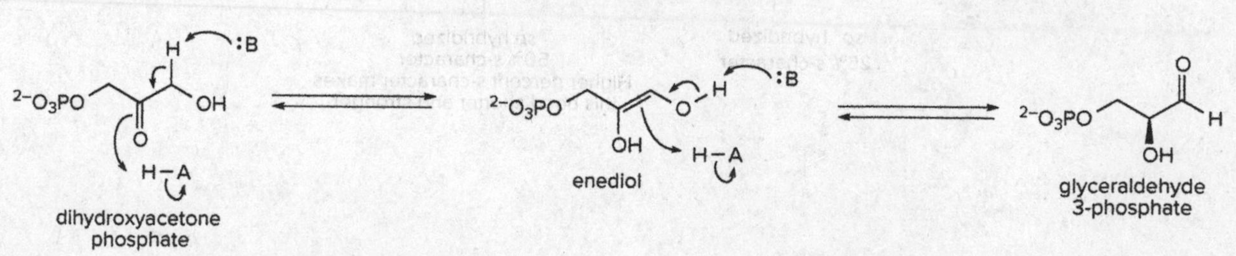

$^{2-}O_3PO$ OH :B H–A
dihydroxyacetone phosphate

enediol

glyceraldehyde 3-phosphate

Chapter 7 Alkyl Halides and Nucleophilic Substitution

Chapter Review

General facts about alkyl halides

- Alkyl halides contain a halogen atom X bonded to an sp^3 hybridized carbon (7.1).
- Alkyl halides are named as halo alkanes, with the halogen as a substituent (7.2).
- Alkyl halides have a polar C–X bond, so they exhibit dipole–dipole interactions but are incapable of intermolecular hydrogen bonding (7.3).
- The polar C–X bond containing an electrophilic carbon makes alkyl halides reactive toward nucleophiles and bases (7.5).

The central theme (7.6)

- Nucleophilic substitution is one of the two main reactions of alkyl halides. A nucleophile replaces a leaving group on an sp^3 hybridized carbon.

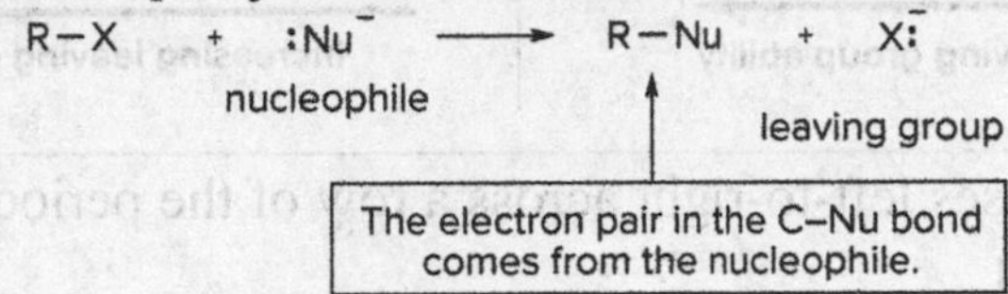

- One σ bond is broken and one σ bond is formed.
- There are two possible mechanisms: S_N1 and S_N2.

S_N1 and S_N2 mechanisms compared

	S_N2 mechanism	S_N1 mechanism
[1] Mechanism	• One step (7.11B)	• Two steps (7.12B)
[2] Alkyl halide	• Order of reactivity: CH_3X > RCH_2X > R_2CHX > R_3CX (7.11D)	• Order of reactivity: R_3CX > R_2CHX > RCH_2X > CH_3X (7.12D)
[3] Rate equation	• rate = k[RX][:Nu⁻]	• rate = k[RX]
	• second-order kinetics (7.11A)	• first-order kinetics (7.12A)
[4] Stereochemistry	• backside attack of the nucleophile (7.11C)	• trigonal planar carbocation intermediate (7.12C)
	• inversion of configuration at a stereogenic center	• racemization at a stereogenic center
[5] Nucleophile	• favored by stronger nucleophiles (7.15B)	• favored by weaker nucleophiles (7.15B)
[6] Leaving group	• better leaving group → faster reaction (7.15C)	• better leaving group → faster reaction (7.15C)
[7] Solvent	• favored by polar aprotic solvents (7.15D)	• favored by polar protic solvents (7.15D)

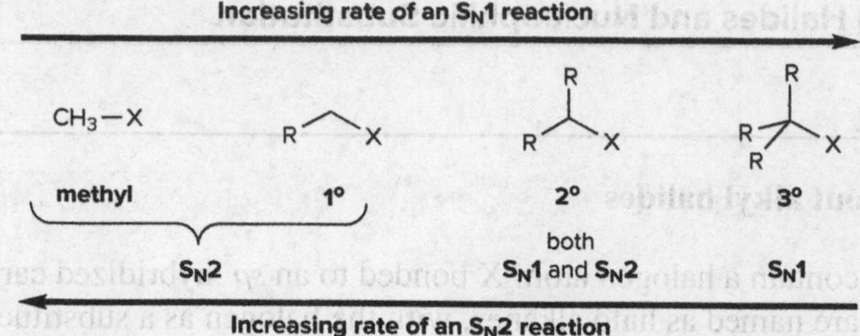

Important trends

- The best leaving group is the weakest base. Leaving group ability increases left-to-right across a row and down a column of the periodic table (7.7).

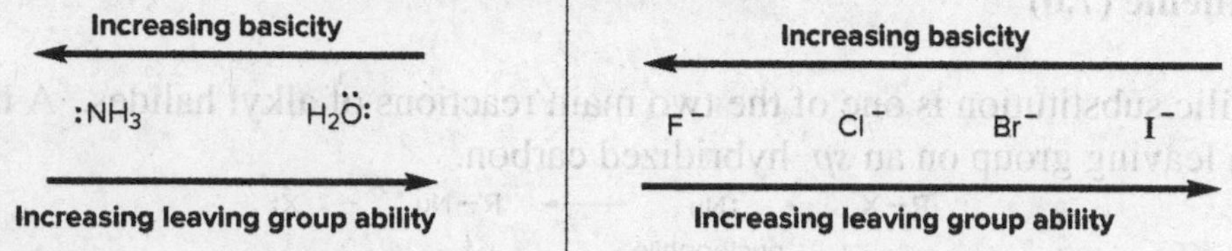

- Nucleophilicity decreases left-to-right across a row of the periodic table (7.8A).

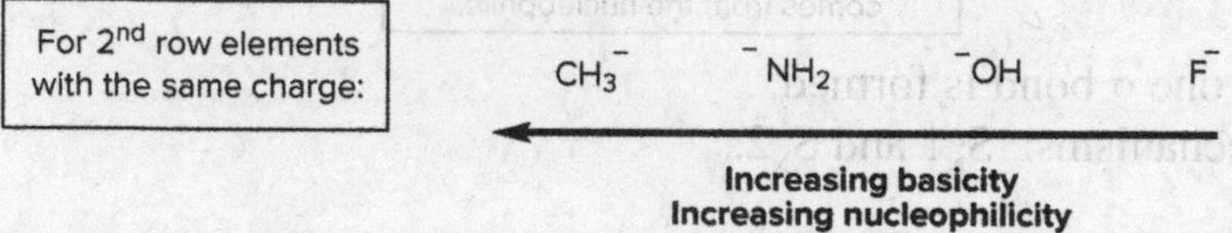

- Nucleophilicity decreases down a column of the periodic table in polar aprotic solvents (7.8C).

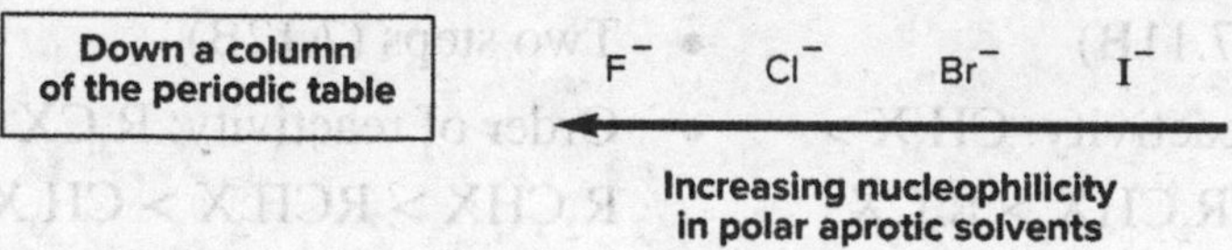

- Nucleophilicity increases down a column of the periodic table in polar protic solvents (7.8C).

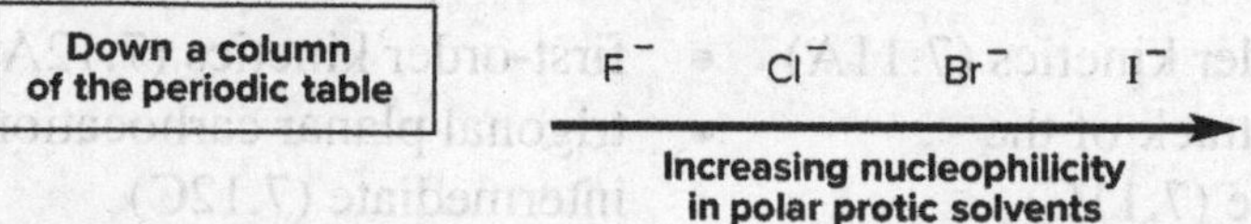

- The stability of a carbocation increases as the number of R groups bonded to the positively charged carbon increases (7.13).

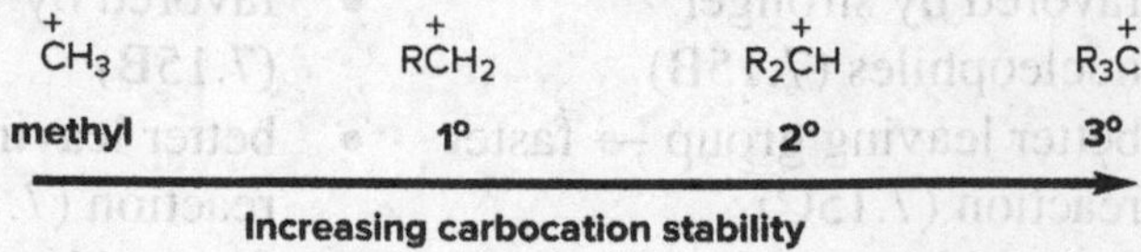

Important principles

Principle	Example
• Electron-donating groups (such as R groups) stabilize a positive charge (7.13A).	• 3° Carbocations (R_3C^+) are more stable than 2° carbocations (R_2CH^+), which are more stable than 1° carbocations (RCH_2^+).
• Steric hindrance decreases nucleophilicity but not basicity (7.8B).	• $(CH_3)_3CO^-$ is a stronger base but a weaker nucleophile than $CH_3CH_2O^-$.
• Hammond postulate: In an endothermic reaction, the more stable product is formed faster. In an exothermic reaction, this fact is not necessarily true (7.14).	• S_N1 reactions are faster when more stable (more substituted) carbocations are formed, because the rate-determining step is endothermic.
• Planar, sp^2 hybridized atoms react with reagents from both sides of the plane (7.12C).	• A trigonal planar carbocation reacts with nucleophiles from both sides of the plane.

Practice Test on Chapter Review

1. Give the IUPAC name for the following compound, including the appropriate *R*,*S* prefix.

Cl

2. a. Which of the following carbocations is the most stable?

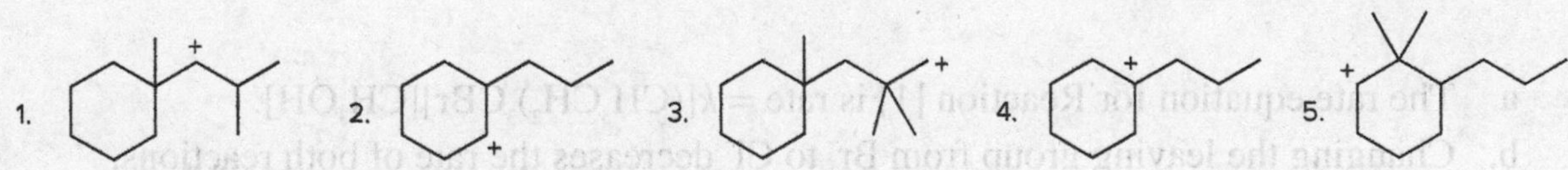

b. Which of the following anions is the best leaving group?

1. CH_3^- 2. ^-OH 3. H^- 4. $^-NH_2$ 5. Cl^-

c. Which species is the strongest nucleophile in polar protic solvents?

1. F^- 2. ^-OH 3. Cl^- 4. H_2O 5. ^-SH

d. Which of the following statements is true about the given reaction?

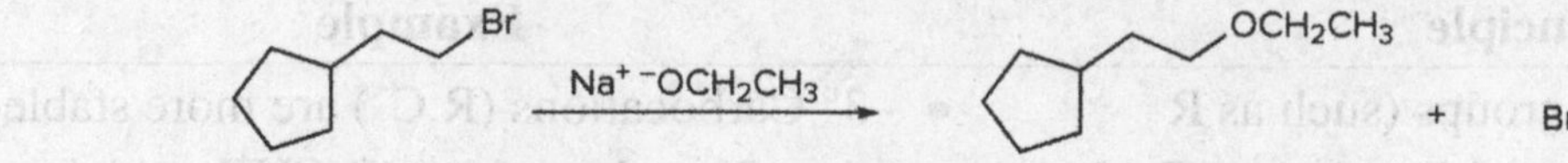

1. The reaction follows second-order kinetics.
2. The rate of the reaction increases when the solvent is changed from CH_3CH_2OH to DMSO.
3. The rate of the reaction increases when the leaving group is changed from Br to F.
4. Statements (1) and (2) are both true.
5. Statements (1), (2), and (3) are all true.

3. Rank the following compounds in order of ***increasing*** **reactivity in an S_N1 reaction.** Rank the ***least reactive*** compound as **1**, the ***most reactive*** compound as **4**, and the compounds of intermediate reactivity as **2** and **3**.

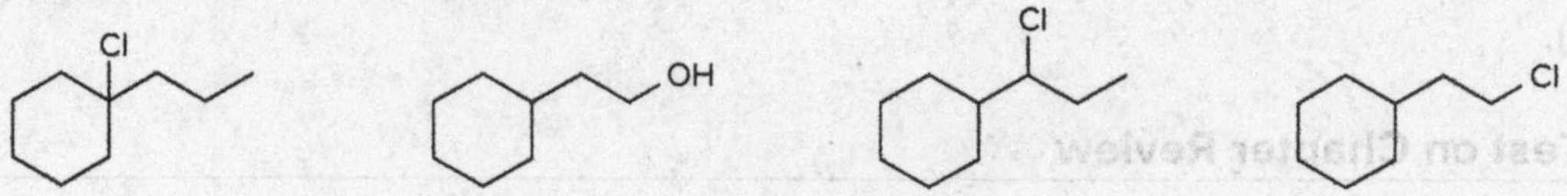

4. Consider the following two nucleophilic substitution reactions, labeled Reaction [1] and Reaction [2]. (Only the starting materials are drawn.) Then answer True (T) or False (F) to each of the following statements.

Reaction [1] $(CH_3CH_2)_3CBr$ + CH_3OH ⟶

Reaction [2] $CH_3CH_2CH_2Br$ + $^-OCH_3$ ⟶

a. The rate equation for Reaction [1] is rate = $k[(CH_3CH_2)_3CBr][CH_3OH]$.
b. Changing the leaving group from Br^- to Cl^- decreases the rate of both reactions.
c. Changing the solvent from CH_3OH to $(CH_3)_2S{=}O$ increases the rate of Reaction [2].
d. Doubling the concentration of both $CH_3CH_2CH_2Br$ and $^-OCH_3$ in Reaction [2] doubles the rate of the reaction.
e. If entropy is ignored and K_{eq} for Reaction [1] is <1, then the reaction is exothermic.
f. If entropy is ignored and $\Delta H°$ is negative for Reaction [1], then the bonds in the product are stronger than the bonds in the starting materials.
g. The energy diagram for Reaction [2] exhibits only one energy barrier.

5. Draw the organic products formed in the following reactions. **Use wedges and dashed wedges to show stereochemistry in compounds with stereogenic centers.**

a.

b. 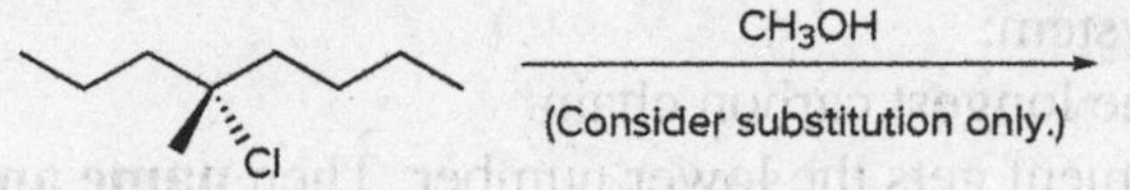

c. 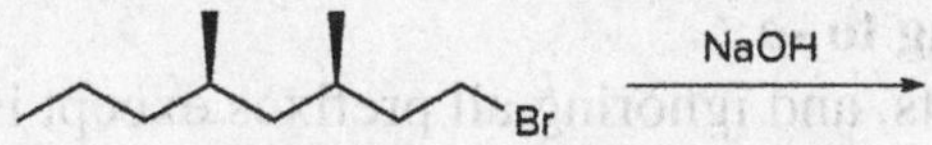

d. 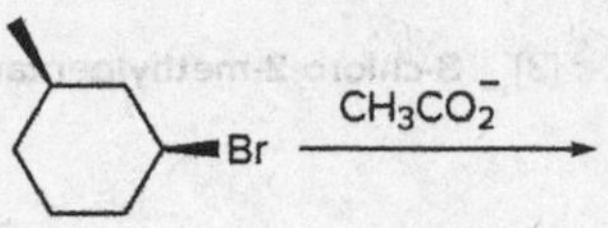

Answers to Practice Test

1. (*R*)-6-chloro-2-methylnonane

2. a. 4
   b. 5
   c. 5
   d. 4

3. A–4
   B–1
   C–3
   D–2

4. a. F
   b. T
   c. T
   d. F
   e. F
   f. T
   g. T

5. a. 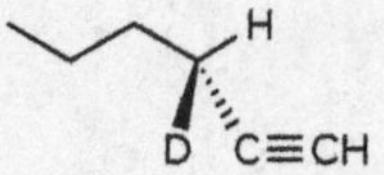

b. 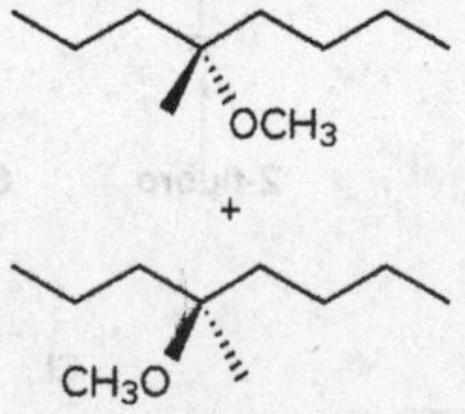

c. 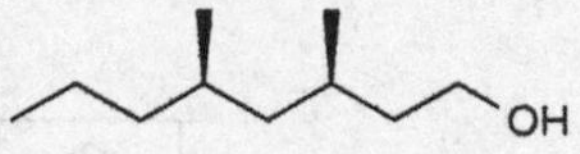

d. 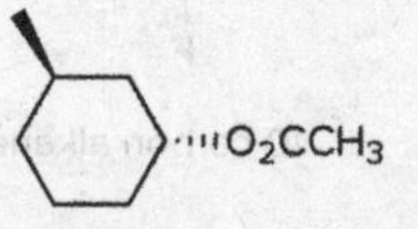

Answers to Problems

7.1 Classify the alkyl halide as 1°, 2°, or 3° **by counting the number of carbons bonded directly to the carbon bonded to the halogen.**

a, b. 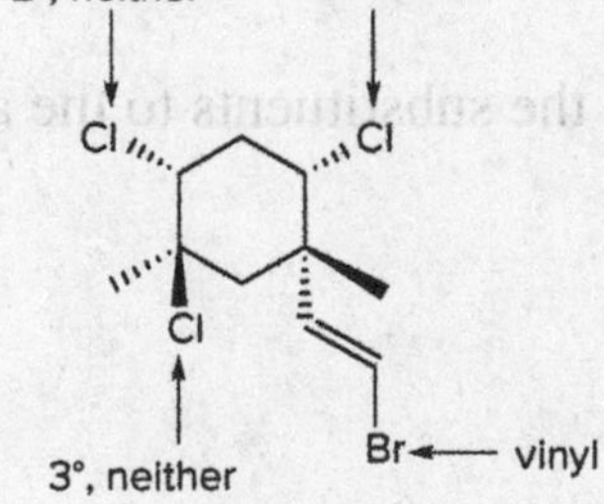

telfairine

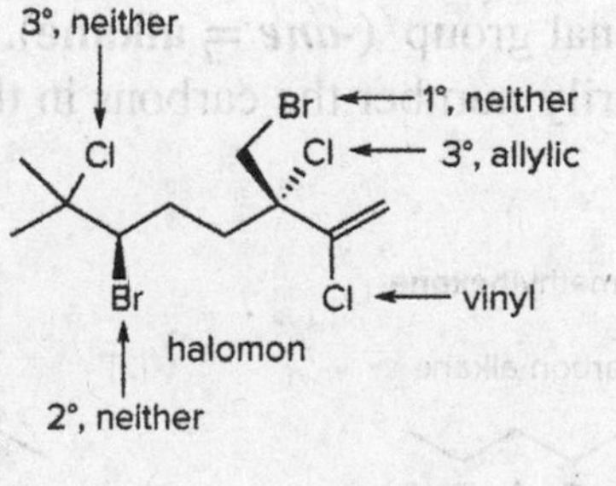

halomon

7.2 To name a compound with the IUPAC system:

[1] **Name the parent** chain by finding the longest carbon chain.

[2] **Number the chain** so the first substituent gets the lower number. Then **name and number all substituents**, giving like substituents a prefix (di, tri, etc.). **To name the halogen substituent, change the *-ine* ending to *-o*.**

[3] **Combine all parts**, alphabetizing substituents, and ignoring all prefixes except iso.

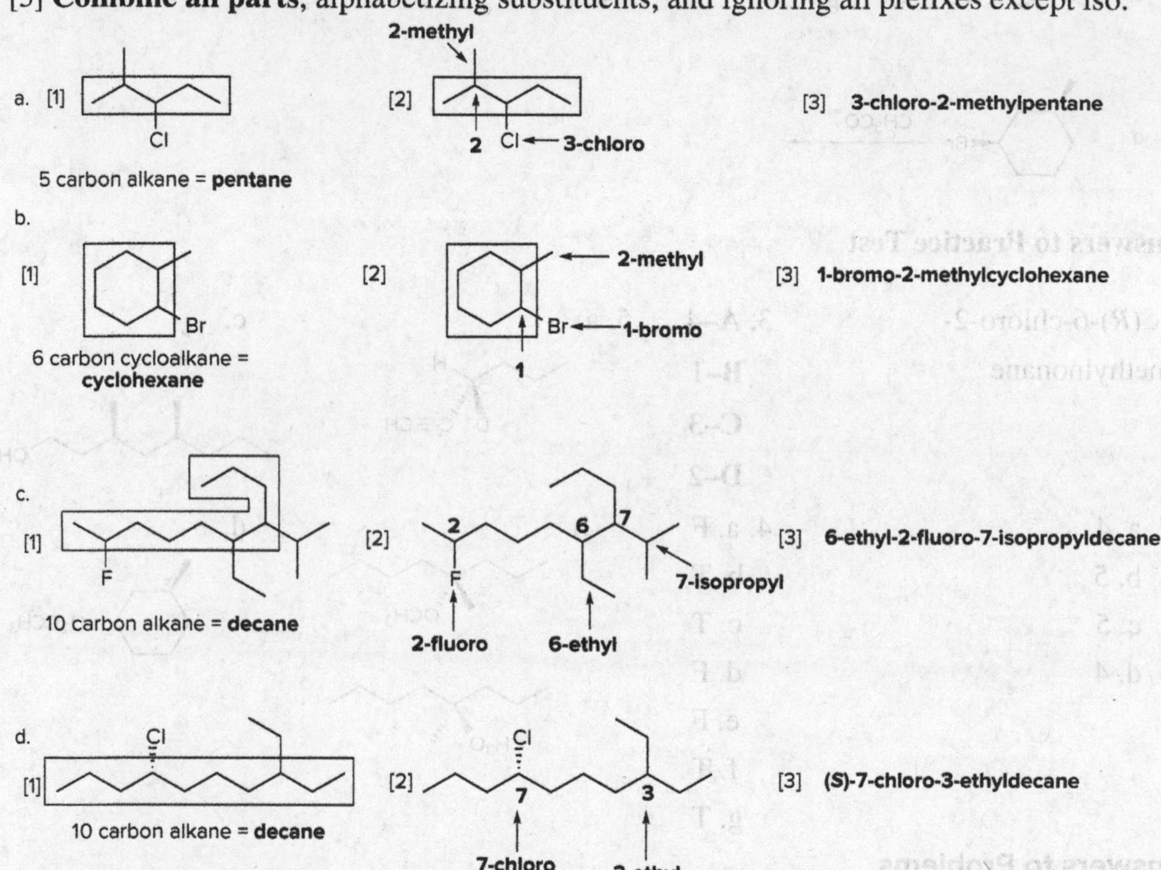

7.3 To work backwards from a name to a structure:

[1] Find the parent name and draw that number of carbons. Use the suffix to identify the functional group **(*-ane* = alkane).**

[2] Arbitrarily number the carbons in the chain. Add the substituents to the appropriate carbon.

a. 3-chloro-2-methyl**hexane**

[1] 6 carbon alkane

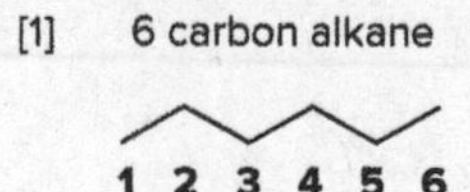

[2] methyl at C2

Cl chloro at C3

b. 4-ethyl-5-iodo-2,2-dimethyl**octane**

[1] 8 carbon alkane

1 2 3 4 5 6 7 8

[2] ethyl at C4

2 methyls at C2

I iodo at C5

c. *cis*-1,3-dichloro**cyclopentane**

[1] 5 carbon cycloalkane [2] chloro groups at C1 and C3, both on the same side

Cl Cl C1 C3

d. 1,1,3-tribromo**cyclohexane**

[1] 6 carbon cycloalkane [2] 3 Br groups

Br C3 Br Br C1

e. 6-ethyl-3-iodo-3,5-dimethylnonane

[1] 9 carbon chain [2] 3 5 6 I

f. (*R*)-1-fluoro-2,6,6-trimethylnonane

[1] 9 carbon chain [2]

1 2 6 F H *R* F

7.4 a. Because an *sp*^{2} hybridized C has a higher percent *s*-character than an *sp*^{3} hybridized C, it holds electron density closer to C. This pulls a little more electron density toward C, away from Cl, and thus a C*sp*^{2}–Cl bond is less polar than a C*sp*^{3}–Cl bond.

b.

Cl — **lowest boiling point**

Cl — *sp*^{3} C–Cl bond **intermediate boiling point**

Br — larger halogen, *sp*^{3} C–Br bond **highest boiling point**

7.5 a. Chondrocole A has 10 C's and only one functional group capable of hydrogen bonding to water (an ether), so it is insoluble in H_2O. Because it is organic, it is soluble in CH_2Cl_2.

b.

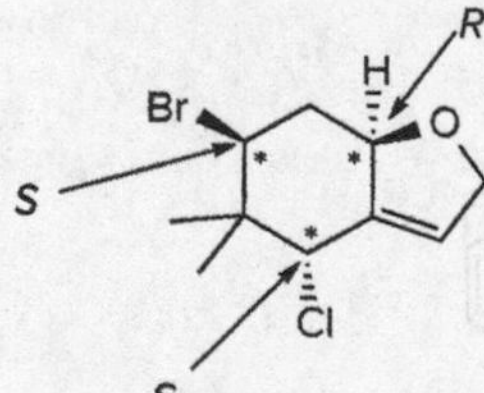

Three stereogenic centers are labeled with (*).

c.

Br O Cl — enantiomer

Br O Cl — constitutional isomer

7.6 To draw the products of a nucleophilic substitution reaction:
[1] **Find the *sp*^{3} hybridized electrophilic carbon** with a leaving group.
[2] **Find the nucleophile** with lone pairs or electrons in π bonds.
[3] **Substitute the nucleophile for the leaving group** on the electrophilic carbon.

a. Br (leaving group) + $^{-}OCH_2CH_3$ (nonbonded e$^-$ pairs, nucleophile) → OCH_2CH_3 + Br^-

b. Cl (leaving group) + $Na^+ {}^-OH$ (nonbonded e$^-$ pairs, nucleophile) → OH + Na^+Cl^-

c. I (leaving group) + N_3^- (nonbonded e$^-$ pairs, nucleophile) → N_3 + I^-

d. Br (leaving group) + $Na^+ {}^-CN$ (nonbonded e$^-$ pairs, nucleophile) → CN + Na^+Br^-

7.7 Use the steps from Answer 7.6 and then draw the proton transfer reaction.

a. Br (leaving group) + $:N(CH_2CH_3)_3$ (nucleophile) —substitution→ $\overset{+}{N}(CH_2CH_3)_3$ + Br^-

b. Cl (leaving group) + $H_2\ddot{O}:$ (nucleophile) —substitution→ $\ddot{O}^+$(H)(H) + $:\ddot{Cl}:^-$ —proton transfer→ $\ddot{O}H$ + HCl

7.8

A → ticlopidine

These atoms come from the nucleophile.

NH

S

7.9 Good leaving groups include Cl^-, Br^-, I^-, and H_2O.

a. Br
Br^- is a **good leaving group.**

b. OH
no good leaving group
^-OH is too strong a base.

c. $\overset{+}{O}H_2$
H_2O is a **good leaving group.**

d.
no good leaving group
H^- is too strong a base.

7.10 Good leaving groups include Cl^-, Br^-, I^-, and H_2O. F^-, ^-OH, $^-NH_2$, H^-, and R^- are *poor* leaving groups.

worst leaving group → H_2N
best leaving group
A

worst leaving group → HO
best leaving group
B

7.11 To decide whether the equilibrium favors the starting material or the products, **compare the nucleophile and the leaving group**. The reaction proceeds toward the weaker base.

a. NH_2 + Br^- ⟶ Br + $^-NH_2$
nucleophile
better leaving group
weaker base
pK_a (HBr) = –9
leaving group
pK_a (NH_3) = 38
Reaction favors starting material.

b. I + ^-CN ⟶ CN + I^-
nucleophile
pK_a (HCN) = 9.1
leaving group
better leaving group
weaker base
pK_a (HI) = –10
Reaction favors product.

7.12 Use these three rules to find the stronger nucleophile in each pair:

[1] Comparing two nucleophiles having the ***same attacking atom*****, the stronger base is a stronger nucleophile.**

[2] **Negatively charged nucleophiles** are always **stronger than their conjugate acids.**

[3] **Nucleophilicity decreases left to right across a row of the periodic table,** when comparing species of similar charge.

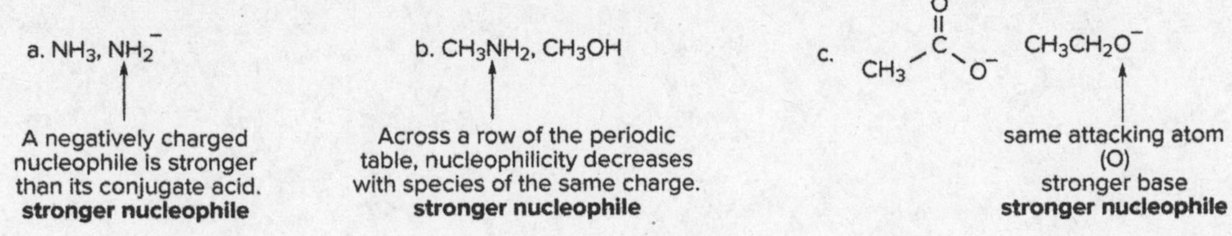
a. NH3, NH2−
A negatively charged nucleophile is stronger than its conjugate acid.
stronger nucleophile
b. CH3NH2, CH3OH
Across a row of the periodic table, nucleophilicity decreases with species of the same charge.
stronger nucleophile
c.
CH3CH2O−
same attacking atom (O)
stronger base
stronger nucleophile

7.13 ***Polar protic solvents*** **are capable of hydrogen bonding,** and therefore must contain a **H bonded to an electronegative O or N.** ***Polar aprotic solvents*** **are incapable of hydrogen bonding,** and therefore do not contain any O–H or N–H bonds.

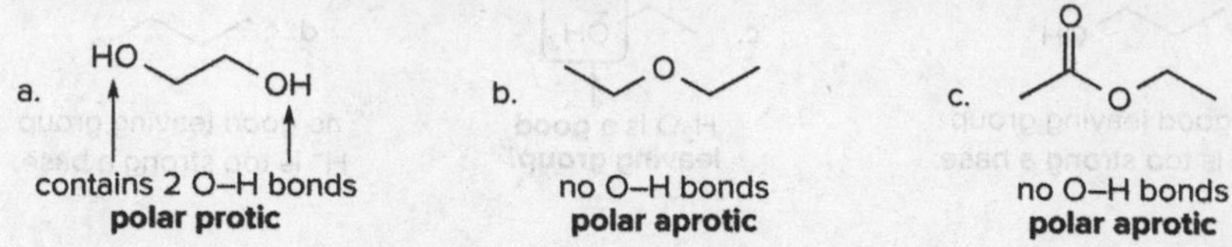

7.14 • In ***polar protic solvents,*** **the trend in nucleophilicity is opposite to the trend in basicity** down a column of the periodic table, so nucleophilicity increases.

• In ***polar aprotic solvents,*** **the trend is identical to basicity**, so nucleophilicity decreases down a column.

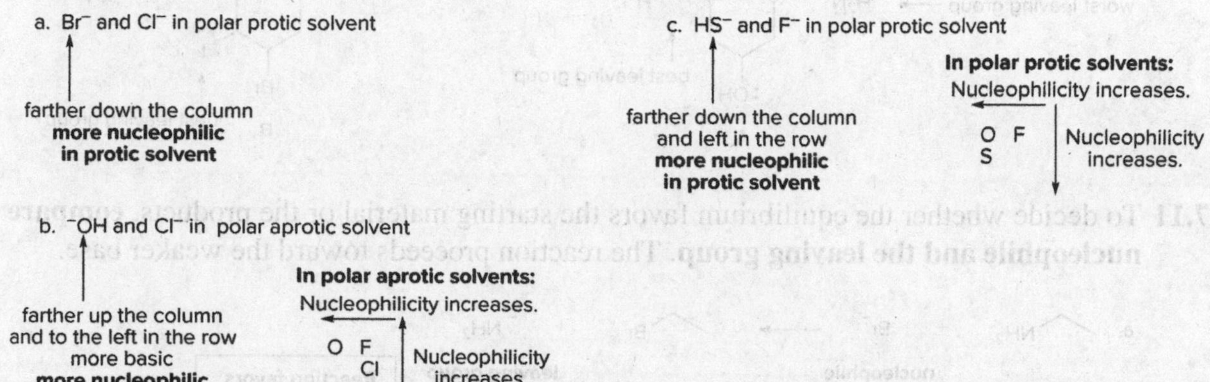

7.15 The stronger base is the stronger nucleophile, except in polar protic solvents, where nucleophilicity increases down a column. For other rules, see Answers 7.12 and 7.14.

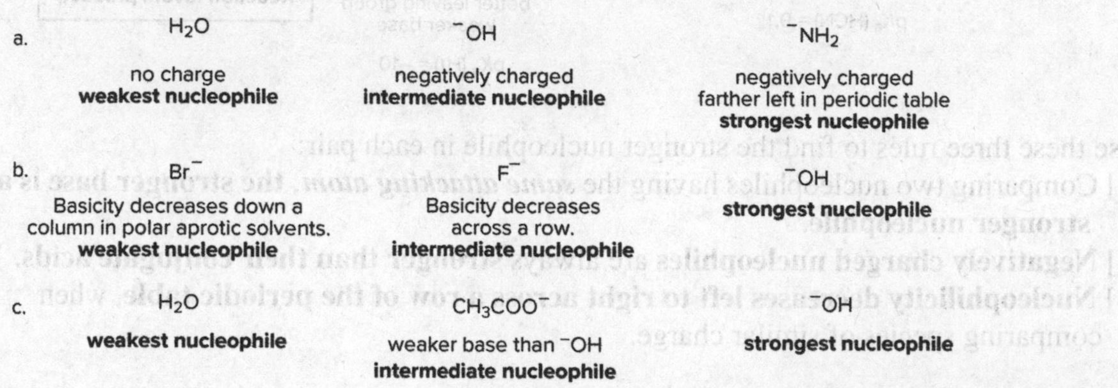

7.16 To determine what nucleophile is needed to carry out each reaction, look at the product to see what has replaced the leaving group.

a. SH replaces Br. **HS^- is needed.**

b. OCH_2CH_3 replaces Br. **$CH_3CH_2O^-$ is needed.**

c. $OCOCH_3$ replaces Br. **CH_3COO^- is needed.**

d. C≡CH replaces Br. **HC≡C⁻ is needed.**

7.17 The general rate equation for an S_N2 reaction is rate = k[RX][:Nu⁻].

a. [RX] is tripled, and [:Nu⁻] stays the same: **rate triples.**
b. Both [RX] and [:Nu⁻] are tripled: **rate increases by a factor of 9 (3 × 3 = 9).**
c. [RX] is halved, and [:Nu⁻] stays the same: **rate halved.**
d. [RX] is halved, and [:Nu⁻] is doubled: **rate stays the same (1/2 × 2 = 1).**

7.18 All S_N2 reactions have one step. The transition state in an S_N2 reaction has **dashed bonds to both the leaving group and the nucleophile,** and must contain partial charges.

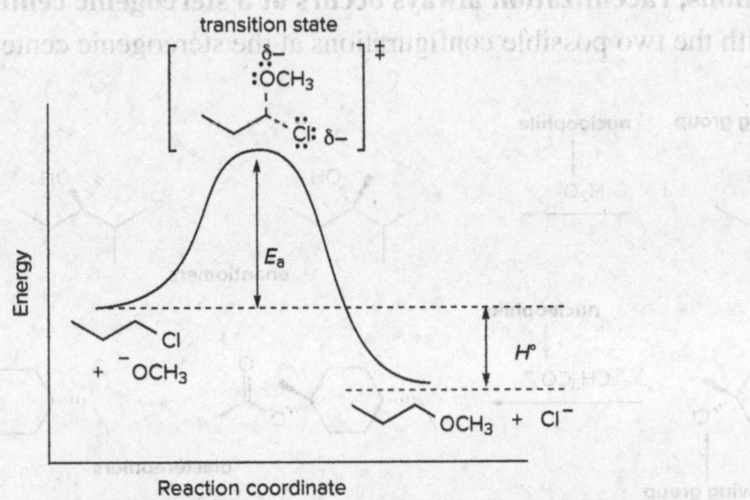

7.19 To draw the products of S_N2 reactions, **replace the leaving group by the nucleophile, and then draw the stereochemistry with *inversion* at the stereogenic center.**

a. (D)(H)C–Br + :Ö⁻ (ethoxide) ⟶ :O–C(D)(H) (inverted)

b. cyclopentyl–I + ⁻:C≡N: ⟶ cyclopentyl····C≡N:

7.20 *Increasing* the number of R groups *increases* crowding of the transition state and *decreases* the rate of an S_N2 reaction.

a. Cl — 2° alkyl halide or Cl — 1° alkyl halide **faster reaction**

b. Br — 2° alkyl halide **faster reaction** or Br — 3° alkyl halide

7.21

N–H + Na^+ H:$^-$ → DMF; :N:$^-$ + Br (leaving group) — NO_2 → NO_2 **X**

7.22 In a first-order reaction, **the rate changes with any change in [RX]**. The rate is independent of any change in [:Nu$^-$].

a. [RX] is tripled, and [:Nu$^-$] stays the same: **rate triples.**
b. Both [RX] and [:Nu$^-$] are tripled: **rate triples.**
c. [RX] is halved, and [:Nu$^-$] stays the same: **rate is halved.**
d. [RX] is halved, and [:Nu$^-$] is doubled: **rate is halved.**

7.23 In S_N1 reactions, racemization always occurs at a stereogenic center. Draw two products, with the two possible configurations at the stereogenic center.

a. **leaving group** Br; **nucleophile** H_2O → OH + OH + HBr — **enantiomers**

b. **leaving group** Cl; **nucleophile** $CH_3CO_2^-$ → O + O + Cl^- — **diastereomers**

7.24 Carbocations are classified by the number of R groups bonded to the carbon: 0 R groups = methyl, 1 R group = 1°, 2 R groups = 2°, and 3 R groups = 3°.

a. 2 R groups **2° carbocation**

b. 1 R group **1° carbocation**

c. 3 R groups **3° carbocation**

d. 2 R groups **2° carbocation**

7.25 For carbocations: Increasing number of R groups = Increasing stability.

1° carbocation **least stable** | 2° carbocation **intermediate stability** | 3° carbocation **most stable**

7.26 The rate of an S_N1 reaction increases with increasing alkyl substitution.

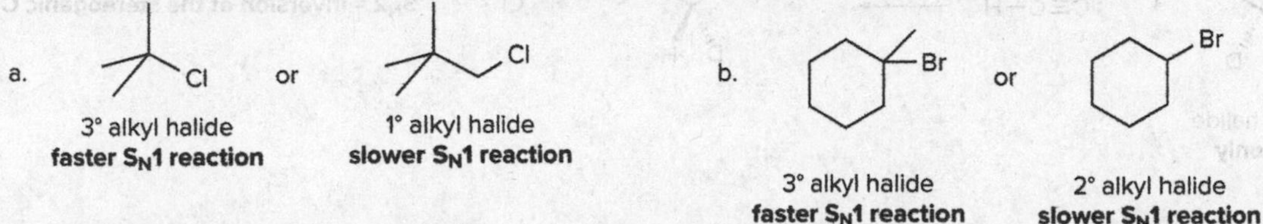

7.27 • For **methyl and 1° alkyl halides**, only S_N2 will occur.
• For **2° alkyl halides**, S_N1 and S_N2 will occur.
• For **3° alkyl halides**, only S_N1 will occur.

a. Br — 2° alkyl halide **S_N1 and S_N2**

b. Br — 1° alkyl halide **S_N2**

c. Br — 2° alkyl halide **S_N1 and S_N2**

d. Br — 3° alkyl halide **S_N1**

7.28 • Draw the product of nucleophilic substitution for each reaction.
• For **methyl and 1° alkyl halides**, only S_N2 will occur.
• For **2° alkyl halides**, S_N1 and S_N2 will occur and other factors determine which mechanism operates.
• For **3° alkyl halides**, only S_N1 will occur.

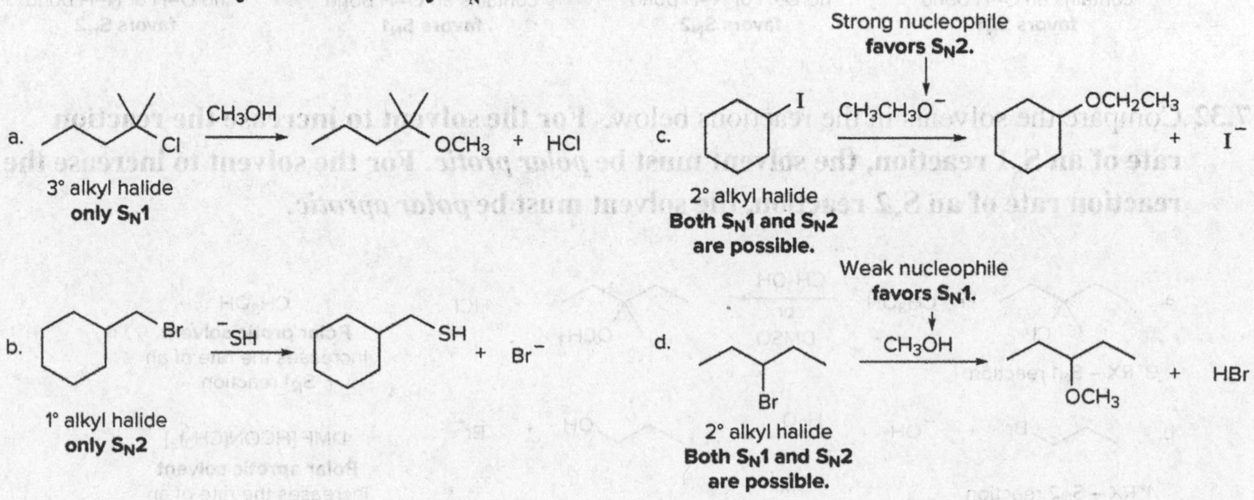

7.29 First decide whether the reaction will proceed via an S_N1 or S_N2 mechanism. Then draw the products with stereochemistry.

a. 2° alkyl halide + H_2O → (H, OH) + (HO, H) + HBr

2° alkyl halide
S_N1 and S_N2

Weak nucleophile
favors S_N1.

enantiomers

S_N1 = racemization at the stereogenic C

b. (Cl; H, D) + $^{-}$:C≡C–H → HC≡C (D, H) + Cl^-

1° alkyl halide
S_N2 only

S_N2 = inversion at the stereogenic C

7.30

HO, Br, Cl, Br

3° bromide
(c)

Most substituted RX reacts fastest.
order: (b) < (a) < (c)

2° bromide
(a)

2° chloride
worst leaving group
(b)

7.31 • **Polar protic solvents** favor the S_N1 mechanism by solvating the intermediate carbocation and halide.

• **Polar aprotic solvents** favor the S_N2 mechanism by making the nucleophile stronger.

a. CH_3CH_2OH	b. CH_3CN	c. CH_3COOH	d. $CH_3CH_2OCH_2CH_3$
polar protic solvent	polar aprotic solvent	polar protic solvent	polar aprotic solvent
contains an O–H bond	no O–H or N–H bond	contains an O–H bond	no O–H or N–H bond
favors S_N1	**favors S_N2**	**favors S_N1**	**favors S_N2**

7.32 Compare the solvents in the reactions below. **For the solvent to increase the reaction rate of an S_N1 reaction, the solvent must be *polar protic*. For the solvent to increase the reaction rate of an S_N2 reaction, the solvent must be *polar aprotic*.**

a. (Cl) + CH_3OH —(CH_3OH or DMSO)→ (OCH_3) + HCl

3° RX – S_N1 reaction

CH_3OH
Polar protic solvent increases the rate of an S_N1 reaction.

b. (Br) + ^{-}OH —(H_2O or DMF)→ (OH) + Br^-

1° RX – S_N2 reaction

DMF [$HCON(CH_3)_2$]
Polar aprotic solvent increases the rate of an S_N2 reaction.

c. (H, Cl) + CH_3O^- → (CH_3OH or HMPA) → (H, OCH_3) + Cl^-

2° RX — strong nucleophile — S_N2 reaction

HMPA $[(CH_3)_2N]_3P{=}O$
Polar aprotic solvent increases the rate of an S_N2 reaction.

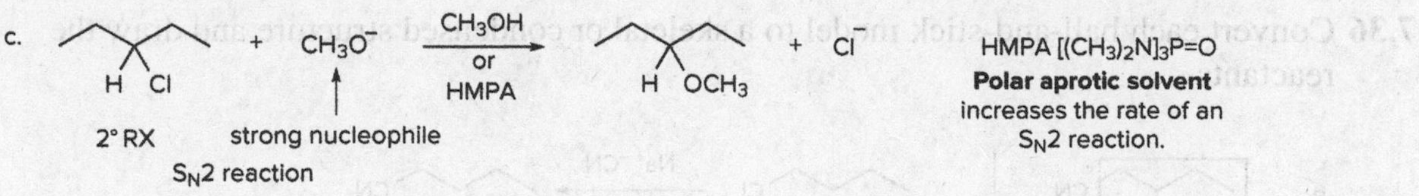

7.33 To predict whether the reaction follows an S_N1 or S_N2 mechanism:

[1] **Classify RX as a methyl, 1°, 2°, or 3° halide.** (Methyl, 1° = S_N2; 3° = S_N1; 2° = either.)
[2] **Classify the nucleophile as strong or weak.** (Strong favors S_N2; weak favors S_N1.)
[3] **Classify the solvent as polar protic or polar aprotic.** (Polar protic favors S_N1; polar aprotic favors S_N2.)

a. (Br) + O^- → (O) + Br^- — S_N2 reaction

1° alkyl halide
S_N2

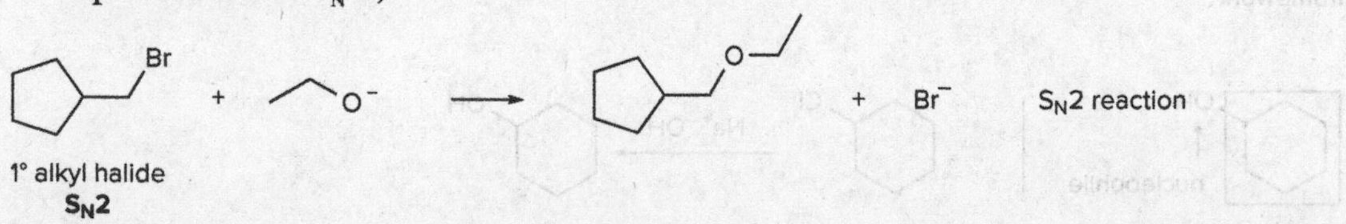

b. (Br) + N_3^- → (N_3) + Br^- — S_N2 reaction = ***inversion*** at the stereogenic center. The leaving group was "up." The nucleophile attacks from below.

2° alkyl halide
S_N1 or S_N2

Strong nucleophile
favors S_N2.

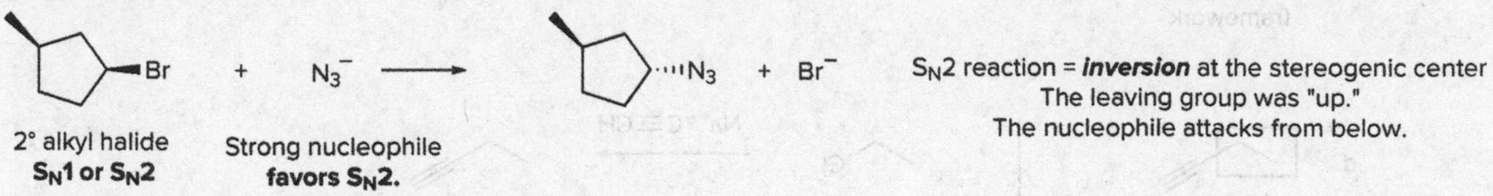

c. (I) + CH_3OH → (OCH_3) + HI — S_N1 reaction

3° alkyl halide
S_N1

Weak nucleophile
favors S_N1.

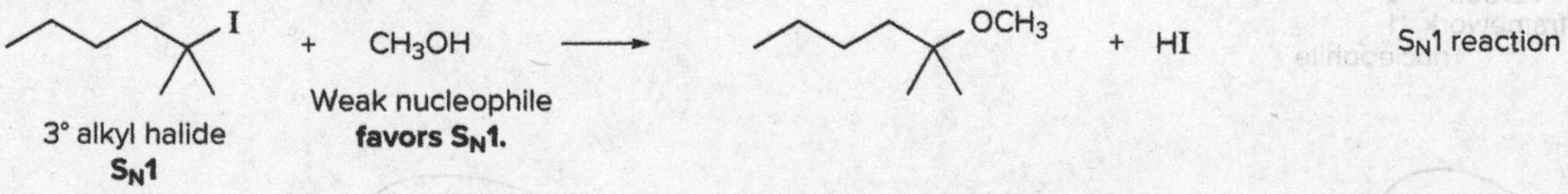

d. (Cl) + H_2O → (OH) + (HO) + HCl — S_N1 reaction forms **two enantiomers.**

3° alkyl halide
S_N1

Weak nucleophile
favors S_N1.

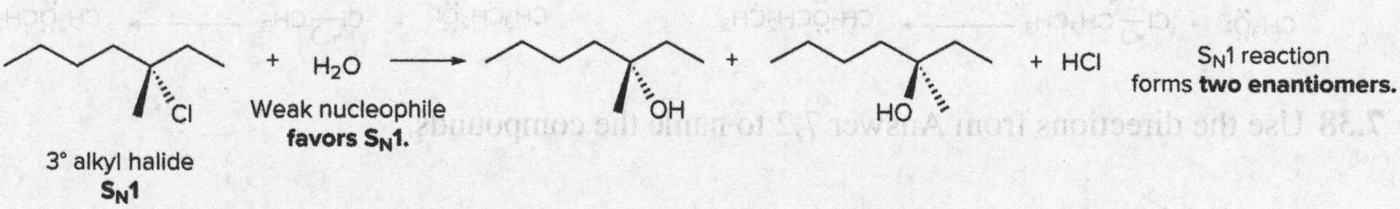

7.34

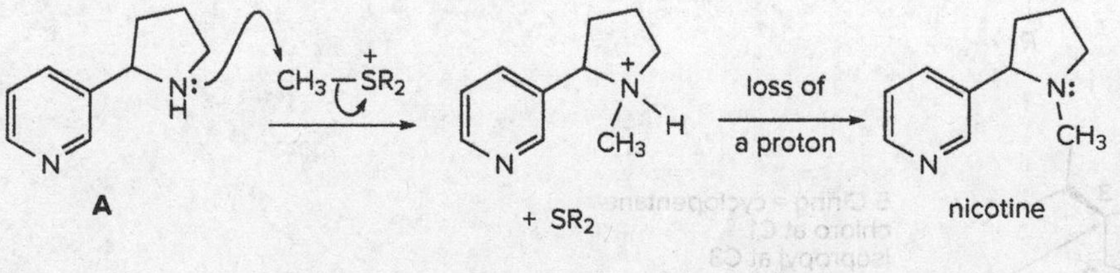

A (N:, H) + CH_3–$\overset{+}{S}R_2$ → (N⁺, H, CH_3) + SR_2 → loss of a proton → (N:, CH_3) nicotine

7.35

(Cl, Br, Cl)

vinyl chloride
nonreactive

2° alkyl bromide
less crowded
better leaving group
most reactive

3° alkyl chloride
intermediate reactivity

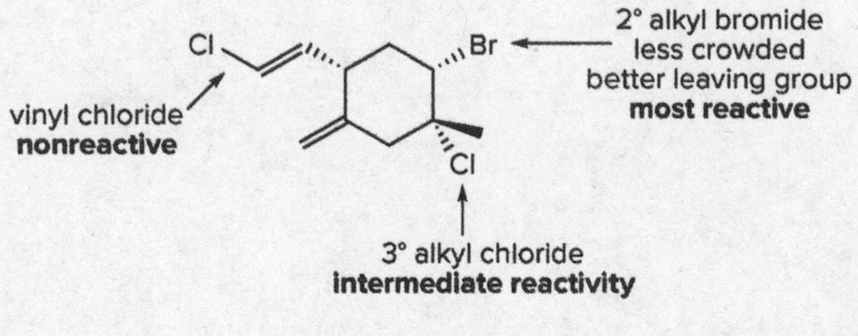

7.36 Convert each ball-and-stick model to a skeletal or condensed structure and draw the reactants.

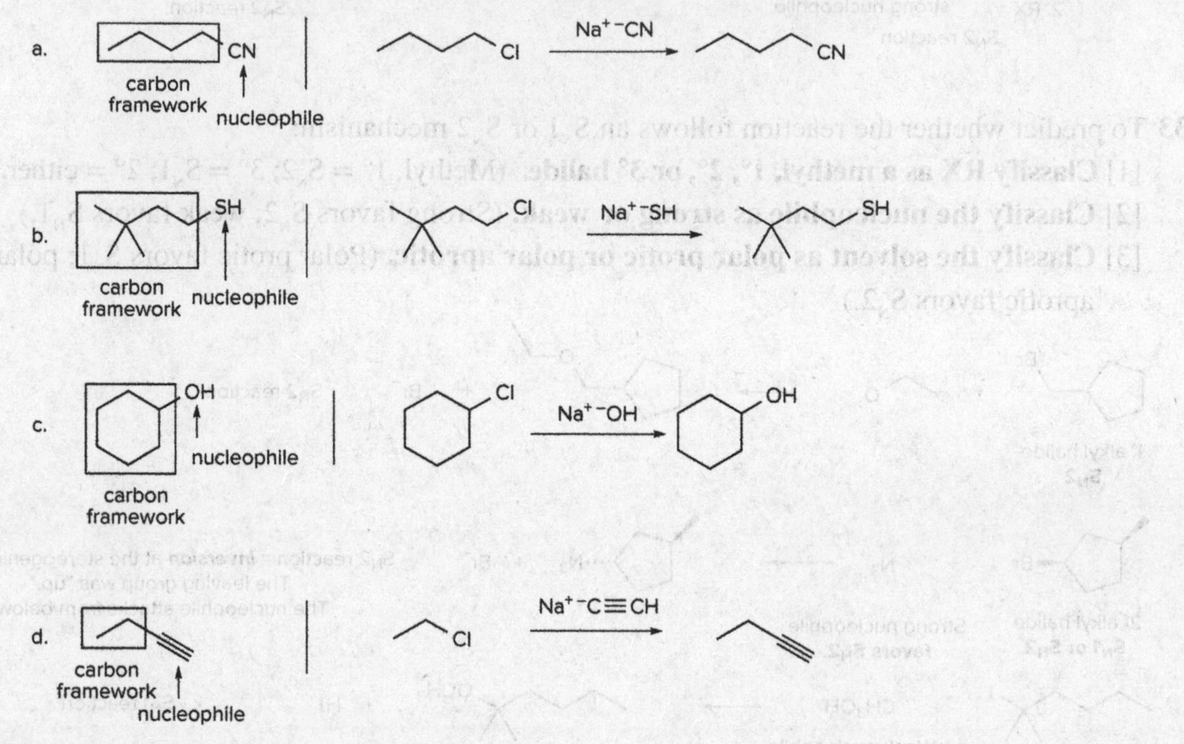

7.37

$CH_3\ddot{O}:^- + Cl-CH_2CH_3 \longrightarrow CH_3\ddot{O}CH_2CH_3$ $\qquad CH_3CH_2\ddot{O}:^- + Cl-CH_3 \longrightarrow CH_3\ddot{O}CH_2CH_3$

7.38 Use the directions from Answer 7.2 to name the compounds.

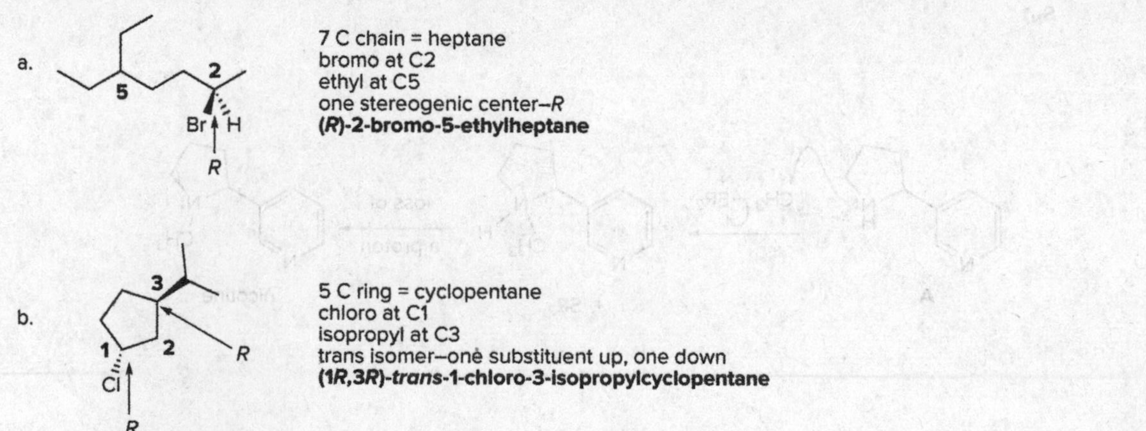

7.39

a. H, Br (eq) $\xrightarrow[\text{acetone}]{^{-}CN}$ CN (axial), H — **inversion** (equatorial to axial)

polar aprotic solvent
S_N2 reaction

Large *tert*-butyl group is in more roomy equatorial position.

b. Br (axial), H $\xrightarrow[\text{acetone}]{^{-}CN}$ H, CN (eq) — **inversion** (axial to equatorial)

polar aprotic solvent
S_N2 reaction

7.40 Use the directions from Answer 7.2 to name the compounds.

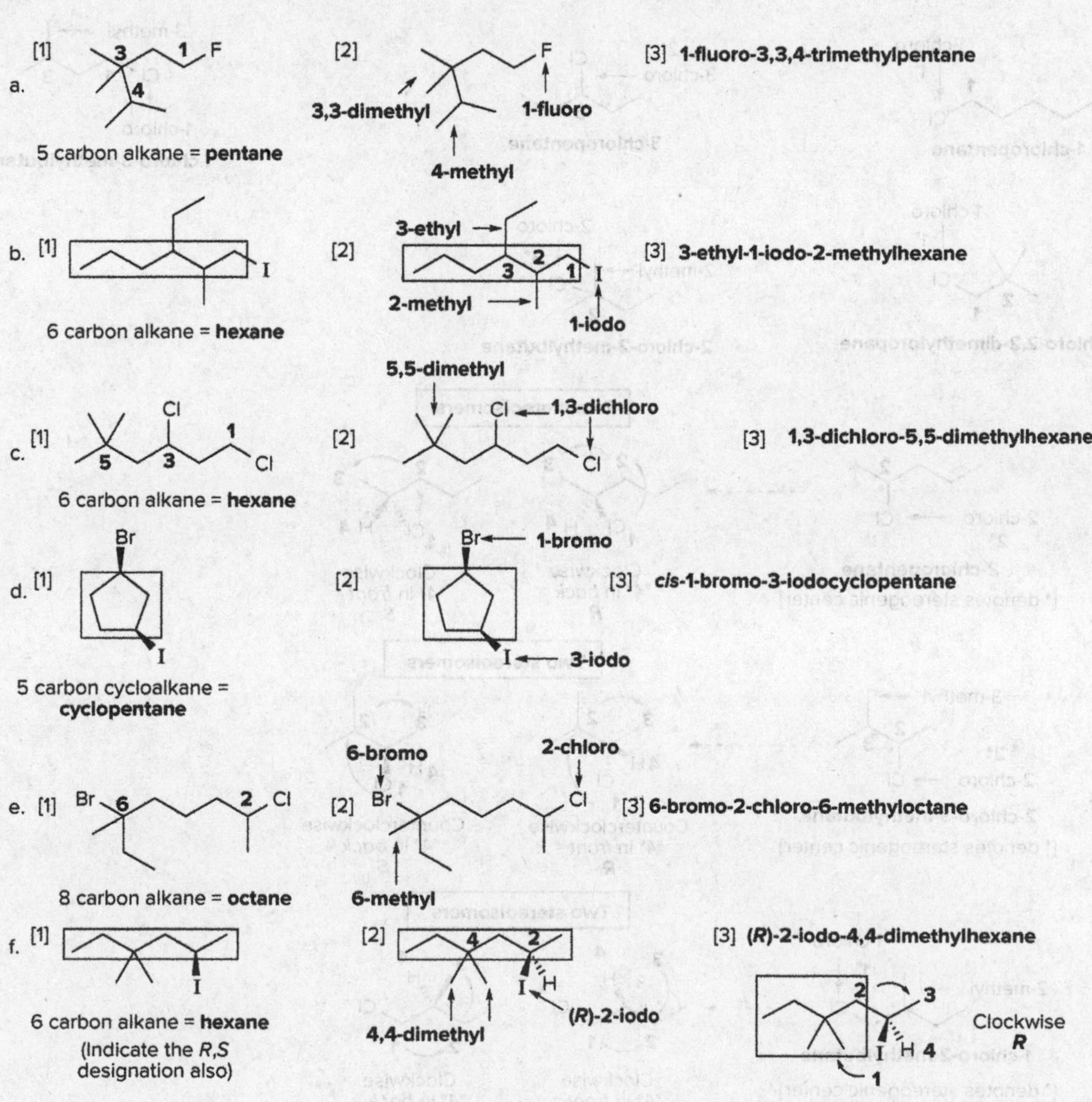

7.41 To work backwards to a structure, use the directions in Answer 7.3.

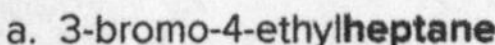

a. 3-bromo-4-ethyl**heptane**

c. 1-bromo-4-ethyl-3-fluoro**octane**

e. (1*R*,2*R*)-*trans*-1-bromo-2-chloro**cyclohexane**

b. 1,1-dichloro-2-methyl**cyclohexane**

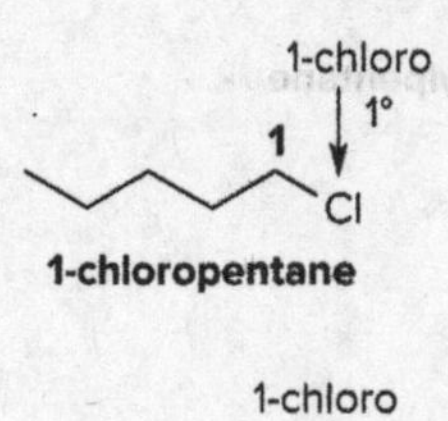

d. (*S*)-3-iodo-2-methyl**nonane**

f. (*R*)-4,4,5-trichloro-3,3-dimethyl**decane**

7.42

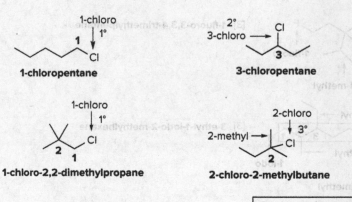

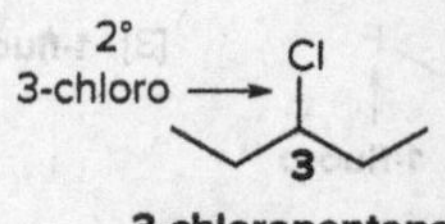

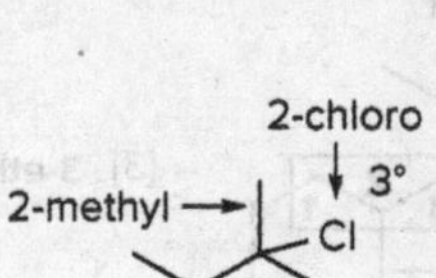

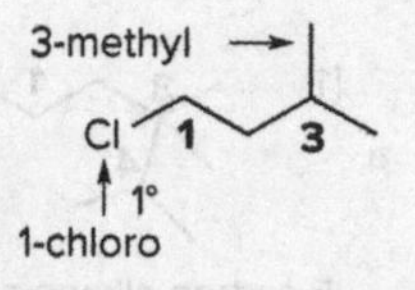

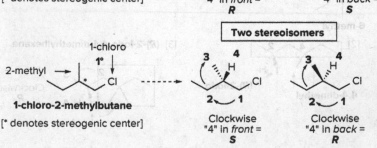

1-chloropentane

1-chloro-2,2-dimethylpropane

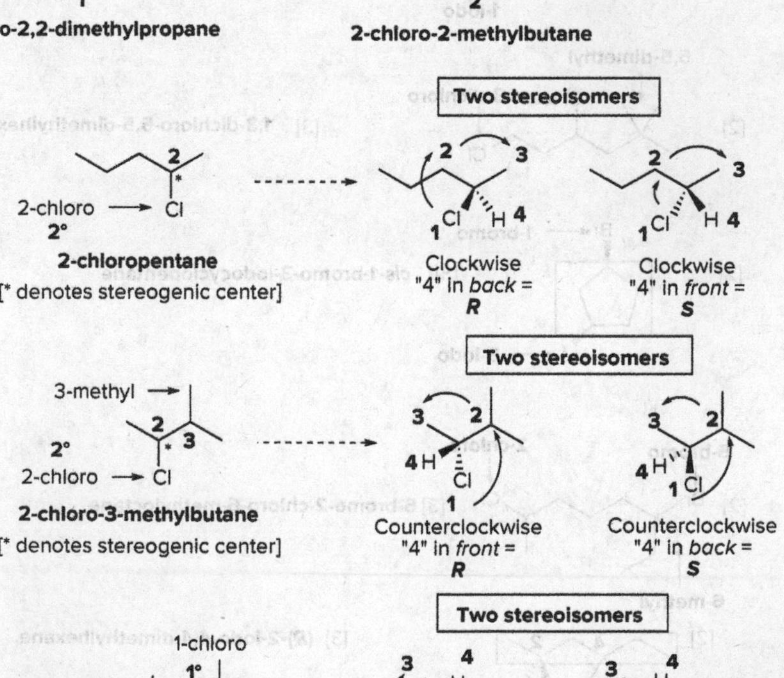

7.43 Use the steps from Answer 7.6 and then draw the proton transfer reaction, when necessary.

a. + O^- → + Cl^-

Cl: leaving group; nucleophile

b. I + $Na^{+}\,^{-}CN$ → CN + NaI

leaving group; nucleophile

c. I + $H_2\ddot{O}$: → OH + HI

leaving group; nucleophile

d. Cl + OH → O + HCl

leaving group; nucleophile

e. Br + $Na^{+}\,^{-}OCH_3$ → OCH_3 + NaBr

leaving group; nucleophile

f. Cl + S → S + Cl^-

leaving group; nucleophile

7.44 A good leaving group is a weak base.

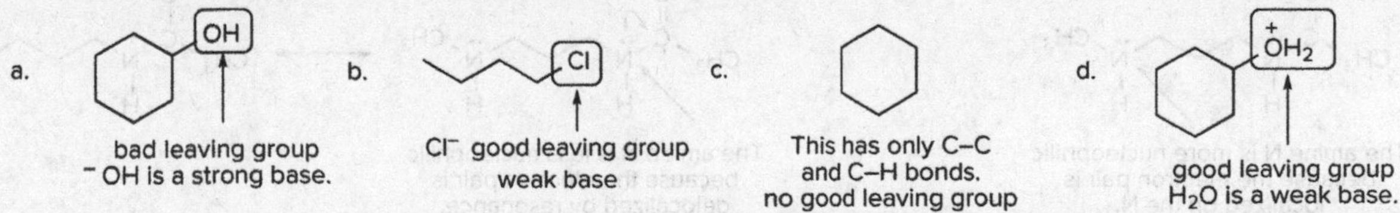

7.45 Leaving group ability increases left-to-right across a row and down a column of the periodic table.

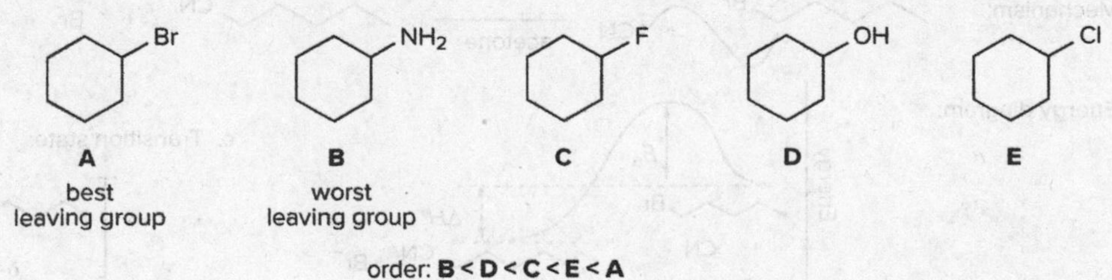

order: **B < D < C < E < A**

7.46 Compare the nucleophile and the leaving group in each reaction. The reaction will occur if it proceeds toward the weaker base. Remember that the stronger the acid (lower pK_a), the weaker the conjugate base.

a. Cyclohexyl-NH_2 + I^- ✗→ cyclohexyl-I + $^-NH_2$ — **Reaction will not occur.**

I^-: weaker base, pK_a (HI) = –10; $^-NH_2$: stronger base, pK_a (NH_3) = 38

b. CH_3CH_2I + $^-O{-}CH_3$ → $CH_3CH_2OCH_3$ + I^- — **Reaction will occur.**

$^-OCH_3$: stronger base, pK_a (CH_3OH) = 15.5; I^-: weaker base, pK_a (HI) = –10

7.47 In acetone, nucleophilicity decreases left-to-right across a row and down a column of the periodic table.

$$I^- < Br^- < CH_3S^- < CH_3O^- < CH_3NH^-$$

7.48 ***Polar protic solvents* are capable of hydrogen bonding**, so they must contain a H bonded to an electronegative O or N. ***Polar aprotic solvents* are incapable of hydrogen bonding**, so they do not contain any O–H or N–H bonds.

a. $(CH_3)_2CHOH$ — contains O–H bond — **protic**

b. CH_3NO_2 — no O–H or N–H bond — **aprotic**

c. CH_2Cl_2 — no O–H or N–H bond — **aprotic**

d. NH_3 — contains N–H bond — **protic**

e. $N(CH_3)_3$ — no O–H or N–H bond — **aprotic**

f. $HCONH_2$ — contains an N–H bond — **protic**

7.49

The amine N is more nucleophilic because the electron pair is localized on the N.

The amide N is less nucleophilic because the electron pair is delocalized by resonance.

7.50

1° alkyl halide
S_N2 reaction

a. Mechanism: $CH_3CH_2CH_2CH_2CH_2Br$ + $:CN^-$ →(acetone) $CH_3CH_2CH_2CH_2CH_2CN$ + Br^-

b. Energy diagram:

Energy; Reaction coordinate; E_a; $\Delta H°$; Br + ^-CN; CN + Br^-

c. Transition state:

$[\,Br\,\delta-\ \cdots\ \delta-\,CN:\,]^‡$

d. Rate equation: one-step reaction with both nucleophile and alkyl halide in the only step:
rate = k[R–Br][$^{-}$CN]

e. [1] The leaving group is changed from Br^- to I^-:
Leaving group becomes less basic → a better leaving group → faster reaction.
[2] The solvent is changed from acetone to CH_3CH_2OH:
Solvent changed to polar protic → decreases reaction rate.
[3] The alkyl halide is changed from $CH_3(CH_2)_4Br$ to $CH_3CH_2CH_2CH(Br)CH_3$:
Changed from 1° to 2° alkyl halide → the alkyl halide gets more crowded, so the reaction rate decreases.
[4] The concentration of $^{-}$CN is increased by a factor of 5.
Reaction rate will increase by a factor of 5.
[5] The concentration of both the alkyl halide and $^{-}$CN are increased by a factor of 5:
Reaction rate will increase by a factor of 25 (5 × 5 = 25).

7.51 All S_N2 reactions proceed with backside attack of the nucleophile. When nucleophilic attack occurs at a stereogenic center, inversion of configuration occurs.

a. + $^{-}OCH_3$ → + Cl^- inversion of configuration

b. + $^{-}OCH_2CH_3$ → OCH_2CH_3 + Cl^-

[No bond to the stereogenic center is broken, because the leaving group is not bonded to the stereogenic center.]

c. + $^{-}$CN → + Br^- inversion of configuration

[* denotes a stereogenic center]

7.52

leaving group

aprepitant

7.53

\+ **M** → NaOH → [These atoms come from the nucleophile.] → [O] → **omeprazole**

M =

During the reaction, NaOH removes the SH proton, to form a stronger nucleophile.

7.54

leaving group

leaving group

→ **N** after loss of proton → KOC(CH_3)$_3$, (CH_3)$_3$COH → **O** after loss of proton → HCl → cetirizine

7.55 For carbocations: **Increasing number of R groups = Increasing stability.**

a.

1° carbocation **least stable** | 2° carbocation **intermediate stablity** | 3° carbocation **most stable**

b.

1° carbocation **least stable** | 2° carbocation **intermediate stablity** | 3° carbocation **most stable**

7.56 Both **A** and **B** are resonance stabilized, but the N atom in **B** is more basic and therefore more willing to donate its electron pair.

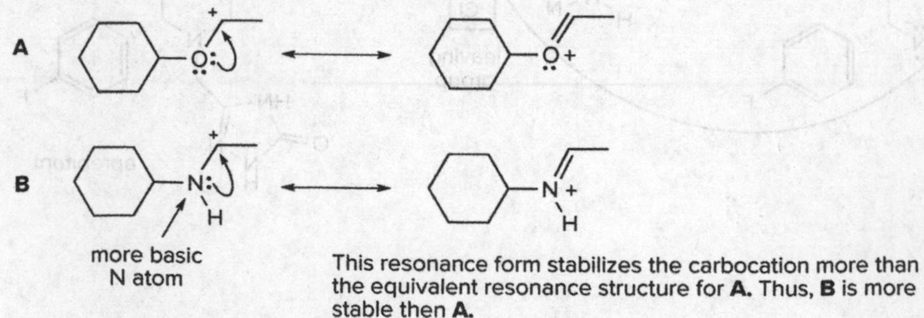

This resonance form stabilizes the carbocation more than the equivalent resonance structure for **A**. Thus, **B** is more stable then **A**.

7.57

a. Mechanism: **S_N1 only**

A + H_2O —Step [1]→ B + I^- ; B + $H_2\ddot{O}$ —Step [2]→ C ($:\overset{+}{O}H_2$)

b. Energy diagram:

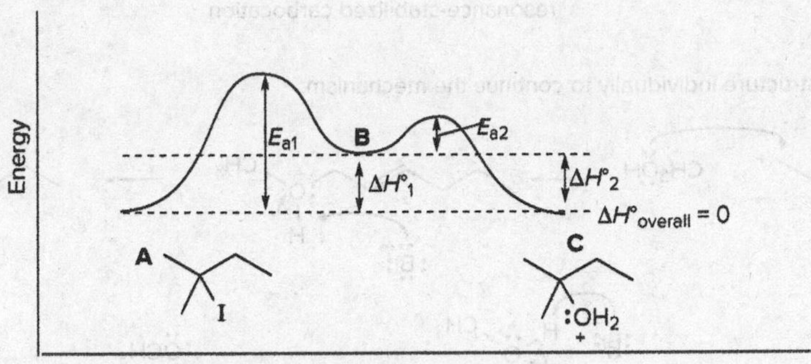

c. Transition states:

$[\delta+ \cdots :\ddot{I}: \delta-]^{\ddagger}$ $[\delta+ \cdots :OH_2\ \delta+]^{\ddagger}$

d. Rate equation: **rate = $k[(CH_3)_2CICH_2CH_3]$**

e. [1] Leaving group changed from I^- to Cl^-: **rate decreases** because I^- is a better leaving group.

[2] Solvent changed from H_2O (polar protic) to DMF (polar aprotic): **rate decreases** because polar protic solvent favors S_N1.

[3] Alkyl halide changed from 3° to 2°: **rate decreases** because 2° carbocations are less stable.

[4] [R–X] and [H_2O] increased by factor of five: **rate increases** by a factor of five. (Only the concentration of R–X affects the rate.)

7.58

a. (Br) + CH_3CH_2OH ⟶ (OCH_2CH_3) + (CH_3CH_2O) + HBr

b. (Br) + H_2O ⟶ (OH) + (OH) + HBr

7.59 The 1° alkyl halide is also allylic, so it forms a resonance-stabilized carbocation. Increasing the stability of the carbocation by resonance increases the rate of the S_N1 reaction.

Br → resonance-stabilized carbocation + Br^-

Use each resonance structure individually to continue the mechanism:

$CH_3\ddot{O}H$ → CH_3 :Br: → $\ddot{O}CH_3$ + HBr

:Br: H CH3 → :ÖCH3 + HBr

$CH_3\ddot{O}H$

7.60

a, b.

Br Cl Br OH OH plocamenol A

3° alkyl bromide most substituted alkyl halide reacts fastest in S_N1

1° alkyl bromide Br better leaving group than Cl reacts fastest in S_N2

7.61

Br A Br B Br C

Vinyl halides like A do not react by either an S_N1 or S_N2 mechanism. With S_N2, the rate is 3° < 2° < 1°, and with S_N1, the rate is 1° < 2° < 3°.

a. **S_N2 reactivity: A < C < B**

b. **S_N1 reactivity: A < B < C**

7.62

a. Br + ^-CN →(acetone) CN + Br^-

1° alkyl halide
S_N2 only

b. Br H + $^-OCH_3$ →(DMSO) H OCH_3 + Br^-

2° alkyl halide
S_N1 and S_N2

strong nucleophile
polar aprotic solvent
Both favor S_N2.

reaction at a stereogenic center
inversion of configuration

c. Br + CH_3OH → OCH_3 + HBr

3° alkyl halide
S_N1 only

d. + CH_3CO_2H → O_2CCH_3 + O_2CCH_3 + HI
I

reaction at a stereogenic center
racemization of product

2° alkyl halide
S_N1 and S_N2

Weak nucleophile
favors S_N1.

e. Br + $^-OCH_2CH_3$ —DMF→ OCH_2CH_3 + Br^-

reaction at a stereogenic center
inversion of configuration

2° alkyl halide
S_N1 and S_N2

strong nucleophile
polar aprotic solvent
Both favor S_N2.

f. Cl + CH_3CH_2OH → OCH_2CH_3 + OCH_2CH_3 + HCl

two products—**diastereomers**
Nucleophile attacks
from above and below.

2° alkyl halide
S_N1 and S_N2

Weak nucleophile
favors S_N1.

7.63 An S_N1 mechanism means the reaction occurs in a stepwise fashion by way of a carbocation.

P–O, O–PP, HO, OH → P–O, HO, OH + ^-O–PP; H, N, O, N, H, O, O, O^- → P–O, O, N^+, H, O, N, H, O, ^-O–PP, HO, OH, O, O^- → P–O, O, N, H, O, N, O, HO, OH, O, O^-

A   B   C + HO–PP

7.64 First decide whether the reaction will proceed via an S_N1 or S_N2 mechanism (Answer 7.33), and then draw the mechanism.

Br → + ; $CH_3CH_2\ddot{O}H$ → $\ddot{O}CH_2CH_3$, H, :Br:$^-$ + H, $^+OCH_2CH_3$, :Br:$^-$ → $\ddot{O}CH_2CH_3$ + $\ddot{O}CH_2CH_3$ + HBr

3° alkyl halide
S_N1 only

can attack from
above or below

7.65

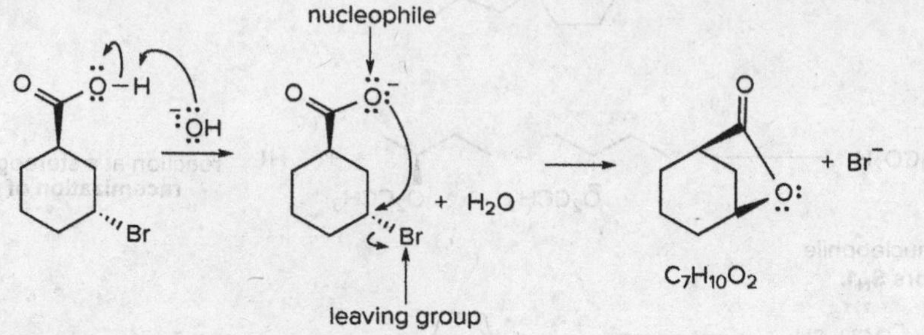

nucleophile

leaving group

+ H_2O

+ Br^-

$C_7H_{10}O_2$

7.66

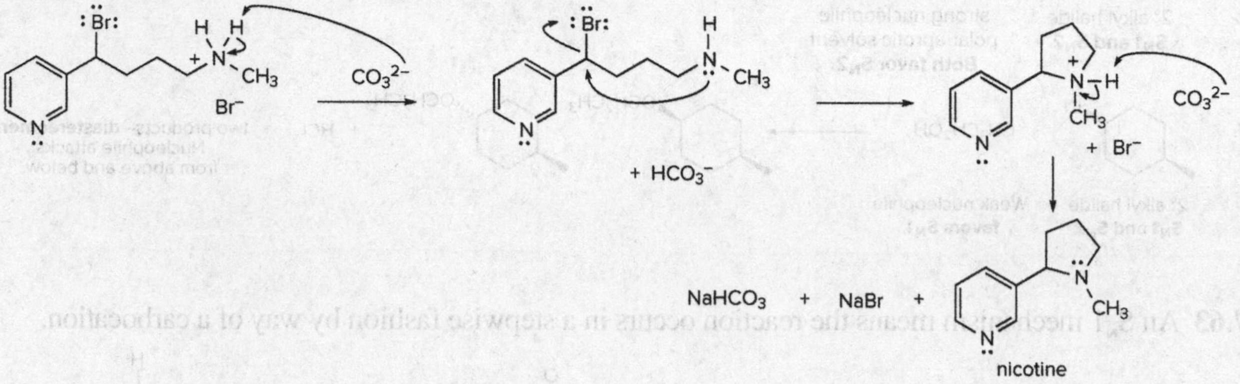

Br^-

CO_3^{2-}

+ HCO_3^-

CO_3^{2-}

+ Br^-

$NaHCO_3$ + NaBr +

nicotine

7.67

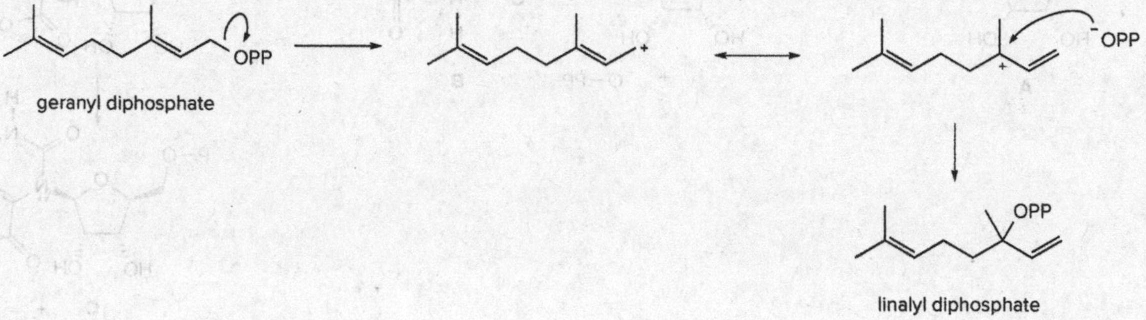

geranyl diphosphate

linalyl diphosphate

7.68

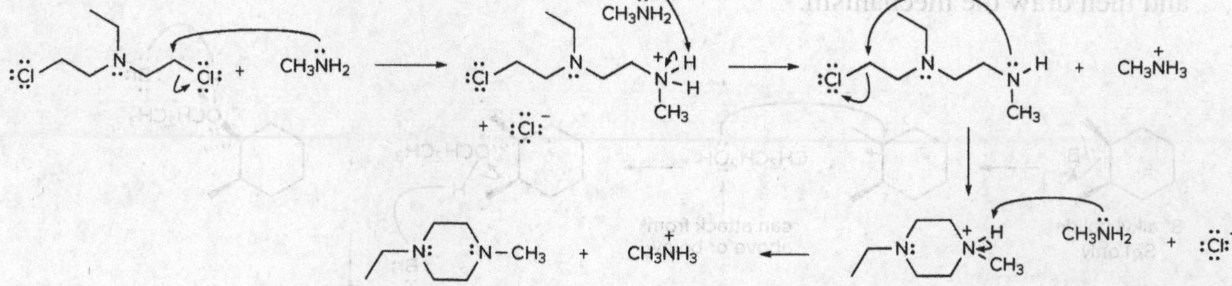

+ CH_3NH_2

CH_3NH_2

+ $:\ddot{Cl}:^-$

+ $CH_3\overset{+}{N}H_3$

CH_3NH_2 + $:\ddot{Cl}:^-$

+ $CH_3\overset{+}{N}H_3$

7.69 In the first reaction, substitution occurs at the stereogenic center. Because an achiral, planar carbocation is formed, the nucleophile can attack from either side, thus generating a racemic mixture.

3° alkyl halide

(R)-6-bromo-2,6-dimethylnonane

$CH_3\ddot{O}H$ / S_N1 → achiral, planar carbocation + Br^- → two steps → racemic mixture **optically inactive**

In the second reaction, the starting material contains a stereogenic center, but the nucleophile does not attack at that carbon. Because a bond to the stereogenic center is not broken, the configuration is retained and a chiral product is formed.

3° alkyl halide

(R)-2-bromo-2,5-dimethylnonane

$CH_3\ddot{O}H$ / S_N1 → + Br^- Reaction does not occur at the stereogenic center. → two steps → configuration retained **optically active**

7.70

a. → The nucleophile has replaced the leaving group. Missing reagent: ^-O-cyclohexyl

b. → The nucleophile has replaced the leaving group. Missing reagent: $^-C{\equiv}CH$

c. N_3^- → The nucleophile has replaced the halide. Starting material: chlorocyclohexane

d. ^-SH → The nucleophile has replaced the halide. Starting material: The leaving group must have the opposite orientation to the position of the nucleophile in the product.

7.71 To devise a synthesis, look for the carbon framework and the functional group in the product. **The carbon framework is from the alkyl halide and the functional group is from the nucleophile.**

a. carbon framework | functional group — SH; Cl + $Na^+\ ^-SH$ → SH

b. carbon framework | functional group — O; Cl + $Na^+\ ^-O$ → O

c. carbon framework | functional group — CN; Cl + $Na^{+-}CN$ → CN

d. carbon framework | functional group — O; Cl (2° halide) + $Na^+\ ^-O$ → O

or

functional group | carbon framework — O; Cl (1° halide) + $O^-\ Na^+$ → O

This path is preferred. The strong nucleophile favors an S_N2 reaction so an unhindered 1° alkyl halide reacts faster.

e. carbon framework | functional group — O, O; Cl + $Na^+\ ^-O$ → O, O

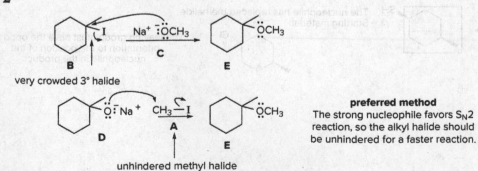

7.72

B (I) + $Na^+\ :\ddot{O}CH_3$ (C) → E ($\ddot{O}CH_3$)

very crowded 3° halide

D ($\ddot{O}:\ Na^+$) + $CH_3{-}I$ (A) → E ($\ddot{O}CH_3$)

unhindered methyl halide

preferred method
The strong nucleophile favors S_N2 reaction, so the alkyl halide should be unhindered for a faster reaction.

7.73

H–C≡C–H —NaH→ H–C≡C⁻ (A) + H_2

A + 1-bromodecane → B

B —NaH→ C + H_2

C + 1-bromotridecane → D

D —addition of H_2 (1 equiv)→ muscalure

7.74

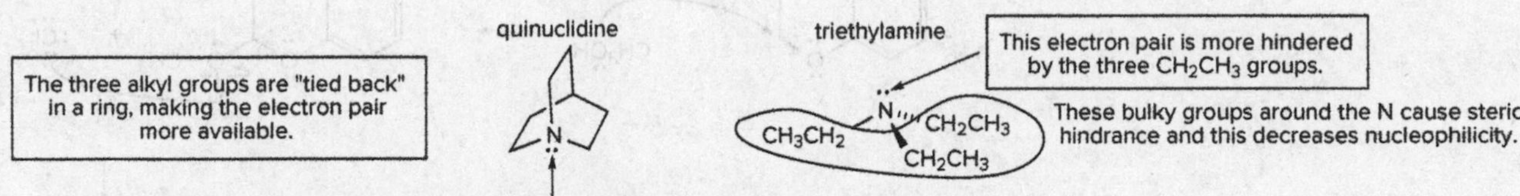

quinuclidine
triethylamine
The three alkyl groups are "tied back" in a ring, making the electron pair more available.
This electron pair is more hindered by the three CH_2CH_3 groups.
These bulky groups around the N cause steric hindrance and this decreases nucleophilicity.
N
CH_3CH_2
CH_2CH_3
CH_2CH_3

This electron pair on quinuclidine is much more available than the one on triethylamine.

less steric hindrance
more nucleophilic

7.75

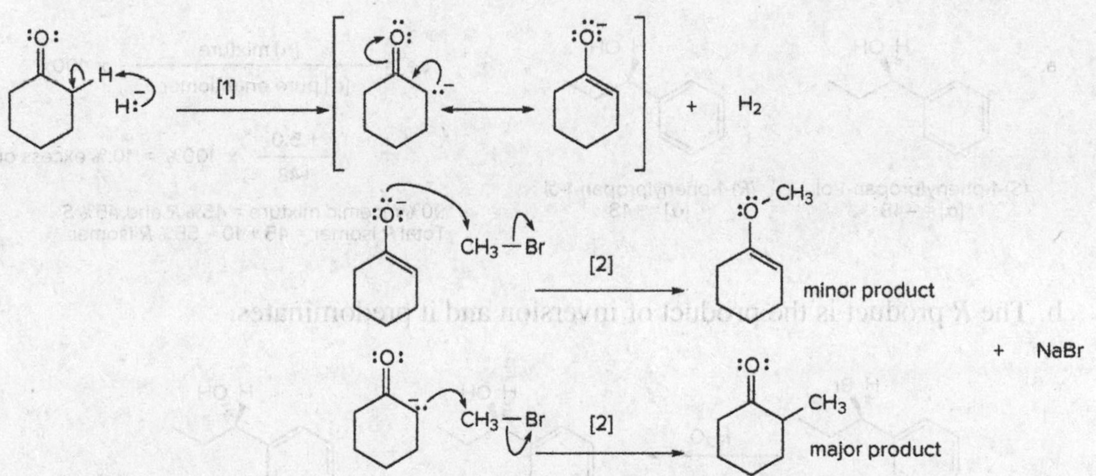

H:⁻
[1]
+ H_2
CH_3–Br
[2]
minor product
+ NaBr
[2]
major product
CH_3

7.76

Cl bonded to sp^2 C cannot undergo S_N1.

Cl bonded to sp^3 C
No resonance stabilization is possible for the carbocation formed here.

Cl bonded to sp^3 C
Resonance-stabilized carbocation forms. best for S_N1

J

CH_3OH (1 equiv)

re-draw

CH_3OH

+ :Cl:⁻

+ HCl

K

7.77

a.

(S)-1-phenylpropan-1-ol
[α] = −48

(R)-1-phenylpropan-1-ol
[α] = +48

$$ee = \frac{[\alpha]\text{ mixture}}{[\alpha]\text{ pure enantiomer}} \times 100\%$$

$$= \frac{+5.0}{+48} \times 100\% = 10.\%\text{ excess of } R \text{ isomer}$$

90% racemic mixture = 45% *R* and 45% *S*
Total *R* isomer = 45 + 10 = 55% *R* isomer

b. The *R* product is the product of inversion and it predominates.

H_2O

S
retention

R
inversion

c. The weak nucleophile favors an S_N1 reaction, which occurs by way of an intermediate carbocation. Perhaps there is more inversion than retention because H_2O attacks the intermediate carbocation while the Br^- leaving group is still in the vicinity of the carbocation. The Br^- would then shield one side of the carbocation and backside attack would be slightly favored.

Chapter 8 Alkyl Halides and Elimination Reactions

Chapter Review

A comparison between nucleophilic substitution and β-elimination

Nucleophilic substitution—A nucleophile attacks a carbon atom (7.6).

X + :Nu⁻ → Nu + :X⁻
substitution product
good leaving group

β-Elimination—A base attacks a proton (8.1).

B: H ... X → + H–B⁺ + :X⁻
elimination product
good leaving group

Similarities	Differences
• In both reactions RX acts as an electrophile, reacting with an electron-rich reagent. • Both reactions require a **good leaving group X:⁻** willing to accept the electron density in the C–X bond.	• In substitution, a nucleophile attacks a single carbon atom. • In elimination, a Brønsted–Lowry base removes a proton to form a π bond, and two carbons are involved in the reaction.

The importance of the base in E2 and E1 reactions (8.9)

The strength of the base determines the mechanism of elimination.

- Strong bases favor E2 reactions.
- Weak bases favor E1 reactions.

Br —[$^{-}$:ÖH, strong base]→ E2 product + H_2O: + :Br:⁻

Br —[H_2O:, weak base]→ E1 product + H_3O^+: + :Br:⁻

E1 and E2 mechanisms compared

	E2 mechanism	E1 mechanism
[1] Mechanism	• one step (8.4B)	• two steps (8.6B)
[2] Alkyl halide	• rate: $R_3CX > R_2CHX > RCH_2X$ (8.4C)	• rate: $R_3CX > R_2CHX > RCH_2X$ (8.6C)
[3] Rate equation	• rate = k[RX][B:] • second-order kinetics (8.4A)	• rate = k[RX] • first-order kinetics (8.6A)
[4] Stereochemistry	• anti periplanar arrangement of H and X (8.8)	• trigonal planar carbocation intermediate (8.6B)
[5] Base	• favored by strong bases (8.4B)	• favored by weak bases (8.6C)
[6] Leaving group	• better leaving group → faster reaction (8.4B)	• better leaving group → faster reaction (Table 8.4)
[7] Solvent	• favored by polar aprotic solvents (8.4B)	• favored by polar protic solvents (Table 8.4)
[8] Product	• more substituted alkene favored (Zaitsev rule, 8.5)	• more substituted alkene favored (Zaitsev rule, 8.6C)

Summary chart on the four mechanisms: S_N1, S_N2, E1, and E2 (8.11)

Alkyl halide type	Conditions	Mechanism
1° RCH_2X	strong nucleophile	**S_N2**
	strong bulky base	**E2**
2° R_2CHX	strong base and nucleophile	**S_N2 + E2**
	strong bulky base	**E2**
	weak base and nucleophile	**S_N1 + E1**
3° R_3CX	weak base and nucleophile	**S_N1 + E1**
	strong base	**E2**

Zaitsev rule

- β-Elimination affords the more stable product having the more substituted double bond.
- Zaitsev products predominate in E2 reactions except when a cyclohexane ring prevents trans diaxial arrangement.

Practice Test on Chapter Review

1. Which of the following is true about an E1 reaction?
   1. The reaction is faster with better leaving groups.
   2. The reaction is fastest with 3° alkyl halides.
   3. The reaction is faster with stronger bases.
   4. Statements (1) and (2) are true.
   5. Statements (1), (2), and (3) are all true.

2. Consider the S_N2 and E1 reaction mechanisms. What effect on the rate of the reaction is observed when each of the following changes is made? Fill in each box of the table with one of the following phrases: **increases, decreases,** or **remains the same.**

Change	S_N2 mechanism	E1 mechanism
a. The alkyl halide is changed from $(CH_3)_3CBr$ to $CH_3CH_2CH_2CH_2Br$.		
b. The solvent is changed from $(CH_3)_2CO$ to CH_3CH_2OH.		
c. The nucleophile/base is changed from ^-OH to H_2O.		
d. The alkyl halide is changed from CH_3CH_2Cl to CH_3CH_2I.		
e. The concentration of the base/nucleophile is increased by a factor of five.		

3. Rank the following compounds in order of ***increasing*** **reactivity in an E2 elimination** reaction. Rank the ***most reactive*** compound as **3**, the ***least reactive*** compound as **1**, and the compound of intermediate reactivity as **2**.

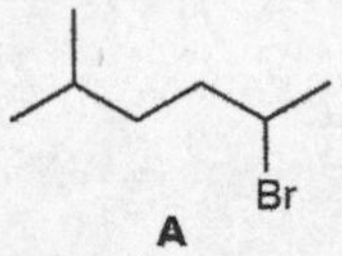

Br Br Br

A 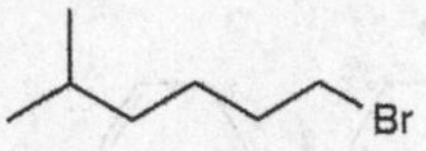 **B** 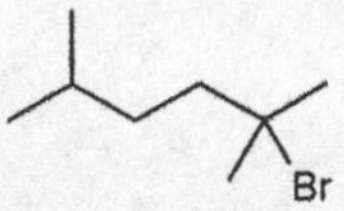 **C**

4. Draw the organic products formed in the following reactions.

a. 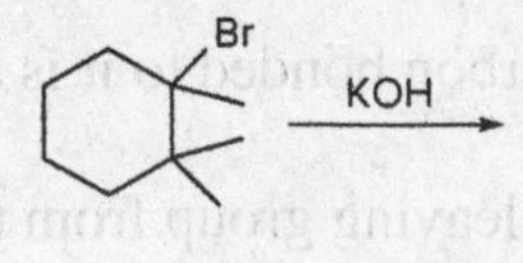 Br KOH

c. 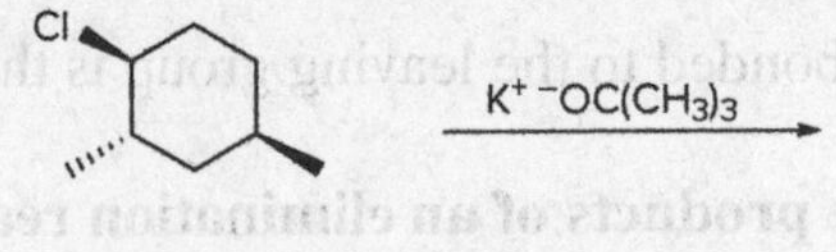 Cl $K^+ {}^-OC(CH_3)_3$

b. 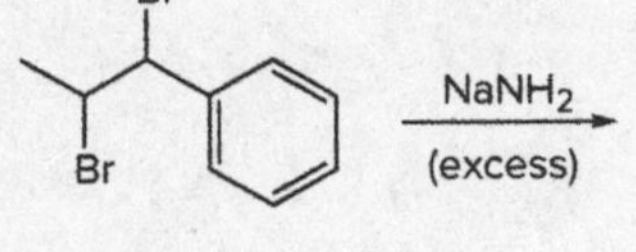 Br Br $NaNH_2$ (excess)

d. 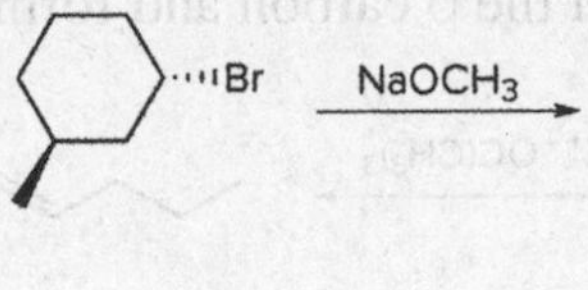 Br $NaOCH_3$

5. a. Fill in the appropriate alkyl halide needed to synthesize the following compound as a single product using the given reagent.

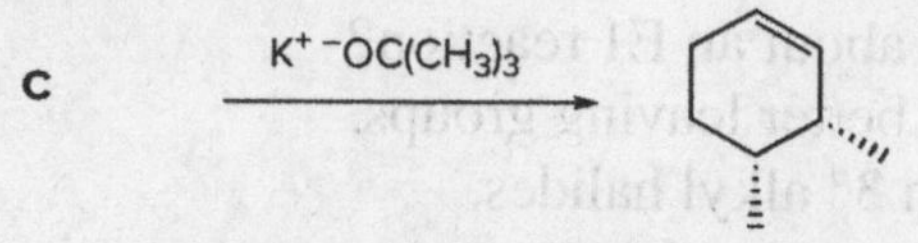

C $\xrightarrow{K^+ \ ^-OC(CH_3)_3}$

b. What starting material is needed for the following reaction? The starting material must yield product cleanly, in one step without any other organic side products.

D $\xrightarrow{K^+ \ ^-OC(CH_3)_3}$ $CH_3CH_2CH{=}CHCH_3$ (cis and trans mixture)

6. Draw all products formed in the following reaction.

Cl $\xrightarrow{CH_3OH}$

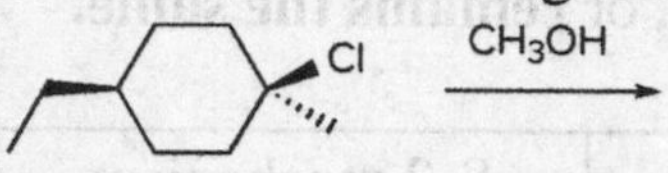

Answers to Practice Test

1\. 4

2\.

	S_N2	E1
a.	increases	decreases
b.	decreases	increases
c.	decreases	same
d.	increases	increases
e.	increases	same

3\. A–2

B–1

C–3

4\. a. 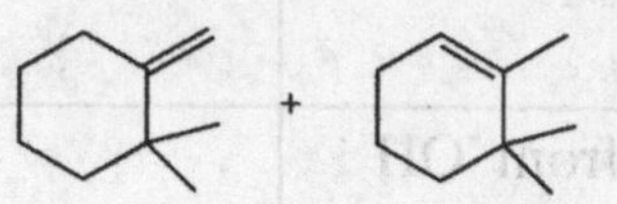 +

b.

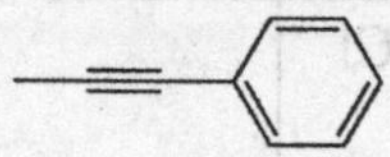

c.

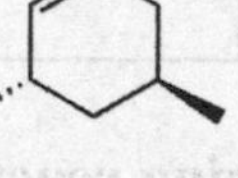

d. 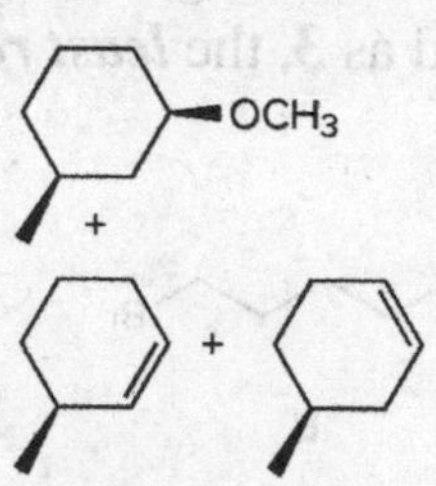 OCH_3 + +

5\. a. 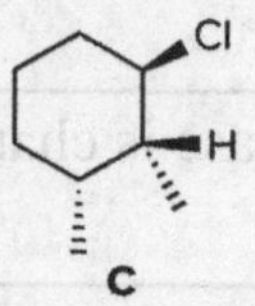 Cl H C

b. 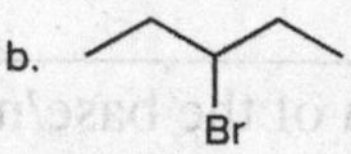 Br (or Cl or I) D

6\. OCH_3 OCH_3

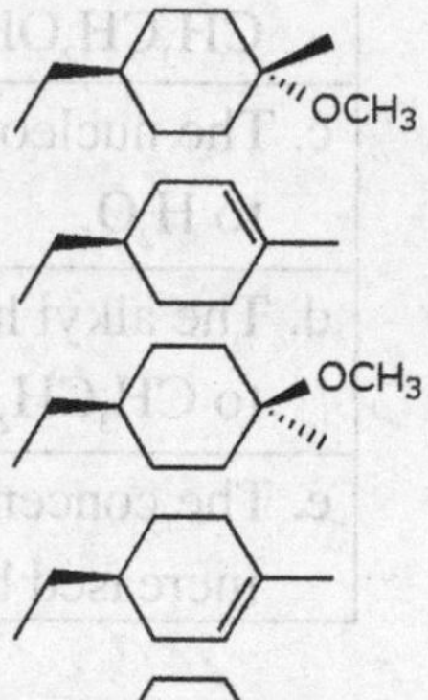

Answers to Problems

8.1 • The carbon bonded to the leaving group is the α **carbon.** Any carbon bonded to it is a β **carbon.**

• **To draw the products of an elimination reaction**: Remove the leaving group from the α carbon and a H from the β carbon and form a π bond.

a. 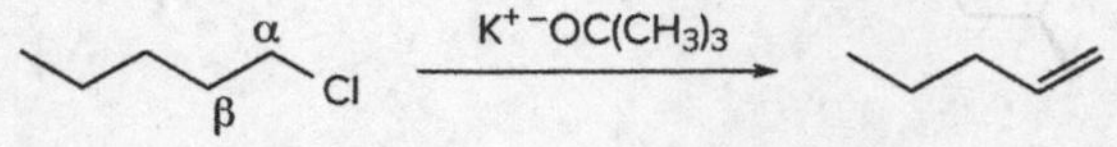 α β Cl $\xrightarrow{K^+ \ ^-OC(CH_3)_3}$

b. β_1 α β_1 β_2 Cl $\xrightarrow{K^{+}\,^{-}OC(CH_3)_3}$ + $CH_3CH{=}C(CH_3)CH_2CH_3$

c. β_1 β_2 α Br β_1 $\xrightarrow{K^{+}\,^{-}OC(CH_3)_3}$ +

8.2 **Alkenes are classified by the number of carbon atoms bonded to the double bond.** A monosubstituted alkene has one carbon atom bonded to the double bond, a disubstituted alkene has two carbon atoms bonded to the double bond, etc.

a. 4 C's bonded to C=C **tetrasubstituted**; 2 C's bonded to each C=C **disubstituted**; OH; 3 C's bonded to each C=C **trisubstituted**; vitamin A

b. vitamin D_3; H; 3 C's bonded to each C=C **trisubstituted**; 2 C's bonded to the C=C **disubstituted**; HO

8.3 To have stereoisomers at a C=C, the two groups on each end of the double bond must be different from each other.

a. OH; two CH_3 groups no stereoisomers possible

b. two CH_3 groups no stereoisomers possible; OH; two different groups on both ends stereoisomers possible

c. two different groups on both ends stereoisomers possible; two identical groups on one end no stereoisomers possible

8.4 Two definitions:

- **Constitutional isomers** differ in the connectivity of the atoms.
- **Stereoisomers** differ only in the 3-D arrangement of the atoms in space.

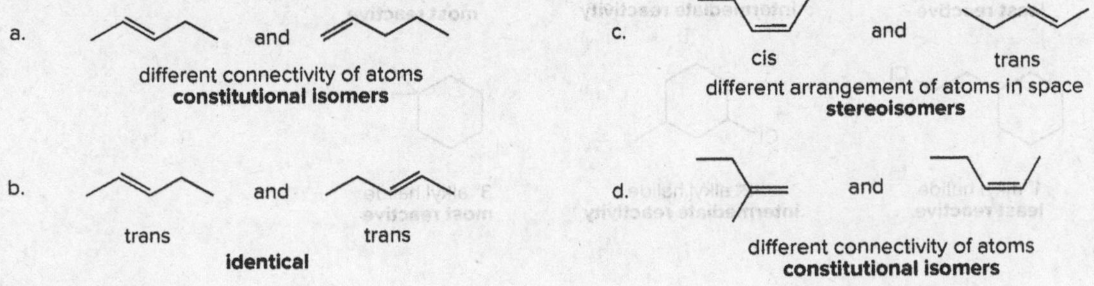

8.5 Two rules to predict the relative stability of alkenes:
[1] Trans alkenes are generally more stable than cis alkenes.
[2] The stability of an alkene increases as the number of R groups on the C=C increases.

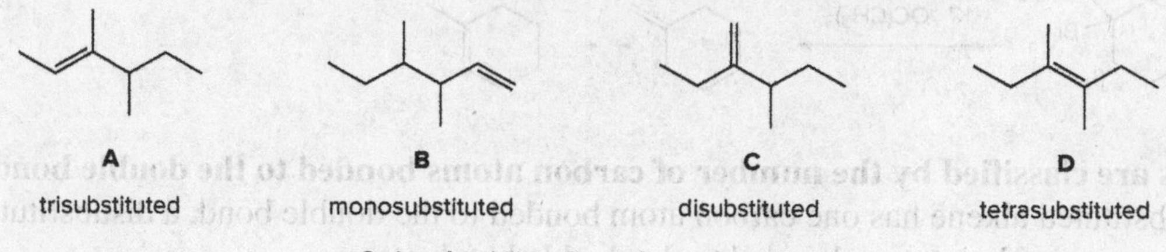

Order of stability: **B < C < A < D**

8.6 In an E2 mechanism, four bonds are involved in the single step. Use curved arrows to show these simultaneous actions:
[1] The base attacks a hydrogen on a β carbon.
[2] A π bond forms.
[3] The leaving group comes off.

Br β H :Ö:⁻ → + ÖH + :Br:⁻

new π bond

transition state: [:Br: δ– H :Ö: δ–]‡

8.7 In both cases, the rate of elimination decreases.

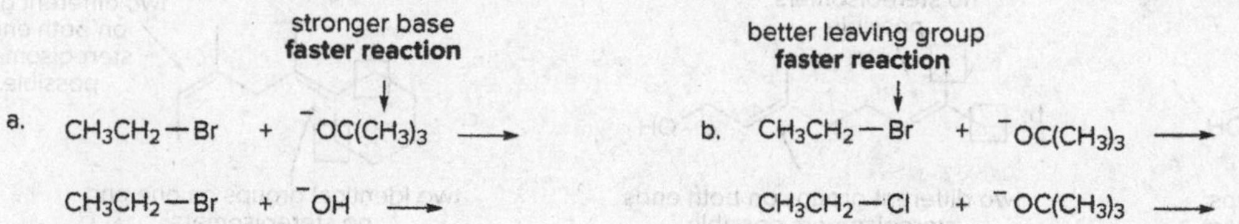

8.8 As the number of R groups on the carbon with the leaving group increases, the rate of an E2 reaction increases.

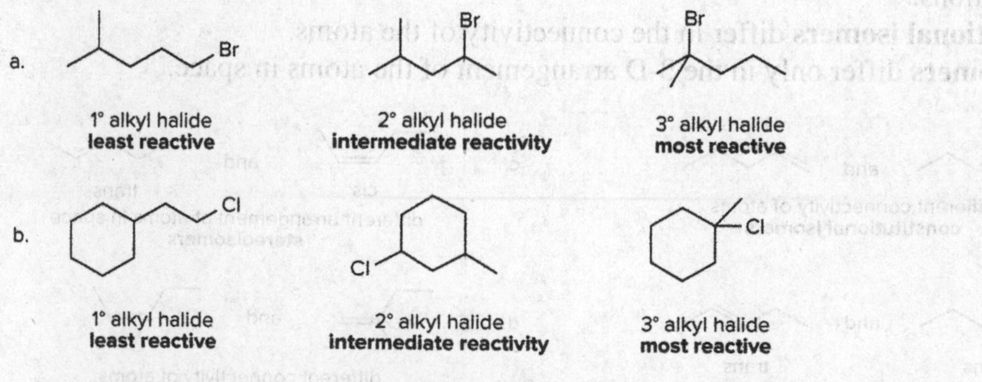

8.9 Use the following characteristics of an E2 reaction to answer the questions:
[1] E2 reactions are second order and one step.
[2] More substituted halides react faster.
[3] Reactions with strong bases or better leaving groups are faster.
[4] Reactions with polar aprotic solvents are faster.

Rate equation: rate = *k*[RX][Base]

a. tripling the concentration of the alkyl halide = **rate triples**
b. halving the concentration of the base = **rate is halved**
c. changing the solvent from CH_3OH to DMSO = **rate increases** (Polar aprotic solvent is better for E2.)
d. changing the leaving group from I^- to Br^- = **rate decreases** (I^- is a better leaving group.)
e. changing the base from ^-OH to H_2O = **rate decreases** (weaker base)
f. changing the alkyl halide from CH_3CH_2Br to $(CH_3)_2CHBr$ = **rate increases** (More substituted halide reacts faster.)

8.10 According to the Zaitsev rule, the major product in a β-elimination reaction has the *more* substituted double bond.

a. Br, α — loss of H and Br → trisubstituted **major product** + (+ stereoisomer) disubstituted **minor product**

b. Br, α — loss of H and Br → trisubstituted **minor product** + tetrasubstituted **major product** + disubstituted **minor product**

c. Cl, α — loss of H and Cl → monosubstituted **minor product** + (+ stereoisomer) disubstituted **major product**

d. α, Cl — loss of H and Cl → trisubstituted **ONLY product**

8.11 An E1 mechanism has two steps:
[1] The leaving group comes off, creating a carbocation.
[2] A base pulls off a proton from a β carbon, and a π bond forms.

c. 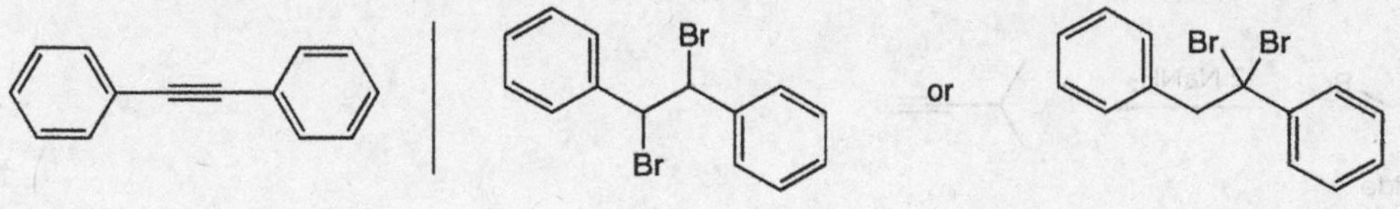 [structure: diphenylacetylene] | [structure: 1,2-dibromo-1,2-diphenylethane, Br, Br] or [structure: 1,1-dibromo-1,2-diphenylethane, Br, Br]

8.47

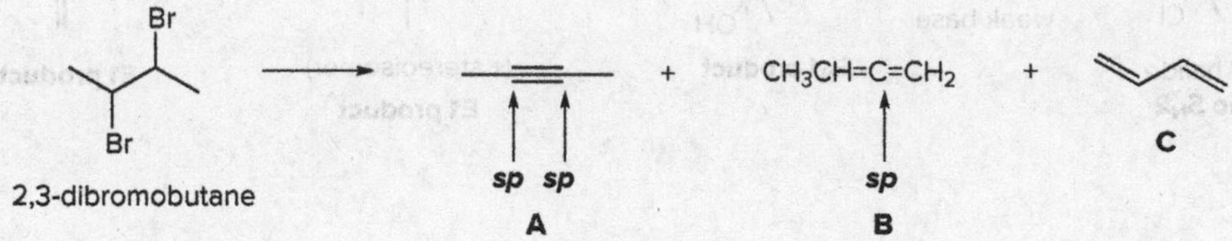

Br, Br

2,3-dibromobutane → [2-butyne] + $CH_3CH{=}C{=}CH_2$ + [1,3-butadiene]

sp *sp* — A; *sp* — B; C

8.48

a. 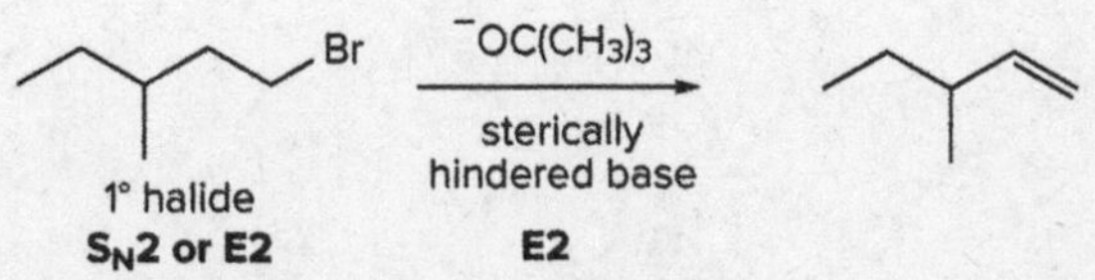 [structure] Br — 1° halide, S_N2 or E2 — $^-OC(CH_3)_3$, sterically hindered base, E2 → [alkene]

b. 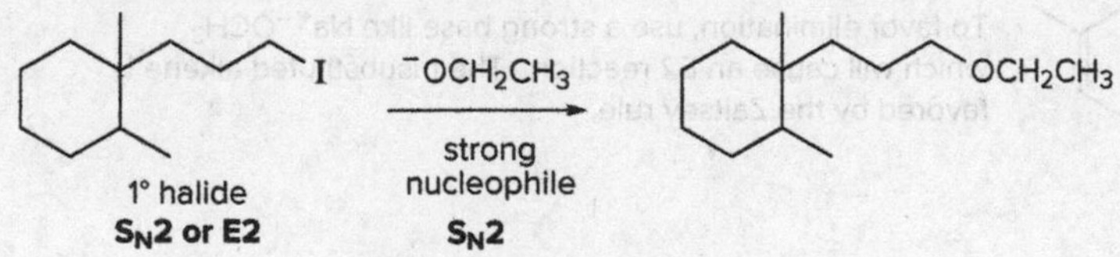 [structure] I — 1° halide, S_N2 or E2 — $^-OCH_2CH_3$, strong nucleophile, S_N2 → [structure] OCH_2CH_3

c. 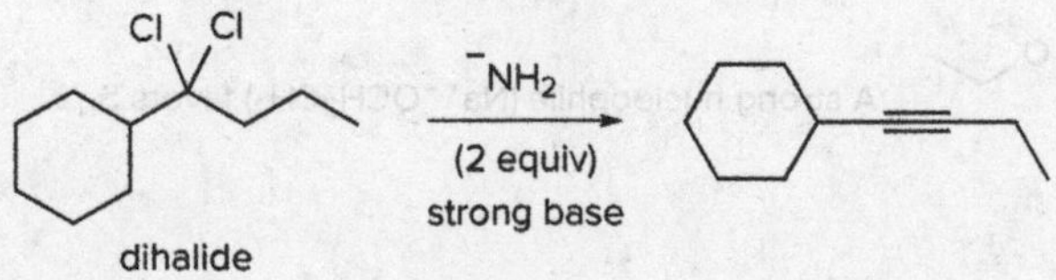 [structure] Cl Cl — dihalide — $^-NH_2$ (2 equiv), strong base → [alkyne]

d. 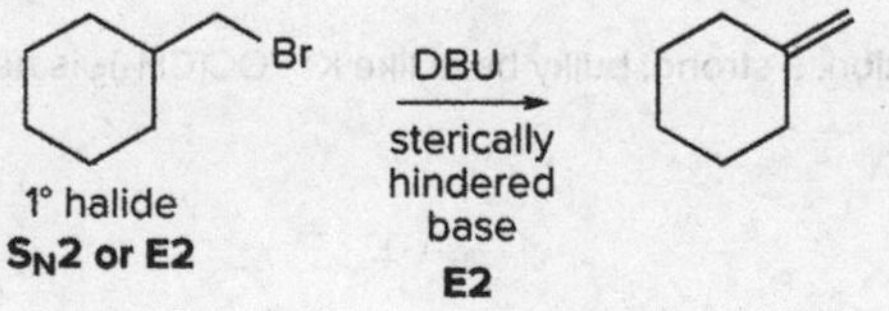 [structure] Br — 1° halide, S_N2 or E2 — DBU, sterically hindered base, E2 → [methylenecyclohexane]

e. 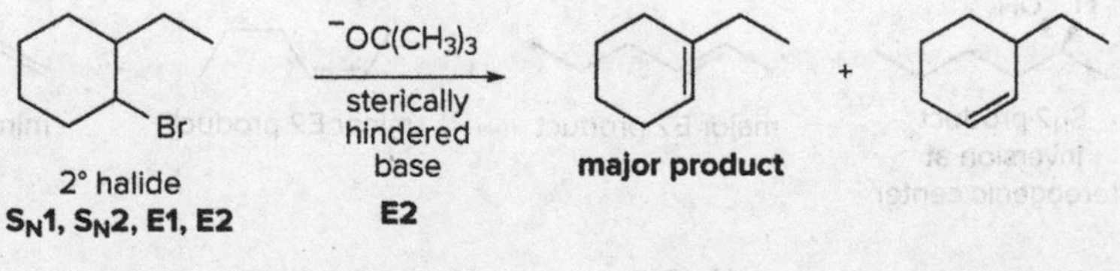 [structure] Br — 2° halide, S_N1, S_N2, E1, E2 — $^-OC(CH_3)_3$, sterically hindered base, E2 → [structure] **major product** + [structure]

f. 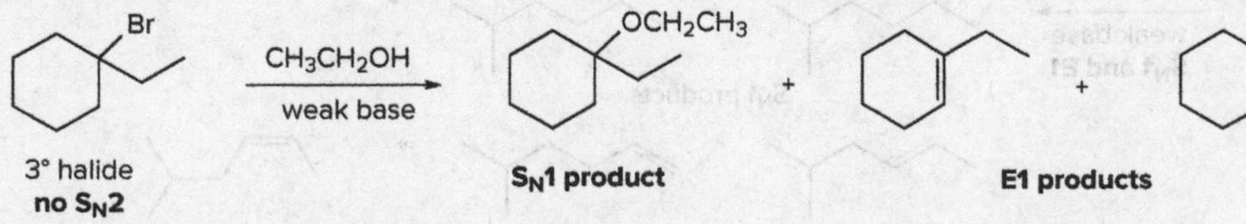 [structure] Br — 3° halide, no S_N2 — CH_3CH_2OH, weak base → [structure] OCH_2CH_3 — S_N1 product + [structure] + [structure] — E1 products

g. Br, Br — diahlide — 2 $NaNH_2$ →

h. Cl — 3° halide **no S_N2** — H_2O weak base → OH **S_N1 product** + (+ stereoisomer) **E1 product** + **E1 product**

8.49

a. Br — 3° halide → To get the substitution product the mechanism must be S_N1, so a weak nucleophile like CH_3CH_2OH must be used.

b. Br — 3° halide → To favor elimination, use a strong base like $Na^+ \ ^-OCH_3$ which will cause an E2 reaction. The trisubstituted alkene is favored by the Zaitsev rule.

c. Cl — 1° halide → A strong nucleophile ($Na^+ \ ^-OCH_2CH_3$) favors S_N2.

d. Cl — 1° halide → To get elimination, a strong, bulky base like $K^+ \ ^-OC(CH_3)_3$ is needed.

8.50

a. Cl H — 2° halide S_N1, S_N2, E1, E2 — ^-OH strong base **S_N2 and E2** → H OH — S_N2 product inversion at stereogenic center + major E2 product + minor E2 product + minor E2 product

b. Cl H — 2° halide **S_N1, S_N2, E1, E2** — H_2O weak base **S_N1 and E1** → HO H + H OH + S_N1 products

major E1 product + minor E1 product + minor E1 product

c. 3° halide **no S_N2** — CH_3OH, weak base, **S_N1 and E1** — S_N1 products (CH_3O, C_6H_5; C_6H_5, OCH_3) + E1 product (C_6H_5)

d. 2° halide **S_N1, S_N2, E1, E2** (Br, D) — KOH, strong base, **S_N2 and E2** — S_N2 product inversion at stereogenic center (OH, D) + E2 product (trans diaxial elimination of D, Br)

e. 2° halide **S_N1, S_N2, E1, E2** (Br) — $NaOCH_3$, strong base, **S_N2 and E2** — S_N2 product inversion (OCH_3) + E2 product + E2 product + E2 product

f. 3° halide **no S_N2** (Br) — $NaOCH_3$, strong base, **E2** — E2 product

g. 2° halide **S_N1, S_N2, E1, E2** (Cl) — NaOH, strong base, **S_N2 and E2** — S_N2 inversion (OH) + E2 product + E2 product + E2 product

h. 2° halide **S_N1, S_N2, E1, E2** (CH_3O, OCH_3, Br) — H_2O, weak base, **S_N1 and E1** — S_N1 (CH_3O, OCH_3, OH) + S_N1 (CH_3O, OCH_3, OH) + E1 (CH_3O, OCH_3) + E1 (CH_3O, OCH_3) + E1 (CH_3O, OCH_3)

8.51

a. CH_3OH, weak base, **S_N1 and E1** → S_N1 (OCH_3 product) + S_N1 (OCH_3 product) + E1 + E1 + E1

b. KOH, strong base, **E2** → products

8.52

a. 3° halide + $^-OC(CH_3)_3$ (strong bulky base, **E2**) — substitution product not formed; elimination gives

major product
more substituted alkene

No substitution occurs with a strong bulky base and a 3° RX. The C with the leaving group is too crowded for an S_N2 substitution to occur. Elimination occurs instead by an E2 mechanism.

b. 1° halide + $^-OCH_3$ (strong nucleophile, **S_N2**) — elimination product not formed; substitution gives product with OCH_3

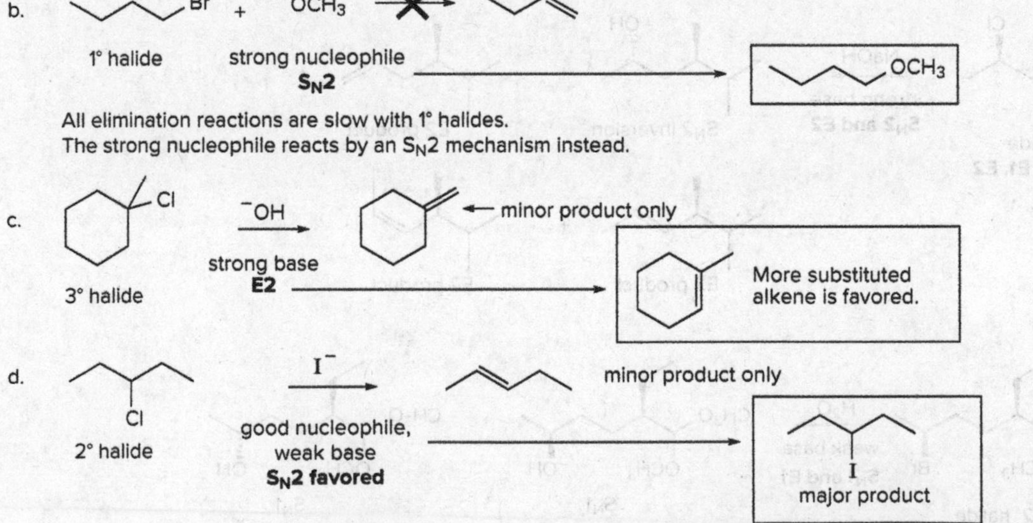

The 2° halide can react by an E2 or S_N2 reaction with a negatively charged nucleophile or base. Since I^- is a weak base, substitution by an S_N2 mechanism is favored.

8.53

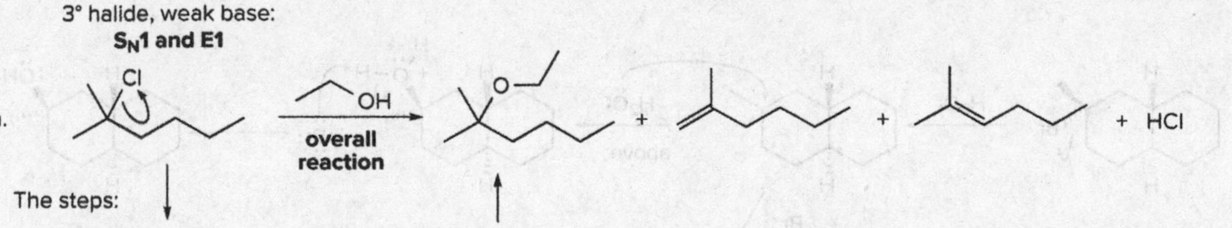

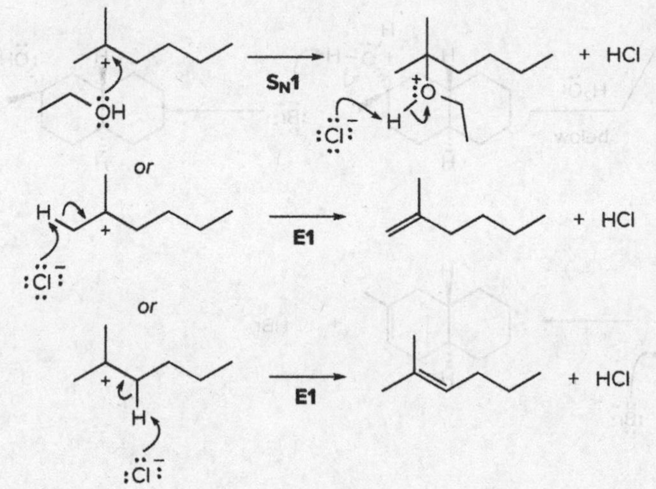

3° halide, weak base:
S_N1 and E1

a. Cl ⟶ (CH3CH2OH, **overall reaction**) ether + alkene + alkene + HCl

The steps:

S_N1 + HCl

Any base (such as CH_3CH_2OH or Cl^-) can be used to remove a proton to form an alkene. If Cl^- is used, HCl is formed as a reaction by-product. If CH_3CH_2OH is used, $(CH_3CH_2OH_2)^+$ is formed instead.

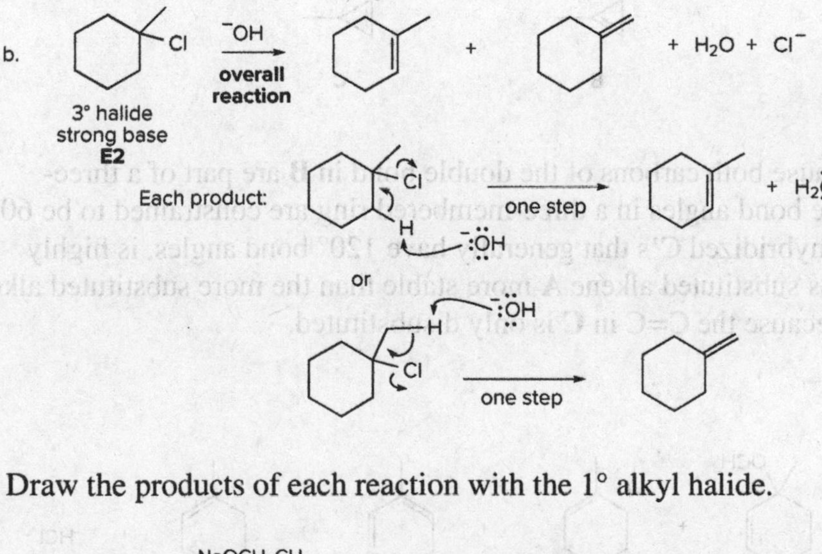

or

E1 + HCl

or

E1 + HCl

b. Cl ⟶ ($^-$OH, **overall reaction**) + H_2O + Cl^-

3° halide
strong base
E2

Each product: one step + H_2O + Cl^-

or

one step

8.54 Draw the products of each reaction with the 1° alkyl halide.

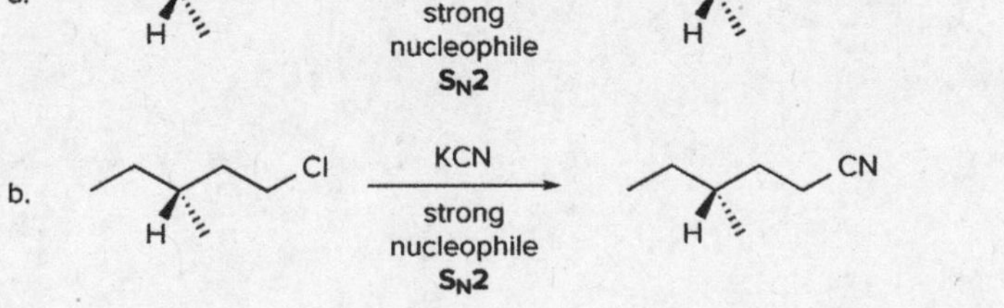

a. Cl ⟶ ($NaOCH_2CH_3$, strong nucleophile, **S_N2**) OCH_2CH_3

b. Cl ⟶ (KCN, strong nucleophile, **S_N2**) CN

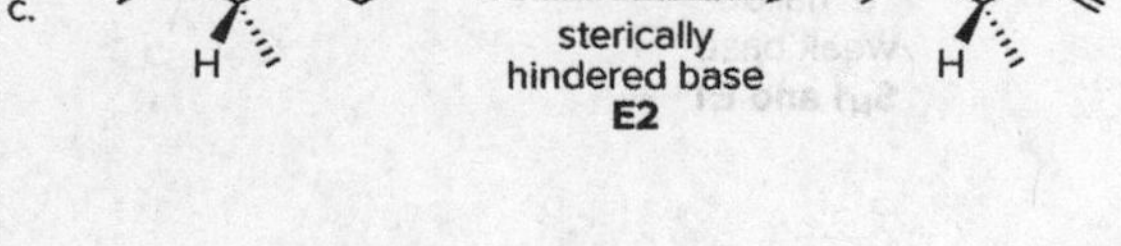

c. Cl ⟶ (DBU, sterically hindered base, **E2**) alkene

8.55

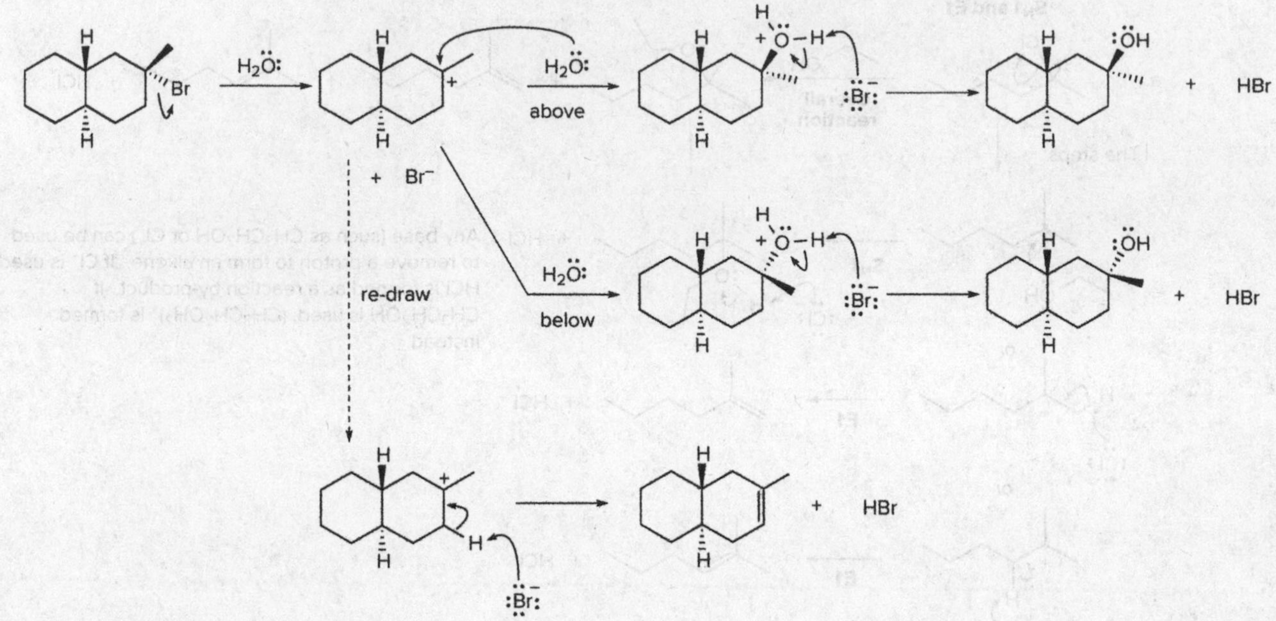

$H_2\ddot{O}$: above; + Br^-; re-draw; $H_2\ddot{O}$: below; :Br:; + HBr

8.56

A B C

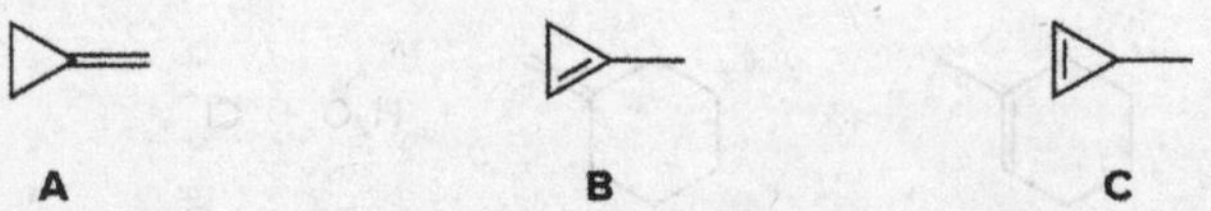

A is more stable than **B** because both carbons of the double bond in **B** are part of a three-membered ring. Because the bond angles in a three-membered ring are constrained to be 60°, the double bond, which has sp^2 hybridized C's that generally have 120° bond angles, is highly strained. This makes the less substituted alkene **A** more stable than the more substituted alkene **B. C** is less stable than **B** because the C=C in **C** is only disubstituted.

8.57

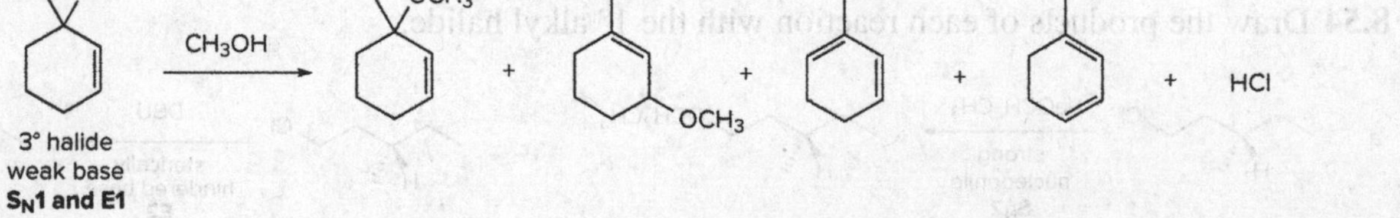

CH_3OH; 3° halide, weak base, S_N1 and E1; OCH_3 + OCH_3 + + + HCl

8.58 E2 elimination needs a leaving group and a hydrogen in the **trans diaxial** position.

Two different conformations:

This conformation has Cl's axial, but no H's axial.

This conformation has no Cl's axial.

For elimination to occur, a cyclohexane must have a H and Cl in the trans diaxial arrangement. Neither conformation of this isomer has both atoms—H and Cl—axial; thus, this isomer only slowly loses HCl by elimination.

8.59

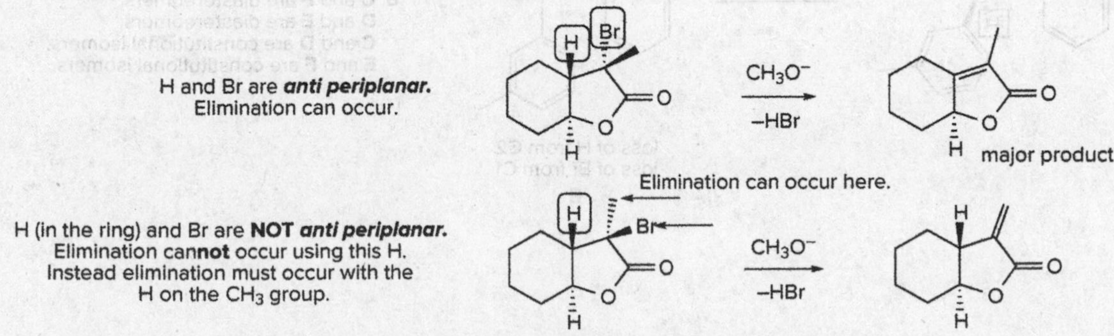
H and Br are ***anti periplanar.***
Elimination can occur.
CH_3O^-
$-HBr$
major product
Elimination can occur here.
H (in the ring) and Br are **NOT *anti periplanar.***
Elimination can**not** occur using this H.
Instead elimination must occur with the
H on the CH_3 group.
Elimination cannot occur in the ring
because the required anti periplanar geometry is not present.

8.60

leaving group

DBN

overall reaction

E2

S_N2

A sequence of two reactions forms the final product: E2 elimination opens the five-membered ring. Then the sulfur nucleophile displaces the Cl^- leaving group to form the six-membered ring.

8.61 One equivalent of $NaNH_2$ removes one mole of HBr in an anti periplanar fashion from each dibromide. Two modes of elimination are possible for each compound.

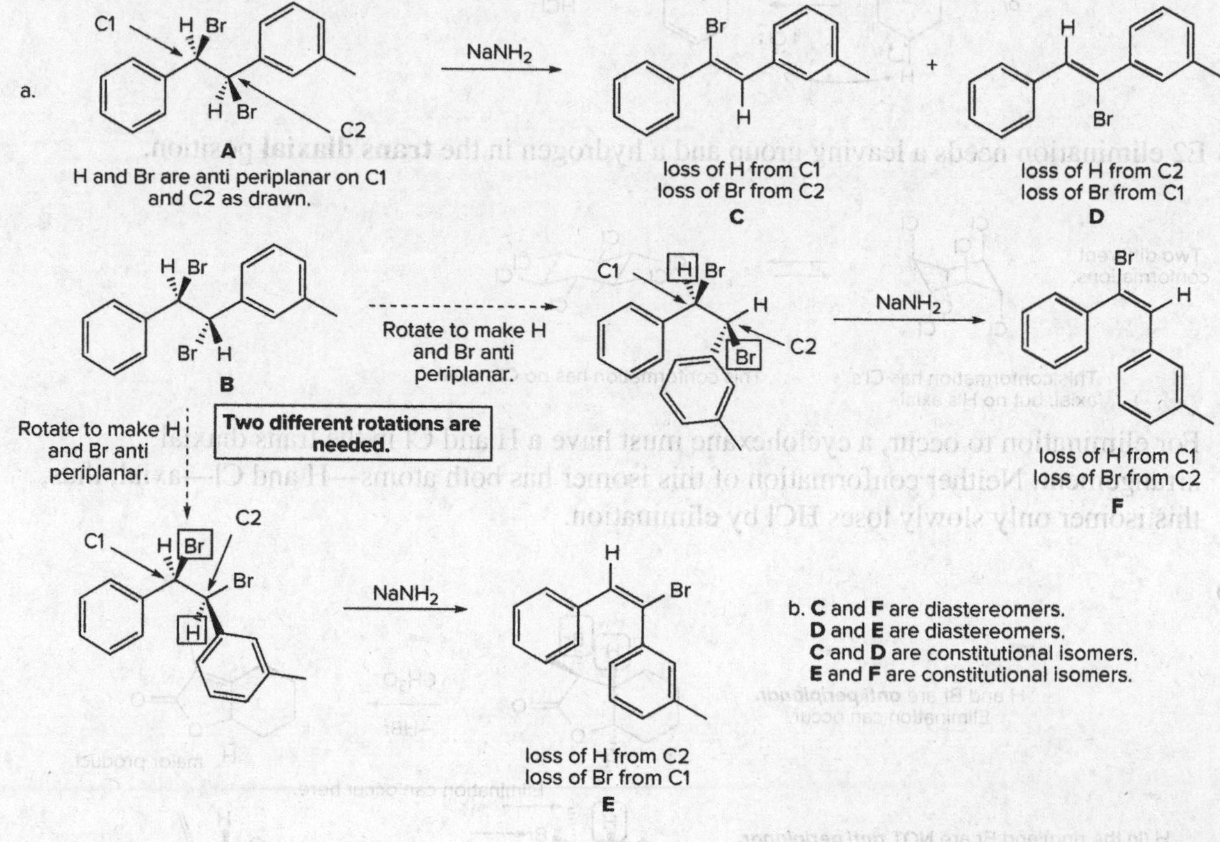

b. **C** and **F** are diastereomers.
D and **E** are diastereomers.
C and **D** are constitutional isomers.
E and **F** are constitutional isomers.

Chapter 9 Alcohols, Ethers, and Related Compounds

Chapter Review

General facts about ROH, ROR, and epoxides

- All three compounds contain an O atom that is sp^3 hybridized and tetrahedral (9.2).

109° an alcohol — 111° an ether — 60° an epoxide

- All three compounds have polar C–O bonds, but only alcohols have an O–H bond for intermolecular hydrogen bonding (9.4).

hydrogen bond

- Alcohols and ethers do not contain a good leaving group. Nucleophilic substitution can occur only after the OH (or OR) group is converted to a better leaving group (9.7A).

$R{-}\ddot{O}H + H{-}Cl \rightleftarrows R{-}\overset{+}{O}H_2 + Cl^-$

strong acid — weak base, good leaving group

- Epoxides have a leaving group located in a strained three-membered ring, making them reactive to strong nucleophiles and acids HZ that contain a nucleophilic atom Z (9.16).

With strong nucleophiles, :Nu⁻ — [1] — leaving group — [2] H–OH — + ⁻OH

A new reaction of carbocations (9.9)

- Less stable carbocations rearrange to more stable carbocations by shift of a hydrogen atom or an alkyl group. Besides rearrangement, carbocations also react with nucleophiles (7.12) and bases (8.6).

R (or H) — 1,2-shift — R (or H)

Preparation of alcohols, ethers, and epoxides (9.6)

[1] Preparation of alcohols

R–X + $^-$OH ⟶ R–OH + X^-

- The mechanism is S_N2.
- The reaction works best for CH_3X and 1° RX.

[2] Preparation of alkoxides (a Brønsted–Lowry acid–base reaction)

R–O–H + Na^+H^- ⟶ R–O^- Na^+ + H_2

alkoxide

[3] Preparation of ethers (Williamson ether synthesis)

R–X + $^-$OR' ⟶ R–OR' + X^-

- The mechanism is S_N2.
- The reaction works best for CH_3X and 1° RX.

[4] Preparation of epoxides (intramolecular S_N2 reaction)

B: H–Ö: C–C X ⟶[1] :Ö:$^-$ C–C X + H–B^+ ⟶[2] epoxide + X^-

halohydrin

- A two-step reaction sequence:
  [1] Removal of a proton with base forms an alkoxide.
  [2] Intramolecular S_N2 reaction forms the epoxide.

Reactions of alcohols

[1] Dehydration to form alkenes

a. Using strong acid (9.8, 9.9)

–C–C– (H, OH) ⟶ (H_2SO_4 or TsOH) C=C + H_2O

- Order of reactivity: $R_3COH > R_2CHOH > RCH_2OH$.
- The mechanism for 2° and 3° ROH is E1; carbocations are intermediates and rearrangements occur.
- The mechanism for 1° ROH is E2.
- The Zaitsev rule is followed.

b. Using $POCl_3$ and pyridine (9.10)

$$-\overset{|}{\underset{H}{C}}-\overset{|}{\underset{OH}{C}}- \xrightarrow[\text{pyridine}]{POCl_3} \boxed{>C=C<} + H_2O$$

- The mechanism is E2.
- No carbocation rearrangements occur.

---

[2] Reaction with HX to form RX (9.11)

$$R-OH + H-X \longrightarrow \boxed{R-X} + H_2O$$

- Order of reactivity: $R_3COH > R_2CHOH > RCH_2OH$.
- The mechanism for 2° and 3° ROH is S_N1; carbocations are intermediates and rearrangements occur.
- The mechanism for CH_3OH and 1° ROH is S_N2.

---

[3] Reaction with other reagents to form RX (9.12)

$$R-OH + SOCl_2 \xrightarrow{\text{pyridine}} \boxed{R-Cl}$$

$$R-OH + PBr_3 \longrightarrow \boxed{R-Br}$$

- Reactions occur with CH_3OH and 1° and 2° ROH.
- The reactions follow an S_N2 mechanism.

---

[4] Reaction with tosyl chloride to form alkyl tosylates (9.13A)

$$R-OH + Cl-SO_2-C_6H_4-CH_3 \xrightarrow{\text{pyridine}} R-O-SO_2-C_6H_4-CH_3$$

$$\boxed{R-OTs}$$

- The C–O bond is not broken, so the configuration at a stereogenic center is retained.

Reactions of alkyl tosylates

Alkyl tosylates undergo either substitution or elimination depending on the reagent (9.13B).

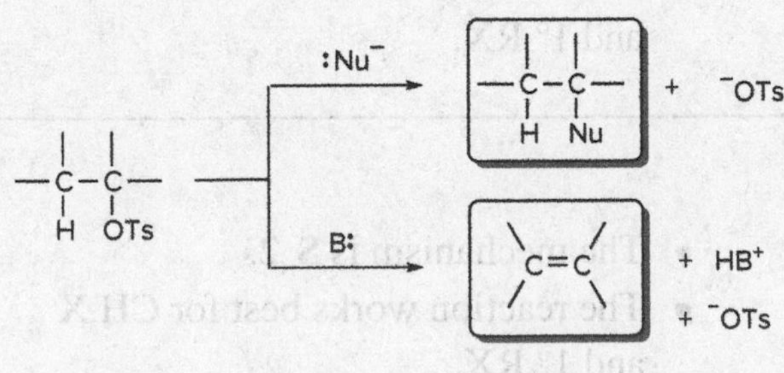

- Substitution is carried out with strong :Nu^-, so the mechanism is S_N2.
- Elimination is carried out with strong bases, so the mechanism is E2.

Reactions of ethers

Only one reaction is useful: Cleavage with strong acids (9.14)

$$\text{R–O–R'} + \underset{\substack{(2\ \text{equiv})\\(\text{X = Br or I})}}{\text{H–X}} \longrightarrow \boxed{\text{R–X}} + \boxed{\text{R'–X}} + H_2O$$

- With 2° and 3° R groups, the mechanism is S_N1.
- With CH_3 and 1° R groups, the mechanism is S_N2.

Reactions involving thiols and sulfides (9.15)

[1] Preparation of thiols

$$\text{R–X} + {}^{-}\text{SH} \longrightarrow \boxed{\text{R–SH}} + X^-$$

- The mechanism is S_N2.
- The reaction works best for CH_3X and 1° RX.

[2] Oxidation and reduction involving thiols

a. Oxidation of thiols to disulfides

$$\text{R–SH} \xrightarrow{Br_2\ \text{or}\ I_2} \boxed{\text{RS–SR}}$$

b. Reduction of disulfides to thiols

$$\text{RS–SR} \xrightarrow[\text{HCl}]{\text{Zn}} \boxed{\text{R–SH}}$$

[3] Preparation of sulfides

$$\text{R–X} + {}^{-}\text{SR'} \longrightarrow \boxed{\text{R–SR'}} + X^-$$

- The mechanism is S_N2.
- The reaction works best for CH_3X and 1° RX.

[4] Reaction of sulfides to form sulfonium ions

$$R'_2S + \text{R–X} \longrightarrow \boxed{R'_2\overset{+}{S}\text{–R}} + X^-$$

- The mechanism is S_N2.
- The reaction works best for CH_3X and 1° RX.

Reactions of epoxides

Epoxide rings are opened with nucleophiles :Nu⁻ and acids HZ (9.16).

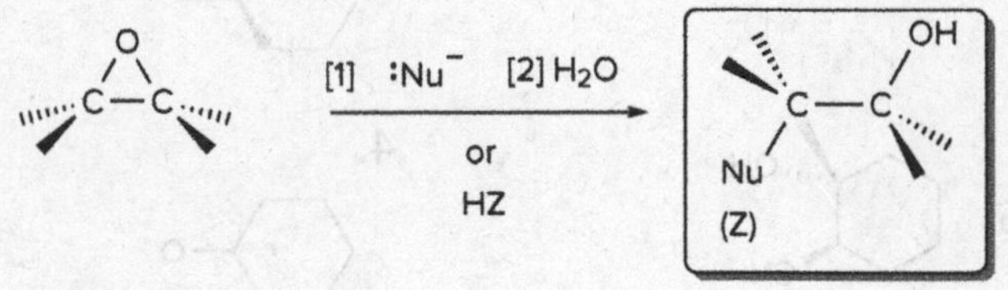

[1] :Nu⁻ [2] H_2O
or
HZ

OH
Nu
(Z)

- The reaction occurs with backside attack, resulting in trans or anti products.
- With :Nu⁻, the mechanism is S_N2, and nucleophilic attack occurs at the *less* substituted C.
- With HZ, the mechanism is between S_N1 and S_N2, and attack of Z⁻ occurs at the *more* substituted C.

Practice Test on Chapter Review

1. Give the IUPAC name for each of the following compounds.

a.

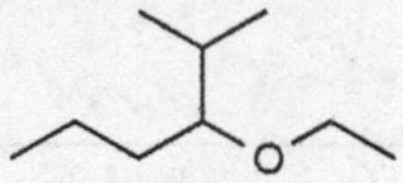

b.

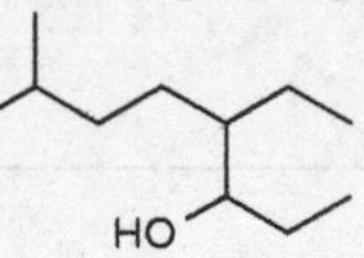

HO

2. Draw the organic products formed in each reaction. Draw all stereogenic centers using wedges and dashed wedges.

a. 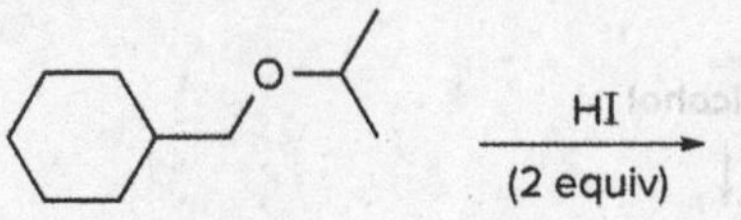 HI (2 equiv)

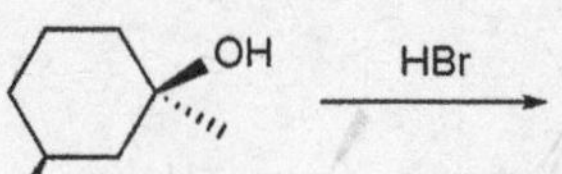

b. 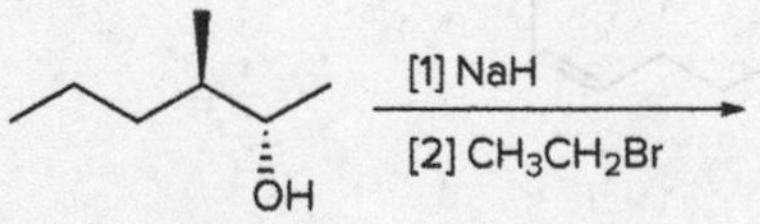 OH [1] NaH [2] CH_3CH_2Br

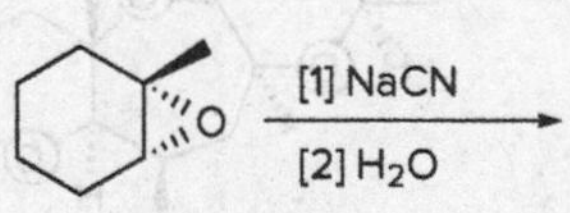

c. 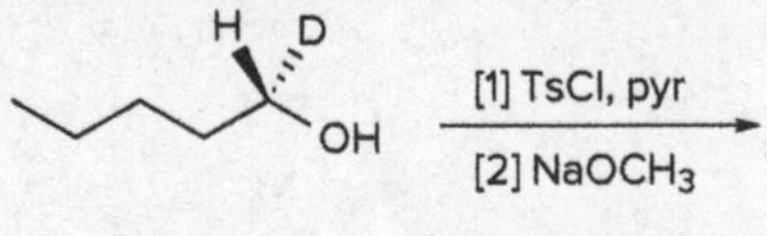 H D OH [1] TsCl, pyr [2] $NaOCH_3$

d. OH HBr

e. O [1] NaCN [2] H_2O

f. 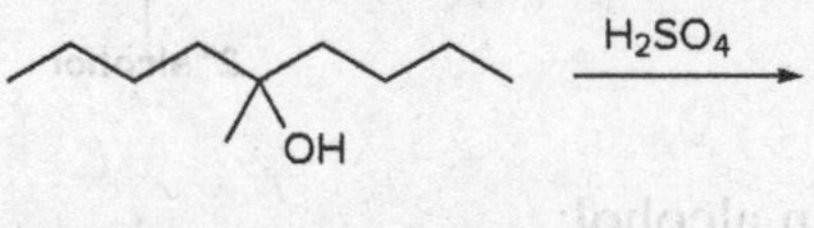 OH H_2SO_4

3. What starting material is needed for the following reaction?

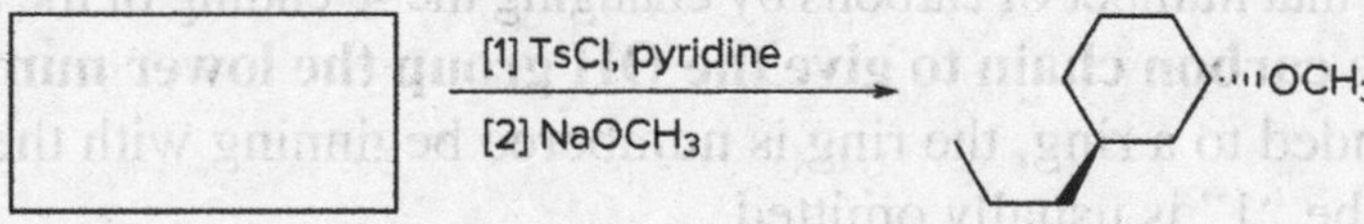

[1] TsCl, pyridine [2] $NaOCH_3$ → OCH_3

4. What alkoxide and alkyl halide are needed to make the following ether?

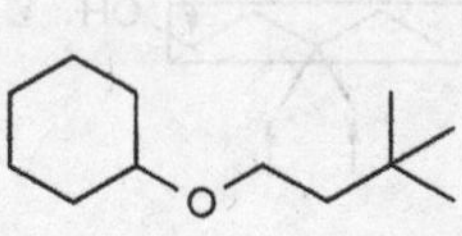

O

Answers to Practice Test

1. a. 3-ethoxy-2-methylhexane

   b. 4-ethyl-7-methyl-octan-3-ol

2. a. I + I

   b. OCH_2CH_3

   c. D H OCH_3

   d. Br + Br

   e. OH CN

   f. + (+ stereoisomer)

3. OH

4. O^- + X

Answers to Problems

9.1 **Alcohols** are classified as 1°, 2°, or 3°, depending on the number of carbon atoms bonded to the carbon with the OH group. Five ether oxygens are circled.

3° alcohol

OH

2° alcohol

9.2 To name an alcohol:

[1] **Find the longest chain that has the OH group as a substituent.** Name the molecule as a derivative of that number of carbons by changing the ***-e*** ending of the alkane to the suffix ***-ol.***

[2] **Number the carbon chain to give the OH group the lower number.** When the OH group is bonded to a ring, the ring is numbered beginning with the C bonded to the OH group, and the "1" is usually omitted.

[3] Apply the other rules of nomenclature to complete the name.

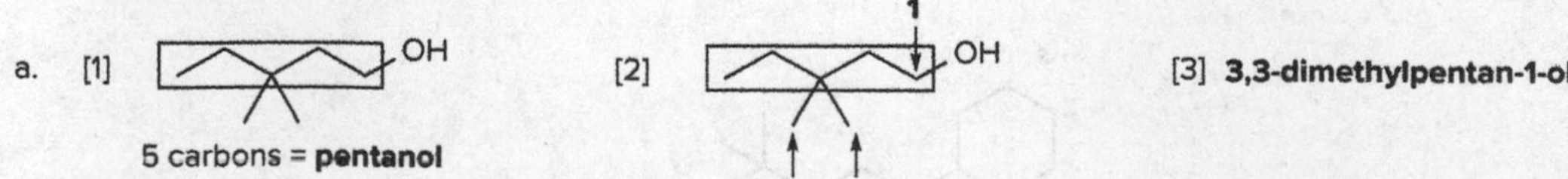

b. [1]

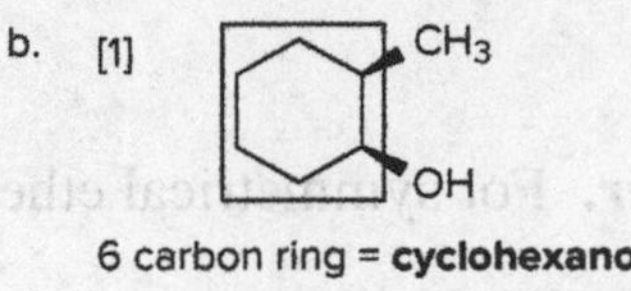

6 carbon ring = **cyclohexanol**

[2]

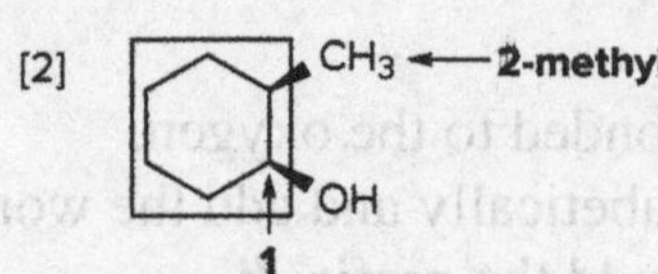

[3] ***cis*-2-methylcyclohexanol**

c. [1]

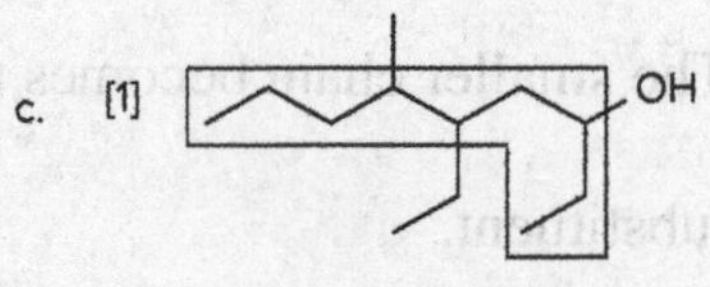

9 carbons = **nonanol**

[2]

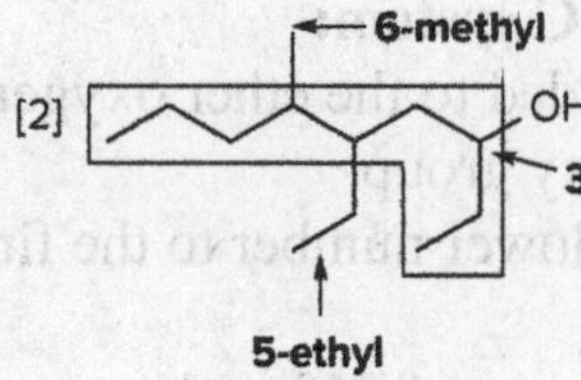

[3] **5-ethyl-6-methylnonan-3-ol**

d. [1]

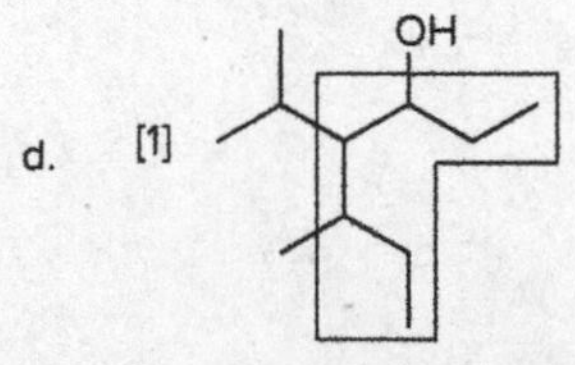

7 carbons = **heptanol**

[2]

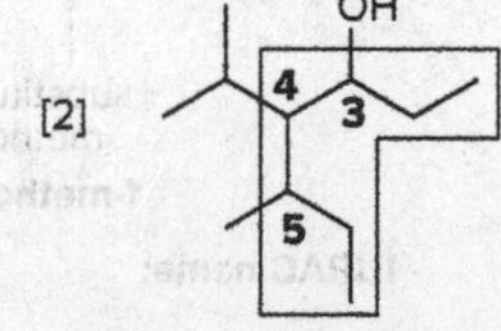

[3] **4-isopropyl-5-methylheptan-3-ol**

e. [1]

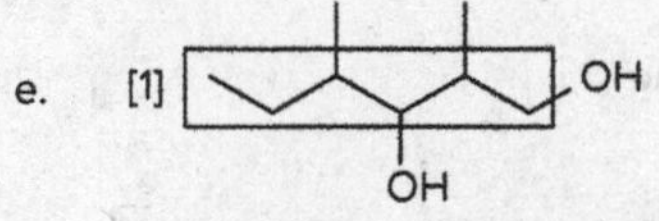

6 carbons = **hexanediol**

[2]

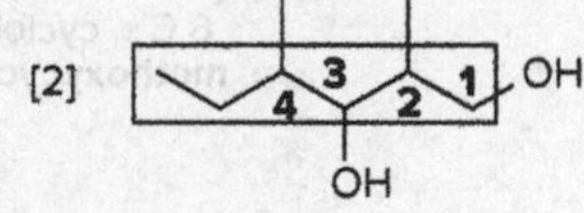

[3] **2,4-dimethylhexane-1,3-diol**

f. [1]

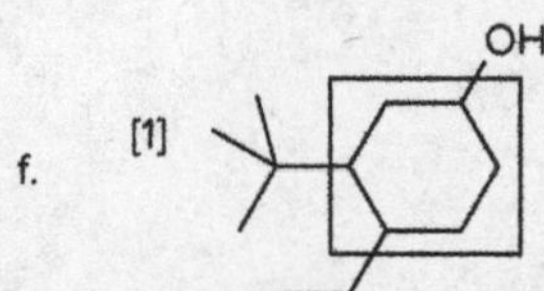

6 carbon ring = **cyclohexanol**

[2]

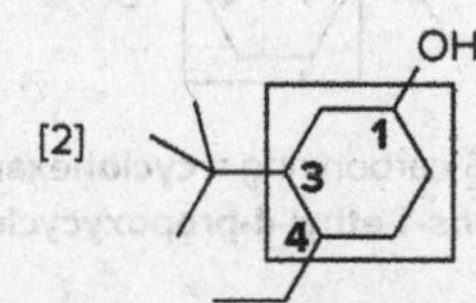

[3] **3-*tert*-butyl-4-ethylcyclohexanol**

9.3 To work backwards from a name to a structure:
[1] Find the parent name and draw its structure.
[2] Add the substituents to the long chain.

a. 7,7-dimethyl**octan-4-ol**

OH
4
7

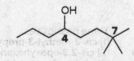

c. 2-*tert*-butyl-3-methyl**cyclohexanol**

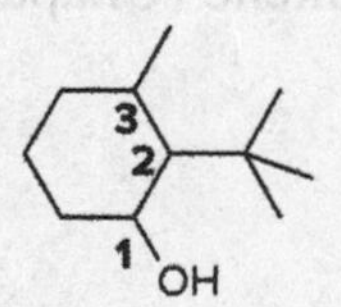

b. 5-methyl-4-propyl**heptan-3-ol**

3
4
5
OH

d. *trans*-**cyclohexane-1,2-diol**

OH
OH
or
OH
OH

9.4 To name simple ethers:

[1] Name both alkyl groups bonded to the oxygen.

[2] Arrange these names alphabetically and add the word ***ether.*** For symmetrical ethers, name the alkyl group and add the prefix ***di-.***

To name ethers using the IUPAC system:

[1] Find the two alkyl groups bonded to the ether oxygen. The smaller chain becomes the substituent, named as an alkoxy group.

[2] Number the chain to give the lower number to the first substituent.

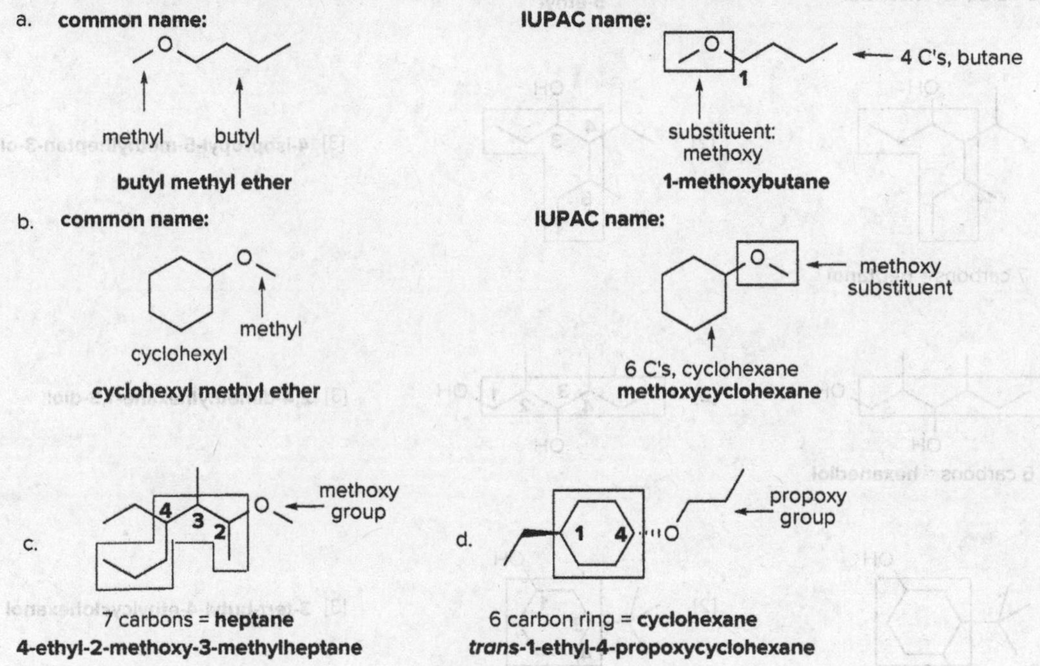

9.5 Three ways to name epoxides:

[1] Epoxides are named as derivatives of oxirane, the simplest epoxide.

[2] Epoxides can be named by considering the oxygen as a substituent called an **epoxy** group, bonded to a hydrocarbon chain or ring. Use two numbers to designate which two atoms the oxygen is bonded to.

[3] Epoxides can be named as **alkene oxides** by mentally replacing the epoxide oxygen by a double bond. Name the alkene (Chapter 10) and add the word ***oxide.***

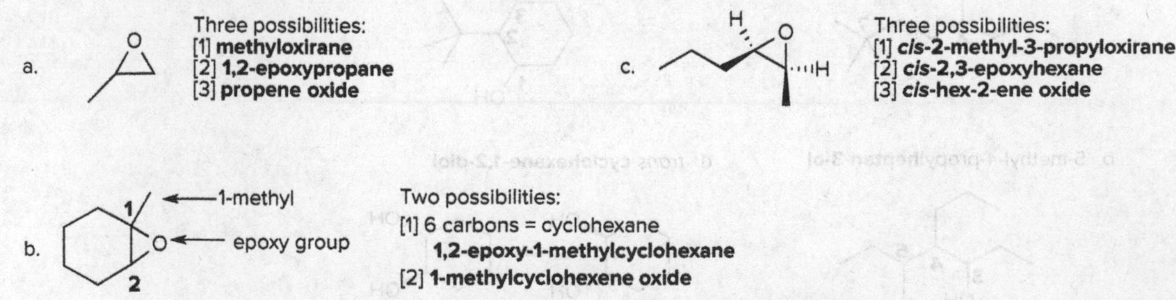

9.6 Two rules for boiling point:
[1] **The stronger the intermolecular forces the higher the bp.**
[2] **Bp increases as the extent of the hydrogen bonding increases.** For alcohols with the same number of carbon atoms: 3° ROH < 2° ROH < 1° ROH.

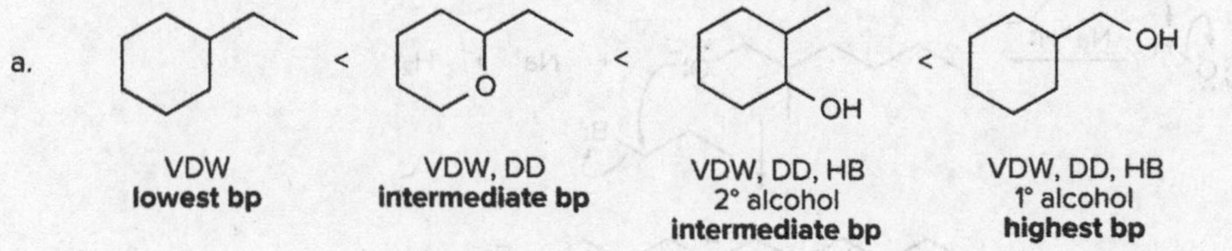

a.
VDW
lowest bp
VDW, DD
intermediate bp
OH
VDW, DD, HB
2° alcohol
intermediate bp
OH
VDW, DD, HB
1° alcohol
highest bp

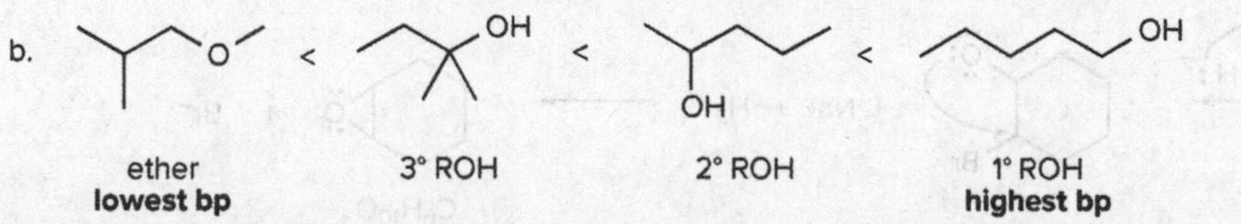

b.
O
ether
lowest bp
OH
3° ROH
OH
2° ROH
OH
1° ROH
highest bp

9.7 Draw the products of substitution in the following reactions by substituting OH or OR for X in the starting material.

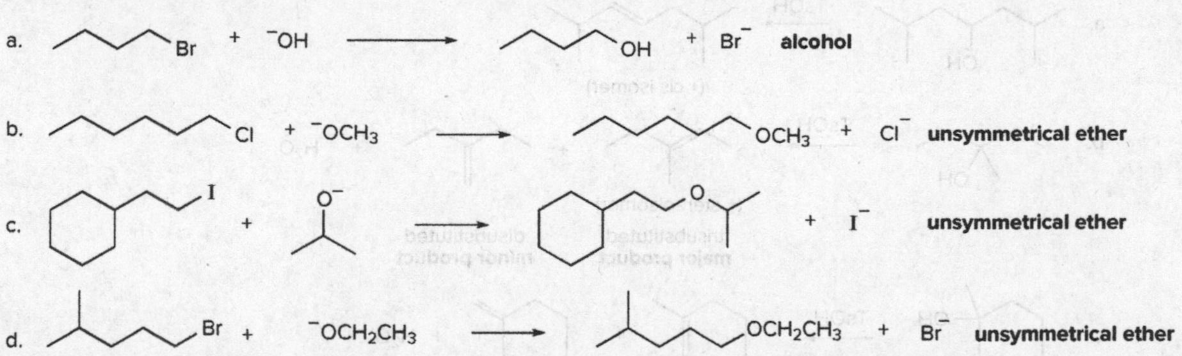

a. Br + ^-OH → OH + Br^- **alcohol**

b. Cl + $^-OCH_3$ → OCH_3 + Cl^- **unsymmetrical ether**

c. I + O^- → O + I^- **unsymmetrical ether**

d. Br + $^-OCH_2CH_3$ → OCH_2CH_3 + Br^- **unsymmetrical ether**

9.8 Two possible routes to **X** are shown. Path [2] with a 1° alkyl halide is preferred. Path [1] cannot occur because the leaving group would be bonded to an sp^2 hybridized C, making it an unreactive aryl halide.

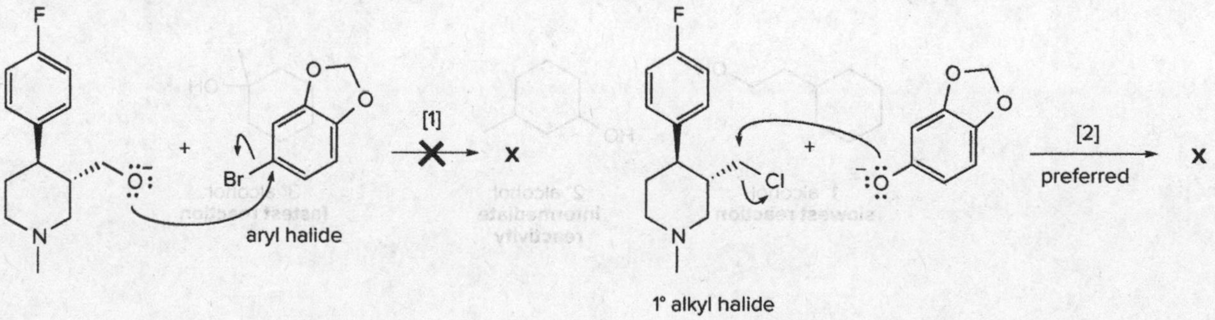

F
O
N
O
O
Br
aryl halide
[1]
X
F
Cl
N
1° alkyl halide
O
O
O
[2]
preferred
X

9.9 **NaH and $NaNH_2$ are strong bases that will remove a proton from an alcohol**, creating a nucleophile.

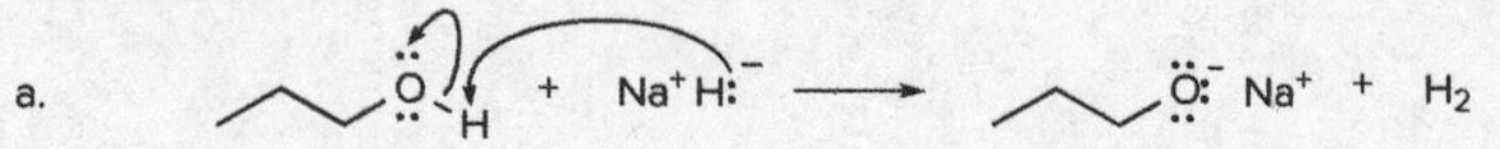

a. O H + $Na^+ H:^-$ → O^- Na^+ + H_2

b. cyclopentyl alcohol + $Na^+ :NH_2^-$ → alkoxide + Na^+ + NH_3

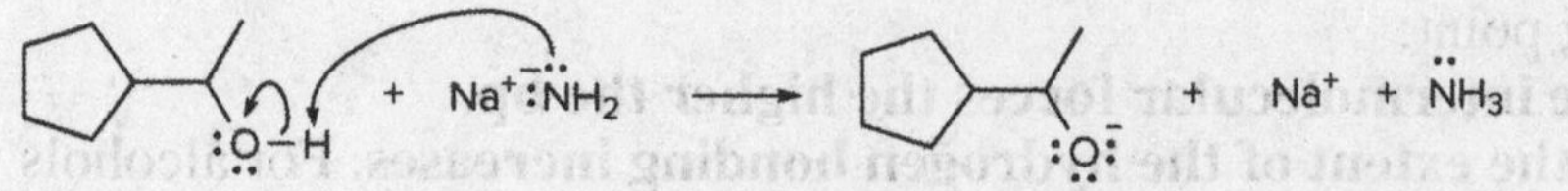

c. alcohol → $Na^+ H:^-$ → alkoxide + Na^+ + H_2; + Br → ether + Br^-

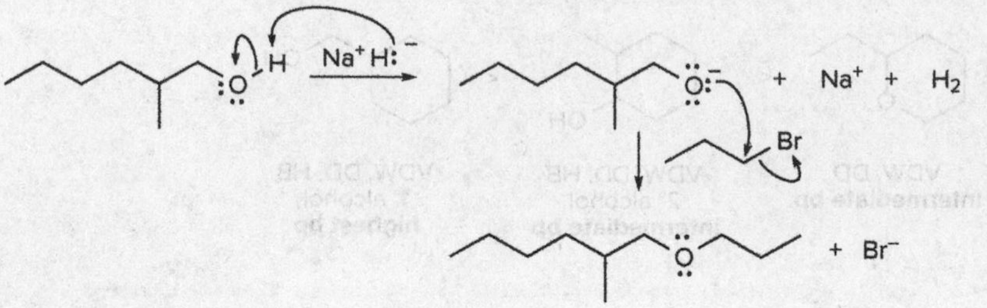

d. $Na^+ H:^-$ → alkoxide + Na^+ + H_2 → epoxide + Br^-

$C_6H_{10}O$

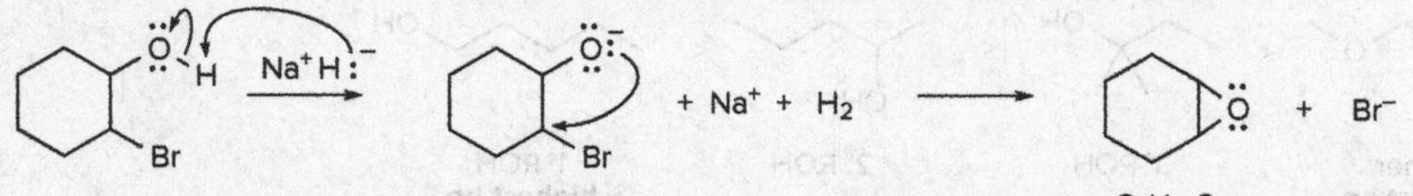

9.10 Dehydration follows the Zaitsev rule, so the more stable, more substituted alkene is the major product.

a. OH —TsOH→ (+ cis isomer)

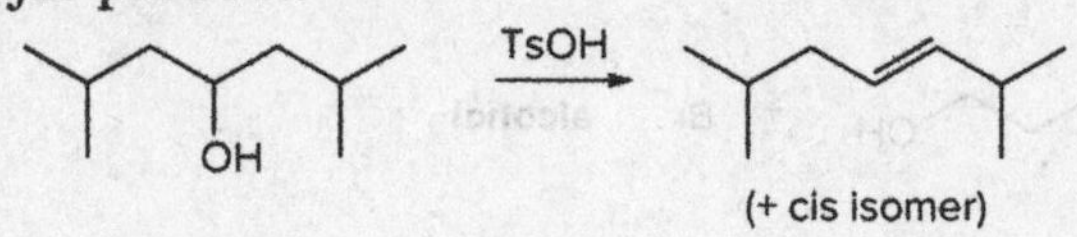

b. OH —TsOH→ + + H_2O

(+ stereoisomer)
trisubstituted
major product

disubstituted
minor product

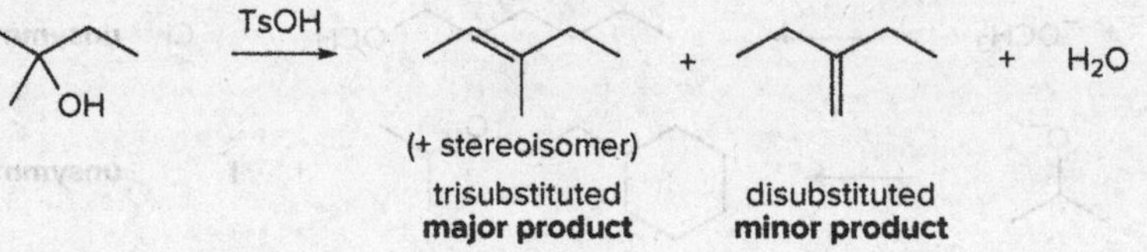

c. OH —TsOH→ + + H_2O

trisubstituted
major product

disubstituted
minor product

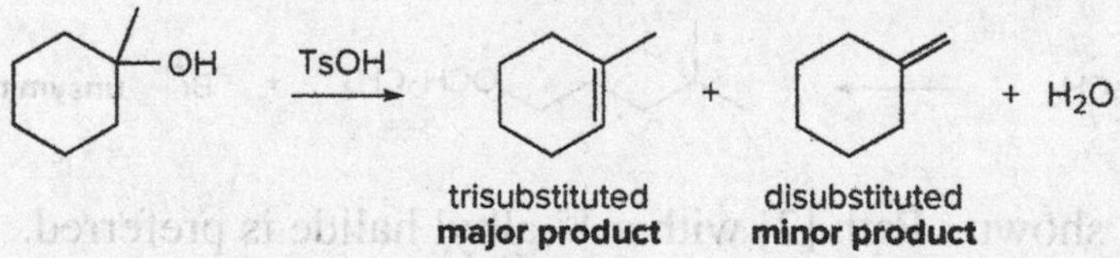

9.11 The rate of dehydration increases as the number of R groups increases.

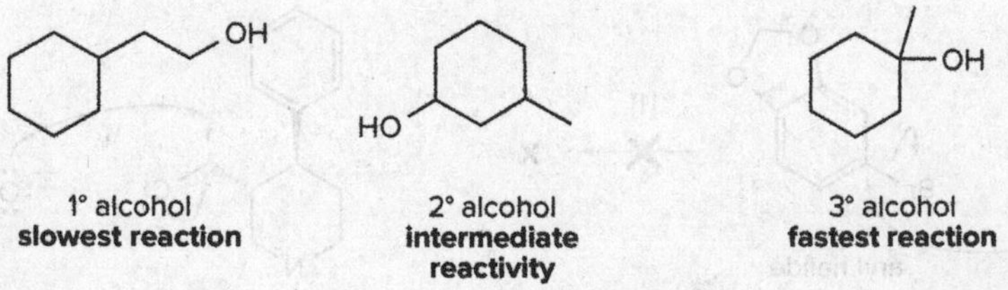

OH — 1° alcohol **slowest reaction**

HO — 2° alcohol **intermediate reactivity**

OH — 3° alcohol **fastest reaction**

9.12

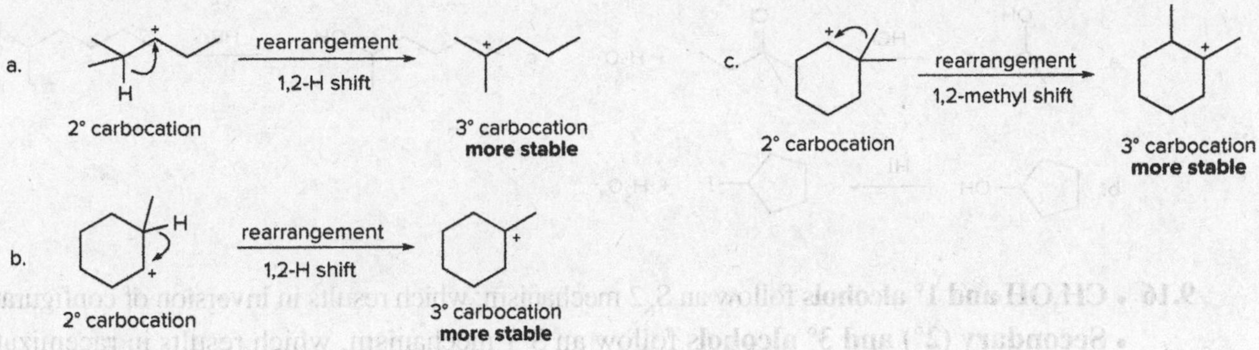

a. 2° carbocation → rearrangement, 1,2-H shift → 3° carbocation **more stable**

b. 2° carbocation → rearrangement, 1,2-H shift → 3° carbocation **more stable**

c. 2° carbocation → rearrangement, 1,2-methyl shift → 3° carbocation **more stable**

9.13

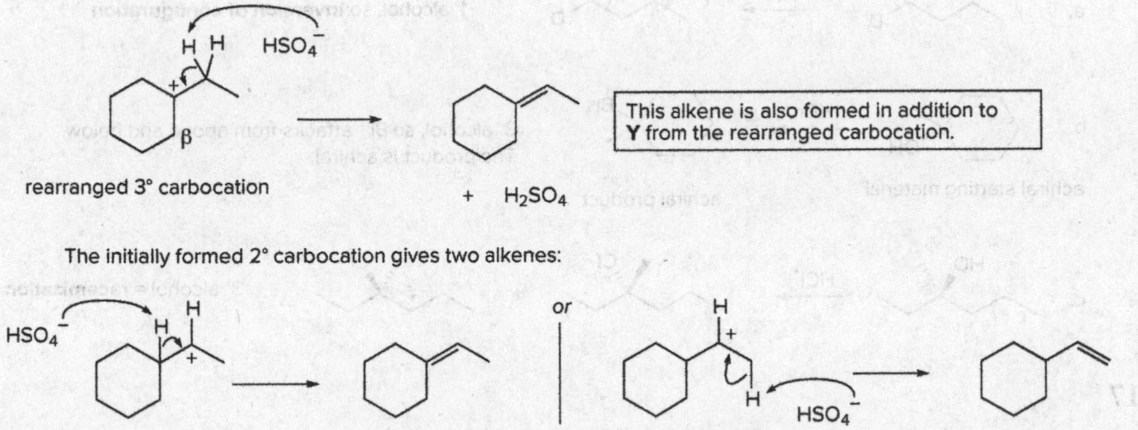

HSO_4^-

rearranged 3° carbocation → + H_2SO_4

This alkene is also formed in addition to **Y** from the rearranged carbocation.

The initially formed 2° carbocation gives two alkenes:

HSO_4^- → or → HSO_4^-

9.14

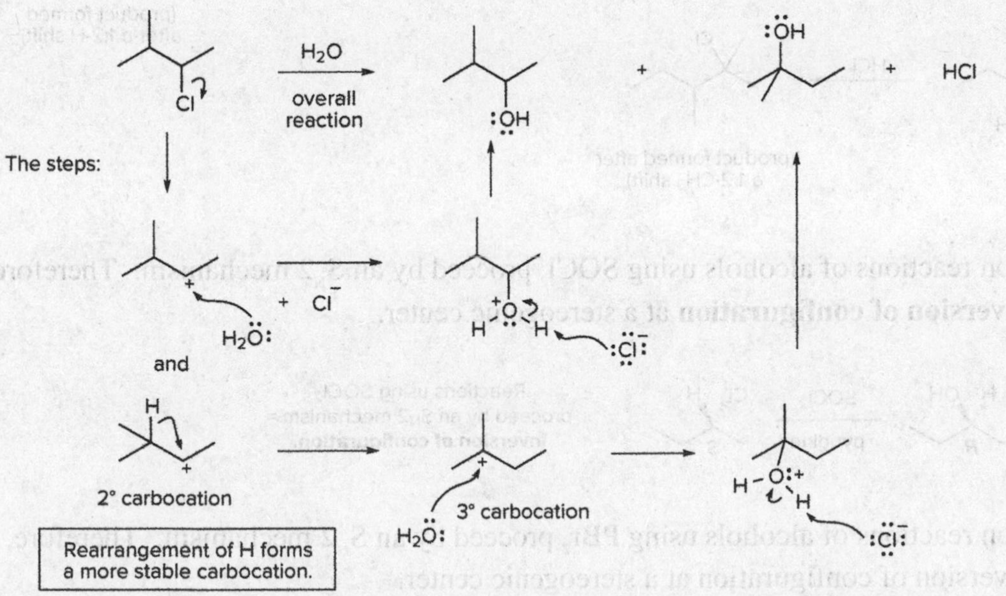

H_2O overall reaction → :OH + :OH + HCl

The steps:

+ Cl^- $H_2\ddot{O}$: :Cl:$^-$

and

2° carbocation → 3° carbocation $H_2\ddot{O}$: → H–O:+ H :Cl:

Rearrangement of H forms a more stable carbocation.

9.15

a. (2-methylbutan-2-ol) $\xrightarrow{HCl}$ (2-chloro-2-methylbutane) + H_2O

b. (cyclopentanol) $\xrightarrow{HI}$ (iodocyclopentane) + H_2O

c. (alcohol) $\xrightarrow{HBr}$ (alkyl bromide) + H_2O

9.16 • **CH_3OH and 1° alcohols** follow an S_N2 mechanism, which results in inversion of configuration.
• **Secondary (2°) and 3° alcohols** follow an S_N1 mechanism, which results in racemization at a stereogenic center.

a. OH, D $\xrightarrow{HI}$ I, D — 1° alcohol, so **inversion of configuration**

b. OH $\xrightarrow{HBr}$ Br — 3° alcohol, so Br^- attacks from above and below. The product is achiral.

achiral starting material — achiral product

c. HO $\xrightarrow{HCl}$ Cl + Cl — 3° alcohol = **racemization**

9.17

a. OH $\xrightarrow{HCl}$ Cl

b. OH $\xrightarrow{HCl}$ Cl (product formed after a 1,2-CH_3 shift)

c. OH $\xrightarrow{HCl}$ Cl (product formed after a 1,2-H shift)

9.18 Substitution reactions of alcohols using $SOCl_2$ proceed by an S_N2 mechanism. Therefore, there is **inversion of configuration** at a stereogenic center.

H OH, *R* $\xrightarrow[\text{pyridine}]{SOCl_2}$ Cl H, *S* — Reactions using $SOCl_2$ proceed by an S_N2 mechanism = **inversion of configuration.**

9.19 Substitution reactions of alcohols using PBr_3 proceed by an S_N2 mechanism. Therefore, there is inversion of configuration at a stereogenic center.

H OH, *R* $\xrightarrow{PBr_3}$ Br H, *S* — Reactions using PBr_3 proceed by an S_N2 mechanism = **inversion of configuration.**

9.20 Stereochemistry for conversion of ROH to RX by reagent:

[1] **HX**—with 1°, S_N2, so inversion of configuration; with 2° and 3°, S_N1, so racemization.

[2] **$SOCl_2$**—S_N2, so inversion of configuration.

[3] **PBr_3**—S_N2, so inversion of configuration.

a. OH —$SOCl_2$, pyridine→ Cl

c. OH —PBr_3→ Br

S_N2 = **inversion**

b. OH —HI→ I + I

3° alcohol, S_N1 = **racemization**

9.21

a. OH + SO_2Cl —pyridine→ O–S(=O)(=O)– + Cl^-

b. OH —TsCl, pyridine→ OTs + Cl^-

9.22

a. OTs + ^-CN —**S_N2**→ CN + ^-OTs

1° tosylate; strong nucleophile

b. OTs + K^+ $^-OC(CH_3)_3$ —**E2**→ + K^+ ^-OTs + $HOC(CH_3)_3$

1° tosylate; strong bulky base

c. OTs + ^-SH —**S_N2**→ SH

2° tosylate; strong nucleophile

S_N2 product (inversion of configuration)

(Substitution is favored over elimination.)

d. OTs —$NaOCH_2CH_3$, **E2**→ + +

major products **enantiomers**; minor product

9.23

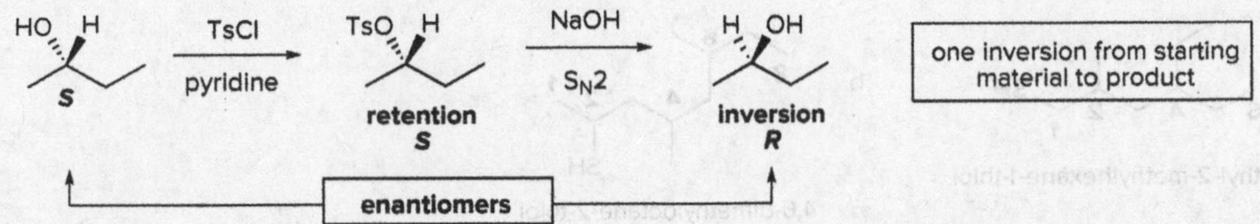

9.24 These reagents can be classified as follows:
[1] $SOCl_2$, PBr_3, HCl, and HBr replace OH with X by a substitution reaction.
[2] Tosyl chloride (TsCl) makes OH a better leaving group by converting it to OTs.
[3] Strong acids (H_2SO_4) and $POCl_3$ (pyridine) result in elimination by dehydration.

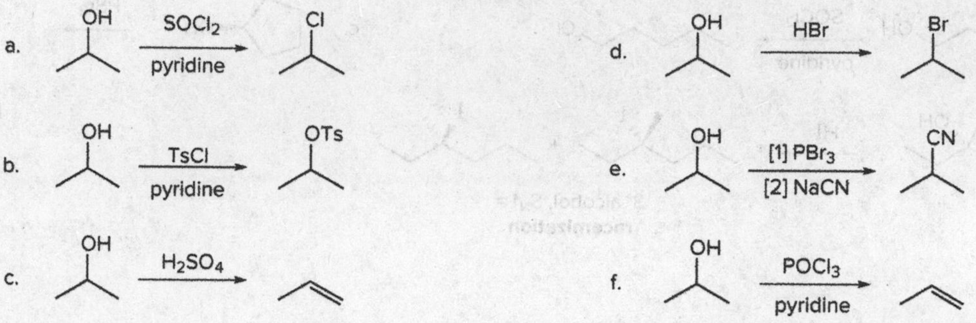

a. OH → ($SOCl_2$, pyridine) → Cl
d. OH → (HBr) → Br
b. OH → (TsCl, pyridine) → OTs
e. OH → ([1] PBr_3; [2] NaCN) → CN
c. OH → (H_2SO_4) →
f. OH → ($POCl_3$, pyridine) →

9.25

a. → (HBr) → 2 Br + H_2O
c. → (HBr) → Br + Br + H_2O
b. → (HBr) → Br + Br + H_2O

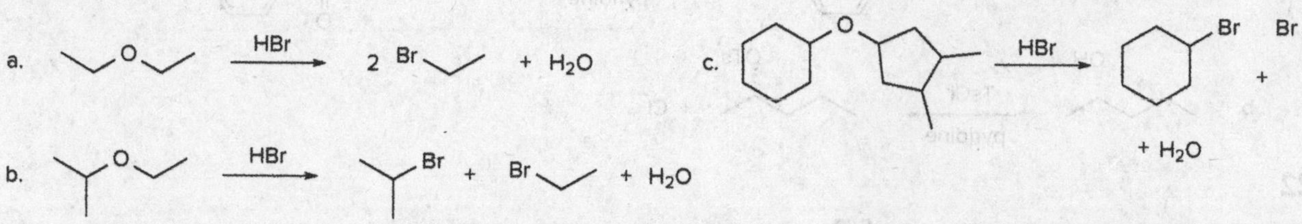

9.26 Ether cleavage can occur by either an S_N1 or S_N2 mechanism, but neither mechanism can occur when the ether O atom is bonded to an aromatic ring. An S_N1 reaction would require formation of a highly unstable carbocation on a benzene ring, a process that does not occur. An S_N2 reaction would require backside attack through the plane of the aromatic ring, which is also not possible. Thus, cleavage of the Ph–OCH_3 bond does not occur.

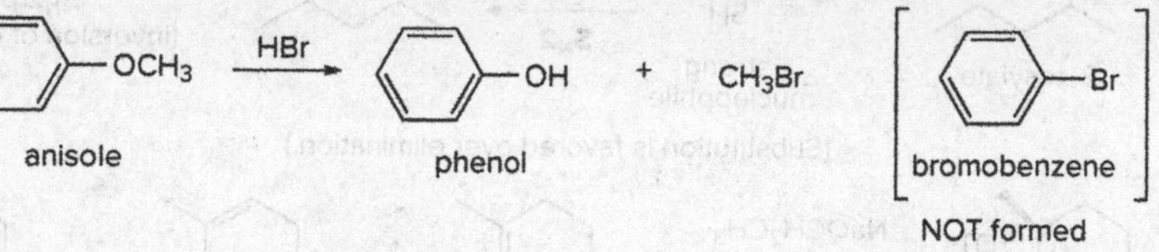

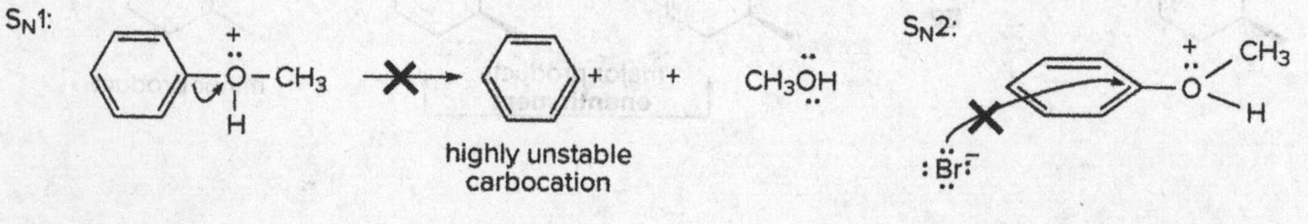

anisole –OCH_3 → (HBr) → phenol –OH + CH_3Br [bromobenzene –Br NOT formed]

S_N1: –$\ddot{O}^+$(CH_3)–H ✗→ highly unstable carbocation + $CH_3\ddot{O}H$

S_N2: :Br:⁻ ✗ → –$\ddot{O}^+$(CH_3)–H

9.27

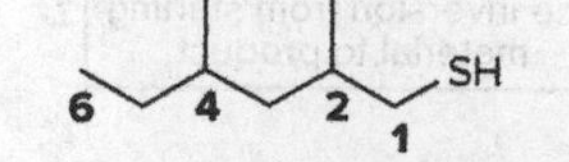

a. 6 4 2 SH 1
4-ethyl-2-methylhexane-1-thiol

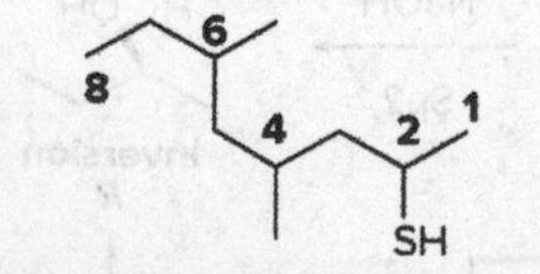

b. 8 6 4 2 1 SH
4,6-dimethyloctane-2-thiol

9.28

a. Br —NaSH→ SH

c. grapefruit mercaptan (SH) —Br_2→ S–S

b. D, Cl —NaSH→ HS, D
backside attack

d. S–S —Zn, HCl→ 2 SH

9.29

a. S, 1
1-ethylthiobutane
or
butyl ethyl sulfide

b. 1, 2, S
2-methyl-1-methylthiocyclopentane

9.30

a. SH —[1] NaH, [2] CH_3Br→ S

b. S + Cl → $\overset{+}{S}$ + Cl^-

9.31 Two rules for the reaction of an epoxide:
[1] Nucleophiles attack from the **back side** of the epoxide.
[2] Negatively charged nucleophiles attack at the **less substituted carbon.**

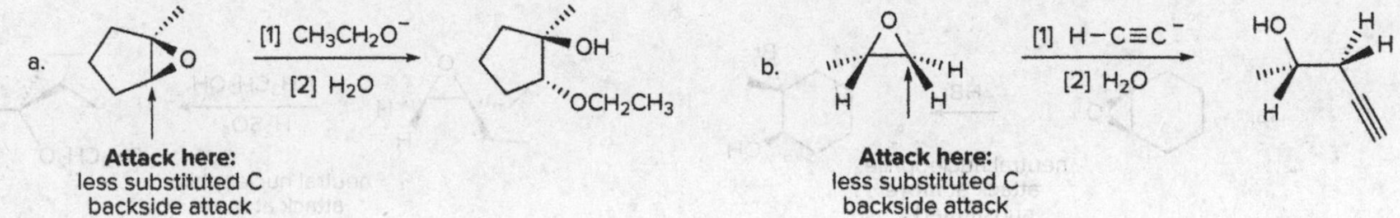

9.32 In both isomers, ^{-}OH attacks from the back side at either C–O bond.

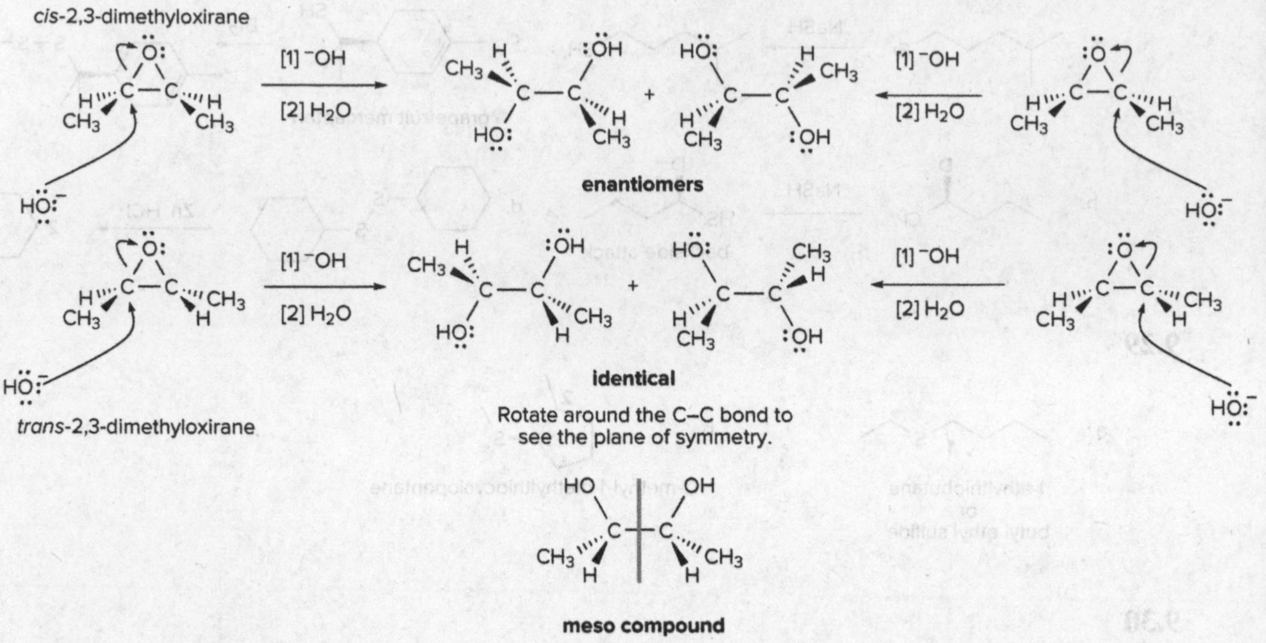

cis-2,3-dimethyloxirane

[1] ^{-}OH [2] H_2O

enantiomers

identical

Rotate around the C–C bond to see the plane of symmetry.

meso compound

trans-2,3-dimethyloxirane

9.33 Remember the difference between negatively charged nucleophiles and neutral nucleophiles:

- **Negatively charged nucleophiles attack first,** followed by protonation, and the nucleophile attacks at the ***less*** **substituted carbon.**
- **Neutral nucleophiles have protonation first,** followed by nucleophilic attack at the ***more*** **substituted carbon.**
- **Trans or anti products are always formed, regardless of the nucleophile.**

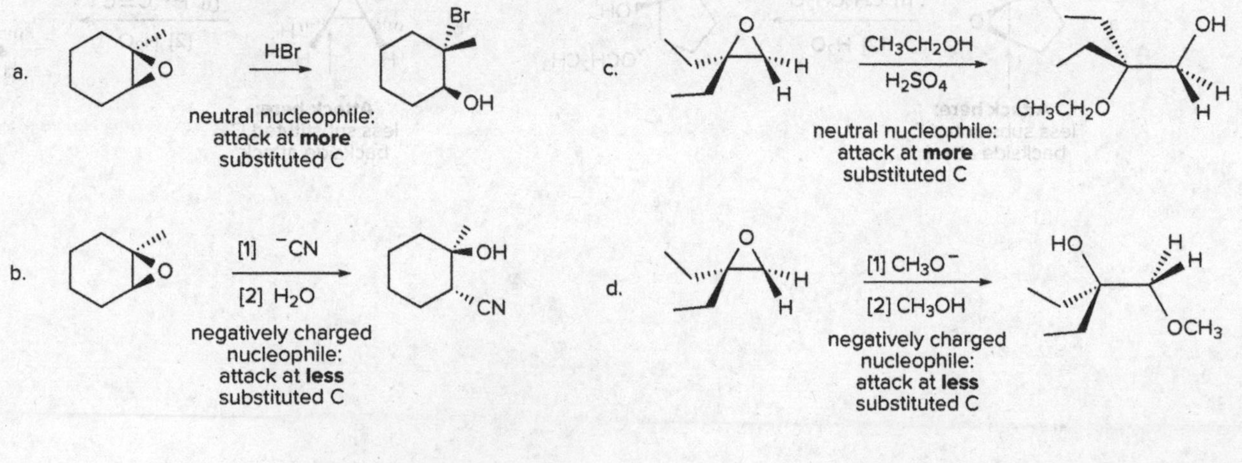

a. HBr — neutral nucleophile: attack at **more** substituted C

b. [1] ^{-}CN [2] H_2O — negatively charged nucleophile: attack at **less** substituted C

c. CH_3CH_2OH, H_2SO_4 — neutral nucleophile: attack at **more** substituted C

d. [1] CH_3O^- [2] CH_3OH — negatively charged nucleophile: attack at **less** substituted C

9.34

a. (1*R*,2*R*)-2-Isobutylcyclopentanol

b. 2° alcohol

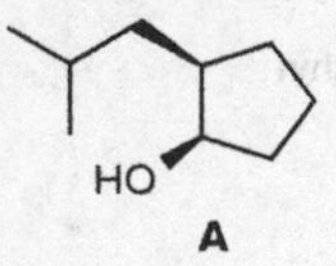

HO

A

c. stereoisomer

HO

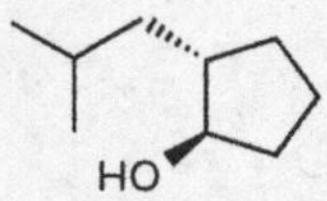

(1*R*,2*S*)-2-Isobutylcyclopentanol

d. constitutional isomer

OH

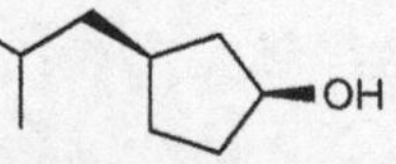

(1*S*,3*S*)-3-Isobutylcyclopentanol

e. constitutional isomer with an ether

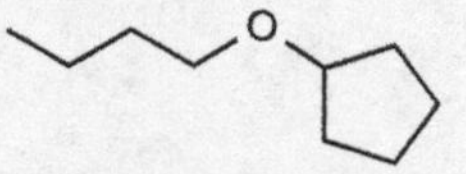

O

butoxycyclopentane

f.

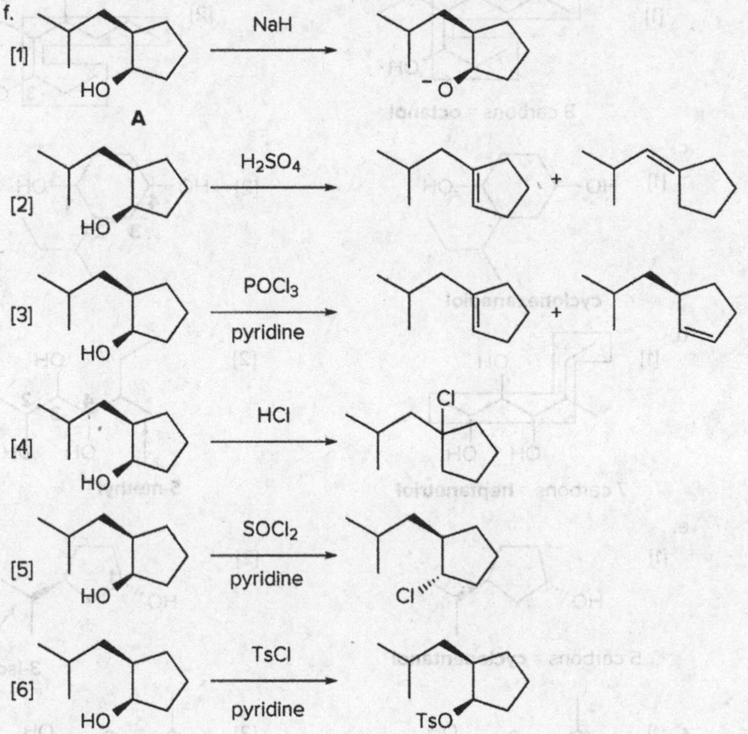

[1] HO, A — NaH → $^-$O

[2] HO — H_2SO_4 → (alkene) + (alkene)

[3] HO — $POCl_3$, pyridine → (alkene) + (alkene)

[4] HO — HCl → Cl

[5] HO — $SOCl_2$, pyridine → Cl

[6] HO — TsCl, pyridine → TsO

9.35

a. HO H — HBr → Br H (*S*) + H Br (*R*) — 2° alcohol, **S_N1 = racemization**

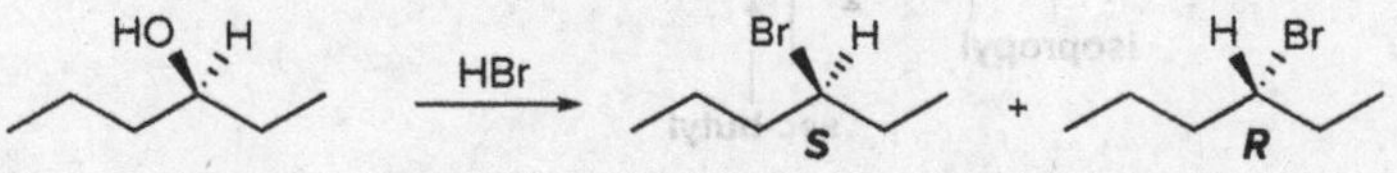

b. HO H — PBr_3 → H Br (*R*) — PBr_3 follows **S_N2 = inversion.**

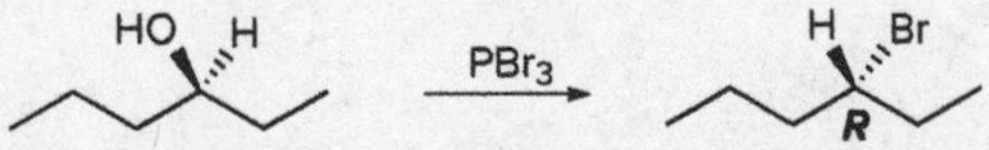

c. HO H — HCl → Cl H (*S*) + H Cl (*R*) — 2° alcohol, **S_N1 = racemization**

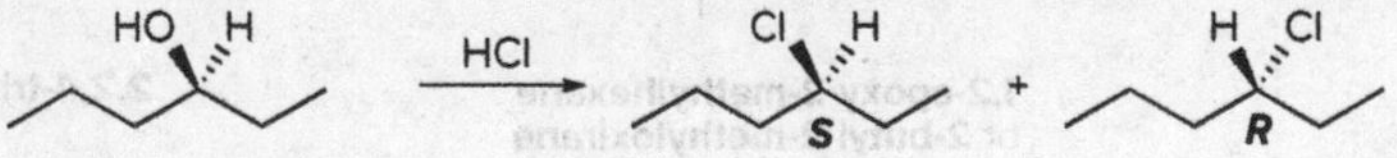

d. HO H — $SOCl_2$, pyridine → H Cl (*R*) — $SOCl_2$ follows **S_N2 = inversion.**

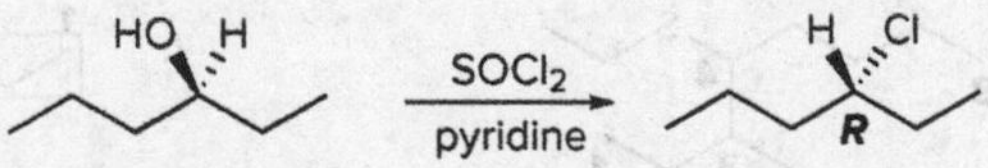

9.36 Use the directions from Answer 9.2.

a.

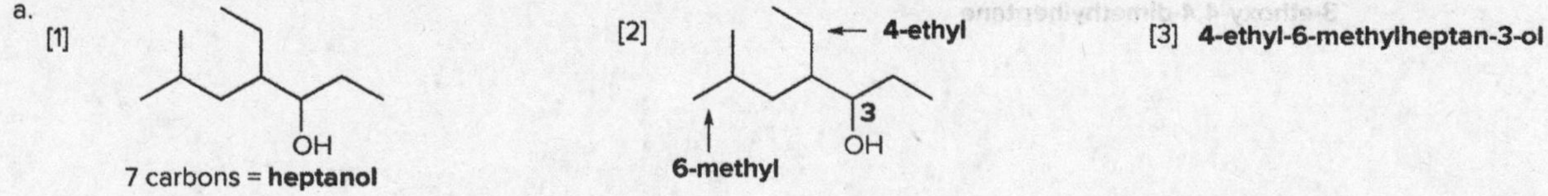

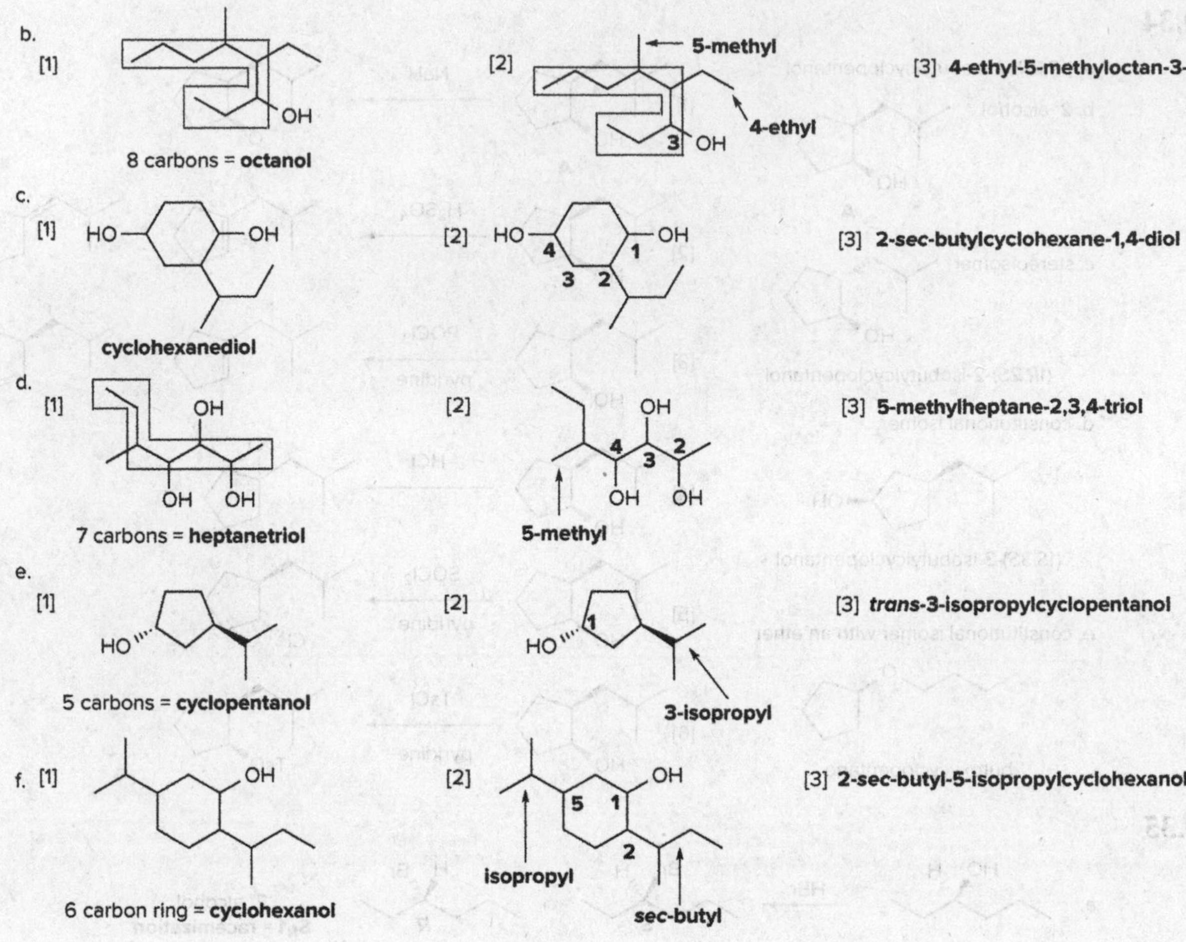

b.
[1]
OH
8 carbons = octanol
[2]
5-methyl
4-ethyl
3 OH
[3] 4-ethyl-5-methyloctan-3-ol
c.
[1] HO OH
cyclohexanediol
[2] HO OH
4 1
3 2
[3] 2-sec-butylcyclohexane-1,4-diol
d.
[1]
OH
OH OH
7 carbons = heptanetriol
[2]
OH
4 2
3
OH OH
5-methyl
[3] 5-methylheptane-2,3,4-triol
e.
[1]
HO
5 carbons = cyclopentanol
[2]
HO 1
3-isopropyl
[3] trans-3-isopropylcyclopentanol
f. [1]
OH
6 carbon ring = cyclohexanol
[2]
OH
5 1
2
isopropyl
sec-butyl
[3] 2-sec-butyl-5-isopropylcyclohexanol

9.37 Use the rules from Answers 9.4 and 9.5.

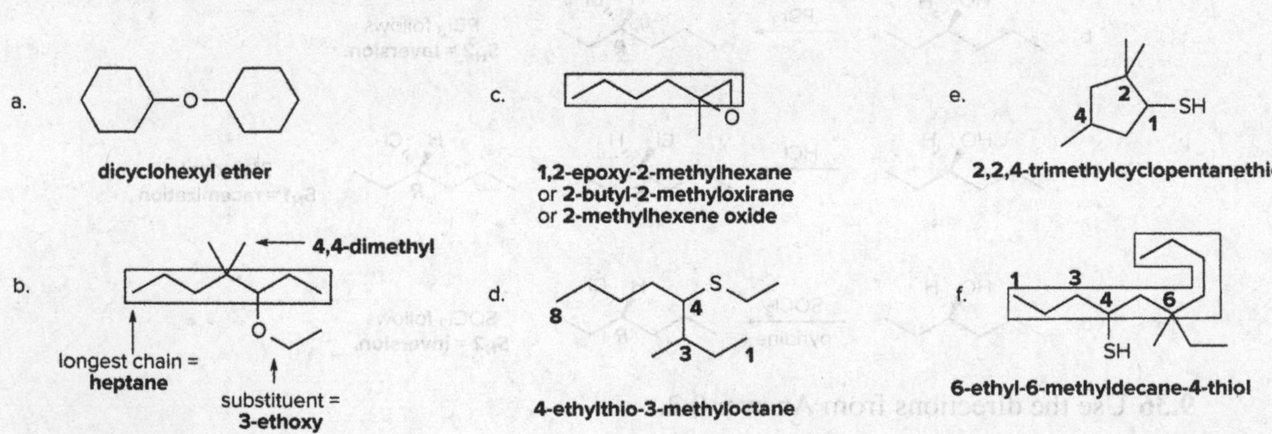

a.
O
dicyclohexyl ether
b.
4,4-dimethyl
O
longest chain =
heptane
substituent =
3-ethoxy
3-ethoxy-4,4-dimethylheptane
c.
O
1,2-epoxy-2-methylhexane
or 2-butyl-2-methyloxirane
or 2-methylhexene oxide
d.
S
8
4
3
1
4-ethylthio-3-methyloctane
e.
2
SH
4
1
2,2,4-trimethylcyclopentanethiol
f.
1
3
4
6
SH
6-ethyl-6-methyldecane-4-thiol

9.38 Use the directions from Answer 9.3.

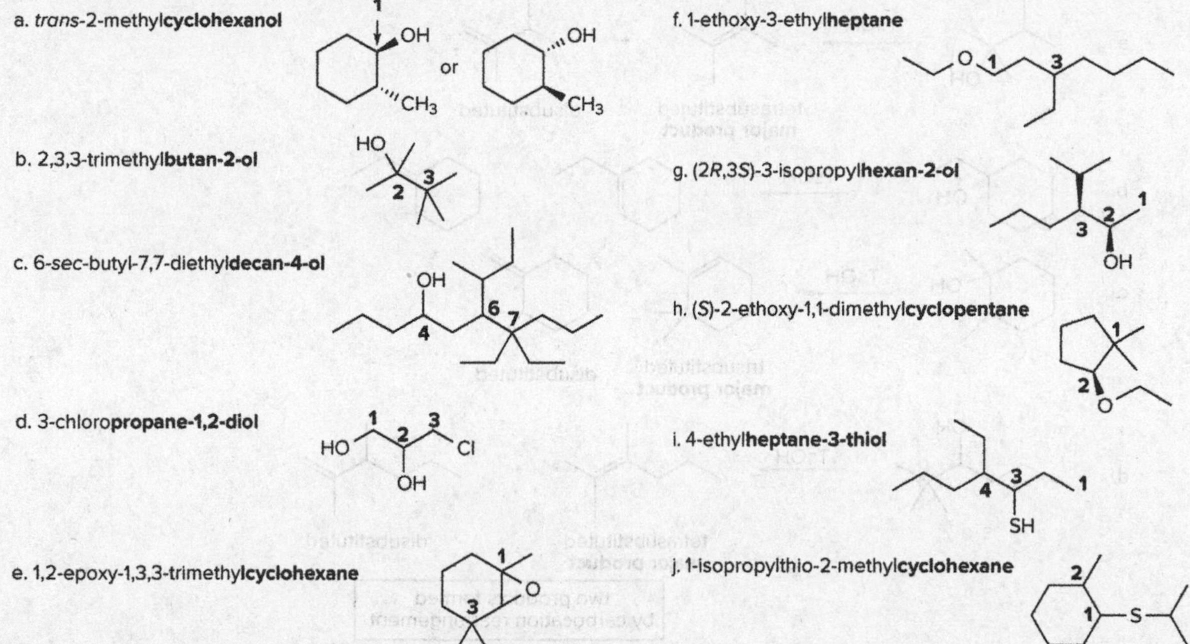

9.39 Stronger intermolecular forces increase boiling point. All of the compounds can hydrogen bond, but both diols have more opportunity for hydrogen bonding because they have two OH groups, making their bp's higher than the bp of butan-1-ol. Propane-1,2-diol can also intramolecularly hydrogen bond. Intramolecular hydrogen bonding decreases the amount of intermolecular hydrogen bonding, so the bp of propane-1,2-diol is somewhat lower.

Increasing boiling point →

butan-1-ol	propane-1,2-diol	propane-1,3-diol
118 °C	187 °C	215 °C

9.40

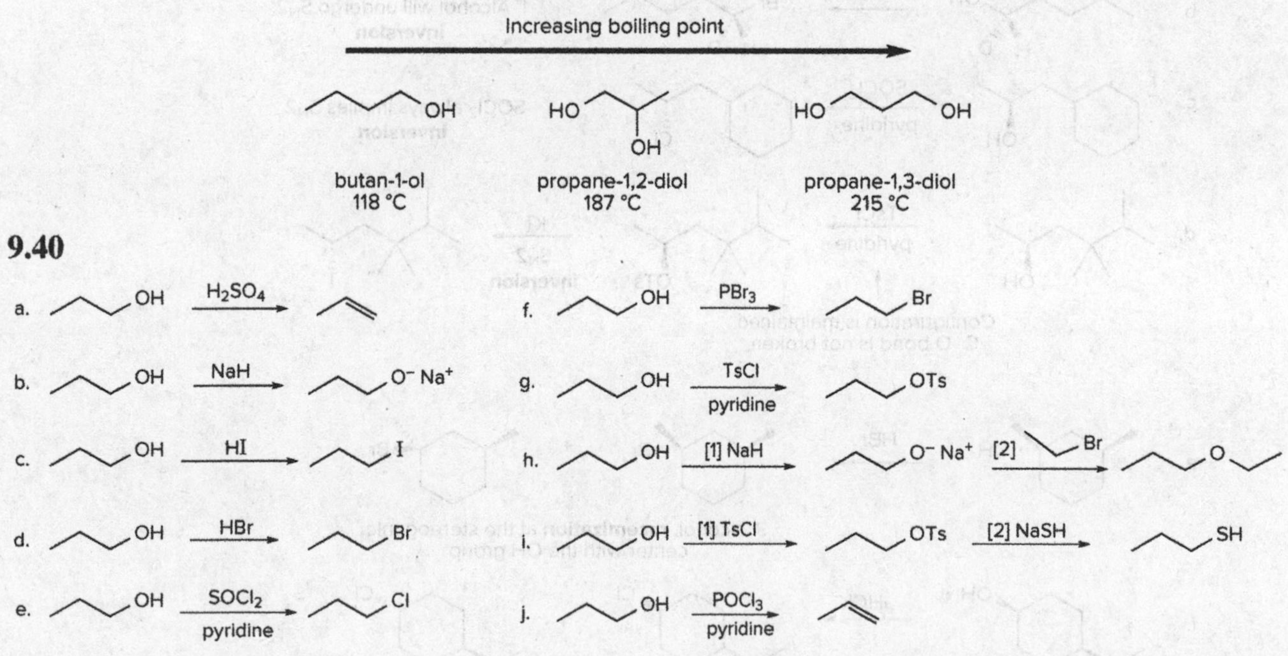

9.41 Dehydration follows the Zaitsev rule, so the more stable, more substituted alkene is the major product.

a. OH — TsOH → +
tetrasubstituted **major product** — disubstituted

b. OH — TsOH → +

c. OH — TsOH → +
trisubstituted **major product** — disubstituted

d. OH — TsOH → +
tetrasubstituted **major product** — disubstituted

two products formed by carbocation rearrangement

9.42

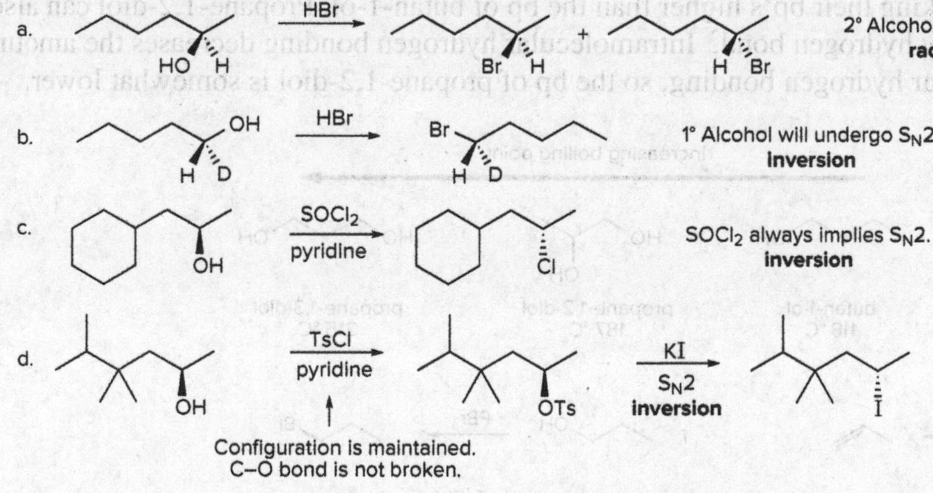

e. OH — HBr → Br + Br

3° alcohol, **racemization** at the stereogenic center with the OH group

f. OH — HCl → Cl + Cl

2° alcohol, **racemization** at the stereogenic center with the OH group

9.43

(R)-hexan-2-ol

(a) NaH → A = ; CH_3I → B =

(b) TsCl, pyridine → C = ; CH_3O^- → D =

(c) PBr_3 → E = ; CH_3O^- → F =

Routes (a) and (c) give identical products, labeled **B** and **F.**

9.44

a. A; H_2SO_4 ($-H_2O$) → 2° carbocation; 1,2-H shift → 3° carbocation → major product +

b. A; $POCl_3$, pyridine, E2 →

H and OH must be trans in the E2 reaction.

The major products are different because the mechanisms are different. With H_2SO_4, the reaction proceeds by an E1 mechanism involving a carbocation rearrangement. With $POCl_3$, the mechanism is E2 and the H and OH must be trans diaxial. No H on the C with the CH_3 group is trans to the OH group, so only one product forms.

9.45 1,2-Shift of a ring carbon leads to ring expansion, so two products with six-membered rings are formed.

H–Cl → 2° carbocation

→ 3° carbocation; $:Cl:^-$ → Cl

or → 3° carbocation; $:Cl:^-$ → Cl

or → 3° carbocation; $:Cl:^-$ → Cl

9.46 Acid-catalyzed dehydration follows an E1 mechanism for 2° and 3° ROH with an added step to make a good leaving group. The three steps are:
[1] Protonate the oxygen to make a good leaving group.
[2] Break the C–O bond to form a carbocation.
[3] Remove a β hydrogen to form the π bond.

overall reaction
The steps:
$H-OSO_3H$ + H_2O
+ HSO_4^- HSO_4^- + H_2O + H_2SO_4
2° carbocation
and
1,2-H shift + HSO_4^- + H_2SO_4
2° carbocation 3° carbocation
and + HSO_4^- + H_2SO_4

9.47 To draw the mechanism:
[1] Protonate the oxygen to make a good leaving group.
[2] Break the C–O bond to form a carbocation.
[3] Look for possible rearrangements to make a more stable carbocation.
[4] Remove a β hydrogen to form the π bond.

Dark and light circles are meant to show where the carbons in the starting material appear in the product.

$H-OSO_3H$ + HSO_4^- 2° carbocation + H_2O : + HSO_4^- 3° carbocation + H_2SO_4

9.48

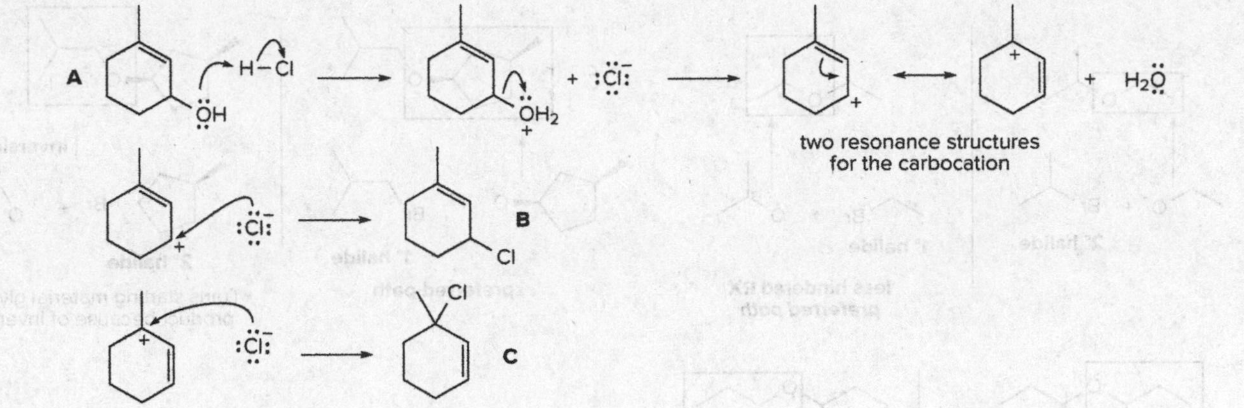

A + H–Cl → $\ddot{O}H_2^+$ + :Cl:⁻ → + $H_2\ddot{O}$

two resonance structures for the carcocation

:Cl:⁻ → B (Cl)

:Cl:⁻ → C (Cl)

9.49

H–OSO$_3$H → +$\ddot{O}H_2$ + HSO_4^- → + $H_2\ddot{O}$ —1,2-shift→ + HSO_4^- → + H_2SO_4

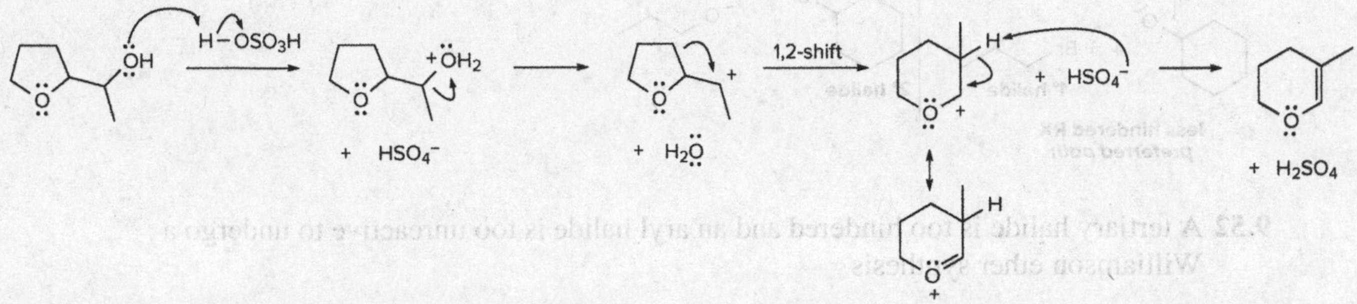

9.50 Elimination of the OH (as H_2O) adjacent to the five-membered ring forms a resonance-stabilized carbocation.

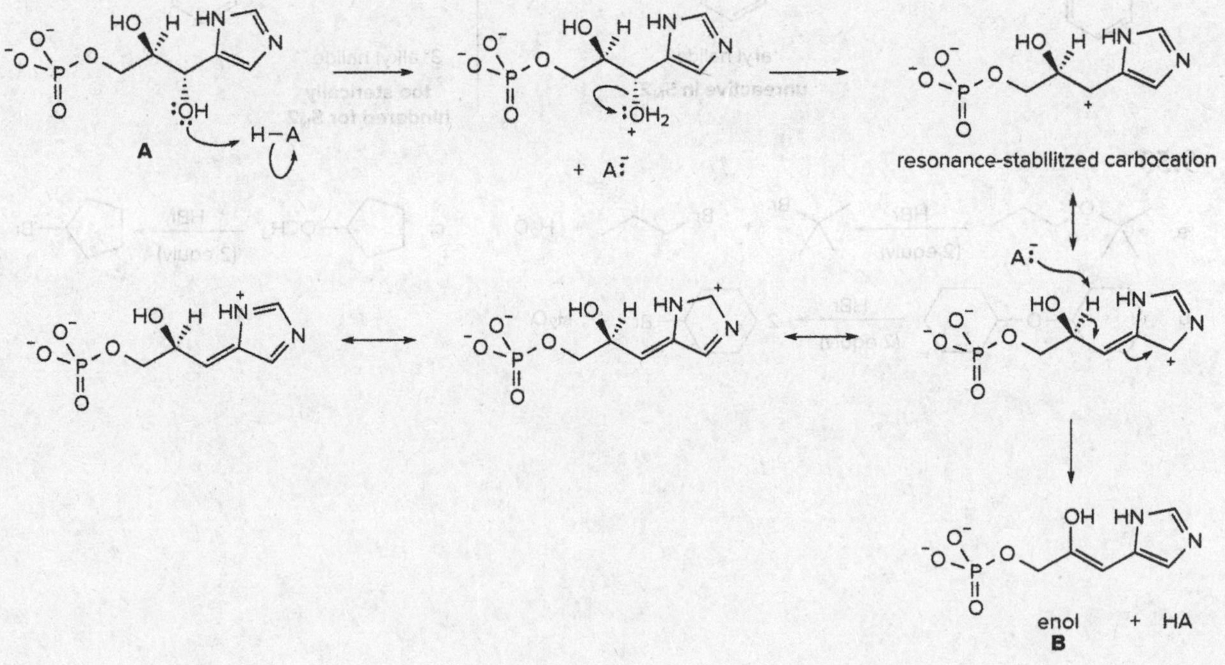

A + H–A → :$\ddot{O}H_2$ + A:⁻ → resonance-stabilitzed carbocation

A:⁻ → enol B + HA

9.51

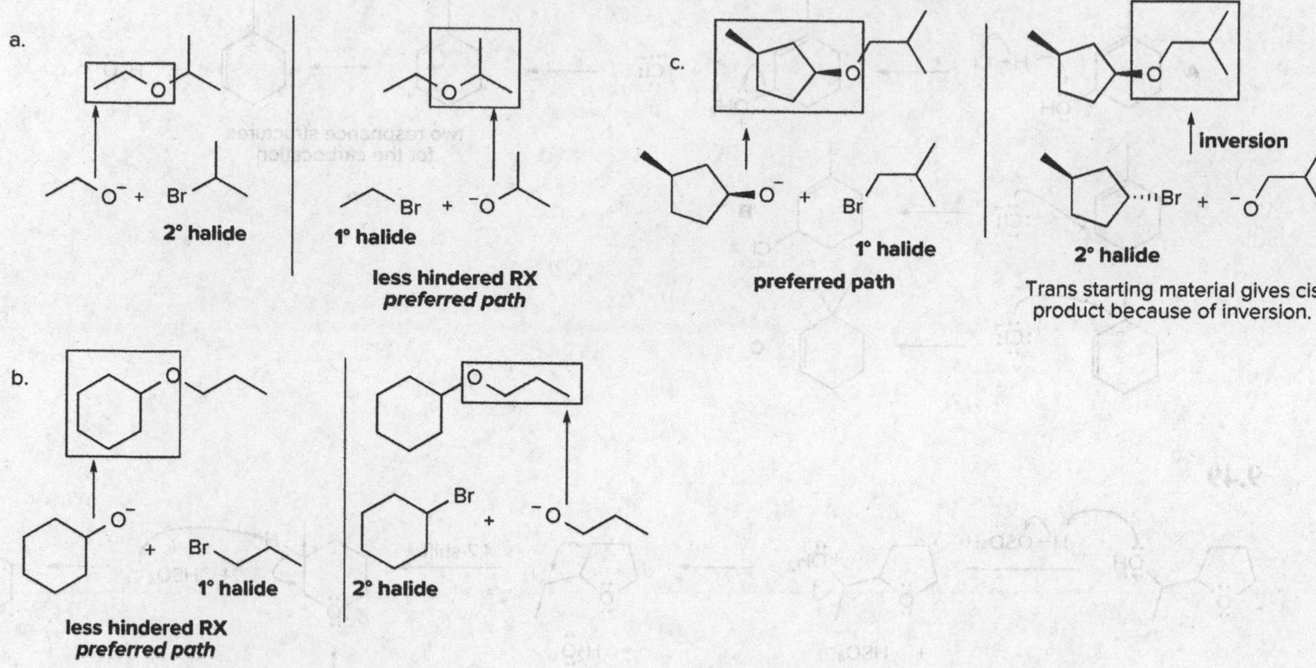

9.52 A tertiary halide is too hindered and an aryl halide is too unreactive to undergo a Williamson ether synthesis.

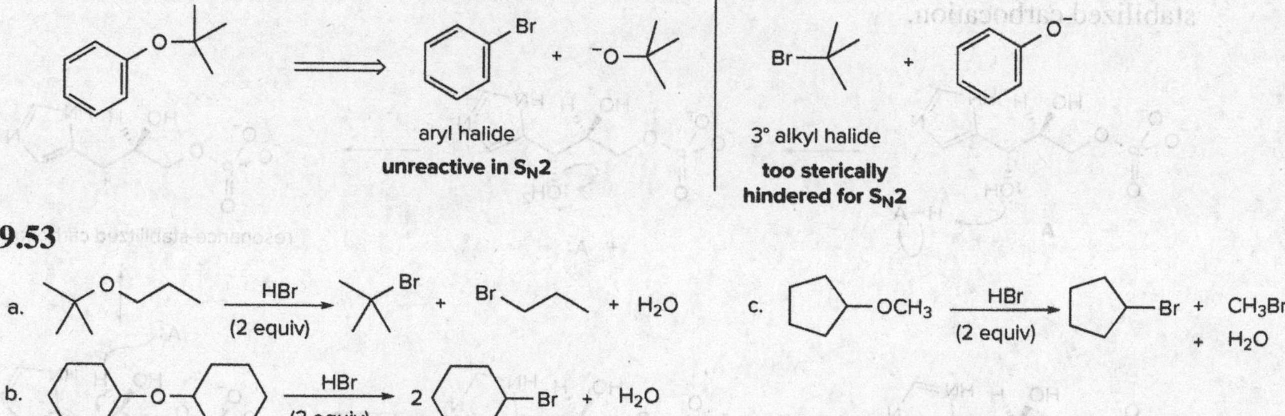

9.53

a. HBr (2 equiv) → tert-butyl Br + Br-propyl + H_2O

b. HBr (2 equiv) → 2 cyclohexyl–Br + H_2O

c. cyclopentyl–OCH_3 HBr (2 equiv) → cyclopentyl–Br + CH_3Br + H_2O

9.54

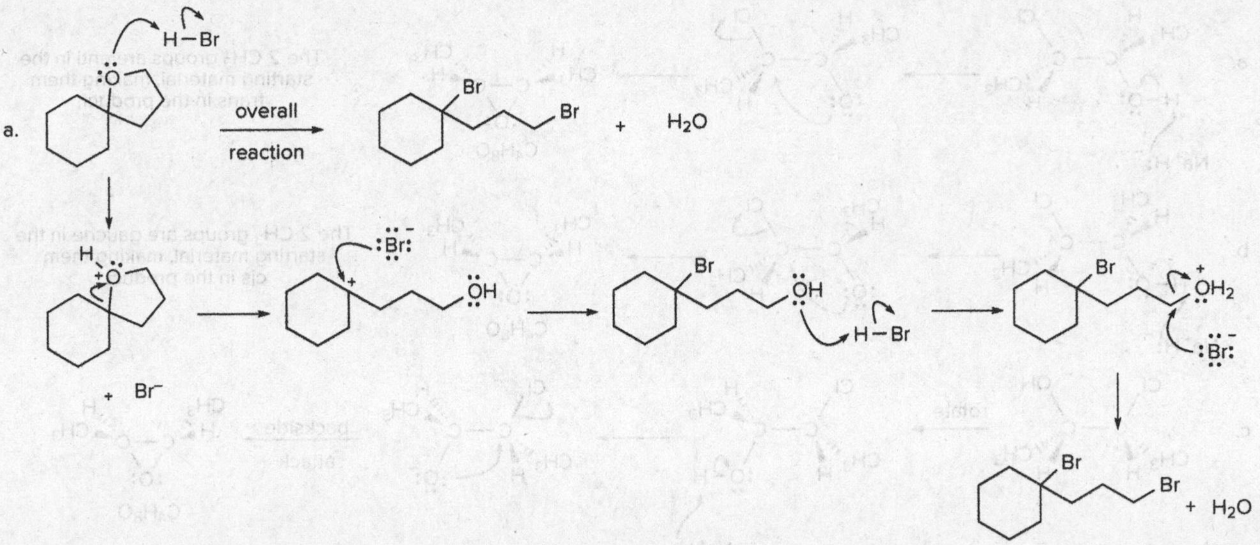

a. overall reaction → + H_2O

Br, Br

H–Br

+ Br^-

:Br:

ÖH

H–Br

OH_2

:Br:

+ H_2O

9.55

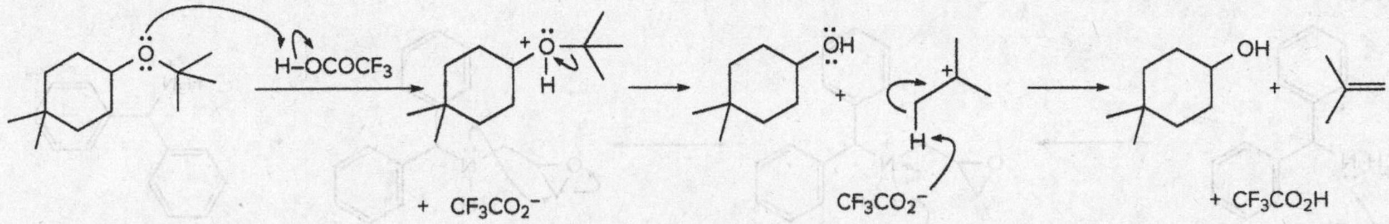

H–OCOCF$_3$

+ $CF_3CO_2^-$

ÖH

$CF_3CO_2^-$

OH

+ CF_3CO_2H

9.56

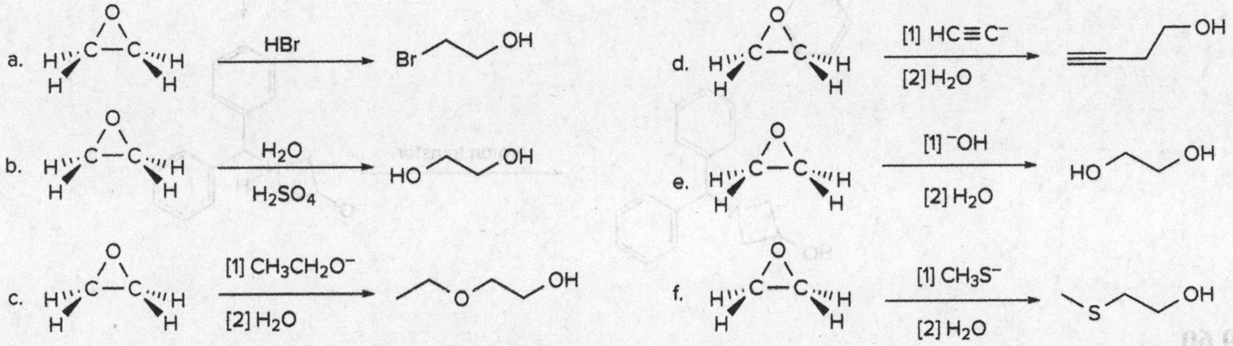

a. HBr → Br–CH$_2$CH$_2$–OH

b. H_2O, H_2SO_4 → HO–CH$_2$CH$_2$–OH

c. [1] $CH_3CH_2O^-$ [2] H_2O → CH$_3$CH$_2$O–CH$_2$CH$_2$–OH

d. [1] HC≡C$^-$ [2] H_2O → HC≡C–CH$_2$CH$_2$–OH

e. [1] $^-$OH [2] H_2O → HO–CH$_2$CH$_2$–OH

f. [1] CH_3S^- [2] H_2O → CH$_3$S–CH$_2$CH$_2$–OH

9.57

a. CH_3CH_2OH, H_2SO_4 → OH

b. [1] $CH_3CH_2O^-$ Na^+ [2] H_2O → OH

c. HBr → OH, Br

d. [1] NaCN [2] H_2O → OH, CN

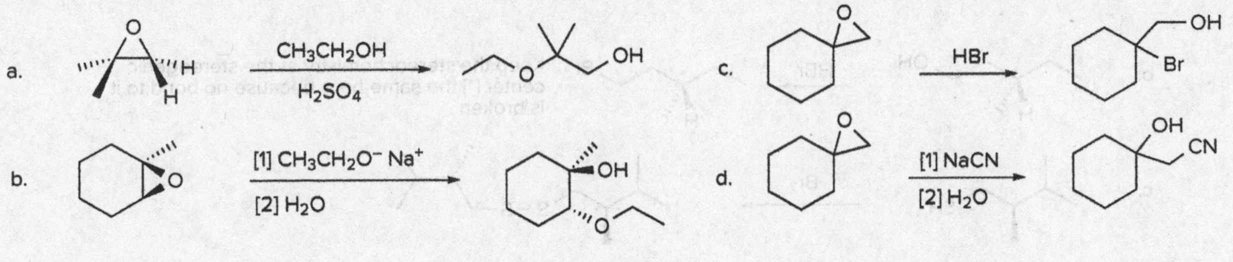

9.58

a. 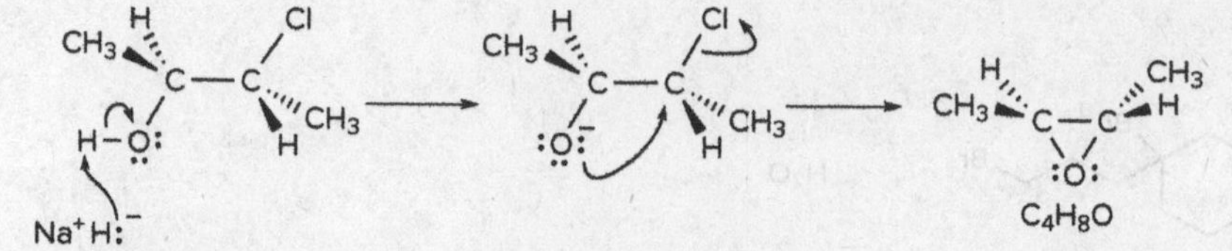 The 2 CH_3 groups are anti in the starting material, making them trans in the product.

C_4H_8O

b. 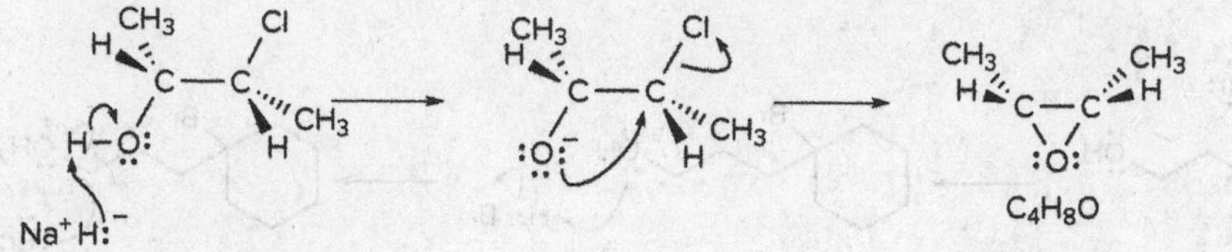 The 2 CH_3 groups are gauche in the starting material, making them cis in the product.

C_4H_8O

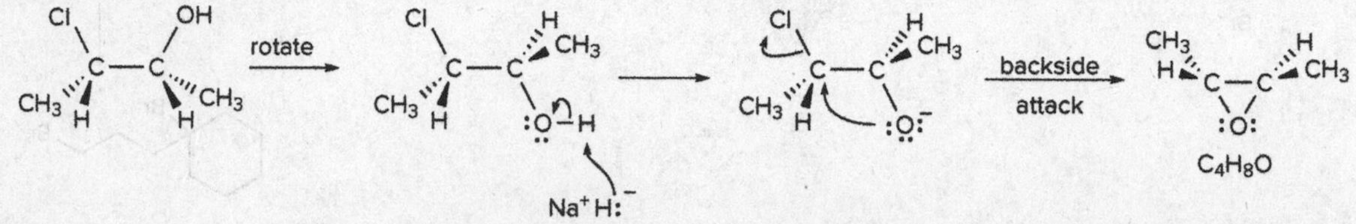

c. rotate → backside attack

C_4H_8O

9.59

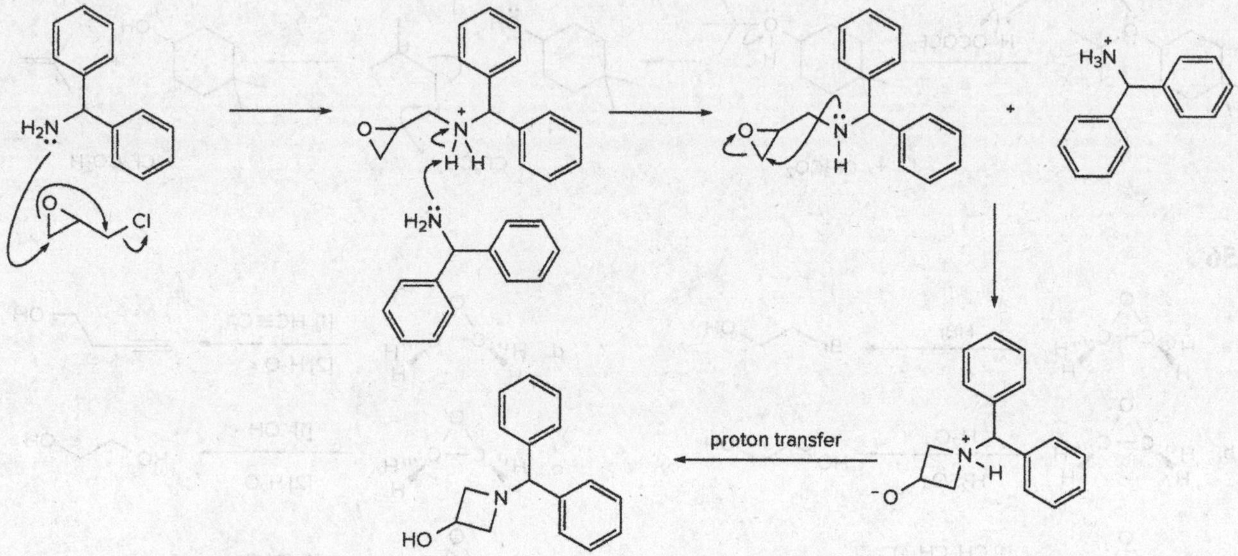

proton transfer

9.60

a. 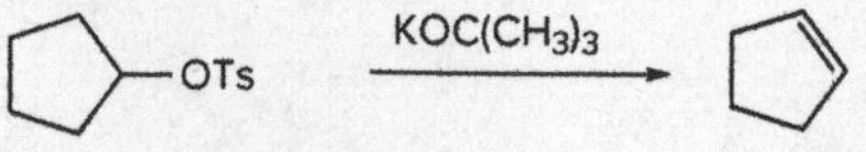 $KOC(CH_3)_3$ — Bulky base favors E2.

b. 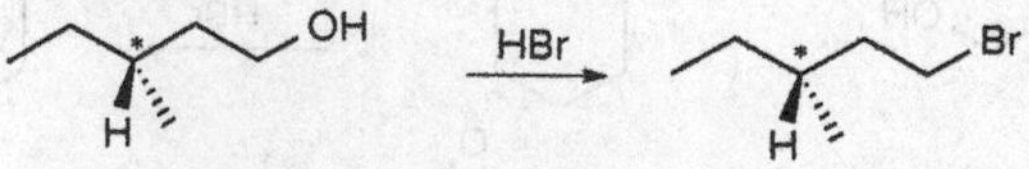 HBr — Keep the stereochemistry at the stereogenic center [*] the same here because no bond to it is broken .

c. 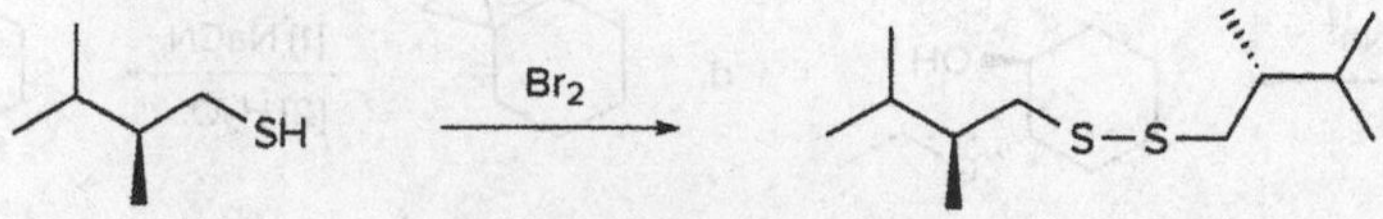 Br_2

d. OTs — KSH → SH

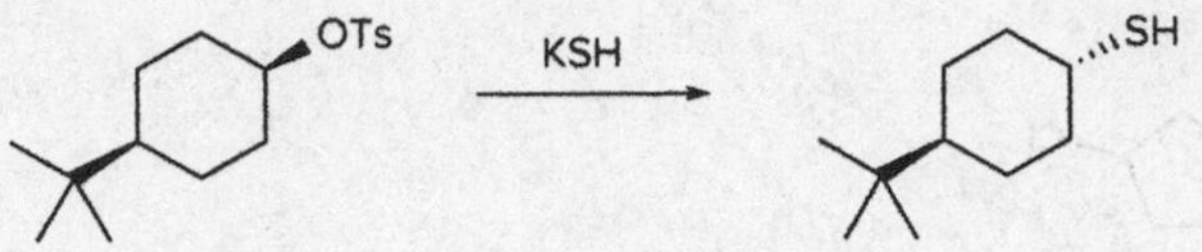

e. OH — PBr_3 → Br **S_N2 inversion**

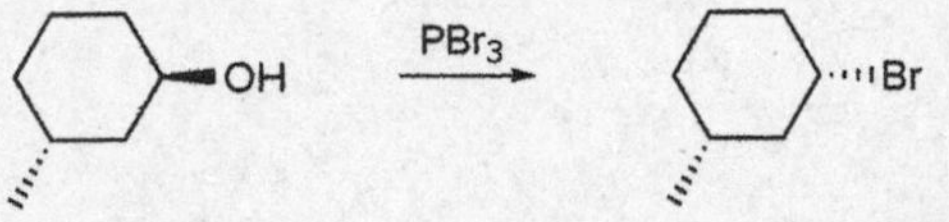

f. OH, H D — TsCl, pyridine → OTs, H D — $CH_3CO_2^-$ → CH_3CO_2, H D

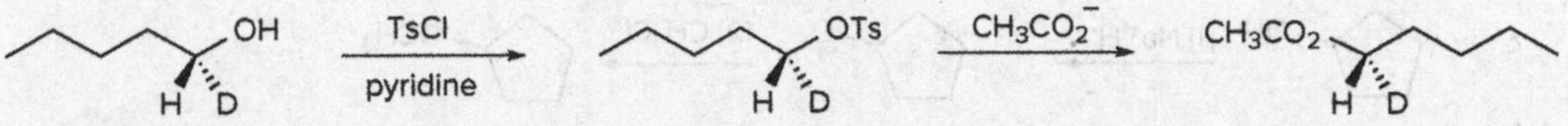

g. O — HBr → OH, Br + Br, OH

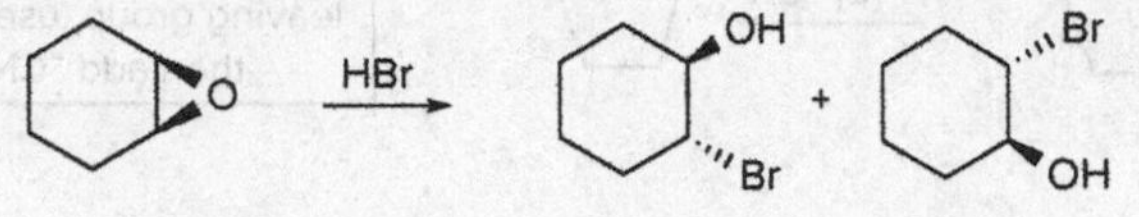

h. O — [1] $NaOCH_3$ [2] H_2O → OH, OCH_3 + OCH_3, OH

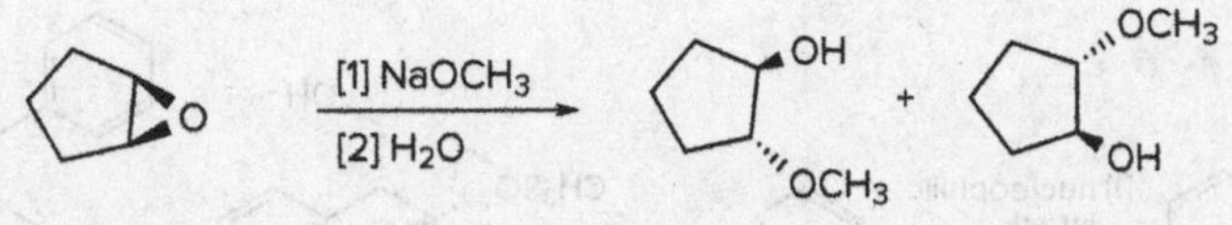

i. OH — NaH → O^- — I → O

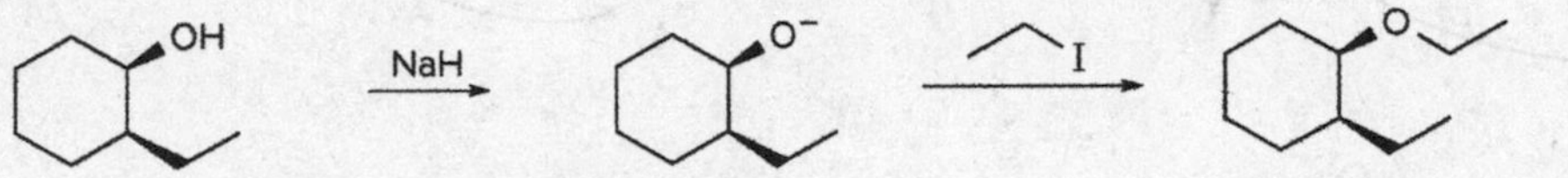

j. O — HI (2 equiv) → I + $I-CH_3$ + H_2O

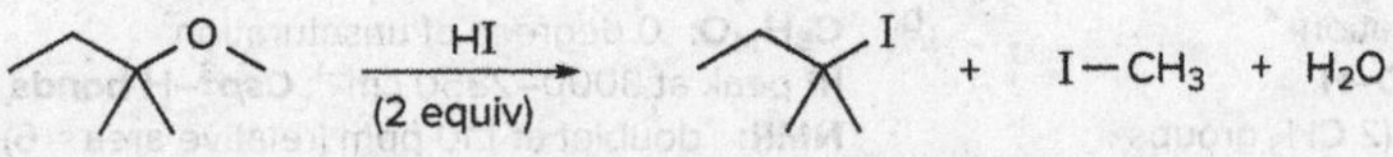

k. D, Br + S: Na^+ → D, S

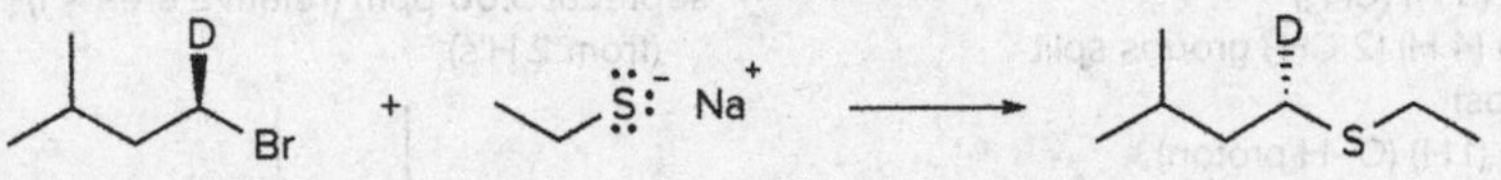

l. S — I → S^+ I^-

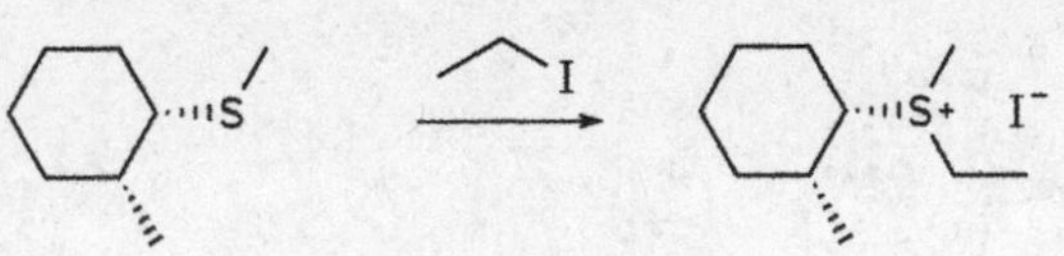

9.61

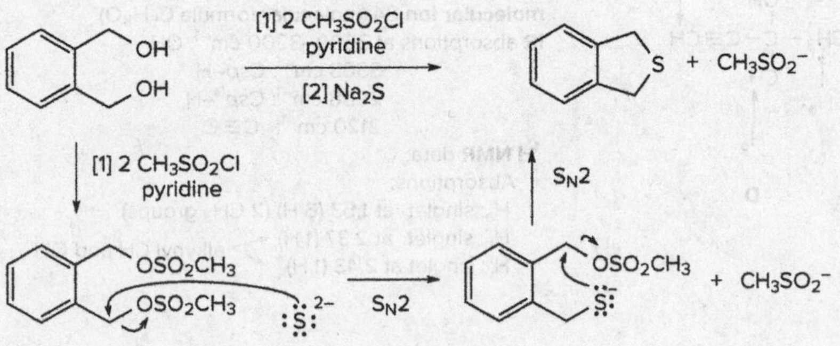

OH, OH — [1] 2 CH_3SO_2Cl pyridine [2] Na_2S → S + $CH_3SO_2^-$

[1] 2 CH_3SO_2Cl pyridine

OSO_2CH_3, OSO_2CH_3 :S:$^{2-}$ — S_N2 → OSO_2CH_3, S: + $CH_3SO_2^-$ — S_N2

9.62

a. Cyclopentanol (OH) → HCl or $SOCl_2$, pyridine → chlorocyclopentane (Cl)

b. Cyclopentanol (OH) → H_2SO_4 → cyclopentene

c. Cyclopentanol (OH) → [1] Na^+H^- → cyclopentoxide (O^-) → [2] CH_3Cl → OCH_3

d. Cyclopentanol (OH) → [1] TsCl, pyridine → OTs → [2] ^-CN → CN

Make OH a good leaving group (use TsCl); then add ^-CN.

9.63

X (CH_3SO_3, OH, Cl) + Y (H_2N) → [1] nucleophilic attack, [2] loss of H^+ → (CH_3SO_3, OH, HN) → Z (N, OH)

9.64

a. **$C_6H_{14}O$:** 0 degrees of unsaturation
IR peak at 3600–3200 cm^{-1}: **O–H**
NMR: triplet at 0.8 ppm (6 H) (2 CH_3 groups split by CH_2 groups)
singlet at 1.0 ppm (3 H) (CH_3)
quartet at 1.5 ppm (4 H) (2 CH_2 groups split by CH_3 groups)
singlet at 1.6 ppm (1 H) (O–H proton)

OH

b. **$C_6H_{14}O$:** 0 degrees of unsaturation
IR peak at 3000–2850 cm^{-1}: **Csp^3–H bonds**
NMR: doublet at 1.10 ppm (relative area = 6) (from 12 H's)
septet at 3.60 ppm (relative area = 1) (from 2 H's)

O

9.65

Acetone (O) → [1] LiC≡CH, [2] H_2O → $CH_3-C(OH)(CH_3)-C\equiv CH$ (D; H_a on CH_3 groups, H_b on OH, H_c on alkynyl CH)

Compound D:
molecular ion 84 (molecular formula C_5H_8O)
IR absorptions at 3600–3200 cm^{-1}: OH
3303 cm^{-1}: Csp–H
2938 cm^{-1}: Csp^3–H
2120 cm^{-1}: C≡C

^{1}H NMR data:
Absorptions:
H_a: singlet at 1.53 (6 H) (2 CH_3 groups)
H_b: singlet at 2.37 (1 H)
H_c: singlet at 2.43 (1 H)
(H_b, H_c: alkynyl CH and OH)

9.66

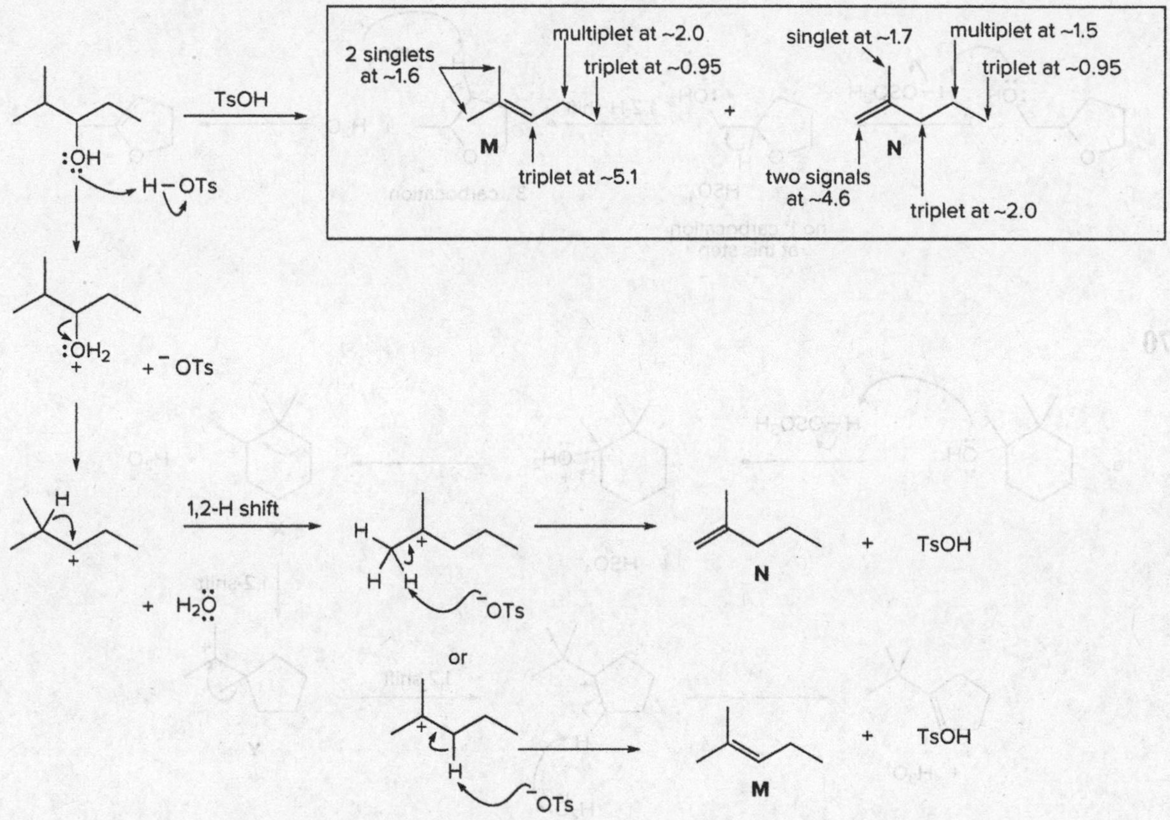

TsOH

multiplet at ~2.0
2 singlets at ~1.6
triplet at ~0.95
M
triplet at ~5.1
singlet at ~1.7
multiplet at ~1.5
triplet at ~0.95
N
two signals at ~4.6
triplet at ~2.0

H–OTs

$+ \ ^{-}OTs$

1,2-H shift

+ H_2O

N + TsOH

or

M + TsOH

9.67

Compound **A:** Molecular formula $C_4H_8O_2$ (one degree of unsaturation)
IR absorptions at 3600–3200 (O–H), 3000–2800 (C–H), and 1700 (C=O) cm^{-1}
^{1}H NMR data:

Absorption	ppm	# of H's	Explanation
singlet	2.2	3	a CH_3 group
singlet	2.55	1	1 H adjacent to none or OH
triplet	2.7	2	2 H's adjacent to 2 H's
triplet	3.8	2	2 H's adjacent to 2 H's

Structure:

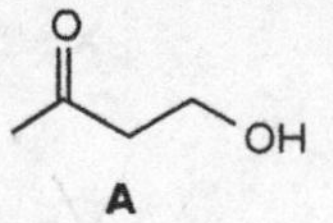

OH
A

9.68 If the base is not bulky, it can react as a nucleophile and open the epoxide ring. The bulky base cannot act as a nucleophile, and will only remove the proton.

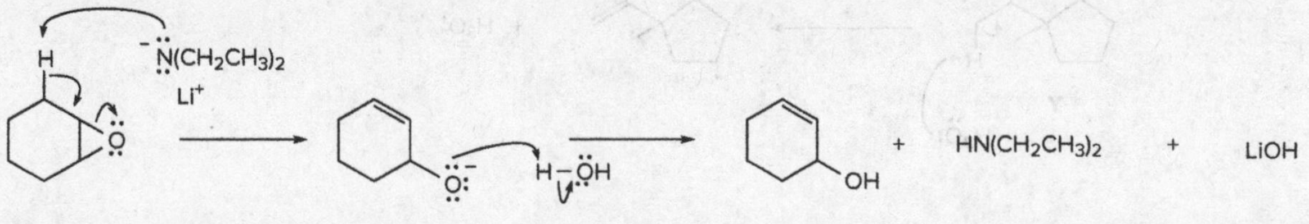

$^{-}N(CH_2CH_3)_2$
Li^+

H–OH

OH + $HN(CH_2CH_3)_2$ + LiOH

9.69

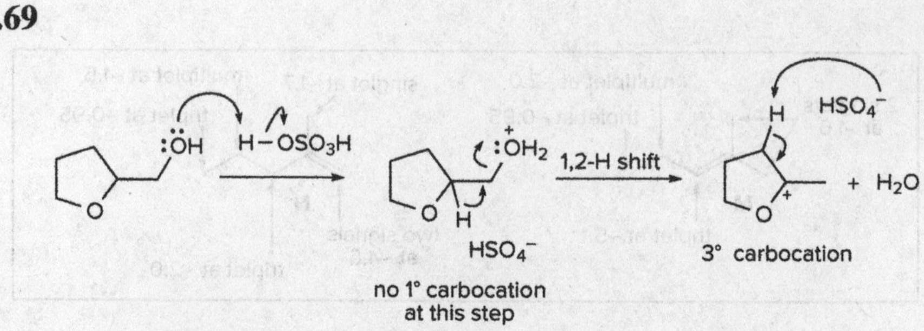

:ÖH H–OSO₃H → (protonated) :ÖH₂ 1,2-H shift → + H₂O →

HSO_4^-

no 1° carbocation at this step

3° carbocation

9.70

a. ÖH H–OSO₃H → ÖH₂⁺ → + H₂O

+ HSO_4^-

X

1,2-shift

Y

1,2-shift

$H_2\ddot{O}$:

+ H_3O^+

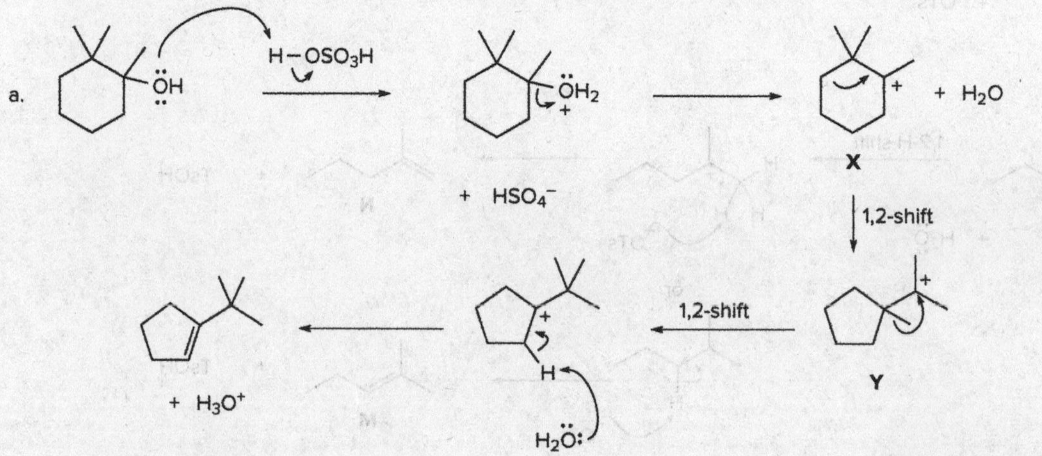

b. **Other elimination products can form from carbocations X and Y.**

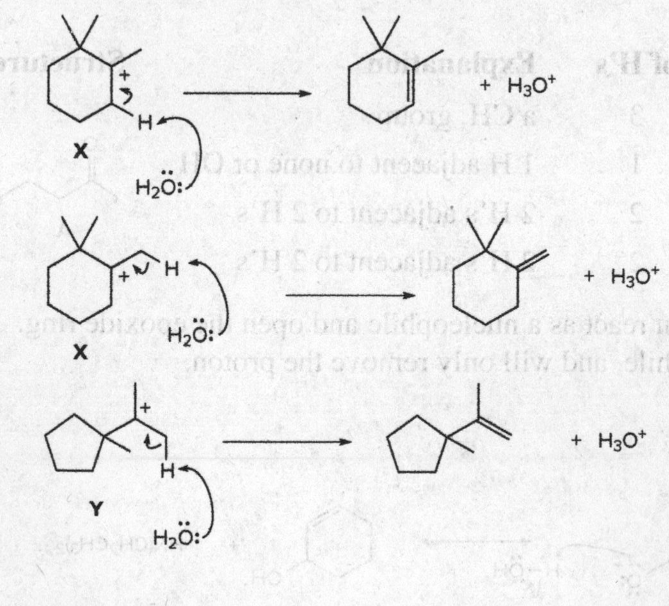

X + H_3O^+

$H_2\ddot{O}$:

X + H_3O^+

$H_2\ddot{O}$:

Y + H_3O^+

$H_2\ddot{O}$:

9.71

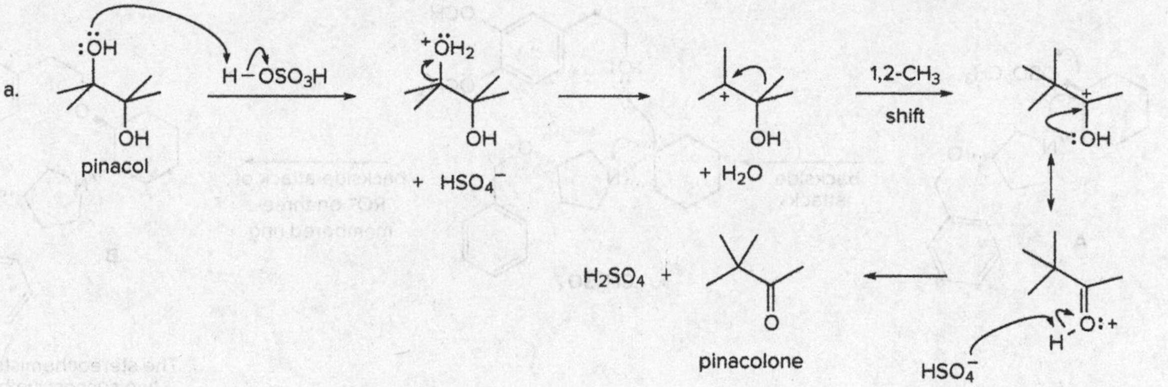

a. :ÖH pinacol OH H–OSO$_3$H → $^+$ÖH$_2$ OH + HSO$_4^-$ → + OH + H$_2$O → 1,2-CH$_3$ shift → + :ÖH ↔ O:+ H HSO$_4^-$ → H$_2$SO$_4$ + O pinacolone

b. Two different carbocations can form. The carbocation with the (+) charge adjacent to the benzene rings (**A**) is more stable, so it is preferred.

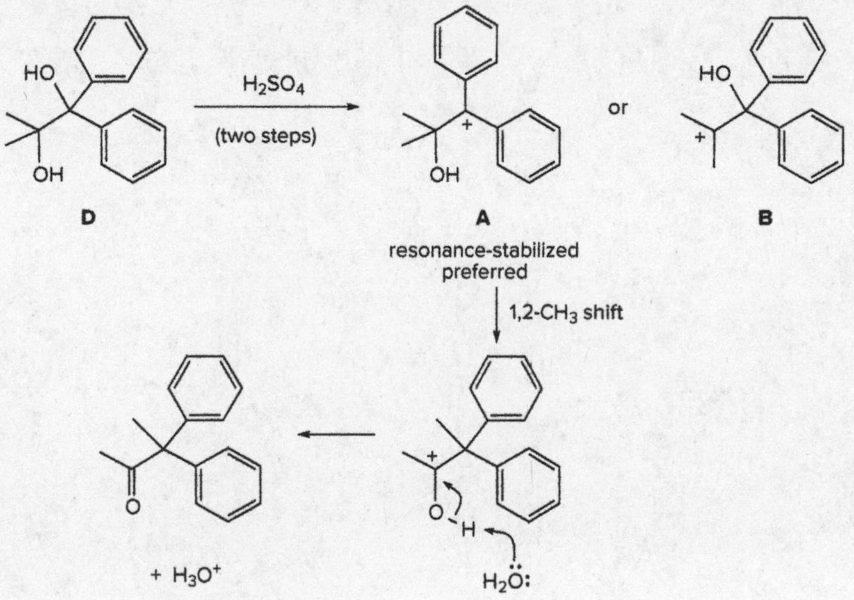

HO OH **D** H$_2$SO$_4$ (two steps) → OH + **A** resonance-stabilized preferred or HO + **B**

1,2-CH$_3$ shift → + O H H$_2$Ö: → O + H$_3$O$^+$

9.72

:Ö–H I O O: Na$^+$ $^-$:ÖH → :Ö:$^-$ I O O: → Ö: I O Ö:$^-$ → :Ö O: I Ö:$^-$ → :Ö O: :Ö: + I$^-$

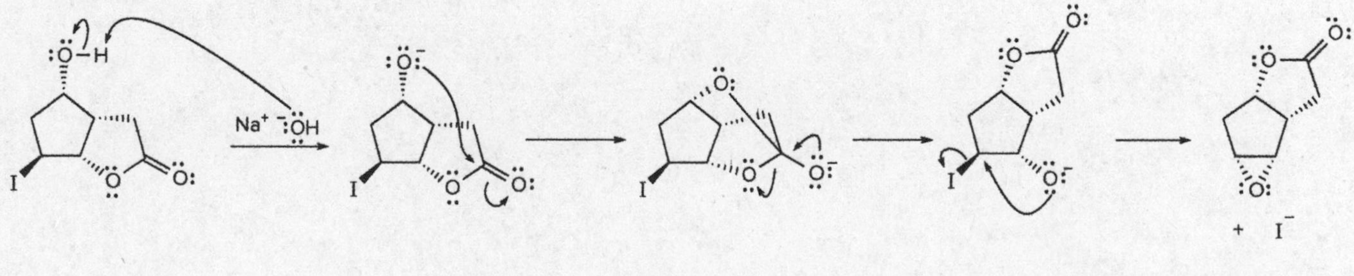

9.73

OSO_2CH_3

A

backside attack

OCH_3

OCH_3

$+ \quad CH_3SO_3^-$

backside attack of RO^- on three-membered ring

OCH_3

OCH_3

B

The stereochemistry in **B** results from two successive backside attacks.

Chapter 10 Alkenes and Alkynes

Chapter Review

General facts about alkenes

- Alkenes contain a carbon–carbon double bond consisting of a stronger σ bond and a weaker π bond. Each carbon is sp^2 hybridized and trigonal planar (10.1).
- Alkenes are named using the suffix ***-ene*** (10.3).
- Alkenes with different groups on each end of the double bond exist as a pair of diastereomers, identified by the prefixes *E* and *Z* (10.3C).

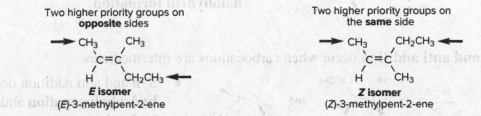

- Alkenes have weak intermolecular forces, giving them low mp's and bp's, and making them water insoluble. A cis alkene is more polar than a trans alkene, giving it a slightly higher boiling point (10.4).

cis-but-2-ene

more polar isomer →

a small net dipole
higher bp

trans-but-2-ene

← **less polar isomer**

no net dipole
lower bp

- A π bond is electron rich and much weaker than a σ bond, so alkenes undergo addition reactions with electrophiles (10.8).

General facts about alkynes

- Alkynes contain a carbon–carbon triple bond consisting of a strong σ bond and two weak π bonds. Each carbon is *sp* hybridized and linear (10.1).

H–C≡C–H =
acetylene
180°
sp hybridized

- Alkynes are named using the suffix ***-yne*** (10.3).
- Alkynes have weak intermolecular forces, giving them low mp's and low bp's, and making them water insoluble (10.4).
- Because its weaker π bonds make an alkyne electron rich, alkynes undergo addition reactions with electrophiles (10.8).

Stereochemistry of alkene addition reactions (10.8)

A reagent XY adds to a double bond in one of three different ways:

- **Syn addition**—X and Y add from the same side.

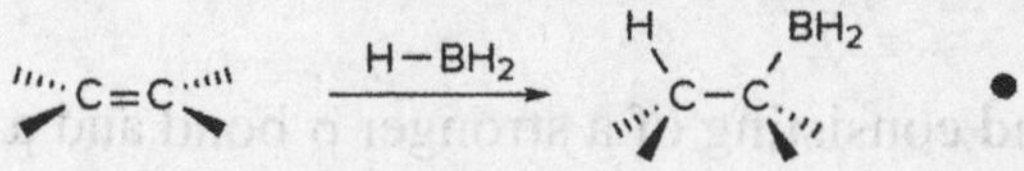

- Syn addition occurs in **hydroboration.**

- **Anti addition**—X and Y add from opposite sides.

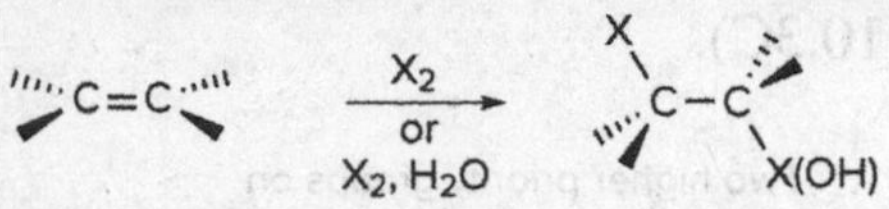

- Anti addition occurs in **halogenation** and **halohydrin formation.**

- **Both syn and anti addition** occur when carbocations are intermediates.

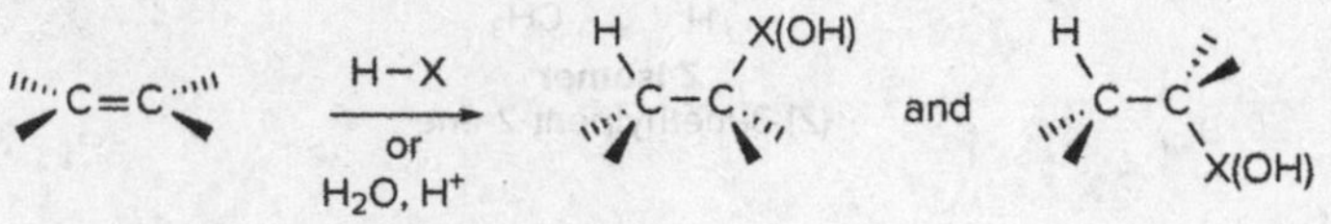

- Syn and anti addition occur in **hydrohalogenation** and **hydration.**

Addition reactions of alkenes

[1] **Hydrohalogenation**—Addition of HX (X = Cl, Br, I) (10.9–10.11)

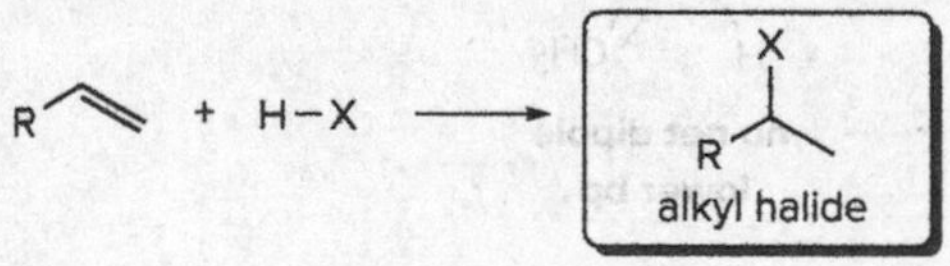

- The mechanism has two steps.
- Carbocations are formed as intermediates.
- Carbocation rearrangements are possible.
- Markovnikov's rule is followed. H bonds to the less substituted C to form the more stable carbocation.
- Syn and anti addition occur.

[2] **Hydration** and related reactions—Addition of H_2O or ROH (10.12)

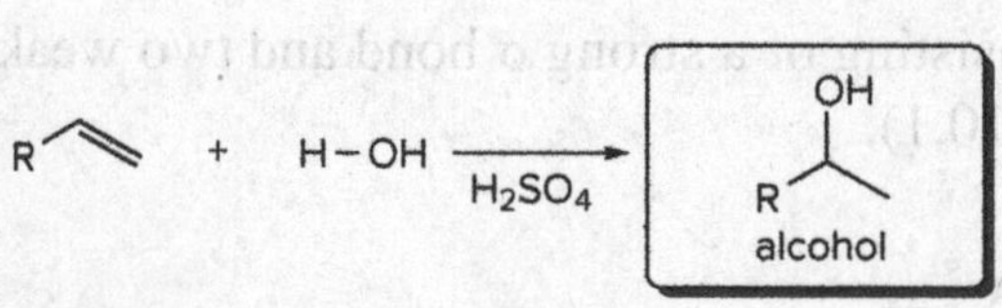

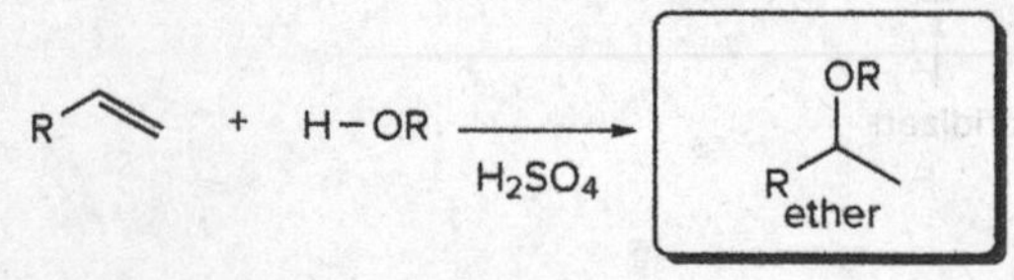

For both reactions:

- The mechanism has three steps.
- Carbocations are formed as intermediates.
- Carbocation rearrangements are possible.
- Markovnikov's rule is followed. H bonds to the less substituted C to form the more stable carbocation.
- Syn and anti addition occur.

[3] **Halogenation**—Addition of X_2 (X = Cl or Br) (10.13–10.14)

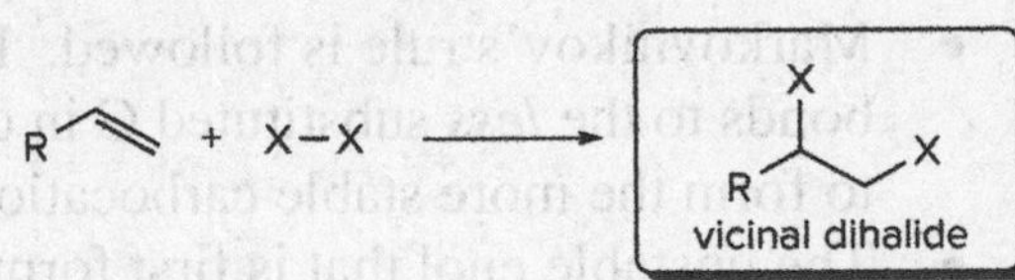

- The mechanism has two steps.
- Bridged halonium ions are formed as intermediates.
- No rearrangements occur.
- Anti addition occurs.

---

[4] **Halohydrin formation**—Addition of OH and X (X = Cl, Br) (10.15)

R + X−X, H_2O → halohydrin

- The mechanism has three steps.
- Bridged halonium ions are formed as intermediates.
- No rearrangements occur.
- X bonds to the less substituted C.
- Anti addition occurs.
- NBS in DMSO and H_2O adds Br and OH in the same fashion.

---

[5] **Hydroboration–oxidation**—Addition of H_2O (10.16)

R → [1] BH_3 or 9-BBN; [2] H_2O_2, HO^- → alcohol

- Hydroboration has a one-step mechanism.
- No rearrangements occur.
- OH bonds to the less substituted C.
- Syn addition of H_2O results.

Addition reactions of alkynes

[1] **Hydrohalogenation**—Addition of HX (X = Cl, Br, or I) (10.17)

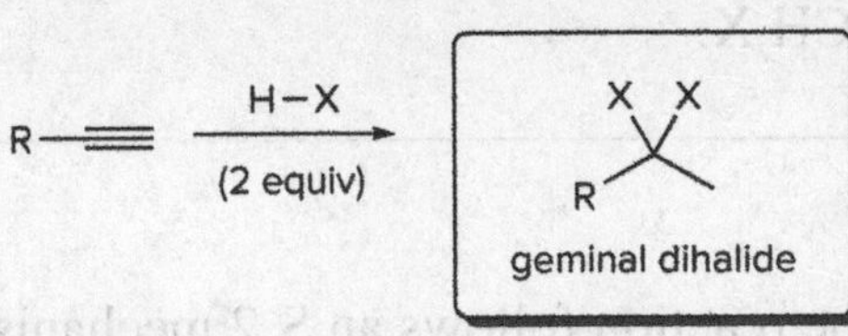

- Markovnikov's rule is followed. H bonds to the *less* substituted C in order to form the more stable carbocation.

---

[2] **Halogenation**—Addition of X_2 (X = Cl or Br) (10.17)

R → X−X (2 equiv) → tetrahalide

- Anti addition of X_2 occurs.

[3] **Hydration**—Addition of H_2O (10.18)

$R-C\equiv C$ $\xrightarrow[H_2SO_4,\ HgSO_4]{H_2O}$ [enol] $\rightleftharpoons$ ketone

- Markovnikov's rule is followed. H bonds to the *less* substituted C in order to form the more stable carbocation.
- The unstable enol that is first formed rearranges to a carbonyl group.

[4] **Hydroboration–oxidation**—Addition of H_2O (10.19)

$R-C\equiv C$ $\xrightarrow[[2]\ H_2O_2,\ HO^-]{[1]\ R_2BH}$ [enol] $\rightleftharpoons$ aldehyde

- The unstable enol, first formed after oxidation, rearranges to a carbonyl group.

Reactions involving acetylide anions

[1] Formation of acetylide anions from terminal alkynes (10.8B)

$R-C\equiv C-H \;+\; :B \rightleftharpoons R-C\equiv C:^- \;+\; HB^+$

- Typical bases used for the reaction are $NaNH_2$ and NaH.

[2] Reaction of acetylide anions with alkyl halides (10.20A)

$H-C\equiv C:^- \;+\; R-X \longrightarrow H-C\equiv C-R \;+\; X^-$

- The reaction follows an S_N2 mechanism.
- The reaction works best with CH_3X and RCH_2X.

[3] Reaction of acetylide anions with epoxides (10.20B)

$H-C\equiv C:^-$ $\xrightarrow[[2]\ H_2O]{[1]\ \text{ethylene oxide}}$ $H-C\equiv C-CH_2CH_2OH$

- The reaction follows an S_N2 mechanism.
- Ring opening occurs from the back side at the *less* substituted end of the epoxide.

Practice Test on Chapter Review

1. Give the IUPAC name for the following compounds.

a.

b.

2. Draw the structure for the compound with the following IUPAC name: 5-*tert*-butyl-6,6-dimethylnon-3-yne.

3. Draw the organic products formed in the following reactions. Draw all stereogenic centers using wedges and dashed wedges.

a. 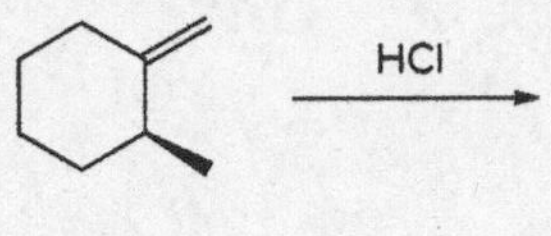 $\xrightarrow{HCl}$

c. 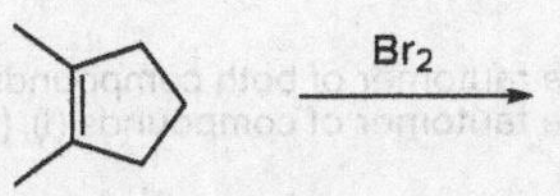 $\xrightarrow{Br_2}$

b. 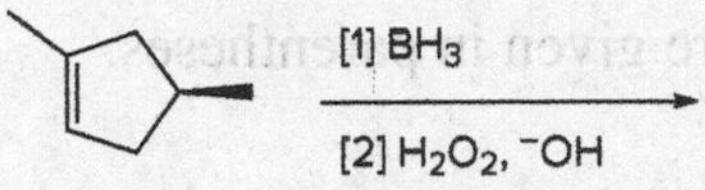 $\xrightarrow{[1]\ BH_3;\ [2]\ H_2O_2,\ ^-OH}$

d. 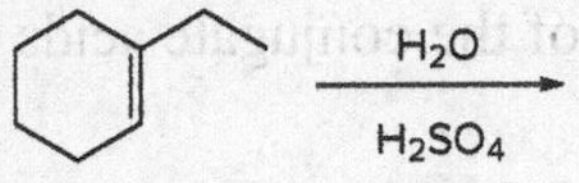 $\xrightarrow{H_2O,\ H_2SO_4}$

4. Draw the organic products formed in the following reactions.

a. 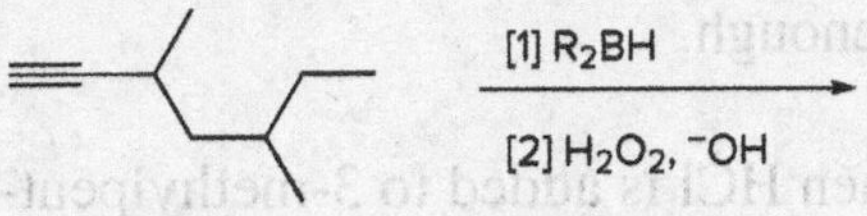 $\xrightarrow{[1]\ R_2BH;\ [2]\ H_2O_2,\ ^-OH}$

d.  $\xrightarrow{H_2O,\ H_2SO_4,\ HgSO_4}$

b. 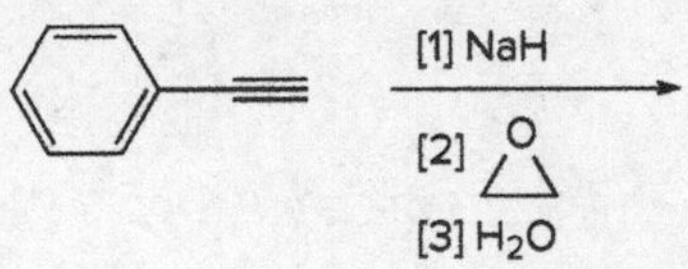 $\xrightarrow{[1]\ NaH;\ [2]\ \text{(epoxide)};\ [3]\ H_2O}$

e. 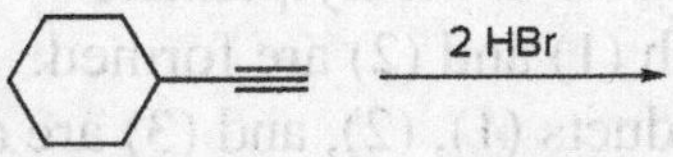 $\xrightarrow{2\ HBr}$

c. 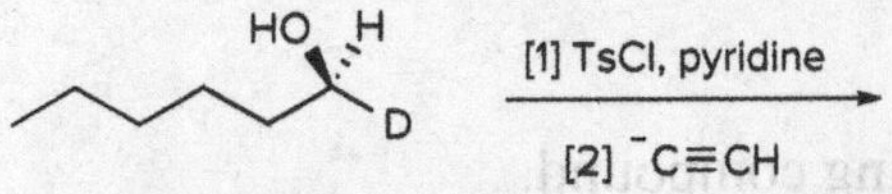 HO H D $\xrightarrow{[1]\ TsCl,\ pyridine;\ [2]\ ^-C{\equiv}CH}$

5. Fill in the table with the stereochemistry observed in the reaction of an alkene with each reagent. Choose from syn, anti, or both syn and anti.

Reagent	Stereochemistry
a. [1] 9-BBN; [2] H_2O_2, ^-OH	
b. H_2O, H_2SO_4	
c. Cl_2, H_2O	
d. HI	
e. Br_2	

6. a. In which of the following reactions are carbocation rearrangements observed?
   1. hydrohalogenation
   2. halohydrin formation
   3. hydroboration–oxidation
   4. Carbocation rearrangements are observed in reactions (1) and (2).
   5. Carbocation rearrangements are observed in reactions (1), (2), and (3).

b. Which of the following compounds is an enol tautomer of compound **A**?

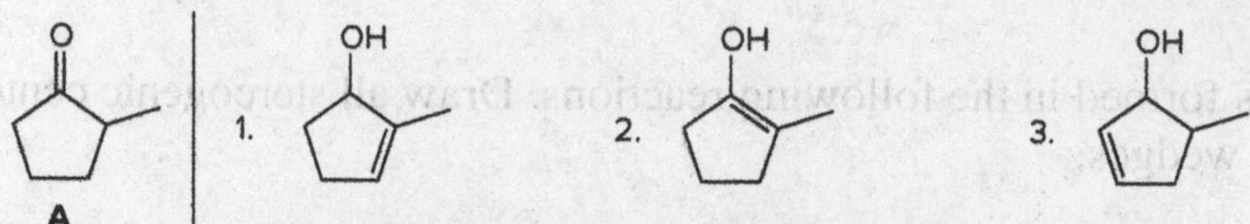

4. **A** can be a tautomer of both compounds (1) and (2).
5. **A** can be a tautomer of compounds (1), (2), and (3).

c. Which of the following bases is strong enough to deprotonate $CH_3C{\equiv}CH$ (propyne, $pK_a = 25$)? The pK_a's of the conjugate acids of the bases are given in parentheses.

1. CH_3Li ($pK_a = 50$)
2. $NaOCH_3$ ($pK_a = 15.5$)
3. $NaOCOCH_3$ ($pK_a = 4.8$)
4. The bases in (1) and (2) are both strong enough.
5. The bases in (1), (2), and (3) are all strong enough.

d. Which of the following products are formed when HCl is added to 3-methylpent-1-ene?

1. 2-chloro-2-methylpentane
2. 3-chloro-3-methylpentane
3. 1-chloro-3-methylpentane
4. Both (1) and (2) are formed.
5. Products (1), (2), and (3) are all formed.

7. Draw two different enol tautomers for the following compound.

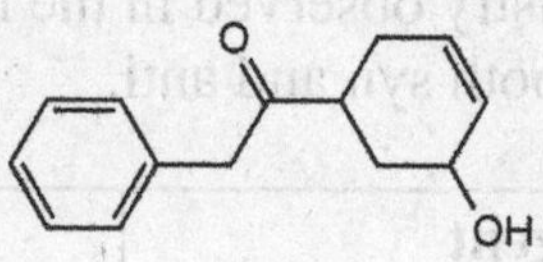

8. What acetylide anion and alkyl halide are needed to make the following alkyne?

Answers to Practice Test

1.

a. (*E*)-4-ethyl-2,5-dimethylnon-3-ene

b. (*E*)-4-isopropyl-2-methyloct-3-ene

2.

3.

a. Cl + Cl

b. HO + HO

c. Br, Br + Br, Br

d. OH

4.

a. H, O

b. OH

c. HC≡C, D, H

d. O

e. Br Br

5. a. syn

b. both

c. anti

d. both

e. anti

6. a. 1

b. 2

c. 1

d. 2

7.

HO, OH (*E* + *Z*)

HO, OH (*E* + *Z*)

8.

CH_3X +

Answers to Problems

10.1

Six alkenes of molecular formula C_5H_{10}:

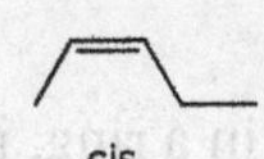

diastereomers

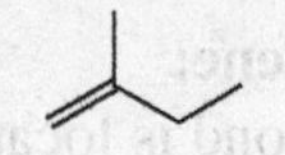

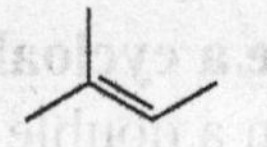

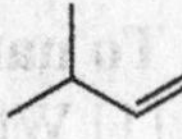

10.2 To determine the number of degrees of unsaturation:

[1] Calculate the maximum number of H's (2*n* + 2).

[2] Subtract the actual number of H's from the maximum number.

[3] Divide by two.

a. C_8H_{12}

[1] maximum number of H's = 2*n* + 2 = 2(8) + 2 = 18

[2] subtract actual from maximum = 18 – 12 = 6

[3] divide by two = 6/2 = **3 degrees of unsaturation**

possible structures:

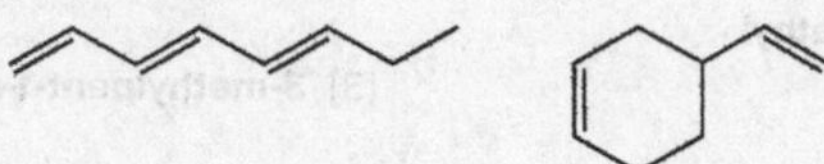

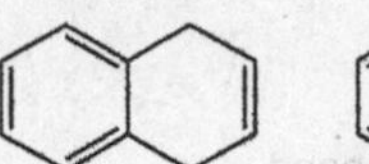

b. $C_{10}H_{10}$

[1] maximum number of H's = 2*n* + 2 = 2(10) + 2 = 22

[2] subtract actual from maximum = 22 – 10 = 12

[3] divide by two = 12/2 = **6 degrees of unsaturation**

possible structures:

10.3 Follow the procedure in Answer 10.2.

a. C_6H_6
[1] maximum number of H's = 2*n* + 2 = 2(6) + 2 = 14
[2] subtract actual from maximum = 14 – 6 = 8
[3] divide by two = 8/2 = **4 degrees of unsaturation**

b. C_8H_{18}
[1] maximum number of H's = 2*n* + 2 = 2(8) + 2 = 18
[2] subtract actual from maximum = 18 – 18 = 0
[3] divide by two = 0/2 = **0 degrees of unsaturation**

c. C_7H_8O
Ignore the O.
[1] maximum number of H's = 2*n* + 2 = 2(7) + 2 = 16
[2] subtract actual from maximum = 16 – 8 = 8
[3] divide by two = 8/2 = **4 degrees of unsaturation**

d. $C_7H_{11}Br$
Because of Br, add one more H (11 + 1 H = 12 H's).
[1] maximum number of H's = 2*n* + 2 = 2(7) + 2 = 16
[2] subtract actual from maximum = 16 – 12 = 4
[3] divide by two = 4/2 = **2 degrees of unsaturation**

e. C_5H_9N
Because of N, subtract one H (9 – 1 H = 8 H's).
[1] maximum number of H's = 2*n* + 2 = 2(5) + 2 = 12
[2] subtract actual from maximum = 12 – 8 = 4
[3] divide by two = 4/2 = **2 degrees of unsaturation**

10.4 Use the directions from Answer 10.2. Ignore O in calculating degrees of unsaturation. Add one more H for each halogen. Subtract one H for each N.

a. $C_{19}H_{21}N_3O$ is equivalent to $C_{19}H_{18}$ when calculating degrees of unsaturation.
For 19 C's, the maximum number of H's = 2*n* + 2 = 2(19) + 2 = 40 H's
Subtract actual from maximum = 40 – 18 = 22 H's fewer
Divide by two = 22/2 = 11 degrees of unsaturation

b. $C_{17}H_{16}F_6N_2O$ is equivalent to $C_{17}H_{20}$ when calculating degrees of unsaturation.
For 17 C's, the maximum number of H's = 2*n* + 2 = 2(17) + 2 = 36 H's
Subtract actual from maximum = 36 – 20 = 16 H's fewer
Divide by two = 16/2 = 8 degrees of unsaturation

10.5 To name an alkene:

[1] Find the longest chain that contains the double bond. Change the ending from ***-ane*** to ***-ene.***
[2] Number the chain to give the double bond the lower number. The alkene is named by the first number.
[3] Apply all other rules of nomenclature.

To name a cycloalkene:

[1] When a double bond is located in a ring, it is always located between C1 and C2. Omit the "1" in the name. Change the ending from ***-ane*** to ***-ene.***
[2] Number the ring clockwise or counterclockwise to give the first substituent the lower number.
[3] Apply all other rules of nomenclature.

To name an alkyne:

[1] Find the longest chain that contains both atoms of the triple bond, change the *-ane* ending of the parent name to *-yne*, and number the chain to give the first carbon of the triple bond the lower number.
[2] Name all substituents following the other rules of nomenclature.

a. [1] 5 C chain with double bond **pentene**

[2] 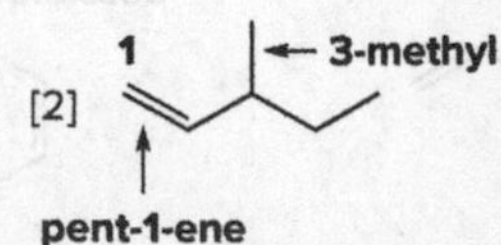

[3] **3-methylpent-1-ene**

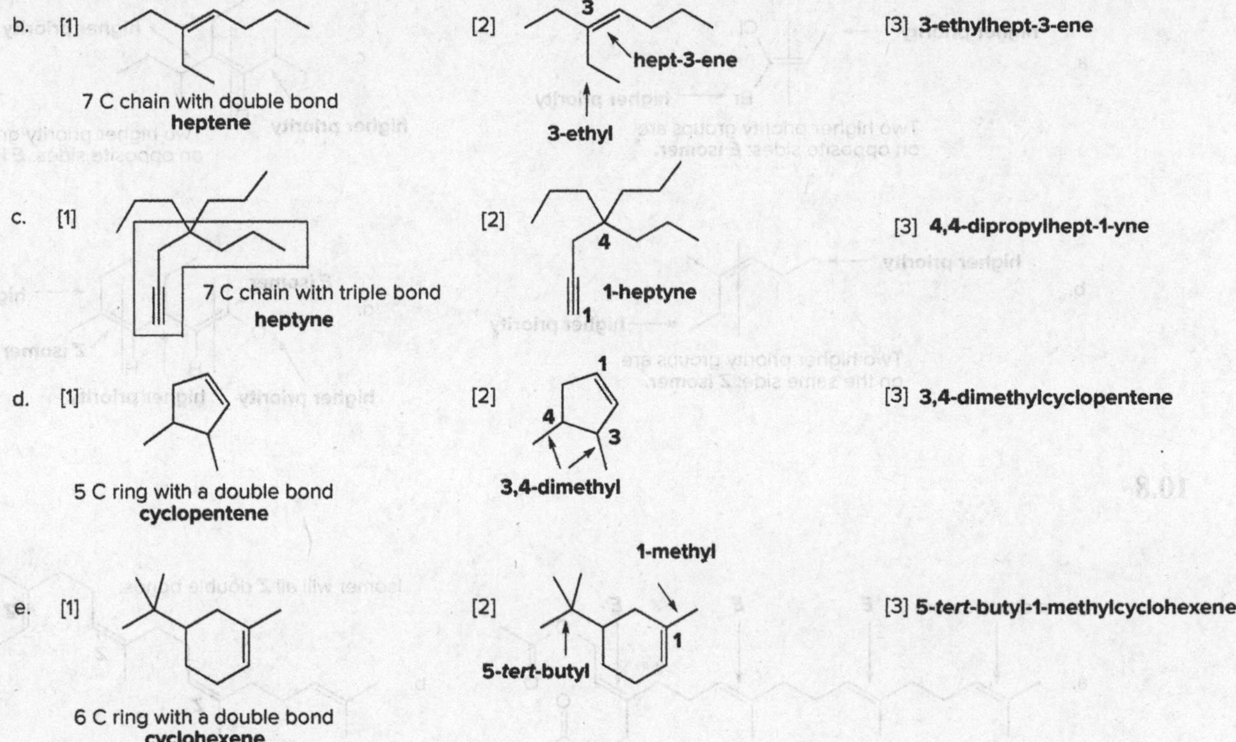

10.6 Use the rules from Answer 10.5 to name the compounds. Enols are named to give the OH the lower number. Compounds with two C=C's are named with the suffix ***-adiene,*** and compounds with two triple bonds are named with the suffix ***-adiyne.***

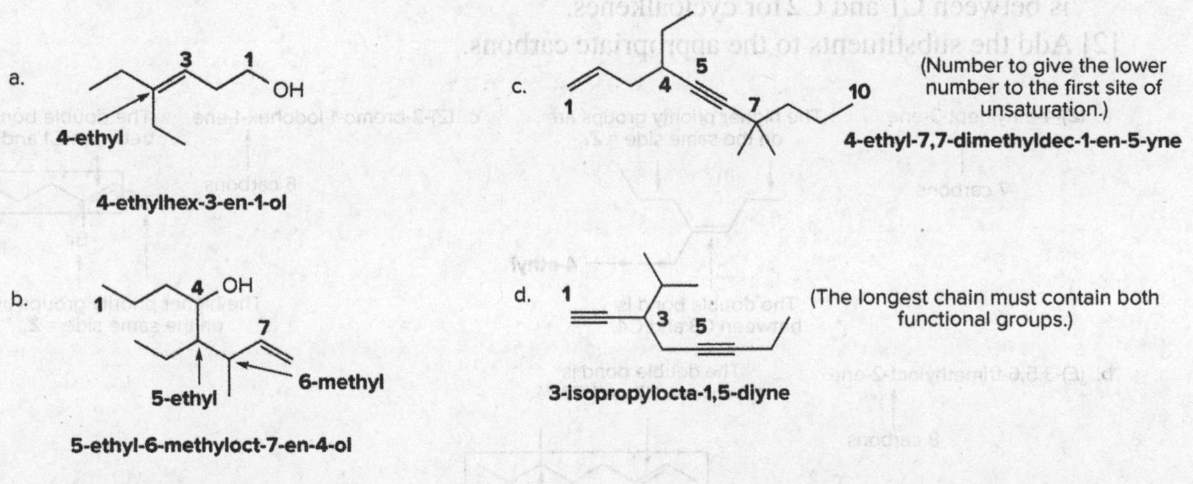

10.7 To label an alkene as *E* or *Z:*

[1] **Assign priorities** to the two substituents ***on each end*** using the rules for *R,S* nomenclature.

[2] **Assign *E* or *Z*** depending on the location of the two higher priority groups.

- The ***E*** prefix is used when the two higher priority groups are on **opposite sides** of the double bond.
- The ***Z*** prefix is used when the two higher priority groups are on the **same side** of the double bond.

a. higher priority → Cl; Br ← higher priority

Two higher priority groups are on opposite sides: ***E*** **isomer.**

b. higher priority →; ← higher priority

Two higher priority groups are on the same side: ***Z*** **isomer.**

c. higher priority; H; higher priority

Two higher priority groups are on opposite sides: ***E*** **isomer.**

d. ***E*** **isomer**; O; H; H; higher priority; ***Z*** **isomer**; higher priority; higher priority

10.8

a. no stereoisomers; ***E***; ***E***; ***E***; ***E***

A

b. Isomer will all ***Z*** double bonds: ***Z***; ***Z***; ***Z***

10.9 To work backwards from a name to a structure:

[1] Find the parent name and functional group and draw, remembering that the double bond is between C1 and C2 for cycloalkenes.

[2] Add the substituents to the appropriate carbons.

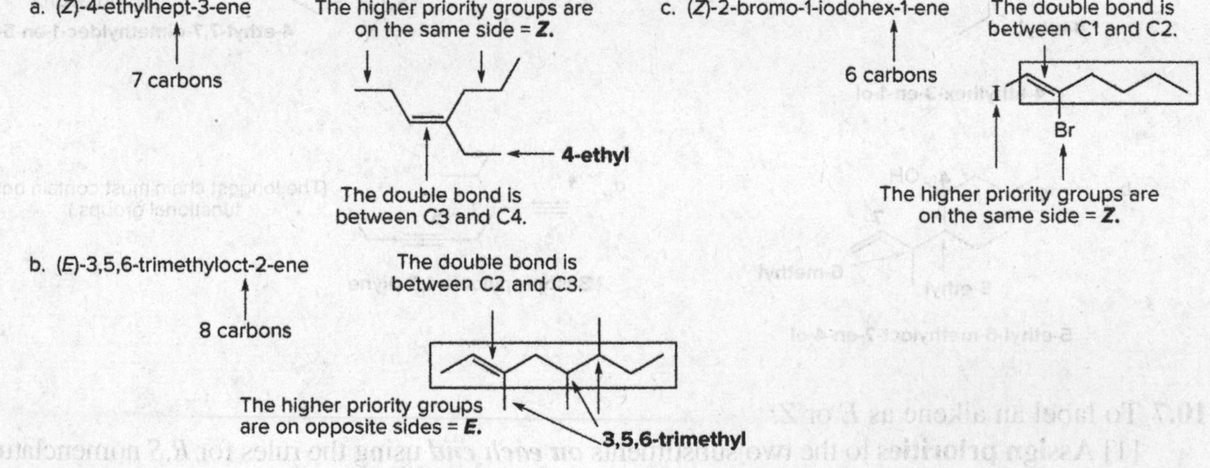
a. (*Z*)-4-ethylhept-3-ene
7 carbons
The higher priority groups are on the same side = ***Z***.
4-ethyl
The double bond is between C3 and C4.
b. (*E*)-3,5,6-trimethyloct-2-ene
8 carbons
The double bond is between C2 and C3.
The higher priority groups are on opposite sides = ***E***.
3,5,6-trimethyl
c. (*Z*)-2-bromo-1-iodohex-1-ene
6 carbons
The double bond is between C1 and C2.
I
Br
The higher priority groups are on the same side = ***Z***.

10.10 To rank the isomers by increasing boiling point:
Look for polarity differences: *small net dipoles* make an alkene more polar, giving it a higher boiling point than an alkene with *no net dipole*. Cis isomers have a higher boiling point than their trans isomers.

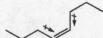

All dipoles cancel.
smallest surface area
no net dipole
lowest bp

Two dipoles cancel.
no net dipole
trans isomer
intermediate bp

Two dipoles reinforce.
net dipole
cis isomer
higher bp

Two dipoles reinforce.
net dipole
cis isomer
more surface area
highest bp

10.11 Increasing number of double bonds = decreasing melting point.

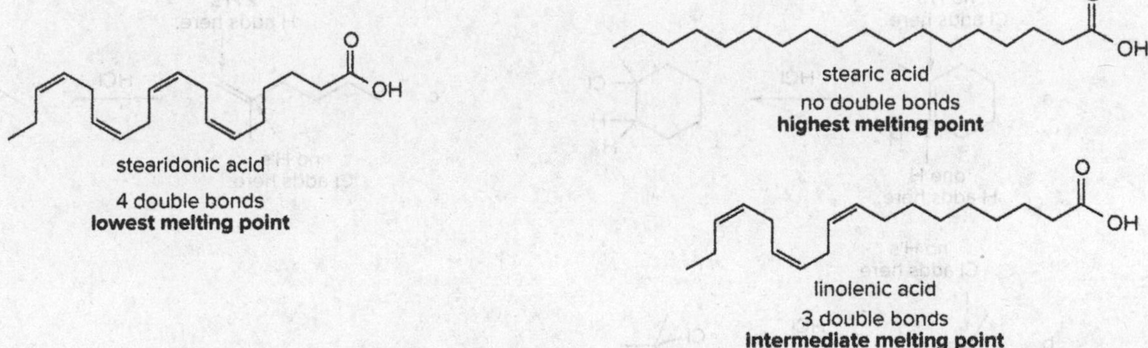

10.12

a. OH $\xrightarrow{H_2SO_4}$ +

c. Br, Br $\xrightarrow{2\ NaNH_2}$

b. Br $\xrightarrow{NaOCH_2CH_3}$ + (+ cis isomer)

10.13 Acetylene has a pK_a of 25, so **bases having a conjugate acid with a pK_a *above* 25** will be able to deprotonate it.

a. CH_3NH^- [pK_a (CH_3NH_2) = 40]
pK_a > 25 = **Can deprotonate acetylene.**

b. CO_3^{2-} [pK_a (HCO_3^-) = 10.2]
pK_a < 25 = **Cannot deprotonate acetylene.**

c. $CH_2{=}CH^-$ [pK_a ($CH_2{=}CH_2$) = 44]
pK_a > 25 = **Can deprotonate acetylene.**

d. $(CH_3)_3CO^-$ [pK_a ($(CH_3)_3COH$) = 18]
pK_a < 25 = **Cannot deprotonate acetylene.**

10.14 To draw the products of an addition reaction:
[1] Locate the two bonds that will be broken in the reaction. Always break the π bond.
[2] Draw the product by forming two new σ bonds.

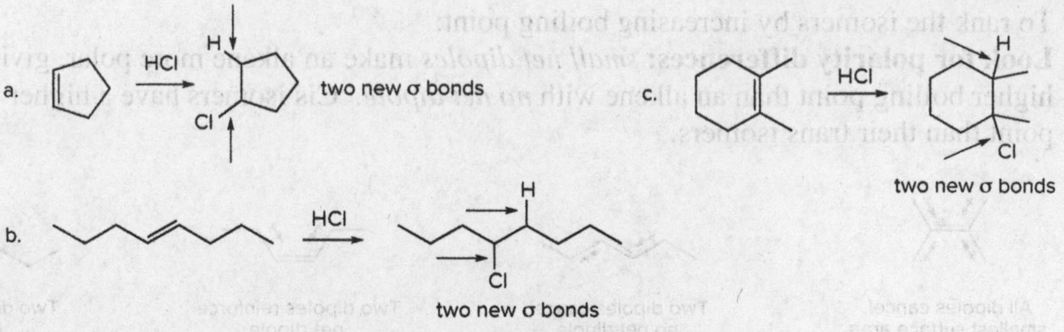

a. HCl → two new σ bonds

c. HCl → two new σ bonds

b. HCl → two new σ bonds

10.15 Addition to alkenes follows Markovnikov's rule: When HX adds to an unsymmetrical alkene, the H bonds to the C that has more H's to begin with.

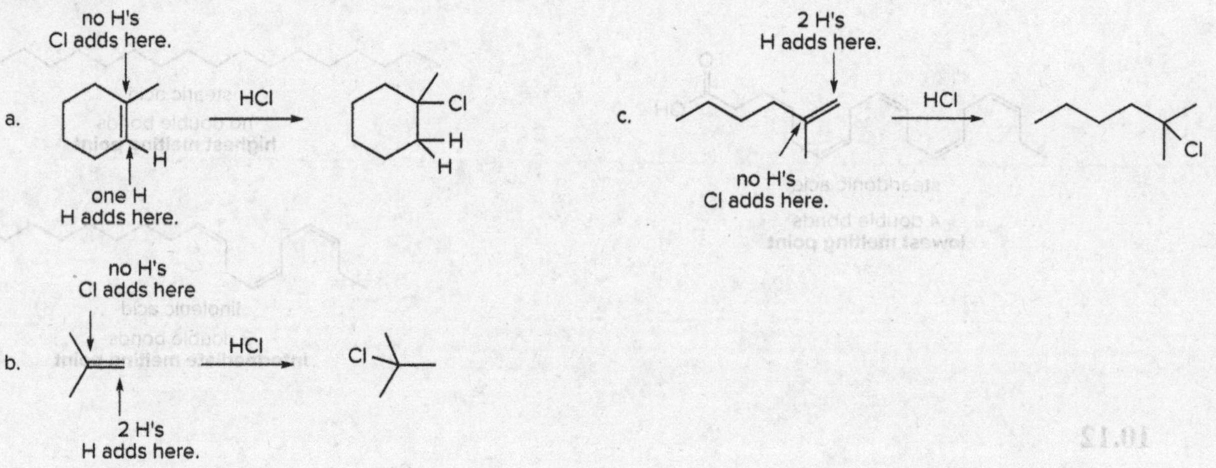

a. no H's Cl adds here. one H H adds here. HCl →

c. 2 H's H adds here. no H's Cl adds here. HCl →

b. no H's Cl adds here 2 H's H adds here. HCl →

10.16 Look for rearrangements of a carbocation intermediate to explain these results.

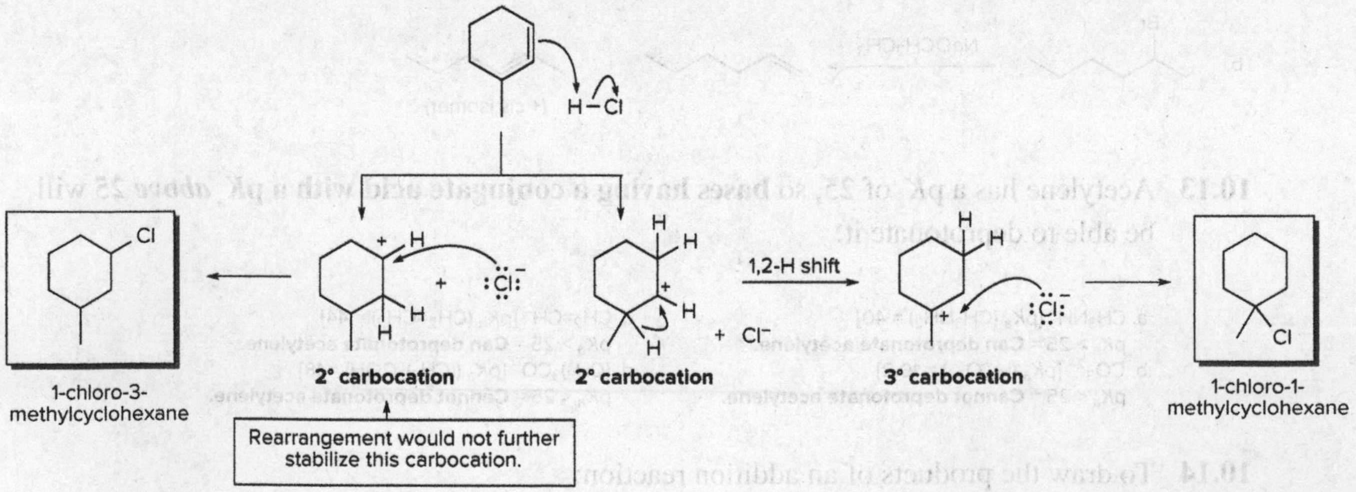

H–Cl

1-chloro-3-methylcyclohexane ← 2° carbocation + :Cl:⁻

Rearrangement would not further stabilize this carbocation.

2° carbocation + Cl⁻ → 1,2-H shift → 3° carbocation + :Cl:⁻ → 1-chloro-1-methylcyclohexane

10.17 Addition of HX to alkenes involves the formation of carbocation intermediates. Rearrangement of the carbocation will occur if it forms a more stable carbocation.

a.

3° carbocation
no rearrangement

b.

2° carbocation 2° carbocation

Rearrangement would not further stabilize either carbocation.
no rearrangement

c.

2° carbocation

Rearrangement would not further stabilize the carbocation.

2° carbocation
rearrangement

1,2-H shift

3° carbocation
more stable

10.18 To draw the products, remember that addition of HX proceeds via a carbocation intermediate.

a. HBr

new stereogenic center

enantiomers

b.

Addition of H^+ (from HCl) from above and below by Markovnikov's rule forms an achiral 3° carbocation.

achiral, trigonal planar 3° carbocation

Cl^- attacks from above and below.

diastereomers

10.19

a.

or

H^+ would add here to form a 3° carbocation.

H^+ would add here to form a 3° carbocation.

b.

or

H^+ would add here to form a 3° carbocation.

H^+ would add here to form a 3° carbocation.

c. OH ⟹ or
H⁺ would add here to form a 2° carbocation. (+ cis isomer)

10.20

pent-1-ene $\xrightarrow[H_2SO_4]{H_2O}$ H OH + HO H
enantiomers

10.21 Halogenation of an alkene adds two elements of X in an anti fashion.

a. $\xrightarrow{Br_2}$ Br, Br + Br, Br

b. $\xrightarrow{Cl_2}$ Cl, Cl + Cl, Cl

10.22 To draw the products of halogenation of an alkene, remember that the halogen adds to both ends of the double bond but only anti addition occurs.

a. $\xrightarrow{Cl_2}$ Cl, Cl + Cl, Cl
enantiomers

c. $\xrightarrow{Br_2}$ Br, Br + Br, Br
diastereomers

b. $\xrightarrow{Br_2}$ Br, Br
achiral meso compound

10.23 Halohydrin formation adds the elements of X and OH across the double bond in an anti fashion. The reaction is regioselective, so X ends up on the carbon that had more H's to begin with.

a. $\xrightarrow[DMSO, H_2O]{NBS}$ Br, HO + HO, Br

b. $\xrightarrow[H_2O]{Cl_2}$ OH, Cl + OH, Cl
Cl bonds to the carbon with more H's to begin with.

10.24 In hydroboration the boron atom is the electrophile and becomes bonded to the carbon atom that had more H's to begin with.

a. BH_3 → BH_2
C with more H's. B will add here.

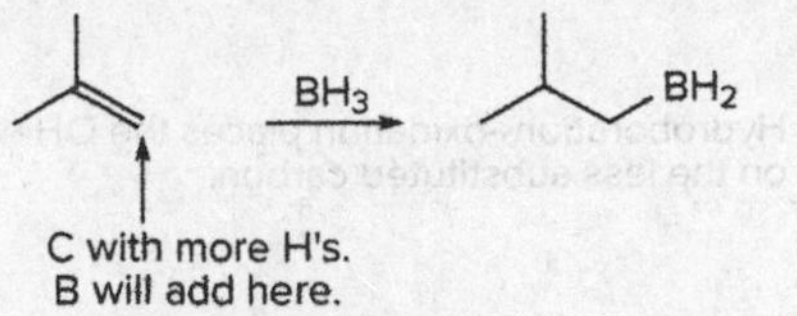

c. BH_3 → BH_2
C with more H's. B will add here.

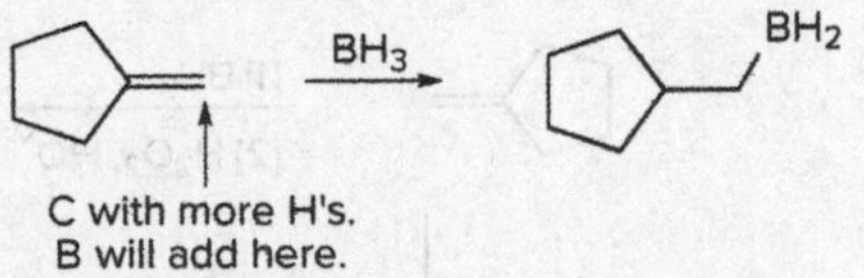

b. BH_3 → BH_2, H
C with more H's. B will add here.

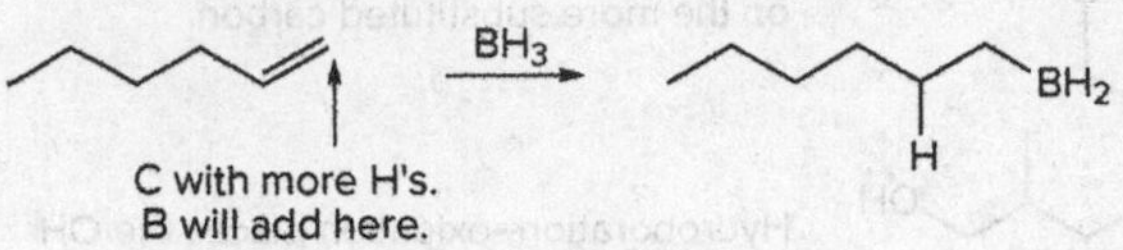

10.25 The hydroboration–oxidation reaction occurs in two steps:
[1] Syn addition of BH_3, with the boron on the less substituted carbon atom
[2] OH replaces the BH_2 with retention of configuration.

a. BH_3 → BH_2 → H_2O_2, ^-OH → OH

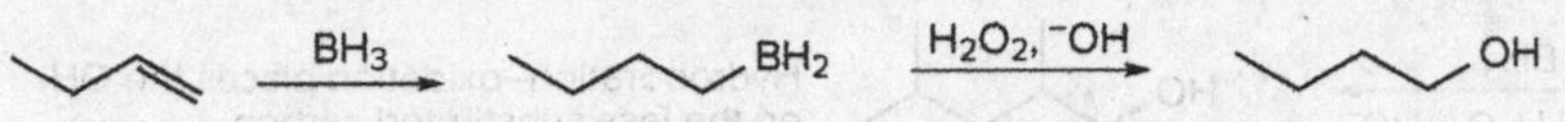

b. → H, BH_2 + H, BH_2 → H_2O_2, ^-OH → H, OH + H, OH

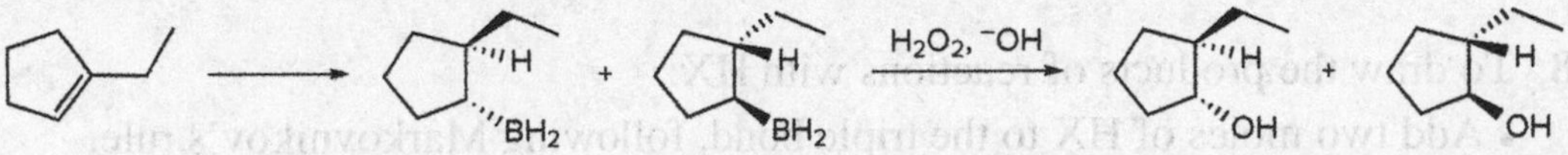

c. BH_3 → H, BH_2 + H, BH_2
H_2O_2, ^-OH
H, OH + H, OH

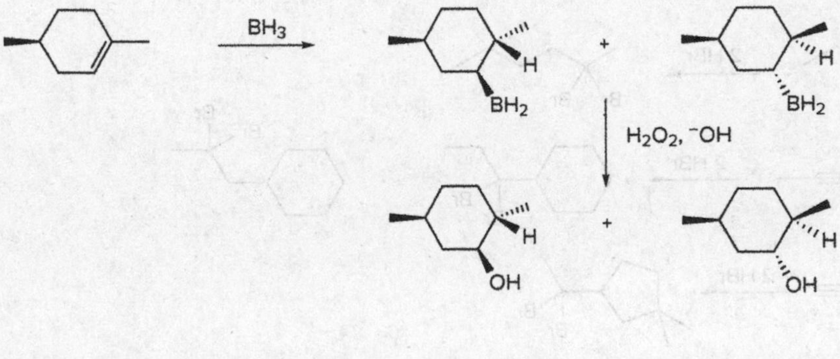

10.26 Remember that hydroboration results in addition of OH on the less substituted C.

a. HO ←

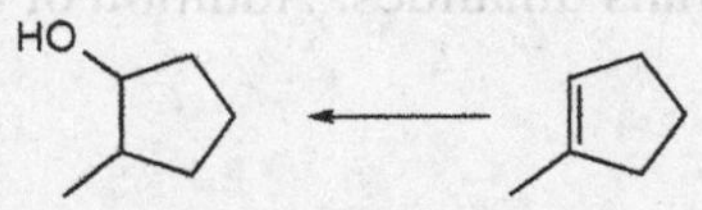

c. OH ←

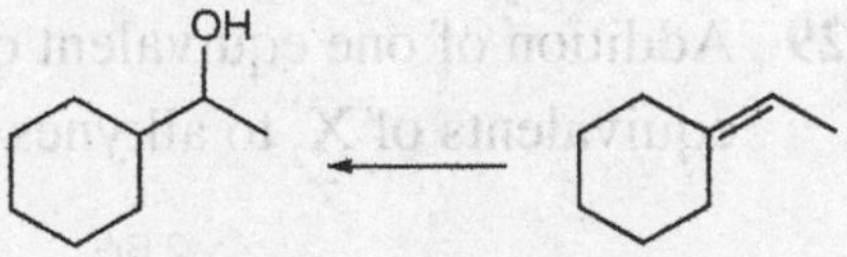

b. OH ← (*E* or *Z* isomer can be used.)

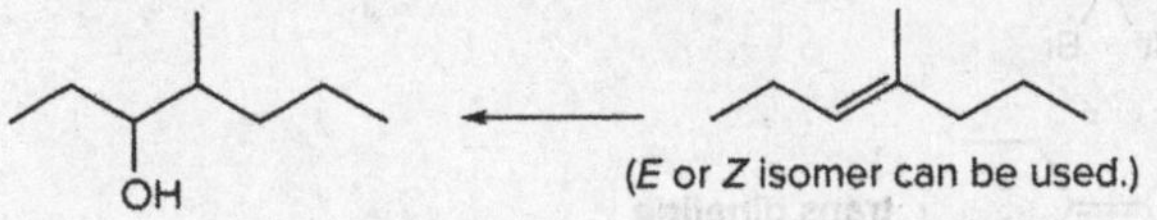

10.27

a. $\xrightarrow[H_2SO_4]{H_2O}$ Hydration places the OH on the more substituted carbon.

$\xrightarrow[[2]\ H_2O_2,\ HO^-]{[1]\ BH_3}$ Hydroboration–oxidation places the OH on the less substituted carbon.

b. $\xrightarrow[H_2SO_4]{H_2O}$ Hydration places the OH on the more substituted carbon.

$\xrightarrow[[2]\ H_2O_2,\ HO^-]{[1]\ BH_3}$ Hydroboration–oxidation places the OH on the less substituted carbon.

c. $\xrightarrow[H_2SO_4]{H_2O}$ Hydration places the OH on the more substituted carbon.

$\xrightarrow[[2]\ H_2O_2,\ HO^-]{[1]\ BH_3}$ HO Hydroboration–oxidation places the OH on the less substituted carbon.

10.28 To draw the products of reactions with HX:

- Add two moles of HX to the triple bond, following Markovnikov's rule.
- Both X's end up on the more substituted C.

a. $\xrightarrow{2\ HBr}$ Br Br

b. $\xrightarrow{2\ HBr}$ Br Br + Br Br

c. $\xrightarrow{2\ HBr}$ Br Br

10.29 Addition of one equivalent of X_2 to alkynes forms trans dihalides. Addition of two equivalents of X_2 to alkynes forms tetrahalides.

a. $\xrightarrow{2\ Br_2}$ Br Br Br Br

b. $\xrightarrow{Cl_2}$ Cl Cl **trans dihalide**

10.30 To draw the keto form of each enol:

[1] Change the C–OH to a C=O at one end of the double bond.

[2] Add a proton at the other end of the double bond.

a. OH — O, H, new C–H bond

c. OH — O, H, new C–H bond

b. OH, H — O, H H, new C–H bond

d. OH — O, H, new C–H bond

10.31 The treatment of alkynes with H_2O, H_2SO_4, and $HgSO_4$ yields ketones.

H_2O / H_2SO_4, $HgSO_4$ → OH + HO ⇌ O + O

Two enols form.
(*E* and *Z* isomers)

two ketones after tautomerization

10.32 Reaction with H_2O, H_2SO_4, and $HgSO_4$ adds the oxygen to the *more* substituted carbon. Reaction with [1] R_2BH, [2] H_2O_2, ^-OH adds the oxygen to the *less* substituted carbon.

a. H_2O / H_2SO_4, $HgSO_4$ → O

Forms a **ketone.** H_2O is added with the O atom on the *more* substituted carbon.

[1] R_2BH / [2] H_2O_2, HO^- → H, O

Forms an **aldehyde.** H_2O is added with the O atom on the *less* substituted carbon.

b. H_2O / H_2SO_4, $HgSO_4$ → O

Forms a **ketone.** H_2O is added with the O atom on the *more* substituted carbon.

[1] R_2BH / [2] H_2O_2, HO^- → O, H

Forms an **aldehyde.** H_2O is added with the O atom on the *less* substituted carbon.

10.33

a. H−C≡C−H —[1] NaH→ H−C≡C:⁻ + H_2 —[2] (Cl)→ + NaCl

b. (cyclopentyl)−C≡CH —[1] NaH→ (cyclopentyl)−C≡C:⁻ + H_2 —[2] (Br)→ + NaBr

1° alkyl halide **substitution product**

—[1] $NaNH_2$→ (cyclopentyl)−C≡C:⁻ + NH_3 —[2] (Cl)→ + NaCl +

3° alkyl halide **elimination product**

10.34

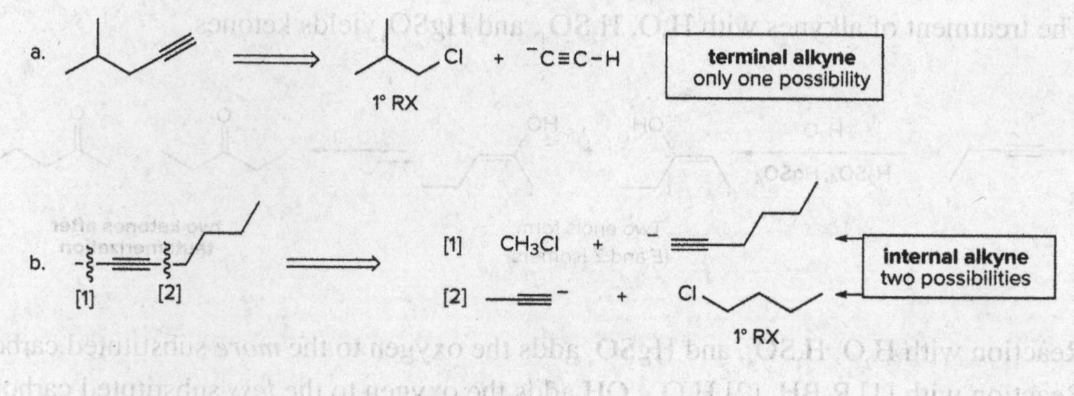

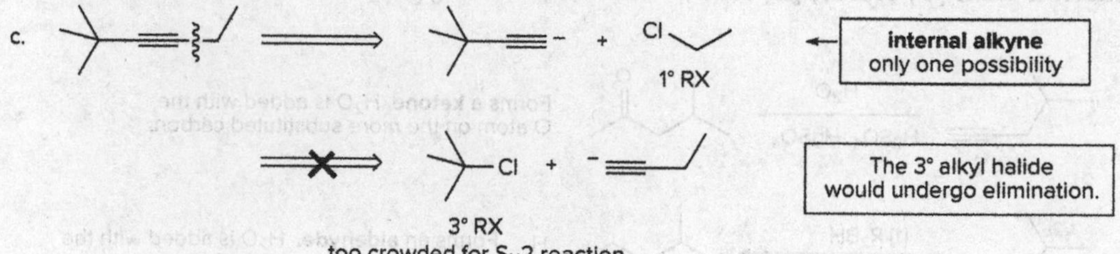

10.35

HC≡C−H + $Na^+H:^-$ → HC≡C:⁻ + H_2 + Na^+ → (Br) → H−C≡C− + $Na^+H:^-$ → :⁻C≡C− + H_2 ; + Na^+Br^-

(Br) → + Br^-

10.36

a. 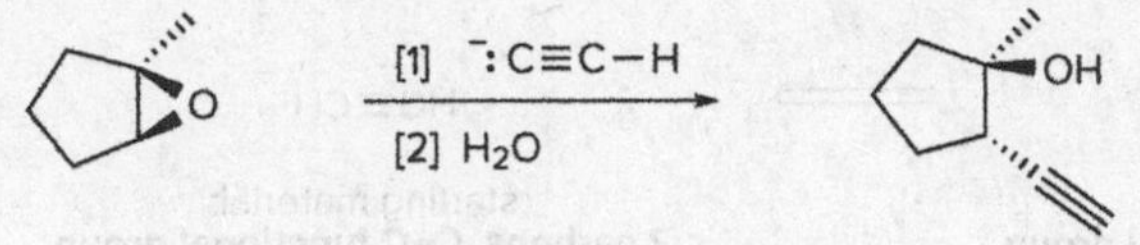 [1] $^{-}$:C≡C–H [2] H_2O

Epoxide is drawn up, so the **acetylide anion attacks from *below* at *less* substituted C.**

b. 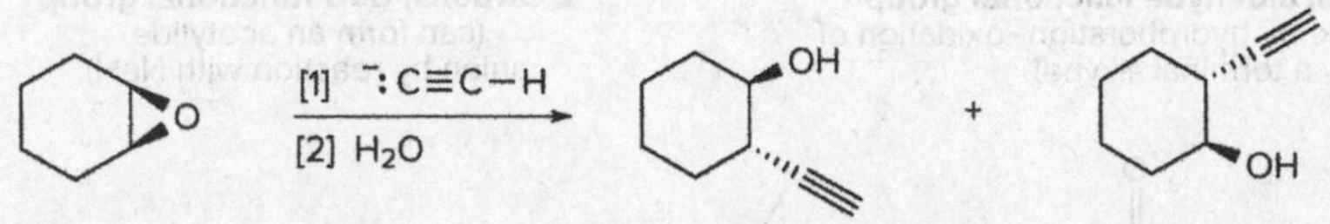 [1] $^{-}$:C≡C–H [2] H_2O

enantiomers

Backside attack of the nucleophile ($^{-}$C≡CH) occurs at either C because both ends are equally substituted.

10.37 To identify the acetylide anion and epoxide starting materials, locate the carbon bonded to the OH group and the carbon adjacent to it bonded to the C≡C.

a. 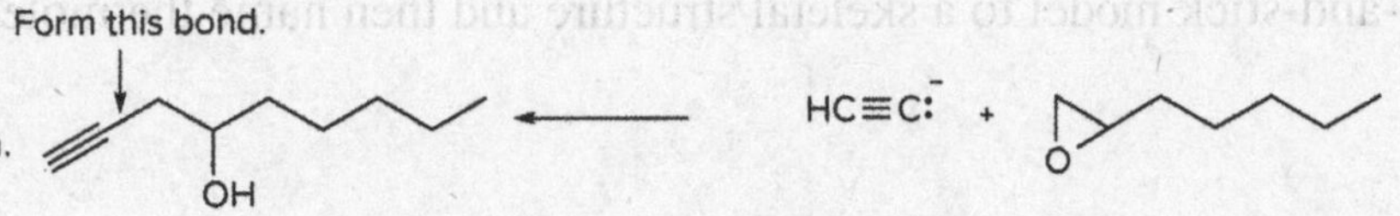 Form this bond. OH ⟵ HC≡C:$^{-}$ + [epoxide]

b. 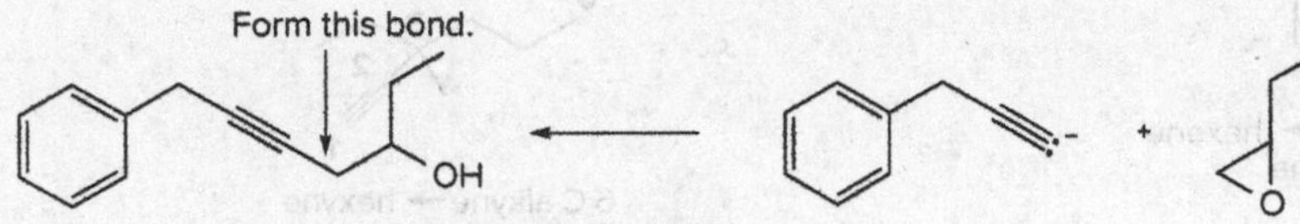 Form this bond. OH ⟵ [acetylide anion] + [epoxide]

c. Form this bond. OH ⟵ [epoxide] + [acetylide anion]

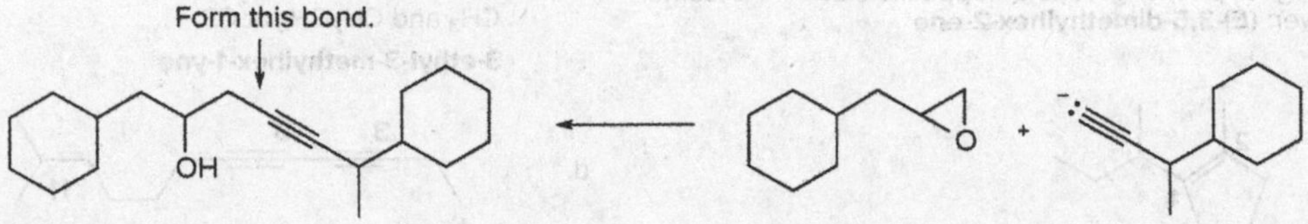

10.38 To use a retrosynthetic analysis:

[1] **Count the number of carbon atoms** in the starting material and product.

[2] **Look at the functional groups** in the starting material and product.

a. Determine what types of reactions can form the product.

b. Determine what types of reactions the starting material can undergo.

[3] **Work backwards** from the product to make the starting material.

[4] Write out the synthesis in the synthetic direction.

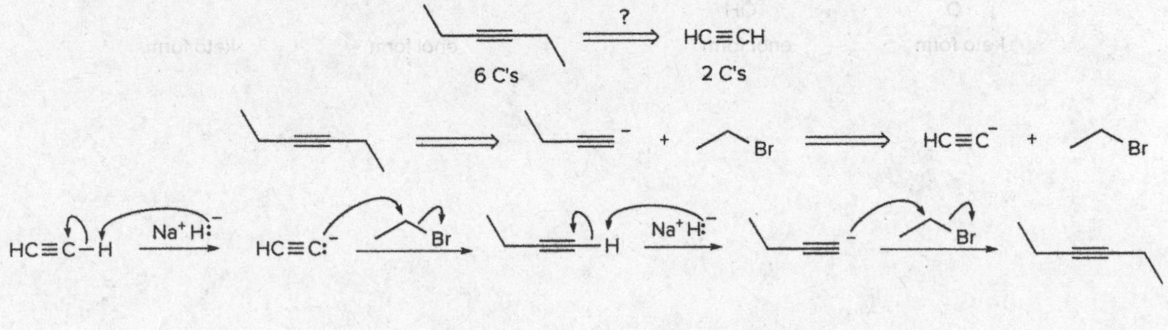

6 C's ⟹ ? HC≡CH 2 C's

⟹ [acetylide anion] + [bromoethane, Br] ⟹ HC≡C^{-} + [bromoethane, Br]

HC≡C–H → ($Na^+H:^-$) HC≡C:$^{-}$ → (Br) [but-1-yne]–H → ($Na^+H:^-$) [acetylide anion] → (Br) [hex-3-yne]

10.39

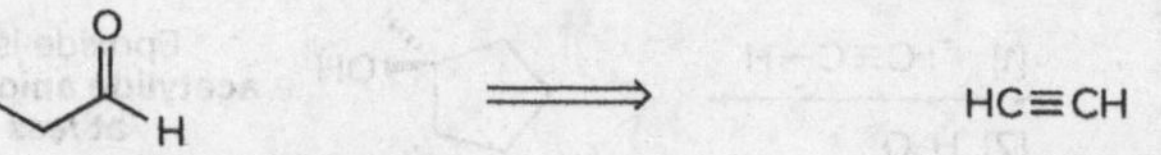

product:
4 carbons, aldehyde functional group
(can be made by hydroboration–oxidation of a terminal alkyne)

$HC\equiv CH$

starting material:
2 carbons, C≡C functional group
(can form an acetylide anion by reaction with NaH)

Retrosynthetic analysis: H–C≡C–H

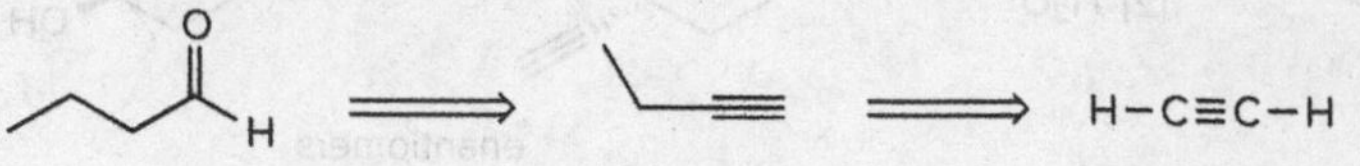

Forward direction: H–C≡C–H, $Na^+ H:^-$, $^-:C\equiv C–H$, Br, [1] R_2BH, [2] H_2O_2, HO^-

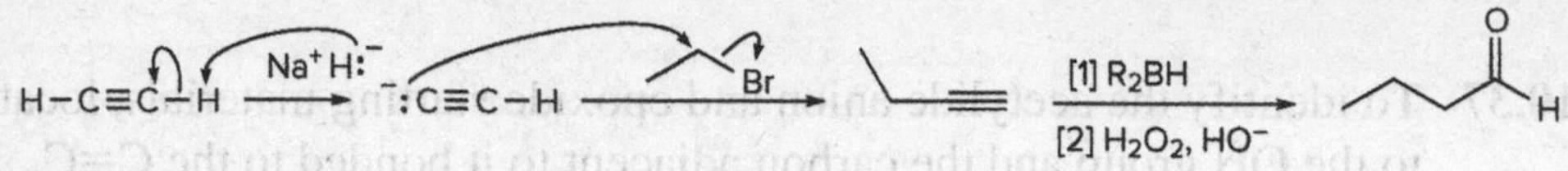

10.40 Convert each ball-and-stick model to a skeletal structure and then name the molecule.

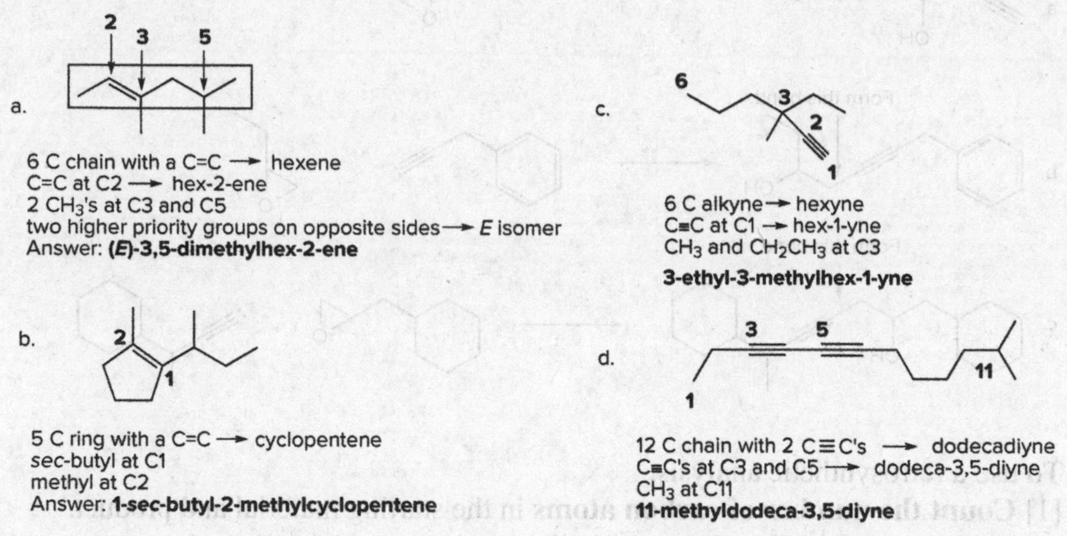

a.
2 3 5
6 C chain with a C=C → hexene
C=C at C2 → hex-2-ene
2 CH3's at C3 and C5
two higher priority groups on opposite sides → E isomer
Answer: (E)-3,5-dimethylhex-2-ene
c.
6 3 2 1
6 C alkyne → hexyne
C≡C at C1 → hex-1-yne
CH3 and CH2CH3 at C3
3-ethyl-3-methylhex-1-yne
b.
2 1
5 C ring with a C=C → cyclopentene
sec-butyl at C1
methyl at C2
Answer: 1-sec-butyl-2-methylcyclopentene
d.
3 5 1 11
12 C chain with 2 C≡C's → dodecadiyne
C≡C's at C3 and C5 → dodeca-3,5-diyne
CH3 at C11
11-methyldodeca-3,5-diyne

10.41

a. O, keto form; OH, enol form

b. OH, enol form; O, H, keto form

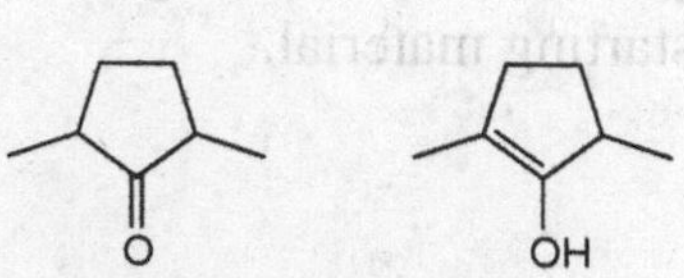

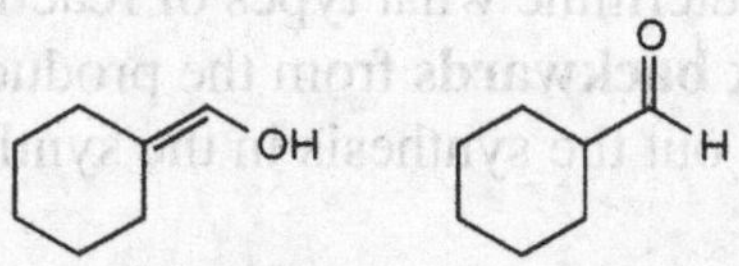

10.42 Use the directions from Answer 10.2 to calculate degrees of unsaturation.

a. C_6H_8
[1] maximum number of H's = 2*n* + 2 = 2(6) + 2 = 14
[2] subtract actual from maximum = 14 – 8 = 6
[3] divide by 2 = 6/2 = **3 degrees of unsaturation**

b. $C_{40}H_{56}$
[1] maximum number of H's = 2*n* + 2 = 2(40) + 2 = 82
[2] subtract actual from maximum = 82 – 56 = 26
[3] divide by 2 = 26/2 = **13 degrees of unsaturation**

c. $C_{10}H_{16}O_2$
Ignore both O's.
[1] maximum number of H's = 2*n* + 2 = 2(10) + 2 = 22
[2] subtract actual from maximum = 22 – 16 = 6
[3] divide by 2 = 6/2 = **3 degrees of unsaturation**

d. C_8H_9Br
Because of Br, add one H (9 + 1 = 10 H's).
[1] maximum number of H's = 2*n* + 2 = 2(8) + 2 = 18
[2] subtract actual from maximum = 18 – 10 = 8
[3] divide by 2 = 8/2 = **4 degrees of unsaturation**

e. C_8H_9ClO
Ignore the O; count Cl as one more H (9 + 1 = 10 H's).
[1] maximum number of H's = 2*n* + 2 = 2(8) + 2 = 18
[2] subtract actual from maximum = 18 – 10 = 8
[3] divide by 2 = 8/2 = **4 degrees of unsaturation**

f. $C_7H_{11}N$
Because of N, subtract one H (11 – 1 = 10 H's).
[1] maximum number of H's = 2*n* + 2 = 2(7) + 2 = 16
[2] subtract actual from maximum = 16 – 10 = 6
[3] divide by 2 = 6/2 = **3 degrees of unsaturation**

g. C_4H_8BrN
Because of Br, add one H, but subtract one for N (8 + 1 – 1 = 8 H's).
[1] maximum number of H's = 2*n* + 2 = 2(4) + 2 = 10
[2] subtract actual from maximum = 10 – 8 = 2
[3] divide by 2 = 2/2 = **1 degree of unsaturation**

h. $C_{10}H_{18}ClNO$
Add one H because of Cl and subtract 1 H for N (18 + 1 – 1 = 18).
[1] maximum number of H's = 2*n* + 2 = 2(10) + 2 = 22
[2] subtract actual from maximum = 22 – 18 = 4
[3] divide by 2 = 4/2 = **2 degrees of unsaturation**

10.43 Name the compounds using the rules in Answers 10.5 and 10.7.

a.
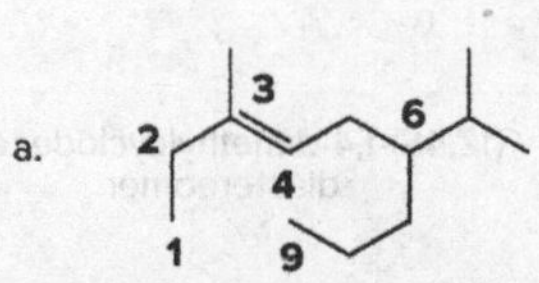

9 C chain with a double bond = **nonene**
(*E*)-6-isopropyl-3-methylnon-3-ene

b.
2-isopropyl
pent-1-ene
4-methyl
2-isopropyl-4-methylpent-1-ene

c.
3
2
5
1
6 C ring with a double bond = **hexene**
5-*sec*-butyl-1,3,3-trimethylcyclohexene

d.
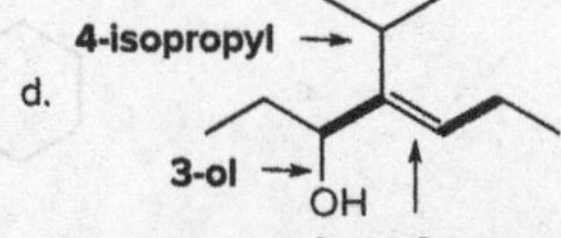

(*E*)-4-isopropylhept-4-en-3-ol

e.
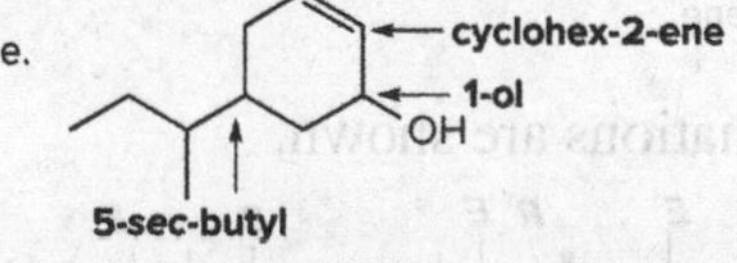

5-*sec*-butylcyclohex-2-enol

f.
non-4-yne
6-ethyl
3-ethyl
6
3
1
7-methyl
7
3,6-diethyl-7-methylnon-4-yne

g.
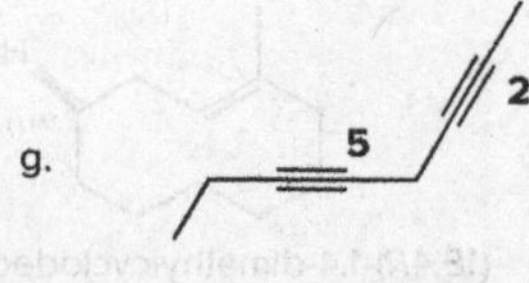

octa-2,5-diyne

h.
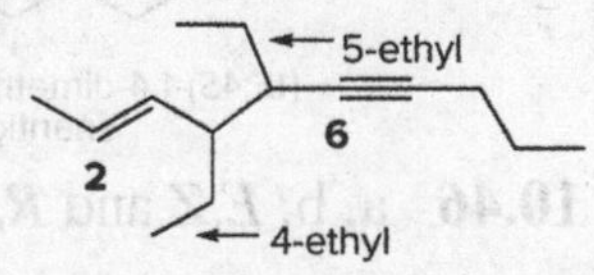

(*E*)-4,5-diethyldec-2-en-6-yne

10.44 Use the directions from Answer 10.9.

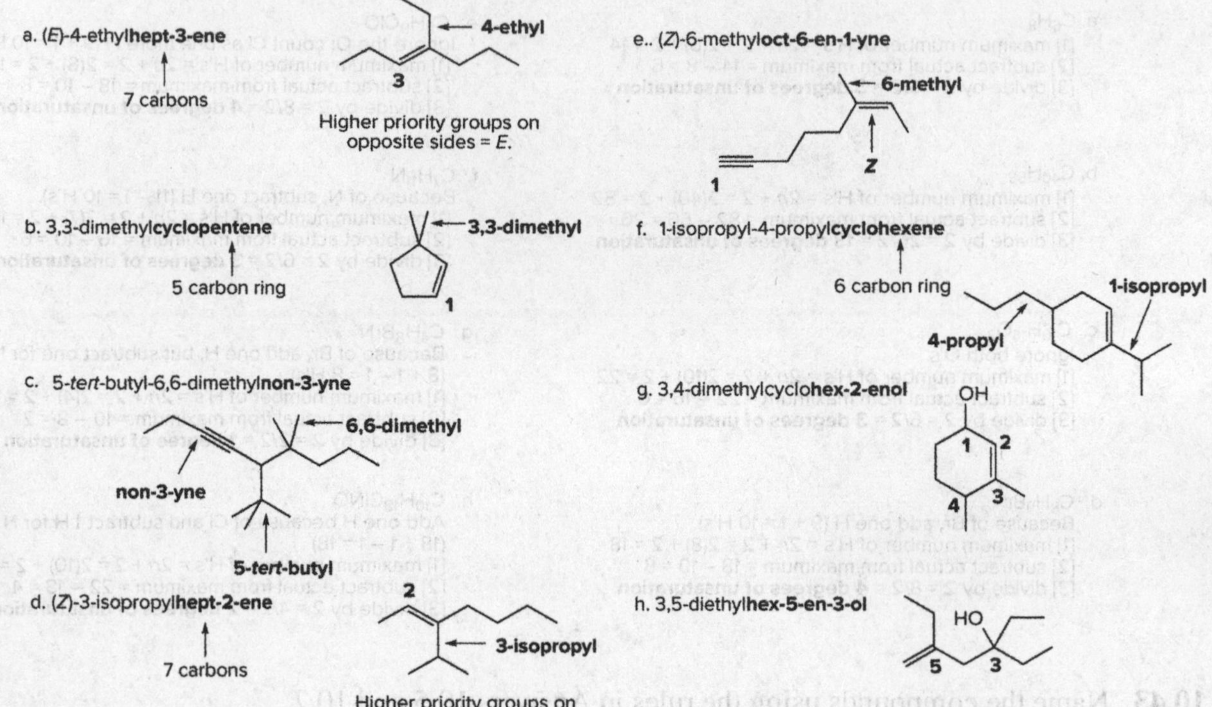

10.45

a. H

(1E,4R)-1,4-dimethylcyclodecene

b. H

(1E,4S)-1,4-dimethylcyclodecene enantiomer

c. (1Z,4S)-1,4-dimethylcyclodecene diastereomer

(1Z,4R)-1,4-dimethylcyclodecene diastereomer

10.46 a, b. *E*,*Z* and *R*,*S* designations are shown.

E E R E S E S
CH_3O S
O O H N CHO OH
OCH_3
E Z S E S

c. Nine double bonds that can be *E* or *Z*
Six tetrahedral stereogenic centers
Maximum possible number of stereoisomers = 2^{15}

10.47

a, b.

erlotinib

most acidic C–H proton

shortest C–C single bond $Csp–Csp^2$

c.

terbinafine

(1) $Csp^2–Csp^3$
(2) $Csp^2–Csp$
(3) $Csp–Csp^3$

Increasing bond strength:
(1) < (3) < (2)

d.

10.48

eleostearic acid

a.

all trans double bonds
higher melting point

b.

all cis double bonds
lower melting point

10.49 Keto–enol tautomers are constitutional isomers in equilibrium that differ in the location of a double bond and a hydrogen. The OH in an enol must be bonded to a C=C.

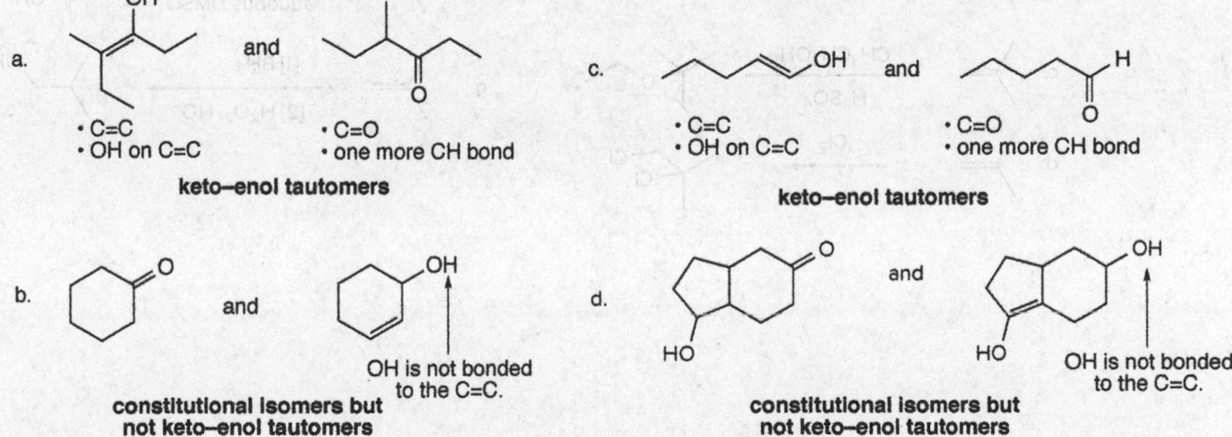

10.50 To draw the enol form of each keto form: [1] Change the C=O to a C–OH. [2] Change one single C–C bond to a double bond, making sure the OH group is bonded to the C=C. Use the directions from Answer 10.30 to draw each keto form.

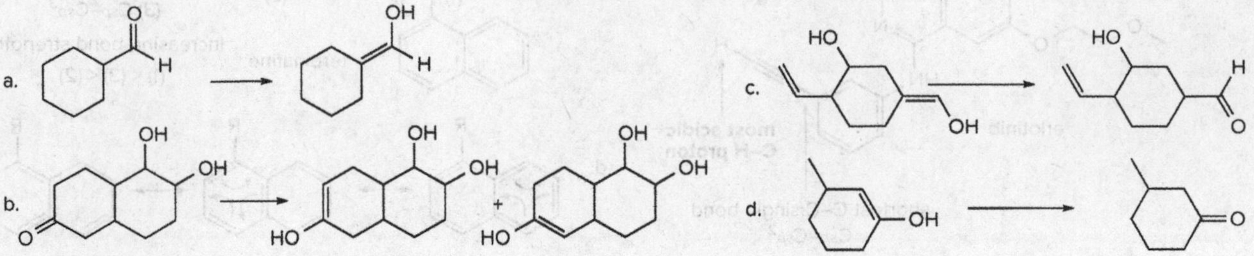

a. →

c. →

b. → +

d. →

10.51 Tautomers are constitutional isomers that are in equilibrium and differ in the location of a double bond and a hydrogen atom.

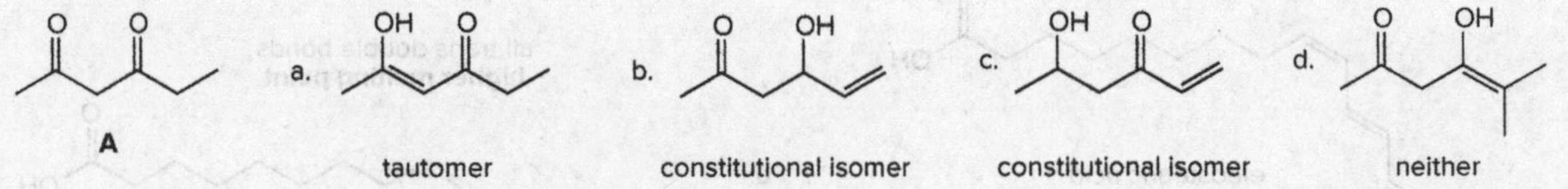

A

a. tautomer

b. constitutional isomer

c. constitutional isomer

d. neither

10.52

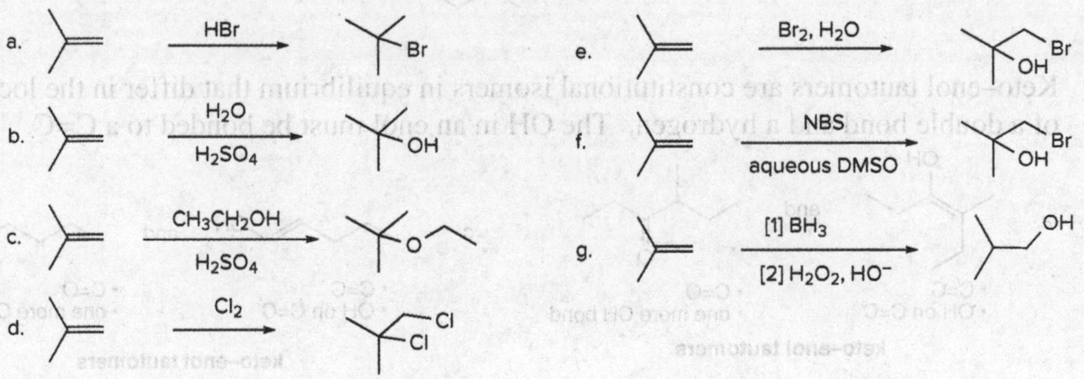

a. HBr

b. H_2O, H_2SO_4

c. CH_3CH_2OH, H_2SO_4

d. Cl_2

e. Br_2, H_2O

f. NBS, aqueous DMSO

g. [1] BH_3, [2] H_2O_2, HO^-

10.53

a. HCl (2 equiv) → Cl Cl

b. HBr (2 equiv) → Br Br

c. Cl_2 (2 equiv) → Cl Cl Cl Cl

d. H_2O, H_2SO_4, $HgSO_4$ → O

e. [1] R_2BH [2] H_2O_2, HO^- → O H

f. NaH → Na^+

g. [1] $^-NH_2$ [2] CH_3CH_2Br

h. [1] $^-NH_2$ [2] O (epoxide) [3] H_2O → HO

10.54

a. O ⟹ OH ⟹

b. O ⟹ OH ⟹

c. O ⟹ OH ⟹

10.55 Hydroboration–oxidation results in addition of an OH group on the *less* substituted carbon, whereas acid-catalyzed addition of H_2O results in the addition of an OH group on the *more* substituted carbon.

a. OH ← hydroboration–oxidation

b. OH ← or; acid-catalyzed addition

c. OH ← hydroboration–oxidation

d. OH — Both methods would give product mixtures.

10.56

a. HCl

H adds here to less substituted C.

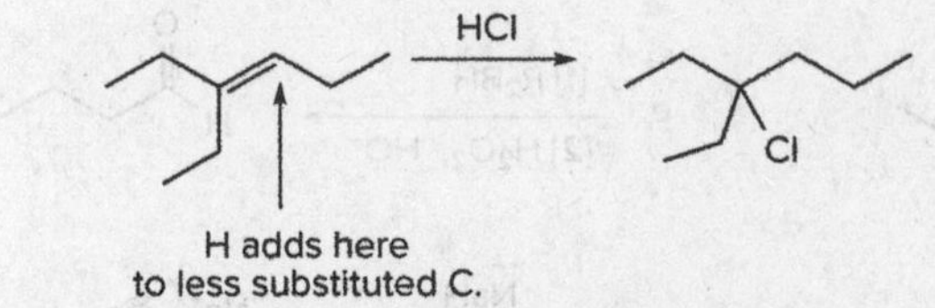

d. Br_2, H_2O

Br adds here to less substituted C.

Br OH

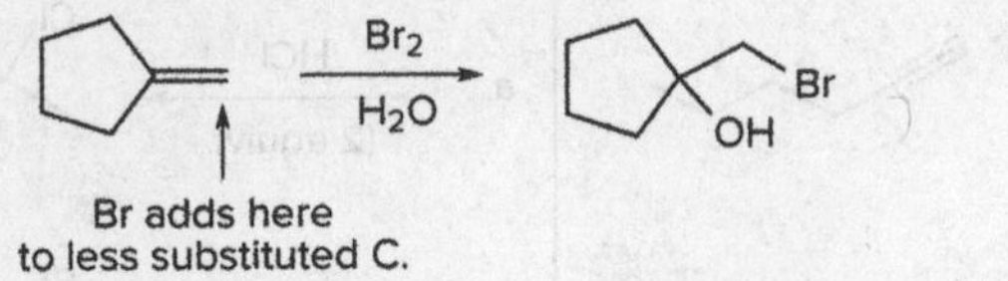

b. H_2O, H_2SO_4

OH

H adds here to less substituted C.

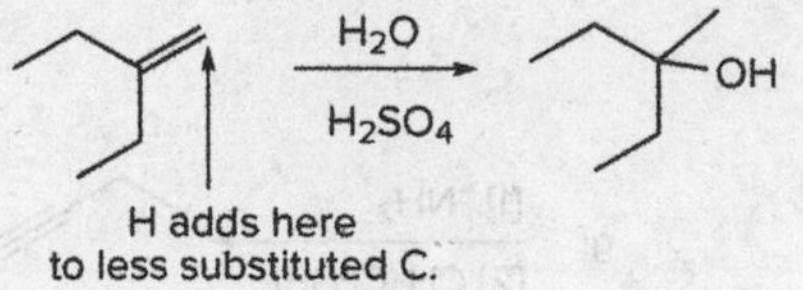

e. [1] 9-BBN [2] H_2O_2, HO^-

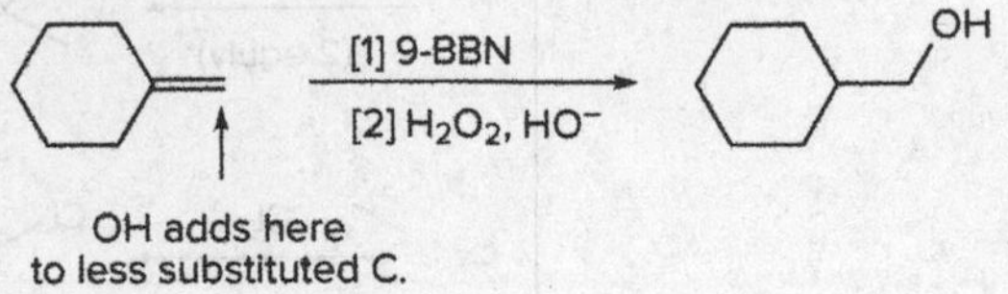

OH

OH adds here to less substituted C.

c. [1] BH_3 [2] H_2O_2, HO^-

OH

BH_3 adds here to less substituted C.

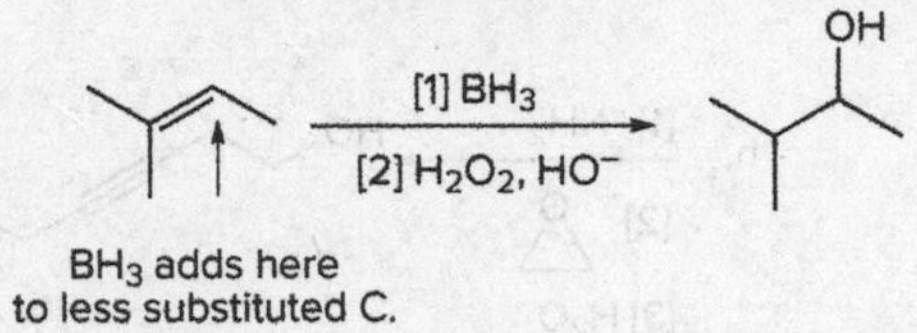

f. Br_2

Br Br

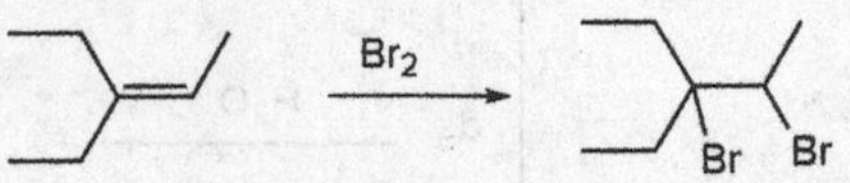

10.57

a. H_2O, H_2SO_4 → OH + OH

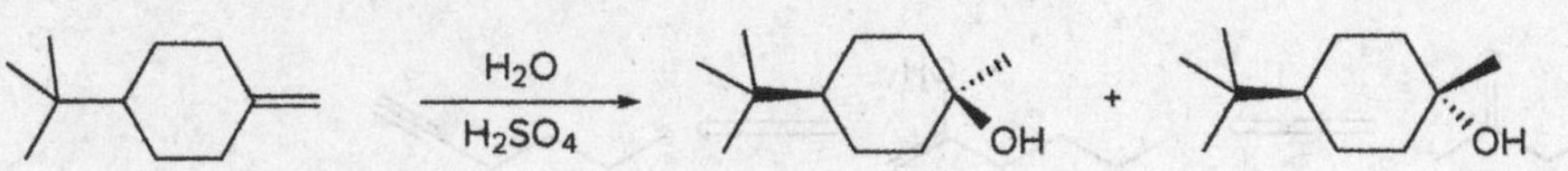

b. HI → I H + H I

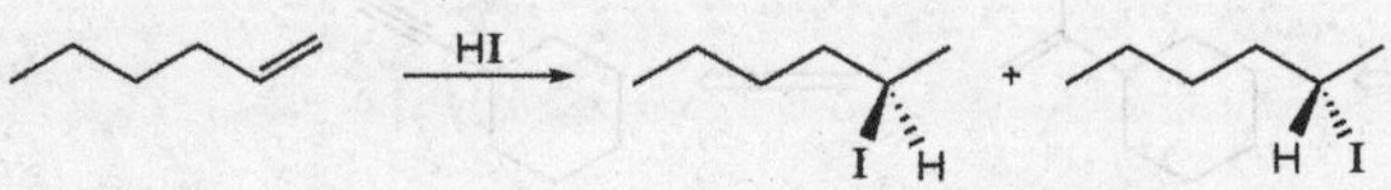

c. Cl_2 → Cl Cl + Cl Cl only anti addition

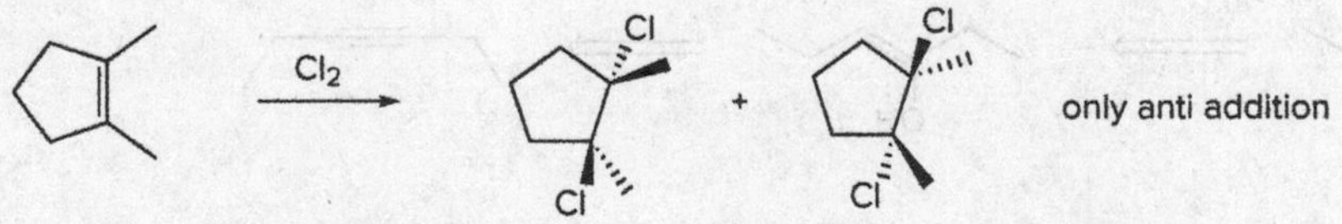

d. [1] BH_3 [2] H_2O_2, HO^- → H OH + H OH only syn addition

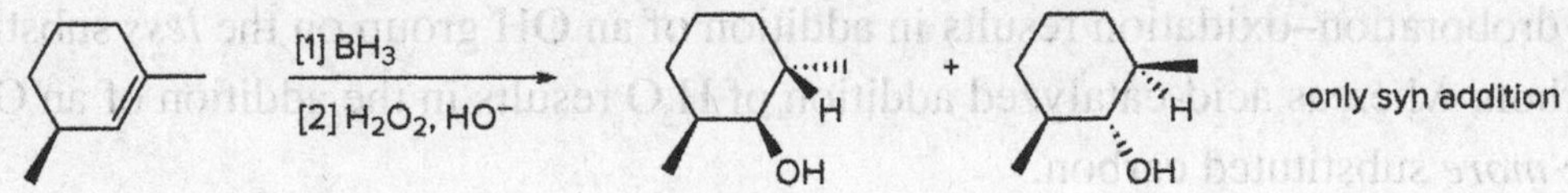

e. Cl_2, H_2O → Cl OH + OH Cl only anti addition

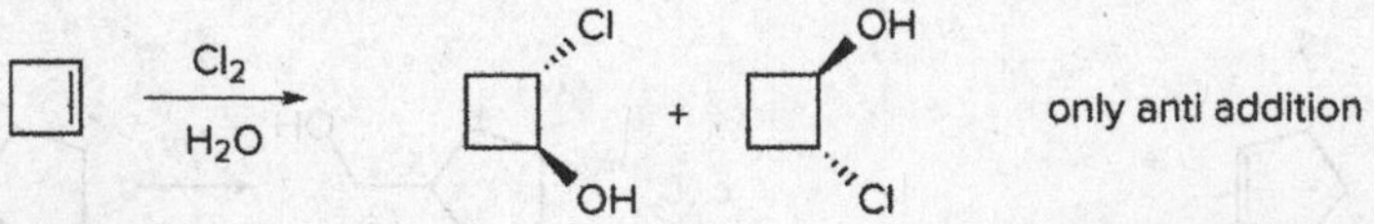

f. H_2O, H_2SO_4 → H OH + HO H + OH

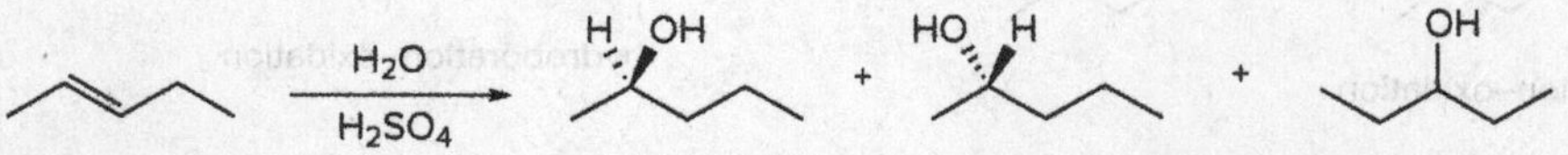

g. Br_2 → Br Br + Br Br

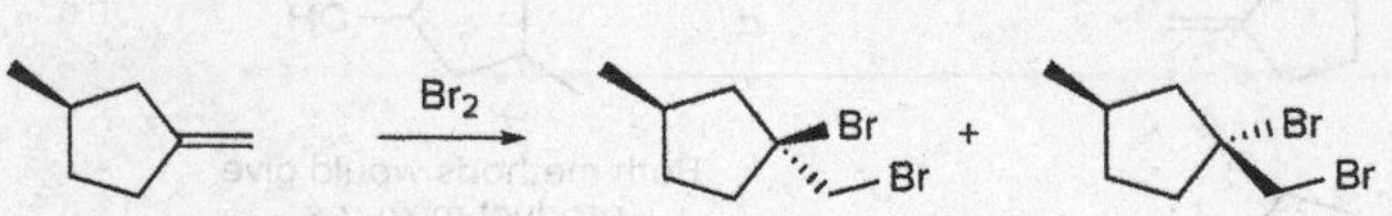

h. CH_3O CH_3O HCl → CH_3O CH_3O Cl + CH_3O CH_3O Cl

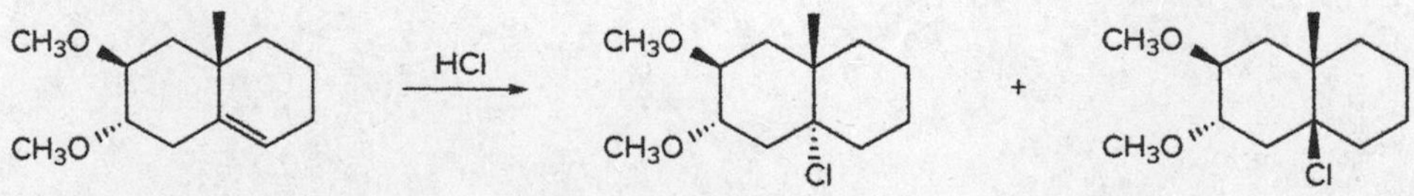

10.58

a. The trans isomer gives **A** and **B**, which are drawn with 2 Br's anti, the stereochemistry that results during Br_2 addition.

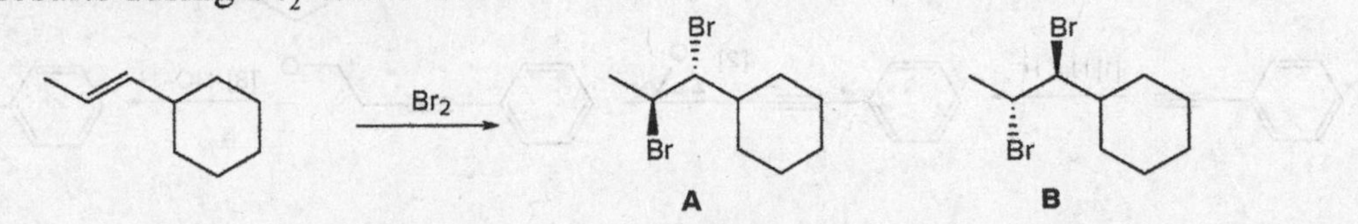

Br_2

Br Br **A** Br Br **B**

b. Because **C** and **D** are drawn with 2 Br's on the same side of the carbon skeleton, the molecules must be re-drawn with Br's anti to see the stereochemistry of the C=C.

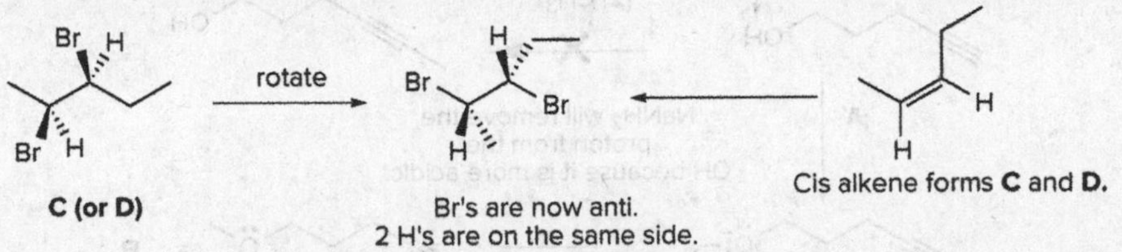

Br H Br H **C (or D)** rotate H Br Br H

Br's are now anti.
2 H's are on the same side.

Cis alkene forms **C** and **D.**

10.59

a. 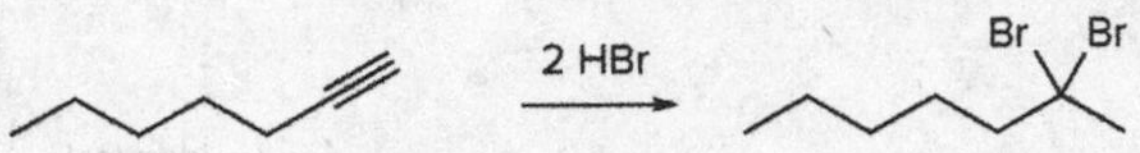 2 HBr → Br Br

b. 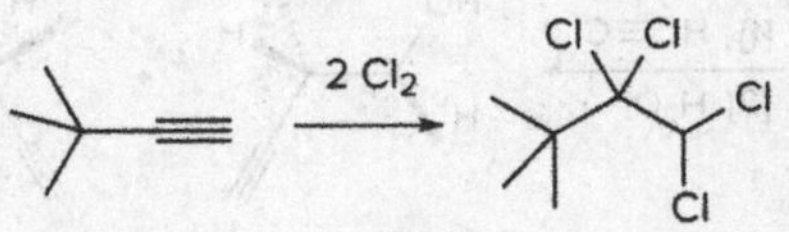 2 Cl_2 → Cl Cl Cl Cl

c. 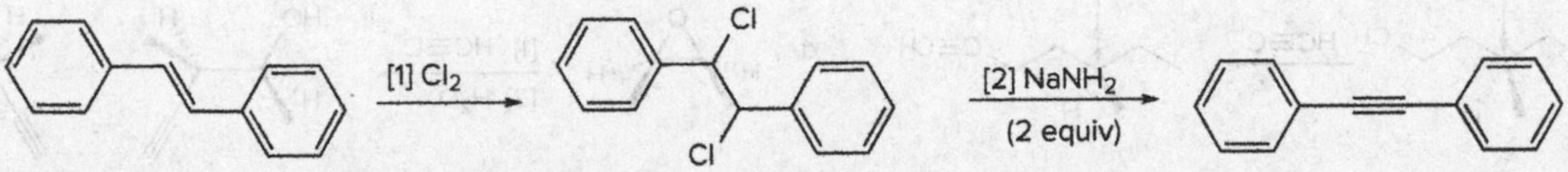 [1] Cl_2 → Cl Cl [2] $NaNH_2$ (2 equiv) →

d. 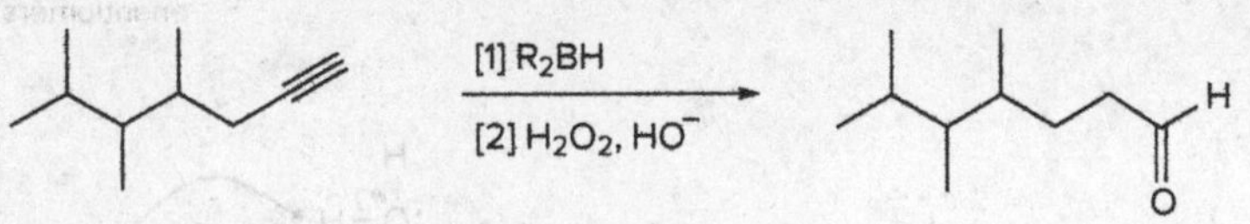 [1] R_2BH [2] H_2O_2, HO^- → H O

e. $HC{\equiv}C^-$ + D_2O → $HC{\equiv}CD$ + DO^-

f. 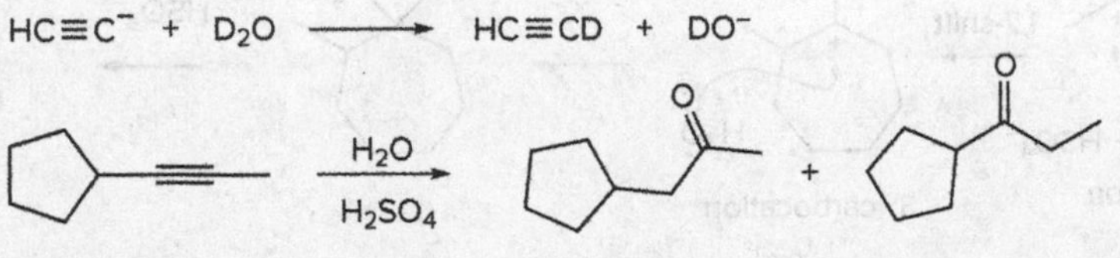 H_2O, H_2SO_4 → O + O

g. 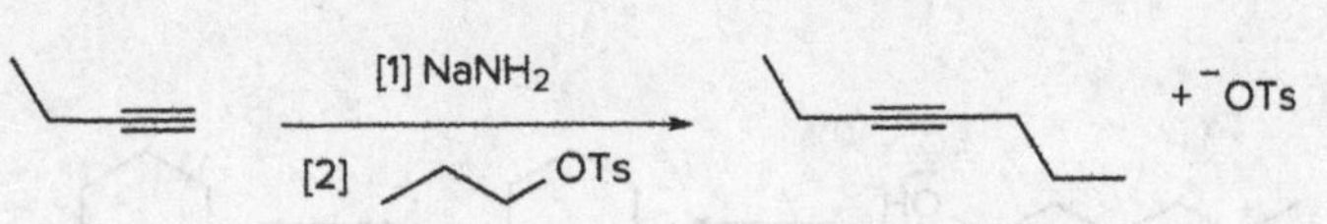 [1] $NaNH_2$ [2] OTs → + ^-OTs

h. O [1] $HC{\equiv}C^-$ → O^- [2] HO–H → OH

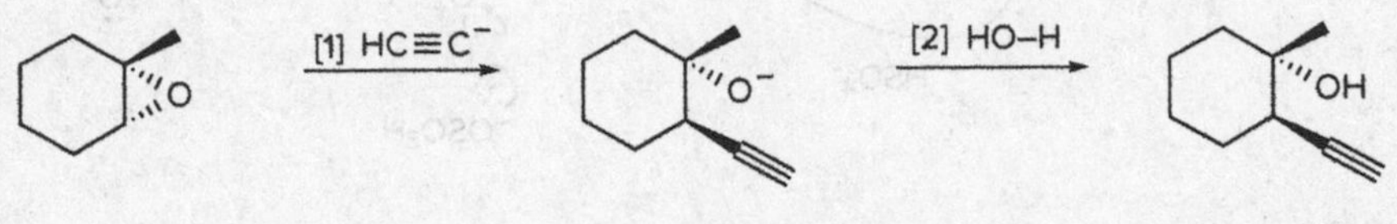

i. [1] $NaNH_2$ → [2] (iodomethyl)cyclohexane, I

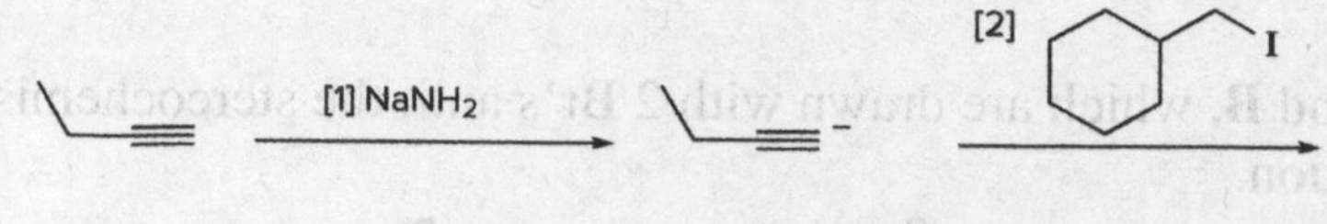

j. 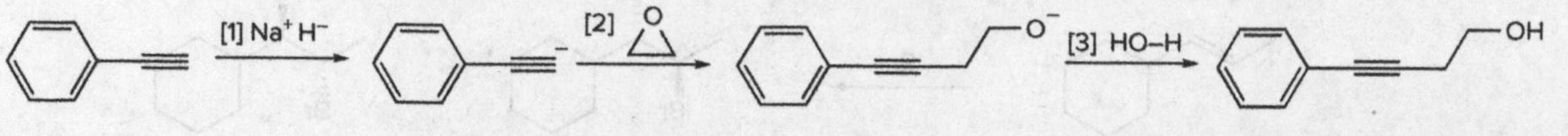 [1] $Na^+ H^-$ → [2] epoxide → [3] HO–H → OH

10.60

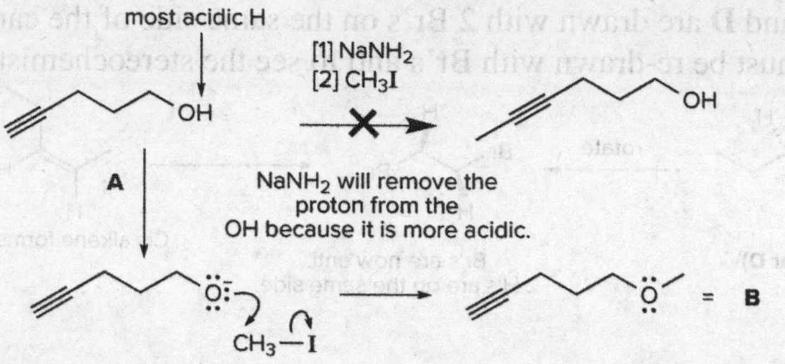

most acidic H

OH

A

[1] $NaNH_2$
[2] CH_3I

✕ → OH

NaNH₂ will remove the proton from the OH because it is more acidic.

CH_3—I

= B

10.61

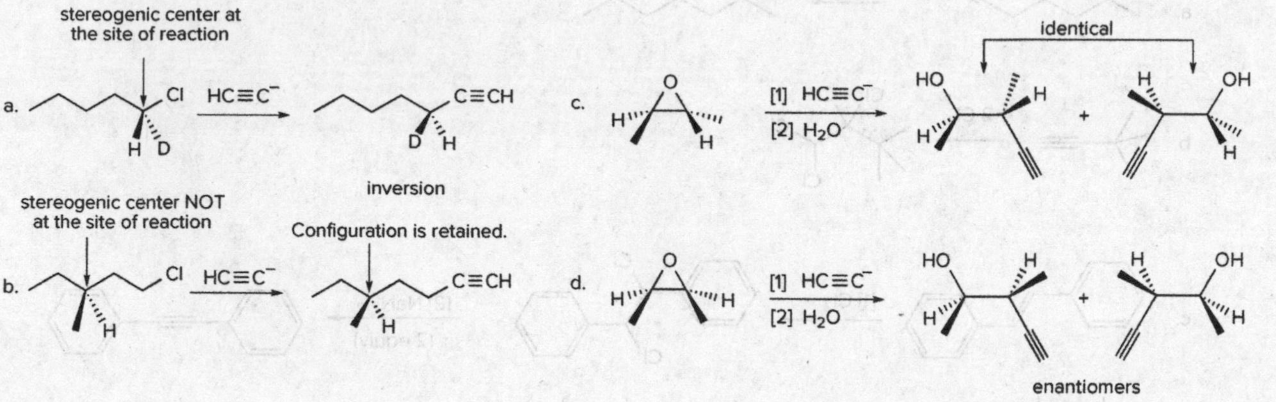

a. stereogenic center at the site of reaction

Cl, H D, $HC{\equiv}C^-$ → C≡CH, D H

inversion

b. stereogenic center NOT at the site of reaction

Cl, H, $HC{\equiv}C^-$ → C≡CH, H

Configuration is retained.

c. O, H, H; [1] $HC{\equiv}C^-$ [2] H_2O → HO, H, H + H, OH, H

identical

d. O, H, H; [1] $HC{\equiv}C^-$ [2] H_2O → HO, H, H + H, OH, H

enantiomers

10.62

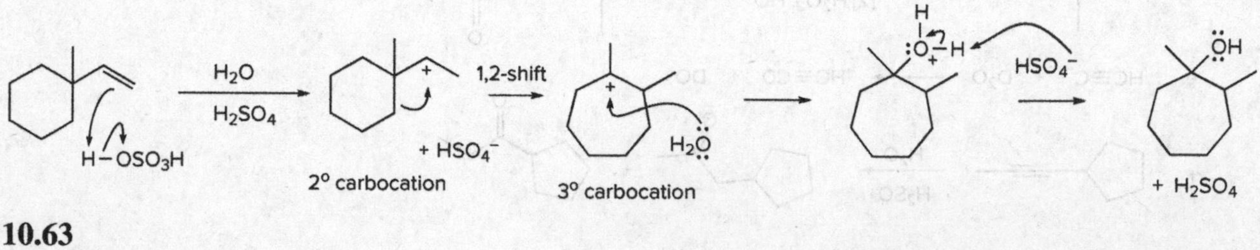

H–OSO_3H; H_2O, H_2SO_4 → + HSO_4^-, 2° carbocation → 1,2-shift → 3° carbocation, H_2O → H, :O–H (+), HSO_4^- → OH + H_2SO_4

10.63

H–OSO_3H; OH, H_2SO_4 → OH + HSO_4^- → O+, H, $^-OSO_3H$ → O + H_2SO_4

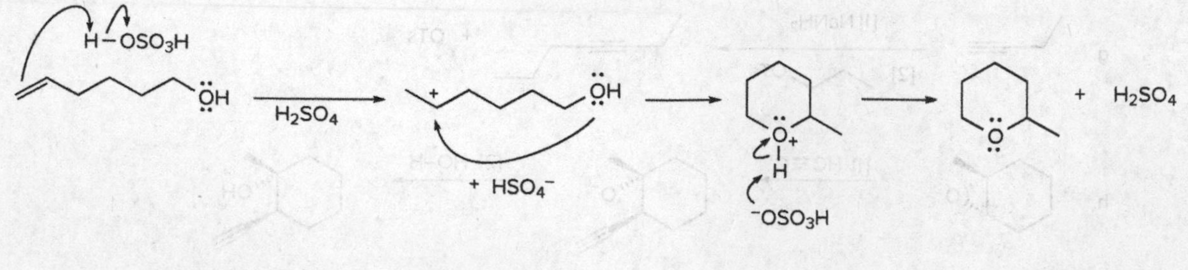

10.64

$H-\ddot{O}H_2^+$

1,2-shift

1,2-shift

$H_2\ddot{O}:$

+ H_3O^+

+ H_3O^+

+ H_3O^+

10.65

A

H_2O

H_2SO_4

X

Y

B

Both carbocations **X** and **Y** have additional resonance stabilization because they are located adjacent to a benzene ring. This explains why only one ketone is formed from the internal alkyne.

10.66

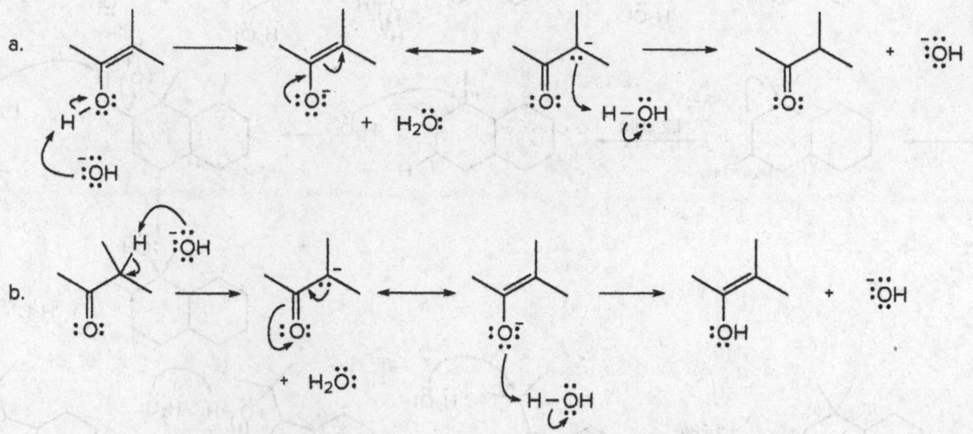

a. :O:, H, :ÖH⁻, :O:⁻, + $H_2\ddot{O}$:, :O:, H–ÖH, :O:, + :ÖH⁻

b. H, :ÖH⁻, :O:, :O:, + $H_2\ddot{O}$:, :O:⁻, H–ÖH, :ÖH, + :ÖH⁻

10.67

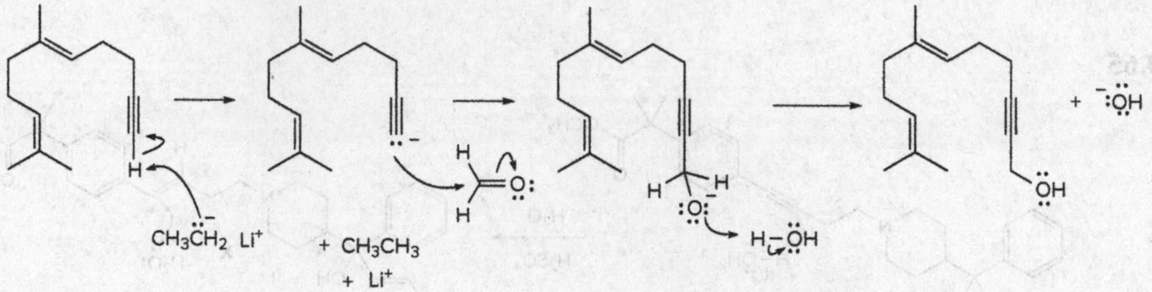

H, $CH_3\ddot{C}H_2$ Li^+, + CH_3CH_3 + Li^+, H, H, :O:, H, H, :O:⁻, $H_2\ddot{O}H$, ÖH, + :ÖH⁻

10.68

a. Br; $K^+\ {}^-OC(CH_3)_3$; [1] BH_3 [2] H_2O_2, HO^-; OH; NaH; O^- + H_2; CH_3I; OCH_3

b. Br_2; Br, Br; $NaNH_2$ excess

c. Br; $K^+\ {}^-OC(CH_3)_3$; Br_2; Br, Br; $NaNH_2$ excess

d. Cl; $K^+\ {}^-OC(CH_3)_3$; Cl_2, H_2O; OH, Cl; NaH; O

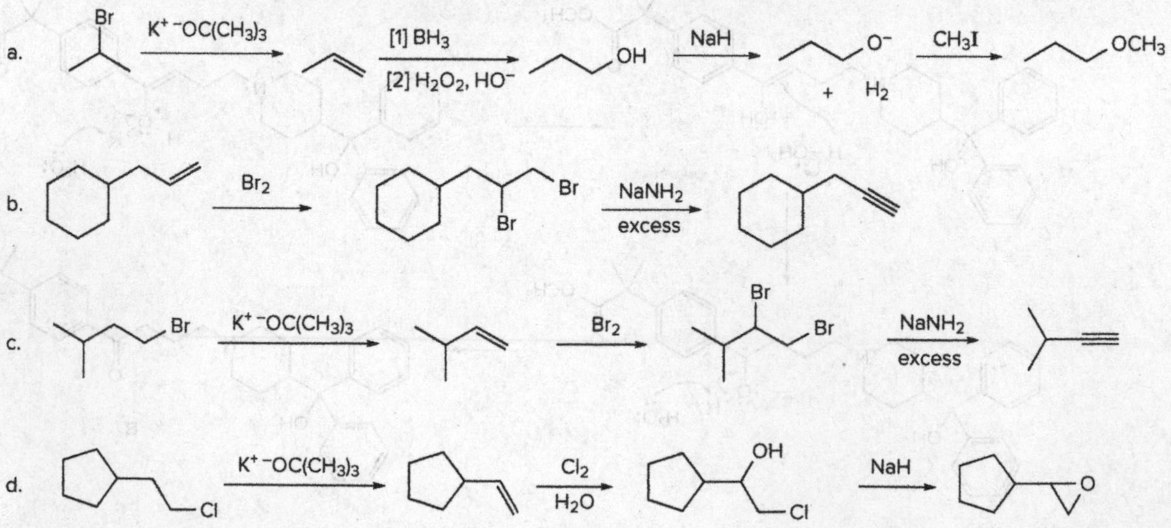

10.69 The alkyl halides must be methyl or 1°.

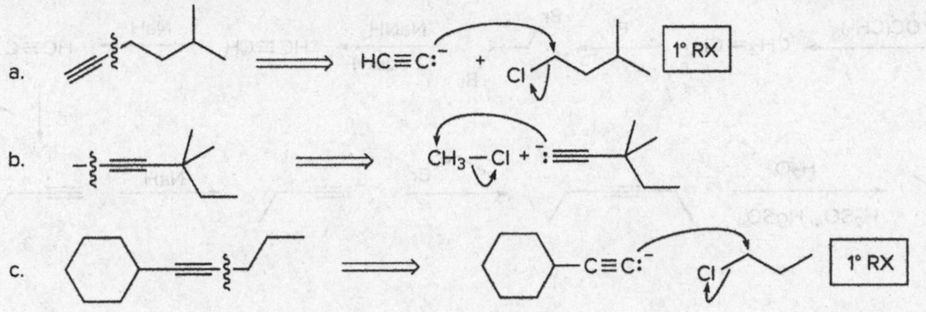

a. ⟹ $HC{\equiv}C:^-$ + Cl (1° RX)

b. ⟹ $CH_3{-}Cl$ + $^-:{\equiv}$

c. ⟹ $-C{\equiv}C:^-$ Cl (1° RX)

10.70 Follow the directions in Problem 10.37.

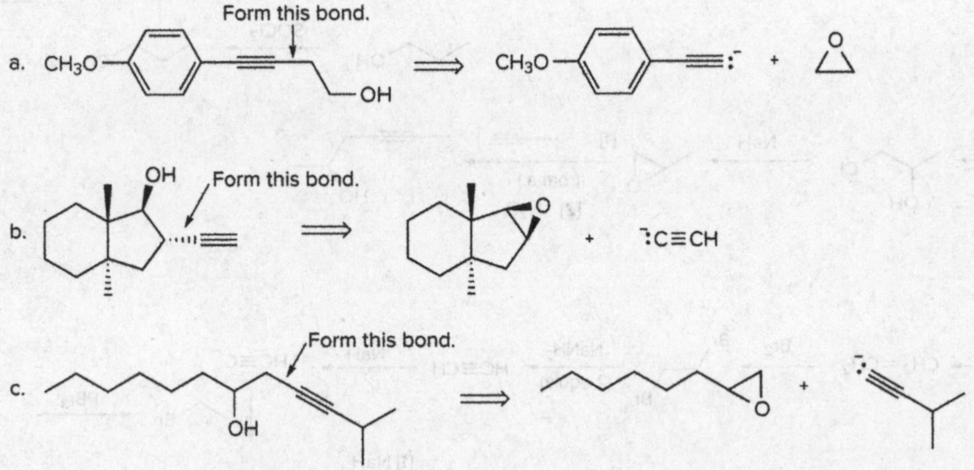

a. CH_3O — Form this bond. — OH ⟹ CH_3O — $\equiv:^-$ + O

b. OH Form this bond. ⟹ O + $^-:C{\equiv}CH$

c. Form this bond. OH ⟹ O +

10.71

a. $HC{\equiv}C{-}H \xrightarrow{Na^+H^-} HC{\equiv}C^-$ → Cl → H → NaH → $^-$ → Cl

b. H (from a.) [1] R_2BH [2] H_2O_2, HO^- → O H

c. (from a.) → Cl → H_2O / H_2SO_4, $HgSO_4$ → O

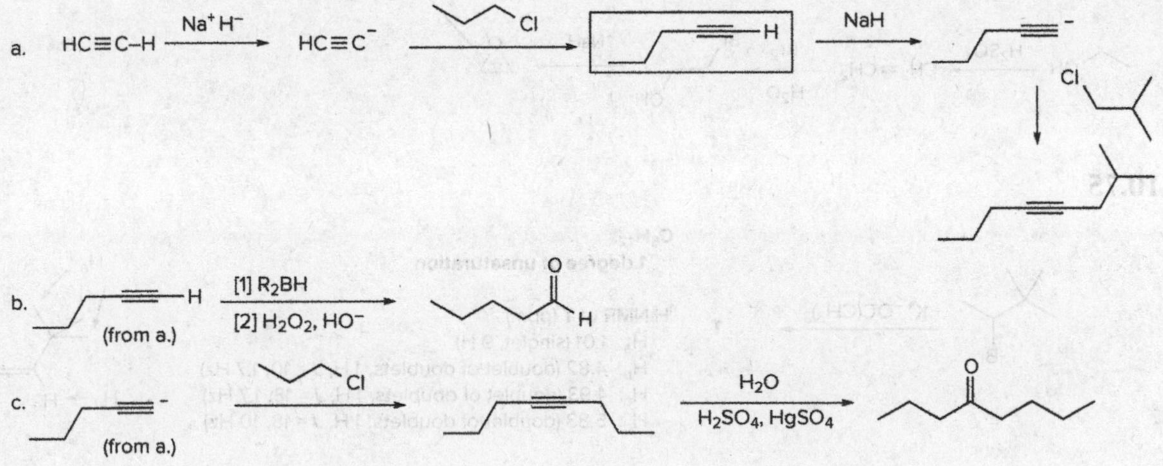

10.72

Br → ($K^+ \, ^-OC(CH_3)_3$) → $CH_2{=}CH_2$ → (Br_2) → Br, Br → ($NaNH_2$ (2 equiv)) → $HC{\equiv}CH$ → (NaH) → $HC{\equiv}C^-$ → (Br) → NaH → ⁻ → (Br) → (H_2O, H_2SO_4, $HgSO_4$) → O

10.73

a. OH → (H_2SO_4) → (Br_2) → Br, Br → ($2\ ^-NH_2$) → (NaH) →

OH → ($SOCl_2$) → Cl

b. (from a.) → (Cl_2, H_2O) → OH, Cl → (NaH) → O → ([1] ⁻ (from a.) [2] H_2O) → HO

10.74

OH → (H_2SO_4) → $CH_2{=}CH_2$ → (Br_2) → Br, Br → ($NaNH_2$ (2 equiv)) → $HC{\equiv}CH$ → (NaH) → $HC{\equiv}C^-$

OH → (PBr_3) → Br

$HC{\equiv}C$— → ([1] NaH [2] O [3] H_2O) → HO—

OH → (H_2SO_4) → $CH_2{=}CH_2$ → (Br_2, H_2O) → Br, OH → (NaH) → O

10.75

Br → ($K^+ \, ^-OC(CH_3)_3$) → **T**

C_6H_{12}:
1 degree of unsaturation

^{1}H NMR of **T** (ppm):
H_a: 1.01 (singlet, 9 H)
H_b: 4.82 (doublet of doublets, 1 H, J = 10, 1.7 Hz)
H_c: 4.93 (doublet of doublets, 1 H, J = 18, 1.7 Hz)
H_d: 5.83 (doublet of doublets, 1 H, J = 18, 10 Hz)

H_a, H_c → H, H_d → H, H ← H_b

OH → (H^+) → **U** ^{1}H NMR of **U:** 1.60 (singlet) ppm

All H's are identical, so there is only one singlet in the NMR.

10.76

C_6H_{10}:
2 degrees of unsaturation
IR peak at 3000 cm^{-1}: **Csp^3–H bonds**
peak at 3300 cm^{-1}: **C$_{sp}$–H bond**
peak at ~2150 cm^{-1}: **C≡C bond**
^{13}C NMR: 4 signals: 4 kinds of C's

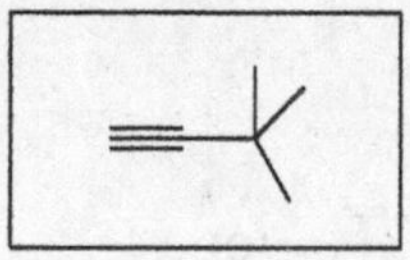

10.77

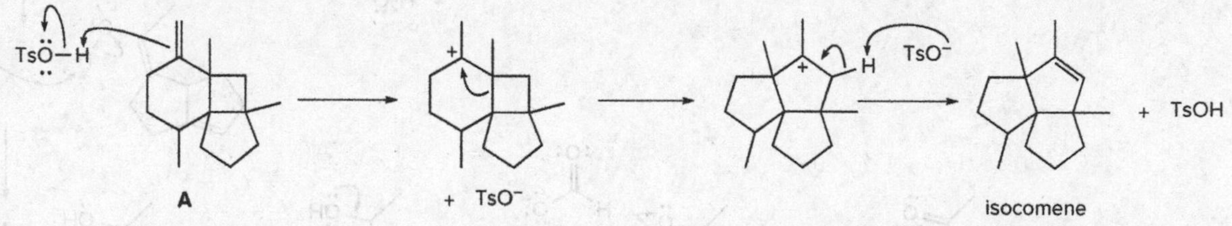

TsÖ–H
A
+ TsO$^-$
TsO$^-$
+ TsOH
isocomene

10.78

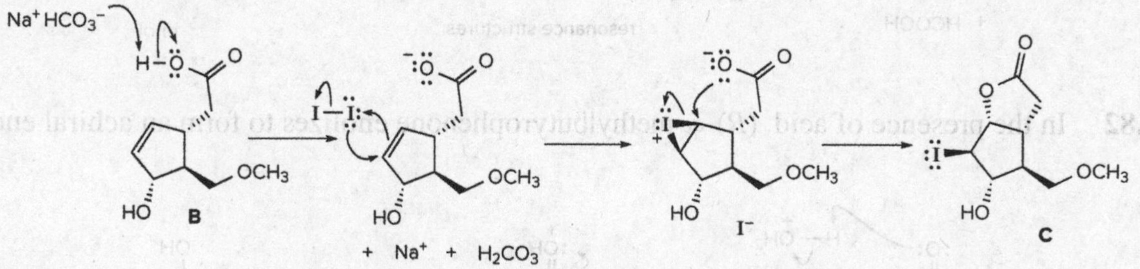

Na$^+$ HCO$_3^-$
H–Ö
OCH$_3$
HO B
+ Na$^+$ + H$_2$CO$_3$
I$^-$
HO C

10.79 Two resonance structures can be drawn for an enol.

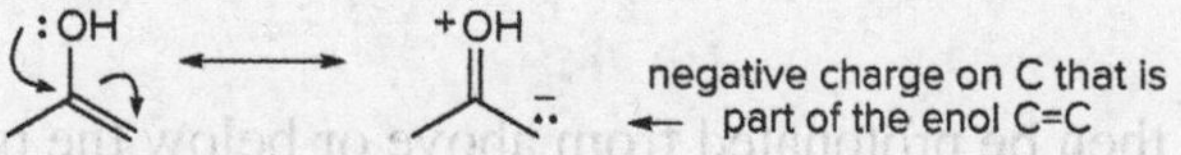

:ÖH
+ÖH
negative charge on C that is part of the enol C=C

Because the second resonance structure places an electron pair (and therefore a negative charge) on an enol carbon, the C=C is more nucleophilic than the C=C of an alkene for which no additional resonance forms can be drawn. Thus, the OH group donates electron density to the C=C by a resonance effect.

10.80

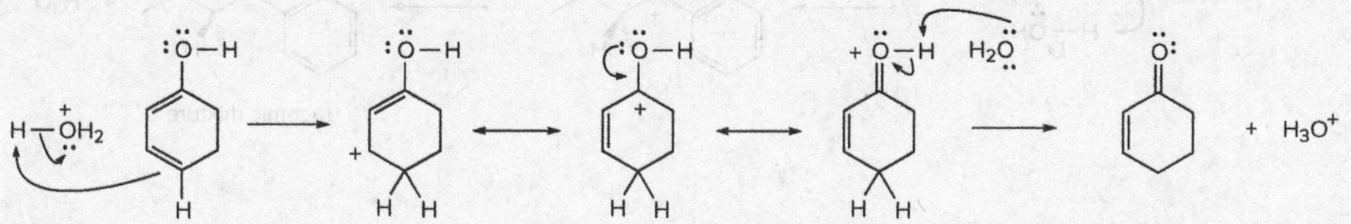

H–ÖH$_2$ (+)
:Ö–H
H H
+ Ö–H H$_2$Ö
+ H$_3$O$^+$

10.81

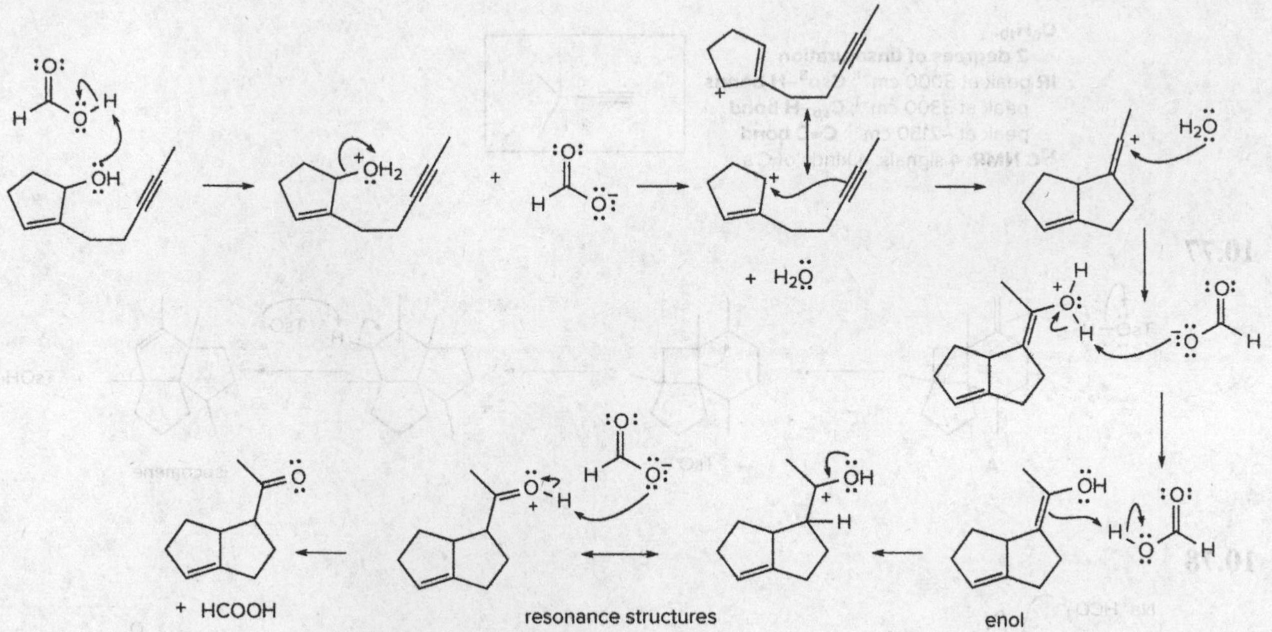

:O: H–C–Ö–H; ÖH; +ÖH2; H–C(=O)–Ö:⁻; + H2Ö; H2Ö; :O:; H; enol; ÖH; H; resonance structures; + HCOOH

10.82 In the presence of acid, (*R*)-α-methylbutyrophenone enolizes to form an achiral enol.

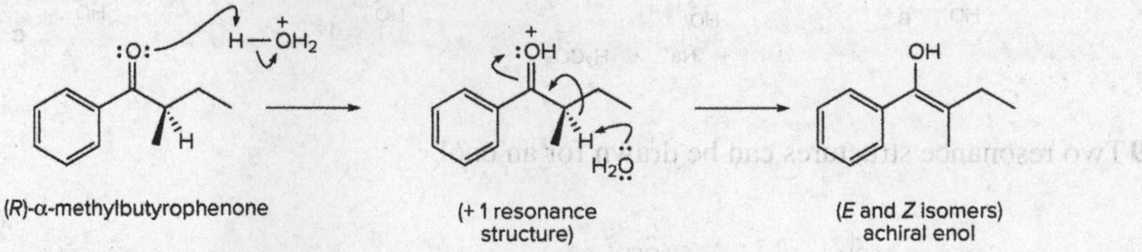

:O:; H–ÖH2 +; (*R*)-α-methylbutyrophenone; :ÖH +; H; H2Ö; (+ 1 resonance structure); OH; (*E* and *Z* isomers) achiral enol

The achiral enol can then be protonated from above or below the plane to form a racemic mixture that is optically inactive.

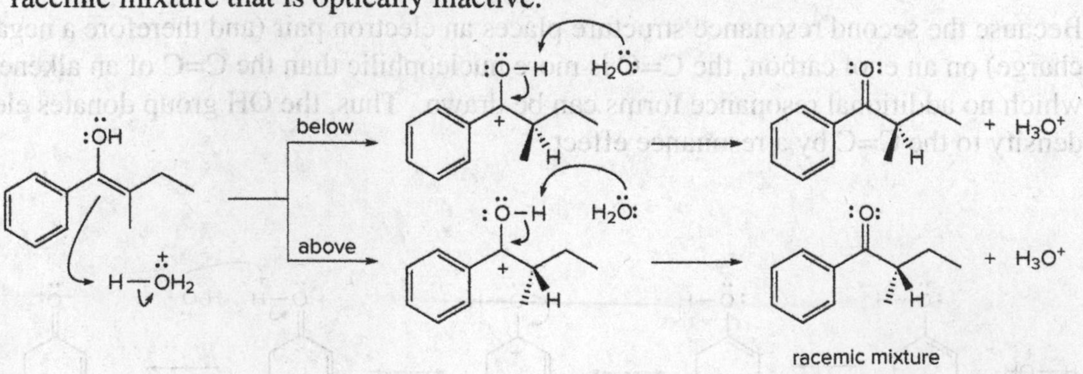

:ÖH; H–ÖH2 +; below; above; :Ö–H; H2Ö:; H; :O:; + H_3O^+; racemic mixture

10.83

H–OSO3H

+ HSO_4^-

re-draw

HSO_4^-

A

1,2-H shift

A

B

HSO_4^-

+ H_2SO_4

+ H_2SO_4

1,2-CH_3 shift

B

HSO_4^-

+ H_2SO_4

10.84

H–OSO3H

+ HSO_4^-

OH

H_2O:

HSO_4^-

HO

OH

+ H_2SO_4

Chapter 11 Oxidation and Reduction

Chapter Review

Summary: Terms that describe reaction selectivity

- A **regioselective reaction** forms predominately or exclusively one constitutional isomer (Section 8.5).

$^-$OH

major product
trisubstituted alkene

minor product
disubstituted alkene

- A **stereoselective reaction** forms predominately or exclusively one stereoisomer (Section 8.5).

Br

$Na^{+}\ ^{-}OCH_2CH_3$

trans alkene
major product

cis alkene
minor product

- An **enantioselective reaction** forms predominately or exclusively one enantiomer (Section 11.14).

OH

allylic alcohol

Sharpless reagent

O

OH

or

O

OH

One enantiomer is favored.

Definitions of oxidation and reduction

Oxidation reactions result in:
- an increase in the number of C–Z bonds, or
- a decrease in the number of C–H bonds.

Reduction reactions result in:
- a decrease in the number of C–Z bonds, or
- an increase in the number of C–H bonds.

[Z = an element more electronegative than C]

Reduction reactions

[1] Reduction of alkenes—Catalytic hydrogenation (11.3)

R R H H

H_2

Pd, Pt, or Ni

H H R R H H

- **Syn addition** of H_2 occurs.
- Increasing alkyl substitution on the C=C decreases the rate of reaction.

[2] Reduction of alkynes

a. R—≡—R $\xrightarrow[\text{Pd-C}]{2\ H_2}$ [alkane: RCH$_2$CH$_2$R]

- Two equivalents of H_2 are added and four new C–H bonds are formed (11.5A).

b. R—≡—R $\xrightarrow[\text{Lindlar catalyst}]{H_2}$ [cis alkene]

- **Syn addition** of H_2 occurs, forming a **cis** alkene (11.5B).
- The Lindlar catalyst is deactivated so that reaction stops after one equivalent of H_2 has been added.

c. R—≡—R $\xrightarrow[NH_3]{Na}$ trans alkene

- **Anti addition** of H_2 occurs, forming a **trans** alkene (11.5C).

[3] Reduction of alkyl halides (11.6)

R–X $\xrightarrow[\text{[2] } H_2O]{\text{[1] } LiAlH_4}$ R–H (alkane)

- The reaction follows an S_N2 mechanism.
- CH_3X and RCH_2X react faster than more substituted RX.

[4] Reduction of epoxides (11.6)

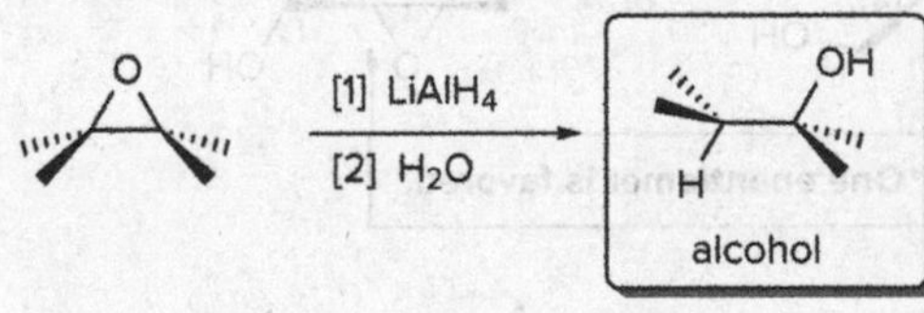

- The reaction follows an S_N2 mechanism.
- In unsymmetrical epoxides, H^- (from $LiAlH_4$) attacks at the less substituted carbon.

Oxidation reactions

[1] Oxidation of alkenes

a. Epoxidation (11.8)

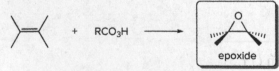

alkene + RCO_3H ⟶ epoxide

- The mechanism has **one step.**
- **Syn addition** of an O atom occurs.
- The reaction is stereospecific.

b. Anti dihydroxylation (11.9A)

[1] RCO_3H; [2] H_2O (H^+ or HO^-) → 1,2-diol

- Ring opening of an epoxide intermediate with ^-OH or H_2O forms a 1,2-diol with two OH groups added in an **anti** fashion.

c. Syn dihydroxylation (11.9B)

[1] OsO_4; [2] $NaHSO_3$, H_2O or [1] OsO_4, NMO; [2] $NaHSO_3$, H_2O or $KMnO_4$, H_2O, HO^- → 1,2-diol

- Each reagent adds two new C–O bonds to the C=C in a **syn** fashion.

d. Oxidative cleavage (11.10)

[1] O_3; [2] Zn, H_2O or CH_3SCH_3 → ketone + aldehyde

- Both the σ and π bonds of the alkene are cleaved to form two carbonyl groups.

[2] Oxidative cleavage of alkynes (11.11)

a. internal alkyne: R–C≡C–R' [1] O_3; [2] H_2O → RCOOH + R'COOH (carboxylic acids)

- The σ bond and both π bonds of the alkyne are cleaved.

b. terminal alkyne: R–C≡C–H [1] O_3; [2] H_2O → RCOOH + CO_2

[3] Oxidation of alcohols (11.12, 11.13)

a. RCH_2OH (1° alcohol) PCC → RCHO (aldehyde)

- Oxidation of a 1° alcohol with PCC stops at the aldehyde stage. Only one C–H bond is replaced by a C–O bond.

b. R–CH₂–OH (1° alcohol) $\xrightarrow[H_2SO_4,\ H_2O]{CrO_3}$ RCOOH (carboxylic acid)

- Oxidation of a 1° alcohol under harsher reaction conditions—CrO_3 (or $Na_2Cr_2O_7$ or $K_2Cr_2O_7$) + H_2O + H_2SO_4—affords a RCO_2H. Two C–H bonds are replaced by two C–O bonds.

c. R₂CHOH (2° alcohol) $\xrightarrow[\text{or } CrO_3]{PCC}$ R₂C=O (ketone)

- A 2° alcohol has only one C–H bond on the carbon bearing the OH group, so all Cr^{6+} reagents—PCC, CrO_3, $Na_2Cr_2O_7$, or $K_2Cr_2O_7$—oxidize a 2° alcohol to a ketone.

[4] Asymmetric epoxidation of allylic alcohols (11.14)

Allylic alcohol $\xrightarrow[Ti[OCH(CH_3)_2]_4]{(CH_3)_3C{-}OOH}$ epoxy alcohol (with (–)-DET) or epoxy alcohol (with (+)-DET)

Practice Test on Chapter Review

1.a. Compound **X** has a molecular formula of C_9H_{12} and contains no triple bonds. **X** is hydrogenated to a compound of molecular formula C_9H_{14} with excess H_2 and a palladium catalyst. What can be said about **X**?

1. **X** has four rings.
2. **X** has three rings and one double bond.
3. **X** has two rings and two double bonds.
4. **X** has one ring and three double bonds.
5. **X** has four double bonds.

b. Syn addition to an alkene occurs exclusively with which reagents?

1. OsO_4
2. $KMnO_4$, H_2O, ^{-}OH
3. mCPBA, then H_2O, ^{-}OH
4. Both reagents (1) and (2) give syn addition exclusively.
5. Reagents (1), (2), and (3) give syn addition exclusively.

c. Which of the following reagents adds to an alkene exclusively in an anti fashion?

1. Br_2
2. H_2, Pd-C
3. BH_3, then H_2O_2, ^{-}OH
4. Reagents (1) and (2) both add in an anti fashion.
5. Reagents (1), (2), and (3) all add in an anti fashion.

2. Label each statement as True (T) or False (F).

a. PCC oxidizes 1° alcohols to aldehydes.
b. CrO_3 oxidizes 2° alcohols to ketones.
c. Treatment of hex-2-yne with Na in NH_3 forms *cis*-hex-2-ene.
d. Reduction of propene oxide with $LiAlH_4$ forms propan-1-ol.
e. Ozonolysis of 2-methyloct-2-ene forms one ketone and one aldehyde.
f. mCPBA is an oxidizing agent that converts alkenes to trans diols.
g. Oct-1-en-5-yne reacts with H_2 and Pd-C, but does not react with H_2 and Lindlar catalyst.
h. Treatment of cyclohexene with OsO_4 affords an optically inactive product mixture that contains two enantiomers.

3. Label each reagent as an oxidizing agent, reducing agent, or neither.

a. O_3
b. $LiAlH_4$
c. mCPBA
d. H_2O, H_2SO_4
e. PCC
f. Na, NH_3

4. Draw the organic products formed in each reaction and indicate stereochemistry when necessary.

a. Na / NH_3

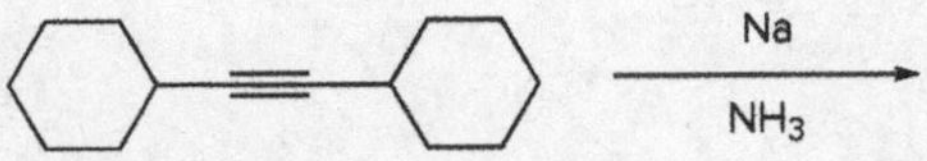

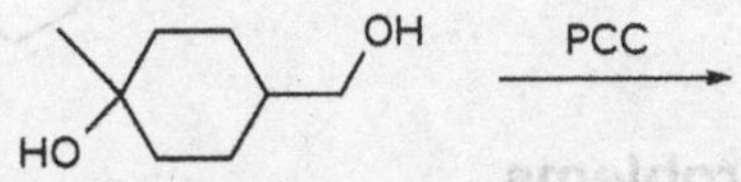

b. H_2 / Pd-C

c. OH HO PCC

d. OH [1] $SOCl_2$ [2] $LiAlH_4$

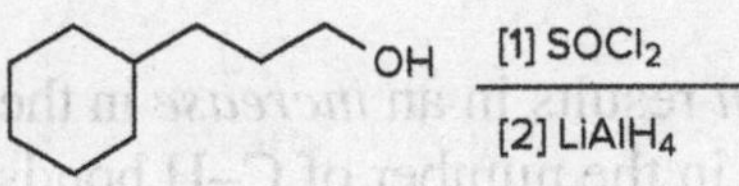

5 a. Fill in the appropriate starting material (including any needed stereochemistry) in the following reaction.

[1] OsO_4
[2] H_2O, $NaHSO_3$

HO H C_6H_5 C—C H C_6H_5 OH + enantiomer

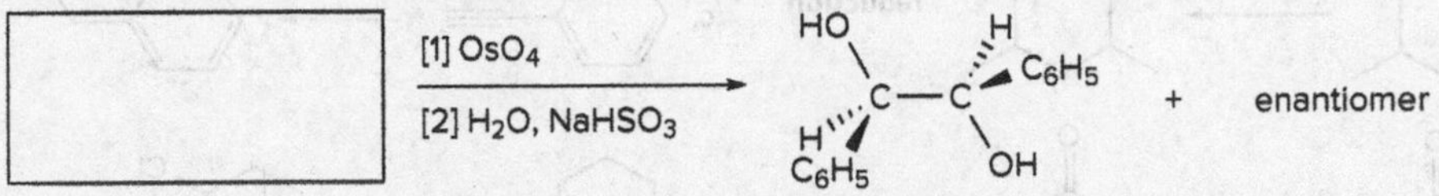

b. Fill in the appropriate reagent in the following reaction.

OH

O H OH

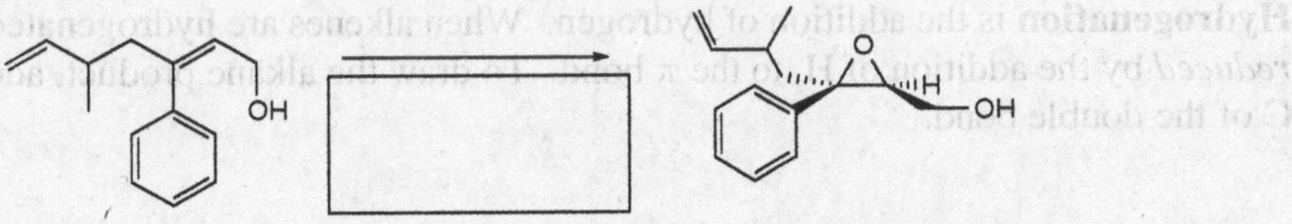

c. What starting material is needed for the following reaction?

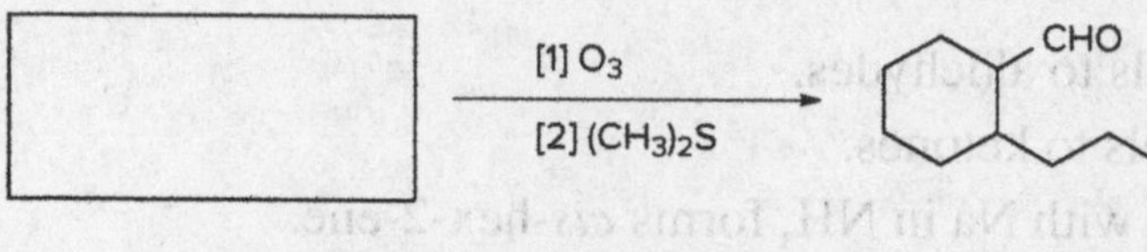

[1] O_3, [2] $(CH_3)_2S$ → CHO

Answers to Practice Test

1.	2.	3.
a. 2	a. T	a. oxidizing
b. 4	b. T	b. reducing
c. 1	c. F	c. oxidizing
	d. F	d. neither
	e. T	e. oxidizing
	f. F	f. reducing
	g. F	
	h. F	

4.

a.

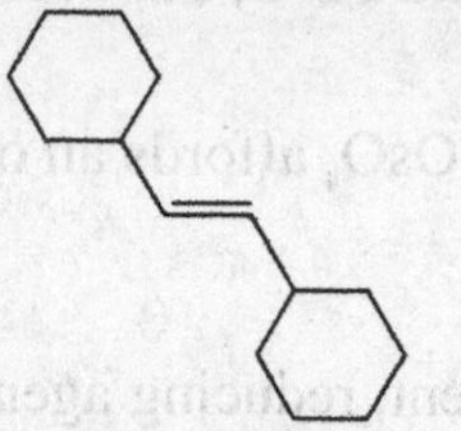

b. 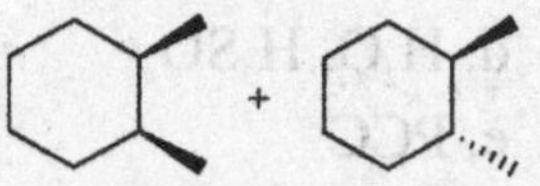 +

c. 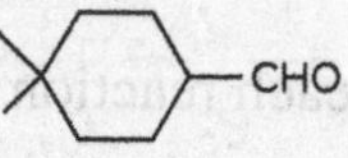 HO CHO

d.

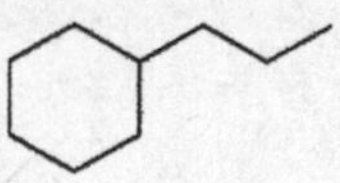

5.

a.

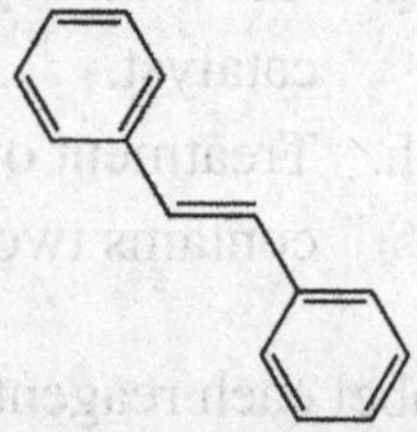

b. Sharpless reagent (+)-DET

c.

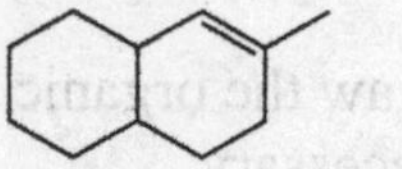

Answers to Problems

11.1 *Oxidation* results in an *increase* in the number of C–Z bonds (usually C–O bonds) *or* a *decrease* in the number of C–H bonds.
Reduction results in a *decrease* in the number of C–Z bonds (usually C–O bonds) *or* an *increase* in the number of C–H bonds.

a. 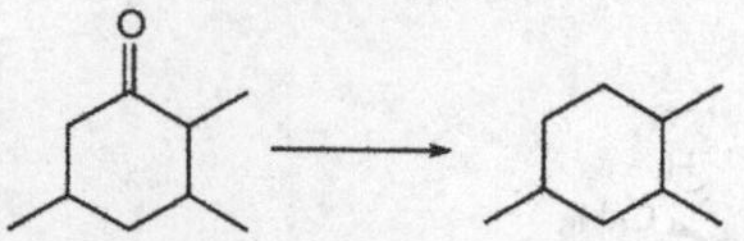 reduction

b. 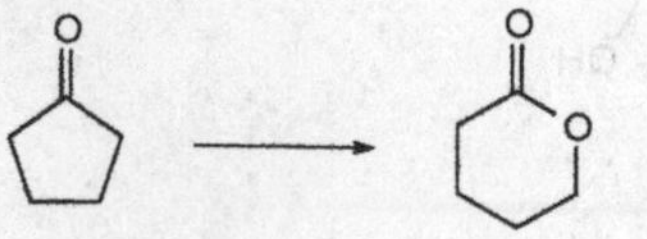 oxidation

c. 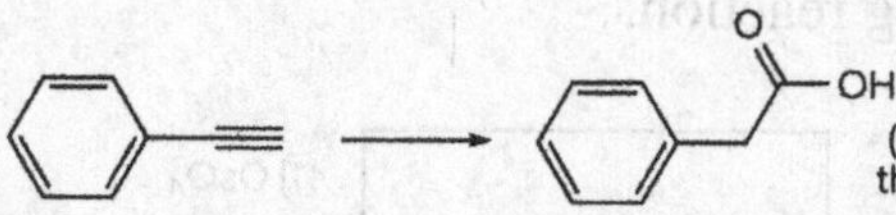 OH oxidation (two new C–H bonds, three new C–O bonds)

d. 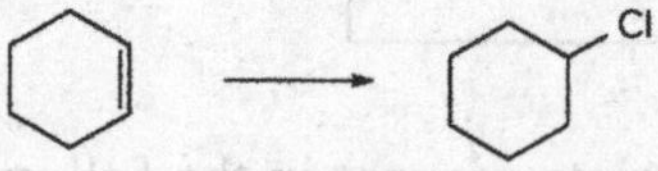 Cl neither (one new C–H bond and one new C–Cl bond)

11.2 Hydrogenation is the addition of hydrogen. When alkenes are hydrogenated, they are *reduced* by the addition of H_2 to the π bond. To draw the alkane product, add a H to each C of the double bond.

a.

c.

b.

11.3 Draw the alkenes that form each alkane when hydrogenated.

a. or or $\xrightarrow[\text{Pd-C}]{H_2}$

c. or or or or or $\xrightarrow[\text{Pd-C}]{H_2}$

b. or $\xrightarrow[\text{Pd-C}]{H_2}$

11.4 Cis alkenes are less stable than trans alkenes, so they have larger heats of hydrogenation. Increasing alkyl substitution increases the stability of a C=C, thus decreasing the heat of hydrogenation.

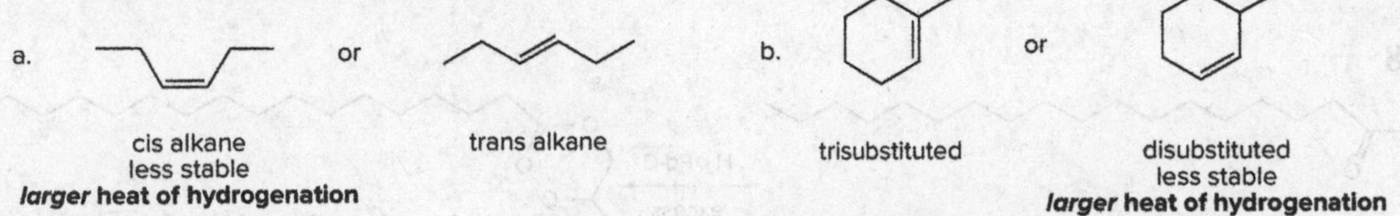

11.5 Hydrogenation products must be identical to use hydrogenation data to evaluate the relative stability of the starting materials.

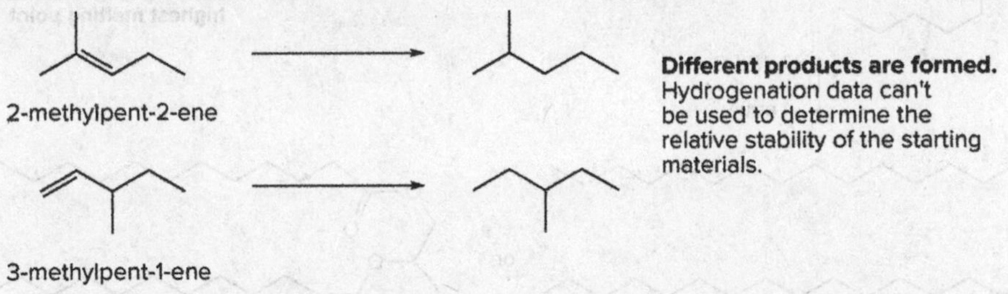

11.6

new stereogenic center

Two enantiomers are formed in equal amounts:

a.

H

=

+

b.

diastereomers

c.

diastereomers

d.

Configuration here stays the same.

new stereogenic center

Two diastereomers are formed.

11.7

Compound	Molecular formula before hydrogenation	Molecular formula after hydrogenation	Number of rings	Number of π bonds
A	$C_{10}H_{12}$	$C_{10}H_{16}$	3	2
B	C_4H_8	C_4H_{10}	0	1
C	C_6H_8	C_6H_{12}	1	2

11.8

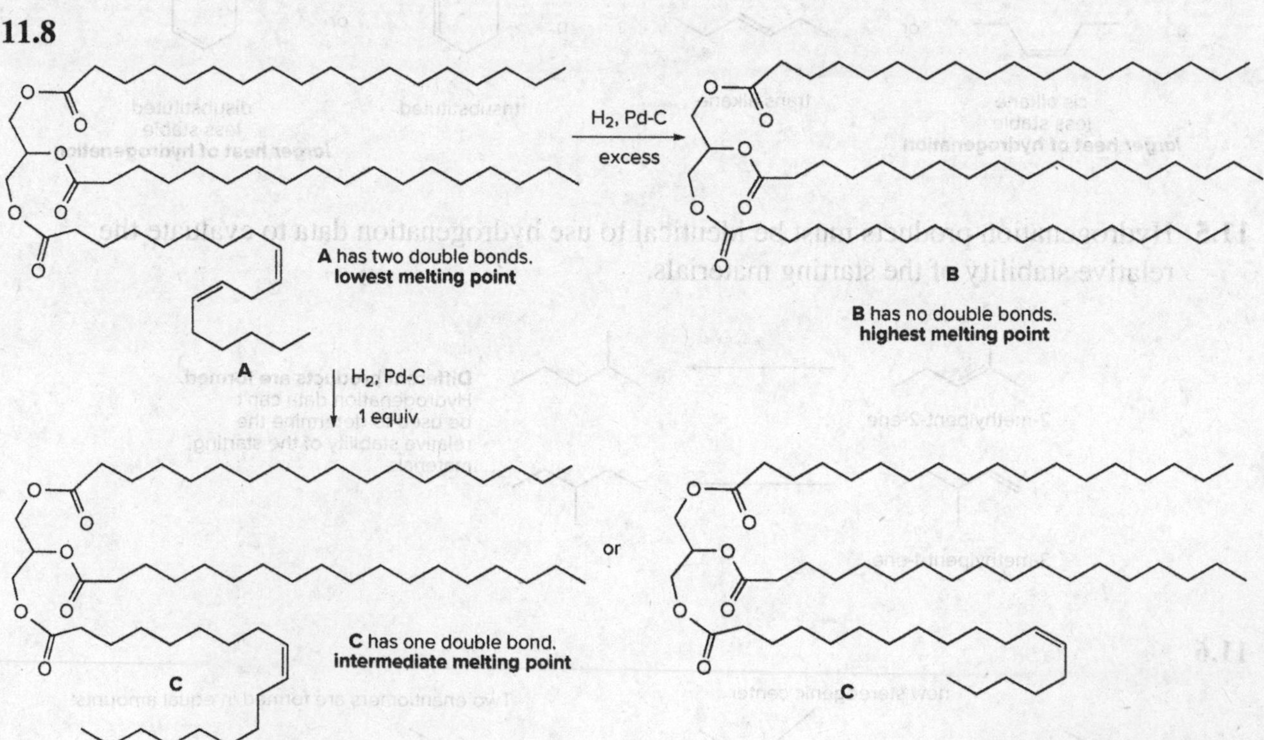

11.9 Hydrogenation of $HC{\equiv}CCH_2CH_2CH_3$ and $CH_3C{\equiv}CCH_2CH_3$ yields the same compound. The heat of hydrogenation is larger for $HC{\equiv}CCH_2CH_2CH_3$ than for $CH_3C{\equiv}CCH_2CH_3$ because internal alkynes are more stable (lower in energy) than terminal alkynes.

11.10 In the presence of the Lindlar catalyst, alkynes react with H_2 to form cis alkenes. Alkenes do not react with the Lindlar catalyst.

A — H_2, Lindlar catalyst → ***cis*-jasmone** (perfume component isolated from jasmine flowers)

11.11 In the presence of Pd-C, H_2 adds to alkenes and alkynes to form alkanes. In the presence of the Lindlar catalyst, alkynes react with H_2 to form cis alkenes. Alkenes do not react with the Lindlar catalyst.

a. C_6H_{10} — H_2, Pd-C → cyclohexane; H_2, Lindlar catalyst → no reaction

b. C_6H_{10} — H_2, Pd-C → hexane; H_2, Lindlar catalyst → cis-3-hexene

11.12 Use the directions from Answer 11.11.

a. H_2 (excess), Pd-C

b. H_2 (1 equiv), Lindlar catalyst

c. H_2 (excess), Lindlar catalyst

d. Na, NH_3

11.13

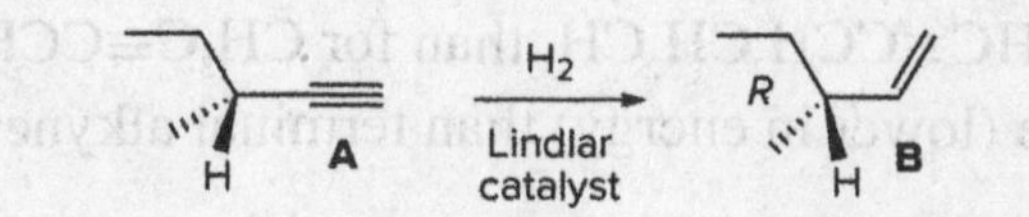

A —[H_2, Lindlar catalyst]→ B (R)

11.14 $LiAlH_4$ reduces alkyl halides to alkanes and epoxides to alcohols.

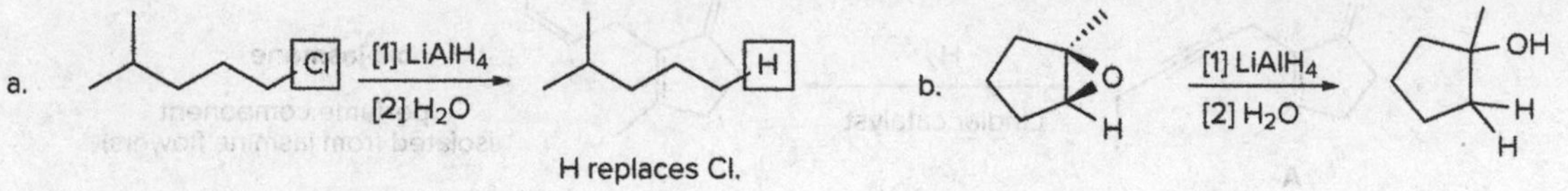

a. [1] $LiAlH_4$ [2] H_2O — H replaces Cl.

b. [1] $LiAlH_4$ [2] H_2O

11.15 To draw the product, add an O atom across the π bond of the C=C.

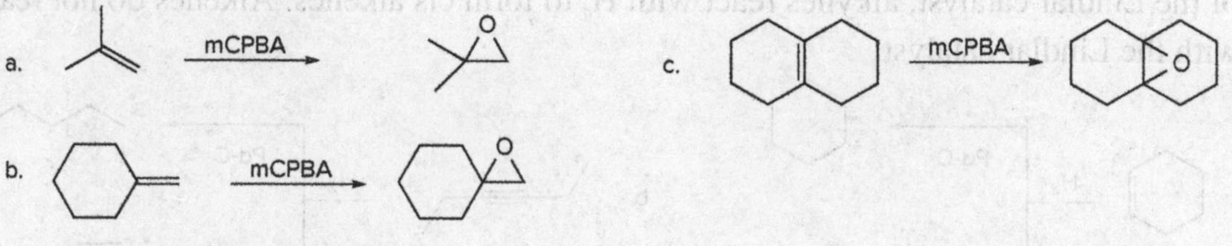

a. mCPBA

b. mCPBA

c. mCPBA

11.16 For epoxidation reactions:

- There are two possible products: O adds from above and below the double bond.
- Substituents on the C=C retain their original configuration in the products.

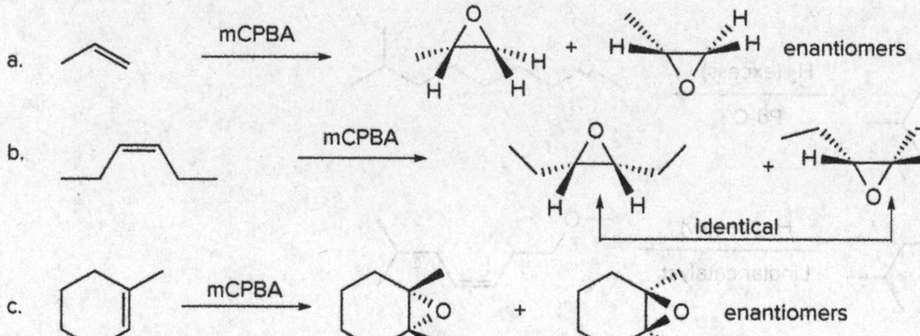

a. mCPBA — enantiomers

b. mCPBA — identical

c. mCPBA — enantiomers

11.17 Treatment of an alkene with a peroxyacid followed by H_2O, HO^- adds two hydroxy groups in an **anti** fashion. *cis*-But-2-ene and *trans*-but-2-ene yield different products of dihydroxylation. *cis*-But-2-ene gives a mixture of two enantiomers and *trans*-but-2-ene gives a meso compound. The reaction is stereospecific because two stereoisomeric starting materials give different products that are also stereoisomers of each other.

cis-but-2-ene —[1] RCO_3H [2] H_2O, HO^-→ enantiomers

trans-but-2-ene —[1] RCO_3H [2] H_2O, HO^-→ identical meso compound

11.18 Treatment of an alkene with OsO_4 adds two hydroxy groups in a **syn** fashion. *cis*-But-2-ene and *trans*-but-2-ene yield different stereoisomers in this dihydroxylation, so the reaction is stereospecific.

cis-but-2-ene —[1] OsO_4 [2] $NaHSO_3$, H_2O→ identical meso compound

trans-but-2-ene —[1] OsO_4 [2] $NaHSO_3$, H_2O→ enantiomers

11.19 To draw the oxidative cleavage products:

- **Locate all the π bonds** in the molecule.
- **Replace all C=C's with *two* C=O's.**

Replace this π bond with two C=O's.

a. —[1] O_3 [2] Zn, H_2O→ ketone + aldehyde. One **ketone** and one **aldehyde** are formed.

b. —[1] O_3 [2] Zn, H_2O→ Two **aldehydes** are formed.

c. —[1] O_3 [2] Zn, H_2O→ A **dicarbonyl** compound is formed.

11.20 To find the alkene that yields the oxidative cleavage products:

- **Find the two carbonyl groups** in the products.
- **Join the two carbonyl carbons** together with a double bond. This is the double bond that was broken during ozonolysis.

a. Join these two C's.

b. Join these two C's.

c. only — With only one product, the alkene must be symmetrical around the double bond. Join this C to the same C in another identical molecule.

11.21

cembrene A; [1] O_3 [2] CH_3SCH_3; CHO chiral; achiral; achiral; achiral

11.22 To draw the products of oxidative cleavage of alkynes:

- **Locate the triple bond.**
- For internal alkynes, **convert the *sp* hybridized C to COOH.**
- For terminal alkynes, the ***sp* hybridized C–H becomes CO_2.**

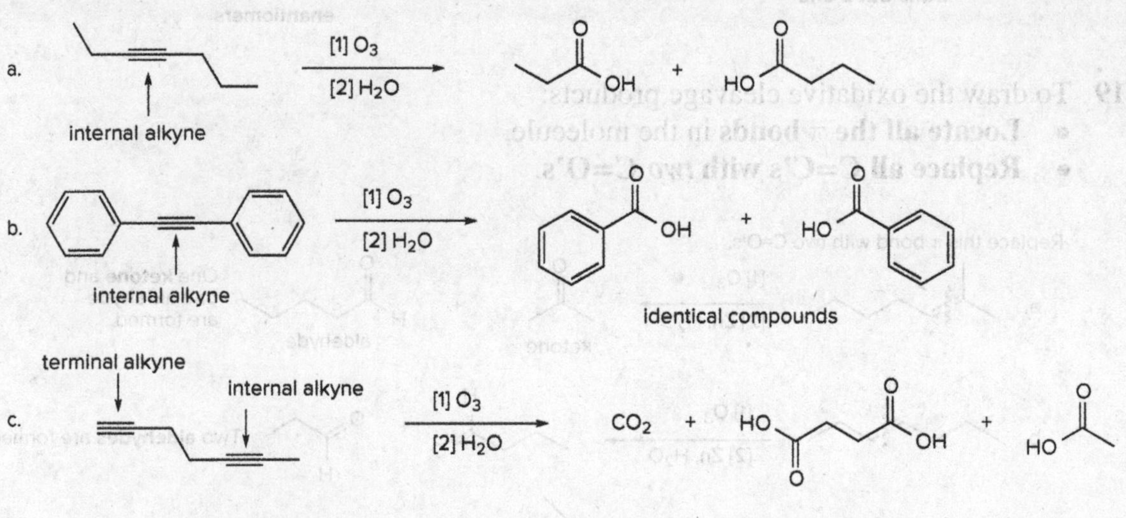

11.23

a. CO_2 + OH

b. OH

c. 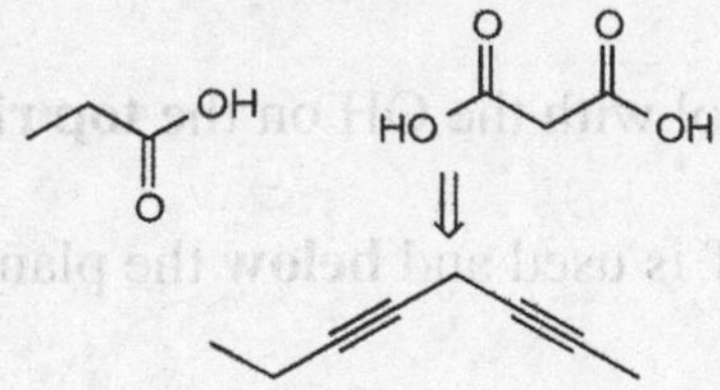 d.

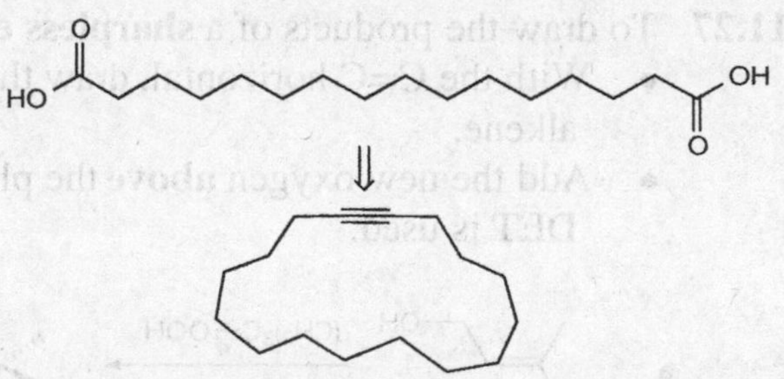

11.24 For the **oxidation of alcohols,** remember:

- **1° Alcohols** are oxidized to aldehydes with PCC.
- **1° Alcohols** are oxidized to carboxylic acids with oxidizing agents like CrO_3 or $Na_2Cr_2O_7$.
- **2° Alcohols** are oxidized to ketones with all Cr^{6+} reagents.

a. 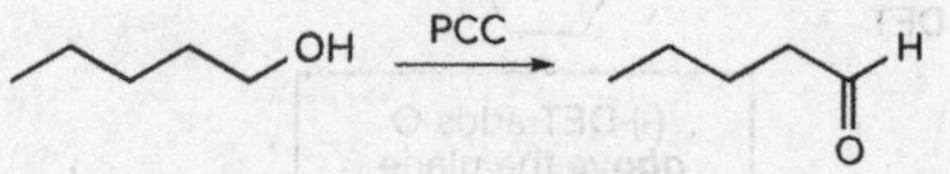 PCC c. 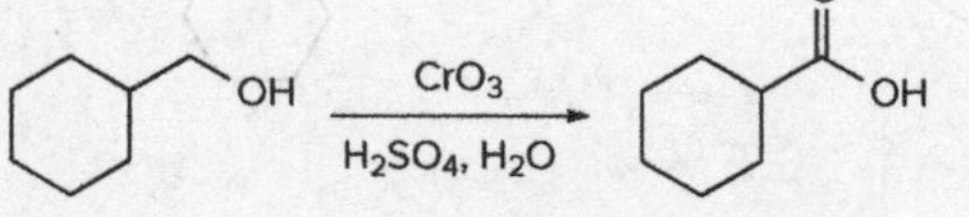 CrO_3, H_2SO_4, H_2O

b. 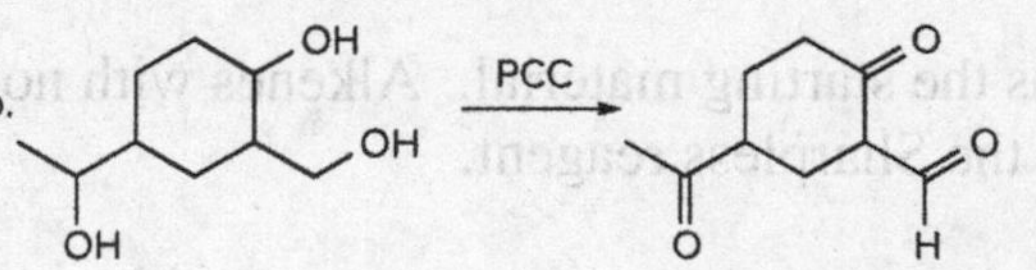 PCC d. 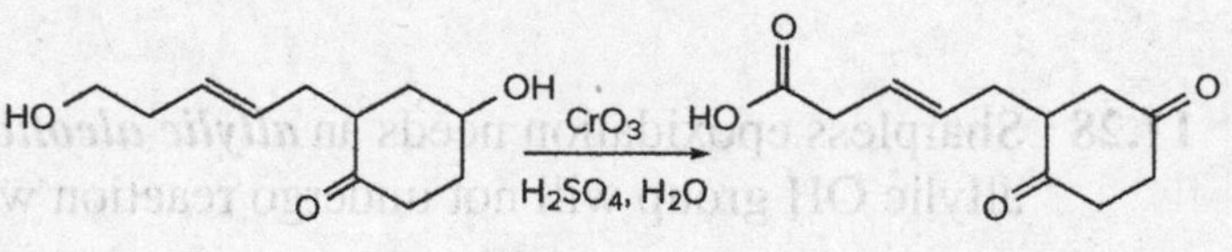 CrO_3, H_2SO_4, H_2O

11.25 An H atom of a CH_2 group is *pro-R* if replacement of H by D forms a stereogenic center with the *R* configuration. An H atom of a CH_2 group is *pro-S* if replacement of H by D forms a stereogenic center with the *S* configuration.

a. 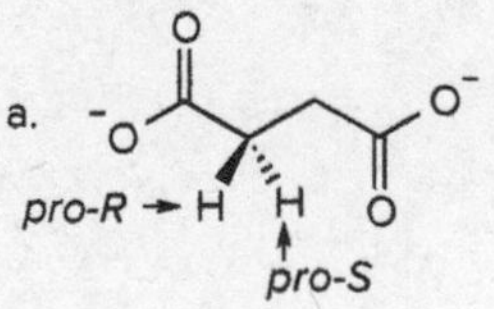 *pro-R* → H, H ← *pro-S* b. 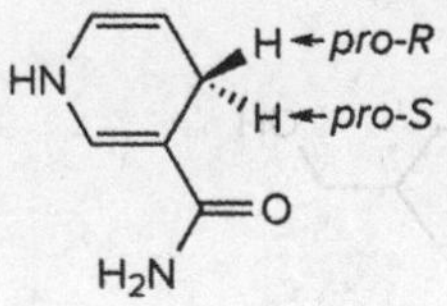 H ← *pro-R*, H ← *pro-S* c. 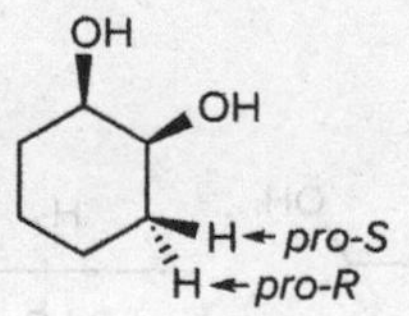 H ← *pro-S*, H ← *pro-R*

11.26 The *pro-R* H's are labeled as H_R; the *pro-S* H's are labeled as H_S.

a. 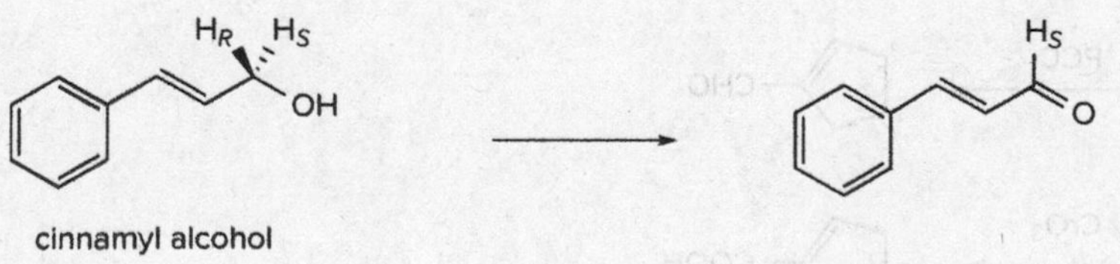 cinnamyl alcohol

b. geraniol

11.27 To draw the products of a **sharpless epoxidation:**

- With the C=C horizontal, draw the allylic alcohol with the OH on the **top right** of the alkene.
- Add the new oxygen **above** the plane if (–)-DET is used and **below** the plane if (+)-DET is used.

a. OH —[$(CH_3)_3C-OOH$ / $Ti[OCH(CH_3)_2]_4$ / (+)-DET]→ OH, H, O

(+)-DET adds O *below* the plane.

b. OH - - re-draw - -→ OH —[$(CH_3)_3C-OOH$ / $Ti[OCH(CH_3)_2]_4$ / (–)-DET]→ H O OH

(–)-DET adds O *above* the plane.

11.28 Sharpless epoxidation needs an ***allylic alcohol*** as the starting material. Alkenes with no allylic OH group will not undergo reaction with the Sharpless reagent.

This alkene is part of an **allylic alcohol** and will be epoxidized.

geraniol OH

This alkene is **not** part of an allylic alcohol and will not be epoxidized.

11.29

a. A (OH) —[H_2 / Pd-C]→ OH

b. A (OH) —[mCPBA]→ O, OH + O, OH

c. A (OH) —[PCC]→ CHO

d. A (OH) —[CrO_3 / H_2SO_4 / H_2O]→ COOH

e. A (OH) —[Sharpless reagent / (+)-DET]→ O, OH

11.30

$HC{\equiv}CH$ —[1] NaH; [2] CH_3CH_2Br→ —[1] NaH; [2] epoxide; [3] H_2O→ OH —Na, NH_3→ OH

11.31 Use the rules from Answer 11.1.

a. → OH reduction

b. HO—OH → O=O oxidation

c. OCH_3, OCH_3 → O, H, Br **oxidation** (one new C–Br bond)

d. OCH_3, OCH_3 → OCH_3 **neither** (one fewer C–O bond and one fewer C–H bond)

11.32 Use the principles from Answer 11.2 and draw the products of syn addition of H_2 from above and below the C=C.

a. —H_2, Pd-C→ +

b. —H_2, Pd-C→ H + H

c. —H_2, Pd-C→ +

diastereomers

d. O —H_2, Pd-C→ H, OH + OH, H

enantiomers

11.33

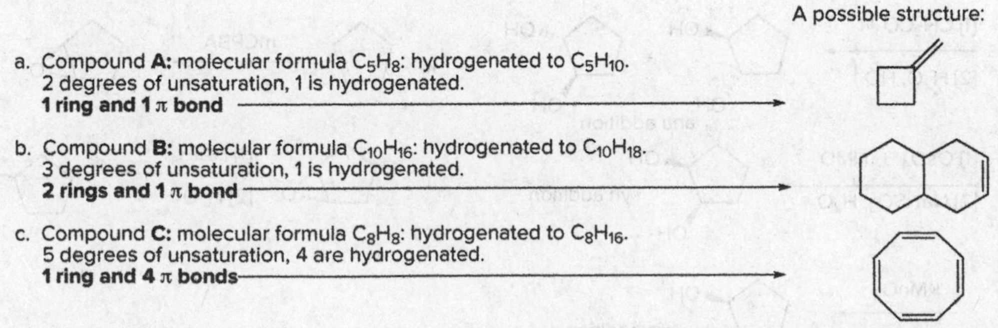

A possible structure:

a. Compound **A**: molecular formula C_5H_8: hydrogenated to C_5H_{10}.
2 degrees of unsaturation, 1 is hydrogenated.
1 ring and 1 π bond

b. Compound **B**: molecular formula $C_{10}H_{16}$: hydrogenated to $C_{10}H_{18}$.
3 degrees of unsaturation, 1 is hydrogenated.
2 rings and 1 π bond

c. Compound **C**: molecular formula C_8H_8: hydrogenated to C_8H_{16}.
5 degrees of unsaturation, 4 are hydrogenated.
1 ring and 4 π bonds

11.34

a. stearidonic acid $\xrightarrow[\text{Pd-C}]{H_2 \text{ (excess)}}$ stearic acid

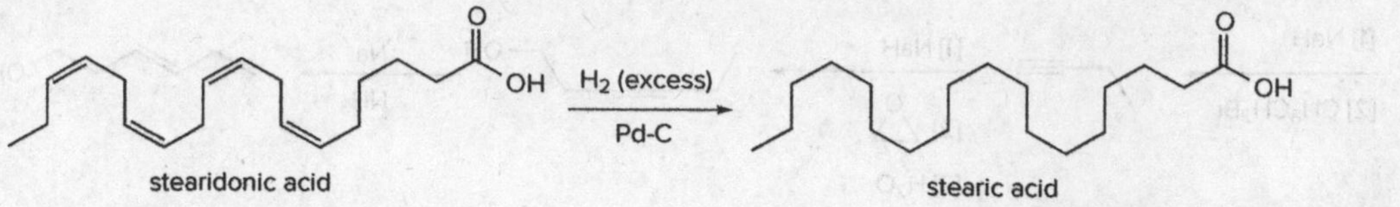

b. $\xrightarrow[\text{Pd-C}]{H_2 \text{ (1 equiv)}}$ + + +

c. trans

one possibility

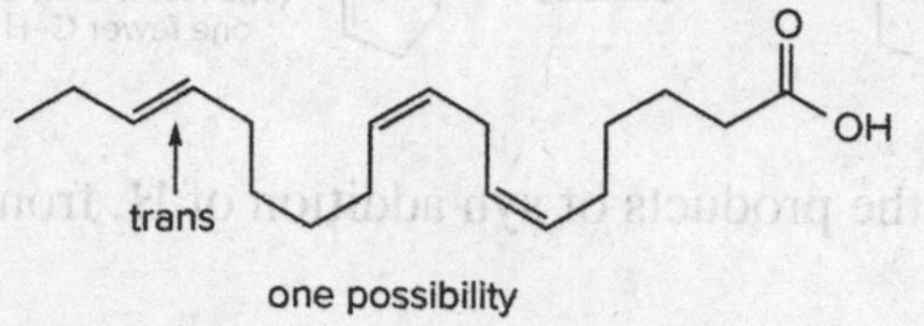

d. stearidonic acid 4 cis C=C's < product in (b) 3 cis C=C's < product in (c) 2 cis and one trans C=C's

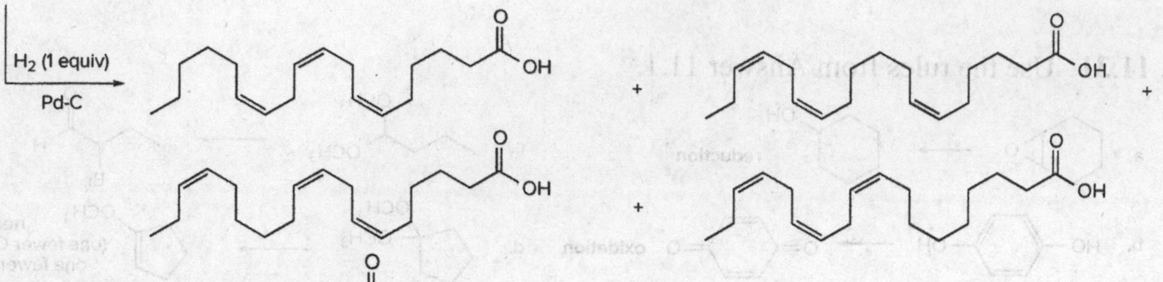

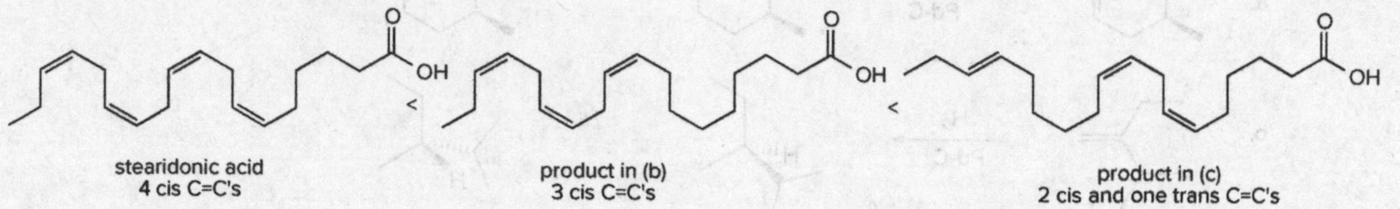

11.35

a. $\xrightarrow[\text{Pd-C}]{H_2}$

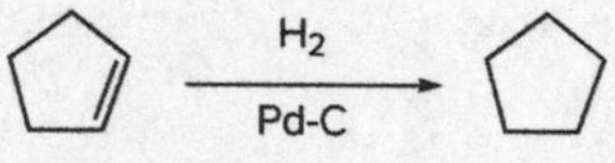

b. $\xrightarrow[\text{Lindlar catalyst}]{H_2}$ no reaction

c. $\xrightarrow[NH_3]{Na}$ no reaction

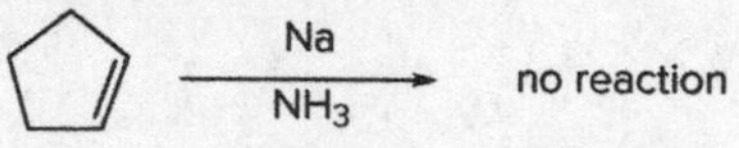

d. $\xrightarrow{CH_3CO_3H}$ O

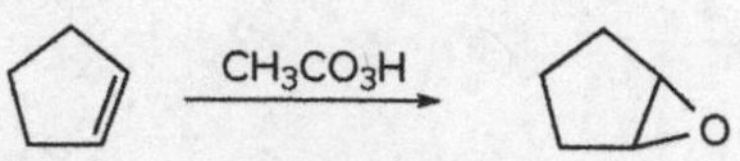

e. [1] CH_3CO_3H [2] H_2O, HO^- → OH, OH + OH, OH anti addition

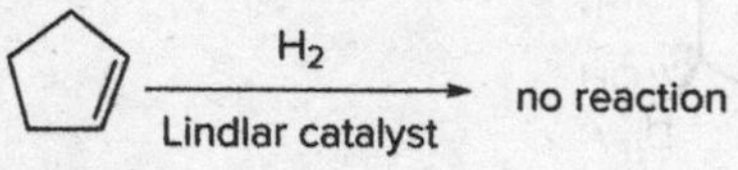

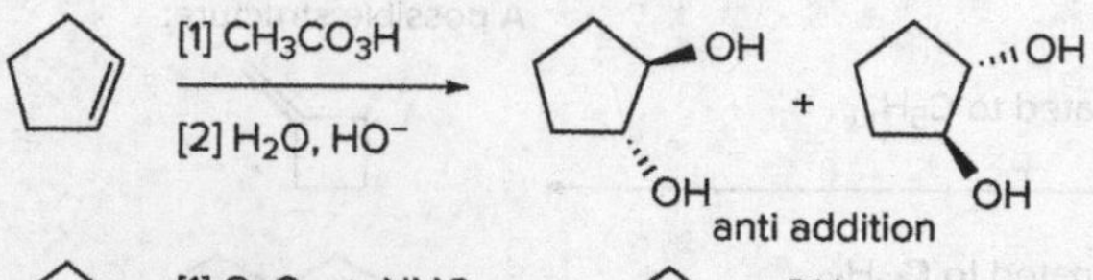

f. [1] OsO_4 + NMO [2] $NaHSO_3$, H_2O → OH, OH syn addition

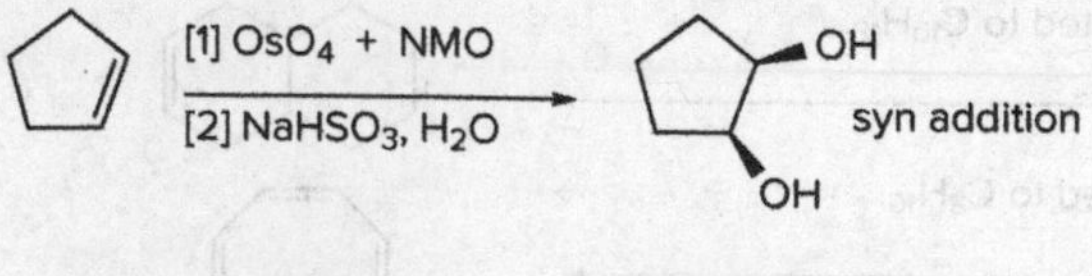

g. $\xrightarrow[H_2O,\ HO^-]{KMnO_4}$ OH, OH syn addition

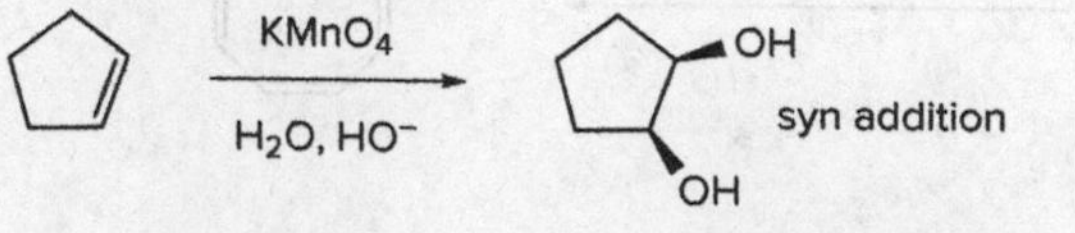

h. [1] $LiAlH_4$ [2] H_2O → no reaction

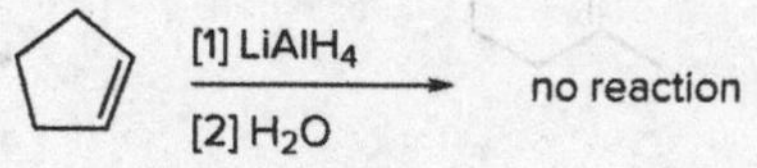

i. [1] O_3 [2] CH_3SCH_3 → H, O, O, H

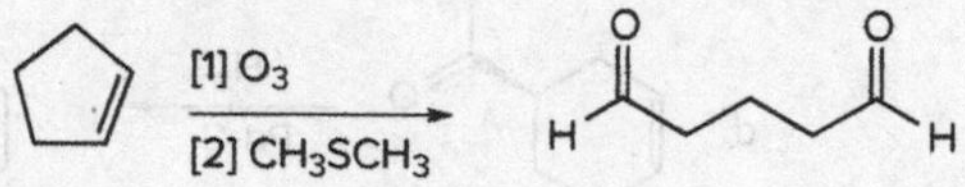

j. $(CH_3)_3COOH$, $Ti[OCH(CH_3)_2]_4$, (–)-DET → no reaction

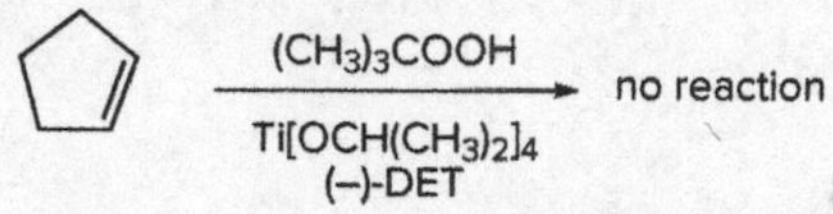

k. $\xrightarrow{\text{mCPBA}}$ O

l. O [1] $LiAlH_4$ [2] H_2O → OH

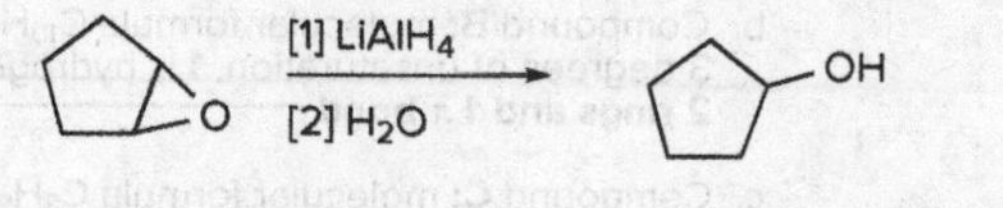

11.36

a. H2 / Pd-C

b. mCPBA

c. PCC

d. CrO_3 / H_2SO_4, H_2O

e. $(CH_3)_3COOH$ / $Ti[OCH(CH_3)_2]_4$ / (+)-DET

f. $(CH_3)_3COOH$ / $Ti[OCH(CH_3)_2]_4$ / (−)-DET

g. [1] PBr_3; [2] $LiAlH_4$; [3] H_2O

11.37

A disubstituted | B monosubstituted | C tetrasubstituted | D trisubstituted

a. The stability of an alkene increases with increasing alkyl substitution. The more stable the alkene, the smaller the heat of hydrogenation.

Heat of hydrogenation: C < D < A < B

most stable, smallest heat of hydrogenation (C) — least stable, largest heat of hydrogenation (B)

b. The rate of hydrogenation increases with decreasing alkyl substitution.

Rate of hydrogenation: C < D < A < B

c.

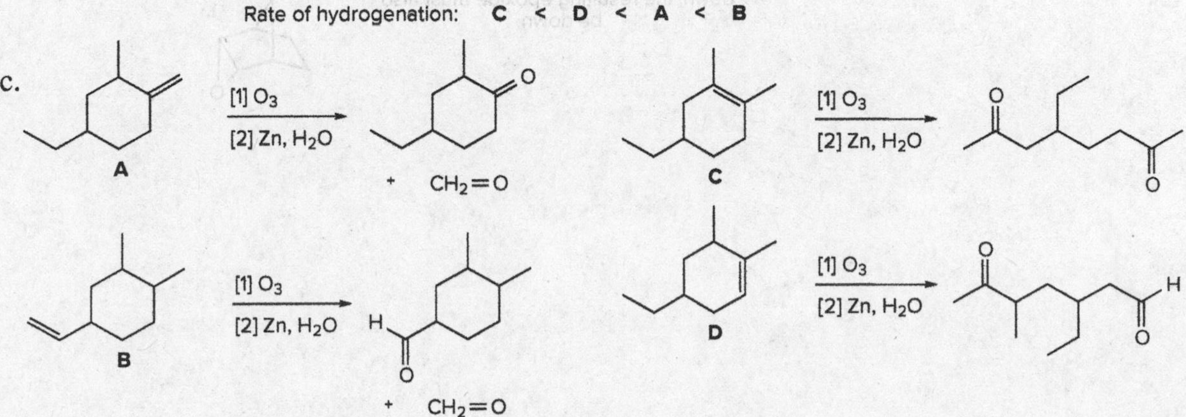

11.38

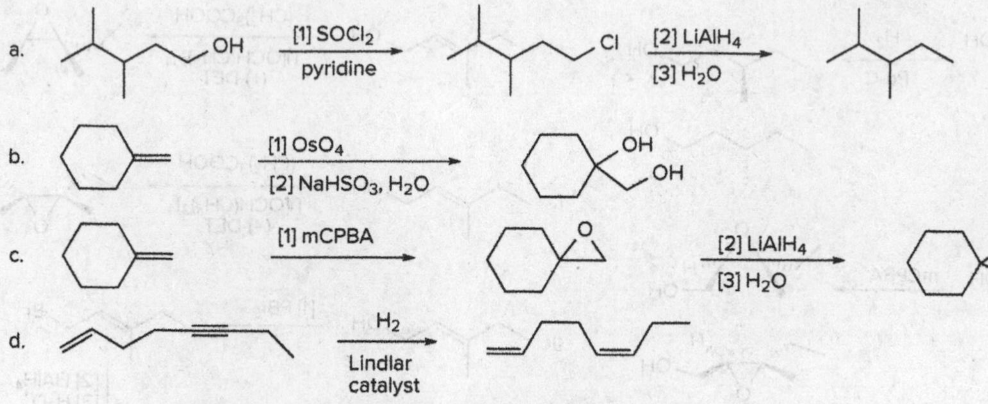

a. [1] $SOCl_2$, pyridine; [2] $LiAlH_4$, [3] H_2O

OH → Cl

b. [1] OsO_4, [2] $NaHSO_3$, H_2O

OH, OH

c. [1] mCPBA; [2] $LiAlH_4$, [3] H_2O

O; OH

d. H_2, Lindlar catalyst

11.39

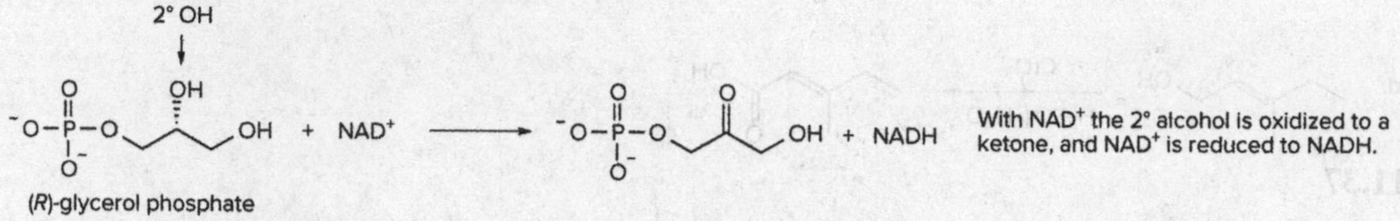

2° OH

OH

$^{-}O-P(=O)(O^{-})-O$ ... OH + NAD^+ → $^{-}O-P(=O)(O^{-})-O$ ... (C=O) ... OH + NADH

(*R*)-glycerol phosphate

With NAD^+ the 2° alcohol is oxidized to a ketone, and NAD^+ is reduced to NADH.

11.40 The two sides of the C=C of **A** are different. D_2 adds only from above, so this side must be less sterically hindered. Other reagents will also add from the same side.

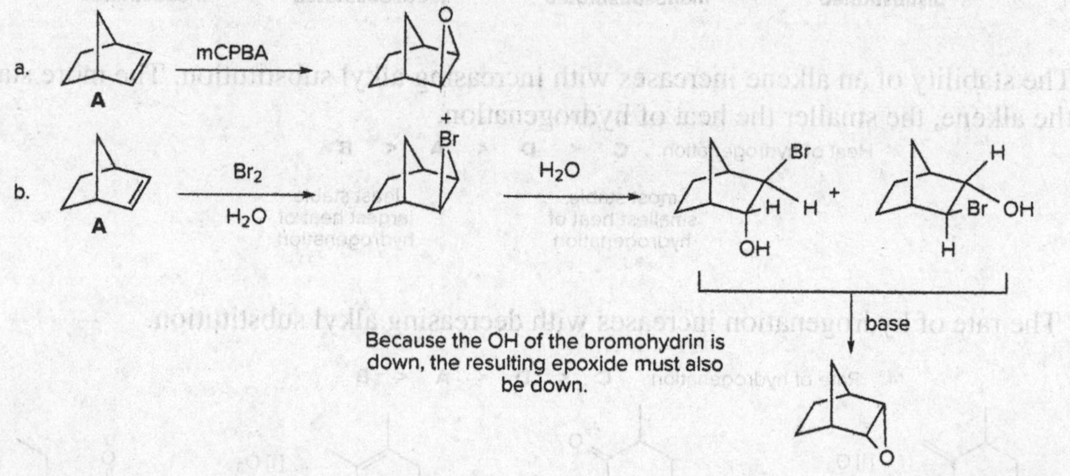

a. A —mCPBA→ (epoxide, O)

b. A —Br_2, H_2O→ (Br$^+$) —H_2O→ (Br, H, H, OH) + (H, Br, OH, H) —base→ (epoxide, O)

Because the OH of the bromohydrin is down, the resulting epoxide must also be down.

11.41 Alkenes treated with [1] OsO_4 followed by $NaHSO_3$ in H_2O will undergo **syn** addition, whereas alkenes treated with [2] CH_3CO_3H followed by ^-OH in H_2O will undergo **anti** addition.

a. [1] [1] OsO_4 [2] $NaHSO_3$, H_2O → HO OH

[2] [1] CH_3CO_3H [2] ^-OH, H_2O anti addition → HO OH → rotate → HO OH

b. [1] [1] OsO_4 [2] $NaHSO_3$, H_2O syn addition → OH OH → rotate → OH OH + enantiomer

[2] [1] CH_3CO_3H [2] ^-OH, H_2O → OH OH + enantiomer

c. [1] [1] OsO_4 [2] $NaHSO_3$, H_2O syn addition → HO OH → rotate → OH, $CH_3CH_2CH_2$, H, H, $CH_2CH_2CH_3$, OH

[2] [1] CH_3CO_3H [2] ^-OH, H_2O → OH, $CH_3CH_2CH_2$, H, H, $CH_2CH_2CH_3$, OH

11.42

a. A (O, H, OH) → CH_3SO_2Cl → O, H, $O-S(=O)_2-CH_3$

b. CH_3S^- + O, H, $O-S(=O)_2-CH_3$ → CH_3S, O^-, H, $O-S(=O)_2-CH_3$ → CH_3S, H, O = O, SCH_3, H **B**

c. To form chiral **A,** Sharpless reagent with (+)-DET could be used.

11.43

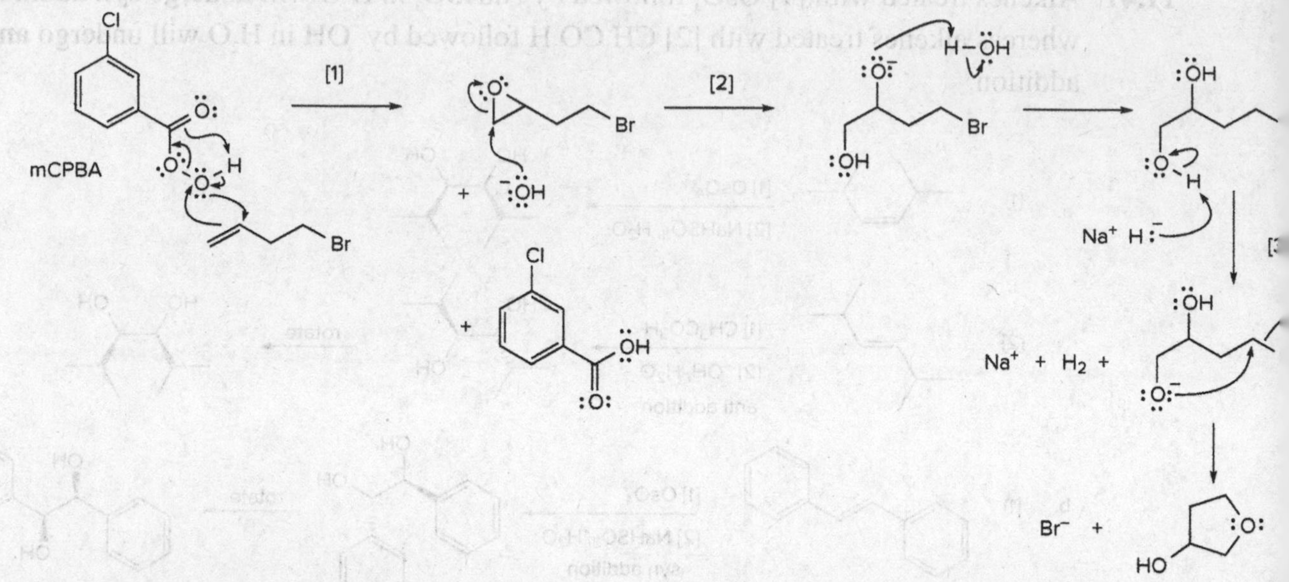

mCPBA

[1]

[2]

Br

Na⁺ H:⁻

Na⁺ + H₂ +

Br⁻ +

HO

11.44 Use the directions from Answers 11.19 and 11.22.

a. 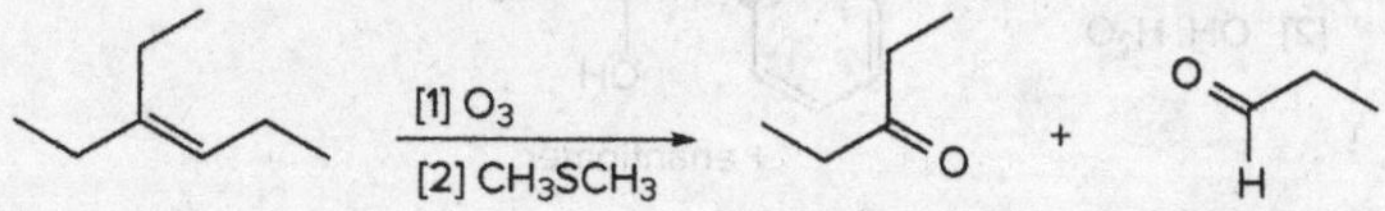 [1] O_3 [2] CH_3SCH_3 +

b. 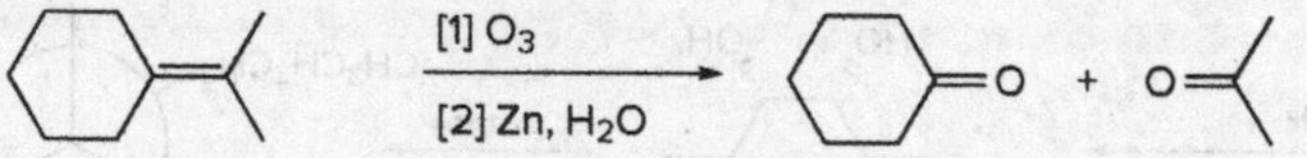 [1] O_3 [2] Zn, H_2O +

c. 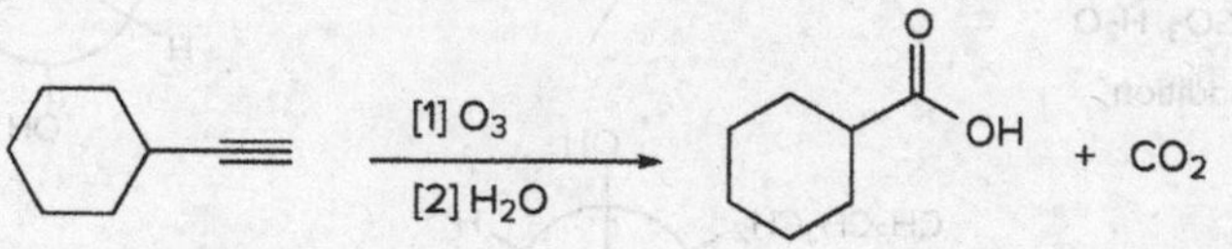 [1] O_3 [2] H_2O OH + CO_2

d. 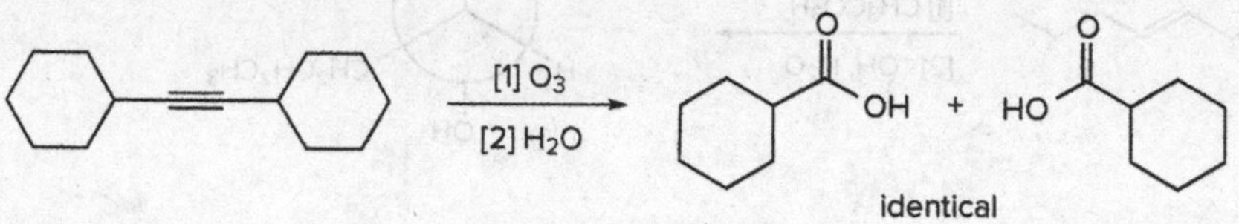 [1] O_3 [2] H_2O OH + HO

identical

11.45

a. 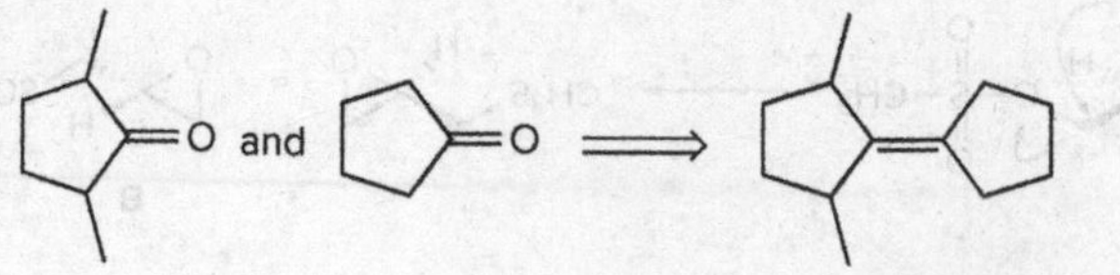 and ⟹

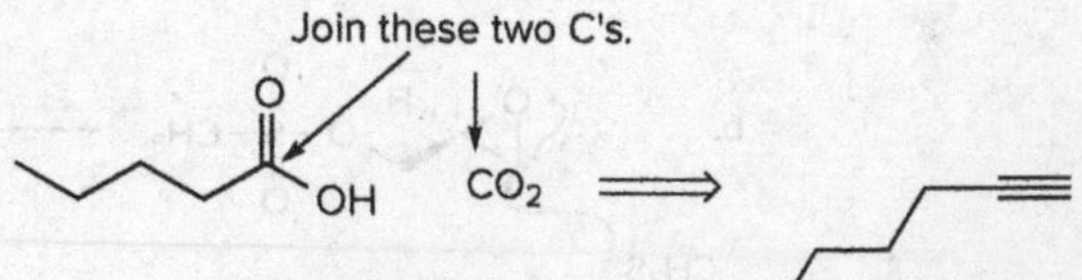

b. 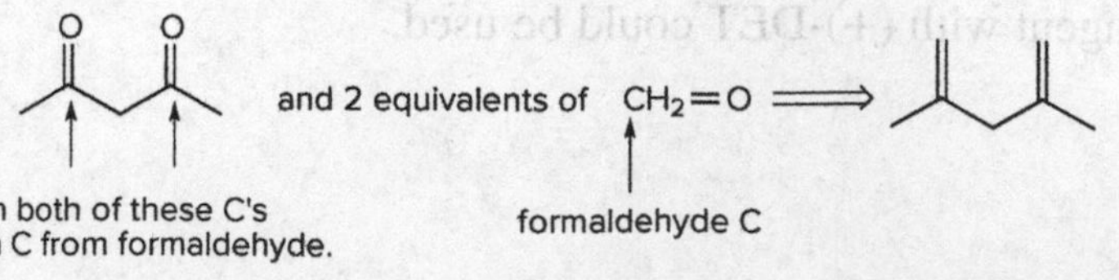 and 2 equivalents of $CH_2{=}O$ ⟹

Join both of these C's to a C from formaldehyde.

formaldehyde C

c. Join these two C's. OH CO_2 ⟹

d. 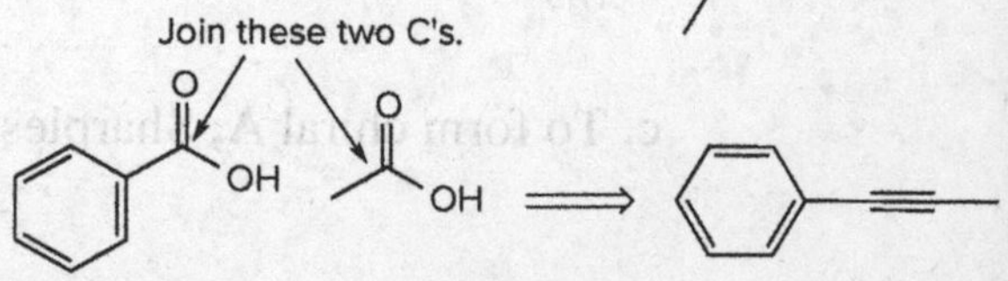 Join these two C's. OH OH ⟹

11.46 Use the directions from Answer 11.20.

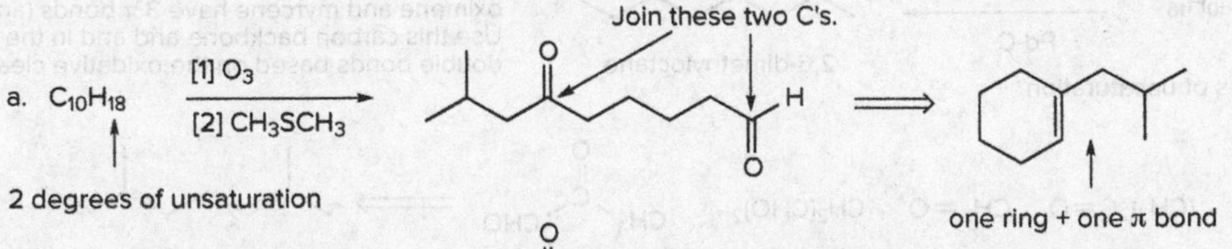

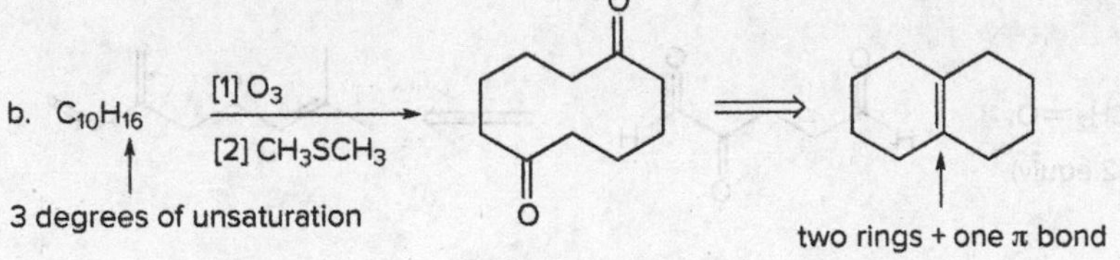

a. $C_{10}H_{18}$ → [1] O_3 [2] CH_3SCH_3

Join these two C's.

2 degrees of unsaturation

one ring + one π bond

b. $C_{10}H_{16}$ → [1] O_3 [2] CH_3SCH_3

3 degrees of unsaturation

two rings + one π bond

11.47

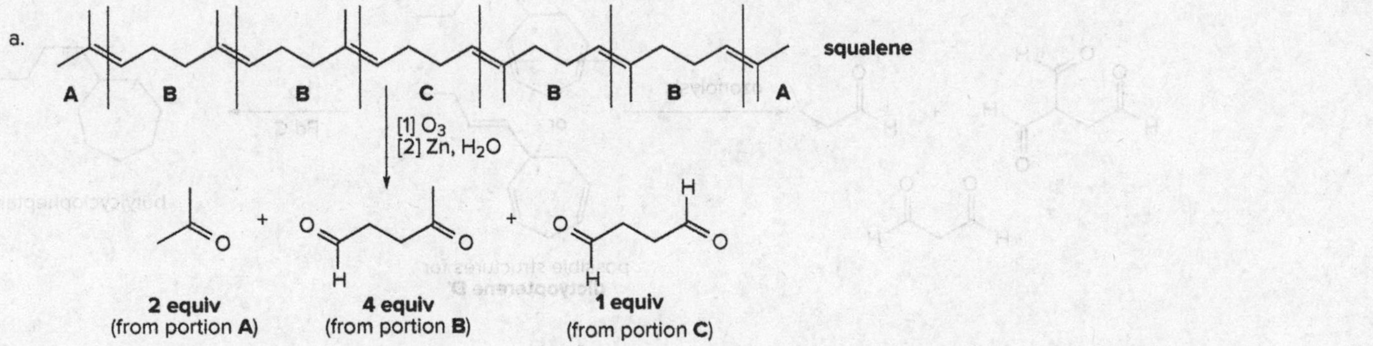

a. squalene

A B B C B B A

[1] O_3 [2] Zn, H_2O

2 equiv (from portion **A**) + 4 equiv (from portion **B**) + 1 equiv (from portion **C**)

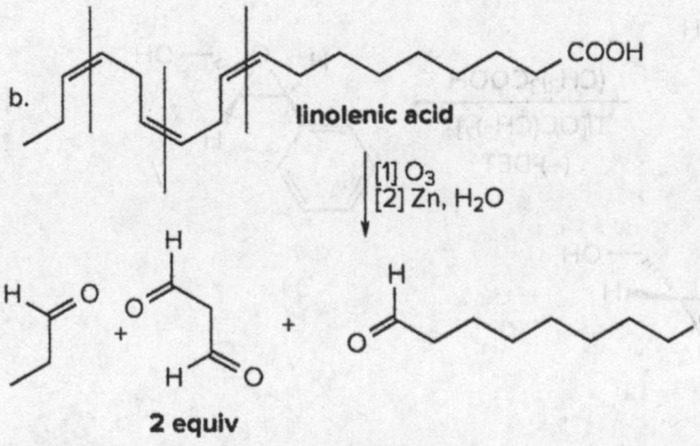

b. linolenic acid

COOH

[1] O_3 [2] Zn, H_2O

2 equiv

COOH

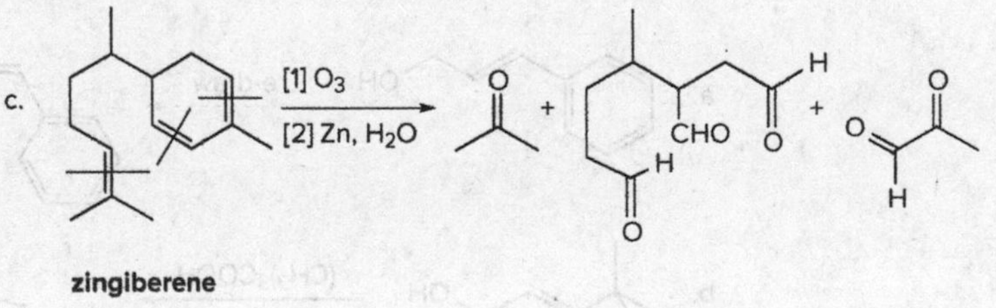

c. zingiberene

[1] O_3 [2] Zn, H_2O

CHO

11.48

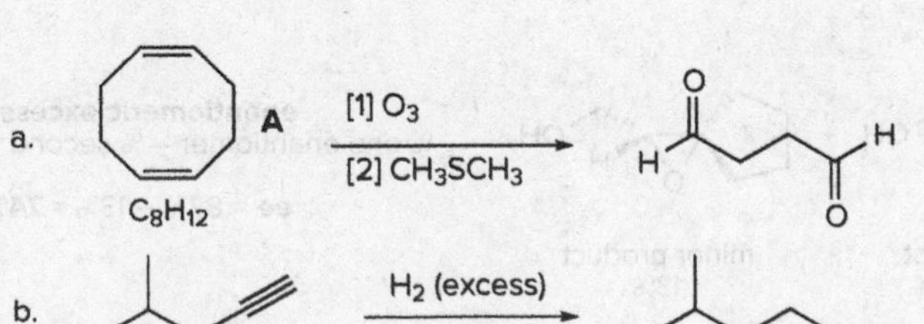

a. A C_8H_{12} → [1] O_3 [2] CH_3SCH_3

b. C_6H_{10} **B** → H_2 (excess) Pd-C

→ [1] $NaNH_2$ [2] CH_3I → C_7H_{12} **C**

11.49

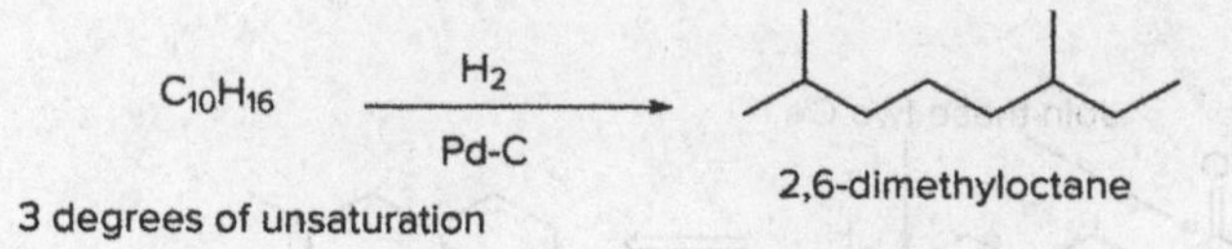

$C_{10}H_{16}$ —H_2, Pd-C→ 2,6-dimethyloctane

3 degrees of unsaturation

The hydrogenation reaction tells you that both oximene and myrcene have 3 π bonds (and no rings). Use this carbon backbone and add in the double bonds based on the oxidative cleavage products.

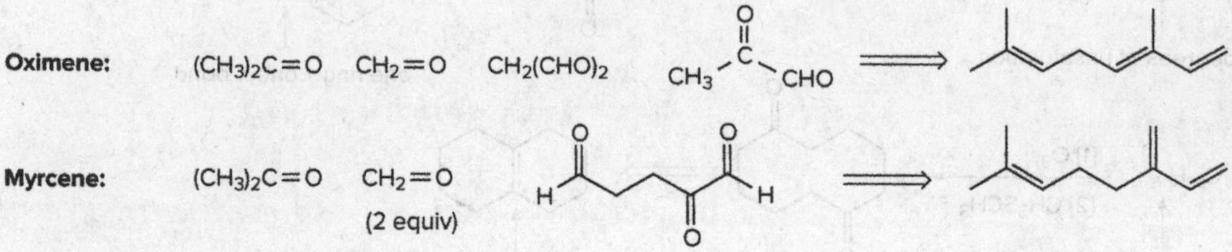

Oximene: $(CH_3)_2C{=}O$ $CH_2{=}O$ $CH_2(CHO)_2$ CH_3COCHO ⟹

Myrcene: $(CH_3)_2C{=}O$ $CH_2{=}O$ (2 equiv) ⟹

11.50 The stereogenic center (labeled with *) in both structures can be *R* or *S*.

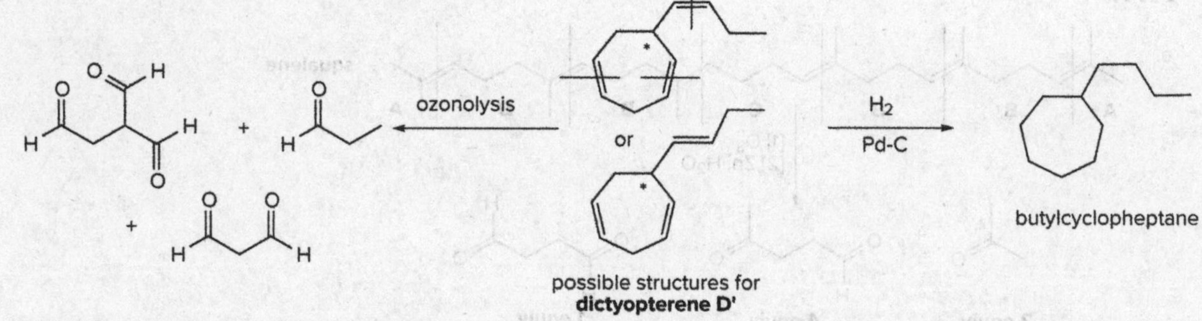

ozonolysis ← possible structures for **dictyopterene D'** (or) —H_2, Pd-C→ butylcyclopheptane

11.51

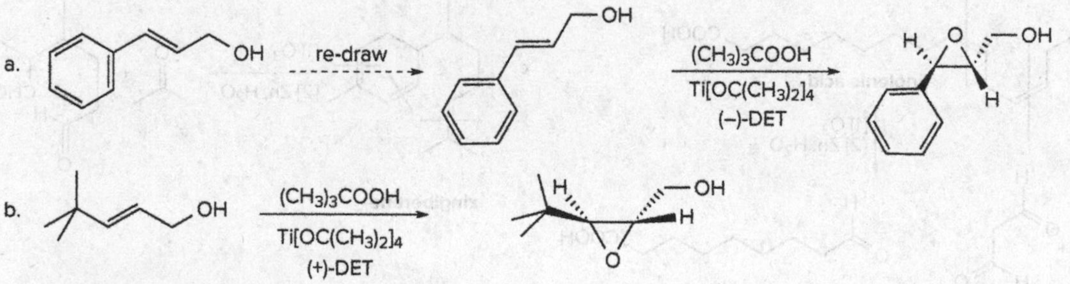

a. re-draw → $(CH_3)_3COOH$, $Ti[OC(CH_3)_2]_4$, (–)-DET

b. OH —$(CH_3)_3COOH$, $Ti[OC(CH_3)_2]_4$, (+)-DET→ OH

11.52

OH re-draw

OH —$(CH_3)_3COOH$, $Ti[OC(CH_3)_2]_4$, (–)-DET→ OH + OH

major product 87%

minor product 13%

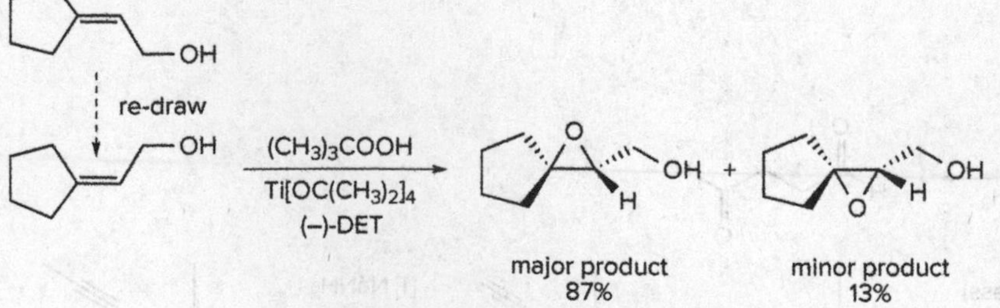

enantiomeric excess = % one enantiomer – % second enantiomer

ee = 87% – 13% = **74%**

11.53

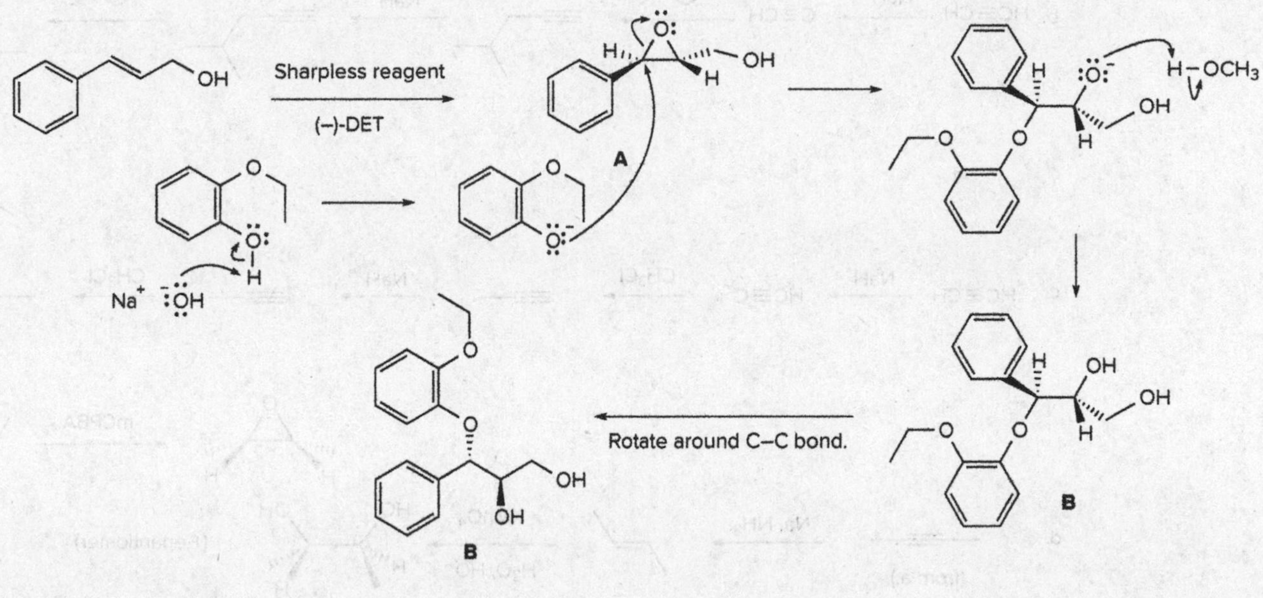

Sharpless reagent

(–)-DET

A

Na^+ :ÖH

H–OCH_3

Rotate around C–C bond.

B

B

11.54

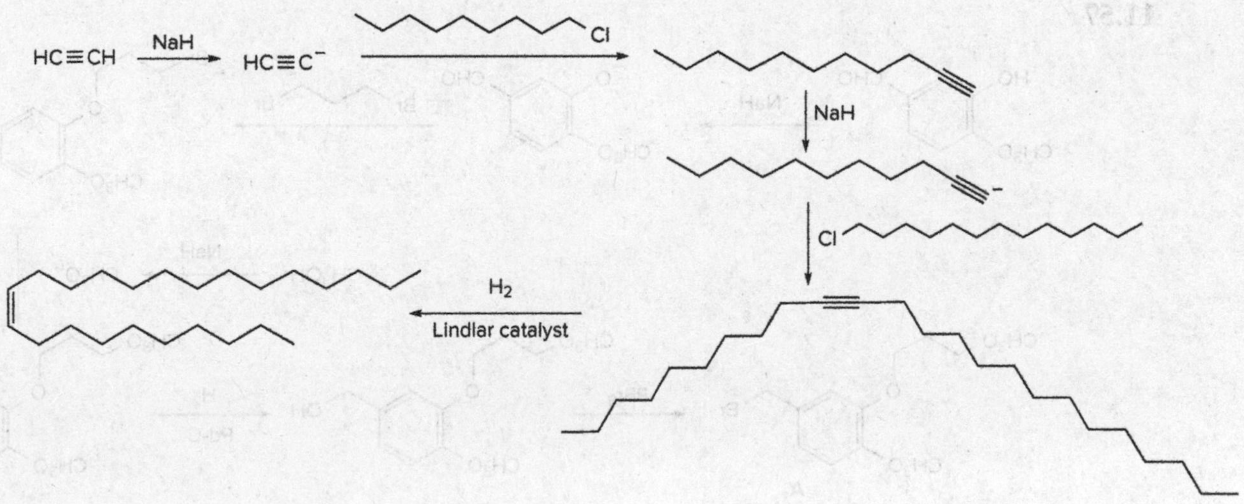

HC≡CH $\xrightarrow{NaH}$ HC≡C^-

Cl

NaH

Cl

H_2

Lindlar catalyst

11.55

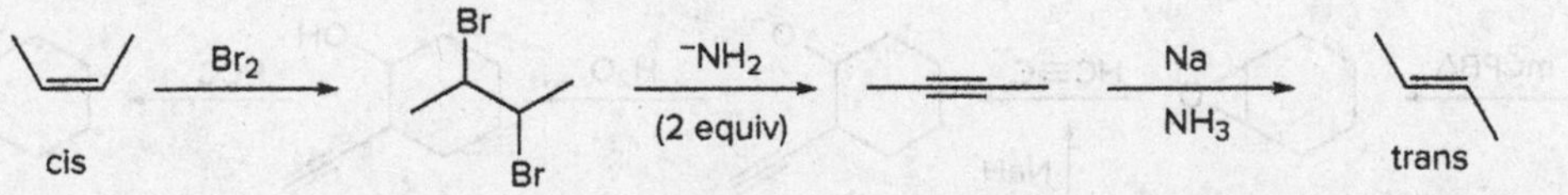

cis

Br_2

Br

Br

$^-NH_2$

(2 equiv)

Na

NH_3

trans

11.56

a. HC≡CH $\xrightarrow{NaH}$ $^-$C≡CH $\xrightarrow{CH_3Cl}$ $\xrightarrow{NaH}$ $\xrightarrow{CH_3Cl}$ $\xrightarrow[NH_3]{Na}$

b. 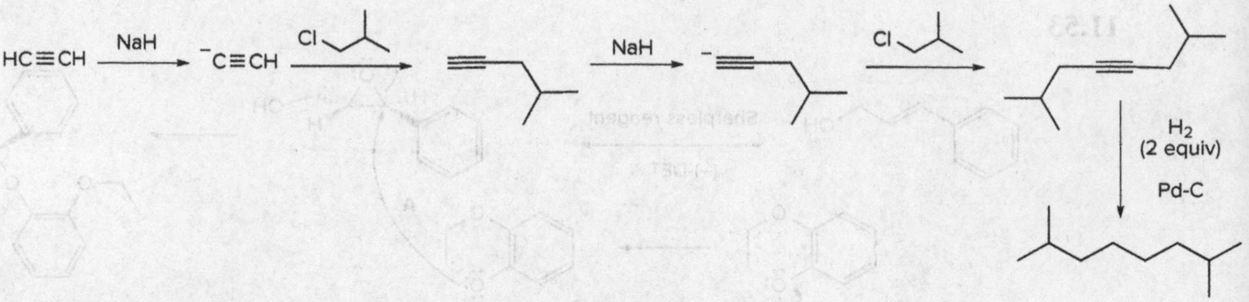 $HC\equiv CH$ —NaH→ $^{-}C\equiv CH$ —Cl(isobutyl)→ … —NaH→ … —Cl(isobutyl)→ … —H_2 (2 equiv), Pd-C→ …

c. $HC\equiv CH$ —NaH→ $HC\equiv C^{-}$ —CH_3Cl→ … —NaH→ … —CH_3Cl→ … —H_2, Lindlar catalyst→ … —mCPBA→ H H

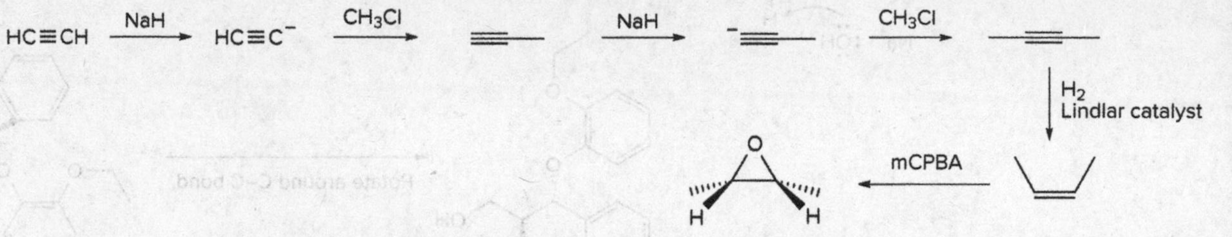

d. 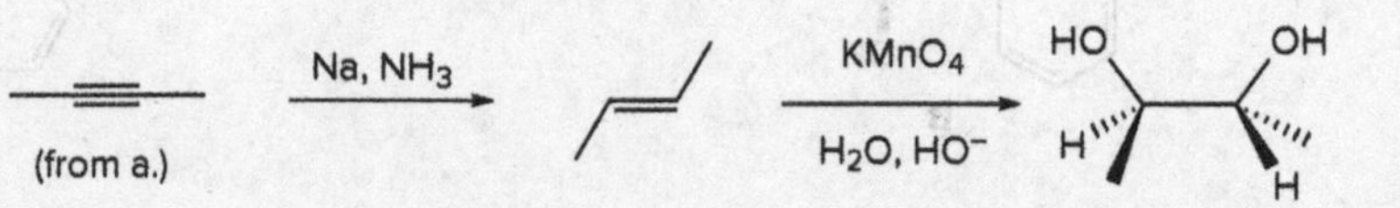 (from a.) —Na, NH_3→ … —$KMnO_4$, H_2O, HO^{-}→ HO OH H H (+ enantiomer)

11.57

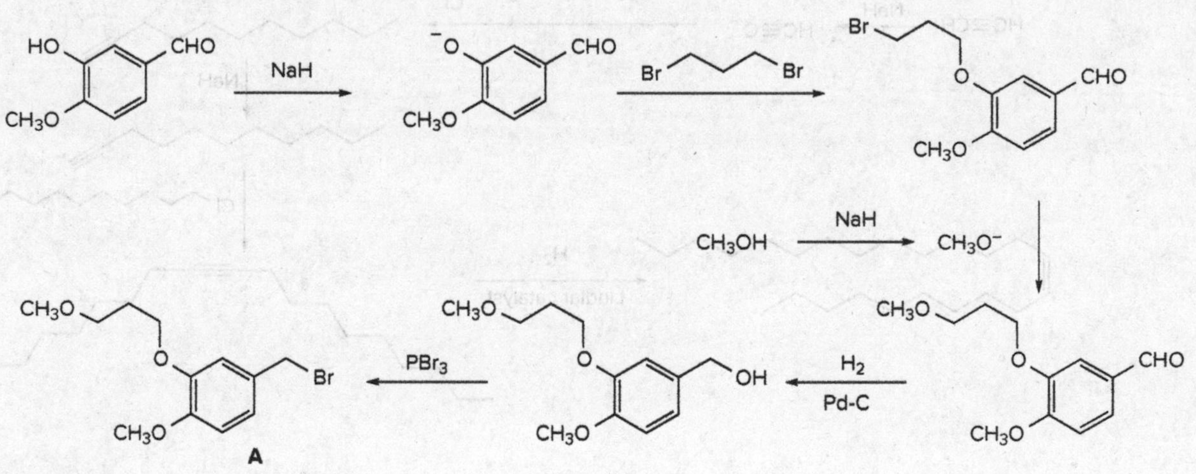

HO CHO CH_3O —NaH→ ^{-}O CHO CH_3O —Br Br→ Br O CHO CH_3O

CH_3OH —NaH→ CH_3O^{-}

CH_3O O CHO CH_3O —H_2, Pd-C→ CH_3O O OH CH_3O —PBr_3→ CH_3O O Br CH_3O **A**

11.58

a. 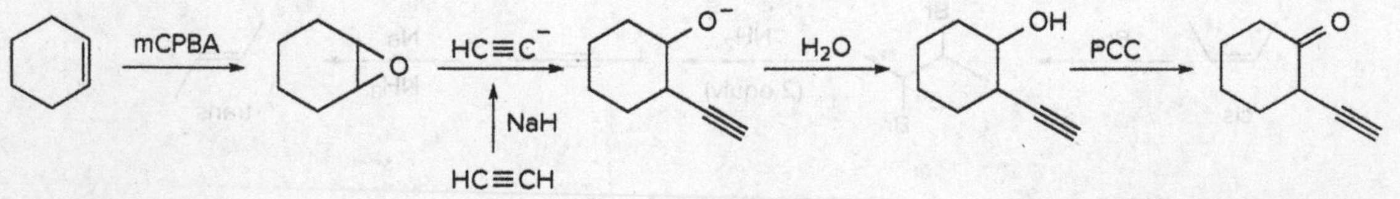 —mCPBA→ O —$HC\equiv C^{-}$ (from $HC\equiv CH$, NaH)→ O^{-} —H_2O→ OH —PCC→ O

b. 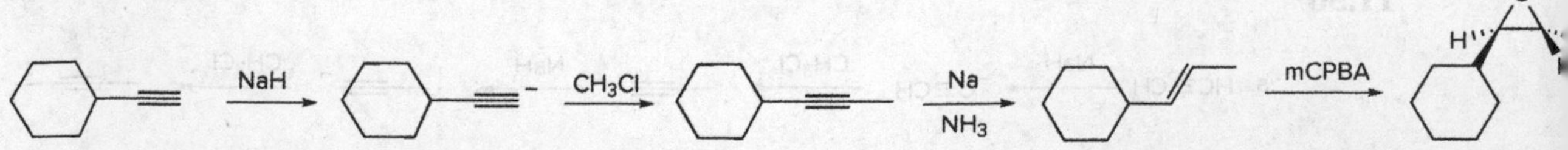 —NaH→ … $^{-}$ —CH_3Cl→ … —Na, NH_3→ … —mCPBA→ O H (+ enantiomer)

c. 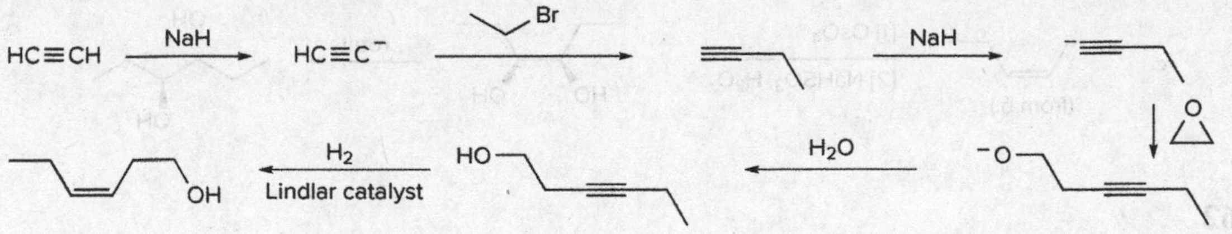 $HC\equiv CH$ $\xrightarrow{NaH}$ $HC\equiv C^-$ → (ethyl bromide) → 1-butyne $\xrightarrow{NaH}$ butynide anion → (ethylene oxide) → ^-O–alkoxide alkyne $\xrightarrow{H_2O}$ HO–(hex-3-yn-1-ol) $\xrightarrow[\text{Lindlar catalyst}]{H_2}$ (Z)-alkene OH

d. 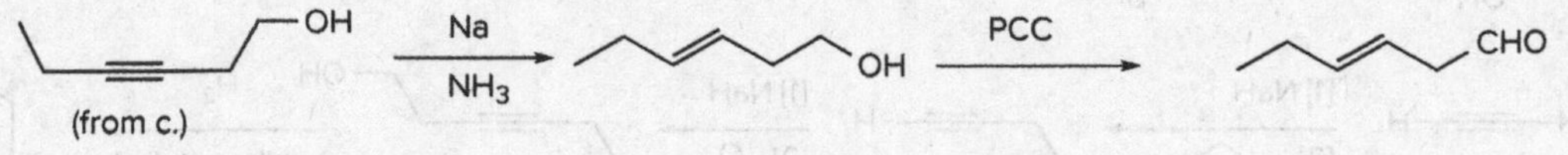 alkyne –OH (from c.) $\xrightarrow[NH_3]{Na}$ (E)-alkene OH $\xrightarrow{PCC}$ (E)-alkene CHO

11.59

a. 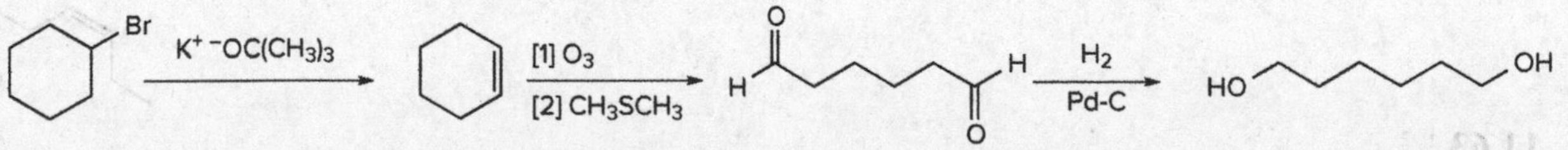 bromocyclohexane (Br) $\xrightarrow{K^+ \ ^-OC(CH_3)_3}$ cyclohexene $\xrightarrow[\text{[2] } CH_3SCH_3]{\text{[1] } O_3}$ H–(dialdehyde, O, O)–H $\xrightarrow[\text{Pd-C}]{H_2}$ HO–(chain)–OH

b. 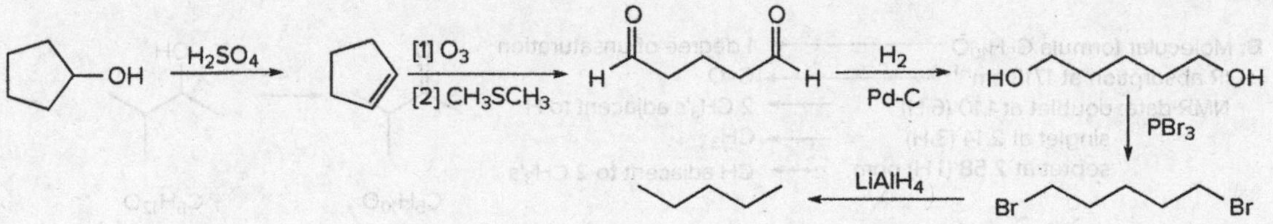 cyclopentanol (OH) $\xrightarrow{H_2SO_4}$ cyclopentene $\xrightarrow[\text{[2] } CH_3SCH_3]{\text{[1] } O_3}$ H–(dialdehyde, O, O)–H $\xrightarrow[\text{Pd-C}]{H_2}$ HO–(chain)–OH $\xrightarrow{PBr_3}$ Br–(chain)–Br $\xrightarrow{LiAlH_4}$ pentane

11.60

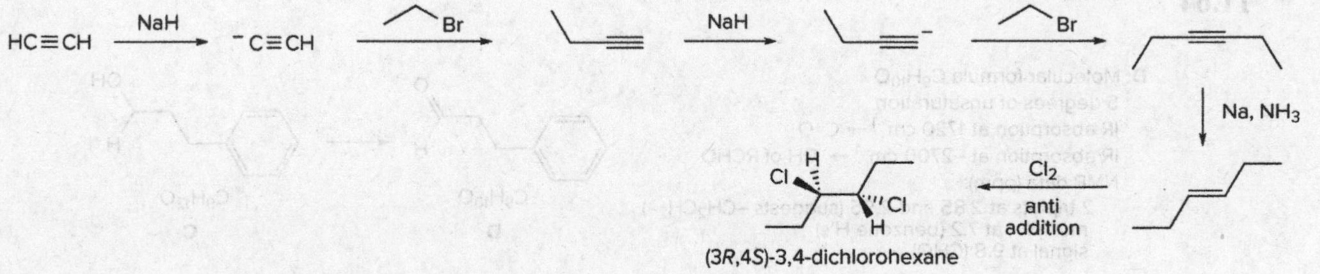

$HC\equiv CH$ $\xrightarrow{NaH}$ $^-C\equiv CH$ → (ethyl bromide, Br) → 1-butyne $\xrightarrow{NaH}$ butynide anion → (ethyl bromide, Br) → 3-hexyne $\xrightarrow{Na,\ NH_3}$ (E)-3-hexene $\xrightarrow[\text{anti addition}]{Cl_2}$ Cl, H, Cl, H

(3*R*,4*S*)-3,4-dichlorohexane

11.61

a. 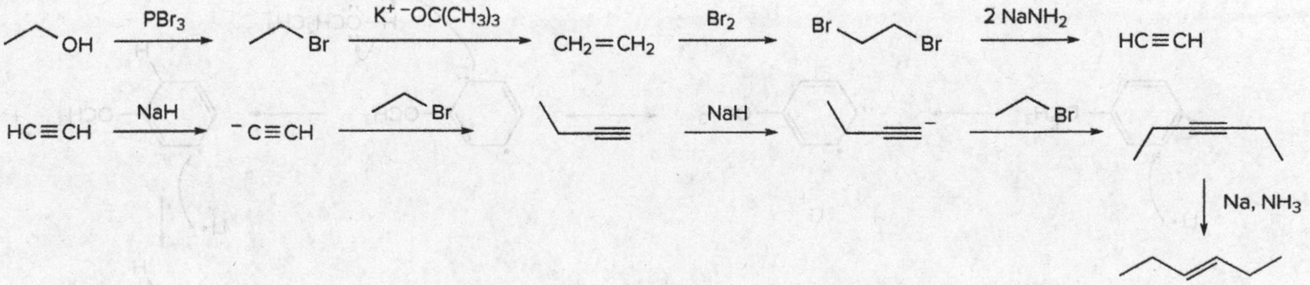 ethanol (OH) $\xrightarrow{PBr_3}$ bromoethane (Br) $\xrightarrow{K^+ \ ^-OC(CH_3)_3}$ $CH_2{=}CH_2$ $\xrightarrow{Br_2}$ Br–CH₂CH₂–Br $\xrightarrow{2\ NaNH_2}$ $HC\equiv CH$

$HC\equiv CH$ $\xrightarrow{NaH}$ $^-C\equiv CH$ → (ethyl bromide, Br) → 1-butyne $\xrightarrow{NaH}$ butynide anion → (ethyl bromide, Br) → 3-hexyne $\xrightarrow{Na,\ NH_3}$ (E)-3-hexene

b. 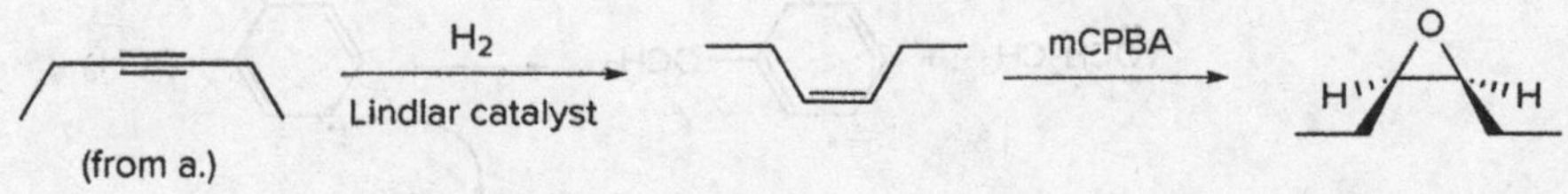 3-hexyne (from a.) $\xrightarrow[\text{Lindlar catalyst}]{H_2}$ (Z)-3-hexene $\xrightarrow{mCPBA}$ epoxide (O, H, H)

c. (from b.) [1] OsO_4 [2] $NaHSO_3$, H_2O → HO OH → rotate → OH OH

11.62

OH → PBr_3 → Br

H—≡—H → [1] NaH [2] Br → ≡—H → [1] NaH [2] O (epoxide) [3] H_2O → ≡ OH → H_2 Lindlar catalyst → OH → [1] NaH [2] Br → O **A**

11.63

B: Molecular formula $C_5H_{10}O$ → 1 degree of unsaturation
IR absorption at 1718 cm^{-1} → C=O
NMR data: doublet at 1.10 (6 H) → 2 CH_3's adjacent to H
singlet at 2.14 (3 H) → CH_3
septet at 2.58 (1 H) ppm → CH adjacent to 2 CH_3's

O ← OH

$C_5H_{10}O$ **B** $C_5H_{12}O$ **A**

11.64

D: Molecular formula $C_9H_{10}O$
5 degrees of unsaturation
IR absorption at 1720 cm^{-1} → C=O
IR absorption at ~2700 cm^{-1} → CH of RCHO
NMR data (ppm):
2 triplets at 2.85 and 2.95 (suggests $-CH_2CH_2-$)
multiplet at 7.2 (benzene H's)
signal at 9.8 (CHO)

O H ← OH H

$C_9H_{10}O$ **D** $C_9H_{12}O$ **C**

11.65

OCH_3 + Li• → OCH_3 + Li^+ ⟷ OCH_3 + $H-\ddot{O}CH_2CH_3$ → H OCH_3 + $^-:\ddot{O}CH_2CH_3$

Li• ↓

$:\ddot{O}CH_2CH_3$ + OCH_3 ← H OCH_3 + Li^+ ; $H-\ddot{O}CH_2CH_3$

11.66

$C_6H_{11}Br$ → [1] OsO_4, [2] NaH

$C_6H_{11}Br$ → OsO_4 → NaH →

11.67

The favored conformation for both molecules places the *tert*-butyl group equatorial.

A = A: This OH is axial and will react **faster** because the OH group is more hindered.

B: This OH is equatorial and will react **more slowly** because the OH group is less hindered. = B

11.68 The two OH's are added to opposite faces of the C=C, so anti addition occurs.

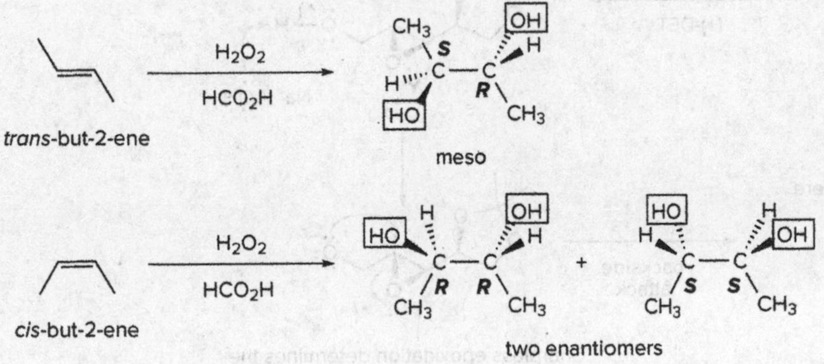

11.69

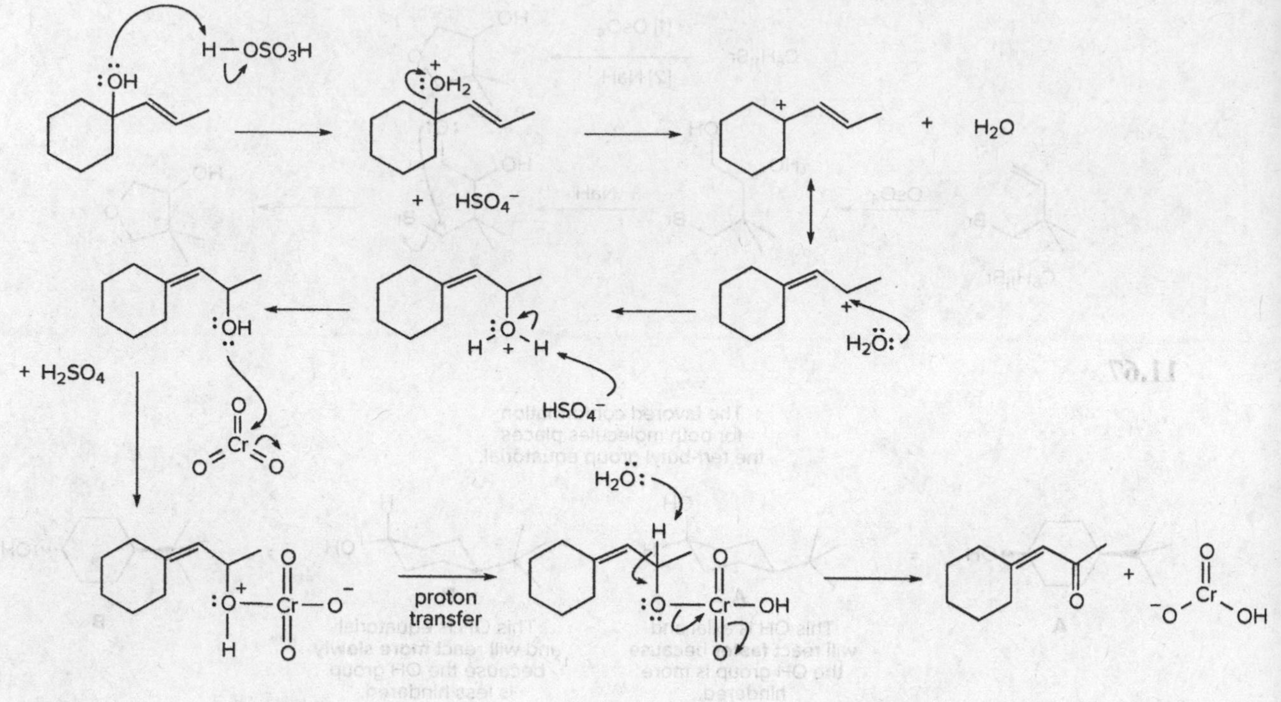

$H-OSO_3H$:OH → :$\overset{+}{O}H_2$ + HSO_4^- → + H_2O

$H_2\ddot{O}$:

:OH + H_2SO_4 ← H + H ← HSO_4^-

O=Cr(=O)=O

$H_2\ddot{O}$: H

:$\overset{+}{O}$—Cr—O⁻ H → proton transfer → :O—Cr—OH → A + ⁻O—Cr(=O)—OH

11.70

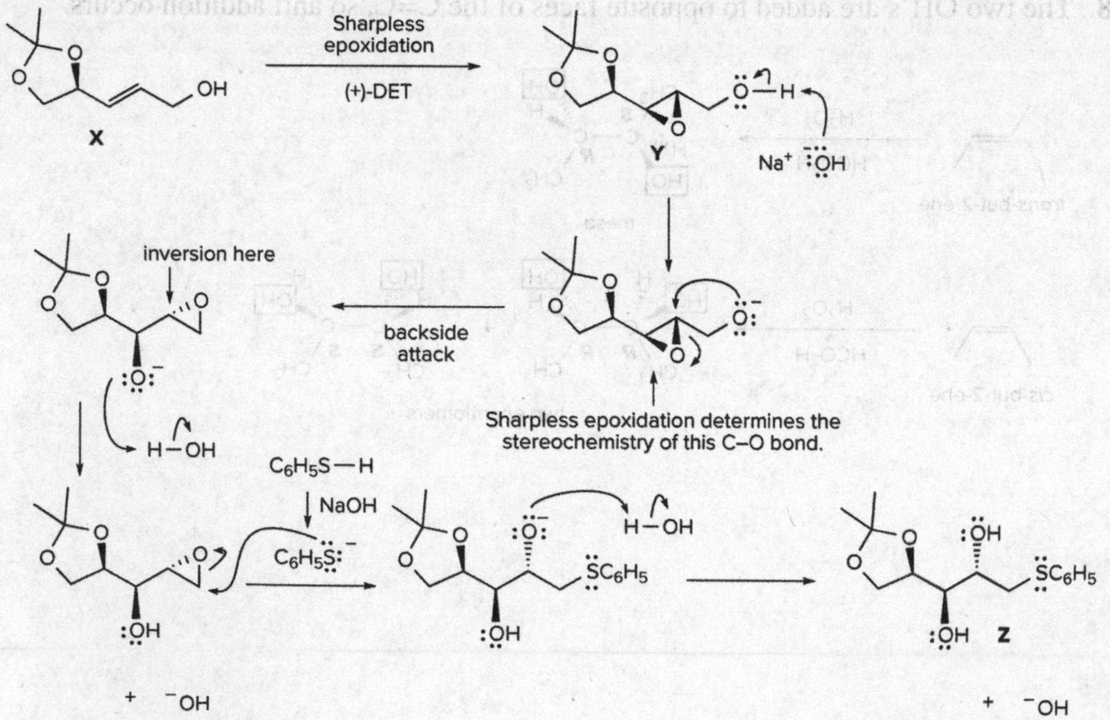

X → Sharpless epoxidation (+)-DET → Y OH—H Na⁺ :ÖH

inversion here ← backside attack ← O:⁻

Sharpless epoxidation determines the stereochemistry of this C–O bond.

:O:⁻ H—OH

C_6H_5S-H NaOH $C_6H_5S:^-$

:OH + ⁻OH → :O:⁻ SC_6H_5 :OH H—OH → :OH SC_6H_5 :OH Z + ⁻OH

Chapter 12 Conjugation, Resonance, and Dienes

Chapter Review

Conjugation and delocalization of electron density

- The overlap of *p* orbitals on three or more adjacent atoms allows electron density to delocalize, thus adding stability (12.1).
- An allyl carbocation ($CH_2{=}CHCH_2^+$) is more stable than a 1° carbocation because of *p* orbital overlap (12.2).
- In a system X=Y–Z:, Z is generally sp^2 hybridized to allow the lone pair to occupy a *p* orbital, making the system conjugated (12.5).

Four common examples of resonance (12.3)

[1] The three atom "allyl" system: X=Y–Z ⟷ X–Y=Z   • = +, –, •, or ••

[2] Conjugated double bonds:

[3] Cations having a positive charge adjacent to a lone pair: $\ddot{X}$–$\overset{+}{Y}$ ⟷ $\overset{+}{X}$=Y

[4] Double bonds having one atom more electronegative than the other: X=Y ⟷ $\overset{+}{X}$–$\overset{-}{Y}$: [Electronegativity of Y > X]

Rules on evaluating the relative "stability" of resonance structures (12.4)

[1] Structures with more bonds and fewer charges are better.

all neutral atoms
one more bond
bettter resonance structure

[2] Structures in which every atom has an octet are better.

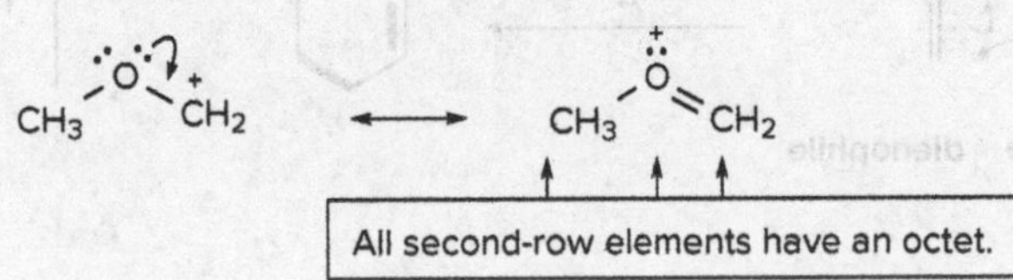

[3] Structures that place a negative charge on a more electronegative element are better.

The (–) charge is on the more electronegative O atom.

better resonance structure

The unusual properties of conjugated dienes

[1] The C–C σ bond joining the two double bonds is unusually short (12.8).

[2] Conjugated dienes are more stable than similar isolated dienes. $\Delta H°$ of hydrogenation is smaller for a conjugated diene than for an isolated diene converted to the same product (12.9).

[3] The reactions are unusual:

- Electrophilic addition affords products of 1,2-addition and 1,4-addition (12.10, 12.11).
- Conjugated dienes undergo the Diels–Alder reaction, a reaction that does not occur with isolated dienes (12.12–12.14).

Reactions of conjugated dienes

[1] Electrophilic addition of HX (X = halogen) (12.10, 12.11)

HX (1 equiv)

1,2-product kinetic product + **1,4-product** thermodynamic product

- The mechanism has two steps.
- Markovnikov's rule is followed. Addition of H^+ forms the *more* stable allylic carbocation.
- The 1,2-product is the kinetic product. When H^+ adds to the double bond, X^- adds to the end of the allylic carbocation to which it is closer (C2 not C4). The kinetic product is formed faster at low temperature.
- The thermodynamic product has the more substituted, more stable double bond. The thermodynamic product predominates at equilibrium. With buta-1,3-diene, the thermodynamic product is the 1,4-product.

[2] Diels–Alder reaction (12.12–12.14)

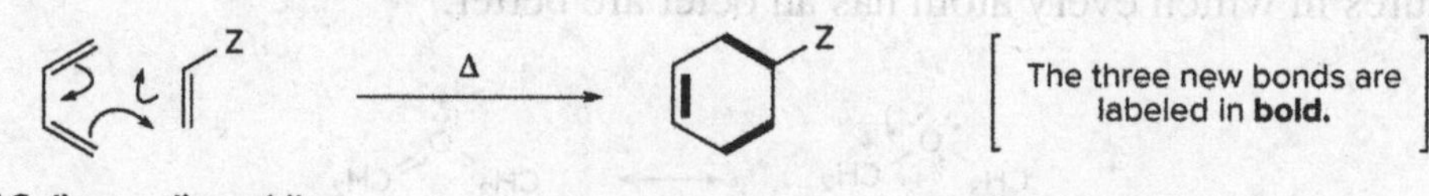

- The reaction forms two σ bonds and one π bond in a six-membered ring.
- The reaction is initiated by heat.
- The mechanism is concerted: all bonds are broken and formed in a single step.
- The diene must react in the *s*-cis conformation (12.13A).
- Electron-withdrawing groups in the dienophile increase the reaction rate (12.13B).
- The stereochemistry of the dienophile is retained in the product (12.13C).
- Endo products are preferred (12.13D).

Practice Test on Chapter Review

1.a. Which of the following statements is true about the Diels–Alder reaction?

1. The reaction is faster with electron-donating groups in the dienophile.
2. The reaction is endothermic.
3. The diene must adopt the *s*-cis conformation.
4. Statements (1) and (2) are true.
5. Statements (1), (2), and (3) are all true.

b. Which of the following compounds contains a labeled carbon atom that is sp^2 hybridized?

$\overset{-}{C}H_2$ $\overset{+}{C}H_2$ $\overset{-}{C}H_2$

A   **B**   **C**

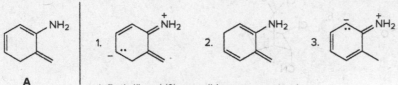

1. **A** only
2. **B** only
3. **C** only
4. **A** and **B**
5. **A, B,** and **C**

c. Which of the following represent valid resonance structures for **A**?

A: NH_2 1. $\overset{+}{N}H_2$ 2. NH_2 3. $\overset{+}{N}H_2$

4. Both (1) and (2) are valid resonance structures.
5. Structures (1), (2), and (3) are all valid.

2. Name the following compounds and indicate the conformation around the σ bond that joins the two double bonds.

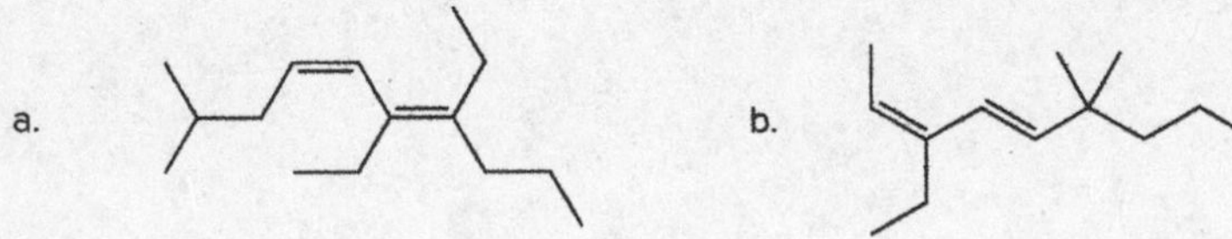

3. Consider the four dienes (A–D) drawn below. [1] Which diene is *most* reactive in the Diels–Alder reaction? [2] Which diene is the *least* reactive in the Diels–Alder reaction?

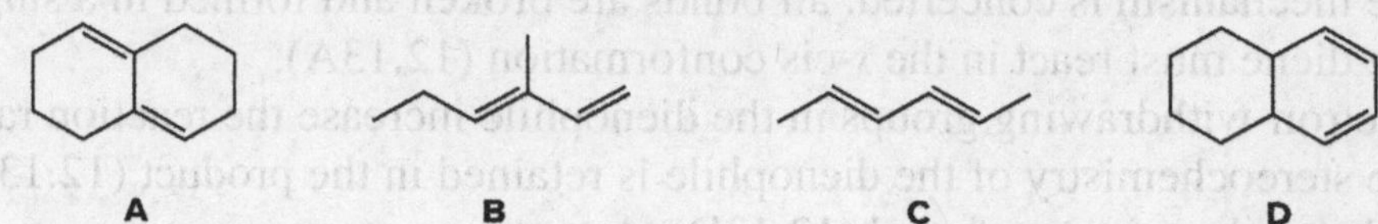

4. Draw the organic products formed in each reaction. In part (b), label the kinetic and thermodynamic products.

a. 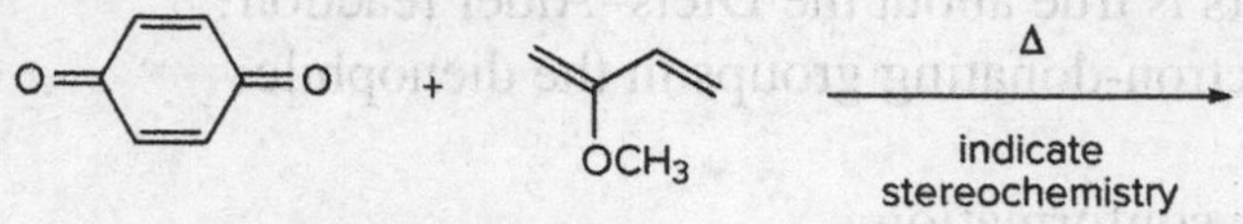 O=⟨⟩=O + (2-methoxy-1,3-butadiene, OCH_3) —Δ, indicate stereochemistry→

b. 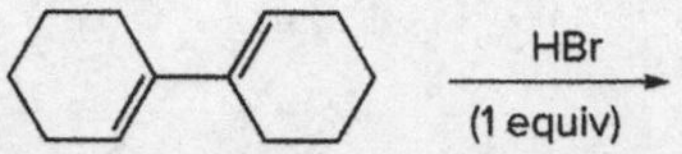 (bicyclohexenyl) —HBr (1 equiv)→

c. 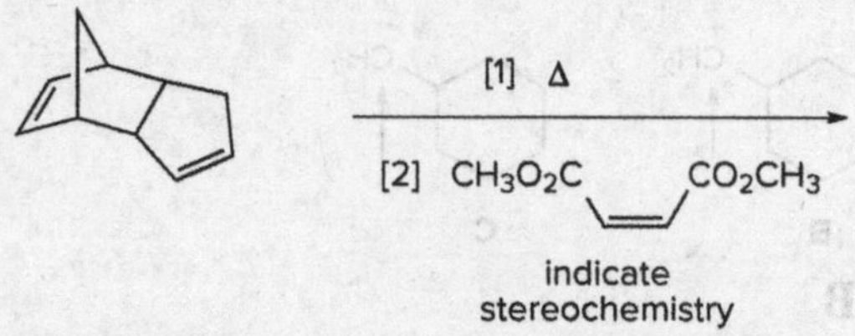 —[1] Δ; [2] CH_3O_2C–CH=CH–CO_2CH_3 (cis), indicate stereochemistry→

5. What diene and dienophile are needed to synthesize the following Diels–Alder adducts?

a. 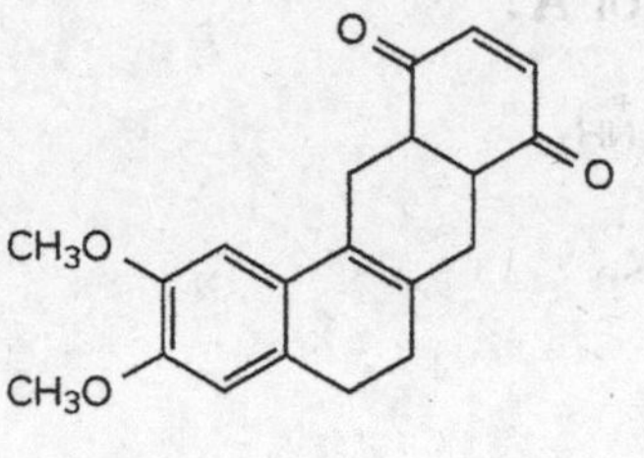 (CH_3O, CH_3O, O, O)

b. 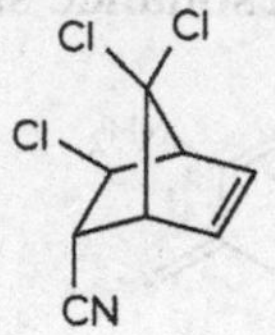 (Cl, Cl, Cl, CN)

Answers to Practice Test

1.a. 3
 b. 4
 c. 1

2.a. (4*Z*,6*E*)-6,7-diethyl-2-methyldeca-4,6-diene, *s*-trans
 b. (2*Z*,4*E*)-3-ethyl-6,6-dimethylnona-2,4-diene, *s*-trans

3. [1] **D;** [2] **A**

4.a

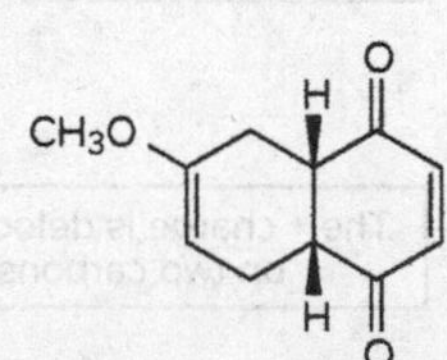

(both H's up or both H's down)

b.

kinetic
+
thermodynamic

c.

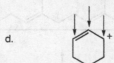

5.a.

b.

Answers to Problems

12.1 **Isolated dienes** have two double bonds separated by two or more σ bonds. **Conjugated dienes** have two double bonds separated by only one σ bond.

a.

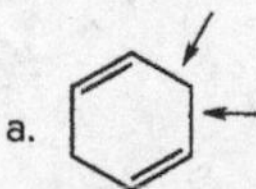

Two σ bonds separate two double bonds = **isolated diene.**

b.

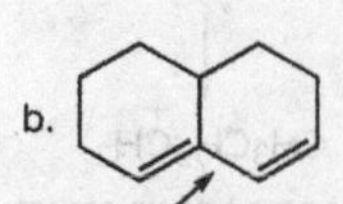

One σ bond separates two double bonds = **conjugated diene.**

c.

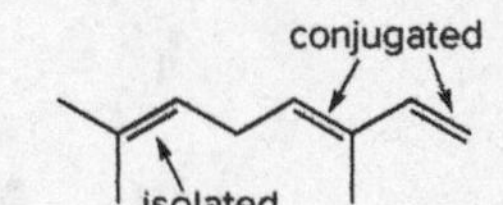

d. isolated, conjugated

e. isolated, conjugated, conjugated

12.2 **Conjugation** occurs when there are overlapping *p* orbitals on three or more adjacent atoms. Double bonds separated by two σ bonds are not conjugated.

a.

Six of the carbon atoms are sp^2 hybridized. Each π bond is separated by only one σ bond.
conjugated

b.

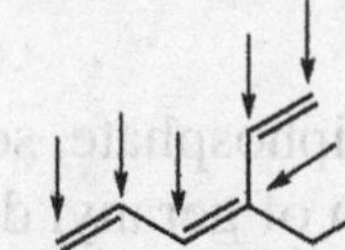

The three π bonds are separated by two or three σ bonds.
NOT conjugated

c.

The two π bonds are separated by only one σ bond.
conjugated

d.

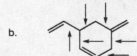

Three adjacent carbon atoms are sp^2 hybridized and have an unhybridized *p* orbital.
conjugated

e.

This carbon is not sp^2 hybridized.
NOT conjugated

12.3 Two resonance structures differ only in the placement of electrons. All σ bonds stay in the same place. Nonbonded electrons and π bonds can be moved. To draw the hybrid:

- Use a dashed line between atoms that have a π bond in one resonance structure and not the other.
- Use a δ symbol for atoms with a charge or radical in one structure but not the other.

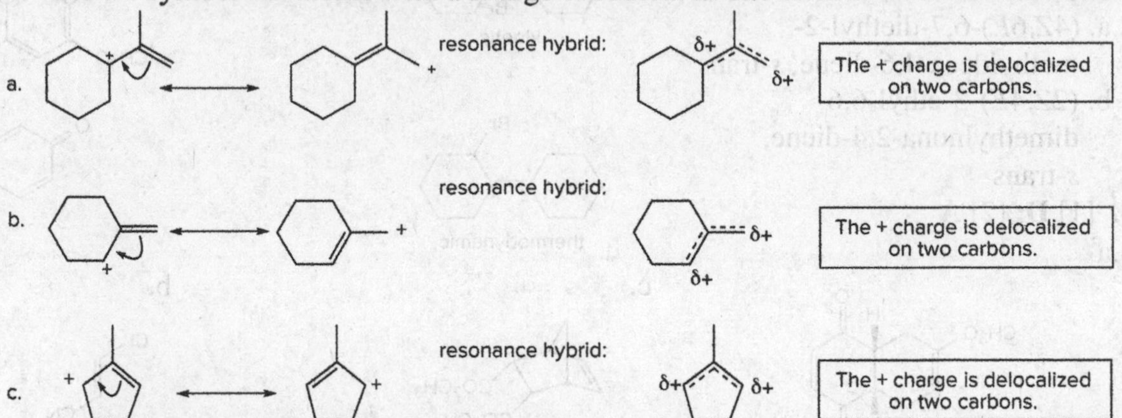
a.
resonance hybrid:
δ+
δ+
The + charge is delocalized on two carbons.
b.
resonance hybrid:
δ+
δ+
The + charge is delocalized on two carbons.
c.
resonance hybrid:
δ+
δ+
The + charge is delocalized on two carbons.

12.4 S_N1 reactions proceed via a carbocation intermediate. Draw the carbocation formed on loss of Cl and compare. The more stable the carbocation, the faster the S_N1 reaction.

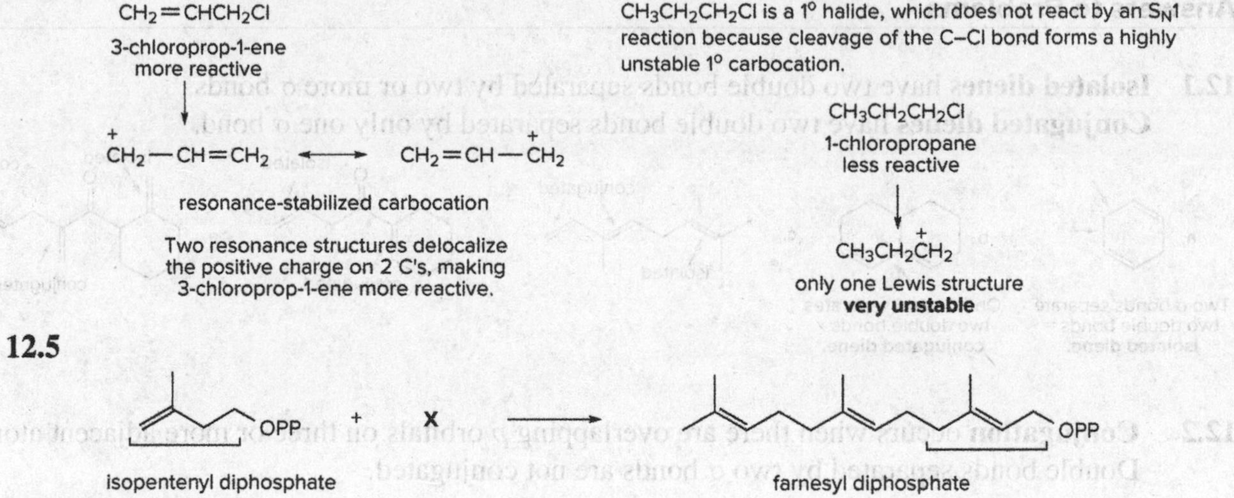
CH2=CHCH2Cl
3-chloroprop-1-ene
more reactive
CH2—CH=CH2 ⟷ CH2=CH—CH2
resonance-stabilized carbocation
Two resonance structures delocalize the positive charge on 2 C's, making 3-chloroprop-1-ene more reactive.
CH3CH2CH2Cl is a 1° halide, which does not react by an SN1 reaction because cleavage of the C–Cl bond forms a highly unstable 1° carbocation.
CH3CH2CH2Cl
1-chloropropane
less reactive
CH3CH2CH2
only one Lewis structure
very unstable
OPP
X
isopentenyl diphosphate
OPP
farnesyl diphosphate

12.5

Five C's of farnesyl diphosphate come from isopentenyl diphosphate, so the remaining 10 C's come from **X**. If the reaction is analogous to the formation of geranyl diphosphate, **X** must contain 10 C's and have an allylic diphosphate.

OPP

OPP = **X**

12.6

a.

Move the charge and the double bond.

b.

Move the charge and the double bond.

c.

Move the lone pair.

d.

Move the charge and the double bond.

12.7 To compare the resonance structures remember the following:

- Resonance structures with **more bonds** are better.
- Resonance structures in which **every atom has an octet** are better.
- Resonance structures with **neutral atoms** are better than those with charge separation.
- Resonance structures that place a **negative charge on a more electronegative atom** are better.

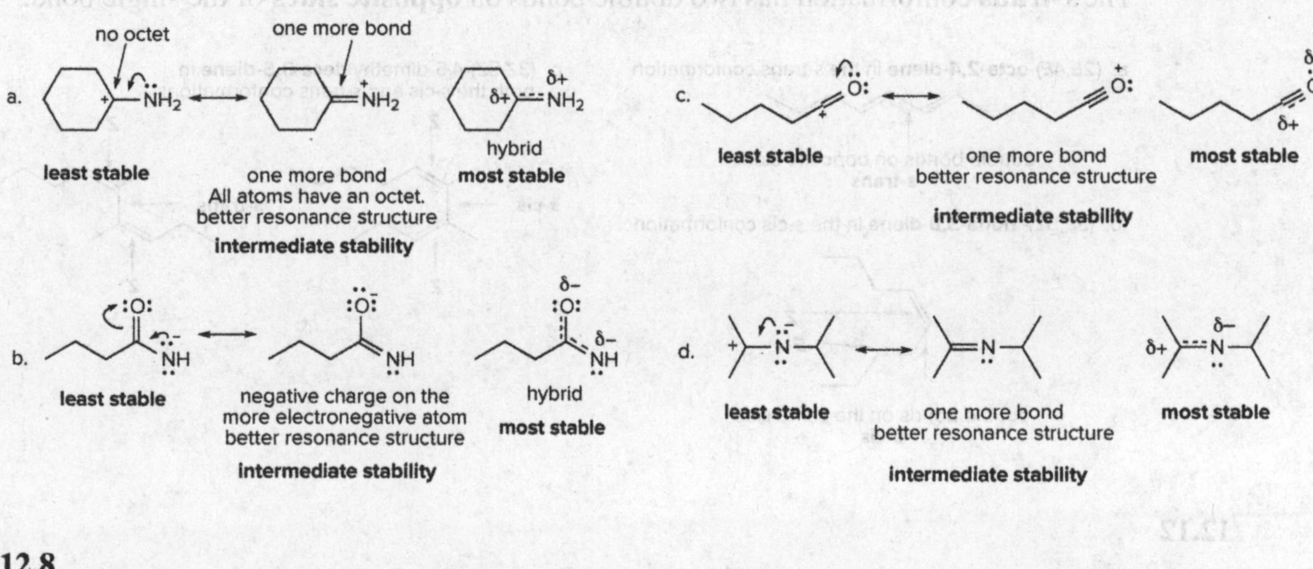

12.8

largest contribution
more bonds
All atoms have octets.

12.9 Remember that in any allyl system, there must be *p* orbitals to delocalize the lone pair.

a. *sp*^{2} b. *sp*^{2} c. *sp*^{3} d. :O: *sp*^{2}

12.10 Constitutional isomers differ in the way the atoms are bonded to each other. Stereoisomers are compounds that differ in 3-D arrangement of groups. Conformations are identical compounds that interconvert.

A: *Z*, *s*-trans — B: *s*-cis — C: bonded to a different C — D: *E*

A and B: conformations
A and C: constitutional isomers
A and D: stereoisomers

12.11 The ***s*-cis** conformation has two double bonds on the **same side** of the single bond. The ***s*-trans** conformation has two double bonds on **opposite sides** of the single bond.

a. (2*E*,4*E*)-**octa-2,4-diene** in the *s*-trans conformation

double bonds on opposite sides
s-trans

b. (3*E*,5*Z*)- **nona-3,5-diene** in the *s*-cis conformation

Z *E*
double bonds on the same side
s-cis

c. (3*Z*,5*Z*)-4,5-dimethyl**deca-3,5-diene** in both the *s*-cis and *s*-trans conformations

Z **s-cis** **Z** — **Z** **s-trans** **Z**

12.12

s-cis conjugated conjugated **s-trans** conjugated OH *E* *E* *Z* *Z* isolated *Z* *Z* OH isolated HO isolated

12.13 Bond length depends on hybridization and percent *s*-character. Bonds with a higher percent *s*-character have smaller orbitals and are shorter.

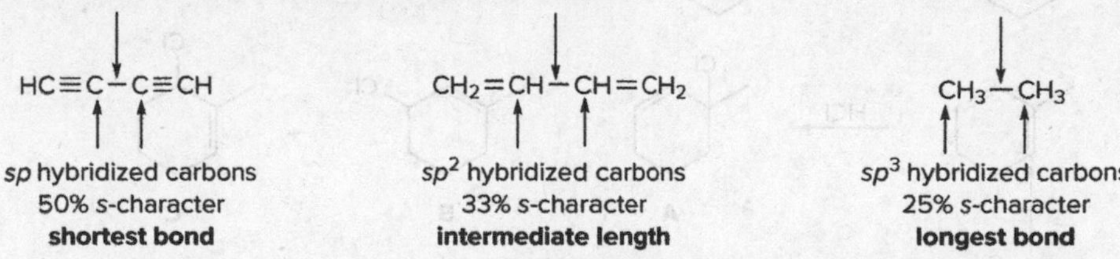

12.14 Two equivalent resonance structures delocalize the π bond and the negative charge.

hybrid: These bond lengths are equal because they are identical.

12.15 The **less stable** (higher energy) **diene** has the **larger heat of hydrogenation.** Isolated dienes are higher in energy than conjugated dienes, so they will have a larger heat of hydrogenation.

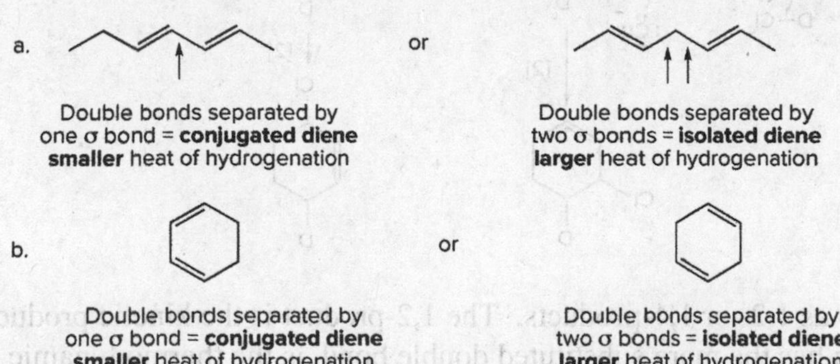

12.16 Isolated dienes are higher in energy than conjugated dienes. Compare the location of the double bonds in the compounds below.

0 conjugated double bonds **least stable**

2 conjugated double bonds **intermediate stability**

3 conjugated double bonds **most stable**

12.17 Conjugated dienes react with HX to form 1,2- and 1,4-products.

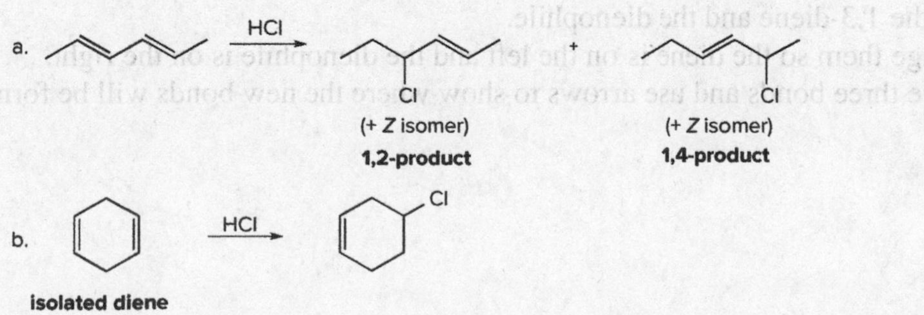

c. [cyclohexa-1,3-diene] $\xrightarrow{HCl}$ [3-chlorocyclohexene]

d. [1-methylcyclohexa-1,3-diene] $\xrightarrow{HCl}$ A + B + C

This double bond is more reactive, so **C** is probably a minor product because it results from HCl addition to the less reactive double bond.

12.18 The mechanism for addition of DCl has two steps:
[1] **Addition of D^+ forms a resonance-stabilized carbocation.**
[2] **Nucleophilic attack of Cl^-** forms 1,2- and 1,4-products.

[1] D–Cl, :Cl:⁻, [2], Cl, D

12.19 Label the products as 1,2- or 1,4-products. The 1,2-product is the kinetic product, and the 1,4-product, which has the more substituted double bond, is the thermodynamic product.

This is C1. The H is added here. The H is added here.

HCl

This is C4.

1,2-product
kinetic product

1,4-product
thermodynamic product

12.20 To draw the products of a Diels–Alder reaction:
[1] Find the 1,3-diene and the dienophile.
[2] Arrange them so the diene is on the left and the dienophile is on the right.
[3] Cleave three bonds and use arrows to show where the new bonds will be formed.

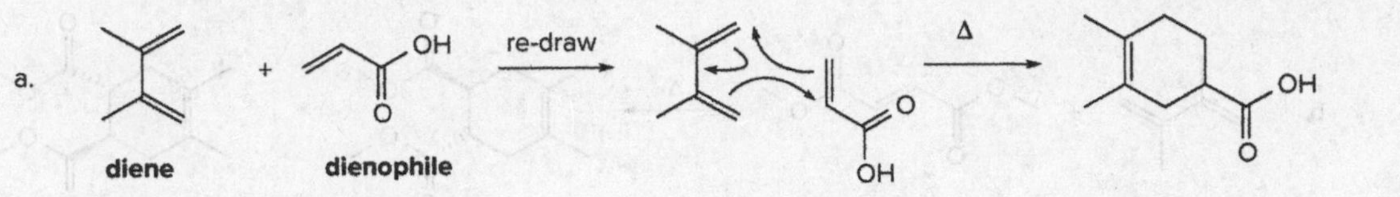

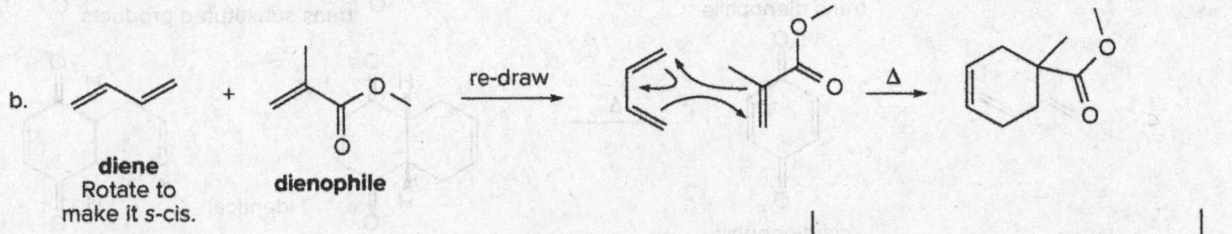

a. diene + dienophile (OH, O) —re-draw→ (OH, O) —Δ→ (OH, O)

b. diene (Rotate to make it *s*-cis.) + dienophile (O) —re-draw→ —Δ→

c. dienophile + diene (Rotate to make it *s*-cis.) —re-draw→ —Δ→

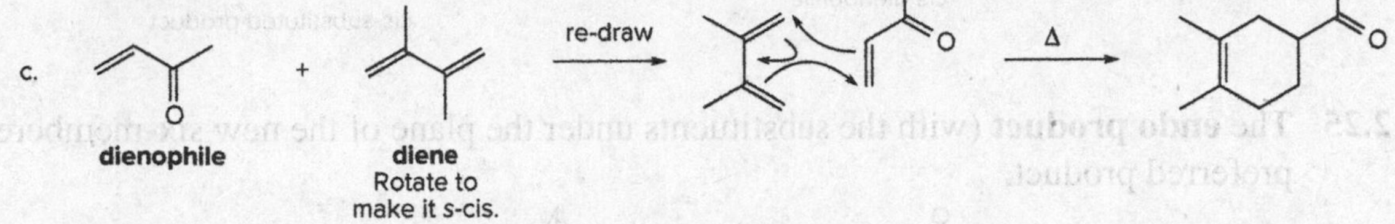

12.21 For a diene to be reactive in a Diels–Alder reaction, **a diene must be able to adopt an *s*-cis conformation.**

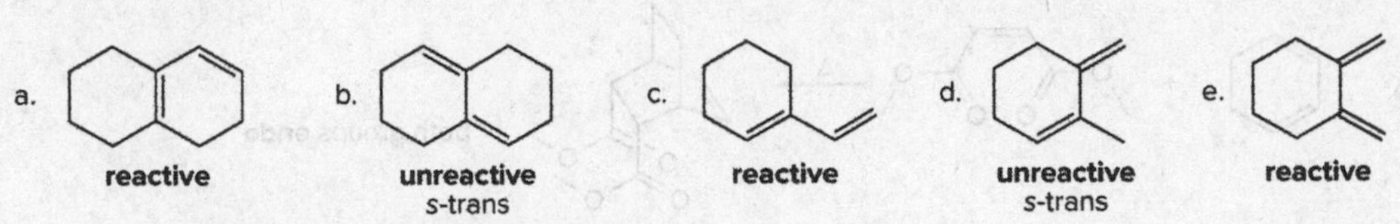

a. reactive

b. unreactive *s*-trans

c. reactive

d. unreactive *s*-trans

e. reactive

12.22 Zingiberene reacts much faster than β-sesquiphellandrene as a Diels–Alder diene because its diene is constrained in the *s*-cis conformation. The diene in β-sesquiphellandrene is constrained in the *s*-trans conformation, so it is unreactive in the Diels–Alder.

12.23 Electron-withdrawing substituents in the dienophile increase the reaction rate.

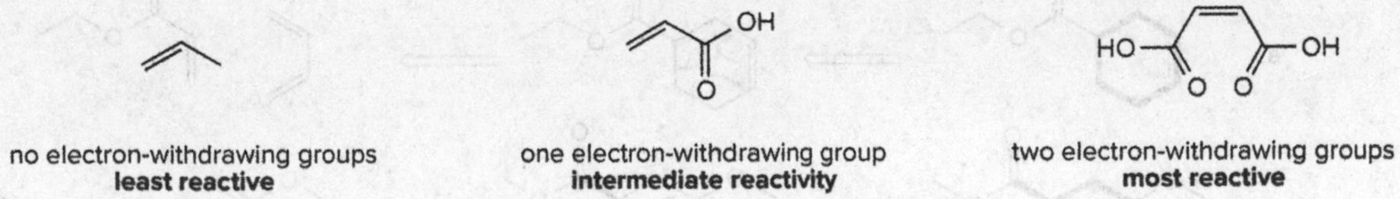

no electron-withdrawing groups **least reactive**

one electron-withdrawing group **intermediate reactivity**

two electron-withdrawing groups **most reactive**

12.24 A cis dienophile forms a cis-substituted cyclohexene.
A trans dienophile forms a trans-substituted cyclohexene.

a. + cis dienophile —Δ→ + cis-substituted products

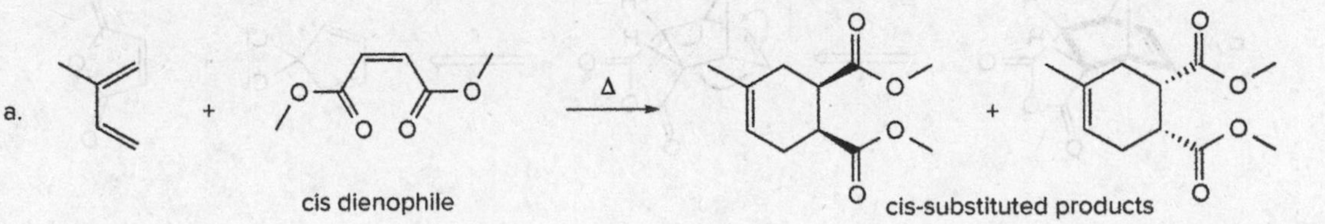

b. + Δ → +

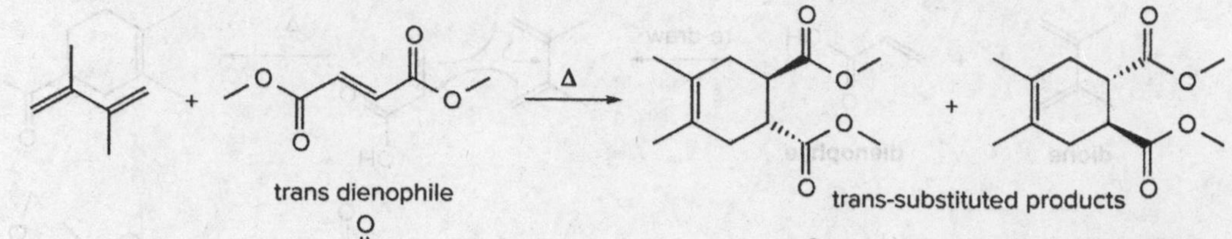

trans dienophile

trans-substituted products

c. + Δ → +

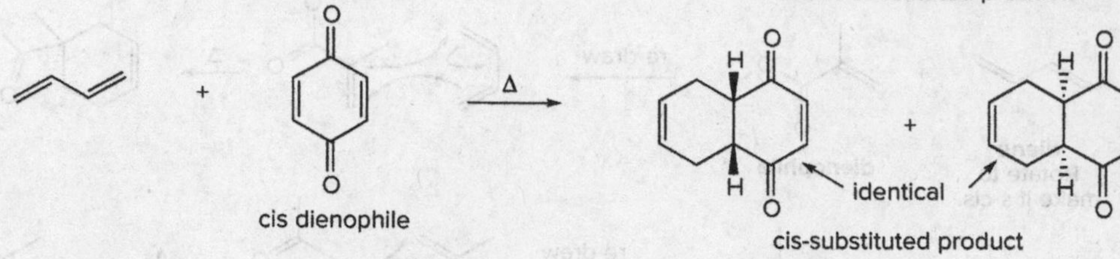

identical

cis dienophile

cis-substituted product

12.25 The **endo product** (with the substituents under the plane of the new six-membered ring) is the preferred product.

a. + Δ →

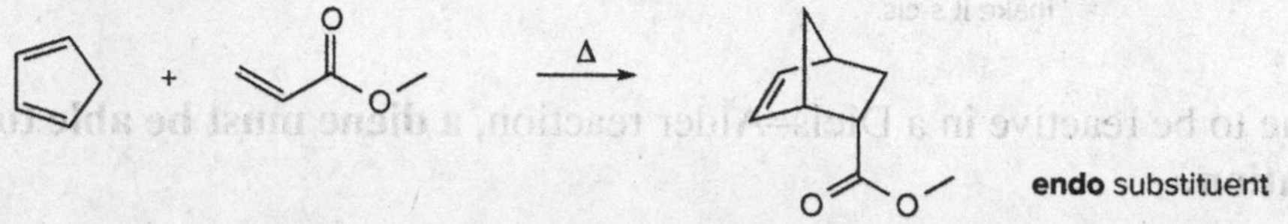

endo substituent

b. + Δ →

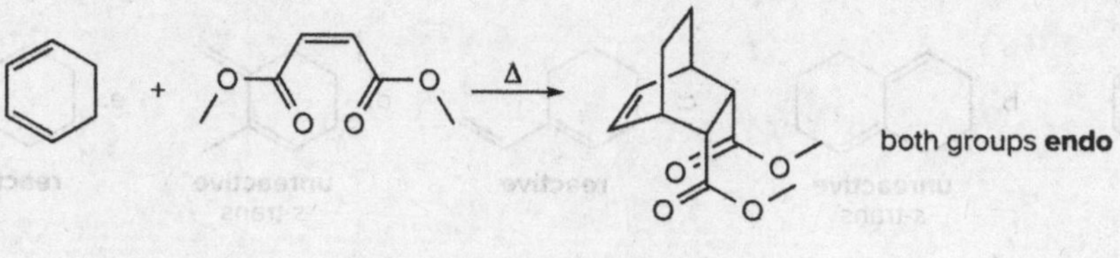

both groups **endo**

12.26 To find the diene and dienophile needed to make each of the products:
[1] Find the six-membered ring with a C–C double bond.
[2] Draw three arrows to work backwards.
[3] Follow the arrows to show the diene and dienophile.

a. ⟹ ⟹ +

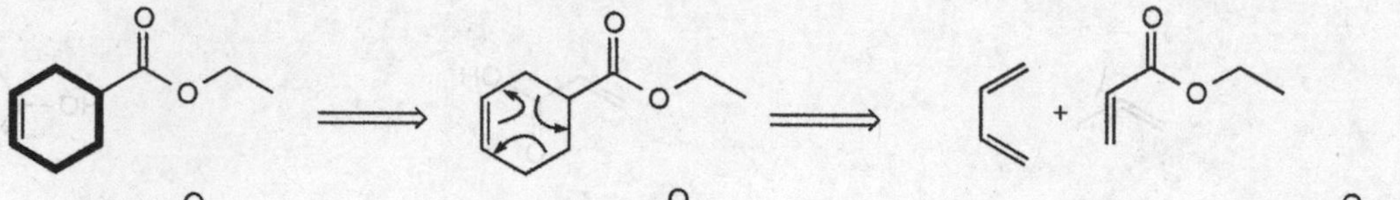

b. ⟹ ⟹ +

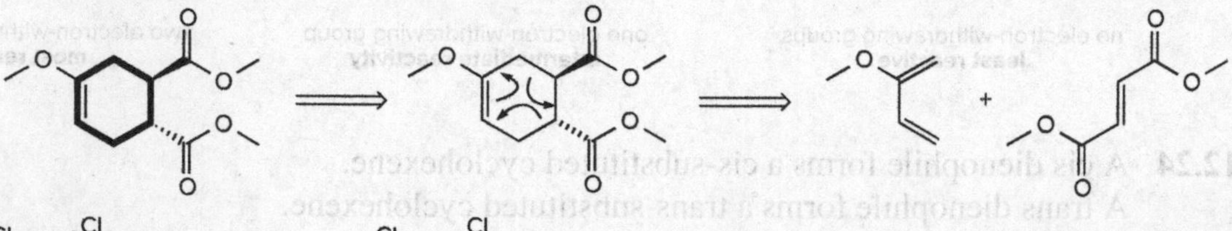

c. ⟹ ⟹ +

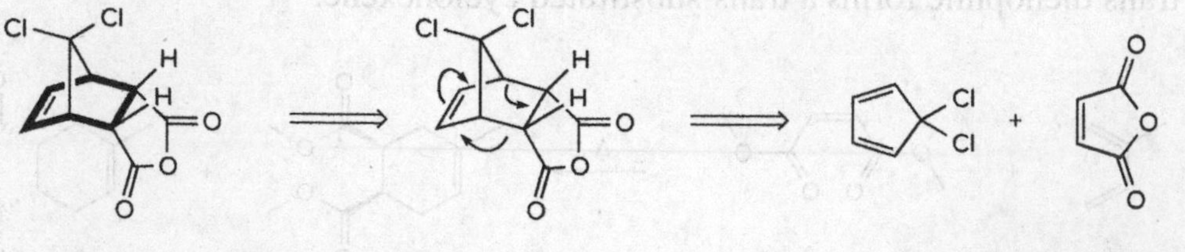

12.27

a. [1] Δ → 2 (cyclopentadiene); [2] → Place this group endo.

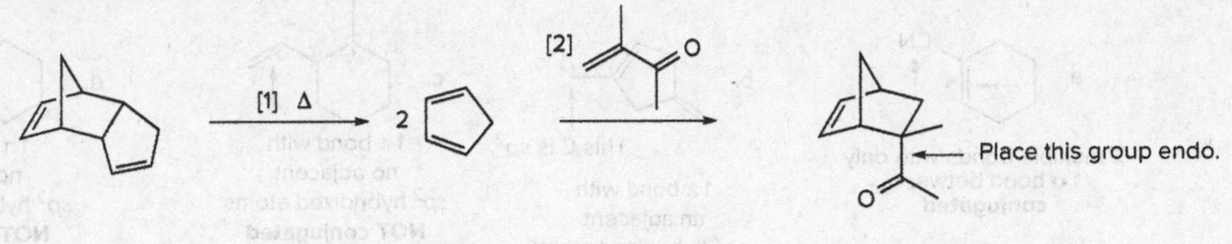

b. [1] Δ → 2 (cyclopentadiene); [2] O=⟨⟩=O → H O H O endo

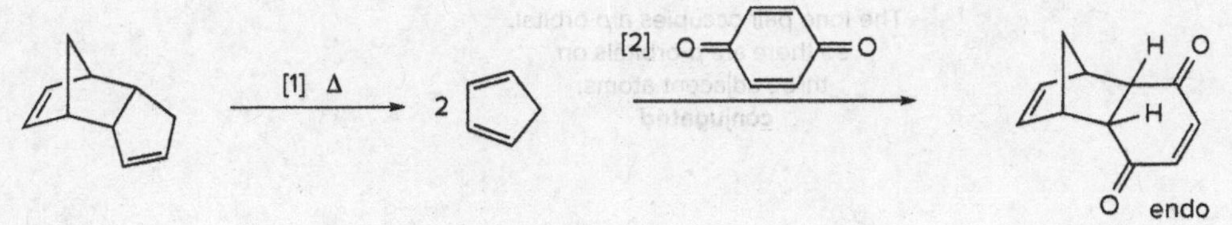

12.28

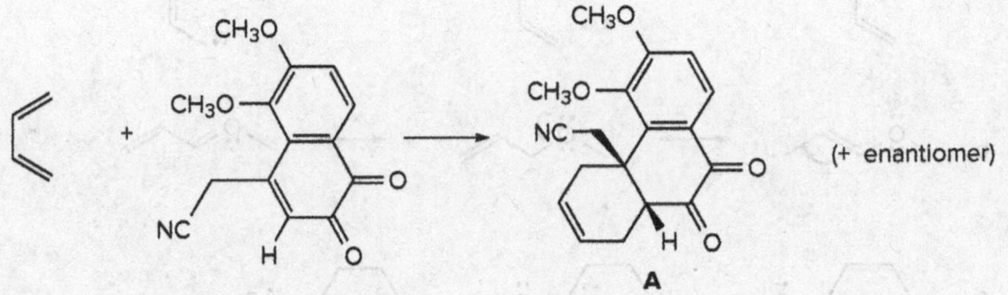

CH_3O CH_3O NC H O O → CH_3O CH_3O NC O H O **A** (+ enantiomer)

12.29

a. 2 3 ← *s*-cis conformation 4

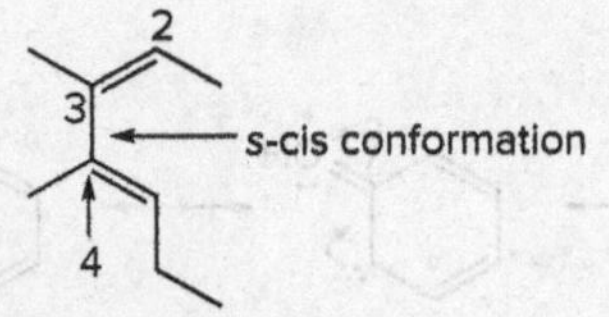

(2*Z*,4*E*)-3,4-dimethylhepta-2,4-diene

b. 4 3 2 *s*-trans conformation

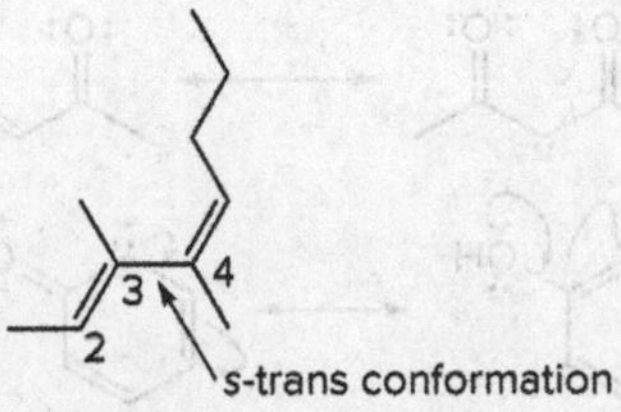

(2*E*,4*Z*)-3,4-dimethylocta-2,4-diene

12.30

a. O OCH_3 H H O OCH_3 ⟹ + CO_2CH_3 CH_3O_2C

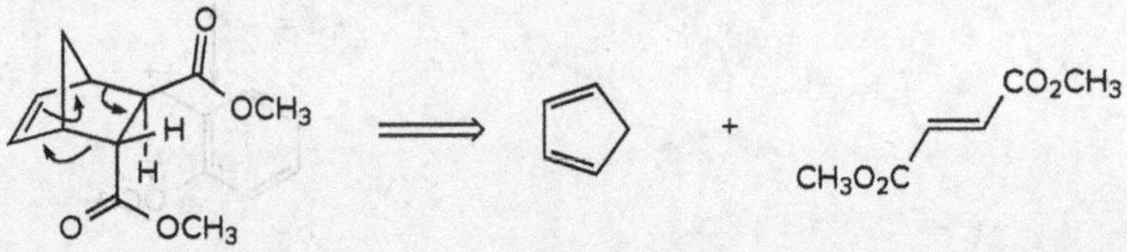

b. H O H O ⟹ + O O

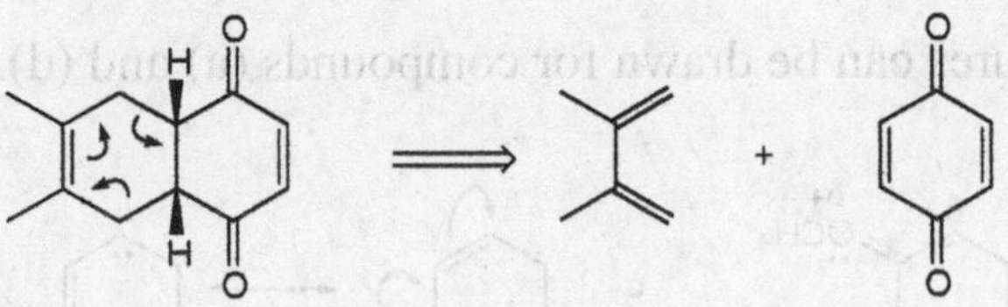

12.31 Use the definition from Answer 12.2.

a. 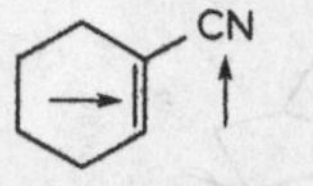 CN

2 multiple bonds with only 1 σ bond between
conjugated

b. 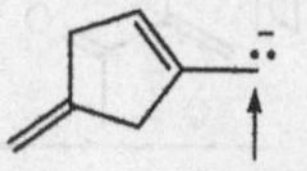 This C is sp^2.

1 π bond with an adjacent sp^2 hybridized atom
The lone pair occupies a *p* orbital, so there are *p* orbitals on three adjacent atoms.
conjugated

c. 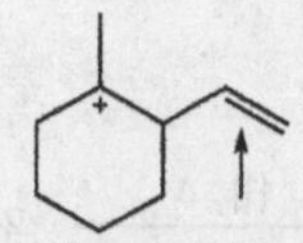 1 π bond with no adjacent sp^2 hybridized atoms
NOT conjugated

d. 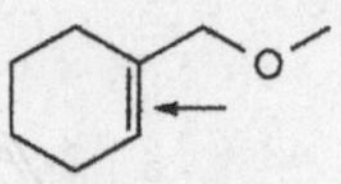 1 π bond with no adjacent sp^2 hybridized atoms
NOT conjugated

12.32

a.

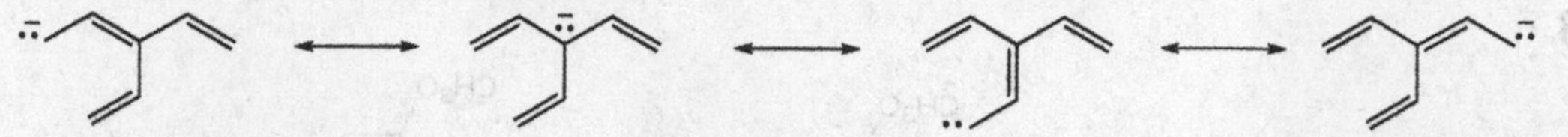

b.

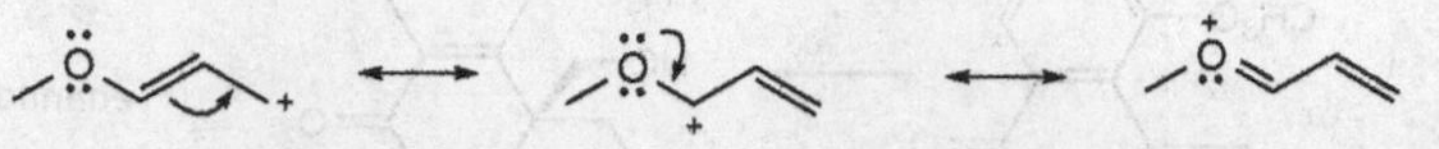

c.

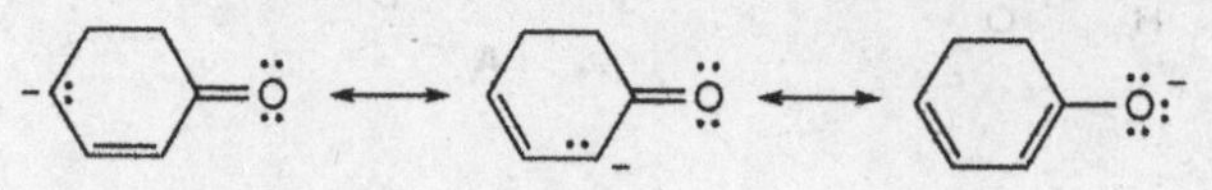

d.

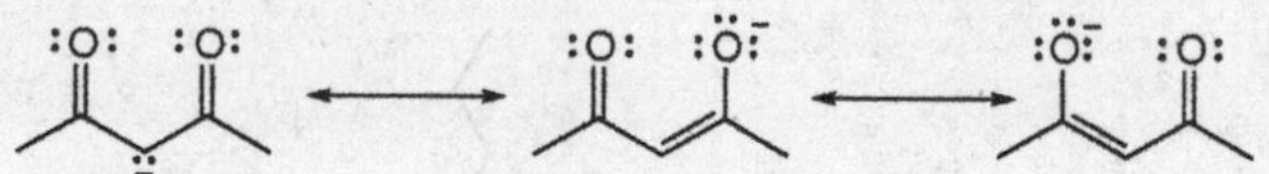

e. 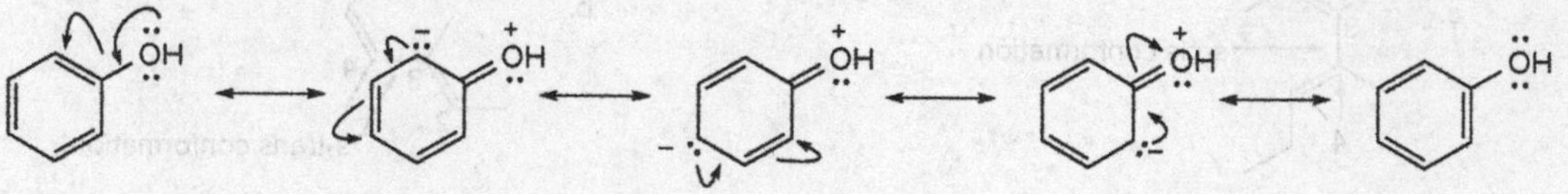 OH OH OH OH OH

f. 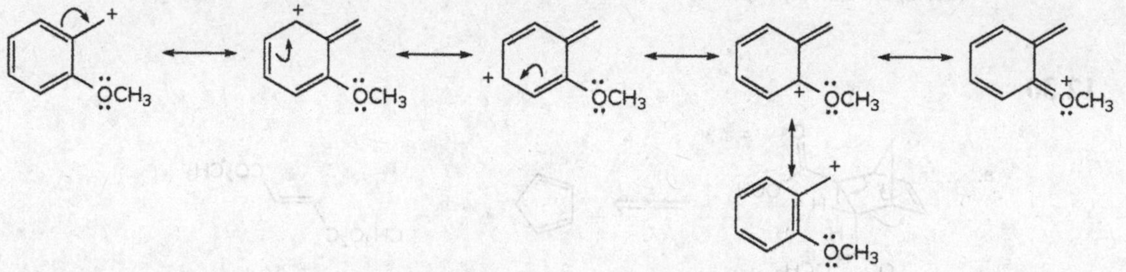 OCH3 OCH3 OCH3 OCH3 OCH3 OCH3

12.33 No additional resonance structures can be drawn for compounds (a) and (d).

b. 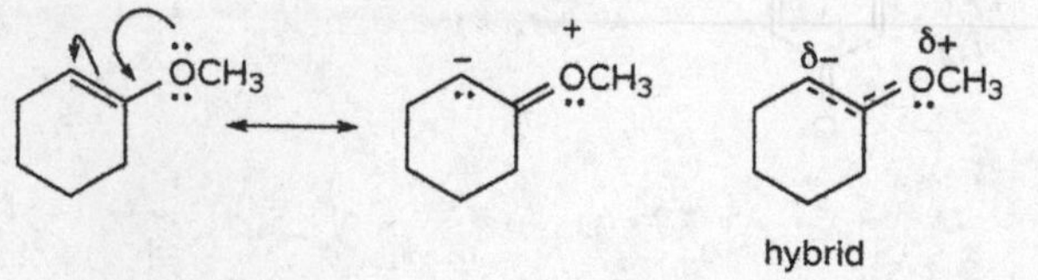 OCH3 OCH3 δ− δ+ OCH3
hybrid

c. 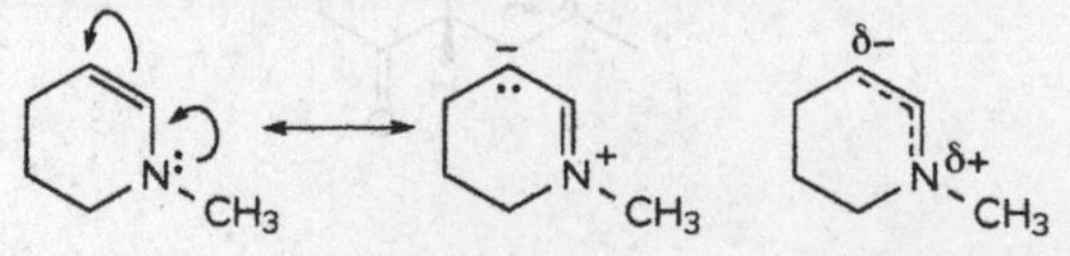 N CH3 N+ CH3 δ− Nδ+ CH3
hybrid

12.34

a.

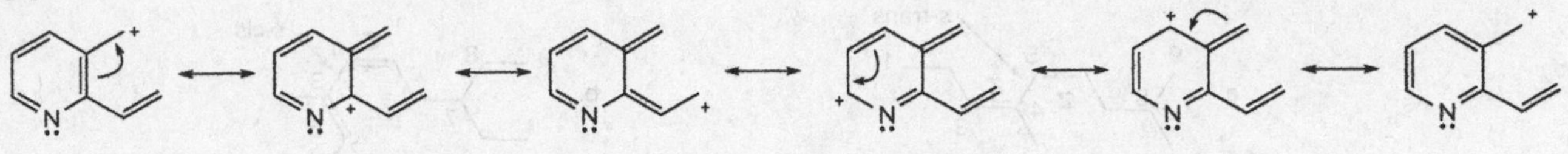

b.

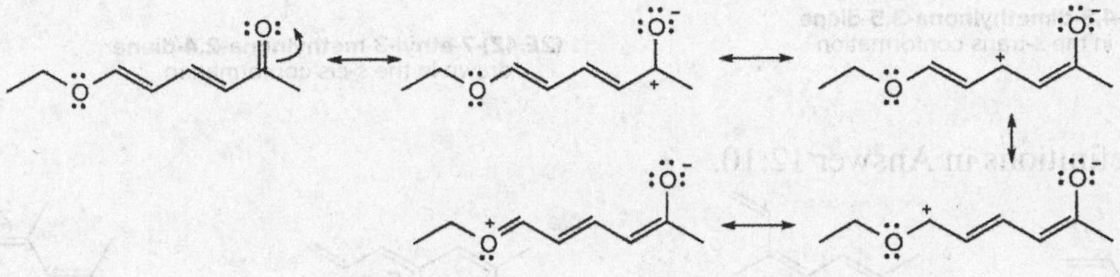

c.

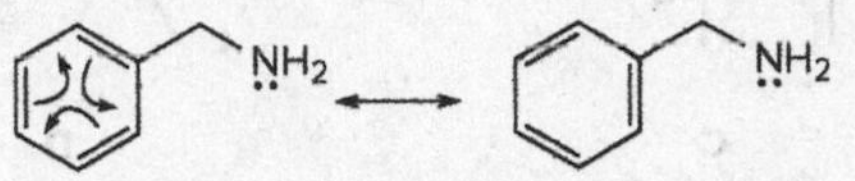

d.

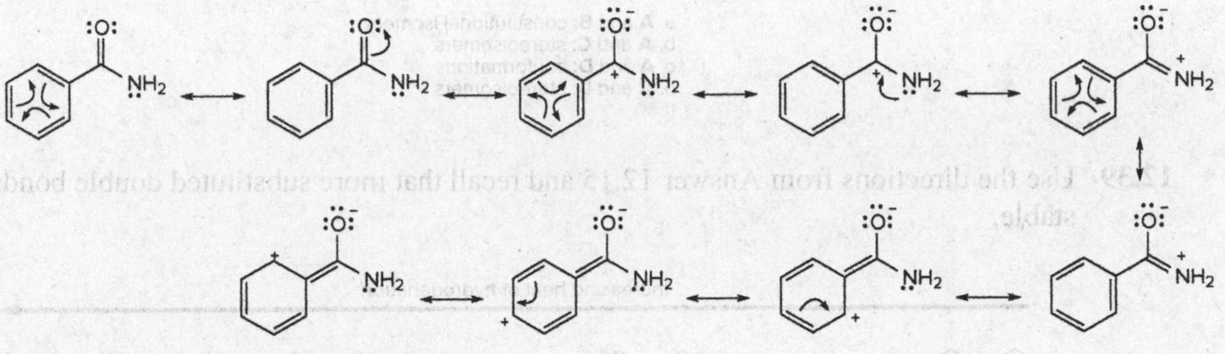

12.35

Draw the products of cleavage of the bond.

ethane CH_3—CH_3 → $\cdot CH_3$ + $\cdot CH_3$

Two unstable radicals form.

but-1-ene → $\cdot CH_3$ +

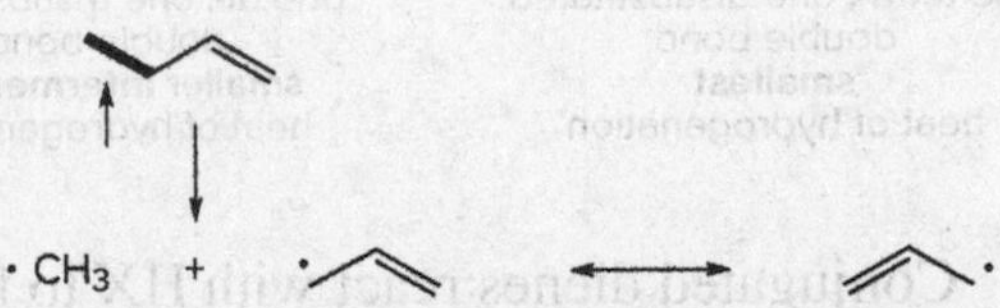

One resonance-stabilized radical forms. This makes the bond dissociation energy lower because a more stable radical is formed.

12.36 Use the directions from Answer 12.11.

a. (*Z*)-penta-1,3-diene in the *s*-trans conformation

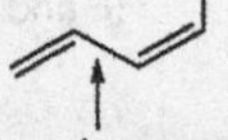

double bonds on opposite sides
s-trans

b. (2*E*,4*Z*)-1-bromo-3-methylhexa-2,4-diene

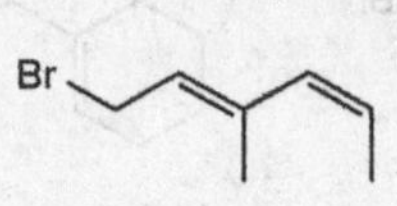

Br

c. (2*E*,4*E*,6*E*)-octa-2,4,6-triene

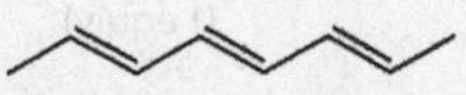

d. (2*E*,4*E*)-3-methylhexa-2,4-diene in the *s*-cis conformation

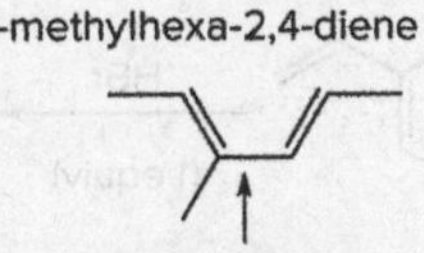

s-cis

12.37

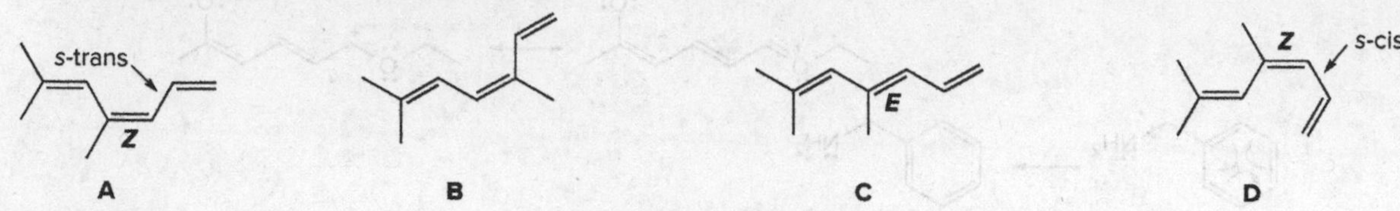

12.38 Use the definitions in Answer 12.10.

a. **A** and **B:** constitutional isomers
b. **A** and **C:** stereoisomers
c. **A** and **D:** conformations
d. **C** and **D:** stereoisomers

12.39 Use the directions from Answer 12.15 and recall that more substituted double bonds are more stable.

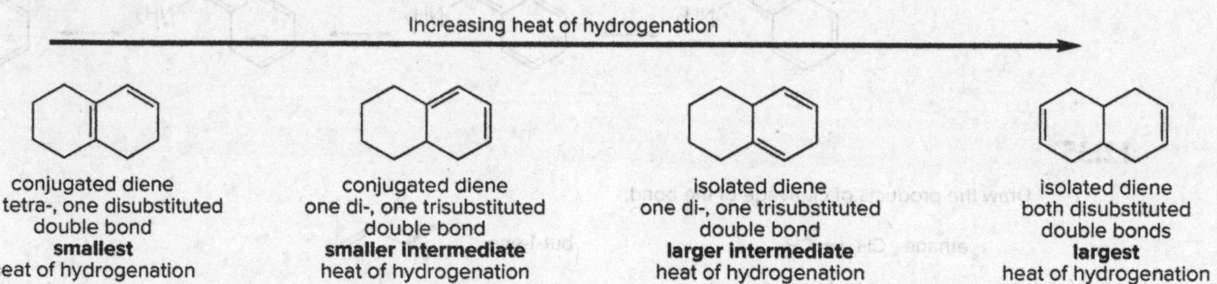

12.40 Conjugated dienes react with HX to form 1,2- and 1,4-products.

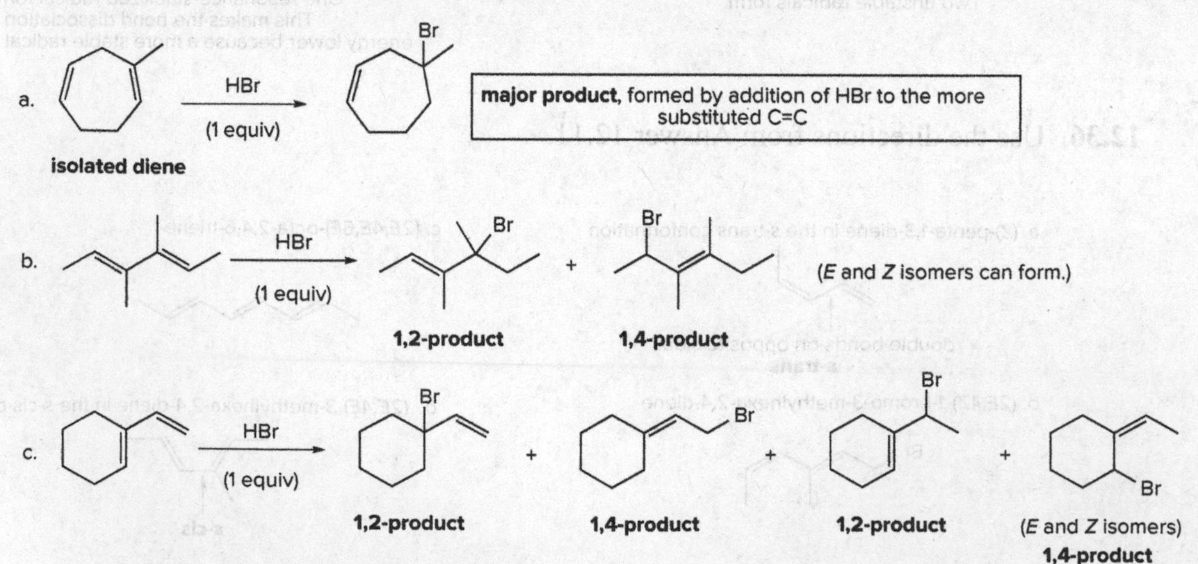

12.41

H−Br

:Br:−

Br

:Br:−

Br

:Br:−

Br

12.42

This cation forms because it is benzylic and resonance stabilized.

A

H−Br

:Br:−

Br

C

B

H−Br

H

2° carbocation

1,2-
H shift

This 2° carbocation
is also benzylic, making it
resonance stabilized, as above.

Br−

Br

C

12.43

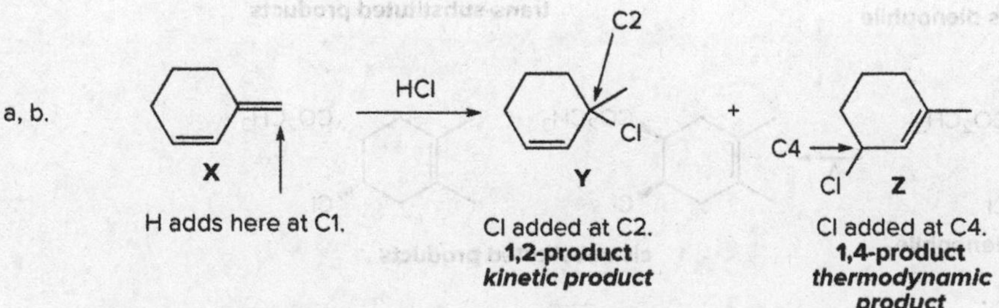

Y is the kinetic product because of the proximity effect. H and Cl add across two adjacent atoms.
Z is the thermodynamic product because it has a more stable trisubstituted double bond.

Addition occurs at the labeled double bond due to the stability of the carbocation intermediate.

c.

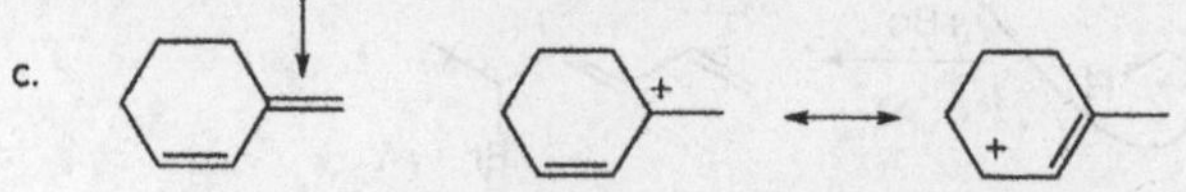

The two resonance structures for this allylic cation are 3° and 2° carbocations.

more stable intermediate
Addition occurs here.

If addition occurred at the other C=C, the following allylic carbocation would form:

The two resonance structures for this allylic cation are 1° and 2° carbocations.

less stable

12.44 There are two possible products:

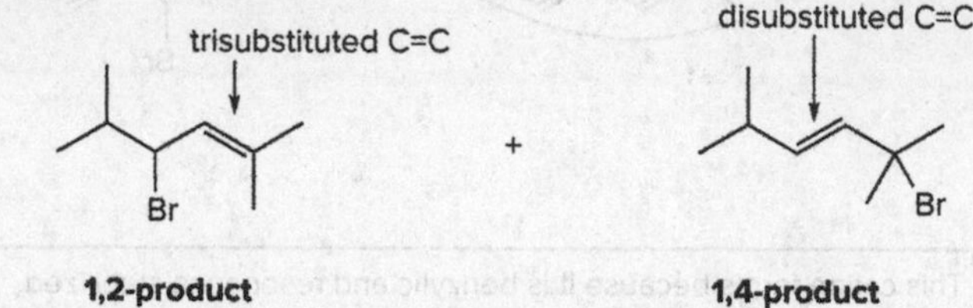

The 1,2-product is always the kinetic product because of the proximity effect. In this case, it is also the thermodynamic (more stable) product because it contains a more highly substituted C=C (trisubstituted) than the 1,4-product (disubstituted). Thus, the 1,2-product is the major product at high and low temperature.

12.45 Draw 1,2- and 1,4-addition products for each reaction.

a. $\xrightarrow[H_2SO_4]{H_2O}$ (OH) 1,2-product + (OH) 1,4-product

b. $\xrightarrow[H_2O]{Br_2}$ (HO, Br) 1,2-product + (Br, OH) 1,4-product

12.46 Use the directions from Answer 12.20.

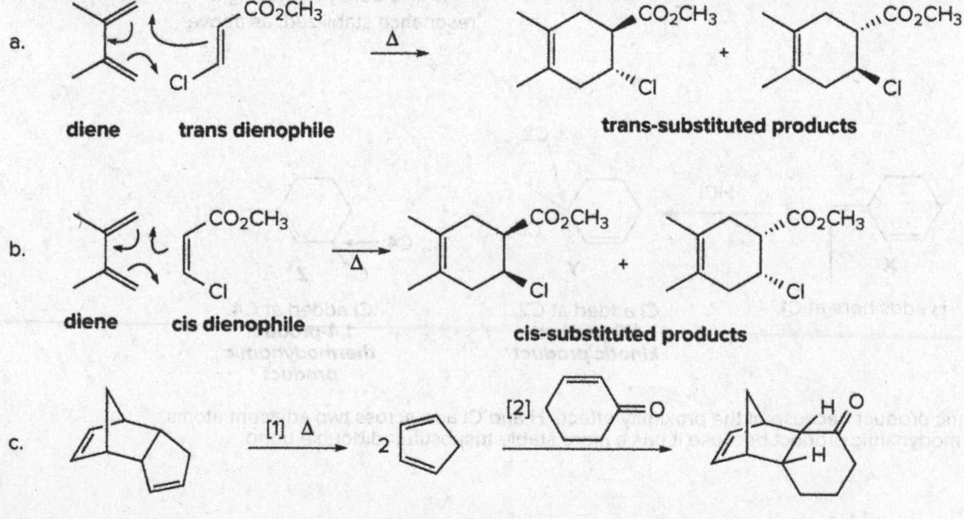

d. 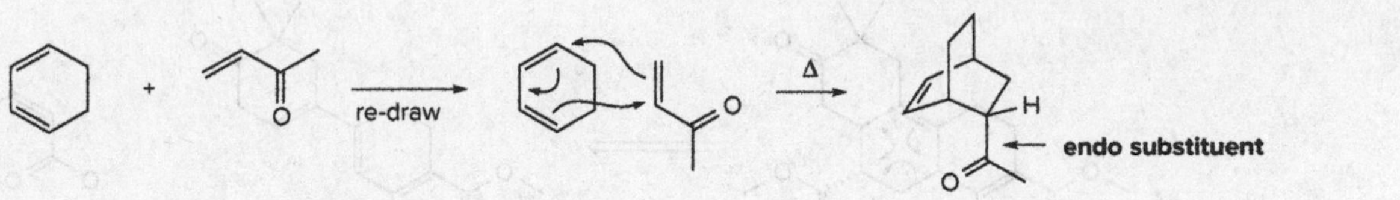 re-draw Δ H ← endo substituent

e. 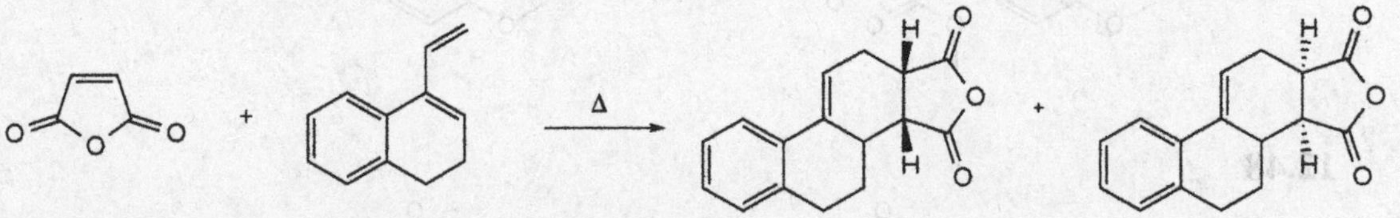 + Δ H H + H H

f. 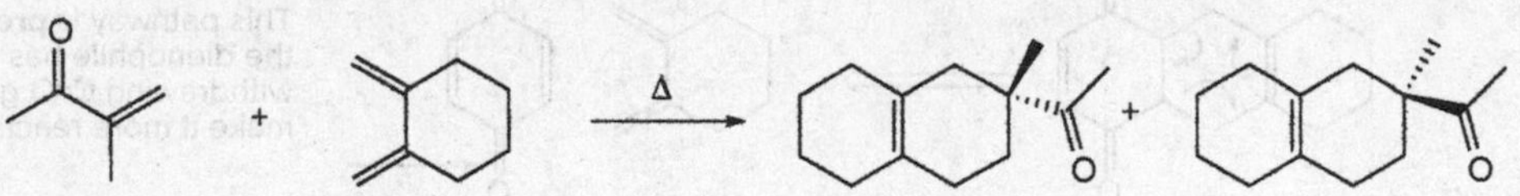 + Δ +

12.47 Use the directions from Answer 12.26.

a. 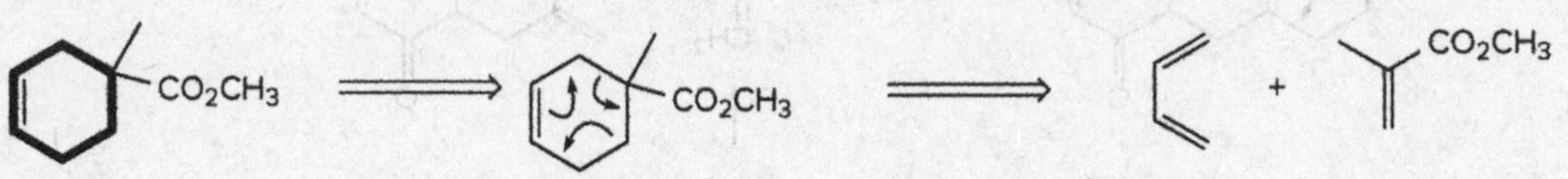 CO_2CH_3 ⟹ CO_2CH_3 ⟹ + CO_2CH_3

b. 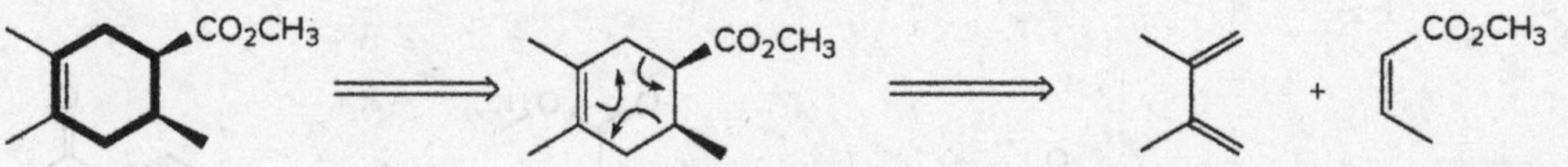 CO_2CH_3 ⟹ CO_2CH_3 ⟹ + CO_2CH_3

c. 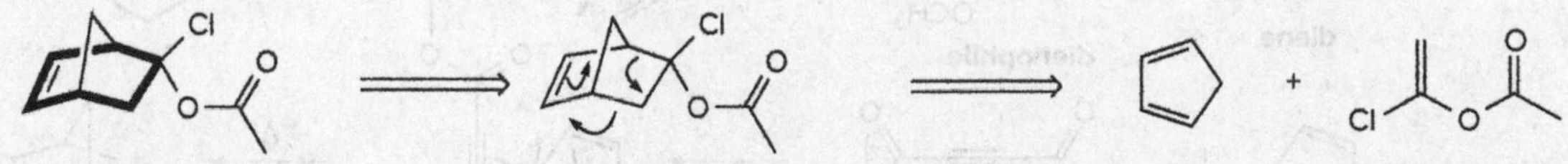 Cl ⟹ Cl ⟹ + Cl

d. 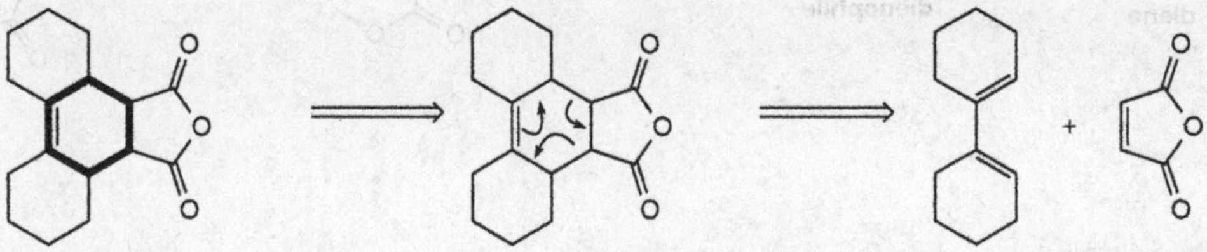 ⟹ ⟹ +

e. 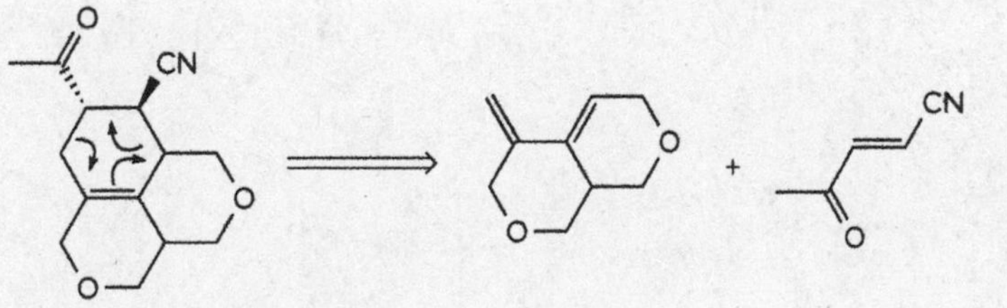 CN ⟹ + CN f.

12.48

This pathway is **preferred** because the dienophile has electron-withdrawing C=O groups that make it more reactive.

CH_2 +
CH_2

no electron-withdrawing groups
less reactive

12.49

a. diene + dienophile (OCH_3) → Δ →

b. diene + dienophile (CH_3O, OCH_3) → Δ →

12.50

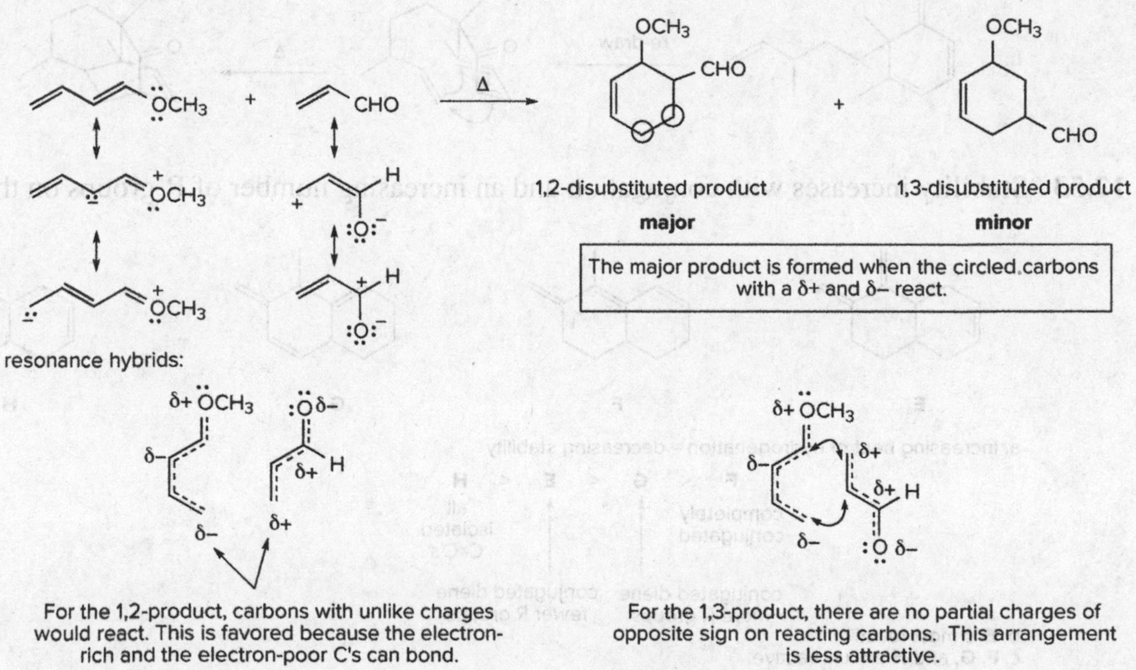

OCH3 + CHO —Δ→ 1,2-disubstituted product **major** + 1,3-disubstituted product **minor**

The major product is formed when the circled carbons with a δ+ and δ– react.

resonance hybrids:

δ+ ÖCH3 :Ö δ– δ– δ+ H δ– δ+

For the 1,2-product, carbons with unlike charges would react. This is favored because the electron-rich and the electron-poor C's can bond.

δ+ ÖCH3 δ– δ+ δ+ H δ– :Ö δ–

For the 1,3-product, there are no partial charges of opposite sign on reacting carbons. This arrangement is less attractive.

12.51

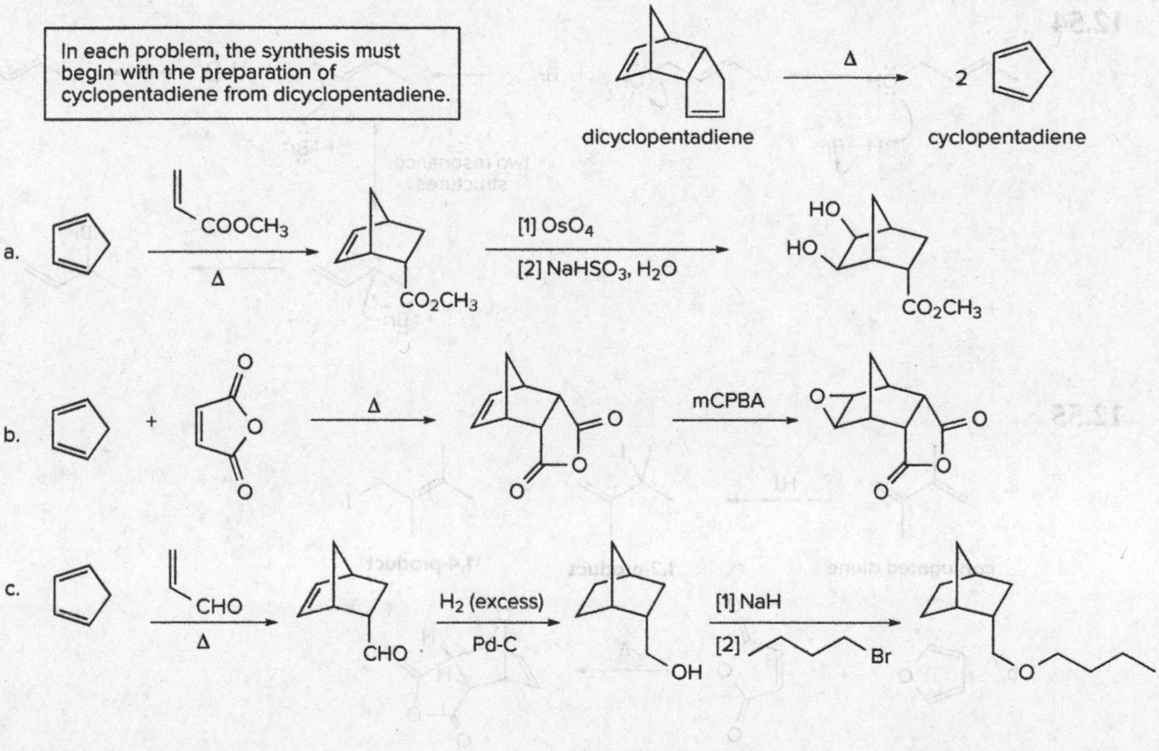

In each problem, the synthesis must begin with the preparation of cyclopentadiene from dicyclopentadiene.

dicyclopentadiene —Δ→ 2 cyclopentadiene

a. COOCH3, Δ → CO_2CH_3 → [1] OsO_4 [2] $NaHSO_3$, H_2O → HO, HO, CO_2CH_3

b. + —Δ→ —mCPBA→

c. CHO, Δ → CHO —H_2 (excess), Pd-C→ OH —[1] NaH [2] Br→

12.52

a. diene dienophile —re-draw→ —Δ→

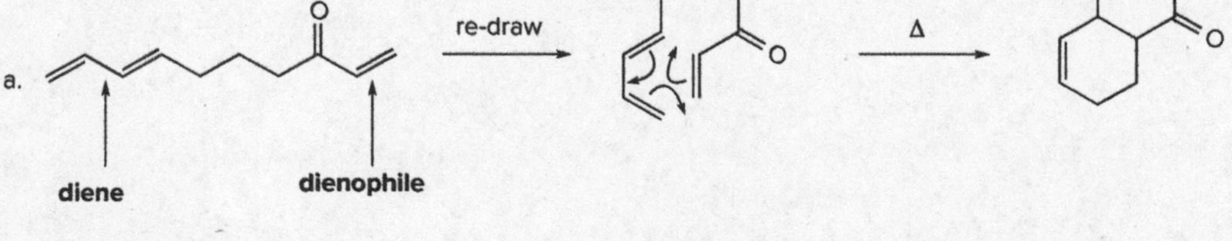

b. re-draw Δ

12.53 Stability increases with conjugation and an increasing number of R groups on the C=C.

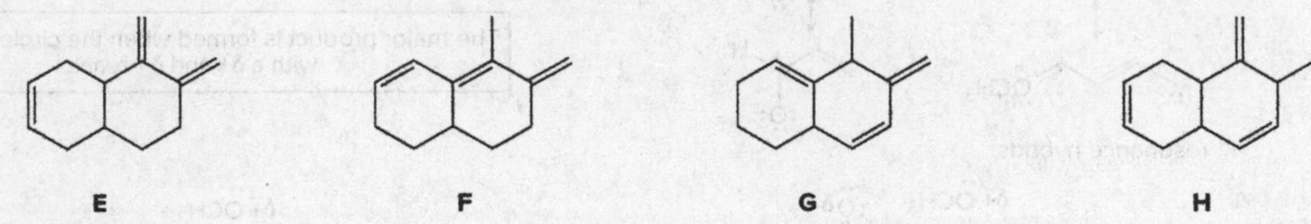

a. Increasing heat of hydrogenation = decreasing stability

F < **G** < **E** < **H**

completely conjugated (F)

conjugated diene, more R groups (G)

conjugated diene, fewer R groups (E)

all isolated C=C's (H)

b. **E** is most reactive.

c. **F**, **G**, and **H** are unreactive.

12.54

ÖH, H–Br → $\overset{+}{O}H_2$ + Br^- → + $H_2\ddot{O}$ → Br

two resonance structures

+ :Br:$^-$ → Br

12.55

a. HI

conjugated diene → 1,2-product + 1,4-product

b. diene + dienophile Δ

c. diene + trans dienophile (CO_2CH_3) $\xrightarrow{\Delta}$ CO_2CH_3 product + CO_2CH_3 product

d. conjugated diene $\xrightarrow{HBr}$ 1,2-product (Br) + 1,4-product (Br)

12.56

OPP → + PP ⟷ re-draw → + HB^+ (H, B:)

12.57 The more stable the carbocation, the faster the S_N1 reaction. The carbocation from **A** is more stable because the lone pairs on the O atom of the OCH_3 afford additional resonance stabilization.

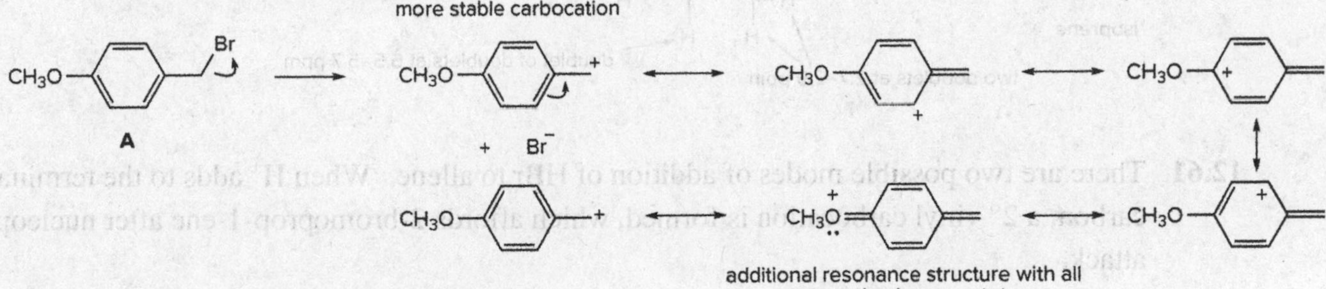

less stable carbocation

B (Br) → + Br^-

This resonance structure is destabilized because the (+) charge is adjacent to the δ+ on the C=O atom.

12.58 The mechanism is E1, with formation of a resonance-stabilized carbocation.

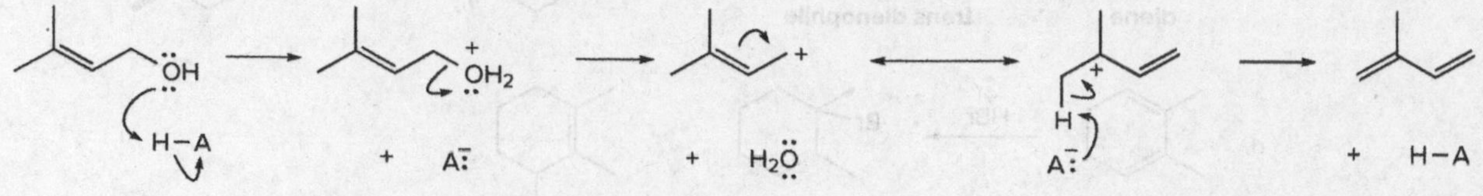

+ A:⁻ + $H_2\ddot{O}$ A:⁻ + H–A

H–A

12.59

a.

Cl β_2 β_1

loss of H (from β_1 C) + Cl
conjugated
more stable
major product

loss of H (from β_2 C) + Cl
more substituted

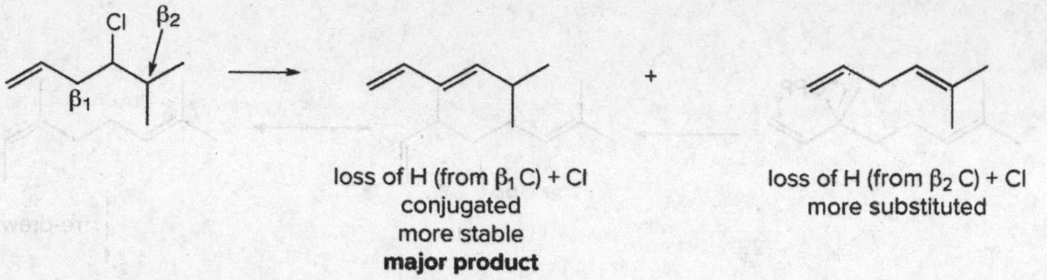

b. Dehydrohalogenation generally forms the more stable product. In this reaction, loss of H from the β_1 carbon forms a more stable conjugated diene, so this product is preferred even though it does not contain the more substituted C=C.

12.60

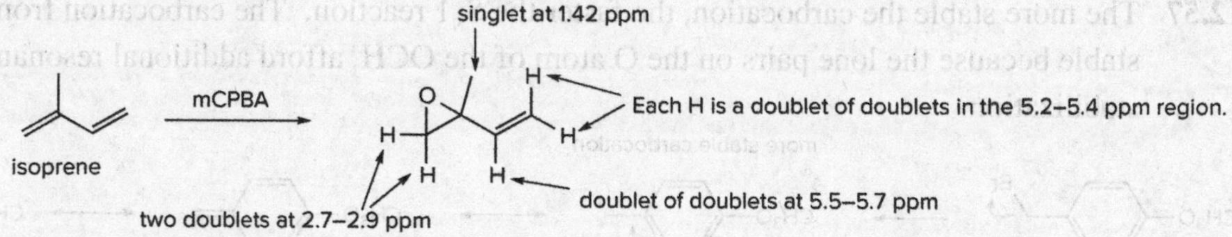

isoprene

mCPBA

singlet at 1.42 ppm

Each H is a doublet of doublets in the 5.2–5.4 ppm region.

doublet of doublets at 5.5–5.7 ppm

two doublets at 2.7–2.9 ppm

12.61 There are two possible modes of addition of HBr to allene. When H^+ adds to the terminal carbon, a 2° vinyl carbocation is formed, which affords 2-bromoprop-1-ene after nucleophilic attack.

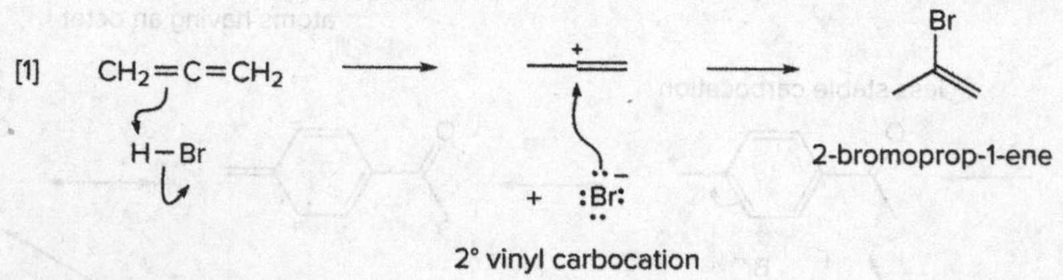

[1] $CH_2{=}C{=}CH_2$ H–Br + :Br:⁻

2° vinyl carbocation

2-bromoprop-1-ene

When H^+ adds to the middle carbon, an intermediate carbocation with a (+) charge adjacent to the C=C is formed. This carbocation is not resonance stabilized (at least initially), because the two C=C's of allene are oriented 90° to each other, a geometry that does not allow for overlap of the C=C with the empty *p* orbital of the carbocation.

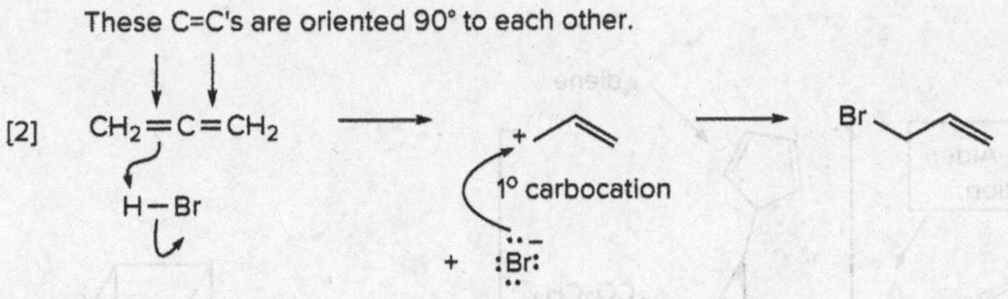

These C=C's are oriented 90° to each other.

[2] $CH_2{=}C{=}CH_2$ + H–Br → 1° carbocation + :Br:⁻ → Br-CH₂CH=CH₂

As a result, path [1] forms a more stable carbocation and is preferred.

12.62

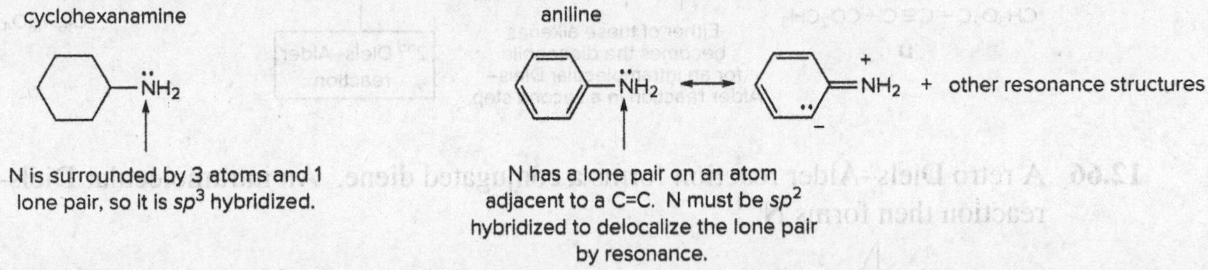

cyclohexanamine

N is surrounded by 3 atoms and 1 lone pair, so it is sp^3 hybridized.

aniline

+ other resonance structures

N has a lone pair on an atom adjacent to a C=C. N must be sp^2 hybridized to delocalize the lone pair by resonance.

Basicity is a measure of how willing an atom is to donate an electron pair. The lone pair on the N in cyclohexanamine is localized on N, whereas the lone pair on the N in aniline is delocalized onto the benzene ring. As a result, the lone pair in cyclohexanamine is much more available for donation to a proton than the lone pair in aniline. This makes cyclohexanamine much more basic than aniline.

12.63

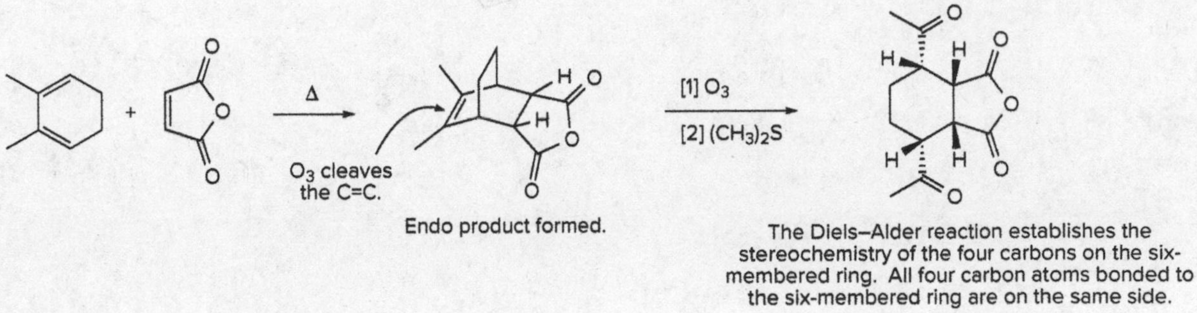

Δ

O_3 cleaves the C=C.

Endo product formed.

[1] O_3

[2] $(CH_3)_2S$

The Diels–Alder reaction establishes the stereochemistry of the four carbons on the six-membered ring. All four carbon atoms bonded to the six-membered ring are on the same side.

12.64

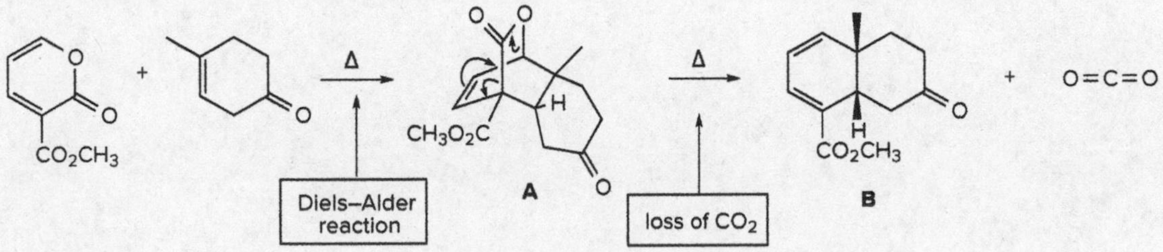

CO_2CH_3 + Δ

Diels–Alder reaction

CH_3O_2C

A

Δ

loss of CO_2

CO_2CH_3

B

+ O=C=O

12.65

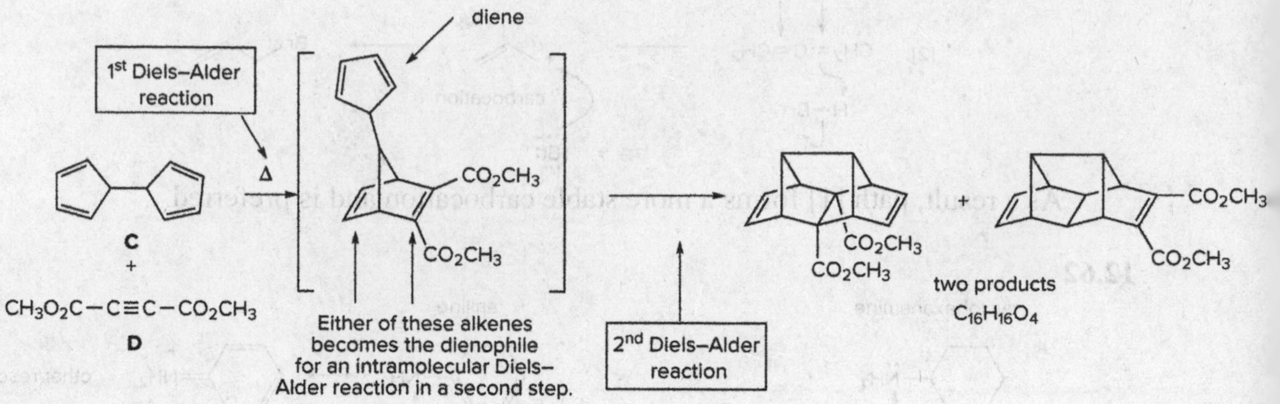

diene

1st Diels–Alder reaction

Δ

CO_2CH_3

CO_2CH_3

C

+

$CH_3O_2C-C\equiv C-CO_2CH_3$

D

Either of these alkenes becomes the dienophile for an intramolecular Diels–Alder reaction in a second step.

2nd Diels–Alder reaction

CO_2CH_3 CO_2CH_3 + CO_2CH_3 CO_2CH_3

two products

$C_{16}H_{16}O_4$

12.66 A retro Diels–Alder reaction forms a conjugated diene. An intramolecular Diels–Alder reaction then forms **N**.

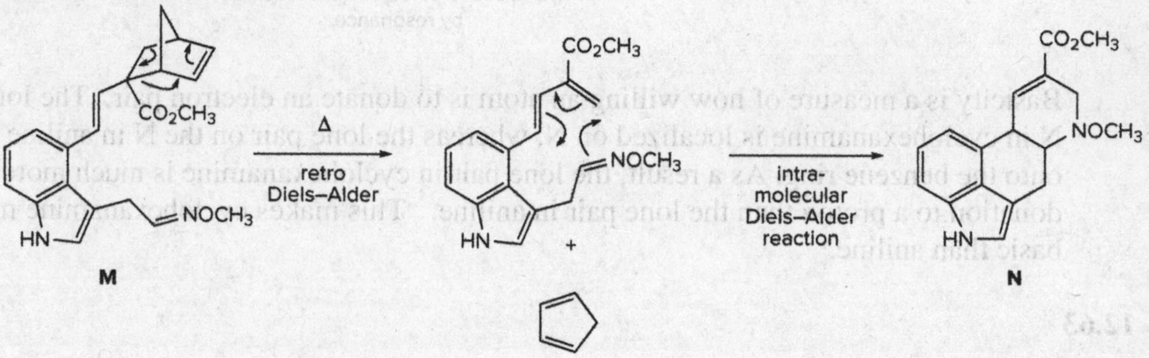

CO_2CH_3 HN $NOCH_3$ **M**

Δ retro Diels–Alder

CO_2CH_3 $NOCH_3$ HN +

intra-molecular Diels–Alder reaction

CO_2CH_3 $NOCH_3$ HN **N**

Spectroscopy A Mass Spectrometry

Chapter Review

Mass spectrometry (MS)

- Mass spectrometry measures the molecular weight of a compound (A.1A).
- The mass of the molecular ion (M) = the molecular weight of a compound. Except for isotope peaks at M + 1 and M + 2, the molecular ion has the highest mass in a mass spectrum (A.1A).
- The base peak is the tallest peak in a mass spectrum (A.1A).
- A compound with an odd number of N atoms gives an odd molecular ion. A compound with an even number of N atoms (including zero) gives an even molecular ion (A.1B).
- Organic chlorides show two peaks for the molecular ion (M and M + 2) in a 3:1 ratio (A.2).
- Organic bromides show two peaks for the molecular ion (M and M + 2) in a 1:1 ratio (A.2).
- The fragmentation of radical cations formed in a mass spectrometer gives lower-molecular-weight fragments, often characteristic of a functional group (A.3, A.4).
- High-resolution mass spectrometry gives the molecular formula of a compound (A.5A).

Practice Test on Chapter Review

1.a. Which compound has a molecular ion at 112 in its mass spectrum?

1. 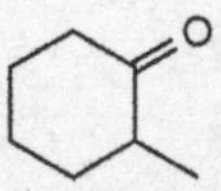 2. 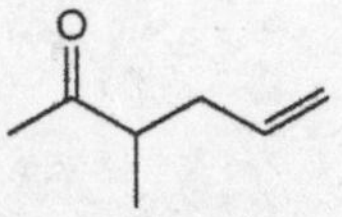 3.

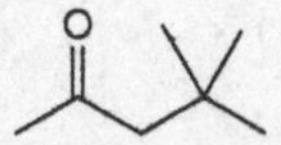

4. Compounds (1) and (2) both fit these criteria.
5. Compounds (1), (2), and (3) all fit these criteria.

b. What is the base peak in a mass spectrum?
1. the peak due to the radical cation formed when a molecule loses an electron
2. the tallest peak in the mass spectrum
3. the peak due to the fragment with the largest *m*/*z* ratio
4. Both (1) and (2) describe the base peak.
5. Statements (1), (2), and (3) all describe the base peak.

c. Which compounds are possible structures for a molecule that has a molecular ion at 150 in its mass spectrum?

1. 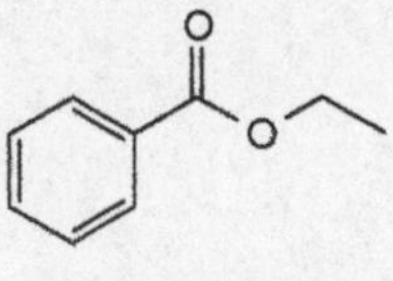 2. 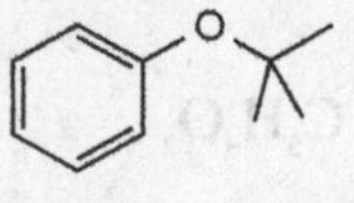 3.

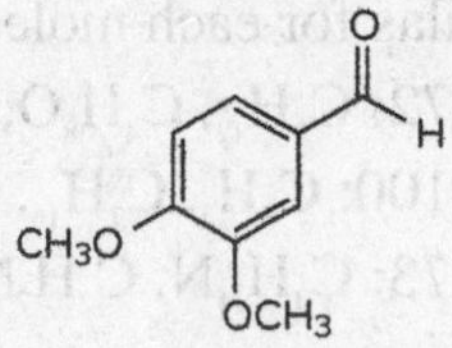

4. Both (1) and (2) are possible structures.
5. Compounds (1), (2), and (3) are all possible structures.

d. Which compounds exhibit prominent M + 2 peaks in their mass spectra?

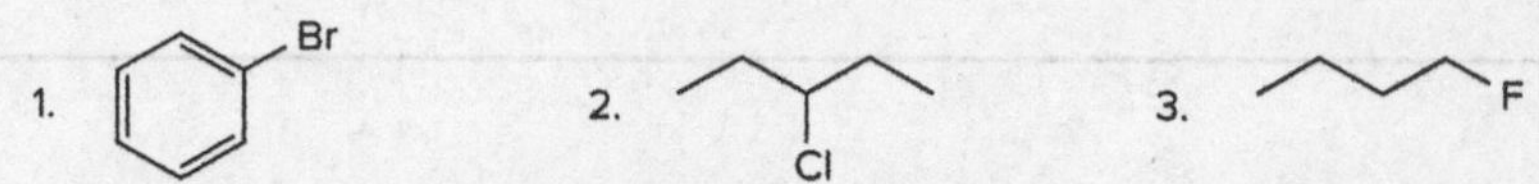

4. Both (1) and (2) show M + 2 peaks.
5. Compounds (1), (2), and (3) all show M + 2 peaks.

2. Answer True (T) or False (F).

a. A compound with a molecular ion at 109 contains a N atom.
b. A compound with a base peak at 57 must contain a N atom.
c. In its mass spectrum, a compound that has a molecular ion with two peaks of approximately equal intensity at 124 and 126 contains chlorine.
d. Fragmentation by alpha (α) cleavage occurs with aldehydes, alcohols, and amines.
e. A high-resolution mass spectrum can be used to distinguish between possible molecular formulas.

Answers to Practice Test

1.a. 4	2.a. T
b. 2	b. F
c. 4	c. F
d. 4	d. T
	e. T

Answers to Problems

A.1 The mass spectrum of pentane has peaks at *m/z* = 72 (molecular ion), 73 (M + 1), and 43 (base peak).

A.2 The molecular ion formed from each compound is equal to its molecular weight.

a. C_3H_6O	b. $C_{10}H_{20}$	c. $C_8H_8O_2$	d. $C_{10}H_{15}N$
molecular weight = **58**	molecular weight = **140**	molecular weight = **136**	molecular weight = **149**
molecular ion (*m/z*) = **58**	molecular ion (*m/z*) = **140**	molecular ion (*m/z*) = **136**	molecular ion (*m/z*) = **149**

A.3 Some possible formulas for each molecular ion:

a. Molecular ion at 72: C_5H_{12}, C_4H_8O, $C_3H_4O_2$
b. Molecular ion at 100: C_8H_4, C_7H_{16}, $C_6H_{12}O$, $C_5H_8O_2$
c. Molecular ion at 73: $C_4H_{11}N$, $C_2H_7N_3$

A.4 Use the molecular ion to propose molecular formulas that contain C, H, and O. Then determine which formula has three degrees of unsaturation.

M at 222: $C_{18}H_6$ (not enough H's)

$C_{17}H_{18}$ $\xrightarrow[+1\,O]{-CH_4}$ $C_{16}H_{14}O$ (not enough H's)

$C_{16}H_{30}$ $\xrightarrow[+1\,O]{-CH_4}$ $C_{15}H_{26}O$ (three degrees of unsaturation)

Answer

A.5 Chlorpheniramine has a molecular formula $C_{16}H_{19}ClN_2$. Calculate the molecular ions by determining the molecular weight using each of the two most common isotopes of Cl (^{35}Cl and ^{37}Cl).

MW of $C_{16}H_{19}{}^{35}ClN_2$ = 274

MW of $C_{16}H_{19}{}^{37}ClN_2$ = 276

Two peaks in a 3:1 ratio at *m/z* = 274 and 276

A.6 To calculate the molecular ions you would expect for compounds with Cl, calculate the molecular weight using each of the two most common isotopes of Cl (^{35}Cl and ^{37}Cl).

a. $C_4H_9{}^{35}Cl$ = **92**
$C_4H_9{}^{37}Cl$ = **94**
Two peaks in 3:1 ratio at *m/z* 92 and 94

b. C_3H_7F = **62**
One peak at *m/z* 62

c. $C_4H_{11}N$ = **73**
One peak at *m/z* 73

d. $C_4H_4N_2$ = **80**
One peak at *m/z* 80

A.7 After calculating the mass of the molecular ion, draw the structure and determine which C–C bond is broken to form fragments of the appropriate mass-to-charge ratio.

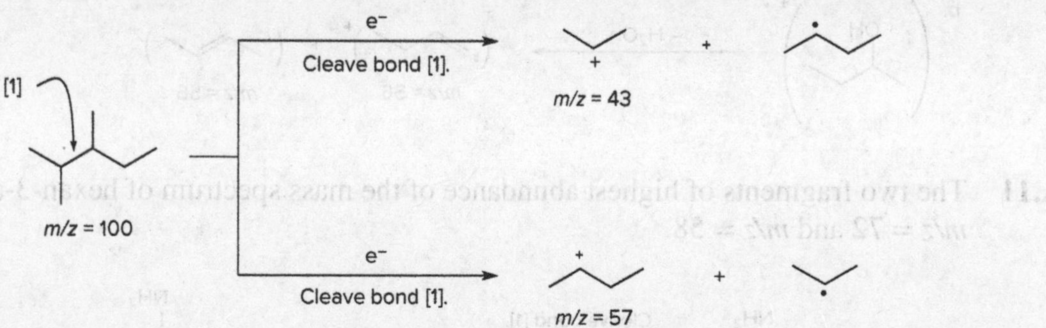

A.8

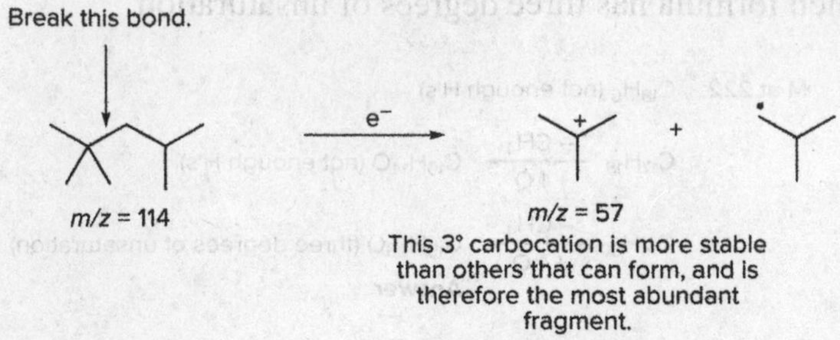

Break this bond.

$m/z = 114$ —e⁻→ $m/z = 57$ (+ radical)

This 3° carbocation is more stable than others that can form, and is therefore the most abundant fragment.

A.9

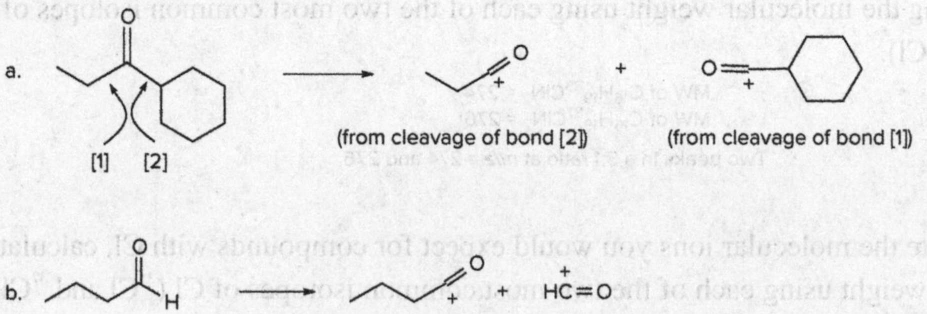

a. [1] [2] → (from cleavage of bond [2]) + (from cleavage of bond [1])

b. → + $\overset{+}{H}C{\equiv}O$

A.10

a.

Cleave bond [1]. → $CH_3\cdot$ + $m/z = 59$

Cleave bond [2]. → $\cdot CH_2CH_3$ + $m/z = 45$

b. —$- H_2O$→ $m/z = 56$ + $m/z = 56$

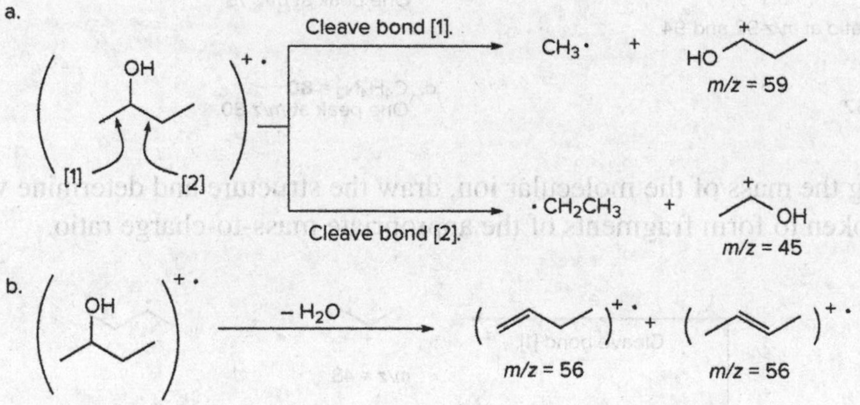

A.11 The two fragments of highest abundance of the mass spectrum of hexan-3-amine occur at $m/z = 72$ and $m/z = 58$.

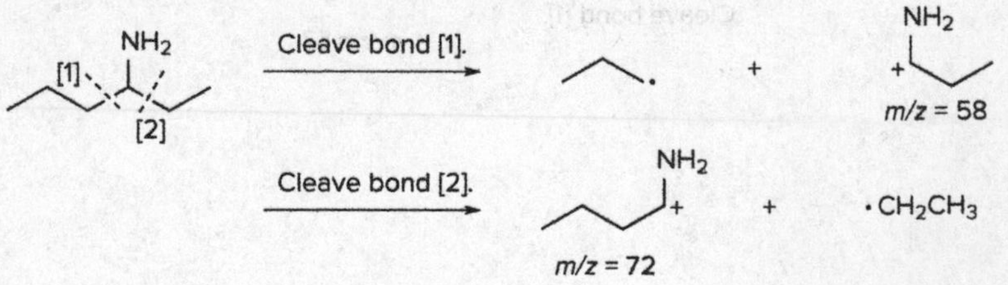

Cleave bond [1]. → + $m/z = 58$

Cleave bond [2]. → $m/z = 72$ + $\cdot CH_2CH_3$

A.12 Use the exact mass values given in Table A.1 to calculate the exact mass of each compound.

$C_7H_5NO_3$	$C_8H_9NO_2$	$C_{10}H_{17}N$
mass: 151.0270	mass: 151.0634 compound **X**	mass: 151.1362

A.13

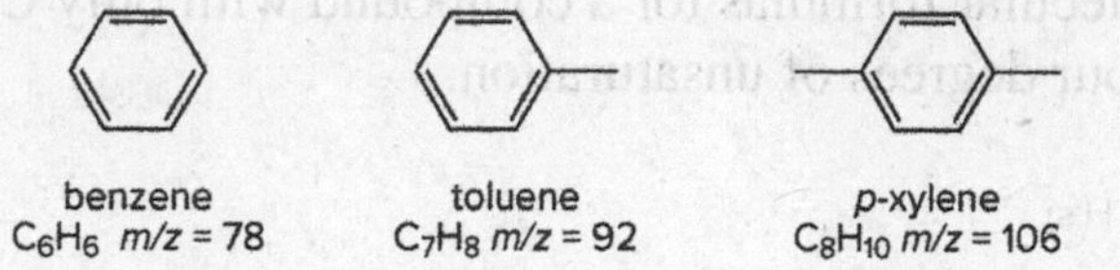

benzene C_6H_6 $m/z = 78$

toluene C_7H_8 $m/z = 92$

p-xylene C_8H_{10} $m/z = 106$

GC–MS analysis:
Three peaks in the gas chromatogram.
Order of peaks: benzene, toluene, *p*-xylene, in order of increasing bp.
Molecular ions observed in the three mass spectra: 78, 92, 106.

A.14

[1] [2] [3]

$m/z = 142$

Cleave bond [1]. → $\cdot CH_3$ + $m/z = 127$

Cleave bond [2]. → + $m/z = 85$

Cleave bond [3]. → $\cdot CH_2CH_3$ + $m/z = 113$

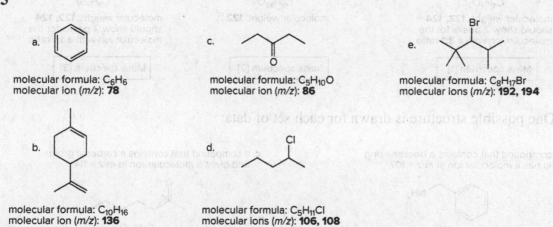

A.15

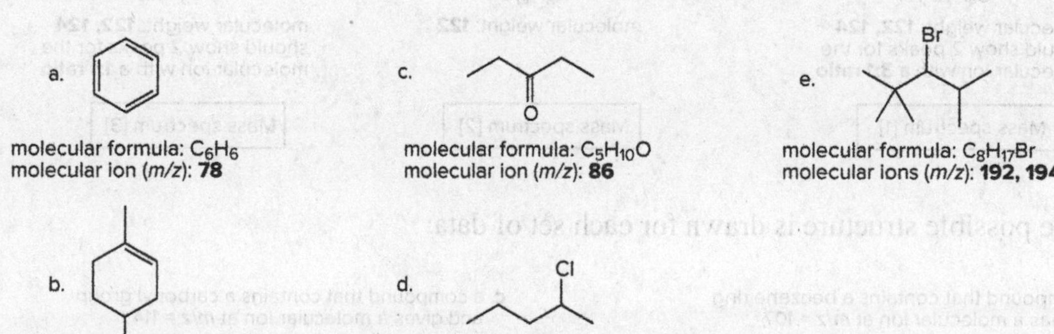

a. molecular formula: C_6H_6
molecular ion (m/z): **78**

c. molecular formula: $C_5H_{10}O$
molecular ion (m/z): **86**

e. Br
molecular formula: $C_8H_{17}Br$
molecular ions (m/z): **192, 194**

b. molecular formula: $C_{10}H_{16}$
molecular ion (m/z): **136**

d. Cl
molecular formula: $C_5H_{11}Cl$
molecular ions (m/z): **106, 108**

A.16 Examples are given for each molecular ion.

a. molecular ion 102: C_8H_6, $C_6H_{14}O$, $C_5H_{10}O_2$, $C_5H_{14}N_2$
b. molecular ion 98: C_8H_2, C_7H_{14}, $C_6H_{10}O$, $C_5H_6O_2$
c. molecular ion 119: C_8H_9N, $C_6H_5N_3$
d. molecular ion 74: C_6H_2, $C_4H_{10}O$, $C_3H_6O_2$

A.17 Likely molecular formula, C_8H_{16} (one degree of unsaturation—one ring or one π bond).

Four structures with m/z = 112

A.18 Use the molecular ion to propose molecular formulas for a compound with only C and H. Then determine which formula has four degrees of unsaturation.

M at 204: C_{17} (no H's)

$C_{16}H_{12}$ (not enough H's)

$C_{15}H_{24}$ (four degress of unsaturation)
Answer

A.19 Use the directions from Answer A.4 and propose a formula with two degrees of unsaturation.

M at 154: $C_{12}H_{10}$ (not enough H's)

$C_{11}H_{22}$ $\xrightarrow[+\,O]{-\,CH_4}$ $C_{10}H_{18}O$ (two degrees of unsaturation)
Answer

A.20

B
$C_4H_7O_2Cl$
molecular weight: **122, 124**
should show 2 peaks for the molecular ion with a **3:1 ratio**
Mass spectrum [1]

C
$C_8H_{10}O$
molecular weight: **122**
Mass spectrum [2]

A
C_3H_7Br
molecular weight: **122, 124**
should show 2 peaks for the molecular ion with a **1:1 ratio**
Mass spectrum [3]

A.21 One possible structure is drawn for each set of data:

a. a compound that contains a benzene ring and has a molecular ion at m/z = 107

NH_2
C_7H_9N

b. a hydrocarbon that contains only sp^3 hybridized carbons and a molecular ion at m/z = 84

C_6H_{12}

c. a compound that contains a carbonyl group and gives a molecular ion at m/z = 114

$C_7H_{14}O$

d. a compound that contains C, H, N, and O and has an exact mass for the molecular ion at 101.0841

$C_5H_{11}NO$

A.22 Use the values given in Table A.1 to calculate the exact mass of each compound. $C_8H_{11}NO_2$ (exact mass 153.0790) is the correct molecular formula.

A.23 In the spectrum of hexan-2-ol ($C_6H_{14}O$, MW = 102), the molecular ion appears at *m/z* = 102, the base peak occurs at *m/z* = 45, and the fragment from loss of water (M – 18) occurs at *m/z* = 84).

[1] OH [2]
Cleave bond [1]. → OH (+) m/z = 87 + •CH3
Cleave bond [2]. → OH (+) m/z = 45 + •

A.24

a. [1] [2] OH → + OH (from cleavage of bond [1]) + OH + (from cleavage of bond [2])

b. [1] O [2] → O+ (from cleavage of bond [1]) + +O (from cleavage of bond [2])

c. [1] [2] O → O+ (from cleavage of bond [1]) + +O (from cleavage of bond [2])

A.25

a. The molecular ion in both spectra is 88. The base peak in spectrum [1] occurs at *m/z* = 59. The base peak in spectrum [2] occurs at *m/z* = 57.

b. One possible fragmentation is shown for each compound.

[1] OH **A** Cleave bond [1]. → + m/z = 57 3° carbocation + •CH2OH

OH [1] **B** Cleave bond [1]. → CH3CH2• + OH + m/z = 59

Thus, spectrum [1], with a base peak at *m/z* = 59 corresponds to **B.** Spectrum [2] with a base peak at *m/z* = 57 corresponds to **A.**

A.26 The mass spectrum of hexan-2-amine ($C_6H_{15}N$) shows a molecular ion at $m/z = 101$ and a base peak at $m/z = 44$. The base peak corresponds to an α cleavage product.

NH2 — Break bond. → NH2 + ·
m/z = 44
base peak

A.27

a. (m/z = 104) ← − H2O — OH [1] (m/z = 122) — e⁻ Cleave bond [1]. → + (resonance-stabilized carbocation) m/z = 91

b. (m/z = 68) ← − H2O — [1] [3] [2] OH (m/z = 86)
e⁻ Cleave bond [1]. → + OH m/z = 71
e⁻ Cleave bond [2]. → + m/z = 41
e⁻ Cleave bond [3]. → $^+CH_2—OH$ m/z = 31

A.28

Ketone **A**
m/z = 128
↓ α cleavage
· CH_2CH_3 + O+
m/z = 99
This is ketone **A** because α cleavage gives a fragment with m/z of 99.

Ketone **B**
m/z = 128
↓ α cleavage
· CH_3 + O+
m/z = 113
This is ketone **B** because α cleavage gives a fragment with m/z of 113.

A.29 An ether fragments by α cleavage because the resulting carbocation is resonance stabilized.

[1] [2] O
Cleave bond [1]. → · + + O ↔ O+
resonance-stabilized carbocations
Cleave bond [2]. → ·CH_3 + + O ↔ O+

A.30 a. There are three molecular ions (M, M + 2, M + 4) for a compound with 1 Cl and 1 Br. The relative peak intensities are determined using the natural occurrence of each isotope. For Cl, there is a 75% probability that a Cl will be ^{35}Cl and a 25% probability that a Cl will be ^{37}Cl. For Br, there is a 50% probability that a Br will be either ^{79}Br or ^{81}Br.

Molecular ions:	M due to	M + 2 due to	M + 4 due to
	$^{35}Cl + ^{79}Br$	$^{35}Cl + ^{81}Br$ or $^{37}Cl + ^{79}Br$	$^{37}Cl + ^{81}Br$
Intensities:	.75 × .5 = .375	(.75 × .5) + (.25 × .5) = .5	.25 × .5 = .125
Ratio of peak heights:	**3** :	**4** :	**1**

b. There are four molecular ions (M, M + 2, M + 4, M + 6) for a compound with 3 Br's. The relative peak intensities are determined as in (a).

M due to	M + 2 due to	M + 4 due to	M + 6 due to
$^{79}Br + ^{79}Br + ^{79}Br$	$^{79}Br + ^{79}Br + ^{81}Br$ $^{79}Br + ^{81}Br + ^{79}Br$ $^{81}Br + ^{79}Br + ^{79}Br$	$^{79}Br + ^{81}Br + ^{81}Br$ $^{81}Br + ^{79}Br + ^{81}Br$ $^{81}Br + ^{81}Br + ^{79}Br$	$^{81}Br + ^{81}Br + ^{81}Br$
Intensities: .5 × .5 × .5 = .125	3(.5 × .5 × .5) = .375	3(.5 × .5 × .5) = .375	3(.5 × .5 × .5) = .125
Ratio of peak heights: **1**:	**3** :	**3** :	**1**

A.31 a. The fragment at m/z = 120 is due to the McLafferty rearrangement.

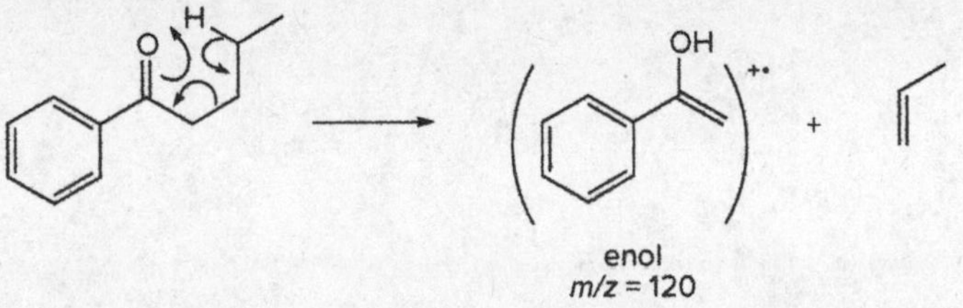

enol
m/z = 120

b.

m/z = 88
fragment due to the
McLafferty rearrangement

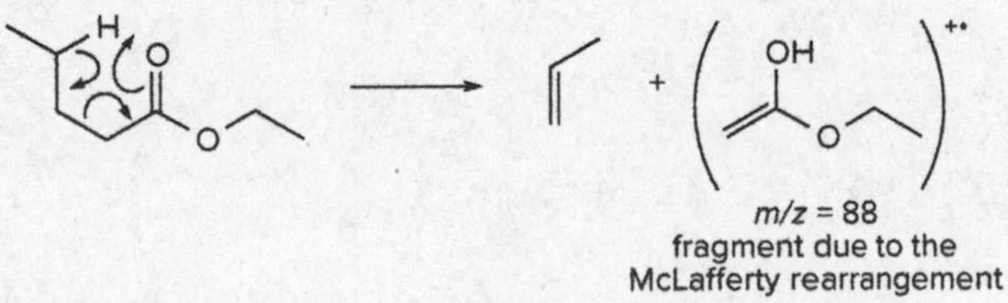

a

c. Compounds need to have a H atom on a carbon three atoms from the C=O to have a McLafferty rearrangement. Only **A** has a H atom for rearrangement.

H atom on this C.

A

Spectroscopy B Infrared Spectroscopy

Chapter Review

Electromagnetic radiation

- The wavelength (λ) and frequency (ν) of electromagnetic radiation are *inversely* related by the following equations, where *c* is the speed of light: $\lambda = c/\nu$ or $\nu = c/\lambda$ (B.1).
- The energy (E) of a photon is proportional to its frequency; the higher the frequency, the higher the energy (h = Planck's constant): $\boldsymbol{E = h\nu}$ (B.1).

Infrared spectroscopy (IR, B.2–B.4)

- Infrared spectroscopy identifies functional groups.
- IR absorptions are reported in wavenumbers:

wavenumber = $\tilde{\nu}$ = 1/λ

- The functional group region, **4000–1500 cm^{-1}**, is the most useful region of an IR spectrum.
- C–H, O–H, and N–H bonds absorb at high frequency, $\geq$ 2500 cm^{-1}.
- As bond strength increases, the wavenumber of an absorption increases; thus triple bonds absorb at higher wavenumber (i.e. higher energy) than double bonds.

C=C ~1650 cm^{-1} → C≡C ~2250 cm^{-1}

Increasing bond strength
Increasing $\tilde{\nu}$

- The higher the percent *s*-character, the stronger the bond, and the higher the wavenumber of an IR absorption (B.3B).

C_{sp^3}–H	C_{sp^2}–H	C_{sp}–H
25% *s*-character	33% *s*-character	50% *s*-character
3000–2850 cm^{-1}	3150–3000 cm^{-1}	3300 cm^{-1}

Increasing percent *s*-character
Increasing $\tilde{\nu}$

- For carbonyl compounds, conjugation shifts a C=O absorption to lower wavenumber (B.3C). For cyclic compounds, a smaller ring size shifts a C=O absorption to higher wavenumber (B.4B).

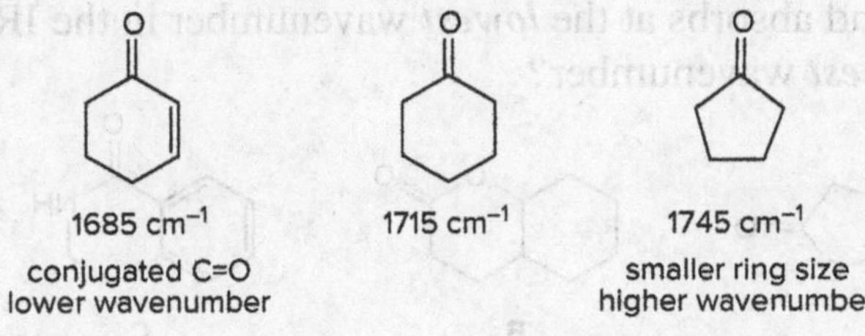

Practice Test on Chapter Review

1. Which of the following compounds has peaks at 3300, 3000, and 2250 cm^{-1} in its IR spectrum?

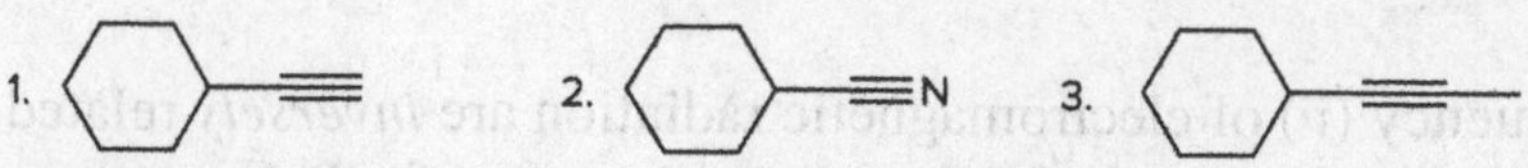

4. Compounds (1) and (2) have these peaks in their IR spectra.
5. Compounds (1), (2), and (3) all contain these peaks in their IR spectra.

2. Answer each question with the number that corresponds to one of the following regions of an IR spectrum.

1. 4000–2500 cm^{-1}
2. 2500–2000 cm^{-1}
3. 2000–1500 cm^{-1}
4. < 1500 cm^{-1}

a. This region is called the fingerprint region of an IR spectrum.
b. The OH group of propan-1-ol absorbs in this region.
c. The C=N of $C_6H_5CH_2CH{=}NCH_3$ absorbs in this region.
d. An unsymmetrical C≡C absorbs in this region.
e. An *sp* hybridized C–H bond absorbs in this region.
f. Ethyl benzoate ($C_6H_5CO_2CH_2CH_3$) absorbs in all regions of the IR except this one.

3. Answer True (T) or False (F).

a. IR spectroscopy is useful for determining the molecular weight of a compound.
b. A C–H bond that absorbs at 3140 cm^{-1} is stronger than a C–H bond that absorbs at 2950 cm^{-1}.
c. But-2-yne shows an IR absorption at 2250 cm^{-1}.
d. Propan-1-ol shows an IR absorption at 3200–3600 cm^{-1}.
e. An ether shows no IR absorptions at 3200–3600 or 1700 cm^{-1}.

4. (a) Considering compounds **A–D,** which compound absorbs at the ***lowest*** wavenumber in its IR spectrum? (b) Which compound absorbs at the ***highest*** wavenumber in its IR spectrum?

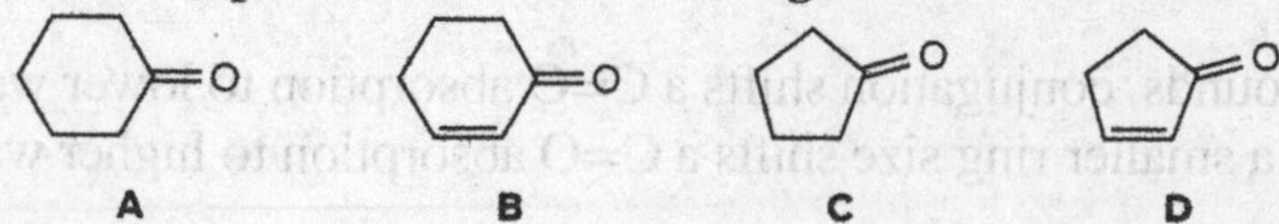

5. (a) Which compound absorbs at the ***lowest*** wavenumber in the IR? (b) Which compound absorbs at the ***highest*** wavenumber?

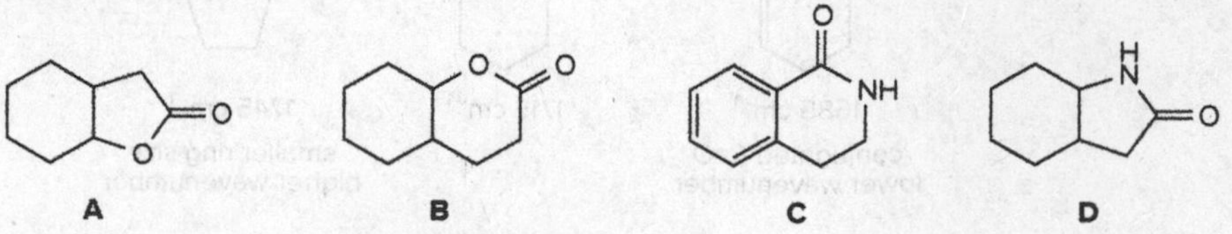

Answers to Practice Test

1. 1	2.a. 4	3.a. F	4.a. **B**
	b. 1	b. T	b. **C**
	c. 3	c. F	
	d. 2	d. T	5.a. **C**
	e. 1	e. T	b. **A**
	f. 2		

Answers to Problems

B.1 **Wavelength and frequency are inversely proportional.** The higher-frequency light will have a shorter wavelength.

a. Light having $\lambda = 10^2$ nm has a higher ν than light with $\lambda = 10^4$ nm.

b. Light having $\lambda = 100$ nm has a higher ν than light with $\lambda = 100$ mm.

c. Blue light has a higher ν than red light.

B.2 **The energy of a photon** is *proportional* to its **frequency** and *inversely* proportional to its wavelength.

a. Light having $\nu = 10^8$ Hz is of higher energy than light having $\nu = 10^4$ Hz.

b. Light having $\lambda = 10$ nm is of higher energy than light having $\lambda = 1000$ nm.

c. Blue light is of higher energy than red light.

B.3 Higher wavenumbers are proportional to higher frequencies and higher energies.

a. IR light with a wavenumber of 3000 cm^{-1} is higher in energy than IR light with a wavenumber of 1500 cm^{-1}.

b. IR light having $\lambda = 10$ mm is higher in energy than IR light having $\lambda = 20$ mm.

B.4 Stronger bonds absorb at a higher wavenumber. Bonds to lighter atoms (H versus D) absorb at higher wavenumber.

B.5 As the percent *s*-character increases, bonds get shorter and stronger, and the wavenumber of the IR absorption increases.

sp hybridized C, 50% *s*-character

[1] [2] [3]

sp^2 hybridized C, 33% *s*-character

sp^3 hybridized C, 25% *s*-character

a. Increasing bond strength: [2] < [3] < [1]
b. Increasing bond length: [1] < [3] < [2]
c. Increasing percent *s*-character: [2] < [3] < [1]
d. Increasing wavenumber: [2] < [3] < [1]

B.6 Conjugation lowers the wavenumber of a C=O absorption. Increased resonance delocalization also lowers the wavenumber of the C=O absorption.

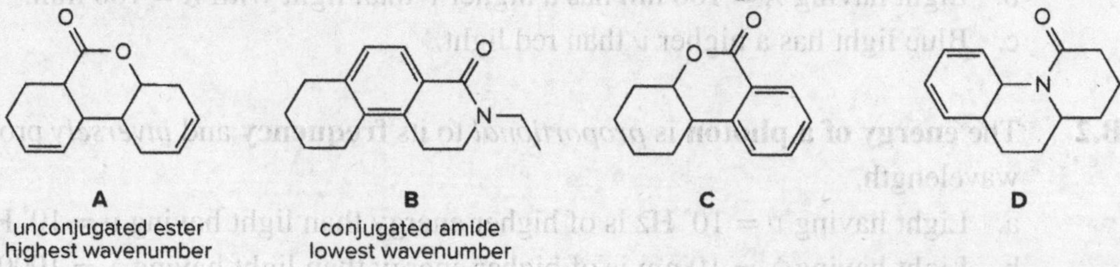

B.7 a. Compound **A** has peaks at ~3150 (sp^2 hybridized C–H), 3000–2850 (sp^3 hybridized C–H), and 1650 (C=C) cm^{-1}.
b. Compound **B** has a peak at 3000–2850 (sp^3 hybridized C–H) cm^{-1}.

B.8 Cyclopentane and pent-1-ene are both composed of C–C and C–H bonds, but pent-1-ene also has a C=C bond. This difference will give the IR of pent-1-ene an additional peak at 1650 cm^{-1} (for the C=C). Pent-1-ene will also show C–H absorptions for sp^2 hybridized C–H bonds at 3150–3000 cm^{-1}.

B.9 Use the principles in Answer B.6.

C < B < A

B.10 Compound **X** has two ethers, so it just has an absorption at 3000–2830 cm^{-1} for its sp^3 hybridized C–H bonds. Compound **B** has a C=O, so in addition to peaks at 3000–2850 cm^{-1}, it should have a strong absorption at ~1700 cm^{-1}.

B.11 Look at the functional groups in each compound below to explain how each IR is different.

A — C=O peak at ~1700 cm^{-1}

B — C=C peak at 1650 cm^{-1}; Csp^2–H at 3150–3000 cm^{-1}

C — O–H peak at 3200–3600 cm^{-1}

B.12 All compounds show an absorption at 3000–2850 cm^{-1} due to the sp^3 hybridized C–H bonds. Additional peaks in the functional group region for each compound are shown.

a. no additional peaks

b. O–H bond at 3600–3200 cm^{-1}

c. Csp^2–H at 3150–3000 cm^{-1}; C=C bond at 1650 cm^{-1}

d. C=O bond at ~1700 cm^{-1}

e. O–H at 3600–3200 cm^{-1}; N–H at 3500–3200 cm^{-1}; Csp^2–H at 3150–3000 cm^{-1}; C=O at ~1700 cm^{-1}; C=C at 1650 cm^{-1}; aromatic ring at 1600, 1500 cm^{-1}

B.13 Look at the functional groups that are removed during the reaction and the functional groups that are formed in the product.

H_2O, H_2SO_4

C=C at ~ 1650 cm^{-1} disappears.

OH at 3600–3200 cm^{-1} appears.

B.14 Possible structures are (a) $CH_3COOCH_2CH_3$ and (c) $CH_3CH_2COOCH_3$. Compounds (b) and (d) also have an OH group that would give a strong absorption at ~3600–3200 cm^{-1}, which is absent in the IR spectrum of **X,** thus excluding them as possibilities.

B.15

a. Hydrocarbon with a molecular ion at $m/z = 68$
IR absorptions at
3310 cm^{-1} = C_{sp}–H bond
3000–2850 cm^{-1} = C_{sp^3}–H bonds
2120 cm^{-1} = C≡C bond
Molecular formula: C_5H_8

or

b. Compound with C, H, and O with a molecular ion at $m/z = 60$
IR absorptions at 3600–3200 cm^{-1} = O–H bond
3000–2850 cm^{-1} = C_{sp^3}–H bonds
Molecular formula: C_3H_8O

OH or OH

B.16

a.
C_{sp^2}–H at 3000–3150 cm^{-1}
C_{sp^3}–H at 2850–3000 cm^{-1}
C=C at 1650 cm^{-1}

b. OH
O–H at 3200–3600 cm^{-1}
C_{sp^2}–H at 3000–3150 cm^{-1}
C_{sp^3}–H at 2850–3000 cm^{-1}
C=C at 1650 cm^{-1}

B.17

O
OH

C_{sp^2}–H at 3150–3000 cm^{-1}
C_{sp^3}–H at 3000–2850 cm^{-1}
C=O at ~1700 cm^{-1}
C=C at 1650 cm^{-1}
O–H above 3000 cm^{-1}
[The OH of a COOH is much broader than the OH of an alcohol and occurs at 3500–2500 cm^{-1}.]

B.18

a. =O or –OH
stronger bond
higher $\tilde{\nu}$ absorption

b. =N or N–H
stronger bond
higher $\tilde{\nu}$ absorption

c. –H or –H
stronger bond
higher $\tilde{\nu}$ absorption

B.19 Locate the functional groups in each compound. Use Tables B.1 and B.2 to determine what IR absorptions each would have.

a.
C_{sp}–H at 3300 cm^{-1}
C_{sp^3}–H at 2850–3000 cm^{-1}
C≡C at 2250 cm^{-1}

b. OH
O–H at 3200–3600 cm^{-1}
C_{sp^3}–H at 2850–3000 cm^{-1}

c. O
C_{sp^3}–H at 2850–3000 cm^{-1}
C=O at 1700 cm^{-1}

d. O OH
O–H at > 3000 cm^{-1}
C_{sp^2}–H at 3000–3150 cm^{-1}
C=O at ~1700 cm^{-1}
phenyl group at 1600, 1500 cm^{-1}

[The OH of the RCOOH is even broader than the OH of an alcohol (3500–2500 cm^{-1})]

B.20 a. Use Tables B.1 and B.2 to identify the IR absorptions.

C=O (1740 cm^{-1})
benzene at 1600, 1500 cm^{-1}
HO
OH at 3600–3200 cm^{-1}
estrone

Csp^2–H at 3150–3000 cm^{-1}
C=C at ~1650 cm^{-1}
conjugated C=O at < 1700 cm^{-1}
Csp^2–H at 3150–3000 cm^{-1}
C=C at ~1650 cm^{-1}
Csp–H at 3300 cm^{-1}
C≡C at 2250 cm^{-1}
OH at 3600–3200 cm^{-1}
etonogestrel

b. The C=O in estrone is at a much higher wavenumber than the C=O in etonogestrol, which is both conjugated and located in a larger ring.

B.21 a. Use Tables B.1 and B.2 to identify the IR absorptions.

a. C=C bond 1650 cm^{-1}; Csp^2–H at 3150–3000 cm^{-1} and C≡C bond 2250 cm^{-1}; Csp–H at 3300 cm^{-1}

b. O–H bond OH at 3500–2500 cm^{-1} and no O–H bond

c. no C=O bond and C=O bond ~1700 cm^{-1}

d. Csp–H bond 3300 cm^{-1} and

B.22 The IR absorptions above 1500 cm^{-1} are different for each of the narcotics.

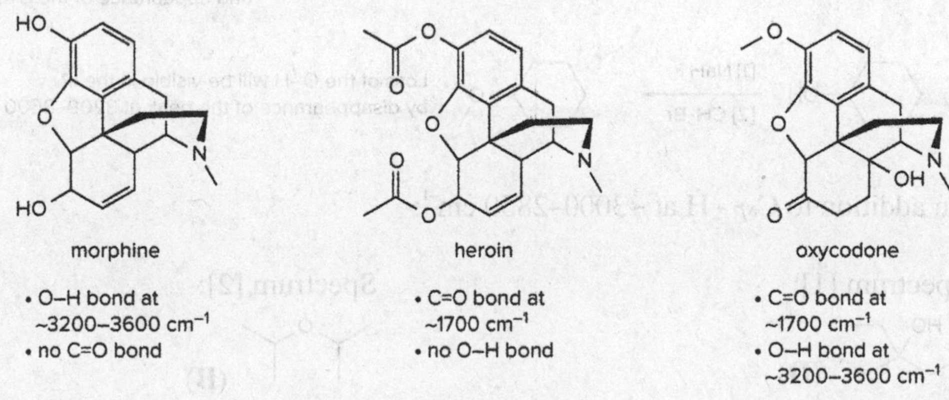

B.23 Use the information in Answer B.6.

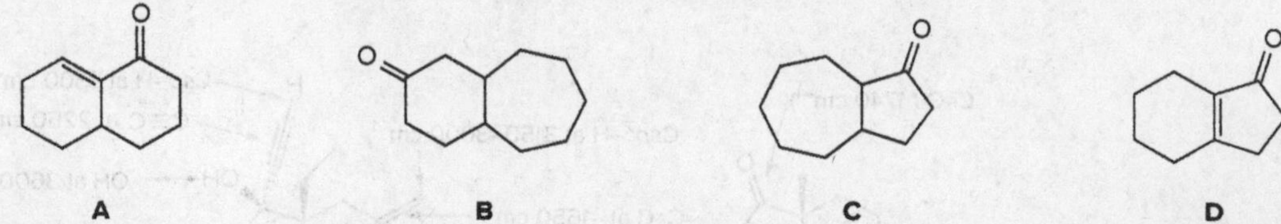

a. **C** absorbs at the highest wavenumber because the C=O is located in a five-membered ring and is not conjugated.
b. **A** absorbs at the lowest wavenumber because the C=O is located in a six-membered ring and is conjugated.

B.24 Use the information in Answer B.6.

A < C < B < D

A: lowest wavenumber — D: highest wavenumber

B.25 Look for a **change in functional groups** from starting material to product to see how IR could be used to determine when the reaction is complete.

a. $\xrightarrow[\text{Pd}]{H_2}$ Loss of the C=C will be visible in the IR by disappearance of the peak at 1650 cm^{-1}.

b. $\xrightarrow{\text{PCC}}$ Loss of the O–H group will be visible in the IR by disappearance of the peak at 3200–3600 cm^{-1} and appearance of the C=O at ~1700 cm^{-1}.

c. $\xrightarrow[\text{[2] } CH_3SCH_3]{\text{[1] } O_3}$ Loss of the C=C will be visible in the IR by disappearance of the peak at 1650 cm^{-1} and appearance of the C=O at ~1700 cm^{-1}.

d. $\xrightarrow[\text{[2] } CH_3Br]{\text{[1] NaH}}$ Loss of the O–H will be visible in the IR by disappearance of the peak at 3200–3600 cm^{-1}.

B.26 In addition to Csp^3–H at ~3000–2850 cm^{-1}:

Spectrum [1]: **(D)**

OH at 3600–3200 cm^{-1}

Spectrum [2]: **(B)**

No other peaks above 1500 cm^{-1}

Spectrum [3]: **(A)**

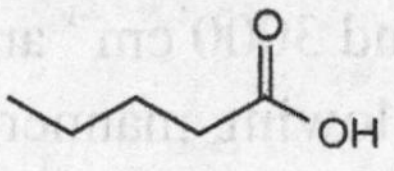

OH at ~3500–2500 cm^{-1}

C=O at ~1700 cm^{-1}

Spectrum [4]: **(C)**

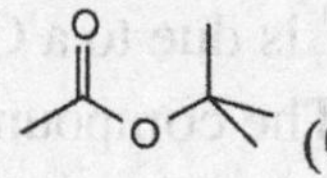

C=O at ~1700 cm^{-1}

B.27

a. Compound with a molecular ion at m/z = 72
IR absorption at 1725 cm^{-1} = C=O bond
Molecular formula: C_4H_8O

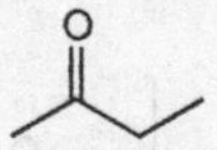

b. Compound with a molecular ion at m/z = 55
The odd molecular ion means an odd number of N's present. Molecular formula: C_3H_5N
IR absorption at 2250 cm^{-1} = C≡N bond

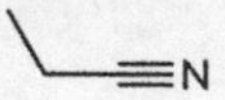

c. Compound with a molecular ion at m/z = 74
IR absorption at 3600–3200 cm^{-1} = O–H bond
Molecular formula: $C_4H_{10}O$

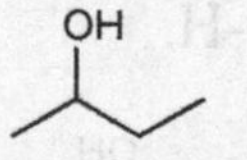

or

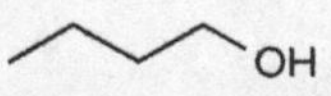

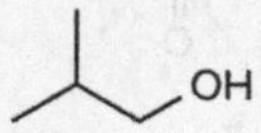

or

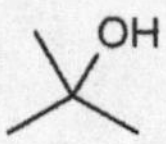

B.28

Chiral hydrocarbon with a molecular ion at m/z = 82
Molecular formula: C_6H_{10}
IR absorptions at 3300 cm^{-1} = C*sp*–H bond
3000–2850 cm^{-1} = C*sp*^{3}–H bonds
2250 cm^{-1} = C≡C bond

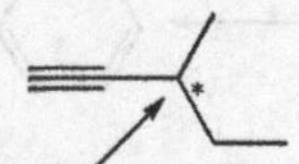

stereogenic center

Two possible enantiomers:

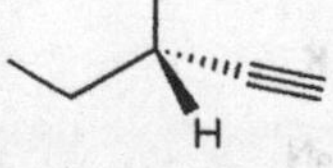

or

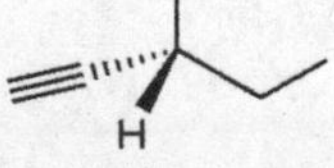

B.29 The chiral compound **Y** has a strong absorption at 2970–2840 cm^{-1} in its IR spectrum due to *sp*^{3} hybridized C–H bonds. The two peaks of equal intensity at 136 and 138 indicate the presence of a Br atom. The molecular formula is C_4H_9Br. Only one constitutional isomer of this molecular formula has a stereogenic center:

Y = and

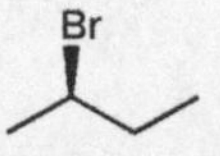

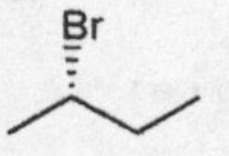

two possible enantiomers

B.30 The molecular ion of 192 suggests $C_{12}H_{16}O_2$ as a possible molecular formula. The IR absorption at 1721 cm^{-1} is due to a C=O, and the absorptions around 3000 cm^{-1} are due to C_{sp^2}–H and C_{sp^3}–H. The compound is an ester, formed in the following manner.

OH + NaOH → I → O

H
$C_{12}H_{16}O_2$

B.31 The molecular ion of 144 suggests $C_8H_{16}O_2$ as a possible molecular formula for **X**. The IR absorption at 1739 cm^{-1} is due to a C=O, and the absorptions at less than 3000 cm^{-1} are due to C_{sp^3}–H.

OH + HO → O

2-methylpropan-1-ol

$C_8H_{16}O_2$
X
possible structure

B.32

J
$C_6H_{12}O$
$m/z = 100$
IR absorption at 2962 cm^{-1} = C_{sp^3}–H bonds
1718 cm^{-1} = C=O bond

Fragments:

a cleavage product
$m/z = 43$

a cleavage product
$m/z = 85$

The fragment at $m/z = 57$ could be due to $(C_4H_9)^+$ or $(C_3H_5O)^+$.

B.33

NH_2
K
C_7H_9N
$m/z = 107$

IR absorptions at 3373 and 3290 cm^{-1} = N–H
3062 cm^{-1} = C_{sp^2}–H bonds
2920 cm^{-1} = C_{sp^3}–H bonds
1600 cm^{-1} = benzene ring

The odd molecular ion indicates the presence of a N atom.

CHO
L
C_7H_6O
$m/z = 106$

$m/z = 105$ + $m/z = 77$

IR absorption at 3068 cm^{-1} = C_{sp^2}–H bonds on ring
2820 cm^{-1} and 2736 cm^{-1} = C–H of RCHO (Appendix G)
1703 cm^{-1} = C=O bond
1600 cm^{-1} = aromatic ring

B.34

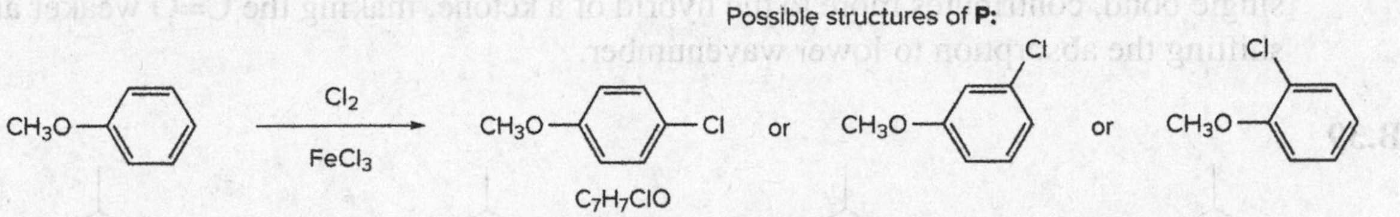

Possible structures of **P**:

CH_3O–C₆H₅ $\xrightarrow[FeCl_3]{Cl_2}$ CH_3O–C₆H₄–Cl or CH_3O–C₆H₄–Cl (meta) or CH_3O–C₆H₄–Cl (ortho)

C_7H_7ClO

m/z = 142, 144

IR absorption at 3096–2837 cm^{-1} = Csp^3–H bonds and Csp^2–H bonds

1582 cm^{-1} and 1494 cm^{-1} = benzene ring

The peak at M + 2 shows the presence of Cl or Br. Since Cl_2 is a reactant, the compound presumably contains Cl.

B.35 The mass spectrum has a molecular ion at 71. The odd mass suggests the presence of an odd number of N atoms; likely formula, C_4H_9N. The IR absorption at ~3300 cm^{-1} is due to N–H and the 3000–2850 cm^{-1} is due to sp^3 hybridized C–H bonds.

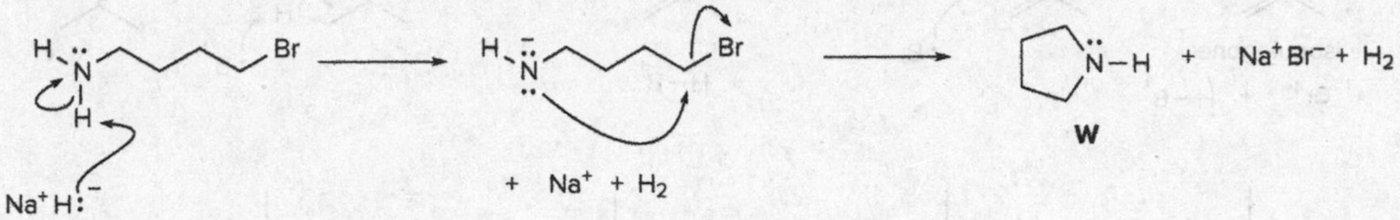

$H_2N(CH_2)_4Br$ + Na^+H^- ⟶ $HN^-(CH_2)_4Br$ + Na^+ + H_2 ⟶ pyrrolidine (N–H) **W** + Na^+Br^- + H_2

B.36 RCOCl should absorb at higher wavenumber than RCO_2R' because Cl is less basic than OR', making RCOCl less resonance stabilized than RCO_2R'.

B.37 Intramolecular hydrogen bonding gives the C=O of methyl salicylate more single-bond character, lowering the wavenumber of the C=O absorption.

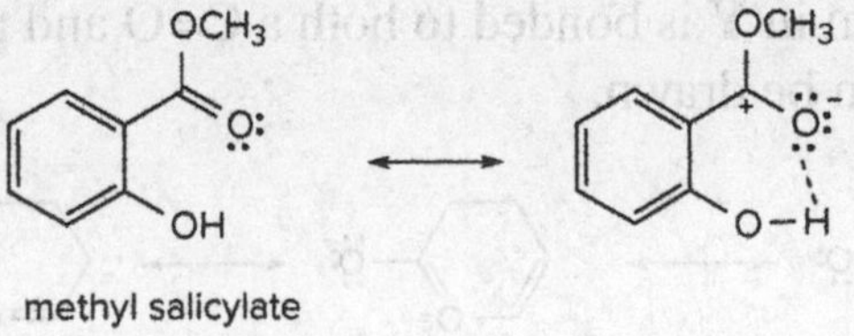

methyl salicylate

$\bar{\nu}$ = 1680 cm^{-1}

B.38 If a ketone carbonyl absorbs at lower wavenumber than an aldehyde carbonyl, the ketone carbonyl is weaker and has more single bond character. This can be explained by the fact that R groups are electron donating and stabilize an adjacent (+) charge.

ketone

The two R groups stabilize the (+) charge, so this form contributes more to the hybrid than the charge-separated resonance form of an aldehyde.

One R group stabilizes the (+) charge less.

As a result, the charge-separated resonance form of a ketone, which contains a C–O single bond, contributes more to the hybrid of a ketone, making the C=O weaker and shifting the absorption to lower wavenumber.

B.39

a, b.

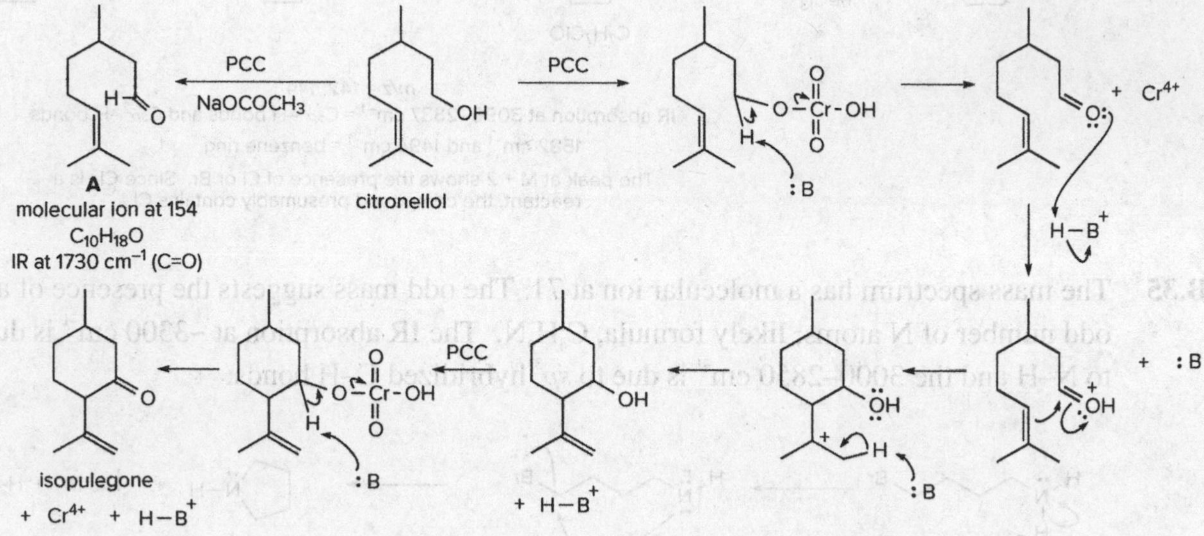

PCC

$NaOCOCH_3$

PCC

$+ Cr^{4+}$

A

molecular ion at 154

$C_{10}H_{18}O$

IR at 1730 cm^{-1} (C=O)

citronellol

PCC

isopulegone

$+ Cr^{4+} + H{-}B^+$

$+ H{-}B^+$

a, c.

isopulegone

$H_2\ddot{O}$:

B

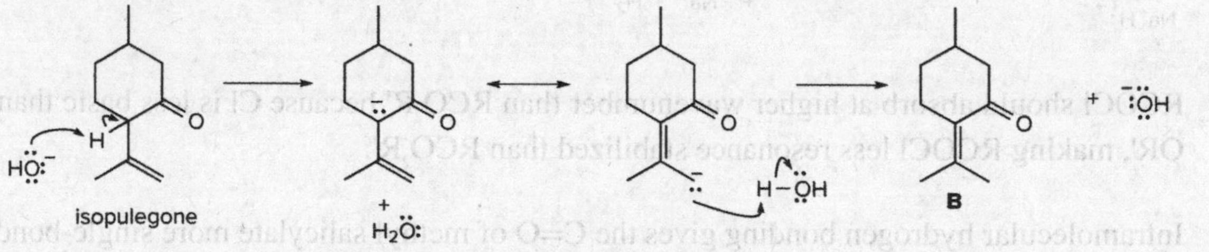

B.40 Because the ring O atom in **Y** is bonded to both a C=O and a C=C, two additional resonance structures can be drawn.

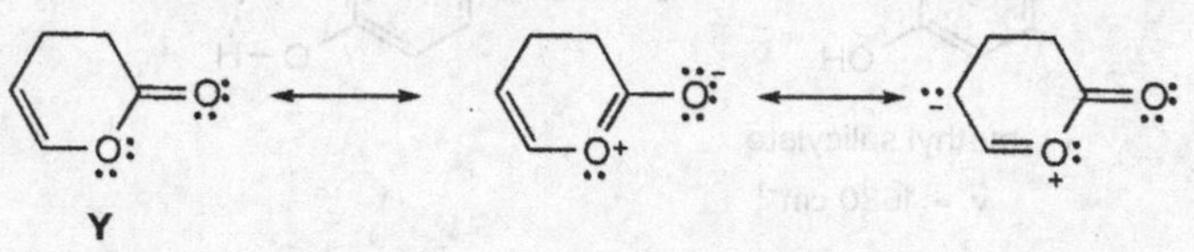

Y

This makes the lone pair on the ring O less available for donation to the C=O, so the C=O has less single-bond character than the C=O in **X.** As a result, **Y** absorbs at higher wavenumber than **X.**

Spectroscopy C Nuclear Magnetic Resonance Spectroscopy

Chapter Review

^{1}H NMR spectroscopy

[1] The **number of signals** equals the number of different types of protons (C.2).

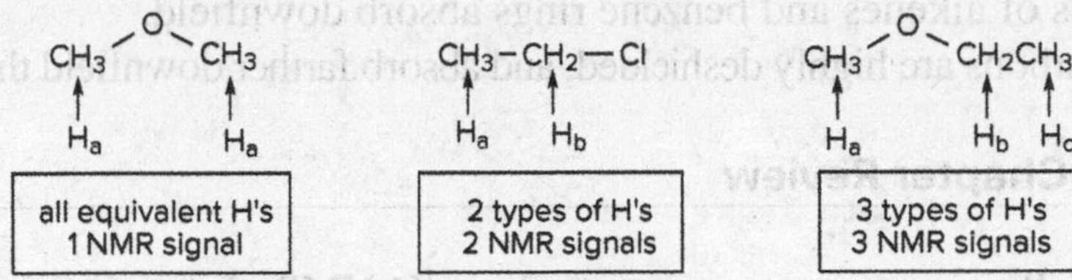

[2] The **position of a signal** (its chemical shift) is determined by shielding and deshielding effects.

- Shielding shifts an absorption upfield; deshielding shifts an absorption downfield.
- Electronegative atoms withdraw electron density, deshield a nucleus, and shift an absorption downfield (C.3).

This proton is shielded. Its absorption is upfield, 0.9–2 ppm.

This proton is deshielded. Its absorption is farther downfield, 2.5–4 ppm.

- Loosely held π electrons can either shield or deshield a nucleus. Protons on benzene rings and double bonds are deshielded and absorb downfield, whereas protons on triple bonds are shielded and absorb upfield (C.4).

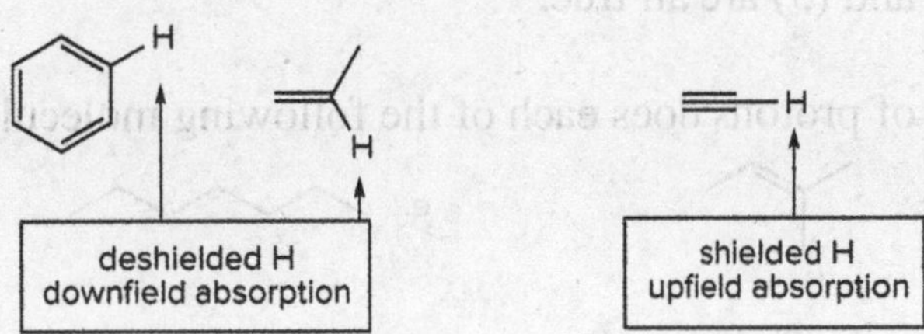

[3] The **area under an NMR signal** is proportional to the number of absorbing protons (C.5).

[4] **Spin–spin splitting** tells about nearby nonequivalent protons (C.6–C.8).

- Equivalent protons do not split each other's signals.
- A set of n nonequivalent protons on the same carbon or adjacent carbons split an NMR signal into $n + 1$ peaks.
- OH and NH protons do not cause splitting (C.9).
- When an absorbing proton has two sets of nearby nonequivalent protons that are equivalent to each other, use the $n + 1$ rule to determine splitting.
- When an absorbing proton has two sets of nearby nonequivalent protons that are not equivalent to each other, the number of peaks in the NMR signal = $(n + 1)(m + 1)$. In flexible alkyl chains, peak overlap often occurs, resulting in $n + m + 1$ peaks in an NMR signal.

^{13}C NMR spectroscopy (C.11)

[1] The number of signals equals the number of different types of carbon atoms. All signals are single lines.

[2] The relative position of ^{13}C signals is determined by shielding and deshielding effects.

- Carbons that are sp^3 hybridized are shielded and absorb upfield.
- Electronegative elements (N, O, and X) shift absorptions downfield.
- The carbons of alkenes and benzene rings absorb downfield.
- Carbonyl carbons are highly deshielded, and absorb farther downfield than other carbon types.

Practice Test on Chapter Review

1.a. Which of the following statements is true about ^{1}H NMR absorptions?
1. A signal that occurs at 1800 Hz on a 300 MHz NMR spectrometer occurs at 3000 Hz on a 500 MHz NMR spectrometer.
2. A signal that occurs at 3.3 ppm on a 60 MHz NMR absorbs at 198 Hz upfield from TMS.
3. A signal that occurs at 600 Hz is downfield from a signal that occurs at 800 Hz.
4. Statements (1) and (2) are both true.
5. Statements (1), (2), and (3) are all true.

b. Which of the following statements is true about ^{1}H NMR spectroscopy?
1. Electronegative elements shield a nucleus so an absorption shifts downfield.
2. A triplet is due to a proton that has four adjacent nonequivalent protons.
3. Circulating π electrons create a magnetic field that reinforces the applied field in the vicinity of the protons in benzene.
4. Statements (1) and (2) are both true.
5. Statements (1), (2), and (3) are all true.

2. How many different types of protons does each of the following molecules contain?

a. c. e. g. Cl Cl

b. O O Cl d. Cl Br Cl Br f. h. Cl H

3. Into how many peaks will each of the circled protons be split in a proton NMR spectrum?

a. O O H c. H e. H g. H Br H

b. H d. HO H OH f. O O H

4. How many lines are present in the ^{13}C NMR spectrum of each compound?

a. CH_3O c. Cl e.

b. d. O f.

5. With reference to the ^1H NMR absorptions in the following compound, (a) which proton absorbs farthest upfield; (b) which proton absorbs farthest downfield?

H_a H_b H_d O O H_c

6. With reference to the ^{13}C NMR absorptions in the following compound, (a) which carbon absorbs farthest downfield; (b) which carbon absorbs farthest upfield?

C_b O C_a F C_c C_d

Answers to Practice Test

1. a. 1	2.a. 5	3.a. 8	4.a. 6	5.a. H_b	6.a. C_c
b. 3	b. 5	b. 3	b. 4	b. H_c	b. C_a
	c. 4	c. 7	c. 6		
	d. 5	d. 4	d. 4		
	e. 4	e. 4	e. 4		
	f. 5	f. 3	f. 5		
	g. 3	g. 8			
	h. 9				

Answers to Problems

C.1 Use the formula δ = [observed chemical shift (Hz)/ν of the NMR (MHz)] to calculate the chemical shifts.

a. **CH_3 protons:**
δ = [1715 Hz] / [500 MHz]
= **3.43 ppm**

OH proton:
δ = [1830 Hz] / [500 MHz]
= **3.66 ppm**

b. The positive direction of the δ scale is downfield from TMS. The CH_3 protons absorb upfield from the OH proton.

C.2 Calculate the chemical shifts as in Answer C.1.

a. one signal:
δ = [1017 Hz] / [300 MHz]
= **3.39 ppm**

second signal:
δ = [1065 Hz] / [300 MHz]
= **3.55 ppm**

b. one signal:
3.39 = [x Hz] / [500 MHz]
x = **1695 Hz**

second signal:
3.55 = [x Hz] / [500 MHz]
x = **1775 Hz**

C.3 To determine if two H's are equivalent replace each by an atom X. If this yields the same compound or mirror images, the two H's are equivalent. Each kind of H will give one NMR signal.

a. 2 kinds of H's
2 NMR signals

c. 2 kinds of H's
2 NMR signals

e. 4 kinds of H's
4 NMR signals

g. 4 kinds of H's
4 NMR signals

b. 2 kinds of H's
2 NMR signals

d. 4 kinds of H's
4 NMR signals

f. 8 kinds of H's
8 NMR signals

h. 5 kinds of H's
5 NMR signals

C.4 Draw in all of the H's and compare them. If two H's are cis and trans to the same group, they are equivalent.

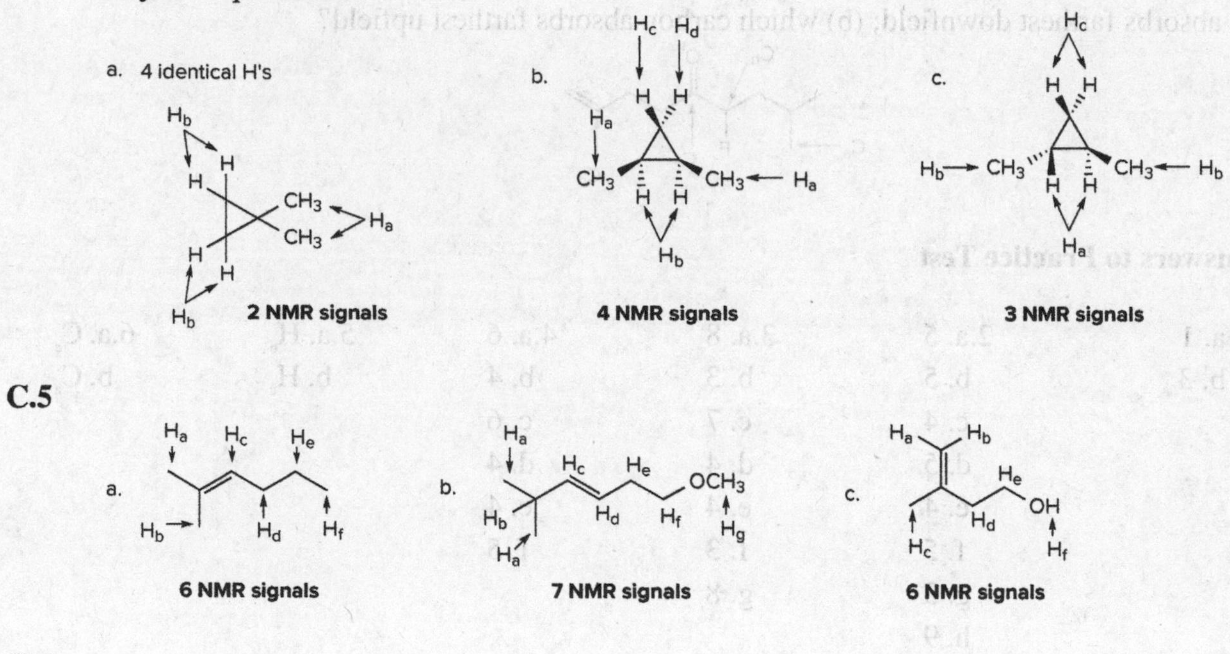

C.6

a. 3 NMR signals

b. 4 NMR signals

c. 6 NMR signals

d. 3 NMR signals

C.7 If replacement of H with X yields the same compound, the protons are **homotopic.** If replacement of H with X yields enantiomers, the protons are **enantiotopic.** If replacement of H with X yields diastereomers, the protons are **diastereotopic.** In general, if the compound has **one stereogenic center,** the protons in a CH_2 group are **diastereotopic.**

a. replacement of H with X

enantiomers = **enantiotopic H's**

c. Pick one configuration at the existing stereogenic center.

replacement of H with X

diastereomers = **diastereotopic H's**

b. replacement of either H with X

no stereogenic center
homotopic H's

d. replacement of H with X

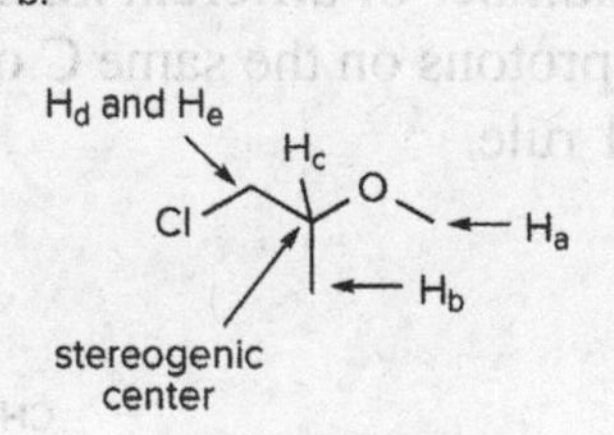
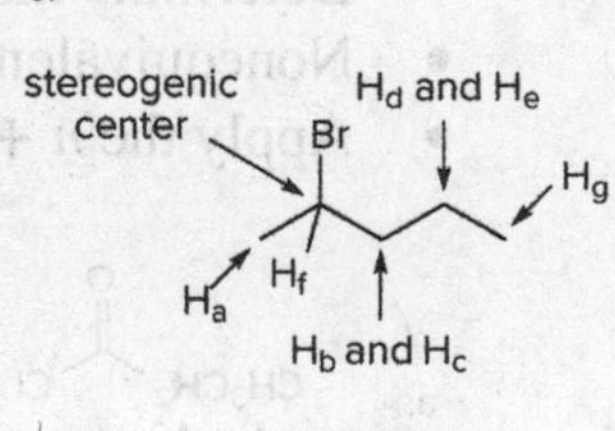

diastereomers = **diastereotopic H's**

C.8 **The two protons of a CH_2 group are different from each other if the compound has one stereogenic center. Replace one proton with X and compare the products.**

a. The stereogenic center makes the H's in the CH_2 group diastereotopic and therefore different from each other.

stereogenic center; H_a; H_b and H_c; H_d; H_e

5 NMR signals

b. H_d and H_e; H_c; H_a; H_b; stereogenic center

5 NMR signals

c. stereogenic center; H_d and H_e; H_g; H_f; H_a; H_b and H_c

7 NMR signals

C.9 **Decreased electron density deshields a nucleus and the absorption goes downfield. Absorption also shifts downfield with increasing alkyl substitution.**

a. $FCH_2CH_2CH_2Cl$ (the CH_2 adjacent to F in bold)
F is more electronegative than Cl. The CH_2 group adjacent to the F is more deshielded and the H's will absorb farther downfield.

b. $CH_3CH_2CH_2CH_2OCH_3$ (the CH_2 adjacent to O in bold)
The CH_2 group adjacent to the O will absorb farther downfield because it is closer to the electronegative O atom.

c. $CH_3OC(CH_3)_3$ (the CH_3 bonded to O in bold)
The CH_3 group bonded to the O atom will absorb farther downfield.

C.10

a. Cl–CH₂–CH₂–CH₂–Br (Ha, Hb, Hc labels)

3 types of protons:
$H_b < H_c < H_a$

b.

3 types of protons:
$H_c < H_a < H_b$

c.

3 types of protons:
$H_c < H_a < H_b$

C.11

a.

H_c protons are shielded because they are bonded to an sp^3 C.
H_a is shielded because it is bonded to an sp C.
H_b protons are deshielded because they are bonded to an sp^2 C.

$H_c < H_a < H_b$

b. CH_3 (Ha) C(=O) OCH_2CH_3 (Hb Hc)

H_c protons are shielded because they are bonded to an sp^3 C.
H_a protons are deshielded slightly because the CH_3 group is bonded to a C=O.
H_b protons are deshielded because the CH_2 group is bonded to an O atom.

$H_c < H_a < H_b$

C.12

An integration ratio of 2:3 means that there are two types of hydrogens in the compound and that the ratio of one type to another type is 2:3.

a. 2 types of H's
3:2 - YES

b. 2 types of H's
6:2 or 3:1 - no

c. 2 types of H's
6:4 or 3:2 - YES

d. 2 types of H's
6:4 or 3:2 - YES

C.13

To determine the **splitting pattern** for a molecule:

- Determine the number of different kinds of protons.
- Nonequivalent protons on the same C or adjacent C's split each other.
- Apply the $n + 1$ rule.

a. CH_3CH_2C(=O)Cl

H_a: 3 peaks: triplet
H_b: 4 peaks: quartet

b. CH_3–C(Br)(Br)–H

H_a: 2 peaks: doublet
H_b: 4 peaks: quartet

c. CH_3C(=O)CH_2CH_2Br

H_a: 1 peak: singlet
H_b: 3 peaks: triplet
H_c: 3 peaks: triplet

d.

H_a: 2 peaks: doublet
H_b: 2 peaks: doublet

e.

H_a: 2 peaks: doublet
H_b: 2 peaks: doublet

f. $ClCH_2$–CH(OCH₃)₂

H_a: 2 peaks: doublet
H_b: 3 peaks: triplet

C.14 Use the directions from Answer C.13.

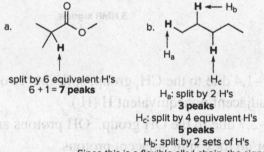

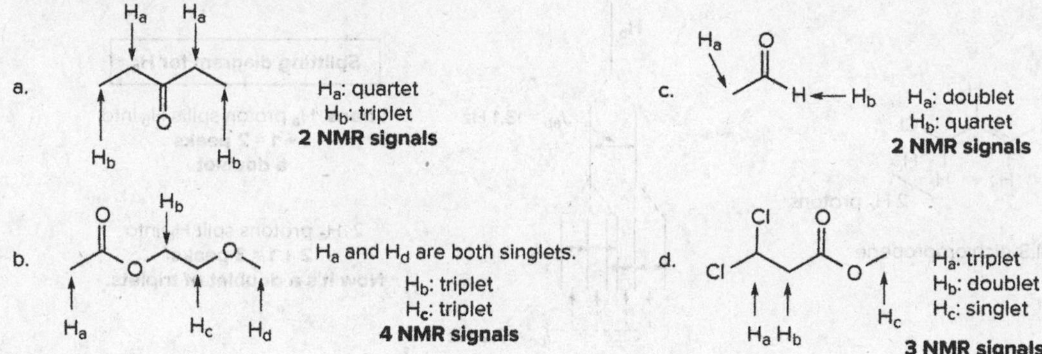

C.15 CH_3CH_2Cl

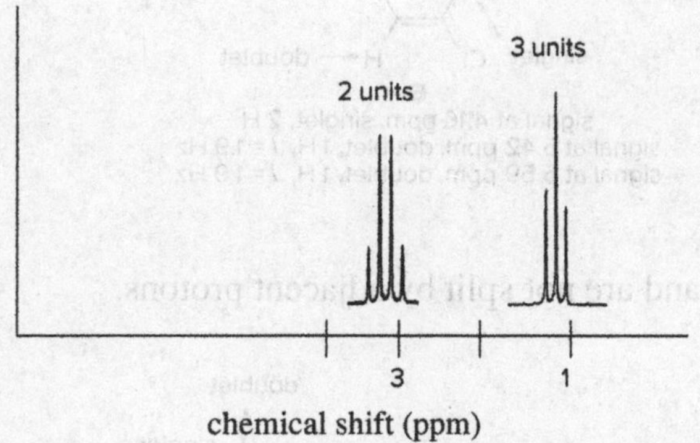

There are two kinds of protons, and they can split each other. The CH_3 signal will be split by the CH_2 protons into $2 + 1 = 3$ peaks. It will be upfield from the CH_2 protons because it is farther from the Cl. The CH_2 signal will be split by the CH_3 protons into $3 + 1 = 4$ peaks. It will be downfield from the CH_3 protons because the CH_2 protons are closer to the Cl. The ratio of integration units will be 3:2.

C.16

a. split by 6 equivalent H's
$6 + 1 = 7$ **peaks**

b. H_a: split by 2 H's
3 peaks
H_c: split by 4 equivalent H's
5 peaks
H_b: split by 2 sets of H's
Since this is a flexible alkyl chain, the signal due to H_b will have peak overlap, and $3 + 2 + 1 =$ **6 peaks** will likely be visible.

c. H_a: split by 1 H
2 peaks
H_b: split by 2 sets of H's
$(1 + 1)(2 + 1) =$ **6 peaks**

d. H_a: split by 2 different H's
$(1+1)(1+1) =$ **4 peaks**
H_b: split by 2 different H's
$(1+1)(1+1) =$ **4 peaks**
H_c: split by 2 different H's
$(1+1)(1+1) =$ **4 peaks**

C.17

a.

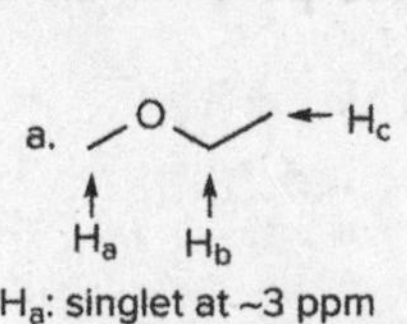

H_a: singlet at ~3 ppm
H_b: quartet at ~3.5 ppm
H_c: triplet at ~1 ppm

b.

H_a: triplet at ~1 ppm
H_b: quartet at ~2 ppm
H_c: septet at ~3.5 ppm
H_d: doublet at ~1 ppm

c.

H_a: singlet at ~3 ppm
H_b: triplet at ~3.5 ppm
H_c: quintet at ~1.5 ppm

d.

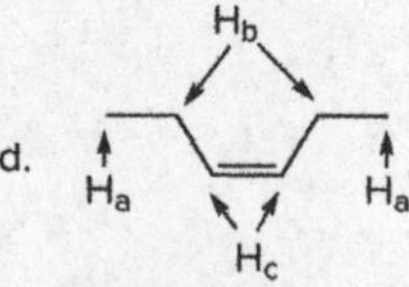

H_a: triplet at ~1 ppm
H_b: multiplet (8 peaks) at ~2.5 ppm
H_c: triplet at ~5 ppm

C.18

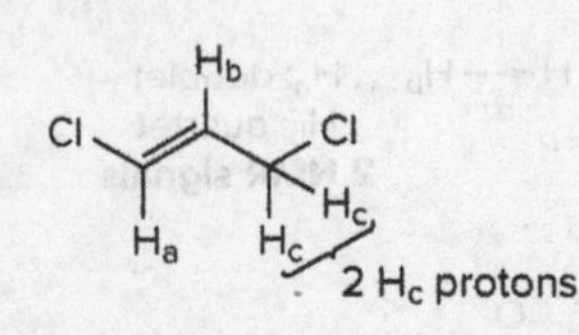

trans-1,3-dichloropropene

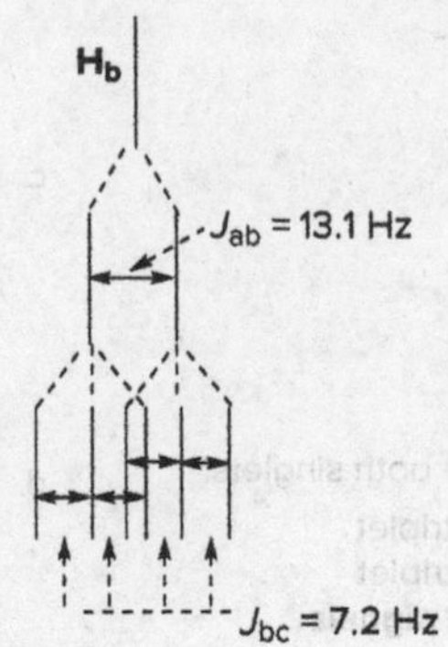

Splitting diagram for H_b

1 trans H_a proton splits H_b into
1 + 1 = 2 peaks
a doublet

2 H_c protons split H_b into
2 + 1 = 3 peaks
Now it's a doublet of triplets.

C.19

Cl, Cl, H ← H_b, ← H_a

$C_3H_4Cl_2$

A

H_a: 1.75 ppm, doublet, 3 H, J = 6.9 Hz
H_b: 5.89 ppm, quartet, 1 H, J = 6.9 Hz

Cl—, singlet, Cl, H ← doublet, H ← doublet

B

signal at 4.16 ppm, singlet, 2 H
signal at 5.42 ppm, doublet, 1 H, J = 1.9 Hz
signal at 5.59 ppm, doublet, 1 H, J = 1.9 Hz

C.20 OH (or NH) protons do not split other signals and are not split by adjacent protons.

a.

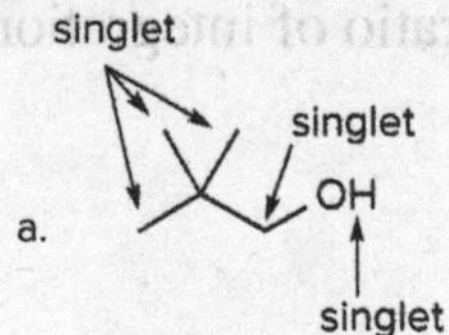

3 NMR signals

b. triplet, triplet, OH, singlet

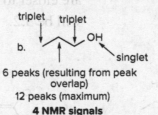

6 peaks (resulting from peak overlap)
12 peaks (maximum)
4 NMR signals

c. doublet, singlet, NH_2

7 peaks
3 NMR signals

C.21

OH ← H_b, ← H_a, H_c, H_d

5 H's on benzene ring

A

H_a: doublet at ~1.4 due to the CH_3 group, split into two peaks by one adjacent nonequivalent H (H_c).

H_b: singlet at ~2.7 due to the OH group. OH protons are not split by nor do they split adjacent protons.

H_c: quartet at ~4.7 due to the CH group, split into four peaks by the adjacent CH_3 group.

H_d: multiplets at ~7.2–7.4 due to five protons on the benzene ring.

C.22

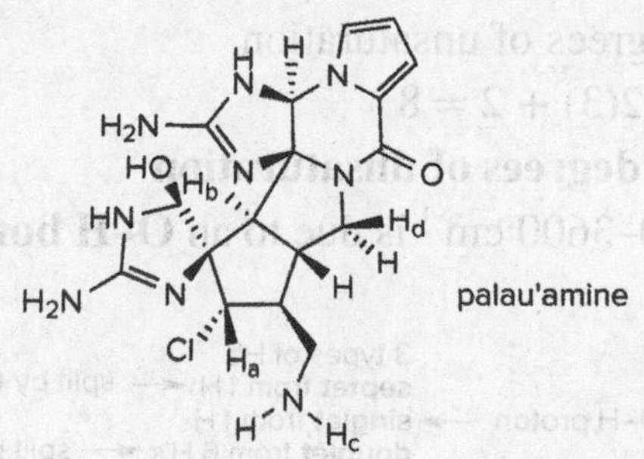

H_a: one adjacent nonequivalent H, so two peaks

H_b: one adjacent nonequivalent H, so two peaks

H_c: H_c is located on a N atom, so there is no splitting and it appears as one peak.

H_d: H_d has one nonequivalent H on the same carbon and one on an adjacent carbon, so it is split into $(1 + 1)(1 + 1) = 4$ peaks (a doublet of doublets).

C.23 Use these steps to propose a structure consistent with the molecular formula, IR, and NMR data.

- Calculate the **degrees of unsaturation.**
- Use the IR data to determine what types of **functional groups** are present.
- Determine the number of different **types of protons.**
- Calculate the **number of H's** giving rise to each signal.
- Analyze the **splitting pattern** and put the molecule together.
- Use the **chemical shift** information to check the structure.

- Molecular formula $C_7H_{14}O_2$

  $2n + 2 = 2(7) + 2 = 16$
  $16 - 14 = 2/2 =$ **1 degree of unsaturation**
  1 π bond or 1 ring

- IR peak at 1740 cm^{-1}

  C=O absorption is around 1700 cm^{-1} (causes the degree of unsaturation).
  No signal at 3200–3600 cm^{-1} means there is no O–H bond.

- NMR data:

absorption	ppm	relative area	
singlet	1.2	9	→ **9 H's**
triplet	1.3	3	→ **3 H's** (probably a CH_3 group)
quartet	4.1	2	→ **2 H's** (probably a CH_2 group)

  - 3 kinds of H's
  - number of H's per signal

    Because the sum of the relative areas equals the number of absorbing H's $(9 + 3 + 2 = 14)$, the relative area shows the actual number of absorbing H's: 9 H's, 3 H's and 2 H's.
  - splitting pattern

    The singlet (9 H) is likely from a *tert*-butyl group: $-C(CH_3)_3$

    The CH_3 and CH_2 groups split each other: CH_3-CH_2-

- Join the pieces together.

  [boxed structure: ethyl ester, $CH_3CH_2O-C(=O)-C(CH_3)_3$] or [$CH_3CH_2-C(=O)-O-C(CH_3)_3$]

  Pick this structure due to the chemical shift data.
  The CH_2 group is shifted downfield (4 ppm), so it is close to the electron-withdrawing O.

C.24

- Molecular formula: C_3H_8O ► Calculate degrees of unsaturation
  $2n + 2 = 2(3) + 2 = 8$
  $8 - 8 =$ **0 degrees of unsaturation**
- IR peak at 3200–3600 cm^{-1} ► Peak at 3200–3600 cm^{-1} is due to an **O–H bond.**
- NMR data:
  - doublet at ~1.2 (6 H)
  - singlet at ~2.2 (1 H)
  - septet at ~4 (1 H)

3 types of H's
septet from 1 H ← split by 6 H's
from the O–H proton → **singlet** from 1 H
doublet from 6 H's ← split by 1 H

► Put information together:

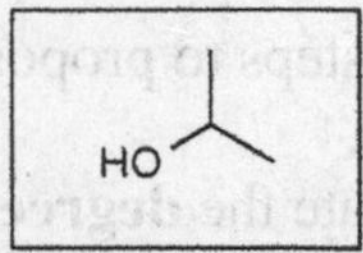

C.25 Each different kind of carbon atom will give a different ^{13}C NMR signal.

a.
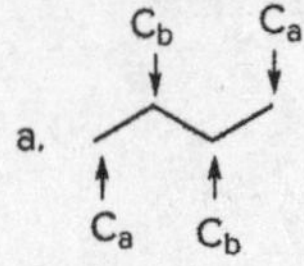

2 kinds of C's
2 ^{13}C NMR signals

b.
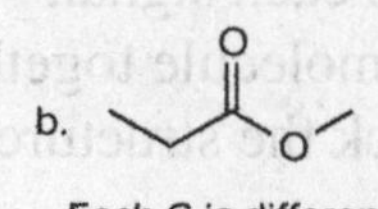

Each C is different.
4 kinds of C's
4 ^{13}C NMR signals

c.
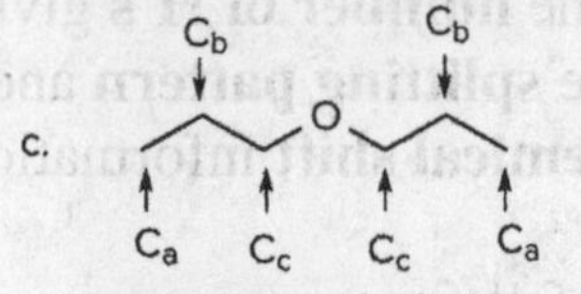

same groups on both sides of O
3 kinds of C's
3 ^{13}C NMR signals

d.
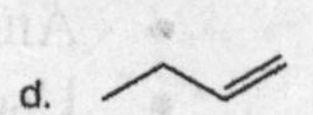

Each C is different.
4 kinds of C's
4 ^{13}C NMR signals

C.26

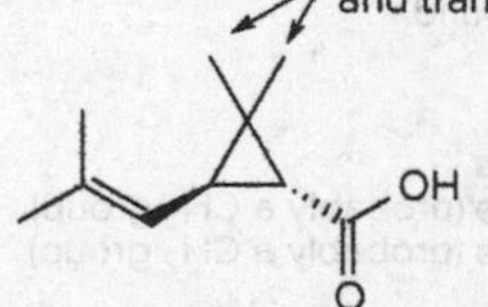

These 2 C's are different because they are cis and trans to different groups.

Every carbon is different so there are 10 lines for the 10 C atoms.

C.27 Electronegative elements shift absorptions downfield. The carbons of alkenes, benzene rings, and carbonyl groups are also shifted downfield.

a. The C closer to the electronegative O will be farther downfield.

b. The C of the CBr_2 group has two bonds to electronegative Br atoms and will be farther downfield.

c. The carbonyl carbon is highly deshielded and will be farther downfield.

d.
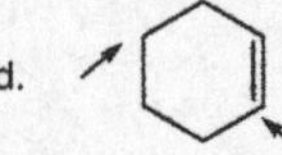

The C atom that is part of the double bond will be farther downfield.

C.28

a. In order of lowest to highest chemical shift:

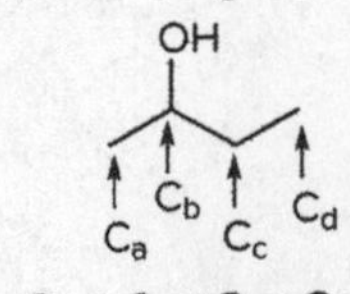

$C_d < C_a < C_c < C_b$

b. In order of lowest to highest chemical shift:

Ca Cb Cc Cb Ca

$C_a < C_b < C_c$

C.29

- molecular formula $C_4H_8O_2$
  $2n + 2 = 2(4) + 2 = 10$
  $10 - 8 = 2/2 =$ **1 degree of unsaturation**
- no IR peaks at 3200–3600 or 1700 cm^{-1}
  no O–H or C=O
- ^1H NMR spectrum at 3.69 ppm
  only one kind of proton
- ^{13}C NMR spectrum at 67 ppm
  only one kind of carbon

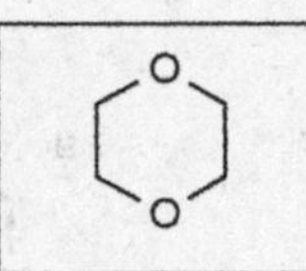

This structure satisifies all the data. The ring is one degree of unsaturation. All carbons and protons are identical.

C.30

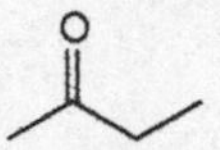

- molecular formula C_4H_8O
  $2n + 2 = 2(4) + 2 = 10$
  $10 - 8 = 2/2 =$ **1 degree of unsaturation**
- ^{13}C NMR signal at > 160 ppm due to C=O

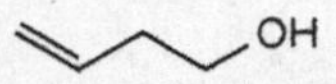

- molecular formula C_4H_8O
  $2n + 2 = 2(4) + 2 = 10$
  $10 - 8 = 2/2 =$ **1 degree of unsaturation**
- all ^{13}C NMR signals at < 160 ppm
  NO C=O

C.31

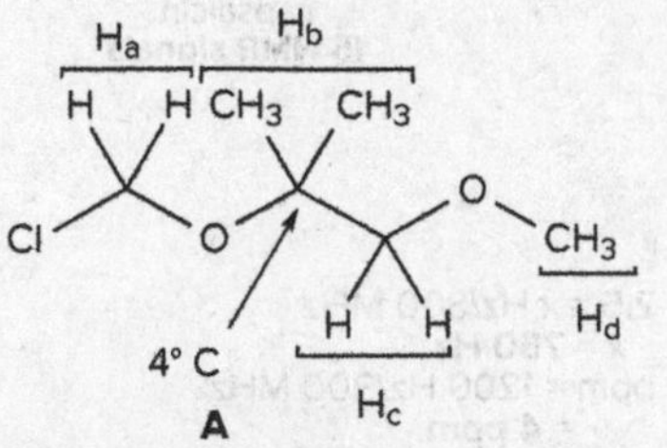

A

a. 4 ^1H NMR signals
b. 5 ^{13}C NMR signals (including the 4° C)

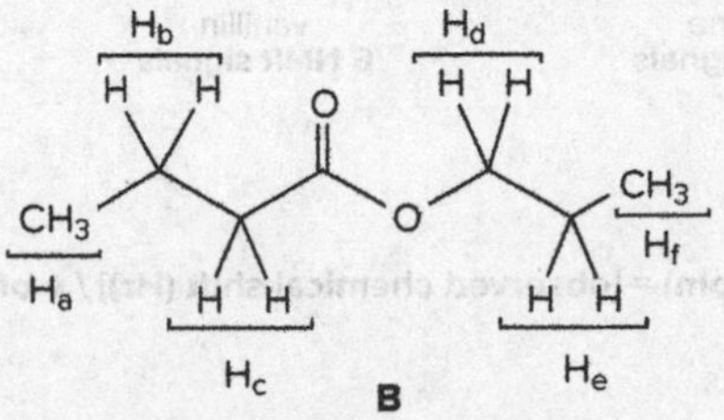

B

a. 6 ^1H NMR signals
b. 7 ^{13}C NMR signals (including the carbonyl C)

C.32

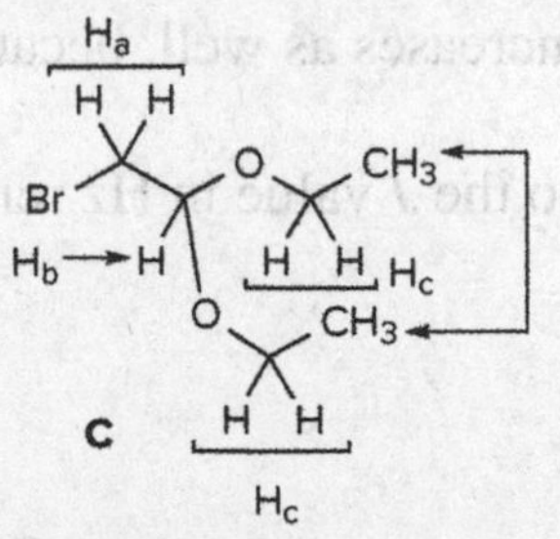

C

a. 4 ^1H NMR signals
b. H_a: 1 adjacent H, so 2 peaks
H_b: 2 adjacent H's, so 3 peaks
H_c: 3 adjacent H's, so 4 peaks
H_d: 2 adjacent H's, so 3 peaks

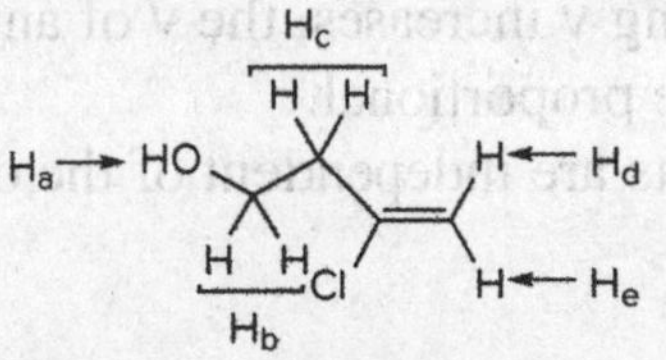

D

a. 5 ^1H NMR signals
b. H_a: singlet
H_b: 2 adjacent H's, so 3 peaks
H_c: 2 adjacent H's, so 3 peaks
H_d: 1 nonequivalent H on the same C, so 2 peaks
H_e: 1 nonequivalent H on the same C, so 2 peaks

C.33 Use the directions from Answer C.3.

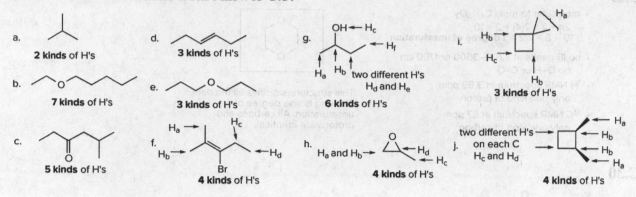

a. 2 kinds of H's

b. 7 kinds of H's

c. 5 kinds of H's

d. 3 kinds of H's

e. 3 kinds of H's

f. H_a, H_b, H_c, H_d, Br — 4 kinds of H's

g. OH ← H_c, H_f, H_a, H_b; two different H's H_d and H_e — 6 kinds of H's

h. H_a and H_b, H_c, H_d — 4 kinds of H's

i. H_a, H_b, H_c, H_b — 3 kinds of H's

j. two different H's on each C H_c and H_d; H_a, H_b, H_b, H_a — 4 kinds of H's

C.34

a. caffeine — **4 NMR signals**

b. vanillin — **6 NMR signals**

c. thymol — equivalent — **7 NMR signals**

d. capsaicin — equivalent — **15 NMR signals**

C.35

δ (in ppm) = [observed chemical shift (Hz)] / ν of the NMR (MHz)]

a. 2.5 = x Hz/300 MHz
x = 750 Hz

b. ppm = 1200 Hz/300 MHz
= 4 ppm

c. 2.0 = x Hz/300 MHz
x = 600 Hz

C.36 a. The chemical shift in δ is independent of the operating frequency, so there is no change in δ when the ν is increased.

b. When the operating ν increases, the ν of an absorption increases as well, because the two quantities are proportional.

c. Coupling constants are independent of the operating ν, so the *J* value in Hz remains the same.

C.37 Use the directions from Answer C.9.

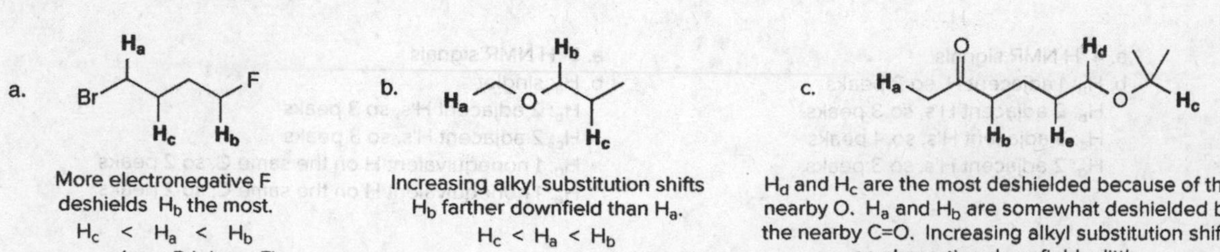

a. More electronegative F deshields H_b the most.
H_c < H_a < H_b
(near Br) (near F)

b. Increasing alkyl substitution shifts H_b farther downfield than H_a.
H_c < H_a < H_b

c. H_d and H_c are the most deshielded because of the nearby O. H_a and H_b are somewhat deshielded by the nearby C=O. Increasing alkyl substitution shifts an absorption downfield a little.
H_e < H_a < H_b < H_d < H_c

C.38

a. H_a protons split by 1 H = **doublet**
H_b proton split by 3 H's = **quartet**

b. both CH_2 groups split each other = **triplets**

c. H_a protons split by 1 H = **doublet**
H_b proton split by 2 H's = **triplet**

d. H_a protons split by 1 H = **doublet**
H_b proton split by 6 H's = **septet**
H_c protons split by 3 H's = **quartet**
H_d protons split by 2 H's = **triplet**

e. H_a protons split by 2 CH_2 groups = **quintet**
H_b protons split by 2 H's = **triplet**

f. H_a protons split by 2 H's = **triplet**
H_b protons split by CH_3 + CH_2 protons = **12 peaks** (maximum)
H_c protons split by 2 different CH_2 groups = **9 peaks** (maximum)
H_d protons split by 2 H's = **triplet**
Since H_b and H_c are located in a flexible alkyl chain, it is likely that peak overlap occurs, so that the following is observed: H_b (3 + 2 + 1 = 6 peaks), H_c (2 + 2 + 1 = 5 peaks).

g. H_a protons split by 2 H's = **triplet**
H_c protons split by 2 H's = **triplet**
H_b protons split by CH_3 + CH_2 protons = **12 peaks** (maximum)
Since H_b is located in a flexible alkyl chain, it is likely that peak overlap occurs, so that only 3 + 2 + 1 = 6 peaks will be observed.

h. H_a: split by CH_3 group + H_b = **8 peaks** (maximum)
H_b: split by 2 H's = **triplet**

i. H_a: split by 1 H = **doublet**
H_b: split by 1 H = **doublet**

j. H_a: split by H_b + H_c = **doublet of doublets** (4 peaks)
H_b: split by H_a + H_c = **doublet of doublets** (4 peaks)
H_c: split by CH_3, H_a + H_b = **16 peaks**

C.39

J_{ab} = 11.8 Hz
J_{bc} = 0.9 Hz
J_{ac} = 18 Hz

H_a: doublet of doublets at 5.7 ppm. Two large *J* values are seen for the H's cis (J_{ab} = 11.8 Hz) and trans (J_{ac} = 18 Hz) to H_a.
H_b: doublet of doublets at ~6.2 ppm. One large *J* value is seen for the cis H (J_{ab} = 11.8 Hz). The geminal coupling (J_{bc} = 0.9 Hz) is hard to see.
H_c: doublet of doublets at ~6.6 ppm. One large *J* value is seen for the trans H (J_{ac} = 18 Hz). The geminal coupling (J_{bc} = 0.9 Hz) is hard to see.

Splitting diagram for H_a

1 trans H_c proton splits H_a into **1 + 1 = 2 peaks a doublet**

1 cis H_b proton splits H_a into **1 + 1 = 2 peaks**
Now it's a doublet of doublets.

J_{ac} = the coupling constant between H_a and H_c

J_{ab} = the coupling constant between H_a and H_b

C.40

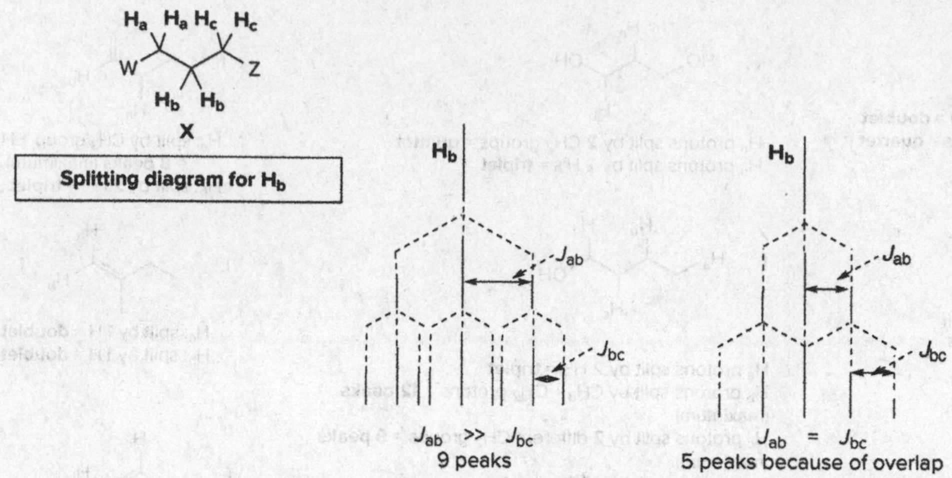

C.41

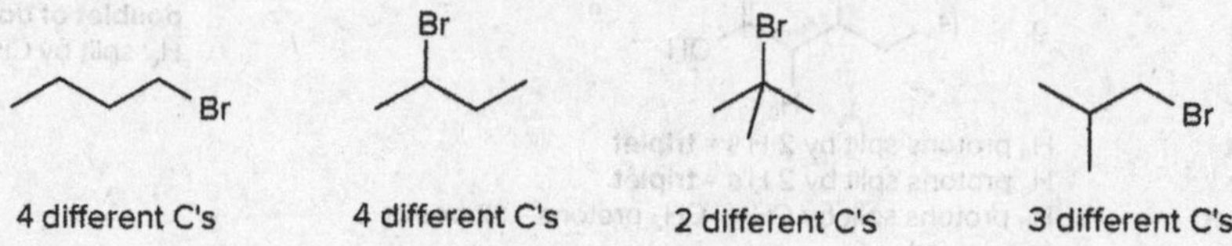

C.42 The O atom of an ester donates electron density, so the carbonyl carbon has less δ+, making it less deshielded than the carbonyl carbon of an aldehyde or ketone. Therefore, the carbonyl carbon of an aldehyde or ketone is more deshielded and absorbs farther downfield.

C.43

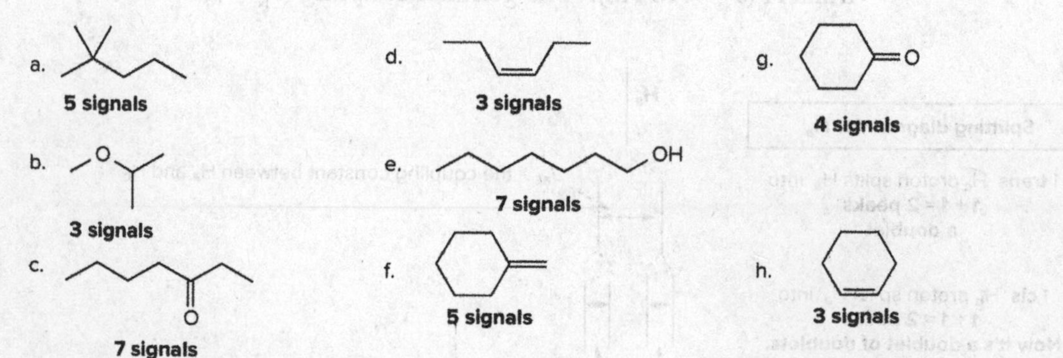

C.44

a. C_a C_b C_c OH

$C_a < C_b < C_c$

b. Br C_a C_b C_c

$C_b < C_c < C_a$

C.45

a. $CH_3CH_2CH_2CH_2OH$ — 14 ppm, 19 ppm, 35 ppm, 62 ppm

b. $(CH_3)_2CHCHO$ — 16 ppm, 41 ppm, 205 ppm

c. CH_2=CHCH(OH)CH_3 — 113 ppm, 143 ppm, 69 ppm, 23 ppm

C.46

H_b H_c H_d H_e H_a **A**

H_c H_d H_e H_b H_a **B**

H_b H_c H_d H_e H_a **C**

The aromatic region for A, B, and C would look similar, so we must concentrate on the signals due to the groups on the benzene ring. The NMR of C would show three singlets for H_a, H_d, and H_e, whereas both A and B have a singlet, triplet, and quartet in their spectra. The locations of these signals differ in A and B. For example, in A, H_a appears as a singlet at 3–4 ppm, whereas in B, H_b appears as a quartet at 3–4 ppm.

C.47 **One distinguishing signal is listed for each pair of compounds.**

a. and

No H's are bonded to C's that are also bonded to Cl's.

This H is downfield because it is bonded to a C with 2 Cl's. It appears as a doublet. There is no similar absorption in the other compound.

b. and

quartet at 3–4 ppm

triplet at 3–4 ppm

c. and

singlet at 3–4 ppm

triplet at 3–4 ppm

C.48

H_a H_b H_c H_d H_e H_f H_g

L

a. 7 ^{1}H NMR signals

b. splitting: H_a: singlet
H_b: singlet
H_c: triplet
H_d: triplet
H_e: singlet
H_f: doublet
H_g: doublet

c. C_a C_b C_c C_d C_e C_f C_g C_h

8 lines in the ^{13}C NMR spectrum.

Br H_a H_a Br H_b H_c

M

a. 3 ^{1}H NMR signals

b. splitting:

H_a: (1 + 1)(1 + 1) = 4 peaks due to nonequivalent protons H_b and H_c

H_b: (1 + 1)(2 + 1) = 6 peaks due to two types of nonequivalent H's, H_c and H_a

H_c: (1 + 1)(2 + 1) = 6 peaks due to two types of nonequivalent H's, H_b and H_a

c. Br C_a C_a Br C_b

2 lines in the ^{13}C NMR spectrum.

H_a H_b H_c Cl H_d H_e H_f

N

a. 6 ^{1}H NMR signals

b. splitting: H_a: 3 peaks
H_b: 4 peaks
H_c: singlet
H_d: 2 peaks
H_e: 2 peaks
H_f: singlet

c. All C's are differnt, so there are 10 lines in the ^{13}C NMR spectrum.

C.49

1-hydroxybutan-2-one
A

4-hydroxybutan-2-one
B

The answers for parts (a)–(d) are the same for both compounds.

a. molecular ion for $C_4H_8O_2$ = 88

b. IR absorptions at 3200–3600 (OH), ~3000 (CH), and ~1700 (C=O) cm^{-1}.

c. Four lines in ^{13}C NMR spectrum

d. Four signals in ^{1}H NMR spectrum

e.

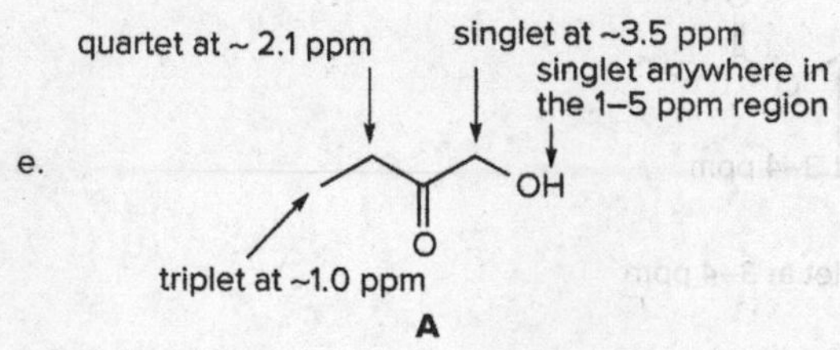

A

triplet at ~3.5 ppm

singlet anywhere in the 1–5 ppm region

singlet at ~2.0 ppm

HO

triplet at ~2.1 ppm

B

C.50 Use the directions from Answer C.23.

a. **$C_4H_8Br_2$:** 0 degrees of unsaturation
IR peak at 3000–2850 cm^{-1}: **Csp^3–H bonds**
NMR: singlet at 1.87 ppm (6 H) (2 CH_3 groups)
singlet at 3.86 ppm (2 H) (CH_2 group)

Br
Br

b. **$C_3H_6Br_2$:** 0 degrees of unsaturation
IR peak at 3000–2850 cm^{-1}: **Csp^3–H bonds**
NMR: quintet at 2.4 ppm (split by 2 CH_2 groups)
triplet at 3.5 ppm (split by 2 H's)

Br Br

c. **$C_5H_{10}O_2$: 1 degree of unsaturation**
IR peak at 1740 cm^{-1}: **C=O**
NMR: triplet at 1.15 ppm (3 H) (CH_3 split by 2 H's)
triplet at 1.25 ppm (3 H) (CH_3 split by 2 H's)
quartet at 2.30 ppm (2 H) (CH_2 split by 3 H's)
quartet at 4.72 ppm (2 H) (CH_2 split by 3 H's)

O
O

d. **C_3H_6O: 1 degree of unsaturation**
IR peak at 1730 cm^{-1}: **C=O**
NMR: triplet at 1.11 ppm
multiplet at 2.46 ppm
triplet at 9.79 ppm

O
H

C.51 IR absorptions:

3088–2897 cm^{-1}: sp^2 and sp^3 hybridized C–H
1740 cm^{-1}: C=O
1606 cm^{-1}: benzene ring

triplet at 2.91
5 H multiplet 7.20–7.35
CH_3 ← singlet at 2.02
triplet at 4.25

$C_{10}H_{12}O_2$
W

C.52 IR absorption at 1713 cm^{-1} is due to C=O.

doublet at 1.09
triplet at 2.43
CH_3 ← triplet at 0.91
septet at 2.60
multiplet at 1.6

$C_7H_{14}O$
V

C.53

Compound C:
molecular ion 146 (molecular formula $C_6H_{10}O_4$)
IR absorption at 1762 cm^{-1}: **C=O**
^{1}H NMR data:
H_a: doublet at 1.47 (3 H) (CH_3 group adjacent to CH)
H_b: singlet at 2.07 (6 H) (2 CH_3 groups)
H_c: quartet at 6.84 (1 H adjacent to CH_3)

H_c
CH_3–C(=O)–O–C(H)(CH_3)–O–C(=O)–CH_3
H_b H_a H_b

C.54

HO–CH₂CH₂–C≡N

Compound **D:**
Molecular ion at m/z = 71: C_3H_5NO (possible formula)
IR absorption at 3600–3200 cm^{-1} → OH
2263 cm^{-1} → CN
^{1}H NMR signals at (ppm):
- 2.5 (triplet, 2 H) CH_2 adjacent to 2 H's
- 3.1 (singlet, 1 H) OH
- 3.8 (triplet, 2 H) CH_2 adjacent to 2 H's

HO–CH₂CH₂CH₂–NH₂

Compound **E:**
Molecular ion at m/z = 75: C_3H_9NO (possible formula)
IR absorption at 3600–3200 cm^{-1} → OH
3636 cm^{-1} → N–H of amine
^{1}H NMR signals at (ppm):
- 1.6 (quintet, 2 H) CH_2 split by 2 CH_2's
- 2.5 (singlet, 3 H) NH_2 and OH
- 2.8 (triplet, 2 H) CH_2 split by CH_2
- 3.7 (triplet, 2 H) CH_2 split by CH_2

C.55

a. **Compound E:**
$C_4H_8O_2$:
1 degree of unsaturation
IR absorption at 1743 cm^{-1}: **C=O**
NMR data:
- H_a: quartet at 4.1 (**2 H**)
- H_b: singlet at 2.0 (**3 H**)
- H_c: triplet at 1.4 (**3 H**)

CH_3–C(=O)–OCH_2CH_3 (CH_3 = H_b; OCH_2 = H_a; CH_3 = H_c)

b. **Compound F:**
$C_4H_8O_2$:
1 degree of unsaturation
IR absorption at 1730 cm^{-1}: **C=O**
NMR data:
- H_a: singlet at 4.1 (**2 H**)
- H_b: singlet at 3.4 (**3 H**)
- H_c: singlet at 2.1 (**3 H**)

CH_3–C(=O)–CH_2OCH_3 (CH_3 = H_c; CH_2 = H_a; OCH_3 = H_b)

C.56

a. **Compound H:**
$C_8H_{11}N$:
4 degrees of unsaturation
IR absorptions at 3365 cm^{-1}: N–H
3284 cm^{-1}: N–H
3026 cm^{-1}: Csp^2–H
2932 cm^{-1}: Csp^3–H
1603 cm^{-1}: due to benzene
1497 cm^{-1}: due to benzene

NMR data:
- multiplet at 7.2–7.4 ppm, **5 H** on a benzene ring
- H_a: triplet at 2.9 ppm, **2 H,** split by 2 H's
- H_b: triplet at 2.8 ppm, **2 H,** split by 2 H's
- H_c: singlet at 1.1 ppm, **2 H,** no splitting (on NH_2)

C_6H_5–CH_2(H_b)–CH_2(H_a)–NH_2(H_c)

b. **Compound I:**
$C_8H_{11}N$:
4 degrees of unsaturation
IR absorptions at 3367 cm^{-1}: N–H
3286 cm^{-1}: N–H
3027 cm^{-1}: Csp^2–H
2962 cm^{-1}: Csp^3–H
1604 cm^{-1} due to benzene
1492 cm^{-1}: due to benzene

NMR data:
- multiplet at 7.2–7.4 ppm, **5 H** on a benzene ring
- H_a: quartet at 4.1 ppm, **1 H,** split by 3H's
- H_b: singlet at 1.45 ppm, **2 H,** no splitting (NH_2)
- H_c: doublet at 1.4 ppm, **3 H,** split by 1 H

C_6H_5–CH(H_a)(CH_3, H_c)–NH_2(H_b)

C.57

a. $C_9H_{10}O_2$:
5 degrees of unsaturation
IR absorption at 1718 cm^{-1}: **C=O**
NMR data:

multiplet at 7.4–8.1 ppm, **5 H** on a benzene ring
quartet at 4.4 ppm, **2 H**, split by 3 H's
triplet at 1.3 ppm, **3 H**, split by 2 H's

$C_6H_5C(=O)OCH_2CH_3$

downfield due to the O atom

b. C_9H_{12}:
4 degrees of unsaturation
IR absorption at 2850–3150 cm^{-1}: **C–H bonds**
NMR data:

singlet at 7.1–7.4 ppm, **5 H**, benzene
septet at 2.8 ppm, **1 H**, split by 6 H's
doublet at 1.3 ppm, **6 H**, split by 1 H

C.58 IR absorption at 1730 cm^{-1} is due to a C=O. Eight lines in the ^{13}C NMR spectrum means there are eight different types of C.

triplet at 4.20

singlet at 9.97

$(CH_3)_2N$

singlet at 2.32

triplet at 3.05

H_a and H_b appear as two doublets.

C.59

The IR shows an OH absorption at 3200–3600 cm^{-1}.

1 H: doublet of doublets at 6.0 ppm

Br

H_2O

B

OH peak at 1.5 ppm

6 H: singlet at 1.3 ppm

Each H is a doublet of doublets in the 4.9–5.2 ppm region.

C.60

D: Molecular ion at m/z = 150: $C_9H_{10}O_2$ (possible molecular formula)
5 degrees of unsaturation
IR absorption at 1692 cm^{-1} → C=O
NMR data (ppm):

triplet at 1.5 (3 H's – CH_3CH_2)
quartet at 4.1 (2 H's – CH_3CH_2)
doublet at 7.0 (2 H's – on benzene ring)
doublet at 7.8 (2 H's – on benzene ring)
singlet at 9.9 (1 H – on aldehyde)

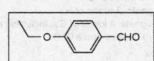

CHO

C.61

Compound **L** has a molecular ion at 90: molecular formula $\mathbf{C_4H_{10}O_2}$

0 degrees of unsaturation

IR absorptions at 2992 and 2941 cm^{-1}: Csp^3–H

^{1}H NMR data (ppm):

H_a: 1.2 (doublet, 3 H), split by 1 H

H_b: 3.3 (singlet, 6 H), due to 2 CH_3 groups

H_c: 4.8 (quartet, 1 H), split by 3 adjacent H's

+ CH_3—OH → H^+ → H_c, H_a, H_b, H_b

L

C.62

Compound **O** has a molecular formula $\mathbf{C_{10}H_{12}O}$.

5 degrees of unsaturation

IR absorption at 1687 cm^{-1}

^{1}H NMR data (ppm):

H_a: 1.0 (triplet, 3 H), due to CH_3 group, split by 2 adjacent H's

H_b: 1.7 (sextet, 2 H), split by CH_3 and CH_2 groups

H_c: 2.9 (triplet, 2 H), split by 2 H's

7.4–8.0 (multiplet, 5 H), benzene ring

O H_c H_b H_a

O

C.63

Compound **P** has a molecular formula $\mathbf{C_5H_9ClO_2}$.

1 degree of unsaturation

^{13}C NMR shows 5 different C's, including a C=O.

^{1}H NMR data (ppm):

H_a: 1.3 (triplet, 3 H), split by 2 H's

H_b: 2.8 (triplet, 2 H), split by 2 H's

H_c: 3.7 (triplet, 2 H), split by 2 H's

H_d: 4.2 (quartet, 2 H), split by CH_3 group

Cl H_c H_b O O H_d H_a

P

C.64

Compound Q: Molecular ion at 86.

Molecular formula: $\mathbf{C_5H_{10}O}$:

1 degree of unsaturation

IR absorption at ~1700 cm^{-1}: **C=O**

NMR data:

H_a: doublet at 1.1 ppm, 2 CH_3 groups split by 1 H

H_b: singlet at 2.1 ppm, CH_3 group

H_c: septet at 2.6 ppm, 1 H split by 6 H's

[1] strong base

[2] CH_3I

H_b O H_c H_a H_a = **Q**

MW = 86

C.65

$\mathbf{C_6H_{12}O_2}$:

1 degree of unsaturation

IR peak at 1740 cm^{-1}: **C=O**

^{1}H NMR 2 signals: 2 types of H's

^{13}C NMR: 4 signals: 4 kinds of C's, including one at ~170 ppm due a C=O

C.66 A second resonance structure for *N,N*-dimethylformamide places the two CH_3 groups in different environments. One CH_3 group is cis to the O atom, and one is cis to the H atom. This gives rise to two different absorptions for the CH_3 groups.

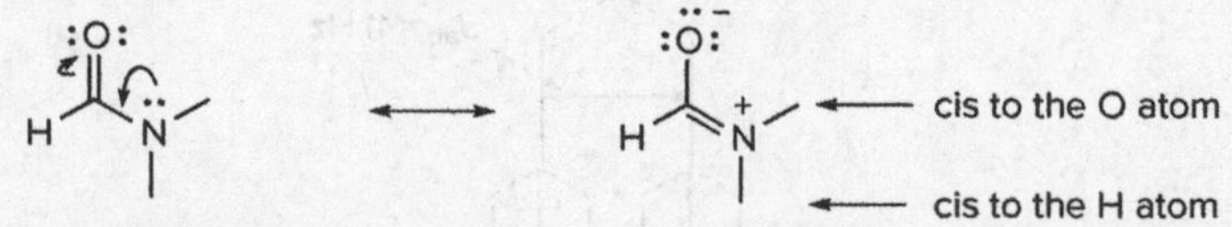

cis to the O atom

cis to the H atom

N,N-dimethylformamide

C.67

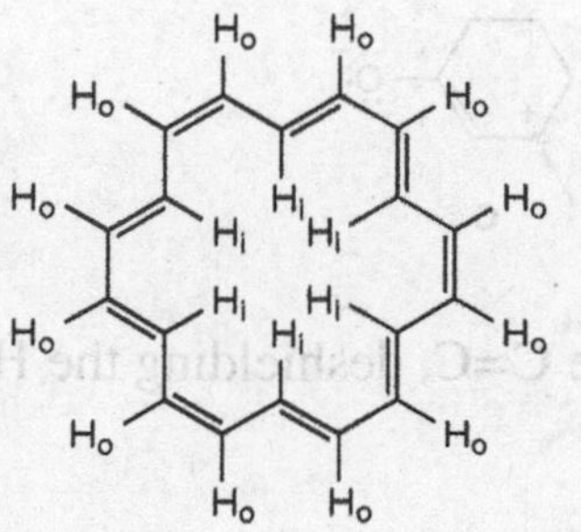

18-Annulene has 18 π electrons that create an **induced magnetic field** similar to the 6 π electrons of benzene. 18-Annulene has 12 protons that are oriented on the outside of the ring (labeled H_o), and 6 protons that are oriented inside the ring (labeled H_i). The induced magnetic field reinforces the external field in the vicinity of the protons on the outside of the ring. These H_o protons are deshielded, so they absorb downfield (8.9 ppm). In contrast, the induced magnetic field is opposite in direction to the applied magnetic field in the vicinity of the protons on the inside of the ring. This shields the H_i protons, so they absorb very far upfield, at –1.8 ppm, which is even higher than TMS.

C.68

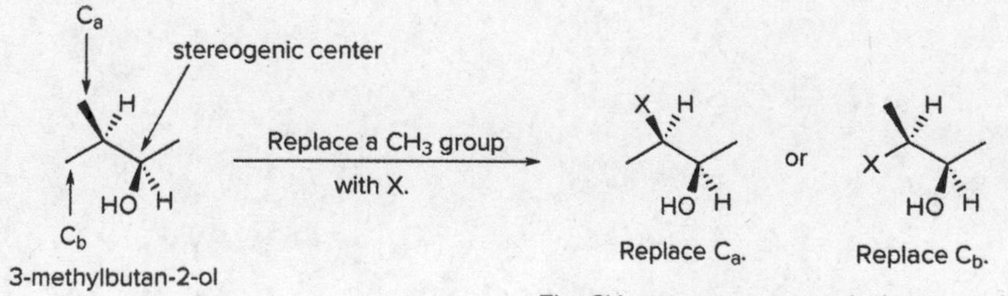

stereogenic center

Replace a CH_3 group with X.

3-methylbutan-2-ol

Replace C_a. or Replace C_b.

The CH_3 groups are not equivalent to each other, because replacement of each by X forms two diastereomers.

Thus, every C in this compound is different and there are five ^{13}C signals.

C.69

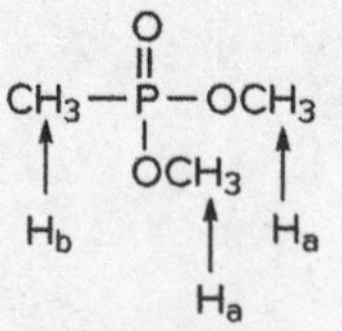

One P atom splits each nearby CH_3 into a doublet by the *n* + 1 rule, making two doublets.

All 6 H_a protons are equivalent.

C.70 a. **Splitting pattern:**

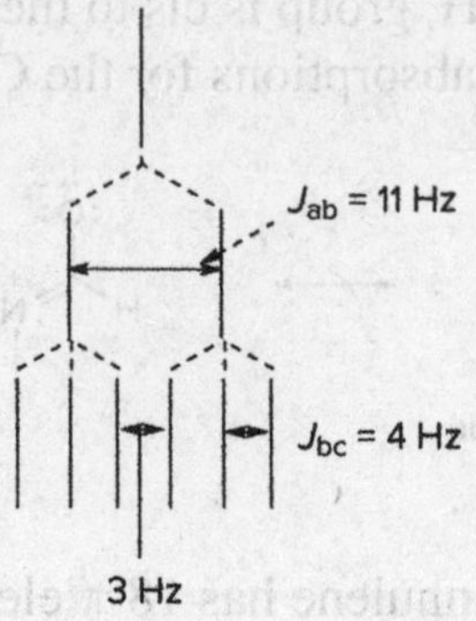

J_{ab} = 11 Hz

J_{bc} = 4 Hz

3 Hz

b. **Three resonance structures can be drawn for cyclohex-2-enone.**

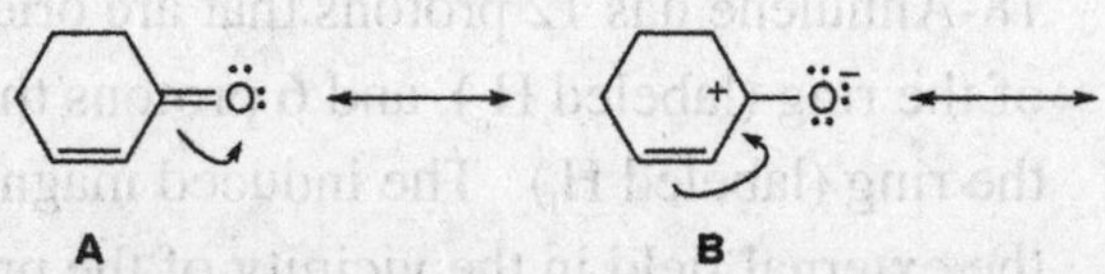

A B C

Resonance structure C places a (+) charge on one C of the C=C, deshielding the H attached to it and shifting the absorption downfield.

Chapter 13 Introduction to Carbonyl Chemistry; Organometallic Reagents; Oxidation and Reduction

Chapter Review

Reduction reactions

[1] Reduction of aldehydes and ketones to 1° and 2° alcohols (13.4)

R' = H or alkyl

$NaBH_4$, CH_3OH
or
[1] $LiAlH_4$ [2] H_2O
or
H_2, Pd-C

1° or 2° alcohol

[2] Reduction of acid chlorides (13.7A)

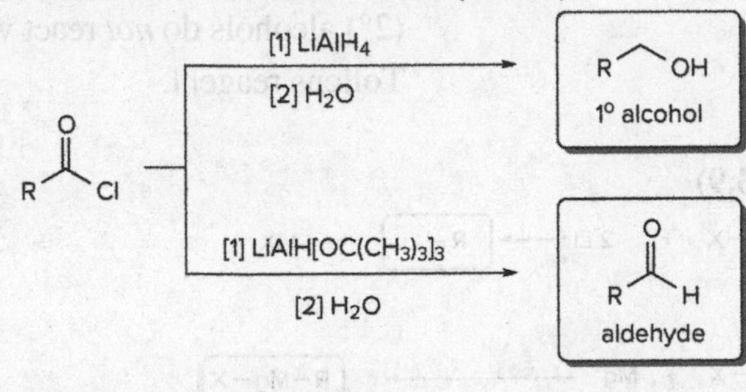

- $LiAlH_4$, a strong reducing agent, reduces an acid chloride to a 1° alcohol.
- With $LiAlH[OC(CH_3)_3]_3$, a milder reducing agent, reduction stops at the aldehyde stage.

[3] Reduction of esters (13.7A)

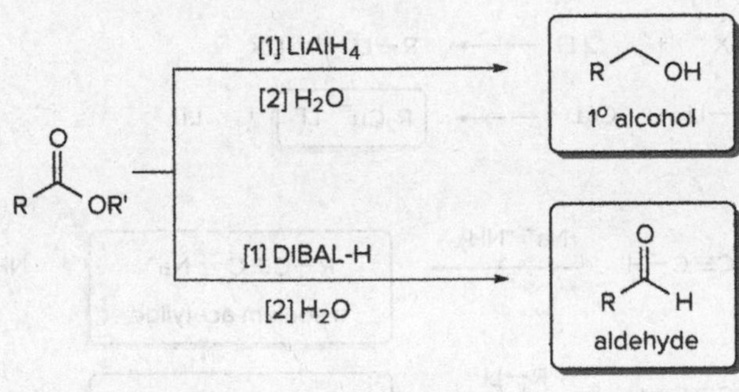

- $LiAlH_4$, a strong reducing agent, reduces an ester to a 1° alcohol.
- With DIBAL-H, a milder reducing agent, reduction stops at the aldehyde stage.

[4] Reduction of carboxylic acids to 1° alcohols (13.7B)

[1] $LiAlH_4$
[2] H_2O

1° alcohol

[5] Reduction of amides to amines (13.7B)

R–C(=O)–N(CH$_3$)$_2$ —[1] $LiAlH_4$, [2] H_2O→ R–CH$_2$–N(CH$_3$)$_2$ (amine)

Oxidation reactions

Oxidation of aldehydes to carboxylic acids (13.8)

RCHO —CrO_3, $Na_2Cr_2O_7$, $K_2Cr_2O_7$, $KMnO_4$ or Ag_2O, NH_4OH→ RCOOH (carboxylic acid)

- All Cr^{6+} reagents except PCC oxidize RCHO to RCOOH.
- Tollens reagent (Ag_2O + NH_4OH) oxidizes RCHO only. Primary (1°) and secondary (2°) alcohols do *not* react with Tollens reagent.

Preparation of organometallic reagents (13.9)

[1] Organolithium reagents: R–X + 2 Li ⟶ R–Li + LiX

[2] Grignard reagents: R–X + Mg —$(CH_3CH_2)_2O$→ R–Mg–X

[3] Organocuprate reagents:

R–X + 2 Li ⟶ R–Li + LiX

2 R–Li + CuI ⟶ R_2Cu^- Li^+ + LiI

[4] Lithium and sodium acetylides:

R–C≡C–H —Na^+ $^-NH_2$→ R–C≡C$^-$ Na^+ (a sodium acetylide) + NH_3

R–C≡C–H —R–Li→ R–C≡C–Li (a lithium acetylide) + R–H

Reactions with organometallic reagents

[1] Reaction as a base (13.9C)

R–M + H–Ö–R ⟶ R–H + M^+ + $^-$:Ö–R

- RM = RLi, RMgX, R_2CuLi
- This acid–base reaction occurs with H_2O, ROH, RNH_2, R_2NH, RSH, RCOOH, $RCONH_2$, and RCONHR.

[2] Reaction with aldehydes and ketones to form 1°, 2°, and 3° alcohols (13.10)

R(C=O)R' —[1] R"MgX or R"Li, [2] H_2O→ RR'R"C–OH (1°, 2°, or 3° alcohol)

R' = H or alkyl

[3] Reaction with esters to form 3° alcohols (13.13A)

R(C=O)OR' —[1] R"Li or R"MgX (2 equiv), [2] H_2O→ RR"R"C–OH (3° alcohol)

[4] Reaction with acid chlorides (13.13)

R(C=O)Cl —[1] R"Li or R"MgX (2 equiv), [2] H_2O→ RR"R"C–OH (3° alcohol)

R(C=O)Cl —[1] R'_2CuLi, [2] H_2O→ R(C=O)R' (ketone)

- More reactive organometallic reagents—R"Li and R"MgX—add two equivalents of R" to an acid chloride to form a 3° alcohol with two identical R" groups.
- Less reactive organometallic reagents—R'_2CuLi—add only one equivalent of R' to an acid chloride to form a ketone.

[5] Reaction with carbon dioxide—Carboxylation (13.14A)

R–MgX —[1] CO_2, [2] H_3O^+→ R–COOH (carboxylic acid)

[6] Reaction with epoxides (13.14B)

ethylene oxide —[1] RLi, RMgX, or R_2CuLi, [2] H_2O→ R–CH_2CH_2–OH (alcohol)

[7] Reaction with α, β-unsaturated aldehydes and ketones (13.15B)

[1] R'Li or R'MgX
[2] H_2O
allylic alcohol

[1] R'_2CuLi
[2] H_2O
ketone

- More reactive organometallic reagents—R'Li and R'MgX—react with α,β-unsaturated carbonyls by 1,2-addition.
- Less reactive organometallic reagents—R'_2CuLi— react with α,β-unsaturated carbonyls by 1,4-addition.

Protecting groups (13.12)

[1] Protecting an alcohol as a *tert*-butyldimethylsilyl ether

[Cl—TBDMS]

[R—O—TBDMS]
tert-butyldimethylsilyl ether

[2] Deprotecting a *tert*-butyldimethylsilyl ether to re-form an alcohol

[R—O—TBDMS]
$Bu_4N^+ F^-$
[F—TBDMS]

Practice Test on Chapter Review

1. Which compounds undergo nucleophilic addition and which undergo substitution?

a. b. c. d.

2. What product is formed when $CH_3CH_2CH_2Li$ reacts with each compound, followed by quenching with water and acid?

a. CH_3CH_2CHO
b. $(CH_3)_2CO$
c. $CH_3CH_2CO_2CH_3$
d. CH_3CH_2COCl
e. CO_2
f. $CH_2{=}CHCOCH_3$
g. ethylene oxide
h. CH_3COOH

3. What product is formed when $HO(CH_2)_4CHO$ is treated with each reagent?
   a. $NaBH_4$, CH_3OH
   b. PCC
   c. Ag_2O, NH_4OH
   d. $Na_2Cr_2O_7$, H_2SO_4, H_2O

4. What reagent is needed to convert $(CH_3CH_2)_2CHCOCl$ into each compound?
   a. $(CH_3CH_2)_2CHCOCH_2CH_3$
   b. $(CH_3CH_2)_2CHCHO$
   c. $(CH_3CH_2)_2CHC(OH)(CH_2CH_3)_2$
   d. $(CH_3CH_2)_2CHCH_2OH$

5. Draw the organic products formed in the following reactions.

a. [1] $LiAlH_4$; [2] H_2O

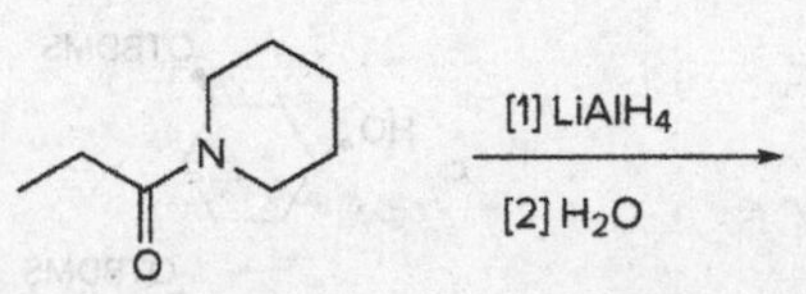

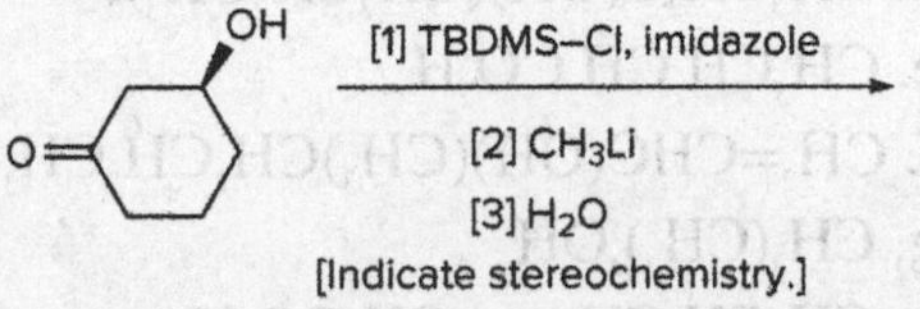

b. [1] Mg; [2] (epoxide); [3] H_3O^+

c. [1] TBDMS–Cl, imidazole; [2] CH_3Li; [3] H_2O [Indicate stereochemistry.]

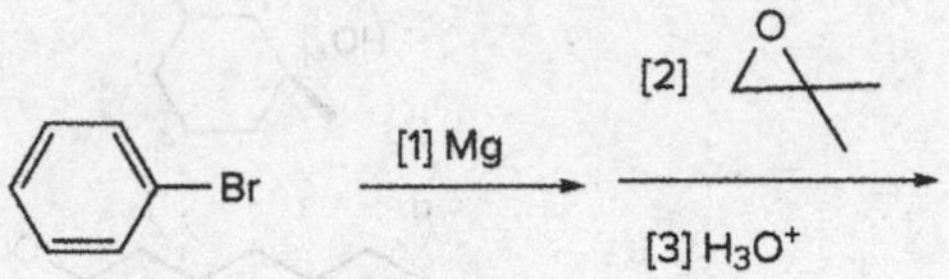

d. $NaBH_4$, CH_3OH

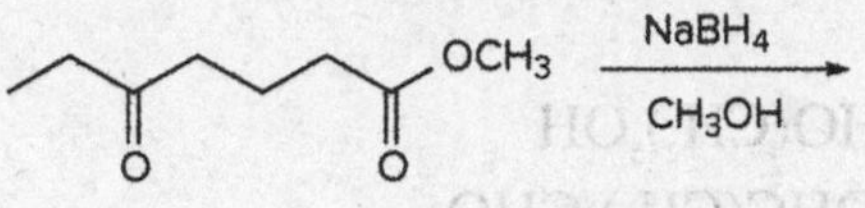

6. What starting materials are needed to synthesize each compound using the indicated reagent or functional group?

a. Synthesize: from an ester

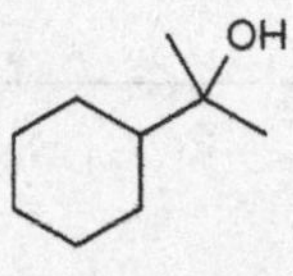

b. Synthesize: using an organocuprate reagent

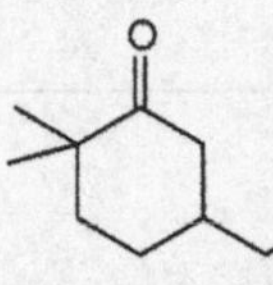

c. Synthesize: using a Grignard reagent

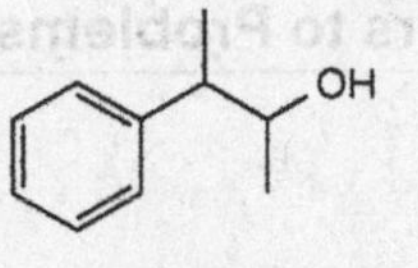

Answers to Practice Test

1.a. addition
   b. substitution
   c. substitution
   d. addition

4.a. $(CH_3CH_2)_2CuLi$
   b. $LiAlH[OC(CH_3)_3]_3$
   c. CH_3CH_2MgBr
   d. $LiAlH_4$

5. 6.

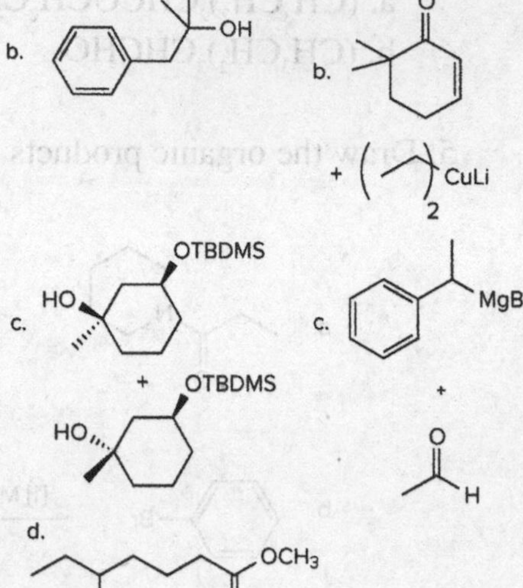

2. a. $CH_3CH_2CH(OH)CH_2CH_2CH_3$
   b. $(CH_3)_2C(OH)CH_2CH_2CH_3$
   c. $CH_3CH_2C(OH)(CH_2CH_2CH_3)_2$
   d. $CH_3CH_2C(OH)(CH_2CH_2CH_3)_2$
   e. $CH_3CH_2CH_2CO_2H$
   f. $CH_2{=}CHC(OH)(CH_3)CH_2CH_2CH_3$
   g. $CH_3(CH_2)_4OH$
   h. $CH_3CH_2CH_3$ + CH_3CO_2H

3.a. $HO(CH_2)_5OH$
   b. $OHC(CH_2)_3CHO$
   c. $HO(CH_2)_4CO_2H$
   d. $HO_2C(CH_2)_3CO_2H$

Answers to Problems

13.1

a. [1] Csp^2–Csp^2
H
← [2] s: Csp^2–Osp^2
p: Cp–Op
:O:
[3]
α-sinensal
Csp^2–Csp^2

b. The O is sp^2 hybridized.
Both lone pairs occupy sp^2 hybrid orbitals.

13.2 A carbonyl compound with a leaving group (NR_2 or OR bonded to the C=O) undergoes substitution reactions. Those without leaving groups undergo addition.

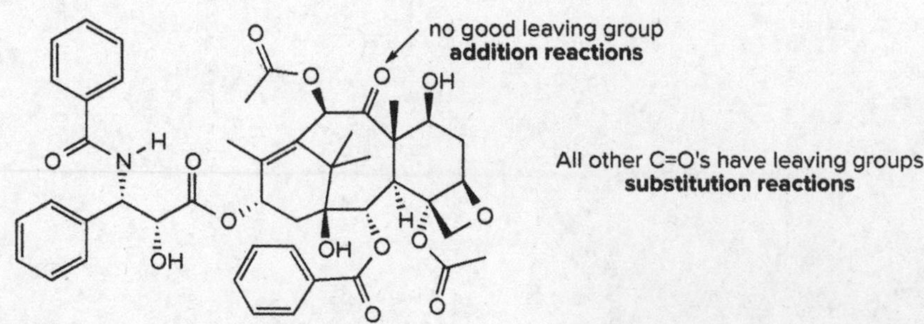

13.3 Aldehydes are more reactive than ketones. In carbonyl compounds with leaving groups, the better the leaving group, the more reactive the carbonyl compound.

a. 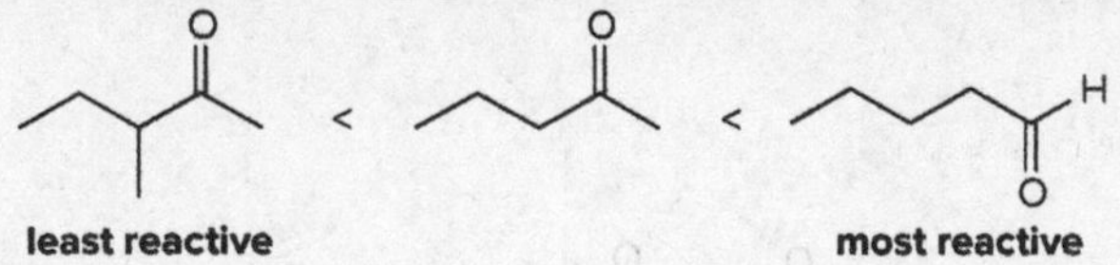 < <

least reactive most reactive

b. 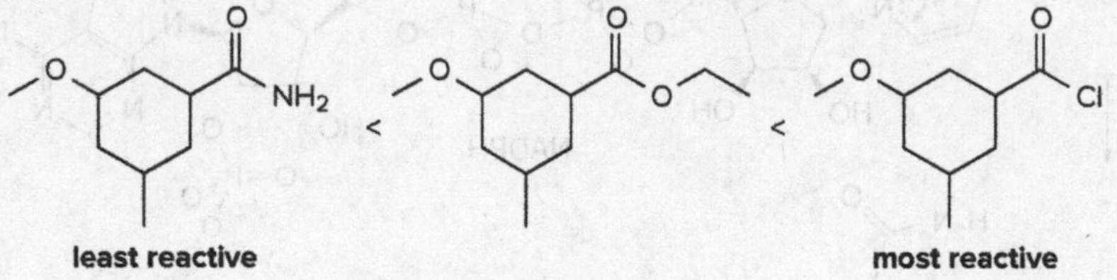 NH_2 < < Cl

least reactive most reactive

13.4 $NaBH_4$ reduces aldehydes to 1° alcohols and ketones to 2° alcohols.

a.  H $\xrightarrow[CH_3OH]{NaBH_4}$ OH

c. 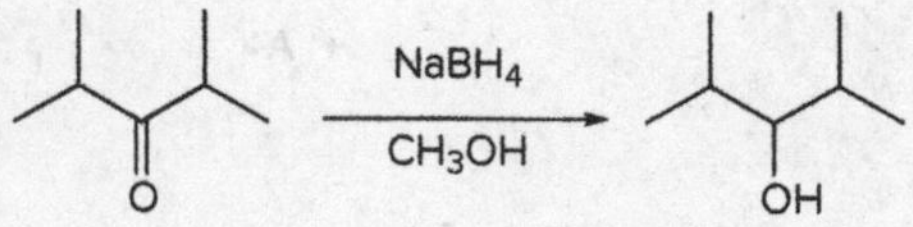 O $\xrightarrow[CH_3OH]{NaBH_4}$ OH

b. 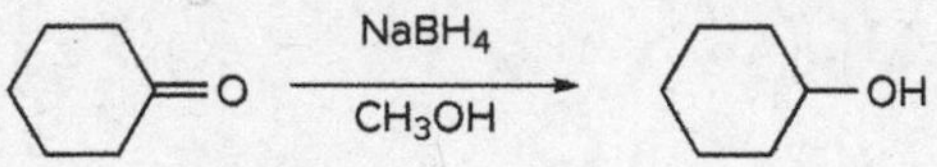 =O $\xrightarrow[CH_3OH]{NaBH_4}$ OH

13.5 1° Alcohols are prepared from aldehydes and 2° alcohols are from ketones.

a. 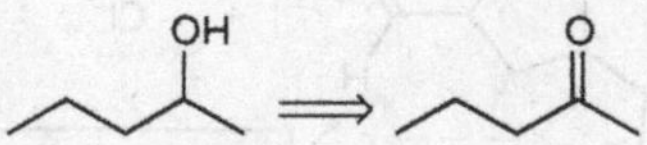 OH ⟹ O

b. 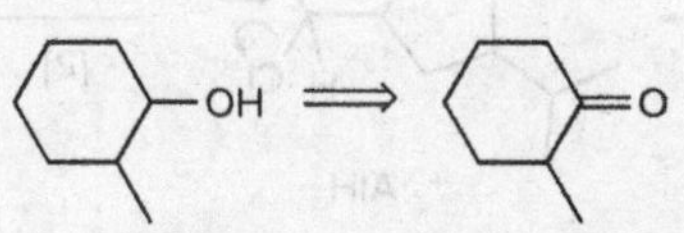 OH ⟹ O

c. 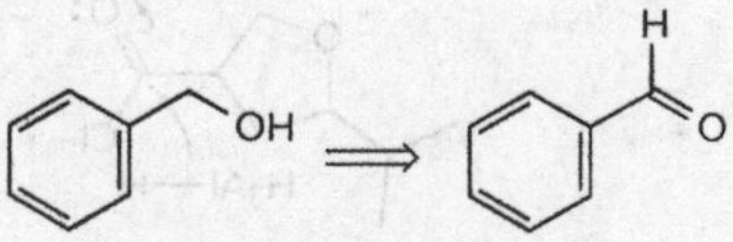 OH ⟹ H O

13.6

a. 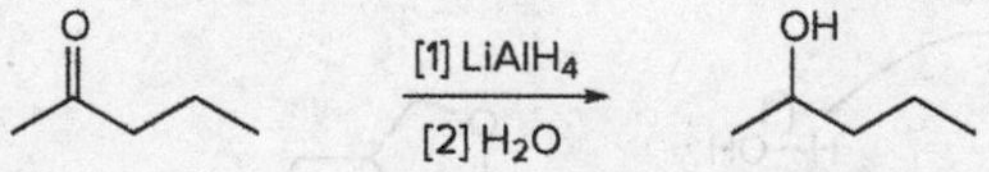 O $\xrightarrow[[2]\ H_2O]{[1]\ LiAlH_4}$ OH

c. 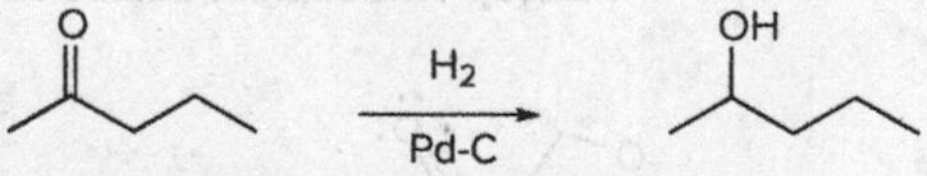 O $\xrightarrow[Pd\text{-}C]{H_2}$ OH

b. 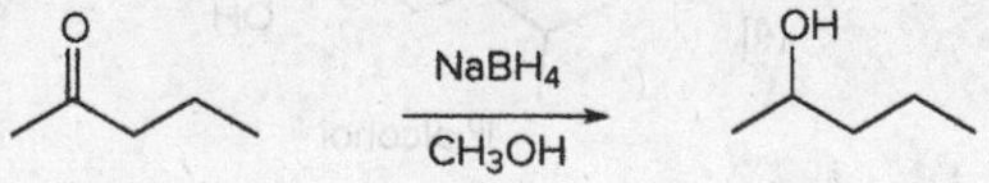 O $\xrightarrow[CH_3OH]{NaBH_4}$ OH

d. 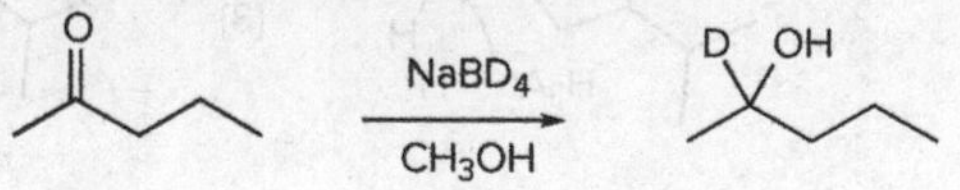 O $\xrightarrow[CH_3OH]{NaBD_4}$ D OH

13.7

a. 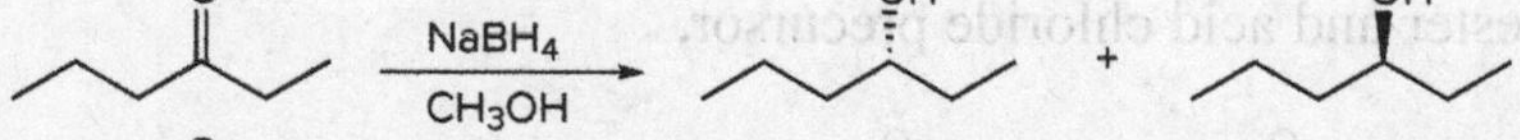 O $\xrightarrow[CH_3OH]{NaBH_4}$ OH + OH

b. 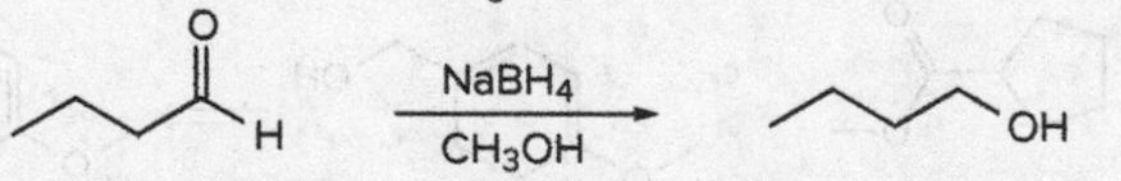 O H $\xrightarrow[CH_3OH]{NaBH_4}$ OH

c. 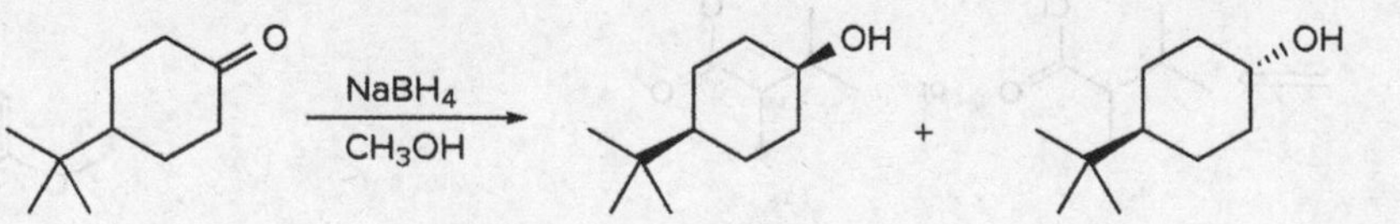 O $\xrightarrow[CH_3OH]{NaBH_4}$ OH + OH

13.49

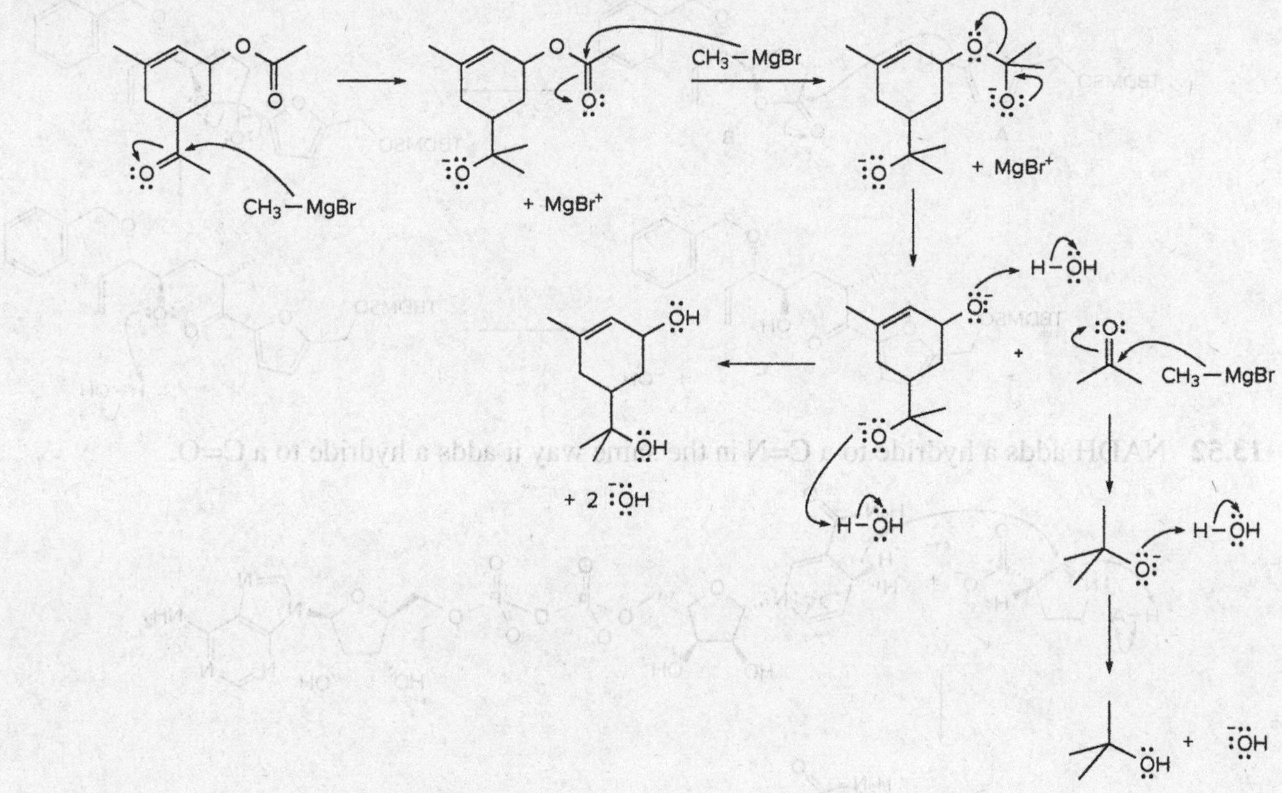

CH3—MgBr

+ MgBr+

CH3—MgBr

+ MgBr+

H—OH

CH3—MgBr

+ 2 −OH

H—OH

H—OH

+ −OH

13.50

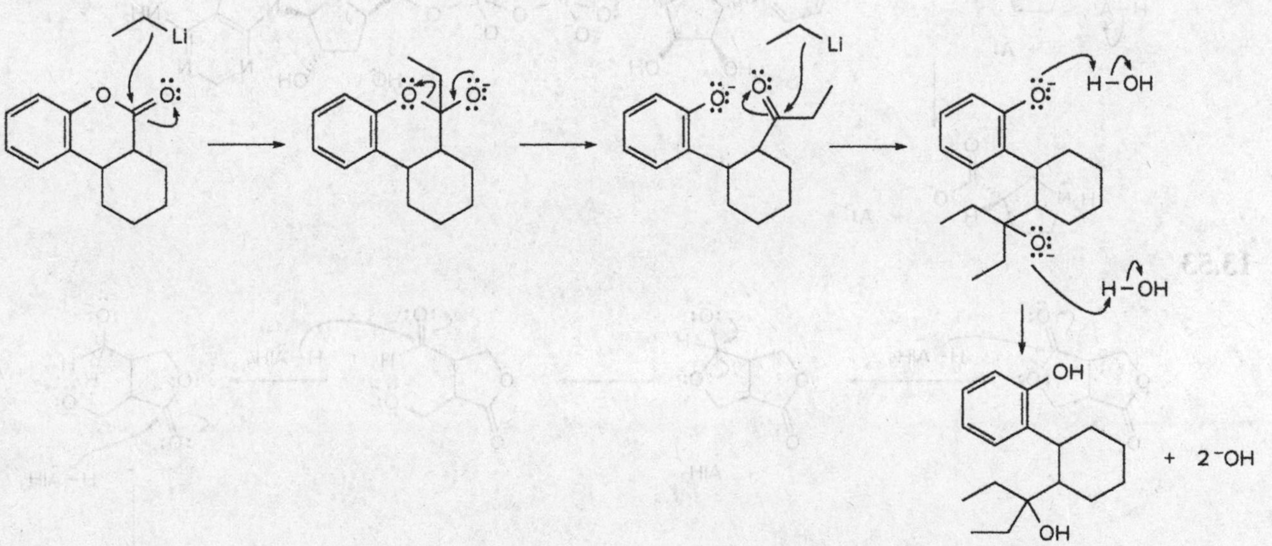

Li

Li

H—OH

H—OH

+ 2 −OH

13.51

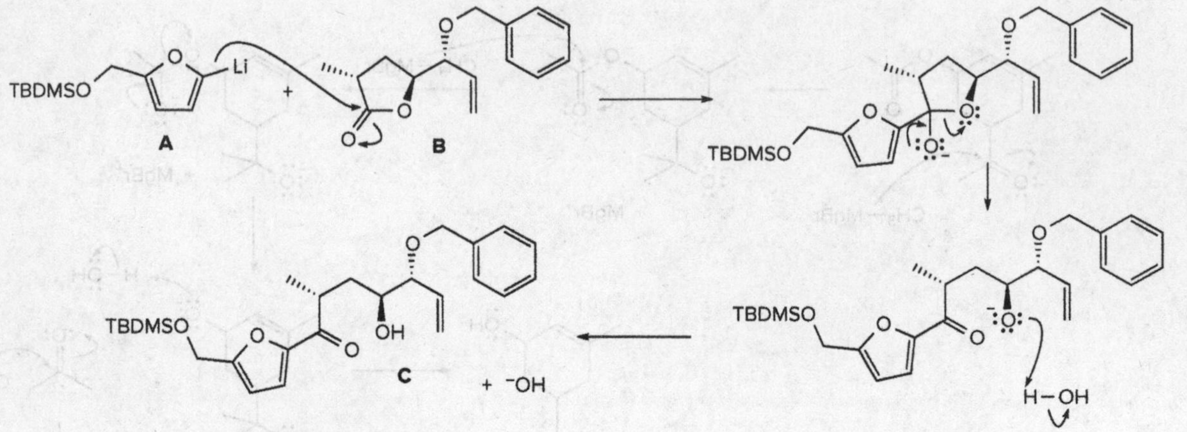

TBDMSO A Li + B → TBDMSO … C + ⁻OH

13.52 NADH adds a hydride to a C=N in the same way it adds a hydride to a C=O.

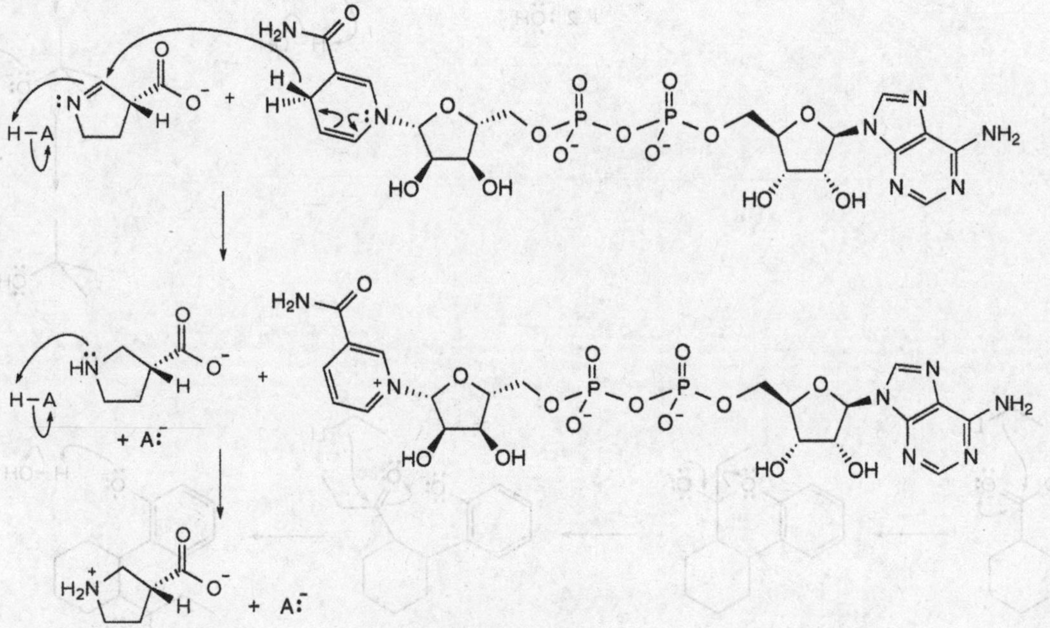

H–A + A:⁻

13.53

H–AlH₃ + AlH₃

H₂O

Protonate four alkoxides.

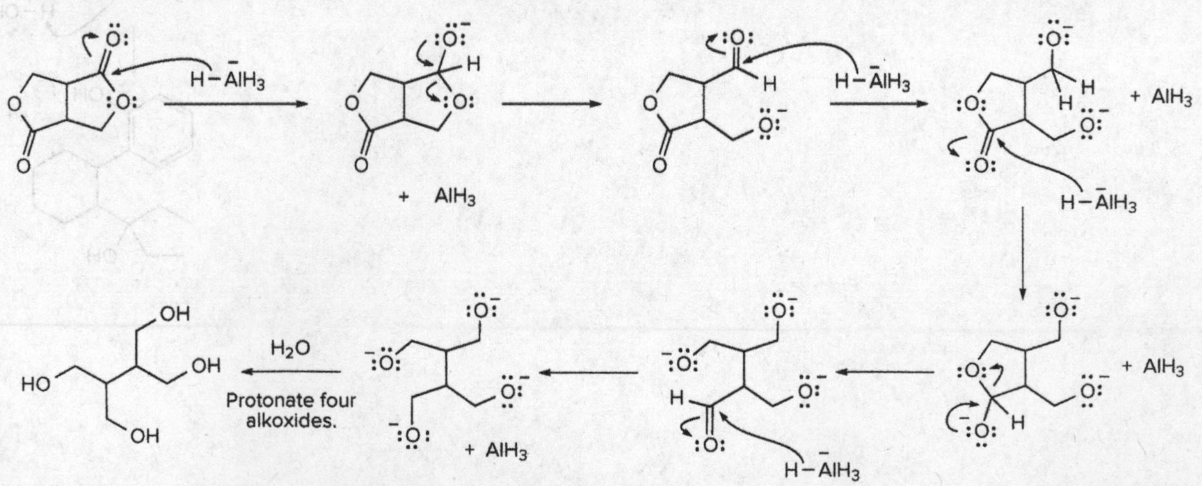

13.54 Any CH_2 bonded to the amine N can be formed by reduction of a C=O.

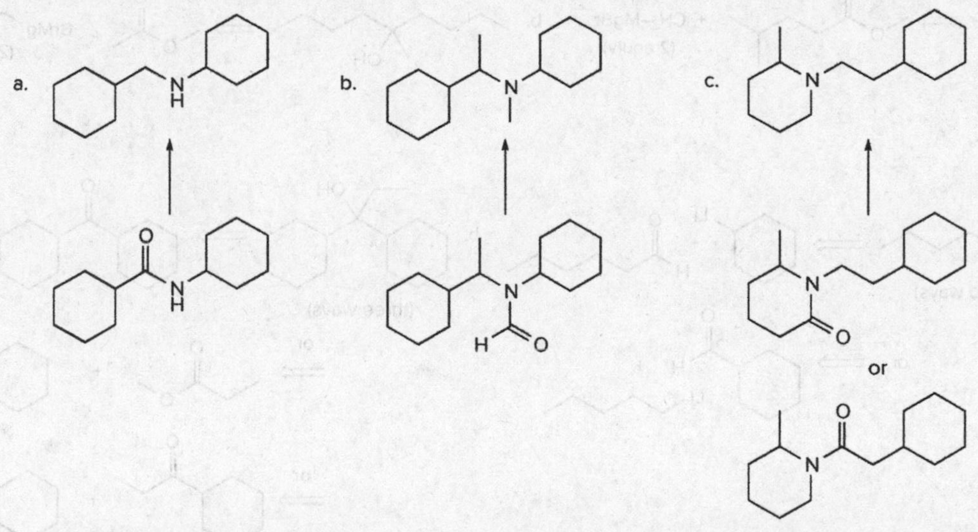

a. b. c.

N H

N

N

N H O

N H O

N O

or

N O

13.55

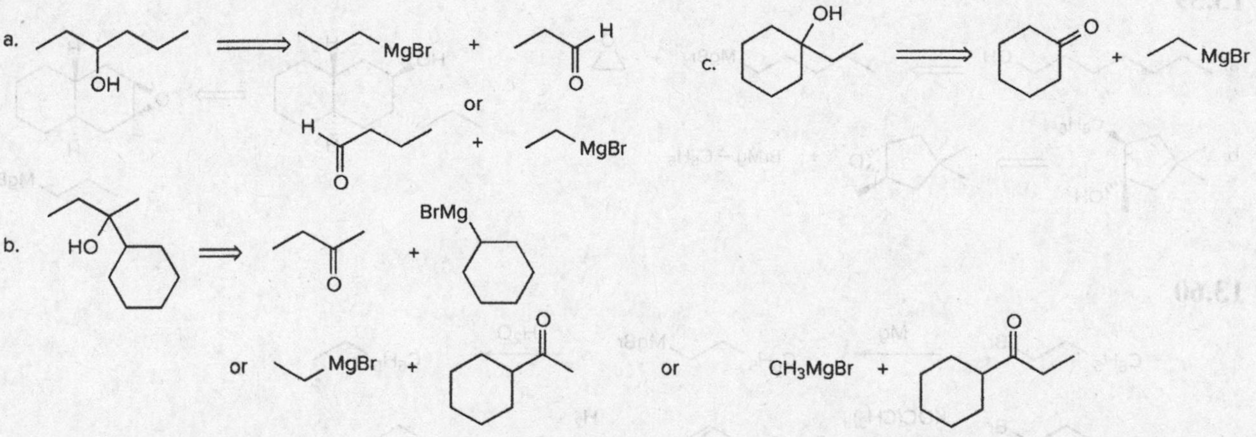

a. OH ⟹ MgBr + H O

or

H O + MgBr

c. OH ⟹ O + MgBr

b. HO ⟹ O + BrMg

or MgBr + O or CH_3MgBr + O

13.56

(a) (b) (c) OH N

(a) ⟹ O N + MgBr

(b) ⟹ O + BrMg N

(c) ⟹ O N + MgBr

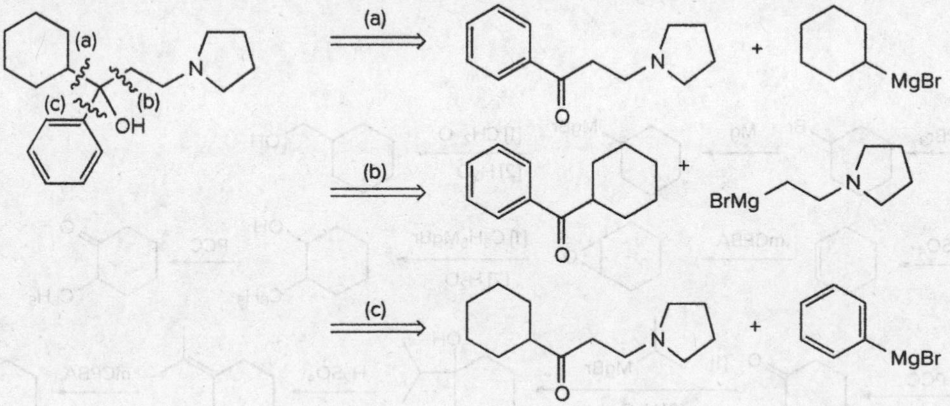

13.57

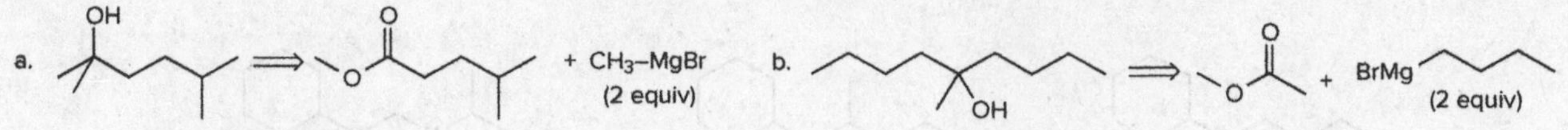

a. [structure] ⟹ [ester] + CH_3–MgBr (2 equiv)

b. [structure with OH] ⟹ [ester] + BrMg [chain] (2 equiv)

13.58

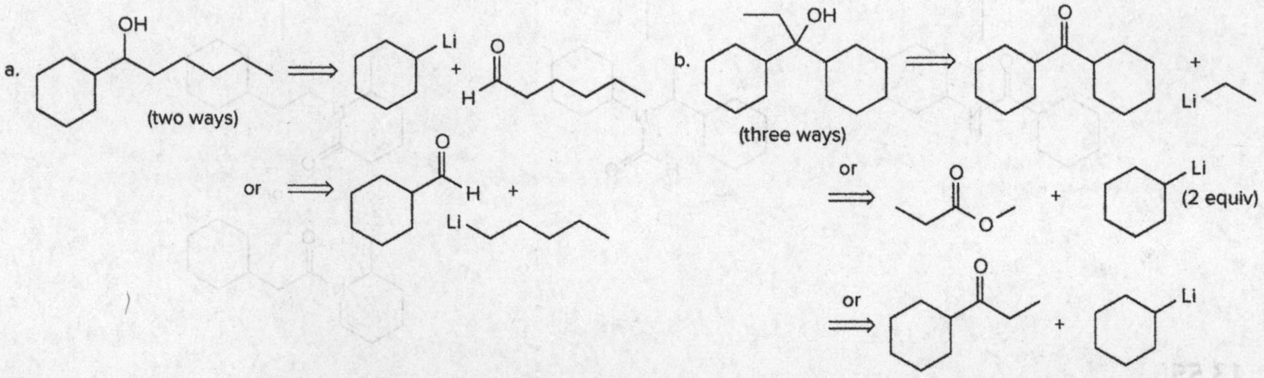

a. [structure with OH] (two ways) ⟹ [cyclohexyl-Li] + [aldehyde, H]

or ⟹ [cyclohexanecarbaldehyde, H] + Li [pentyl chain]

b. [structure with OH] (three ways) ⟹ [dicyclohexyl ketone] + Li [ethyl]

or ⟹ [ester] + [cyclohexyl-Li] (2 equiv)

or ⟹ [cyclohexyl ethyl ketone] + [cyclohexyl-Li]

13.59

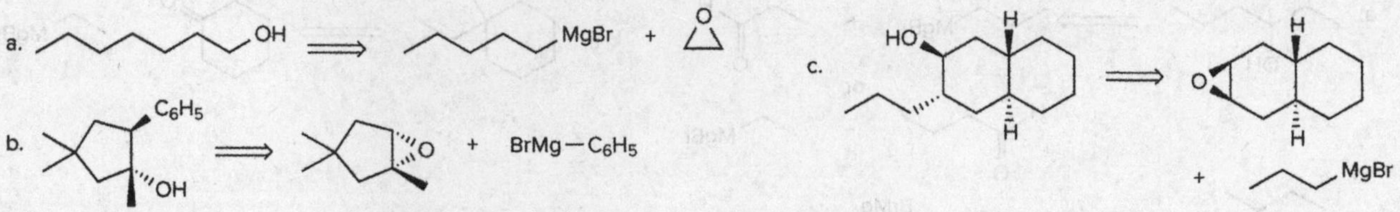

a. [structure] OH ⟹ [chain] MgBr + [epoxide, O]

b. [structure with C_6H_5, OH] ⟹ [epoxide, O] + BrMg – C_6H_5

c. HO [structure, H, H] ⟹ O [epoxide, H, H] + [propyl] MgBr

13.60

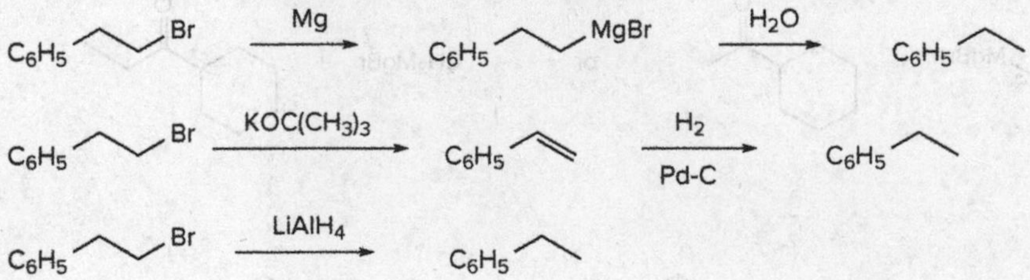

C_6H_5 [chain] Br —Mg→ C_6H_5 [chain] MgBr —H_2O→ C_6H_5 [chain]

C_6H_5 [chain] Br —$KOC(CH_3)_3$→ C_6H_5 [alkene] —H_2, Pd-C→ C_6H_5 [chain]

C_6H_5 [chain] Br —$LiAlH_4$→ C_6H_5 [chain]

13.61

a. [cyclohexanol, OH] —PBr_3→ [bromocyclohexane, Br] —Mg→ [cyclohexyl MgBr] —[1] CH_2=O, [2] H_2O→ [cyclohexylmethanol, OH]

b. [cyclohexanol, OH] —H_2SO_4→ [cyclohexene] —mCPBA→ [epoxide, O] —[1] C_6H_5MgBr, [2] H_2O→ [OH, C_6H_5] —PCC→ [ketone, O, C_6H_5]

c. [cyclohexanol, OH] —PCC→ [cyclohexanone, O] —[1] [isopropyl] MgBr, [2] H_2O→ [OH] —H_2SO_4→ [alkene] major product —mCPBA→ [epoxide, O]

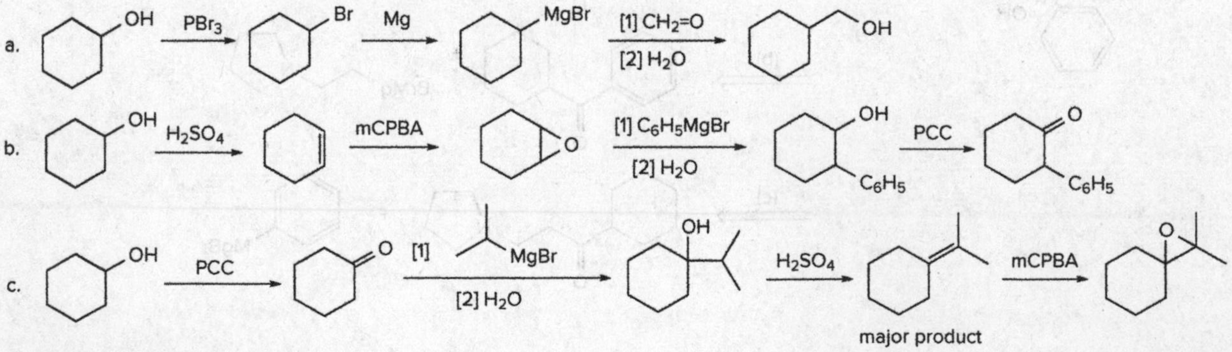

d. (from a.) PCC; [1] MgBr (from a.) [2] H_2O; OH; PCC; O; H

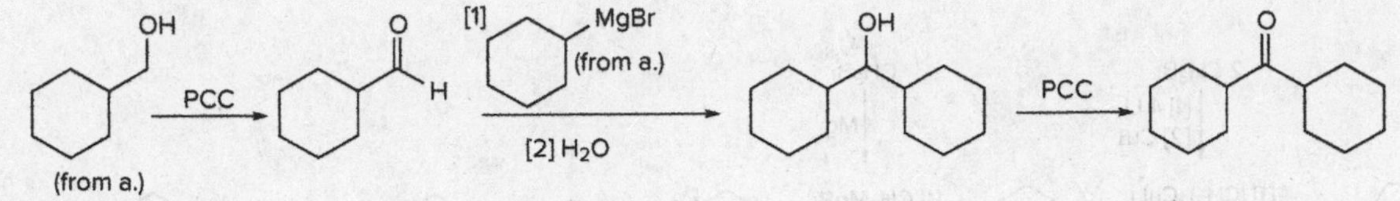

13.62

a. OH PCC H O [1] MgBr [2] H_2O OH PCC O

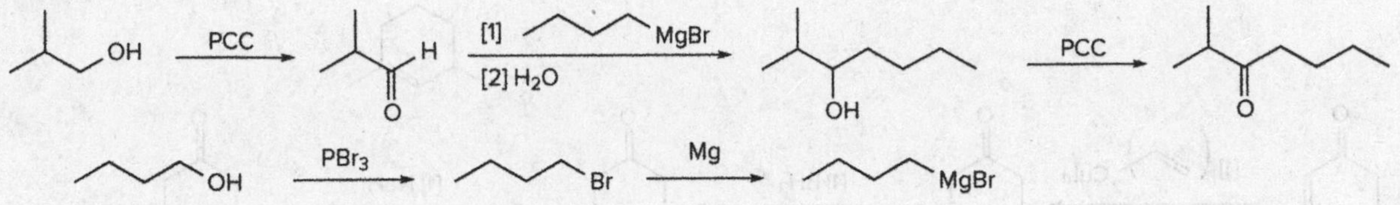

OH PBr_3 Br Mg MgBr

b. H O (from a.) [1] MgBr [2] H_2O OH PBr_3 Br Mg MgBr [1] CO_2 [2] H_3O^+ CO_2H

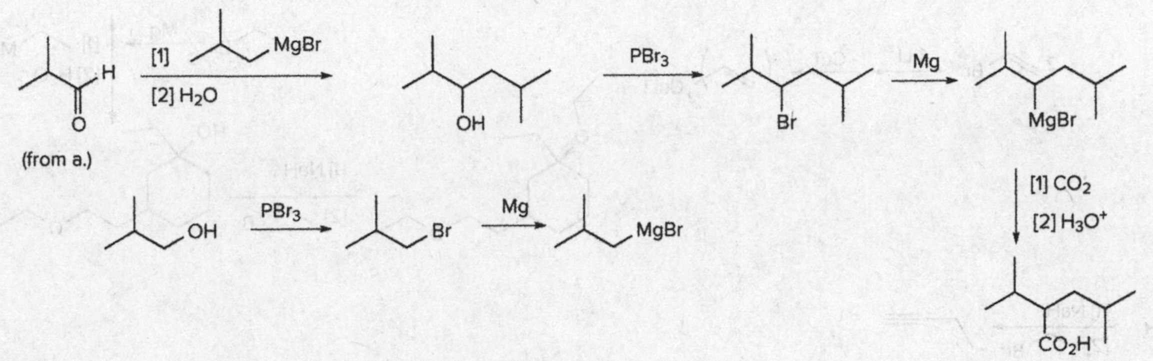

OH PBr_3 Br Mg MgBr

c. OH HBr Br Mg MgBr + H O (from a.) H_2O OH PBr_3 Br $K^+ {}^-OC(CH_3)_3$

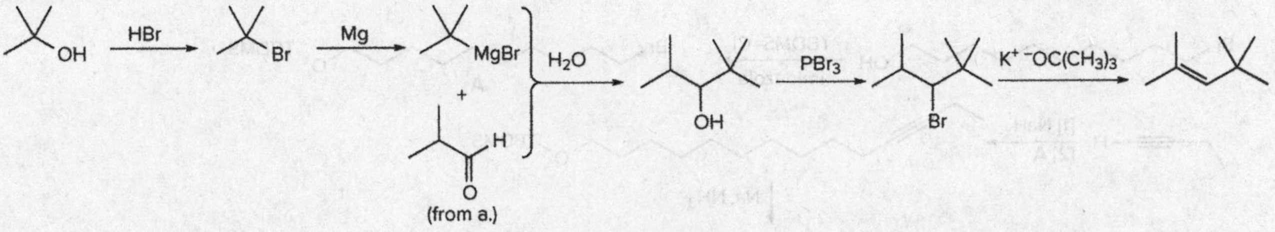

d. HC≡CH [1] NaH [2] $CH_3CH=O$ [3] H_2O HO TBDMS–Cl imidazole TBDMSO NaH TBDMSO – [1] O [2] H_2O TBDMSO OH $(Bu)_4NF$ HO OH

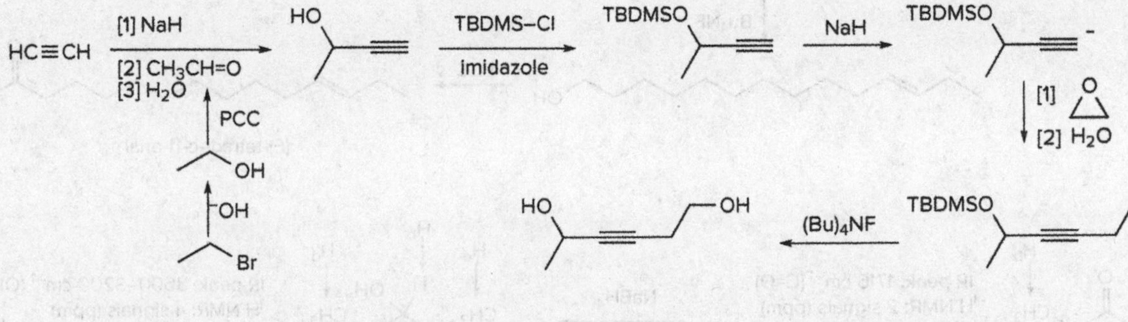

PCC OH OH Br

13.63

a. Cyclohex-2-enone → [1] $(CH_3)_2CuLi$ [2] H_2O (reagent from 2 CH_3Br: [1] 4 Li [2] CuI) → 3-methylcyclohexanone → [1] CH_3MgBr [2] H_2O (reagent from CH_3Br, Mg) → 1,3-dimethylcyclohexanol → H_2SO_4 → 1,5-dimethylcyclohexene + 1,3-dimethylcyclohexene → H_2, Pd-C → 1,3-dimethylcyclohexane

b. Cyclohex-2-enone → [1] $(CH_2{=}CH)_2CuLi$ [2] H_2O → 3-vinylcyclohexanone → [1] BH_3 [2] H_2O_2, ^-OH → 3-(2-hydroxyethyl)cyclohexanone → [1] NaH [2] CH_3CH_2Br → 3-(2-ethoxyethyl)cyclohexanone

2 $CH_2{=}CHBr$ → 4 Li → CuI → $(CH_2{=}CH)_2CuLi$

CH_3CH_2Br → Mg → CH_3CH_2MgBr

3-(2-ethoxyethyl)cyclohexanone → [1] CH_3CH_2MgBr [2] H_2O → 1-ethyl-3-(2-ethoxyethyl)cyclohexanol (HO) → [1] NaH [2] CH_3CH_2Br → 1-ethoxy-1-ethyl-3-(2-ethoxyethyl)cyclohexane

13.64

HC≡CH → [1] NaH [2] CH_3CH_2Br → but-1-yne

Br(CH₂)₁₀OH → TBDMS–Cl, imidazole → Br(CH₂)₁₀O–TBDMS (**A**)

$CH_3CH_2C{\equiv}C{-}H$ → [1] NaH [2] **A** → $CH_3CH_2C{\equiv}C(CH_2)_{10}O$–TBDMS → Na, NH_3 → (E)-$CH_3CH_2CH{=}CH(CH_2)_{10}O$–TBDMS → Bu_4NF → (E)-$CH_3CH_2CH{=}CH(CH_2)_{10}OH$ → PCC → (E)-$CH_3CH_2CH{=}CH(CH_2)_9CHO$

(*E*)-tetradec-11-enal

13.65

A: $C_7H_{14}O$ — $(CH_3)_2CH$–CO–$CH(CH_3)_2$ (CH₃ = H_b; CH = H_a)

IR peak: 1716 cm^{-1} (C=O)
^{1}H NMR: 2 signals (ppm)
doublet 1.2 (H_b)
septet 2.7 (H_a)

→ $NaBH_4$, CH_3OH →

B: $C_7H_{16}O$ — $(CH_3)_2CH$–CH(OH)–$CH(CH_3)_2$ (CH₃ = H_d; CH of isopropyl = H_c; CH–O = H_b; OH = H_a)

IR peak: 3600–3200 cm^{-1} (OH)
^{1}H NMR: 4 signals (ppm)
doublet 0.9 (H_d)
singlet 1.5 (H_a)
multiplet 1.7 (H_c)
triplet 3.0 (H_b)

13.66 Molecular ion at m/z = 86: $C_5H_{10}O$ (possible molecular formula).

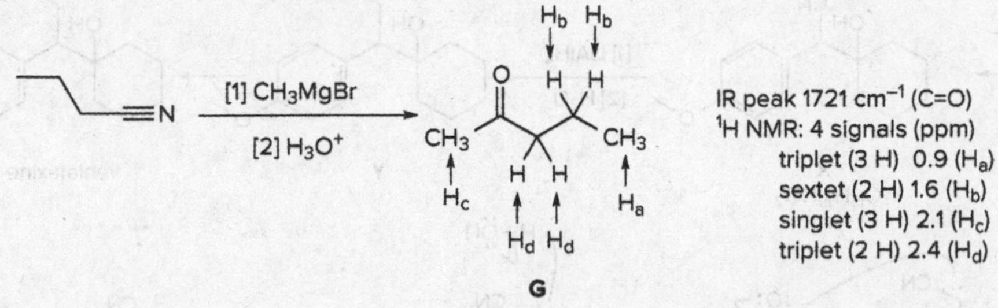
[1] CH3MgBr
[2] H3O+
Hb Hb
H H
CH3
CH3
H H
Hc
Ha
Hd Hd
G
IR peak 1721 cm−1 (C=O)
1H NMR: 4 signals (ppm)
triplet (3 H) 0.9 (Ha)
sextet (2 H) 1.6 (Hb)
singlet (3 H) 2.1 (Hc)
triplet (2 H) 2.4 (Hd)

13.67 Molecular ion at m/z = 86: $C_5H_{10}O$ (possible molecular formula).

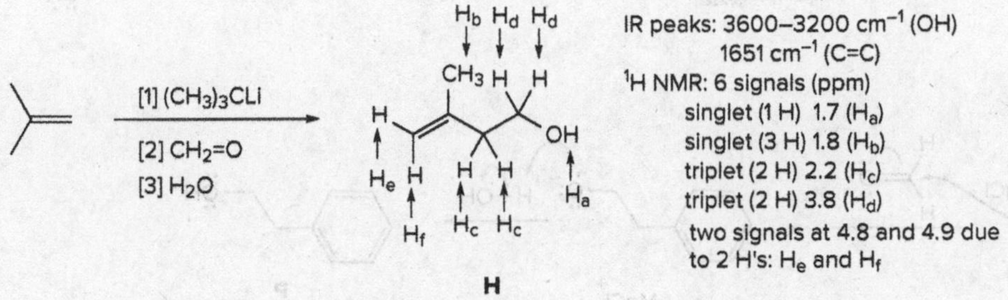
[1] (CH3)3CLi
[2] CH2=O
[3] H2O
Hb Hd Hd
CH3 H H
OH
He H
H H
Ha
Hf
Hc Hc
H
IR peaks: 3600–3200 cm−1 (OH)
1651 cm−1 (C=C)
1H NMR: 6 signals (ppm)
singlet (1 H) 1.7 (Ha)
singlet (3 H) 1.8 (Hb)
triplet (2 H) 2.2 (Hc)
triplet (2 H) 3.8 (Hd)
two signals at 4.8 and 4.9 due to 2 H's: He and Hf

13.68

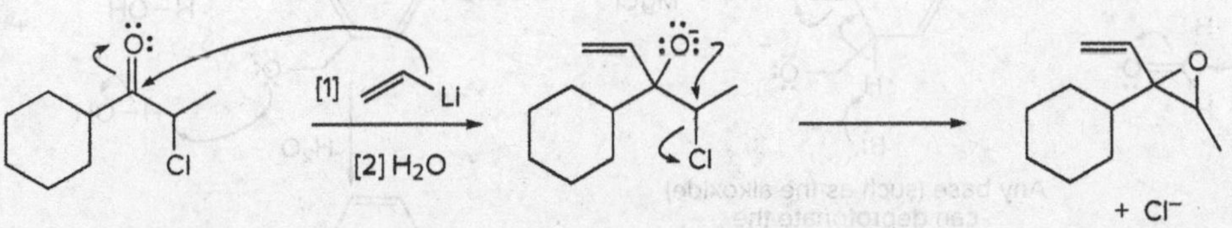
Cl
[1]
Li
[2] H2O
Cl
+ Cl−

13.69 The β carbon of an α,β-unsaturated carbonyl compound absorbs farther downfield in the ^{13}C NMR spectrum than the α carbon, because the β carbon is deshielded and bears a partial positive charge as a result of resonance. Because three resonance structures can be drawn for an α,β-unsaturated carbonyl compound, one of which places a positive charge on the β carbon, the decrease in electron density at this carbon deshields it, shifting the ^{13}C absorption downfield. This is not the case for the α carbon.

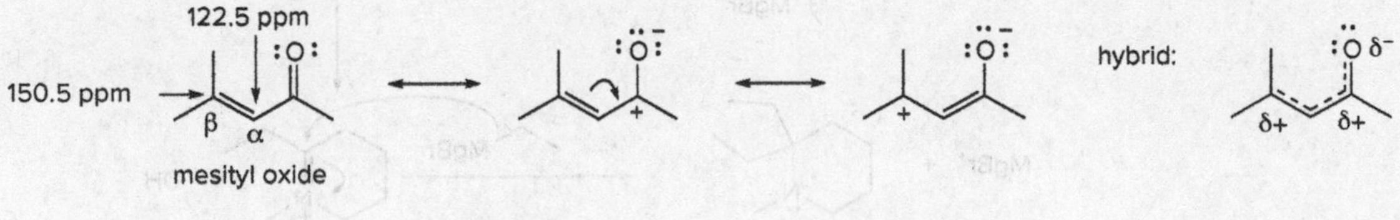
122.5 ppm
150.5 ppm
β
α
mesityl oxide
hybrid:
δ−
δ+
δ+

13.70

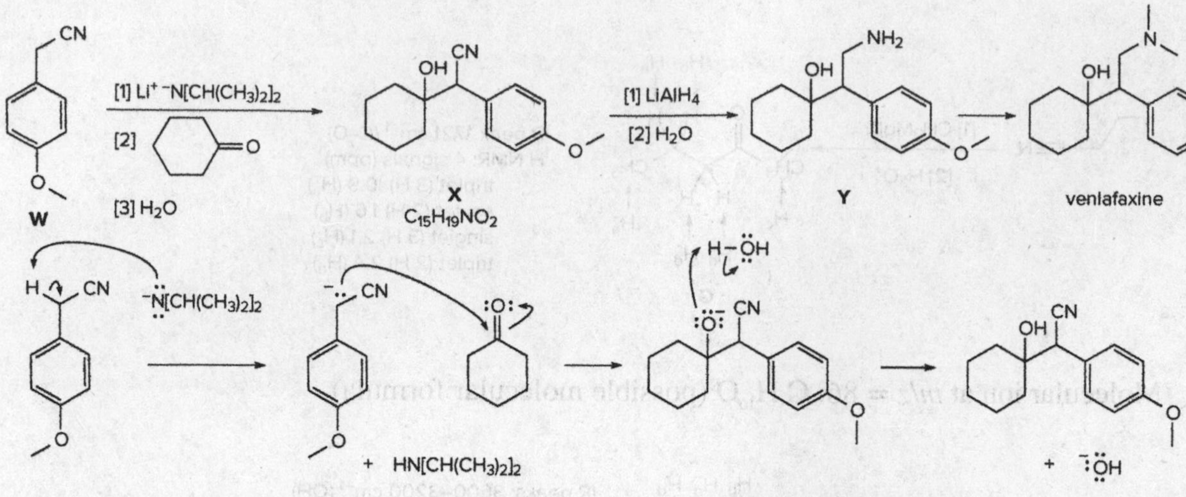

[1] Li$^+$ $^-$N[CH(CH$_3$)$_2$]$_2$
[2] cyclohexanone
[3] H_2O

W

X
$C_{15}H_{19}NO_2$

[1] $LiAlH_4$
[2] H_2O

Y

venlafaxine

+ HN[CH(CH$_3$)$_2$]$_2$

+ $^-$OH

13.71

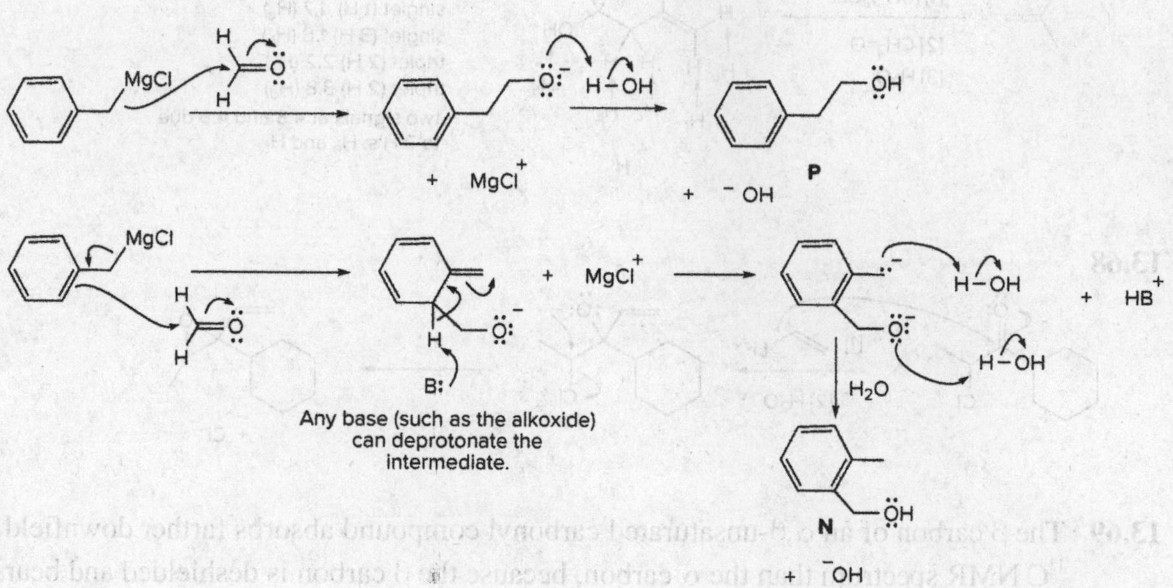

+ $MgCl^+$

P

+ $^-$OH

+ $MgCl^+$

+ HB^+

B:

Any base (such as the alkoxide) can deprotonate the intermediate.

H_2O

N

+ $^-$OH

13.72

MgBr

+ $MgBr^+$

(any proton source)

+ A^-

MgBr

$MgBr^+$ +

+ $^-$OH

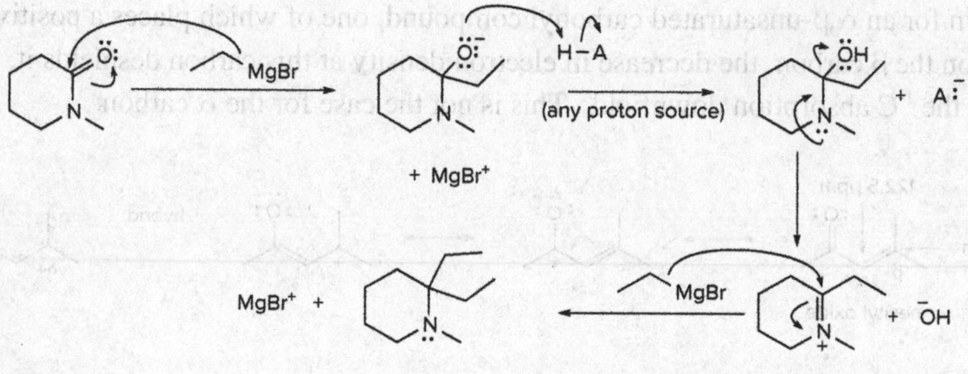

Chapter 14 Aldehydes and Ketones—Nucleophilic Addition

Chapter Review

General facts

- Aldehydes and ketones contain a carbonyl group bonded to only H atoms or R groups. The carbonyl carbon is sp^2 hybridized and trigonal planar (14.1).
- Aldehydes are identified by the suffix *-al,* whereas ketones are identified by the suffix *-one* (14.2).
- Aldehydes and ketones are polar compounds that exhibit dipole–dipole interactions (14.3).

Summary of spectroscopic absorptions of RCHO and R_2CO (14.3)

IR absorptions	C=O	~1700 cm^{-1} • increasing frequency with decreasing ring size for cyclic ketones • For both RCHO and R_2CO, the frequency decreases with conjugation.
	C_{sp^2}–H of CHO	~2700–2830 cm^{-1} (one or two peaks)
^{1}H NMR absorptions	CHO	9–10 ppm (highly deshielded proton)
	C–H α to C=O	2–2.5 ppm (somewhat deshielded C_{sp^3}–H)
^{13}C NMR absorption	C=O	190–215 ppm

Nucleophilic addition reactions

[1] Addition of hydride (H^-) (14.7)

RCOR' (R' = H or alkyl) —[NaBH₄, CH₃OH or [1] LiAlH₄ [2] H₂O]→ RR'CH(OH) (1° or 2° alcohol)

$$\text{RC(=O)R'} \xrightarrow[\text{[1] LiAlH}_4\text{ [2] H}_2\text{O}]{\text{NaBH}_4\text{, CH}_3\text{OH or}} \text{RR'C(H)OH}$$

R' = H or alkyl; 1° or 2° alcohol

- The mechanism has two steps.
- $H:^-$ adds to the planar C=O from both sides.

[2] Addition of organometallic reagents (R^-) (14.7)

$$\text{RC(=O)R'} \xrightarrow[\text{[2] H}_2\text{O}]{\text{[1] R''MgX or R''Li}} \text{RR'C(R'')OH}$$

R' = H or alkyl

- The mechanism has two steps.
- $R:^-$ adds to the planar C=O from both sides.

[3] Addition of cyanide (^{-}CN) (14.8)

R' = H or alkyl

NaCN, HCl

cyanohydrin

- The mechanism has two steps.
- ^{-}CN adds to the planar C=O from both sides.

[4] Wittig reaction (14.9)

R' = H or alkyl

Ph_3P=

Wittig reagent

- The reaction forms a new C–C σ bond and a new C–C π bond.
- $Ph_3P=O$ is formed as by-product.

[5] Addition of 1° amines (14.10)

R' = H or alkyl

$R''NH_2$, mild acid

imine

- The reaction is fastest at pH 4–5.
- The intermediate carbinolamine is unstable, and loses H_2O to form the C=N.

[6] Addition of 2° amines (14.11)

R' = H or alkyl

R_2NH, mild acid

NR_2

- The reaction is fastest at pH 4–5.
- The intermediate carbinolamine is unstable, and loses H_2O to form the C=C.

[7] Addition of H_2O—Hydration (14.14)

R' = H or alkyl

H_2O, H^+ or ^{-}OH

- The reaction is reversible. Equilibrium favors the product only with less stable carbonyl compounds (e.g., H_2CO and Cl_3CCHO).
- The reaction is catalyzed with either H^+ or ^{-}OH.

[8] Addition of alcohols (14.15)

R' = H or alkyl

+ R"OH (2 equiv), H^+ + H_2O

- The reaction is reversible.
- The reaction is catalyzed with acid.
- Removal of H_2O drives the equilibrium to favor the products.

Other reactions

[1] Synthesis of Wittig reagents (14.9A)

R–CH$_2$–X → [1] Ph_3P: [2] Bu—Li → Ph_3P=CH–R

- Step [1] is best with CH_3X and RCH_2X because the reaction follows an S_N2 mechanism.
- A strong base is needed for proton removal in Step [2].

[2] Conversion of cyanohydrins to aldehydes and ketones (14.8)

R(R')C(OH)CN → ^{-}OH → aldehyde or ketone (RCOR') + H_2O + ^{-}CN

- This reaction is the reverse of cyanohydrin formation.

[3] Hydrolysis of nitriles (14.8)

R(R')C(HO)–C≡N → H_2O, H^+ or ^{-}OH, Δ → α-hydroxy carboxylic acid

R' = H or alkyl

[4] Hydrolysis of imines and enamines (14.12)

imine or enamine → H_2O, H^+ → aldehyde or ketone + RNH_2 or R_2NH

[5] Hydrolysis of acetals (14.15)

R(R')C(OR")$_2$ + H_2O ⇌ (H^+) aldehyde or ketone (RCOR') + R"OH (2 equiv)

R' = H or alkyl

- The reaction is acid catalyzed and is the reverse of acetal synthesis.
- A large excess of H_2O drives the equilibrium to favor the products.

Practice Test on Chapter Review

1. Give the IUPAC name for the following compounds.

a.

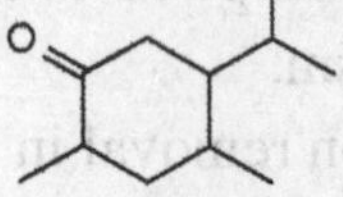

b.

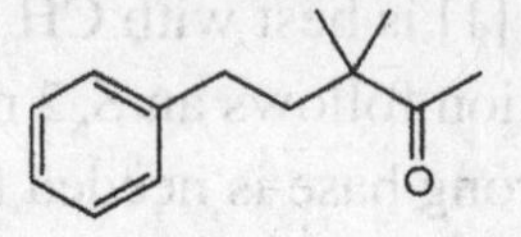

c.

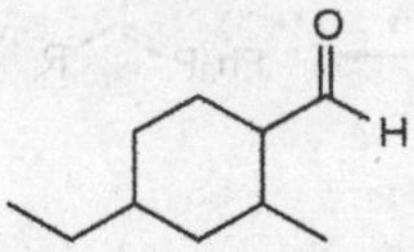

2. (a) Considering compounds **A–D,** which compound forms the smallest amount of hydrate? (b) Which compound forms the largest amount of hydrate?

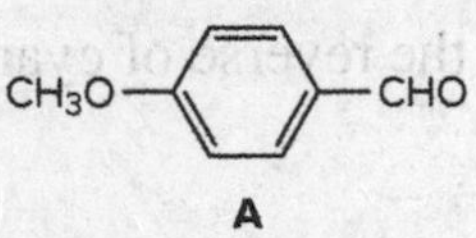

CH_3O–C₆H₄–CHO **A**; Cl–C₆H₄–COCH₃ **B**; O_2N–C₆H₄–CHO **C**; CH_3O–C₆H₄–COCH₃ **D**

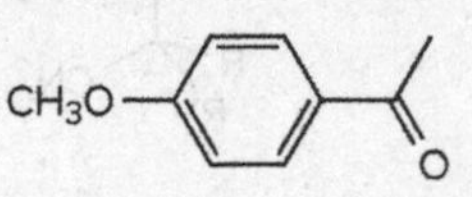

3. (a) Considering compounds **A–D,** which compound absorbs at the ***lowest*** wavenumber in its IR spectrum? (b) Which compound absorbs at the ***highest*** wavenumber in its IR spectrum?

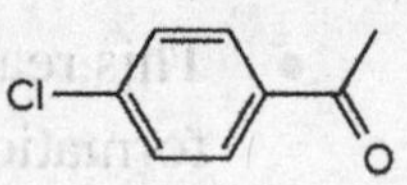

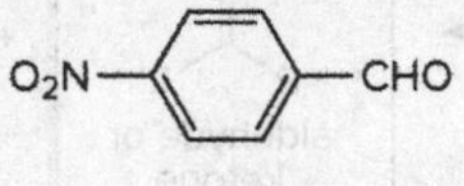

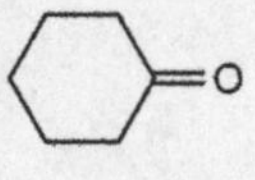

A 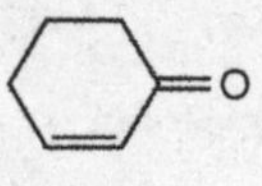 **B** 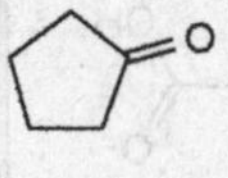 **C** 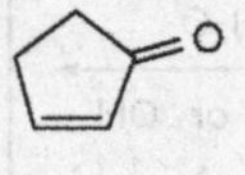 **D**

4. Fill in the lettered reagents (**A–G**) in the following reaction scheme.

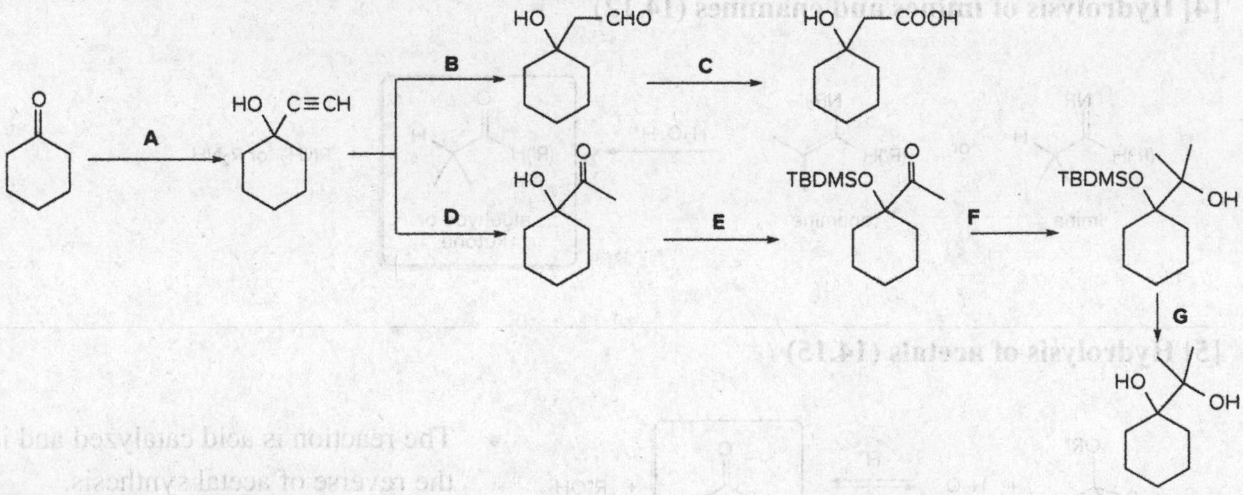

A, B, C, D, E, F, G

HO C≡CH; HO CHO; HO COOH; HO; TBDMSO; TBDMSO OH; HO OH

5. Draw the organic products formed in the following reactions.

a. 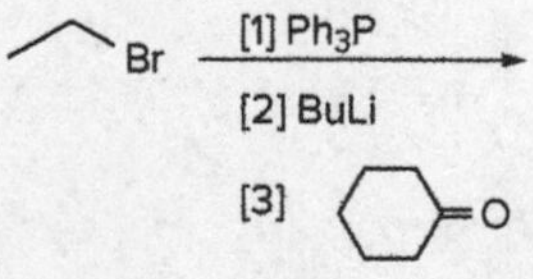 Br → [1] Ph_3P [2] BuLi [3] cyclohexanone

b. 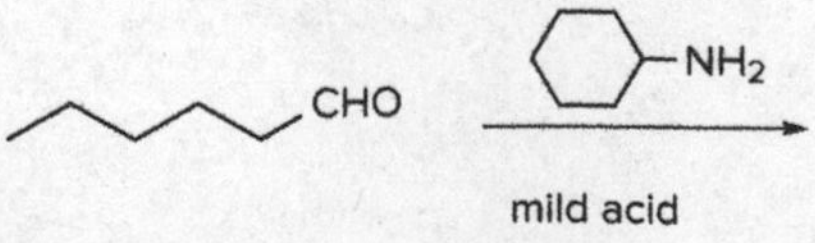 CHO → cyclohexyl–NH_2, mild acid

c. [1] NaCN, HCl; [2] H_2O, H^+, Δ

e. + HO⁀OH, TsOH

d. mild acid

f. CH_3CH_2OH, H^+ [Indicate stereochemistry.]

Answers to Practice Test

1.a. 5-isopropyl-2,4-dimethylcyclohexanone
b. 3,3-dimethyl-5-phenylpentan-2-one
c. 4-ethyl-2-methylcyclohexanecarbaldehyde

2.a. **D**
b. **C**

3.a. **B**
b. **C**

4.
A = [1] LiC≡CH; [2] H_2O
B = [1] R_2BH; [2] H_2O_2, ^-OH
C = Ag_2O, NH_4OH
D = H_2O, H_2SO_4, $HgSO_4$
E = TBDMS–Cl, imidazole
F = [1] CH_3Li; [2] H_2O
G = $Bu_4N^+F^-$

5.
a.
b. N
c. OH, CO_2H
d. N (E + Z)
e. O, O
f. O, OH, +, O, OH

Answers to Problems

14.1 As the number of R groups bonded to the carbonyl C increases, reactivity toward nucleophilic attack decreases. Steric hindrance decreases reactivity as well.

Increasing reactivity
decreasing steric hindrance

14.2 More stable aldehydes are less reactive toward nucleophilic attack.

CHO

benzaldehyde
Several resonance structures delocalize the partial positive charge on the carbonyl carbon, making it more stable and less reactive toward nucleophilic attack.

CHO

cyclohexanecarbaldehyde
This aldehyde has no added resonance stabilization.

14.3 • To name an aldehyde with a chain of atoms: [1] Find the longest chain with the CHO group and change the *-e* ending to *-al*. [2] Number the carbon chain to put the CHO at C1, but omit this number from the name. Apply all other nomenclature rules.

• To name an aldehyde with the CHO bonded to a ring: [1] Name the ring and add the suffix *-carbaldehyde*. [2] Number the ring to put the CHO group at C1, but omit this number from the name. Apply all other nomenclature rules.

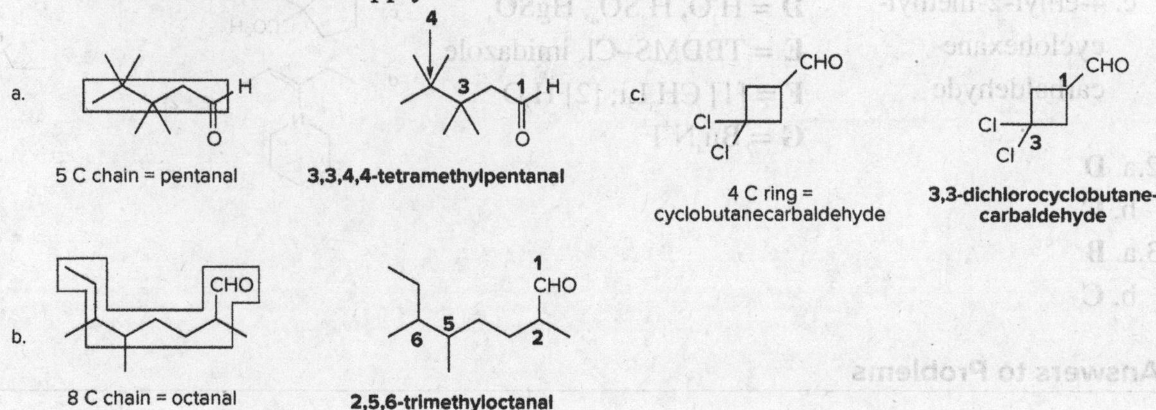

14.4 Work backwards from the name to the structure, referring to the nomenclature rules in Answer 14.3.

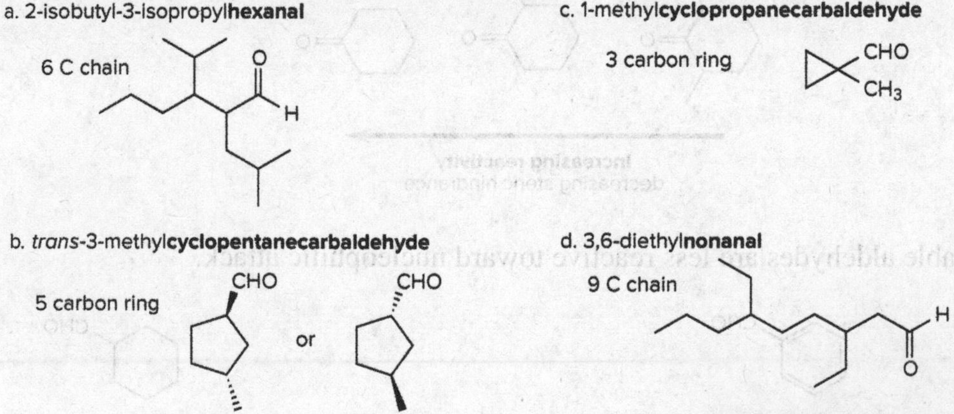

14.5 • To name an acyclic ketone: [1] Find the longest chain with the carbonyl group and change the *-e* ending to *-one*. [2] Number the carbon chain to give the carbonyl C the lower number. Apply all other nomenclature rules.

• To name a cyclic ketone: [1] Name the ring and change the *-e* ending to *-one*. [2] Number the C's to put the carbonyl C at C1 and give the next substituent the lower number. Apply all other nomenclature rules.

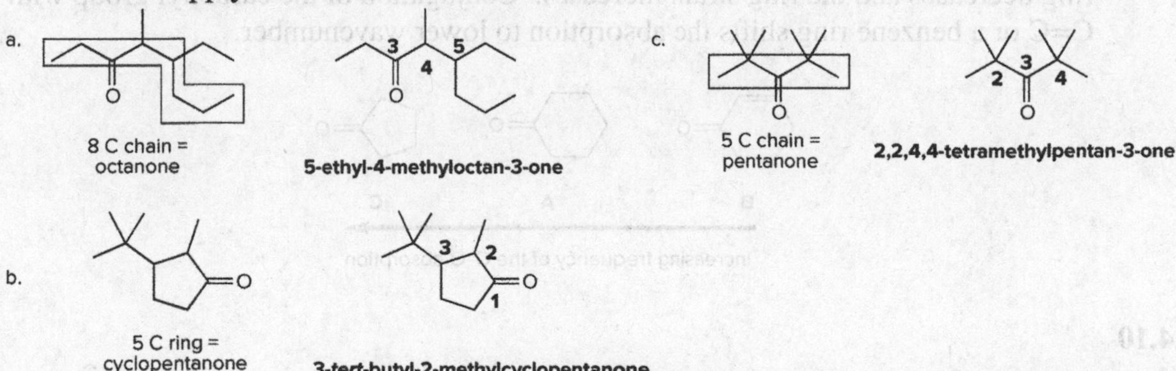

14.6 Most common names are formed by naming both alkyl groups on the carbonyl C, arranging them alphabetically, and adding the word ketone.

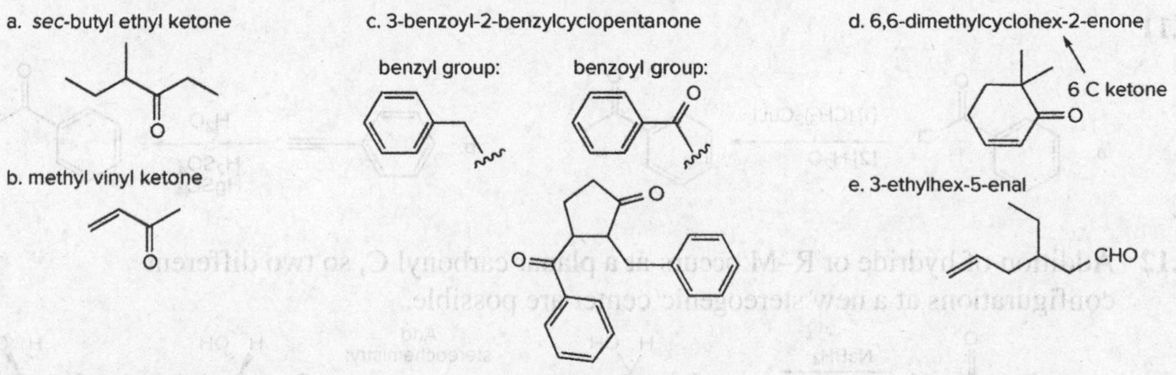

14.7 Compounds with both a C–C double bond and an aldehyde are named as enals.

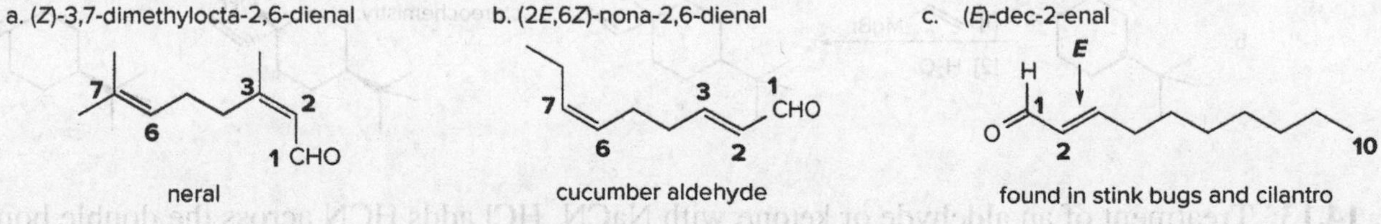

14.8 Even though both compounds have polar C–O bonds, the electron pairs around the sp^3 hybridized O atom of diethyl ether are more crowded and less able to interact with electron-deficient sites in other diethyl ether molecules. The O atom of the carbonyl group of butan-2-one extends out from the carbon chain, making it less crowded. The lone pairs of electrons on the O atom can more readily interact with the electron-deficient sites in the other molecules, resulting in stronger forces and a higher boiling point.

butan-2-one diethyl ether

14.9 For cyclic ketones, the carbonyl absorption shifts to higher wavenumber as the size of the ring decreases and the ring strain increases. Conjugation of the carbonyl group with a C=C or a benzene ring shifts the absorption to lower wavenumber.

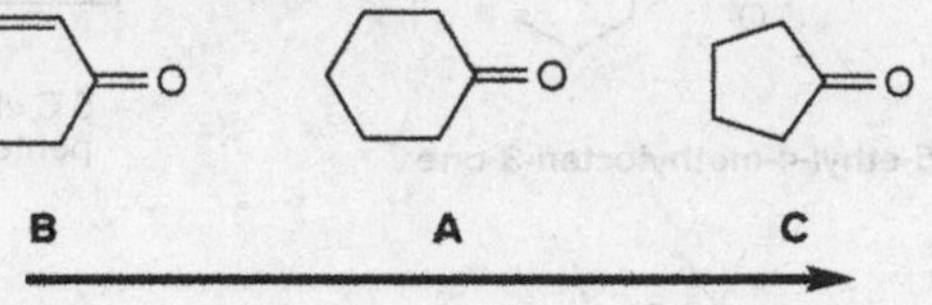

B A C

increasing frequency of the C=O absorption

14.10

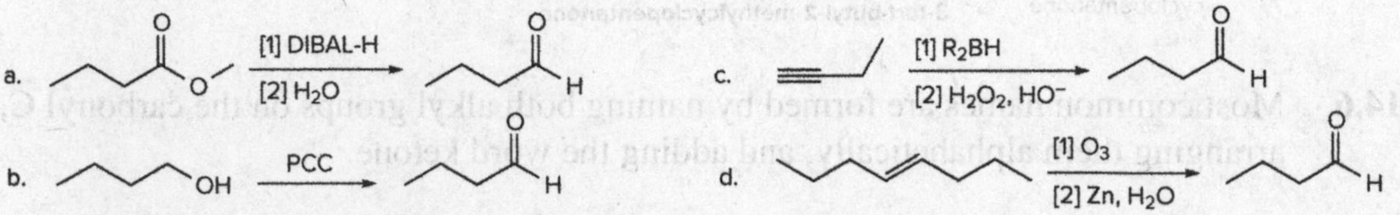

a. [1] DIBAL-H [2] H_2O

b. PCC

c. [1] R_2BH [2] H_2O_2, HO^-

d. [1] O_3 [2] Zn, H_2O

14.11

a. [1] $(CH_3)_2CuLi$ [2] H_2O

b. H_2O H_2SO_4 $HgSO_4$

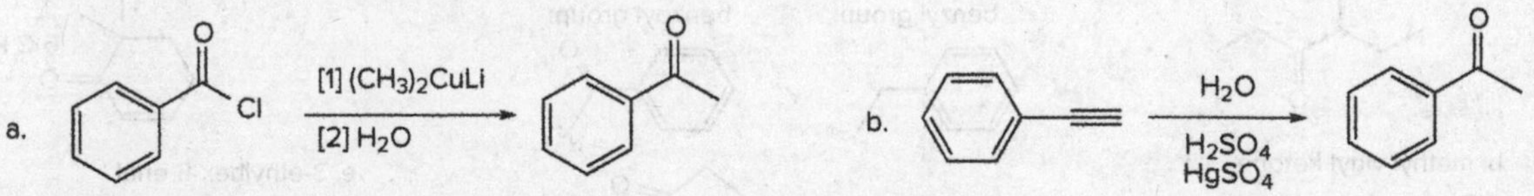

14.12 Addition of hydride or R–M occurs at a planar carbonyl C, so two different configurations at a new stereogenic center are possible.

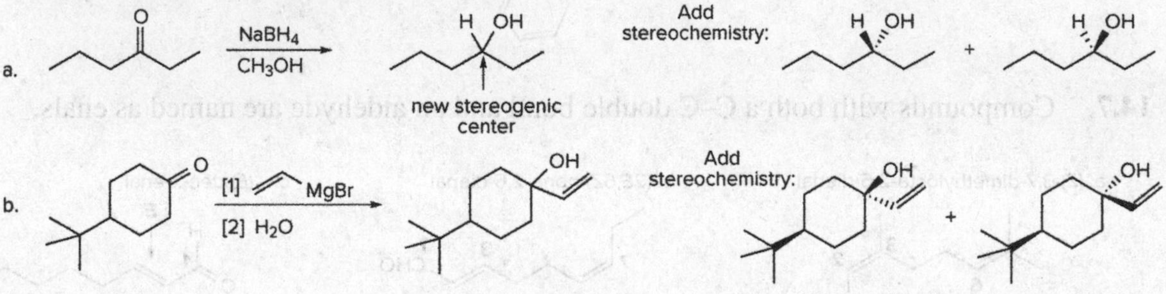

a. $NaBH_4$ CH_3OH — new stereogenic center — Add stereochemistry:

b. [1] MgBr [2] H_2O — Add stereochemistry:

14.13 Treatment of an aldehyde or ketone with NaCN, HCl adds HCN across the double bond. Cyano groups are hydrolyzed by H_3O^+ to replace the three C–N bonds with three C–O bonds.

a. CHO NaCN HCl → OH CN

b. OH CN H_3O^+, Δ → OH COOH

14.14

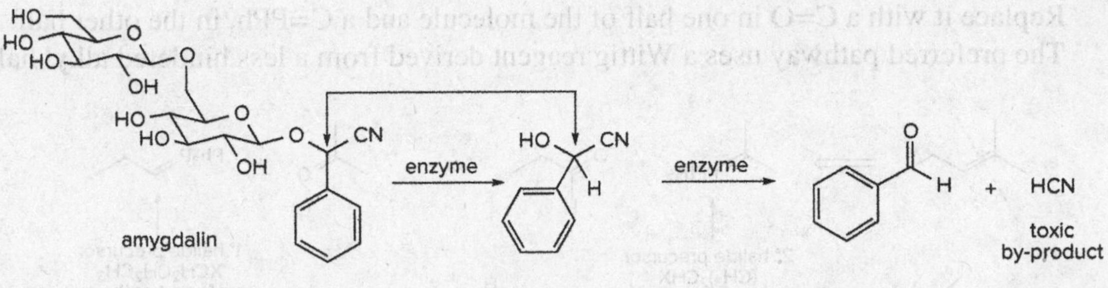

HO, HO, HO, OH, O, O, HO, HO, OH, O, O, CN — enzyme → HO, CN, H — enzyme → O, H + HCN

amygdalin

toxic by-product

14.15

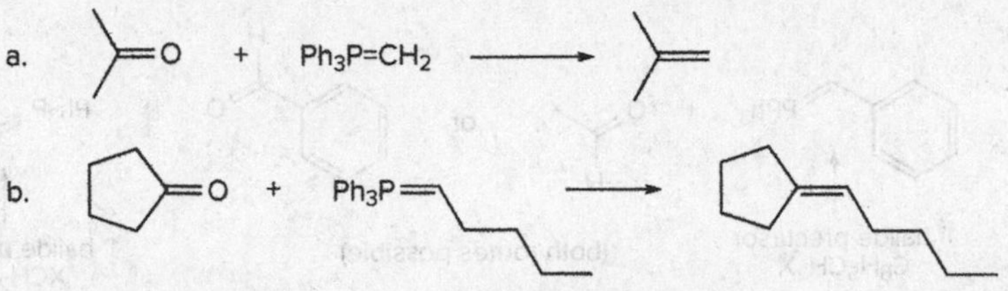

a. (CH$_3$)$_2$C=O + $Ph_3P{=}CH_2$ ⟶

b. cyclopentanone + $Ph_3P{=}$ ⟶

14.16

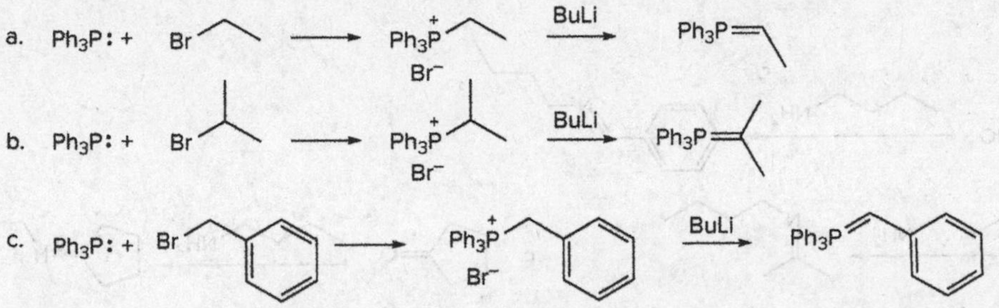

a. Ph_3P: + Br ⟶ $Ph_3\overset{+}{P}$ Br^- —BuLi→ $Ph_3P{=}$

b. Ph_3P: + Br ⟶ $Ph_3\overset{+}{P}$ Br^- —BuLi→ $Ph_3P{=}$

c. Ph_3P: + Br ⟶ $Ph_3\overset{+}{P}$ Br^- —BuLi→ $Ph_3P{=}$

14.17

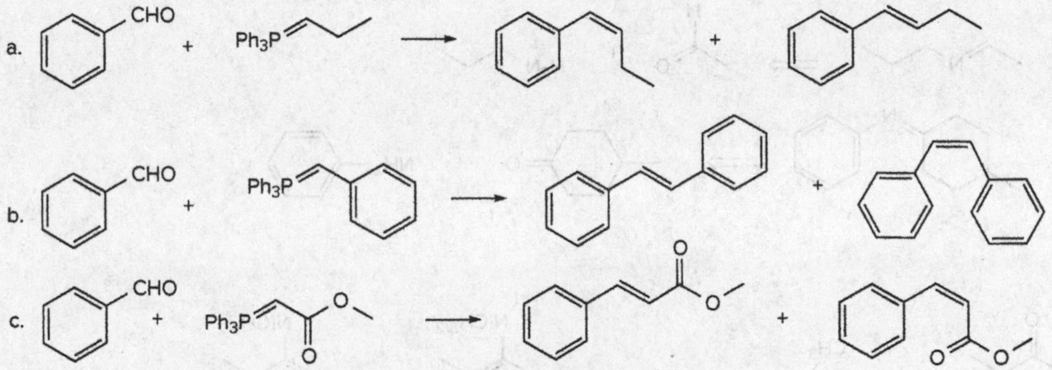

a. CHO + $Ph_3P{=}$ ⟶ +

b. CHO + $Ph_3P{=}$ ⟶ +

c. CHO + $Ph_3P{=}$ O, O ⟶ O, O + O, O

14.18 To draw the starting materials of the Wittig reactions, find the C=C and cleave it. Replace it with a C=O in one half of the molecule and a $C{=}PPh_3$ in the other half. The preferred pathway uses a Wittig reagent derived from a less hindered alkyl halide.

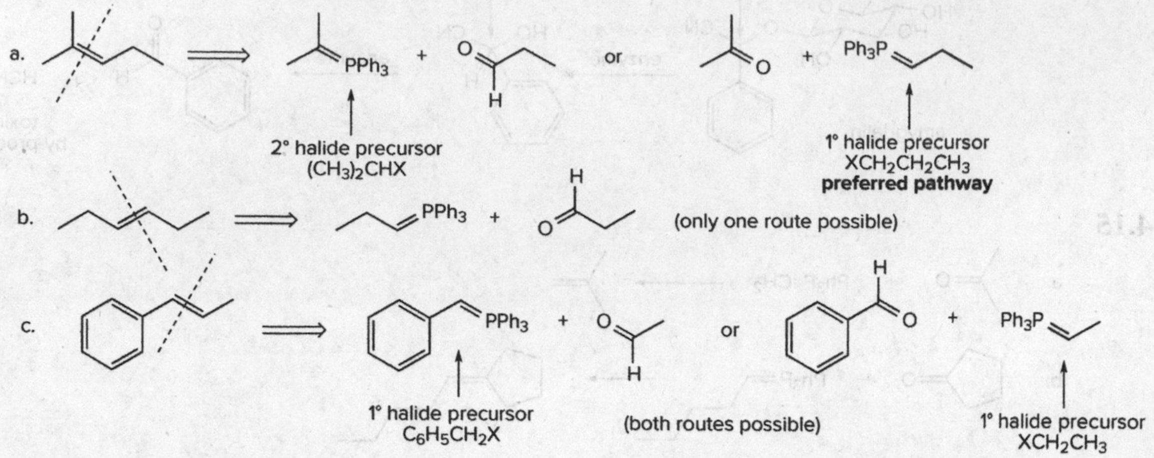

a. ⟹ PPh_3 + O, H or O + Ph_3P

2° halide precursor $(CH_3)_2CHX$

1° halide precursor $XCH_2CH_2CH_3$ **preferred pathway**

b. ⟹ PPh_3 + O (only one route possible)

c. ⟹ PPh_3 + O, H or H, O + Ph_3P

1° halide precursor $C_6H_5CH_2X$

(both routes possible)

1° halide precursor XCH_2CH_3

14.19 When a 1° amine reacts with an aldehyde or ketone, the C=O is replaced by C=NR.

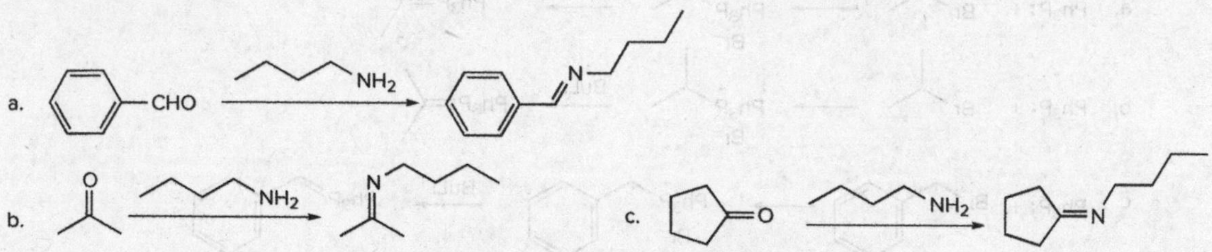

a. CHO —NH_2→ N

b. O —NH_2→ N

c. O —NH_2→ N

14.20 The C=NR is formed from a C=O and an NH_2 group of a 1° amine.

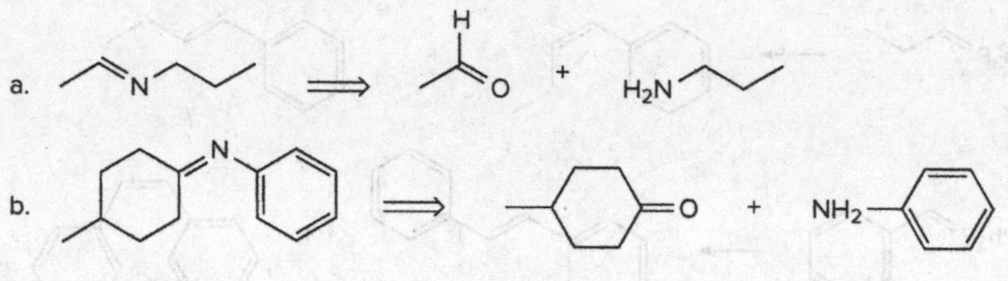

a. N ⟹ H, O + H_2N

b. N ⟹ O + NH_2

14.21

O + CH_3–N–H (CH_3) ⟶ $N(CH_3)_2$ + $N(CH_3)_2$

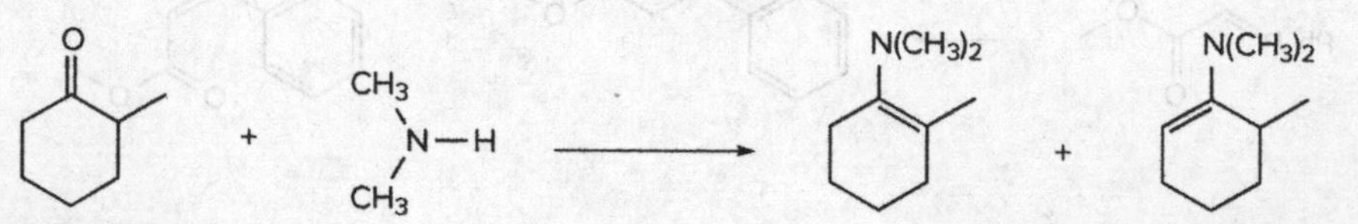

14.22 • Imines are hydrolyzed to 1° amines and a carbonyl compound.
• Enamines are hydrolyzed to 2° amines and a carbonyl compound.

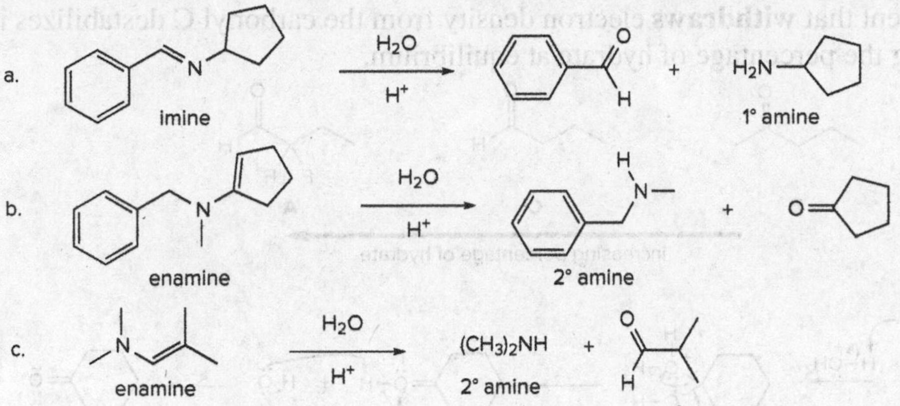

a. imine $\xrightarrow[H^+]{H_2O}$ + H_2N– (1° amine)

b. enamine $\xrightarrow[H^+]{H_2O}$ 2° amine +

c. enamine $\xrightarrow[H^+]{H_2O}$ $(CH_3)_2NH$ (2° amine) +

14.23

$H-\ddot{O}H_2^+$ → $H_2\ddot{O}$: → $H_2\ddot{O}$: → + H_3O^+

proton transfer

$H-\ddot{O}H_2^+$ ← NH_2 ← $H_2\ddot{O}$: + $\overset{+}{N}H_3$

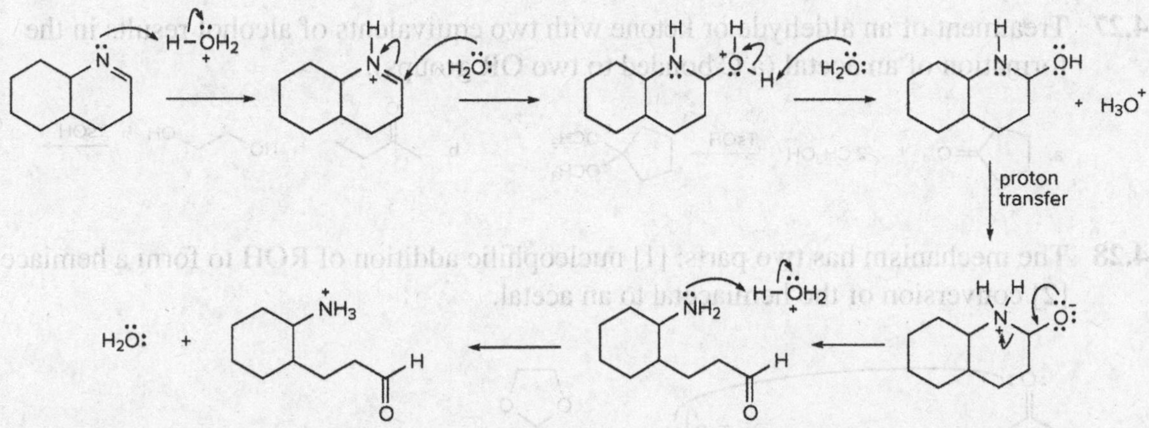

14.24 The amine of the amino acid reacts with the aldehyde carbonyl of pyridoxal phosphate to form an imine. The α carbon of the amino acid becomes the ketone carbonyl of the α-keto acid.

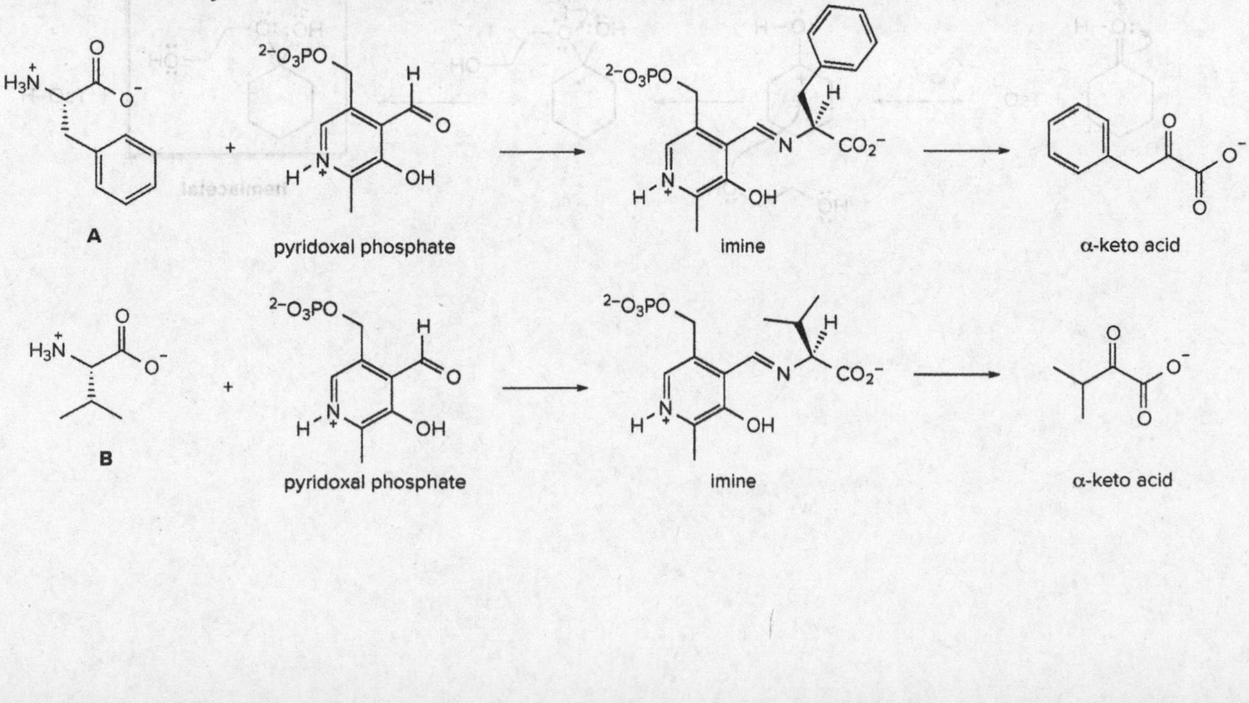

A + pyridoxal phosphate → imine → α-keto acid

B + pyridoxal phosphate → imine → α-keto acid

14.25

- A substituent that **donates** electron density to the carbonyl C stabilizes it, **decreasing** the percentage of hydrate at equilibrium.
- A substituent that **withdraws** electron density from the carbonyl C destabilizes it, **increasing** the percentage of hydrate at equilibrium.

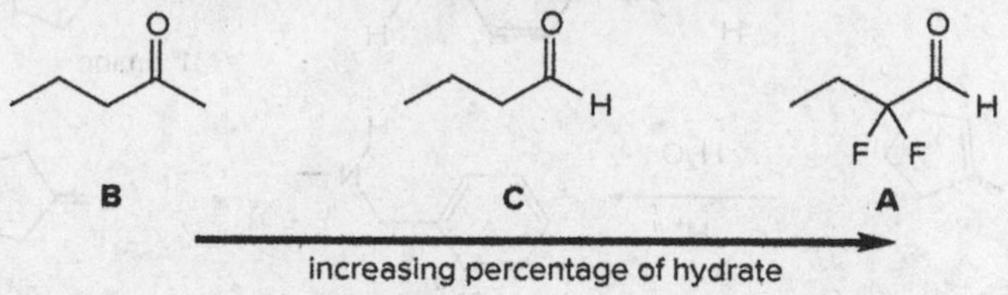

B C A

increasing percentage of hydrate

14.26

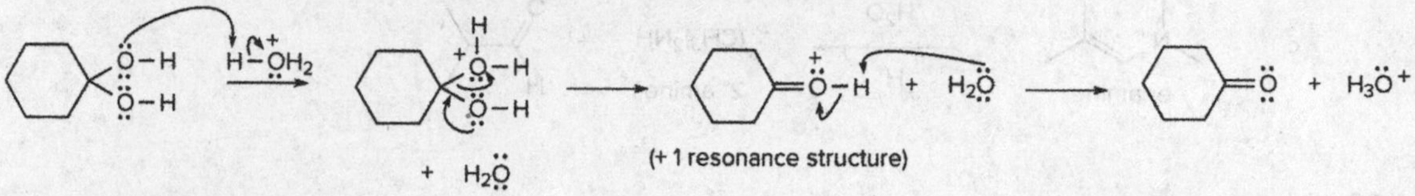

$H-\ddot{O}H_2^+$

(+ 1 resonance structure)

$+ H_2\ddot{O}$ $+ H_3\ddot{O}^+$

14.27 Treatment of an aldehyde or ketone with two equivalents of alcohol results in the formation of an acetal (a C bonded to two OR groups).

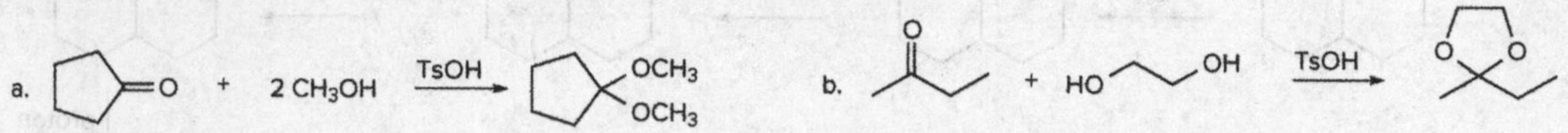

a. + 2 CH_3OH TsOH OCH_3 OCH_3 b. + HO OH TsOH

14.28 The mechanism has two parts: [1] nucleophilic addition of ROH to form a hemiacetal; [2] conversion of the hemiacetal to an acetal.

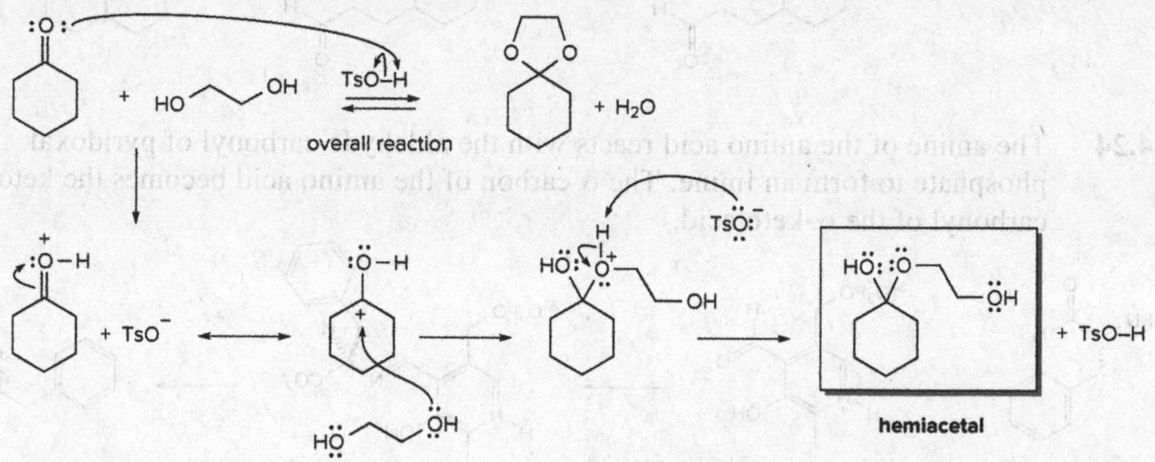

+ HO OH TsO–H + H_2O

overall reaction

+ TsO^- TsÖ:⁻ + TsO–H

hemiacetal

HO: :O—/\—OH TsO—H → H—O⁺: :O—/\—OH + TsO⁻ → :O—/\—OH (carbocation) ↔ ⁺O=/\—OH + H_2O:

hemiacetal

resonance-stabilized carbocation

carbocation re-drawn → (cyclic oxonium ion) TsO:⁻ → acetal + TsO–H

acetal

14.29

a. CH_3O OCH_3 (2,2-dimethoxypropane) $\xrightarrow[H_2SO_4]{H_2O}$ acetone + 2 CH_3OH

b. (cyclic acetal) $\xrightarrow[H_2SO_4]{H_2O}$ acetone + cyclohexane-1,2-diol (OH, OH)

c. (diethyl acetal) $\xrightarrow[H_2SO_4]{H_2O}$ ethanol (OH) + ketone

d. (spiro acetal) $\xrightarrow[H_2SO_4]{H_2O}$ cyclohexanone + HO–(CH₂)₃–HO

e. CH_3O, CH_3O, OCH_3, OCH_3 $\xrightarrow[H_2SO_4]{H_2O}$ CH_3O, CH_3O (ketone) + 2 CH_3OH

14.30

HO, O, O, acetal, oleandrin, H, H, OH, H $\xrightarrow{H_3O^+}$ HO, OH, H, O + HO, H, H, OH, O

14.31 Use an acetal protecting group to carry out the reaction.

(ethyl 3-oxocyclohexanecarboxylate) $\xrightarrow[TsOH]{HO\text{—}OH}$ (acetal-protected ester) $\xrightarrow[[2]\ H_2O]{[1]\ CH_3Li\ (2\ equiv)}$ (acetal, tertiary alcohol OH) $\xrightarrow{H_3O^+}$ (ketone, tertiary alcohol OH)

14.32

a. 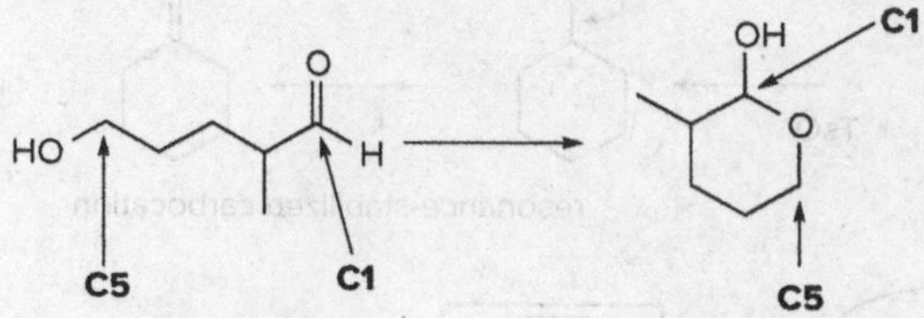 HO–(C5)…–C(=O)H (C1) → cyclic hemiacetal (OH on C1; ring O; C5)

b. 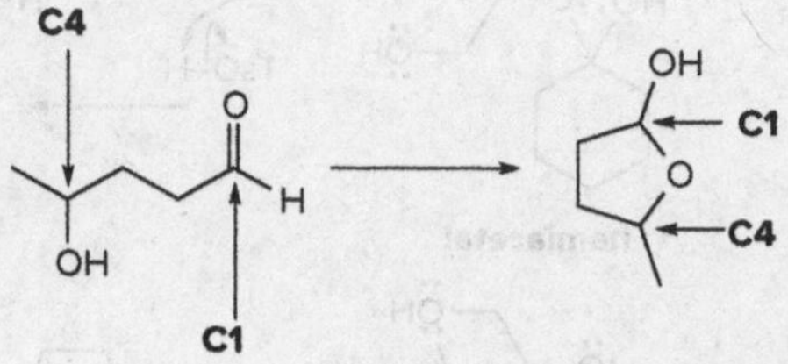 (C4)…–C(=O)H (C1), OH on C4 → cyclic hemiacetal (OH on C1; ring O; C4)

14.33

monensin: acetal; hemiacetal

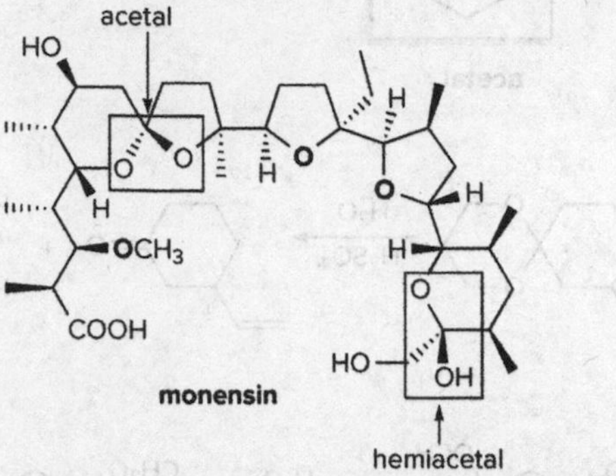

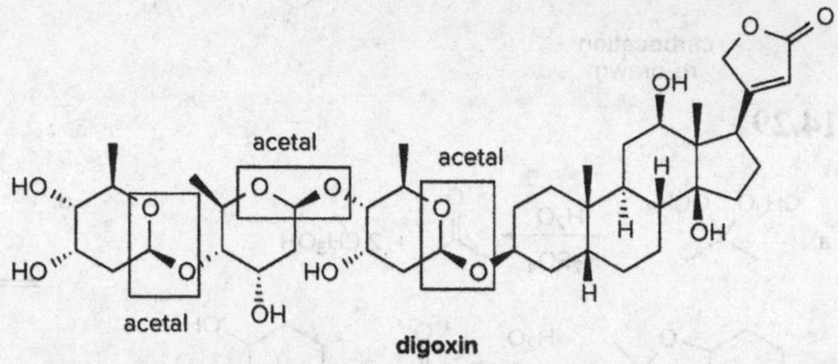

Ether **O** atoms are indicated in **bold.**

digoxin: acetal; acetal; acetal

14.34 The hemiacetal OH is replaced by an OR group to form an acetal.

a. 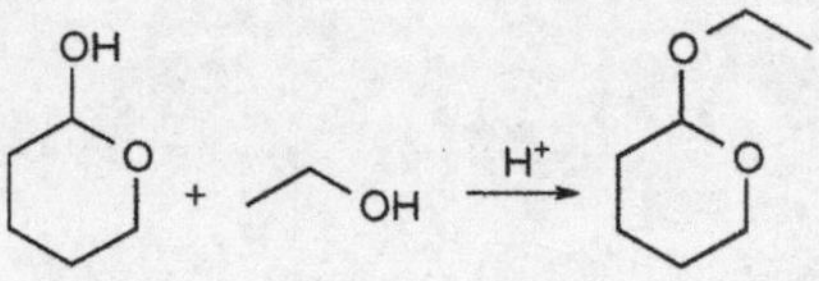 cyclic hemiacetal + ethanol $\xrightarrow{H^+}$ ethyl acetal

b. 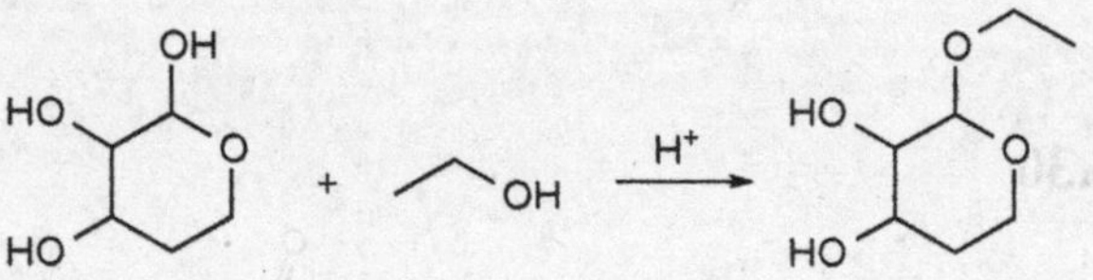 HO, HO substituted cyclic hemiacetal + ethanol $\xrightarrow{H^+}$ ethyl acetal

14.35

a. 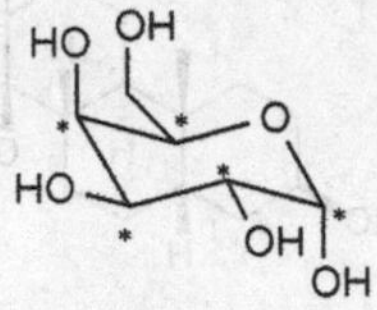 5 stereogenic centers (labeled with *)

b. 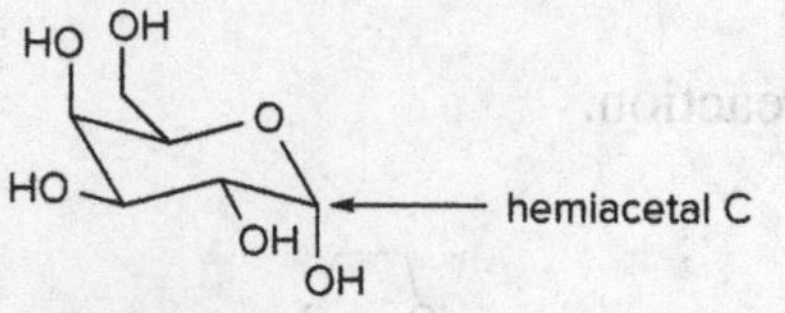 hemiacetal C

α-D-galactose

c. 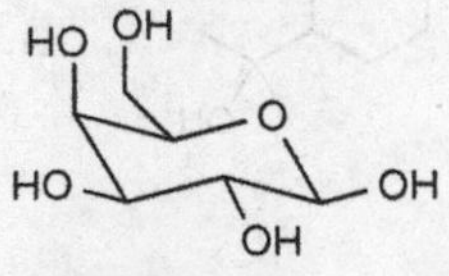 β-D-galactose

d. 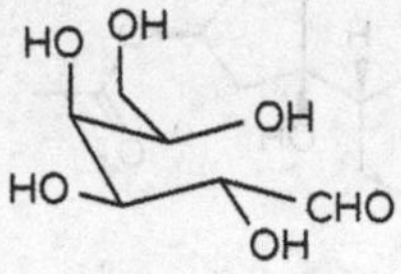 HO, OH, OH, HO, CHO, OH (open-chain form)

e. 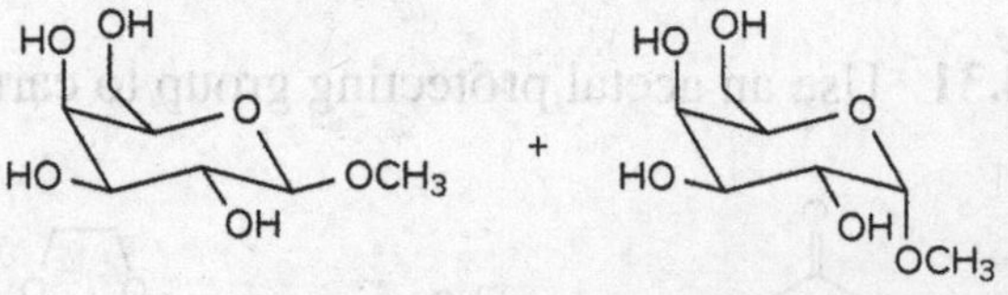 OCH$_3$ + OCH$_3$

14.36

a. 3,3-dimethylbutanal **A**; *cis*-5-isopropyl-2-methylcyclohexanone **B**

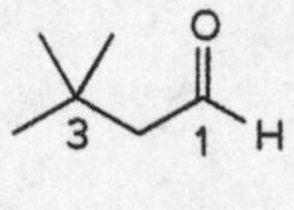

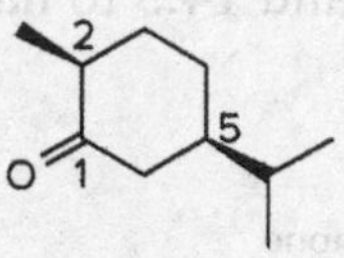

b. [1] **A** $\xrightarrow[CH_3OH]{NaBH_4}$

[2] **A** $\xrightarrow{CH_3MgBr}$ $\xrightarrow{H_2O}$ +

[3] **A** $\xrightarrow{Ph_3P{=}CHOCH_3}$ OCH_3 (+ cis isomer)

[4] **A** $\xrightarrow[\text{mild } H^+]{}$

[5] **A** $\xrightarrow[H^+]{HOCH_2CH_2CH_2OH}$

[1] **B** $\xrightarrow[CH_3OH]{NaBH_4}$ +

[2] **B** $\xrightarrow{CH_3MgBr}$ $\xrightarrow{H_2O}$ +

[3] **B** $\xrightarrow{Ph_3P{=}CHOCH_3}$ +

[4] **B** $\xrightarrow[\text{mild } H^+]{}$

[5] **B** $\xrightarrow[H^+]{HOCH_2CH_2CH_2OH}$

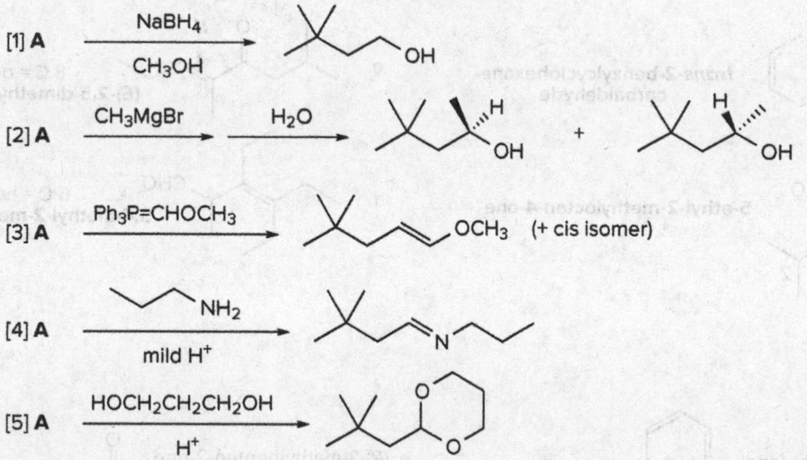

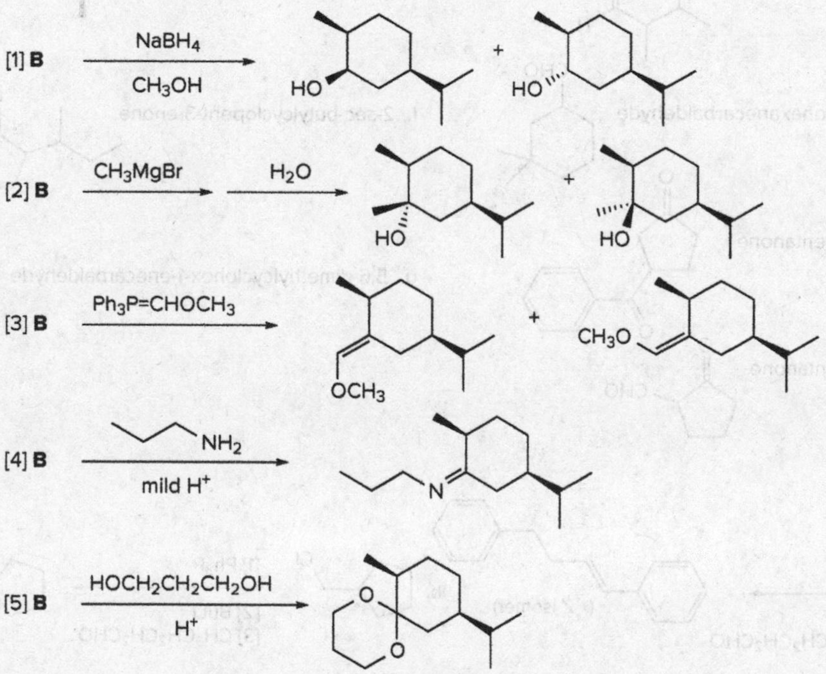

14.37

a. ⟹ CHO + HO OH

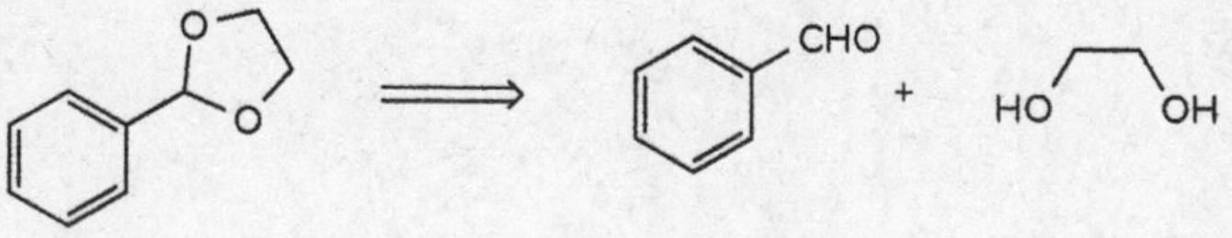

b. ⟹ OH OH +

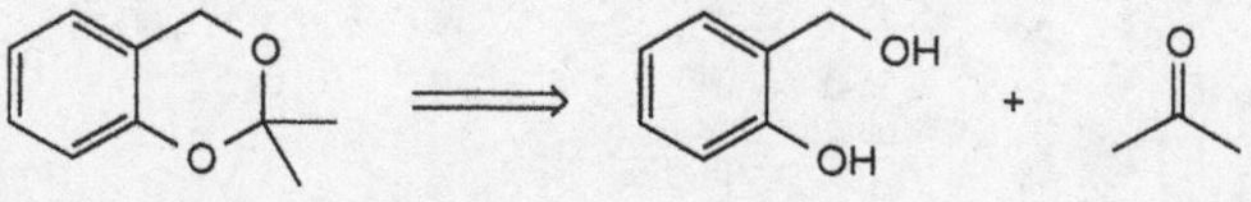

14.38 Use the rules from Answers 14.3 and 14.5 to name the aldehydes and ketones.

a. 6 C ring = cyclohexanone
5-ethyl-2-methyl-cyclohexanone

b. CHO
***trans*-2-benzylcyclohexane-carbaldehyde**

c. 5 4 O 2
5-ethyl-2-methyloctan-4-one

d. CHO
6 C = hexanal
3,4-diethylhexanal

e. O *E*
8 C = octenone
(*E*)-2,5-dimethyloct-5-en-4-one

f. CHO
6 C = hexenal
3,4-diethyl-2-methylhex-3-enal

14.39

a. 2-methyl-3-phenylbutanal
O H

b. 3,3-dimethylcyclohexanecarbaldehyde
CHO

c. 3-benzoylcyclopentanone
O O

d. 2-formylcyclopentanone
O CHO

e. (*R*)-3-methylheptan-2-one
O

f. 2-*sec*-butylcyclopent-3-enone
O

g. 5,6-dimethylcyclohex-1-enecarbaldehyde
CHO

14.40

a. Br [1] Ph_3P [2] BuLi [3] $C_6H_5CH_2CH_2CHO$ → (+ *Z* isomer)

b. Cl [1] Ph_3P [2] BuLi [3] $CH_3CH_2CH_2CHO$ → (+ *Z* isomer)

14.41

a. 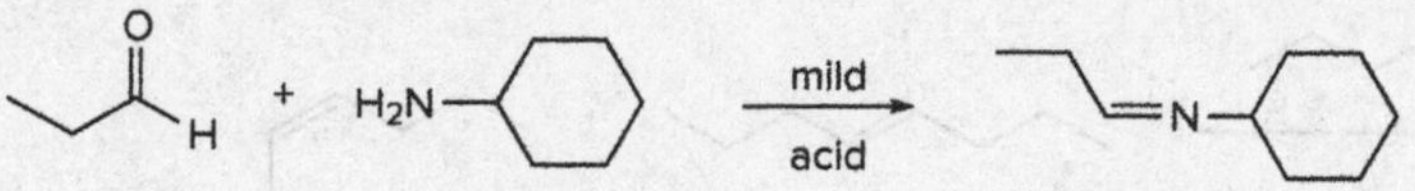 + H_2N– —mild acid→ =N–

b. 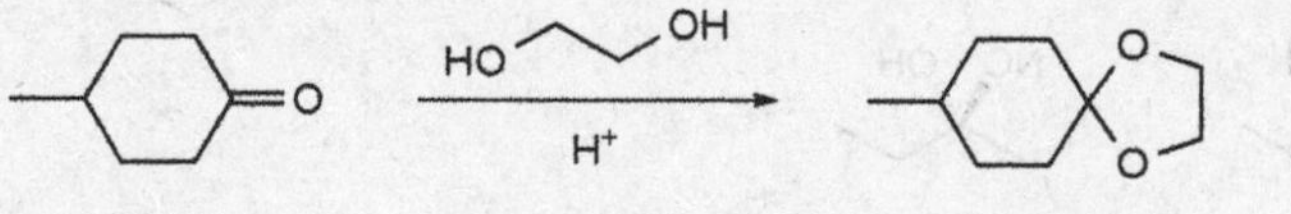 —HO⁀OH, H^+→

c.  —H_3O^+→ + H_2N

d. 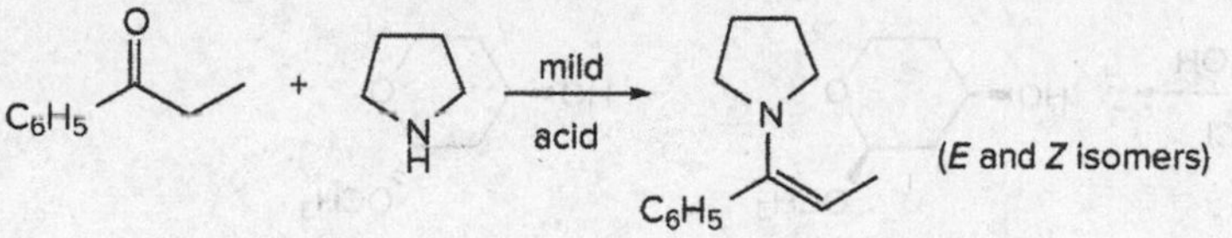 C_6H_5 + N H —mild acid→ N C_6H_5 (*E* and *Z* isomers)

e. 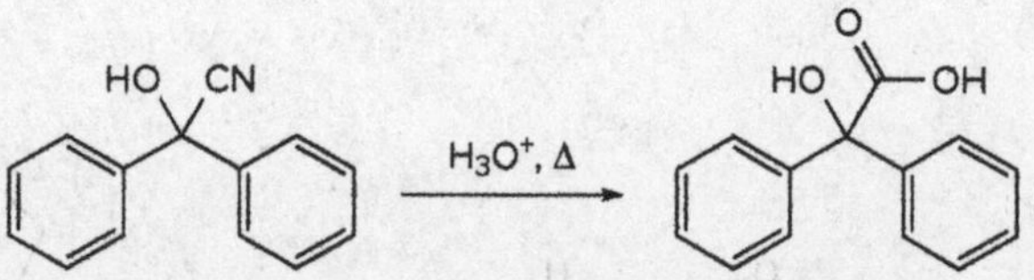 HO CN —H_3O^+, Δ→ HO OH O

f. OH O —CH_3CH_2OH, H^+→ O O

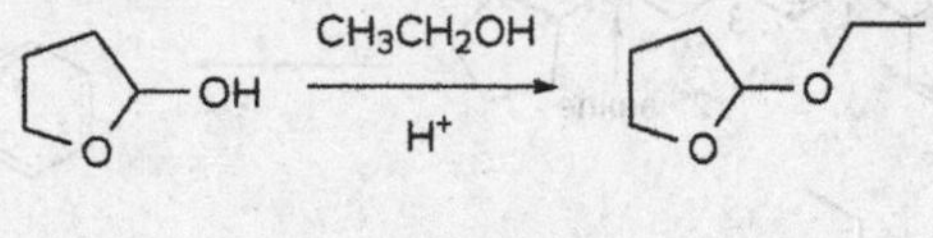

g. 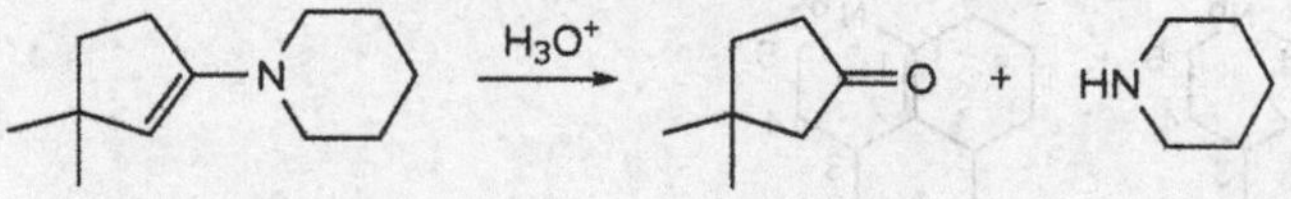 N —H_3O^+→ =O + HN

h. 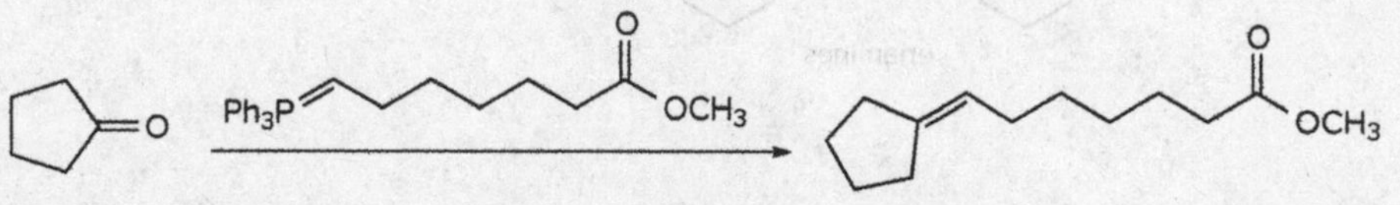 =O —Ph_3P= OCH₃ O→ OCH₃ O

14.42

a. 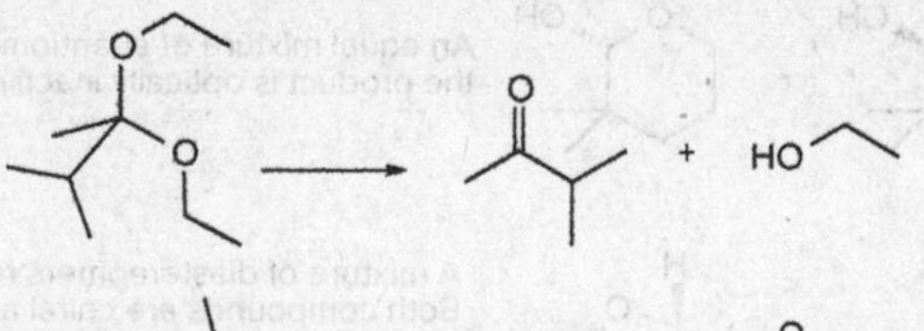 O O → O + HO

b. O O O O → O O O + HOCH₃

c. O O → HO H O + $HOCH_2CH_3$

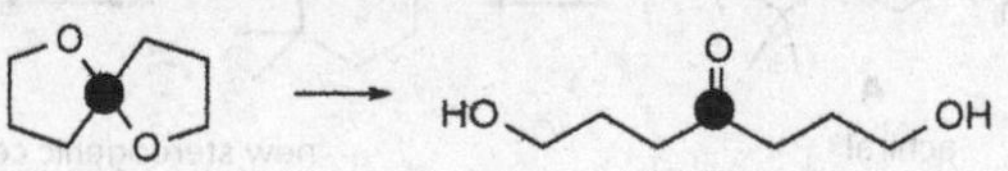

d. O O → HO O OH

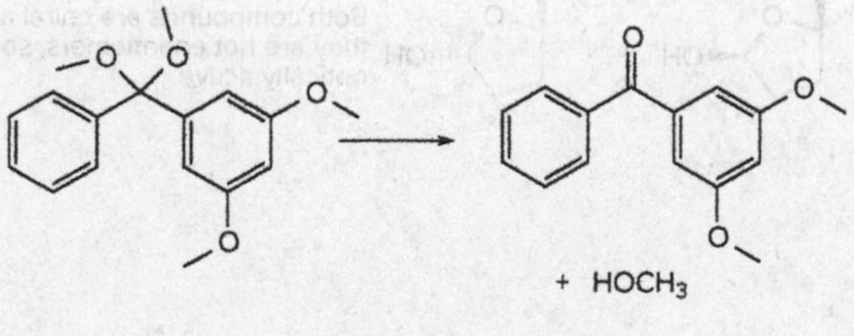

e. 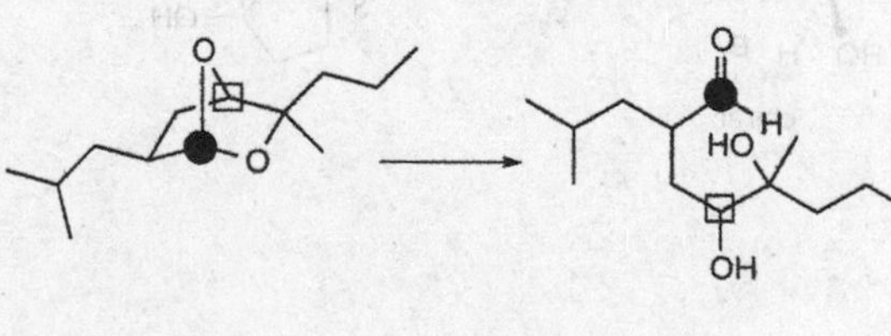 O O O → O H HO OH

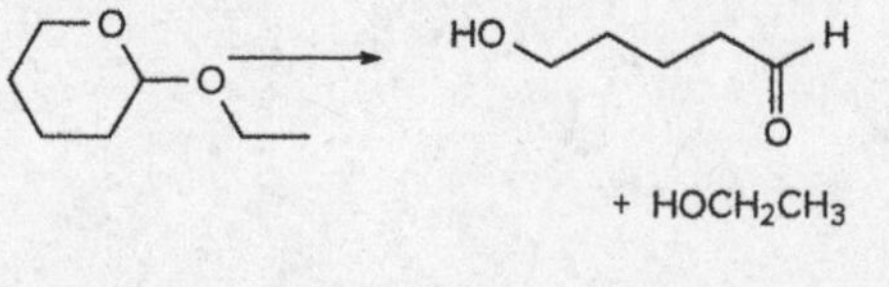

14.43

a. Ph_3P= → (E and Z alkenes) +

b. NaCN, HCl → HO CN + NC OH

c. $NaBH_4$, CH_3OH → OH + OH

d. HO, O, OH → CH_3OH, HCl → HO, O, OCH_3 + HO, O, OCH_3

14.44

a. O, 1, 2, 3, 4, 5, NH_2, 1° amine → 5N, 4, 3, 1, 2

b. O, 1, 2, 3, 4, 5, H, N6, 2° amine → N6, 1, 2, 3, 4, 5 + N6, 1, 2, 3, 4, 5

enamines

c. O, 1, 2, 3, 4, H, N5, 2° amine → 5N, 4, 3, 2, 1

14.45

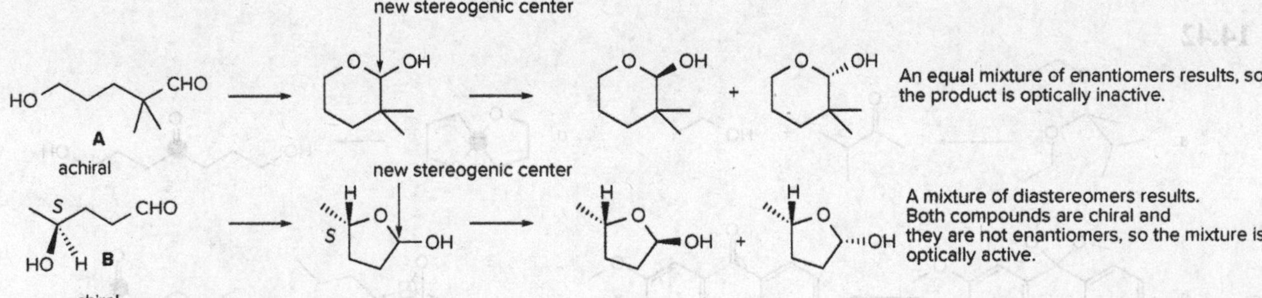
new stereogenic center
HO
CHO
A
achiral
OH
OH
OH
An equal mixture of enantiomers results, so the product is optically inactive.
new stereogenic center
S
CHO
HO H B
chiral
H
S
OH
H
OH
H
OH
A mixture of diastereomers results. Both compounds are chiral and they are not enantiomers, so the mixture is optically active.

14.46

a, b.

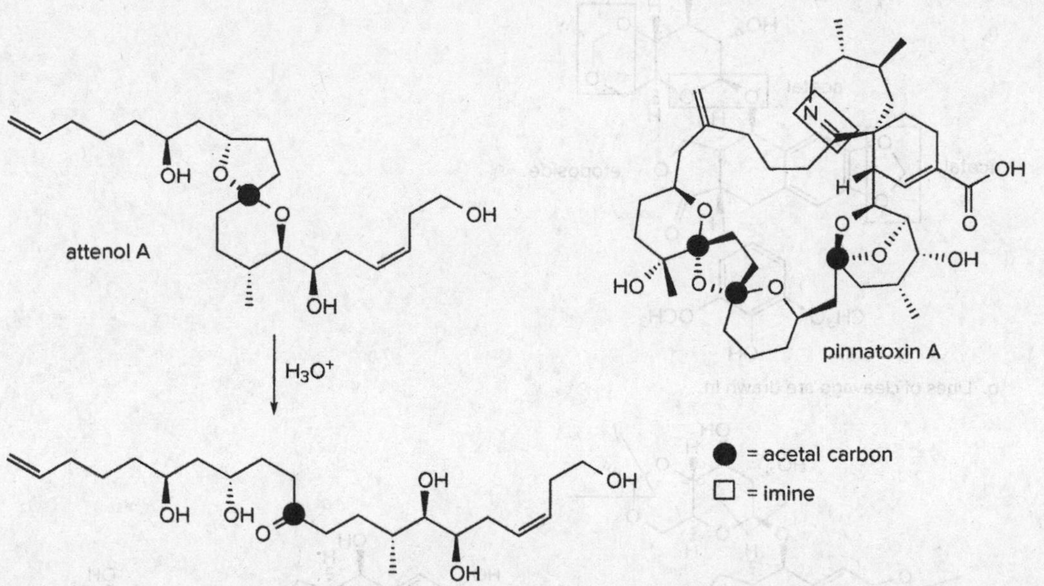

attenol A

H_3O^+

pinnatoxin A

● = acetal carbon
□ = imine

14.47

a. imine → H_3O^+ → 1° amine

b. enamine → H_3O^+ → 2° amine

c. enamine → H_3O^+ → 2° amine

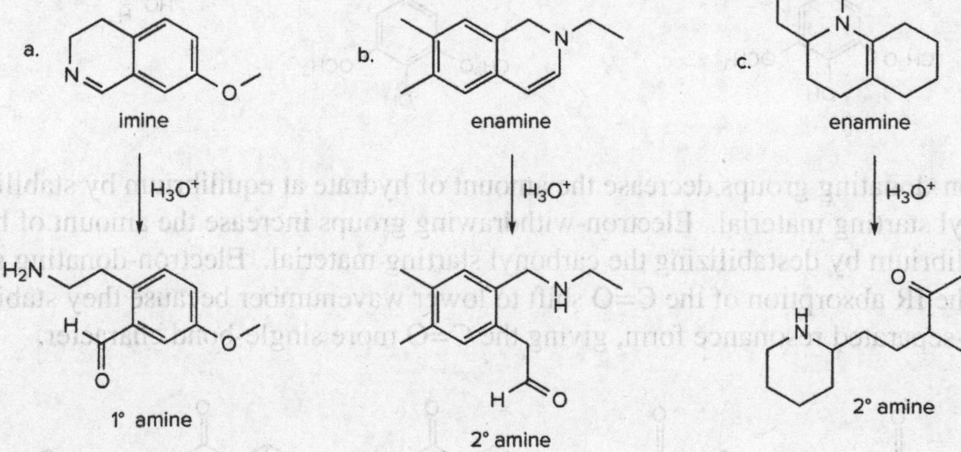

14.48 The ketone carbonyl is formed from the α carbon of the α-amino acid, which is bonded to NH_3^+ and H.

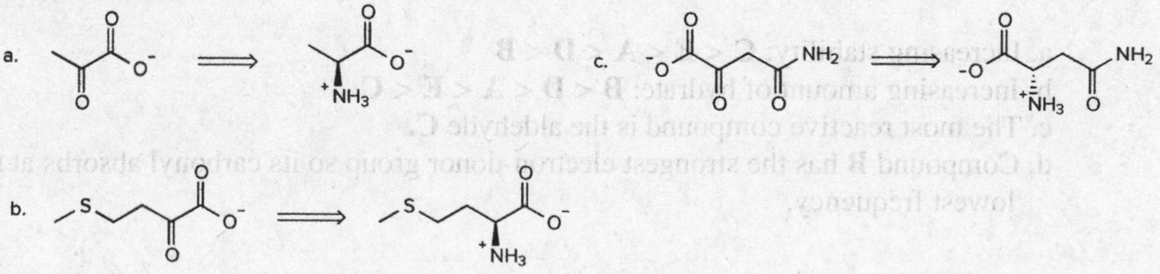

a.

b.

c.

14.49

a.

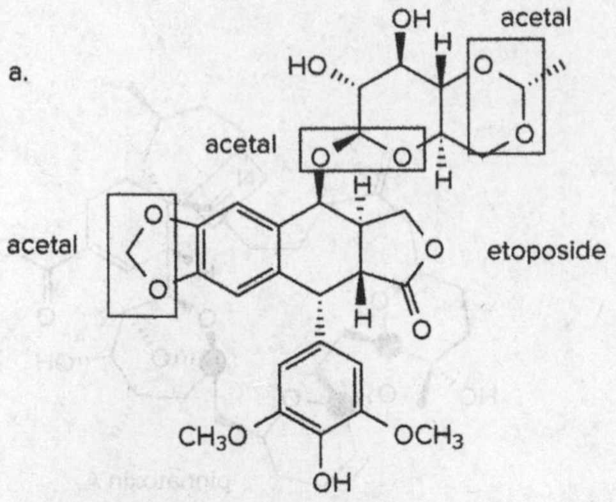
OH
acetal
HO
H
acetal
H
H
acetal
etoposide
H
CH3O
OCH3
OH

b. Lines of cleavage are drawn in.

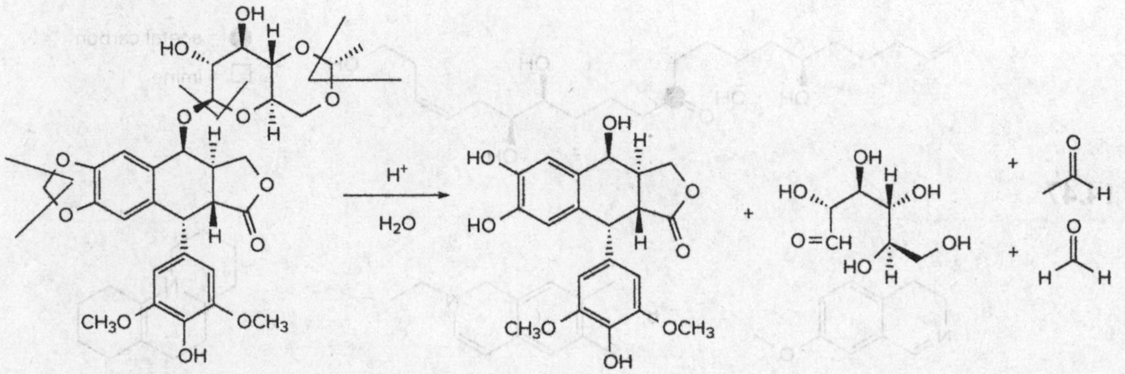
OH
HO
H
H
H
H
CH3O
OCH3
OH
H+
H2O
HO
HO
OH
H
H
CH3O
OCH3
OH
OH
HO
H
OH
O=CH
OH
HO
H
H
H
H

14.50 Electron-donating groups decrease the amount of hydrate at equilibrium by stabilizing the carbonyl starting material. Electron-withdrawing groups increase the amount of hydrate at equilibrium by destabilizing the carbonyl starting material. Electron-donating groups make the IR absorption of the C=O shift to lower wavenumber because they stabilize the charge-separated resonance form, giving the C=O more single-bond character.

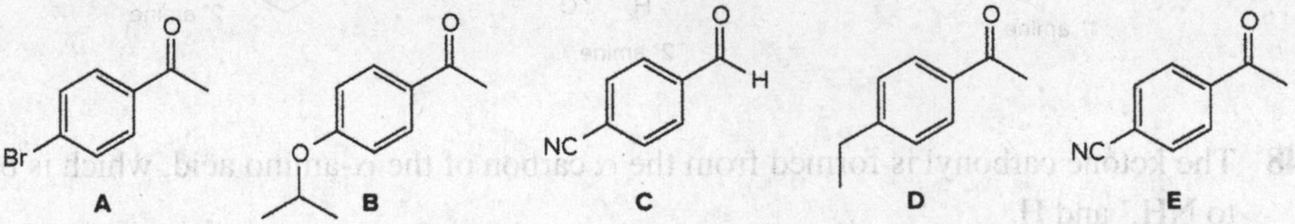
Br
A
B
NC
H
C
D
NC
E

a. Increasing stability: **C < E < A < D < B**
b. Increasing amount of hydrate: **B < D < A < E < C**
c. The most reactive compound is the aldehyde **C.**
d. Compound **B** has the strongest electron-donor group so its carbonyl absorbs at the lowest frequency.

14.51 Use the principles from Answer 14.18.

a. [alkene] ⟹ [ketone] + Ph_3P=[ylide] or [ylide]PPh_3 + [aldehyde, H, O]

1° alkyl halide precursor
($XCH_2CH_2CH_2CH_3$)
preferred pathway

2° alkyl halide precursor
[$(CH_3CH_2)_2CHX$]

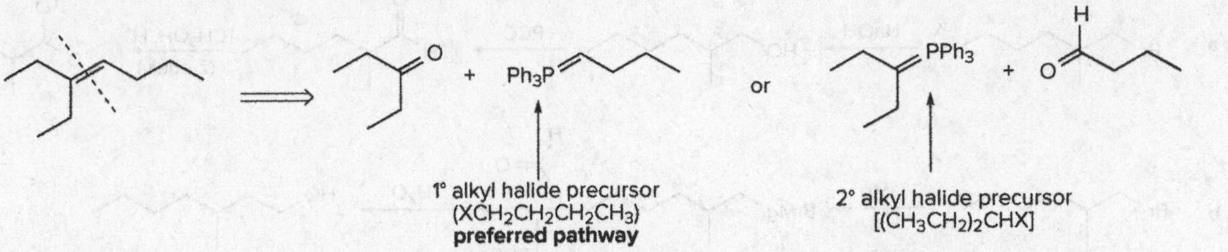

b. [alkene] ⟹ [ketone, O] + $Ph_3P{=}CH_2$ or [ylide, PPh_3] + $CH_2{=}O$

methyl halide precursor
(CH_3X)
preferred pathway

2° alkyl halide precursor
($CH_3CH_2CH_2CHXCH_3$)

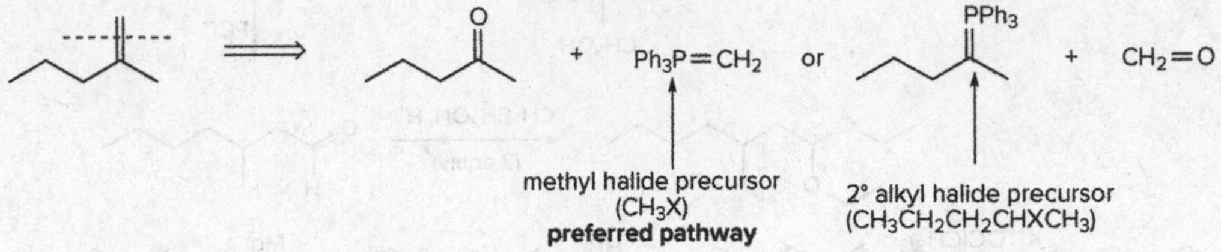

c. [alkene, C_6H_5] ⟹ [cyclohexanone]=O + Ph_3P=[ylide, C_6H_5] or [cyclohexylidene]=PPh_3 + O=[aldehyde, C_6H_5, H]

1° alkyl halide precursor
($C_6H_5CH_2X$)
preferred pathway

2° alkyl halide precursor
[cyclohexyl]–X

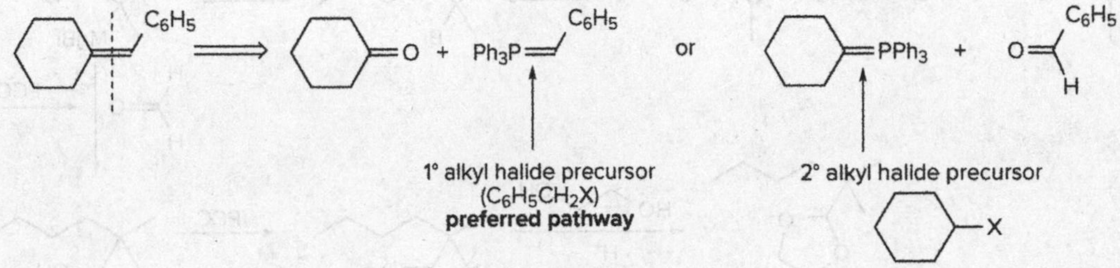

14.52

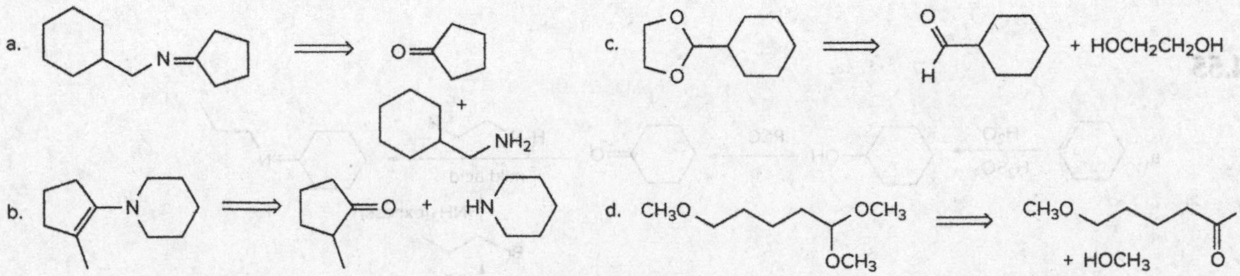

a. [imine, N=] ⟹ O=[cyclopentanone] + [amine, NH_2]

b. [enamine, N] ⟹ [ketone]=O + HN[piperidine]

c. [acetal, O, O] ⟹ [aldehyde, O, H] + $HOCH_2CH_2OH$

d. CH_3O[chain]OCH_3, OCH_3 ⟹ CH_3O[chain]H, O + $HOCH_3$

14.53

a. [alkene]

[benzyl alcohol] OH $\xrightarrow{PBr_3}$ [benzyl bromide] Br $\xrightarrow{[1]\ Ph_3P,\ [2]\ BuLi}$ [ylide] PPh_3 $\xrightarrow{CHO}$ [alkene] (+ *Z* isomer)

[alcohol] OH $\xrightarrow{PCC}$ CHO

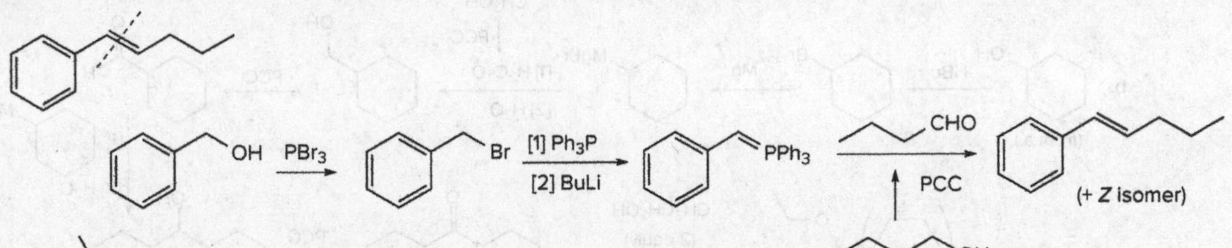

b. [alkene]

[alcohol]–OH $\xrightarrow{PCC}$ [ketone]=O

[alcohol] OH $\xrightarrow{PBr_3}$ [alkyl bromide] Br $\xrightarrow{[1]\ Ph_3P,\ [2]\ BuLi}$ Ph_3P=[ylide] $\xrightarrow{\text{ketone}=O}$ [alkene]

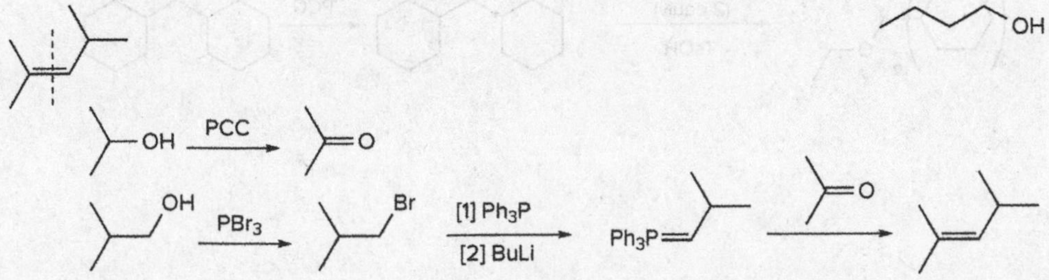

14.54

a. Br–(alkyl bromide) $\xrightarrow{NaOH}$ HO–(alcohol) $\xrightarrow{PCC}$ H–C(=O)–(aldehyde) $\xrightarrow[(2\ equiv)]{CH_3OH,\ H^+}$ (dimethyl acetal)

b. Br–(alkyl bromide) $\xrightarrow{Mg}$ BrMg–(Grignard) $\xrightarrow{H_2C=O}$ $\xrightarrow{H_2O}$ HO–(alcohol) $\xrightarrow{PCC}$ (aldehyde, H) $\xrightarrow[(2\ equiv)]{CH_3CH_2OH,\ H^+}$ (diethyl acetal)

H–C(=O)–H prepared from CH_3OH with PCC

c. Br–(alkyl bromide) $\xrightarrow{K^+\,{}^-OC(CH_3)_3}$ (alkene) $\xrightarrow{HBr}$ (tertiary bromide, Br) $\xrightarrow{Mg}$ (MgBr) $\xrightarrow{H_2C=O}$ (–OH) $\xrightarrow{PCC}$ (aldehyde, H, O) $\xrightarrow[H^+]{HO\quad OH}$ (1,3-dioxolane, O, O)

H–C(=O)–H prepared from CH_3OH with PCC

14.55

a. (cyclohexene) $\xrightarrow[H_2SO_4]{H_2O}$ (cyclohexanol, –OH) $\xrightarrow{PCC}$ (cyclohexanone, =O) $\xrightarrow[\text{mild acid}]{H_2N\text{–propyl}}$ (imine, =N–propyl)

HO–propyl $\xrightarrow{PBr_3}$ Br–propyl $\xrightarrow{NH_3\ (excess)}$ H_2N–propyl

b. (cyclohexanol, OH) (from a.) $\xrightarrow{HBr}$ (Br) $\xrightarrow{Mg}$ (MgBr) $\xrightarrow[\text{[2] } H_2O]{\text{[1] } H_2C=O}$ (OH) $\xrightarrow{PCC}$ (aldehyde, O, H) $\xrightarrow[\text{[2] } H_2O]{\text{[1] cyclohexyl-MgBr}}$ (OH) $\xrightarrow{PCC}$ (ketone, O) $\xrightarrow[TsOH]{CH_3CH_2OH\ (2\ equiv)}$ (cyclohexyl)$_2$C(OEt)$_2$ (O, O, 2)

$H_2C=O$ prepared from CH_3OH with PCC

14.56

a. PCC; PBr_3; Mg; [1] (acetaldehyde) [2] H_2O; PCC; (ethanol) TsOH

b. H_2SO_4 → $CH_2{=}CH_2$; Br_2; 2 $NaNH_2$ → HC≡CH; NaH → $HC{\equiv}C^-\ Na^+$; [1] OsO₄ [2] $NaHSO_3$, H_2O → HO⁄⁄OH; [1] (acetaldehyde) (from a.) [2] H_2O; PCC; TsOH

14.57

HO⁄⁄OH, TsOH; Mg; [1] (cyclopentanone) [2] H_2O; H_2O, H^+

Br⁄⁄⁄CHO; BrMg; CHO

A

14.58

(+ 1 resonance structure)

H_2O + ... $\overset{+}{N}H_3$

(+ 1 resonance structure)

14.59

H_2SO_4

$H-\overset{+}{O}SO_3H$

$+ HSO_4^-$

(+ 1 resonance structure)

HSO_4^-

$H-\overset{+}{O}SO_3H$

(+ 1 resonance structure)

$+ HSO_4^-$

$+ H_2SO_4$

HSO_4^-

14.60

Ar

$NHNH_2$

H_2NSO_2

Ar'

$Ar'NHNH_2$

proton transfer

$Ar'NHNH$

$H-Cl$

$+ Cl^-$

H_2O

$+ H_3O^+$

proton transfer

$+ Cl^-$

$+ H_3O^+$

celecoxib

14.61 The OH groups react with the C=O in an intramolecular reaction, first to form a hemiacetal, and then to form an acetal.

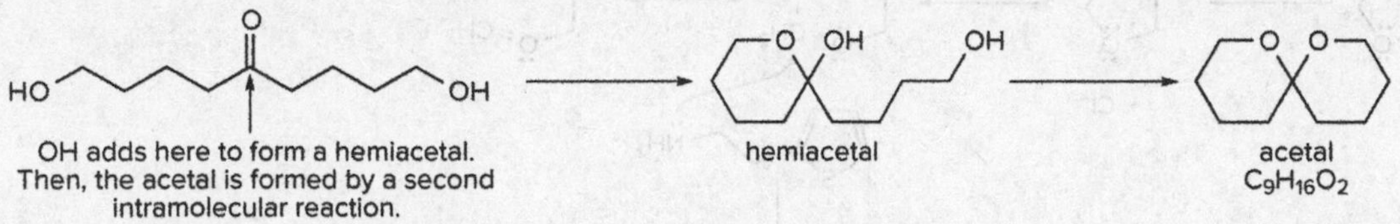

OH adds here to form a hemiacetal. Then, the acetal is formed by a second intramolecular reaction.

hemiacetal

acetal
$C_9H_{16}O_2$

14.62

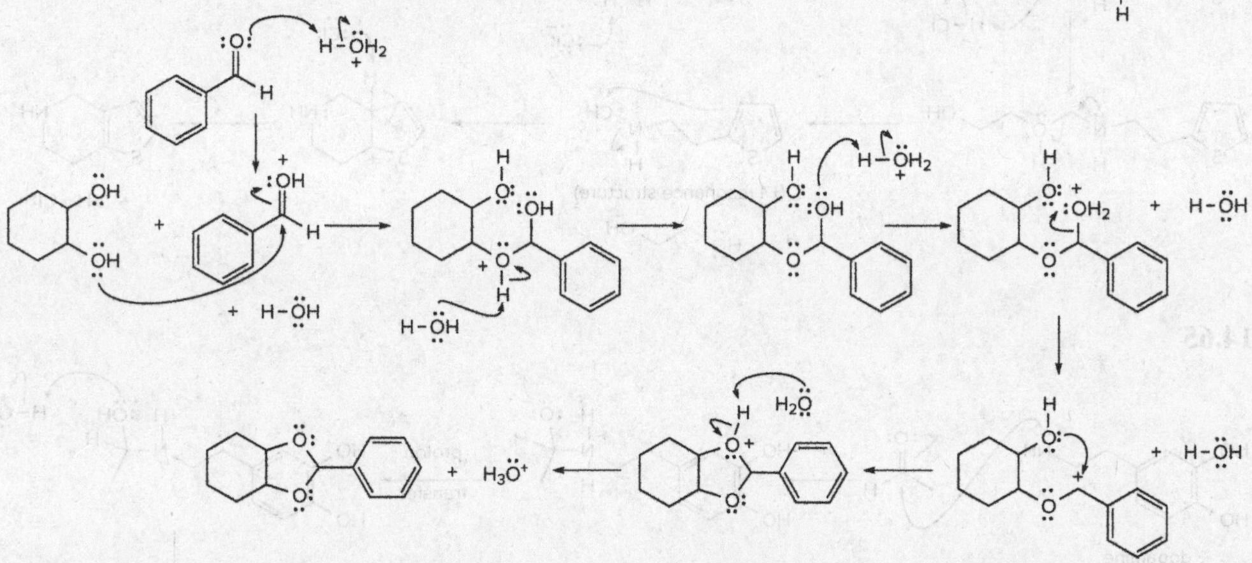

14.63 Nucleophilic addition of H_2O to the protonated imine forms a tetrahedral intermediate. After two proton transfers the amine comes off as a leaving group to form the α-keto acid and pyridoxamine phosphate.

A

carbinolamine

pyridoxamine phosphate

α-keto acid

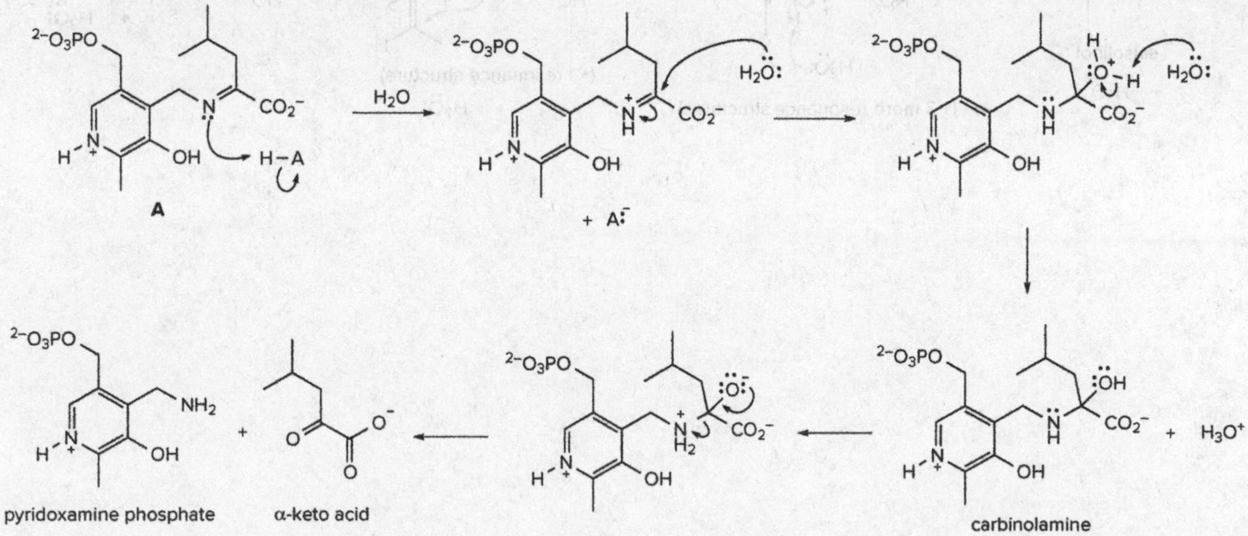

14.64

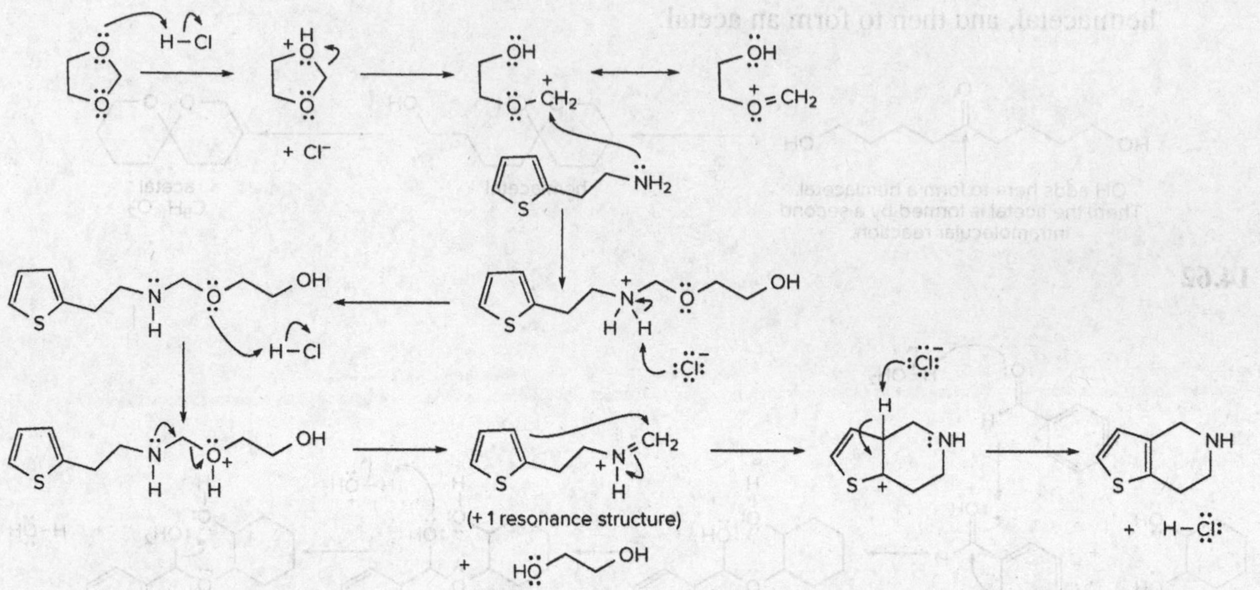

H–Cl
+ Cl⁻
(+ 1 resonance structure)
+ HO–CH₂CH₂–OH
+ H–Cl

14.65

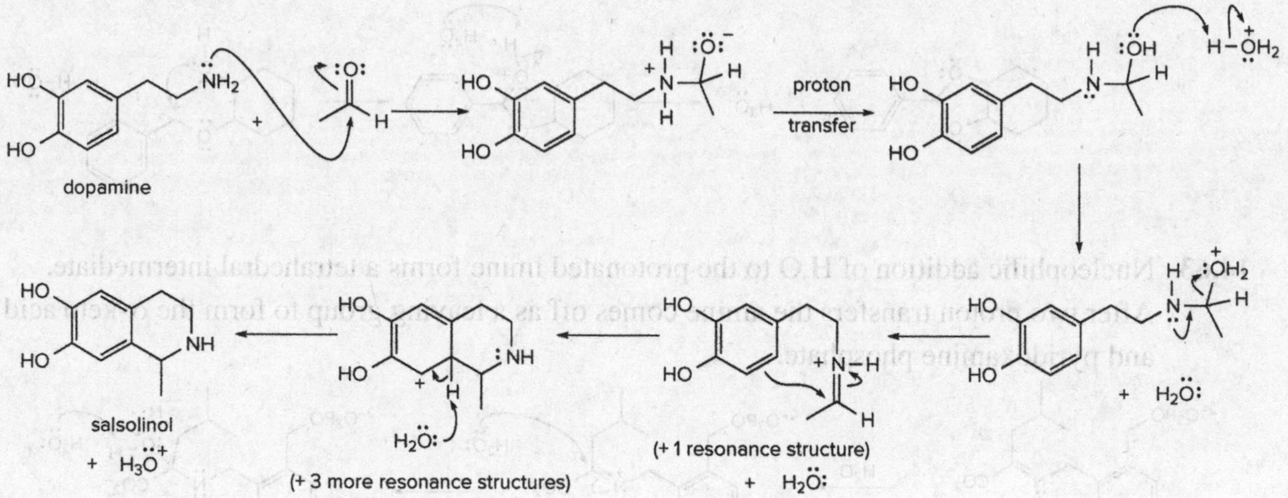

dopamine
proton transfer
(+ 1 resonance structure)
+ H_2O
(+ 3 more resonance structures)
H_2O
salsolinol
+ H_3O^+

14.66

CH_3–MgI

5,5-dimethoxypentan-2-one

(+ 1 resonance structure)

(+ 1 resonance structure)

Y

14.67

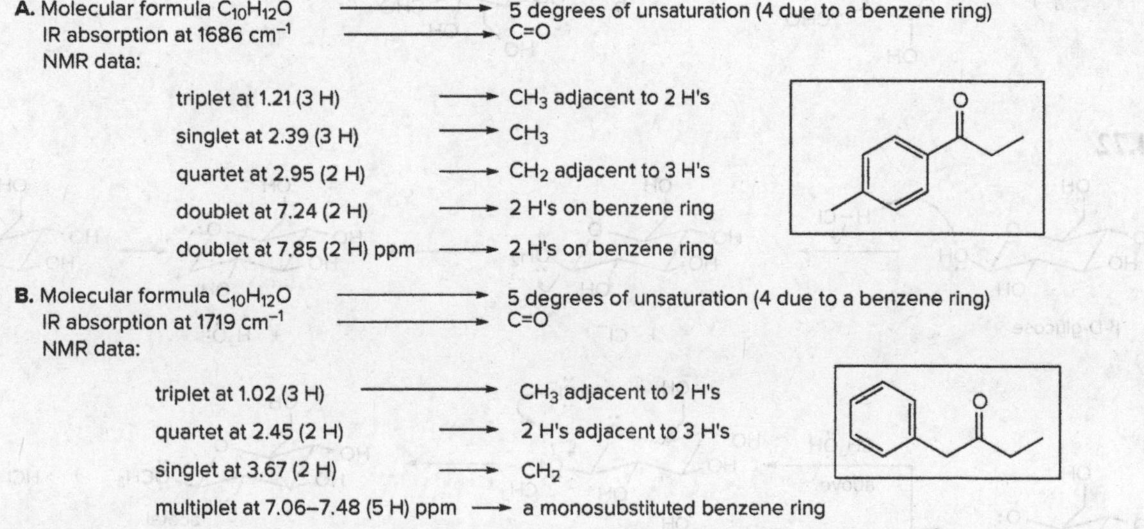

A. Molecular formula $C_{10}H_{12}O$ → 5 degrees of unsaturation (4 due to a benzene ring)

IR absorption at 1686 cm^{-1} → C=O

NMR data:

- triplet at 1.21 (3 H) → CH_3 adjacent to 2 H's
- singlet at 2.39 (3 H) → CH_3
- quartet at 2.95 (2 H) → CH_2 adjacent to 3 H's
- doublet at 7.24 (2 H) → 2 H's on benzene ring
- doublet at 7.85 (2 H) ppm → 2 H's on benzene ring

B. Molecular formula $C_{10}H_{12}O$ → 5 degrees of unsaturation (4 due to a benzene ring)

IR absorption at 1719 cm^{-1} → C=O

NMR data:

- triplet at 1.02 (3 H) → CH_3 adjacent to 2 H's
- quartet at 2.45 (2 H) → 2 H's adjacent to 3 H's
- singlet at 3.67 (2 H) → CH_2
- multiplet at 7.06–7.48 (5 H) ppm → a monosubstituted benzene ring

14.68

$C_7H_{16}O_2$: 0 degrees of unsaturation

IR: 3000 cm^{-1}: **C–H bonds**

NMR data (ppm):

H_a: quartet at 3.8 (**4 H**), split by 3 H's

H_b: singlet at 1.5 (**6 H**)

H_c: triplet at 1.2 (**6 H**), split by 2 H's

$CH_3CH_2–O–C(CH_3)_2–O–CH_2CH_3$

H_a, H_b, H_c

14.69

A: Molecular formula $C_9H_{10}O$
5 degrees of unsaturation
IR absorption at 1700 cm^{-1} → C=O
IR absorption at ~2700 cm^{-1} → CH of RCHO
NMR data (ppm):
- triplet at 1.2 (2 H's adjacent)
- quartet at 2.7 (3 H's adjacent)
- doublet at 7.3 (2 H's on benzene)
- doublet at 7.7 (2 H's on benzene)
- singlet at 9.9 (CHO)

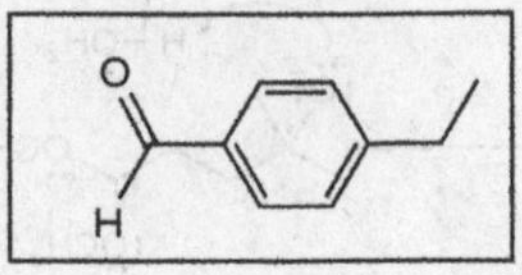

14.70

C. Molecular formula $C_6H_{12}O_3$
1 degree of unsaturation
IR absorption at 1718 cm^{-1} → C=O
NMR data (ppm):
- singlet at 2.1 (3 H's)
- doublet at 2.7 (2 H's)
- singlet at 3.3 (6 H's – 2 OCH_3 groups)
- triplet at 4.8 (1 H)

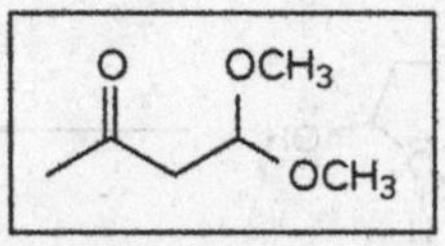

14.71

a.

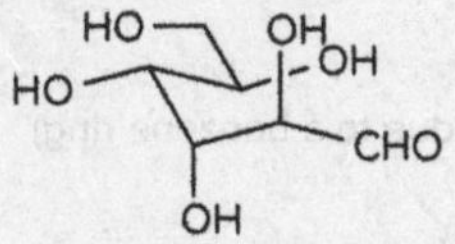

b.

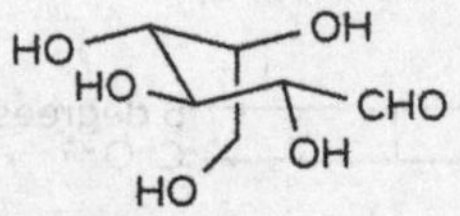

14.72

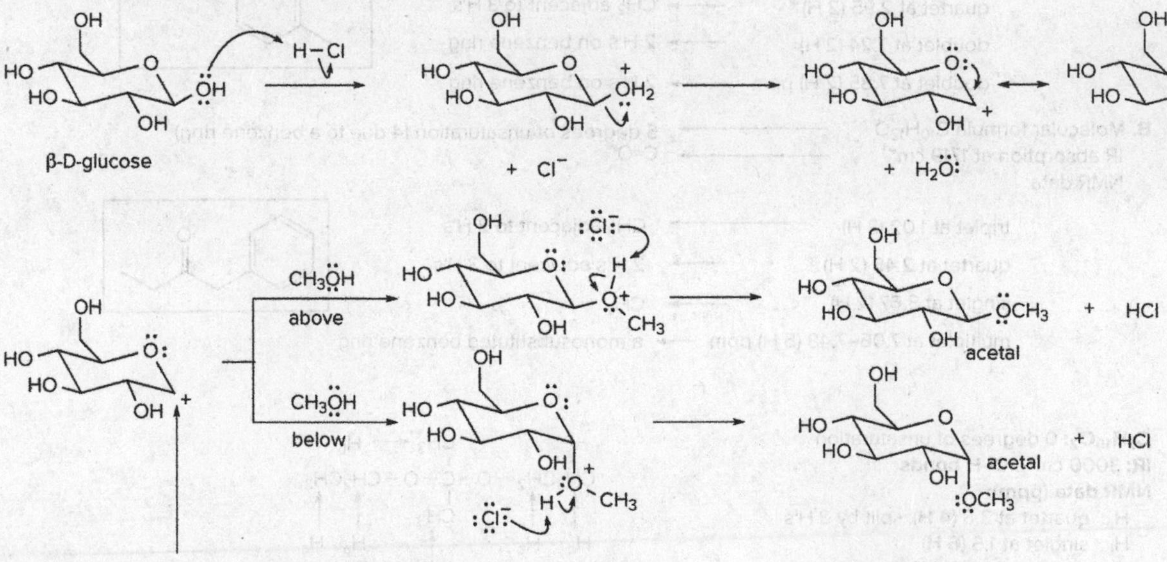

The carbocation is trigonal planar, so CH_3OH attacks from two different directions, and two different acetals are formed.

14.73

a. brevicomin

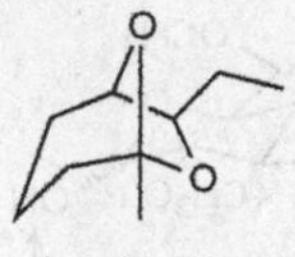

b. [1] Ph_3P [2] BuLi

PPh_3

PCC

[1] OsO_4 [2] $NaHSO_3$, H_2O

(+ *Z* isomer)

H_3O^+

brevicomin

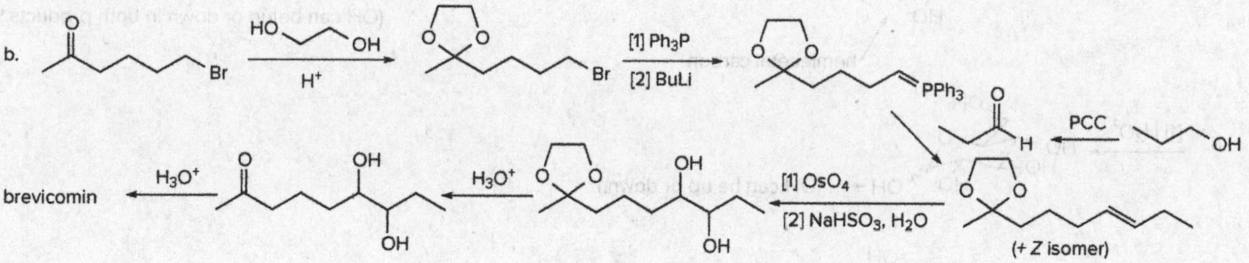

14.74

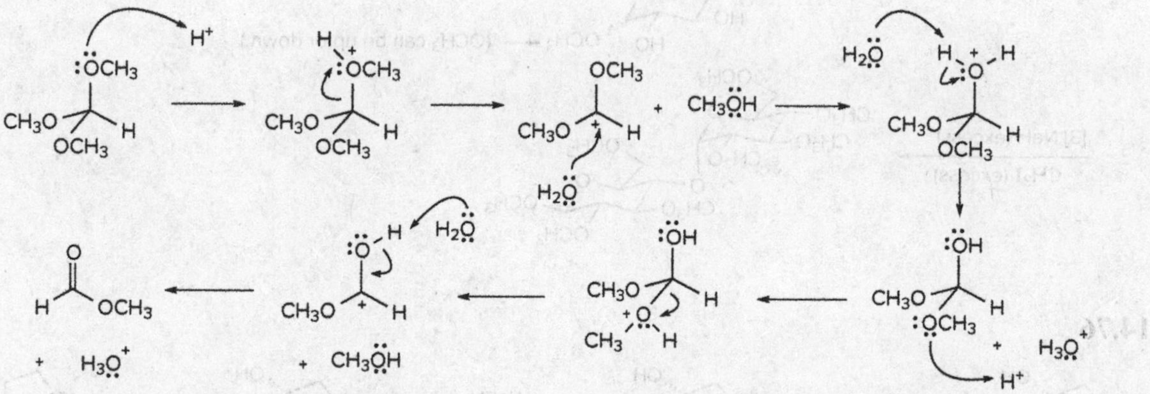

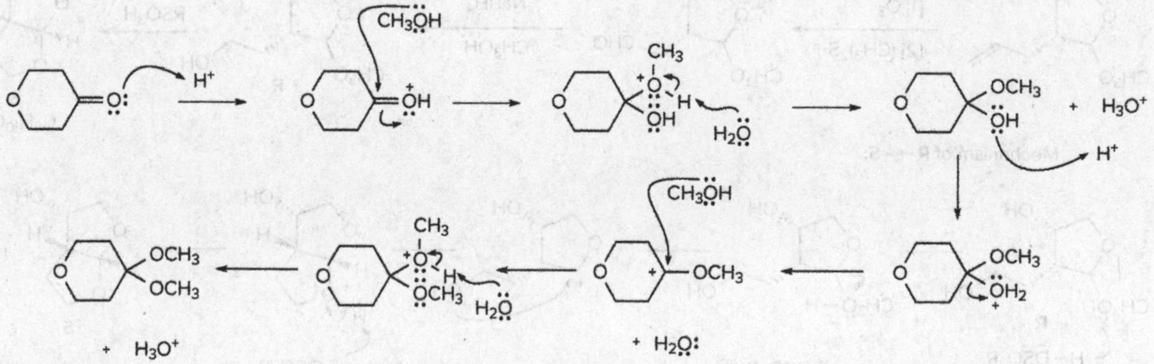

H^+ CH_3OH H_2O H_3O^+ H^+

CH_3OH H_2O H_3O^+ H^+

+ H_2O

+ H_3O^+

14.75

a. acetal carbon

hemiacetal carbon

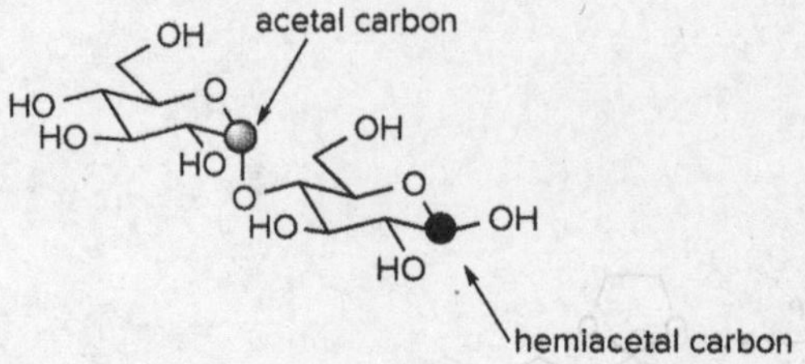

c. (OH can be up or down in both products.)

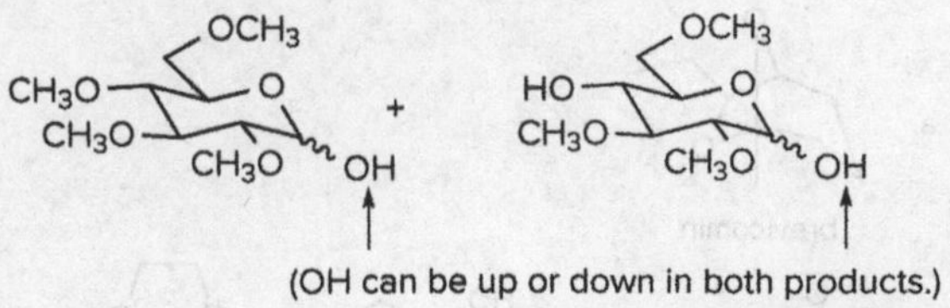

b. [1] H_3O^+ → (OH can be up or down.)

[2] CH_3OH, HCl → (OCH_3 can be up or down.)

[3] NaH (excess), CH_3I (excess) →

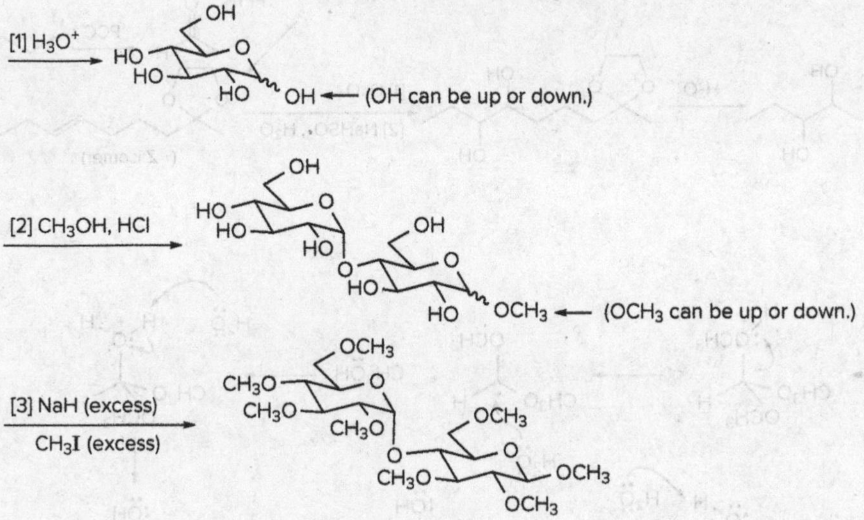

14.76

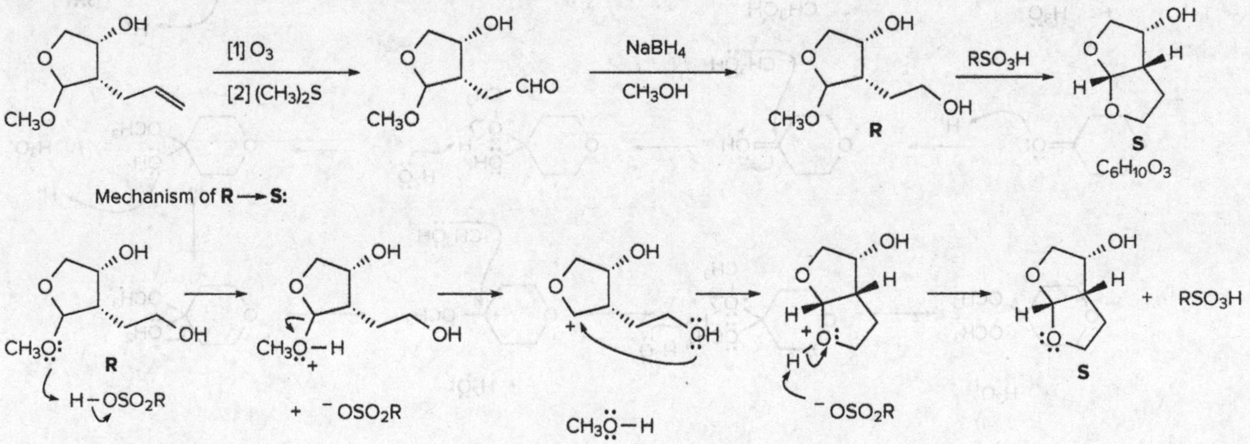

[1] O_3, [2] $(CH_3)_2S$ → CHO; $NaBH_4$, CH_3OH → **R**; RSO_3H → **S** $C_6H_{10}O_3$

Mechanism of **R** → **S**:

R + H–OSO_2R → + $^-OSO_2R$ → + $CH_3\ddot{O}$–H → + $^-OSO_2R$ → **S** + RSO_3H

14.77 The mechanism involves S_N2 displacement of Br, followed by intramolecular enamine formation.

S_N2

+ HCO_3^-

CO_3^{2-}

+ Br^-

+ HCO_3^-

proton transfer

(any acid)

+ HCO_3^- + H_2O

conivaptan

Chapter 15 Carboxylic Acids and Nitriles

Chapter Review

General facts

- Carboxylic acids contain a carboxy group (COOH). The central carbon is sp^2 hybridized and trigonal planar (15.1).
- Nitriles contain a cyano group (CN). The C atom of the C≡N is *sp* hybridized and linear (15.1).

Summary of spectroscopic absorptions (15.3B)

Carboxylic Acids		
IR absorptions	C=O	~1710 cm^{-1}
	O–H	3500–2500 cm^{-1} (very broad and strong)
^1H NMR absorptions	O–H	10–12 ppm (highly deshielded proton)
	C–H α to COOH	2–2.5 ppm (somewhat deshielded Csp^3–H)
^{13}C NMR absorption	C=O	170–210 ppm (highly deshielded carbon)
Nitriles		
IR absorption	–C≡N	2500 cm^{-1}
^{13}C NMR absorption	–C≡N	115–120 ppm

General acid–base reaction of carboxylic acids (15.7)

R–COOH (pK_a ≈ 5) + :B ⇌ R–COO⁻ (carboxylate anion) + H–B⁺

- Carboxylic acids are especially acidic because carboxylate anions are resonance stabilized.
- For equilibrium to favor the products, the base must have a conjugate acid with a pK_a > 5. Common bases are listed in Table 15.4.

Factors that affect acidity

Resonance effects. A carboxylic acid is more acidic than an alcohol or phenol because its conjugate base is more effectively stabilized by resonance (15.7).

ROH (pK_a = 16–18) → phenol (pK_a = 10) → RCOOH (pK_a ≈ 5)

Increasing acidity

Inductive effects. Acidity increases with the presence of electron-withdrawing groups (like the electronegative halogens) and decreases with the presence of electron-donating groups (like polarizable alkyl groups) (15.9).

Nitrile synthesis (15.13)

Nitriles are prepared by S_N2 substitution using unhindered alkyl halides as starting materials.

$$R{-}X + {}^{-}CN \xrightarrow{S_N2} R{-}C{\equiv}N + X^{-}$$

$R = CH_3$, 1°

Reactions of nitriles

[1] Hydrolysis (15.13)

$$R{-}C{\equiv}N \xrightarrow[(H^+ \text{ or } {}^{-}OH)]{H_2O} RCOOH \text{ (with acid) or } RCOO^{-} \text{ (with base)}$$

[2] Reduction (15.13)

$$R{-}C{\equiv}N \xrightarrow[\text{[2] } H_2O]{\text{[1] } LiAlH_4} RCH_2NH_2$$

1° amine

$$R{-}C{\equiv}N \xrightarrow[\text{[2] } H_2O]{\text{[1] DIBAL-H}} RCHO$$

aldehyde

[3] Reaction with organometallic reagents (15.13)

$$R{-}C{\equiv}N \xrightarrow[\text{[2] } H_2O]{\text{[1] R'MgX or R'Li}} RCOR'$$

ketone

Other facts

- The Henderson–Hasselbalch equation tells whether a compound will exist in its acidic form (HA) or as its conjugate base ($A:^{-}$) at a particular pH. An acid exists in its protonated form HA in solutions that are more acidic than its pK_a. An acid exists as its conjugate base $A:^{-}$ in solutions that are more basic than its pK_a (15.8).
- Extraction is a useful technique for separating compounds having different solubility properties. Carboxylic acids can be separated from other organic compounds by extraction, because aqueous base converts a carboxylic acid into a water-soluble carboxylate anion (15.10).
- Phosphoric acid esters [$ROPO(OH)_2$] exist as monophosphate dianions ($ROPO_3^{2-}$) at physiological pH (15.11), because the pK_a values for the two OH groups are less than 7.4.

- Amino acids have an amino group on the α carbon to the carboxy group [$RCH(NH_2)COOH$]. Amino acids exist as zwitterions at pH ≈ 6. Adding acid forms a species with a net (+1) charge [$RCH(NH_3)COOH]^+$. Adding base forms a species with a net (–1) charge [$RCH(NH_2)COO]^-$ (15.12).

Practice Test on Chapter Review

1. Give the IUPAC name for each of the following compounds.

[1] CO_2H [2] CN

2. a. Which of the labeled atoms is least acidic?

H_a O O $O–H_d$ H_e H_c H_b

1. H_a 2. H_b 3. H_c 4. H_d 5. H_e

b. Which compound(s) can be converted to **A** by an oxidation reaction?

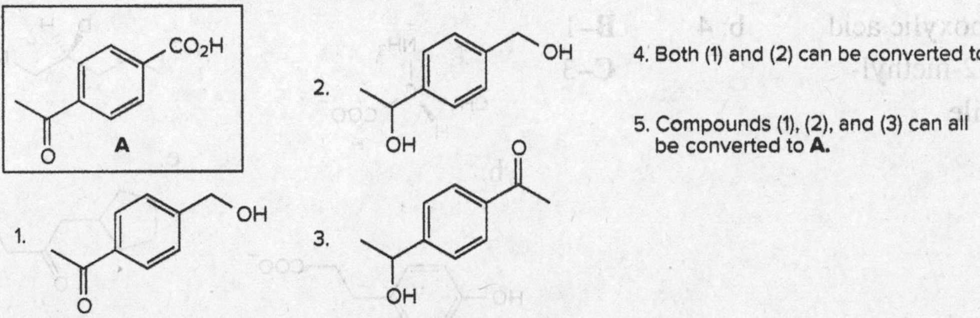
CO2H
A
OH
2.
OH
4. Both (1) and (2) can be converted to A.
5. Compounds (1), (2), and (3) can all be converted to A.
OH
1.
3.
OH

3. Rank the following compounds in order of increasing basicity. Label the ***least*** basic compound as **1,** the ***most*** basic compound as **3,** and the compound of ***intermediate*** basicity as **2.**

COO^- O_2N COO^- O^-

A **B** **C**

4. Draw the organic products formed in each of the following reactions.

a. 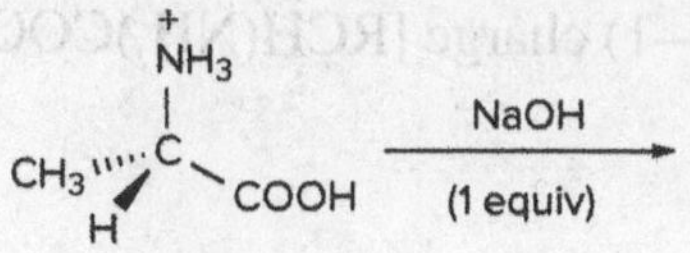 $CH_3C(H)(\overset{+}{N}H_3)COOH$ —NaOH (1 equiv)→

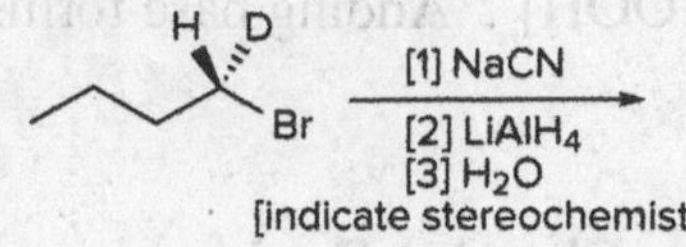

b. 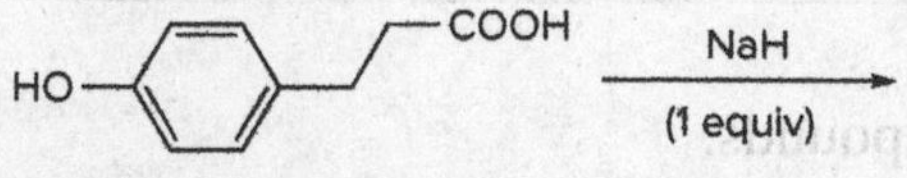 HO–C_6H_4–CH_2CH_2COOH —NaH (1 equiv)→

c. (2-hydroxyphenyl)–$CH_2CH_2N(CH_3)_2$ —[1] NaH; [2] CH_3I→

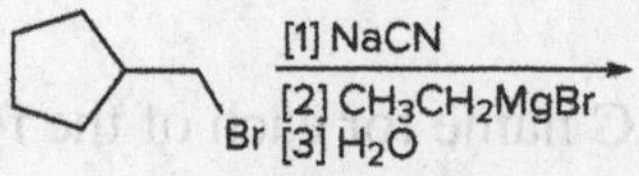

d. $CH_3CH_2CH_2CH(D)Br$ —[1] NaCN; [2] $LiAlH_4$; [3] H_2O [indicate stereochemistry]→

e. (cyclopentyl)CH_2Br —[1] NaCN; [2] CH_3CH_2MgBr; [3] H_2O→

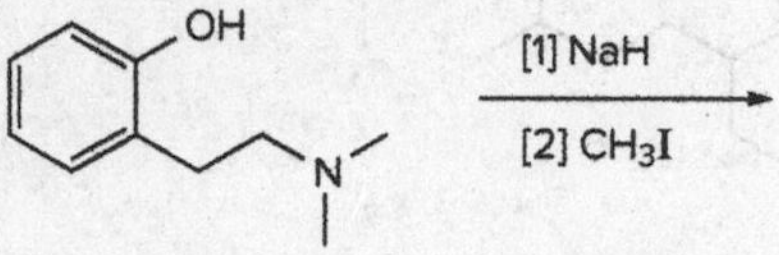

5. What reagent is needed to convert $CH_3CH_2CH_2CN$ to each compound?

a. $CH_3CH_2CH_2COOH$

b. $CH_3CH_2CH_2CH_2NH_2$

c. $CH_3CH_2CH_2COCH_2CH_3$

d. $CH_3CH_2CH_2CHO$

Answers to Practice Test

1. [1] *cis*-2-methylcyclopentanecarboxylic acid
[2] 5-ethyl-2-methylheptanenitrile

2.a. 1
b. 4

3. **A**–2
B–1
C–3

4.a. $CH_3C(H)(\overset{+}{N}H_3)COO^-$

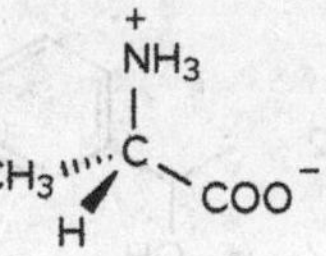

b. 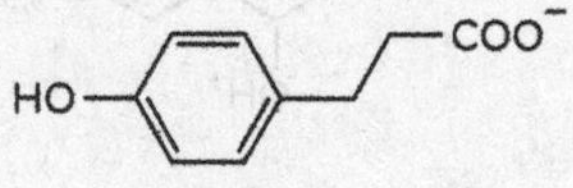 HO–C_6H_4–$CH_2CH_2COO^-$

c. (2-methoxyphenyl)–$CH_2CH_2N(CH_3)_2$ (OCH_3)

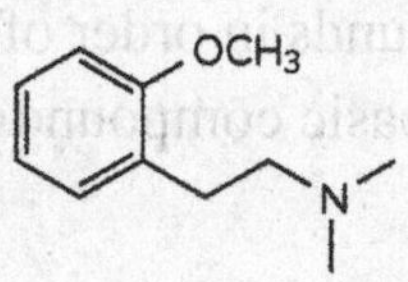

d. 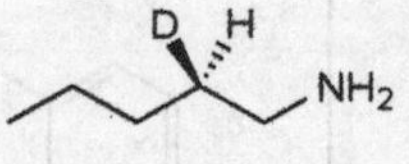 $CH_3CH_2CH_2CH(D)CH_2NH_2$

e. 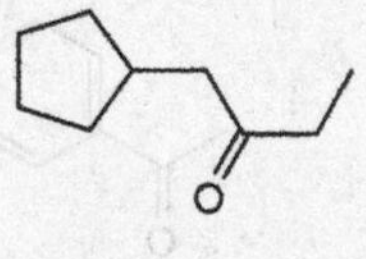 (cyclopentyl)$CH_2COCH_2CH_3$

5.a. H_3O^+
b. [1] $LiAlH_4$; [2] H_2O
c. [1] CH_3CH_2Li; [2] H_2O
d. [1] DIBAL-H; [2] H_2O

Answers to Problems

15.1 To name a carboxylic acid:
[1] Find the longest chain containing the COOH group and change the *-e* ending to *-oic acid.*
[2] Number the chain to put the COOH carbon at C1, but omit the number from the name.
[3] Follow all other rules of nomenclature.

a.

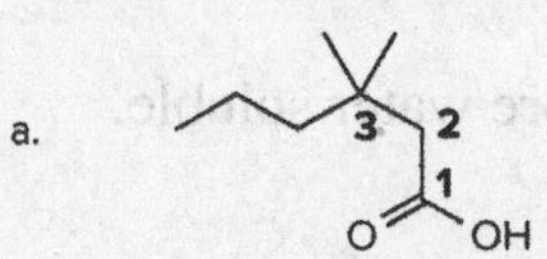

Number the chain to put COOH at C1.
6 carbon chain = **hexanoic acid**
3,3-dimethylhexanoic acid

b.

Number the chain to put COOH at C1.
6 carbon chain = **hexanoic acid**
2,4-diethylhexanoic acid

c.

Number the chain to put COOH at C1.
9 carbon chain = **nonanoic acid**
4-isopropyl-6,8-dimethylnonanoic acid

15.2

a. 2-bromo**butanoic acid**

b. 2,3-dimethyl**pentanoic acid**

c. 3,3,4-trimethyl**heptanoic acid**

d. 2-*sec*-butyl-4,4-diethyl**nonanoic acid**

e. 3,4-diethyl**cyclohexanecarboxylic acid**

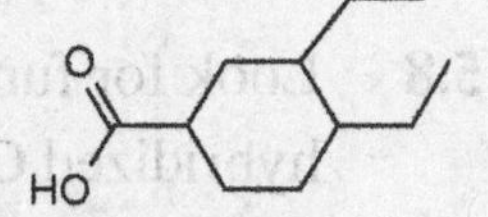

f. 1-isopropyl**cyclobutane-carboxylic acid**

15.3

a. α-methoxy**valeric acid**

b. β-phenyl**propionic acid**

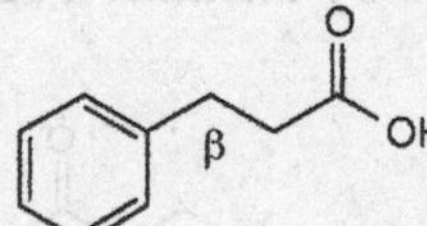

c. α,β-dimethyl**caproic acid**

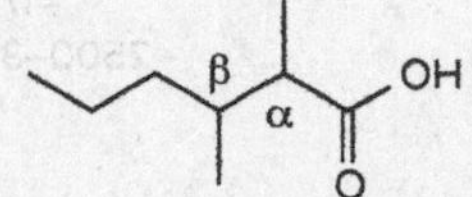

d. α-chloro-β-methyl**butyric acid**

15.4

a. **lithium benzoate**

b. **sodium formate**

c. **potassium 2-methylpropanoate**

d. **sodium 4-bromo-6-ethyl-octanoate**

15.5

2-propyl**pentanoic acid**

sodium 2-propyl**pentanoate**

15.6

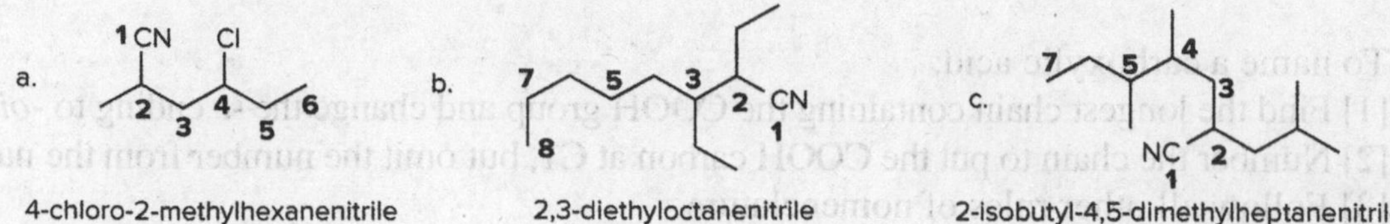

15.7 More polar molecules have a higher boiling point and are more water soluble.

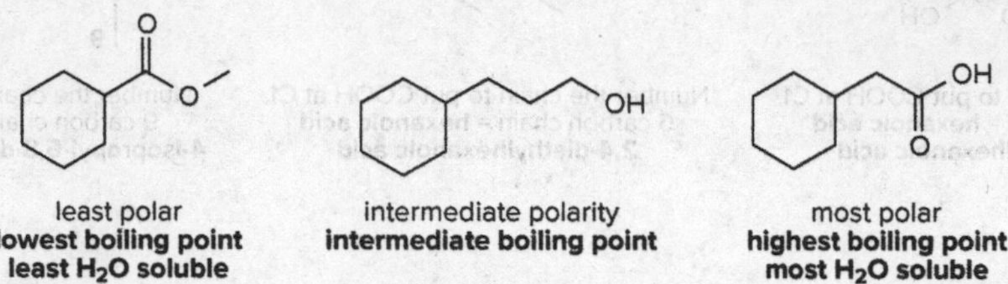

15.8 Look for functional group differences to distinguish the compounds by IR. Besides sp^3 hybridized C–H bonds at 3000–2850 cm^{-1} (which all three compounds have), the following functional group absorptions are seen:

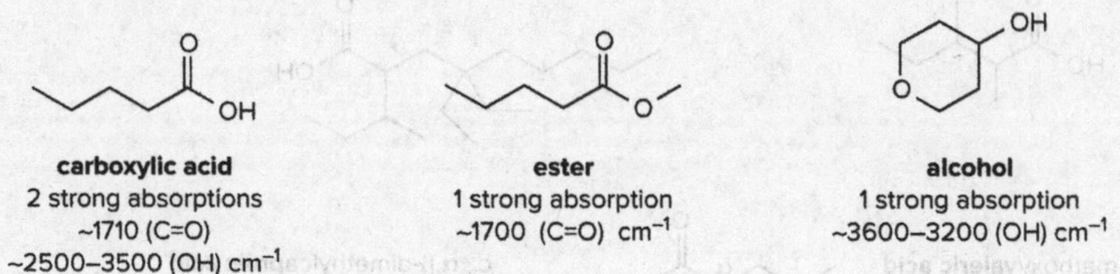

15.9 1° Alcohols are converted to carboxylic acids by oxidation reactions.

15.10

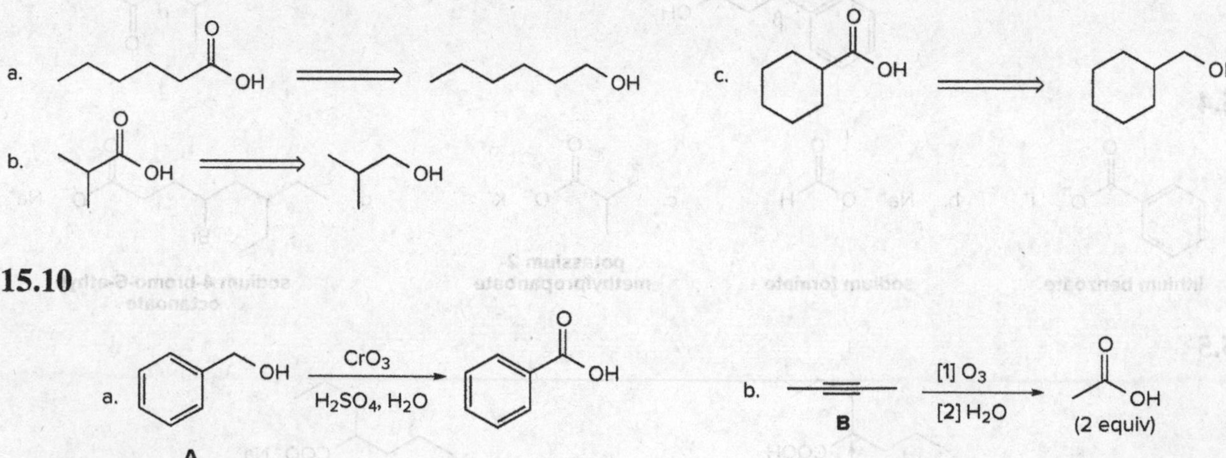

c. C ($2°$ OH, $1°$ OH) $\xrightarrow[H_2SO_4,\ H_2O]{CrO_3}$ (keto acid, OH)

d. D (Br) $\xrightarrow{[1]\ Mg;\ [2]\ CO_2;\ [3]\ H_3O^+}$ (carboxylic acid, OH)

e. E (O, OH, H) $\xrightarrow[NH_4OH]{Ag_2O}$ (O, OH, OH)

15.11

a. (cyclohexyl)–COOH $\xrightarrow{NaOH}$ (cyclohexyl)–COO^- Na^+ + H_2O

b. (4-methylphenyl)–OH $\xrightarrow{NaOCH_3}$ (4-methylphenyl)–O^- Na^+ + $HOCH_3$

c. ($(CH_3)_3C$)–OH $\xrightarrow{NaH}$ ($(CH_3)_3C$)–O^- Na^+ + H_2

d. C_6H_5–COOH $\xrightarrow{NaHCO_3}$ C_6H_5–COO^- Na^+ + H_2CO_3

15.12 CH_3COOH has a pK_a of 4.8. Any base having a conjugate acid with a pK_a higher than 4.8 can deprotonate it.

a. F^- pK_a(HF) = 3.2 **not strong enough**
b. $(CH_3)_3CO^-$ pK_a[$(CH_3)_3COH$] = 18 **strong enough**
c. CH_3^- pK_a(CH_4) = 50 **strong enough**
d. $^-NH_2$ pK_a(NH_3) = 38 **strong enough**
e. Cl^- pK_a(HCl) = −7.0 **not strong enough**

15.13

Increasing acidity: $H_a < H_b < H_c$

mandelic acid (H_a, H OH_b, OH_c)

$-H_a$ → negative charge on C
unstable conjugate base

$-H_b$ → negative charge on O
more stable conjugate base

$-H_c$ → negative charge on O, resonance stabilized
most stable conjugate base

15.14 When the pH of the solution is higher (more basic) than the pK_a of the acid, the compound will exist primarily as its conjugate base. When the pH of the solution is lower (more acidic) than the pK_a of the acid, the compound will exist primarily as the acid and retain its proton.

a. The pH is higher than the pK_a, so the acid is deprotonated.

b. The pH = pK_a of the acid, so equal amounts of the acid and its conjugate base are present.

c. The pH is lower than the pK_a, so the acid retains its proton.

15.15

a. The pH of the stomach is lower than the pK_a values of all three carboxy groups, so all of the carboxy groups retain their protons.

b. The pH of the intestines is higher than the pK_a's of all three carboxy groups, so all three protons are removed and a trianion is formed.

citric acid

predominates in the stomach

predominates in the intestines

15.16 Electron-withdrawing groups make an acid more acidic, lowering its pK_a.

least acidic
pK_a = 4.9

one electron-withdrawing group
intermediate acidity
pK_a = 3.2

three electron-withdrawing F's
most acidic
pK_a = 0.2

15.17

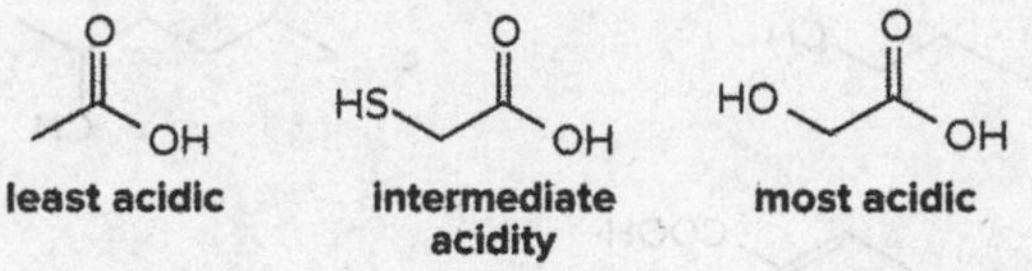

least acidic — intermediate acidity — most acidic

15.18 To separate compounds by an extraction procedure, they must have different solubility properties.

a. $CH_3(CH_2)_6COOH$ and $CH_3CH_2CH_2CH_2CH{=}CH_2$: **YES.** The acid can be extracted into aqueous base, while the alkene will remain in the organic layer.

b. $CH_3CH_2CH_2CH_2CH{=}CH_2$ and $(CH_3CH_2CH_2)_2O$: **NO.** Both compounds are soluble in organic solvents and insoluble in water. Neither is acidic enough to be extracted into aqueous base.

c. $CH_3(CH_2)_6COOH$ and NaCl: one carboxylic acid, one salt: **YES.** The carboxylic acid is soluble in an organic solvent, whereas the salt is soluble in water.

d. NaCl and KCl: two salts: **NO.**

15.19 Because the pK_a values for loss of the first three protons from pyrophosphoric acid (0.9, 2.1, and 6.7) are less than physiological pH (7.4), all three protons of ADP will be removed. The predominant form is a trianion.

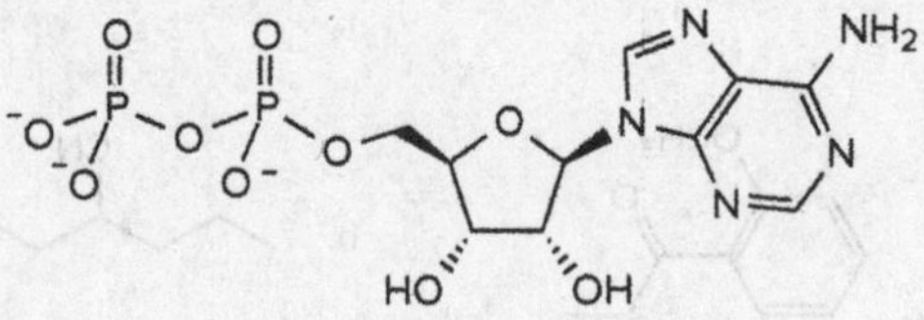

ADP at physiological pH

15.20

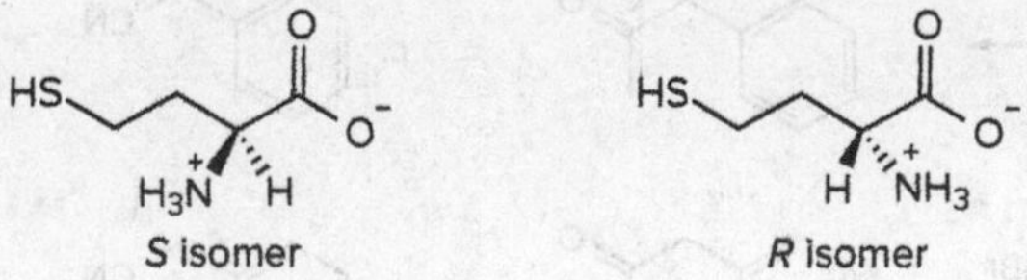

S isomer — *R* isomer

15.21

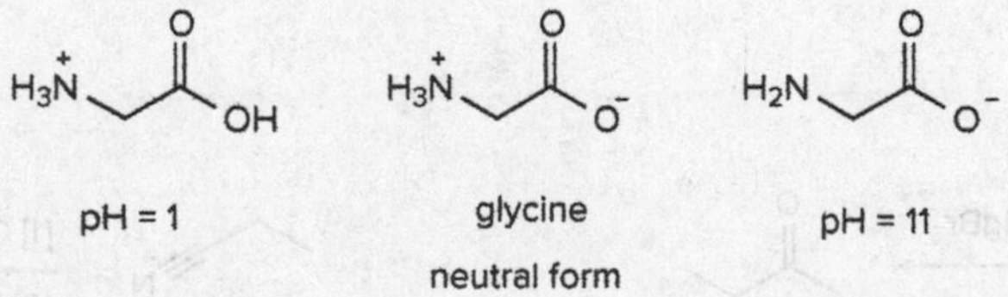

pH = 1 — glycine neutral form — pH = 11

15.22

$$pI = \frac{pK_a(COOH) + pK_a(NH_3^+)}{2} = \frac{(2.29) + (9.72)}{2} = 6.01$$

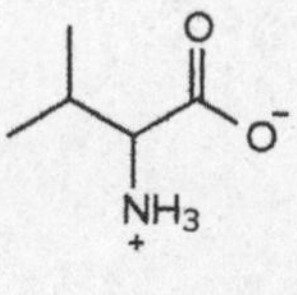

15.23

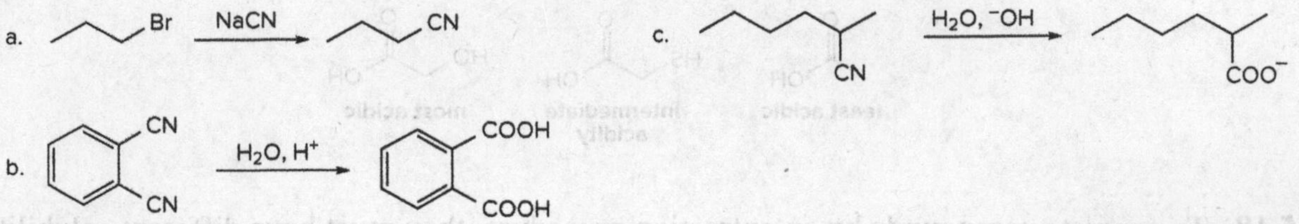

a. Br → NaCN → CN

c. CN → H_2O, ^-OH → COO^-

b. CN, CN → H_2O, H^+ → COOH, COOH

15.24

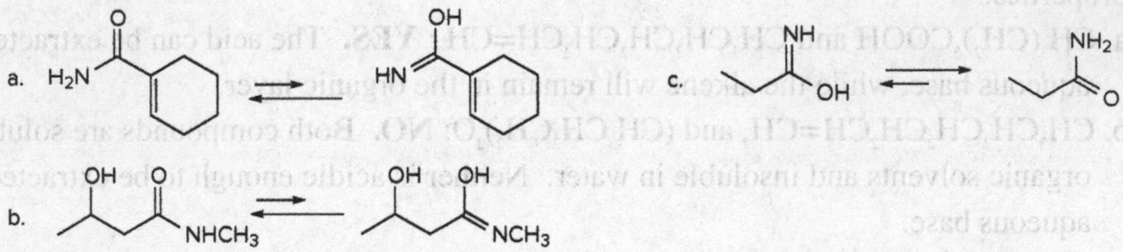

a. H_2N, O ⇄ OH, HN

c. NH, OH ⇄ NH_2, O

b. OH, O, $NHCH_3$ ⇄ OH, OH, NCH_3

15.25

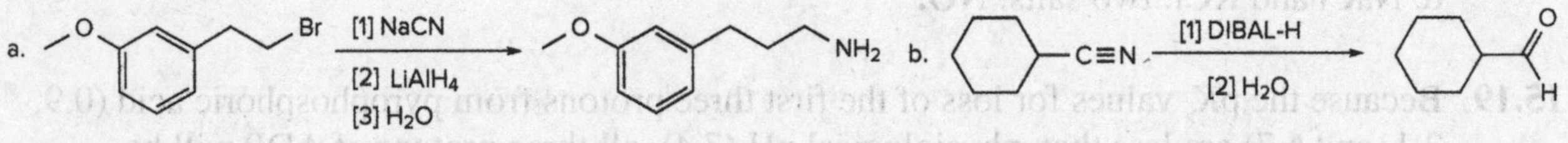

a. O, Br → [1] NaCN [2] $LiAlH_4$ [3] H_2O → O, NH_2

b. C≡N → [1] DIBAL-H [2] H_2O → O, H

15.26

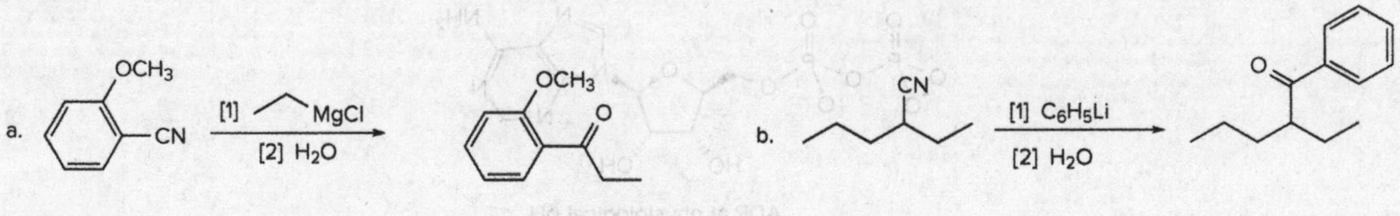

a. OCH_3, CN → [1] MgCl [2] H_2O → OCH_3, O

b. CN → [1] C_6H_5Li [2] H_2O → O

15.27

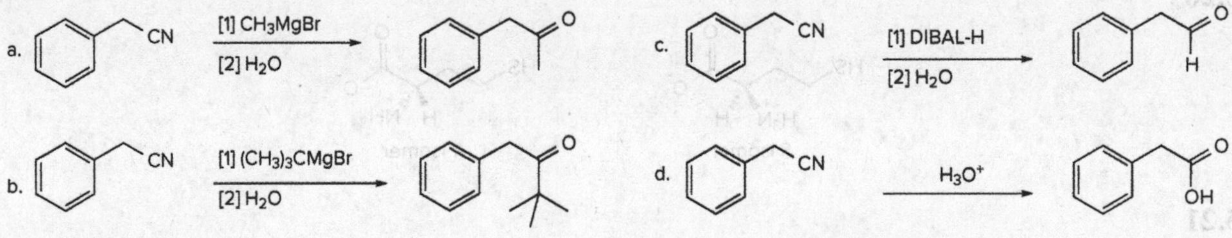

a. CN → [1] CH_3MgBr [2] H_2O → O

c. CN → [1] DIBAL-H [2] H_2O → O, H

b. CN → [1] $(CH_3)_3CMgBr$ [2] H_2O → O

d. CN → H_3O^+ → O, OH

15.28

$CH_3-C{\equiv}N$ → [1] MgBr [2] H_2O → O

N → [1] CH_3MgBr [2] H_2O → O

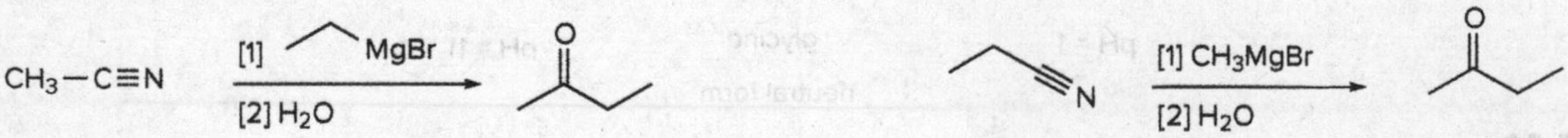

15.29

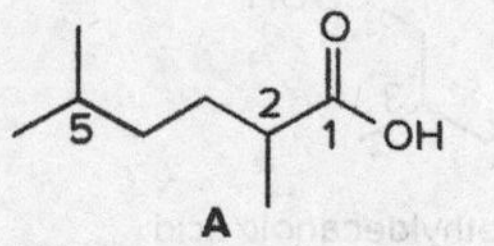

5
2
1
OH
O

A

a. 2,5-dimethylhexanoic acid

NaOH

b.

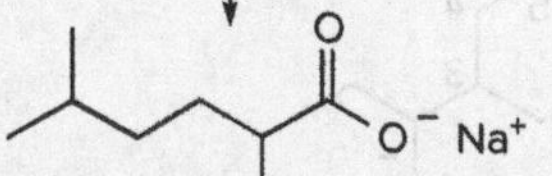

O
O⁻ Na⁺

c. sodium 2,5-dimethylhexanoate

d. An alcohol or ether would have a much higher pK_a than a carboxylic acid.

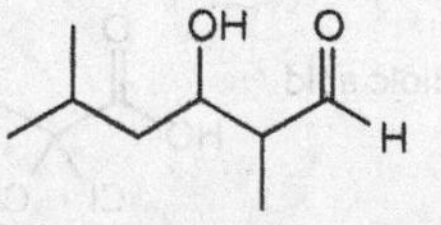

OH
O
H

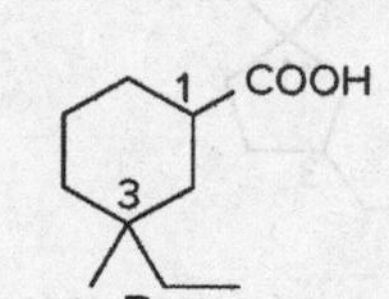

1
COOH
3

B

a. 3-ethyl-3-methylcyclohexanecarboxylic acid

NaOH

b.

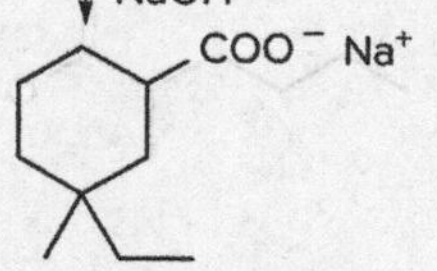

COO⁻ Na⁺

c. sodium 3-ethyl-3-methylcyclohexanecarboxylate

d.

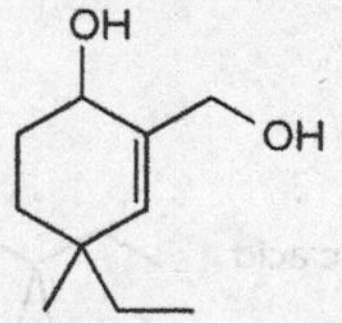

OH
OH

15.30

a.

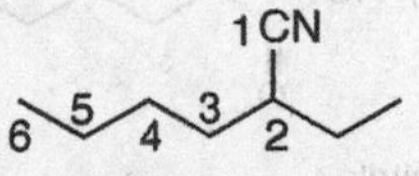

1 CN
6
5
4
3
2

C

2-ethylhexanenitrile

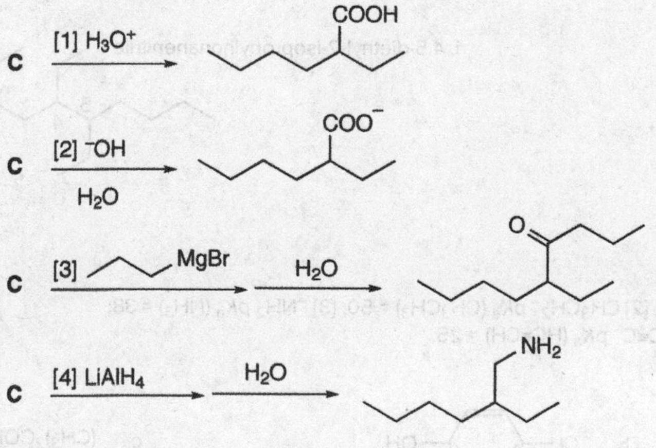

b. C
[1] H₃O⁺
COOH
C
[2] ⁻OH
H₂O
COO⁻
C
[3]
MgBr
H₂O
O
C
[4] LiAlH₄
H₂O
NH₂

15.31

a.

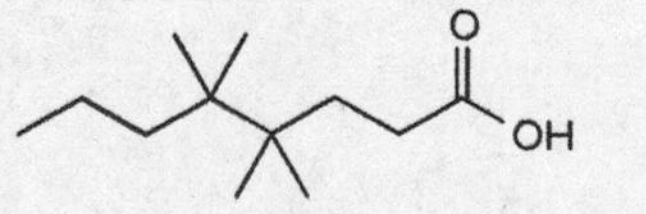

O
OH

4,4,5,5-tetramethyloctanoic acid

b.

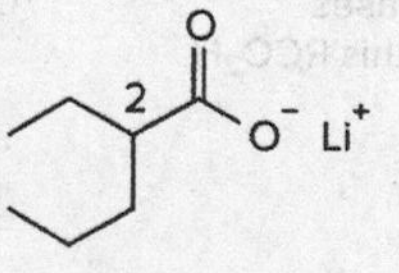

2
O
O⁻ Li⁺

lithium 2-ethylpentanoate

c. 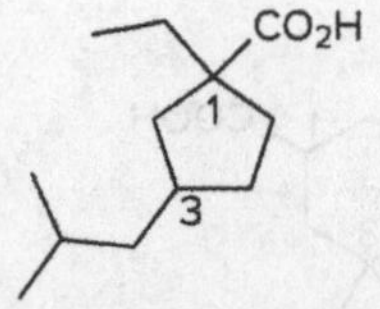 1-ethyl-3-isobutylcyclopentanecarboxylic acid

e. 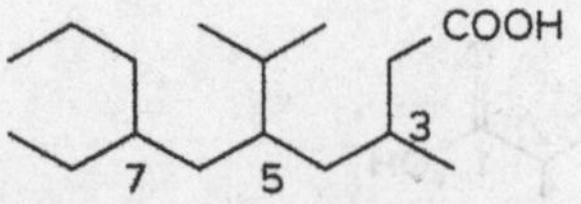 7-ethyl-5-isopropyl-3-methyldecanoic acid

d. 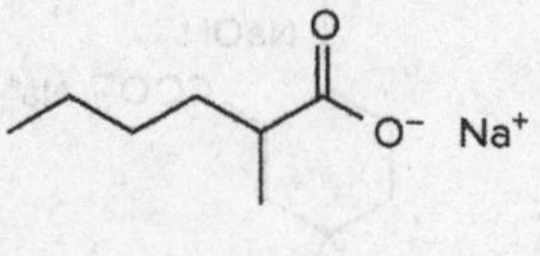 sodium 2-methylhexanoate

f. 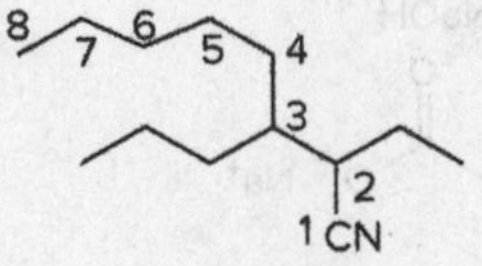 2-ethyl-3-propyloctanenitrile

15.32

a. 3,3-dimethylpentanoic acid

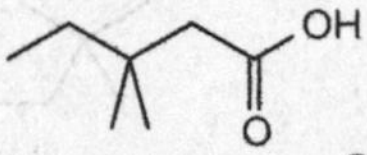

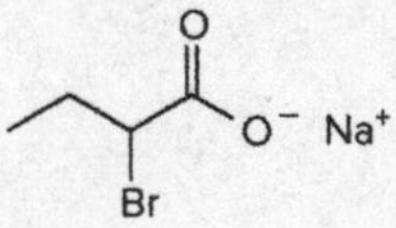

f. 2,2-dichloropentanedioic acid

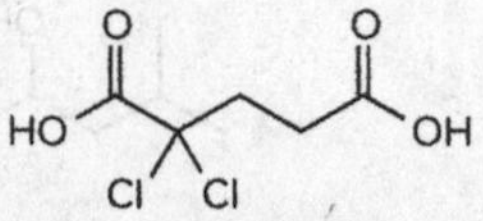

b. 4-chloro-3-phenylheptanoic acid

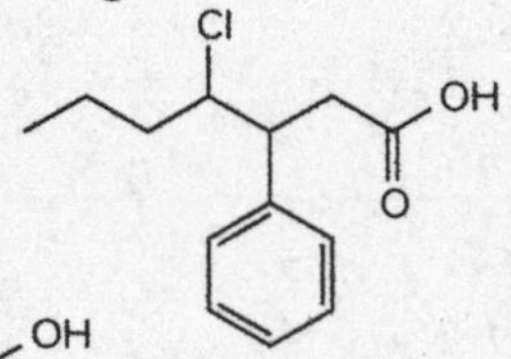

g. 4-isopropyl-2-methyloctanedioic acid

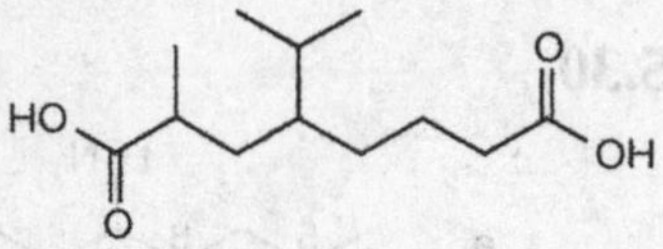

c. (*R*)-2-chloropropanoic acid

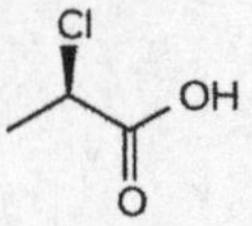

h. 3,3-dimethylpentanenitrile

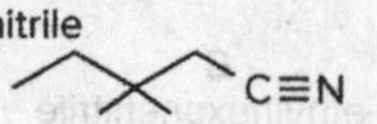

d. potassium acetate

i. 4,5-diethyl-2-isopropylnonanenitrile

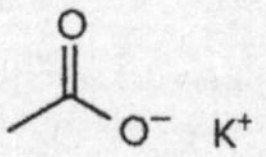

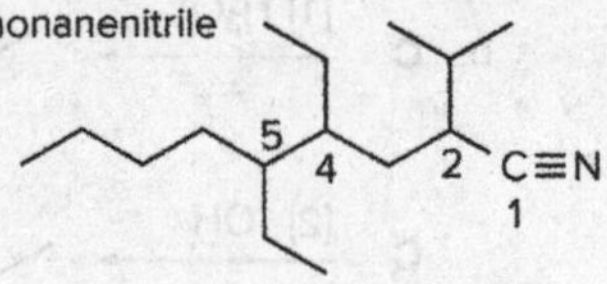

e. sodium α-bromobutyrate

15.33

Bases: [1] ^-OH pK_a (H_2O) = 15.7; [2] $CH_3CH_2^-$ pK_a (CH_3CH_3) = 50; [3] $^-NH_2$ pK_a (NH_3) = 38; [4] NH_3 pK_a (NH_4^+) = 9.4; [5] $HC{\equiv}C^-$ pK_a ($HC{\equiv}CH$) = 25.

a. 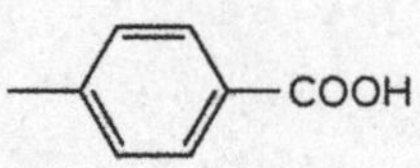 p-$CH_3C_6H_4COOH$

pK_a = 4.3
All of the bases can deprotonate this RCO_2H.

b. 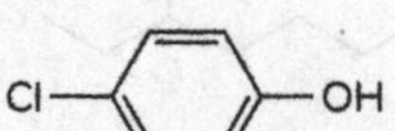 p-ClC_6H_4OH

pK_a = 9.4
^-OH, $CH_3CH_2^-$, $^-NH_2$, and $HC{\equiv}C^-$ can deprotonate this phenol.

c. $(CH_3)_3COH$

pK_a = 18
$CH_3CH_2^-$, $^-NH_2$, and $HC{\equiv}C^-$ can deprotonate this ROH.

15.34

a. $CH_3CH_2CH_2CH_2CH_2OH + NH_3 \rightleftharpoons CH_3CH_2CH_2CH_2CH_2O^- + NH_4^+$

$pK_a \approx 16$ $\quad$ $pK_a = 9.4$

Reaction favors reactants.

b. $C_6H_5OH + Na^+\ ^-NH_2 \rightleftharpoons C_6H_5O^- + NH_3 + Na^+$

$pK_a = 10$ $\quad$ $pK_a = 38$

Reaction favors products.

c. (2-methylphenyl)COOH $+ CH_3^-\ Li^+ \rightleftharpoons$ (2-methylphenyl)$COO^-\ Li^+ + CH_4$

$pK_a \approx 4$ $\quad$ $pK_a = 50$

Reaction favors products.

d. (4-methylphenyl)OH $+ Na_2CO_3 \rightleftharpoons$ (4-methylphenyl)$O^-\ Na^+ + Na^+\ HCO_3^-$

$pK_a = 10.2$ $\quad$ $pK_a = 10.2$

With the same pK_a for the starting acid and the conjugate acid, an equal amount of starting materials and products is present.

15.35

caftaric acid

H_b: C–H, least acidic, unstable conjugate base with a (–) charge on carbon

OH_d: most acidic H; part of a carboxylic acid; most highly stabilized conjugate base

OH_c: alcohol OH

OH_a: phenolic OH

Increasing acidity: $H_b < H_c < H_a < H_d$

15.36

D ($C_6H_5COO^-$) < B ($C_6H_5O^-$) < A ($C_6H_5NH^-$) < C ($C_6H_5CH_2^-$)

increasing basicity

15.37 The OH of the phenol group in morphine is more acidic than the OH of the alcohol ($pK_a \approx 10$ versus $pK_a \approx 16$). KOH is basic enough to remove the phenolic OH, the most acidic proton.

most acidic proton → HO
The OH is part of a phenol.
Methylation occurs here.

an alcohol → HO

morphine

[1] KOH

Many resonance structures stabilize the conjugate base.

[2] CH_3I

codeine

15.38

The O in **A** is more electronegative than the N in **C,** so there is a stronger electron-withdrawing inductive effect. This stabilizes the conjugate base of **A,** making **A** more acidic than **C.**

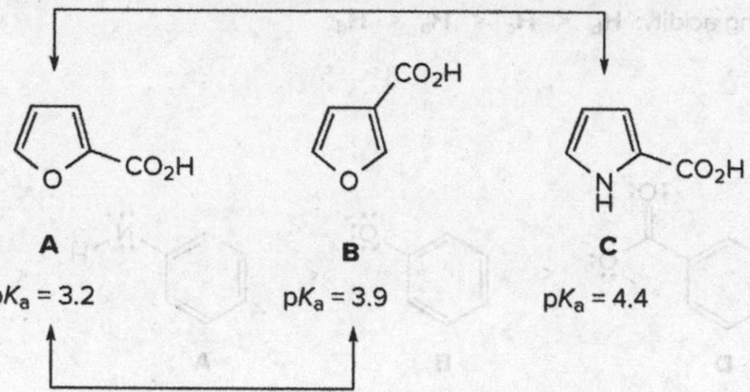

Since the O in **A** is closer to the COOH group than the O atom in **B,** there is a stronger electron-withdrawing inductive effect. This makes **A** more acidic than **B.**

15.39 a. The pK_{a1} of phthalic acid is lower than the pK_{a1} of isophthalic acid because the electron-withdrawing CO_2H is closer to the negatively charged CO_2^- of the conjugate base.

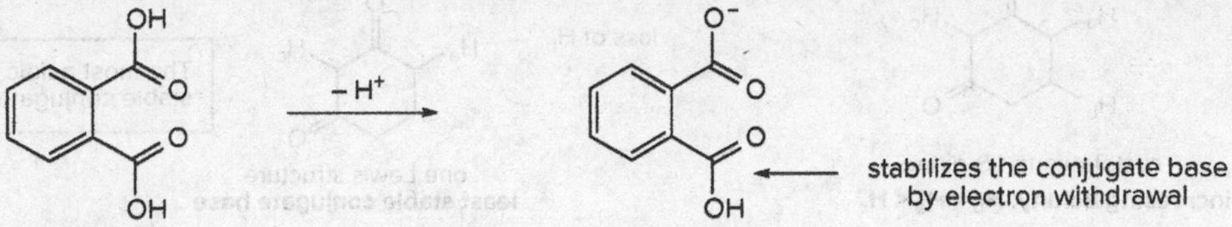

OH
O
O
OH
– H⁺
O⁻
O
O
OH
stabilizes the conjugate base
by electron withdrawal

phthalic acid
stronger acid
lower pK_a

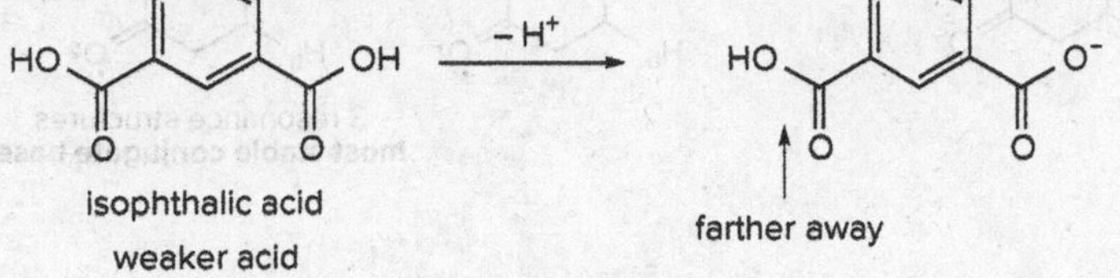

HO
OH
O
O
– H⁺
HO
O⁻
O
O
farther away

isophthalic acid
weaker acid

b. After loss of the first proton, each compound now has one CO_2H group and one CO_2^-. The CO_2^- is electron donating, so the closer it is located to the CO_2H, the more it destabilizes the resulting conjugate base.

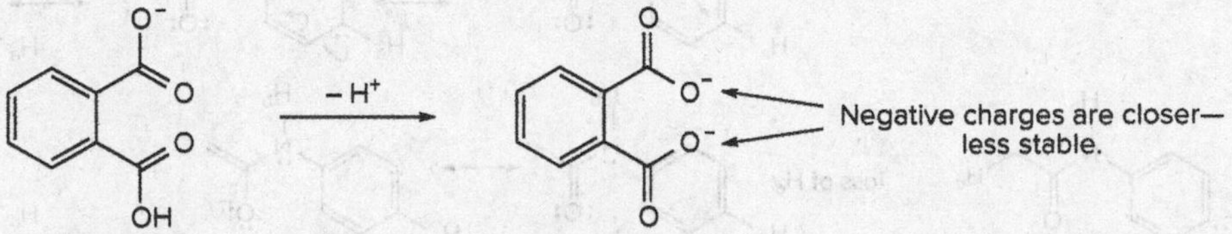

O⁻
O
O
OH
– H⁺
O
O⁻
O⁻
O
Negative charges are closer—
less stable.

weaker acid

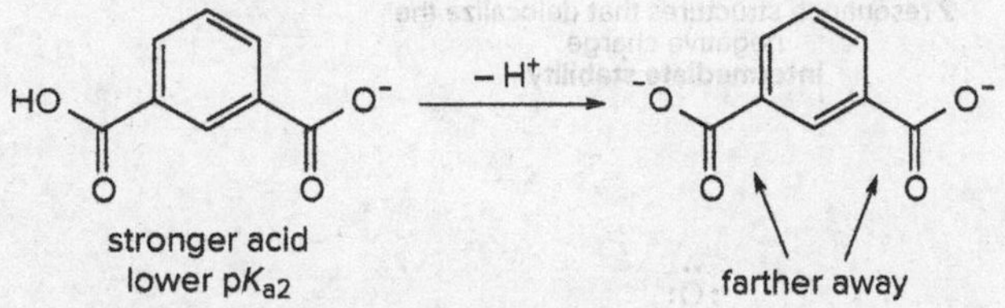

HO
O
O⁻
O
– H⁺
⁻O
O
O⁻
O
farther away

stronger acid
lower pK_{a2}

15.40

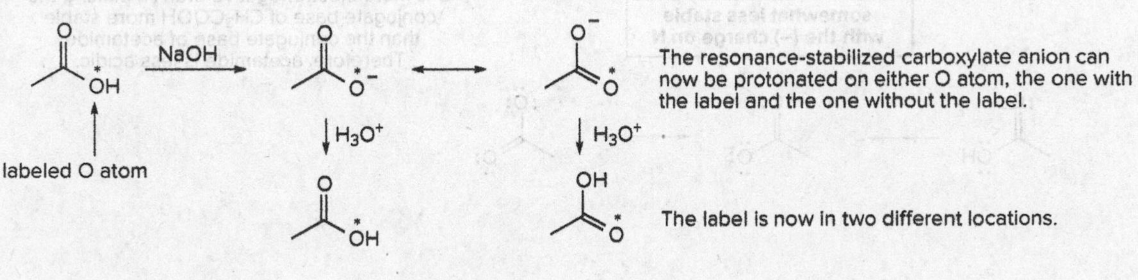

NaOH
H₃O⁺
H₃O⁺
labeled O atom
The resonance-stabilized carboxylate anion can now be protonated on either O atom, the one with the label and the one without the label.
The label is now in two different locations.

15.41

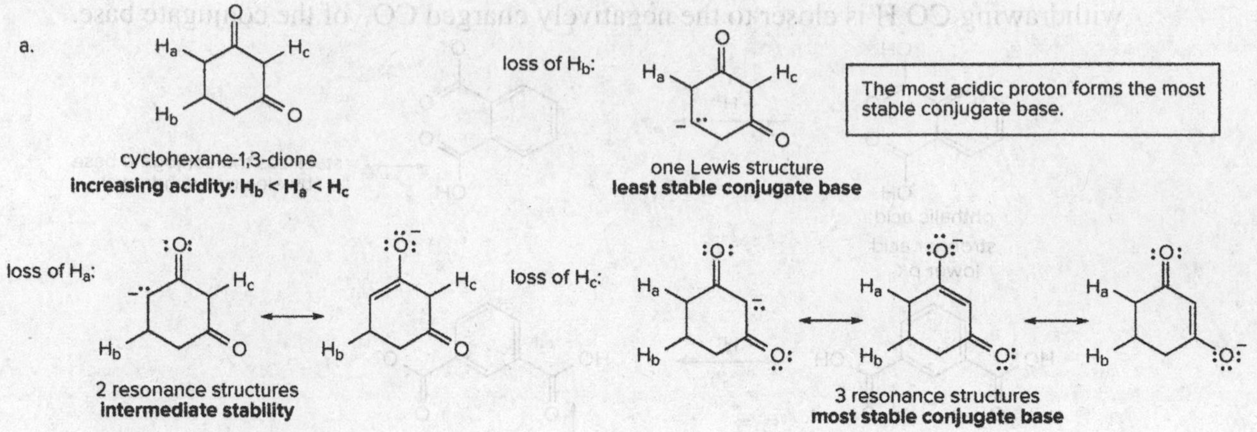

a.

cyclohexane-1,3-dione

increasing acidity: $H_b < H_a < H_c$

loss of H_b:

one Lewis structure
least stable conjugate base

The most acidic proton forms the most stable conjugate base.

loss of H_a:

2 resonance structures
intermediate stability

loss of H_c:

3 resonance structures
most stable conjugate base

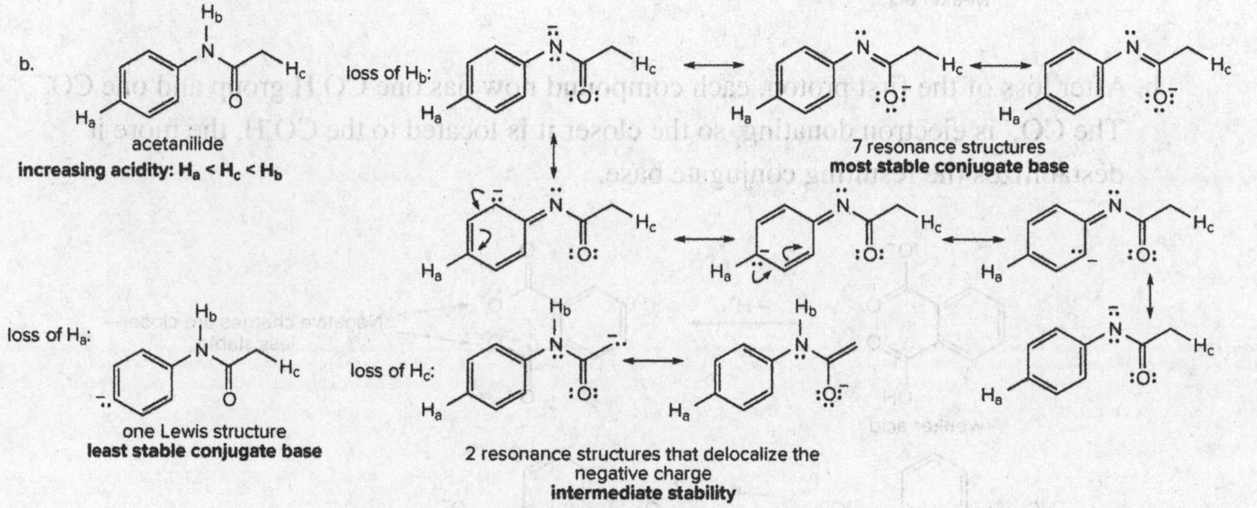

b.

acetanilide

increasing acidity: $H_a < H_c < H_b$

loss of H_b:

7 resonance structures
most stable conjugate base

loss of H_a:

one Lewis structure
least stable conjugate base

loss of H_c:

2 resonance structures that delocalize the negative charge
intermediate stability

15.42

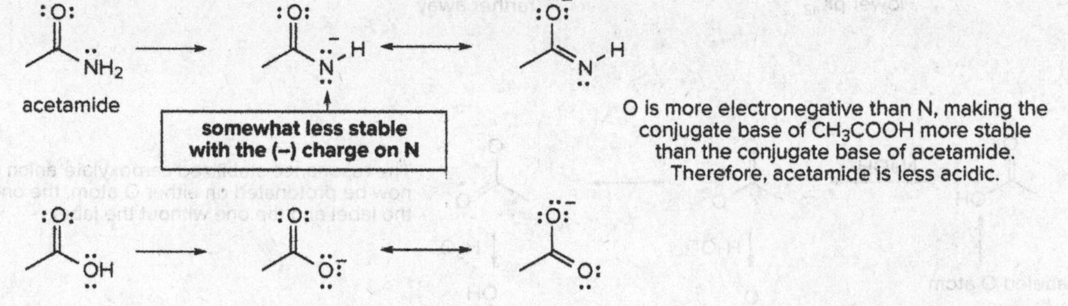

acetamide
somewhat less stable
with the (–) charge on N
O is more electronegative than N, making the
conjugate base of CH3COOH more stable
than the conjugate base of acetamide.
Therefore, acetamide is less acidic.

15.43

a. 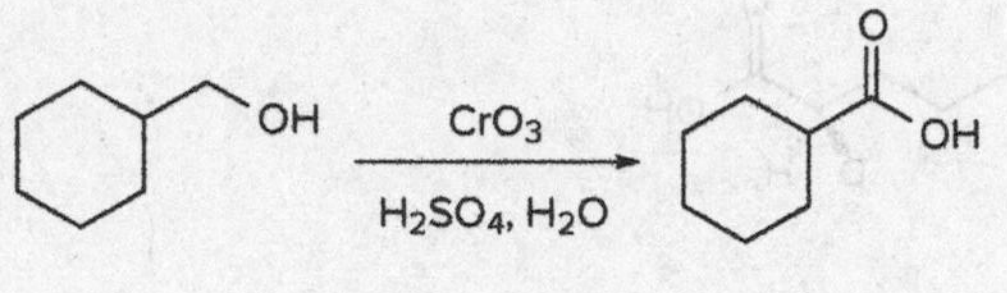 OH —[CrO_3 / H_2SO_4, H_2O]→ OH (O)

b. 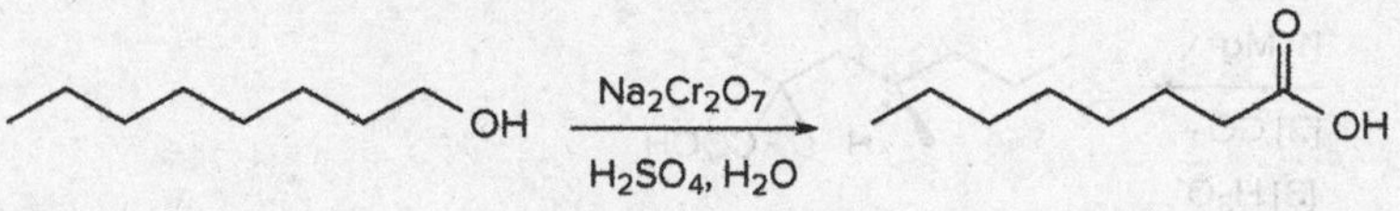 OH —[$Na_2Cr_2O_7$ / H_2SO_4, H_2O]→ OH (O)

c. 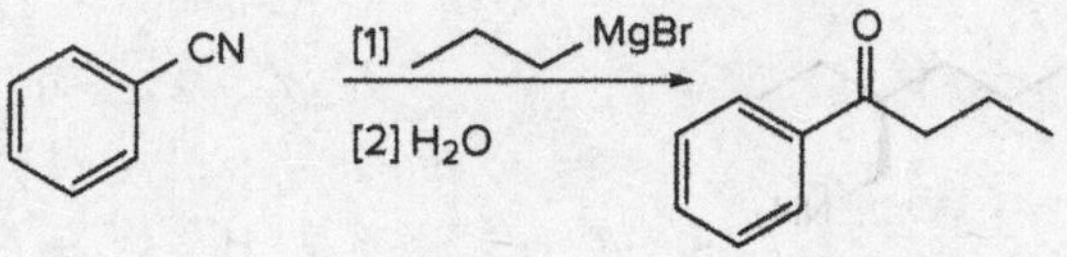 CN —[[1] MgBr / [2] H_2O]→ O

d. 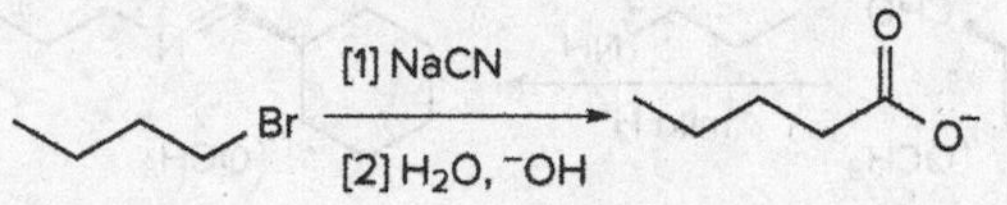 Br —[[1] NaCN / [2] H_2O, ^-OH]→ O^- (O)

15.44

a. 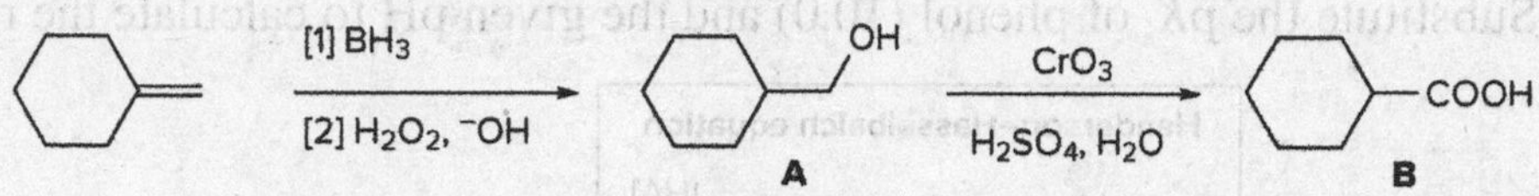 —[[1] BH_3 / [2] H_2O_2, ^-OH]→ OH **A** —[CrO_3 / H_2SO_4, H_2O]→ COOH **B**

b. HC≡CH —[[1] $NaNH_2$ / [2] CH_3I]→ **C** —[[1] $NaNH_2$ / [2] CH_3CH_2I]→ **D** —[[1] O_3 / [2] H_2O]→ OH (O) + OH (O) **E** + **F**

c. Br, Br —[NaCN]→ CN, Br **I** —[[1] DIBAL-H / [2] H_2O]→ CHO, Br **J** —[Ag_2O, NH_4OH]→ CO_2H, Br **K**

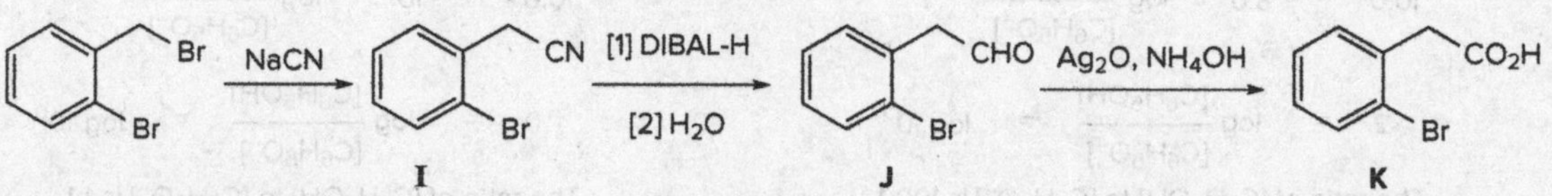

15.45

CN

a. —[H_3O^+]→ COOH

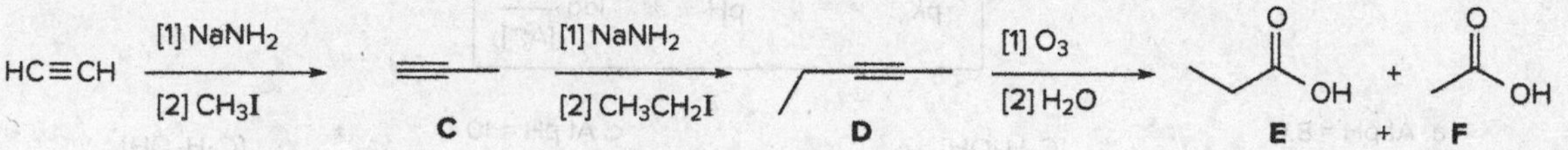

b. —[H_2O, ^-OH]→ COO^-

c. —[[1] CH_3MgBr / [2] H_2O]→ O

d. —[[1] CH_3CH_2Li / [2] H_2O]→ O

e. —[[1] DIBAL-H / [2] H_2O]→ H, O

f. —[[1] $LiAlH_4$ / [2] H_2O]→ NH_2

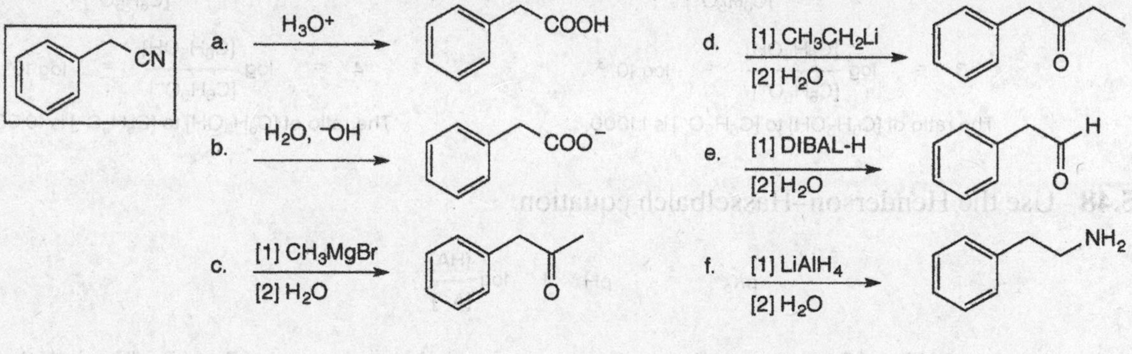

15.46

a. 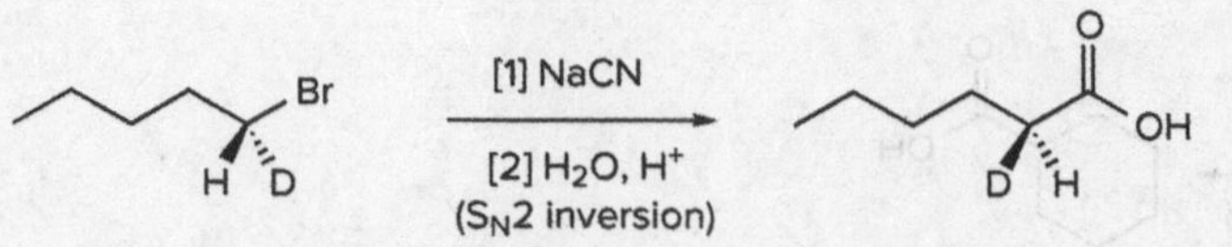 [1] NaCN; [2] H_2O, H^+ (S_N2 inversion)

b. 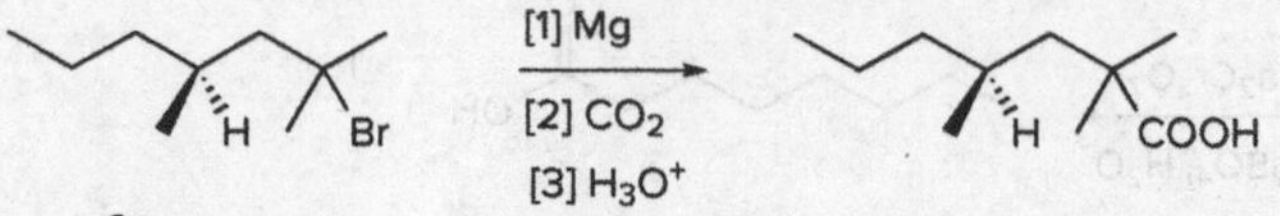 [1] Mg; [2] CO_2; [3] H_3O^+

c. 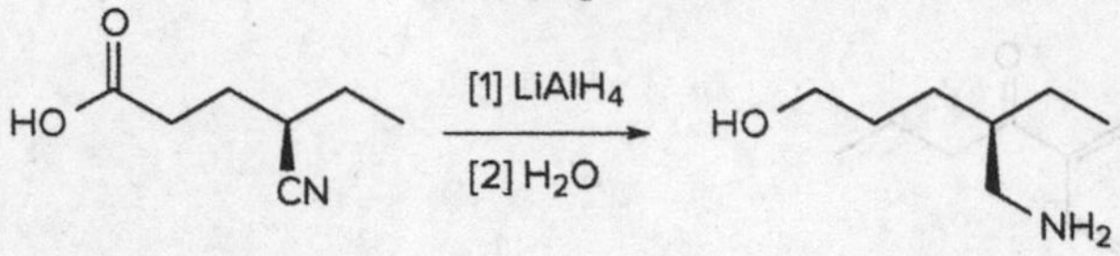 [1] $LiAlH_4$; [2] H_2O

d. 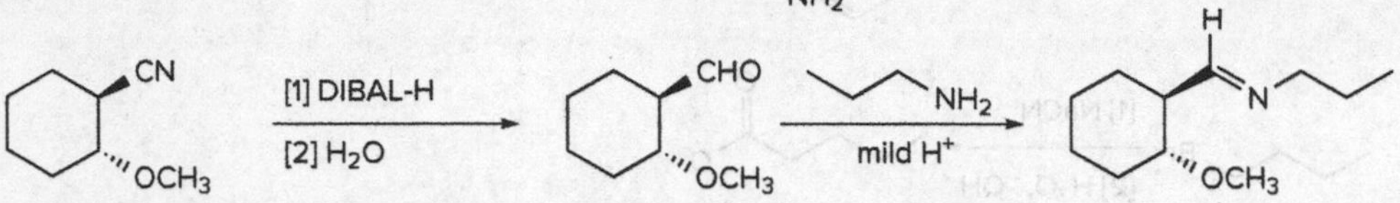 [1] DIBAL-H; [2] H_2O; then $CH_3CH_2CH_2NH_2$, mild H^+

15.47 Use the Henderson–Hasselbalch equation and the principles in Answer 15.14 to answer the question. Substitute the pK_a of phenol (10.0) and the given pH to calculate the ratio.

Henderson–Hasselbalch equation

$$pK_a = pH + \log\frac{[HA]}{[A:^-]}$$

a. At pH = 8.0

$$10.0 = 8.0 + \log\frac{[C_6H_5OH]}{[C_6H_5O^-]}$$

$$2 = \log\frac{[C_6H_5OH]}{[C_6H_5O^-]} = \log 10^2$$

The ratio of $[C_6H_5OH]$ to $[C_6H_5O^-]$ is 100:1.

b. At pH = 13

$$10.0 = 13 + \log\frac{[C_6H_5OH]}{[C_6H_5O^-]}$$

$$-3 = \log\frac{[C_6H_5OH]}{[C_6H_5O^-]} = \log 10^{-3}$$

The ratio of $[C_6H_5OH]$ to $[C_6H_5O^-]$ is 1:1000.

c. At pH = 10

$$10.0 = 10 + \log\frac{[C_6H_5OH]}{[C_6H_5O^-]}$$

$$0 = \log\frac{[C_6H_5OH]}{[C_6H_5O^-]} = \log 10^0$$

The ratio of $[C_6H_5OH]$ to $[C_6H_5O^-]$ is 1:1.

d. At pH = 6.0

$$10.0 = 6 + \log\frac{[C_6H_5OH]}{[C_6H_5O^-]}$$

$$4 = \log\frac{[C_6H_5OH]}{[C_6H_5O^-]} = \log 10^4$$

The ratio of $[C_6H_5OH]$ to $[C_6H_5O^-]$ is 10,000:1.

15.48 Use the Henderson–Hasselbalch equation.

$$pK_a = pH + \log\frac{[HA]}{[A:^-]}$$

a. An equal amount of **A** and **B** will be present when pH = pK_a (**A**) = 4.3 because log [**A**]/[**B**] = log 1/1 = 0.

$$pK_a = 4.3 = x + \log 1$$

$$4.3 = x + 0 = 4.3 \text{ pH}$$

b. An equal amount of **B** and **C** will be present when pH = pK_a (**B**) = 5.4 because log [**A**]/[**C**] = log 1/1 = 0.

$$pK_a = 5.4 = x + \log 1$$

$$5.4 = x + 0 = 5.4 \text{ pH}$$

15.49 Use the principles in Answer 15.14 to answer the questions.

$$RCO_2H \xrightarrow{-H^+} RCO_2^- \quad pK_a = 4.8 \qquad RNH_3^+ \xrightarrow{-H^+} RNH_2 \quad pK_a = 9$$

a. At pH = 8, RCO_2^- and RNH_3^+ predominate.
b. At pH = 3, RCO_2H and RNH_3^+ predominate.
c. At pH = 10, RCO_2^- and RNH_2 predominate.

15.50

A B COOH

- Dissolve both compounds in CH_2Cl_2.
- Add 10% $NaHCO_3$ solution. This makes a carboxylate anion ($C_{10}H_7COO^-$) from **B**, which dissolves in the aqueous layer. The other compound (**A**) remains in the CH_2Cl_2.
- Separate the layers.

15.51

OH + OH + OH

CH_2Cl_2

10% $NaHCO_3$

aqueous: O^- Na^+

CH_2Cl_2: OH + OH

10% NaOH

aqueous: O^- Na^+

CH_2Cl_2: OH

15.52

A B OH C OH

A and **C** are neutral organic compounds that are soluble in CH_2Cl_2 and insoluble in any aqueous solution. **B** is a carboxylic acid that is soluble in CH_2Cl_2 and insoluble in H_2O. In basic aqueous solution (10% NaOH or 10% $NaHCO_3$), however, the acid is deprotonated to form a carboxylate anion that is soluble in the aqueous medium.

	CH_2Cl_2	H_2O
a.	CH_2Cl_2	H_2O
	A, B, C	(none)
b.	CH_2Cl_2	10% NaOH
	A, C	**B**
c.	CH_2Cl_2	10% NaOH
	A, C	**B**

15.53

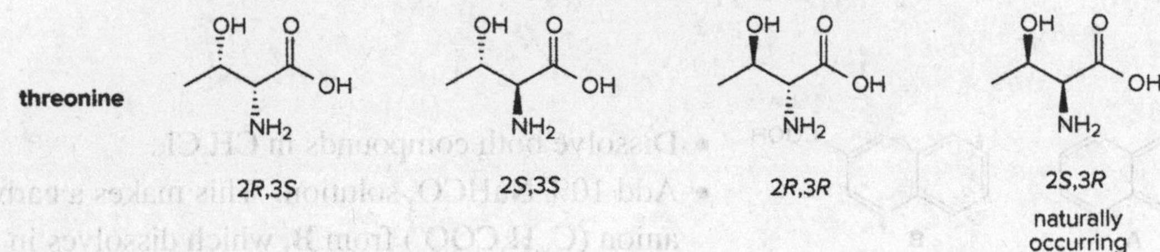

15.54

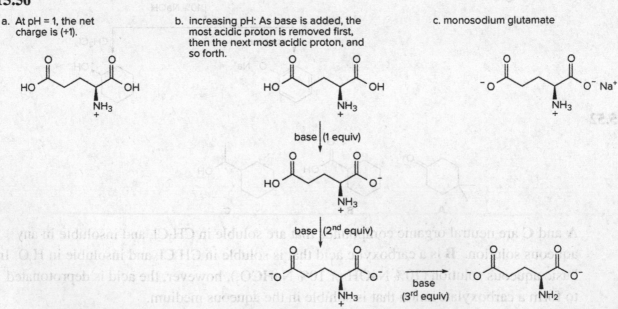

neutral
pH = 6
form at isoelectric point

(+1) form
pH = 1

(–1) form
pH = 11

15.55

a. asparagine $pI = \dfrac{pK_a(COOH) + pK_a(NH_3^+)}{2} = (2.02 + 8.80) / 2 = \mathbf{5.41}$

b. methionine $pI = \dfrac{pK_a(COOH) + pK_a(NH_3^+)}{2} = (2.28 + 9.21) / 2 = \mathbf{5.75}$

15.56

a. At pH = 1, the net charge is (+1).

b. increasing pH: As base is added, the most acidic proton is removed first, then the next most acidic proton, and so forth.

base (1 equiv)

base (2nd equiv)

base (3rd equiv)

c. monosodium glutamate

15.57 Use the principles in Answer 15.14 to answer the questions.

$H_3\overset{+}{N}$–$CH_2CH_2CH_2$–CO_2^- **A**

$pK_a = 10.43$ $pK_a = 4.23$

a. [1] $H_3\overset{+}{N}CH_2CH_2CH_2COOH$

predominant form at pH = 1.21

[2] $H_2NCH_2CH_2CH_2CO_2^-$

predominant form at pH = 12.1

[3] An equal mixture of **B** and **C** is present at pH = 10.43.

[4] An equal mixture of product **A** and **B** is at pH = 4.23.

b. Isoelectric point = p*I* = $\dfrac{pK_a(CO_2H) + pK_a(NH_3^+)}{2}$

$= \dfrac{4.23 + 10.43}{2}$

$= 7.33$

c. The NH_3^+ group is electron withdrawing. The closer it is located to the carboxy group, the more acidic the CO_2H group is. Because the NH_3^+ group is farther from the CO_2H group in γ–aminobutyric acid compared to an α-amino acid, γ-aminobutyric acid is less acidic and has a higher pK_a.

15.58 Use the Henderson–Hasselbalch equation and the principles in Answer 15.14 to answer the questions.

$HSCH_2CH(\overset{+}{N}H_3)COOH$ (**A**, $pK_a = 2.05$) ⟶ $HSCH_2CH(\overset{+}{N}H_3)CO_2^-$ (**B**, $pK_a = 8.00$) ⟶ $^-SCH_2CH(\overset{+}{N}H_3)CO_2^-$ (**C**, $pK_a = 10.25$) ⟶ $^-SCH_2CH(NH_2)CO_2^-$ (**D**)

a. At pH = 12, **D** predominates.
b. At pH = 1, **A** predominates.
c. At pH = 8, an equal mixture of **B** and **C** is present.
d. At pH = 3, **B** predominates.

15.59

a. $CH_3Cl + NaCN \longrightarrow CH_3{-}CN \xrightarrow{H_3O^+} CH_3{-}COOH$

$CH_3{-}Cl + Mg \longrightarrow CH_3{-}MgCl \xrightarrow[\text{[2] } H_3O^+]{\text{[1] } CO_2} CH_3{-}COOH$

b. Bromobenzene + NaCN → This method can't be used because an S_N2 reaction can't be done on an sp^2 hybridized C.

Bromobenzene + Mg → phenylmagnesium bromide (MgBr) → [1] CO_2, [2] H_3O^+ → benzoic acid (COOH)

c. HO(CH₂)₄Br + NaCN → HO(CH₂)₄CN → H_3O^+ → HO(CH₂)₄COOH

HO(CH₂)₄Br + Mg → This method can't be used because you can't make a Grignard reagent with an acidic OH group.

15.60

a. Benzonitrile (CN) → [1] $LiAlH_4$, [2] H_2O → benzylamine (NH_2)

Benzonitrile → [1] DiBAL-H, [2] H_2O → benzaldehyde (CHO)

Benzylamine + benzaldehyde → mild H^+ → imine (N-benzylidenebenzylamine)

b. Benzaldehyde (CHO) (from a.) → $NaBH_4$, CH_3OH → benzyl alcohol (OH) → PBr_3 → benzyl bromide (Br) → Mg → benzylmagnesium bromide (MgBr) → [1] benzaldehyde (CHO), [2] H_2O → 1,2-diphenylethanol (OH) → H_2SO_4 → stilbene

major stereisomer

c. 1,2-Diphenylethanol (OH) (from b.) → PCC → deoxybenzoin (ketone) → [1] benzylmagnesium bromide (MgBr) (from b.), [2] H_2O → tertiary alcohol (OH)

15.61

a. 1-Bromobutane (Br) → NaCN → pentanenitrile (CN)

1-Bromobutane → Mg → butylmagnesium bromide (MgBr)

Pentanenitrile + butylmagnesium bromide → H_2O → 5-nonanone (O)

b. [Reaction scheme: cyclohexylmethyl bromide → [1] Mg, [2] $H_2C{=}O$, [3] H_2O → alcohol → PCC → aldehyde → [1] cyclohexylmethyl MgBr, [2] H_2O → secondary alcohol → PCC → ketone → ethylMgBr (from ethyl bromide + Mg) → H_2O → tertiary alcohol → HBr → tertiary bromide → [1] Mg, [2] CO_2, [3] H_3O^+ → carboxylic acid]

15.62

a. Molecular formula: $C_3H_5ClO_2$ ⟶ one double bond or ring
IR: 3500–2500 cm^{-1}, 1714 cm^{-1} ⟶ C=O and O–H
NMR data: 2.87 (triplet, 2 H), 3.76 (triplet, 2 H), and 11.8 (singlet, 1 H) ppm

b. Molecular formula: $C_8H_8O_3$ ⟶ 5 double bonds or rings
IR: 3500–2500 cm^{-1}, 1710 cm^{-1} ⟶ C=O and O–H
NMR data: 4.7 (singlet, 2 H), 6.9–7.3 (multiplet, 5 H), and 11.3 (singlet, 1 H) ppm
monosubstituted benzene ring

c. Molecular formula: C_4H_7N
IR: 2250 cm^{-1} → triple bond
NMR: 1.08 (triplet, 3 H), 1.70 (multiplet, 2 H), 2.34 (triplet, 2 H) ppm

15.63

Compound **B:** Molecular formula $C_4H_8O_2$ (one degree of unsaturation)
IR absorptions at 3500–2500 (O–H) and 1700 (C=O) cm^{-1}
^{1}H NMR data:

Absorption	ppm	# of H's	Explanation	Structure:
doublet	1.0	6	6 H's adjacent to 1 H	
septet	2.3	1	1 H adjacent to 6 H's	
singlet (very broad)	10.6	1	OH of RCOOH	**B**

15.64

Compound **C:** Molecular formula $C_4H_8O_3$ (one degree of unsaturation)
IR absorptions at 3600–2500 (O–H) and 1734 (C=O) cm^{-1}
^1H NMR data:

Absorption	ppm	# of H's	Explanation
triplet	1.3	3	CH_3 group adjacent to 2 H's
quartet	3.6	2	2 H's adjacent to 3 H's
singlet	4.2	2	2 H's
singlet	11.3	1	OH of COOH

Structure:

C

15.65

Compound **D:** Molecular formula $C_9H_9ClO_2$ (five degrees of unsaturation)
^{13}C NMR data: 30, 36, 128, 130, 133, 139, 179 = 7 different types of C's
^1H NMR data:

Absorption	ppm	# of H's	Explanation
triplet	2.7	2	2 H's adjacent to 2 H's
triplet	2.9	2	2 H's adjacent to 2 H's
two signals	7.2	4	disubstituted benzene ring
singlet	11.7	1	OH of COOH

Structure:

Cl COOH

D

15.66 The first equivalent of NaH removes the most acidic proton—that is, the OH proton on the phenol. The resulting phenoxide can then act as a nucleophile to displace I to form a substitution product. With two equivalents, both OH protons are removed. In this case the more nucleophilic O atom is the stronger base—that is, the alkoxide derived from the alcohol (not the phenoxide), so this negatively charged O atom reacts first in a nucleophilic substitution reaction.

most acidic proton

HO OH

NaH (1 equiv) → nucleophile ^-O OH → [1] CH_3I [2] H_2O → O OH

NaH (2 equiv) → ^-O O^- nucleophile → [1] CH_3I [2] H_2O → HO O

15.67 A CH_3O group has an electron-withdrawing inductive effect and an electron-donating resonance effect. In 2-methoxyacetic acid, the OCH_3 group is bonded to an sp^3 hybridized C, so there is no way to donate electron density by resonance. The CH_3O group withdraws electron density because of the electronegative O atom, stabilizing the conjugate base, and making CH_3OCH_2COOH a stronger acid than CH_3COOH.

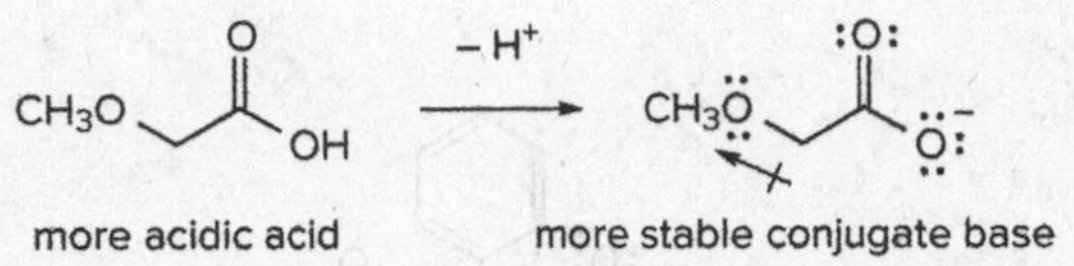

more acidic acid — more stable conjugate base

In *p*-methoxybenzoic acid, the CH_3O group is bonded to an sp^2 hybridized C, so it can donate electron density by a resonance effect. This destabilizes the conjugate base, making the starting material less acidic than C_6H_5COOH.

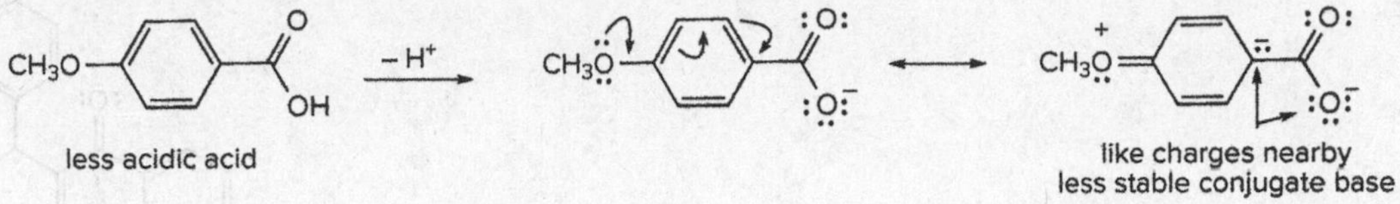

less acidic acid — like charges nearby, less stable conjugate base

15.68

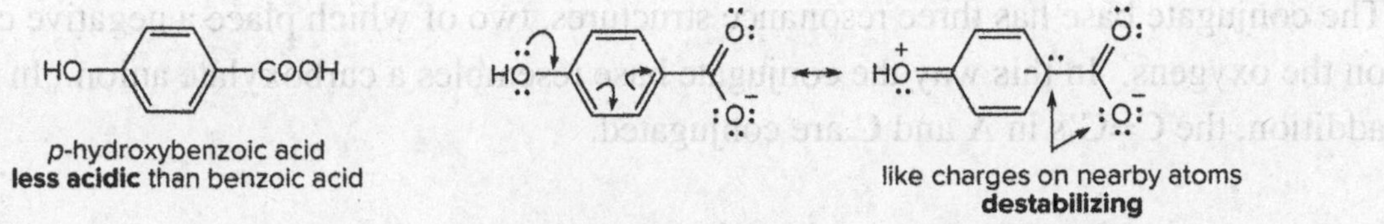

p-hydroxybenzoic acid
less acidic than benzoic acid

like charges on nearby atoms
destabilizing

The OH group donates electron density by its resonance effect and this destabilizes the conjugate base, making the acid less acidic than benzoic acid.

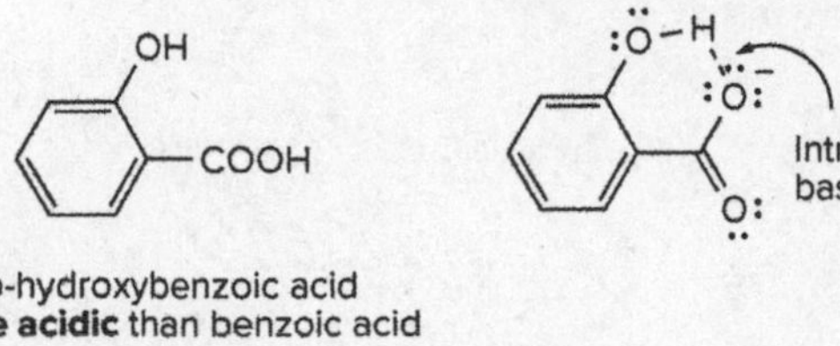

Intramolecular hydrogen bonding stabilizes the conjugate base, making the acid more acidic than benzoic acid.

o-hydroxybenzoic acid
more acidic than benzoic acid

15.69

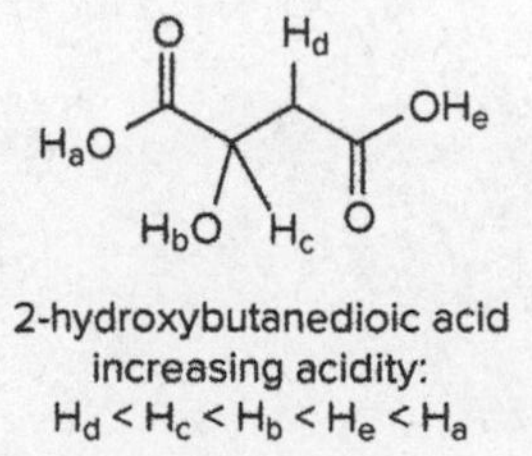

H_a and H_e must be the two most acidic protons because they are part of a carboxy group. Loss of a proton forms a resonance-stabilized carboxylate anion that has the negative charge delocalized on two O atoms. H_a is more acidic than H_e because the nearby OH group on the α carbon increases acidity by an electron-withdrawing inductive effect. H_b is the next most acidic proton because the conjugate base places a negative charge on the electronegative O atom, but it is not resonance stabilized.

2-hydroxybutanedioic acid
increasing acidity:
$H_d < H_c < H_b < H_e < H_a$

The least acidic H's are H_c and H_d because these H's are bonded to C atoms. The electronegative O atom further acidifies H_c by an electron-withdrawing inductive effect.

15.70

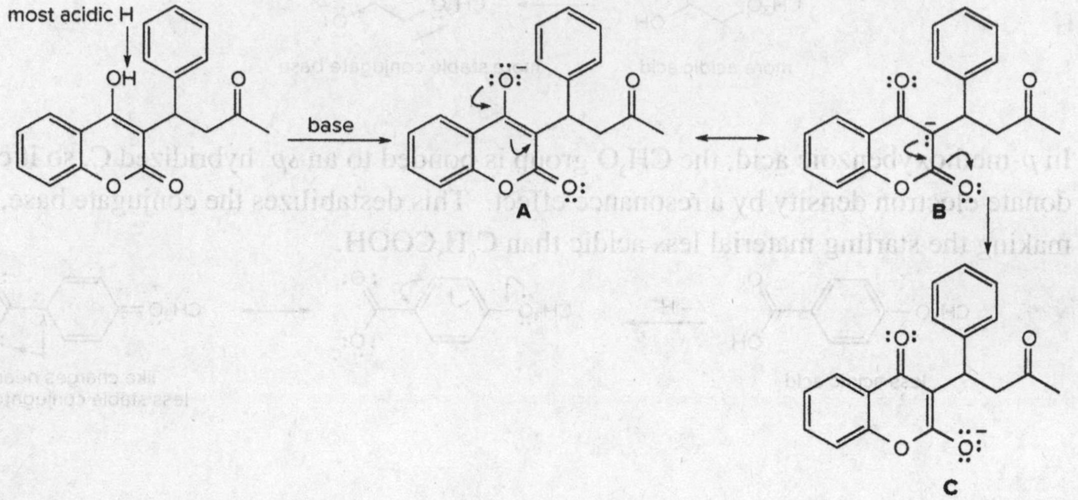

most acidic H

OH

base

A

B

C

The conjugate base has three resonance structures, two of which place a negative charge on the oxygens. In this way the conjugate base resembles a carboxylate anion. In addition, the C=C's in **A** and **C** are conjugated.

Chapter 16 Carboxylic Acids and Their Derivatives—Nucleophilic Acyl Substitution

Chapter Review

Summary of spectroscopic absorptions of RCOZ (16.4)

IR absorptions

- All RCOZ compounds have a C=O absorption in the region 1600–1850 cm^{-1}.
  - RCOCl: 1800 cm^{-1}
  - $(RCO)_2O$: 1820 and 1760 cm^{-1} (two peaks)
  - RCOOR': 1735–1745 cm^{-1}
  - $RCO_2PO_3^{2-}$: 1700–1730 cm^{-1}
  - RCOSR': 1690–1720 cm^{-1}
  - $RCONR'_2$: 1630–1680 cm^{-1}
- Additional amide absorptions occur at 3200–3400 cm^{-1} (N–H stretch) and 1640 cm^{-1} (N–H bending).
- Decreasing the ring size of a cyclic lactone, lactam, or anhydride increases the frequency of the C=O absorption.
- Conjugation shifts the C=O to lower wavenumber.

^{1}H NMR absorptions

- C–H α to the C=O absorbs at 2–2.5 ppm.
- N–H of an amide absorbs at 7.5–8.5 ppm.

^{13}C NMR absorption

- C=O absorbs at 160–180 ppm.

Summary: The relationship between the basicity of Z^- and the properties of RCOZ

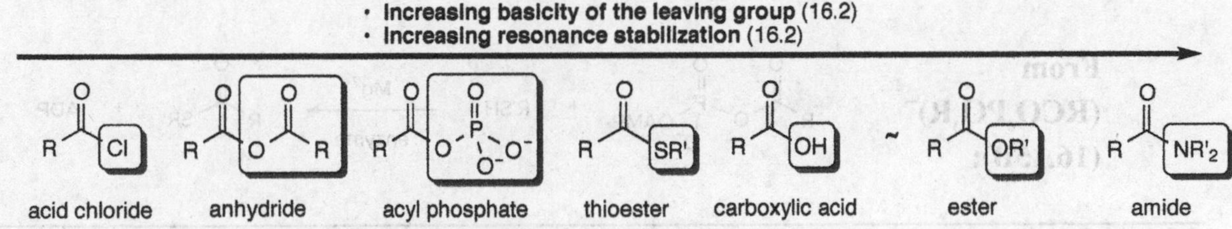

General features of nucleophilic acyl substitution

- The characteristic reaction of compounds having the general structure RCOZ is nucleophilic acyl substitution (16.1).
- The mechanism consists of two steps (16.6A):
  [1] Addition of a nucleophile to form a tetrahedral intermediate
  [2] Elimination of a leaving group
- More reactive acyl compounds can be used to prepare less reactive acyl compounds. The reverse is not necessarily true (16.6B).

Nucleophilic acyl substitution reactions

[1] Reaction that synthesizes acid chlorides (RCOCl)

From RCOOH (16.9A):

$$RCOOH + SOCl_2 \longrightarrow RCOCl + SO_2 + HCl$$

[2] Reactions that synthesize anhydrides [$(RCO)_2O$]

a. From RCOCl (16.7):

$$RCOCl + {}^{-}OCOR' \longrightarrow RCOOCOR' + Cl^-$$

b. From dicarboxylic acids (16.9B):

$$HOOC\text{–}CH_2CH_2\text{–}COOH \xrightarrow{\Delta} \text{cyclic anhydride} + H_2O$$

[3] Reaction that produces acyl phosphates ($RCO_2PO_3^{2-}$)

From RCO_2^- (16.15A):

$$RCO_2^- + ATP \xrightarrow[\text{enzyme}]{Mg^{2+}} RCO\text{–}O\text{–}PO_3^{2-} + ADP$$

[4] Reaction that produces thioesters (RCOSR')

From $(RCO_2PO_3R)^-$ (16.15B):

$$RCO\text{–}O\text{–}P(O)(O^-)\text{–}OAMP + R'SH \xrightarrow[\text{enzyme}]{Mg^{2+}} RCOSR' + ADP$$

[5] Reactions that synthesize carboxylic acids (RCOOH) and carboxylates ($RCOO^-$)

a. From RCOCl (16.7):

$$RCOCl + H_2O \xrightarrow{\text{pyridine}} RCOOH + C_5H_5NH^+\ Cl^-$$

b. From $(RCO)_2O$ (16.8):

$$RCOOCOR + H_2O \longrightarrow 2\ RCOOH$$

c. From RCOOR' (16.10):

$$RCOOR' + H_2O \xrightarrow{(H^+ \text{ or } {}^-OH)} RCOOH \text{ (with acid) or } RCOO^- \text{ (with base)} + R'OH$$

d. From RCOSR' (16.16): RCOSR' + H_2O —(enzyme)→ $RCOO^-$ + HSR'

e. From RCONR'$_2$ (R' = H or alkyl, 16.12): $RCONR'_2$ (R' = H or alkyl) —(H_2O, H^+)→ RCOOH + $R'_2\overset{+}{N}H_2$; —(H_2O, ^-OH)→ $RCOO^-$ + R'_2NH

[6] Reactions that synthesize esters (RCOOR')

a. From RCOCl (16.7): RCOCl + R'OH —(pyridine)→ RCOOR' + pyridinium ($C_5H_5NH^+$) Cl^-

b. From (RCO)$_2$O (16.8): RCOOCOR + R'OH → RCOOR' + RCOOH

c. From RCOSR' (16.16): RCOSR' + R'OH —(enzyme)→ RCOOR' + HSR'

d. From RCOOH (16.9C): RCOOH + R'OH —(H_2SO_4)→ RCOOR' + H_2O

[7] Reactions that synthesize amides (RCONH$_2$) [The reactions are written with NH$_3$ as the nucleophile to form RCONH$_2$. Similar reactions occur with R'NH$_2$ to form RCONHR' and with R'$_2$NH to form RCONR'$_2$.]

a. From RCOCl (16.7): RCOCl + NH_3 (2 equiv) → $RCONH_2$ + $NH_4^+Cl^-$

b. From (RCO)$_2$O (16.8): RCOOCOR + NH_3 (2 equiv) → $RCONH_2$ + $RCOO^-\,NH_4^+$

c. From RCO$_2$PO$_3^{2-}$ (16.15B): $RCOOPO_3^{2-}$ + $R''NH_2$ —(Mg^{2+}, enzyme)→ RCONHR'' + $HOPO_3^{2-}$

d. From RCOOH (16.9D):

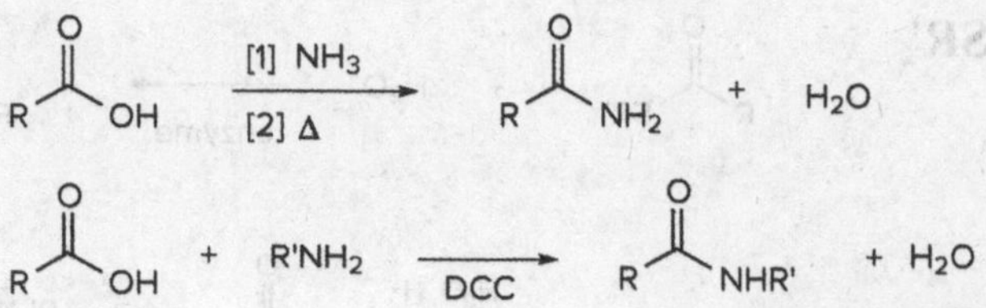

$$RCOOH \xrightarrow[\text{[2] } \Delta]{\text{[1] } NH_3} RCONH_2 + H_2O$$

$$RCOOH + R'NH_2 \xrightarrow{DCC} RCONHR' + H_2O$$

e. From RCOOR' (16.10):

$$RCOOR' + NH_3 \longrightarrow RCONH_2 + R'OH$$

Practice Test on Chapter Review

1. Give the IUPAC name for each of the following compounds.

a. 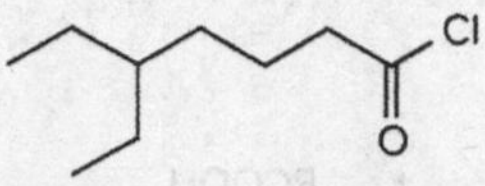 b. 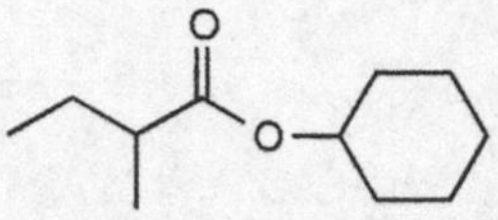 c.

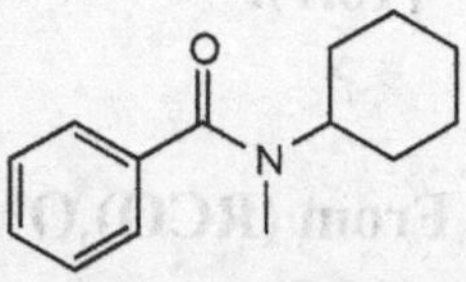

2. (a) Which compound absorbs at the *lowest* wavenumber in the IR? (b) Which compound absorbs at the *highest* wavenumber?

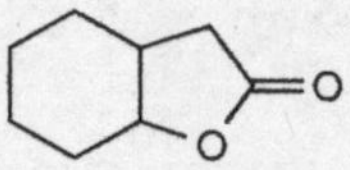

A 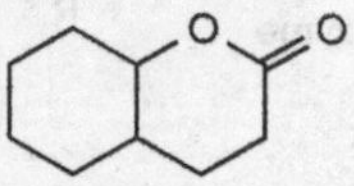 B 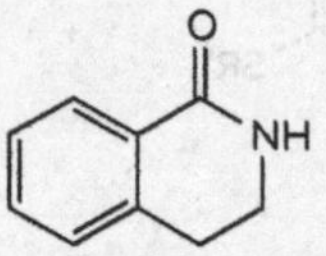 C 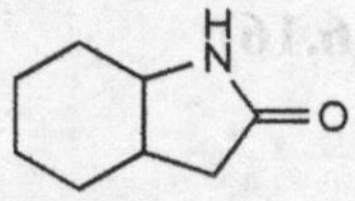 D

3.a. Which of the following reaction conditions can be used to synthesize an ester?

1. RCOCl + R'OH + pyridine
2. RCOOH + R'OH + H_2SO_4
3. RCOOH + R'OH + NaOH
4. Both methods (1) and (2) can be used to synthesize an ester.
5. Methods (1), (2), and (3) can all be used to synthesize an ester.

b. Which of the following compounds is most reactive in nucleophilic acyl substitution?

1. CH_3COCl
2. CH_3COOCH_3
3. $CH_3CON(CH_3)_2$
4. $(CH_3CO)_2O$
5. CH_3COOH

4. What reagent is needed to convert $(CH_3CH_2)_2CHCOOH$ into each compound?

a. $(CH_3CH_2)_2CHCOO^-Na^+$
b. $(CH_3CH_2)_2CHCOCl$
c. $(CH_3CH_2)_2CHCON(CH_3)_2$
d. $(CH_3CH_2)_2CHCO_2CH_2CH_3$
e. $[(CH_3CH_2)_2CHCO]_2O$

5. Draw the organic products formed in the following reactions.

a. 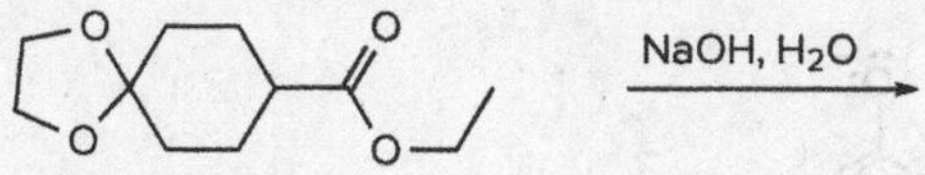 NaOH, H_2O

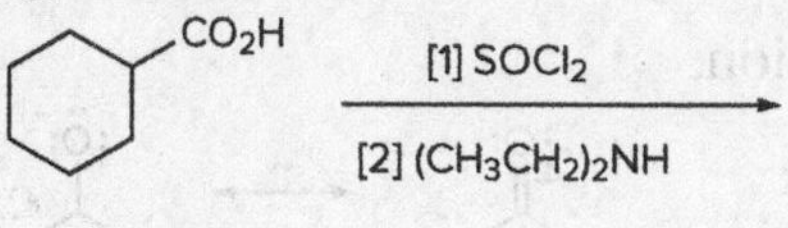

b. 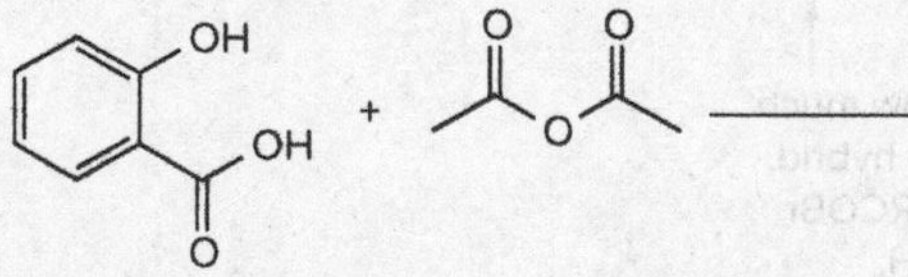 +

c. [1] $SOCl_2$ [2] $(CH_3CH_2)_2NH$

d. 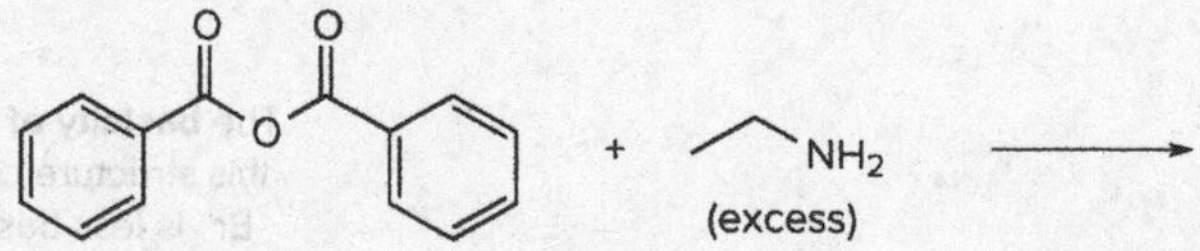 + NH_2 (excess)

Answers to Practice Test

1.a. 5-ethyl-heptanoyl chloride
b. cyclohexyl 2-methylbutanoate
c. *N*-cyclohexyl-*N*-methylbenzamide

2.a. **C**
b. **A**

3.a. 4
b. 1

4.a. NaOH
b. $SOCl_2$
c. $HN(CH_3)_2$, DCC
d. CH_3CH_2OH, H_2SO_4
e. heat

5. a. 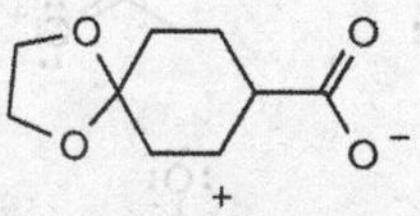 b. 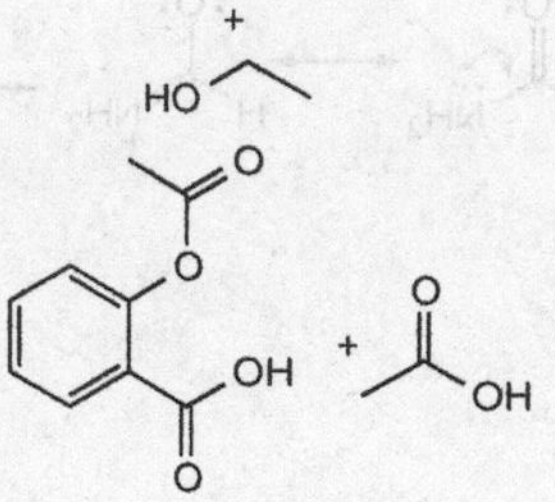 c. d.

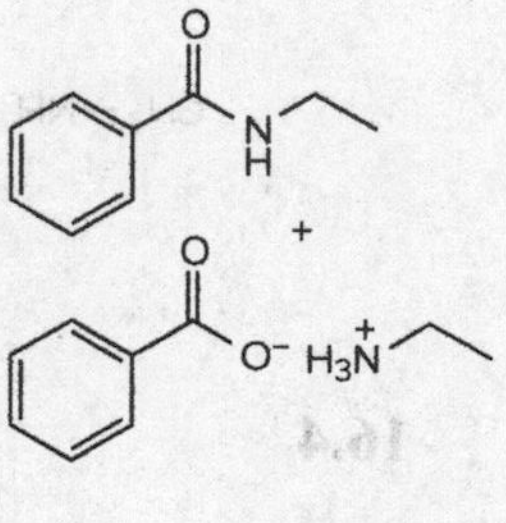

Answers to Problems

16.1 The number of C–N bonds determines the classification as a 1°, 2°, or 3° amide.

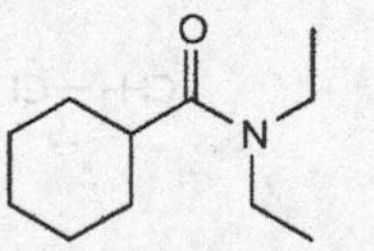

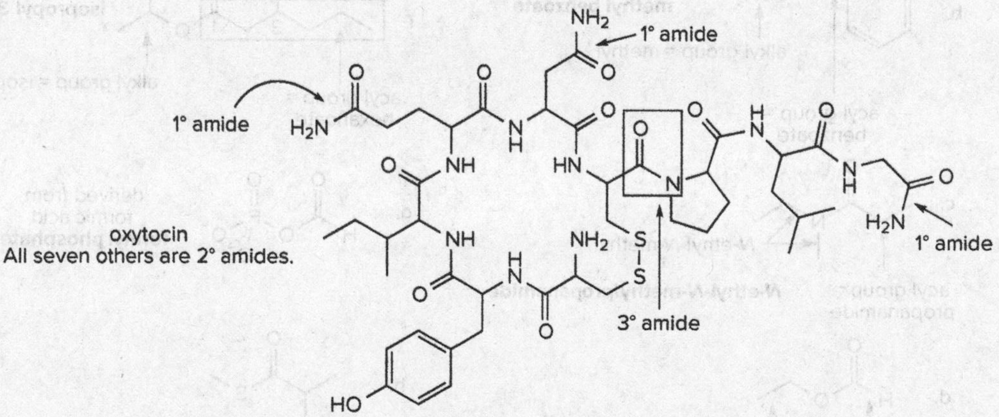

1° amide
1° amide
1° amide
3° amide
oxytocin
All seven others are 2° amides.

16.2 As the basicity of Z increases, the stability of RCOZ increases because of added resonance stabilization.

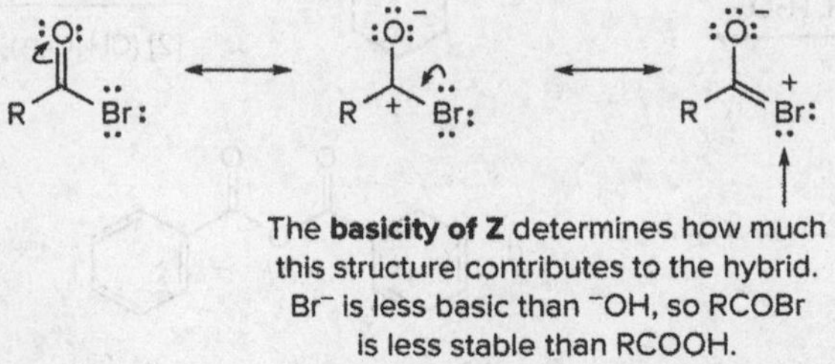

The **basicity of Z** determines how much this structure contributes to the hybrid. Br^- is less basic than ^-OH, so RCOBr is less stable than RCOOH.

16.3

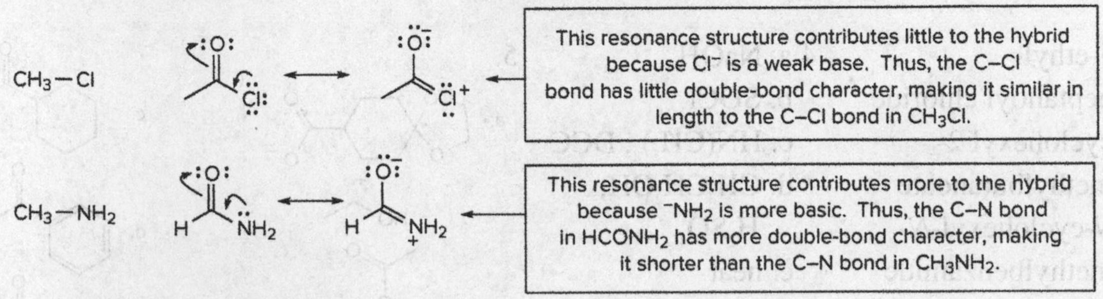

CH_3-Cl

This resonance structure contributes little to the hybrid because Cl^- is a weak base. Thus, the C–Cl bond has little double-bond character, making it similar in length to the C–Cl bond in CH_3Cl.

CH_3-NH_2

This resonance structure contributes more to the hybrid because $^-NH_2$ is more basic. Thus, the C–N bond in $HCONH_2$ has more double-bond character, making it shorter than the C–N bond in CH_3NH_2.

16.4

a. **2-ethylbutanoyl chloride** ← 2-ethyl

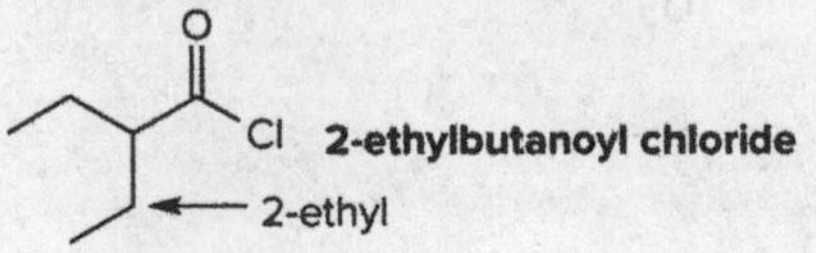

e. **benzoic propanoic anhydride** — acyl group = propanoic; acyl group = benzoic

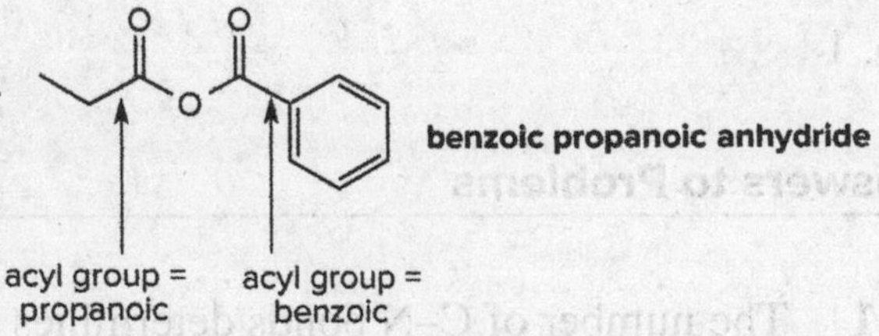

b. **methyl benzoate** — alkyl group = methyl; acyl group = benzoate

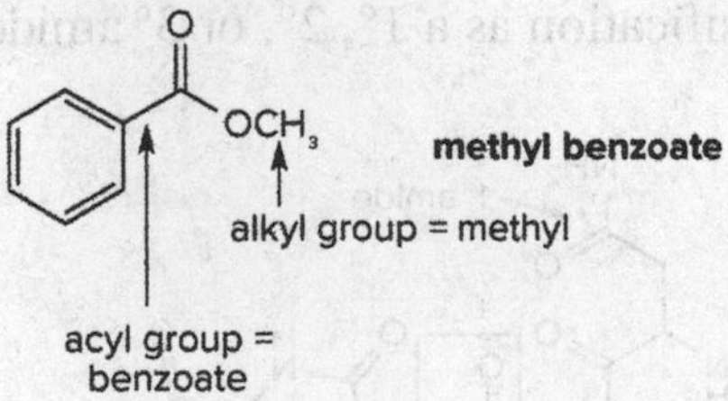

f. **isopropyl 3-ethylhexanoate** — acyl group = hexanoate; alkyl group = isopropyl

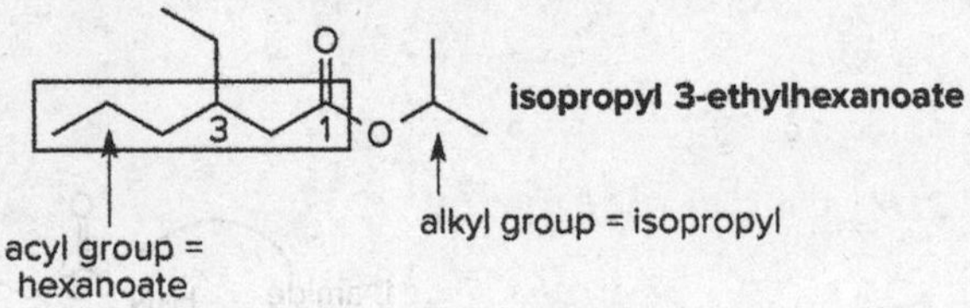

c. **N-ethyl-N-methylpropanamide** — acyl group = propanamide; N-ethyl-N-methyl

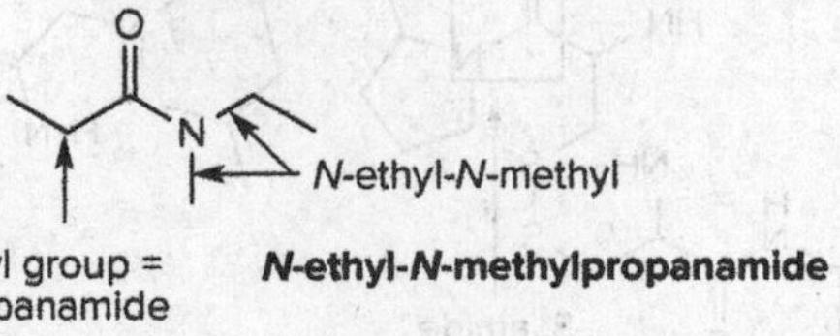

g. derived from formic acid **formyl phosphate**

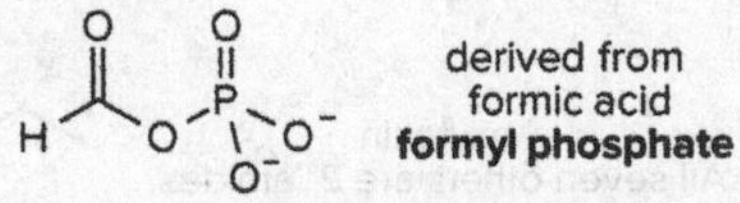

d. **ethyl formate** — acyl group = formate; alkyl group = ethyl

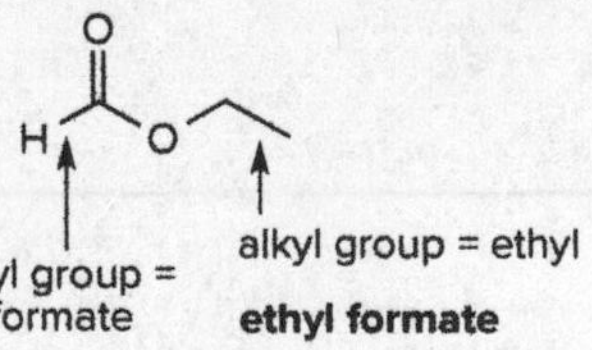

h. **S-methyl 2-methylpropanethioate** — derived from 2-methylpropanoic acid

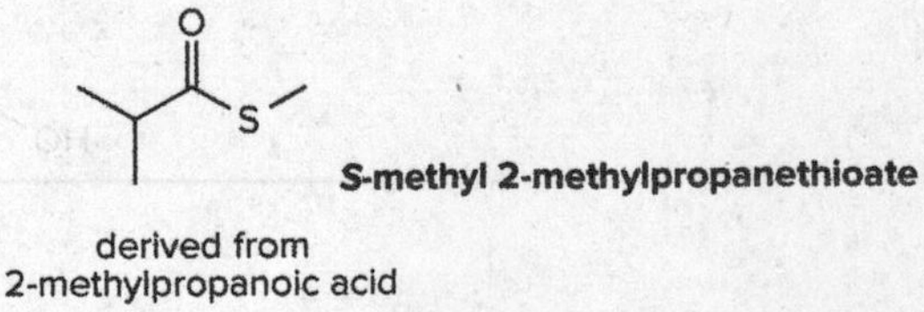

16.5

a. 5-methylheptanoyl chloride

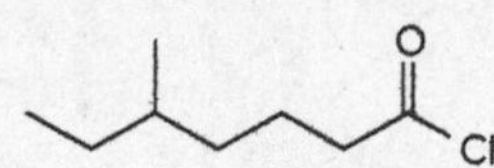

c. acetic formic anhydride

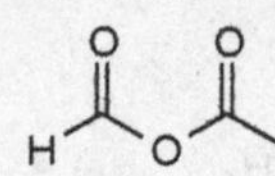

e. sec-butyl 2-methylhexanoate

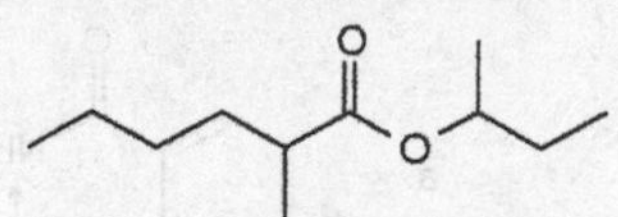

b. *S*-isopropyl 3-ethylcyclobutanecarbothioate

d. *N*-isobutyl-*N*-methylbutanamide

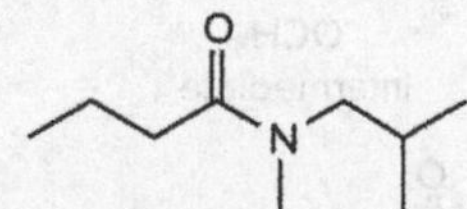

f. 3-ethyl-2-methylpentanoyl phosphate

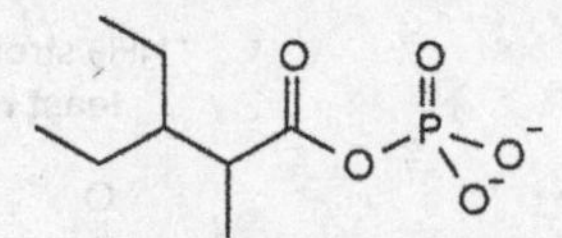

16.6 CH_3CONH_2 has a higher boiling point than CH_3CO_2H because it has more opportunities for hydrogen bonding. Each CH_3CONH_2 can hydrogen bond to three other molecules (at its O atom and two N–H bonds), whereas CH_3CO_2H has only two intermolecular hydrogen bonds possible (at its carbonyl O atom and OH bond).

16.7

- Amides are more resonance stabilized than esters.
- Conjugation *decreases* the frequency of C=O absorption.
- Decreasing ring size *increases* the frequency of C=O absorption.

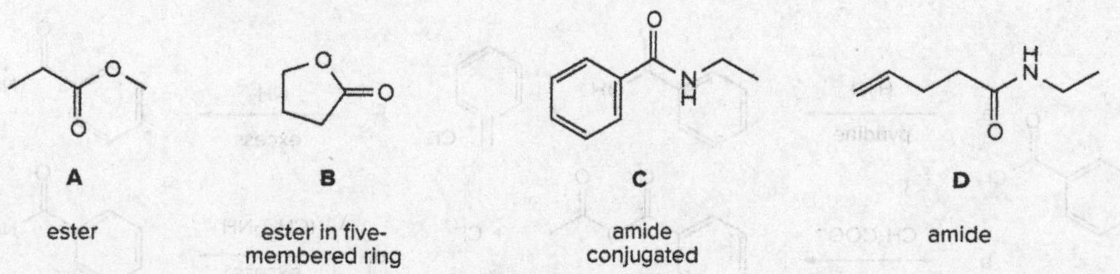

16.8 More reactive acyl compounds can be converted to less reactive acyl compounds.

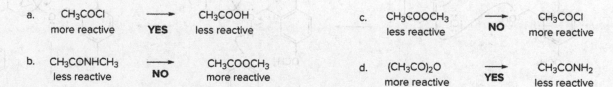

16.9 The better the leaving group is, the more reactive the carboxylic acid derivative. The weakest base is the best leaving group.

a.

$^{-}NH_2$ strongest base
least reactive

$^{-}OCH_3$
intermediate

Cl^{-} weakest base
most reactive

b.

$^{-}NHCH_3$ strongest base
least reactive

^{-}OH
intermediate

$^{-}O_2CCH_2CH_3$ weakest base
most reactive

16.10

acetic anhydride

trichloroacetic anhydride

The Cl atoms are electron withdrawing, which makes the conjugate base (the leaving group, CCl_3COO^{-}) weaker and more stable.

16.11

a. H_2O, pyridine → benzoic acid + pyridinium Cl^{-}

b. CH_3COO^{-} → mixed anhydride + Cl^{-}

c. NH_3 excess → benzamide + $NH_4^{+}Cl^{-}$

d. $(CH_3)_2NH$ excess → $N(CH_3)_2$ amide + $(CH_3)_2\overset{+}{N}H_2$ Cl^{-}

16.12 The mechanism has three steps: [1] nucleophilic attack by O; [2] proton transfer; and [3] elimination of the Cl^{-} leaving group to form the product.

A

+ :Cl:⁻

16.13

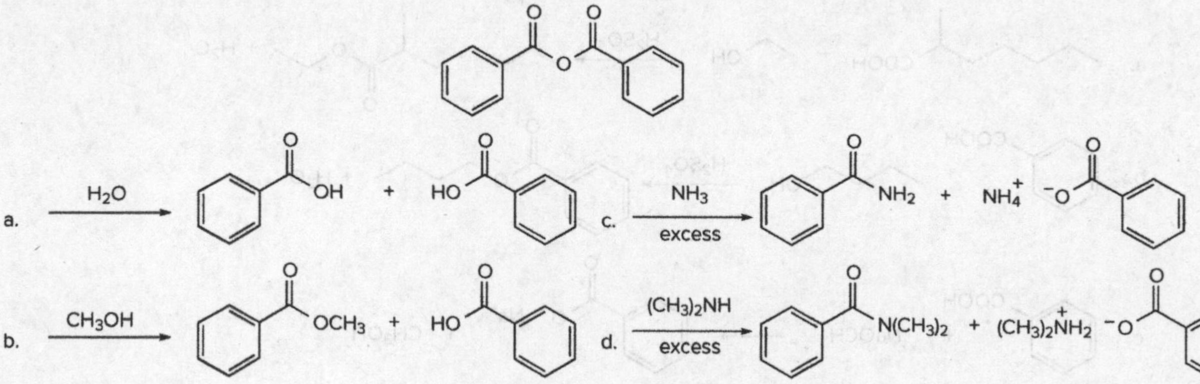

a. H_2O → OH + HO

b. CH_3OH → OCH_3 + HO

c. NH_3 excess → NH_2 + NH_4^+ ^-O

d. $(CH_3)_2NH$ excess → $N(CH_3)_2$ + $(CH_3)_2\overset{+}{N}H_2$ ^-O

16.14

a. [1] CH_3MgBr (2 equiv) [2] H_2O → OH + OH

b. [1] $LiAlH_4$ [2] H_2O → OH + OH

c. [1] $LiAlH[OC(CH_3)_3]_3$ [2] H_2O → H + OH

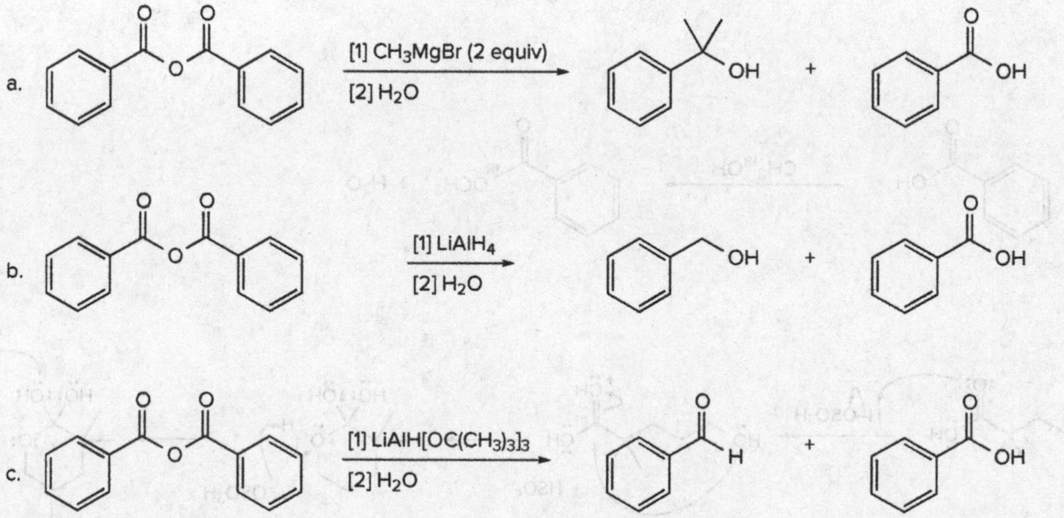

16.15 Reaction of a carboxylic acid with thionyl chloride converts it to an acid chloride.

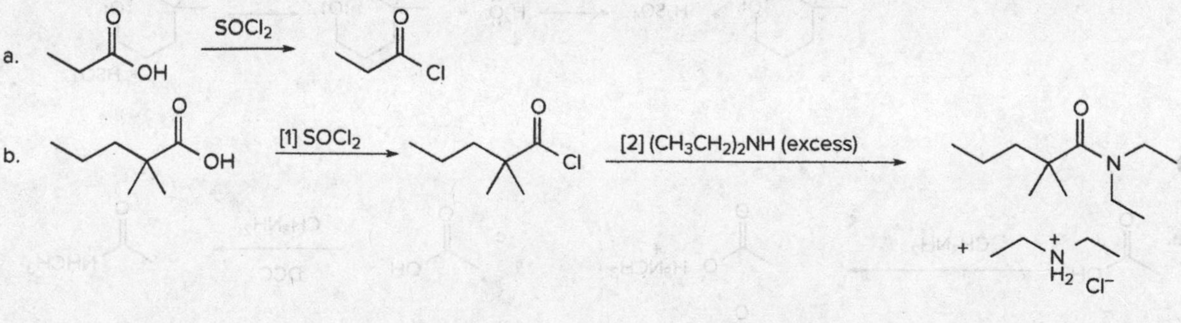

a. OH $SOCl_2$ → Cl

b. OH [1] $SOCl_2$ → Cl [2] $(CH_3CH_2)_2NH$ (excess) → N + $\overset{+}{N}H_2$ Cl^-

16.16

a. (2-methylheptanoic acid, COOH) + (ethanol, OH) —H_2SO_4→ ethyl 2-methylheptanoate + H_2O

b. (benzoic acid, COOH) + (1-butanol, OH) —H_2SO_4→ butyl benzoate + H_2O

c. (benzoic acid, COOH) + $NaOCH_3$ → sodium benzoate (O^- Na^+) + CH_3OH

d. HO(CH$_2$)$_5$C(=O)OH —H_2SO_4→ six-membered lactone + H_2O

16.17

(benzoic acid, OH) —$CH_3{}^{18}OH$→ $^{18}OCH_3$ ester + H_2O

16.18

HÖ ... C(=:O:)ÖH + H–OSO$_3$H → HÖ ... C(=$\overset{+}{Ö}$H)ÖH + HSO_4^- → HÖ: :ÖH ... +O–H + $^-OSO_3H$ → HÖ: :ÖH ... :O: + H–ÖSO$_3$H → HÖ: :ÖH$_2$ ... :O: + HSO_4^- → :$\overset{+}{O}$–H ... :O: + HSO_4^- → :O: ... Ö: + H_2SO_4 ← $H_2Ö$ +

16.19

a. (acetic acid, OH) —CH_3NH_2→ (acetate, O^-) $H_3\overset{+}{N}CH_3$

b. (acetic acid, OH) —[1] CH_3NH_2; [2] Δ→ (NHCH$_3$ amide) + H_2O

c. (acetic acid, OH) —CH_3NH_2, DCC→ (NHCH$_3$ amide)

16.20

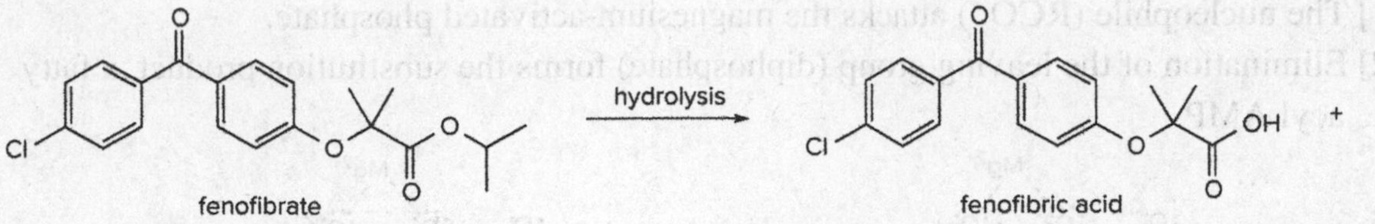

fenofibrate —hydrolysis→ fenofibric acid + HO—isopropyl

16.21

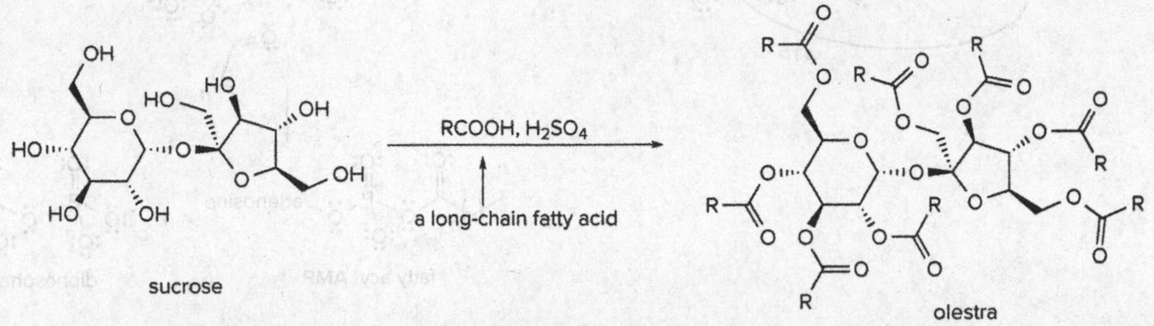

sucrose —$RCOOH, H_2SO_4$ (a long-chain fatty acid)→ olestra

16.22

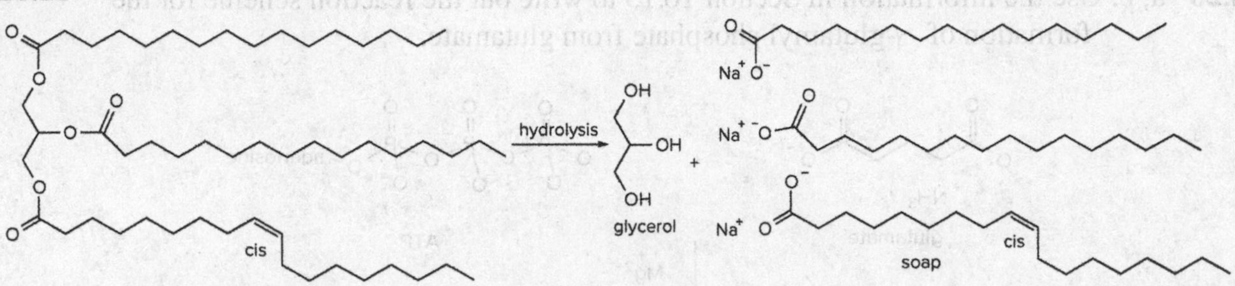

hydrolysis → glycerol + soap ($Na^+\ ^-O$ carboxylates; cis)

16.23 Aspirin has an ester, a more reactive acyl group, whereas acetaminophen has an amide, a less reactive acyl group. The ester makes aspirin more easily hydrolyzed with water from the air than acetaminophen. Therefore, Tylenol can be kept for many years, whereas aspirin decomposes.

ester — acetylsalicylic acid

amide — acetaminophen

16.24

H_3O^+

"Regular" amide is not hydrolyzed.

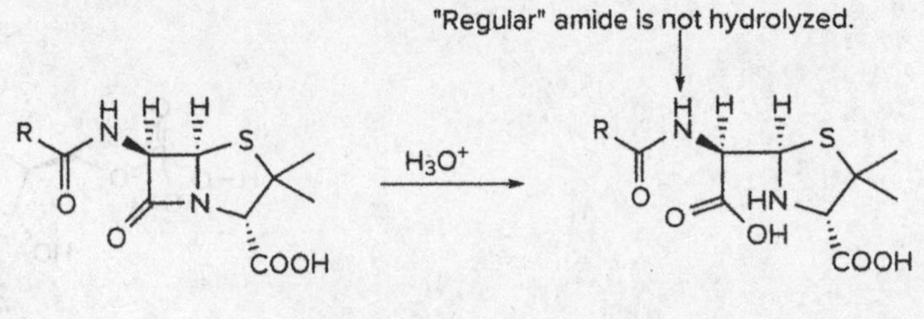

16.25 The mechanism has two steps:

[1] The nucleophile (RCO_2^-) attacks the magnesium-activated phosphate.

[2] Elimination of the leaving group (diphosphate) forms the substitution product, a fatty acyl AMP.

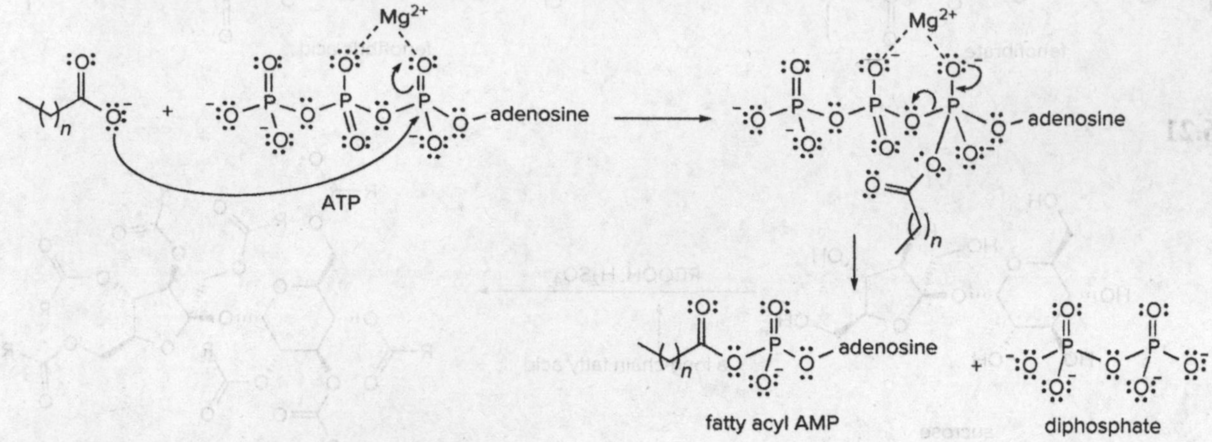

Mg2+
adenosine
ATP
Mg2+
adenosine
adenosine
+
fatty acyl AMP
diphosphate

16.26 a, b. Use the information in Section 16.15 to write out the reaction scheme for the formation of γ-glutamyl phosphate from glutamate.

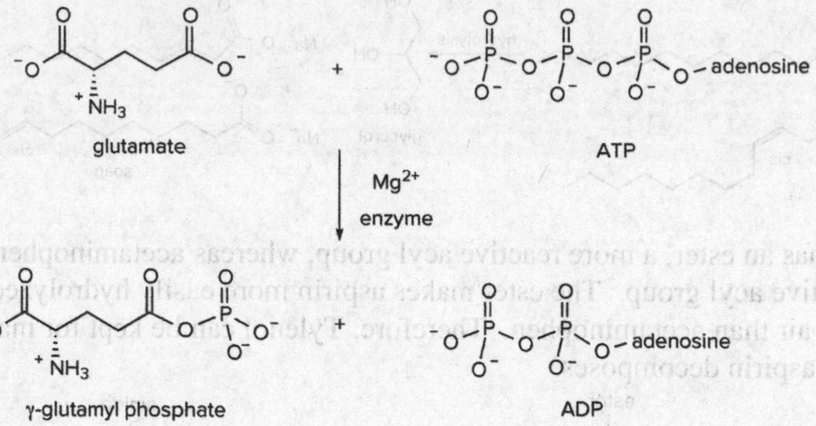

glutamate
+
adenosine
ATP
Mg2+
enzyme
γ-glutamyl phosphate
+
adenosine
ADP

16.27

+ H–NH2
Mg2+
enzyme
asparagine
+
AMP

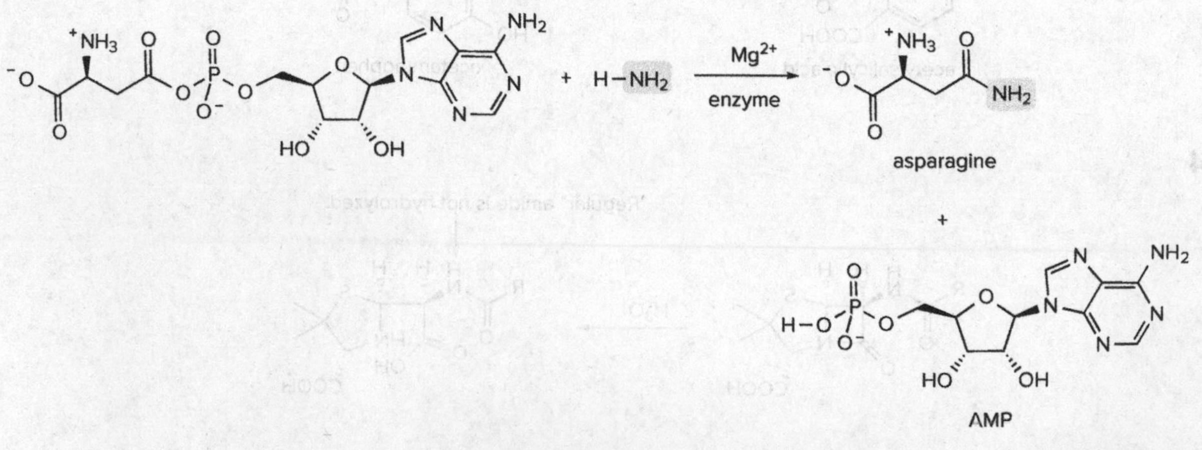

16.28 The mechanism has two steps:

[1] The nucleophile (HSCoA) attacks the magnesium-activated carbonyl, forming an sp^3 hybridized carbon.

[2] Elimination of the leaving group and protonation forms the substitution product, acetyl CoA.

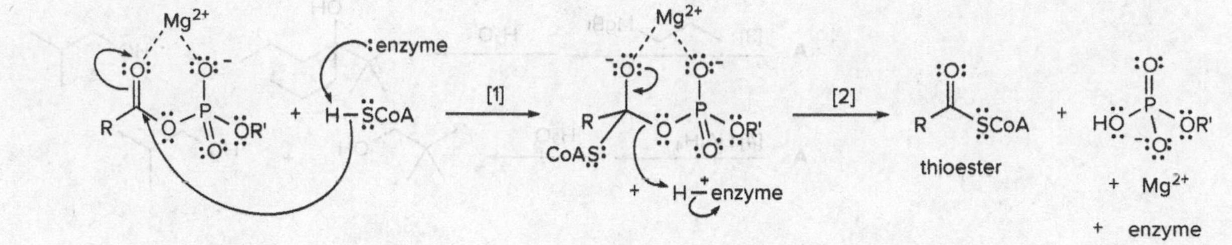

Mg2+ · enzyme · [1] · CoAS · H–enzyme · [2] · thioester · + Mg2+ · + enzyme

16.29 The mechanism involves three operations:

[1] The nucleophile (H_2O) is deprotonated and attacks the enzyme-activated carbonyl, forming an sp^3 hybridized carbon.

[2] Protonation and elimination of the leaving group (HSCoA) forms a carboxylic acid.

[3] The carboxylic acid loses a proton to form citrate.

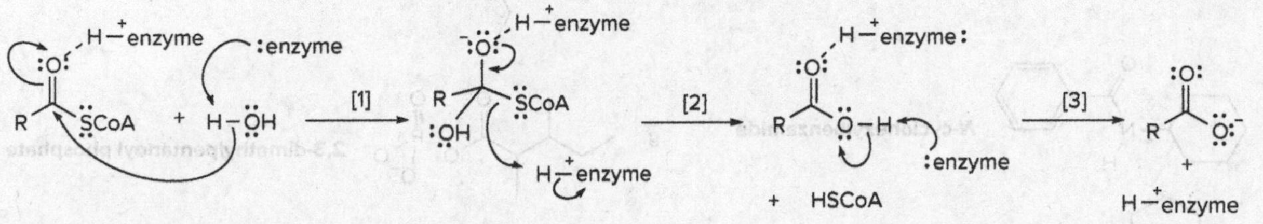

H–enzyme · :enzyme · [1] · [2] · + HSCoA · [3] · H–enzyme

16.30 Acetyl CoA acetylates the NH_2 group of glucosamine, because the NH_2 group is the most nucleophilic site.

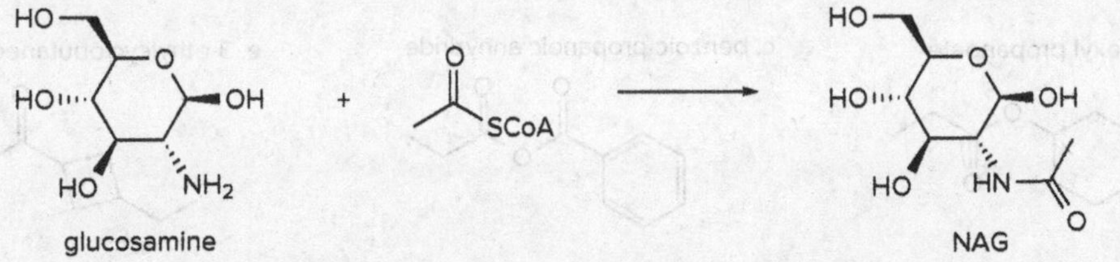

glucosamine · SCoA · NAG

16.31 The better the leaving group, the more reactive the acyl compound.

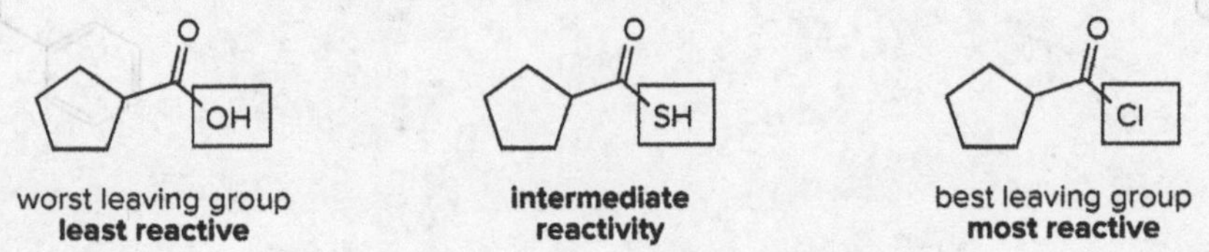

worst leaving group **least reactive** · **intermediate reactivity** · best leaving group **most reactive**

16.32

a. 3 2 1 O O

A
isobutyl 2,2-dimethyl-propanoate

b. **A** $\xrightarrow{[1]\ H_3O^+}$ COOH + HO

A $\xrightarrow[H_2O]{[2]\ ^-OH}$ COO^- + HO

A $\xrightarrow{[3]\ \text{MgBr}}$ $\xrightarrow{H_2O}$ OH + HO

A $\xrightarrow{[4]\ LiAlH_4}$ $\xrightarrow{H_2O}$ OH + HO

16.33

a. O O O
cyclohexanecarboxylic anhydride

b. O O
phenyl phenylacetate

c. O N H
***N*-cyclohexylbenzamide**

d. Cl O OCH_3
methyl 3-chlorocyclohexane-carboxylate

e. 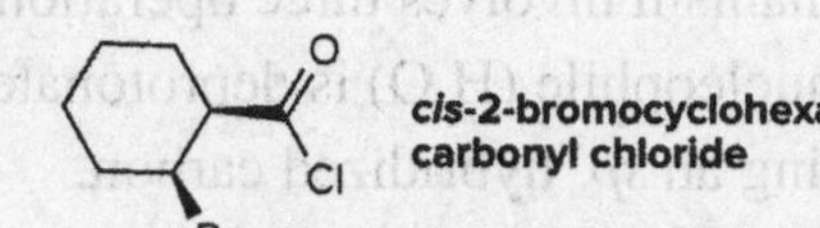
***cis*-2-bromocyclohexane-carbonyl chloride**

f. O N
***N,N*-diethylcyclohexanecarboxamide**

g. O O P O O^- O^-
2,3-dimethylpentanoyl phosphate

h. O S
***S*-ethyl propanethioate**

16.34

a. cyclohexyl propanoate

b. cyclohexanecarboxamide

NH_2

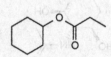

c. benzoic propanoic anhydride

d. 3-methylhexanoyl chloride

Cl

e. 3-ethylcyclobutanecarbonyl phosphate

f. *N,N*-dibenzylformamide

H N

16.35

Reaction as an acid:

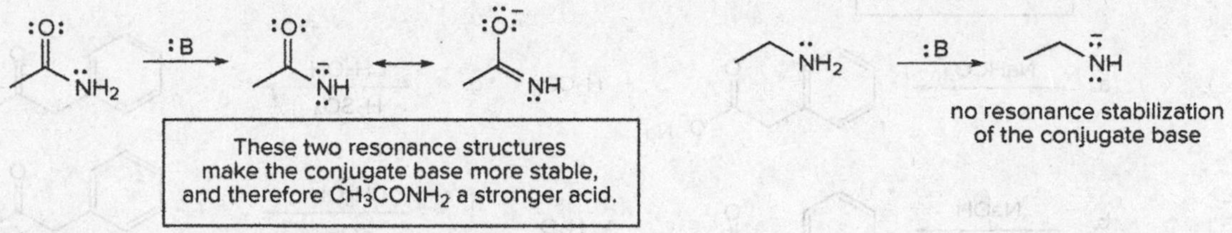
:O:
NH2
:B
:O:
NH
:O:
NH
These two resonance structures
make the conjugate base more stable,
and therefore CH3CONH2 a stronger acid.
NH2
:B
NH
no resonance stabilization
of the conjugate base

Reaction as a base:

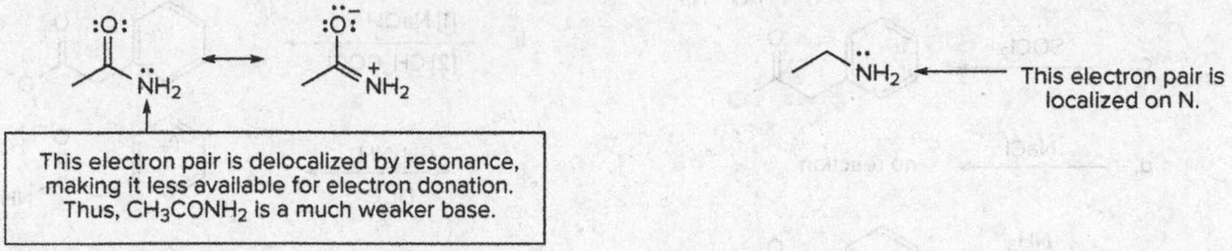
:O:
NH2
:O:
NH2
This electron pair is delocalized by resonance,
making it less available for electron donation.
Thus, CH3CONH2 is a much weaker base.
NH2
This electron pair is
localized on N.

16.36 a. The electron pair on the O atom in phenyl acetate is delocalized on both the carbonyl group and the benzene ring. Thus, the C=O has less single bond character than the C=O of cyclohexyl acetate, so the absorption occurs at higher wavenumber.

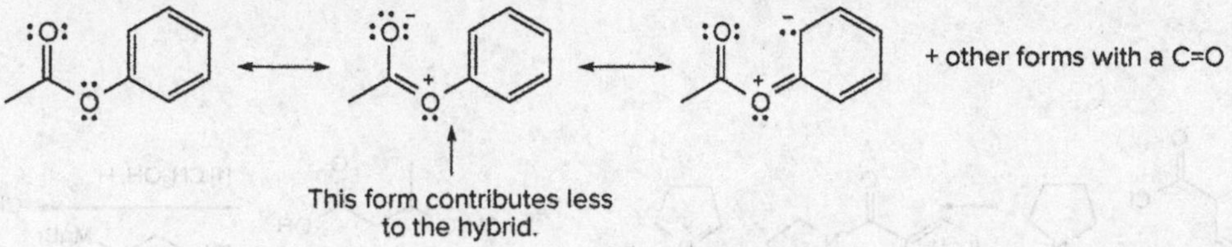
:O:
:O:
:O:
+ other forms with a C=O
This form contributes less
to the hybrid.

b. Because the lone pair on the O atom in cyclohexyl acetate can only be delocalized on the C=O, the C=O of cyclohexyl acetate is more effectively stabilized by resonance.

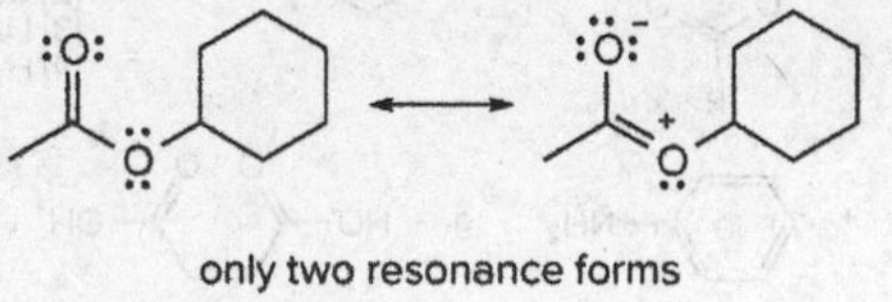
:O:
:O:
only two resonance forms

c. Phenyl acetate has a better leaving group and is more reactive.

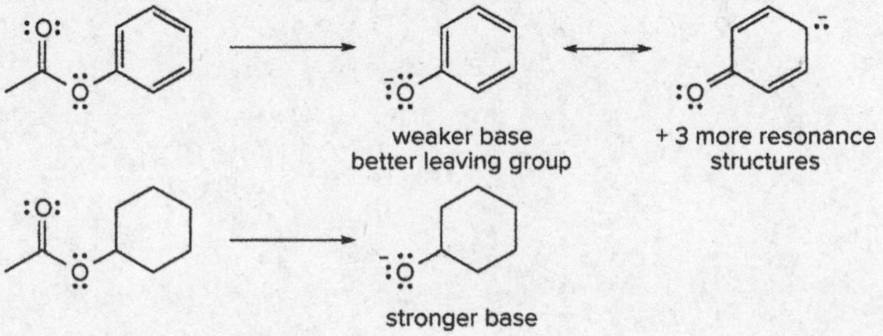
:O:
weaker base
better leaving group
+ 3 more resonance
structures
:O:
stronger base

16.37

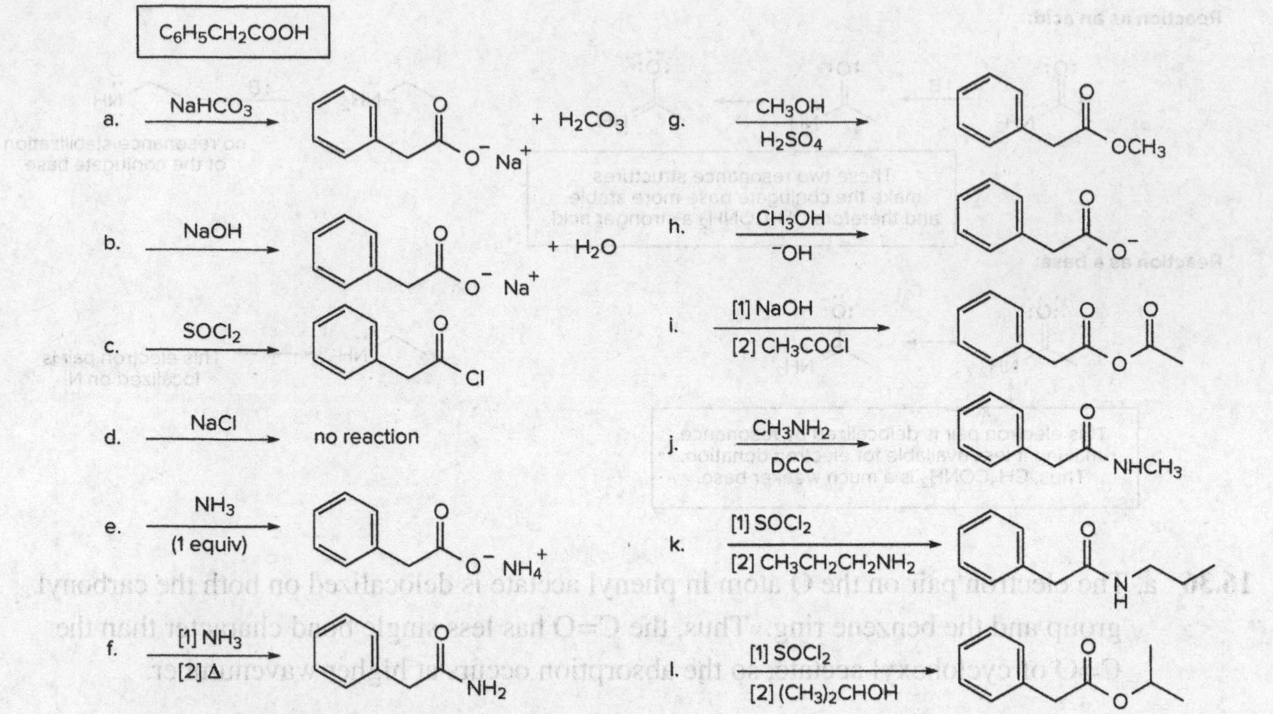

$C_6H_5CH_2COOH$

a. $NaHCO_3$ → O^- Na^+ + H_2CO_3

b. NaOH → O^- Na^+ + H_2O

c. $SOCl_2$ → Cl

d. NaCl → no reaction

e. NH_3 (1 equiv) → O^- NH_4^+

f. [1] NH_3 [2] Δ → NH_2

g. CH_3OH, H_2SO_4 → OCH_3

h. CH_3OH, ^-OH → O^-

i. [1] NaOH [2] CH_3COCl →

j. CH_3NH_2, DCC → $NHCH_3$

k. [1] $SOCl_2$ [2] $CH_3CH_2CH_2NH_2$ → N–H

l. [1] $SOCl_2$ [2] $(CH_3)_2CHOH$ →

16.38

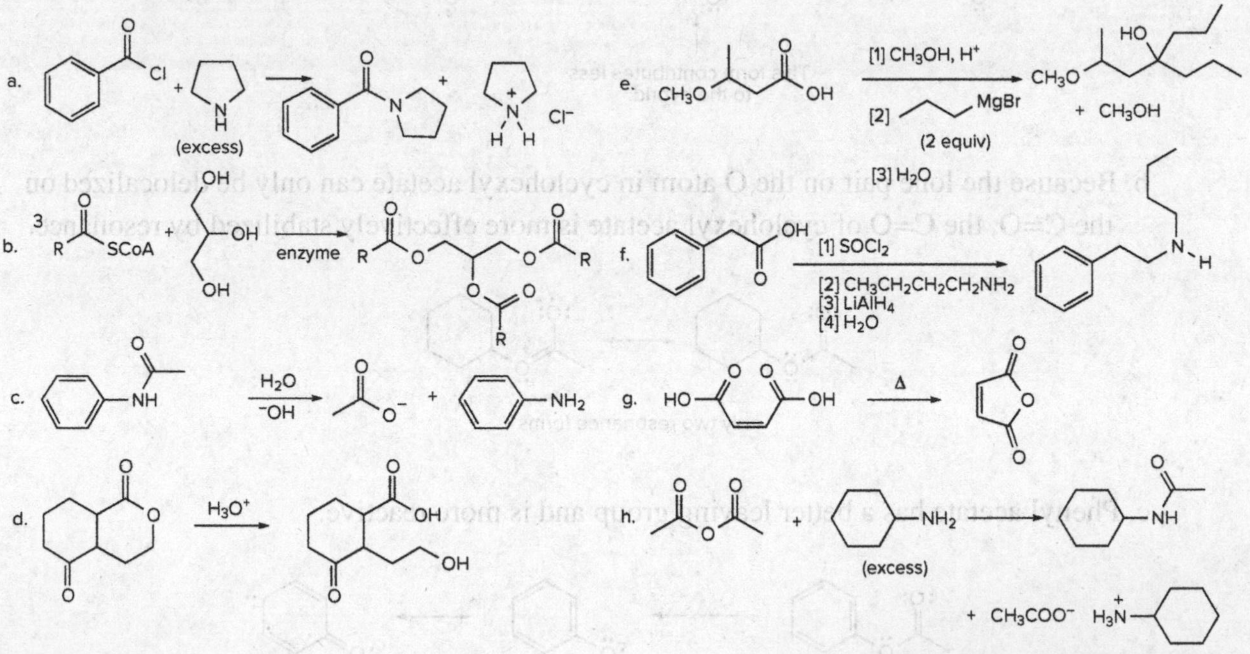

a. Cl + N–H (excess) → N + N^+ H H Cl^-

b. 3 R–C(=O)–SCoA + HO–CH₂–CH(OH)–CH₂–OH → (enzyme) R–C(=O)–O–CH₂–CH(O–C(=O)–R)–CH₂–O–C(=O)–R

c. NH → (H_2O, ^-OH) → O^- + NH_2

d. H_3O^+ → OH, OH

e. CH_3O … OH → [1] CH_3OH, H^+ [2] MgBr (2 equiv) [3] H_2O → CH_3O … HO + CH_3OH

f. OH → [1] $SOCl_2$ [2] $CH_3CH_2CH_2CH_2NH_2$ [3] $LiAlH_4$ [4] H_2O → N–H

g. HO … OH → Δ →

h. + NH_2 (excess) → NH + CH_3COO^- $H_3\overset{+}{N}$

16.39

cinnamoylcocaine
Hydrolyze both esters.

H_3O^+ → X + CH_3OH

Ph–CO–O–CO–Ph → Y

NaOCH$_3$
CH$_3$I

cocaine

16.40

a. H_3O^+

b. H_3O^+

c. H_3O^+

d. H_3O^+

16.41

a. CH_3COCl, pyridine

b. H_3O^+

c. H^+

d. (2 equiv)

Cl^-

16.42 Hydrolyze the amide and ester bonds in both starting materials to draw the products.

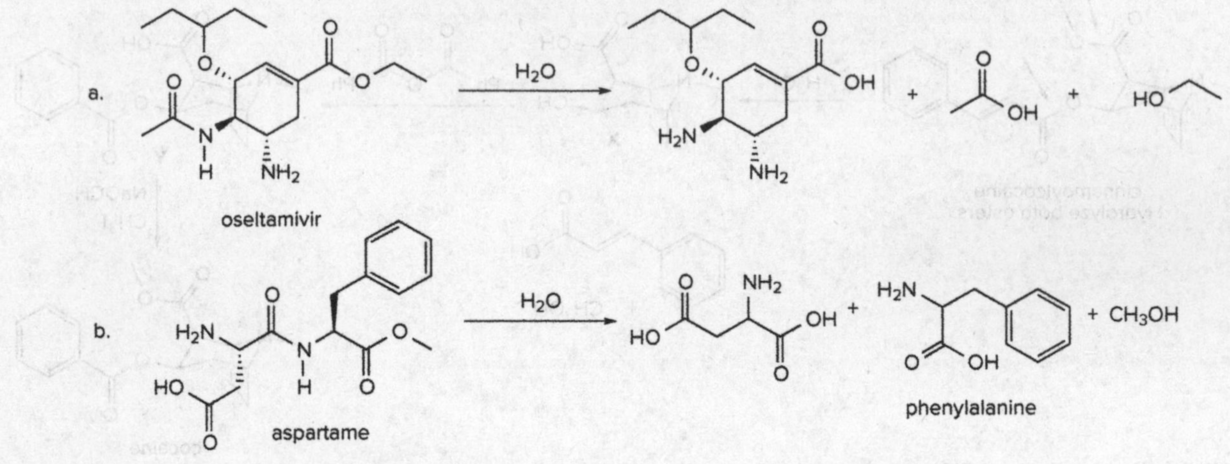

a. oseltamivir $\xrightarrow{H_2O}$ + + HO

b. aspartame $\xrightarrow{H_2O}$ + phenylalanine + CH_3OH

16.43

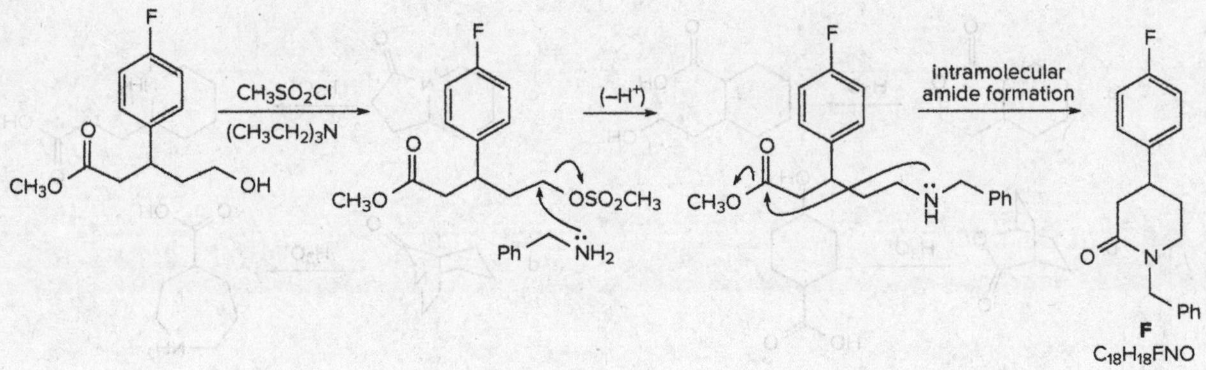

CH_3SO_2Cl, $(CH_3CH_2)_3N$; OSO_2CH_3; Ph–NH_2; $(-H^+)$; intramolecular amide formation

F $C_{18}H_{18}FNO$

16.44

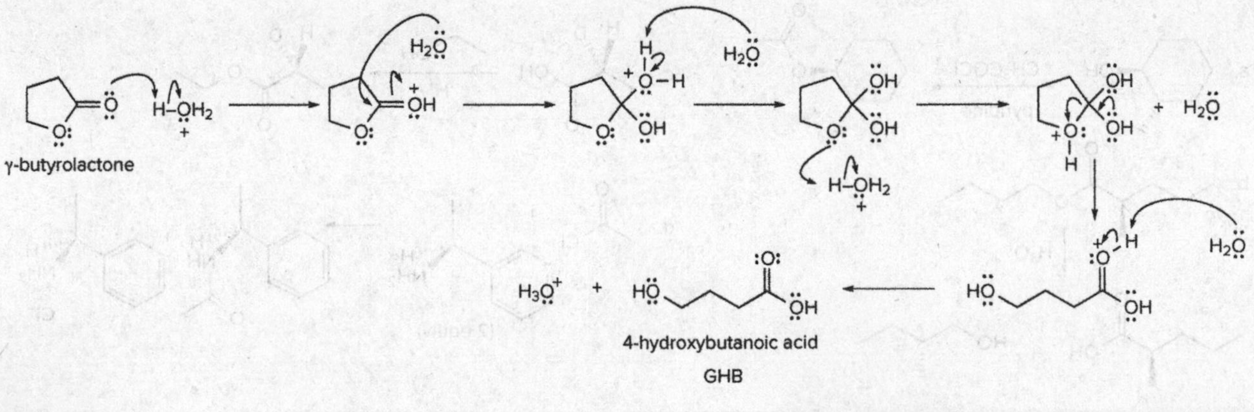

γ-butyrolactone

4-hydroxybutanoic acid

GHB

16.45

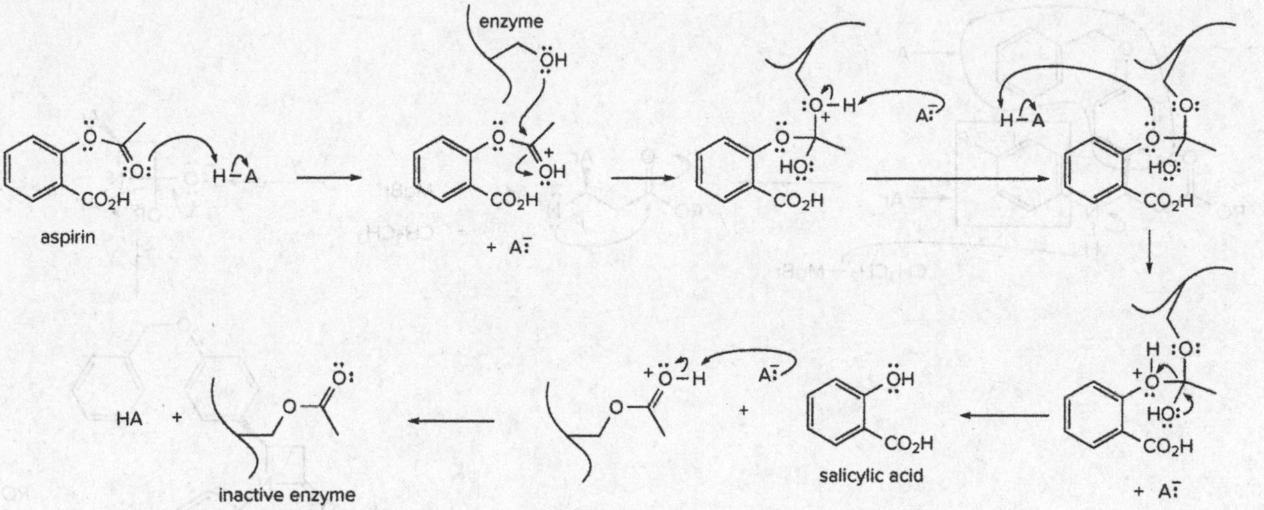

aspirin

enzyme

+ A:⁻

salicylic acid

inactive enzyme

HA +

16.46 The mechanism has two steps:

[1] The nucleophile attacks the magnesium-activated carbonyl, forming an *sp*^{3} hybridized carbon.

[2] Elimination of the leaving group and protonation forms the substitution product, glycinamide ribonucleotide.

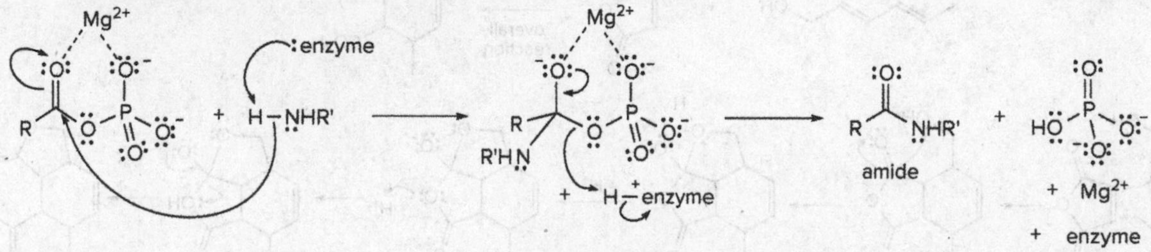

amide

+ Mg^{2+}

+ enzyme

16.47 The mechanism involves three operations:

[1] The nucleophile (H_2O) is deprotonated and attacks the enzyme-activated carbonyl, forming an *sp*^{3} hybridized carbon.

[2] Protonation and elimination of the leaving group (HSCoA) forms a carboxylic acid.

[3] The carboxylic acid loses a proton to form succinate.

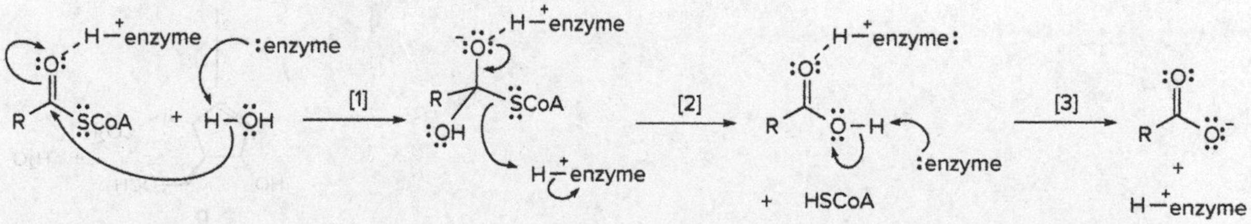

+ HSCoA

Protons are added or removed from the substrate in the active site of the enzyme. A basic electron pair on the enzyme removes protons, whereas acidic functional groups in the enzyme donate protons.

16.48

Ar

Ar'

RO

CH_3CH_2-MgBr

$+ MgBr^+$

$+ CH_3CH_3$

$+ RO^-$

16.49

OH + Δ overall reaction CO_2H

$-H^+$

H^+

OH

16.50

CH_3O

C

$^-:\ddot{O}H$

proton transfer

$H-\overset{+}{O}H_2$

CO_2H + H_2O

HO OCH_3

D

16.51

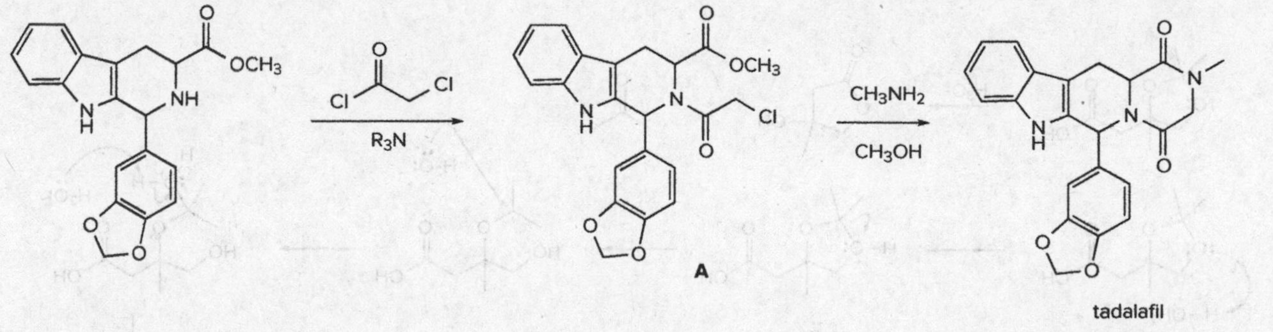

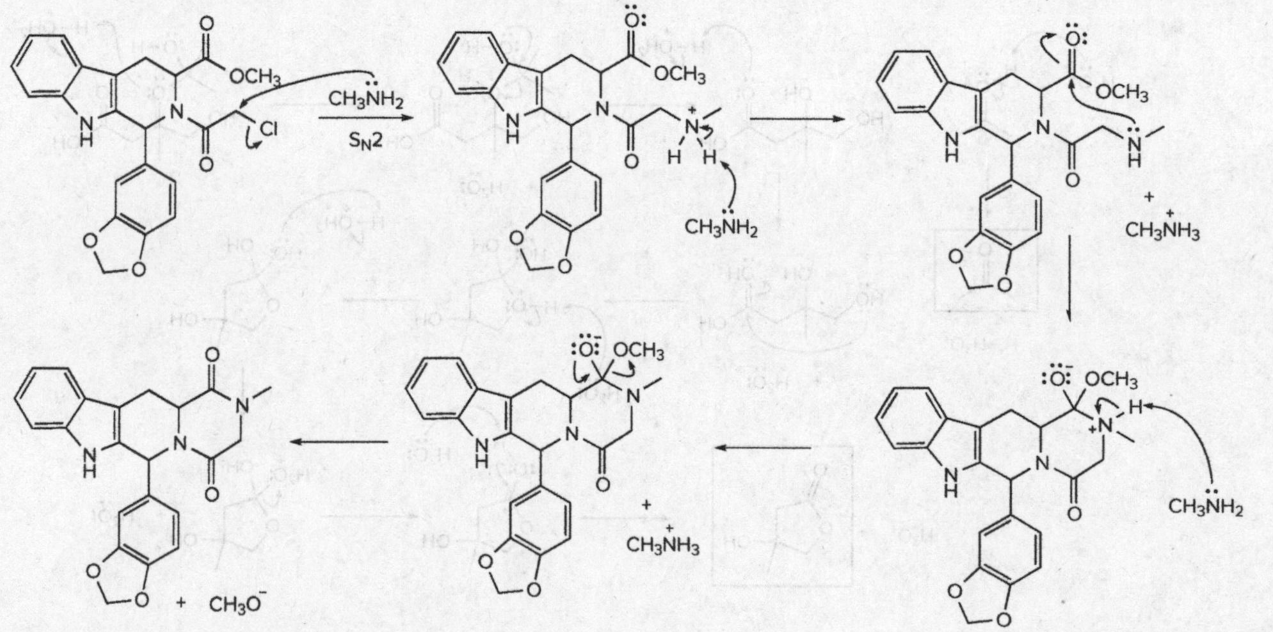

Cl–C(=O)–CH2–Cl, R_3N

A

CH_3NH_2, CH_3OH

tadalafil

$CH_3\ddot{N}H_2$, S_N2

$CH_3\ddot{N}H_2$

$CH_3\overset{+}{N}H_3$

$CH_3\ddot{N}H_2$

$CH_3\overset{+}{N}H_3$

+ CH_3O^-

16.52 The mechanism combines hydrolysis of an acetal and Fischer esterification.

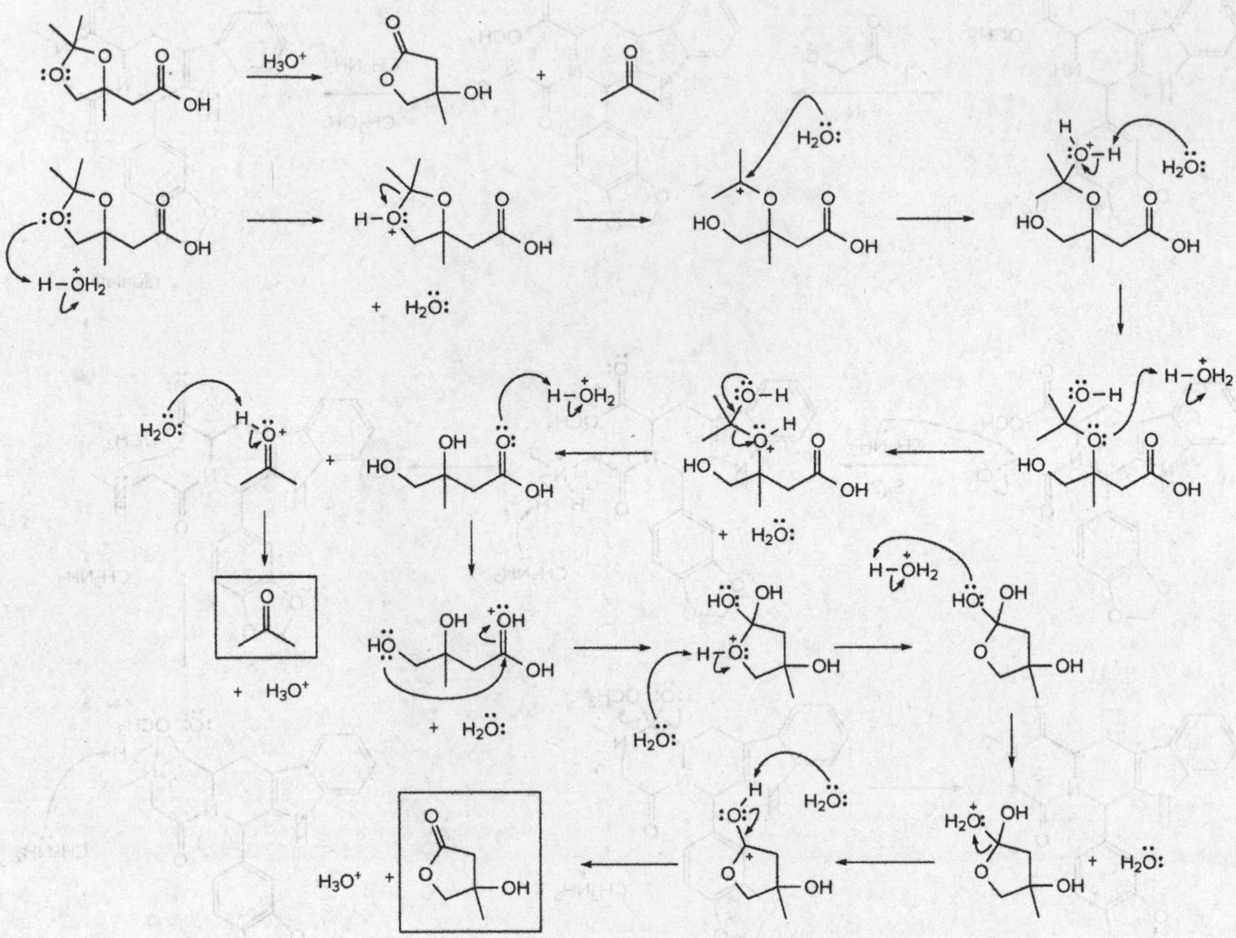

H_3O^+ + $H_2\ddot{O}$: + $H_2\ddot{O}$: H_3O^+ + $H_2\ddot{O}$: $H_2\ddot{O}$: H_3O^+ + $H_2\ddot{O}$:

16.53

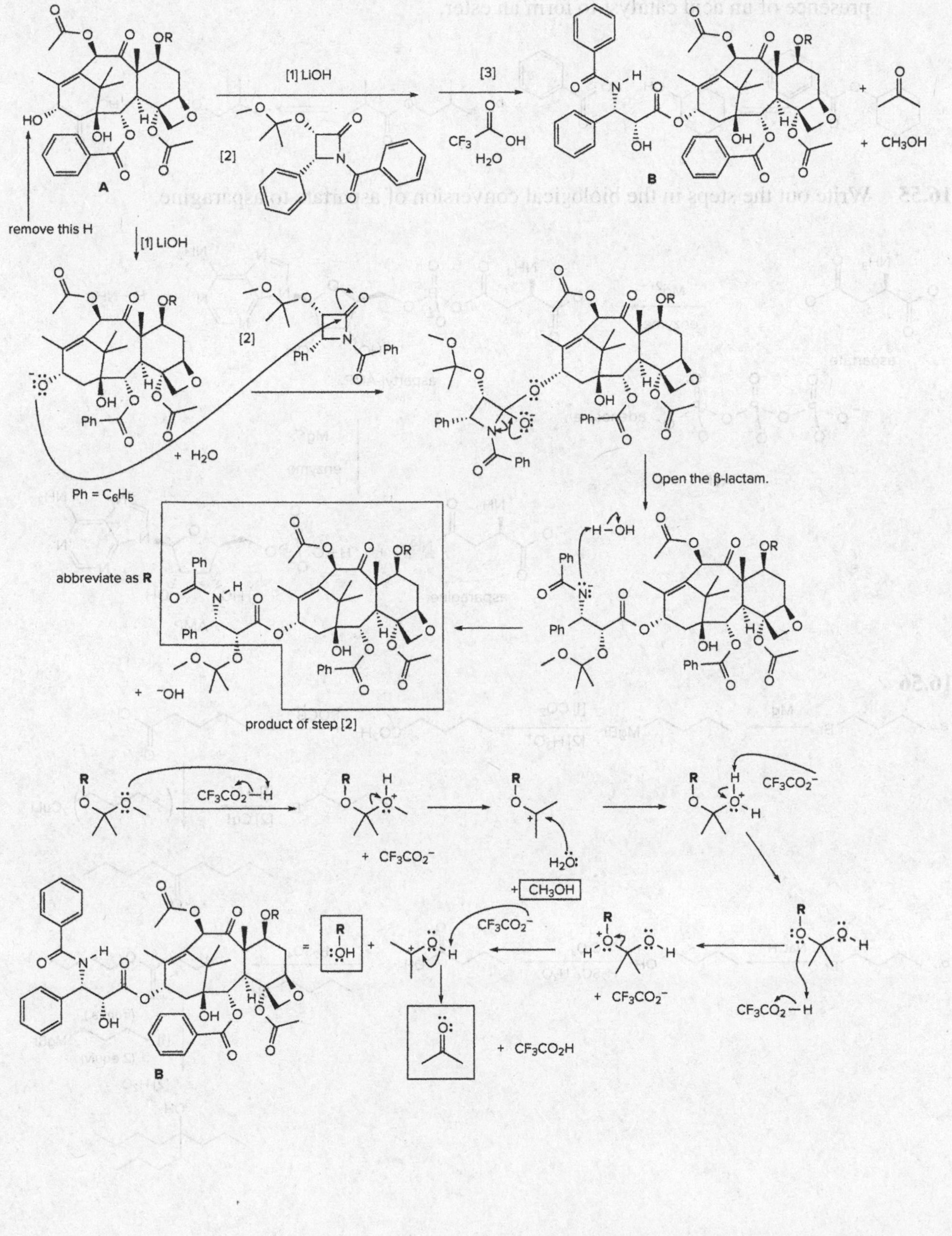

[1] LiOH
[2]
[3] CF_3CO_2H, H_2O

A

B

+ CH_3OH

remove this H

[1] LiOH

[2]

Ph

+ H_2O

$Ph = C_6H_5$

Open the β-lactam.

abbreviate as **R**

+ ^{-}OH

product of step [2]

CF_3CO_2—H

+ $CF_3CO_2^-$

$H_2\ddot{O}$:

+ CH_3OH

$CF_3CO_2^-$

+ $CF_3CO_2^-$

CF_3CO_2—H

B

+ CF_3CO_2H

16.54 Fischer esterification is the treatment of a carboxylic acid with an alcohol in the presence of an acid catalyst to form an ester.

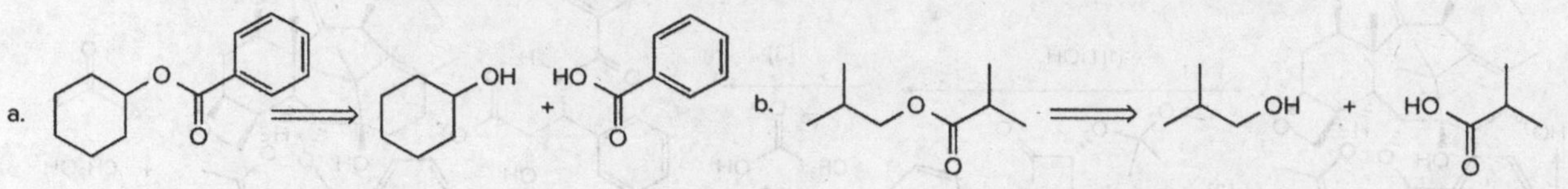

a. O, O, OH, HO, +, O

b. O, O, OH, +, HO, O

16.55 Write out the steps in the biological conversion of aspartate to asparagine.

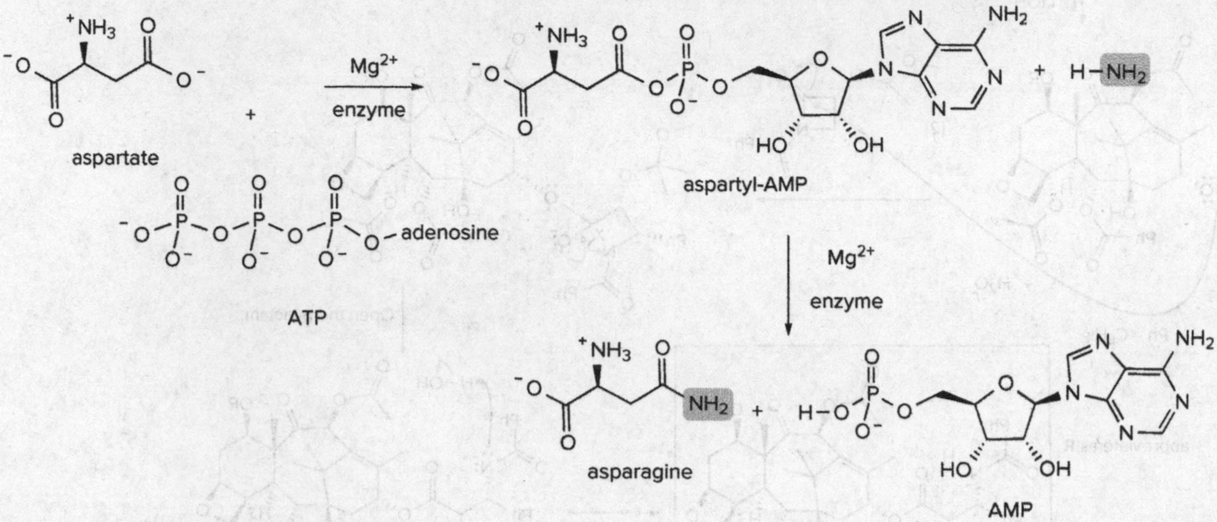

$^+NH_3$, O, ^-O, O^-, O, aspartate, +, ^-O, P, O, P, O, P, O, adenosine, ATP

Mg^{2+}, enzyme

$^+NH_3$, O, O, O, P, O, O, N, N, NH_2, N, N, HO, OH, aspartyl-AMP, + H–NH_2

Mg^{2+}, enzyme

$^+NH_3$, O, ^-O, NH_2, O, asparagine, + H–O, P, O, O, N, N, NH_2, N, N, HO, OH, AMP

16.56

a. Br $\xrightarrow{Mg}$ MgBr $\xrightarrow{[1]\ CO_2,\ [2]\ H_3O^+}$ CO_2H $\xrightarrow{SOCl_2}$ Cl, O

Br $\xrightarrow{[1]\ 4\ Li,\ [2]\ CuI}$ $(\ )_2CuLi$

O

b. Br $\xrightarrow{NaOH}$ OH $\xrightarrow{CrO_3,\ H_2SO_4,\ H_2O}$ O, OH $\xrightarrow{H_2SO_4}$ (OH) O, O

[1] MgBr (from a.) (2 equiv)

[2] H_2O

OH

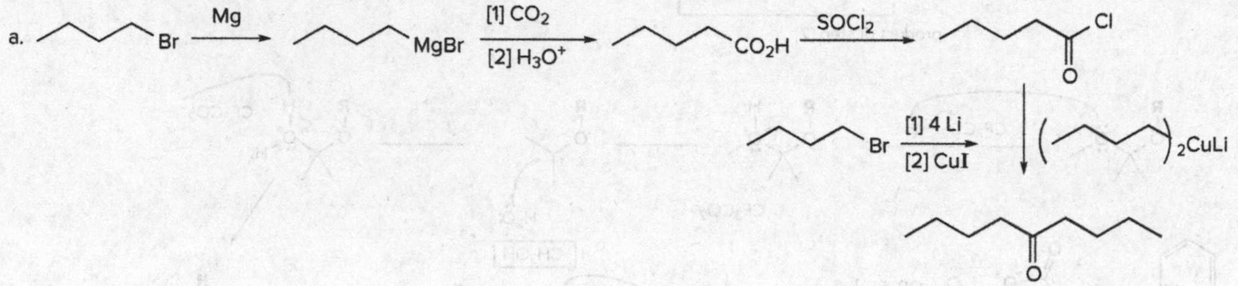

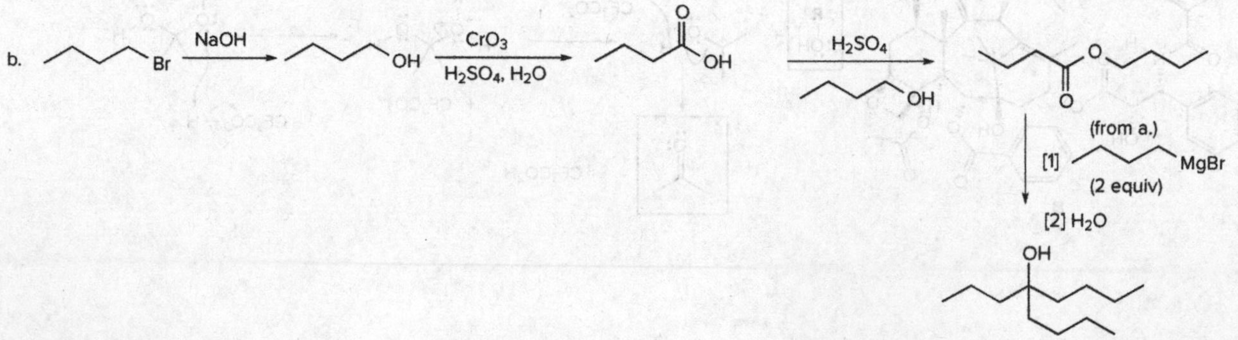

c. Br —NaCN→ CN —[1] $LiAlH_4$ [2] H_2O→ NH_2 → (amide product, N–H)

Br —^{-}OH→ OH —CrO_3→ OH (carboxylic acid) —$SOCl_2$→ Cl (acid chloride)

16.57

a. OH —$SOCl_2$→ Cl —[1] $(\text{isopropyl})_2CuLi$ [2] H_2O→ (ketone product)

b. Cl —[1] $LiAlH[OC(CH_3)_3]_3$ [2] H_2O→ H —HO–CH₂CH₂–OH, H^+→ (acetal product)

[from part (a)]

c. Cl —NH_3 (2 equiv)→ NH_2 —[1] $LiAlH_4$ [2] H_2O→ NH_2 —(acetyl chloride, Cl)→ (amide product, N–H)

[from part (a)]

d. Cl —[1] $CH_3CH_2CH_2MgBr$ (2 equiv) [2] H_2O→ HO (tertiary alcohol product)

[from part (a)]

16.58

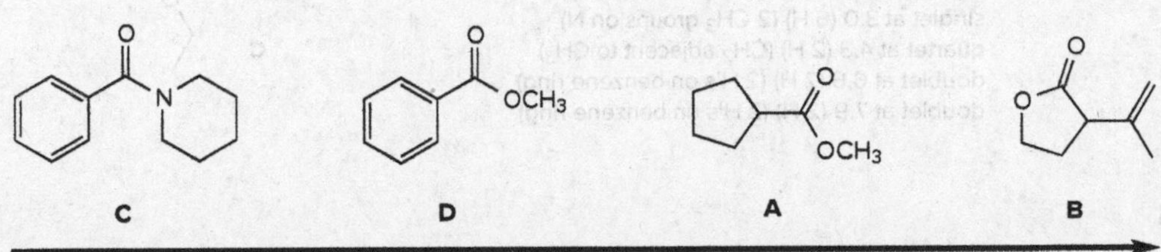

Increasing wavenumber of C=O

16.59

a. $C_6H_{12}O_2$ → one degree of unsaturation
IR: 1738 cm^{-1} → C=O
NMR: 1.12 (triplet, 3 H), 1.23 (doublet, 6 H), 2.28 (quartet, 2 H), 5.00 (septet, 1 H) ppm

b. C_8H_9NO
IR: 3328 (NH), 1639 (conjugated amide C=O) cm^{-1}
NMR: 2.95 (singlet, 3 H), 6.95 (singlet, 1 H), 7.3–7.7 (multiplet, 5 H) ppm

16.60

A. Molecular formula $C_{10}H_{12}O_2$ → five degrees of unsaturation
IR absorption at 1718 cm^{-1} → C=O
NMR data (ppm):
- triplet at 1.4 (CH_3 adjacent to 2 H's)
- singlet at 2.4 (CH_3)
- quartet at 4.4 (CH_2 adjacent to CH_3)
- doublet at 7.2 (2 H's on benzene ring)
- doublet at 7.9 (2 H's on benzene ring)

B. IR absorption at 1740 cm^{-1} → C=O
NMR data (ppm):
- singlet at 2.1 (CH_3)
- triplet at 2.9 (CH_2 adjacent to CH_2)
- triplet at 4.3 (CH_2 adjacent to CH_2)
- multiplet at 7.3 (5 H's, monosubstituted benzene)

16.61

Molecular formula $C_{10}H_{13}NO_2$ → five degrees of unsaturation
IR absorptions at 3300 (NH) and 1680 (C=O, amide or conjugated) cm^{-1}
NMR data (ppm):
- triplet at 1.4 (CH_3 adjacent to CH_2)
- singlet at 2.2 (CH_3C=O)
- quartet at 3.9 (CH_2 adjacent to CH_3)
- doublet at 6.8 (2 H's on benzene ring)
- singlet at 7.1 (NH)
- doublet at 7.3 (2 H's on benzene ring)

phenacetin

16.62

Molecular formula $C_{11}H_{15}NO_2$ → five degrees of unsaturation
IR absorption 1699 (C=O, amide or conjugated) cm^{-1}
NMR data (ppm):
- triplet at 1.3 (3 H) (CH_3 adjacent to CH_2)
- singlet at 3.0 (6 H) (2 CH_3 groups on N)
- quartet at 4.3 (2 H) (CH_2 adjacent to CH_3)
- doublet at 6.6 (2 H) (2 H's on benzene ring)
- doublet at 7.9 (2 H) (2 H's on benzene ring)

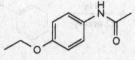

C

16.63

a. Molecular formula $C_6H_{12}O_2$ → one degree of unsaturation
IR absorption at 1743 cm^{-1} → C=O
^1H NMR data (ppm):
- triplet at 0.9 (3 H) – CH_3 adjacent to CH_2
- multiplet at 1.35 (2 H) – CH_2
- multiplet at 1.60 (2 H) – CH_2
- singlet at 2.1 (3 H – from CH_3 bonded to C=O)
- triplet at 4.1 (2 H) – CH_2 adjacent to the electronegative O atom and another CH_2

D

b. Molecular formula $C_6H_{12}O_2$ → one degree of unsaturation
IR absorption at 1746 cm^{-1} → C=O
^1H NMR data (ppm):
- doublet at 0.9 (6 H) – 2 CH_3's adjacent to CH
- multiplet at 1.9 (1 H)
- singlet at 2.1 (3 H) – CH_3 bonded to C=O
- doublet at 3.85 (2 H) – CH_2 bonded to electronegative O and CH

E

16.64

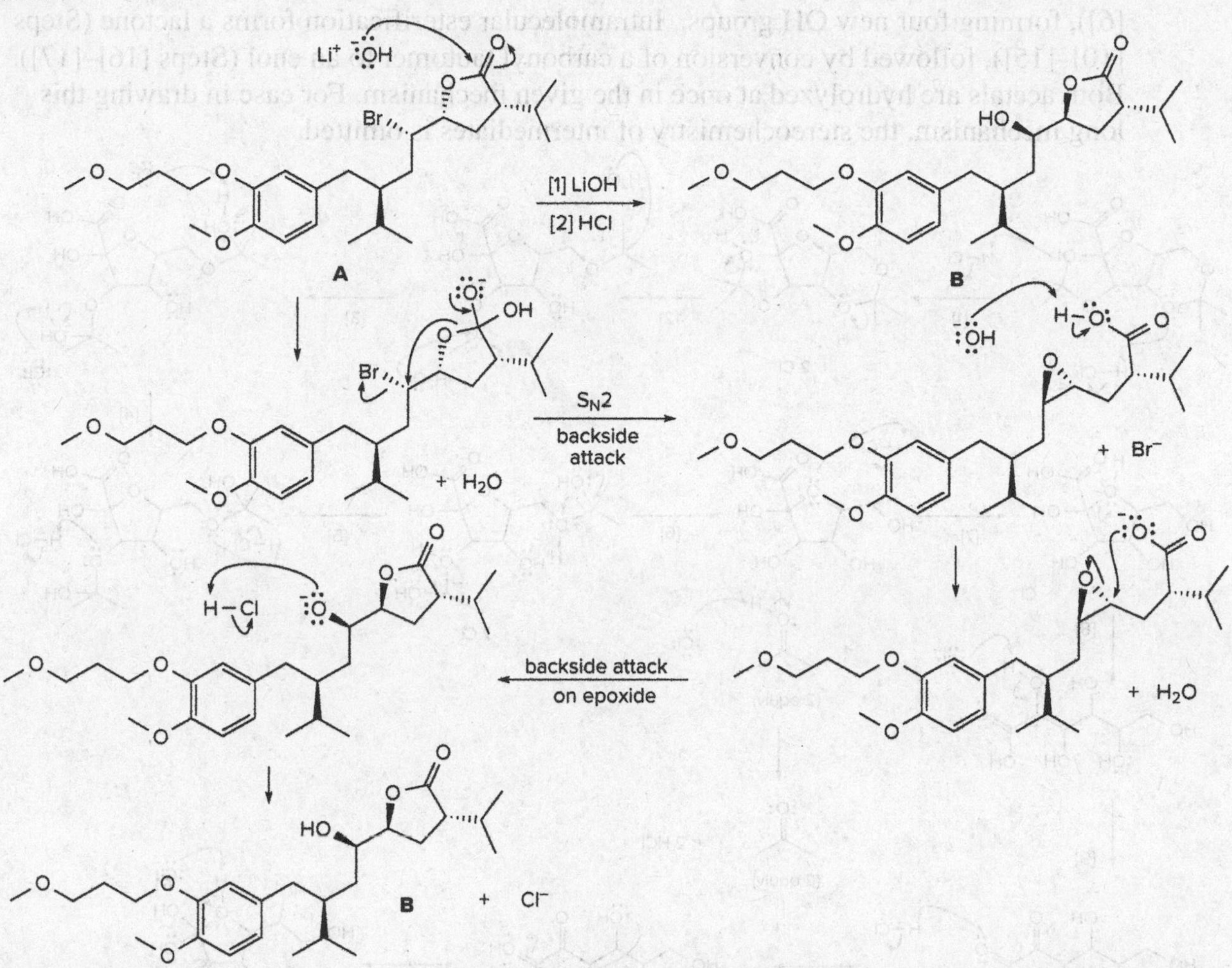

[1] LiOH
[2] HCl

A

B

S_N2
backside attack

+ H_2O

+ Br^-

backside attack on epoxide

+ H_2O

B + Cl^-

16.65 The extent of resonance stabilization affects the position of the C=O absorption in the IR of an amide.

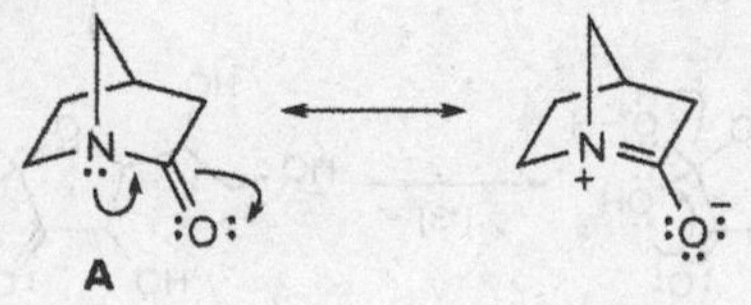

A

This resonance structure does not contribute significantly to the hybrid because it places a double bond at the bridgehead N, an impossible geometry in small rings. The C=O has more double-bond character, so the absorption is at a higher wavenumber.

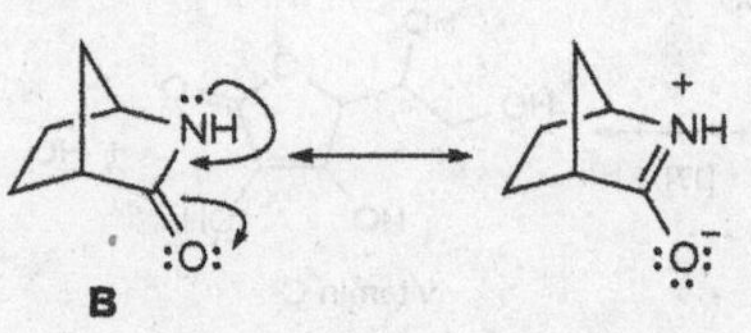

B

This resonance structure contributes significantly to the hybrid, so the C=O has more single-bond character. The absorption shifts to a lower wavenumber.

16.66

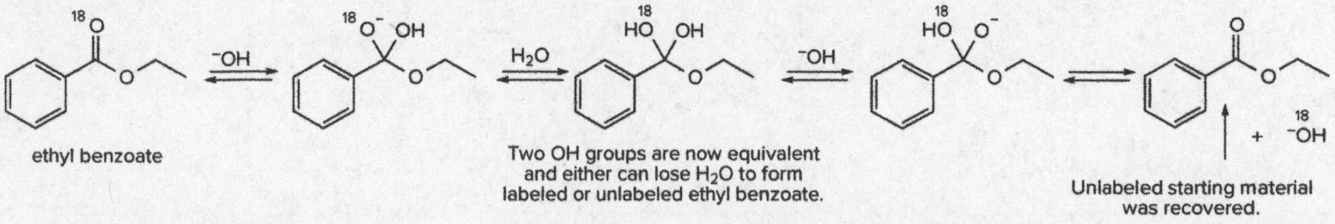

ethyl benzoate

Two OH groups are now equivalent and either can lose H_2O to form labeled or unlabeled ethyl benzoate.

Unlabeled starting material was recovered.

16.67 Both acetals are hydrolyzed by the usual mechanism for acetal hydrolysis (Steps [1]–[6]), forming four new OH groups. Intramolecular esterification forms a lactone (Steps [10]–[15]), followed by conversion of a carbonyl tautomer to an enol (Steps [16]–[17]). Both acetals are hydrolyzed at once in the given mechanism. For ease in drawing this long mechanism, the stereochemistry of intermediates is omitted.

[1] + 2 Cl⁻ [2] H_2O [3] [4] [5] + 2 Cl⁻ [6] [7] + Cl⁻ [8] (2 equiv) → (2 equiv) + 2 HCl [9] [10] + Cl⁻ [11] [12] [13] [14] [15] + H_2O + Cl⁻ [16] + Cl⁻ [17] + HCl

vitamin C

16.68

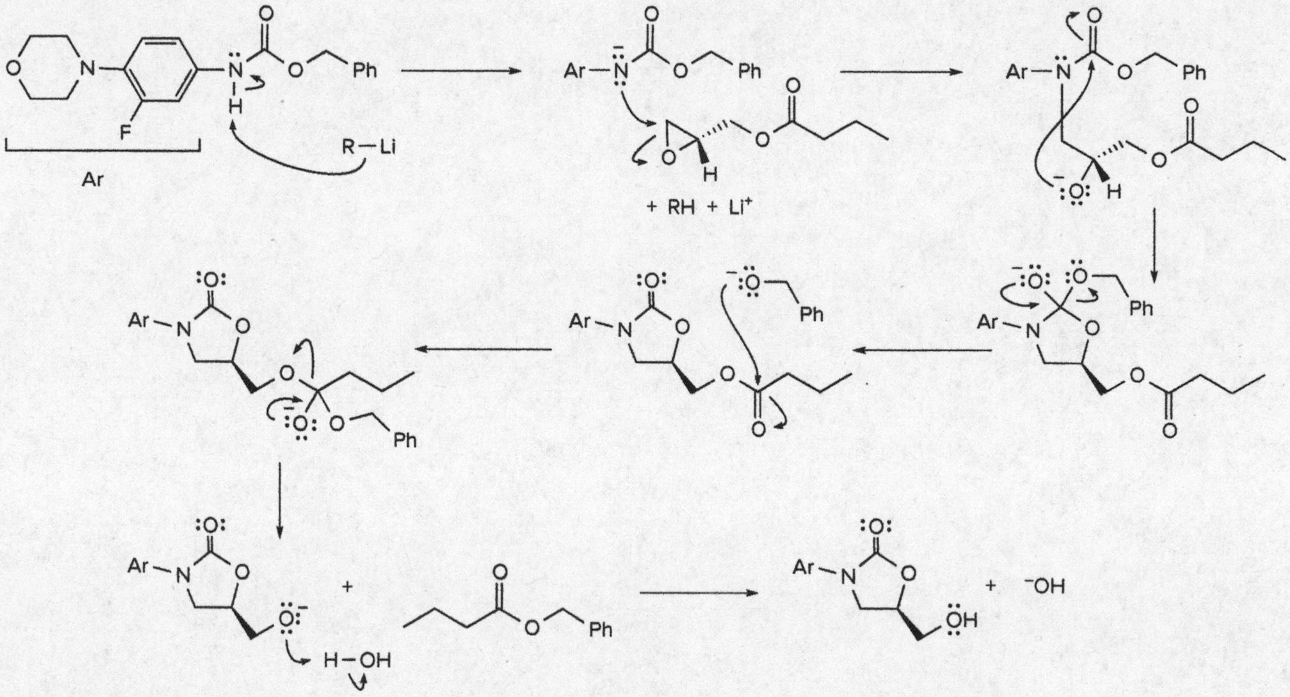

Ar

R—Li

Ar—N—C(=O)—O—CH2—Ph

+ RH + Li+

Ar—N

:O:

Ph

O

H

H—OH

+ ⁻OH

OH

Chapter 17 Substitution Reactions of Carbonyl Compounds at the α Carbon

Chapter Review

Kinetic versus thermodynamic enolates (17.4)

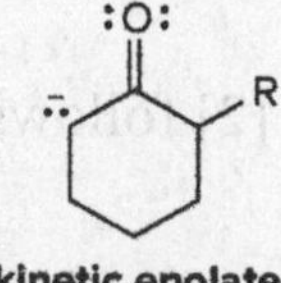

kinetic enolate

Kinetic enolate

- The less substituted enolate
- Favored by strong base, polar aprotic solvent, low temperature: LDA, THF, –78 °C

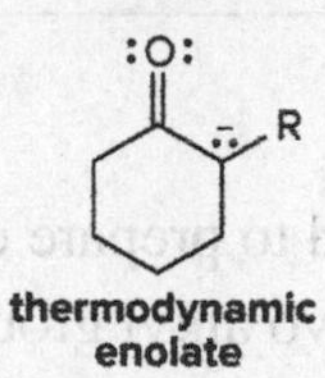

thermodynamic enolate

Thermodynamic enolate

- The more substituted enolate
- Favored by strong base, protic solvent, higher temperature: $NaOCH_2CH_3$, CH_3CH_2OH, room temperature

Halogenation at the α carbon (17.7A)

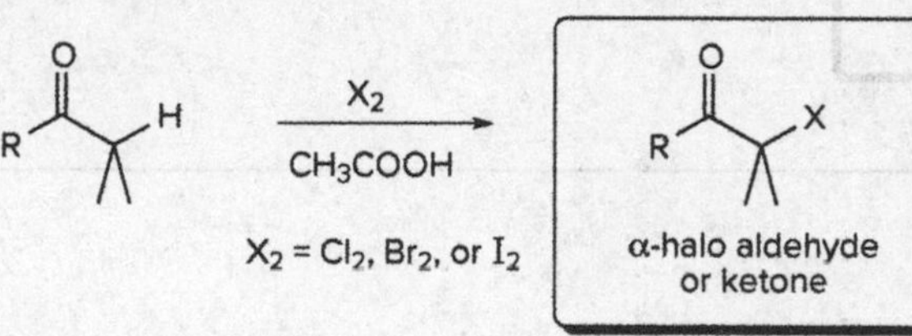

- The reaction occurs via enol intermediates.
- Monosubstitution of X for H occurs on the α carbon.

Reactions of α-halo carbonyl compounds (17.7B)

[1] Elimination to form α,β-unsaturated carbonyl compounds

Li_2CO_3, LiBr, DMF

- Elimination of the elements of Br and H forms a new π bond, giving an α,β-unsaturated carbonyl compound.

[2] Nucleophilic substitution

$:Nu^-$

- The reaction follows an S_N2 mechanism, generating an α-substituted carbonyl compound.

Alkylation reactions at the α carbon

[1] Direct alkylation at the α carbon (17.8)

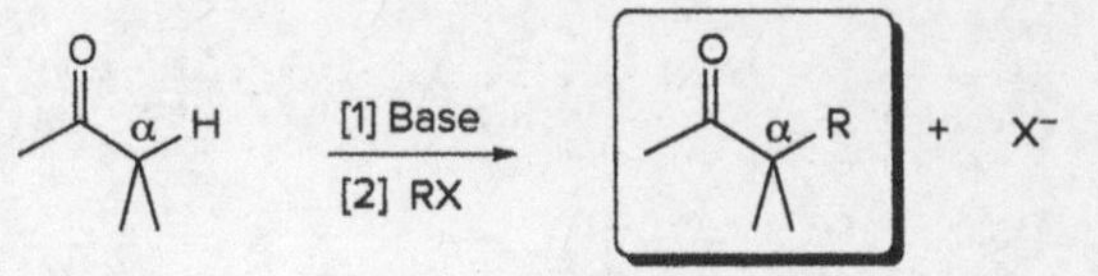

[1] Base
[2] RX

+ X^-

- The reaction forms a new C–C bond to the α carbon.
- LDA is a common base used to form an intermediate enolate.
- The alkylation in Step [2] follows an S_N2 mechanism.

[2] Malonic ester synthesis (17.9)

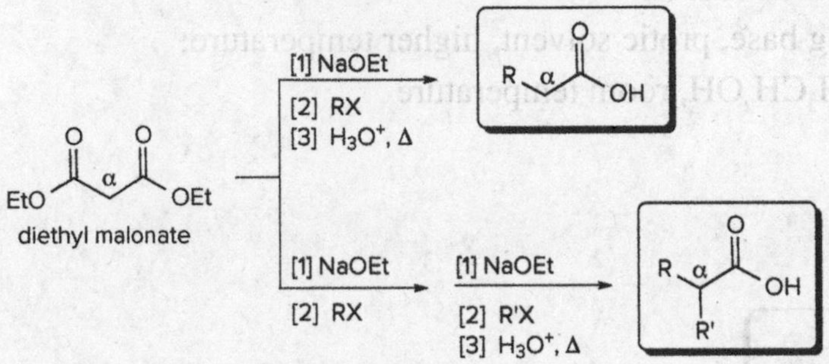

EtO OEt
diethyl malonate

[1] NaOEt
[2] RX
[3] H_3O^+, Δ

OH

[1] NaOEt
[2] RX

[1] NaOEt
[2] R'X
[3] H_3O^+, Δ

R R' OH

- The reaction is used to prepare carboxylic acids with one or two alkyl groups on the α carbon.
- The alkylation in Step [2] follows an S_N2 mechanism.

[3] Acetoacetic ester synthesis (17.10)

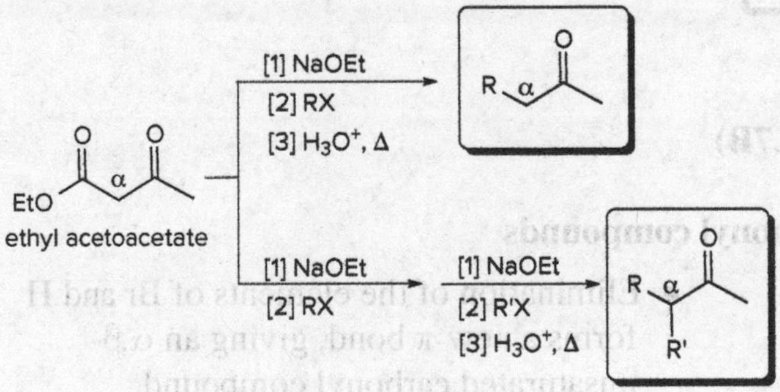

EtO
ethyl acetoacetate

[1] NaOEt
[2] RX
[3] H_3O^+, Δ

[1] NaOEt
[2] RX

[1] NaOEt
[2] R'X
[3] H_3O^+, Δ

R R'

- The reaction is used to prepare ketones with one or two alkyl groups on the α carbon.
- The alkylation in Step [2] follows an S_N2 mechanism.

Practice Test on Chapter Review

1.a. Which of the following compounds can be prepared using either the acetoacetic ester synthesis or the malonic ester synthesis?

1. COOH
2.
3.

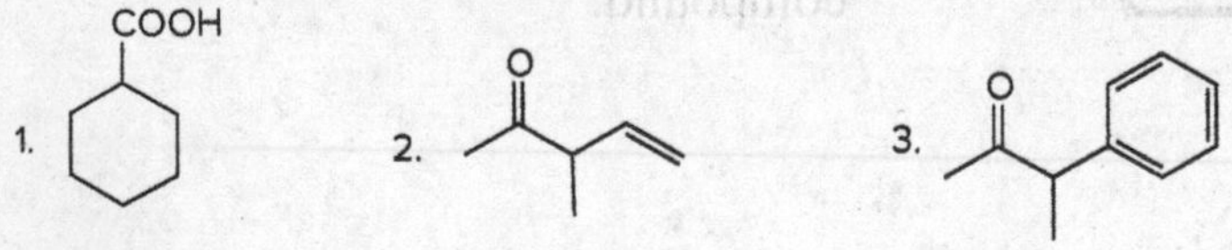

4. Both (1) and (2) can be prepared by one of these routes.
5. Compounds (1), (2), and (3) can all be prepared by these routes.

b. Which of the following compounds is ***not*** an enol form of dicarbonyl compound **A?**

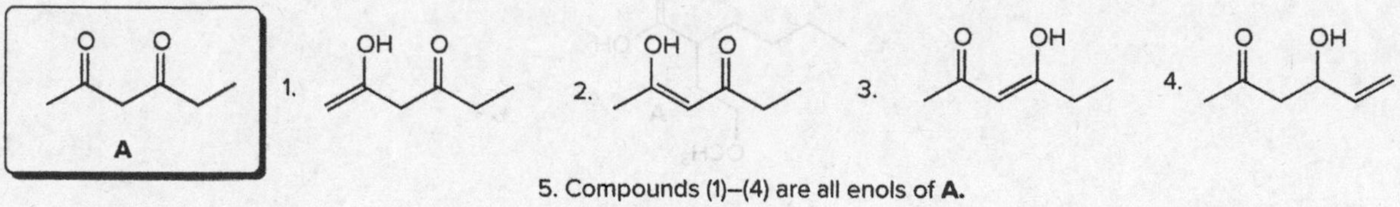

A 1. 2. 3. 4.

5. Compounds (1)–(4) are all enols of **A.**

2.a. Which proton in compound **A** has the *lowest* pK_a?

b. Which proton in compound **A** has the *highest* pK_a?

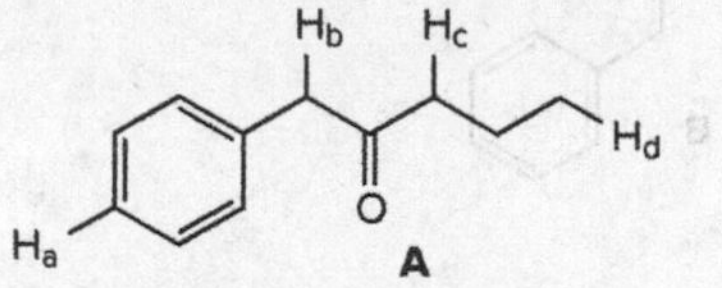

H_b H_c H_d H_a **A**

c. Which proton in compound **B** is the least acidic?

d. Which proton in compound **B** is the most acidic?

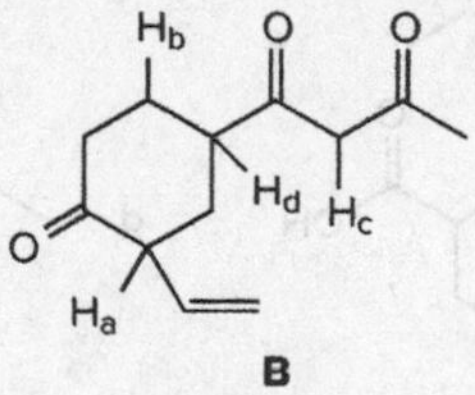

H_b H_d H_c H_a **B**

3. What reagents are needed to convert heptan-4-one to each compound?

a. 3-bromoheptan-4-one

b. 3-methylheptan-4-one

c. 3,5-dimethylheptan-4-one

d. hept-2-en-4-one

e. 2-methylheptan-4-one

4. Draw the organic products formed in each of the following reactions.

a. CO_2Et [1] LDA [2] CH_3CH_2Br [3] H_3O^+, Δ

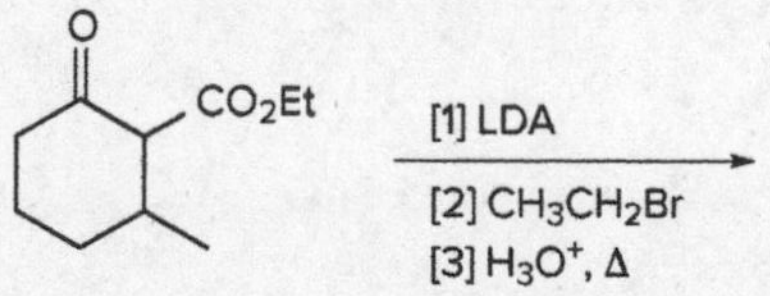

b. $CH_2(CO_2Et)_2$ [1] NaOEt [2] CH_3CH_2Br → [1] NaOEt [2] $C_6H_5CH_2Cl$ [3] H_3O^+, Δ

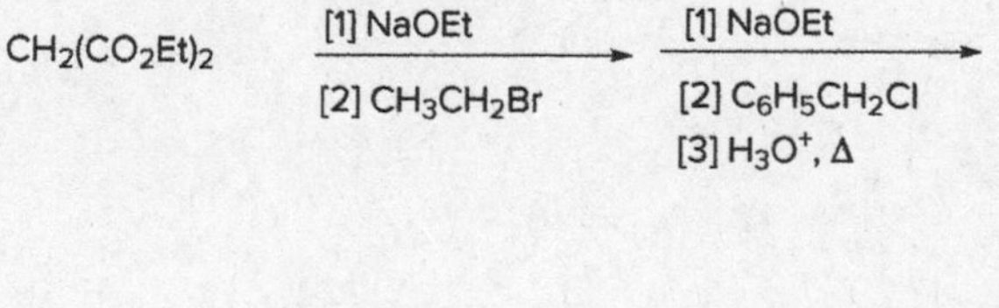

c. OEt [1] NaOEt [2] $CH_3CH_2CH_2CH_2Br$ [3] H_3O^+, Δ

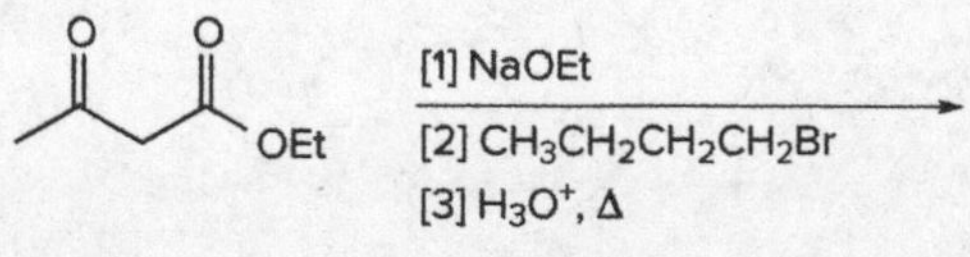

d. [1] LDA [2] CH_3CH_2Br

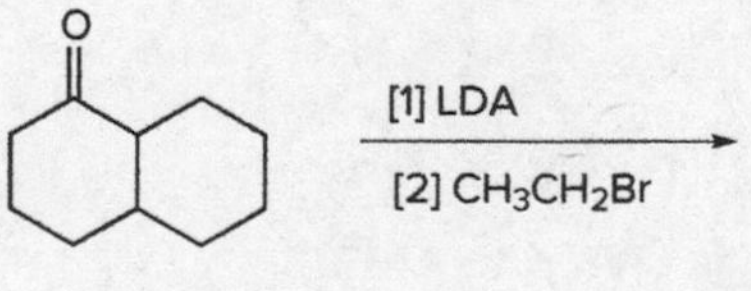

5.a. What starting materials are needed to synthesize carboxylic acid **A** by a malonic ester synthesis?

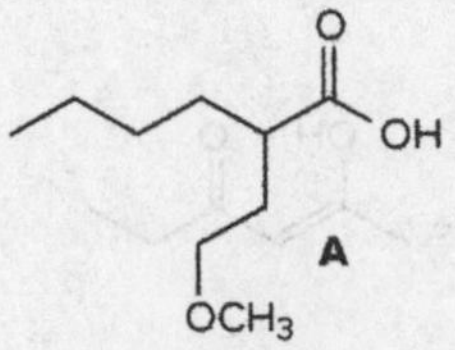

A

b. What starting materials are needed to synthesize ketone **B** by the acetoacetic ester synthesis?

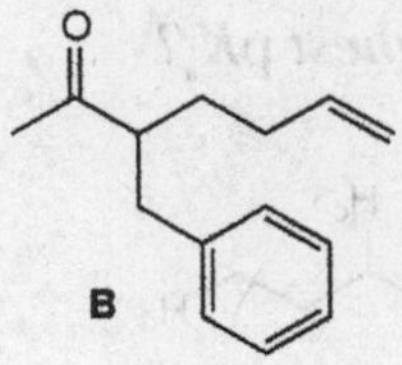

B

Answers to Practice Test

1.a. 1
   b. 4
2.a. H_b
   b. H_d
   c. H_b
   d. H_c
3.a. Br_2, CH_3CO_2H
   b. [1] LDA; [2] CH_3I
   c. []1 LDA; [2] CH_3I; [3] LDA; [4] CH_3I
   d. [1] Br_2, CH_3CO_2H; [2] Li_2CO_3, LiBr, DMF
   e. [1] Br_2, CH_3CO_2H; [2] Li_2CO_3, LiBr, DMF; [3] $(CH_3)_2CuLi$; [4] H_2O

4.
a. 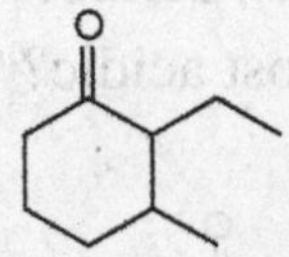
b. 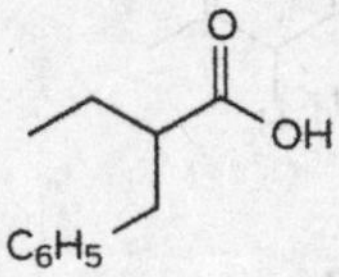 OH C_6H_5
c.
d.

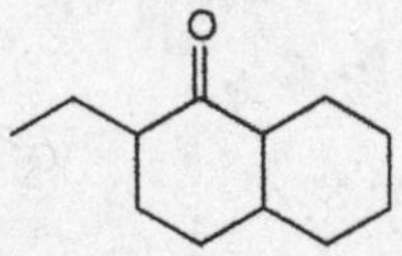

5.
a. 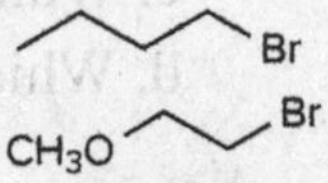 Br
CH_3O Br
$CH_2(CO_2Et)_2$
b. 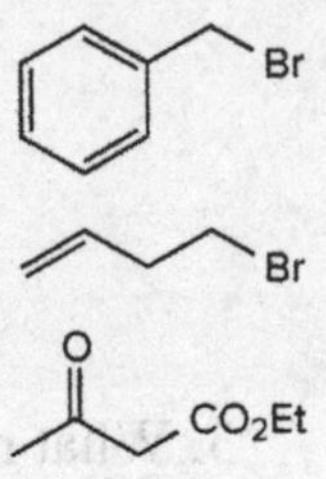 Br
Br
CO_2Et

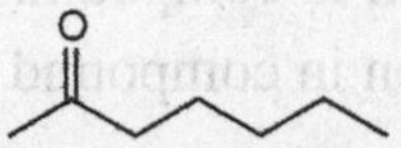

Answers to Problems

17.1

- To convert a ketone to its enol tautomer, change the C=O to C–OH, make a new double bond to an α carbon, and remove a proton at the other end of the C=C.
- To convert an enol to its keto form, find the C=C bonded to the OH. Change the C–OH to a C=O, add a proton to the other end of the C=C, and delete the double bond.

[In cases where *E* and *Z* isomers are possible, only one stereoisomer is drawn.]

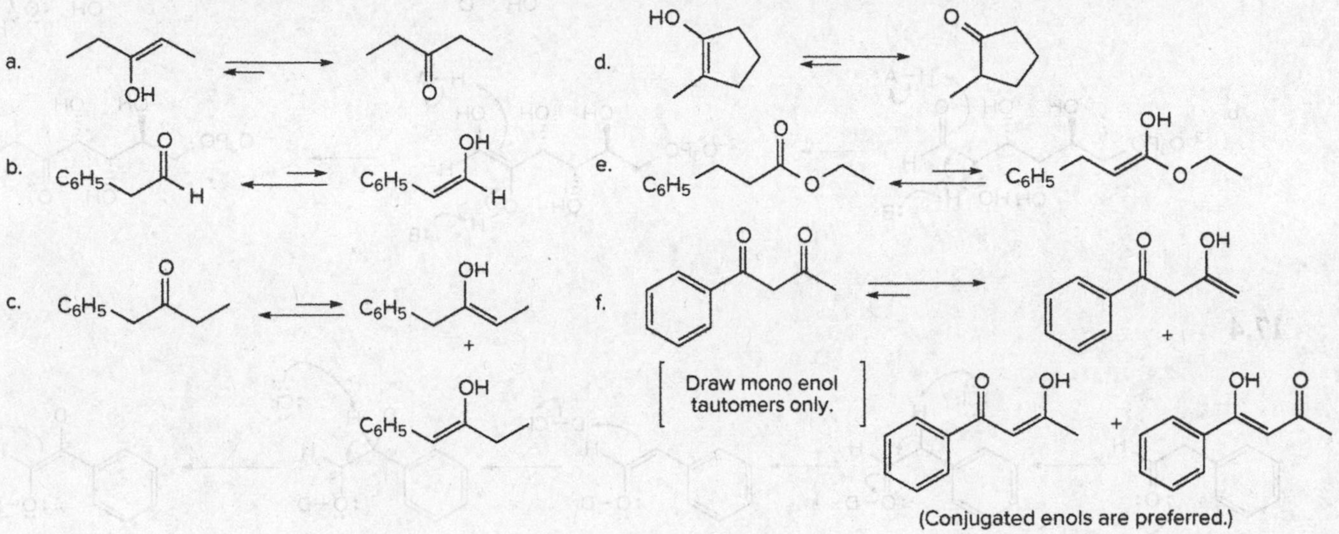

a. OH ⇌ O

b. C_6H_5 O H ⇌ C_6H_5 OH H

c. C_6H_5 O ⇌ C_6H_5 OH + OH C_6H_5

d. HO ⇌ O

e. C_6H_5 O O ⇌ OH C_6H_5 O

f. O O ⇌ O OH +

[Draw mono enol tautomers only.]

O OH + OH O

(Conjugated enols are preferred.)

17.2

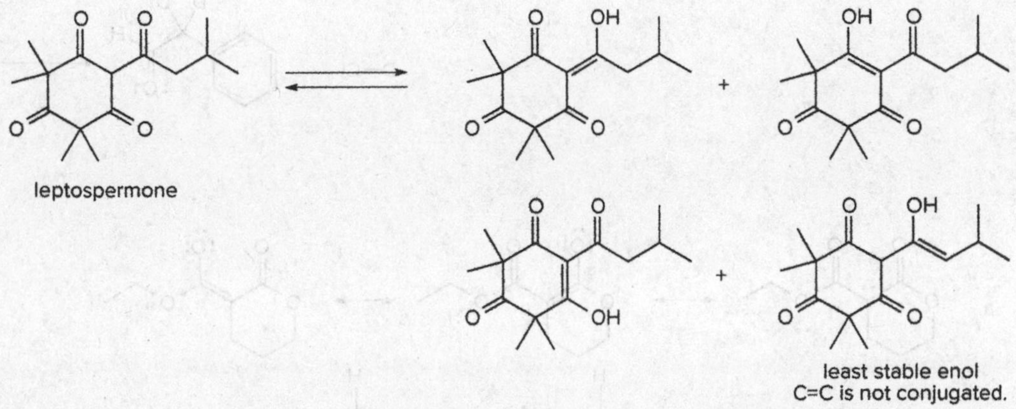

O O O O ⇌ O OH O O + OH O O O

leptospermone

O O O OH + O OH O O

least stable enol
C=C is not conjugated.

17.3

a.

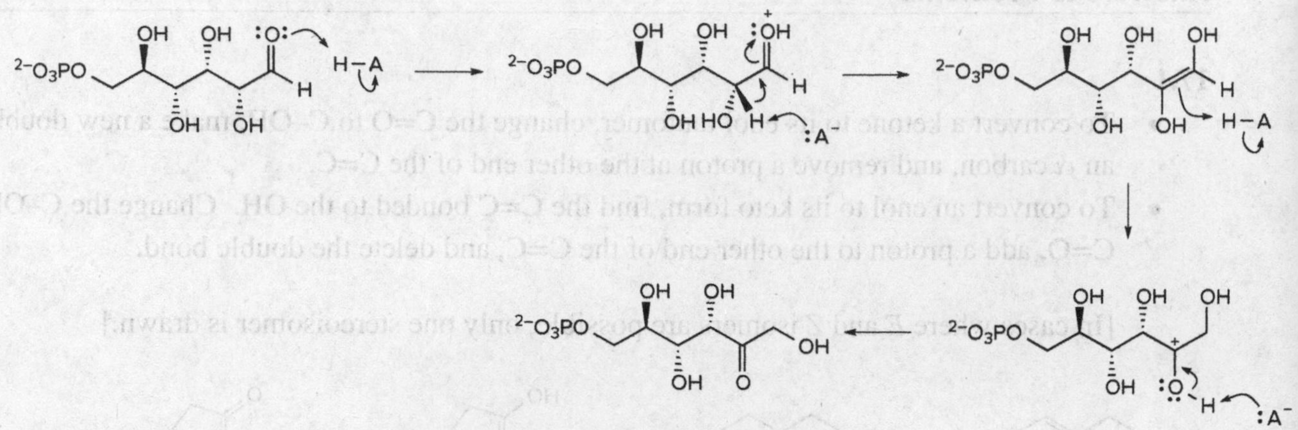

b.

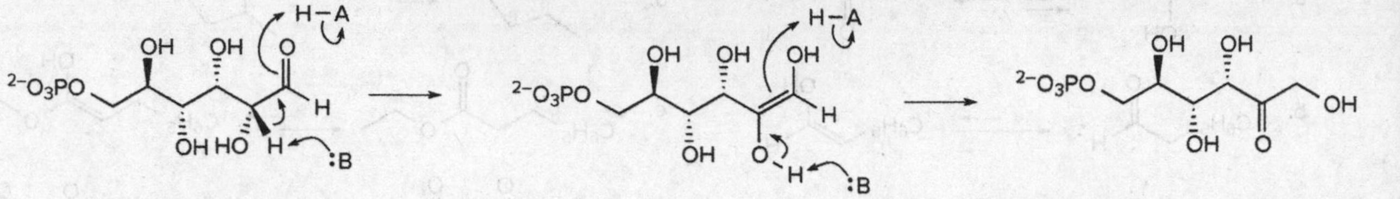

17.4

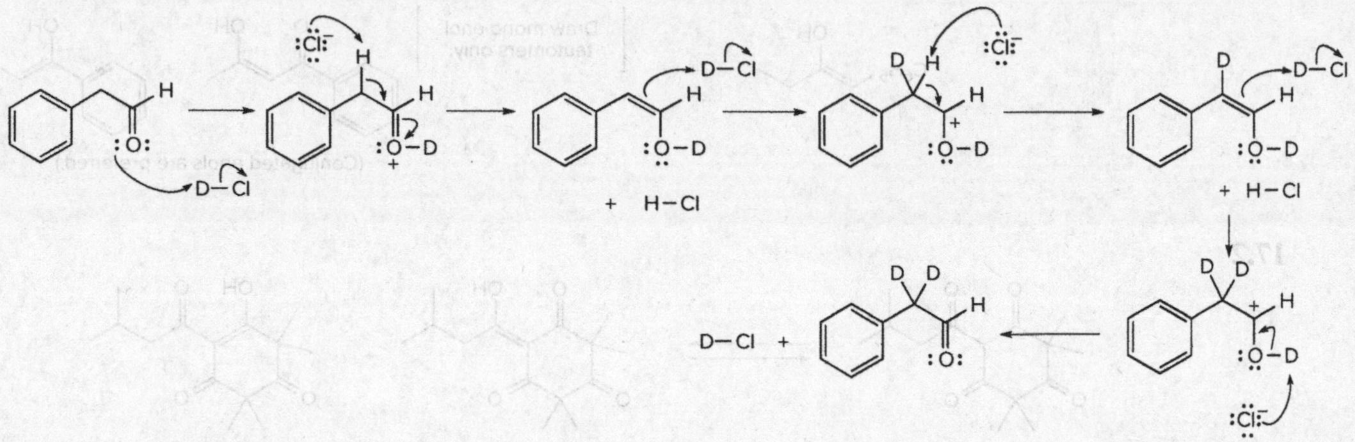

17.5

a.

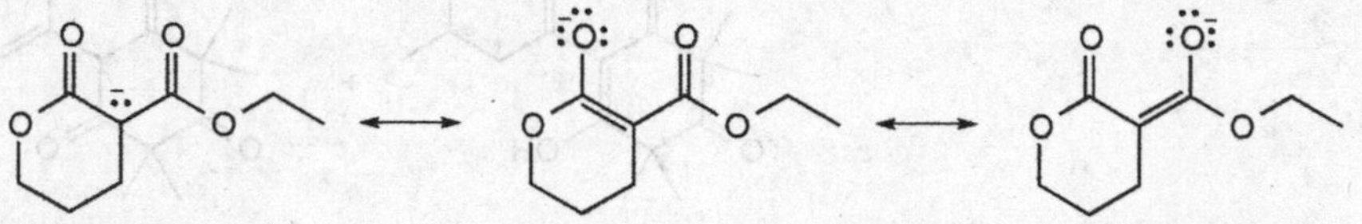

b.

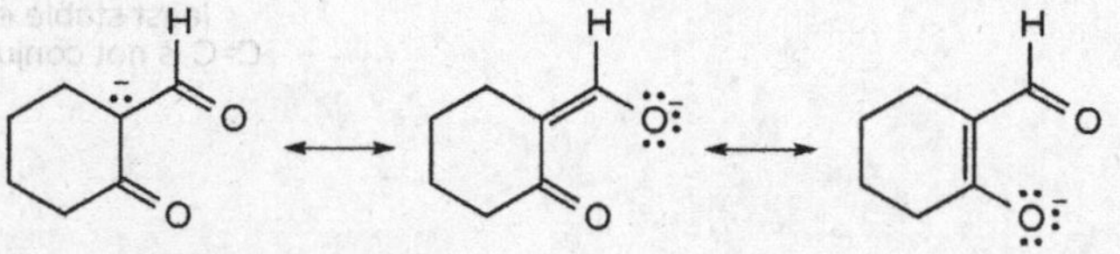

c.

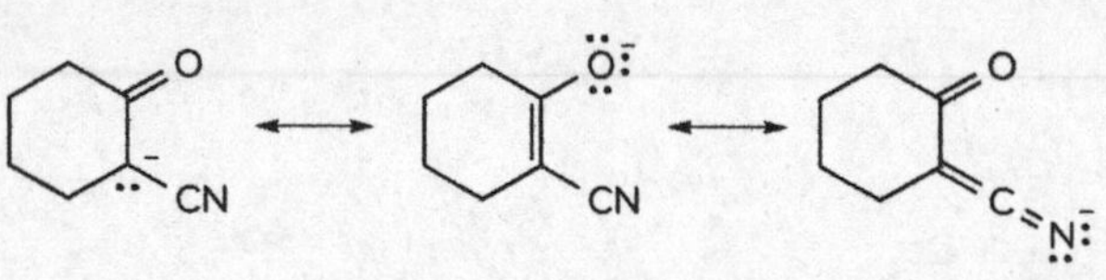

17.6 The indicated H's are α to a C=O or C≡N group, making them more acidic because their removal forms conjugate bases that are resonance stabilized.

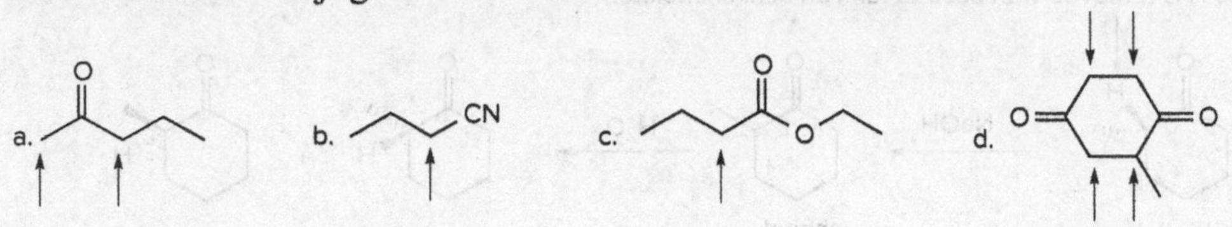

a. b. CN c. O d. O O

17.7

−H⁺ −H⁺ −H⁺

no resonance stabilization
least acidic

Two resonance structures stabilize the conjugate base.
intermediate acidity

Three resonance structures stabilize the conjugate base.
most acidic

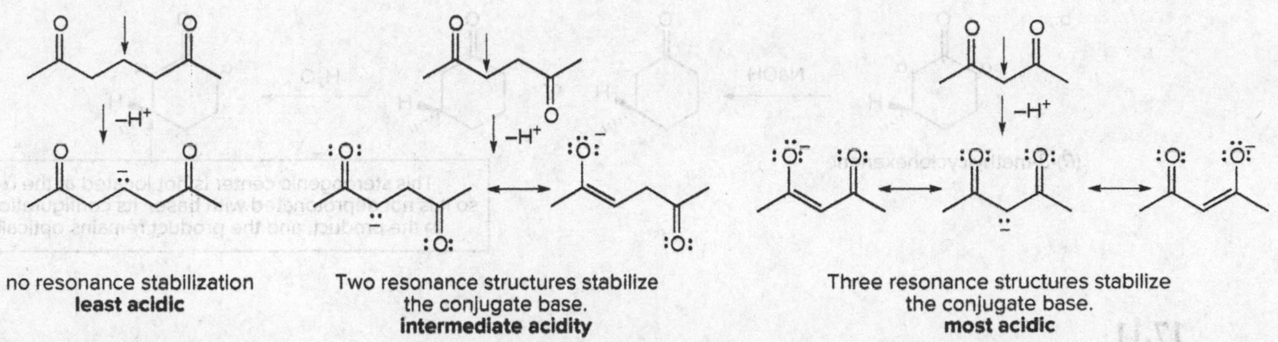

17.8 In each of the reactions, the LDA pulls off the most acidic proton.

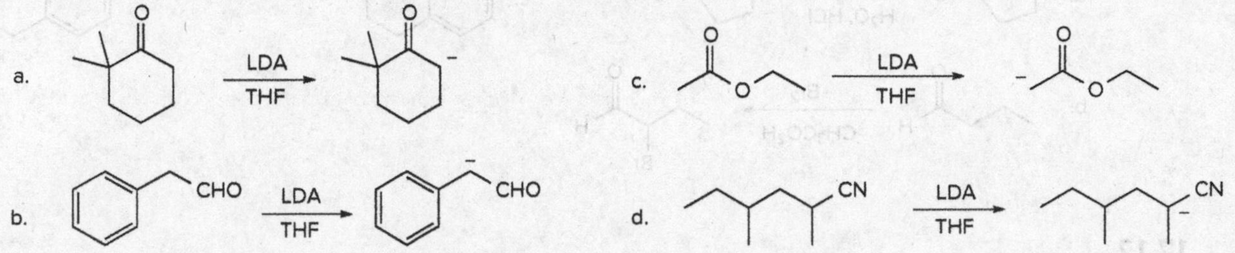

a. LDA THF

b. CHO LDA THF CHO

c. LDA THF

d. CN LDA THF CN

17.9

- LDA, THF forms the kinetic enolate by removing a proton from the less substituted C.
- Treatment with $NaOCH_3$, CH_3OH forms the thermodynamic enolate by removing a proton from the more substituted C.

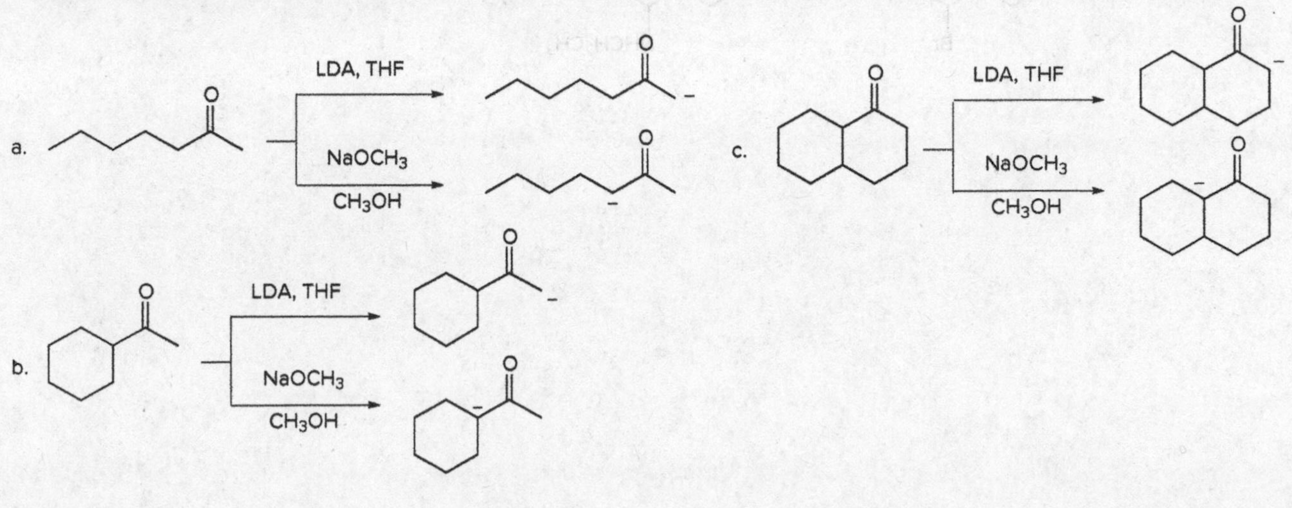

a. LDA, THF; $NaOCH_3$, CH_3OH

b. LDA, THF; $NaOCH_3$, CH_3OH

c. LDA, THF; $NaOCH_3$, CH_3OH

17.10

a. This acidic H is removed with base to form an achiral enolate.

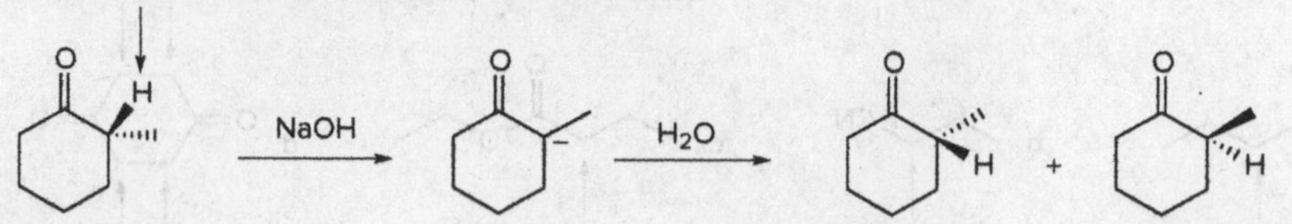

(R)-2-methylcyclohexanone → NaOH → achiral → H_2O → +

Protonation of the planar achiral enolate occurs with equal probability from two sides, so a racemic mixture is formed. The racemic mixture is optically inactive.

b. (R)-3-methylcyclohexanone → NaOH → or → H_2O →

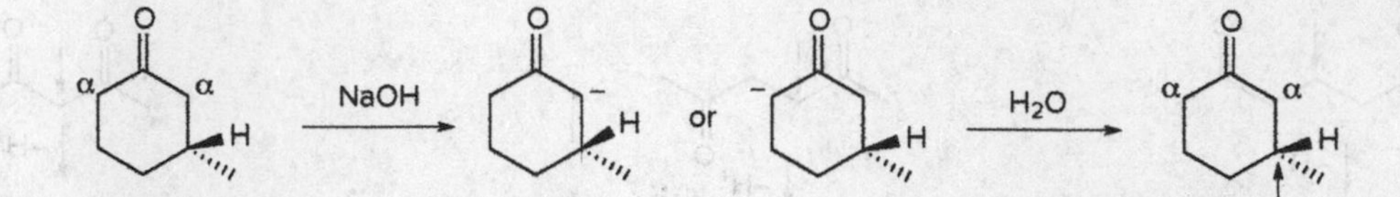

This stereogenic center is not located at the α carbon, so it is not deprotonated with base. Its configuration is retained in the product, and the product remains optically active.

17.11

a. Cl_2, H_2O, HCl

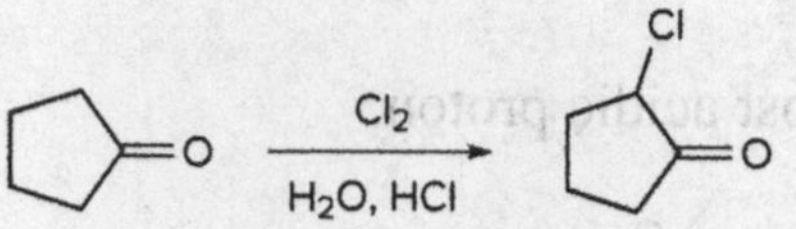

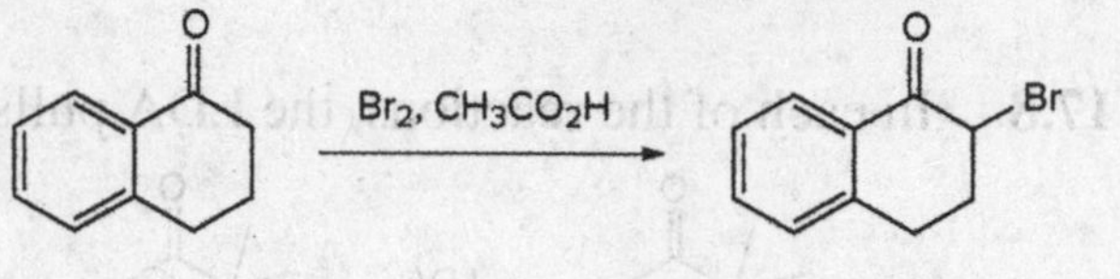

b. Br_2, CH_3CO_2H

c. Br_2, CH_3CO_2H

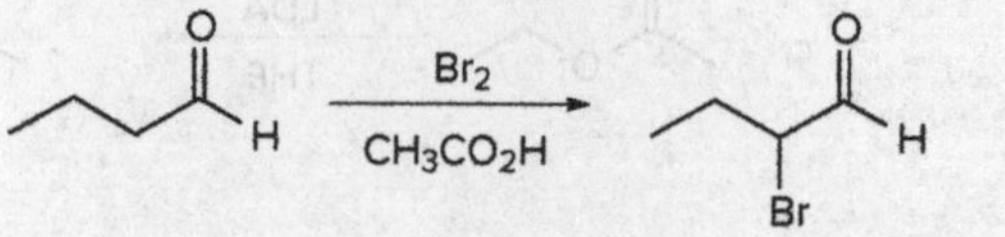

17.12

a. Li_2CO_3, LiBr, DMF

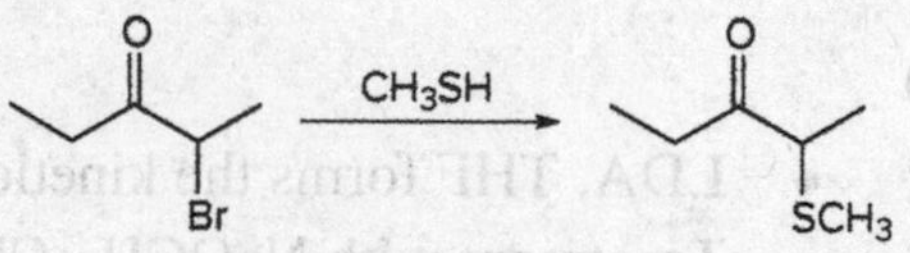

b. $CH_3CH_2NH_2$ → $NHCH_2CH_3$

c. CH_3SH → SCH_3

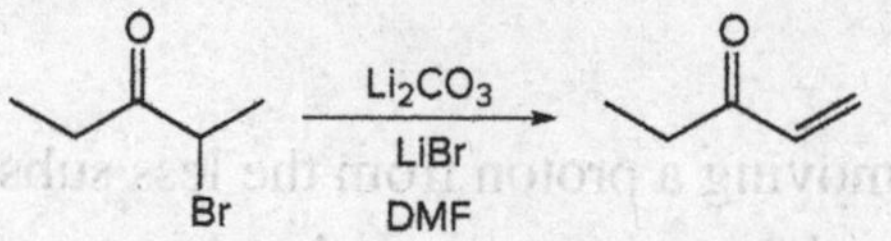

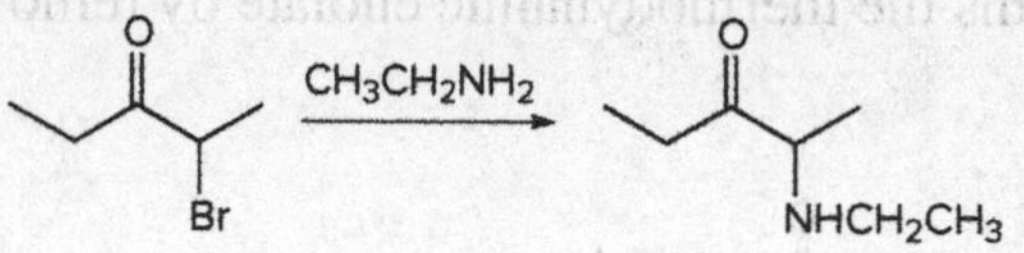

17.13

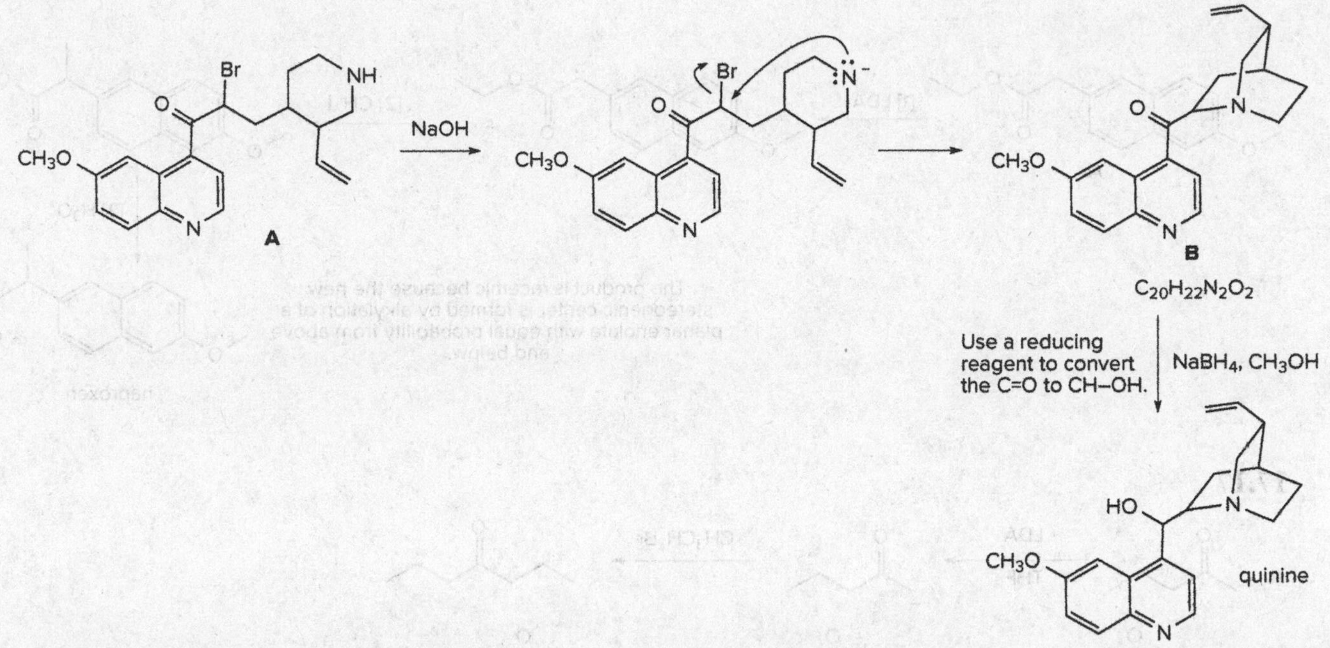

NaOH

A

B

$C_{20}H_{22}N_2O_2$

Use a reducing reagent to convert the C=O to CH–OH.

$NaBH_4$, CH_3OH

quinine

17.14

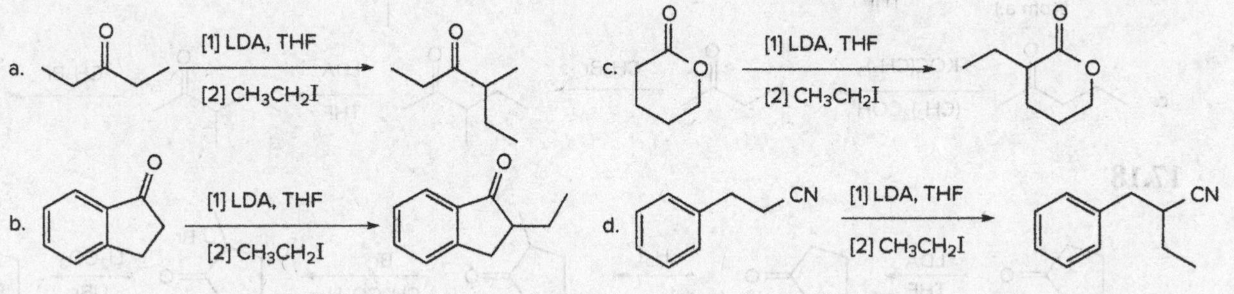

a. [1] LDA, THF [2] CH_3CH_2I

b. [1] LDA, THF [2] CH_3CH_2I

c. [1] LDA, THF [2] CH_3CH_2I

d. [1] LDA, THF [2] CH_3CH_2I

17.15

a. [1] LDA, THF [2] CH_3I

b. [1] LDA, THF [2] CH_3I +

c. [1] LDA, THF [2] CH_3I +

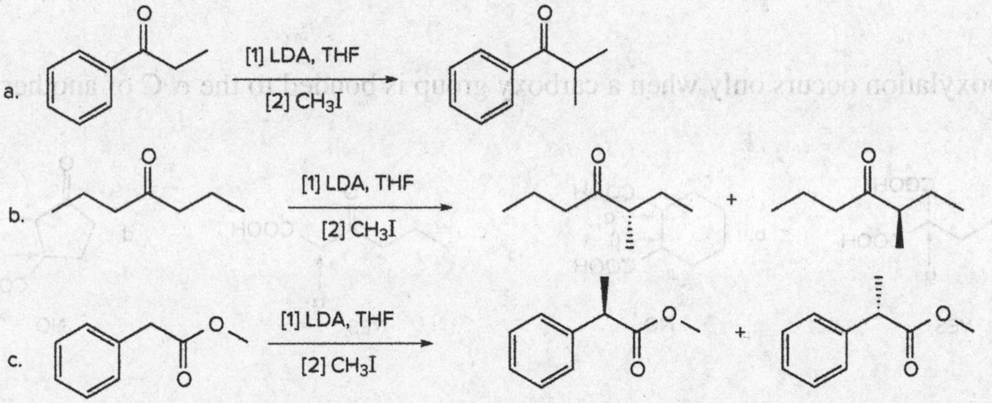

17.16 Three steps are needed: [1] formation of an enolate; [2] alkylation; [3] hydrolysis of the ester.

A

[1] LDA

[2] CH_3I

[3] H_3O^+

OH

The product is racemic because the new stereogenic center is formed by alkylation of a planar enolate with equal probability from above and below.

naproxen

17.17

a. LDA, THF → CH_3CH_2Br →

b. (from a.) LDA, THF → CH_3I →

c. $KOC(CH_3)_3$, $(CH_3)_3COH$ → CH_3Br → LDA, THF → CH_3Br →

17.18

LDA, THF → A → CH_3I → B → Br_2, CH_3CO_2H → C (Br) → Li_2CO_3, LiBr, DMF → α-methylene-γ-butyrolactone

17.19 Decarboxylation occurs only when a carboxy group is bonded to the α C of another carbonyl group.

a. COOH, COOH, α — YES

b. COOH, α, β, COOH — NO

c. COOH, α — YES

d. α, β, COOH — NO

17.20

a. $CH_2(CO_2Et)_2$ → [1] NaOEt, [2] (cyclohexylmethyl bromide, Br) → H_3O^+, Δ → OH

b. $CH_2(CO_2Et)_2$ → [1] NaOEt, [2] CH_3Br → [1] NaOEt, [2] CH_3Br → H_3O^+, Δ → OH

17.21

a. Cl–(CH$_2$)$_3$–Cl → cyclobutane-COOH

b. Br–CH$_2$CH$_2$–O–CH$_2$CH$_2$–Br → tetrahydropyran–COOH

17.22 Locate the α C to the COOH group, and identify all of the alkyl groups bonded to it. These groups are from alkyl halides, and the remainder of the molecule is from diethyl malonate.

a. $CH_2(CO_2Et)_2$ — [1] NaOEt; [2] alkyl bromide → H_3O^+, Δ → OH (carboxylic acid), α

b. $CH_2(CO_2Et)_2$ — [1] NaOEt; [2] Br → [1] NaOEt; [2] Br → H_3O^+, Δ → OH, α

c. $CH_2(CO_2Et)_2$ — [1] NaOEt; [2] Br → [1] NaOEt; [2] Br → H_3O^+, Δ → HO, O, α

17.23 The reaction works best when the alkyl halide is 1° or CH_3X, because this is an S_N2 reaction.

a. OH; $(CH_3)_3CX$ 3° alkyl halide (too crowded)

b. OH; X; aryl halide (leaving group on an *sp*[2] hybridized C). Aryl halides are unreactive in S_N2 reactions.

c. OH; α. This compound has 3 CH_3 groups on the α carbon to the COOH. The malonic ester synthesis can be used to prepare mono- and disubstituted carboxylic acids only: RCH_2COOH and $R_2CHCOOH$, but not R_3CCOOH.

17.24

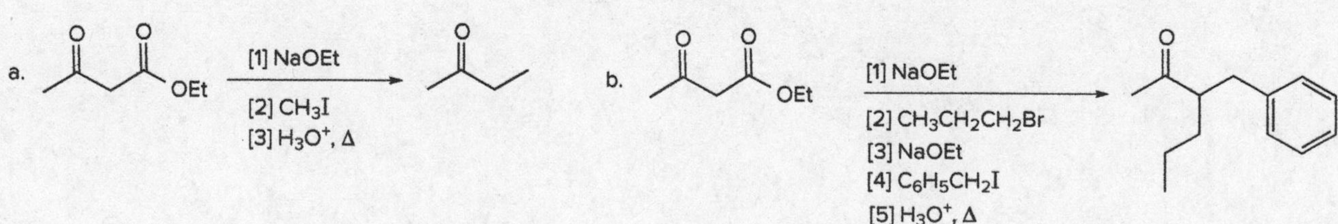

17.25 Locate the α C. All alkyl groups on the α C come from alkyl halides, and the remainder of the molecule comes from ethyl acetoacetate.

a. [1] NaOEt; [2] CH_3CH_2Br → H_3O^+, Δ

b. [1] NaOEt; [2] CH_3CH_2Br → [1] NaOEt; [2] CH_3CH_2Br → H_3O^+, Δ

c. [1] NaOEt; [2] CH_3CH_2Br → [1] NaOEt; [2] $CH_3(CH_2)_3Br$ → H_3O^+, Δ

d. [1] NaOEt; [2] 1-bromobutane (Br) → [1] NaOEt; [2] benzyl bromide (Br) → H_3O^+, Δ

17.26

a. [1] NaOEt; [2] CH_3O-naphthyl-CH_2Br → (product with CO_2Et, CH_3O) → H_3O^+, Δ → nabumetone (CH_3O)

b. [1] LDA, THF; [2] CH_3O-naphthyl-CH_2Br → nabumetone (CH_3O)

17.27

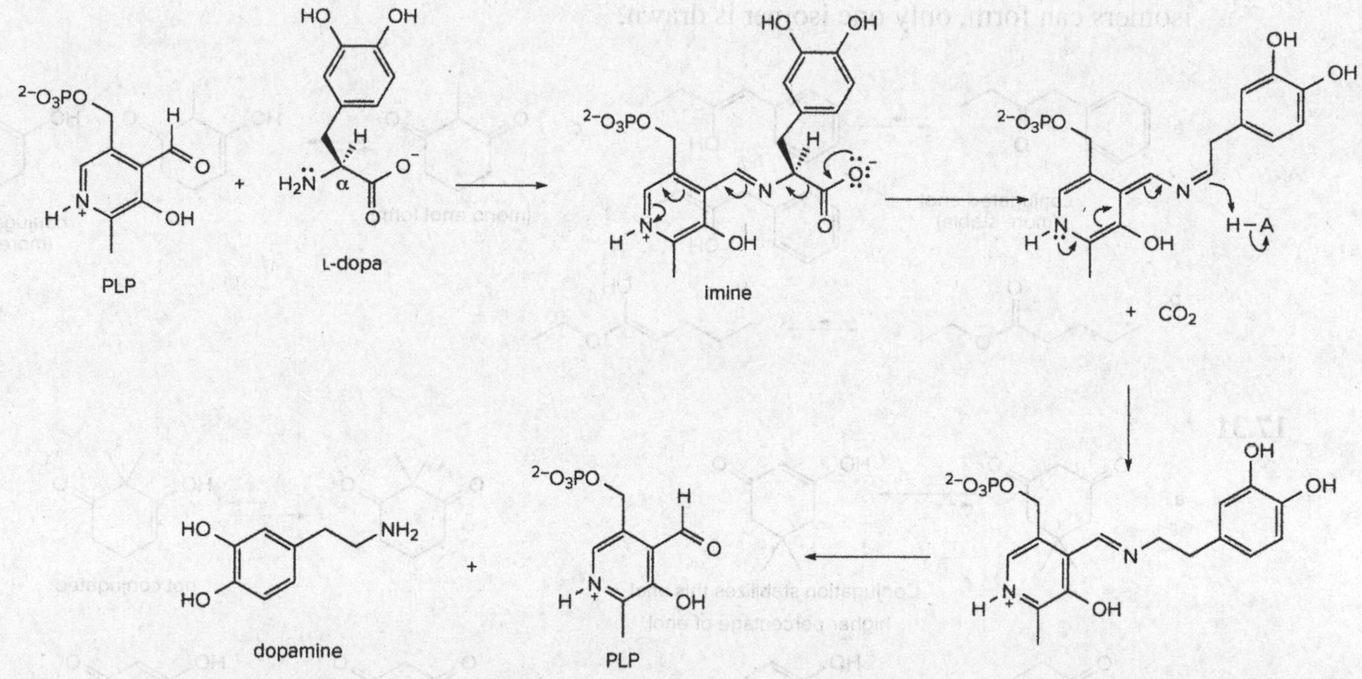

17.28 Use the directions from Answer 17.1 to draw the enol tautomer(s). In cases where *E* and *Z* isomers can form, only one isomer is drawn.

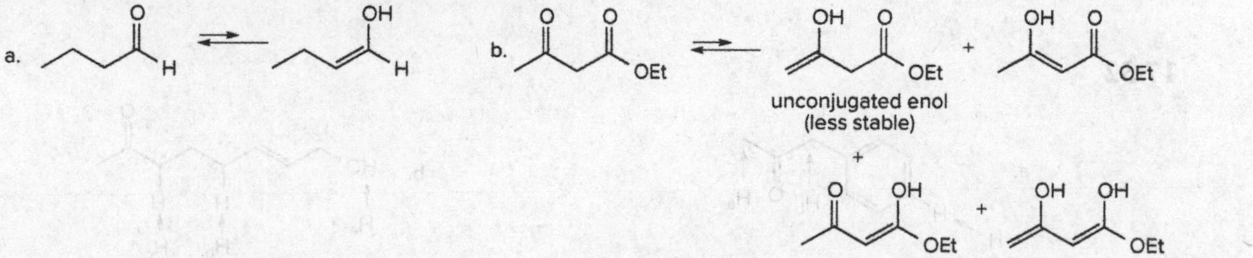

17.29

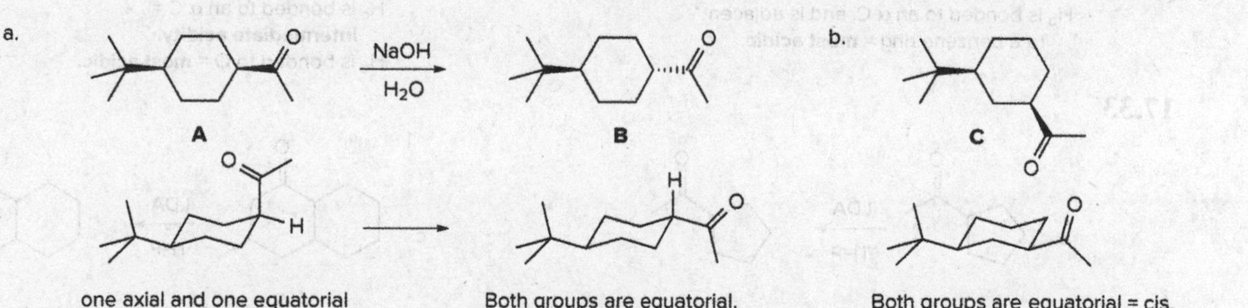

17.30 Use the directions from Answer 17.1 to draw the enol tautomer(s). In cases where *E* and *Z* isomers can form, only one isomer is drawn.

a. conjugated enol (more stable)

OH

\+

OH

c. O O

HO O + HO O

(mono enol form)

conjugated enol (more stable)

b. O

OH

O O

17.31

a. O O

HO O

Conjugation stabilizes this enol.

higher percentage of enol

O O

HO O

not conjugated

b. O O

HO O

not conjugated

O O

HO O

conjugated C=C

higher percentage of enol

17.32

a. H_b O H_a H H_c

H_c is bonded to an sp^2 hybridized C = **least acidic.**
H_a is bonded to an α C = **intermediate acidity.**
H_b is bonded to an α C, and is adjacent to a benzene ring = **most acidic.**

b. HO H_a O H H H_b H_c

H_b is bonded to an sp^3 hybridized C = **least acidic.**
H_c is bonded to an α C = **intermediate acidity.**
H_a is bonded to O = **most acidic.**

17.33

a. O —LDA, THF→ O −

c. O —LDA, THF→ O −

b. CN —LDA, THF→ CN −

d. O O —LDA, THF→ O O −

17.34

H_a H_a H_b H_b remove H_a → H_a H_b H_b ⟷ H_a H_b H_b

Removal of H_a gives two resonance structures. The negative charge is never on O.

H_a H_a H_b H_b remove H_b → H_a H_a H_b ⟷ H_a H_a H_b ⟷ H_a H_a H_b

Removal of H_b gives three resonance structures. The negative charge is on O in one resonance structure, making the conjugate base more stable and H_b more acidic (lower pK_a).

17.35

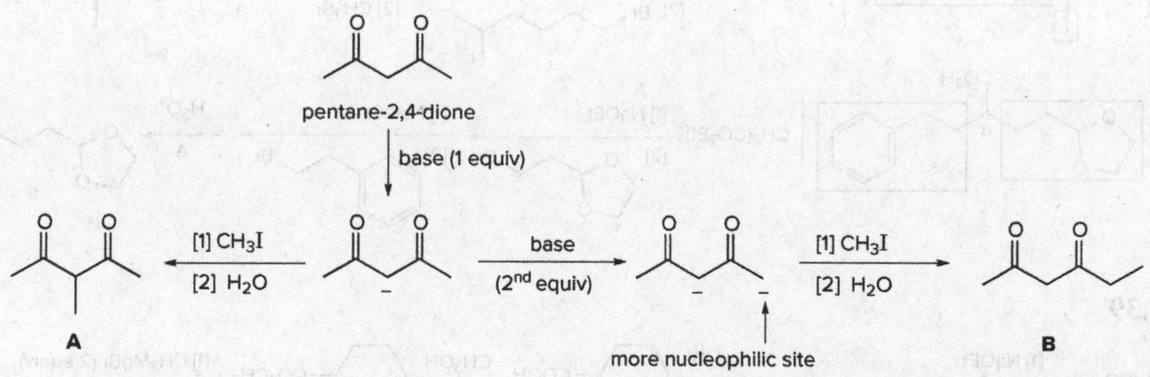

One equivalent of base removes the most acidic proton between the two C=O's, to form **A** on alkylation with CH_3I.

With a second equivalent of base a dianion is formed. Since the second enolate is less resonance stabilized, it is more nucleophilic and reacts first in an alkylation with CH_3I, forming **B** after protonation with H_2O.

17.36

HO HO H HO OH O O — enediol

HO HO H HO O O O — keto tautomer **A**

HO HO H O OH O O — keto tautomer **B**

The enediol is more stable than either of the two keto tautomers because it is more highly conjugated. More resonance structures can be drawn, which delocalize the lone pairs of the two OH groups bonded to the C=C. Such delocalization is not possible with either keto tautomer.

17.37 Use the directions from Answer 17.22.

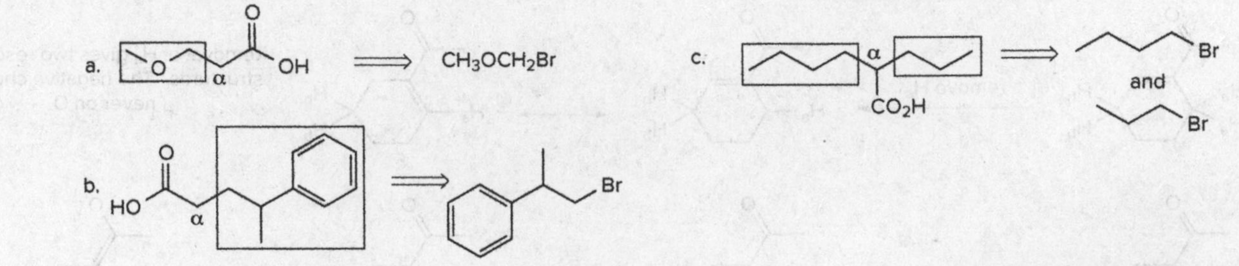

a. (structure) ⟹ CH_3OCH_2Br

b. (structure) ⟹ (structure) Br

c. (structure) CO_2H ⟹ Br and Br

17.38

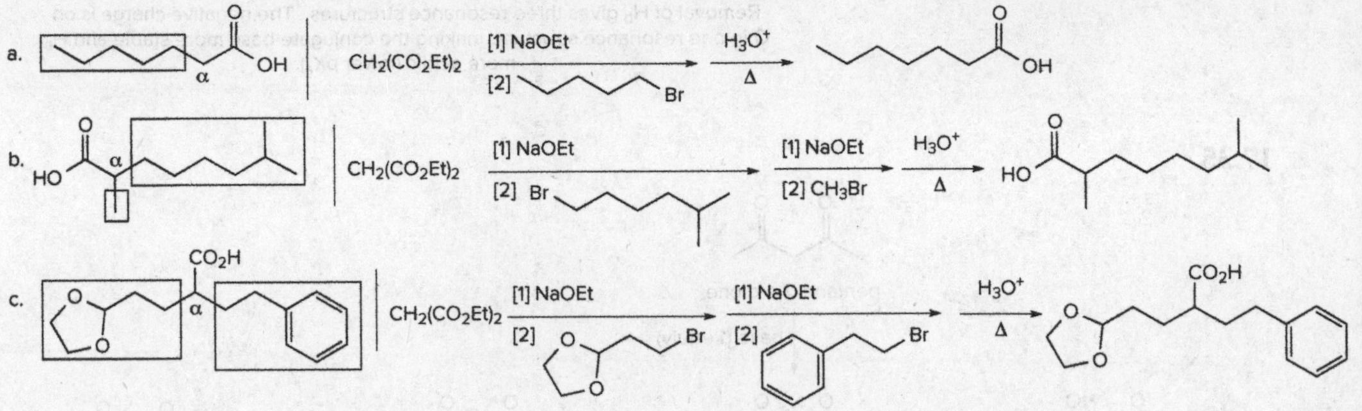

a. OH | $CH_2(CO_2Et)_2$ —[1] NaOEt, [2] Br→ —H_3O^+, Δ→ OH

b. HO | $CH_2(CO_2Et)_2$ —[1] NaOEt, [2] Br→ —[1] NaOEt, [2] CH_3Br→ —H_3O^+, Δ→ HO

c. CO_2H | $CH_2(CO_2Et)_2$ —[1] NaOEt, [2] Br→ —[1] NaOEt, [2] Br→ —H_3O^+, Δ→ CO_2H

17.39

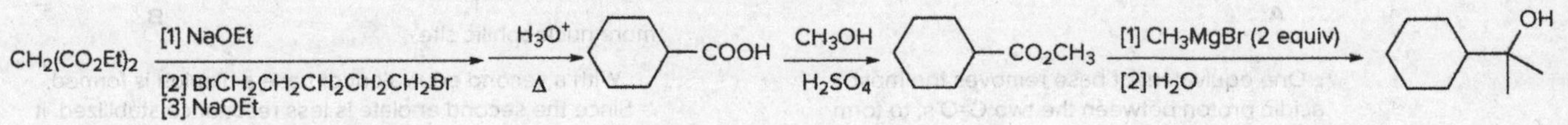

$CH_2(CO_2Et)_2$ —[1] NaOEt, [2] $BrCH_2CH_2CH_2CH_2CH_2Br$, [3] NaOEt→ —H_3O^+, Δ→ COOH —CH_3OH, H_2SO_4→ CO_2CH_3 —[1] CH_3MgBr (2 equiv), [2] H_2O→ OH

17.40

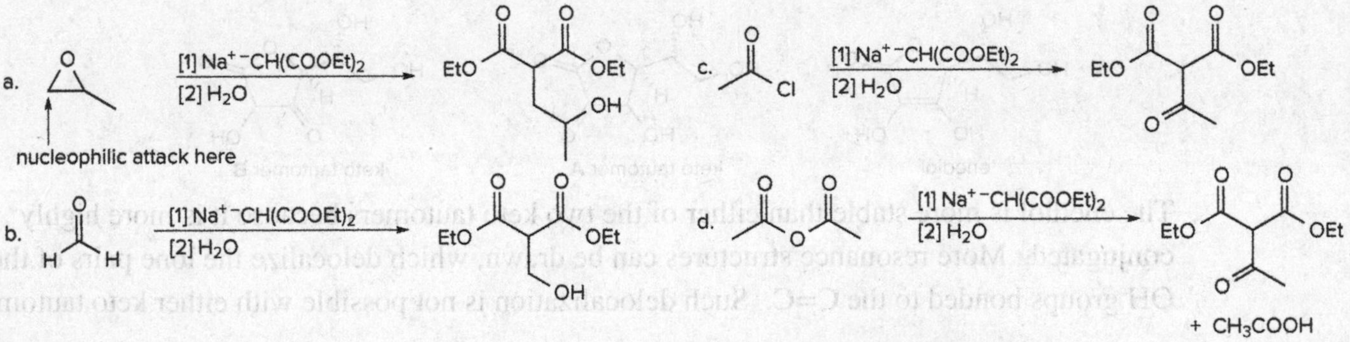

a. (epoxide) —[1] Na^+ $^-CH(COOEt)_2$, [2] H_2O→ EtO OEt OH

nucleophilic attack here

b. H H —[1] Na^+ $^-CH(COOEt)_2$, [2] H_2O→ EtO OEt OH

c. Cl —[1] Na^+ $^-CH(COOEt)_2$, [2] H_2O→ EtO OEt

d. —[1] Na^+ $^-CH(COOEt)_2$, [2] H_2O→ EtO OEt + CH_3COOH

17.41 Use the directions from Answer 17.25.

a. [1] NaOEt, [2] Br–CH₂CH₂CH₂CH(CH₃)₂; H_3O^+, Δ

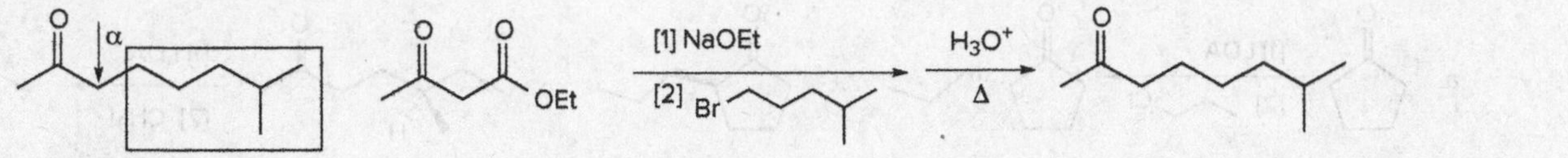

b. [1] NaOEt, [2] CH_3CH_2Br; [1] NaOEt, [2] Br; H_3O^+, Δ

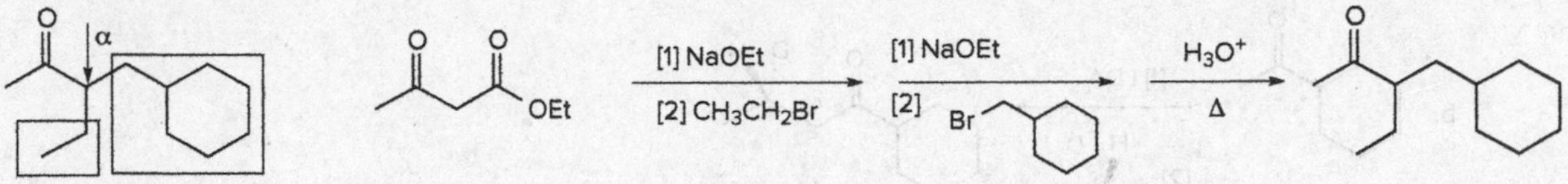

c. [1] NaOEt, [2] Br, Br, [3] NaOEt; H_3O^+, Δ

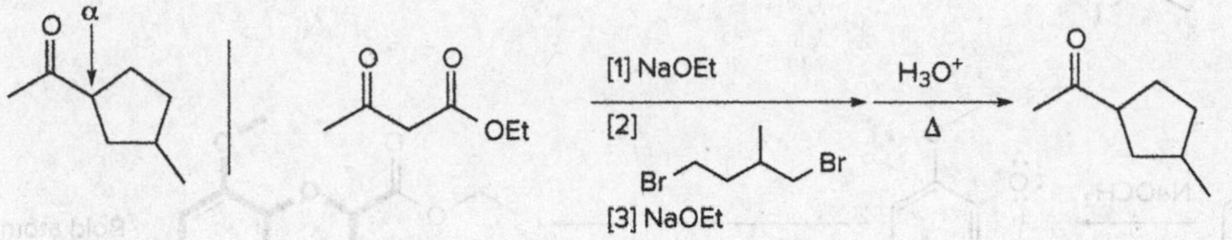

17.42

a. OEt; [1] NaOEt, [2] CH_3Br; [1] NaOEt, [2] CH_3Br; H_3O^+, Δ

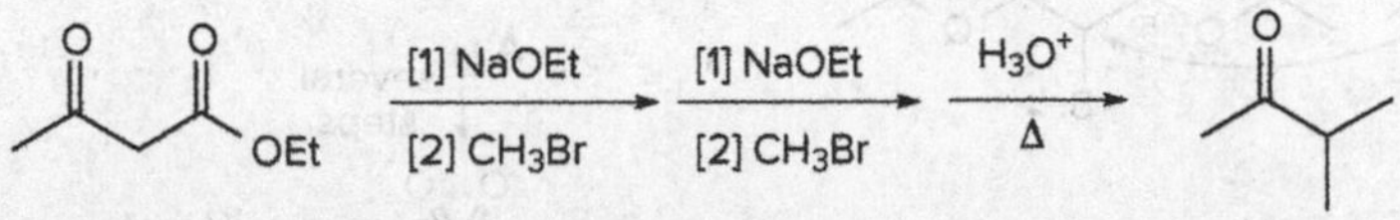

b. (from a.) LDA, THF; CH_3I

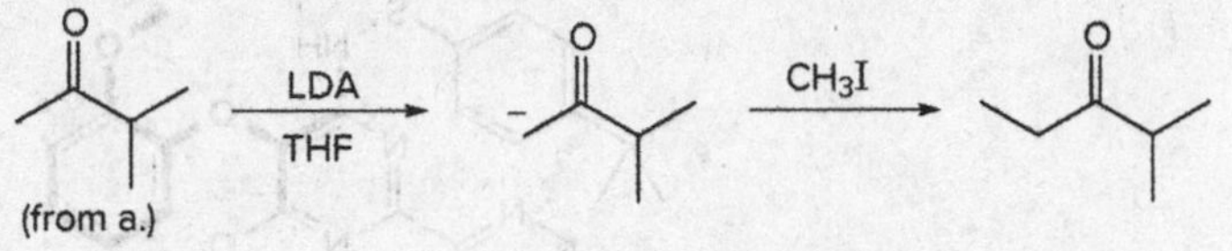

c. OEt; [1] NaOEt, [2] Br; [1] NaOEt, [2] Br; H_3O^+, Δ

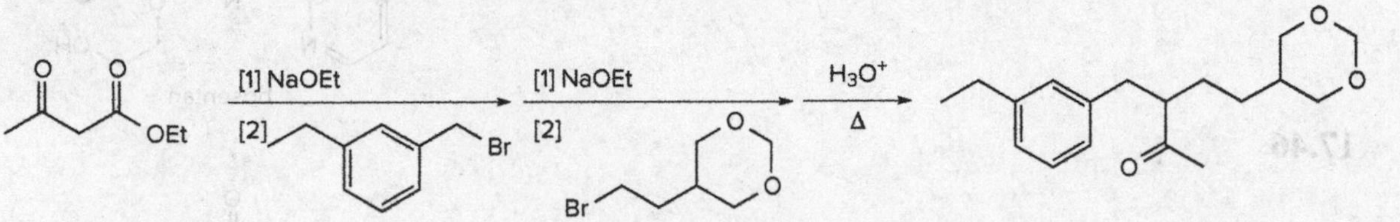

17.43

a. OH, HO; Δ; HO

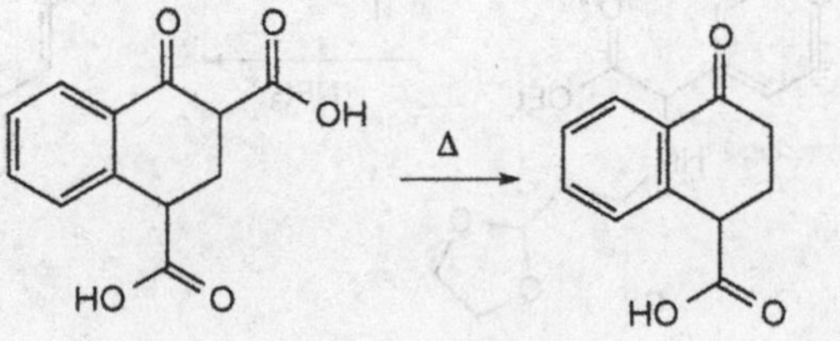

b. OEt; [1] LDA, [2] CH_3CH_2I; OEt

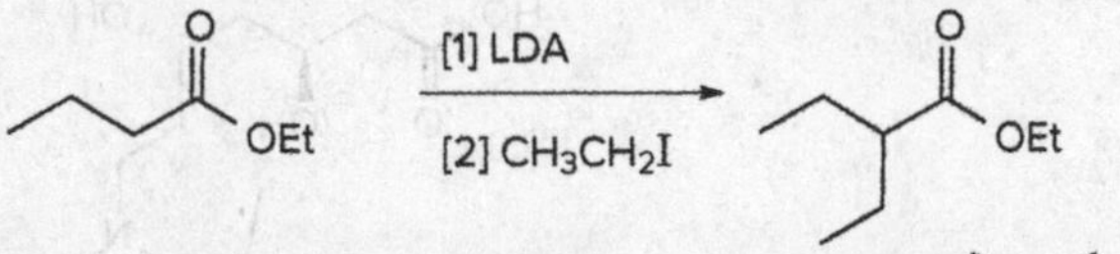

c. Br; NH_2; NH

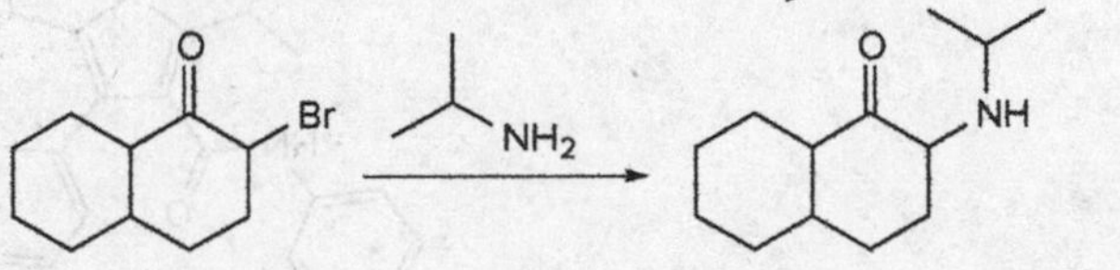

d. [1] LDA, [2] CH_3CH_2I

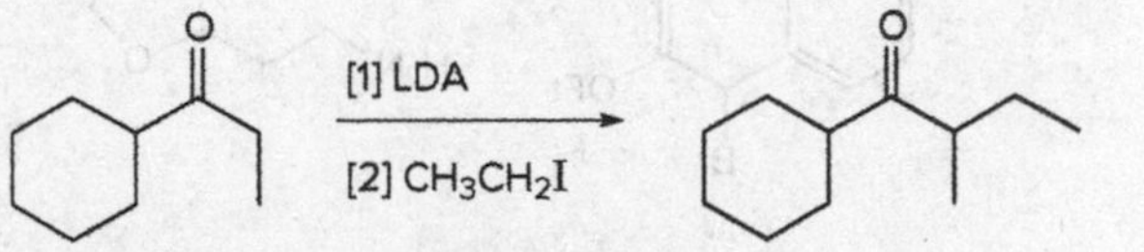

e. [1] Br_2, CH_3CO_2H, [2] Li_2CO_3, LiBr, DMF

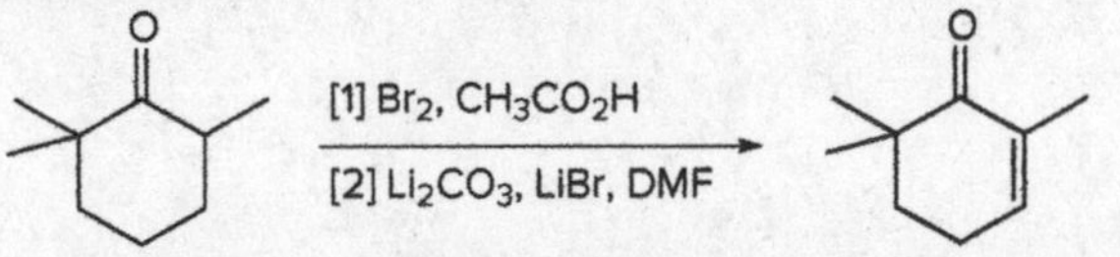

f. Cl, CN; NaH; C≡N

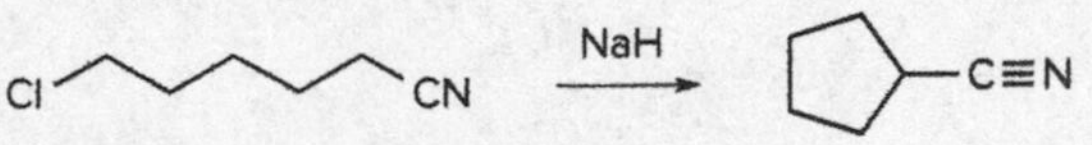

17.44

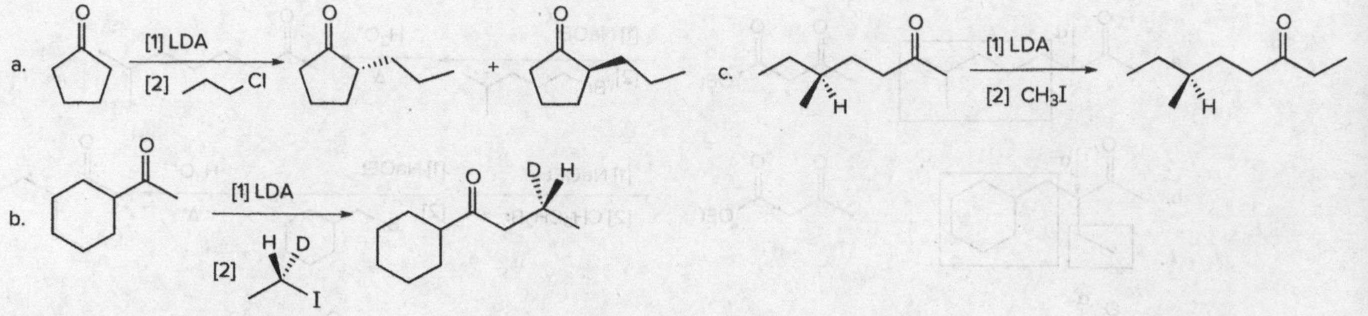

a. [1] LDA [2] Cl

+

c. [1] LDA [2] CH_3I

b. [1] LDA [2]

17.45

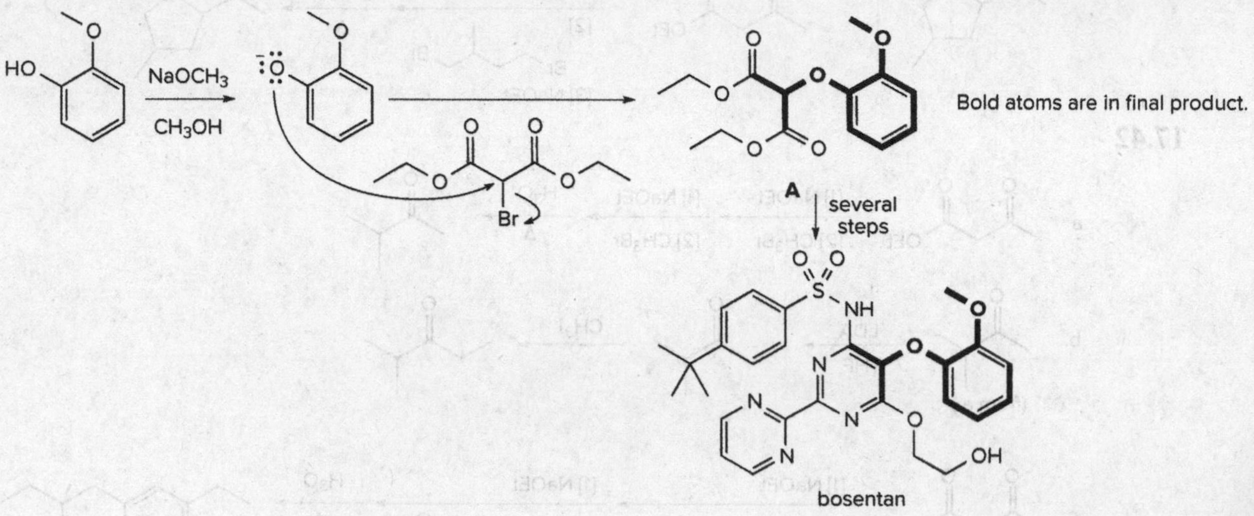

$NaOCH_3$ CH_3OH

Bold atoms are in final product.

A several steps

bosentan

17.46

+ NEt_3 C NEt_3 D several steps

atorvastatin

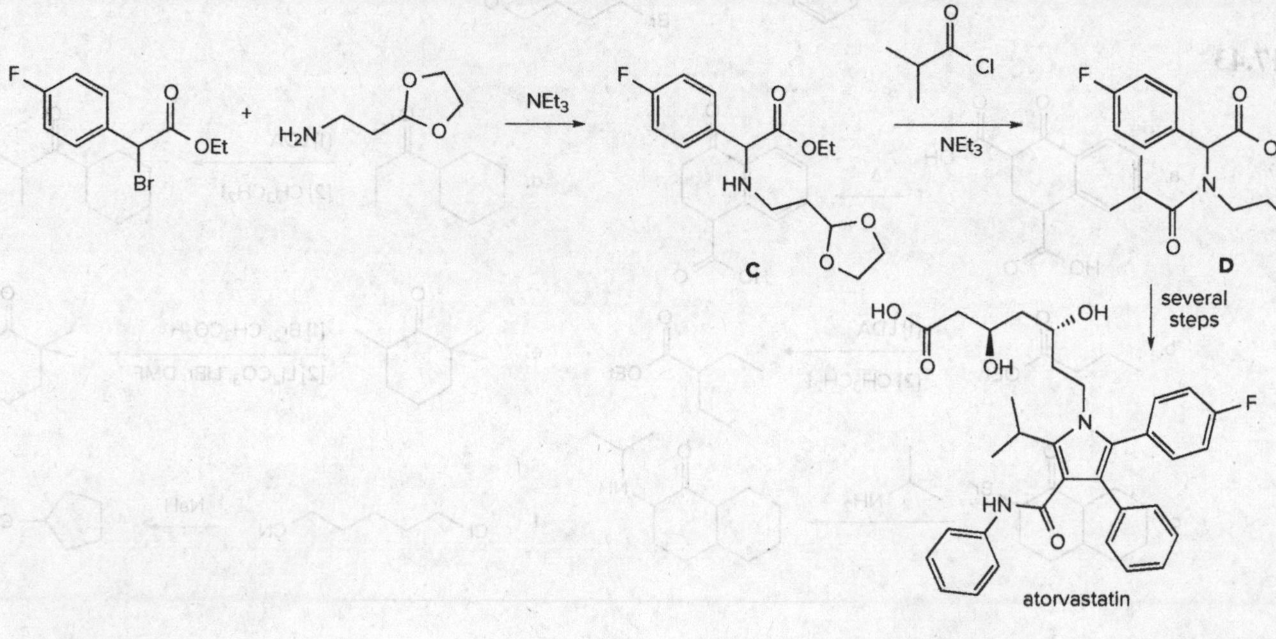

17.47

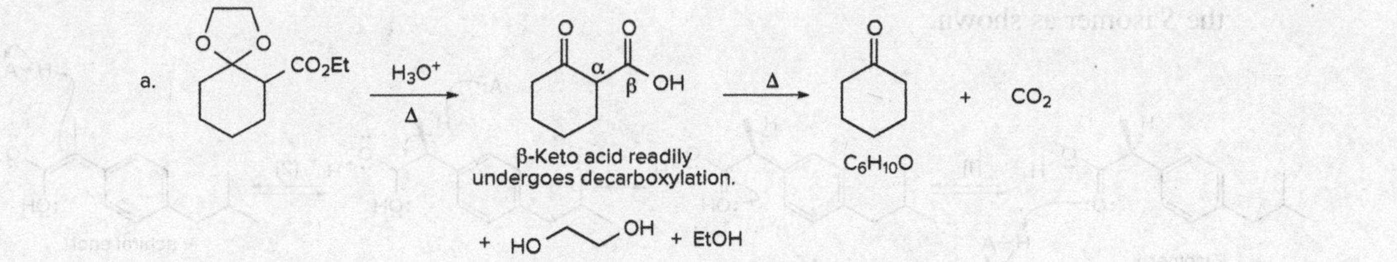

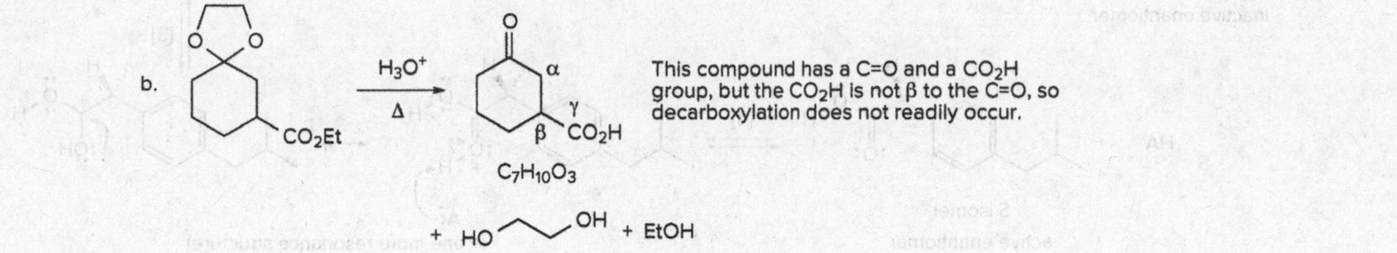

a. $\xrightarrow[\Delta]{H_3O^+}$ β-Keto acid readily undergoes decarboxylation. + HO–CH₂CH₂–OH + EtOH $\xrightarrow{\Delta}$ $C_6H_{10}O$ + CO_2

b. $\xrightarrow[\Delta]{H_3O^+}$ $C_7H_{10}O_3$ + HO–CH₂CH₂–OH + EtOH

This compound has a C=O and a CO_2H group, but the CO_2H is not β to the C=O, so decarboxylation does not readily occur.

17.48

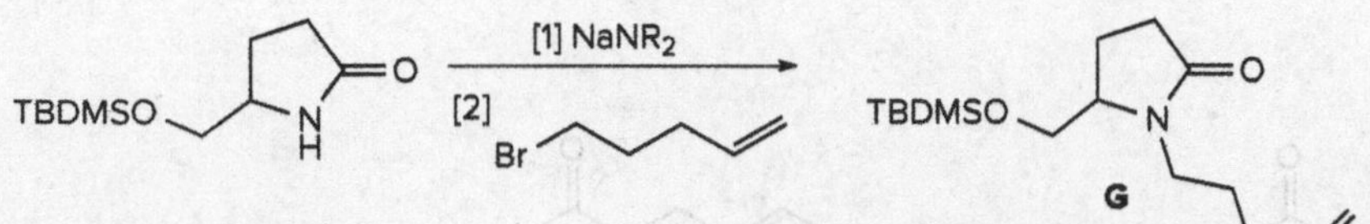

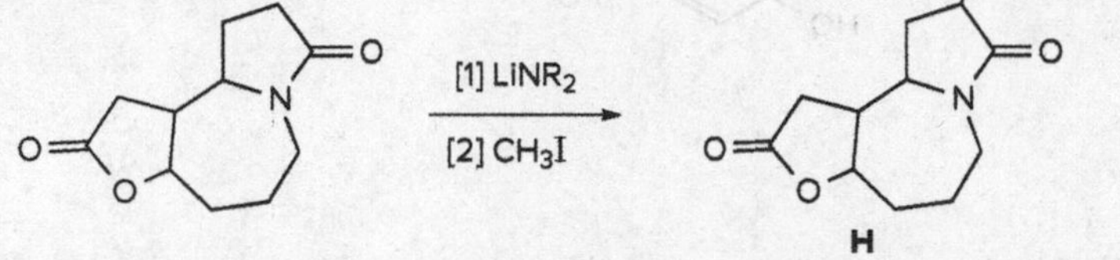

TBDMSO $\xrightarrow[\text{[2] Br}]{\text{[1] NaNR}_2}$ TBDMSO **G**

$\xrightarrow[\text{[2] CH}_3\text{I}]{\text{[1] LiNR}_2}$ **H**

17.49

a. $\xrightarrow{\text{NaH}}$ $\xrightarrow{}$ $\xrightarrow{\text{NaH}}$ **X**

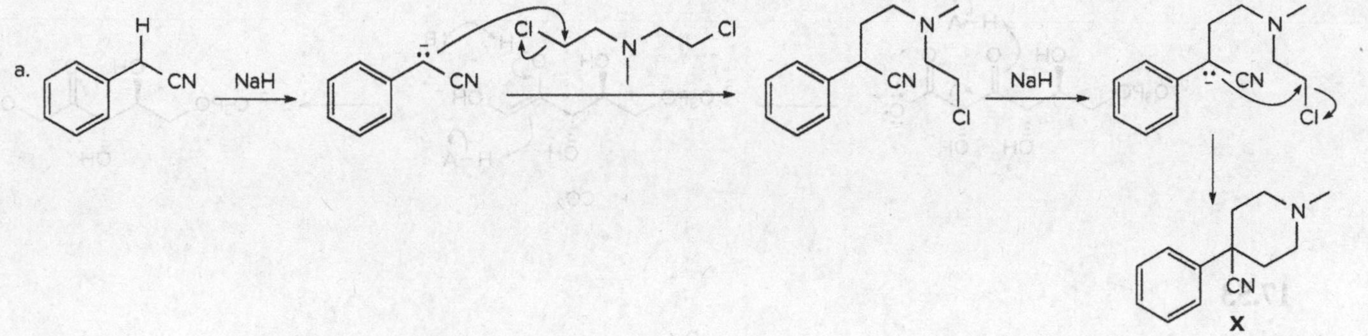

b. **X** can be converted to meperidine by hydrolysis of the nitrile and esterification.

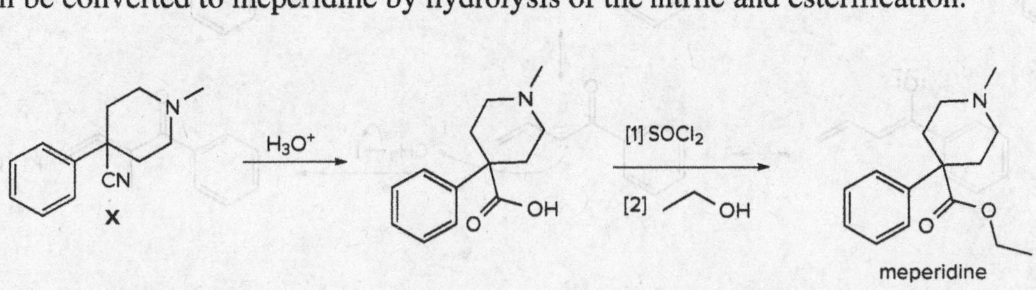

X $\xrightarrow{H_3O^+}$ $\xrightarrow[\text{[2] EtOH}]{\text{[1] SOCl}_2}$ meperidine

17.50 Protonation in Step [3] can occur from below (to re-form the *R* isomer) or from above to form the *S* isomer as shown.

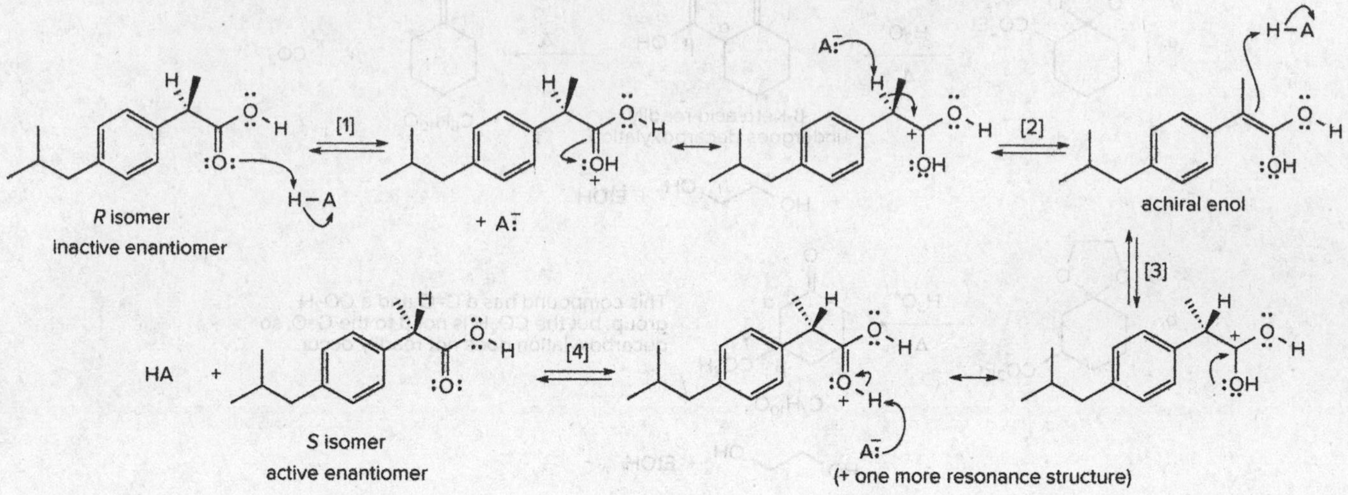

[1] [2] [3] [4]

H–A

+ A:⁻

R isomer
inactive enantiomer

achiral enol

HA +

S isomer
active enantiomer

(+ one more resonance structure)

17.51

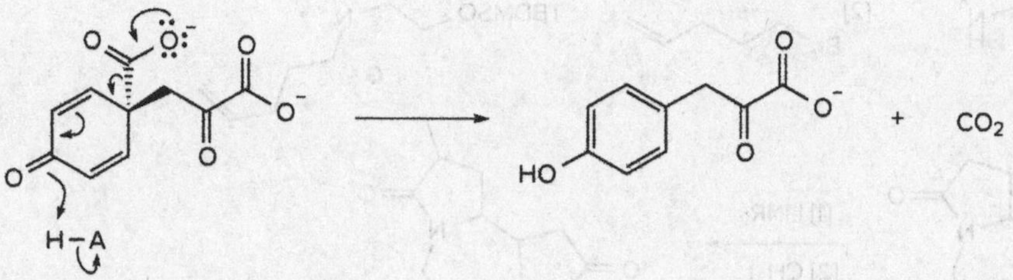

HO

+ CO_2

H–A

17.52

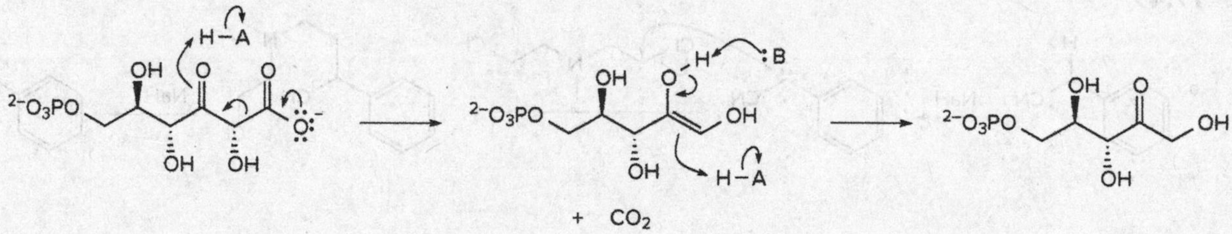

$^{2-}O_3PO$

OH OH OH

H–A :B

+ CO_2

17.53

LDA

CH_3–I

+ I^-

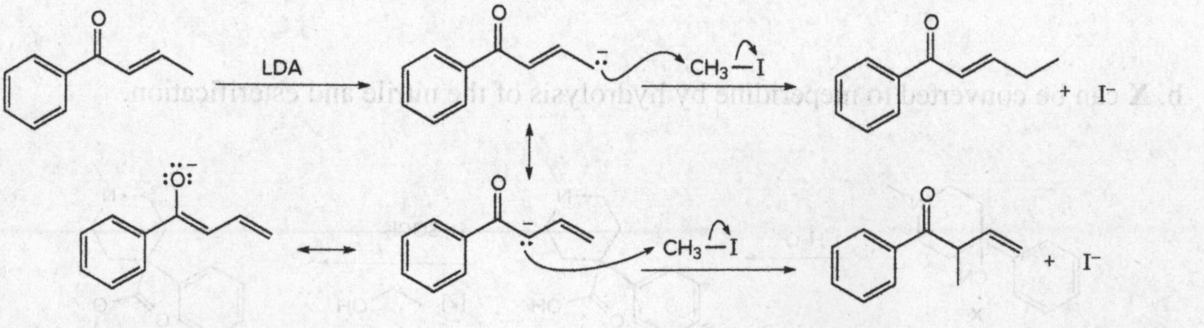

17.54 Protons on the γ carbon of an α,β-unsaturated carbonyl compound are acidic because of resonance.

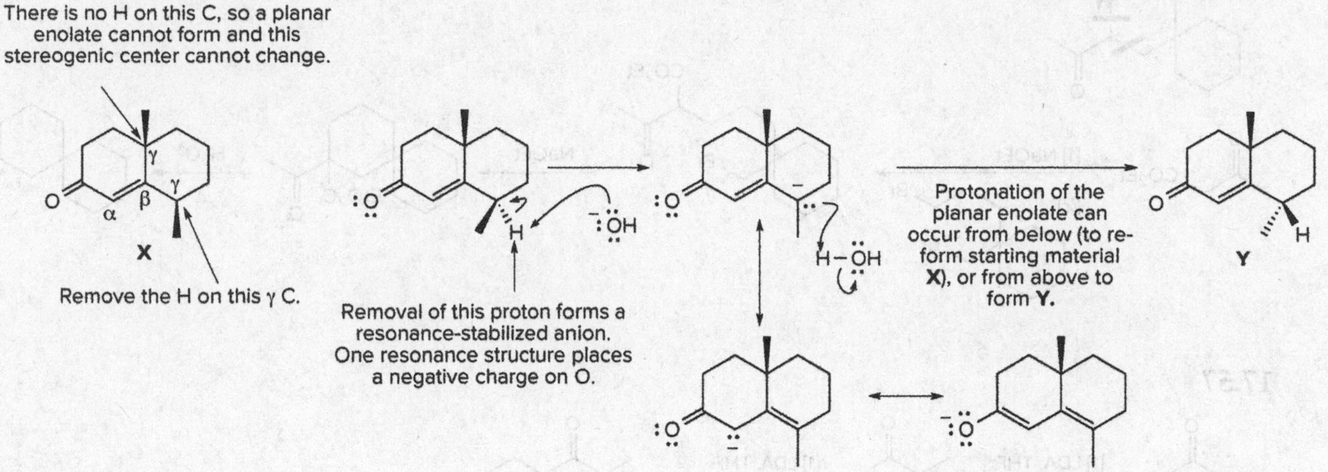

There is no H on this C, so a planar enolate cannot form and this stereogenic center cannot change.

X

Remove the H on this γ C.

Removal of this proton forms a resonance-stabilized anion. One resonance structure places a negative charge on O.

Protonation of the planar enolate can occur from below (to re-form starting material **X**), or from above to form **Y**.

Y

17.55

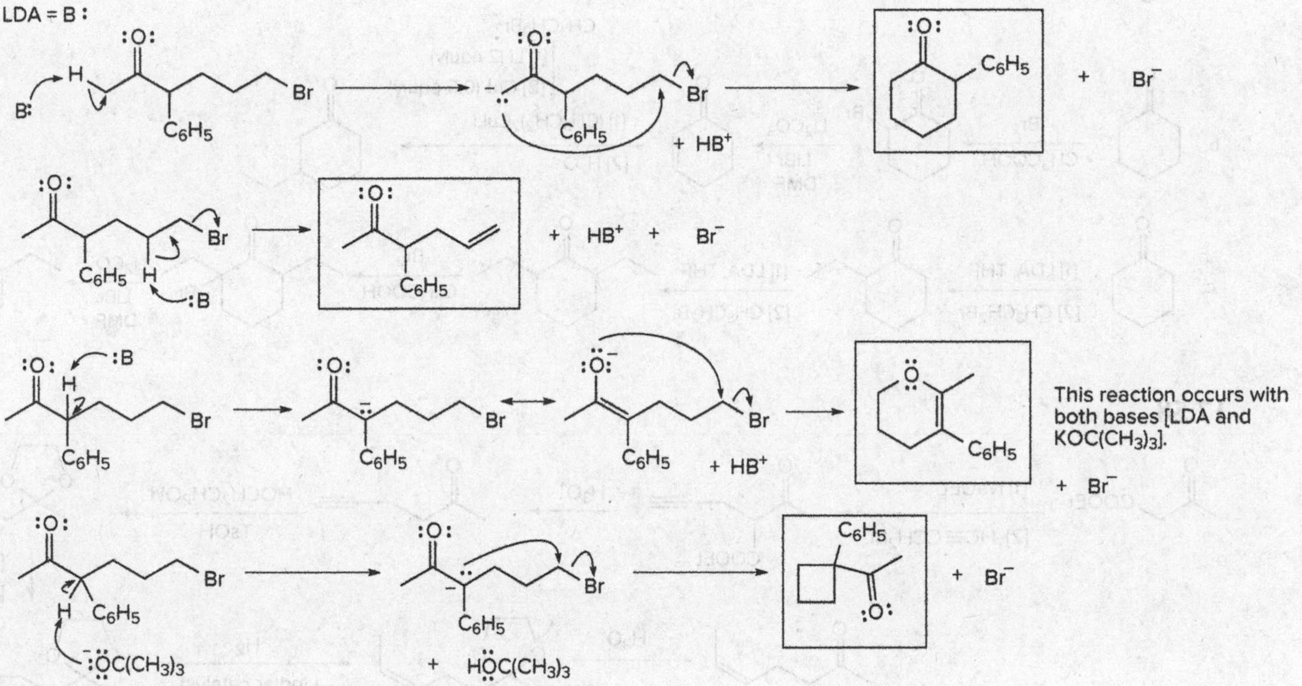

LDA = B:

C_6H_5 + HB^+ → C_6H_5 + Br^-

+ HB^+ + Br^-

This reaction occurs with both bases [LDA and $KOC(CH_3)_3$].

+ Br^-

+ $HOC(CH_3)_3$ ... + Br^-

17.56

a. Because there are two C's bonded to the α carbon, there are two possible intramolecular alkylation reactions.

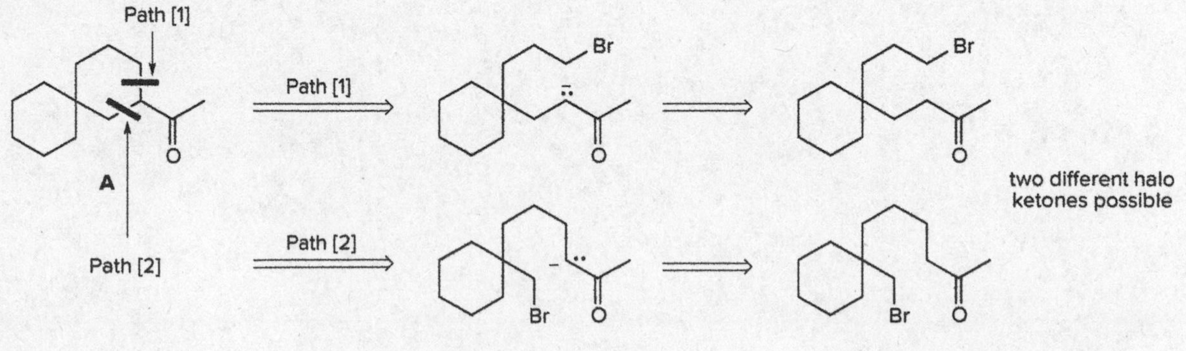

Path [1]

Path [2]

A

two different halo ketones possible

b.

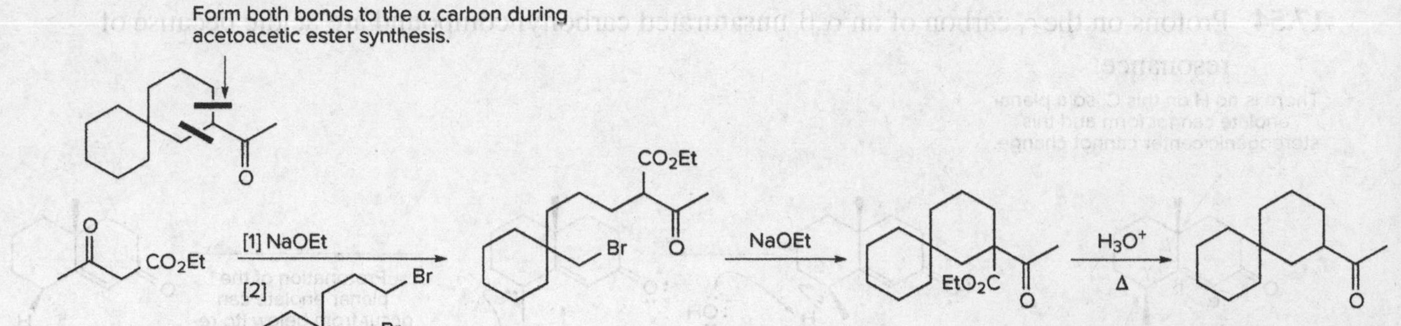

Form both bonds to the α carbon during acetoacetic ester synthesis.

[1] NaOEt
[2] Br, Br
NaOEt
CO_2Et
EtO_2C
H_3O^+
Δ

17.57

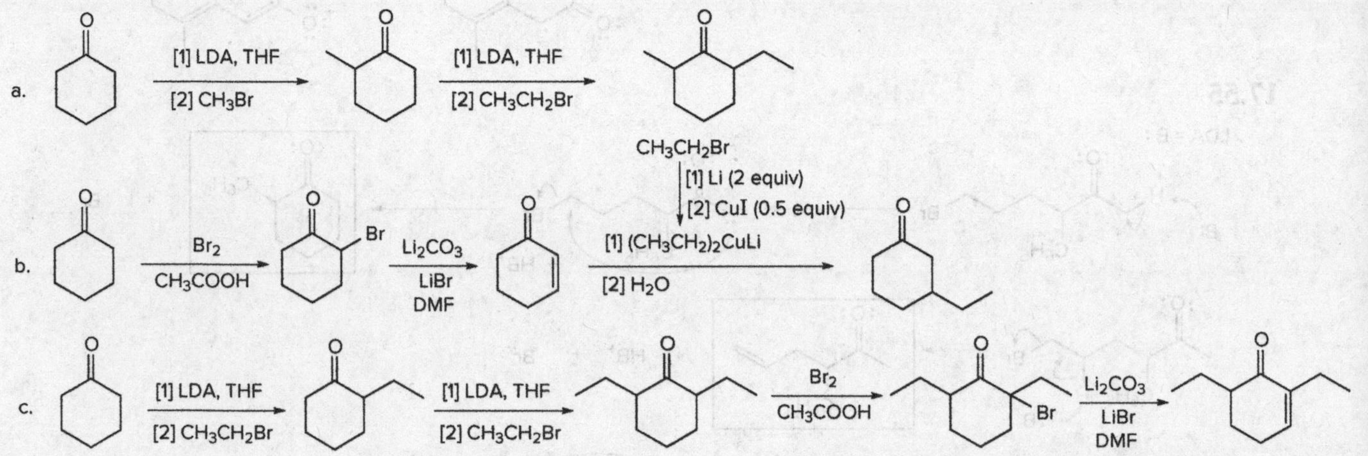

a. [1] LDA, THF [2] CH_3Br → [1] LDA, THF [2] CH_3CH_2Br

b. Br_2, CH_3COOH → Li_2CO_3, LiBr, DMF → [1] $(CH_3CH_2)_2CuLi$ [2] H_2O

CH_3CH_2Br [1] Li (2 equiv) [2] CuI (0.5 equiv)

c. [1] LDA, THF [2] CH_3CH_2Br → [1] LDA, THF [2] CH_3CH_2Br → Br_2, CH_3COOH → Li_2CO_3, LiBr, DMF

17.58

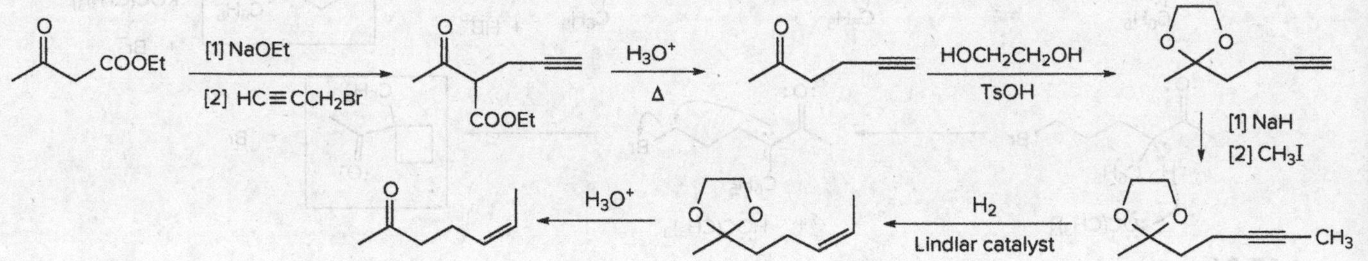

COOEt → [1] NaOEt [2] $HC\equiv CCH_2Br$ → COOEt → H_3O^+, Δ → $HOCH_2CH_2OH$, TsOH → [1] NaH [2] CH_3I → CH_3 → H_2, Lindlar catalyst → H_3O^+

17.59

a. W → [1] CH_3Li [2] CH_3I → Y

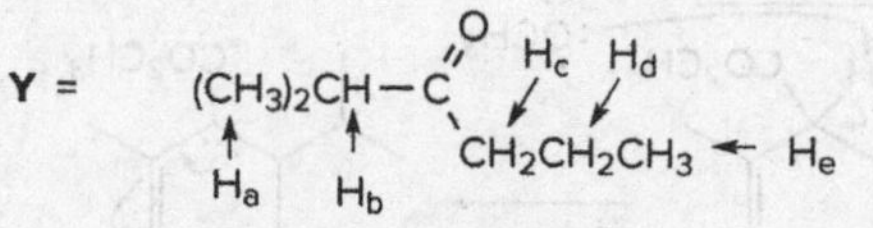

Y = $(CH_3)_2CH-C(=O)-CH_2CH_2CH_3$ (labels: H_a, H_b, H_c, H_d, H_e)

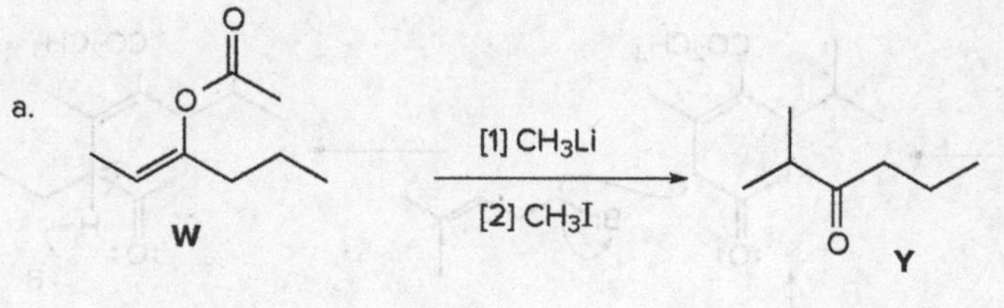

$C_7H_{14}O$ → one degree of unsaturation

IR peak at 1713 cm^{-1} → C=O

^1H NMR signals at (ppm)

H_e: triplet at 0.7 (3 H)

H_a: doublet at 0.9 (6 H)

H_d: sextet at 1.3 (2 H)

H_c: triplet at 1.9 (2 H)

H_b: septet at 2.1 (1 H)

b. CH_3-Li → … + Y + I^- (CH_3-I)

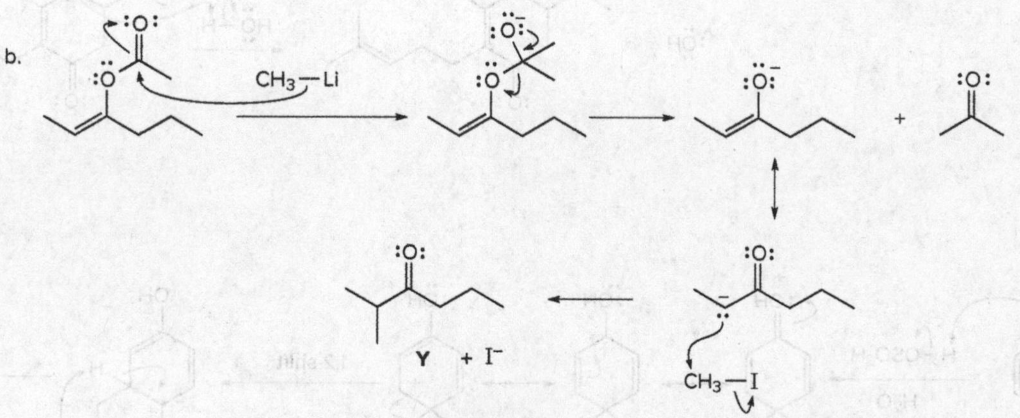

17.60 Removal of H_a with base does not generate an anion that can delocalize onto the carbonyl O atom, whereas removal of H_b generates an enolate that is delocalized on O.

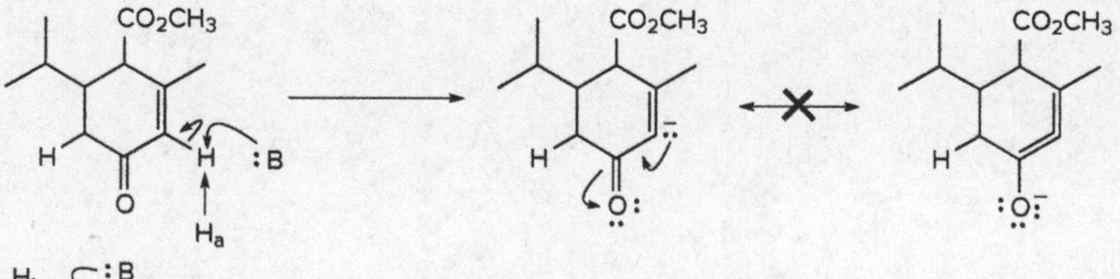

CO_2CH_3 … H … :B … H_a

Delocalization of this sort can't occur by removal of H_a, making H_a less acidic.

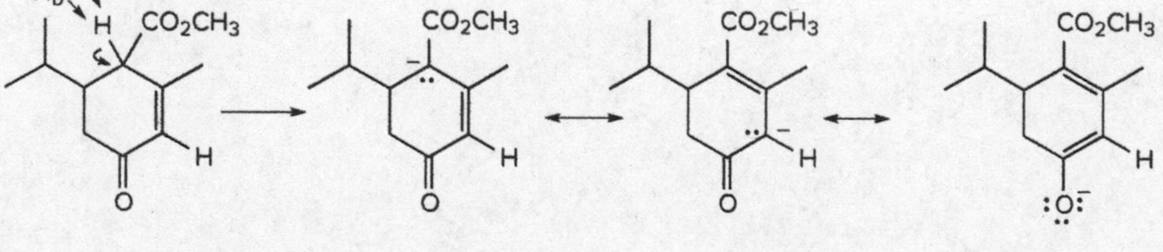

H_b … :B … CO_2CH_3

Removal of H_b gives an anion that is resonance stabilized, so H_b is more acidic.

Mechanism:

CO2CH3 :OCH3 H :O: CO2CH3 :O: + CH3OH CO2CH3 :O: Br CO2CH3 :O: H :B + Br⁻ CO2CH3 :O:⁻ CO2CH3 :O: + HB⁺ CO2CH3 :O: HO–H CO2CH3 :O: ⁻:OH +

17.61

:O: H–OSO3H H2O :OH + HSO4 :OH :OH 1,2-shift OH H HSO4⁻ OH

17.62

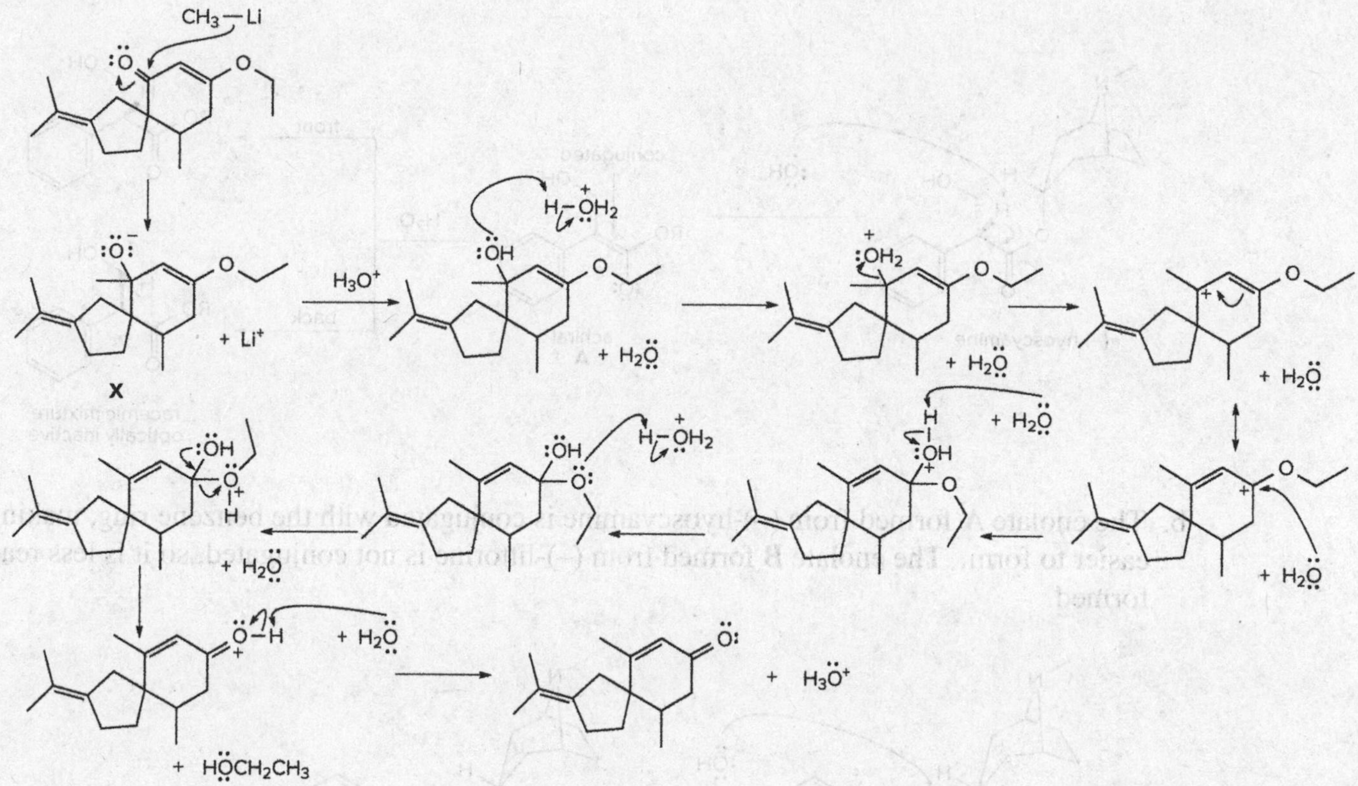

CH_3—Li

X

+ Li^+

H_3O^+

+ $H_2\ddot{O}$

+ $H_3\ddot{O}^+$

+ $H\ddot{O}CH_2CH_3$

17.63

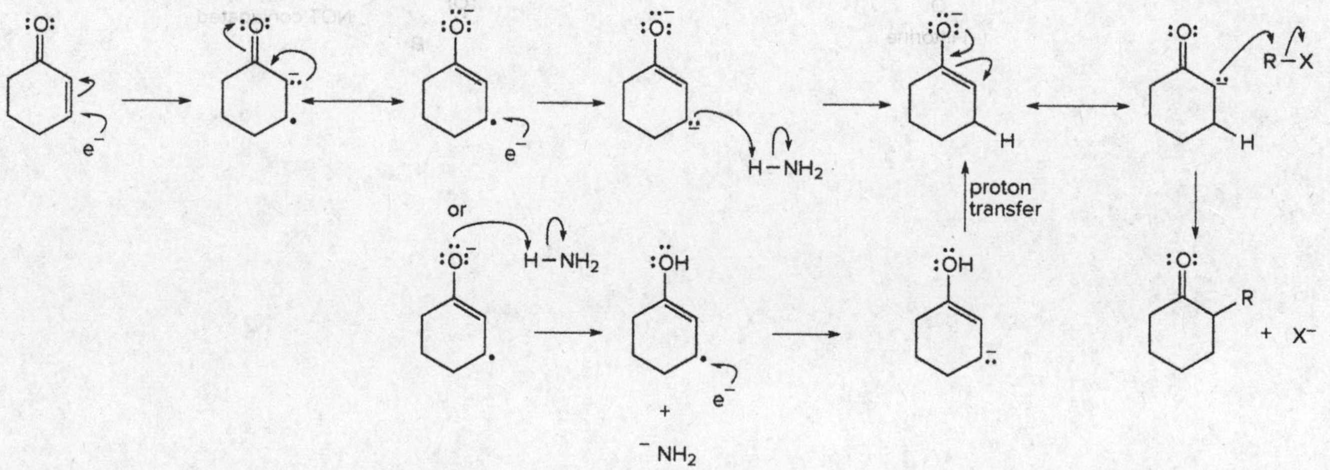

proton transfer

or

+ $^-NH_2$

+ X^-

17.64 a. In the presence of base an achiral enolate that can be protonated from both sides is formed.

(–)-hyoscyamine

conjugated

achiral
A

H_2O

front

back

racemic mixture
optically inactive

b. The enolate **A** formed from (–)-hyoscyamine is conjugated with the benzene ring, making it easier to form. The enolate **B** formed from (–)-littorine is not conjugated, so it is less readily formed.

(–)-littorine

NOT conjugated

B

Chapter 18 Carbonyl Condensation Reactions

Chapter Review

The four major carbonyl condensation reactions

Reaction type	Reaction
[1] Aldol reaction (18.1)	2 aldehyde (or ketone) ⇌ (^-OH, H_2O) β-hydroxy carbonyl compound → (^-OH or H_3O^+) (*E* and *Z*) α,β-unsaturated carbonyl compound
[2] Claisen reaction (18.5)	2 ester → ([1] NaOR' [2] H_3O^+) β-keto ester
[3] Michael reaction (18.9)	α,β-unsaturated carbonyl compound + carbonyl compound → ($^-OR'$ or ^-OH; H_2O) 1,5-dicarbonyl compound
[4] Robinson annulation (18.10)	α,β-unsaturated carbonyl compound + carbonyl compound → (^-OH, H_2O) cyclohex-2-enone

Useful variations

[1] Retro-aldol reaction (18.1B)

β-hydroxy carbonyl compound ⇌ (^-OH, H_2O) 2 aldehyde

[2] Directed aldol reaction (18.3)

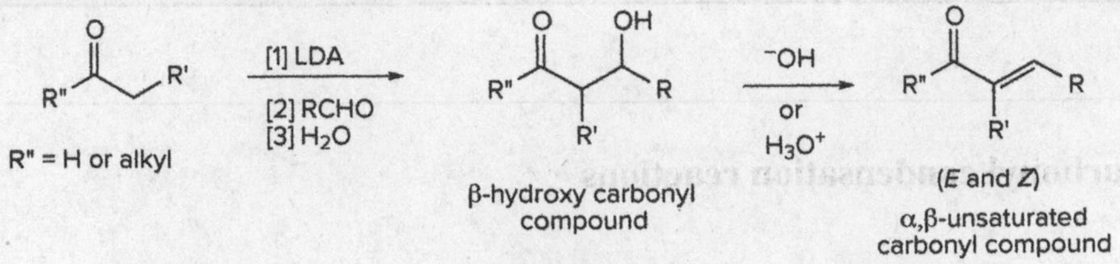

R" = H or alkyl

[1] LDA
[2] RCHO
[3] H_2O

β-hydroxy carbonyl compound

^-OH or H_3O^+

(*E* and *Z*)
α,β-unsaturated carbonyl compound

[3] Intramolecular aldol reaction (18.4)

a. With 1,4-dicarbonyl compounds:

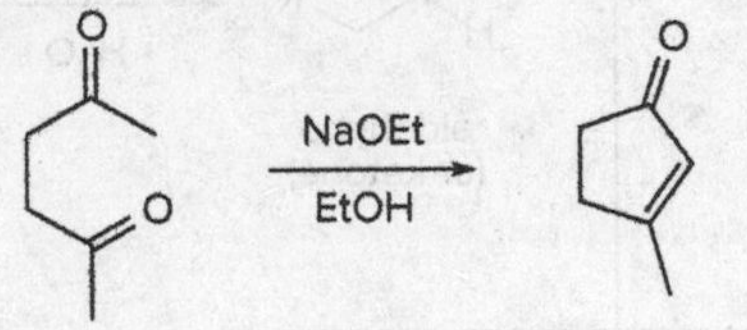

NaOEt
EtOH

b. With 1,5-dicarbonyl compounds:

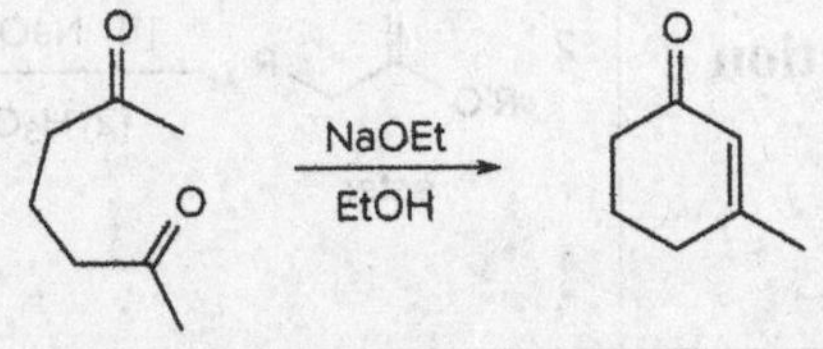

NaOEt
EtOH

[4] Dieckmann reaction (18.7)

a. With 1,6-diesters:

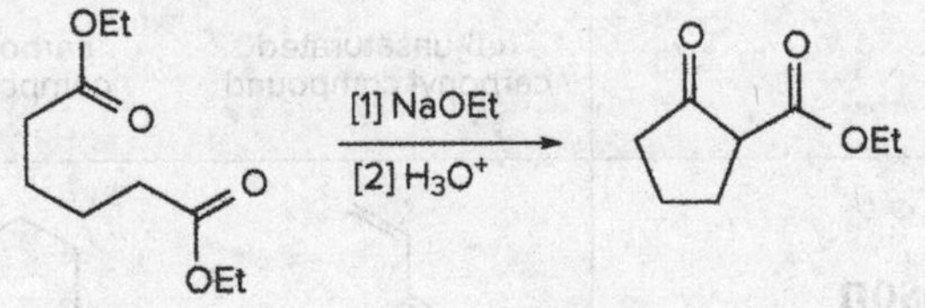

OEt
[1] NaOEt
[2] H_3O^+
OEt

b. With 1,7-diesters:

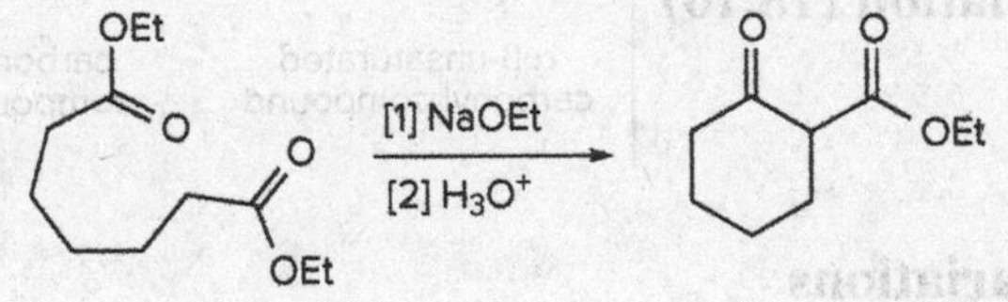

OEt
[1] NaOEt
[2] H_3O^+
OEt

Practice Test on Chapter Review

1.a. Which compounds are possible Michael acceptors?

1.

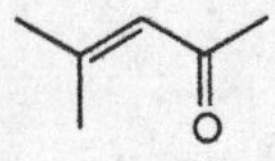

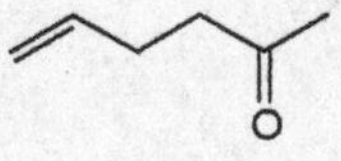

2. 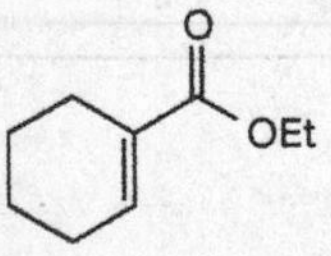 OEt

3.

4. Both compounds (1) and (2) are Michael acceptors.

5. Compounds (1), (2), and (3) are all Michael acceptors.

b. Which of the following compounds can be formed by an aldol reaction?

1. HO CHO
2. OH
3. HO

4. Both (1) and (2) can be formed by aldol reactions.
5. Compounds (1), (2), and (3) can all be formed by aldol reactions.

c. Which compounds can be formed in a Robinson annulation?

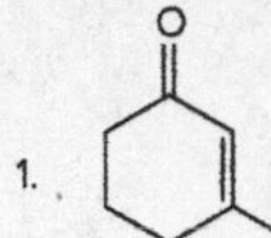

1.
2. CO_2Et
3.

4. Compounds (1) and (2) can be formed by Robinson annulations.
5. Compounds (1), (2), and (3) can all be formed by Robinson annulations.

d. What compounds can be used to form **A** by a condensation reaction?

A

1. and CO_2Et
2. OEt
3. and CHO

4. Compounds (1) and (2) can be used to form **A.**
5. Compounds (1), (2), and (3) can all be used to form **A.**

2. Give the reagents required for each step.

(a) H (b) HO (c) EtO (d) (e)

3. Draw the organic products formed in the following reactions.

a. [1] LDA [2] CH_3CH_2CHO [3] H_2O

b. + HCO_2Et [1] NaOEt, EtOH [2] H_3O^+

c. + NaOEt EtOH

d. CHO + ^-OH, H_2O

4.a. What organic starting materials are needed to synthesize **D** by a Robinson annulation reaction?

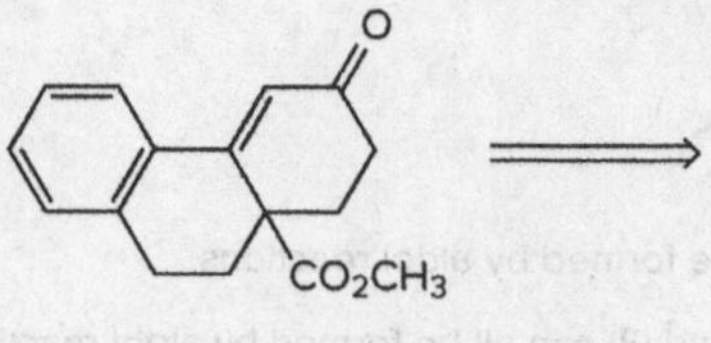

CO_2CH_3

D

b. What organic starting materials are needed to synthesize β-keto ester **B** by a Dieckmann reaction?

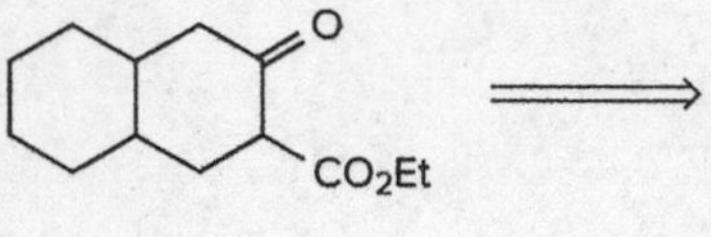

CO_2Et

B

c. What starting materials are needed to synthesize **A** by an aldol reaction?

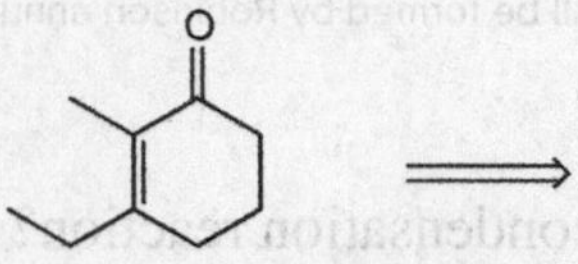

A

Answers to Practice Test

1.a. 4
b. 2
c. 1
d. 4

2.a. [1] O_3; [2] $(CH_3)_2S$
b. CrO_3, H_2SO_4, H_2O
c. CH_3CH_2OH, H_2SO_4
d. [1] NaOEt, EtOH; [2] H_3O^+
e. [1] LDA; [2] CH_3I

3.
a. 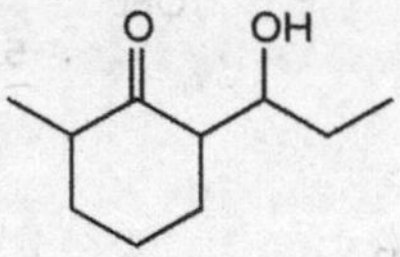 OH
b. 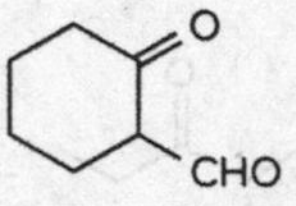 CHO
c. 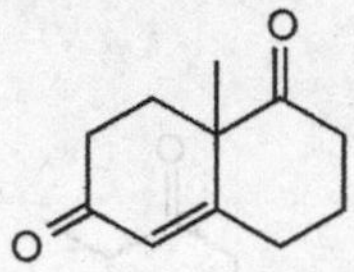
d. 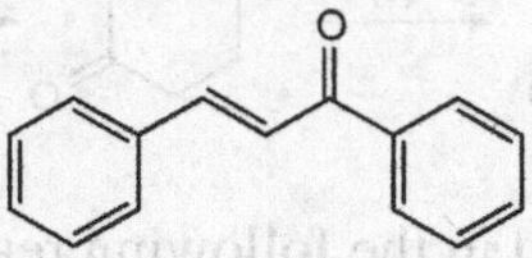 (*E* and *Z* isomers)

4.
a. 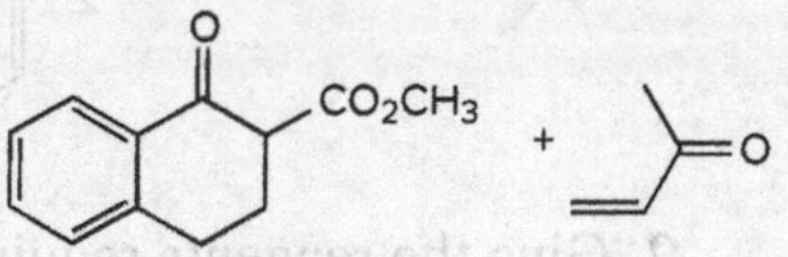 CO_2CH_3 +
b. 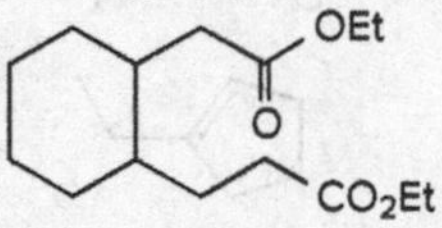 OEt, CO_2Et
c.

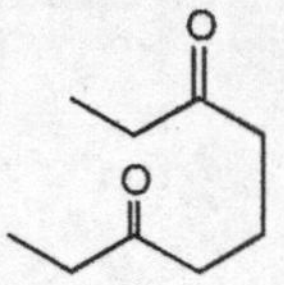

Answers to Problems

18.1

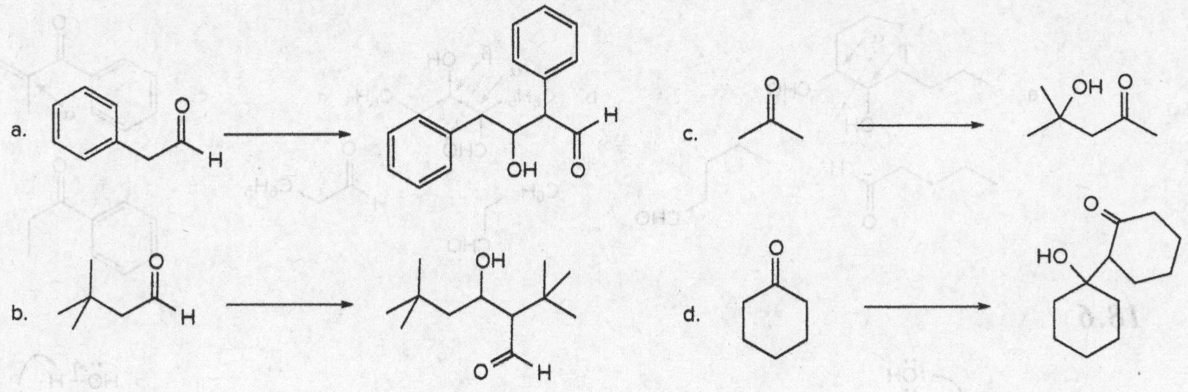

a. H

OH O H

c. OH O

b. O H

OH

O H

d. O

HO O

18.2

a. CHO

no α H
no aldol reaction

b. O

α H

yes

c. O H

no α H
no aldol reaction

d. CHO α H

yes

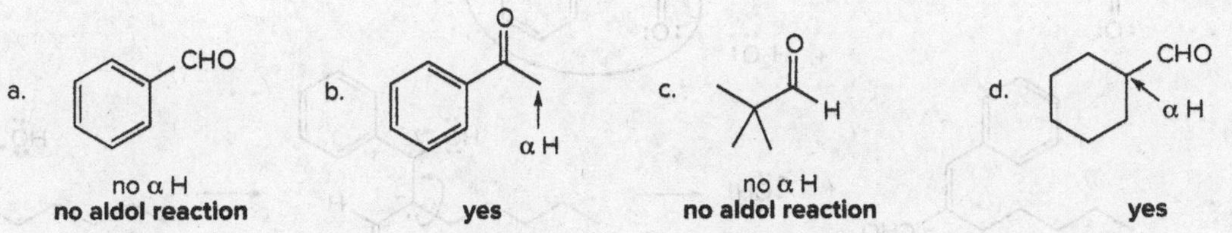

18.3 To draw the product of a retro-aldol reaction, break the bond between the α carbon and the β carbon that has the OH group. This forms two carbonyl compounds.

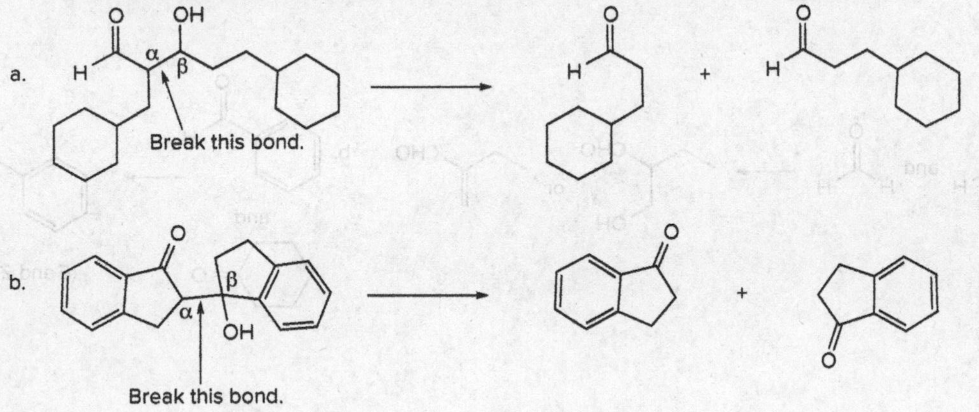

a. O OH H α β Break this bond. → O H + O H

b. O β α OH Break this bond. → O + O

18.4

a. HO O —base→ O

c. O OH —base→ O

b. HO CHO —base→ CHO

(*E* and *Z* isomers)

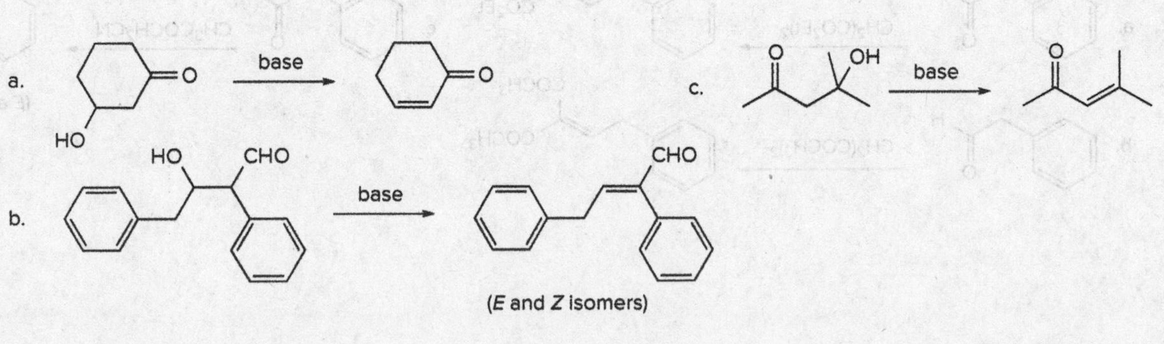

18.5 Locate the α and β C's to the carbonyl group, and break the molecule into two halves at this bond. The α C and all of the atoms bonded to it belong to one carbonyl component. The β C and all of the atoms bonded to it belong to the other carbonyl component.

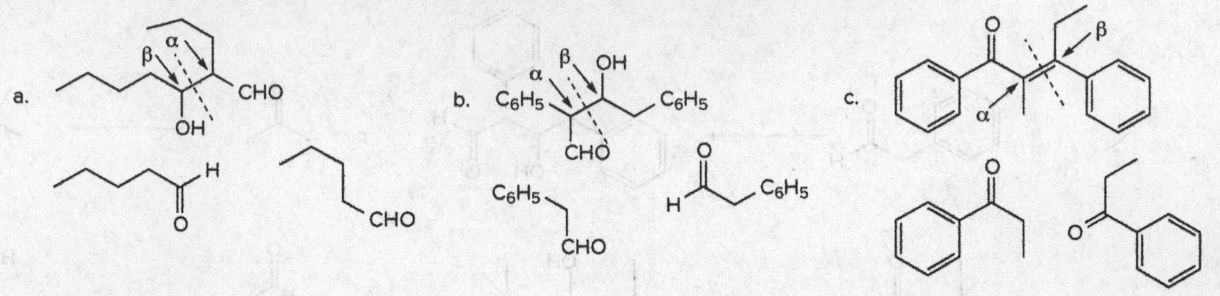

a. β α CHO OH

H O CHO

b. α β OH C_6H_5 C_6H_5 CHO

C_6H_5 CHO H O C_6H_5

c. O β α

O O

18.6

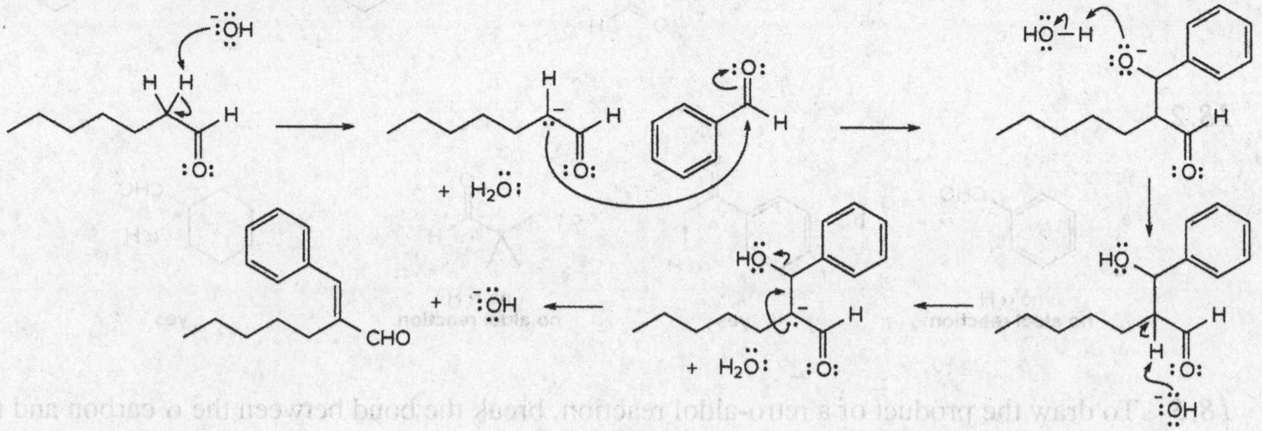

:ÖH H H H :O: → H – H :O: + $H_2\ddot{O}$: :O: H → HÖ–H :Ö⁻ H :O:

HÖ H H :O: :ÖH ← HÖ – H + $H_2\ddot{O}$: :O: ← + :ÖH CHO

18.7

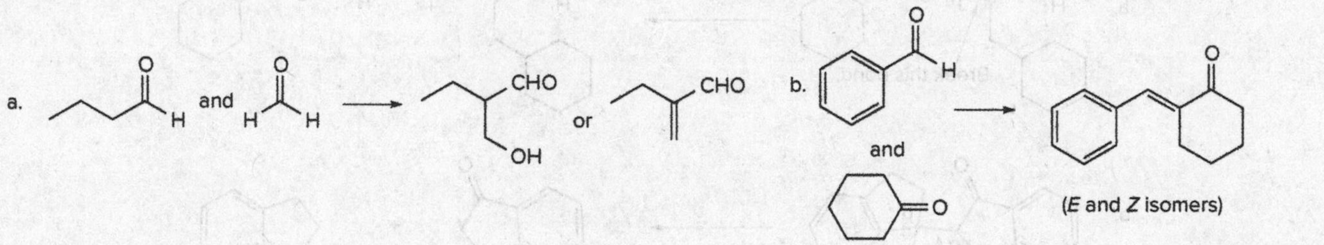

a. O H and O H H → CHO OH or CHO

b. O H and O → O (*E* and *Z* isomers)

18.8

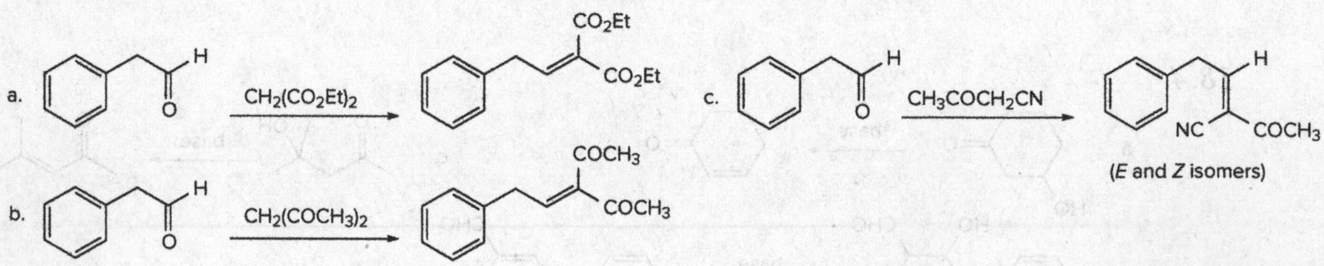

a. H O $CH_2(CO_2Et)_2$ → CO_2Et CO_2Et

b. H O $CH_2(COCH_3)_2$ → $COCH_3$ $COCH_3$

c. H O CH_3COCH_2CN → H NC $COCH_3$ (*E* and *Z* isomers)

18.9 Form the enolate of the β-dicarbonyl compound **A** and react this enolate with the aldehyde **B.**

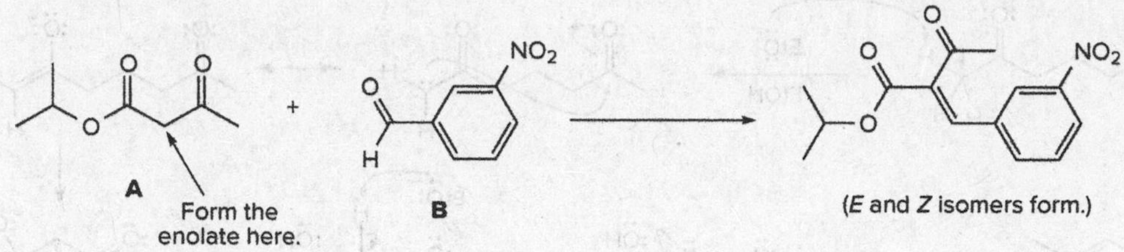

A
Form the enolate here.
+
NO_2
H
B
NO_2
(*E* and *Z* isomers form.)

18.10

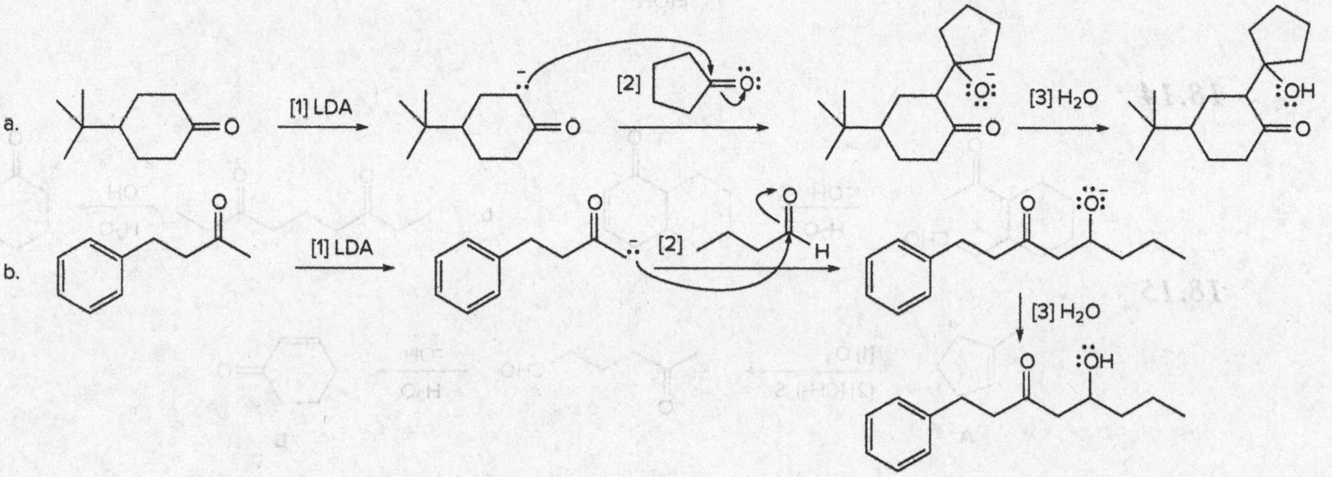

a. [1] LDA [2] [3] H_2O :OH

b. [1] LDA [2] H [3] H_2O :OH

18.11 Find the α and β C's to the carbonyl group and break the bond between them.

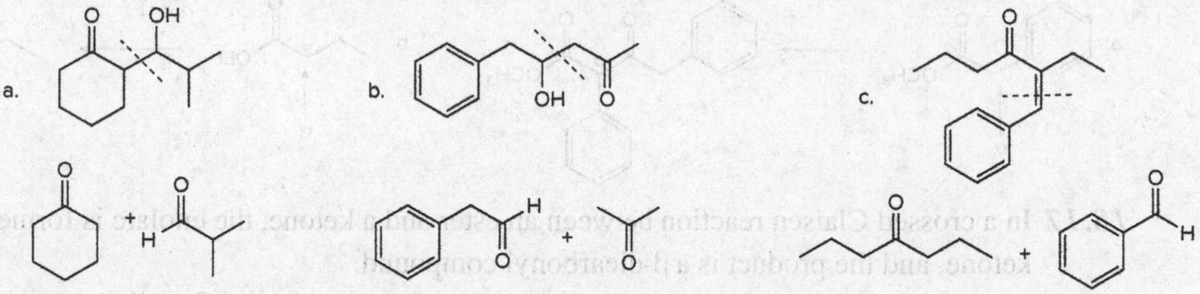

a. OH
b. OH
c.

+ H
+ H
+ H

18.12

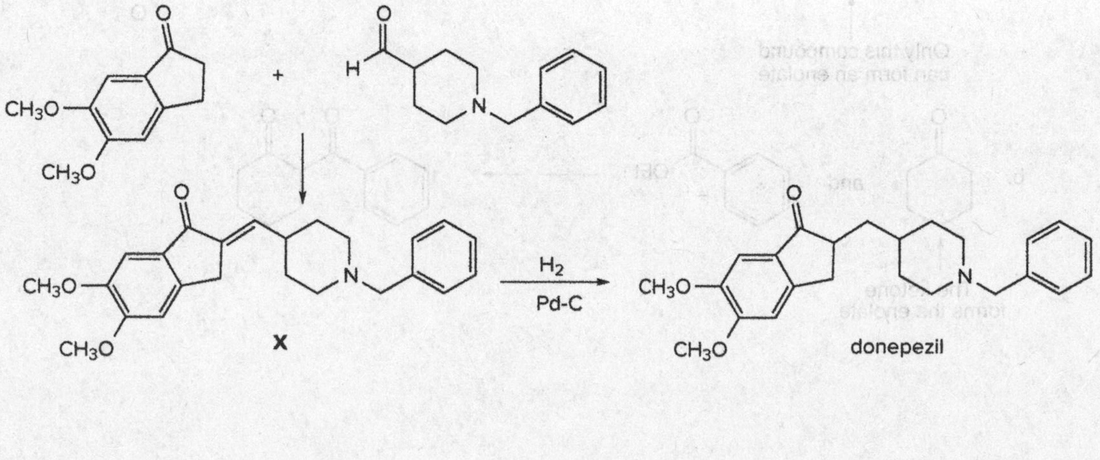

CH_3O
CH_3O
+ H N
CH_3O
CH_3O
N
X
H_2
Pd-C
CH_3O
CH_3O
N
donepezil

18.13 All enolates have a second resonance structure with a negative charge on O.

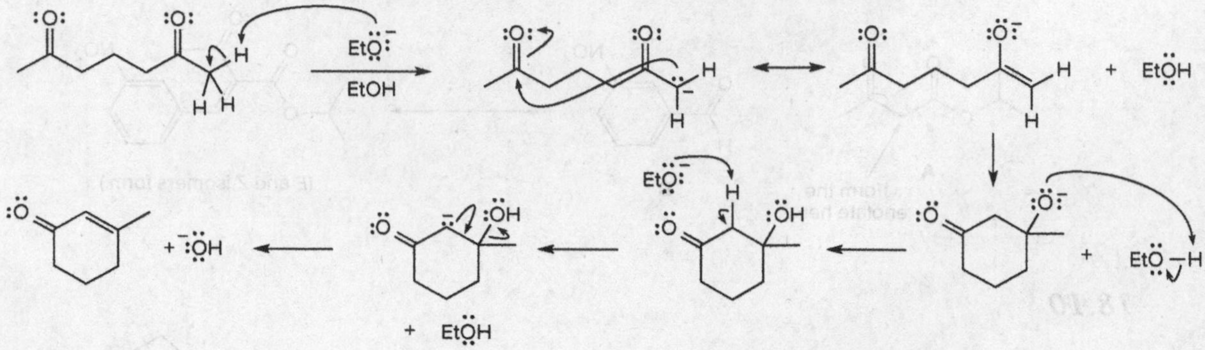

EtÖ:⁻ EtOH + EtÖH + EtÖ–H + ÖH + EtÖH

18.14

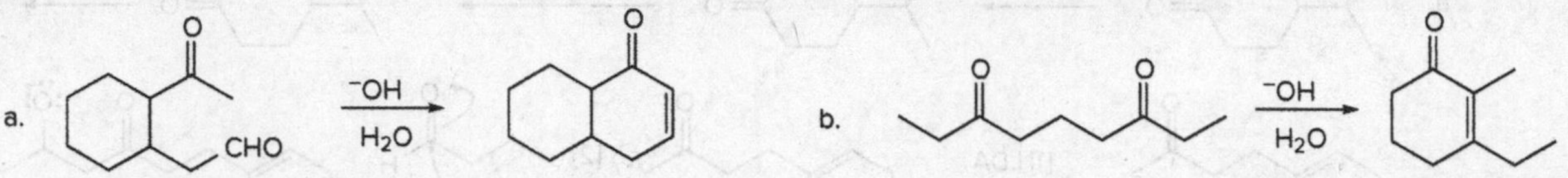

a. CHO ^{-}OH H_2O

b. ^{-}OH H_2O

18.15

A [1] O_3 [2] $(CH_3)_2S$ CHO ^{-}OH H_2O B

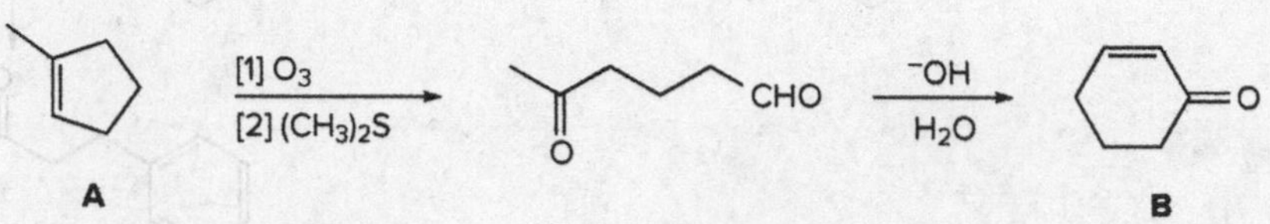

18.16 Join the α C of one ester to the carbonyl C of the other ester to form the β-keto ester.

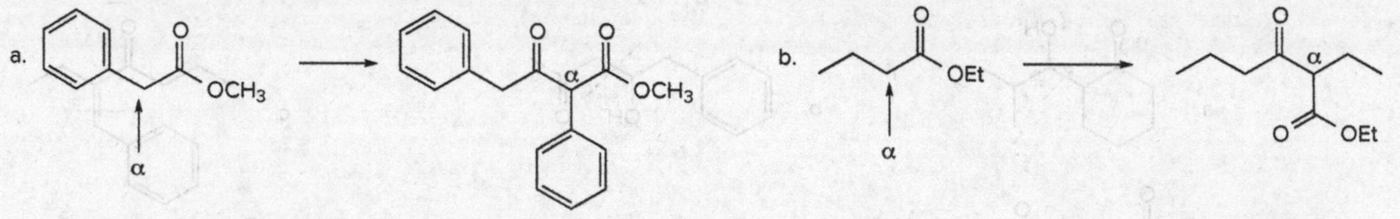

a. OCH_3 α OCH_3 α

b. OEt α α OEt

18.17 In a crossed Claisen reaction between an ester and a ketone, the enolate is formed from the ketone, and the product is a β-dicarbonyl compound.

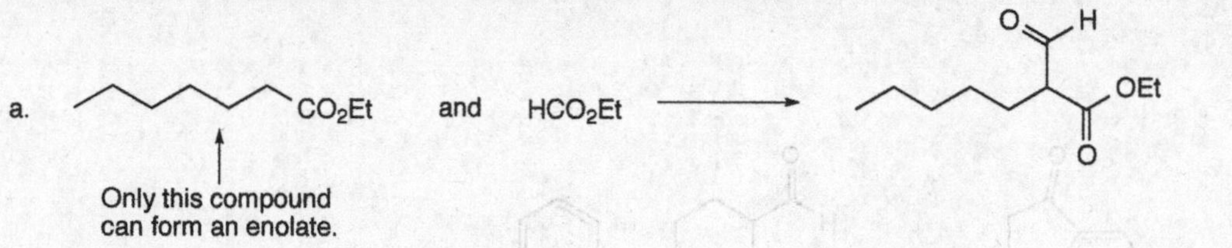

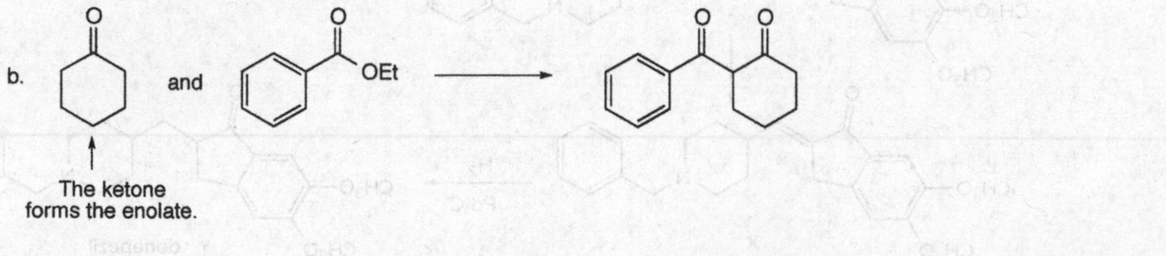

a. CO_2Et and HCO_2Et → OEt

Only this compound can form an enolate.

b. and OEt →

The ketone forms the enolate.

18.18 A β-dicarbonyl compound like avobenzone is prepared by a crossed Claisen reaction between a ketone and an ester.

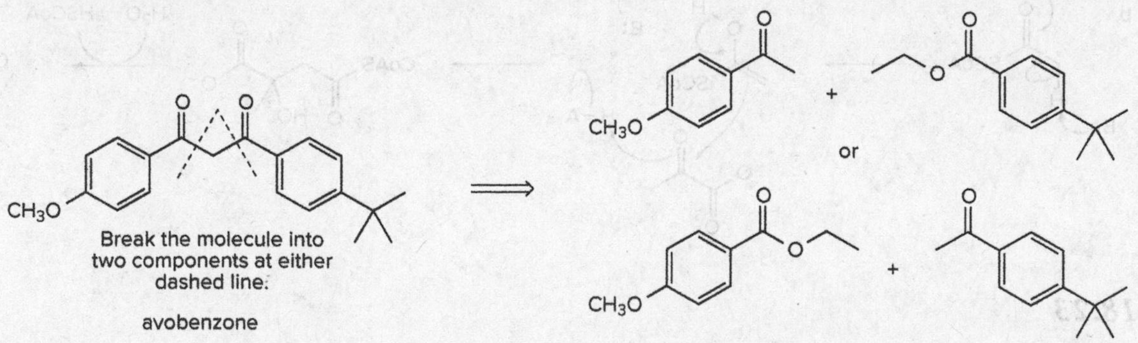

18.19

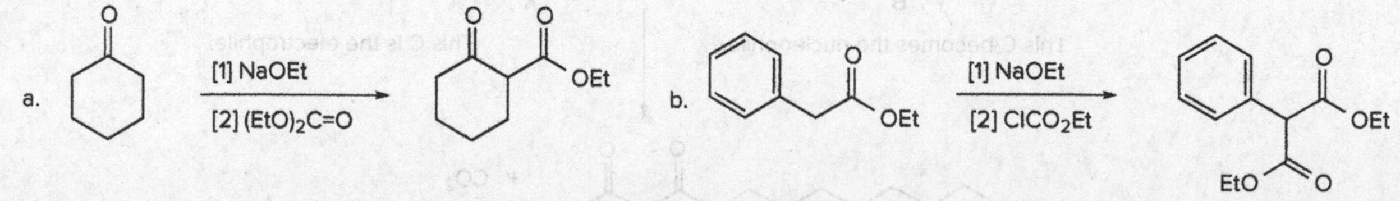

18.20

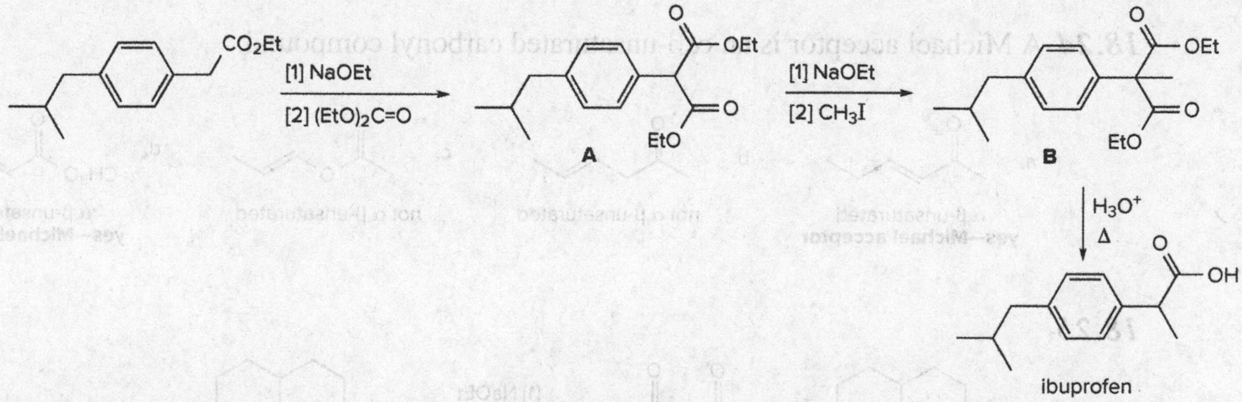

18.21

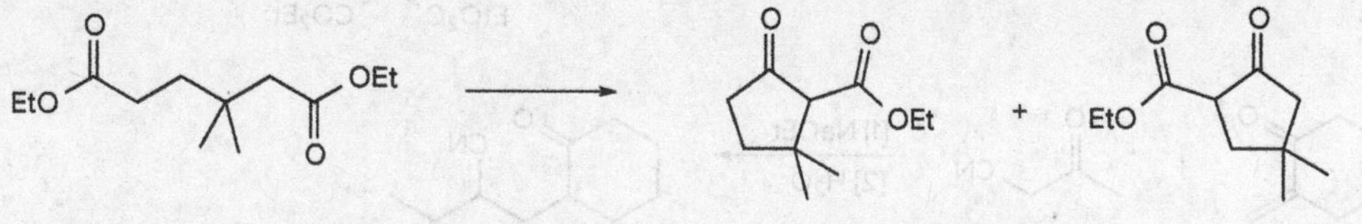

18.22

a. In a biological aldol, the α carbon of one component adds to the carbonyl carbon of the second component. The carbons that are joined are labeled in the reactants and product.

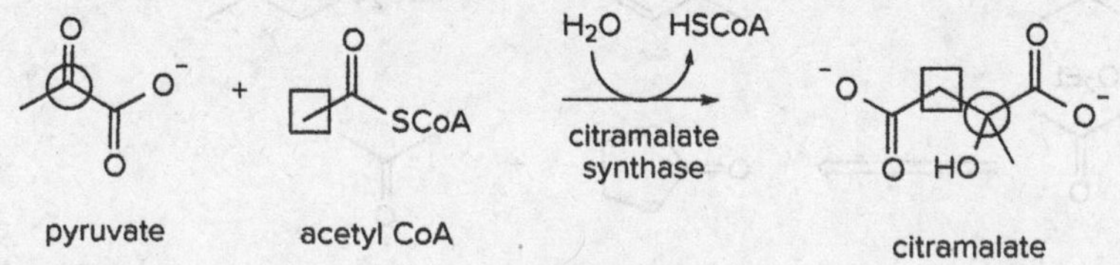

b.

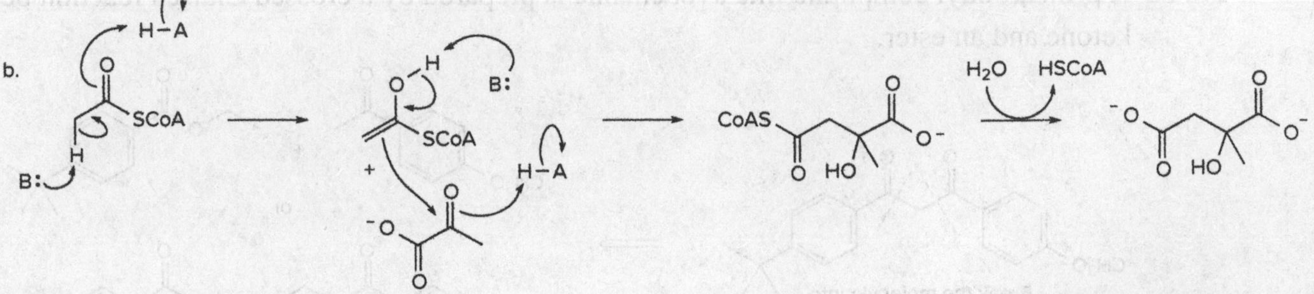

H–A

B:

H_2O HSCoA

SCoA

CoAS

HO

H–A

18.23

^-O_2C α SR' **B** + SR' **X**

This C becomes the nucleophile.

This C is the electrophile.

SR' + CO_2

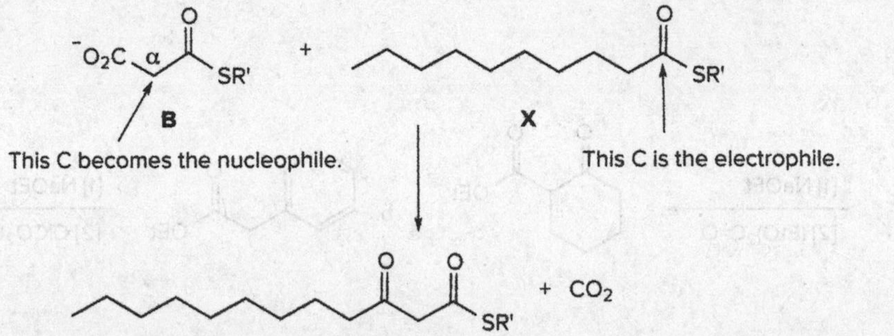

18.24 A Michael acceptor is an α,β-unsaturated carbonyl compound.

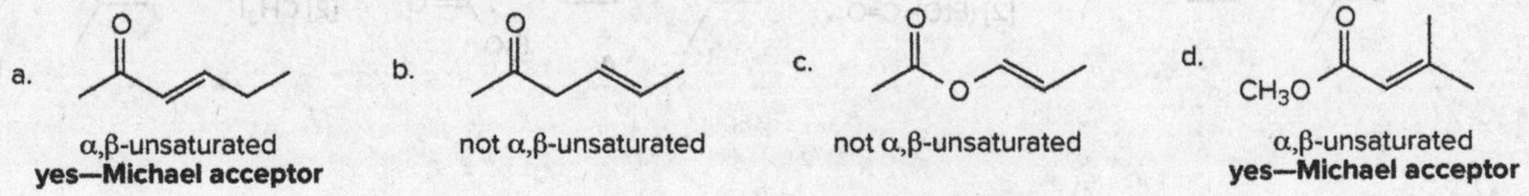

a. α,β-unsaturated **yes—Michael acceptor**

b. not α,β-unsaturated

c. not α,β-unsaturated

d. CH_3O α,β-unsaturated **yes—Michael acceptor**

18.25

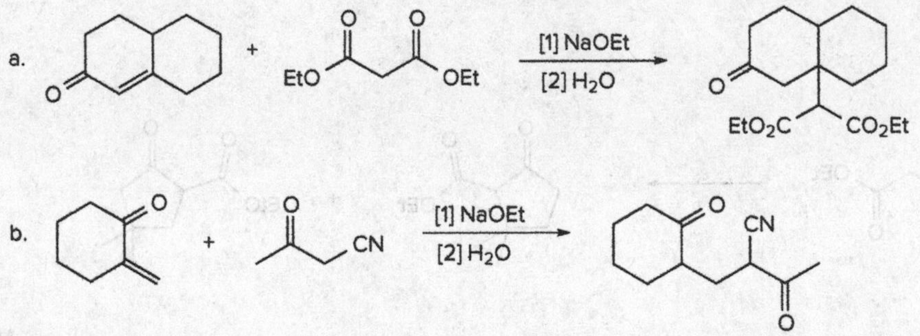

a. + EtO OEt [1] NaOEt [2] H_2O → EtO_2C CO_2Et

b. + CN [1] NaOEt [2] H_2O → CN

18.26

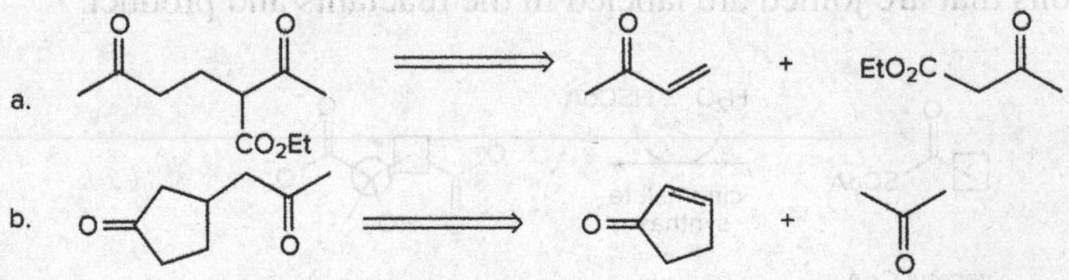

a. CO_2Et ⟹ + EtO_2C

b. ⟹ +

18.27 **The Robinson annulation forms a six-membered ring and three new carbon–carbon bonds: two σ bonds and one π bond.**

a. re-draw

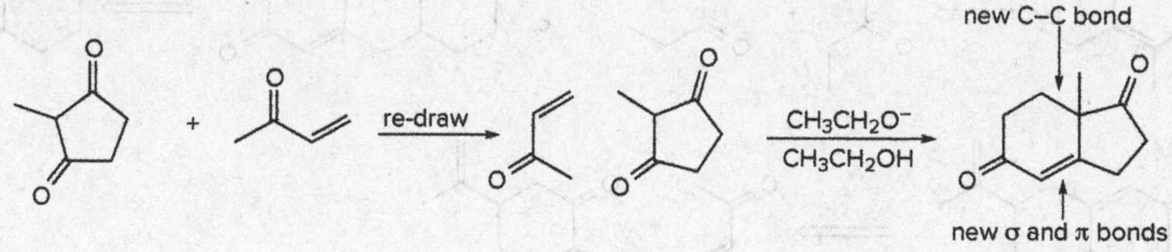

$CH_3CH_2O^-$
CH_3CH_2OH

new C–C bond

new σ and π bonds

b. COOEt + re-draw

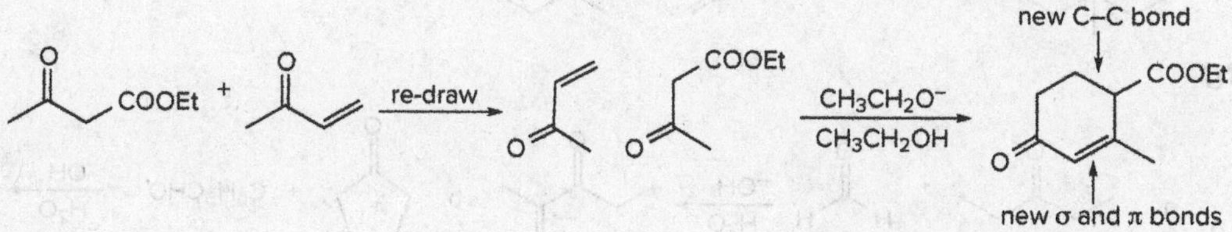

COOEt

$CH_3CH_2O^-$
CH_3CH_2OH

new C–C bond

COOEt

new σ and π bonds

c. + re-draw

$CH_3CH_2O^-$
CH_3CH_2OH

new C–C bond

new σ and π bonds

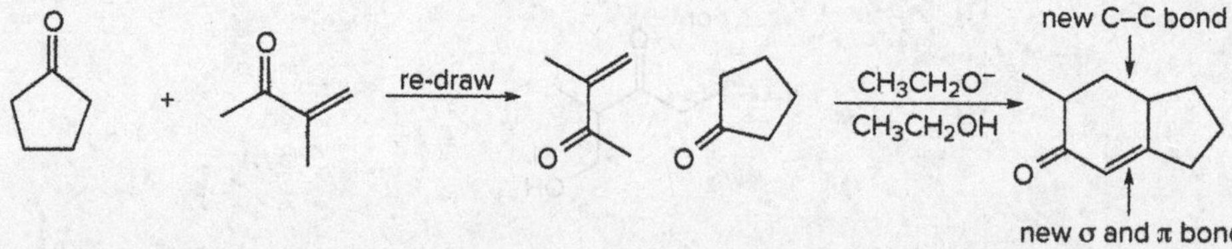

d. EtO_2C + re-draw

EtOOC

$CH_3CH_2O^-$
CH_3CH_2OH

new C–C bond

COOEt

new σ and π bonds

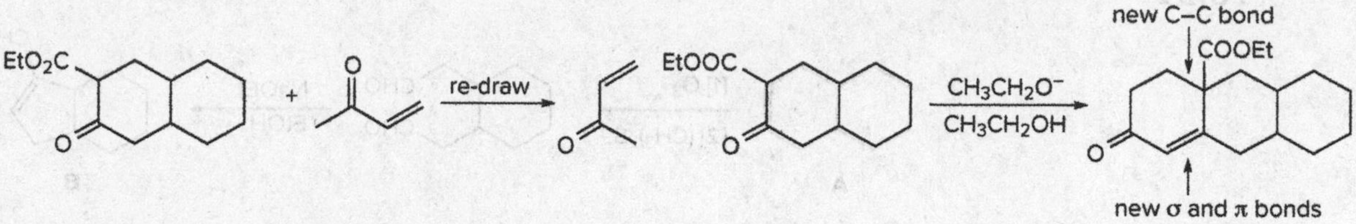

18.28 **A Robinson annulation forms a conjugated cyclohex-2-enone. In a bicyclic product, one carbon of the C=C must be shared by both rings.**

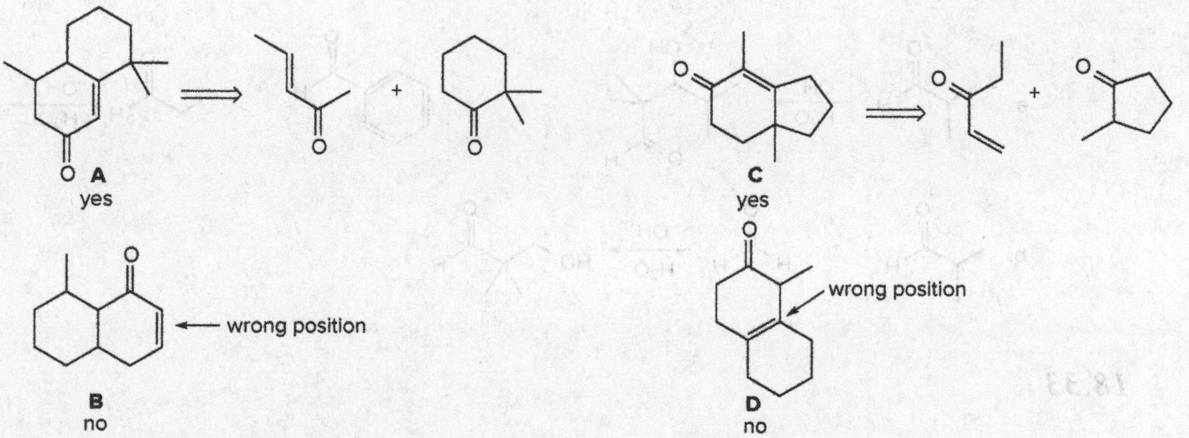

A
yes

C
yes

wrong position

B
no

wrong position

D
no

18.29

a. ⟹ +

c. ⟹ +

b. ⟹ +

18.30

a. + $\xrightarrow[H_2O]{^-OH}$ or

b. + C_6H_5CHO $\xrightarrow[H_2O]{^-OH}$ (*E* and *Z* isomers)

18.31

A $\xrightarrow[\text{[2] }(CH_3)_2S]{\text{[1] }O_3}$ CHO CHO $\xrightarrow[EtOH]{NaOEt}$ CHO B

18.32 The product of an aldol reaction is a β-hydroxy carbonyl compound or an α,β-unsaturated carbonyl compound. The α,β-unsaturated carbonyl compound is drawn as product unless elimination of H_2O cannot form a conjugated system.

a. $\xrightarrow[H_2O]{^-OH}$

c. + $\xrightarrow[H_2O]{^-OH}$ CHO (*E* and *Z* isomers)

b. + $\xrightarrow[H_2O]{^-OH}$ HO

18.33

a. [1] LDA [2] $CH_3CH_2CH_2CHO$ [3] H_2O →

b. OEt [1] LDA [2] CHO [3] H_2O → OH OEt

18.34

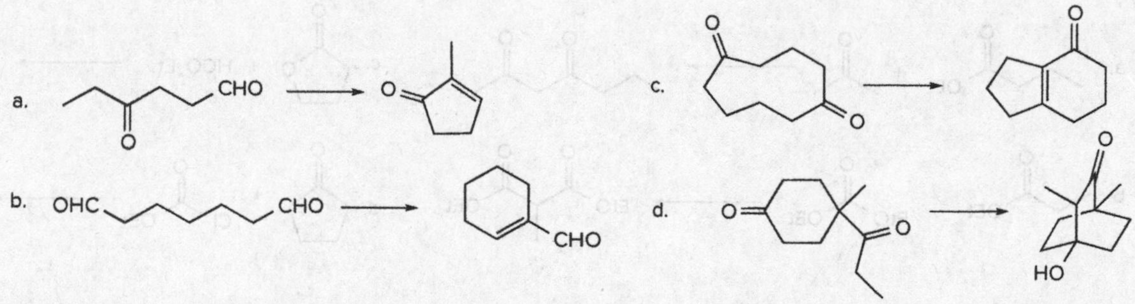

a. CHO
b. OHC CHO → CHO
c.
d. HO

18.35 Locate the α and β C's to the carbonyl group, and break the molecule into two halves at this bond. The α C and all of the atoms bonded to it belong to one carbonyl component. The β C and all the atoms bonded to it belong to the other carbonyl component.

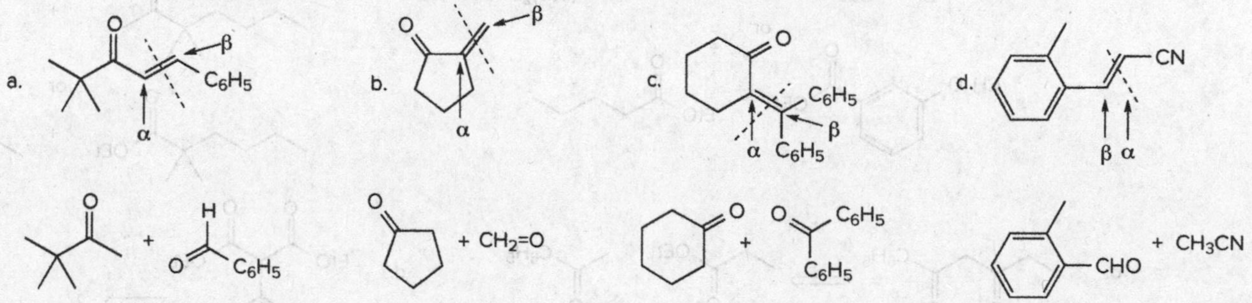

a. β, α, C_6H_5
b. β, α
c. C_6H_5, β, α, C_6H_5
d. CN, β α

a. + H C_6H_5
b. + CH_2=O
c. + C_6H_5 C_6H_5
d. CHO + CH_3CN

18.36

a. H
b. H
c. H
d. HO

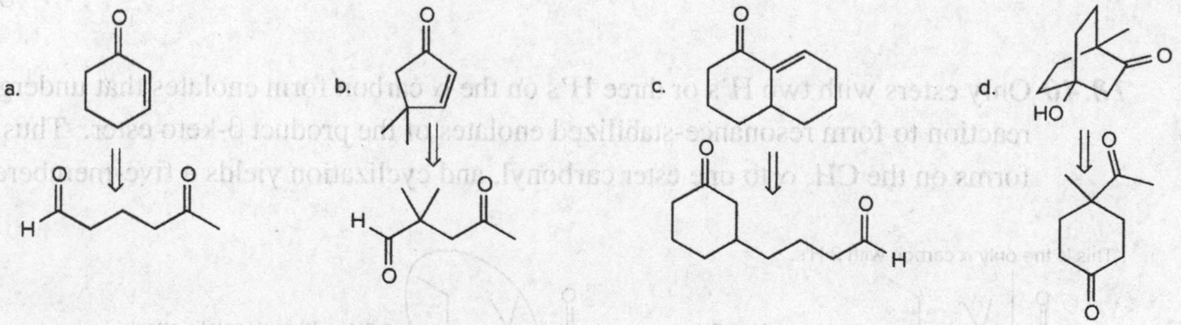

18.37 Ozonolysis cleaves the C=C, and base catalyzes an intramolecular aldol reaction.

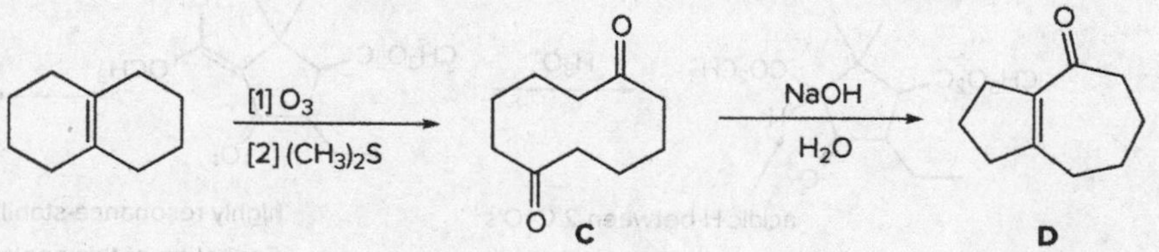

[1] O_3
[2] $(CH_3)_2S$
→ **C** → NaOH, H_2O → **D**

$C_{10}H_{14}O$

18.38

a. [ester] + [ketone] → [β-diketone]

b. [ester] + EtO–CO–OEt → EtO, OEt [β-keto diester]

c. [lactone] + HCO_2Et → H [formyl lactone]

d. [cyclopentanone] + Cl–CO–OEt → EtO [β-keto ester]

18.39

a. CH_3O [aryl β-keto ester] OEt ⟹ CH_3O [aryl ketone] + $(EtO)_2C{=}O$

or

CH_3O [aryl ester] OEt + EtO [ester]

b. [β-diketone] C_6H_5 ⟹ [ester] OEt + [ketone] C_6H_5

c. [aldehyde] CHO ⟹ [ketone] + H–CO–OEt

or

[ester] OEt + [aldehyde] H

d. EtO, OEt [phenyl malonate] ⟹ [phenylacetate] OEt + EtO–CO–OEt

18.40 Only esters with two H's or three H's on the α carbon form enolates that undergo Claisen reaction to form resonance-stabilized enolates of the product β-keto ester. Thus, the enolate forms on the CH_2 α to one ester carbonyl, and cyclization yields a five-membered ring.

This is the only α carbon with 2 H's.

CH_3O [B] OCH_3, CH_3O —NaOCH₃ / CH₃OH→ CH_3O [enolate] OCH_3, CH_3O —[1] nucleophilic attack [2] loss of CH_3O^- [3] deprotonation→ CH_3O_2C, CO_2CH_3 [cyclopentanone enolate] ↔ CH_3O_2C, CO_2CH_3, $:\ddot{O}:^-$ ↔ CH_3O_2C, OCH_3, $:\ddot{O}:$

B

highly resonance-stabilized enolate

Formation of this enolate drives the reaction to completion.

—H_3O^+→ CH_3O_2C, CO_2CH_3, H [β-keto ester]

acidic H between 2 C=O's

18.41 Join the α C of acetyl CoA to the ketone carbonyl of 2-oxoisovalerate.

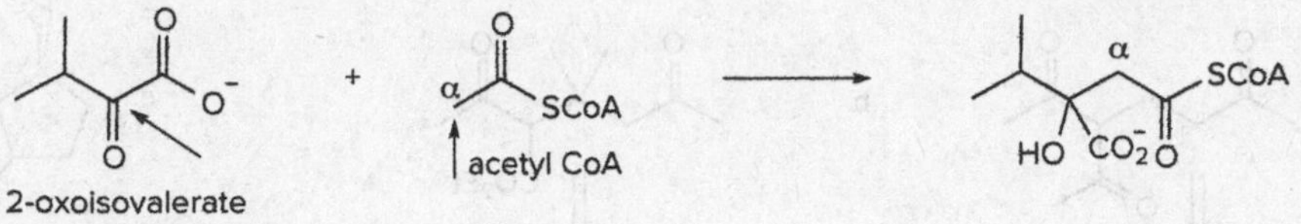

18.42 Cleave the bond between the α carbon to the ketone carbonyl and the β carbon that has the OH group.

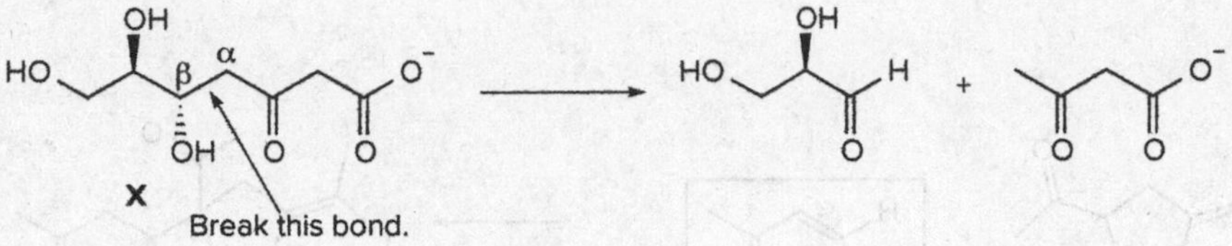

18.43 This bond cleavage is a retro-aldol reaction.

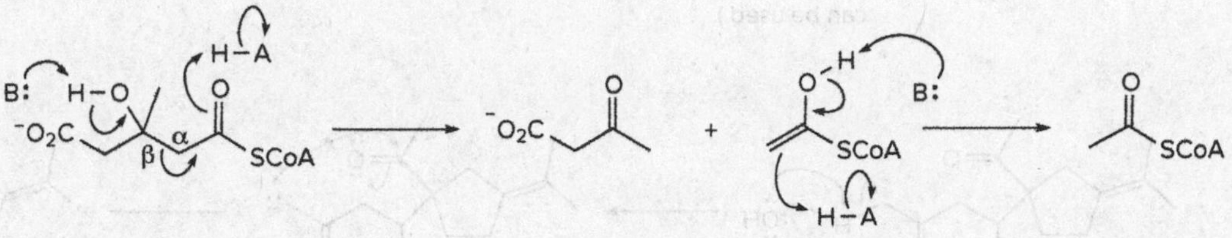

18.44 The mechanism follows the steps of a Claisen condensation.

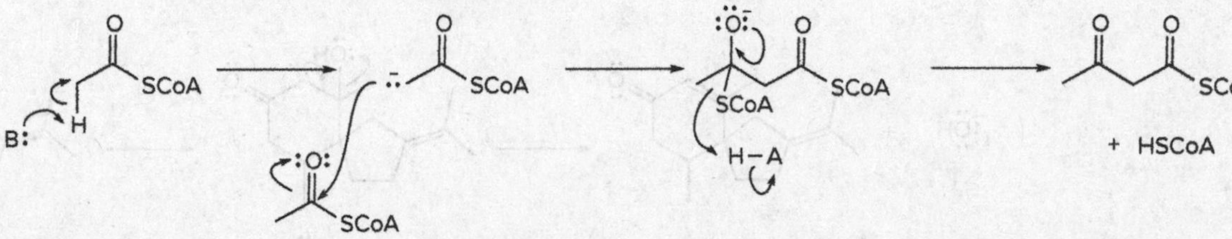

18.45 To draw the steps for a retro-Claisen, reverse the steps of the Claisen mechanism.

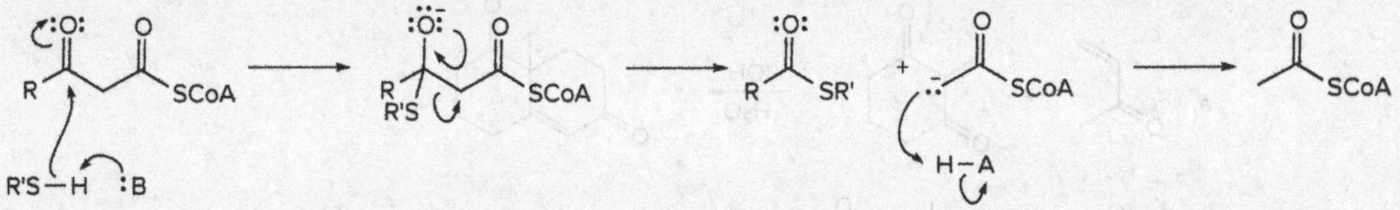

18.46

a.

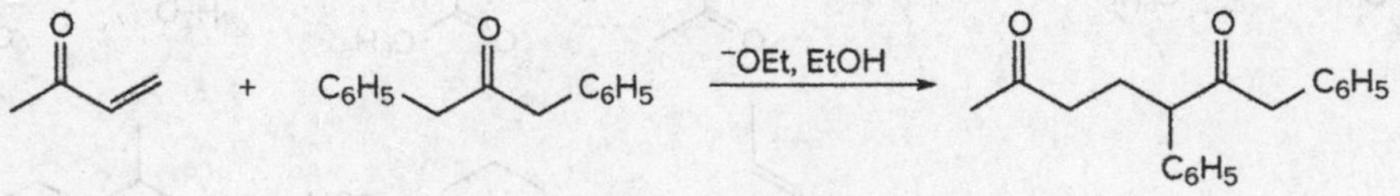

b.

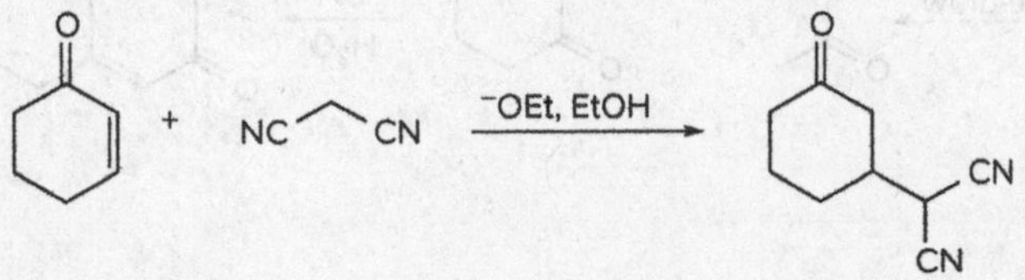

18.47

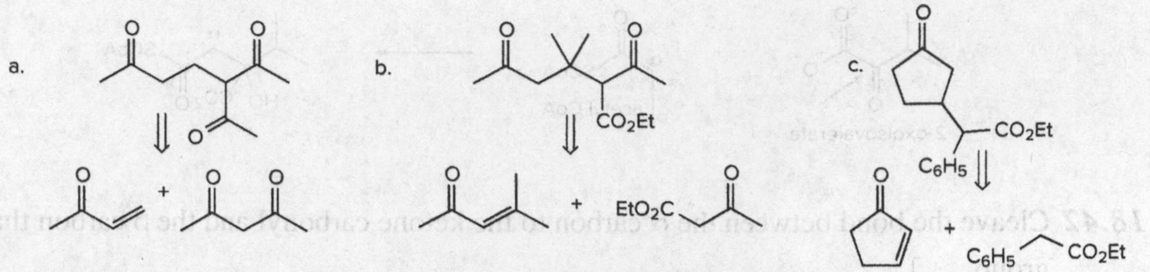

a. b. c.

EtO_2C CO_2Et C_6H_5 CO_2Et

18.48

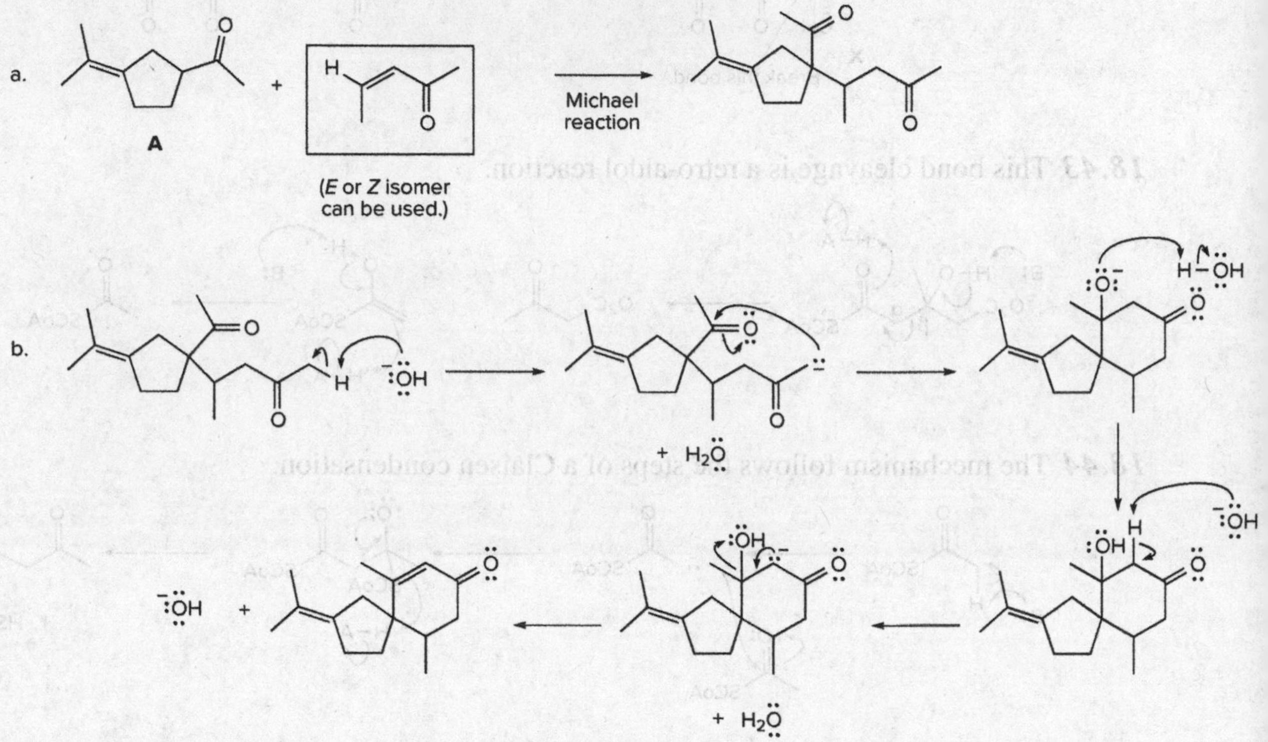

a. A + → Michael reaction

(*E* or *Z* isomer can be used.)

b. + $H_2\ddot{O}$

+ $H_2\ddot{O}$

18.49

a. + ^{-}OH / H_2O →

b. + C_6H_5 re-draw → + C_6H_5 ^{-}OH / H_2O → C_6H_5

c. + re-draw → + ^{-}OH / H_2O →

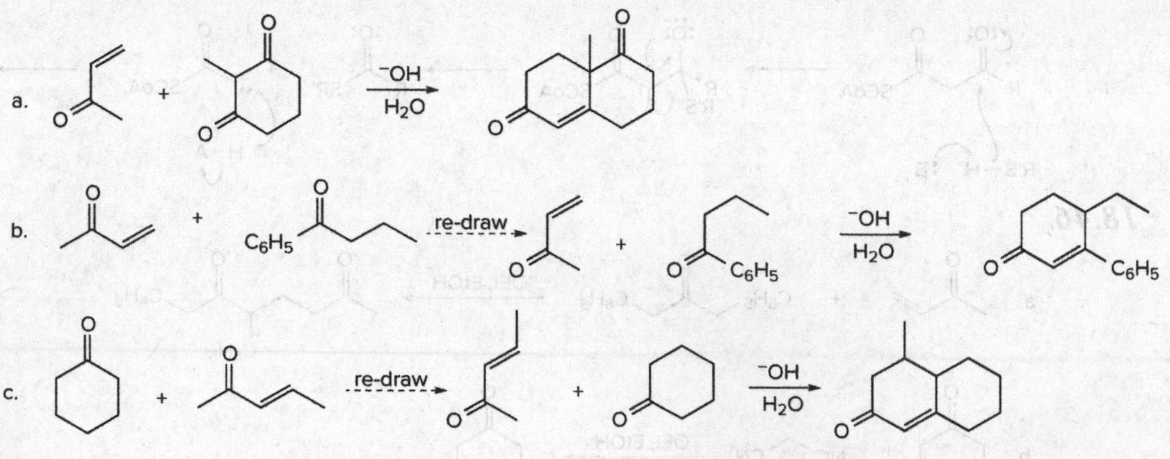

d. 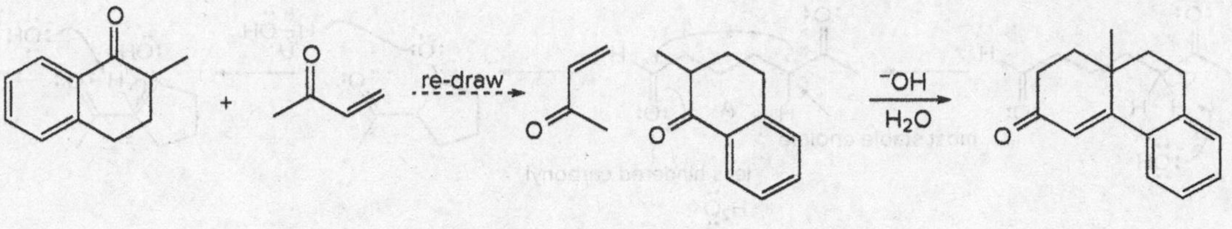 + re-draw ^{-}OH H_2O

18.50

a. 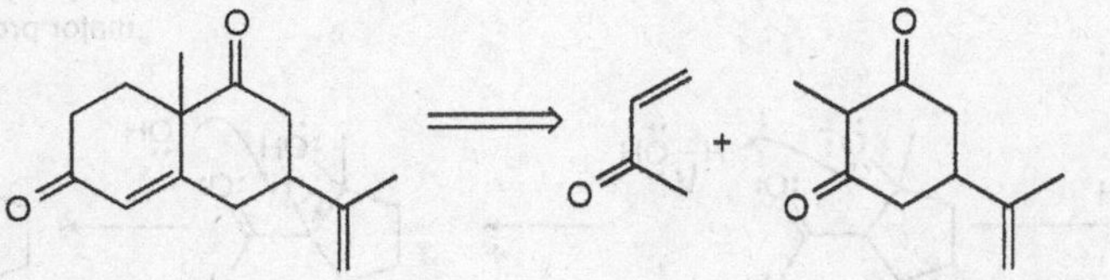 ⟹ +

c. 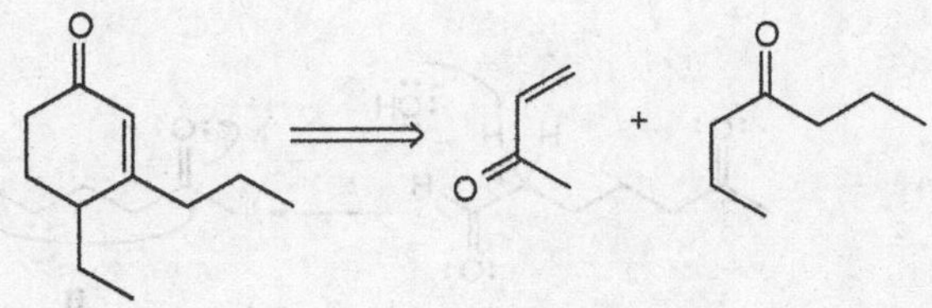 ⟹ +

b. 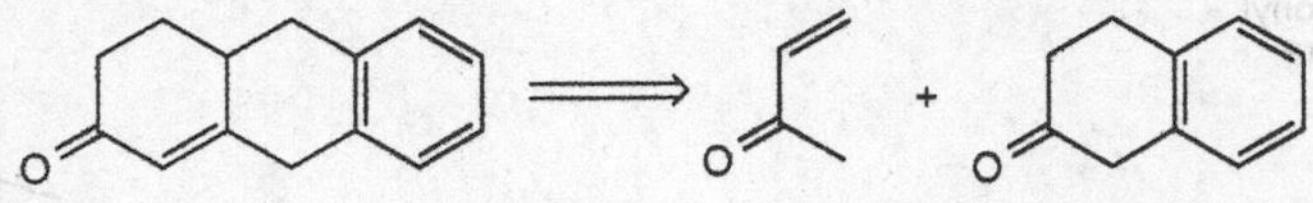 ⟹ +

18.51

a. 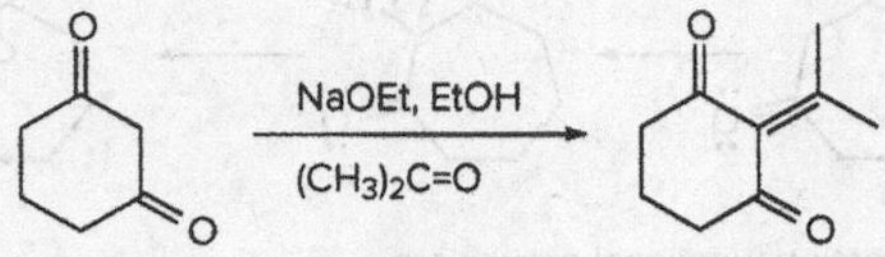 NaOEt, EtOH $(CH_3)_2C{=}O$

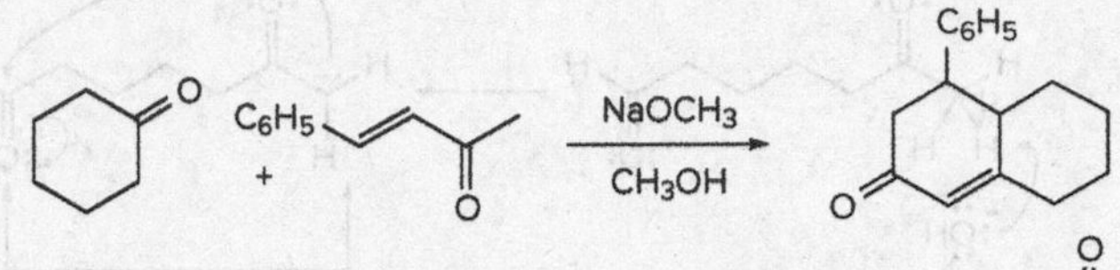

b. 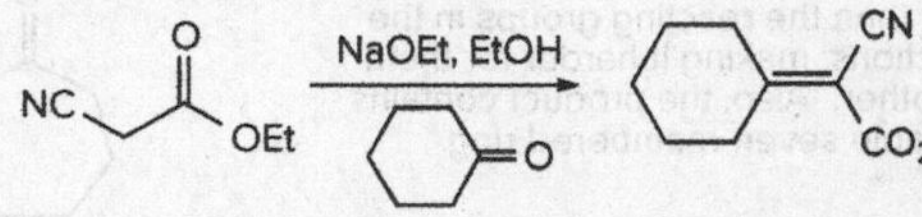 NC OEt NaOEt, EtOH CN CO_2Et

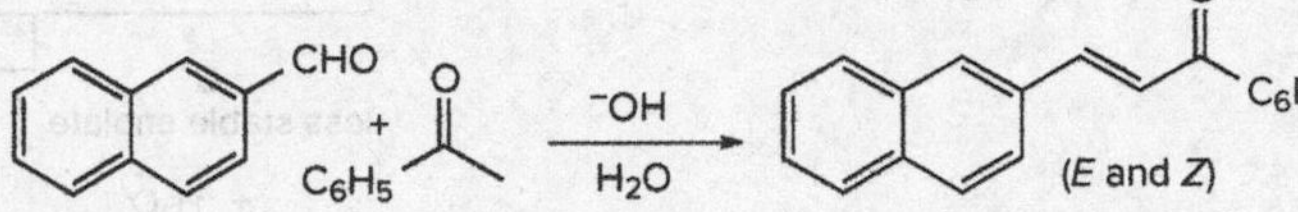

c. 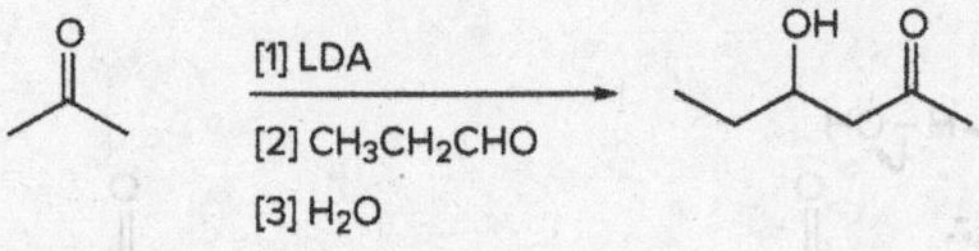 [1] LDA [2] CH_3CH_2CHO [3] H_2O OH

d. C_6H_5 + NaOCH₃ CH_3OH C_6H_5

e. CHO + C_6H_5 ^{-}OH H_2O C_6H_5 (*E* and *Z*)

f. 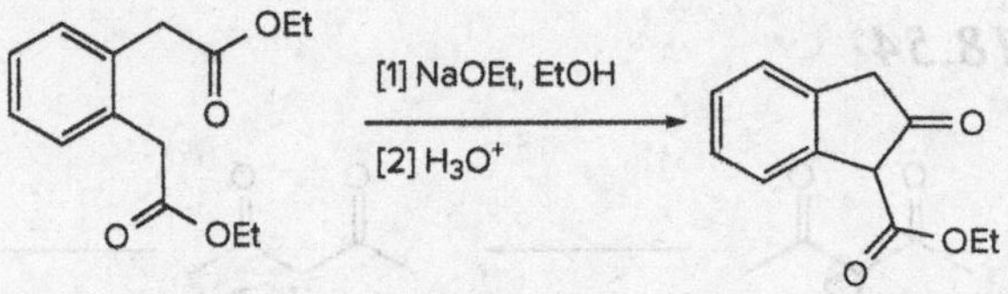 OEt OEt [1] NaOEt, EtOH [2] H_3O^+ OEt

18.52

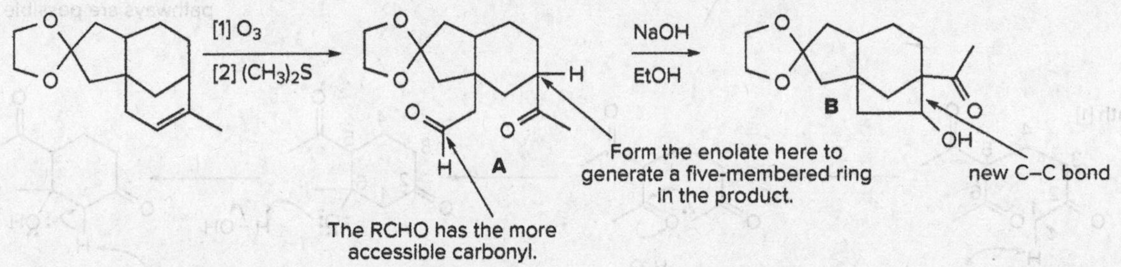

[1] O_3 [2] $(CH_3)_2S$ H A NaOH EtOH B OH

Form the enolate here to generate a five-membered ring in the product.

The RCHO has the more accessible carbonyl.

new C–C bond

18.53 Enolate **A** is more substituted (and more stable) than either of the other two possible enolates and attacks an aldehyde carbonyl group, which is sterically less hindered than a ketone carbonyl. The resulting ring size (five-membered) is also quite stable. That is why 1-acetylcyclopentene is the major product.

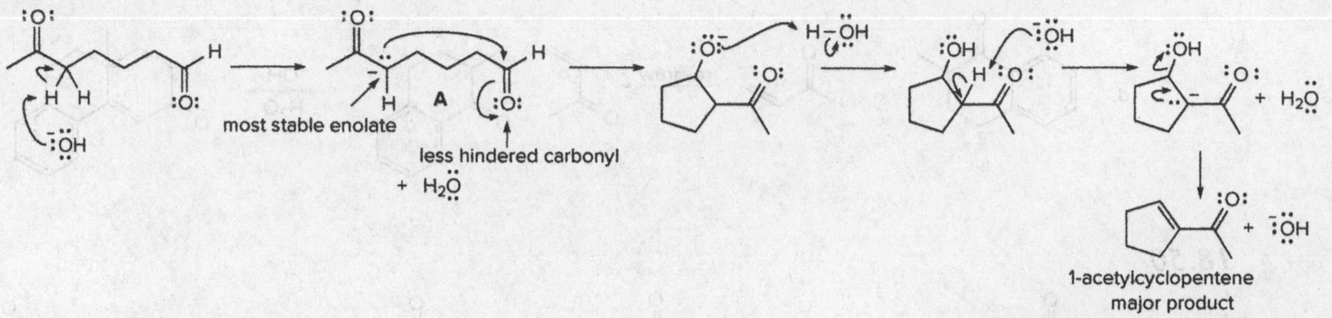

most stable enolate

A

less hindered carbonyl

+ H_2O

+ H_2O

+ ^-OH

1-acetylcyclopentene
major product

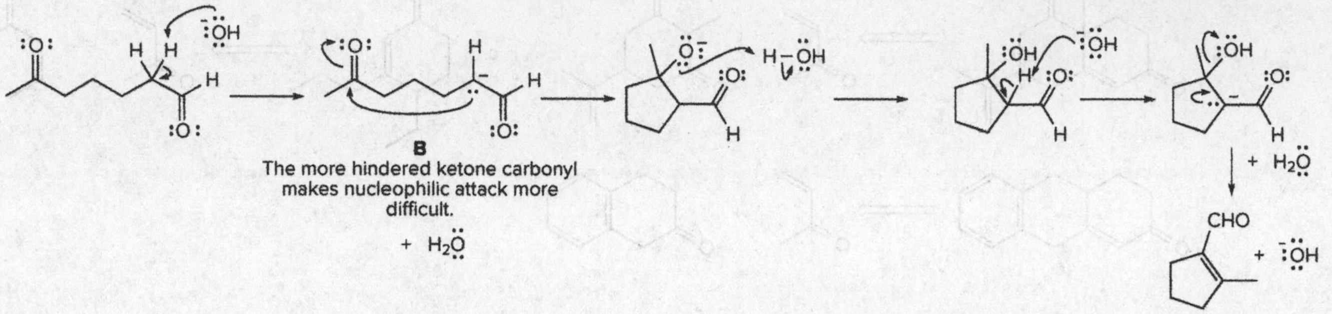

B

The more hindered ketone carbonyl makes nucleophilic attack more difficult.

+ H_2O

+ H_2O

CHO

+ ^-OH

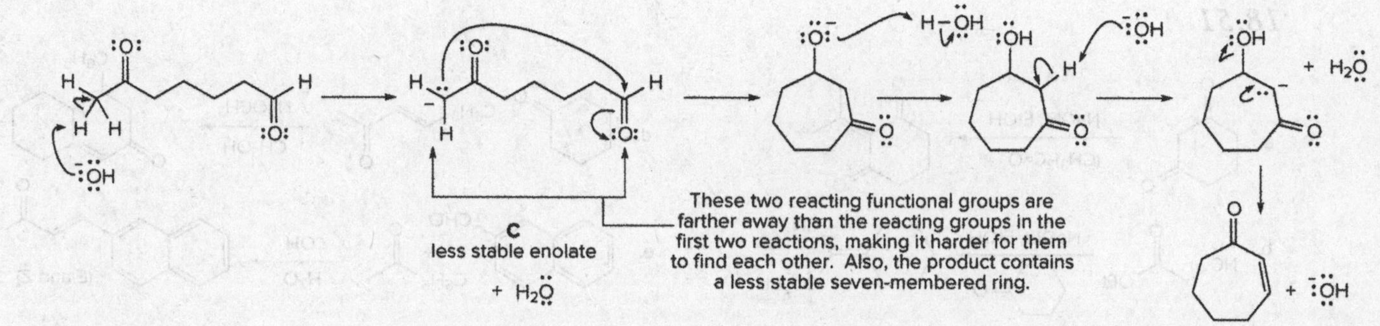

C

less stable enolate

+ H_2O

These two reacting functional groups are farther away than the reacting groups in the first two reactions, making it harder for them to find each other. Also, the product contains a less stable seven-membered ring.

+ H_2O

+ ^-OH

18.54

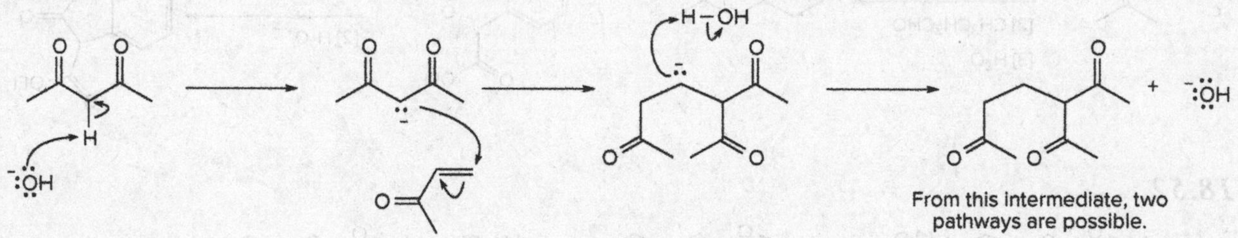

+ ^-OH

From this intermediate, two pathways are possible.

Path [1]

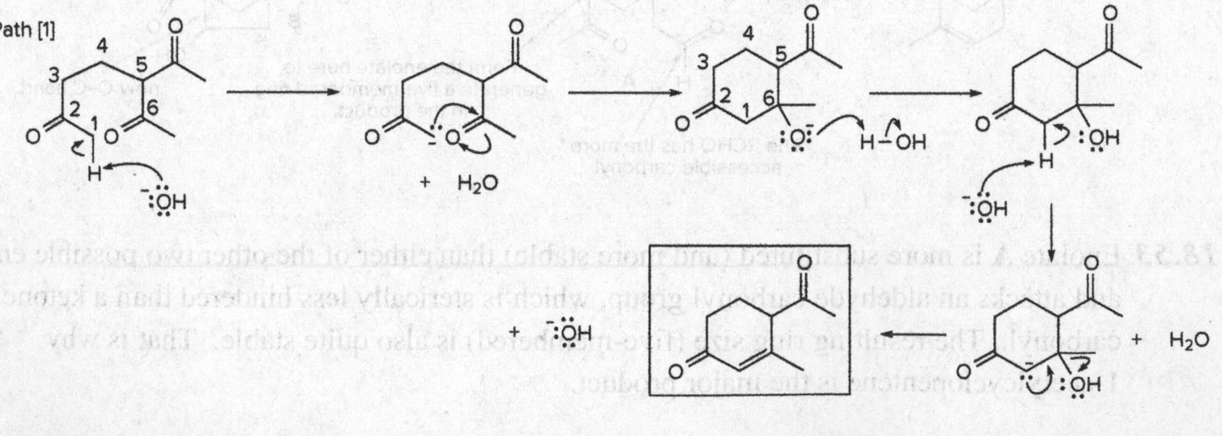

+ H_2O

+ ^-OH

+ H_2O

Path [2]

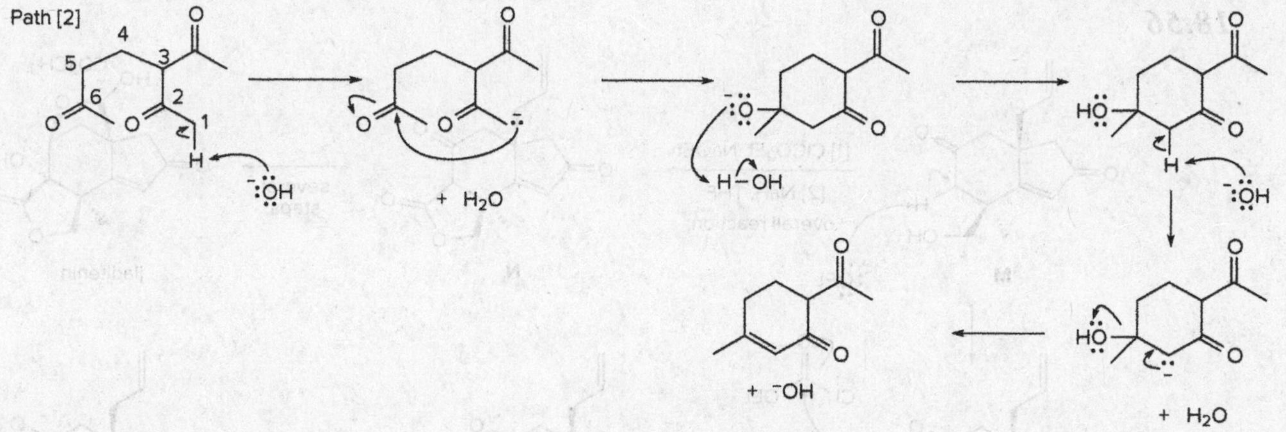

4 5 3 6 2 1

+ H_2O

+ ^{-}OH

+ H_2O

18.55

KOH, H_2O, THF
overall reaction

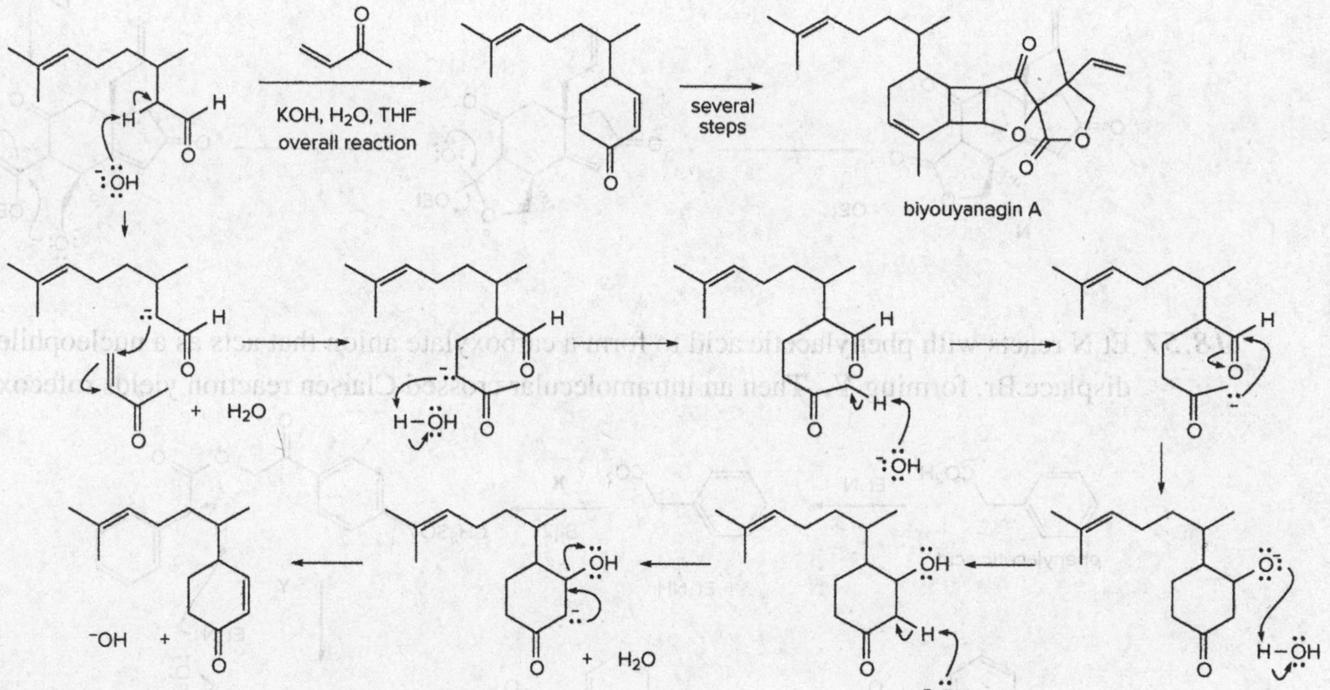

several steps

biyouyanagin A

+ H_2O

^{-}OH +

+ H_2O

18.56

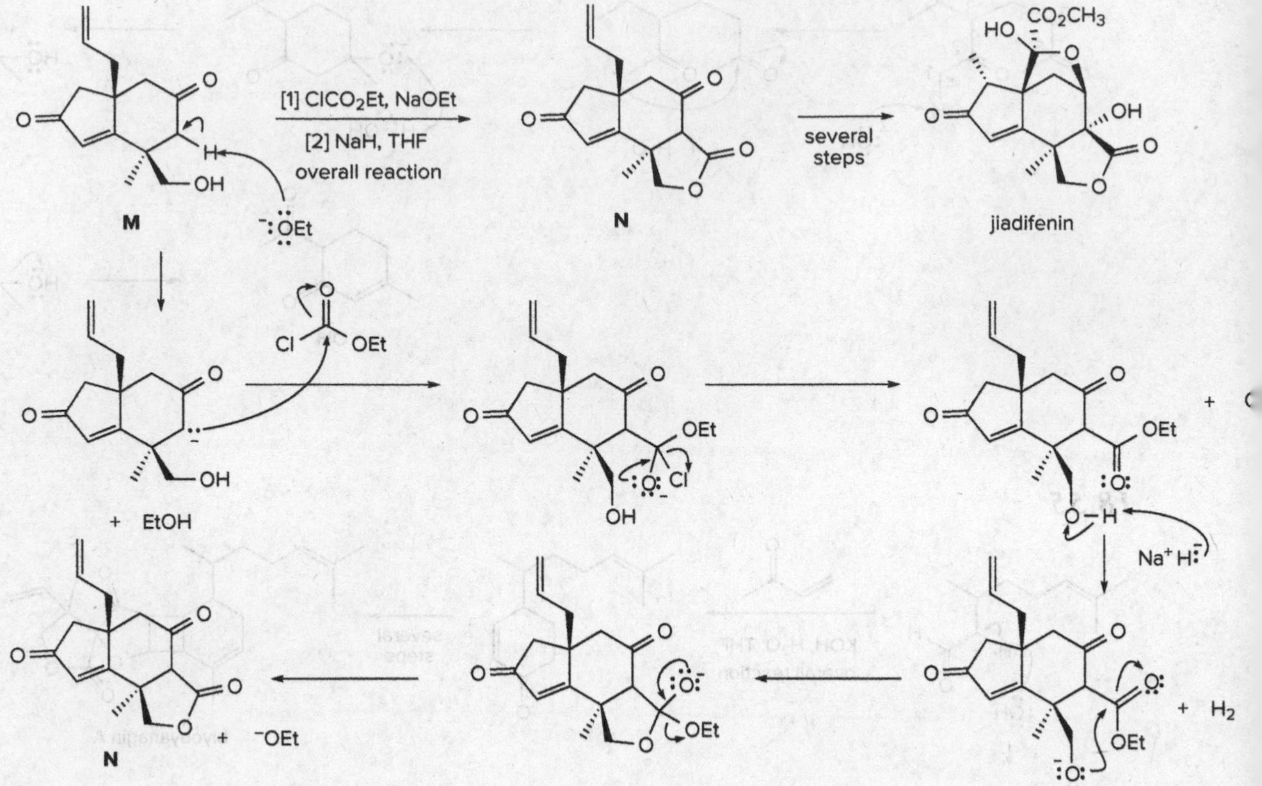

[1] ClCO$_2$Et, NaOEt
[2] NaH, THF
overall reaction

M → N → several steps → jiadifenin

+ EtOH

+ $^{-}$OEt

+ H$_2$

Na$^+$ H$^{:-}$

18.57 Et$_3$N reacts with phenylacetic acid to form a carboxylate anion that acts as a nucleophile to displace Br, forming **Y.** Then an intramolecular crossed Claisen reaction yields rofecoxib.

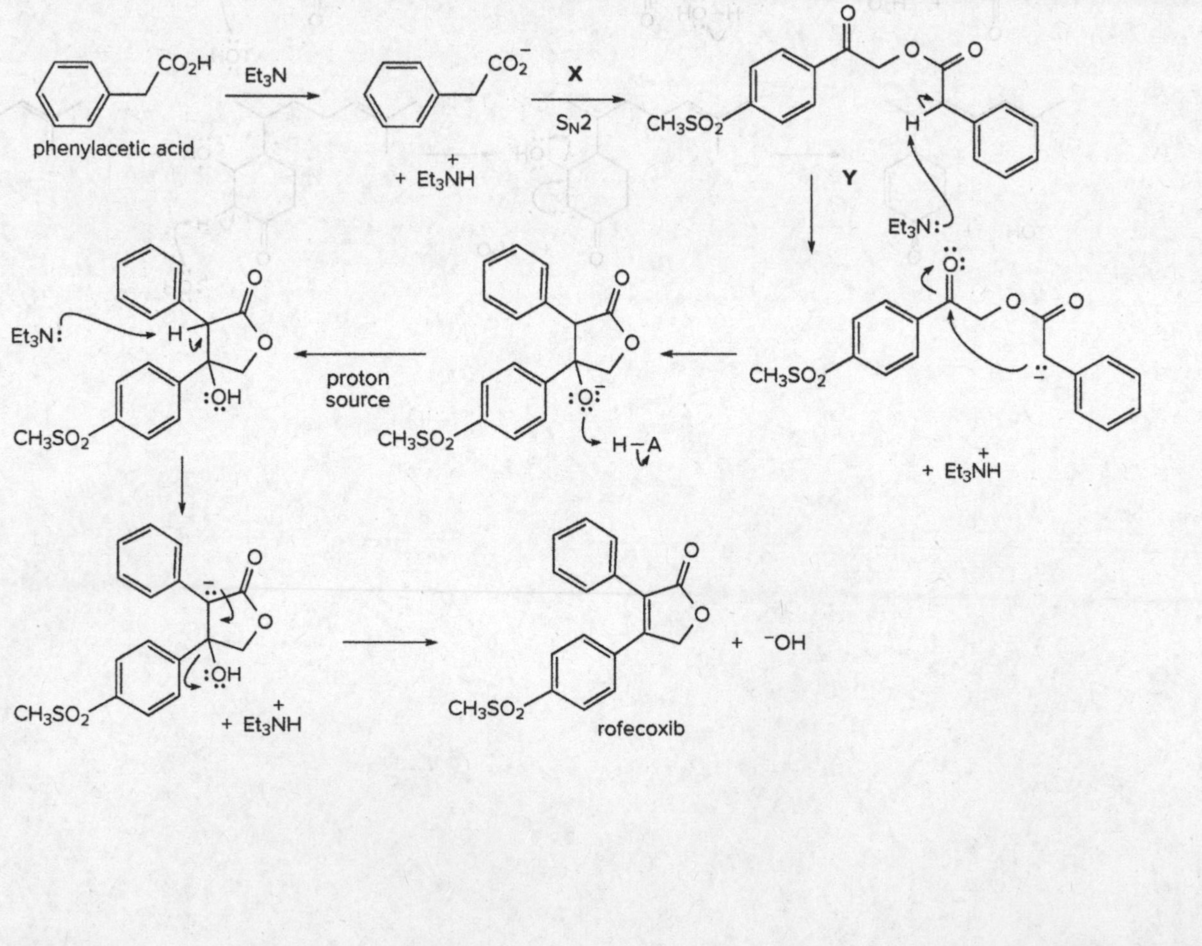

phenylacetic acid → Et$_3$N → + Et$_3\overset{+}{N}$H → X, S$_N$2 → Y

proton source

+ $^{-}$OH

rofecoxib

18.58

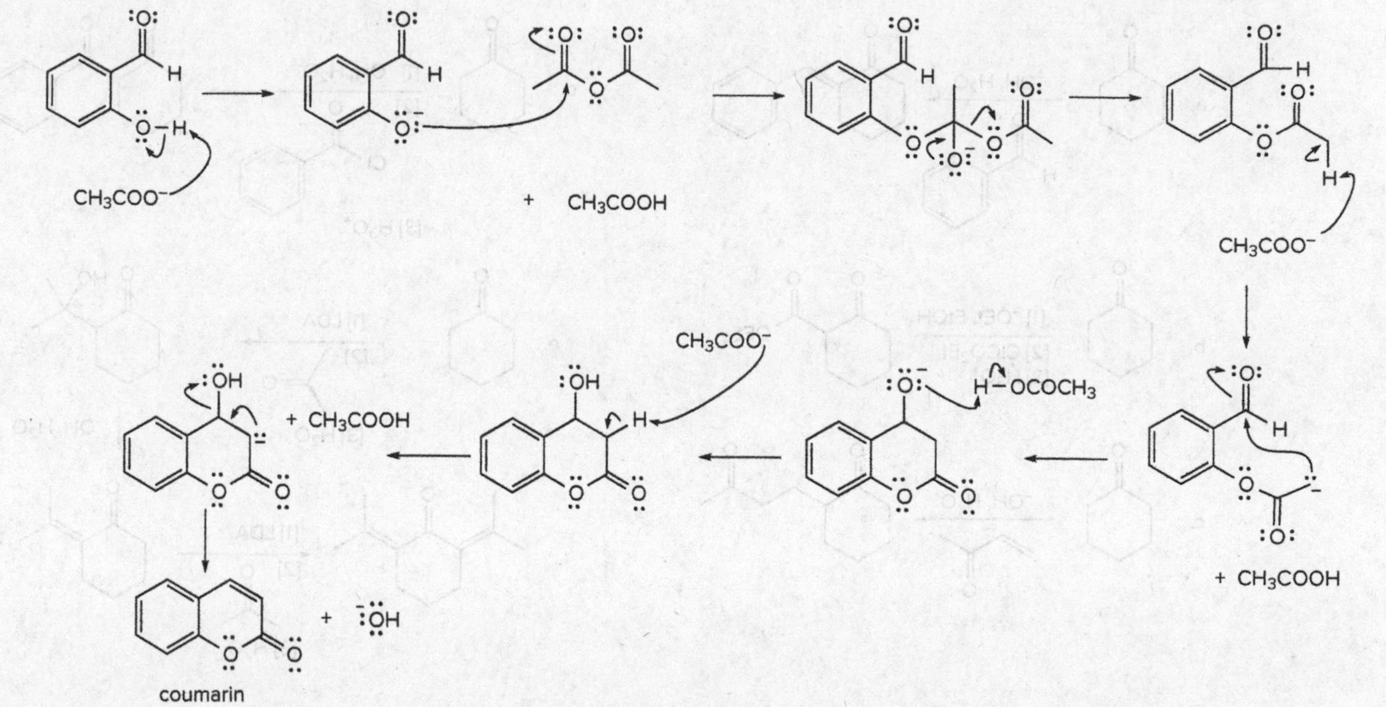

CH_3COO^-

$+ \quad CH_3COOH$

CH_3COO^-

CH_3COO^-

$H-OCOCH_3$

$+ \quad CH_3COOH$

$+ \quad CH_3COOH$

$+ \quad :\ddot{O}H$

coumarin

18.59

a. [1] LDA; [2] CH_3CO_2Et — [1] NaOEt; [2] CH_3Br

OH, H_2SO_4 — PBr3, CH_3OH

OH, CrO_3, H_2SO_4, H_2O — OH

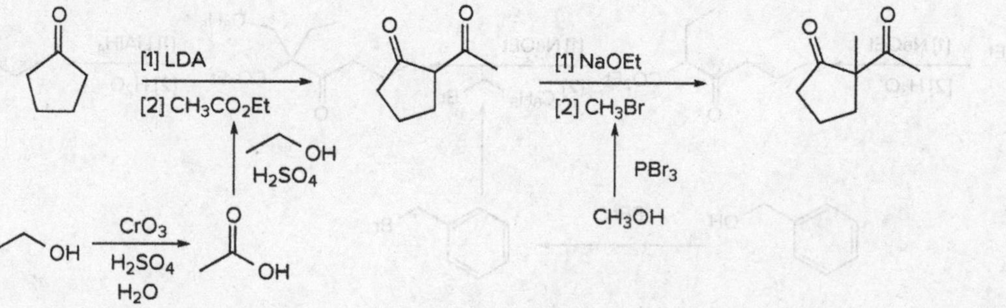

b. [1] CH_3MgBr; [2] H_2O — OH — H_2SO_4 — major product — [1] O_3; [2] $(CH_3)_2S$ — CHO — ^-OH, H_2O

CH_3OH — PBr_3 — CH_3Br — Mg

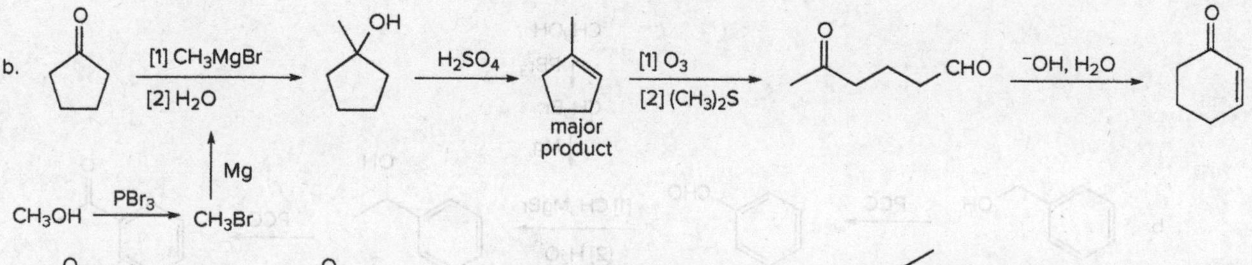

c. [1] LDA; [2] Br — [1] MgBr; [2] H_2O — HO — H_2SO_4 — major product

PBr_3 — OH — Mg — Br

[1] O_3; [2] $(CH_3)_2S$

NaOH, H_2O

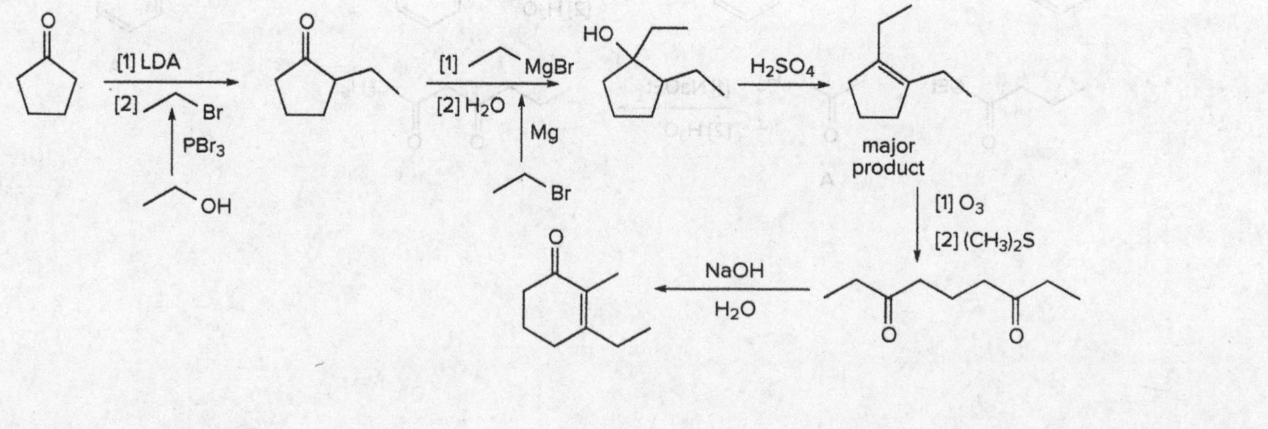

18.60

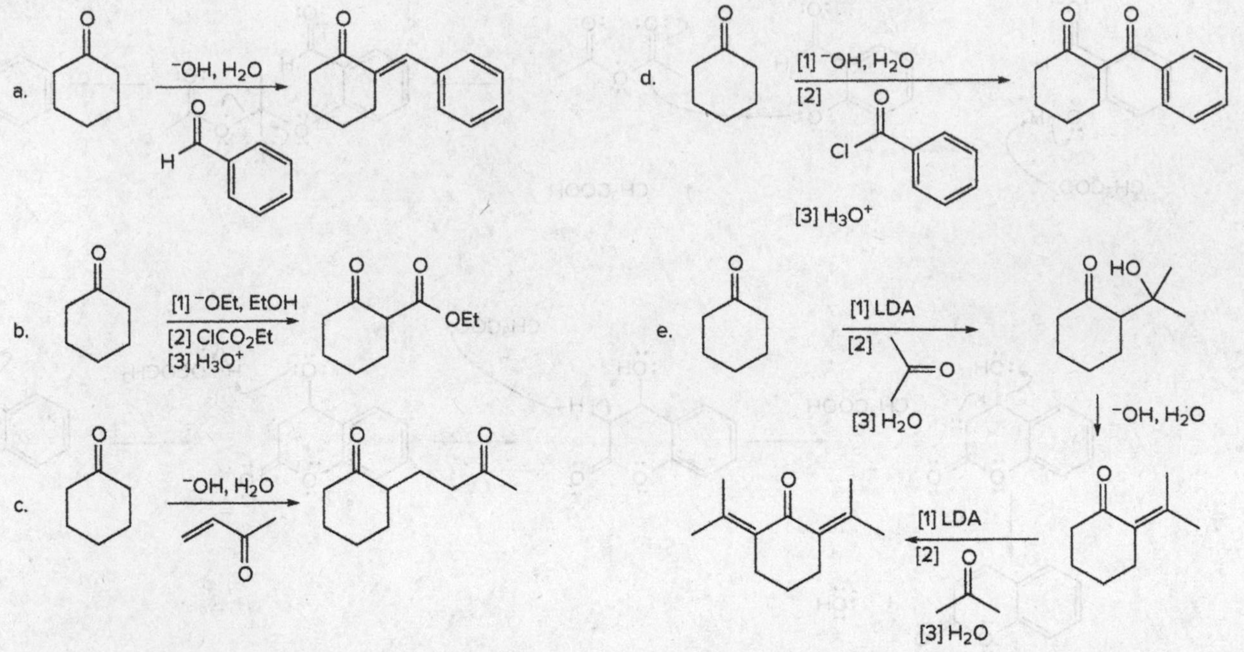

a. ^{-}OH, H_2O

d. [1] ^{-}OH, H_2O
[2] Cl
[3] H_3O^+

b. [1] ^{-}OEt, EtOH
[2] $ClCO_2Et$
[3] H_3O^+
OEt

e. [1] LDA
[2]
[3] H_2O
HO
^{-}OH, H_2O
[1] LDA
[2]
[3] H_2O

c. ^{-}OH, H_2O

18.61

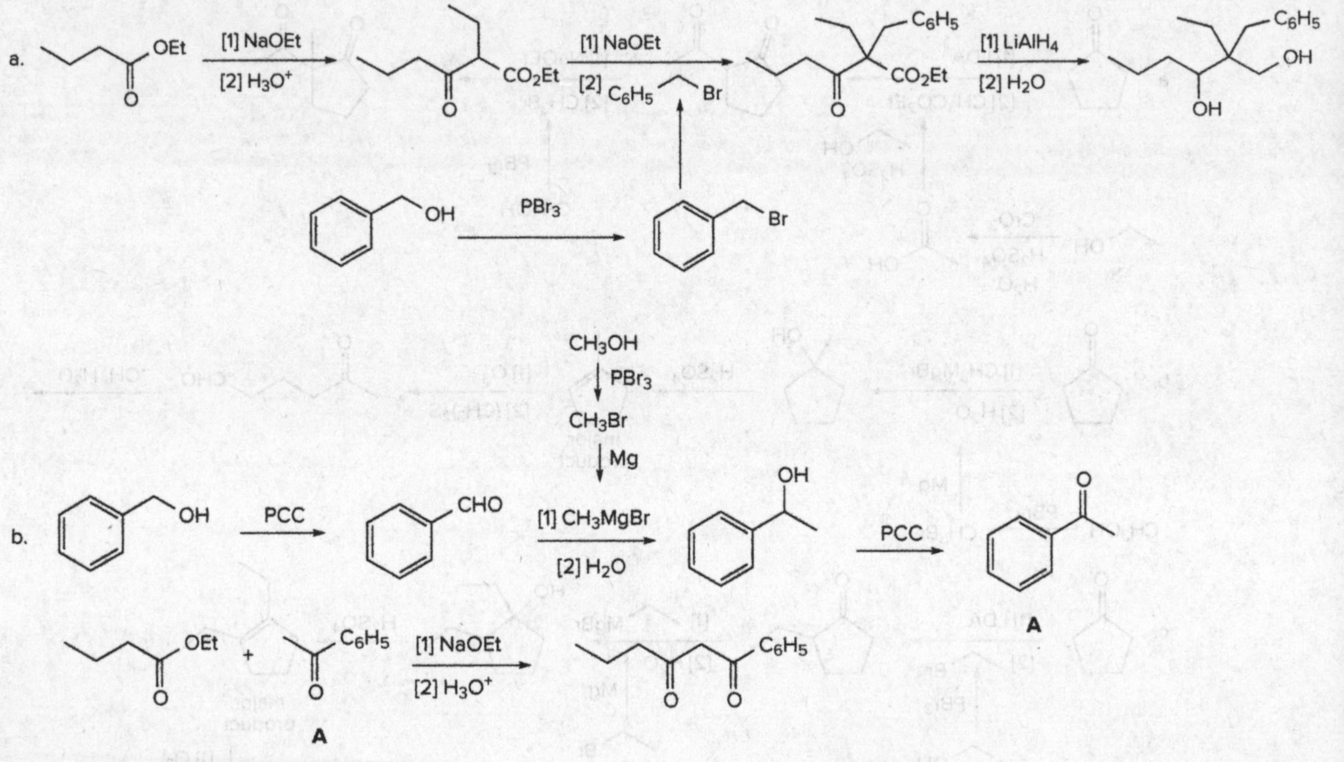

a. OEt
[1] NaOEt
[2] H_3O^+
CO_2Et
[1] NaOEt
[2] C_6H_5 Br
C_6H_5
CO_2Et
[1] $LiAlH_4$
[2] H_2O
C_6H_5
OH
OH

OH
PBr_3
Br

CH_3OH
PBr_3
CH_3Br
Mg

b. OH
PCC
CHO
[1] CH_3MgBr
[2] H_2O
OH
PCC
A

OEt + C_6H_5 **A**
[1] NaOEt
[2] H_3O^+
C_6H_5

18.62

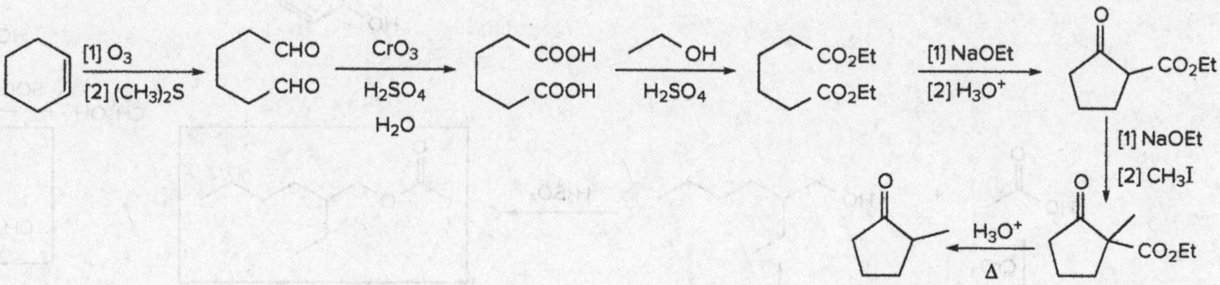

CHO CHO COOH COOH CO2Et CO2Et CO2Et CO2Et

[1] O_3 [2] $(CH_3)_2S$; CrO_3, H_2SO_4, H_2O ; OH, H_2SO_4 ; [1] NaOEt [2] H_3O^+ ; [1] NaOEt [2] CH_3I ; H_3O^+, Δ

18.63

a. ^-OH, H_2O ; $NaBH_4$, CH_3OH ; H^+

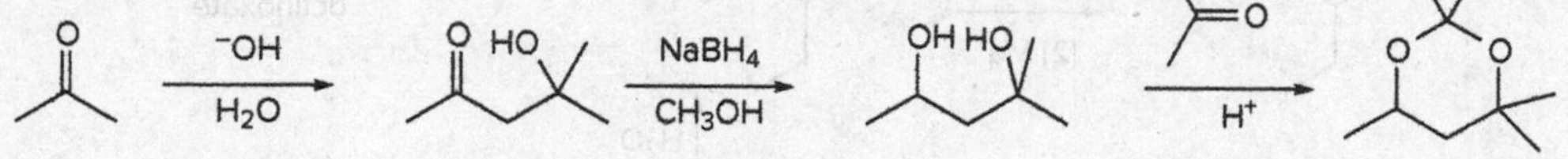

b. [from part (a)] ; ^-OH, H_2O

$NaBH_4$, CH_3OH ; OH ; PBr_3 ; Br ; Mg ; [1] MgBr [2] H_2O ; OH ; H_2SO_4

(*E* and *Z* isomers)

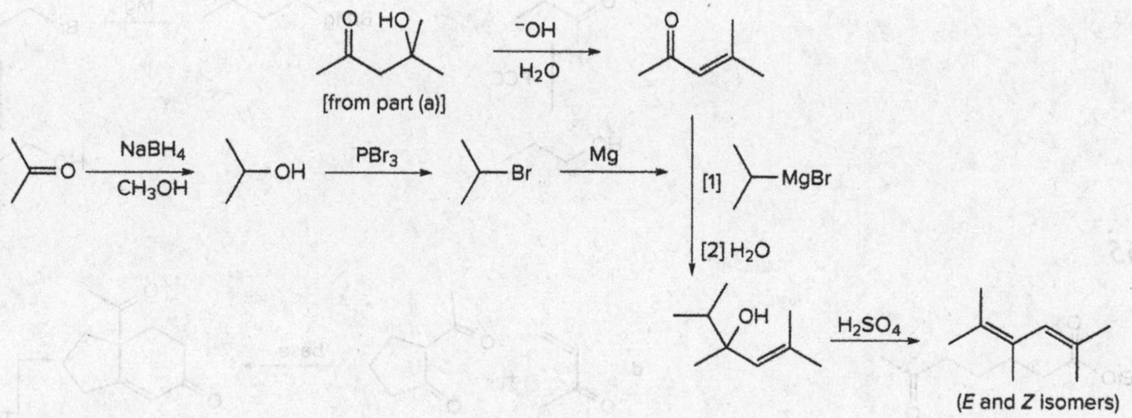

c. [from part (b)] ; $NaBH_4$, CH_3OH ; OH ; H^+

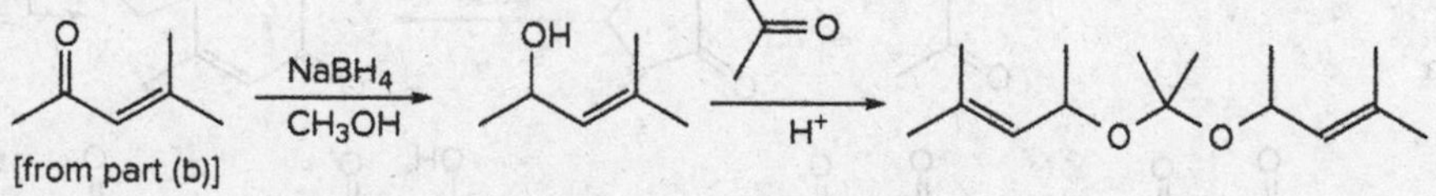

18.64

a. CH_3O octinoxate ⟹ CHO CH_3O +

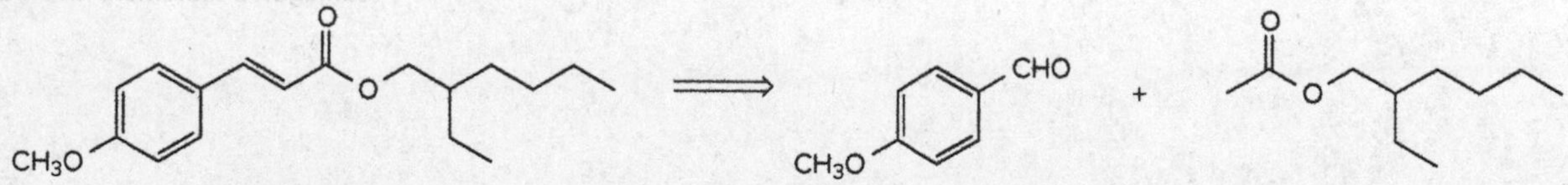

b.

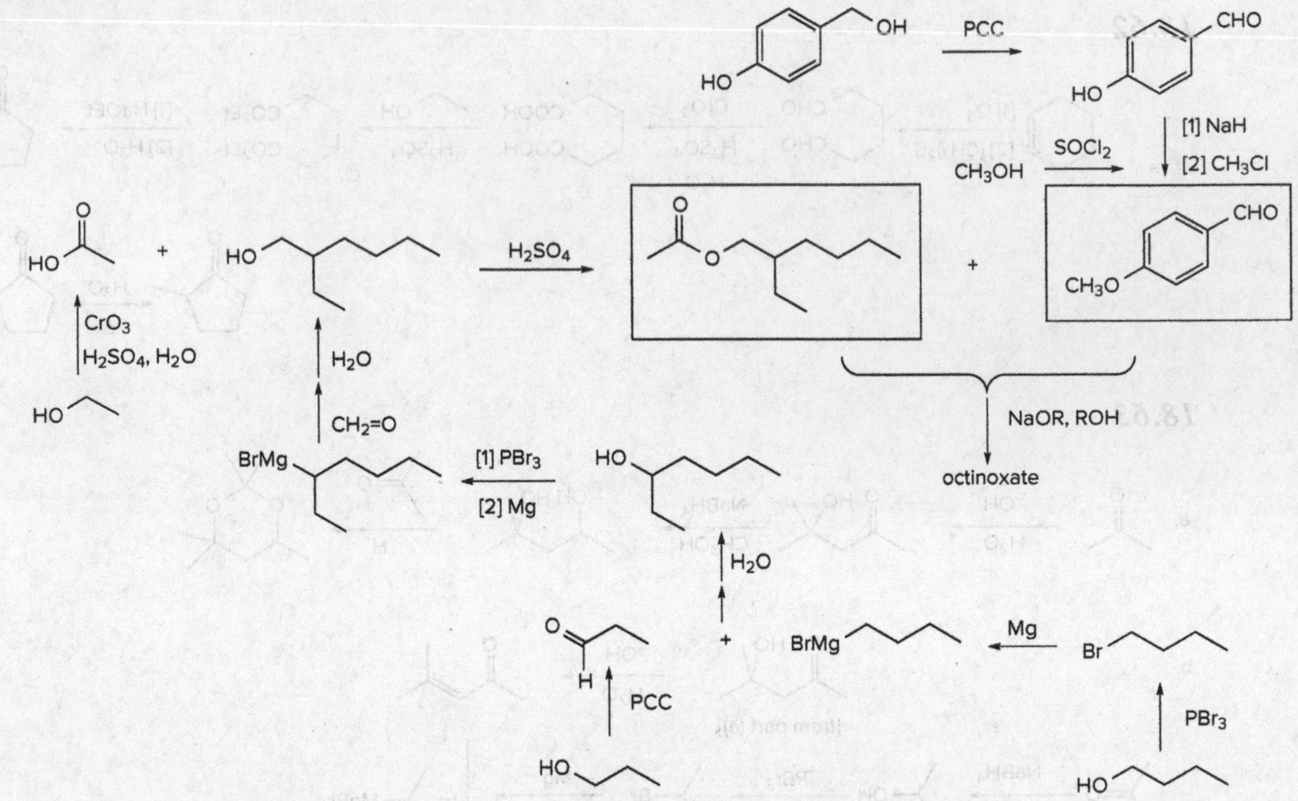

PCC

CHO

HO

[1] NaH
[2] CH_3Cl

$SOCl_2$
CH_3OH

H_2SO_4

+

CHO

CH_3O

CrO_3
H_2SO_4, H_2O

H_2O

$CH_2{=}O$

BrMg

[1] PBr_3
[2] Mg

HO

NaOR, ROH

octinoxate

H_2O

+ BrMg

Mg

Br

PCC

PBr_3

HO

18.65

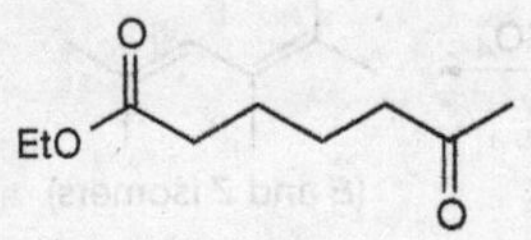

a. EtO

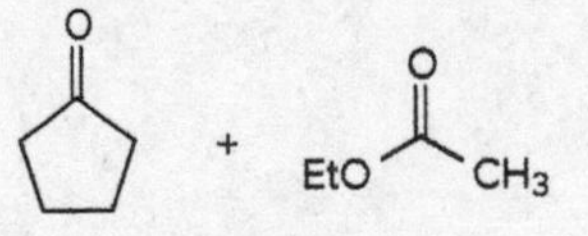

d. + base

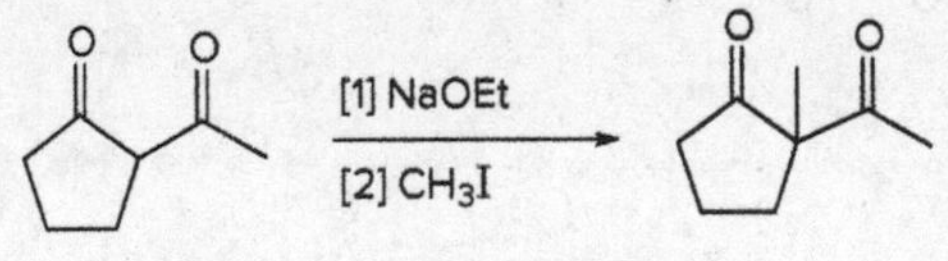

Two products are possible.

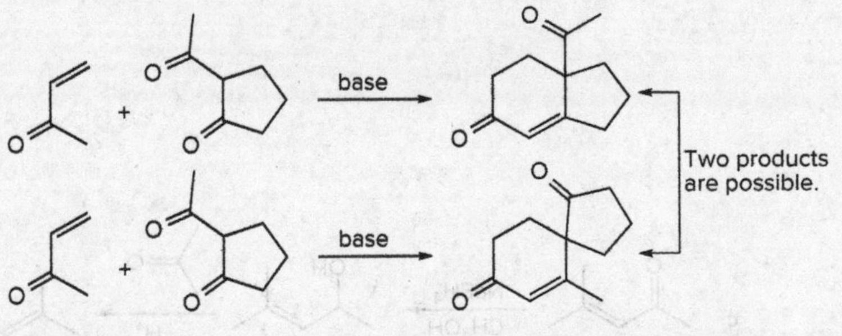

b. + EtO CH_3

base

c. [1] NaOEt
[2] CH_3I

e. OH or OH

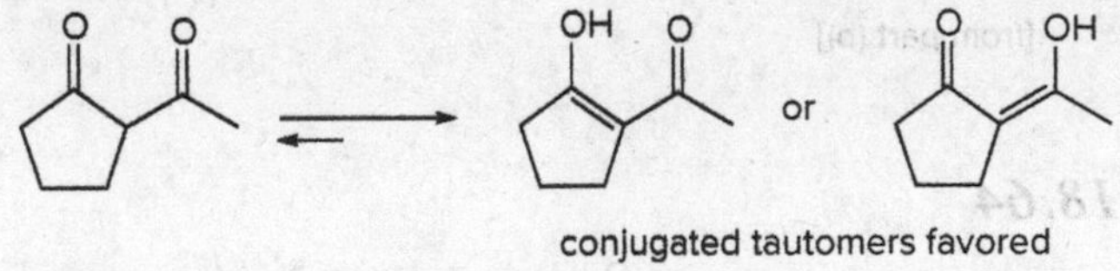

conjugated tautomers favored

18.66

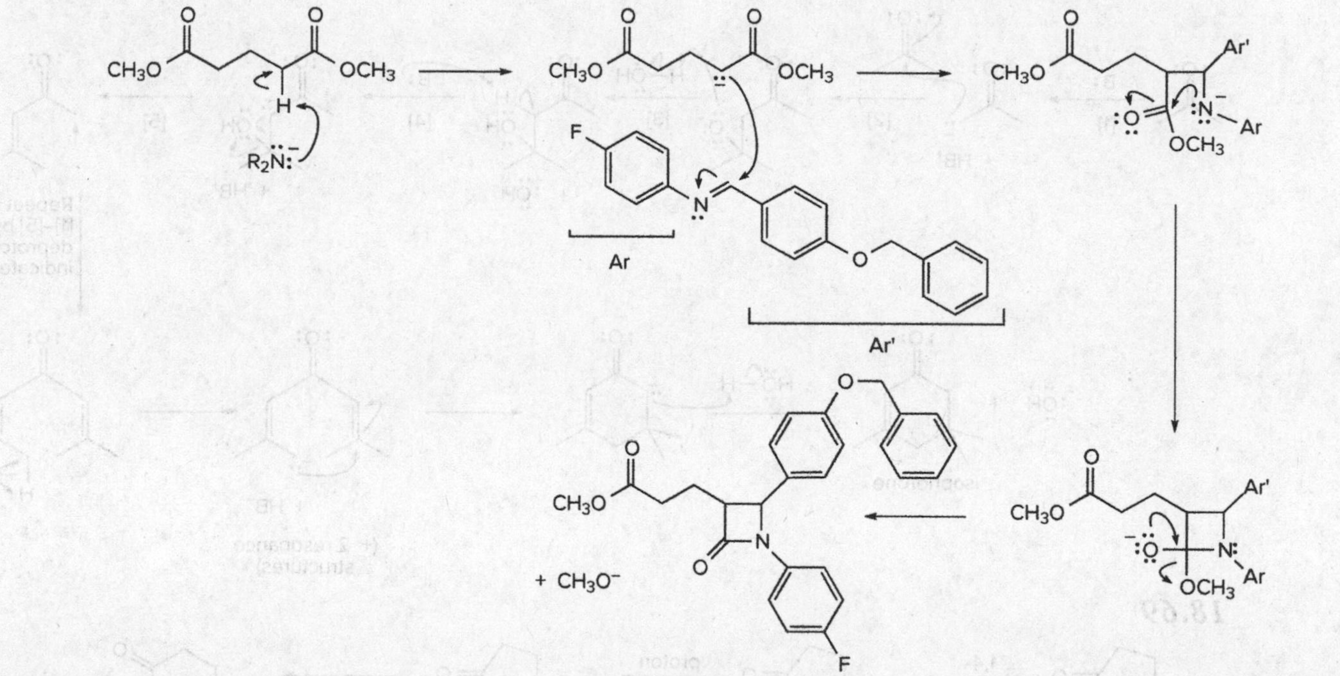

CH_3O OCH_3 H $R_2\ddot{N}{:}^-$ → CH_3O OCH_3 F Ar Ar' → CH_3O Ar' N Ar OCH_3

↓

CH_3O Ar' N Ar OCH_3 → CH_3O O N F + CH_3O^-

18.67

SO_2 Br N CHO H **A** CO_3^{2-} → SO_2 Br N CHO + HCO_3^- → N SO_2 CHO H CO_3^{2-}

↓

N SO_2 H + HCO_3^- → N SO_2 $H{-}OCO_2^-$ → N SO_2 HO H CO_3^{2-}

↓

N SO_2 HO + HCO_3^- → N SO_2 **B** + ^-OH

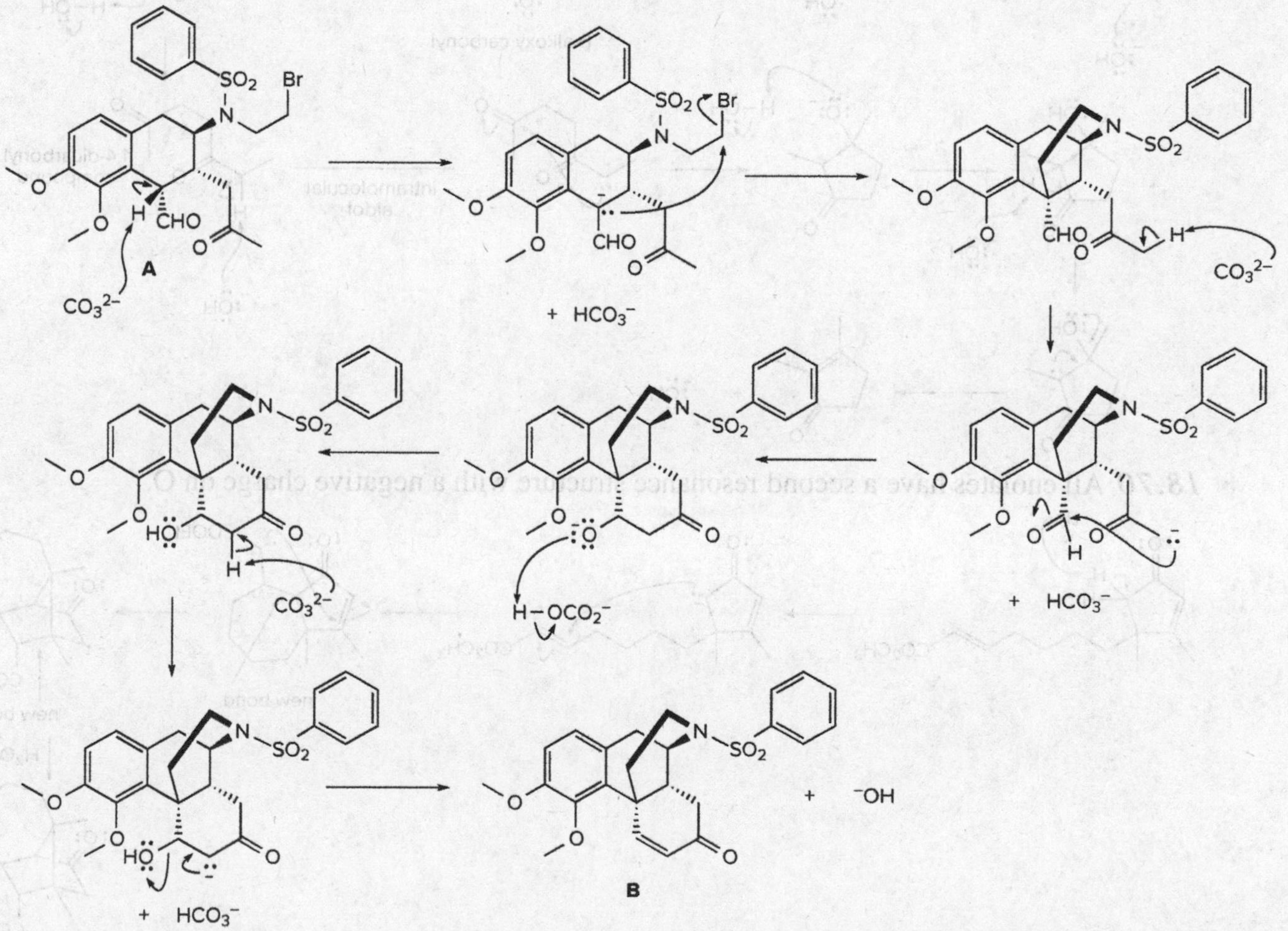

18.68 All enolates have a second resonance structure with a negative charge on O.

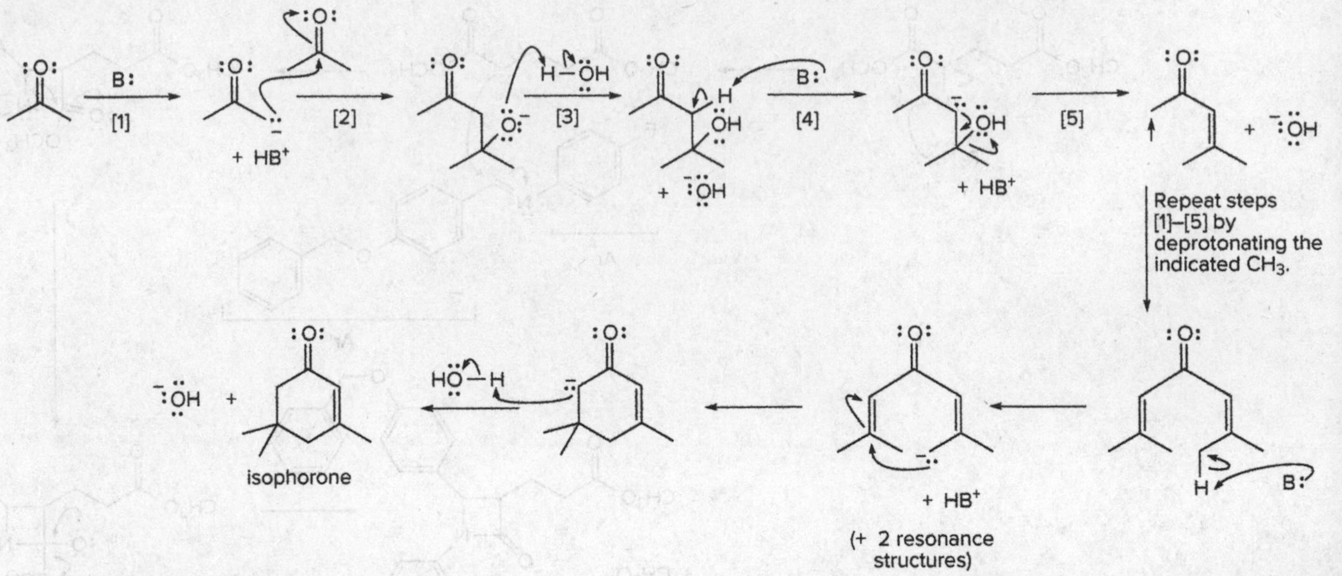

B: [1] + HB⁺ [2] [3] + :ÖH [4] + HB⁺ [5] + :ÖH

Repeat steps [1]–[5] by deprotonating the indicated CH_3.

+ HB⁺ (+ 2 resonance structures)

:ÖH + isophorone

18.69

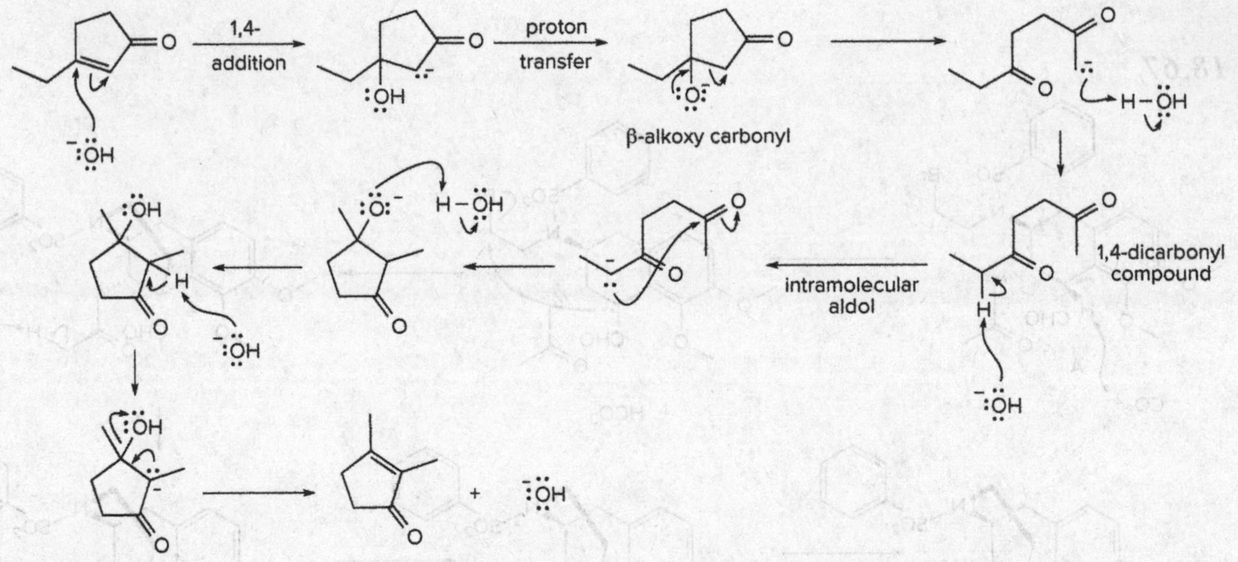

1,4-addition

proton transfer

β-alkoxy carbonyl

1,4-dicarbonyl compound

intramolecular aldol

18.70 All enolates have a second resonance structure with a negative charge on O.

B: CO_2CH_3 COOEt new bond new bond H_2O + ⁻OH CO_2CH_3

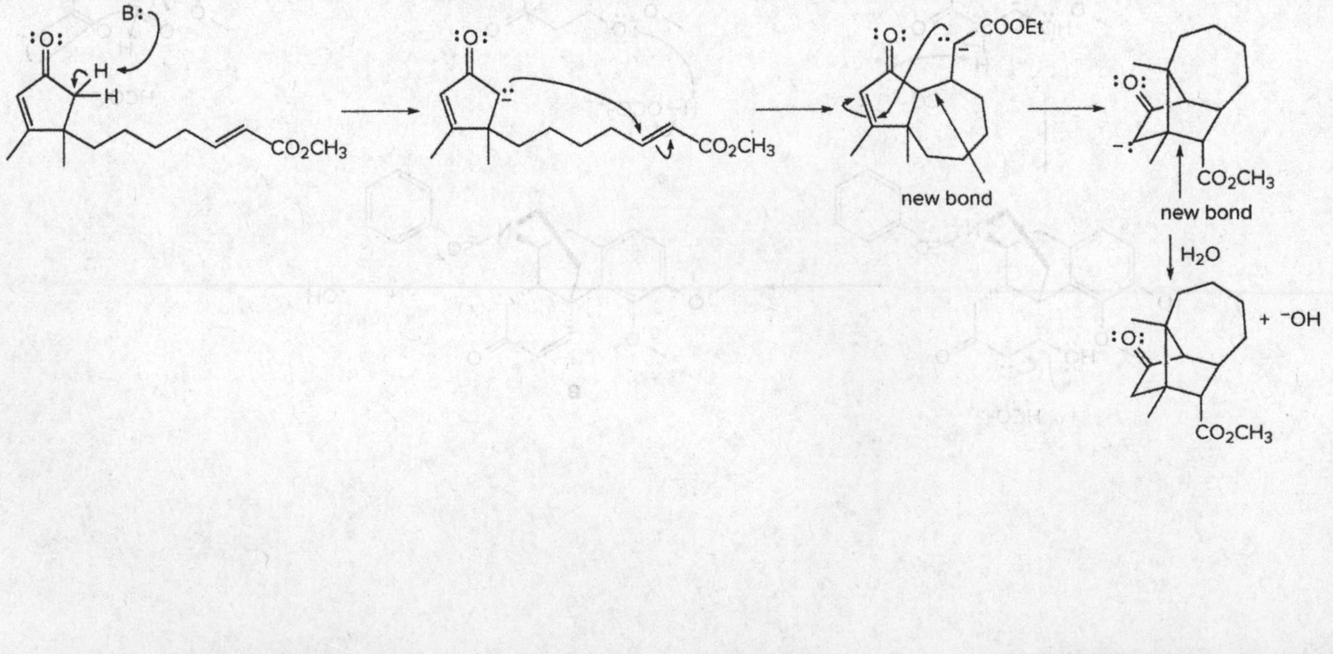

18.71 The reaction takes place in acid, so enols are involved. After the initial condensation reaction, the NH_2 and C=O groups form an imine by an intramolecular reaction.

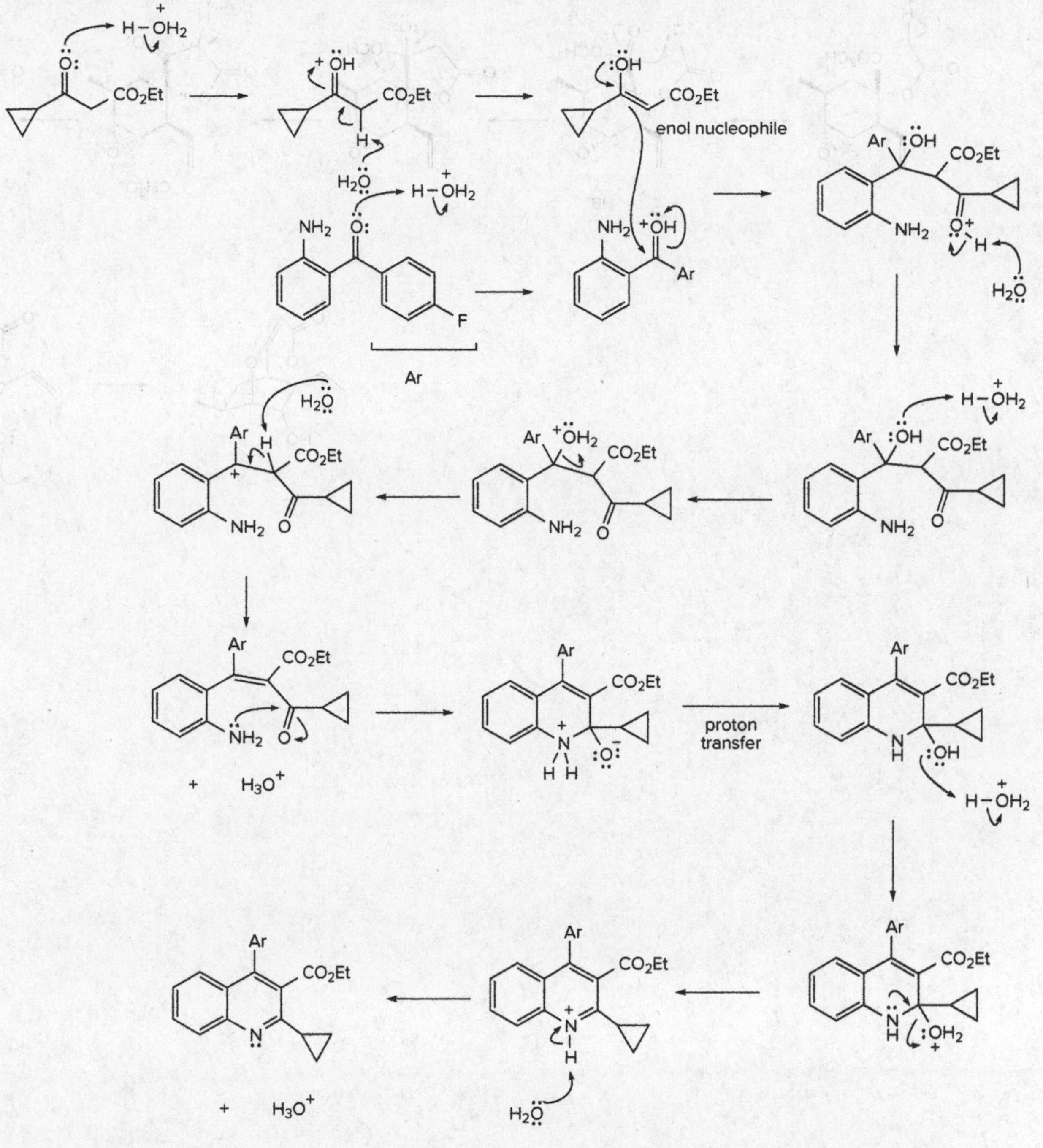

enol nucleophile

Ar

proton transfer

+ H_3O^+

+ H_3O^+

18.72

:B
H
CO_2CH_3
O
O
O
OCH₃
+ HB⁺
:Ö:⁻
OCH₃
:B
H
O
+ CH_3O^-
+ HB⁺
The enolate opens the epoxide ring.
HO
$^-$:Ö:
H⁺

Chapter 19 Benzene and Aromatic Compounds

Chapter Review

Comparing aromatic, antiaromatic, and nonaromatic compounds (19.7)

- **Aromatic compound**
  - A cyclic, planar, completely conjugated compound that contains $4n + 2$ π electrons ($n = 0, 1, 2, 3$, and so forth).
  - An aromatic compound is more stable than a similar acyclic compound having the same number of π electrons.
- **Antiaromatic compound**
  - A cyclic, planar, completely conjugated compound that contains $4n$ π electrons ($n = 0, 1, 2, 3$, and so forth).
  - An antiaromatic compound is less stable than a similar acyclic compound having the same number of π electrons.
- **Nonaromatic compound**
  - A compound that lacks one (or more) of the requirements to be aromatic or antiaromatic.

Properties of aromatic compounds

- Every carbon in an aromatic ring has a *p* orbital to delocalize electron density (19.2).
- Aromatic compounds are unusually stable. $\Delta H°$ for hydrogenation is much less than expected, given the number of degrees of unsaturation (19.6).
- Aromatic compounds do not undergo the usual addition reactions of alkenes (19.6).
- ^{1}H NMR spectra show highly deshielded protons because of ring currents (19.4).

Examples of aromatic compounds with 6 electrons (19.8, 19.9)

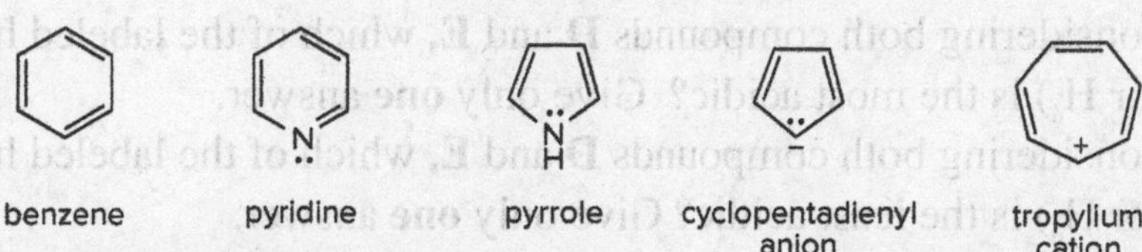

Examples of compounds that are not aromatic (19.8)

Practice Test on Chapter Review

1. Give the IUPAC name for each of the following compounds.

a. b. c.

2. Label each compound as aromatic, nonaromatic, or antiaromatic. Choose only **one** possibility. Assume all completely conjugated rings are planar.

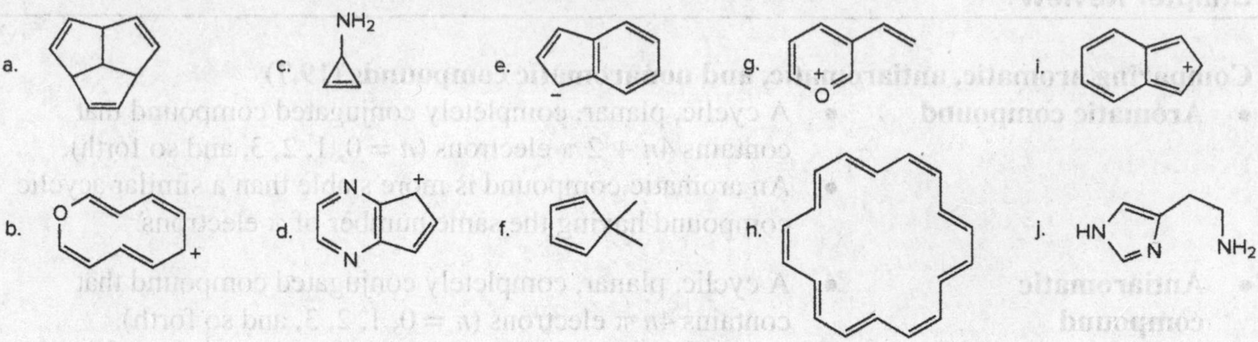

3. Answer the following questions about compounds A–E drawn below.

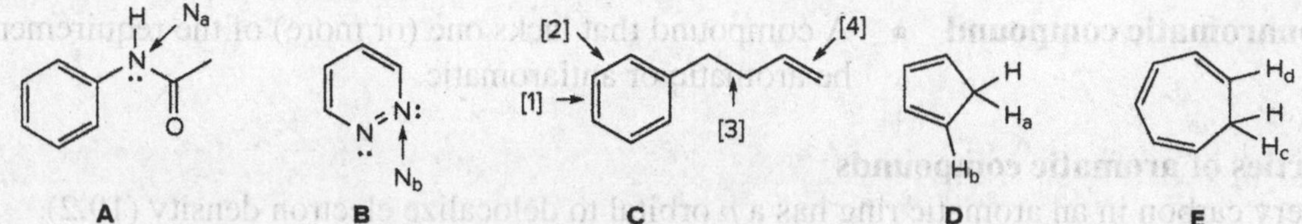

a. How is nitrogen N_a in compound **A** hybridized?
b. In what type of orbital does the lone pair on N_a reside?
c. How is nitrogen N_b in compound **B** hybridized?
d. In what type of orbital does the lone pair on N_b reside?
e. Which of the labeled bonds in compound **C** is the shortest?
f. Which of the labeled bonds in compound **C** is the longest?
g. When considering both compounds **D** and **E,** which of the labeled hydrogen atoms (H_a, H_b, H_c, or H_d) is the most acidic? Give only **one** answer.
h. When considering both compounds **D** and **E,** which of the labeled hydrogen atoms (H_a, H_b, H_c, or H_d) is the least acidic? Give only **one** answer.

Answers to Practice Test

1.a. 2-*sec*-butyl-5-nitrophenol
b. *o*-isobutyltoluene
c. 2-*tert*-butyl-4-nitrophenol

2.a. nonaromatic
b. aromatic
c. nonaromatic
d. antiaromatic
e. aromatic
f. nonaromatic
g. aromatic
h. aromatic
i. antiaromatic
j. aromatic

3.a. sp^2
b. *p*
c. sp^2
d. sp^2
e. 4
f. 3
g. H_a
h. H_c

Answers to Problems

19.1 Move the electrons in the π bonds to draw all major resonance structures.

diphenhydramine

19.2 Look at the hybridization of the atoms involved in each bond. Carbons in a benzene ring are surrounded by three groups and are sp^2 hybridized.

Csp^2–Csp^3

H 1 2 3 4 5

Csp^2–$H1s$

Csp^2–Csp^2
Cp–Cp
shortest of all
the indicated bonds

Csp^2–Csp^2

Csp^2–Csp^2
Cp–Cp

19.3

- To name a benzene ring with **one substituent,** name the substituent and add the word *benzene*.
- To name a **disubstituted ring,** select the correct prefix (ortho = 1,2; meta = 1,3; para = 1,4) and alphabetize the substituents. Use a common name if it is a derivative of that monosubstituted benzene.
- To name a **polysubstituted ring,** number the ring to give the lowest possible numbers and then follow other rules of nomenclature.

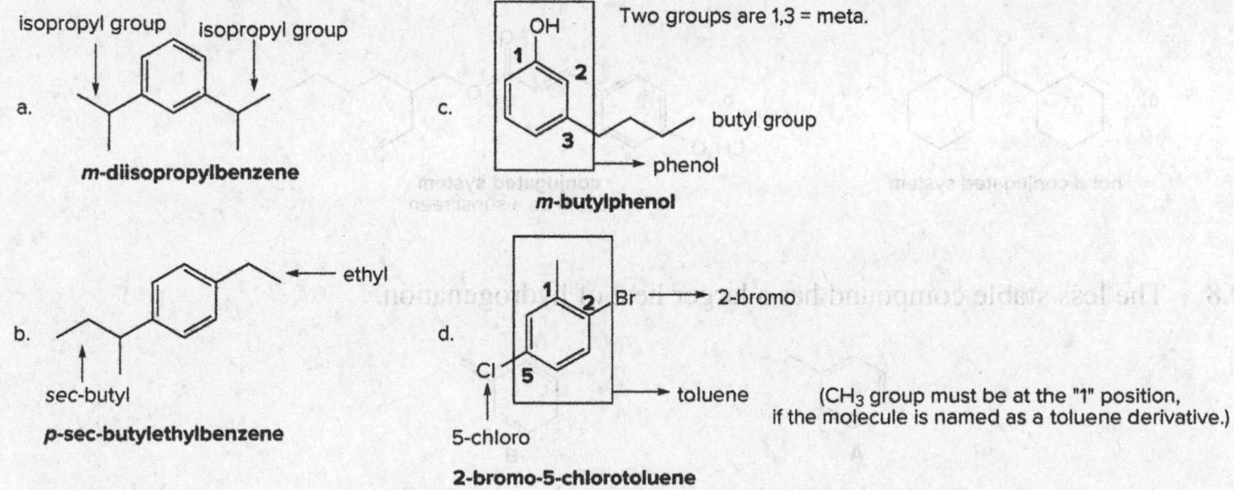

19.4 Work backwards to draw the structures from the names.

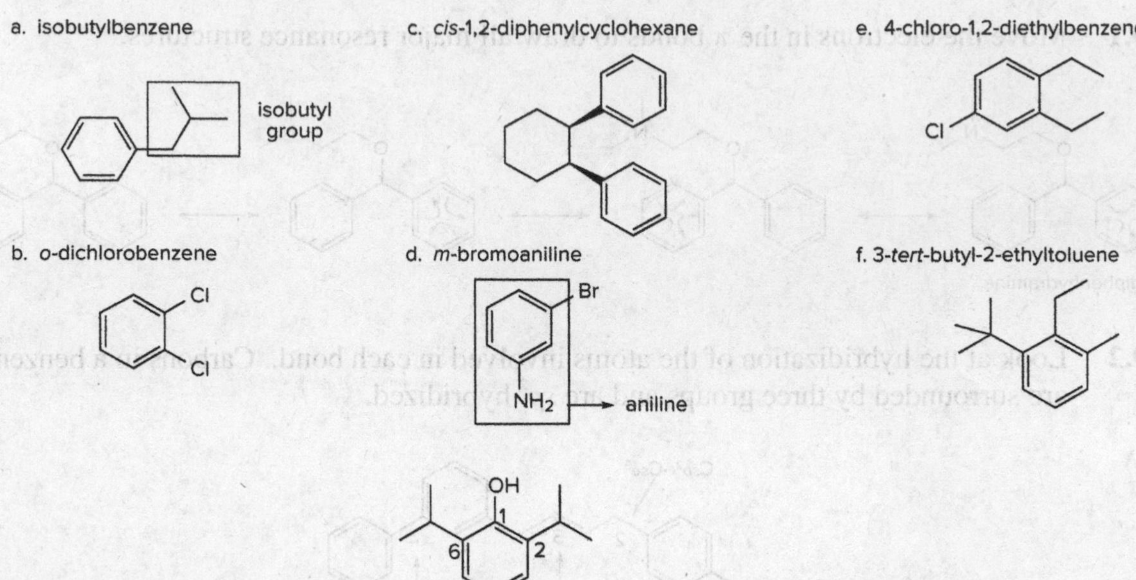

19.5

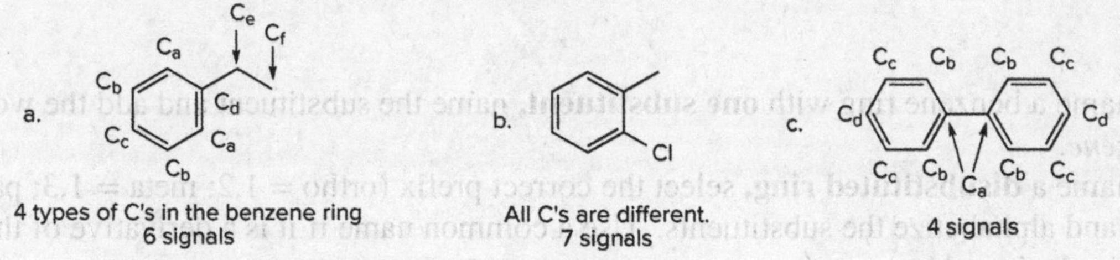

19.6 Count the different types of carbons to determine the number of ^{13}C NMR signals.

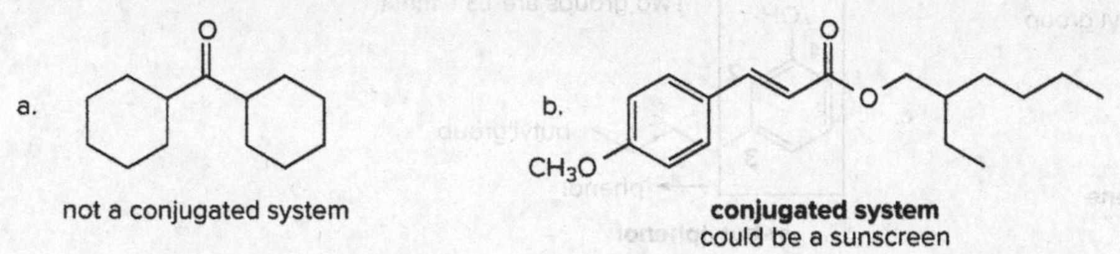

19.7 Sunscreens contain conjugated systems to absorb UV radiation from sunlight. Look for conjugated systems in the compounds below.

19.8 The less stable compound has a larger heat of hydrogenation.

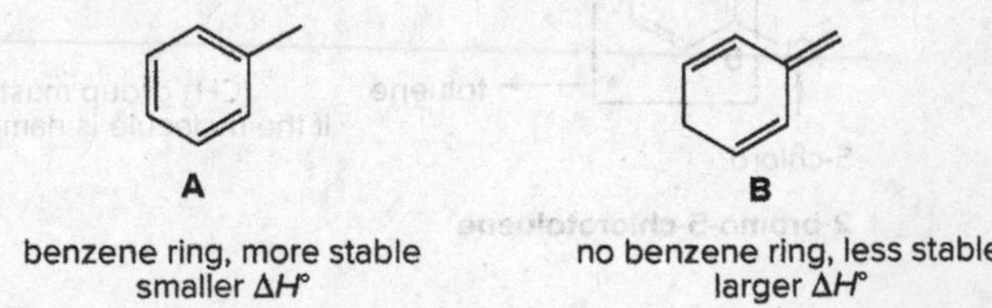

19.9 To be aromatic, a ring must have $4n + 2$ π electrons.

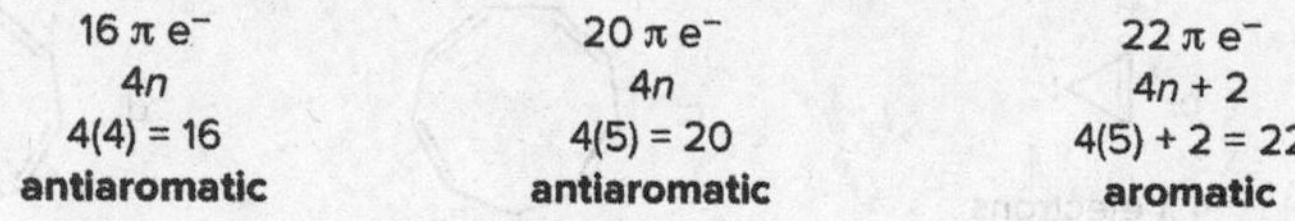

16 π e⁻
4n
4(4) = 16
antiaromatic
20 π e⁻
4n
4(5) = 20
antiaromatic
22 π e⁻
4n + 2
4(5) + 2 = 22
aromatic

19.10

19.11 Compare the conjugate base of cyclohepta-1,3,5-triene with the conjugate base of cyclopentadiene. The compound with the more stable conjugate base will have a lower pK_a.

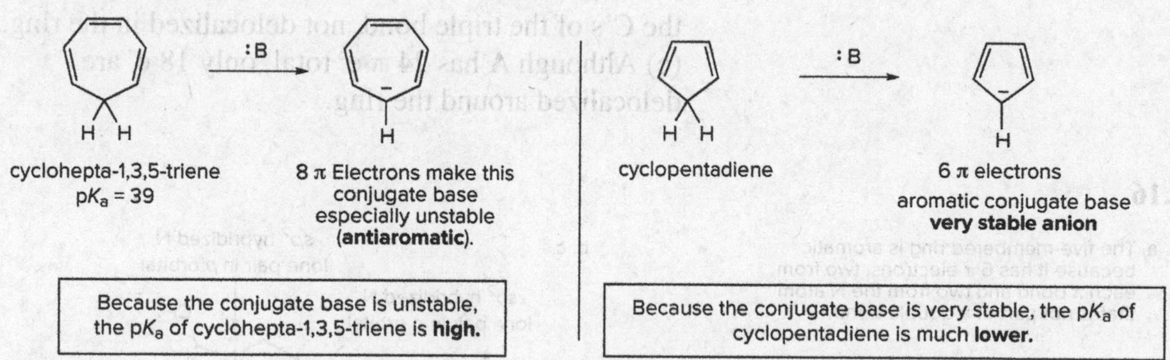

:B
H H
H
cyclohepta-1,3,5-triene
pKa = 39
8 π Electrons make this conjugate base especially unstable (antiaromatic).
Because the conjugate base is unstable, the pKa of cyclohepta-1,3,5-triene is high.
:B
H H
H
cyclopentadiene
6 π electrons
aromatic conjugate base
very stable anion
Because the conjugate base is very stable, the pKa of cyclopentadiene is much lower.

19.12 The compound with the most stable conjugate base is the most acidic.

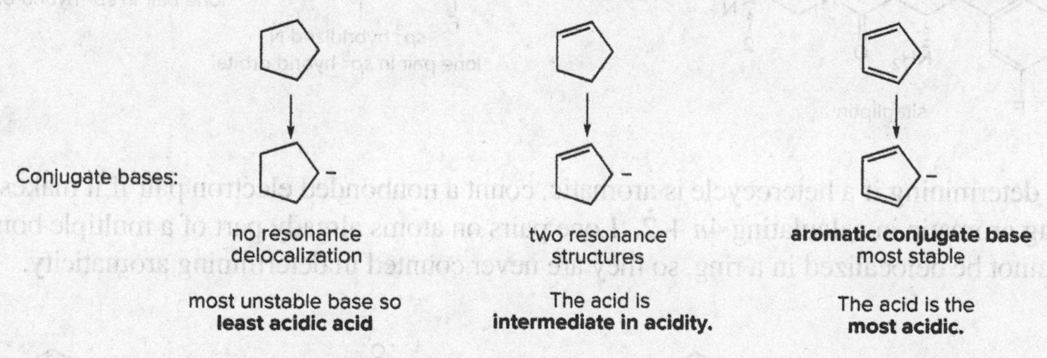

Conjugate bases:
no resonance delocalization
most unstable base so least acidic acid
two resonance structures
The acid is intermediate in acidity.
aromatic conjugate base
most stable
The acid is the most acidic.

19.13

19.14 To be aromatic, the ions must have $4n + 2$ π electrons. Ions in (b) and (c) do not have the right number of π electrons to be aromatic.

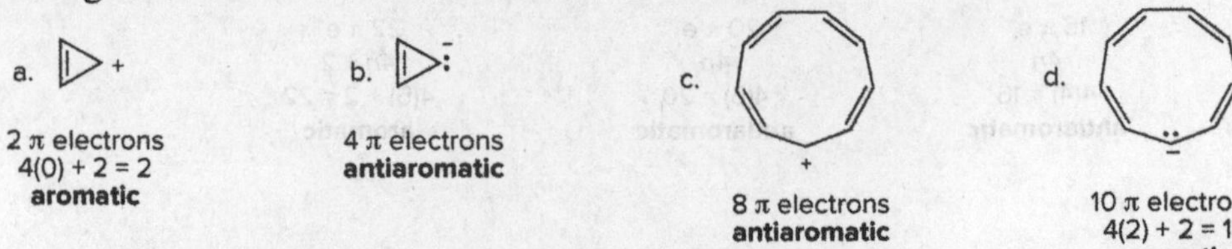

a. 2 π electrons
4(0) + 2 = 2
aromatic

b. 4 π electrons
antiaromatic

c. 8 π electrons
antiaromatic

d. 10 π electrons
4(2) + 2 = 10
aromatic

19.15

A =

The NMR indicates that **A** is aromatic. (a) The C's of the triple bond are *sp* hybridized. (b) Each triple bond has one set of electrons in *p* orbitals that overlap with other *p* orbitals on adjacent atoms in the ring. This overlap allows electrons to delocalize. Each C of the triple bonds also has a *p* orbital in the plane of the ring. The electrons in these *p* orbitals are localized between the C's of the triple bond, not delocalized in the ring. (c) Although **A** has 24 π e⁻ total, only 18 e⁻ are delocalized around the ring.

19.16

a. The five-membered ring is aromatic because it has 6 π electrons, two from each π bond and two from the N atom that is not part of a double bond.

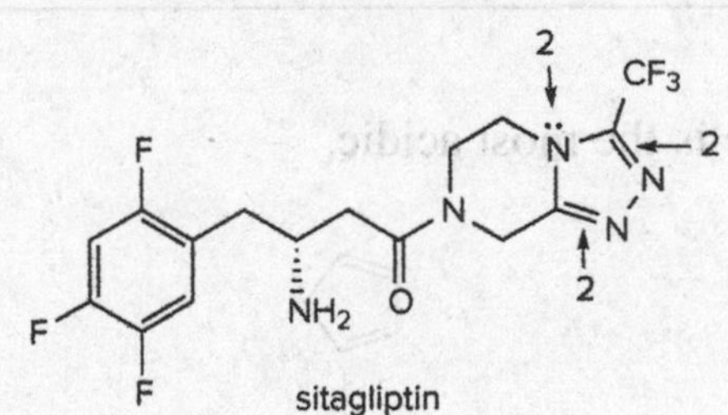

sitagliptin

b, c.

*sp*² hybridized N
lone pair in *p* orbital

*sp*² hybridized N
lone pair in *p* orbital

*sp*² hybridized N
lone pair in *sp*² hybrid orbital

*sp*² hybridized N
lone pair in *sp*² hybrid orbital

*sp*³ hybridized N
lone pair in *sp*³ hybrid orbital

19.17 In determining if a heterocycle is aromatic, count a nonbonded electron pair if it makes the ring aromatic in calculating $4n + 2$. Lone pairs on atoms already part of a multiple bond cannot be delocalized in a ring, so they are never counted in determining aromaticity.

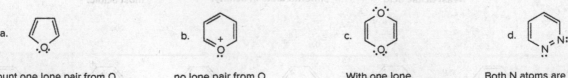

a. Count one lone pair from O.
$4n + 2 = 4(1) + 2 = 6$
aromatic

b. no lone pair from O
$4n + 2 = 4(1) + 2 = 6$
aromatic

c. With one lone pair from each O there would be 8 π electrons.
If O's are *sp*³ hybridized, the ring is not completely conjugated.
not aromatic

d. Both N atoms are part of a double bond, so the lone pairs cannot be counted: there are 6 π electrons.
$4n + 2 = 4(1) + 2 = 6$
aromatic

19.18

a. 8 π e⁻ from C=C's, 2 π e⁻ from O, 10 π e⁻, aromatic

b. 8 π e⁻ from C=C's, 8 π e⁻, antiaromatic

c. sp^3; not completely conjugated, not aromatic

d. 8 π e⁻ from double bonds, 2 π e⁻ from O, 10 π e⁻, aromatic

e. 2 π e⁻; 10 π e⁻, aromatic

19.19 **Answer each question for the four bases that compose DNA. All N atoms are sp^2 hybridized.**

cytosine: electron pair in p orbital (:NH_2); electron pair in sp^2 hybrid orbital (:N); electron pair in p orbital (N–H)

thymine: electron pair in p orbital (HN); electron pair in p orbital (N–H)

adenine: electron pair in p orbital (:NH_2); electron pair in sp^2 hybrid orbital (:N); electron pair in p orbital (N–H); electron pair in sp^2 hybrid orbital; electron pair in sp^2 hybrid orbital

guanine: electron pair in p orbital (HN); electron pair in p orbital (H_2N); electron pair in p orbital (N–H); electron pair in sp^2 hybrid orbital; electron pair in sp^2 hybrid orbital

19.20 **Draw resonance structures for the six-membered thymine and guanine rings that clearly show the π electrons delocalized within the aromatic rings.**

thymine ⟷ 6 π electrons

guanine ⟷ 10 π electrons

19.21 **In using the inscribed polygon method, always draw the vertex pointing down.**

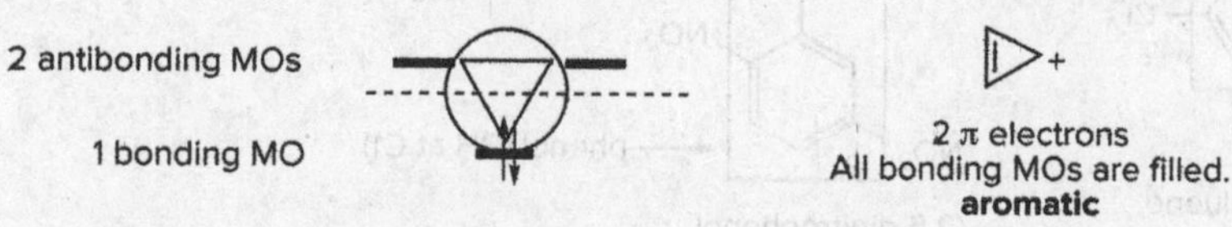

19.22 Draw the inscribed pentagons with the vertex pointing down. Then draw the molecular orbitals (MOs) and add the electrons.

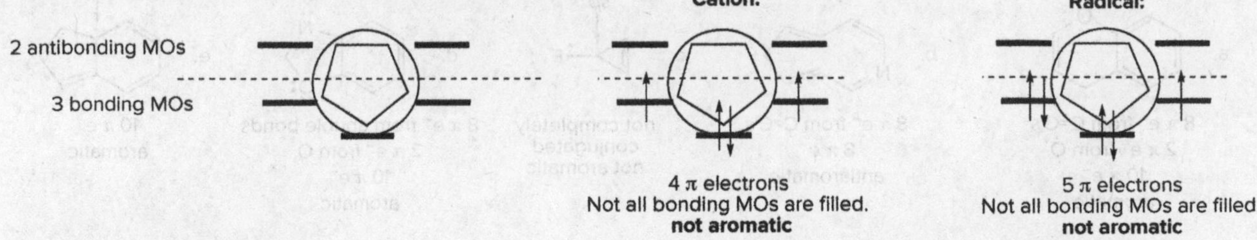

19.23

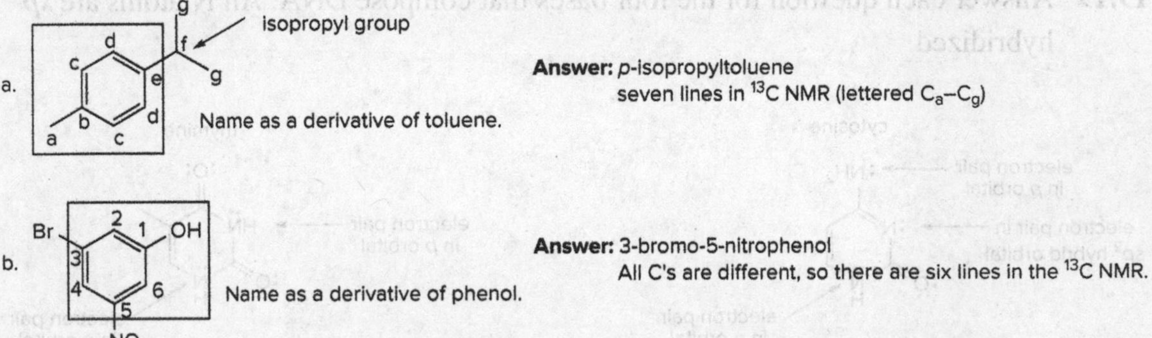

19.24

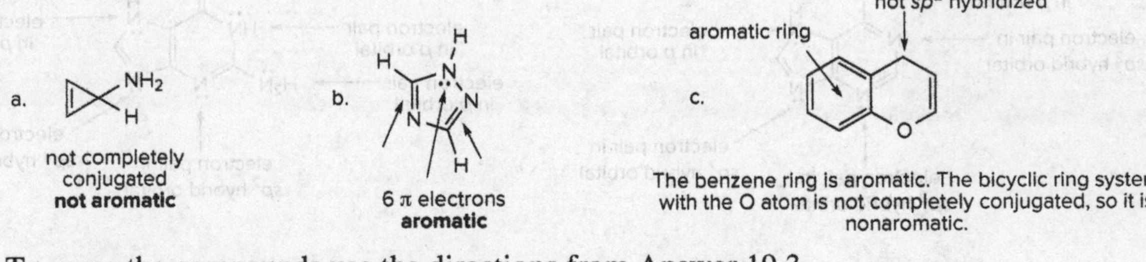

19.25 To name the compounds use the directions from Answer 19.3.

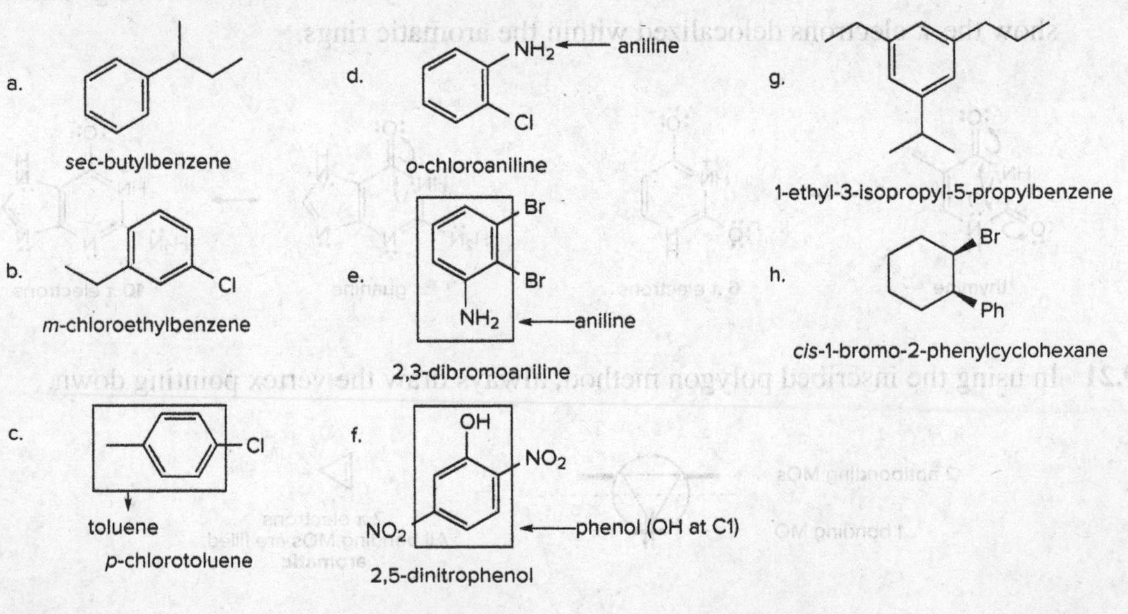

19.26

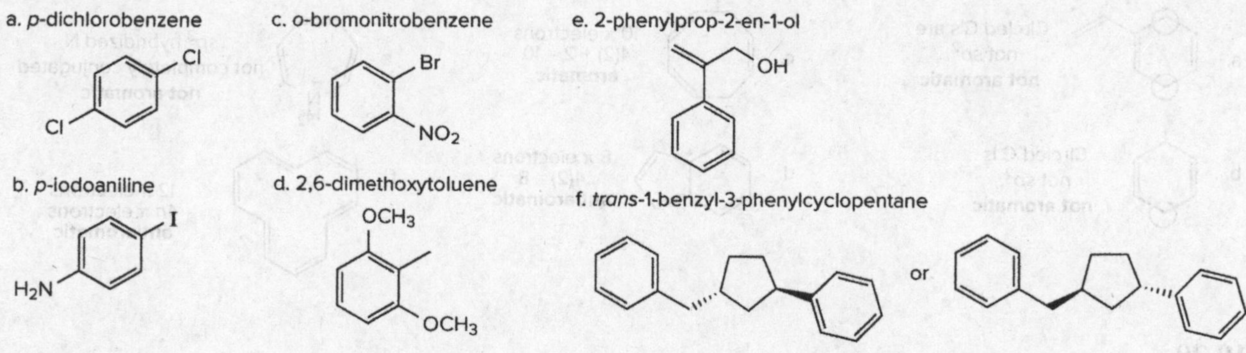

19.27

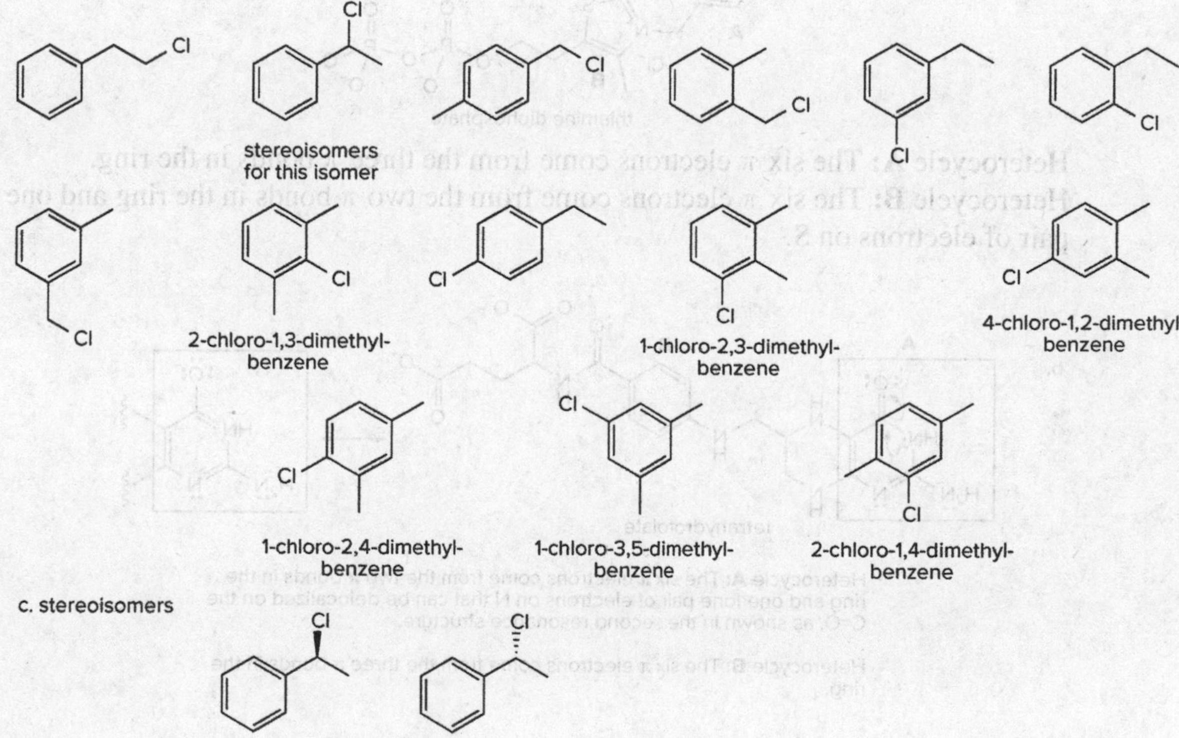

19.28 In determining if a heterocycle is aromatic, count a nonbonded electron pair if it makes the ring aromatic in calculating $4n + 2$. Lone pairs on atoms already part of a multiple bond cannot be delocalized in a ring, so they are never counted in determining aromaticity.

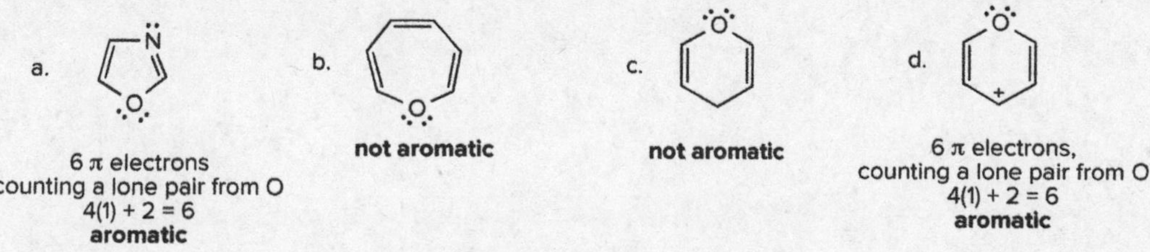

19.29

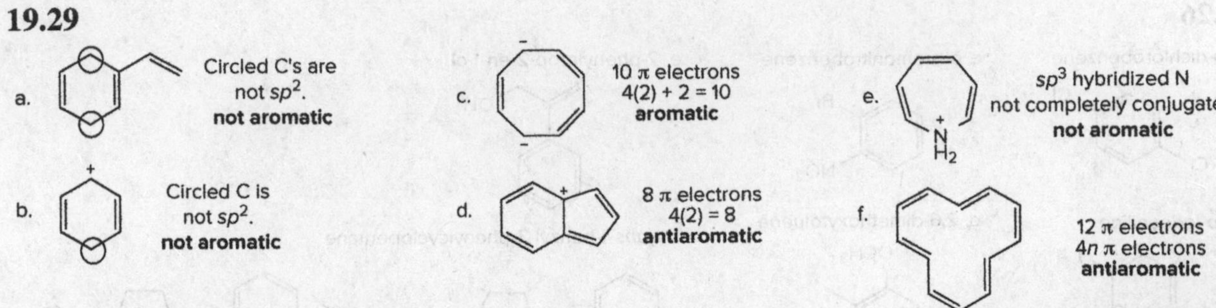

19.30

a. Both heterocycles in thiamine diphosphate are aromatic.

thiamine diphosphate

Heterocycle **A:** The six π electrons come from the three π bonds in the ring.
Heterocycle **B:** The six π electrons come from the two π bonds in the ring and one lone pair of electrons on S.

b.

tetrahydrofolate

Heterocycle **A:** The six π electrons come from the two π bonds in the ring and one lone pair of electrons on N that can be delocalized on the C=O, as shown in the second resonance structure.

Heterocycle **B:** The six π electrons come from the three π bonds in the ring.

19.31 To draw each enol form, replace the C=O by C–OH, draw a double bond between C and N, and remove a proton on the N atom.

keto forms | enol forms

cytosine

thymine

guanine

19.32

6 π electrons in this ring | 6 π electrons in this ring

A

A resonance structure can be drawn for **A** that places a negative charge in the five-membered ring and a positive charge in the seven-membered ring. This resonance structure shows that each ring has six π electrons, making it aromatic. The molecule possesses a dipole such that the seven-membered ring is electron deficient and the five-membered ring is electron rich.

19.33 Each compound is completely conjugated. A compound with $4n + 2$ π electrons is especially stable, whereas a compound with $4n$ π electrons is especially unstable.

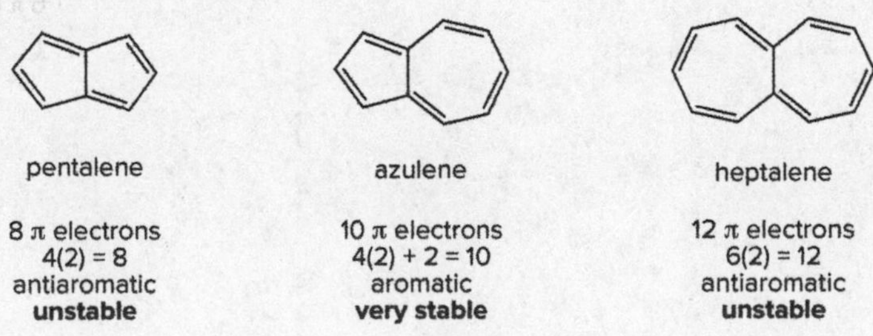

19.34

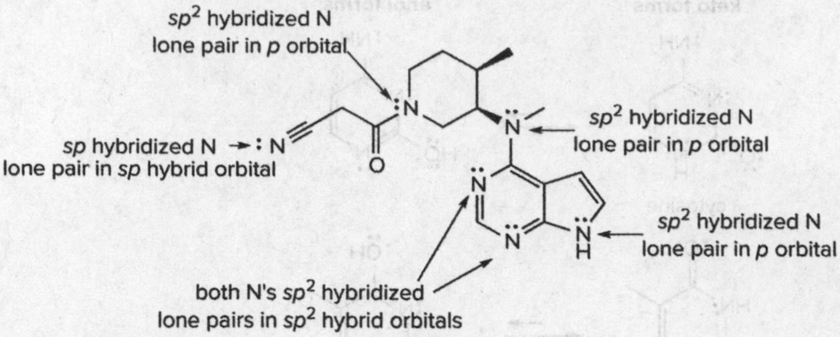

sp² hybridized N
lone pair in p orbital

sp hybridized N
lone pair in sp hybrid orbital

sp² hybridized N
lone pair in p orbital

sp² hybridized N
lone pair in p orbital

both N's sp² hybridized
lone pairs in sp² hybrid orbitals

19.35

a, b.

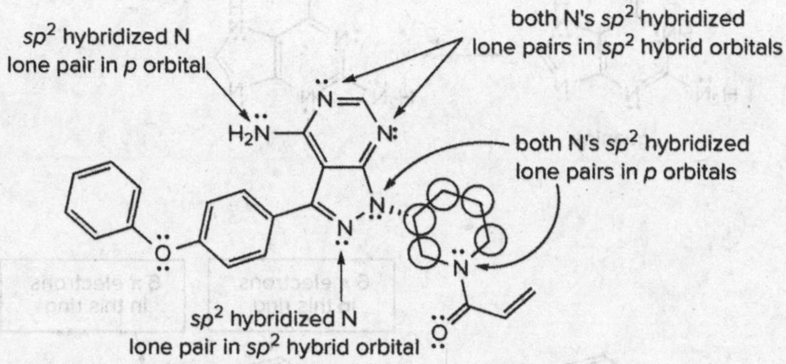

sp² hybridized N
lone pair in p orbital

both N's sp² hybridized
lone pairs in sp² hybrid orbitals

both N's sp² hybridized
lone pairs in p orbitals

sp² hybridized N
lone pair in sp² hybrid orbital

c. Ibrutinib has 28 sp^2 hybridized atoms. Circled atoms are sp^3 hybridized and all remaining C's, N's, and O's are sp^2 hybridized.

19.36 A second resonance structure can be drawn for the six-membered ring that gives it three π bonds, thus making it aromatic with six π electrons.

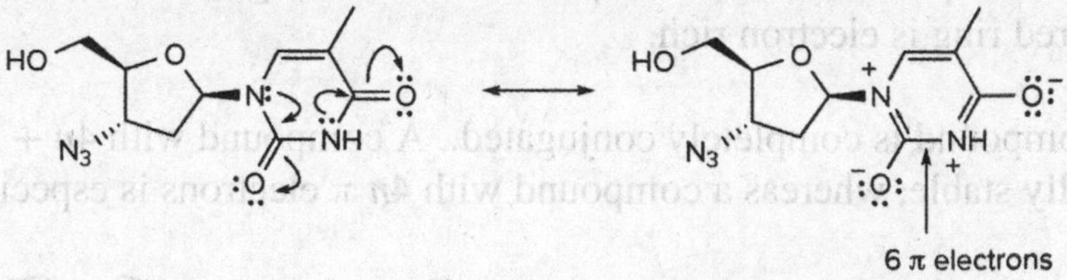

6 π electrons

19.37 The rate of an S_N1 reaction increases with increasing stability of the intermediate carbocation.

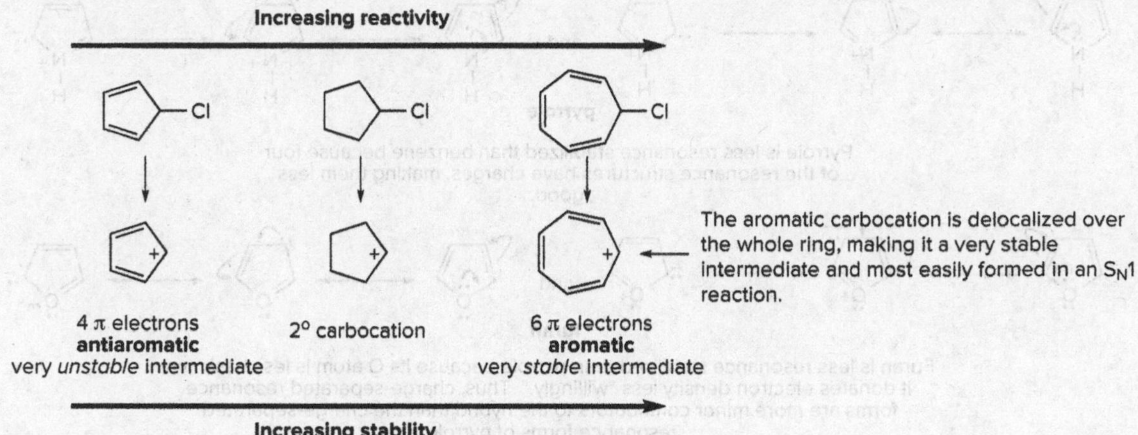
Increasing reactivity
Cl
Cl
Cl
The aromatic carbocation is delocalized over the whole ring, making it a very stable intermediate and most easily formed in an S_N1 reaction.
4 π electrons
antiaromatic
very unstable intermediate
2° carbocation
6 π electrons
aromatic
very stable intermediate
Increasing stability

19.38

Na⁺H⁻ ; H–OH ; + H–D ; + Na⁺ ⁻OH ; HO–H

[Two additional resonance structures are not drawn.]

19.39

Naphthalene can be drawn as three resonance structures:

(a) (b)

In two of the resonance structures bond (a) is a double bond, and bond (b) is a single bond. Therefore, bond (b) has more single bond character, making it longer.

19.40

a.

pyrrole

Pyrrole is less resonance stabilized than benzene because four of the resonance structures have charges, making them less good.

b.

furan

Furan is less resonance stabilized than pyrrole because its O atom is less basic, so it donates electron density less "willingly." Thus, charge-separated resonance forms are more minor contributors to the hybrid than the charge-separated resonance forms of pyrrole.

19.41 The compound with the most stable conjugate base is the strongest acid. Draw and compare the stability of the conjugate bases.

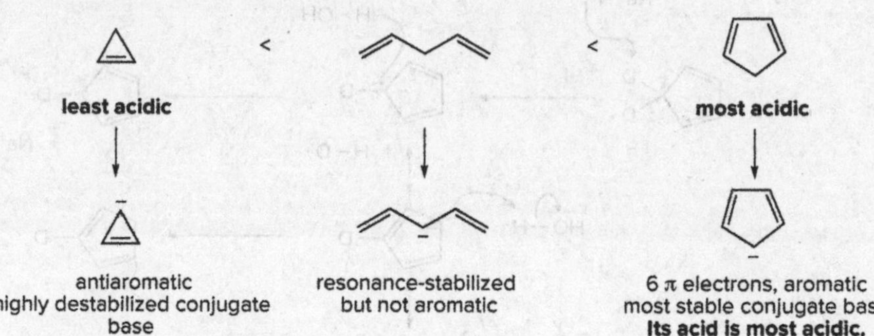

19.42

indene + $NaNH_2$ → Na^+ + NH_3

The conjugate base of indene has 10 π electrons, making it aromatic and very stable. Therefore, indene is more acidic than many hydrocarbons.

19.43

pyrrole — conjugate base →

Both pyrrole and the conjugate base of pyrrole have 6 π electrons in the ring, making them both aromatic. Thus, deprotonation of pyrrole does not result in a gain of aromaticity because the starting material is aromatic to begin with.

cyclopentadiene — conjugate base →

more acidic → H H

Cyclopentadiene is not aromatic, but the conjugate base has 6 π electrons and is therefore aromatic. This makes the C–H bond in cyclopentadiene more acidic than the N–H bond in pyrrole, because deprotonation of cyclopentadiene forms an aromatic conjugate base.

19.44

a. pyrrole + H^+ → **A** (protonation at C2) ⟷ ⟷ ; or + H^+ → **B**

Protonation at C2 forms conjugate acid **A** because the positive charge can be delocalized by resonance. There is no resonance stabilization of the positive charge in **B**.

b. **A**

pK_a = 0.4

Loss of a proton from **A** (which is not aromatic) gives two electrons to N, and forms pyrrole, which has six π electrons that can then delocalize in the five-membered ring, making it aromatic. This makes deprotonation a highly favorable process, and **A** more acidic.

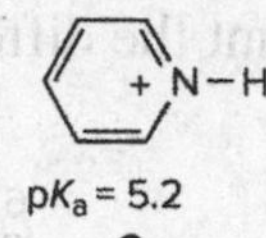

Both **C** and its conjugate base pyridine are aromatic. Because **C** has six π electrons, it is already aromatic to begin with, so there is less to be gained by deprotonation, and **C** is thus less acidic than **A**.

19.45

a.

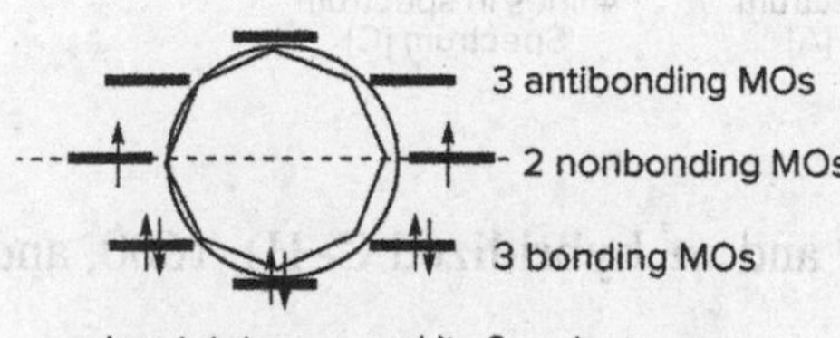

cyclooctatetraene and its 8 π electrons

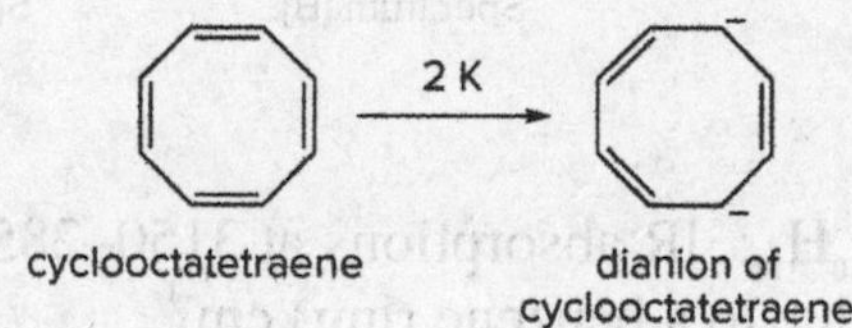

b. Even if cyclooctatetraene were flat, it has two unpaired electrons in its HOMOs (nonbonding MOs), so it cannot be aromatic.

c. The dianion has 10 π electrons.

d. The two additional electrons fill the nonbonding MOs; that is, all the bonding and nonbonding MOs are filled with electrons in the dianion.

e. The dianion is aromatic because its HOMOs are completely filled, and it has no electrons in antibonding MOs.

19.46

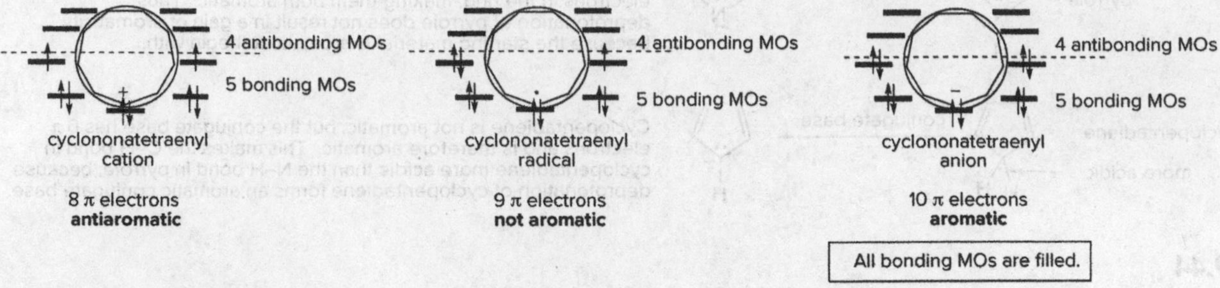

19.47 The number of different types of C's = the number of signals.

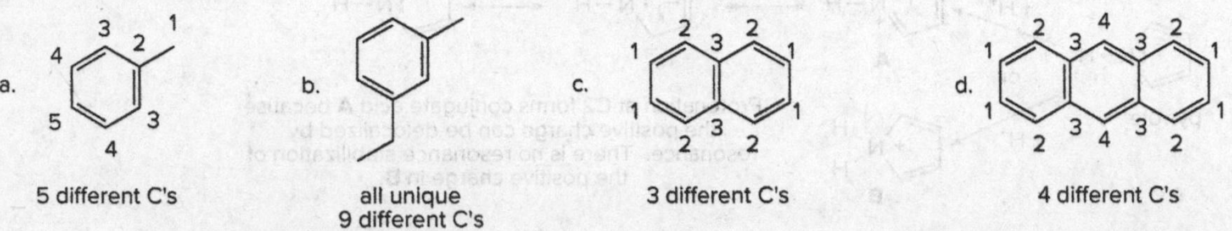

19.48 Draw the three isomers and count the different types of carbons in each. Then match the structures with the data.

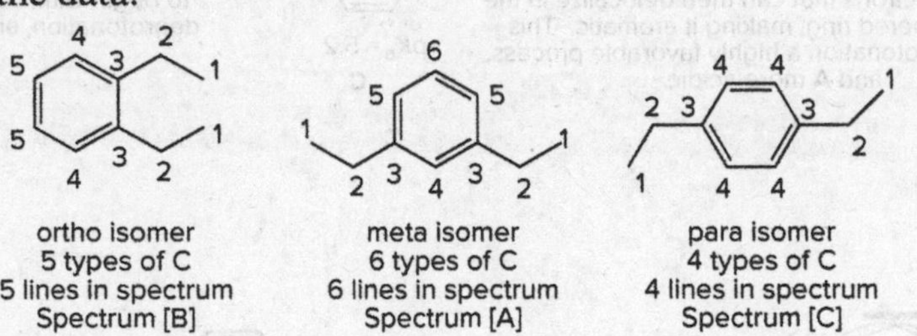

19.49

a. $C_{10}H_{14}$: IR absorptions at 3150–2850 (sp^2 and sp^3 hybridized C–H), 1600, and 1500 (due to a benzene ring) cm^{-1}

^1H NMR data:

Absorption	ppm	# of H's	Explanation	
doublet	1.2	6	6 H's adjacent to 1 H	$(CH_3)_2CH–$
singlet	2.3	3	CH_3	
septet	3.1	1	1 H adjacent to 6 H's	$(CH_3)_2CH–$
multiplet	7–7.4	4	a disubstituted benzene ring	

You can't tell from these data where the two groups are on the benzene ring. They are not para, because the para arrangement usually gives two sets of distinct peaks (resembling two doublets), so there are two possible structures—ortho and meta isomers.

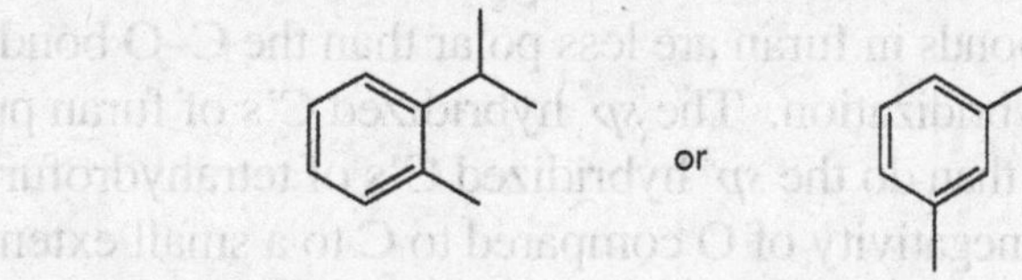

or

b. C_9H_{12}: ^{13}C NMR signals at 21, 127, and 138 ppm → means three different types of C's. ^1H NMR shows two types of H's: 9 H's probably means 3 CH_3 groups; the other 3 H's are very deshielded so they are bonded to a benzene ring.
Only one possible structure fits:

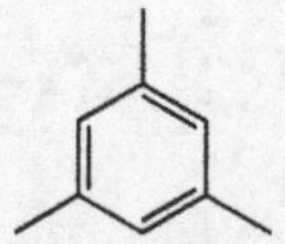

c. C_8H_{10}: IR absorptions at 3108–2875 (sp^2 and sp^3 hybridized C–H), 1606, and 1496 (due to a benzene ring) cm^{-1}
^1H NMR data:

Absorption	ppm	# of H's	Explanation
triplet	1.3	3	3 H's adjacent to 2 H's
quartet	2.7	2	2 H's adjacent to 3 H's
multiplet	7.3	5	a monosubstituted benzene ring

Structure:

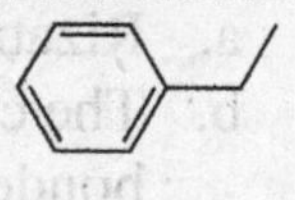

19.50

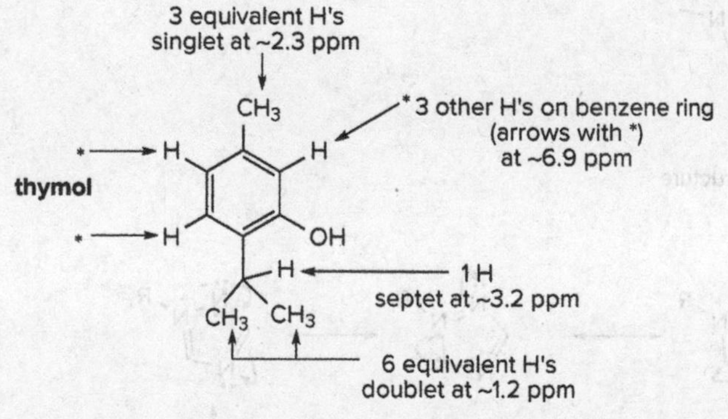

3 equivalent H's singlet at ~2.3 ppm

CH_3

*3 other H's on benzene ring (arrows with *) at ~6.9 ppm

thymol

H H H OH

H ← 1 H septet at ~3.2 ppm

CH_3 CH_3

6 equivalent H's doublet at ~1.2 ppm

IR absorptions:
3500–3200 cm^{-1} (O–H)
3150–2850 cm^{-1} (C–H bonds)
1621 and 1585 cm^{-1} (benzene ring)

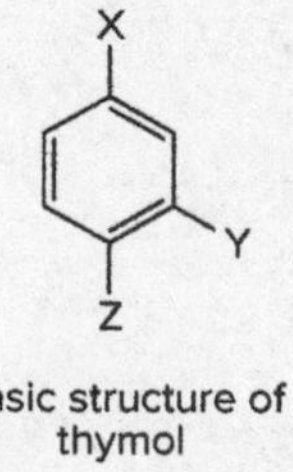

X Y Z

basic structure of thymol

Thymol must have this basic structure, given the NMR and IR data because it is a trisubstituted benzene ring with one singlet and two doublets in the NMR at ~6.9 ppm. However, which group [OH, CH_3, or $CH(CH_3)_2$] corresponds to X, Y, and Z is not readily distinguished with the given data. The correct structure for thymol is given.

19.51 Because tetrahydrofuran has a higher boiling point and is more water soluble, it must be more polar and have stronger intermolecular forces than furan. There are two contributing factors. One lone pair on furan's O atom is delocalized on the five-membered ring to make it aromatic. This makes it less available for H-bonding with water and other intermolecular interactions. Also, the C–O bonds in furan are less polar than the C–O bonds in tetrahydrofuran because of hybridization. The sp^2 hybridized C's of furan pull a little more electron density toward them than do the sp^3 hybridized C's of tetrahydrofuran. This counteracts the higher electronegativity of O compared to C to a small extent.

sp^3 C → more polar C–O bond

sp^2 C → less polar C–O bond

delocalized

19.52

rizatriptan

a. Rizatriptan contains three aromatic rings.

b. The circled N atom is sp^3 hybridized. All other N atoms are either part of a π bond or bonded to a π bond, making them sp^2 hybridized.

c. The lone pair on the circled N atom occupies an sp^3 hybrid orbital. The lone pairs on the boxed-in N atoms are in a *p* orbital. The lone pairs on the unlabeled N atoms occupy an sp^2 hybrid orbital.

d.

one additional resonance structure

e.

19.53

a, b.

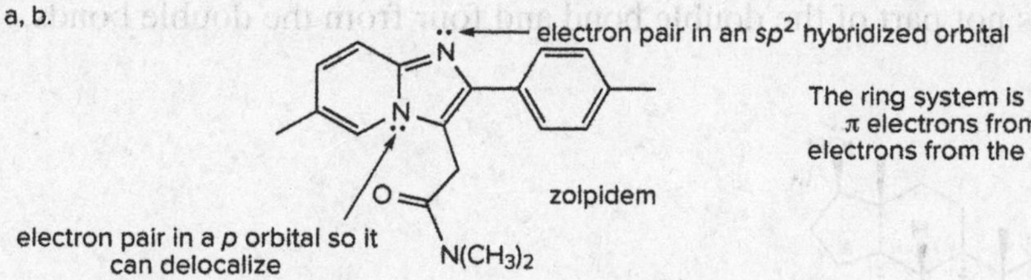

electron pair in an sp^2 hybridized orbital

The ring system is aromatic with 10 π electrons, 8 π electrons from the double bonds and 2 π electrons from the N atom common to both rings.

zolpidem

electron pair in a *p* orbital so it can delocalize

$N(CH_3)_2$

c.

$N(CH_3)_2$ $N(CH_3)_2$ $N(CH_3)_2$

$N(CH_3)_2$ $N(CH_3)_2$ $N(CH_3)_2$

$N(CH_3)_2$ $N(CH_3)_2$ $N(CH_3)_2$

$N(CH_3)_2$ $N(CH_3)_2$

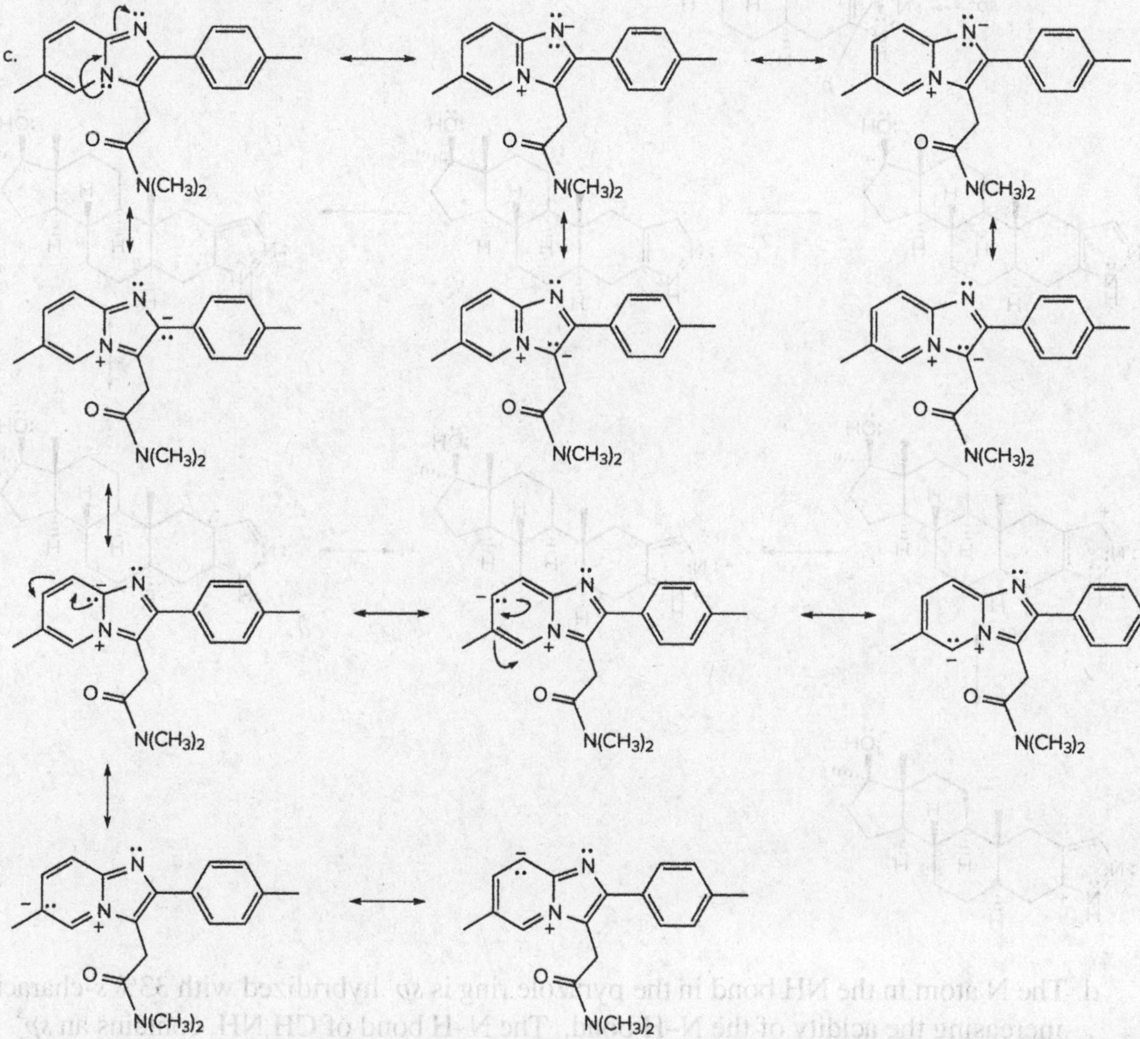

19.54

a. Pyrazole rings are aromatic because they have six π electrons—two from the lone pair on the N atom that is not part of the double bond and four from the double bonds.

b.

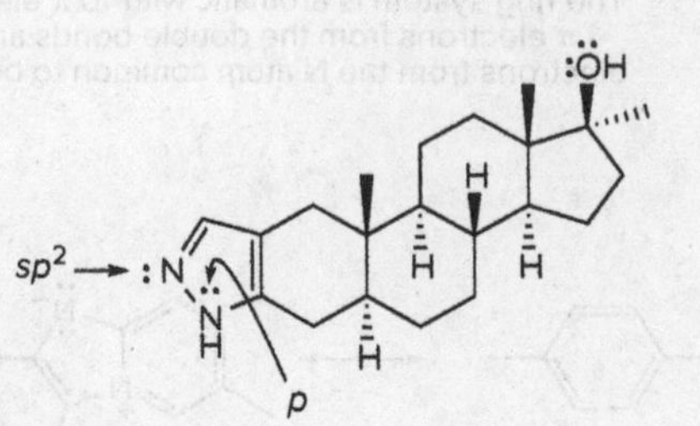

sp2 → :N
p

c.

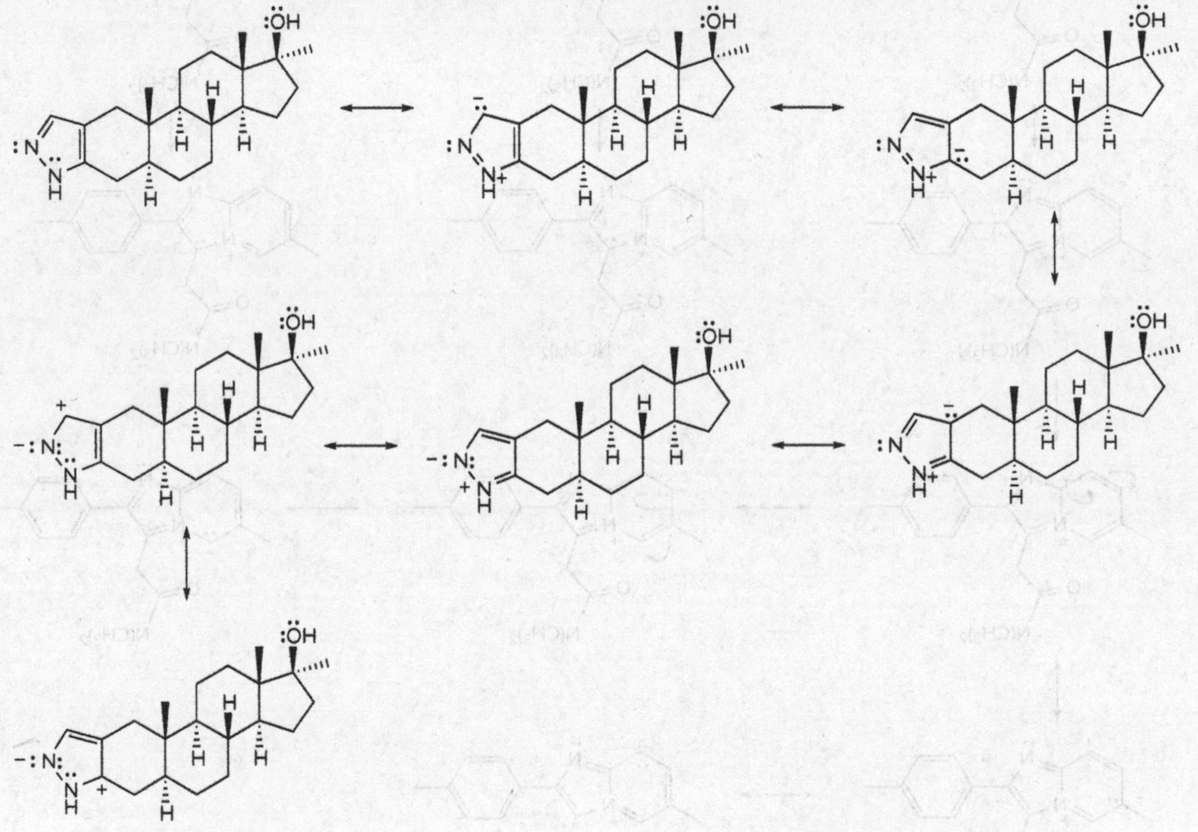

d. The N atom in the NH bond in the pyrazole ring is sp^2 hybridized with 33% *s*-character, increasing the acidity of the N–H bond. The N–H bond of CH_3NH_2 contains an sp^3 hybridized N atom (25% *s*-character)

19.55 Both **A** and **B** are cyclic, and if the lone pair of electrons on N is in a *p* orbital, they are completely conjugated with 10 π electrons, a number that satisfies Hückel's rule. To be aromatic, **A** and **B** must be planar, and the internal bond angles of **A** and **B** would be much larger than 120°, the theoretical bond angle of sp^2 hybridized C's. The fact that **A** is aromatic means that the lone pair on N occupies a *p* orbital, so it can delocalize onto the nine-membered ring. The stabilization gained by being aromatic is greater than any angle strain. With **B,** the lone pair on N is also delocalized onto the C=O, making it less available for donation to the ring, so the ring is not aromatic.

:NH

A

2 electrons in a p orbital

10 π electrons

:O:

:N

OCH_2CH_3

:O:

+N

OCH_2CH_3

B

Electron pair is less available, so the ring is not aromatic.

19.56

a. Loss of the most acidic proton forms the most stable conjugate base. When a conjugate base is both resonance stabilized and aromatic, it is especially stable.

A H most acidic H 1 2 3 4 5 6

The boxed-in H is the most acidic because the conjugate base has six resonance structures, five of which (**2–6**) have an aromatic five-membered ring with six π electrons.

b. **A** is more acidic than **B,** because loss of a proton in **B** generates a resonance-stabilized conjugate base, but it is **not** aromatic.

B: H B

No resonance structure has 6 π electrons in the ring.

19.57 With 14 π electrons in the double bonds, the system is aromatic [4(3) + 2 = 14 π electrons]. The ring current generated by the circulating π electrons deshields the protons on the C=C's, so they absorb downfield (8.14–8.67 ppm). The CH_3 groups, however, are very shielded because they lie above and below the plane, so they absorb far upfield (–4.25 ppm). The dianion of **C** now has 16 π electrons, making it antiaromatic, so the position of the absorptions reverses. The C=C protons are now shielded (–3 ppm), and the CH_3 protons are now deshielded (21 ppm).

–4.75 ppm

~ 8 ppm

C

21 ppm

2–

–3 ppm

dianion

19.58 A second resonance structure for **A** shows that the ring is completely conjugated and has six π electrons, making it aromatic and especially stable. A similar charge-separated resonance structure for **B** makes the ring completely conjugated, but gives the ring four π electrons, making it antiaromatic and especially unstable.

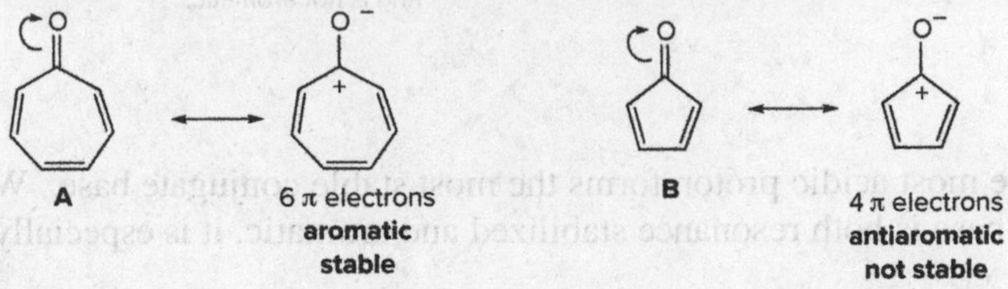

A
6 π electrons
aromatic
stable

B
4 π electrons
antiaromatic
not stable

19.59 The conversion of carvone to carvacrol involves acid-catalyzed isomerization of two double bonds and tautomerization of a ketone to an enol tautomer. In this case the enol form is part of an aromatic phenol. Each isomerization of a C=C involves Markovnikov addition of a proton, followed by deprotonation.

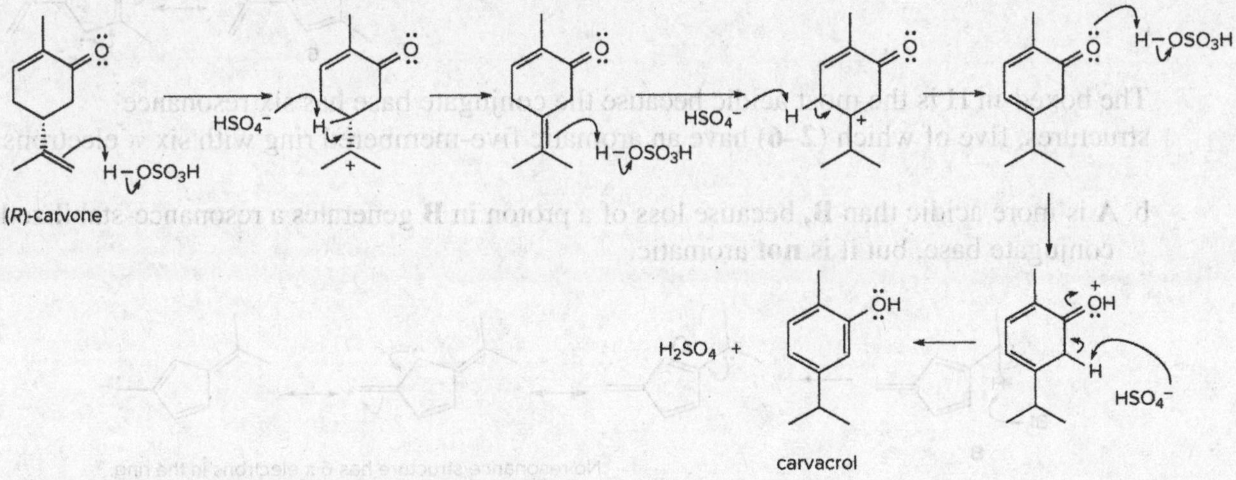

H–OSO$_3$H
(*R*)-carvone
HSO_4^-
H–OSO$_3$H
HSO_4^-
H–OSO$_3$H
H_2SO_4 +
carvacrol
HSO_4^-

19.60 Resonance structures for triphenylene:

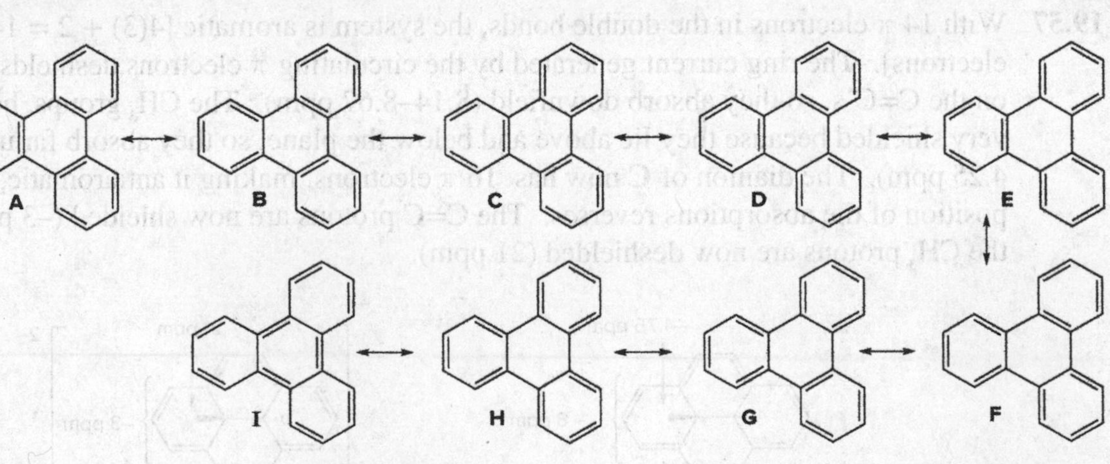

A B C D E
I H G F

Resonance structures **A–H** all keep three double and three single bonds in the three six-membered rings on the periphery of the molecule. This means that each ring behaves like an isolated benzene ring undergoing substitution rather than addition because the π electron density is delocalized within each six-membered ring. Only resonance structure **I** does not have this form. Each C–C bond of triphenylene has four (or five) resonance structures in which it is a single bond and four (or five) resonance structures in which it is a double bond.

Resonance structures for phenanthrene:

A B C D E

With phenanthrene, however, four of the five resonance structures keep a double bond at the labeled C's. (Only **C** does not.) This means that these two C's have more double bond character than other C–C bonds in phenanthrene, making them more susceptible to addition rather than substitution.

19.61

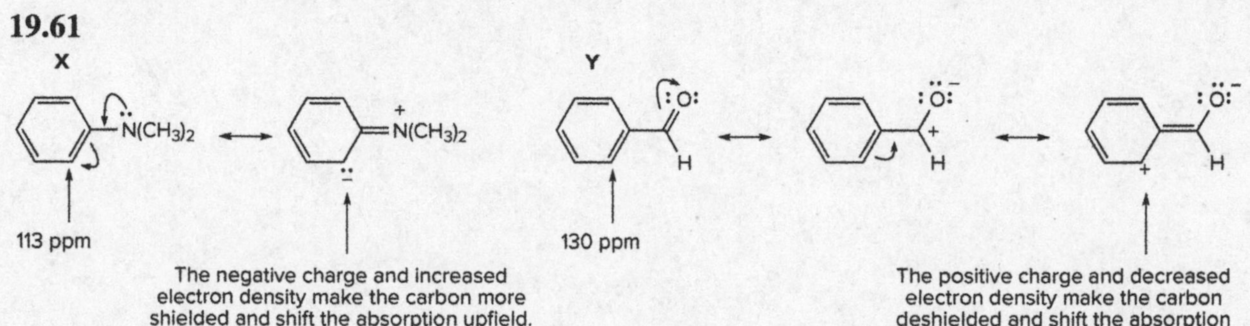
X
N(CH3)2
N(CH3)2
113 ppm
The negative charge and increased electron density make the carbon more shielded and shift the absorption upfield.
Y
H
H
H
130 ppm
The positive charge and decreased electron density make the carbon deshielded and shift the absorption downfield.

Chapter 20 Reactions of Aromatic Compounds

Chapter Review

Mechanism of electrophilic aromatic substitution (20.2)

- Electrophilic aromatic substitution follows a two-step mechanism. Reaction of the aromatic ring with an electrophile forms a carbocation, and loss of a proton regenerates the aromatic ring.
- The first step is rate-determining.
- The intermediate carbocation is stabilized by resonance; a minimum of three resonance structures can be drawn. The positive charge is always located ortho or para to the new C–E bond.

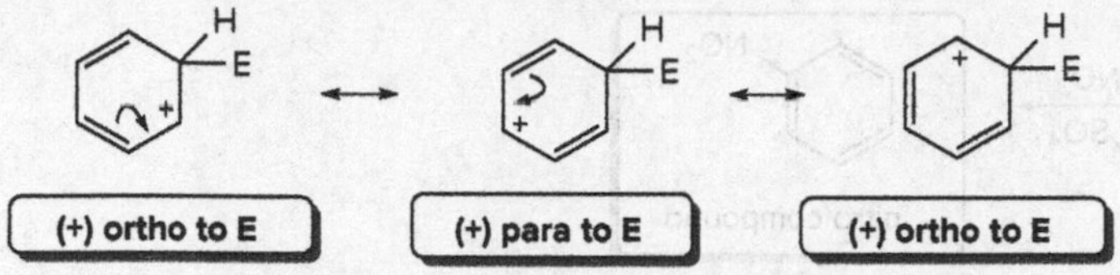

(+) ortho to E ⟷ (+) para to E ⟷ (+) ortho to E

Three rules describing the reactivity and directing effects of common substituents (20.7–20.9)

[1] All ortho, para directors except the halogens activate the benzene ring.
[2] All meta directors deactivate the benzene ring.
[3] The halogens deactivate the benzene ring.

Summary of substituent effects in electrophilic aromatic substitution (20.6–20.9)

	Substituent	Inductive effect	Resonance effect	Reactivity	Directing effect
[1]	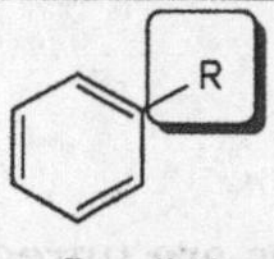 R, R = alkyl	**donating**	**none**	**activating**	**ortho, para**
[2]	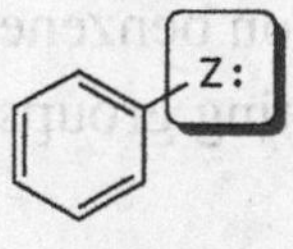 Z:, Z = N or O	**withdrawing**	**donating**	**activating**	**ortho, para**
[3]	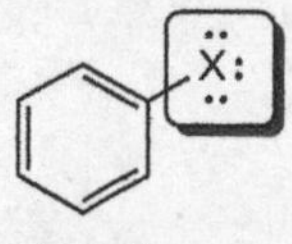 X:, X = halogen	**withdrawing**	**donating**	**deactivating**	**ortho, para**
[4]	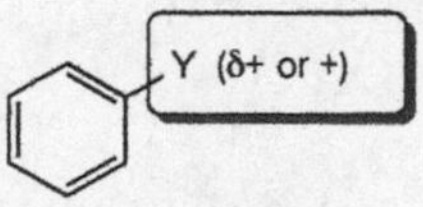 Y (δ+ or +)	**withdrawing**	**withdrawing**	**deactivating**	**meta**

Five examples of electrophilic aromatic substitution

[1] Halogenation—Replacement of H by Cl or Br (20.3)

Benzene (H) $\xrightarrow[\text{FeX}_3]{X_2}$ aryl chloride (Cl) or aryl bromide (Br)

[X = Cl, Br]

- Polyhalogenation occurs on benzene rings substituted by OH and NH_2 (and related substituents) (20.10A).

[2] Nitration—Replacement of H by NO_2 (20.4)

Benzene (H) $\xrightarrow[\text{H}_2\text{SO}_4]{\text{HNO}_3}$ nitro compound (NO_2)

[3] Sulfonation—Replacement of H by SO_3H (20.4)

Benzene (H) $\xrightarrow[\text{H}_2\text{SO}_4]{\text{SO}_3}$ benzenesulfonic acid (SO_3H)

[4] Friedel–Crafts alkylation—Replacement of H by R (20.5)

Benzene (H) $\xrightarrow[\text{AlCl}_3]{\text{RCl}}$ alkyl benzene (arene) (R)

- Rearrangements can occur.
- Vinyl halides and aryl halides are unreactive.
- The reaction does not occur on benzene rings substituted by meta deactivating groups or NH_2 groups (20.10B).
- Polyalkylation can occur.

Variations:

[1] with alcohols

Benzene (H) $\xrightarrow[\text{H}_2\text{SO}_4]{\text{ROH}}$ R-substituted benzene

[2] with alkenes

Benzene (H) $\xrightarrow[\text{H}_2\text{SO}_4]{\text{CH}_2=\text{CHR}}$ R-substituted ethylbenzene

[5] Friedel–Crafts acylation—Replacement of H by RCO (20.5)

RCOCl, $AlCl_3$ → ketone

- The reaction does not occur on benzene rings substituted by meta deactivating groups or NH_2 groups (20.10B).

Nucleophilic aromatic substitution (20.13)

[1] Nucleophilic substitution by an addition–elimination mechanism

:Nu⁻ → Nu

X = F, Cl, Br, I
A = electron-withdrawing group

- The mechanism has two steps.
- Strong electron-withdrawing groups at the ortho or para position are required.
- Increasing the number of electron-withdrawing groups increases the rate.
- Increasing the electronegativity of the halogen increases the rate.

[2] Nucleophilic substitution by an elimination–addition mechanism

:Nu⁻ → Nu

X = halogen

- Reaction conditions are harsh.
- Benzyne is formed as an intermediate.
- Product mixtures may result.

Other reactions of benzene derivatives

[1] Benzylic halogenation (20.14A)

Br_2, $h\nu$ or Δ → benzylic bromide

[2] Oxidation of alkyl benzenes (20.14B)

$KMnO_4$ → COOH, benzoic acid

- A benzylic C–H bond is needed for reaction.

[3] Reduction of ketones to alkyl benzenes (20.14C)

Zn(Hg), HCl
or
NH_2NH_2, ^-OH

alkyl benzene

[4] Reduction of nitro groups to amino groups (20.14D)

NO_2 → H_2, Pd-C or Fe, HCl or Sn, HCl → NH_2

aniline

Practice Test on Chapter Review

1.a. Which of the following statements is true about an ethoxy substituent ($-OCH_2CH_3$) on a benzene ring?

1. OCH_2CH_3 increases the rates of both electrophilic substitution and nucleophilic substitution.
2. OCH_2CH_3 decreases the rates of both electrophilic substitution and nucleophilic substitution.
3. OCH_2CH_3 increases the rate of electrophilic substitution and decreases the rate of nucleophilic substitution.
4. OCH_2CH_3 decreases the rate of electrophilic substitution and increases the rate of nucleophilic substitution.
5. None of these statements is true.

b. Which of the following statements is true about a $-CO_2CH_3$ group on a benzene ring?

1. CO_2CH_3 increases the rates of both electrophilic substitution and nucleophilic substitution.
2. CO_2CH_3 decreases the rates of both electrophilic substitution and nucleophilic substitution.
3. CO_2CH_3 increases the rate of electrophilic substitution and decreases the rate of nucleophilic substitution.
4. CO_2CH_3 decreases the rate of electrophilic substitution and increases the rate of nucleophilic substitution.
5. None of these statements is true.

c. Which of the following is *not* a valid resonance structure for the carbocation that results from ortho attack of an electrophile on $C_6H_5C(CH_3){=}CH_2$?

1. 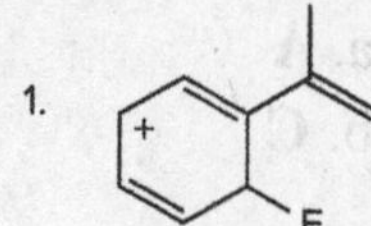

3.

5.

2.

4.

2. Draw the organic products formed in the following reactions.

a. [1] $CH_3CH_2CH_2COCl$, $AlCl_3$; [2] Zn(Hg), HCl

b. HNO_3, H_2SO_4; Fe, HCl

c. $KMnO_4$

d. CH_3O– ; HNO_3, H_2SO_4; H_2, Pd-C

e. Br, OCH_3; SO_3, H_2SO_4

f. Br_2 (1 equiv), $FeBr_3$

g. O_2N–, –Cl; $NaOCH_2CH_3$

h. –Cl; NaOH, Δ

3. (a) Considering the compound drawn below, which ring is *most* reactive in electrophilic aromatic substitution? (b) Which ring is the *least* reactive in electrophilic aromatic substitution?

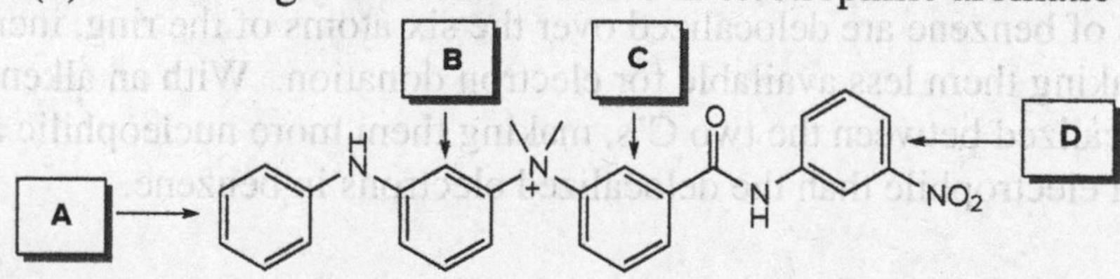

4. Classify each substituent as [1] ortho, para activating, [2] ortho, para deactivating, or [3] meta deactivating.

a. –Br
b. $-CH_2CH_2CH_2Br$
c. –COOH
d. $-NHCOCH_2CH_3$
e. $-N(CH_3)_6COCH_3$
f. $-CCl_3$

5. What reagents are needed to convert toluene ($C_6H_5CH_3$) to each compound?

a. C_6H_5COOH
b. $C_6H_5CH_2Br$
c. *p*-bromotoluene
d. *o*-nitrotoluene
e. *p*-ethyltoluene
f.

Answers to Practice Test

1.a. 3
b. 4
c. 4

2.
a. 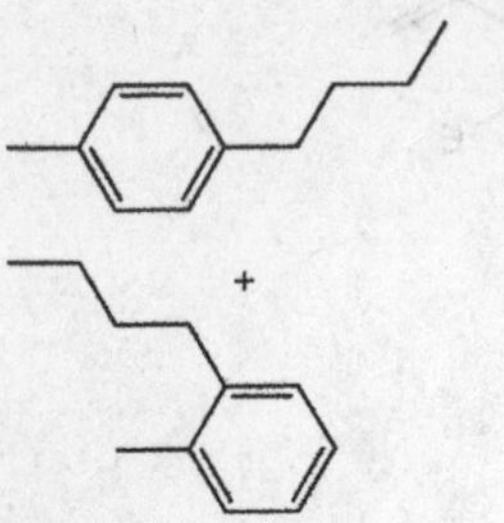 +
b. 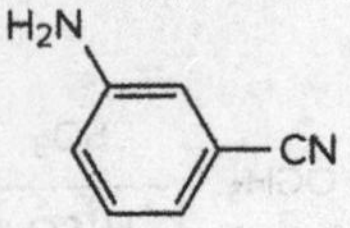 H_2N, CN
c. CO_2H, CO_2H
d. H_2N, CH_3O
e. OCH$_3$, Br, SO$_3$H
f. O, N, H, Br + O, N, H, Br
g. O_2N, O
h. OH + OH

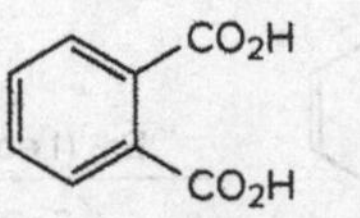

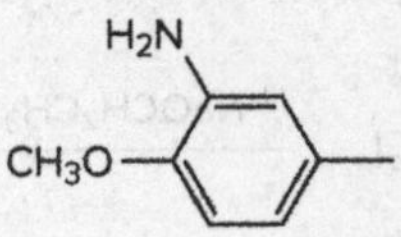

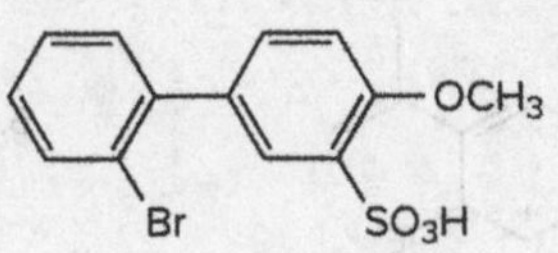

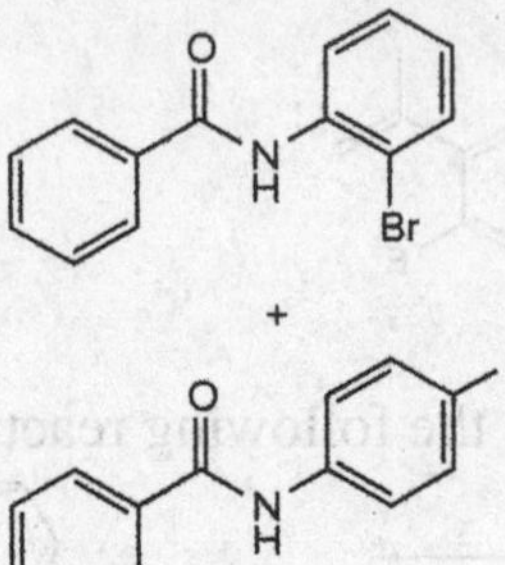

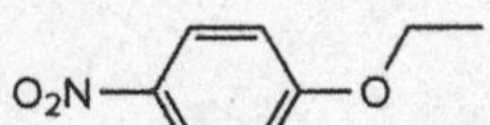

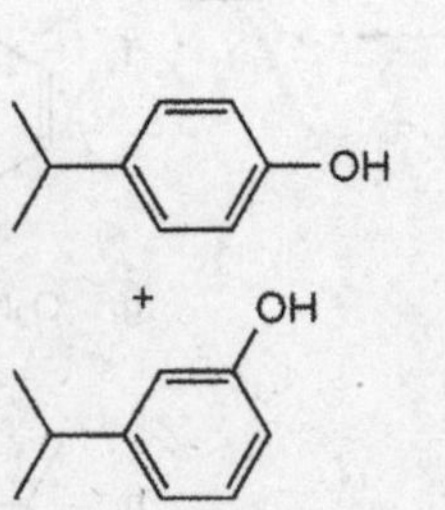

3. a. **A**
b. **C**

4.a. 2
b. 1
c. 3
d. 1
e. 1
f. 3

5.a. $KMnO_4$
b. Br_2, $h\nu$
c. Br_2, $FeBr_3$
d. HNO_3, H_2SO_4
e. CH_3CH_2Cl, $AlCl_3$
f. CH_3CH_2COCl, $AlCl_3$

Answers to Problems

20.1 The π electrons of benzene are delocalized over the six atoms of the ring, increasing benzene's stability and making them less available for electron donation. With an alkene, the two π electrons are localized between the two C's, making them more nucleophilic and thus more reactive with an electrophile than the delocalized electrons in benzene.

20.2

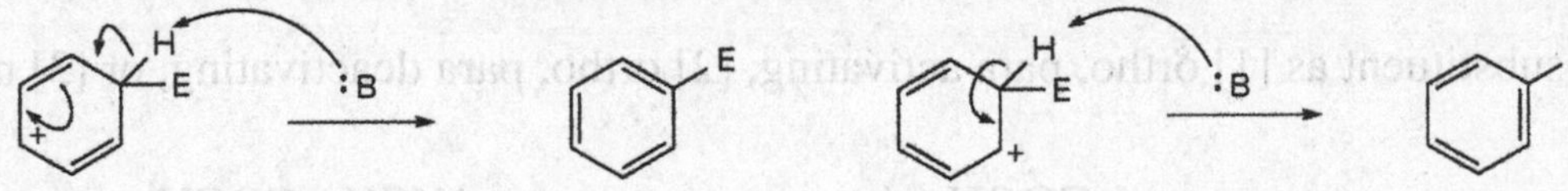

H, E, :B, +, E, H, E, :B, +, E

20.3 Reaction with Cl_2 and $FeCl_3$ as the catalyst occurs in two parts. First is the formation of an electrophile, followed by a two-step substitution reaction.

[1] :Cl–Cl: + $FeCl_3$ → :Cl–Cl–$FeCl_3$ (electrophile)

Lewis base Lewis acid electrophile

[2] resonance-stabilized carbocation + $FeCl_4^-$

[3] → chlorobenzene + HCl + $FeCl_3$

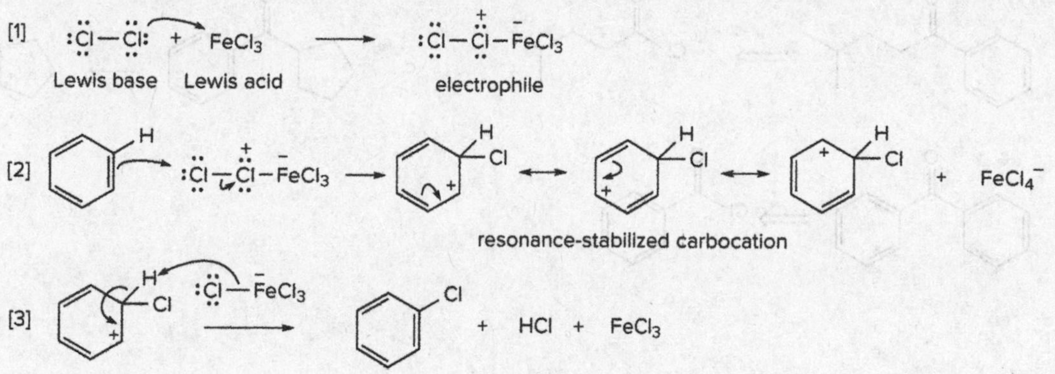

20.4 There are two parts in the mechanism. The first part is formation of an electrophile. The second part is a two-step substitution reaction.

A

[1] SO_3 + H–OSO_3H → $^+SO_3H$ + HSO_4^-

electrophile

[2] resonance-stabilized carbocation

[3] HSO_4^- → SO_3H + H_2SO_4

B

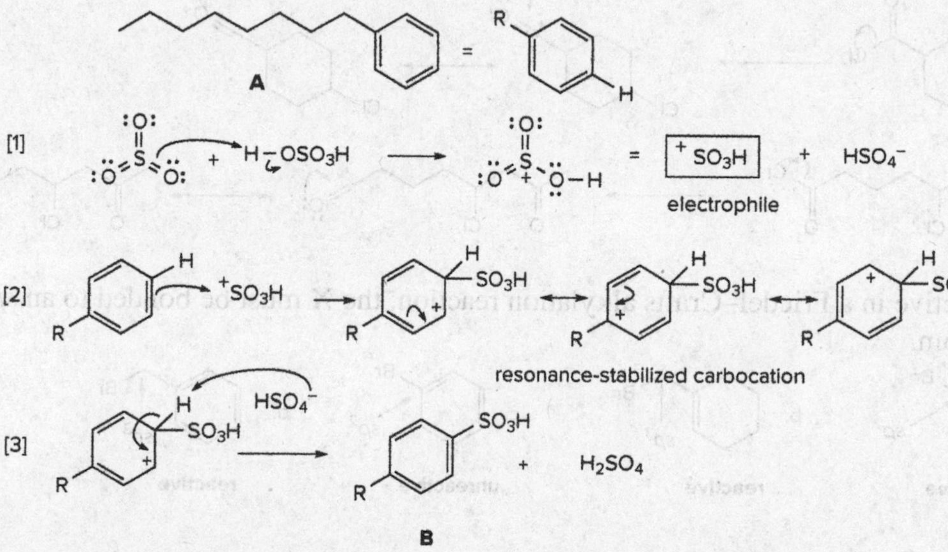

20.5 Friedel–Crafts alkylation results in the transfer of an alkyl group from a halogen to a benzene ring. In Friedel–Crafts acylation an acyl group is transferred from a halogen to a benzene ring.

a. benzene + 2-chloropropane —$AlCl_3$→ isopropylbenzene

b. benzene + chlorocyclohexane —$AlCl_3$→ cyclohexylbenzene

c. benzene + propanoyl chloride —$AlCl_3$→ propiophenone

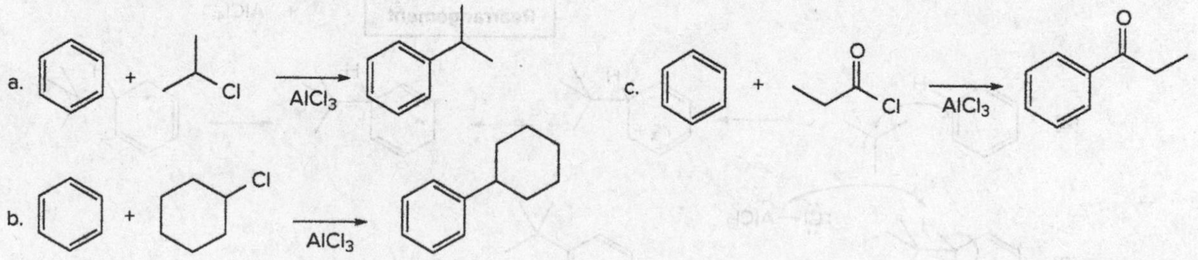

20.6 An acyl group is transferred from a Cl atom to a benzene ring. To draw the acid chloride, substitute a Cl for the benzene ring.

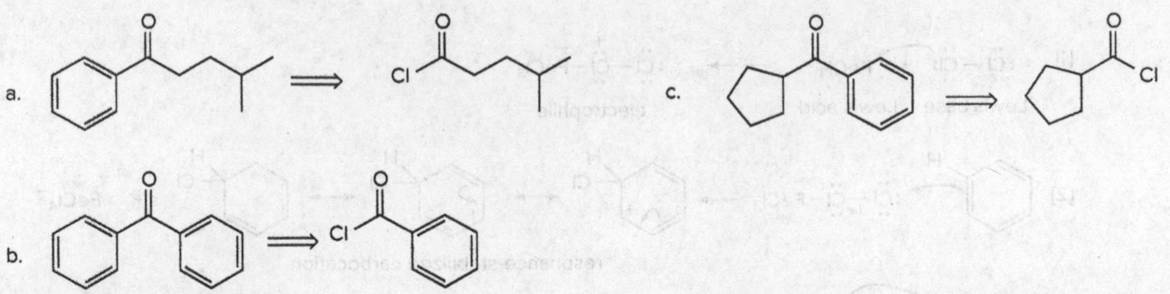

a. ⟹ Cl

c. ⟹ Cl

b. ⟹ Cl

20.7

a. OH, Cl → OH, O: ⟷ OH, O: +

resonance-stabilized acylium ion

b. Cl, Cl → Cl, O: ⟷ Cl, O+

c. Cl, Cl → Cl, O: ⟷ Cl, O: +

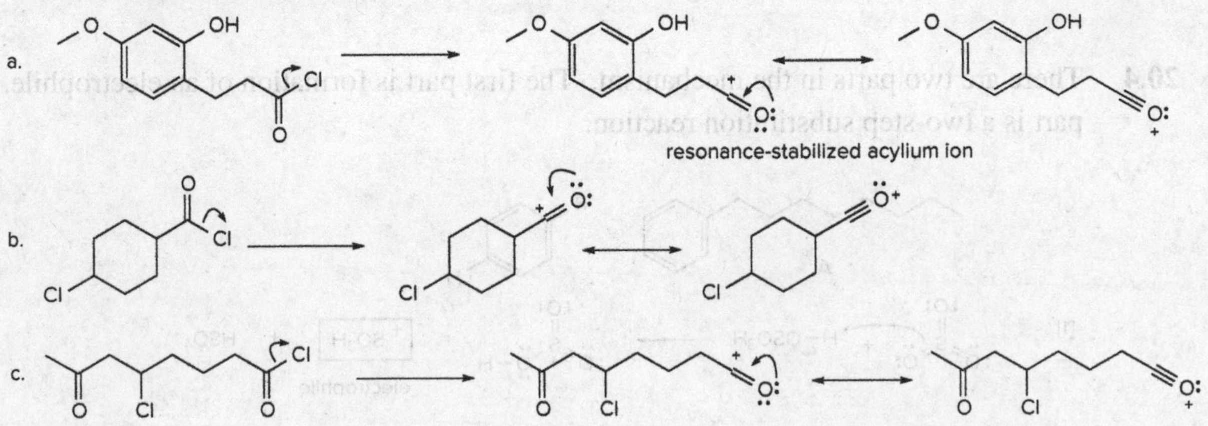

20.8 To be reactive in a Friedel–Crafts alkylation reaction, the X must be bonded to an sp^3 hybridized carbon atom.

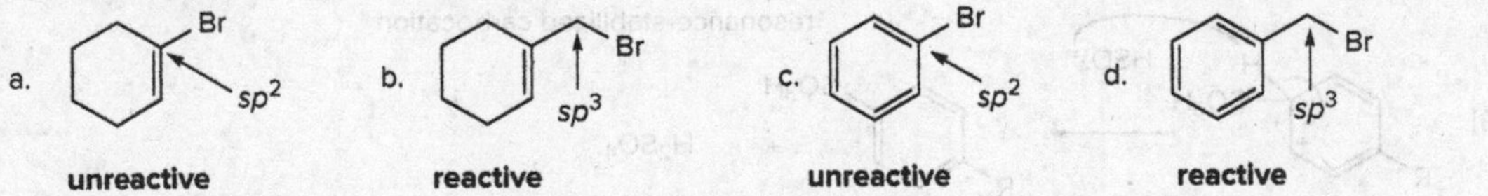

a. Br, sp^2 **unreactive**

b. Br, sp^3 **reactive**

c. Br, sp^2 **unreactive**

d. Br, sp^3 **reactive**

20.9 The product has an "unexpected" carbon skeleton, so rearrangement must have occurred.

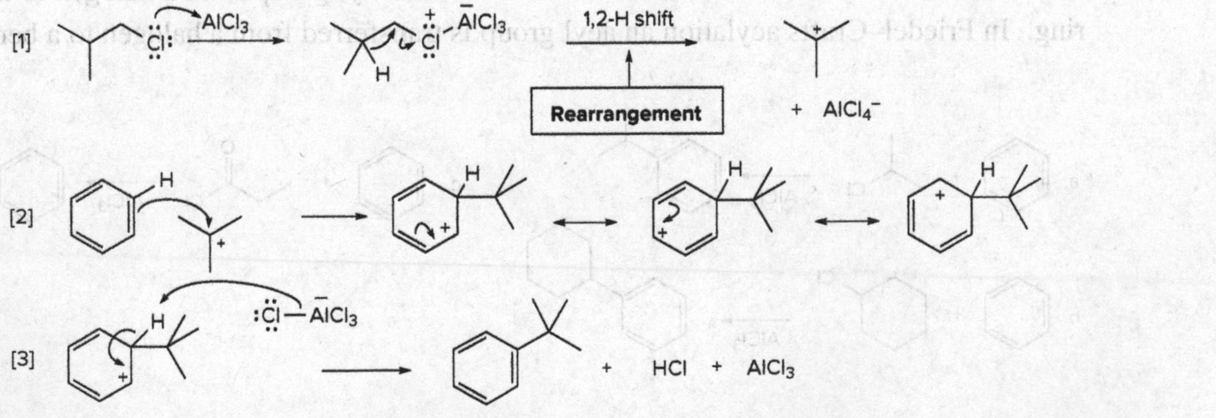

[1] Cl, $AlCl_3$ → $\overset{+}{Cl}$, $\overset{-}{A}lCl_3$, H → 1,2-H shift → + $AlCl_4^-$

Rearrangement

[2] H, + → H, + ⟷ H, + ⟷ H, +

[3] H, + , :Cl—$\overset{-}{A}lCl_3$ → + HCl + $AlCl_3$

20.10 Both alkenes and alcohols can form carbocations for Friedel–Crafts alkylation reactions.

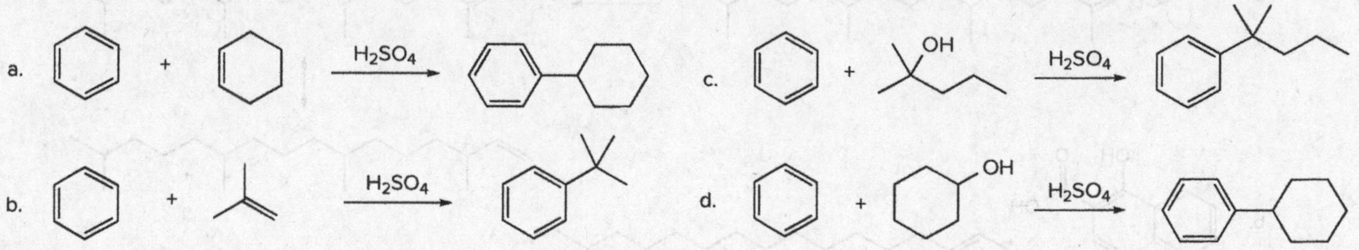

a. + $\xrightarrow{H_2SO_4}$

b. + $\xrightarrow{H_2SO_4}$

c. + OH $\xrightarrow{H_2SO_4}$

d. + OH $\xrightarrow{H_2SO_4}$

20.11

[1] Cl Cl A :Cl: AlCl₃ → Cl Cl Cl₃Al—Cl: Ö: → Cl Cl + $AlCl_4^-$ O: ↔ Cl Cl + O:

[2] Cl Cl H + O → Cl Cl H O ↔ Cl Cl H O ↔ Cl Cl H O

[3] Cl Cl H O :Cl—AlCl₃ → Cl Cl B O + HCl + AlCl₃

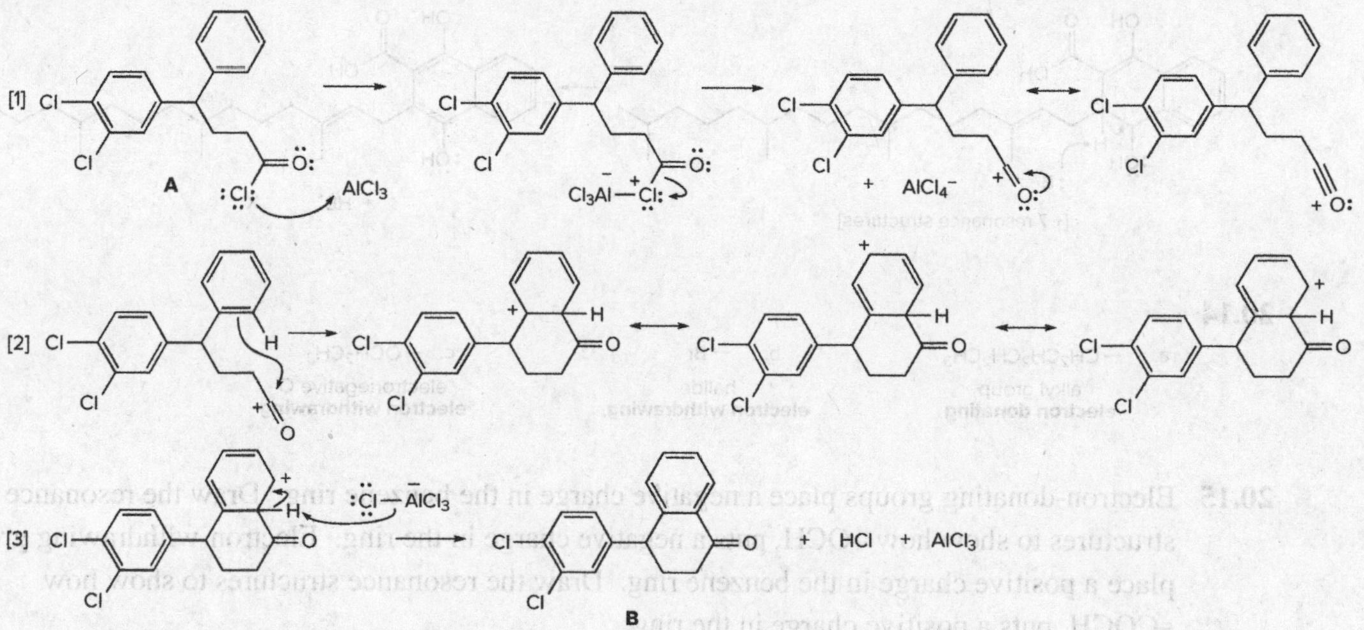

20.12 In parts (b) and (c), a 1,2-shift occurs to afford a rearrangement product.

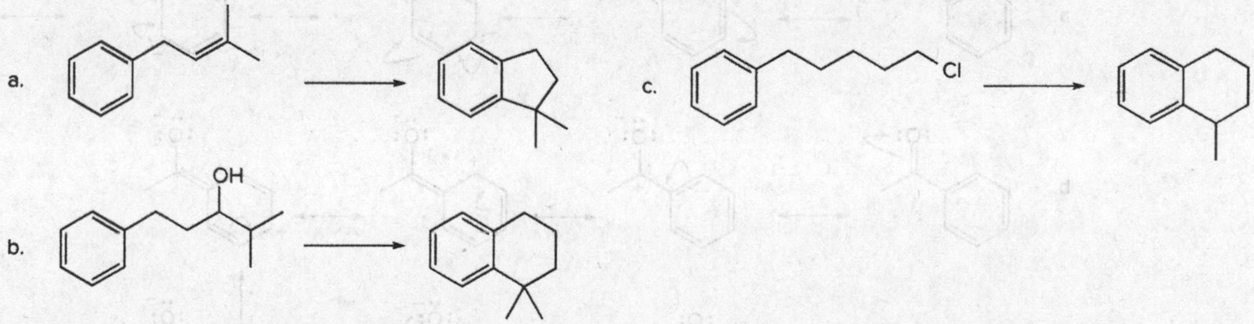

a. →

b. OH →

c. Cl →

20.13

a. PPO [structure] ⟶ [carbocation] ↔ [carbocation]

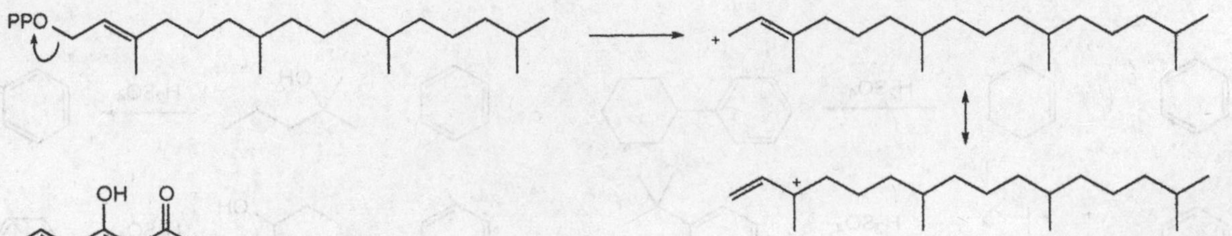

b. [mechanism structures]

OH O OH OH

:OH :B

[+ 7 resonance structures]

OH O OH :OH

+ HB⁺ X

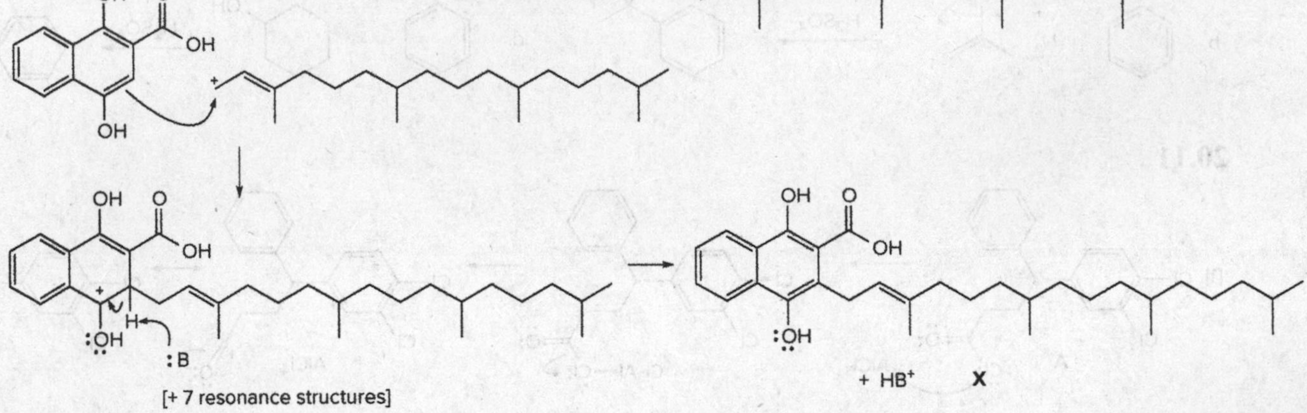

20.14

a. —CH₂CH₂CH₂CH₃ alkyl group **electron donating**

b. —Br halide **electron withdrawing**

c. —OCH₂CH₃ electronegative O **electron withdrawing**

20.15 Electron-donating groups place a negative charge in the benzene ring. Draw the resonance structures to show how –OCH_3 puts a negative charge in the ring. Electron-withdrawing groups place a positive charge in the benzene ring. Draw the resonance structures to show how –$COCH_3$ puts a positive charge in the ring.

a.

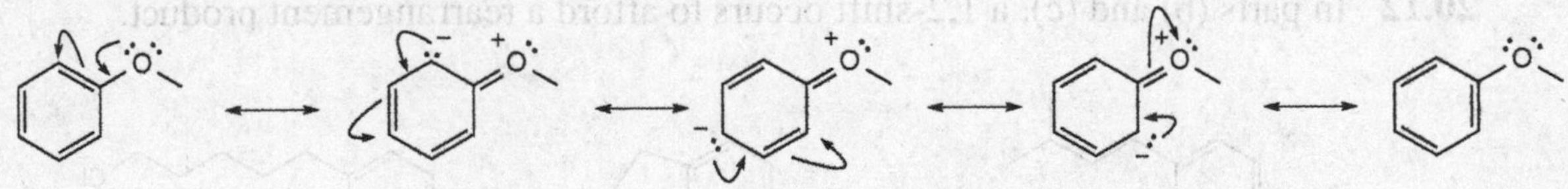

b.

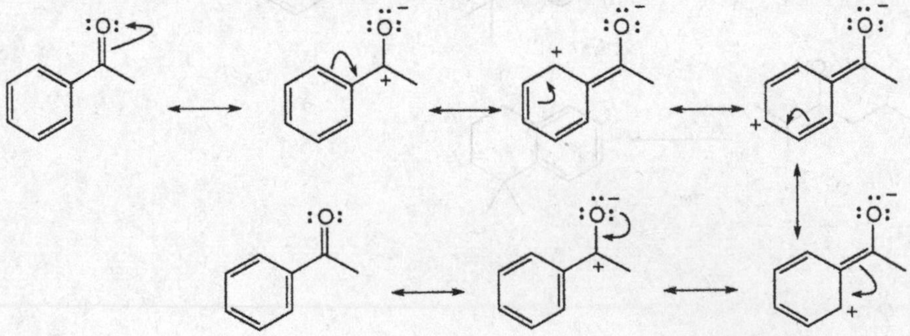

20.16 To classify each substituent, look at the atom bonded directly to the benzene ring. All R groups and Z groups (except halogens) are electron donating. All groups with a positive charge, δ+, or halogens are electron withdrawing.

a. OCH_3 — lone pair on O **electron donating**

b. I — halogen **electron withdrawing**

c. R group **electron donating**

20.17 **Electron-donating groups** make the compound ***react faster*** than benzene in electrophilic aromatic substitution. **Electron-withdrawing groups** make the compound ***react more slowly*** than benzene in electrophilic aromatic substitution.

a. electron withdrawing **reacts slower** → O_2N (meta product)

b. CN — electron withdrawing **reacts slower** → O_2N, CN (meta product)

c. OH — lone pairs on O electron donating **reacts faster** → OH, NO_2 (ortho) + O_2N, OH (para)

d. Cl — halogen electron withdrawing **reacts slower** → Cl, NO_2 (ortho) + Cl, O_2N (para)

e. R group electron donating **reacts faster** → NO_2 (ortho) + O_2N (para)

20.18

a. more electron rich, more reactive; δ+; electron-donating R group; electron-withdrawing group that bears δ+

b. NH_2; H–N; δ+; more electron rich, more reactive; electron-donating N with a lone pair; C with a δ+ electron-withdrawing

20.19

a, b. Look at each substituent individually and classify it as electron donating or withdrawing. The most reactive ring has electron-donor groups (D), and the least reactive ring has electron-withdrawing groups (W).

20.20 **Electron-donating groups** make the compound ***more reactive*** than benzene in electrophilic aromatic substitution. **Electron-withdrawing groups** make the compound ***less reactive*** than benzene in electrophilic aromatic substitution.

20.21 Chlorine inductively withdraws electron density, which decreases the rate of electrophilic aromatic substitution. The closer the Cl is to the ring, the larger the effect it has. The larger the number of Cl's, the larger the effect.

20.22 Especially stable resonance structures have all atoms with an octet. Carbocations with additional electron donor R groups are also more stable structures. Especially unstable resonance structures have adjacent like charges.

c. [Resonance structures: benzaldehyde + $\overset{+}{NO_2}$ → arenium ion intermediates with H and NO_2]

especially unstable
2 adjacent (+) charges

20.23 Polyhalogenation occurs with highly activated benzene rings containing OH, NH_2, and related groups with a catalyst.

a. phenol $\xrightarrow[FeCl_3]{Cl_2}$ 2,4,6-trichlorophenol

b. phenol $\xrightarrow{Cl_2}$ 2-chlorophenol + 4-chlorophenol

c. toluene $\xrightarrow[FeCl_3]{Cl_2}$ 2-chlorotoluene + 4-chlorotoluene

20.24 Friedel–Crafts reactions do not occur with strongly deactivating substituents including NO_2, or with NH_2, NR_2, or NHR groups.

a. benzene–SO_3H $\xrightarrow[AlCl_3]{CH_3Cl}$ **no reaction**

strongly deactivating

b. chlorobenzene $\xrightarrow[AlCl_3]{CH_3Cl}$ o-methyl chlorobenzene + p-methyl chlorobenzene

Cl is an o,p director.

c. N,N-dimethylaniline $\xrightarrow[AlCl_3]{CH_3Cl}$ **no Friedel–Crafts reaction**

d. acetanilide $\xrightarrow[AlCl_3]{CH_3Cl}$ ortho-methyl product + para-methyl product

20.25 To draw the product of a reaction with these disubstituted benzene derivatives and HNO_3, H_2SO_4, remember the following:

- If the two directing effects reinforce each other, the new substituent will be on the position reinforced by both.
- If the directing effects oppose each other, the stronger activator wins.
- No substitution occurs between two meta substituents.

a. methyl 4-methoxybenzoate (o,p; meta) $\xrightarrow[H_2SO_4]{HNO_3}$ product with NO_2 ortho to OCH₃

o,p

meta

b. 2-bromoanisole $\xrightarrow[H_2SO_4]{HNO_3}$ O_2N ortho product + NO_2 para product

o,p (strong)

o,p

opposing effects
stronger OCH_3 wins out

c. o,p; NO_2; meta; HNO_3, H_2SO_4; O_2N, NO_2 + NO_2, NO_2

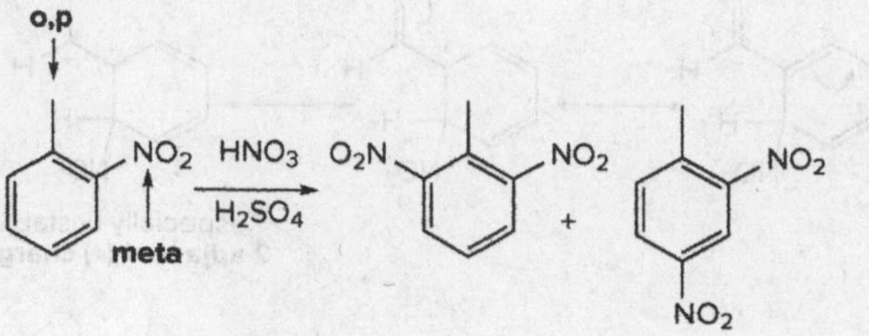

d. o,p; Cl; Br; o,p; HNO_3, H_2SO_4; O_2N, Cl, Br + Cl, Br, NO_2

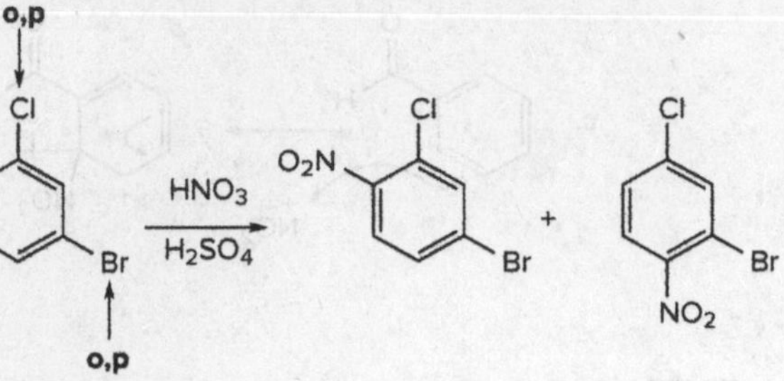

20.26

a. SO_3, H_2SO_4 → SO_3H; Cl_2, $FeCl_3$ → Cl, SO_3H

Put meta director on first.

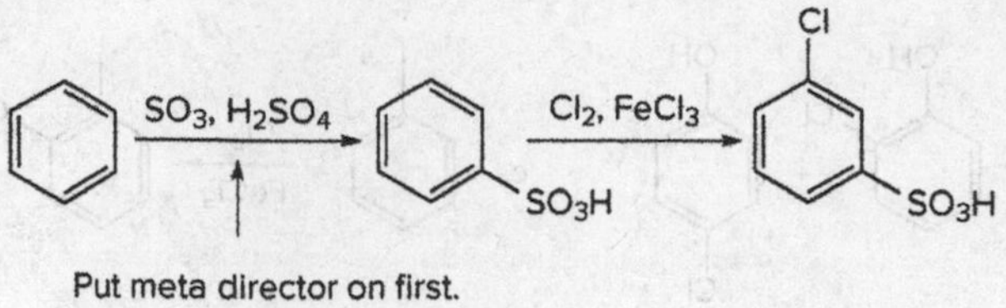

c. OH; CH_3Cl, $AlCl_3$ → OH (+ ortho isomer); Br_2 → OH, Br

Br goes ortho to the stronger activator.

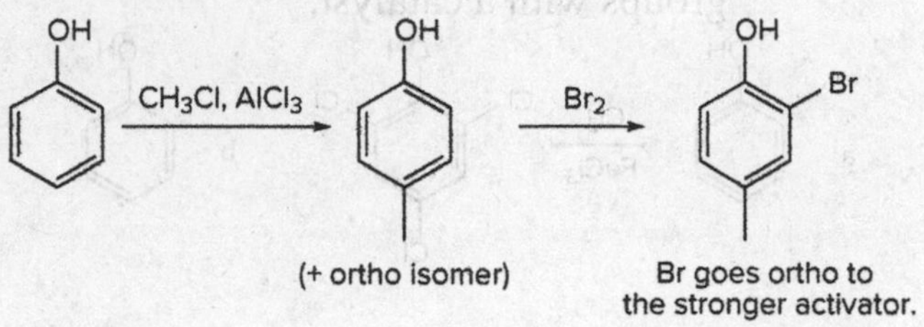

b. $ClCOCH_3$, $AlCl_3$ → O; HNO_3, H_2SO_4 → O, O_2N

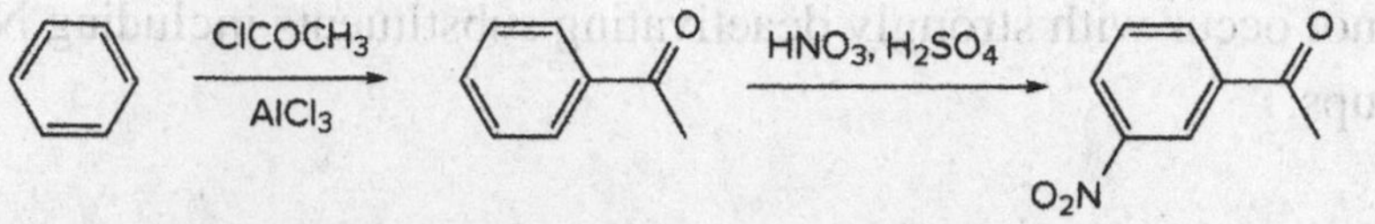

20.27

a. O_2N, Cl, NO_2; $NaOCH_3$ → O_2N, OCH_3, NO_2

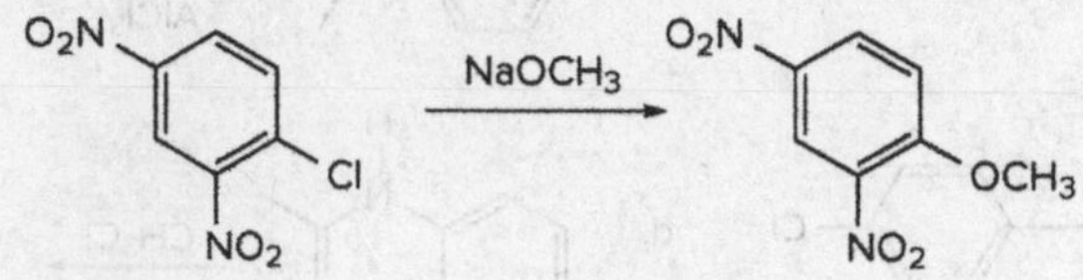

b. F, O; ^-OH → HO, O

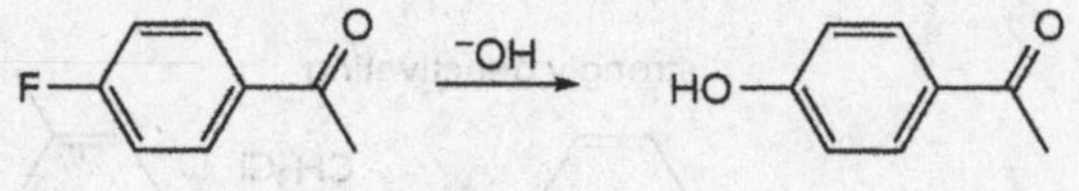

20.28

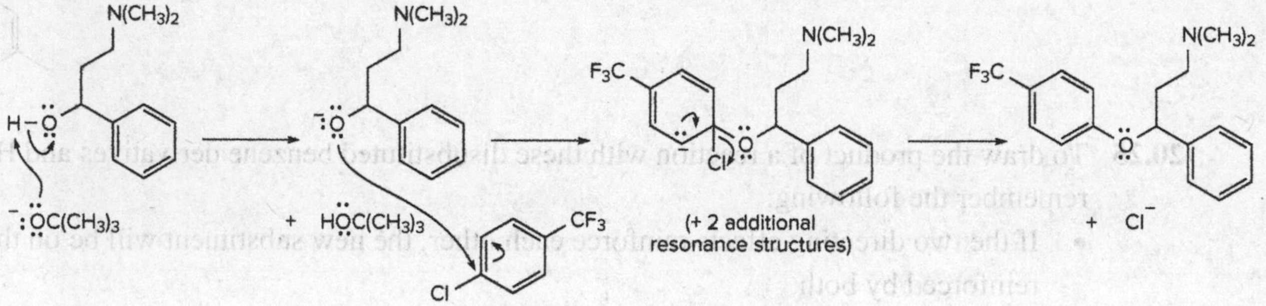

$N(CH_3)_2$; H–Ö; $^-$:ÖC$(CH_3)_3$ → $N(CH_3)_2$; $^-$:Ö: + HÖC$(CH_3)_3$; CF_3, Cl → F_3C, $N(CH_3)_2$, Cl, Ö (+ 2 additional resonance structures) → F_3C, $N(CH_3)_2$, Ö + Cl^-

20.29

a. Cl; $NaNH_2$, NH_3 → NH_2

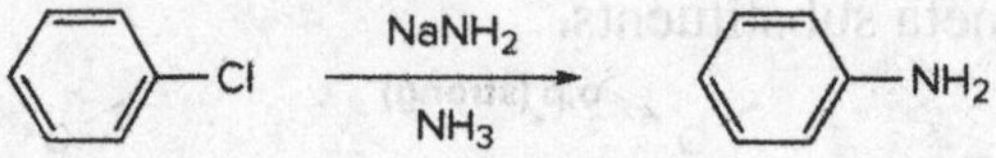

b. CH_3O, Cl; NaOH, H_2O, Δ → CH_3O, OH + CH_3O, OH

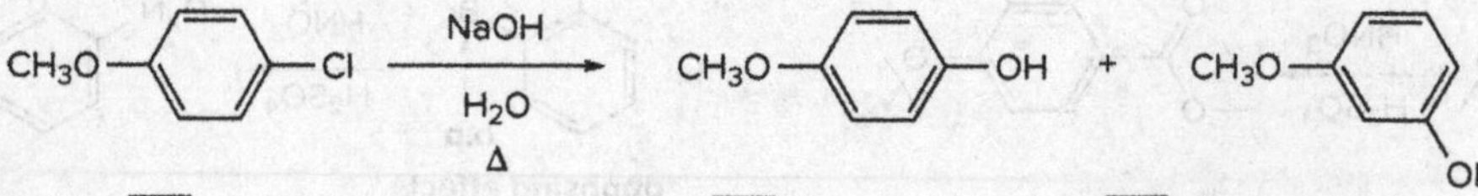

c. Cl; KNH_2, NH_3 → NH_2 + NH_2

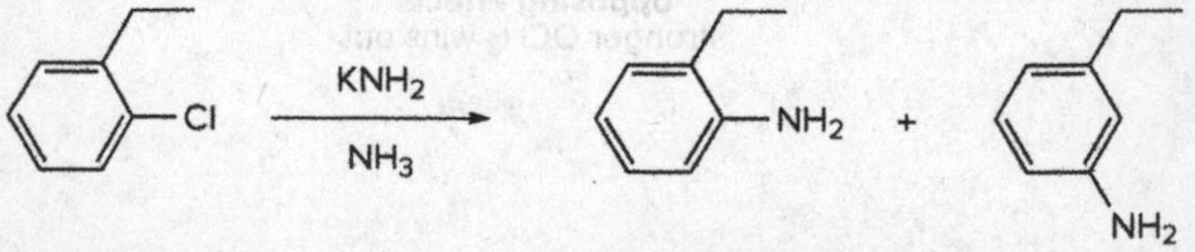

20.30

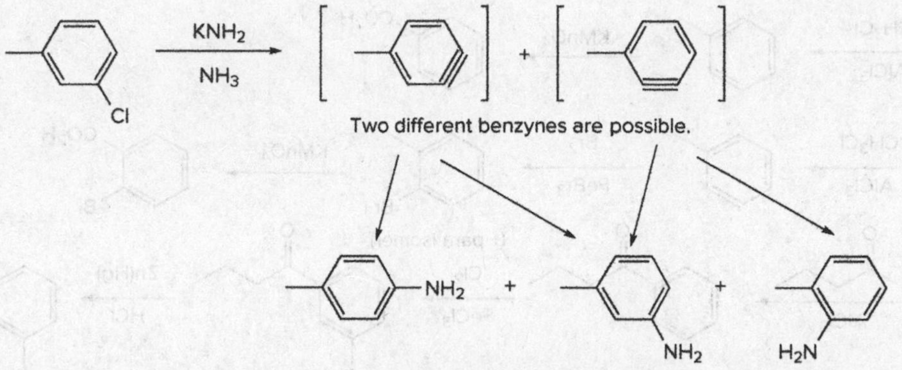

KNH$_2$ / NH$_3$

Two different benzynes are possible.

NH$_2$ + NH$_2$ + H$_2$N

Ortho, meta, and para products are formed.

20.31

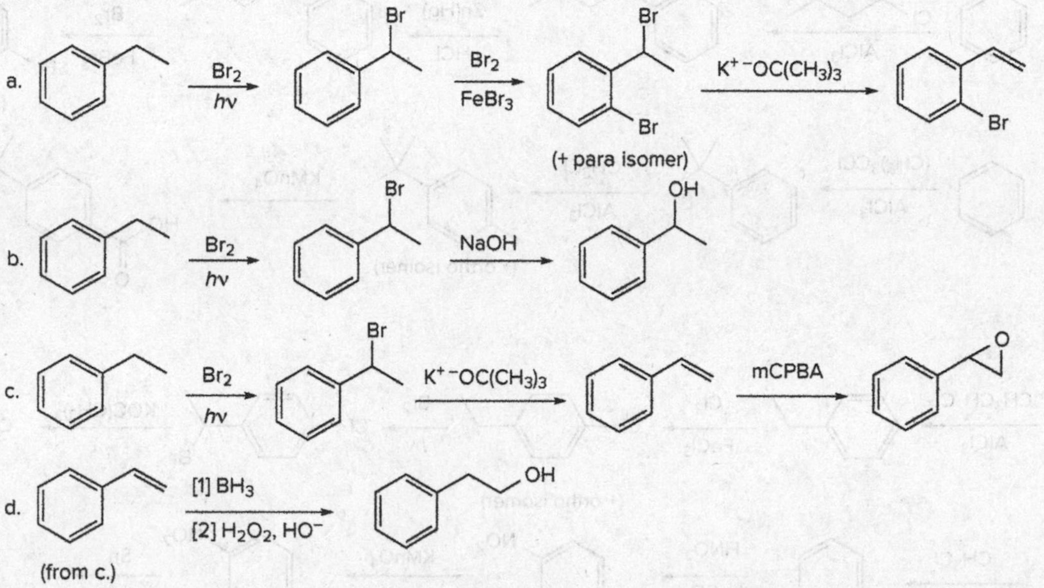

a. Br$_2$ / $h\nu$ → Br → Br$_2$ / FeBr$_3$ → Br, Br (+ para isomer) → K$^+$ $^-$OC(CH$_3$)$_3$ → Br

b. Br$_2$ / $h\nu$ → Br → NaOH → OH

c. Br$_2$ / $h\nu$ → Br → K$^+$ $^-$OC(CH$_3$)$_3$ → mCPBA → O

d. (from c.) [1] BH$_3$ [2] H$_2$O$_2$, HO$^-$ → OH

20.32 First use an acylation reaction, and then reduce the carbonyl group to form the alkyl benzenes.

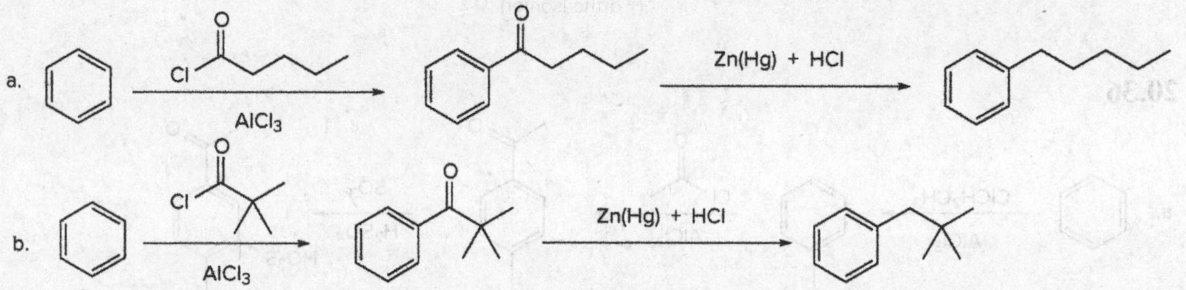

a. Cl, O / AlCl$_3$ → O → Zn(Hg) + HCl →

b. Cl, O / AlCl$_3$ → O → Zn(Hg) + HCl →

20.33

Cl, O / AlCl$_3$ → O → Zn(Hg) / HCl → O, Cl / AlCl$_3$ → O

p-isobutylacetophenone (+ ortho isomer)

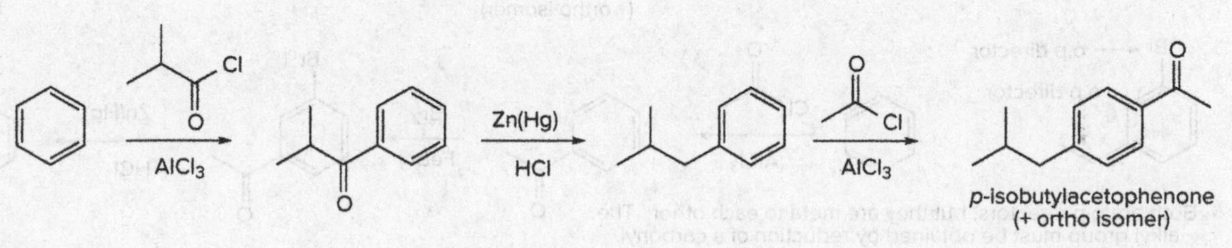

20.34

a. Benzene $\xrightarrow[AlCl_3]{CH_3Cl}$ toluene $\xrightarrow{KMnO_4}$ benzoic acid (CO_2H)

b. Benzene $\xrightarrow[AlCl_3]{CH_3Cl}$ toluene $\xrightarrow[FeBr_3]{Br_2}$ o-bromotoluene (+ para isomer) $\xrightarrow{KMnO_4}$ o-bromobenzoic acid (CO_2H, Br)

c. Benzene + butanoyl chloride (Cl) $\xrightarrow{AlCl_3}$ butyrophenone $\xrightarrow[FeCl_3]{Cl_2}$ m-chlorobutyrophenone (Cl) $\xrightarrow[HCl]{Zn(Hg)}$ m-chlorobutylbenzene (Cl)

d. Benzene + 3-methylbutanoyl chloride (Cl) $\xrightarrow{AlCl_3}$ aryl ketone $\xrightarrow[HCl]{Zn(Hg)}$ isopentylbenzene $\xrightarrow[FeBr_3]{Br_2}$ p-bromo product (Br) (+ ortho isomer)

e. Benzene $\xrightarrow[AlCl_3]{(CH_3)_3CCl}$ tert-butylbenzene $\xrightarrow[AlCl_3]{CH_3Cl}$ p-tert-butyltoluene (+ ortho isomer) $\xrightarrow{KMnO_4}$ p-tert-butylbenzoic acid (HO, O)

20.35

a. Benzene $\xrightarrow[AlCl_3]{CH_3CH_2Cl}$ ethylbenzene $\xrightarrow[FeCl_3]{Cl_2}$ Cl–C₆H₄–CH₂CH₃ (+ ortho isomer) $\xrightarrow[h\nu]{Br_2}$ Cl–C₆H₄–CH(Br)CH₃ $\xrightarrow{KOC(CH_3)_3}$ Cl–C₆H₄–CH=CH₂

b. Benzene $\xrightarrow[AlCl_3]{CH_3Cl}$ toluene $\xrightarrow[H_2SO_4]{HNO_3}$ p-nitrotoluene (NO_2) (+ ortho isomer) $\xrightarrow{KMnO_4}$ HOOC–C₆H₄–NO_2 $\xrightarrow[HCl]{Sn}$ HOOC–C₆H₄–NH_2

20.36

a. Benzene $\xrightarrow[AlCl_3]{ClCH_2CH_3}$ ethylbenzene $\xrightarrow[AlCl_3]{CH_3COCl}$ p-ethylacetophenone (O) (+ ortho isomer) $\xrightarrow[H_2SO_4]{SO_3}$ product with HO_3S ortho to ethyl group

b. Br ← o,p director

o,p director

Benzene + acetyl chloride (Cl) $\xrightarrow{AlCl_3}$ acetophenone $\xrightarrow[FeBr_3]{Br_2}$ m-bromoacetophenone (Br, O) $\xrightarrow[HCl]{Zn(Hg)}$ m-bromoethylbenzene (Br)

Both are o,p directors, but they are meta to each other. The alkyl group must be obtained by reduction of a carbonyl.

c. Benzene $\xrightarrow{Cl_2,\ FeCl_3}$ chlorobenzene $\xrightarrow{CH_3CH_2Cl,\ AlCl_3}$ 4-chloroethylbenzene (+ ortho isomer) $\xrightarrow{Br_2,\ h\nu}$ 1-bromo-1-(4-chlorophenyl)ethane $\xrightarrow{K^+\ ^-OC(CH_3)_3}$ 4-chlorostyrene $\xrightarrow{[1]\ BH_3;\ [2]\ H_2O_2,\ ^-OH}$ 2-(4-chlorophenyl)ethanol $\xrightarrow{PCC}$ (4-chlorophenyl)acetaldehyde

20.37

A (3-isopropoxybenzaldehyde, CHO, O–isopropyl)

a. Br_2, $FeBr_3$ → 4-bromo-3-isopropoxybenzaldehyde (Br, CHO) + 2-bromo-5-isopropoxybenzaldehyde (Br, CHO)

b. HNO_3, H_2SO_4 → 3-isopropoxy-4-nitrobenzaldehyde (NO_2, CHO) + 5-isopropoxy-2-nitrobenzaldehyde (O_2N, CHO)

c. CH_3CH_2COCl, $AlCl_3$ → 3-isopropoxy-4-propanoylbenzaldehyde (CHO) + 5-isopropoxy-2-propanoylbenzaldehyde (CHO)

B (Br–C₆H₄–NHCOCH₃, para)

a. Br_2, $FeBr_3$ → Br–C₆H₃(Br)–NHCOCH₃ (Br ortho to NH)

b. HNO_3, H_2SO_4 → Br–C₆H₃(NO_2)–NHCOCH₃ (NO_2 ortho to NH)

c. CH_3CH_2COCl, $AlCl_3$ → Br–C₆H₃(COCH₂CH₃)–NHCOCH₃ (acyl ortho to NH)

20.38 OH is an ortho, para director.

a. OH (phenol) —HNO_3, H_2SO_4→ OH/NO_2 (ortho) + OH/NO_2 (para) —Sn, HCl→ OH/NH_2 (ortho) + OH/NH_2 (para)

b. OH (phenol) + Cl (acid chloride, O) —$AlCl_3$→ OH, O (ortho ketone) + OH, O (para ketone) —Zn(Hg), HCl→ OH (ortho alkyl) + OH (para alkyl)

c. OH (phenol) + Cl —$AlCl_3$→ OH (ortho ethyl) + OH (para ethyl) —Br_2, $h\nu$→ OH, Br + OH, Br

d. OH (phenol) + Cl —$AlCl_3$→ OH (ortho isopropyl) + OH (para isopropyl) —$KMnO_4$→ OH, O, OH + OH, O, OH

20.39

a. O (ketone) + Cl, O —$AlCl_3$→ No reaction

b. N (dimethylaniline) + Cl, O —$AlCl_3$→ No Friedel–Crafts reaction

c. H, N, O (acetanilide) + Cl, O —$AlCl_3$→ O, H, N, O (ortho) + H, N, O, O (para)

20.40

a. HO, NO_2 (benzene) → HNO_3 / H_2SO_4 → O_2N, NO_2, HO + NO_2, HO, NO_2

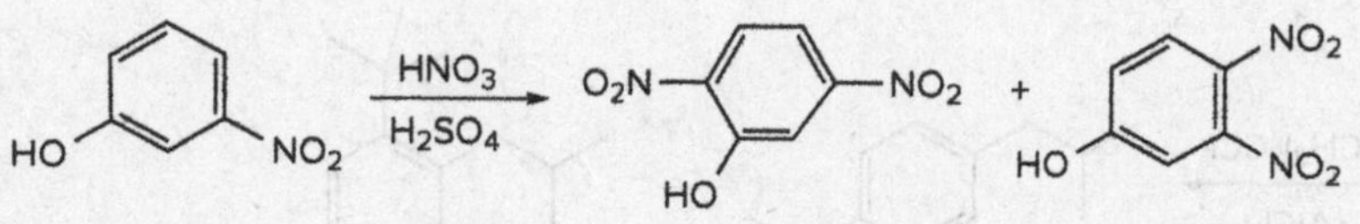

b. Cl (benzene) O, O + Cl / $AlCl_3$ → Cl, O, O

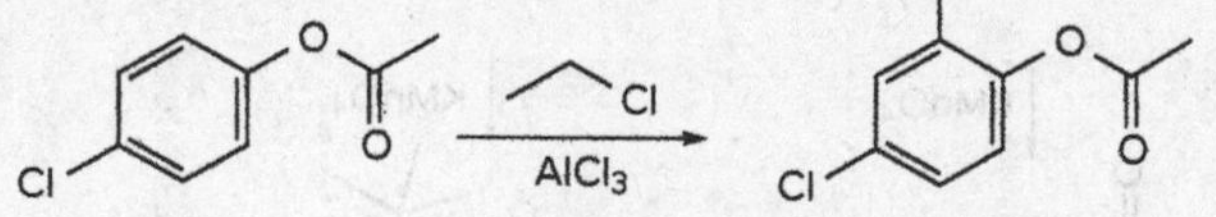

c. O, H → Br_2 / $FeBr_3$ → O, H, Br + Br, O, H

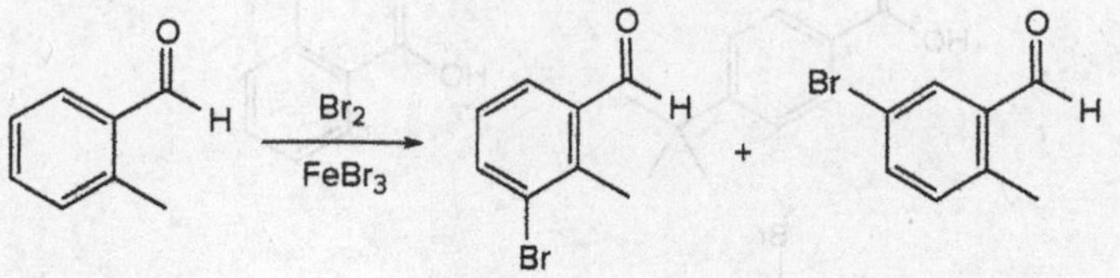

d. O, O, O → Cl_2 / $FeCl_3$ → Cl, O, O, O

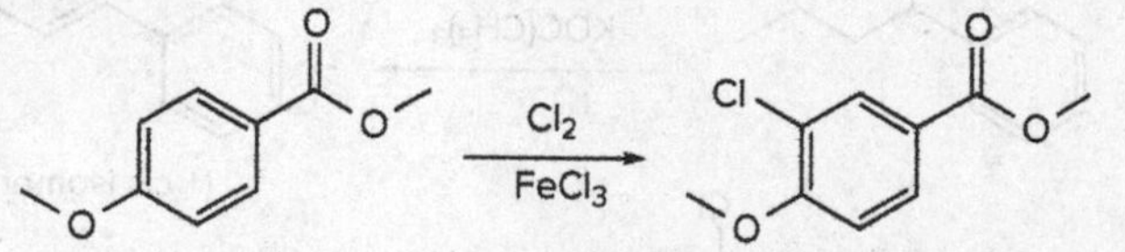

e. Br, O → SO_3 / H_2SO_4 → Br, O, SO_3H + Br, O, HO_3S

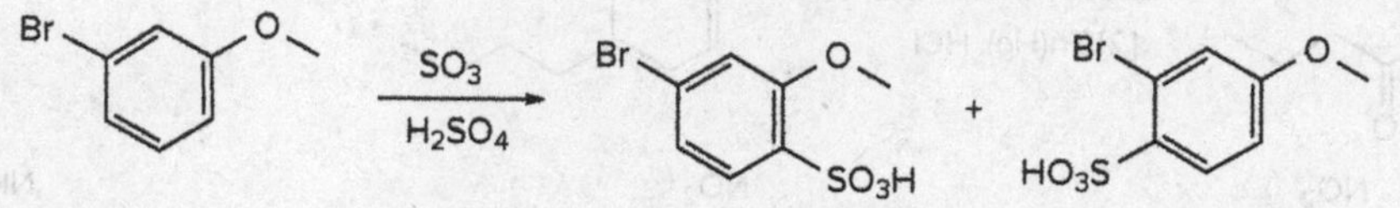

f. F, NO_2 → Na^+ :S: → S, NO_2

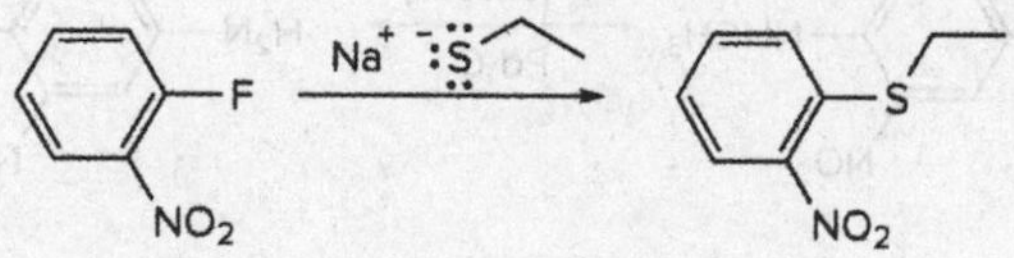

20.41 Watch out for rearrangements.

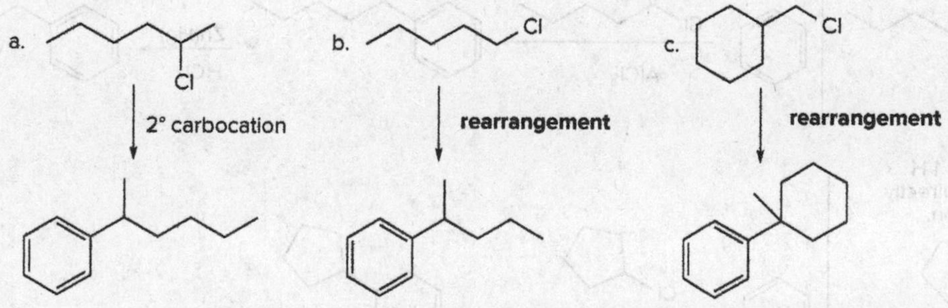

a. Cl → 2° carbocation

b. Cl → rearrangement

c. Cl → rearrangement

20.42

a. $(CH_3)_3CCl$, $AlCl_3$ → + ; $KMnO_4$; $KMnO_4$

HO HO

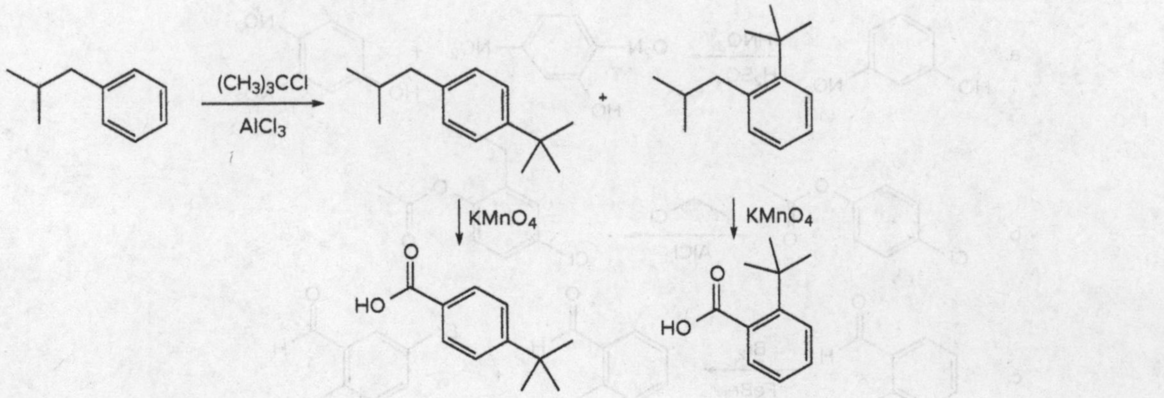

b. Br_2, $h\nu$ → Br → $KOC(CH_3)_3$ → (+ cis isomer)

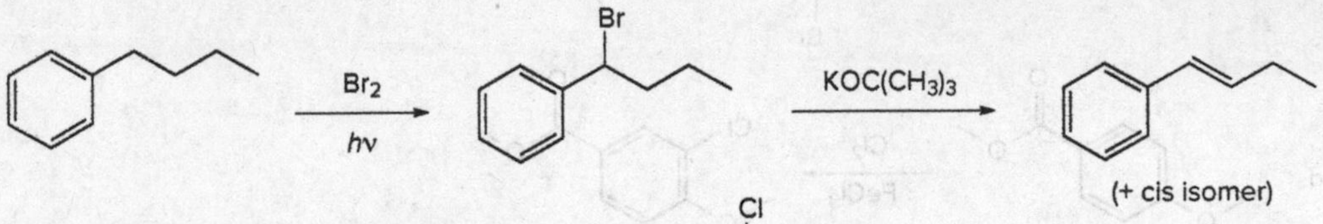

c. [1] Cl_2, $FeCl_3$ [2] Zn(Hg), HCl → Cl

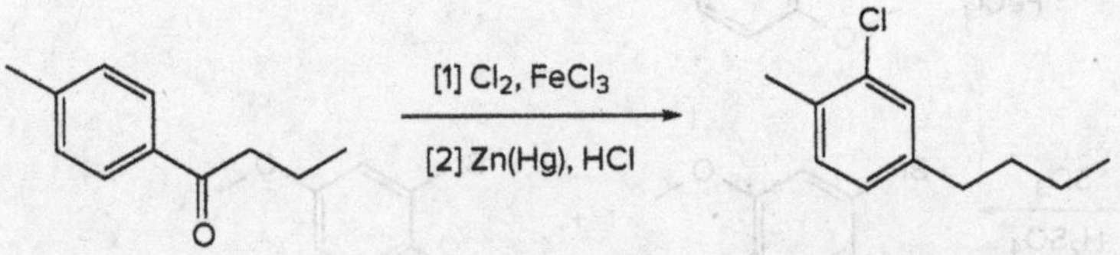

d. NO_2, O_2N, Cl, NO_2 → CH_3NH_2 → NO_2, O_2N, $NHCH_3$, NO_2 → H_2 (excess), Pd-C → NH_2, H_2N, $NHCH_3$, NH_2

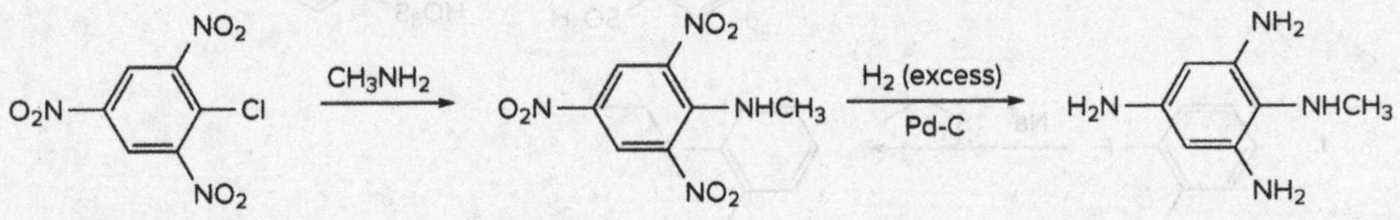

20.43

a. C bonded to 2 H's must use acylation followed by reduction.

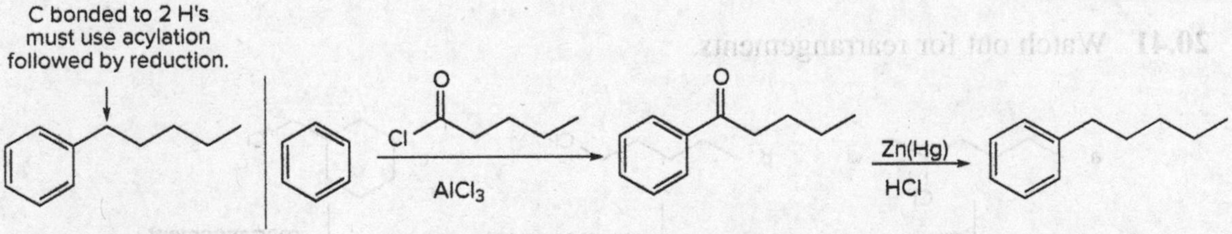

Cl, O, $AlCl_3$ → O → Zn(Hg), HCl →

b. C bonded to 1 H can be added directly by alkylation.

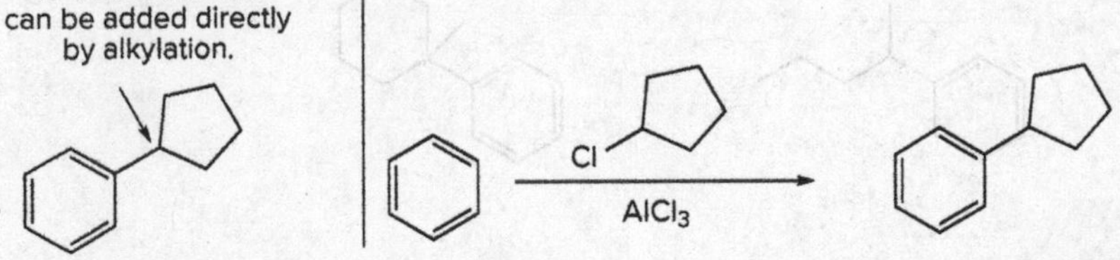

Cl, $AlCl_3$ →

c. Ethyl group can be introduced by two methods.

O, Cl, $AlCl_3$ → O → Zn(Hg), HCl → Method [1]

CH_3CH_2Cl, $AlCl_3$ ← Method [2]

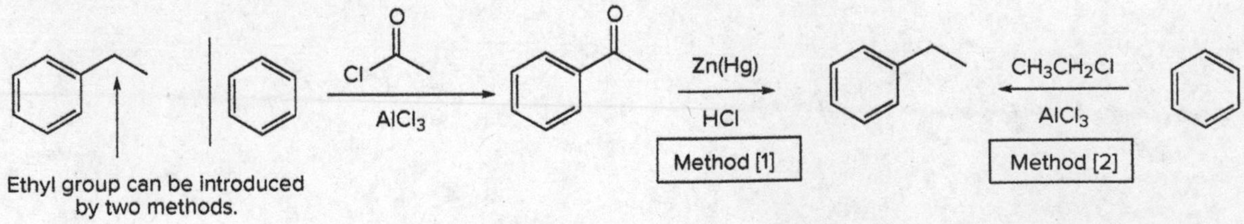

20.44

a. NaH; F–C6H4–CN; X

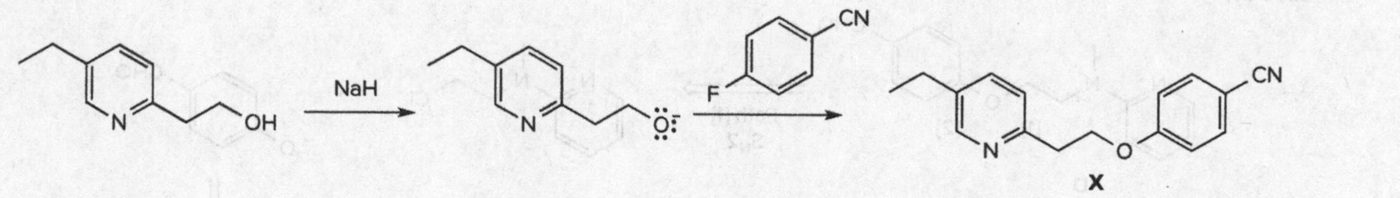

b. [1] TsCl; [2] NaOH, HO–C6H4–CHO; Y

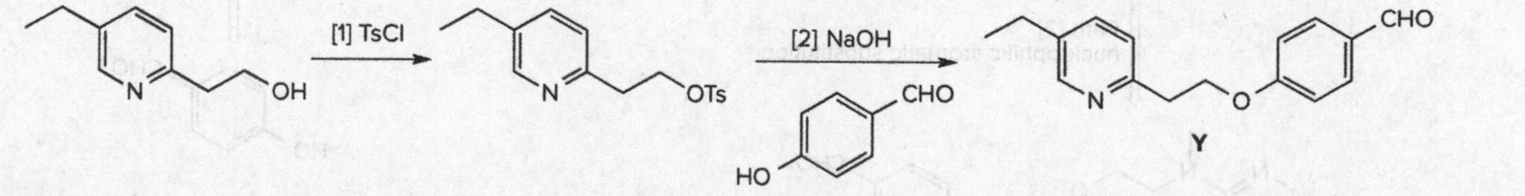

20.45

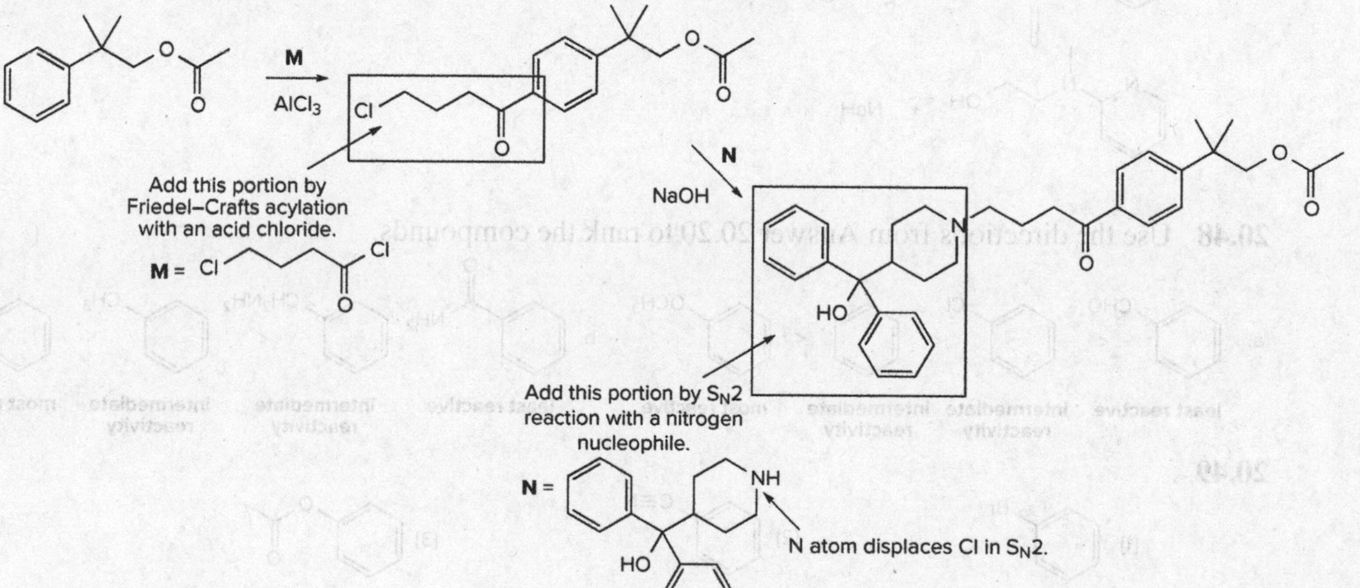

M, $AlCl_3$

Add this portion by Friedel–Crafts acylation with an acid chloride.

M =

N, NaOH

Add this portion by S_N2 reaction with a nitrogen nucleophile.

N =

N atom displaces Cl in S_N2.

20.46

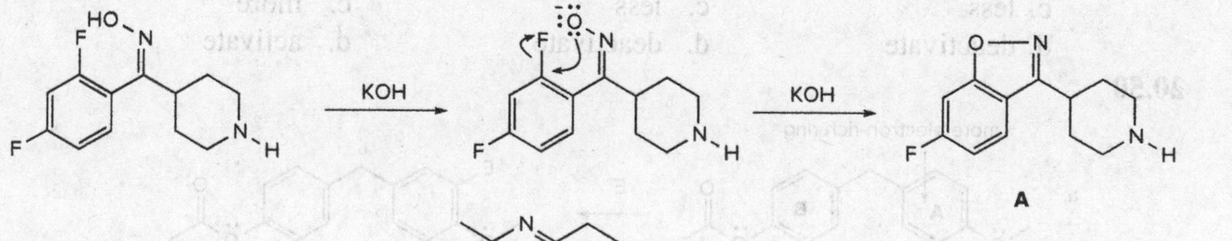

KOH → KOH → A

A second resonance structure can be drawn for the **C** ring, showing that it has six π electrons.

Rings **A** and **B** contain 10 π electrons, eight from the double bonds and two from O, making this ring system aromatic.

6 π electrons

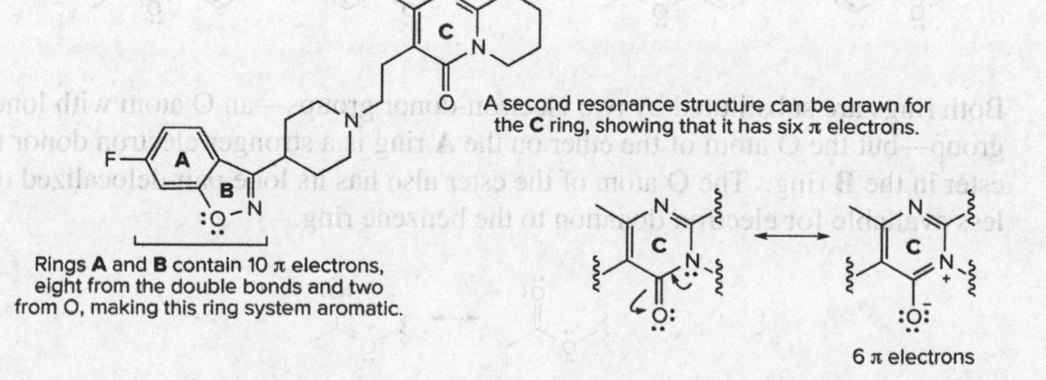

20.47

CHO
[1] [2]
D
Path [1]
S_N2
Cl
+
CHO
$^-$O
CHO
HO
+ NaH
Path [2]
nucleophilic aromatic substitution
O$^-$
+
CHO
F
OH + NaH

20.48 Use the directions from Answer 20.20 to rank the compounds.

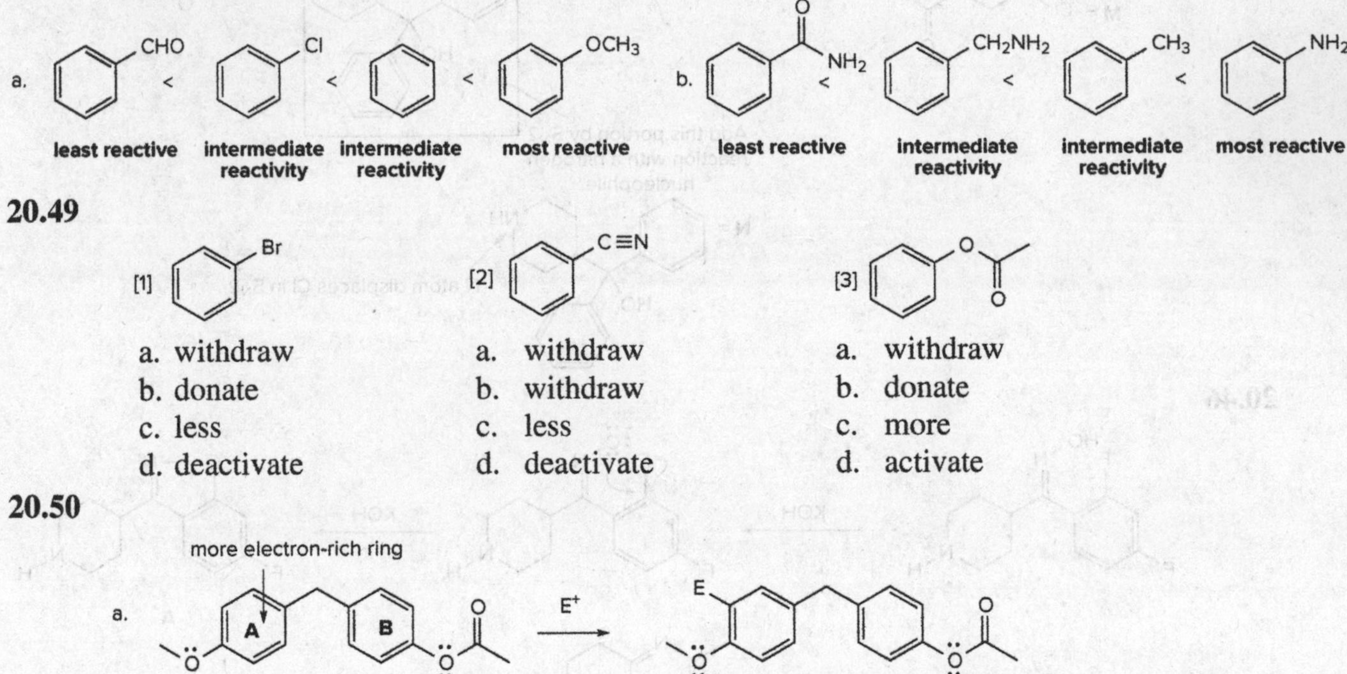

Both rings are substituted by two electron-donor groups—an O atom with lone pairs and an alkyl group—but the O atom of the ether on the **A** ring is a stronger electron donor than the O atom of ester in the **B** ring. The O atom of the ester also has its lone pair delocalized on the C=O, so it is less available for electron donation to the benzene ring.

Thus, A is more electron rich and electrophilic aromatic substitution occurs on the A ring.

b. C D $\xrightarrow{E^+}$ (para product, E) + (ortho product, E)

Both the C and D rings are bonded to N atoms with lone pairs, but the D ring is more electron rich. The N atom on the C ring has its lone pair delocalized on the other N in the six membered ring, so it is less available for donation to the benzene ring.

As a result, electrophilic aromatic substitution occurs on the D ring to afford a mixture of ortho and para products.

20.51

A B C D
NH₂ O Cl

a. **B**
b. **A**
c. (Br products) NH₂ O Cl Br + Cl NH₂ O Br

20.52

OCH_2CH_3

alkyl group on the benzene ring
R stabilizes (+) charges on the o,p positions by an electron-donating inductive effect. This group behaves like any other R group, so that ortho and para products are formed in electrophilic aromatic substitution.

OCH_2CH_3 ⟷ OCH_2CH_3

(+) charge on atom bonded to the benzene ring
Drawing resonance structures in electrophilic aromatic substitution results in especially unstable structures for attack at the o,p positions—two (+) charges on adjacent atoms. This doesn't happen with meta attack, so meta attack is preferred. This is identical to the situation observed with all meta directors.

20.53 Increasing the number of electron-withdrawing groups (especially at the ortho and para positions to the leaving group) increases the rate of nucleophilic aromatic substitution. Increasing the electronegativity of the halogen increases the rate.

a.

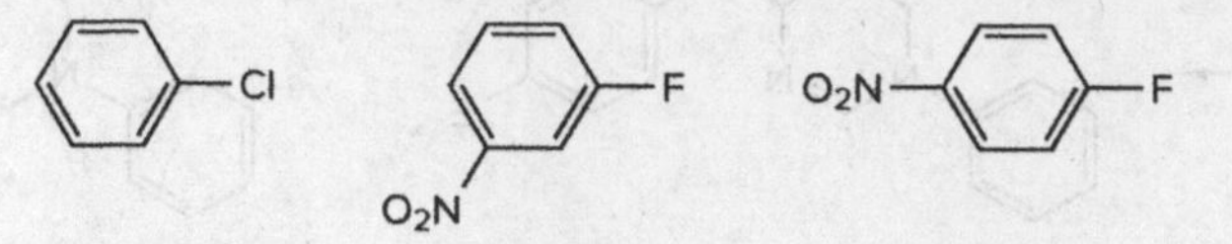

Cl

F

O_2N

O_2N

F

chlorobenzene
least reactive

m-fluoronitro-
benzene

p-fluoronitro-
benzene
most reactive

b.

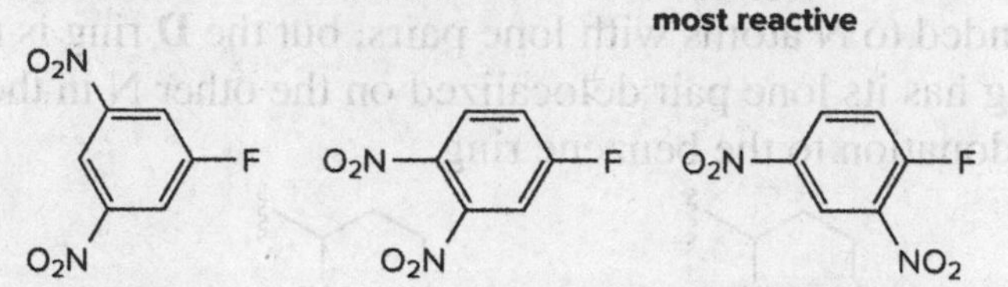

O_2N

F

O_2N

O_2N

F

O_2N

O_2N

F

NO_2

1-fluoro-3,5-dinitro-
benzene
least reactive

1-fluoro-3,4-
dinitrobenzene

1-fluoro-2,4-dinitro-
benzene
most reactive

c.

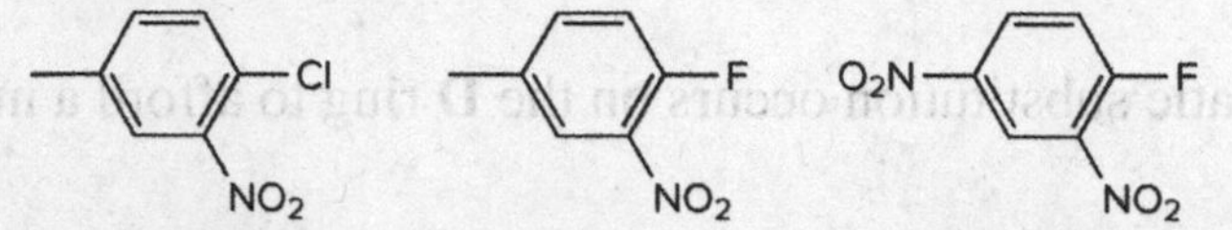

Cl

NO_2

F

NO_2

O_2N

F

NO_2

4-chloro-3-nitro-
toluene
least reactive

4-fluoro-3-
nitrotoluene

1-fluoro-2,4-dinitro-
benzene
most reactive

20.54

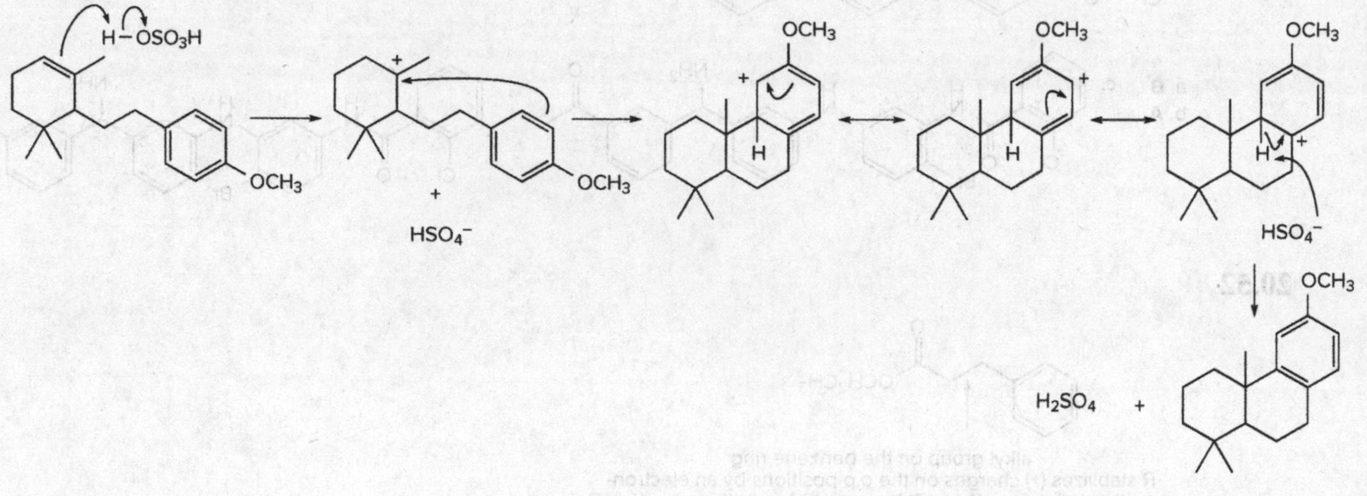

H–OSO_3H

OCH_3

+

OCH_3

+

HSO_4^-

OCH_3

+

H

OCH_3

+

H

OCH_3

+

H

HSO_4^-

OCH_3

H_2SO_4 +

20.55

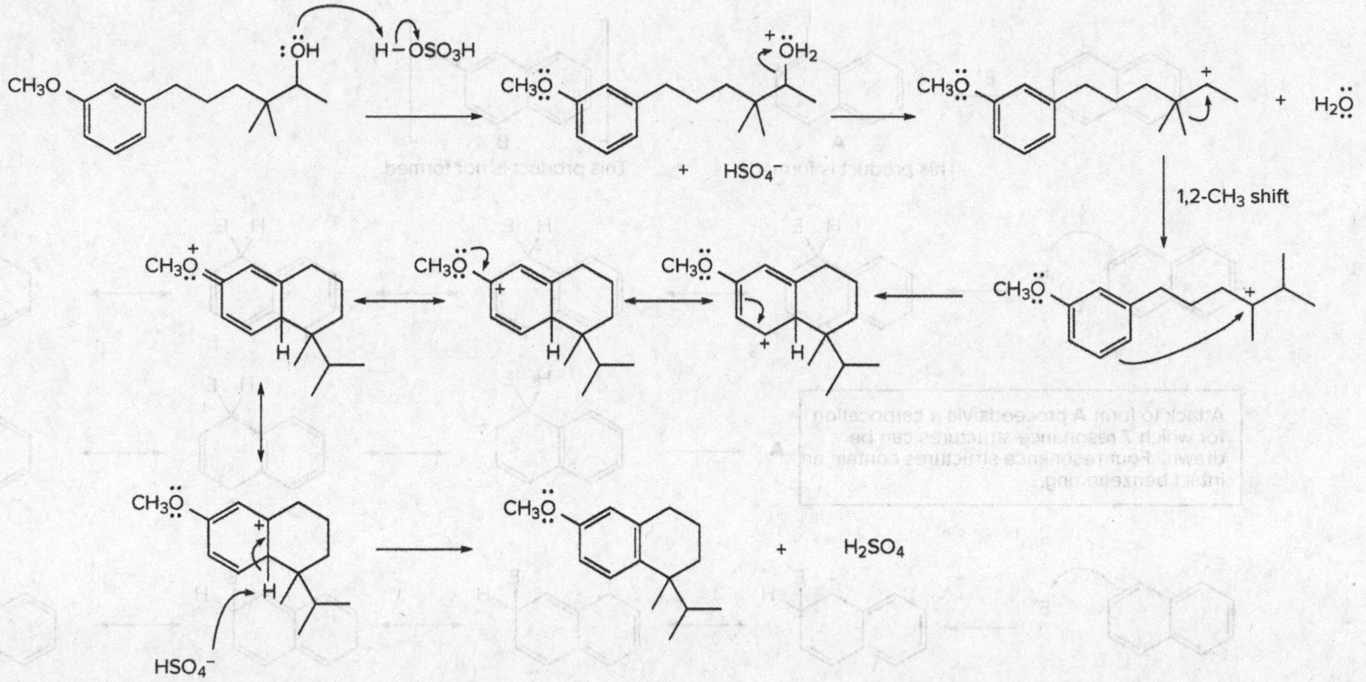

CH_3O :ÖH H–ÖSO$_3$H → $CH_3\ddot{O}$ $\overset{+}{O}H_2$ + HSO_4^- → $CH_3\ddot{O}$ + $H_2\ddot{O}$

1,2-CH_3 shift

$CH_3\overset{+}{O}$ ⟷ $CH_3\ddot{O}$ ⟷ $CH_3\ddot{O}$ ⟵ $CH_3\ddot{O}$

$CH_3\ddot{O}$ → $CH_3\ddot{O}$ + H_2SO_4

HSO_4^-

20.56 The reaction follows the two-step addition–elimination mechanism for nucleophilic aromatic substitution.

2-chloropyridine Na^+ $:\ddot{O}CH_3^-$ → $\ddot{O}CH_3$ Cl ⟷ $\ddot{O}CH_3$ Cl ⟷ $\ddot{O}CH_3$ Cl

A

↓

N OCH_3 + Cl^-

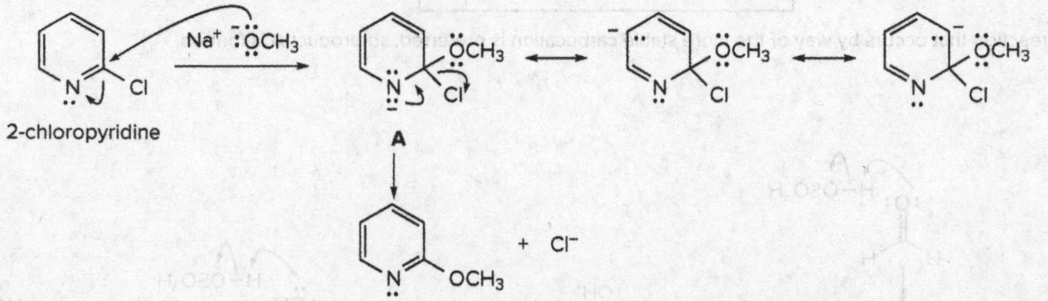

Resonance structure **A** is stabilized because the negative charge is located on an electronegative N. This makes nucleophilic aromatic substitution on 2-chloropyridine faster than a similar reaction with chlorobenzene, which has no N atom to stabilize the intermediate negative charge.

20.57

E⁺
A
This product is formed.
B
This product is *not* formed.

Attack to form **A** proceeds via a carbocation for which **7** resonance structures can be drawn. Four resonance structures contain an intact benzene ring.

Attack to form **B** proceeds via a carbocation for which **6** resonance structures can be drawn. Only two resonance structures contain an intact benzene ring.

A reaction that occurs by way of the more stable carbocation is preferred, so product **A** is formed.

20.58

$H-OSO_3H$

$+ HSO_4^-$

[+ 3 resonance structures]

HSO_4^-

$H-OSO_3H$

$+ HSO_4^-$

[+ 5 resonance structures]

$+ H_2O$

HSO_4^-

[+ 3 resonance structures]

$+ H_2SO_4$

20.59 Intramolecular cyclization of the protonated ketone forms a resonance-stabilized carbocation that loses CO_2 (instead of a proton) to re-form the aromatic ring.

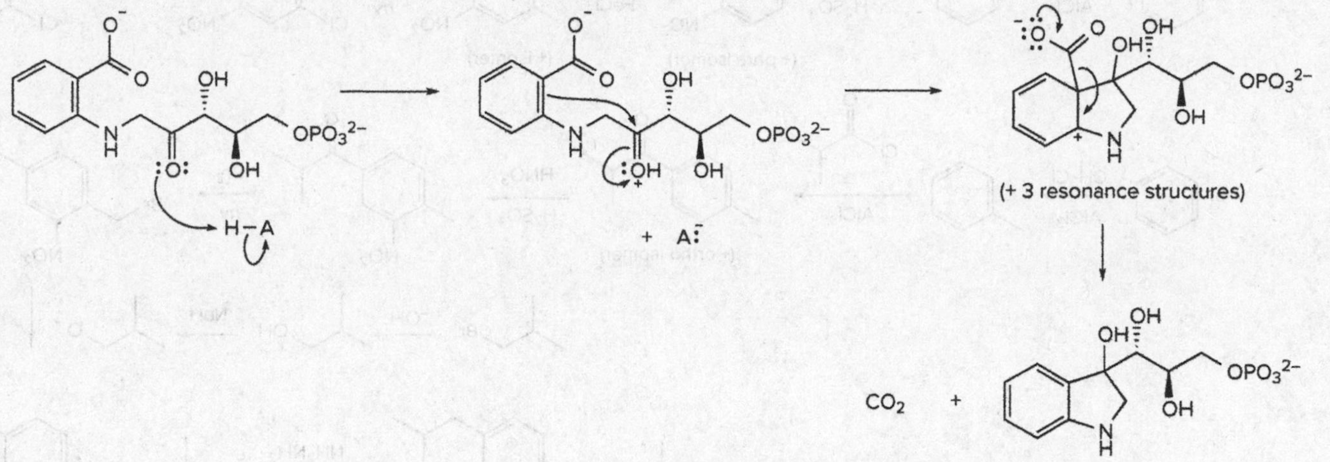

OH OPO$_3^{2-}$ OH H–A + A$^-$ (+ 3 resonance structures) CO_2 +

20.60 Loss of diphosphate forms a resonance-stabilized allylic carbocation, which acts as the electrophile in electrophilic aromatic substitution with tryptophan.

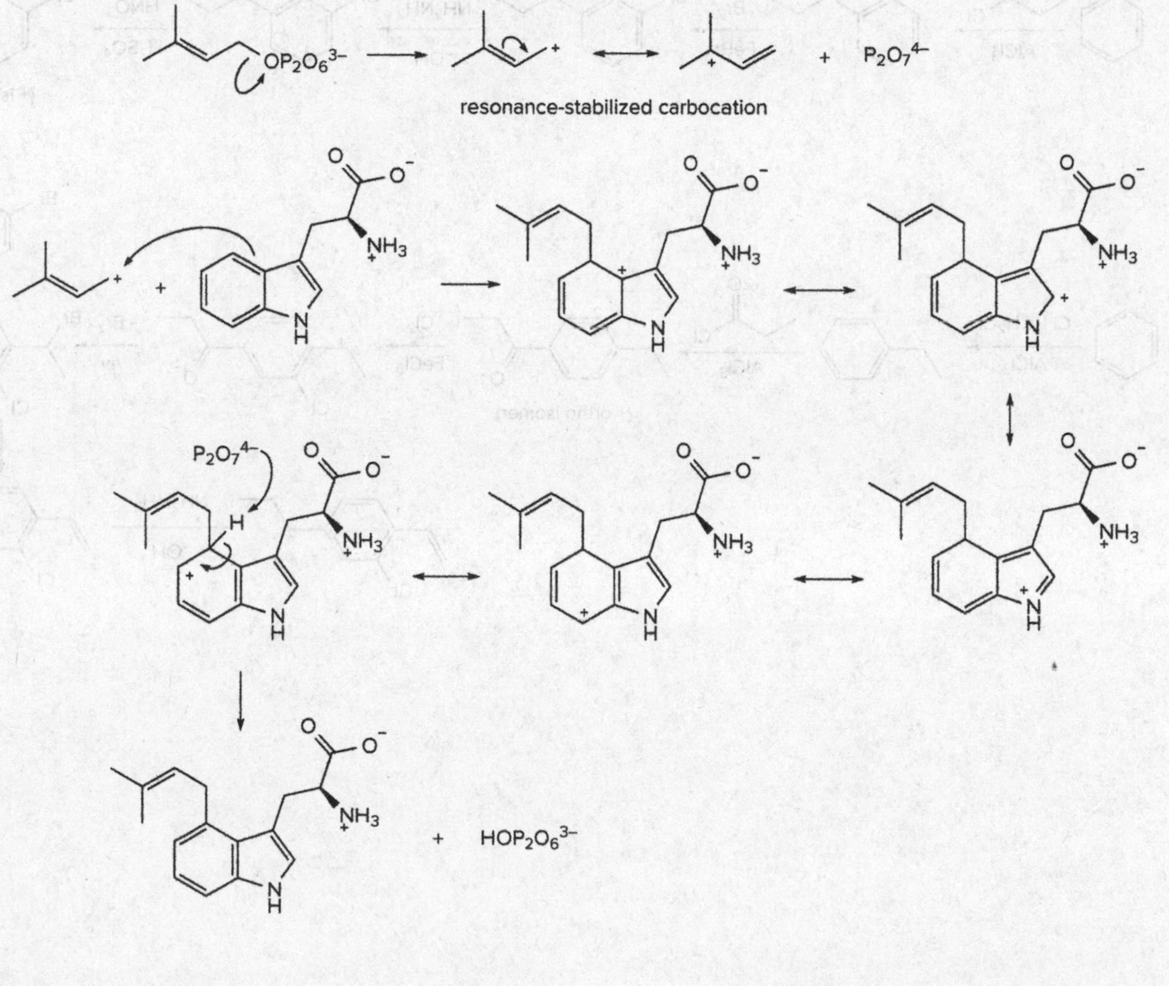

OP$_2$O$_6^{3-}$ + P$_2$O$_7^{4-}$

resonance-stabilized carbocation

$P_2O_7^{4-}$ + HOP$_2$O$_6^{3-}$

20.61

a. (CH$_3$)$_2$CHCl, AlCl$_3$; HNO$_3$, H$_2$SO$_4$ → NO$_2$ (+ para isomer); Cl$_2$, FeCl$_3$ → Cl, NO$_2$ (+ isomer); Br$_2$, $h\nu$ → Br, Cl, NO$_2$; K$^+$ $^-$OC(CH$_3$)$_3$ → Cl, NO$_2$

b. CH$_3$Cl, AlCl$_3$; Cl(C=O)CH(CH$_3$)$_2$, AlCl$_3$ (+ ortho isomer); HNO$_3$, H$_2$SO$_4$ → NO$_2$; Br$_2$, $h\nu$ → Br, NO$_2$

Br; $^-$OH → OH; NaH → O$^-$

NH$_2$NH$_2$, $^-$OH; O, NO$_2$ → O, NO$_2$

c. CH$_3$CH$_2$COCl (O, Cl), AlCl$_3$; Br$_2$, FeBr$_3$ → Br; NH$_2$NH$_2$, $^-$OH → Br; HNO$_3$, H$_2$SO$_4$ → Br, NO$_2$ (+ isomer); H$_2$, Pd-C → Br, NH$_2$

d. CH$_3$CH$_2$Cl, AlCl$_3$; CH$_3$CH$_2$COCl (O, Cl), AlCl$_3$ → O (+ ortho isomer); Cl$_2$, FeCl$_3$ → Cl, O; Br$_2$, $h\nu$ → Br, Cl, O; K$^+$ $^-$OC(CH$_3$)$_3$ → Cl, O; NH$_2$NH$_2$, $^-$OH → Cl

e. CH_3Cl, $AlCl_3$; butanoyl chloride, $AlCl_3$ (+ ortho isomer); Br_2, $FeBr_3$; Br_2, $h\nu$; HO–butyl, NaH → ^-O–butyl

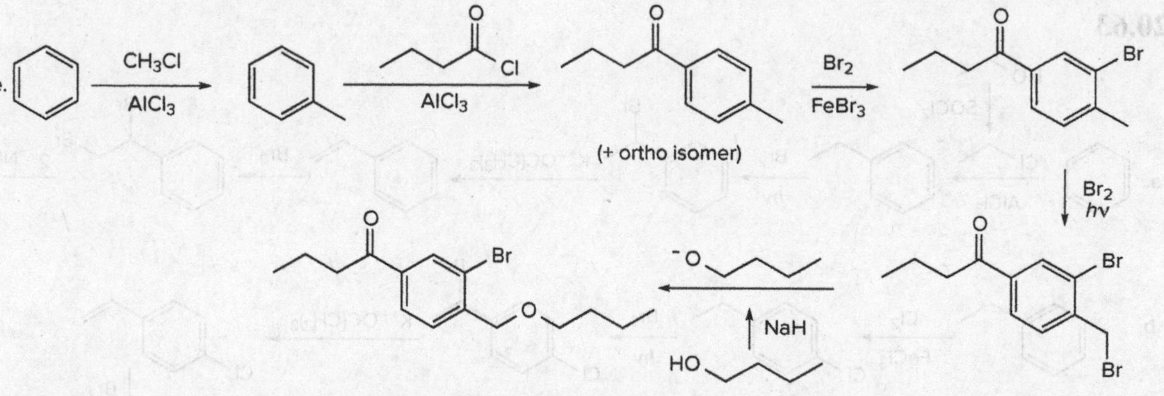

f. chloroethane, $AlCl_3$; isobutyryl chloride, $AlCl_3$ (+ ortho isomer); HNO_3, H_2SO_4; Br_2, $h\nu$; $K^+ {}^-OC(CH_3)_3$; [1] OsO_4 [2] $NaHSO_3$, H_2O

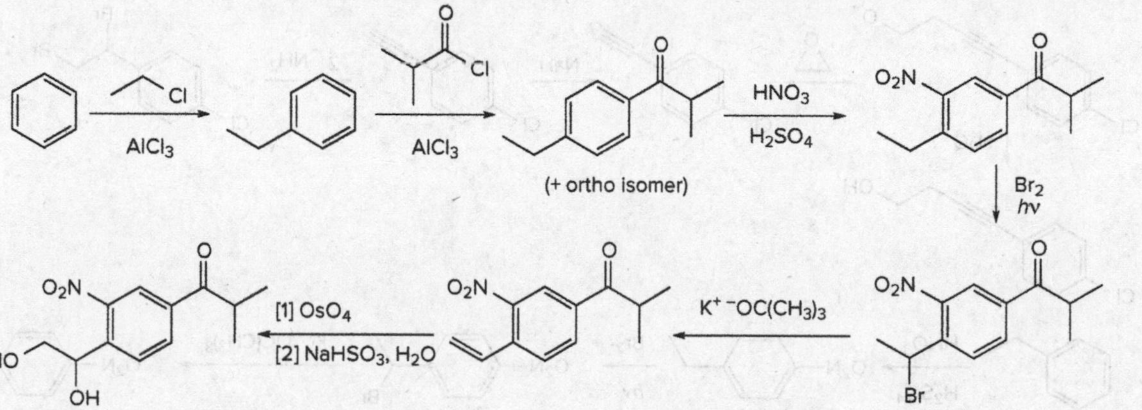

O₂N, HO, OH, Br labels in structures.

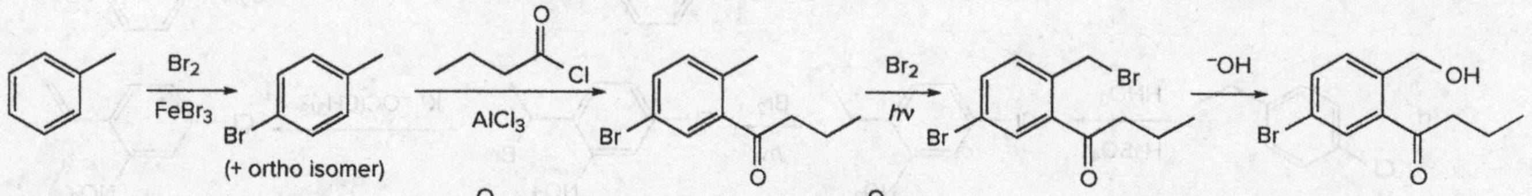

20.62

a. Br_2, $FeBr_3$ (+ ortho isomer); butanoyl chloride, $AlCl_3$; Br_2, $h\nu$; ^-OH

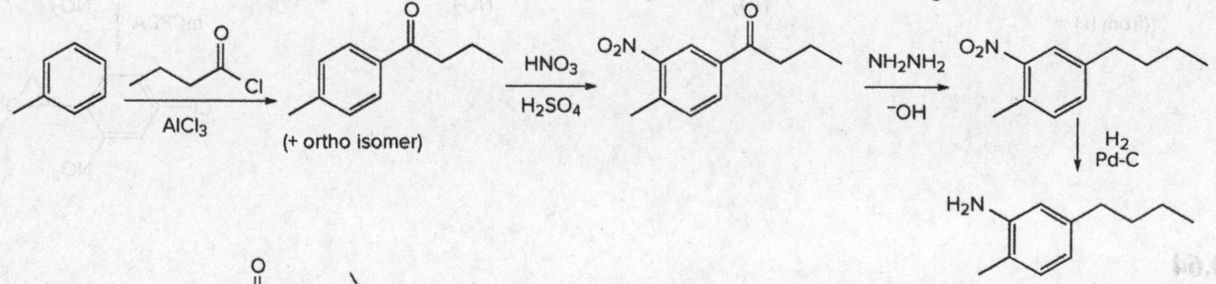

b. butanoyl chloride, $AlCl_3$ (+ ortho isomer); HNO_3, H_2SO_4; NH_2NH_2, ^-OH; H_2, Pd-C → H_2N product

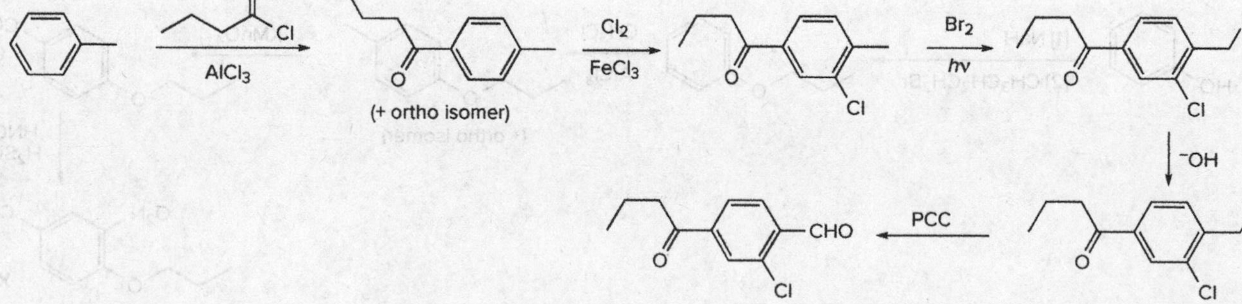

c. butanoyl chloride, $AlCl_3$ (+ ortho isomer); Cl_2, $FeCl_3$; Br_2, $h\nu$; ^-OH; PCC → CHO product

20.63

a. HO (propanoic acid) → $SOCl_2$ → Cl (propanoyl chloride); benzene → Cl, $AlCl_3$ → ethylbenzene → Br_2, $h\nu$ → (1-bromoethyl)benzene (Br) → $K^+ \ ^-OC(CH_3)_3$ → styrene → Br_2 → (1,2-dibromoethyl)benzene (Br, Br) → $2\ ^-NH_2$ → phenylacetylene

b. ethylbenzene (from a.) → Cl_2, $FeCl_3$ → 1-chloro-4-ethylbenzene (Cl) (+ ortho isomer) → Br_2, $h\nu$ → Cl, Br → $K^+ \ ^-OC(CH_3)_3$ → Cl (4-chlorostyrene) → Br_2 → Cl, Br, Br → $2\ ^-NH_2$ → Cl (alkyne) → NaH → Cl (acetylide anion) → ethylene oxide → Cl, O^- → H_2O → Cl, OH

c. ethylbenzene (from a.) → HNO_3, H_2SO_4 → O_2N (4-ethylnitrobenzene) (+ ortho isomer) → Br_2, $h\nu$ → O_2N, Br → $K^+ \ ^-OC(CH_3)_3$ → O_2N (4-nitrostyrene) → HBr, ROOR → O_2N, Br → $^-C{\equiv}CH$ (from $HC{\equiv}CH$, NaH) → O_2N (4-(but-3-ynyl)nitrobenzene)

d. Cl (from b.) → HNO_3, H_2SO_4 → Cl, NO_2 → Br_2, $h\nu$ → Cl, Br, NO_2 → $K^+ \ ^-OC(CH_3)_3$ → Cl, NO_2 → mCPBA → Cl, O (epoxide), NO_2

20.64

HO (phenol) → [1] NaH, [2] $CH_3CH_2CH_2Br$ → propoxybenzene (O) → CH_3Cl, $AlCl_3$ → O (+ ortho isomer) → $KMnO_4$ → O, COOH → HNO_3, H_2SO_4 → O_2N, O, COOH **X**

20.65

a.

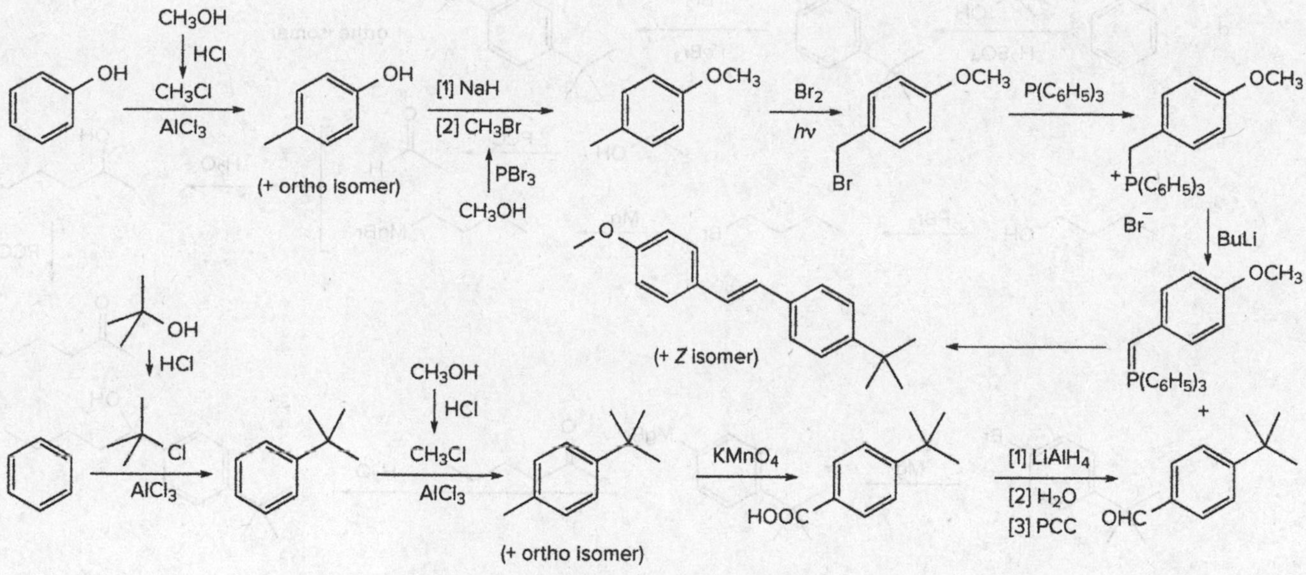

OH

CH_3OH → HCl → CH_3Cl

CH_3Cl, $AlCl_3$ →

OH (+ ortho isomer)

[1] NaH, [2] CH_3Br →

CH_3OH → PBr_3

OCH_3

Br_2, $h\nu$ →

OCH_3, Br

$P(C_6H_5)_3$ →

OCH_3, $^+P(C_6H_5)_3$, Br^-

BuLi

OCH_3, $P(C_6H_5)_3$ +

OH → HCl → Cl

$AlCl_3$ →

CH_3OH → HCl → CH_3Cl

CH_3Cl, $AlCl_3$ → (+ ortho isomer)

$KMnO_4$ → HOOC

[1] $LiAlH_4$, [2] H_2O, [3] PCC → OHC

O (+ *Z* isomer)

b.

OH → PCC → CHO

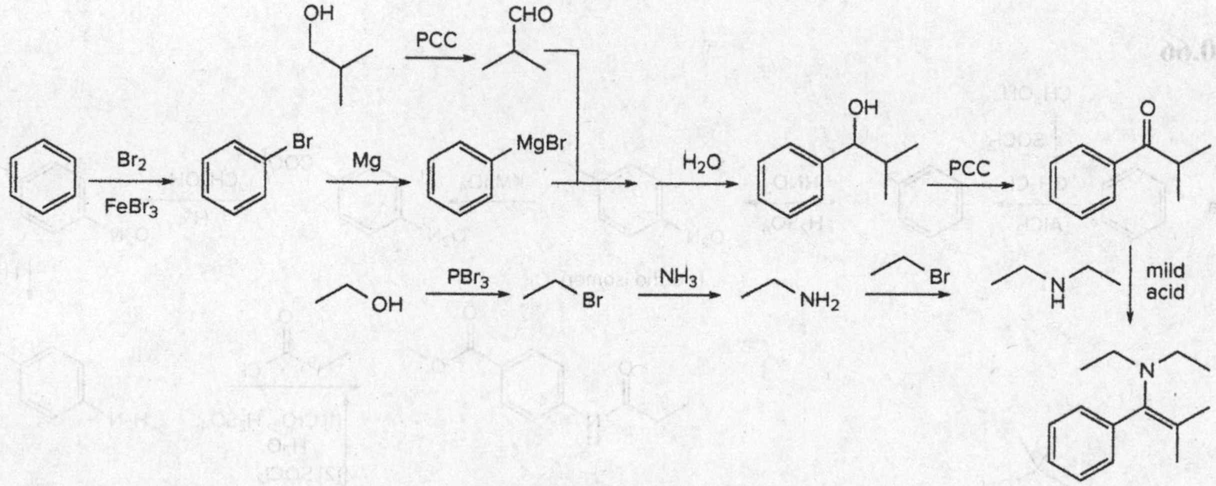

Br_2, $FeBr_3$ → Br → Mg → MgBr → H_2O → OH → PCC → O

OH → PBr_3 → Br → NH_3 → NH_2 → Br → N–H

mild acid

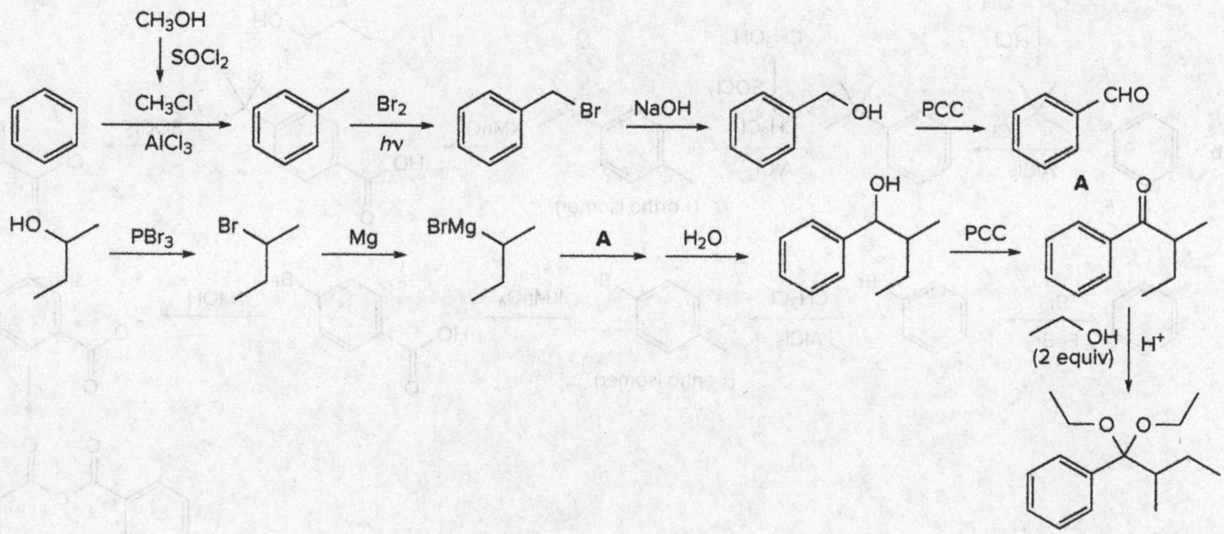

N

c.

CH_3OH → $SOCl_2$ → CH_3Cl

CH_3Cl, $AlCl_3$ → Br_2, $h\nu$ → Br → NaOH → OH → PCC → CHO **A**

HO → PBr_3 → Br → Mg → BrMg → **A** → H_2O → OH → PCC → O

OH (2 equiv), H^+

O O

d. 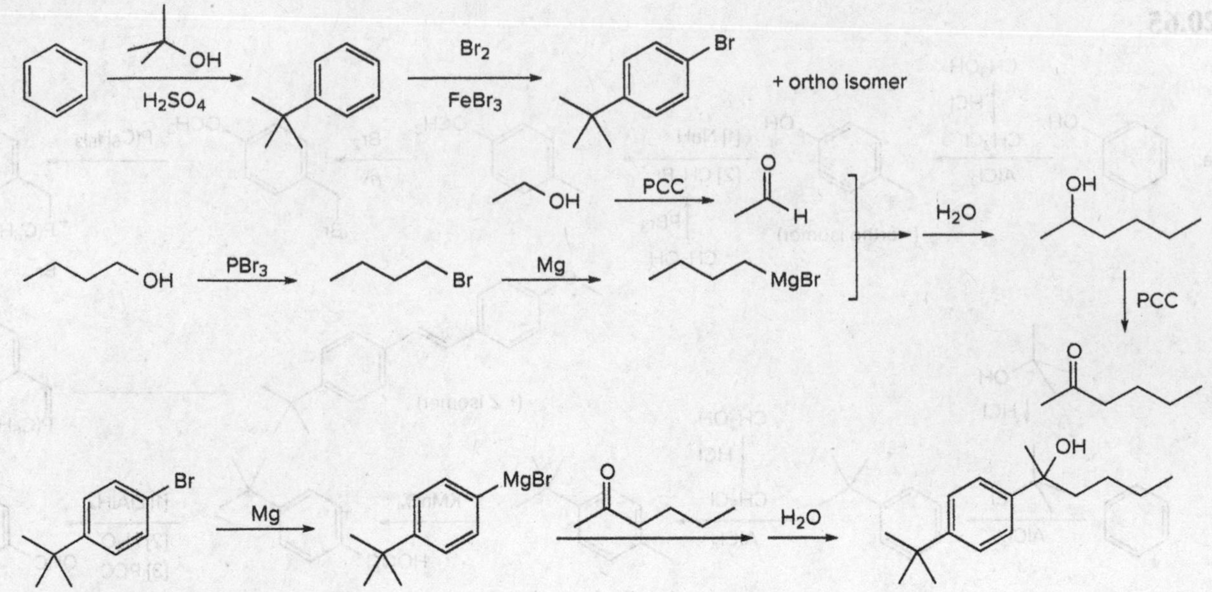 H_2SO_4 (with tert-butyl alcohol, OH); Br_2, $FeBr_3$; Br + ortho isomer

OH → PCC → H; OH → PBr_3 → Br → Mg → MgBr; → H_2O → OH → PCC → O

Br → Mg → MgBr → (ketone) → H_2O → OH

20.66

a. 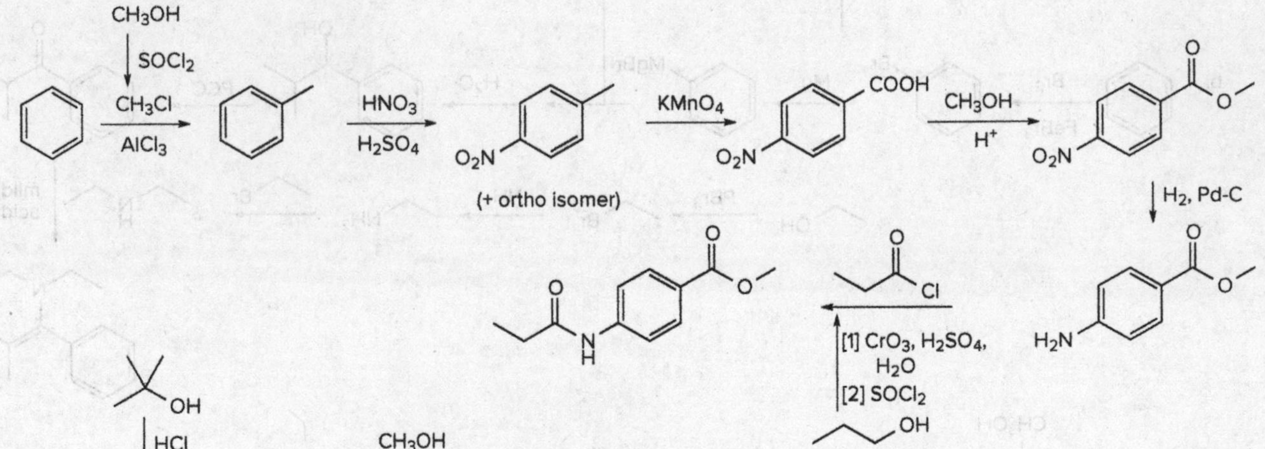 CH_3OH → $SOCl_2$ → CH_3Cl, $AlCl_3$; HNO_3, H_2SO_4; O_2N (+ ortho isomer); $KMnO_4$; O_2N, COOH; CH_3OH, H^+; O_2N; H_2, Pd-C; H_2N; Cl; [1] CrO_3, H_2SO_4, H_2O [2] $SOCl_2$; OH; N H

b. 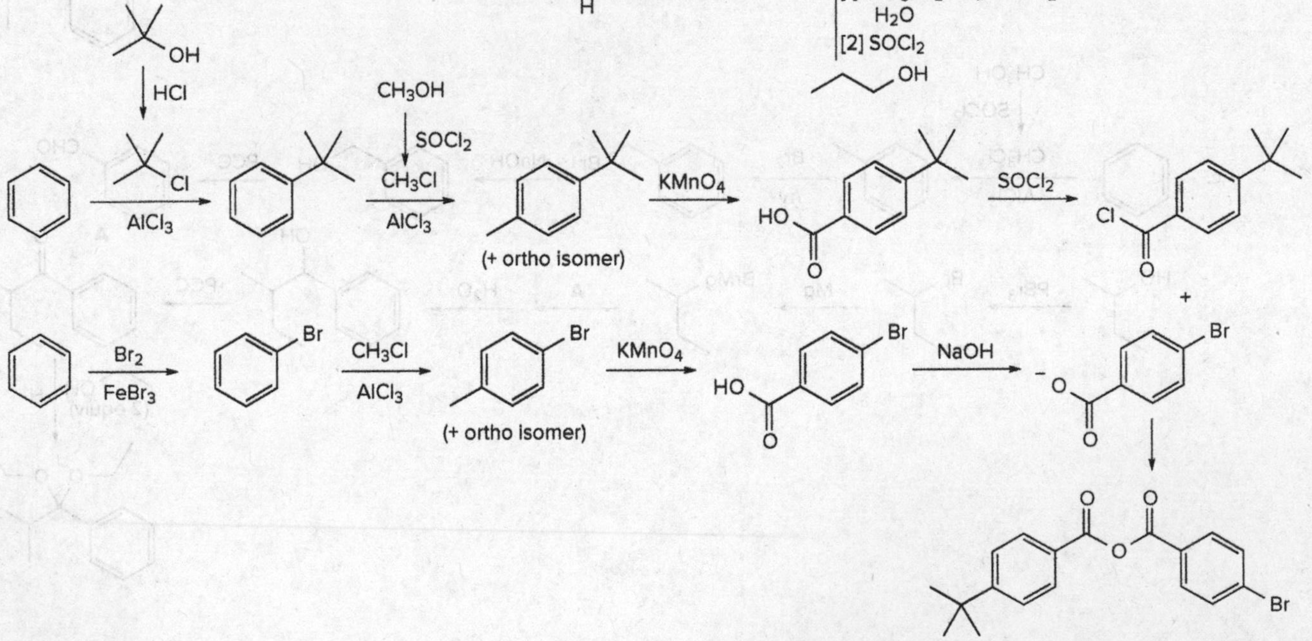 OH → HCl → Cl; $AlCl_3$; CH_3OH → $SOCl_2$ → CH_3Cl, $AlCl_3$; (+ ortho isomer); $KMnO_4$; HO, O; $SOCl_2$; Cl, O; +

Br_2, $FeBr_3$; Br; CH_3Cl, $AlCl_3$; Br (+ ortho isomer); $KMnO_4$; HO, Br, O; NaOH; –O, Br, O

O, O, O, Br

c. [Reaction scheme: 1-butanol →(CrO_3) butanoic acid →($SOCl_2$) butanoyl chloride →(benzene) butyrophenone →(NH_2NH_2, ^-OH) butylbenzene →(Br_2, $FeBr_3$) 1-bromo-4-butylbenzene (+ ortho isomer) →([1] Mg, [2] CO_2, [3] H_3O^+) 4-butylbenzoic acid →(2-methyl-1-propanol, H^+) isobutyl 4-butylbenzoate]

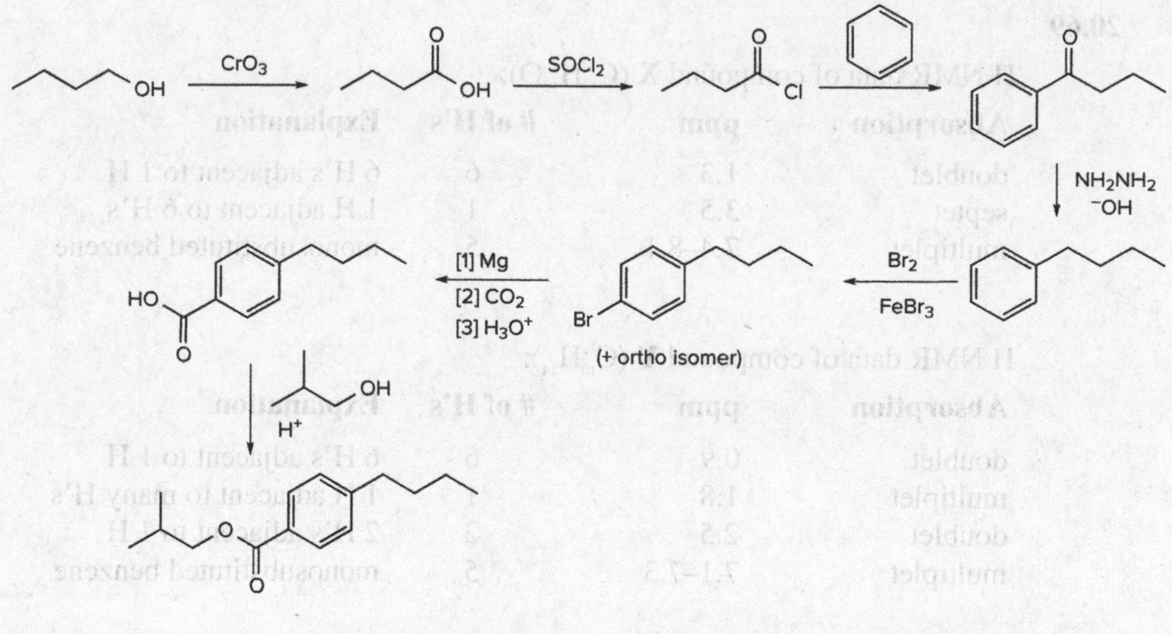

20.67

^{1}H NMR data of compound **A** (C_8H_9Br):

Absorption	ppm	# of H's	Explanation
triplet	1.2	3	3 H's adjacent to 2 H's
quartet	2.6	2	2 H's adjacent to 3 H's
two signals	7.1 and 7.4	4	para disubstituted benzene

Structure: [1-bromo-4-ethylbenzene]

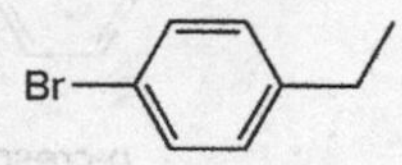

^{1}H NMR data of compound **B** (C_8H_9Br):

Absorption	ppm	# of H's	Explanation
triplet	3.1	2	2 H's adjacent to 2 H's
triplet	3.5	2	2 H's adjacent to 2 H's
multiplet	7.1–7.4	5	monosubstituted benzene

Structure: [(2-bromoethyl)benzene]

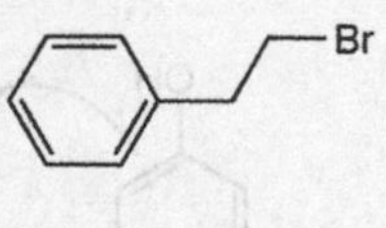

20.68 IR absorption at 1717 cm^{-1} means compound **C** has a C=O.

^{1}H NMR data of compound **C** ($C_{10}H_{12}O$):

Absorption	ppm	# of H's	Explanation
singlet	2.1	3	3 H's
triplet	2.8	2	2 H's adjacent to 2 H's
triplet	2.9	2	2 H's adjacent to 2 H's
multiplet	7.1–7.4	5	monosubstituted benzene

Structure: [4-phenyl-2-butanone]

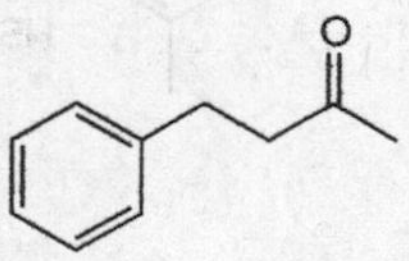

20.69

^{1}H NMR data of compound **X** ($C_{10}H_{12}O$):

Absorption	**ppm**	**# of H's**	**Explanation**
doublet	1.3	6	6 H's adjacent to 1 H
septet	3.5	1	1 H adjacent to 6 H's
multiplet	7.4–8.1	5	monosubstituted benzene

Structure:

^{1}H NMR data of compound **Y** ($C_{10}H_{14}$):

Absorption	**ppm**	**# of H's**	**Explanation**
doublet	0.9	6	6 H's adjacent to 1 H
multiplet	1.8	1	1 H adjacent to many H's
doublet	2.5	2	2 H's adjacent to 1 H
multiplet	7.1–7.3	5	monosubstituted benzene

Structure:

20.70

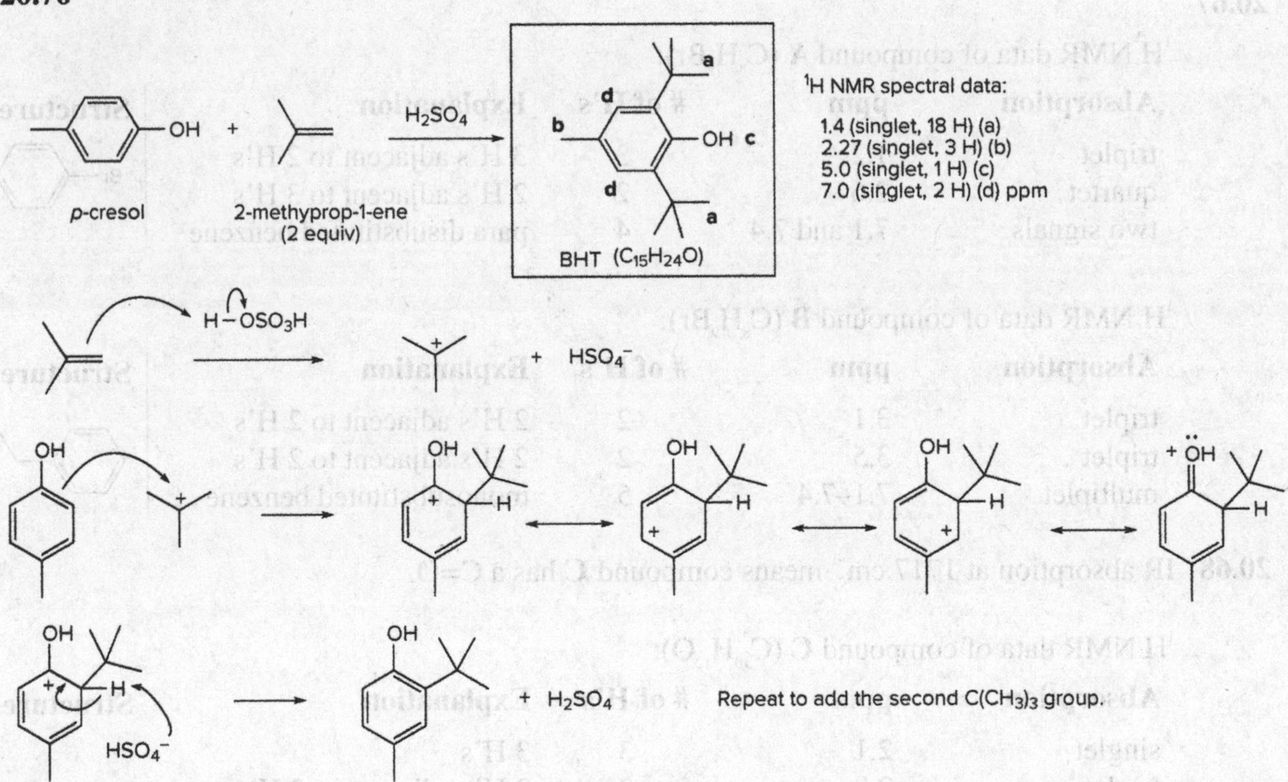

20.71

a. Pyridine: The electron-withdrawing inductive effect of N makes the ring electron poor. Also, electrophiles E^+ can react with N, putting a positive charge on the ring. This makes the ring less reactive with another positively charged species.

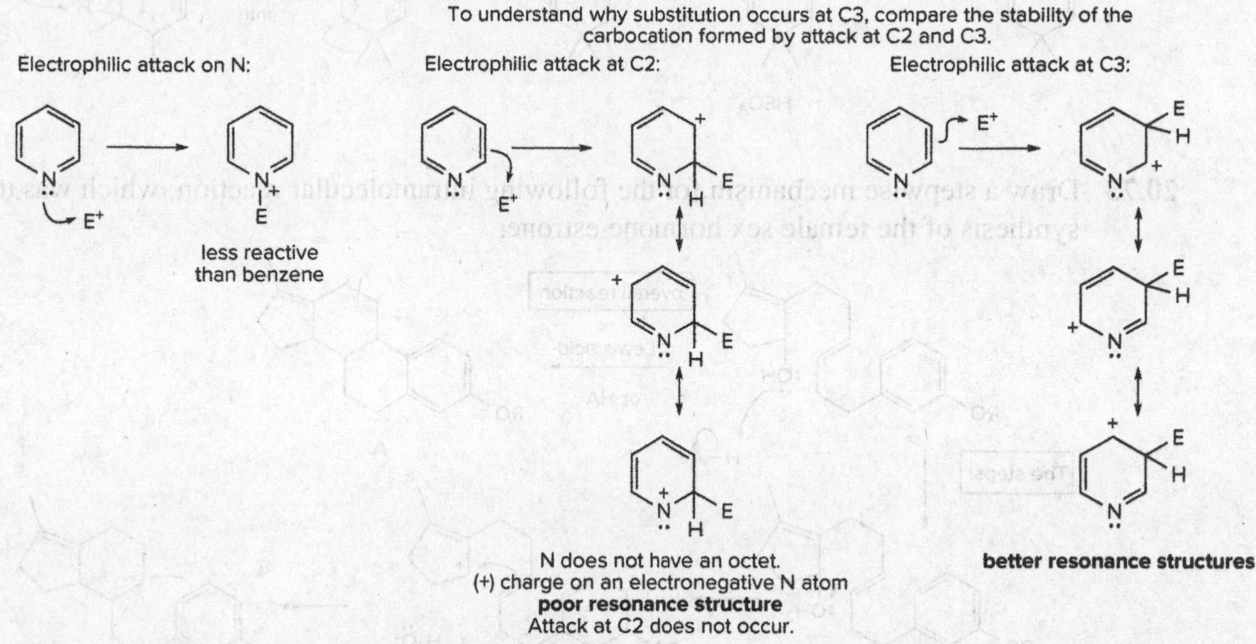

Attack at C3 forms a more stable carbocation, so attack at C3 occurs. Attack at C4 generates a carbocation of similar stability to attack at C2, so attack at C4 does *not* occur.

b. Pyrrole is more reactive than benzene because the C's are more electron rich. The lone pair on N has an electron-donating resonance effect.

Attack at C2:

Attack at C3:

more reactive than benzene

3-position

fewer resonance structures
Attack at C3 does not occur.

2-position
more resonance structures
attack at C2

Attack at C2 forms a more stable carbocation, so electrophilic substitution occurs at C2.

20.72

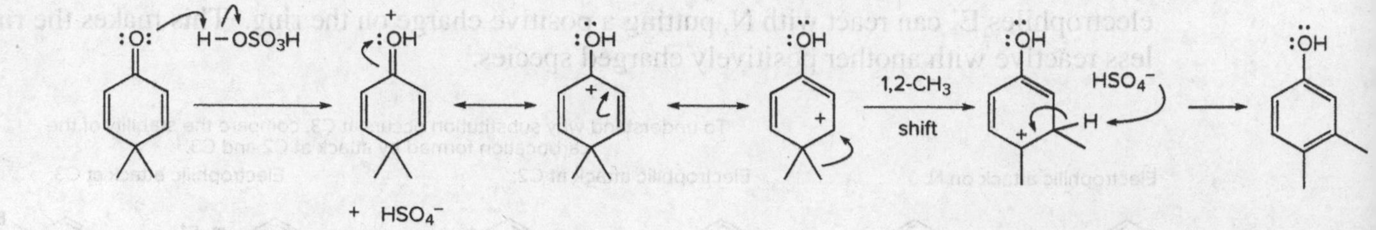

:O: H–OSO3H :OH :OH :OH 1,2-CH3 shift :OH HSO4⁻ :OH

+ HSO4⁻

20.73 Draw a stepwise mechanism for the following intramolecular reaction, which was used in the synthesis of the female sex hormone estrone.

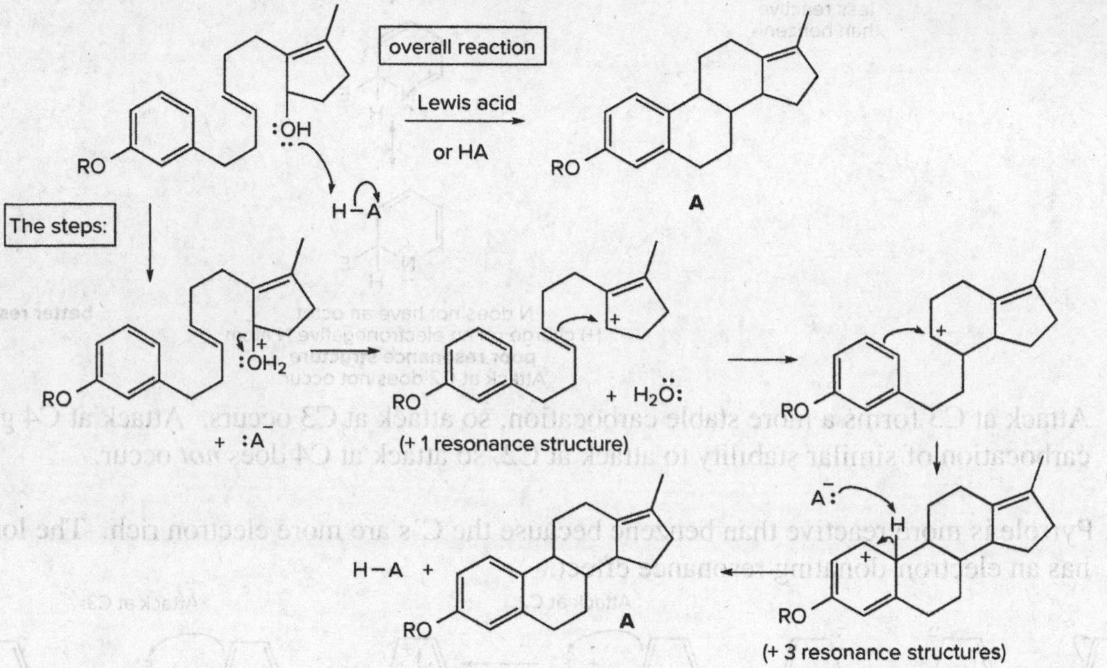

overall reaction

RO :OH Lewis acid or HA RO A

The steps:

H–A

RO :OH2 + :A RO + H2Ö: (+ 1 resonance structure) RO

A:

H–A + RO A RO (+ 3 resonance structures)

20.74

OH BF3 N H N H Ö–H BF3 N H + HOBF3

(+ 3 additional resonance structures)

1,2-shift

H2O + BF3 + N H N H H HO–BF3

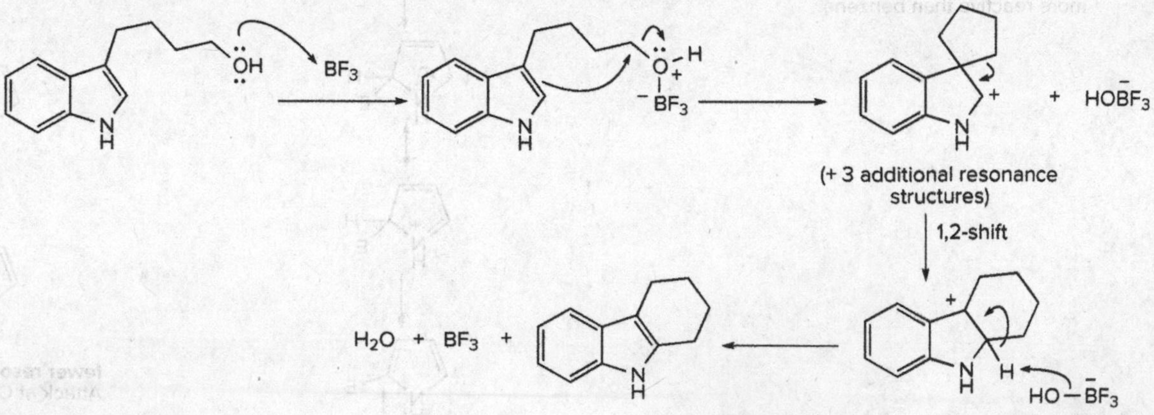

Chapter 21 Radical Reactions

Chapter Review

General features of radicals

- A radical is a reactive intermediate with an unpaired electron (21.1).
- A carbon radical is *sp*^{2} hybridized and trigonal planar (21.1).
- The stability of a radical increases as the number of C's bonded to the radical carbon increases (21.1).

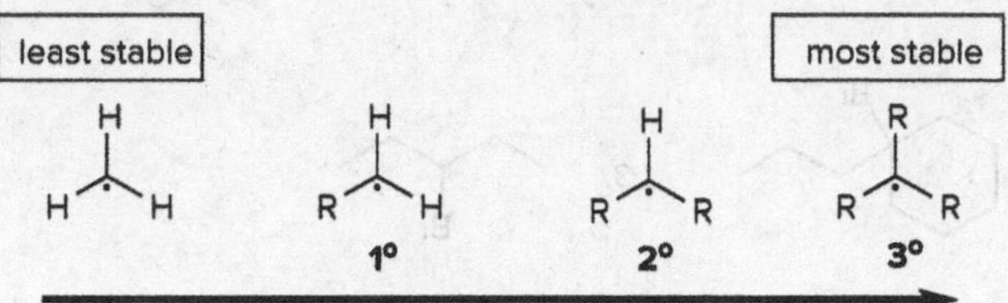

Increasing alkyl substitution
Increasing radical stability

- Allylic radicals are stabilized by resonance, making them more stable than 3° radicals (21.9).

two resonance structures for the allyl radical

Radical reactions

[1] Halogenation of alkanes (21.4)

$R{-}H \xrightarrow[h\nu \text{ or } \Delta,\ X = Cl \text{ or } Br]{X_2} R{-}X$ (alkyl halide)

- The reaction follows a radical chain mechanism.
- The weaker the C–H bond, the more readily the H is replaced by X.
- Chlorination is faster and less selective than bromination (21.6).
- Radical substitution results in racemization at a stereogenic center (21.7).

[2] Allylic halogenation (21.9)

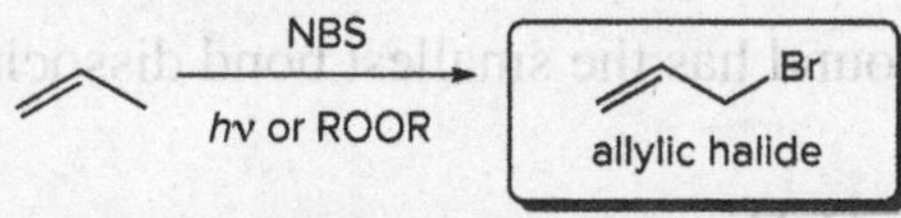

- The reaction follows a radical chain mechanism.

[3] Radical addition of HBr to an alkene (21.12)

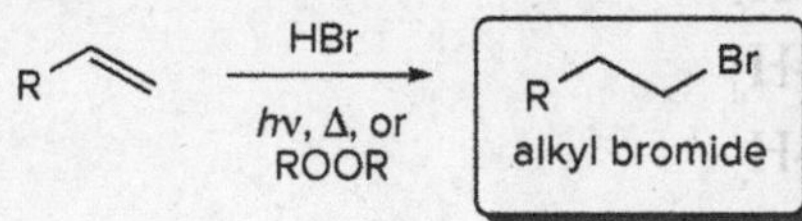

- A radical addition mechanism is followed.
- Br bonds to the less substituted carbon atom to form the more substituted, more stable radical.

[4] Radical polymerization of alkenes (21.13)

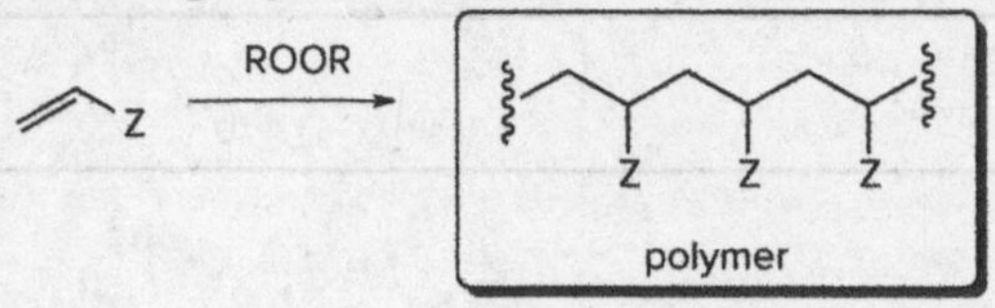

ROOR

polymer

- A radical addition mechanism is followed.

Practice Test on Chapter Review

1.a. Which alkyl halide(s) can be made in good yield by radical halogenation of an alkane?

1. 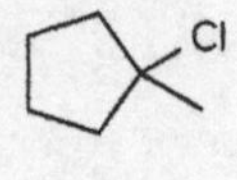 Cl 2. 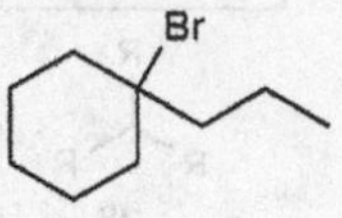 Br 3. 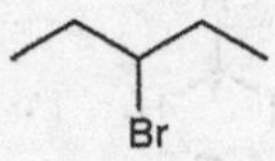 Br

4. Both (1) and (2) can be made in good yield.
5. Compounds (1), (2), and (3) can all be made in good yield.

b. In which of the following reactions will rearrangement *not* occur?

1. halogenation of an alkane with Cl_2 and heat
2. addition of Cl_2 to an alkene
3. addition of HCl to an alkene
4. Rearrangements do not occur in reactions (1) and (2).
5. Rearrangements do not occur in reactions (1), (2), and (3).

c. Which labeled H is most easily abstracted in a radical halogenation reaction?

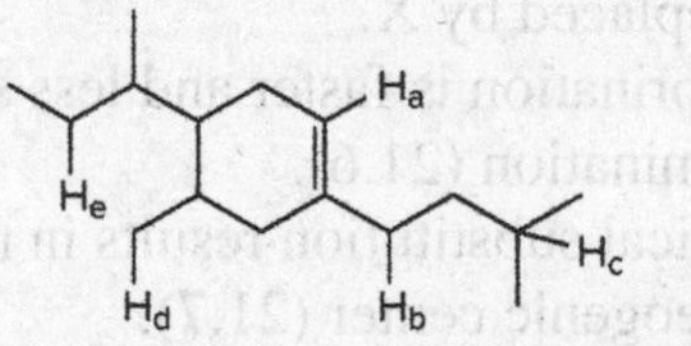

1. H_a
2. H_b
3. H_c
4. H_d
5. H_e

d. Which of the labeled C–H bonds in the following compound has the smallest bond dissociation energy?

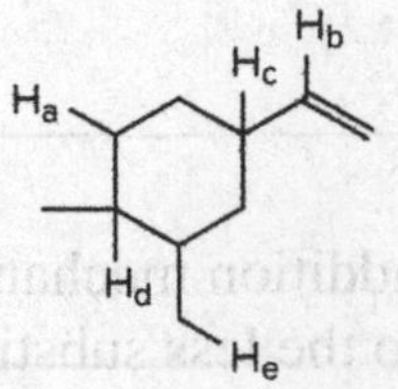

1. C–H_a
2. C–H_b
3. C–H_c
4. C–H_d
5. C–H_e

2. (a) Which radical is the most stable? (b) Which radical is the least stable?

3. Draw all of the organic products formed in each reaction. Indicate stereochemistry in part (c).

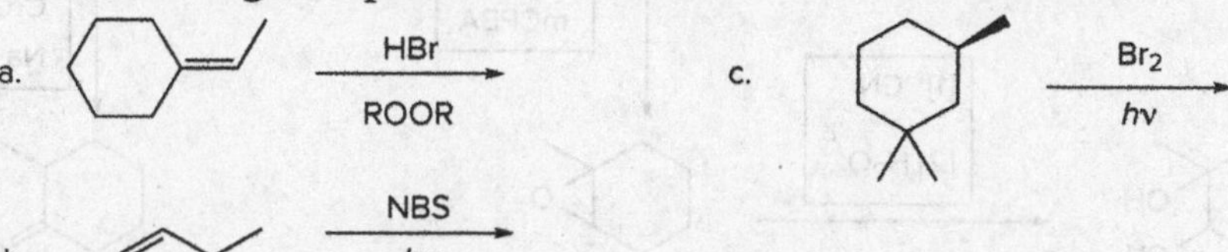

4. What monomer is needed to make the following polymer?

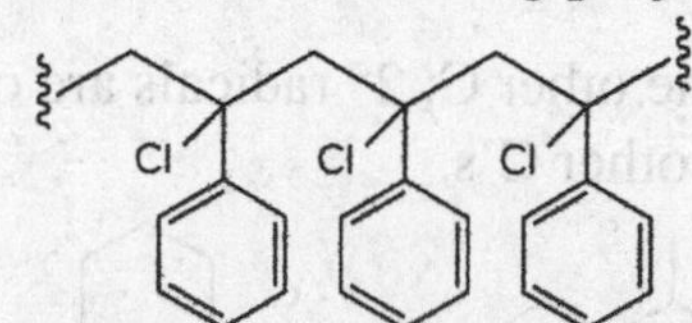

5. In each box, fill in the appropriate reagents needed to carry out the given reaction. This question involves reactions from Chapter 21, as well as previous chapters.

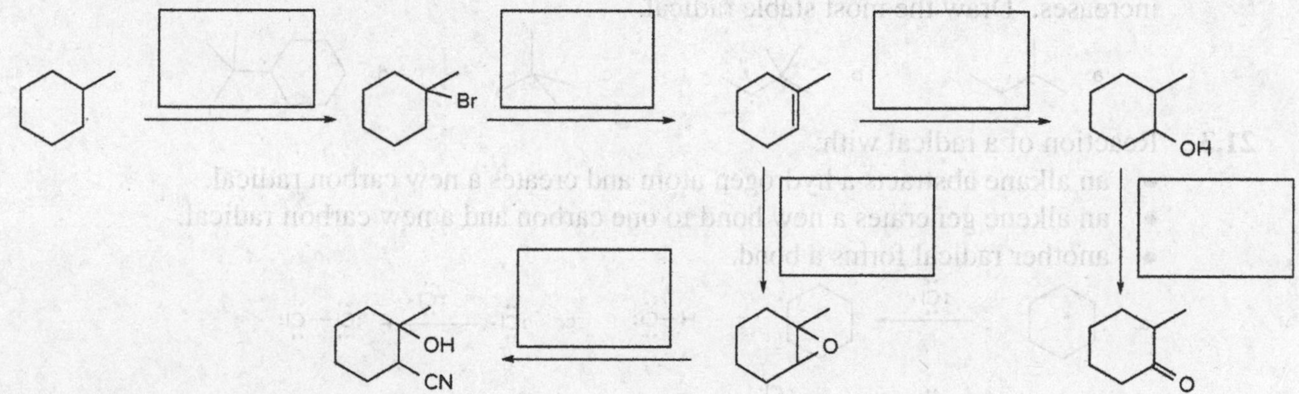

Answers to Practice Test

1.a. 2
b. 4
c. 2
d. 3

2.a. **A**
b. **B**

3.a.

Br

3.b.

Br + Br (+ cis isomer)

3.c.

Br + Br

4.

Cl

5.

[1] Br_2, hν or Δ; KOC$(CH_3)_3$; [1] BH_3 [2] H_2O_2, ^-OH; PCC or CrO_3 or $Na_2Cr_2O_7$; mCPBA; [1] ^-CN [2] H_2O

Answers to Problems

21.1 1° Radicals are on C's bonded to one other C; 2° radicals are on C's bonded to two other C's; 3° radicals are on C's bonded to three other C's.

a. 2° radical b. 3° radical c. 2° radical d. 1° radical

21.2 The stability of a radical increases as the number of alkyl groups bonded to the radical carbon increases. Draw the most stable radical.

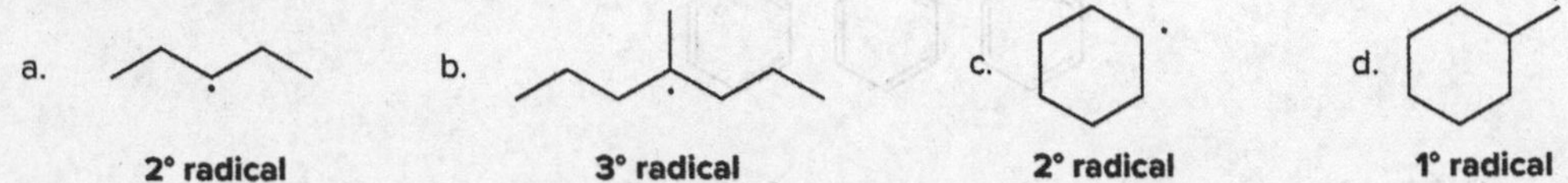

21.3 Reaction of a radical with:

- an alkane abstracts a hydrogen atom and creates a new carbon radical.
- an alkene generates a new bond to one carbon and a new carbon radical.
- another radical forms a bond.

a. cyclohexane + :Cl· → cyclohexyl radical + H–Cl:

b. $CH_2=CH_2$ + :Cl· → $\dot{C}H_2-CH_2$(Cl)

c. :Cl· + :Cl· → :Cl–Cl:

d. :Cl· + ·O–O· → :Cl–O–O·

21.4 **Monochlorination** is a radical substitution reaction in which a Cl replaces a H, thus generating an alkyl halide.

a. cyclobutane + Cl_2 → chlorocyclobutane

b. pentane + Cl_2 → 1-chloropentane + 2-chloropentane + 3-chloropentane

c. isobutane + Cl_2 → 1-chloro-2-methylpropane + 2-chloro-2-methylpropane

21.5

Initiation: $:\ddot{Br}-\ddot{Br}: \xrightarrow[\text{or } \Delta]{h\nu} :\ddot{Br}\cdot + \cdot\ddot{Br}:$

Propagation: $CH_3-H + \cdot\ddot{Br}: \longrightarrow \dot{C}H_3 + H-\ddot{Br}:$

$\dot{C}H_3 + :\ddot{Br}-\ddot{Br}: \longrightarrow CH_3-\ddot{Br}: + \cdot\ddot{Br}:$

Termination: $:\ddot{Br}\cdot + \cdot\ddot{Br}: \longrightarrow :\ddot{Br}-\ddot{Br}:$

or

$\dot{C}H_3 + \dot{C}H_3 \longrightarrow CH_3-CH_3$

or

$\dot{C}H_3 + \cdot\ddot{Br}: \longrightarrow CH_3-\ddot{Br}:$

21.6 The rate-determining step for halogenation reactions is formation of $CH_3^{\cdot}$ + HX.

$CH_3-H + \cdot\ddot{I}: \longrightarrow \cdot CH_3 + H-\ddot{I}:$ $\qquad \Delta H° = +138$ kJ/mol

1 bond broken +435 kJ/mol

1 bond formed −297 kJ/mol

This reaction is more endothermic and has a higher E_a than a similar reaction with Cl_2 or Br_2.

21.7 The **weakest C–H bond** in each alkane is the **most readily cleaved** during radical halogenation.

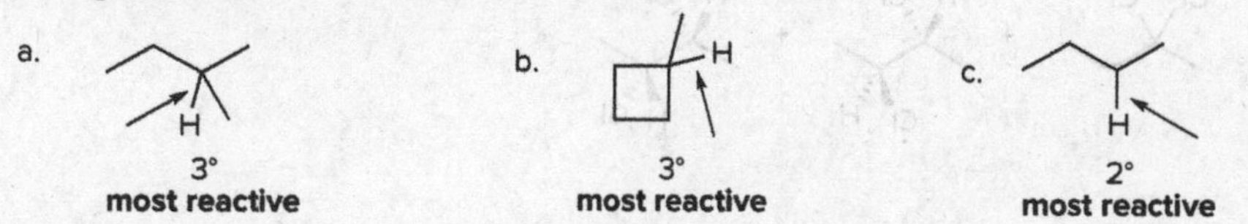

21.8 To draw the product of bromination:

- Draw out the starting material and find the most reactive C–H bond (on the most substituted C).
- The major product is formed by **cleavage of the *weakest* C–H bond.**

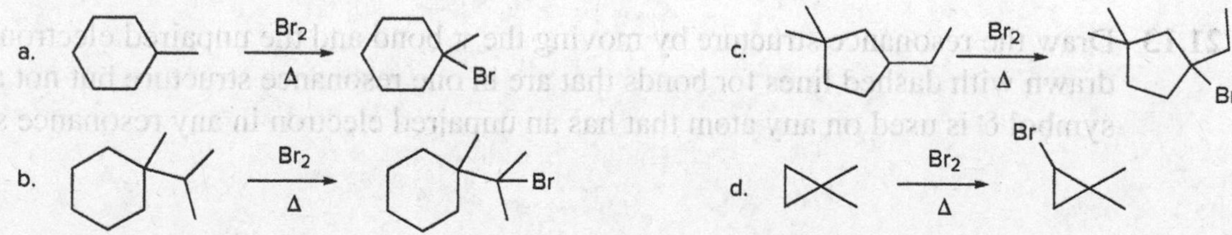

21.9 If 1° C–H and 3° C–H bonds were equally reactive there would be nine times as much $(CH_3)_2CHCH_2Cl$ as $(CH_3)_3CCl$ because the ratio of 1° to 3° H's is 9:1. The fact that the ratio is only 63:37 shows that the 1° C–H bond is less reactive than the 3° C–H bond. $(CH_3)_2CHCH_2Cl$ is still the major product, though, because there are nine 1° C–H bonds and only one 3° C–H bond.

21.10 The reaction does not occur at the stereogenic center, so leave it as is.

C4 C1 Br Cl_2 $h\nu$ Br Cl Br Cl

(*R*)-2-bromobutane *S* *R*

21.11

a. Cl_2 Δ Cl Cl Cl Cl

b. Cl_2 Δ Cl Cl Cl Cl Cl Cl

c. Cl_2 $h\nu$ Cl Cl Cl Cl

d. H Cl Cl_2 $h\nu$ Cl Cl H Cl Cl H H Cl H Cl

(Consider attack at C2 and C3 only.)

21.12

HÖ: H F F F H–Ö–H + F F F

21.13 Draw the resonance structure by moving the π bond and the unpaired electron. The hybrid is drawn with dashed lines for bonds that are in one resonance structure but not another. The symbol δ• is used on any atom that has an unpaired electron in any resonance structure.

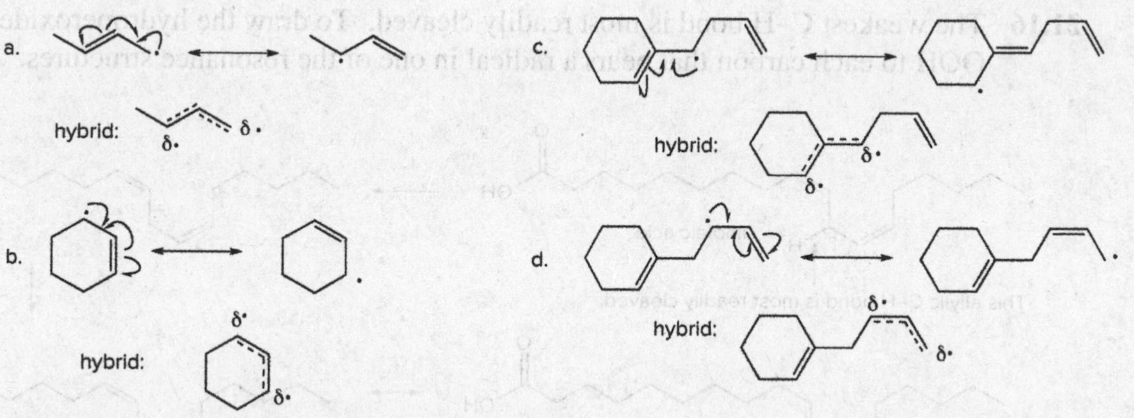

a.

hybrid: δ• δ•

b.

hybrid: δ• δ•

c.

hybrid: δ• δ•

d.

hybrid: δ• δ•

21.14 Reaction of an alkene with NBS or Br_2 + $h\nu$ yields allylic substitution products.

a. NBS, $h\nu$ → Br

b. NBS, $h\nu$ → Br

c. Br_2 → Br, Br

21.15

a. NBS, $h\nu$ → Br + Br (+ *Z* isomer)

b. NBS, $h\nu$ → Br + Br

c. NBS, $h\nu$ → Br + Br (+ *Z* isomer)

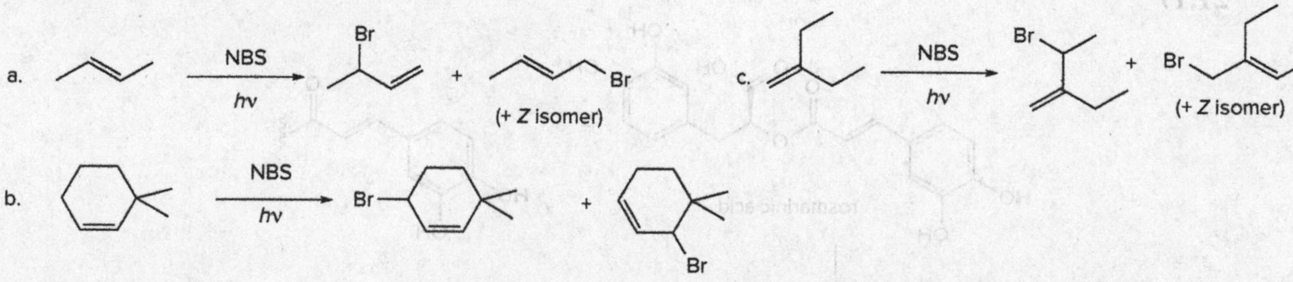

21.16 The weakest C–H bond is most readily cleaved. To draw the hydroperoxide products, add OOH to each carbon that bears a radical in one of the resonance structures.

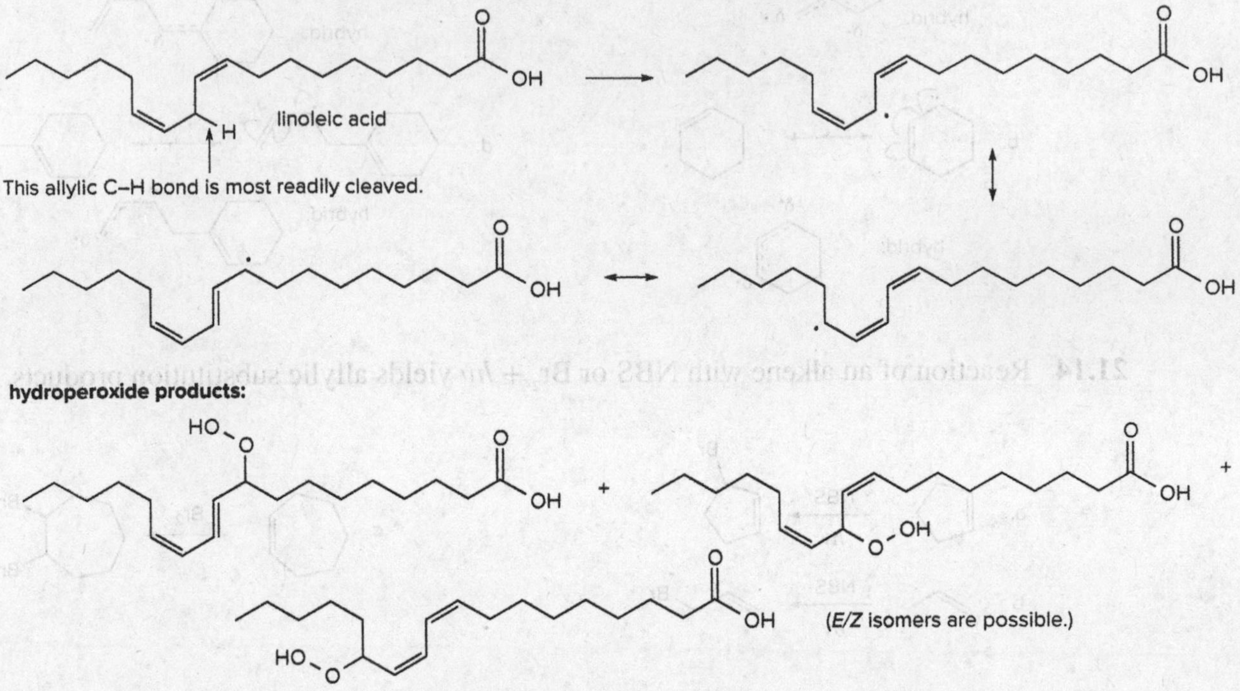

linoleic acid

This allylic C–H bond is most readily cleaved.

hydroperoxide products:

(*E/Z* isomers are possible.)

21.17

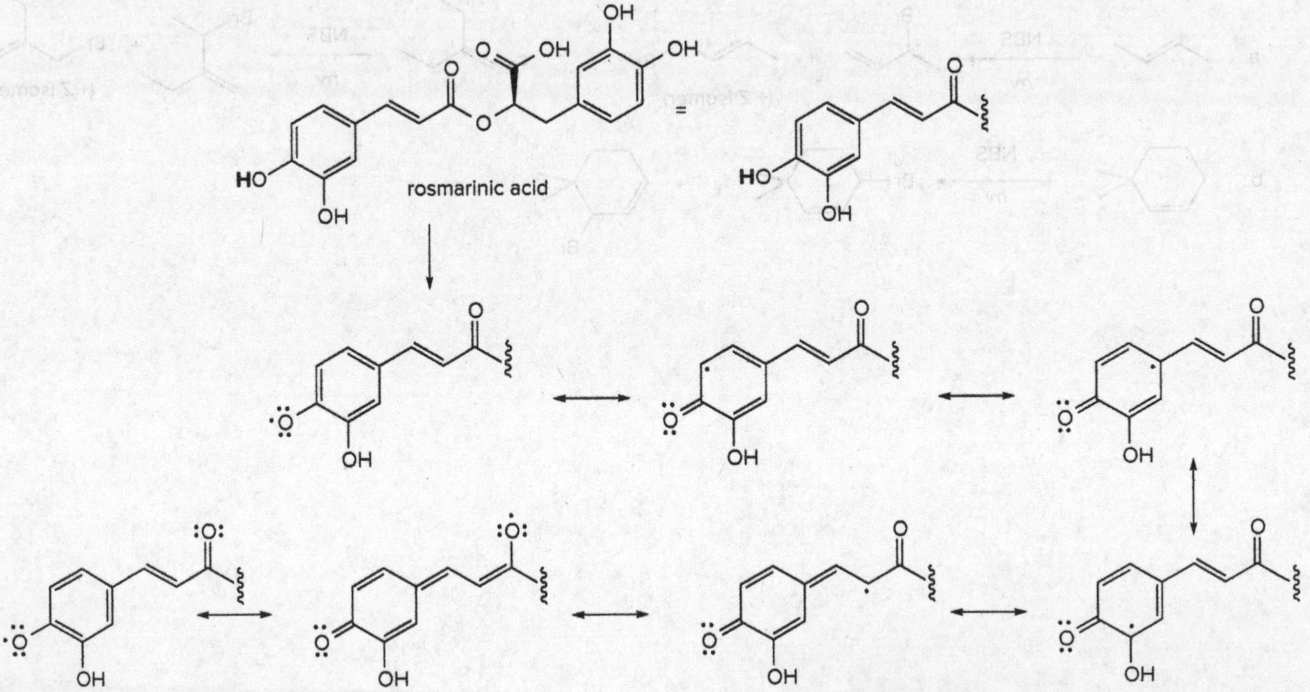

rosmarinic acid

21.18

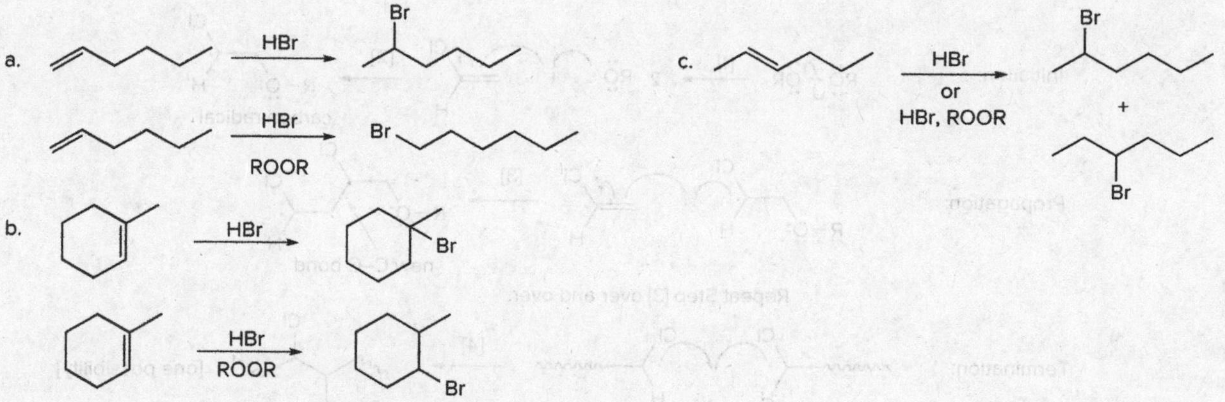

a. HBr

HBr ROOR

Br

b. HBr

HBr ROOR

Br

c. HBr or HBr, ROOR

Br + Br

21.19 In addition of HBr under radical conditions:

- Br$^{\bullet}$ adds first to form the more stable radical.
- Then H$^{\bullet}$ is added to the carbon radical.

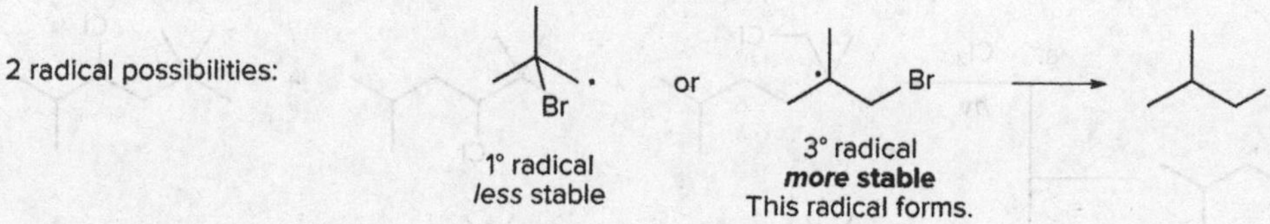

2 radical possibilities:

Br

1° radical
less stable

or

Br

3° radical
***more* stable**
This radical forms.

Br

21.20

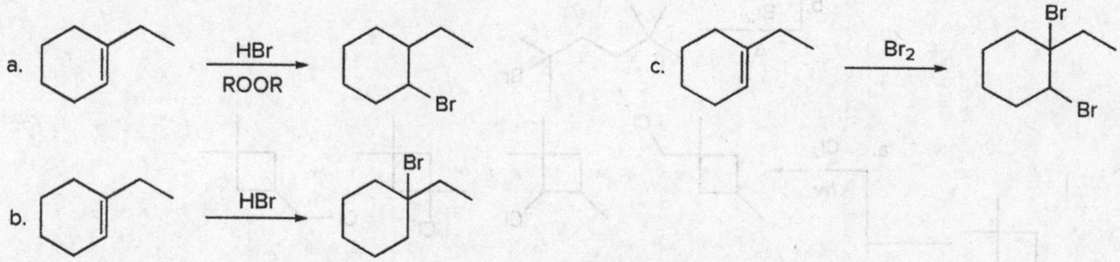

a. HBr ROOR

Br

b. HBr

Br

c. Br_2

Br

Br

21.21

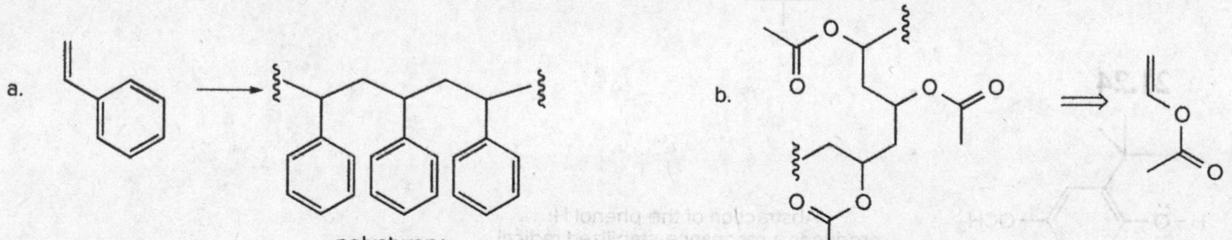

a.

polystyrene

b.

poly(vinyl acetate)

21.22

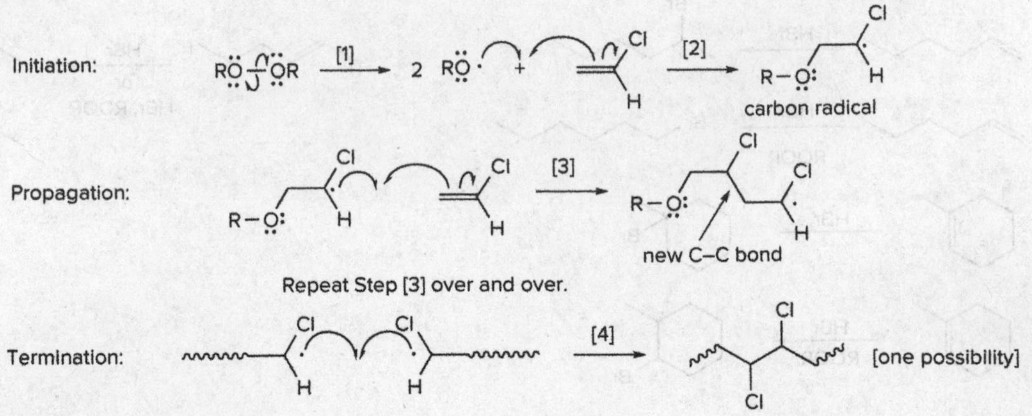

Initiation: RÖ–ÖR →[1] 2 RÖ· + (vinyl chloride) →[2] R–Ö–CH₂–ĊH(Cl) carbon radical

Propagation: R–Ö–CH₂–ĊH(Cl) + vinyl chloride →[3] R–Ö–CH₂–CH(Cl)–CH₂–ĊH(Cl) new C–C bond

Repeat Step [3] over and over.

Termination: ~~~ĊH(Cl) + ĊH(Cl)~~~ →[4] ~~~CH(Cl)–CH(Cl)~~~ [one possibility]

21.23 With Cl_2, each H of the starting material can be replaced by Cl. With Br_2, cleavage of the weakest C–H bond is preferred.

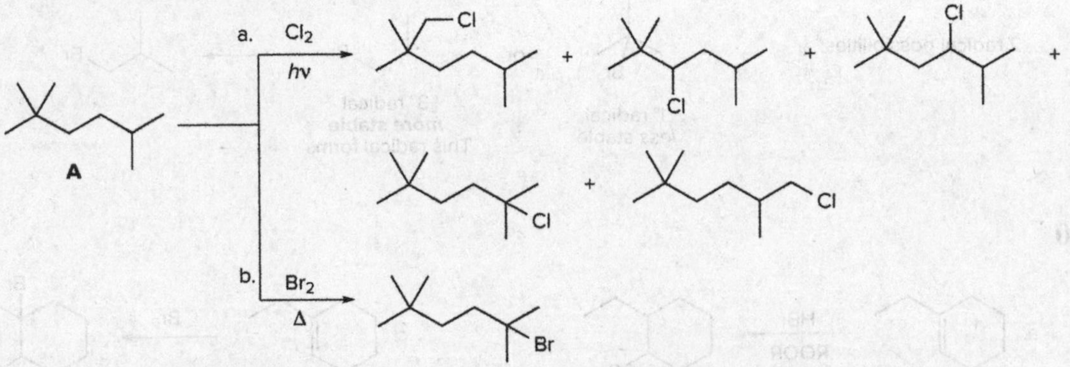

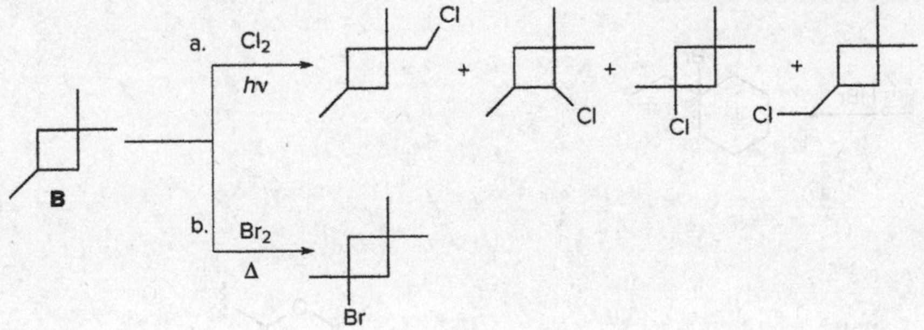

A: a. Cl_2, *hν* → five monochlorinated products (+ + + +)

b. Br_2, Δ → Br product

B: a. Cl_2, *hν* → Cl products (+ + +)

b. Br_2, Δ → Br product

21.24

H–Ö–(aryl)–OCH_3

BHA

Abstraction of the phenol H produces a resonance-stabilized radical.

·Ö–(aryl)–OCH_3 ⟷ :O=(aryl)–OCH_3 ⟷ :O=(aryl)–OCH_3 ⟷ :O=(aryl)–OCH_3 ⟷ ·Ö–(aryl)–OCH_3

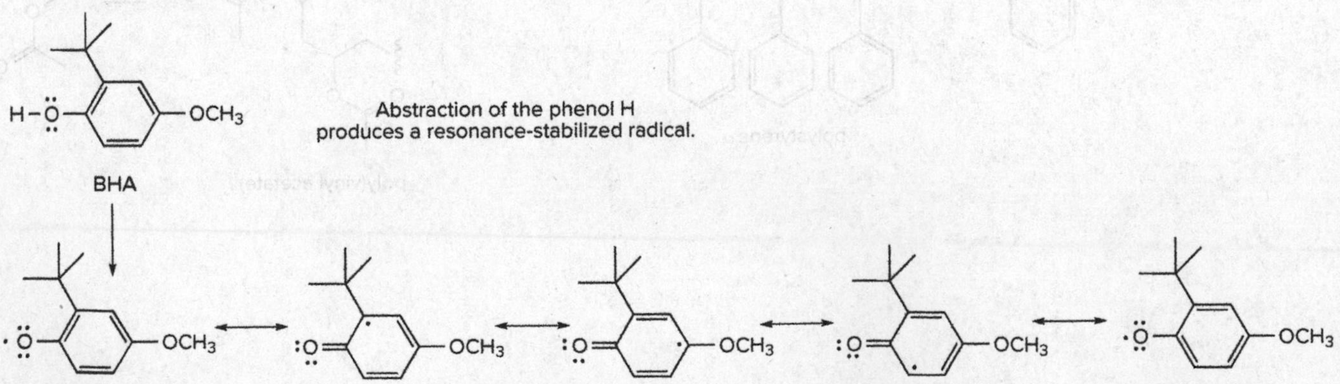

21.25

a. increasing bond strength: 2 < 1 < 3 < 4

b. and c.

alkenyl **least stable** — 1° radical — 2° radical — 2°, allylic **most stable**

increasing stability

d. increasing ease of H abstraction: 4 < 3 < 1 < 2

21.26 Use the directions from Answer 21.2 to rank the radicals.

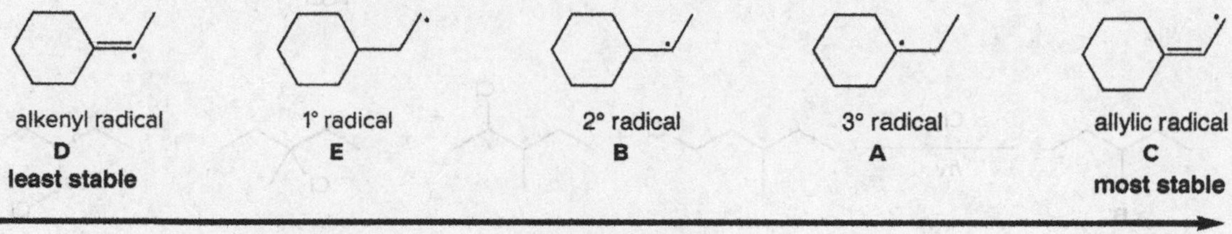

21.27

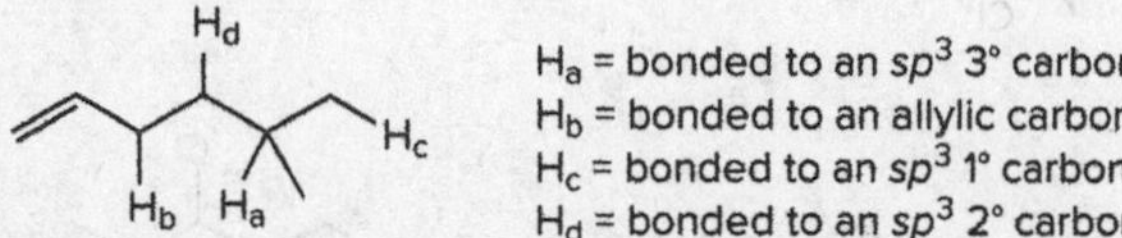

H_a = bonded to an sp^3 3° carbon
H_b = bonded to an allylic carbon
H_c = bonded to an sp^3 1° carbon
H_d = bonded to an sp^3 2° carbon

Increasing ease of abstraction:
$H_c < H_d < H_a < H_b$

21.28

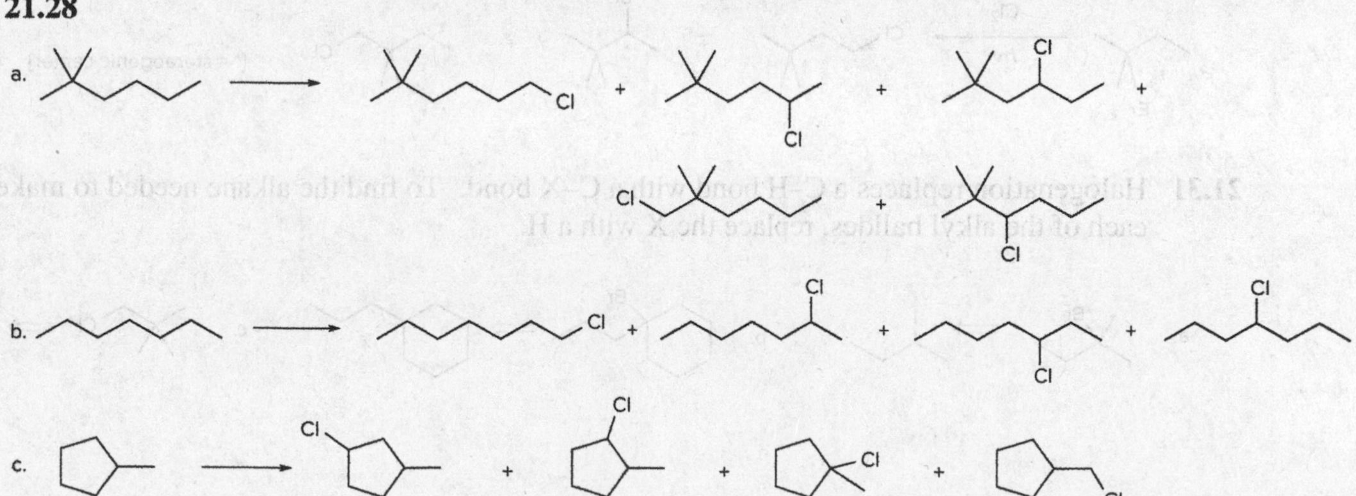

21.29 To draw the product of bromination:

- Draw out the starting material and find the most reactive C–H bond (on the most substituted C).
- The major product is formed by **cleavage of the *weakest* C–H bond.**

a. 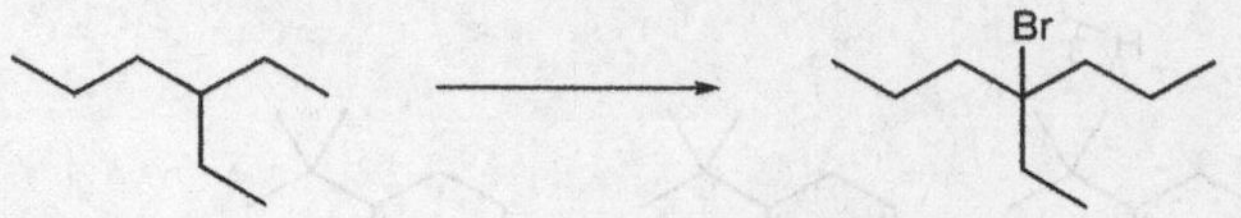 → Br

c. 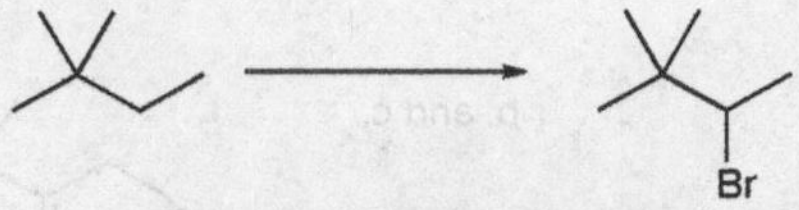 → Br

b. 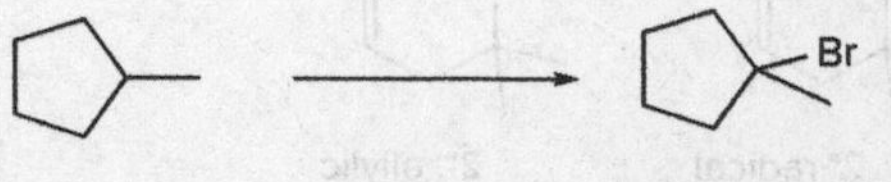 → Br

21.30 Draw all of the alkane isomers of C_6H_{14} and their products from chlorination. Then determine which letter corresponds to which alkane.

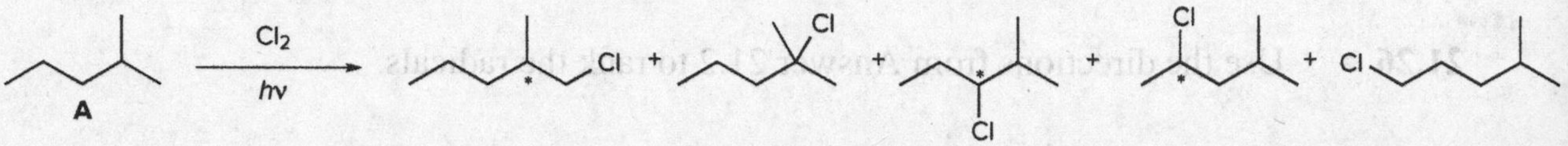

A $\xrightarrow[hv]{Cl_2}$ Cl + Cl + Cl + Cl + Cl

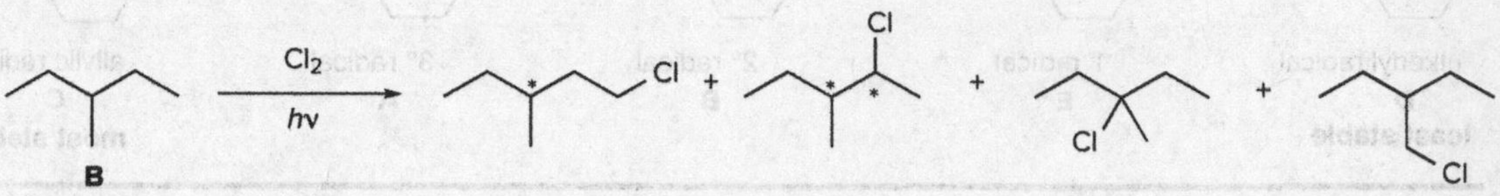

B $\xrightarrow[hv]{Cl_2}$ Cl + Cl + Cl + Cl

C $\xrightarrow[hv]{Cl_2}$ Cl + Cl

D $\xrightarrow[hv]{Cl_2}$ Cl + Cl + Cl

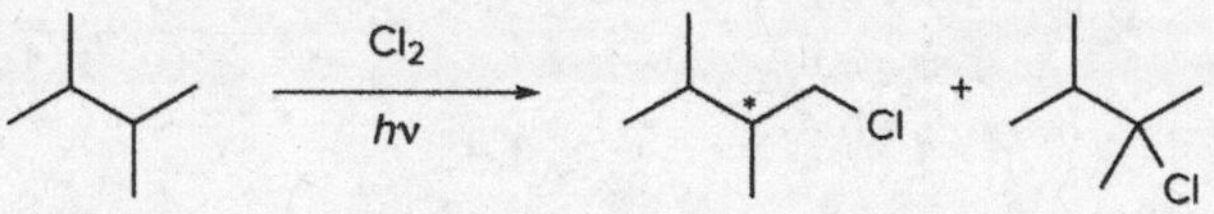

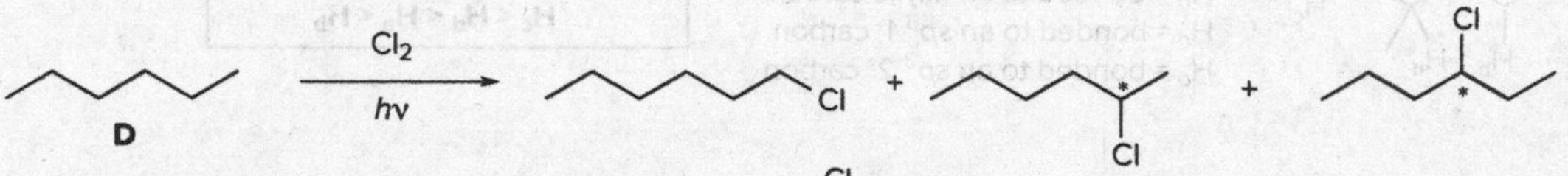

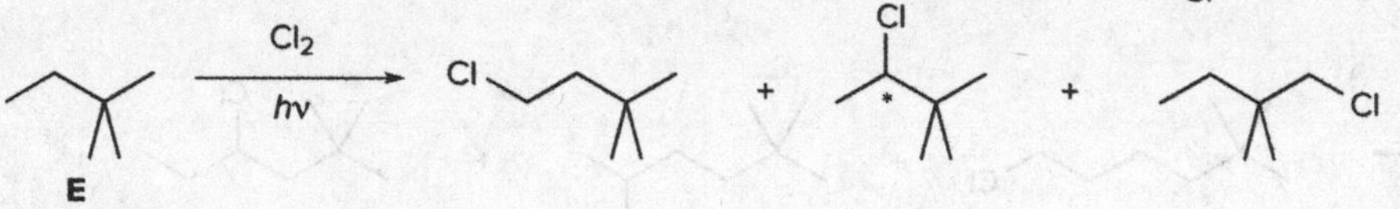

E $\xrightarrow[hv]{Cl_2}$ Cl + Cl + Cl

[* = stereogenic center]

21.31 Halogenation replaces a C–H bond with a C–X bond. To find the alkane needed to make each of the alkyl halides, replace the X with a H.

a. 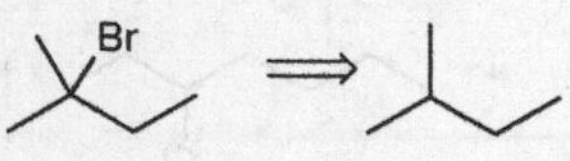 Br ⟹

b. 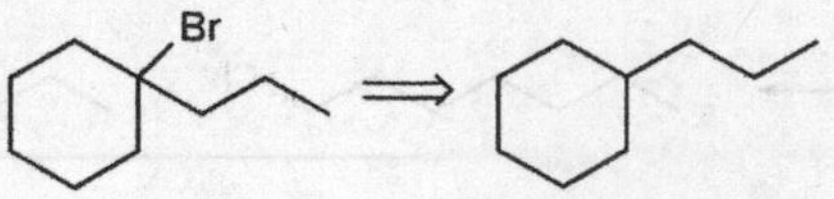 Br ⟹

c. 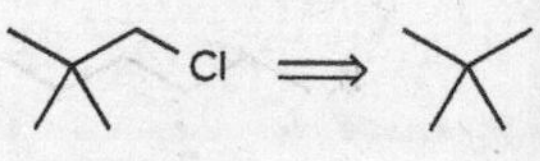 Cl ⟹

21.32 For an alkane to yield one major product on monohalogenation with Cl_2, all of the hydrogens must be identical in the starting material. For an alkane to yield one major product on bromination, it must have a more substituted carbon in the starting material.

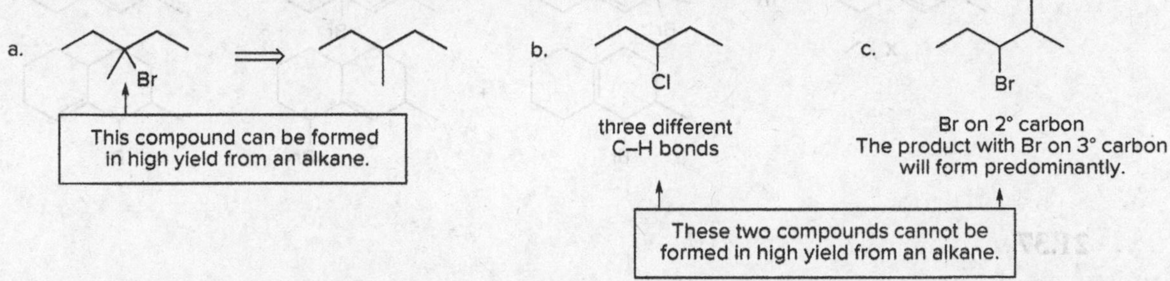

21.33 In bromination, the predominant (or exclusive) product is formed by cleavage of the weaker C–H bond to form the more stable radical intermediate.

H
Br_2
weaker bond Δ
$\Delta H° = 356$ kJ/mol
H
stronger bond
$\Delta H° = 460$ kJ/mol
Br
C
As usual, more product is formed by homolysis of the weaker bond.
Br
D
NOT formed

21.34 Chlorination is not selective, so a mixture of products results. Bromination is selective, and the major product is formed by cleavage of the weakest C–H bond.

a. **Y** $\xrightarrow[\Delta]{Cl_2}$ Cl + Cl + Cl + Cl + Cl + Cl

b. **Y** $\xrightarrow[\Delta]{Br_2}$ Br

c. **Y** $\xrightarrow[\Delta]{Br_2}$ Br $\xrightarrow{K^+ {}^-OC(CH_3)_3}$ **Z**

21.35 Reaction of an alkene with NBS + $h\nu$ yields allylic substitution products.

a. $\xrightarrow[h\nu]{NBS}$ Br

b. $\xrightarrow[h\nu]{NBS}$ Br + Br (+ *Z* isomer) + Br + Br

c. $\xrightarrow[h\nu]{NBS}$ Br + Br + Br + Br

21.36

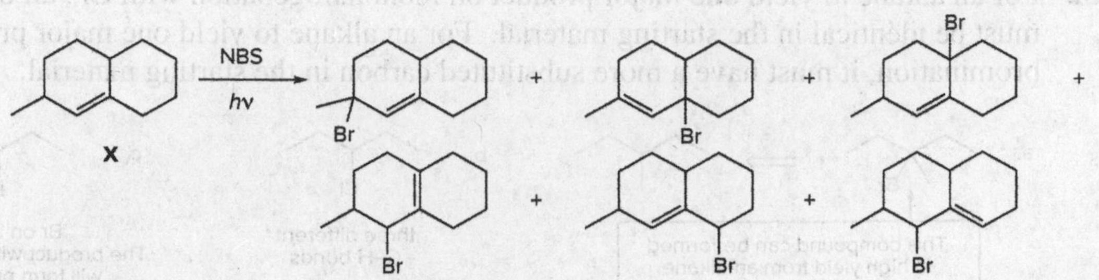

X — NBS, $h\nu$ → Br + Br + Br + Br + Br + Br

21.37

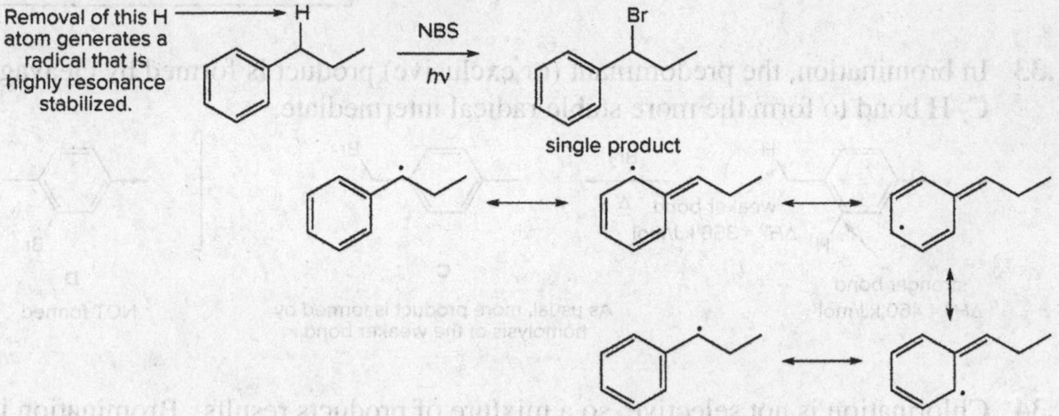

Removal of this H atom generates a radical that is highly resonance stabilized.

NBS, $h\nu$ → single product

21.38

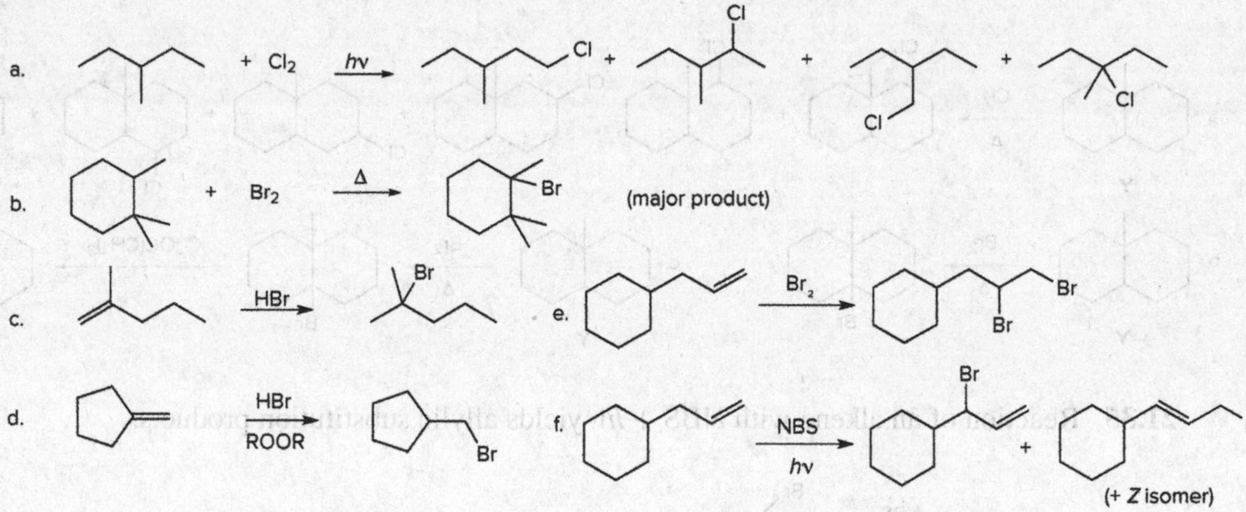

a. + Cl_2 $h\nu$ → Cl + Cl + Cl + Cl

b. + Br_2 Δ → Br (major product)

c. HBr → Br

d. HBr, ROOR → Br

e. Br_2 → Br, Br

f. NBS, $h\nu$ → Br + Br (+ *Z* isomer)

21.39

a. HBr → Br

b. Br_2 → Br, Br + enantiomer

c. NBS, $h\nu$ → Br

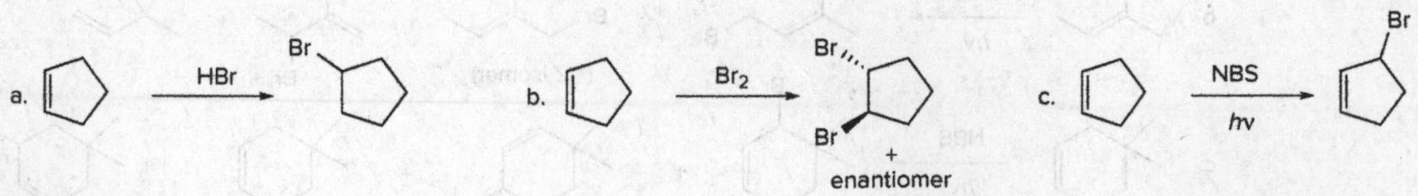

21.40

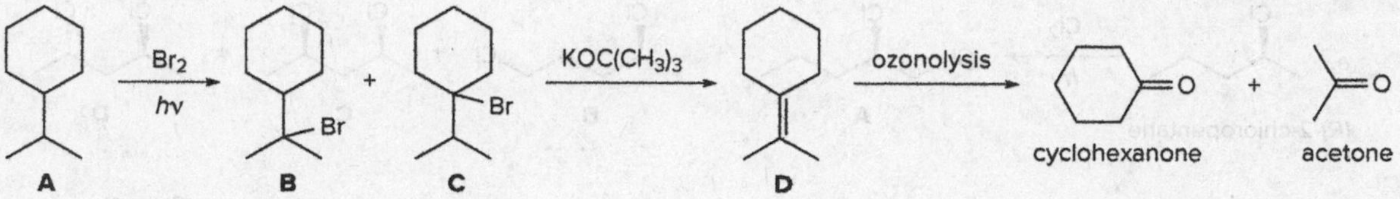

21.41

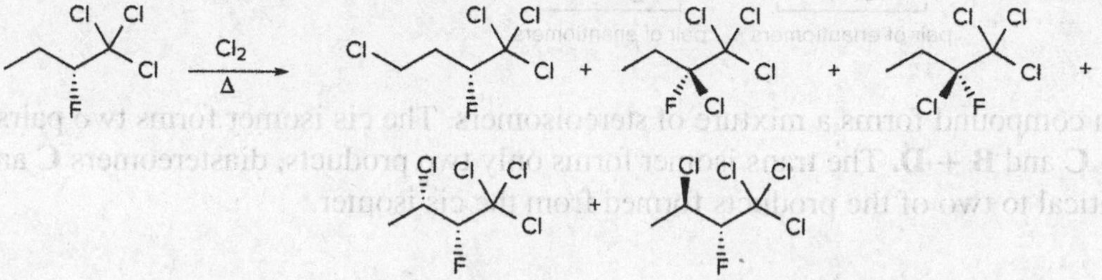

21.42 a.

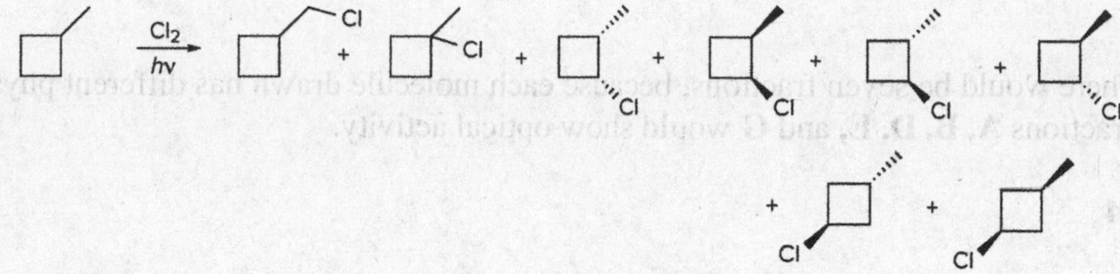

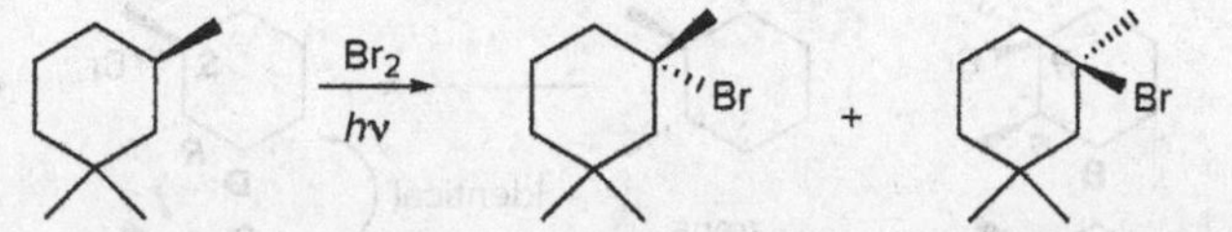

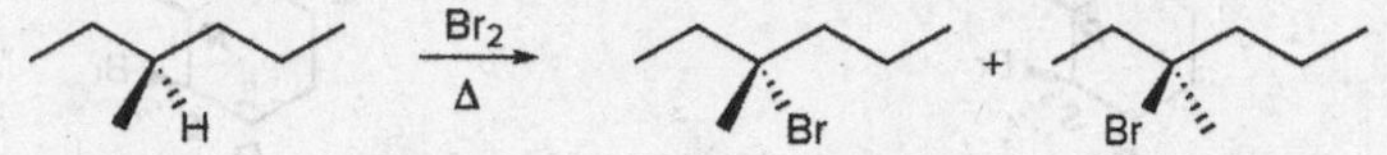

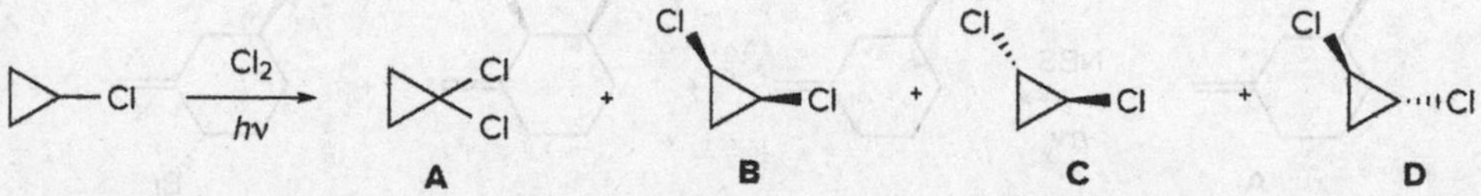

b. Constitutional isomers and diastereomers have different physical properties, so three fractions are obtained:

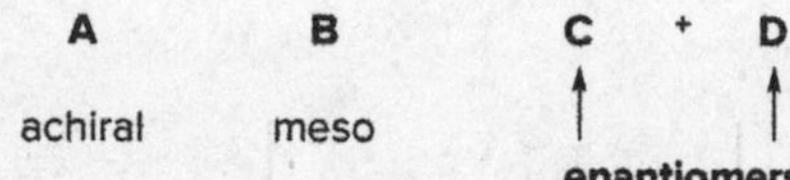

c. No fractions are optically active. One fraction contains achiral **A**, one contains an achiral meso compound **B**, and one contains an equal amount of **C** and **D**.

21.43

a.

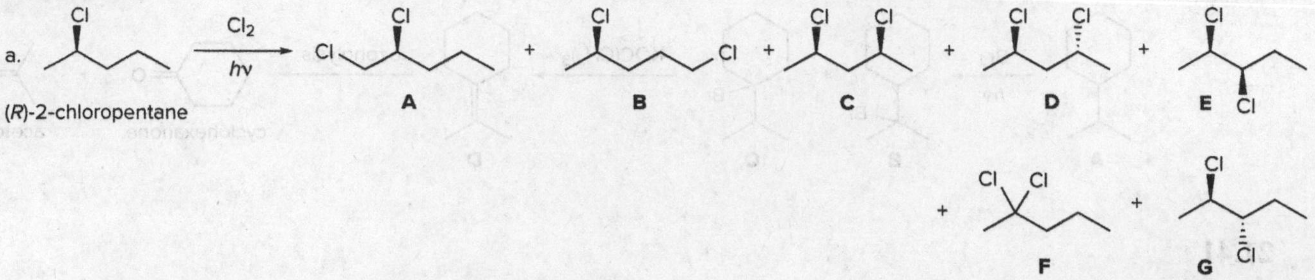

Cl
(R)-2-chloropentane
$\xrightarrow[h\nu]{Cl_2}$
Cl A + B Cl + C + D + E Cl
\+ F Cl Cl + G Cl

b. There would be seven fractions, because each molecule drawn has different physical properties.

c. Fractions **A, B, D, E,** and **G** would show optical activity.

21.44

a.

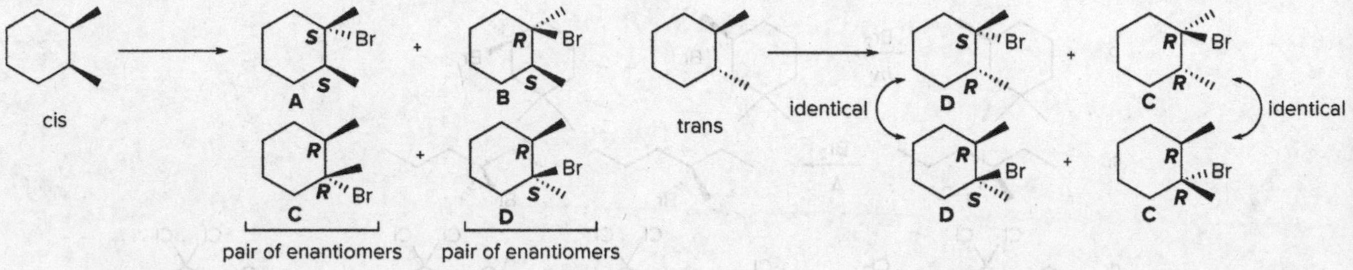

cis → S Br S A + R Br S B
R R Br C + R Br S D
pair of enantiomers pair of enantiomers

trans → S Br R D + R Br R C
identical identical
R Br S D + R Br R C

b. Each compound forms a mixture of stereoisomers. The cis isomer forms two pairs of enantiomers— **A + C** and **B + D.** The trans isomer forms only two products, diastereomers **C** and **D** that are identical to two of the products formed from the cis isomer.

21.45

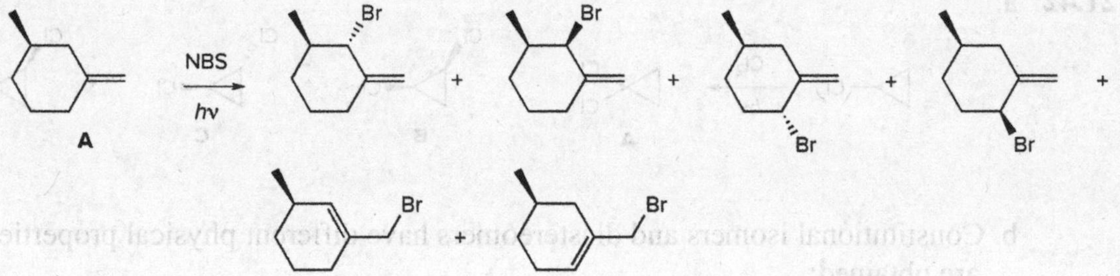

A $\xrightarrow[h\nu]{NBS}$ Br + Br + Br + Br +
Br + Br

21.46

a. $\xrightarrow[ROOR]{HBr}$ Br H + Br H

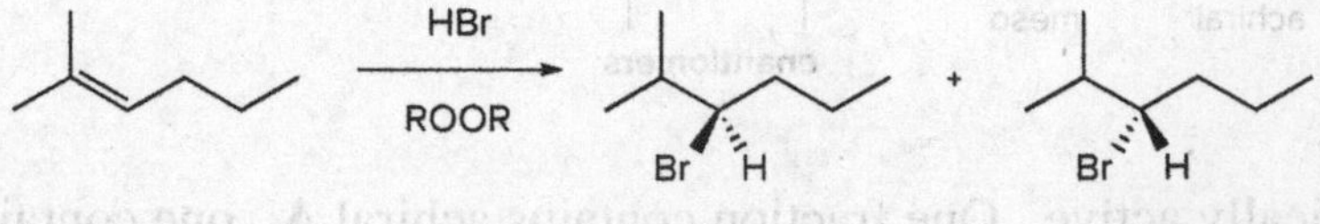

b. $\xrightarrow[ROOR]{HBr}$ Br H + Br H

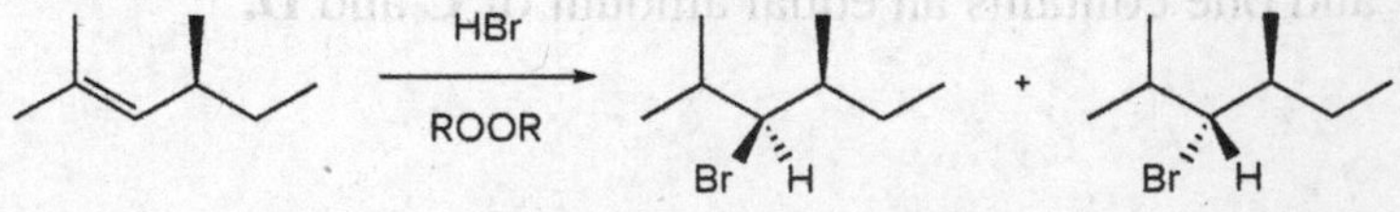

21.47

a. $CH_3-C(CH_3)(H)-CH_3$ + Br_2 $\xrightarrow{\Delta}$ $CH_3-C(CH_3)(Br)-CH_3$ + HBr

C–H bond broken +381 kJ/mol; Br–Br bond broken +192 kJ/mol; C–Br bond formed −272 kJ/mol; H–Br bond formed −368 kJ/mol

total bonds broken = +573 kJ/mol

total bonds formed = −640 kJ/mol

$\Delta H° = -67$ kJ/mol

b. Initiation:

:Br–Br: $\xrightarrow[\text{or } \Delta]{h\nu}$:Br· + ·Br:

Propagation:

$(CH_3)_3C$–H + ·Br: ⟶ $(CH_3)_3C$· + H–Br:

$(CH_3)_3C$· + :Br–Br: ⟶ $(CH_3)_3C$–Br + ·Br:

Termination: (one possibility)

:Br· + ·Br: ⟶ :Br–Br:

c. $\Delta H°$ = (bonds broken) − (bonds formed)
= (+381 kJ/mol) + (−368 kJ/mol)
= +13 kJ/mol

$\Delta H°$ = (bonds broken) − (bonds formed)
= (+192 kJ/mol) + (−272 kJ/mol)
= −80 kJ/mol

d, e.

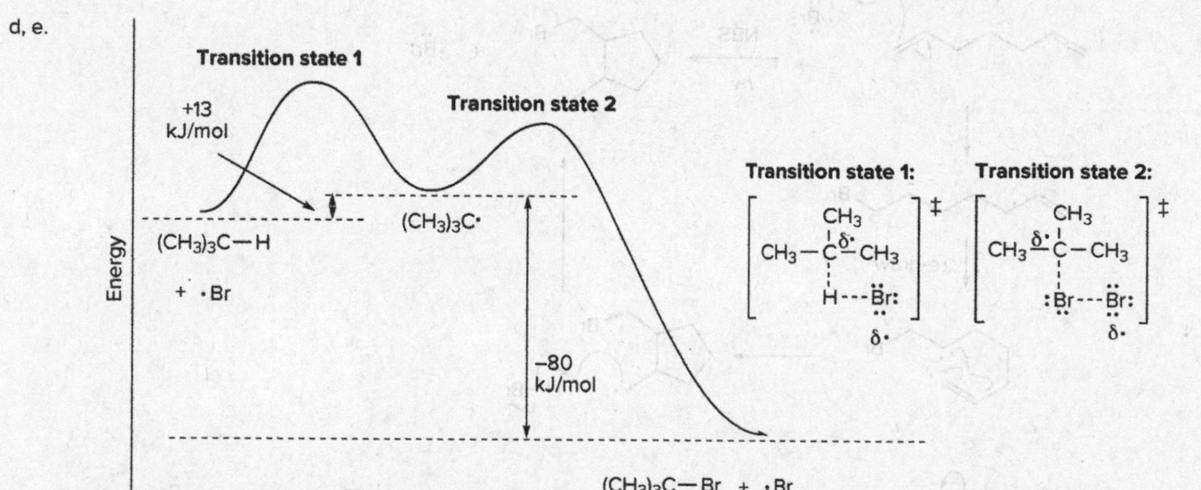

21.48

Initiation:

NBS $\xrightarrow{h\nu}$ N· + ·Br:

Propagation:

+ ·Br: → ↔ ↔ + H–Br:

:Br–Br: (from NBS) :Br–Br: (from NBS)

Br + Br + ·Br:

Termination: :Br· + ·Br: ⟶ :Br–Br:

(one possibility)

21.49

·Br: NBS $h\nu$ Br + ·Br:

Br

re-draw

Br → Br H–Br

21.50

H–Br → + → 1,2-CH$_3$ shift → + Br$^-$ → Br

+ Br$^-$

3,3-dimethylbut-1-ene — 2° carbocation — 3° carbocation — 2-bromo-2,3-dimethylbutane

3,3-dimethylbut-1-ene → Br· (HBr + peroxide) → Br (The 2° radical does NOT rearrange.) → HBr → Br + Br· (1-bromo-3,3-dimethylbutane)

Addition of HBr without added peroxide occurs by an ionic mechanism and forms a 2° carbocation, which rearranges to a more stable 3° carbocation. The addition of H$^+$ occurs first, followed by Br$^-$. Addition of HBr with added peroxide occurs by a radical mechanism and forms a 2° radical that does not rearrange. In the radical mechanism, Br$^\cdot$ adds first, followed by H$^\cdot$.

21.51

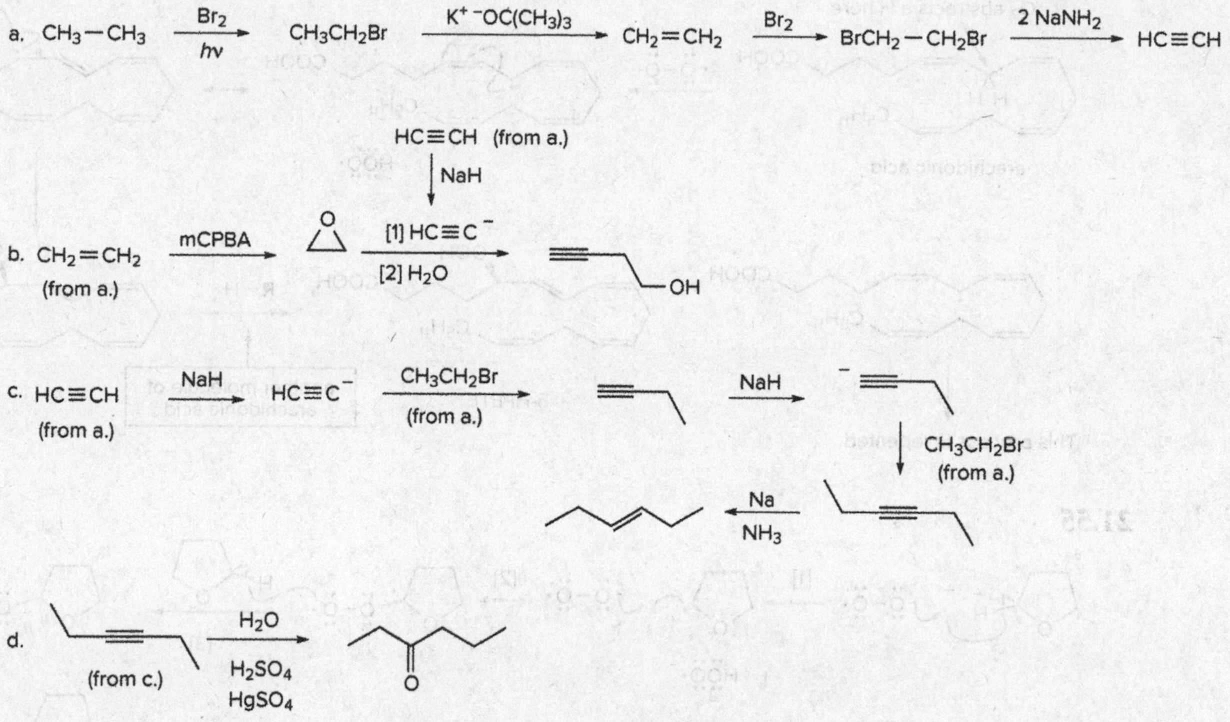

a. $CH_3{-}CH_3$ —(Br_2, $h\nu$)→ CH_3CH_2Br —($K^+\,^-OC(CH_3)_3$)→ $CH_2{=}CH_2$ —(Br_2)→ $BrCH_2{-}CH_2Br$ —(2 $NaNH_2$)→ $HC{\equiv}CH$

$HC{\equiv}CH$ (from a.)

NaH

b. $CH_2{=}CH_2$ (from a.) —(mCPBA)→ epoxide —([1] $HC{\equiv}C^-$, [2] H_2O)→ OH

c. $HC{\equiv}CH$ (from a.) —(NaH)→ $HC{\equiv}C^-$ —(CH_3CH_2Br (from a.))→ —(NaH)→ ⁻ —(CH_3CH_2Br (from a.))→ —(Na, NH_3)→

d. (from c.) —(H_2O, H_2SO_4, $HgSO_4$)→ O

21.52

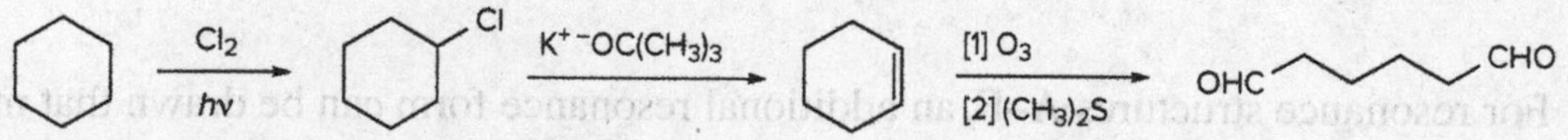

—(Cl_2, $h\nu$)→ Cl —($K^+\,^-OC(CH_3)_3$)→ —([1] O_3, [2] $(CH_3)_2S$)→ OHC CHO

21.53

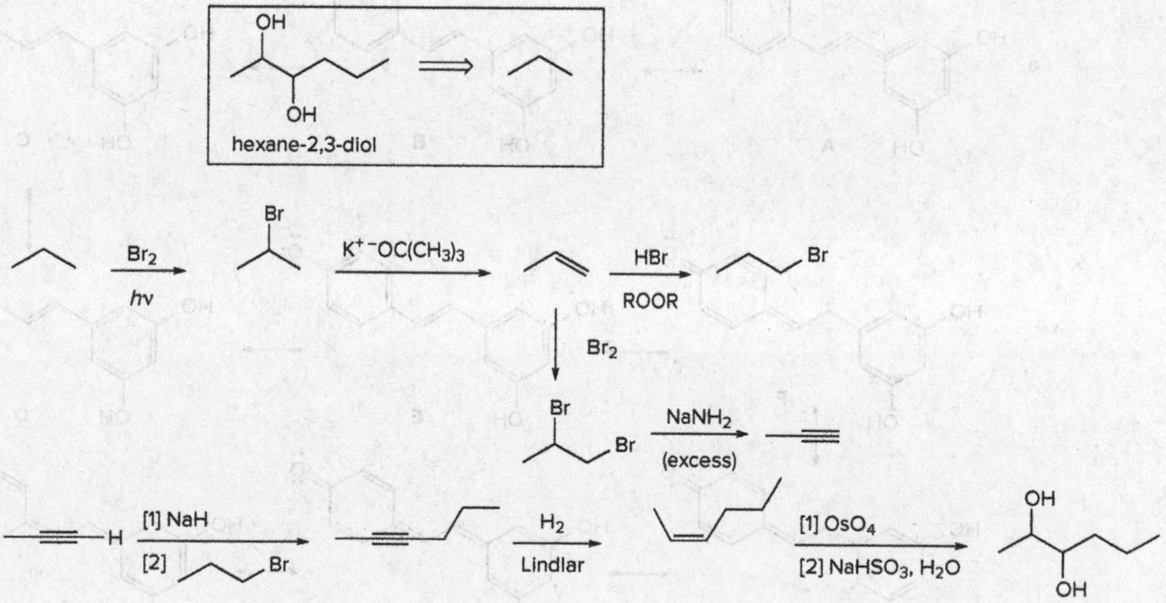

OH
OH
hexane-2,3-diol ⟹

—(Br_2, $h\nu$)→ Br —($K^+\,^-OC(CH_3)_3$)→ —(HBr, ROOR)→ Br

↓ Br_2

Br Br —($NaNH_2$ (excess))→

H —([1] NaH, [2] Br)→ —(H_2, Lindlar)→ —([1] OsO_4, [2] $NaHSO_3$, H_2O)→ OH OH

21.54

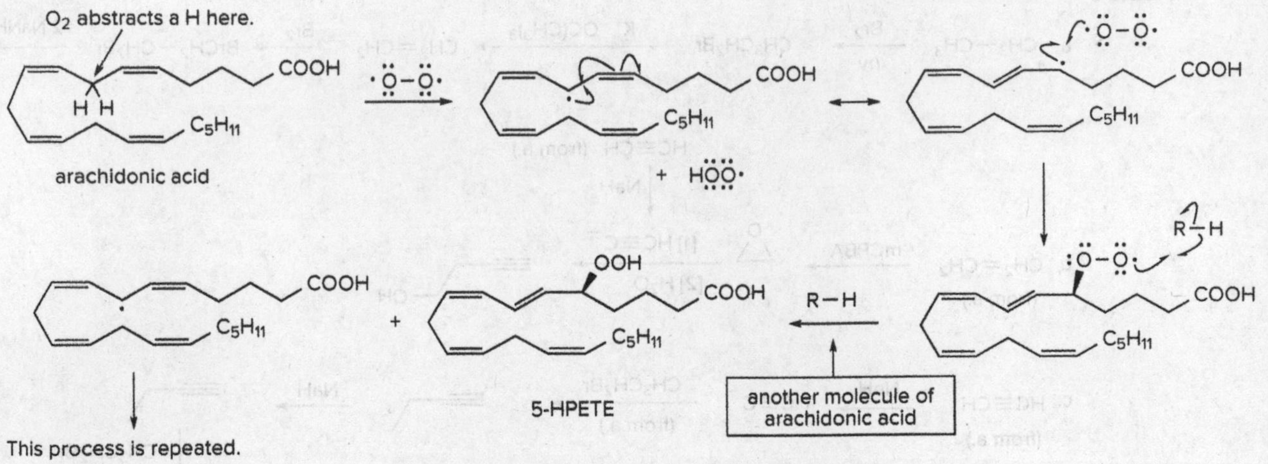

O$_2$ abstracts a H here.

COOH

H H

C_5H_{11}

arachidonic acid

$\cdot\ddot{O}-\ddot{O}\cdot$

COOH

C_5H_{11}

+ $H\ddot{O}\ddot{O}\cdot$

COOH

C_5H_{11}

R–H

COOH

C_5H_{11}

COOH

C_5H_{11}

OOH

COOH

C_5H_{11}

5-HPETE

R–H

another molecule of arachidonic acid

This process is repeated.

21.55

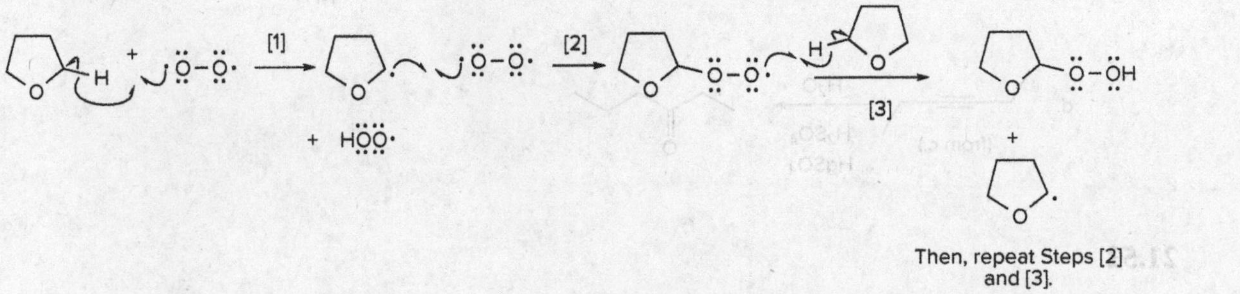

[1] [2] [3]

+ $H\ddot{O}\ddot{O}\cdot$

Then, repeat Steps [2] and [3].

21.56 For resonance structures **A–F,** an additional resonance form can be drawn that moves the position of the three π bonds in the ring bonded to two OH groups.

a. HO OH **A** HO OH **B** HO OH **C**

HO OH **F** HO OH **E** HO OH **D**

HO OH HO OH HO OH

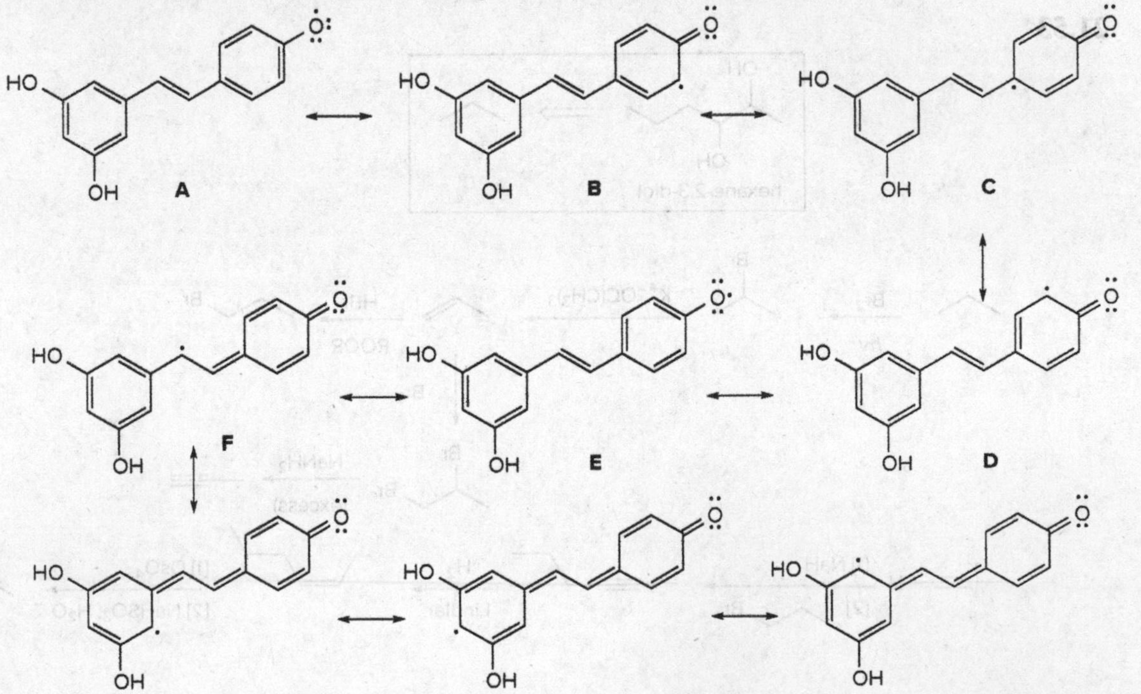

b. Homolysis of the indicated OH bond is preferred because it allows the resulting radical to delocalize over both benzene rings. Cleavage of one of the other OH bonds gives a radical that delocalizes over only one of the benzene rings.

21.57 Abstraction of the labeled H forms a highly resonance-stabilized radical. Four of the possible resonance structures are drawn.

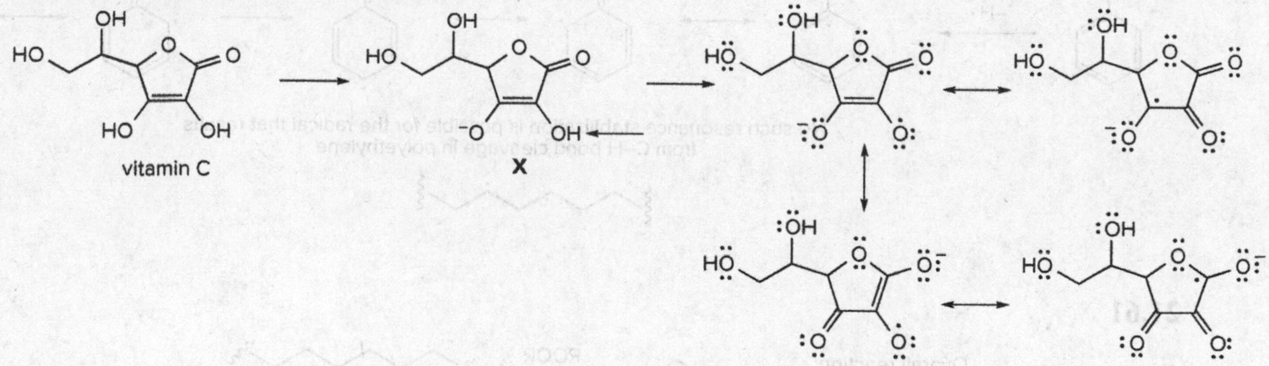

vitamin C

X

21.58 The monomers used in radical polymerization always contain double bonds.

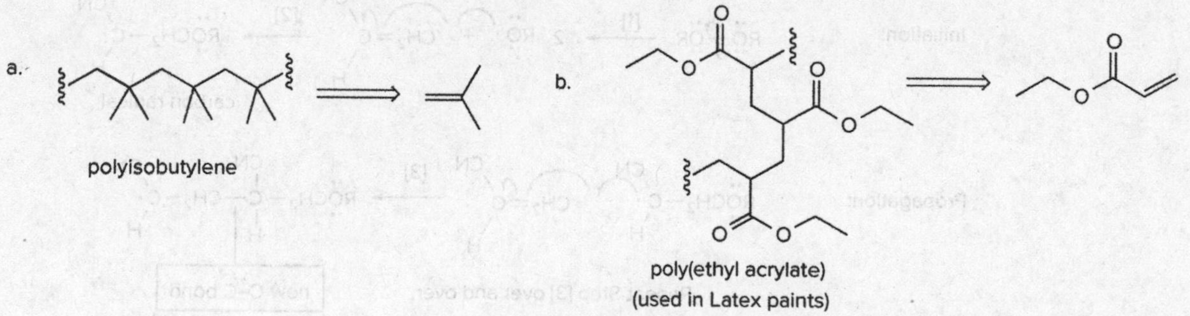

a. polyisobutylene

b. poly(ethyl acrylate) (used in Latex paints)

21.59

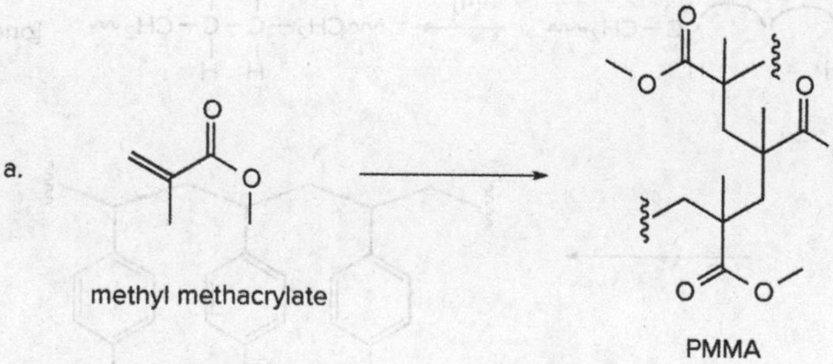

a. methyl methacrylate

PMMA

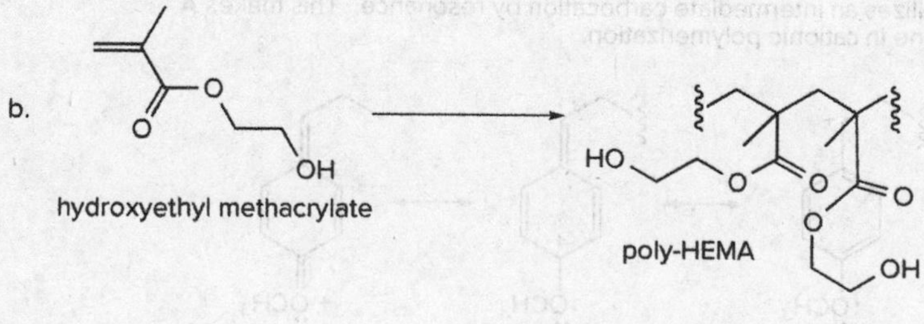

b. hydroxyethyl methacrylate

poly-HEMA

21.60 Polystyrene has H atoms bonded to benzylic carbons—that is, carbons bonded directly to a benzene ring. These C–H bonds are unusually weak because the radical that results from homolysis is resonance stabilized.

–H•

No such resonance stabilization is possible for the radical that results from C–H bond cleavage in polyethylene.

21.61

Overall reaction: CN —ROOR→ CN CN CN

Initiation: RÖ–ÖR —[1]→ 2 RÖ• + $CH_2=C$(CN)(H) —[2]→ $ROCH_2-C•$(CN)(H)

carbon radical

Propagation: $ROCH_2-C•$(CN)(H) + $CH_2=C$(CN)(H) —[3]→ $ROCH_2-C-CH_2-C•$(CN)(H)

new C–C bond

Repeat Step [3] over and over.

Termination: $CH_2-C•$(CN)(H) •$C-CH_2$(NC)(H) —[4]→ $CH_2-C-C-CH_2$ (CN CN, H H) [one possibility]

21.62

a. CH_3O–(A) → OCH_3 OCH_3 OCH_3

A

b. The OCH_3 group stabilizes an intermediate carbocation by resonance. This makes **A** react faster than styrene in cationic polymerization.

+ :OCH_3 ↔ + :OCH_3 ↔ + OCH_3

three of the possible resonance structures

21.63

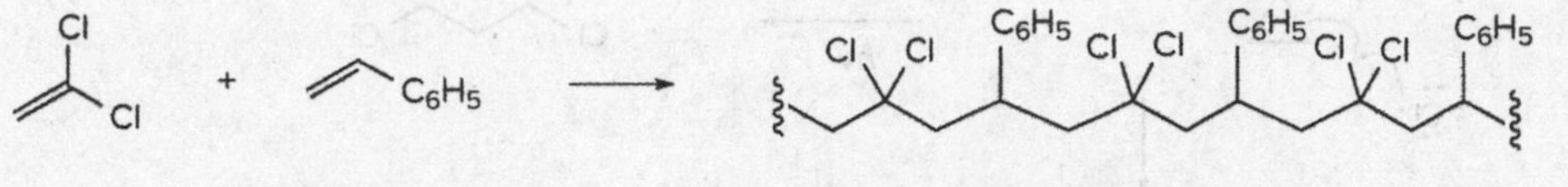

Cl, Cl + C6H5 ⟶ alternating copolymer

21.64

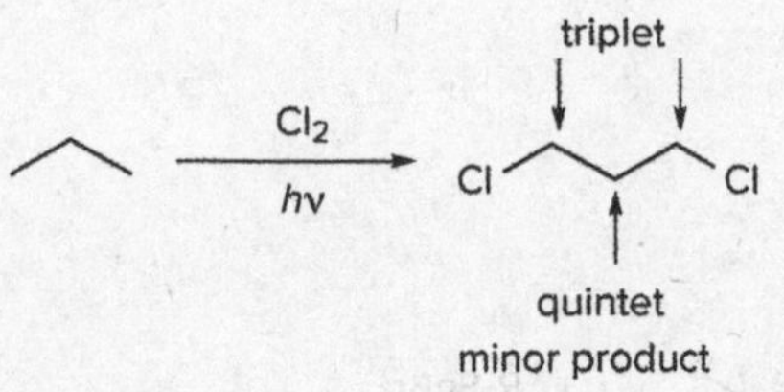

Cl2, hν; triplet; quintet; minor product

Molecular formula $C_3H_6Cl_2$
Integration: relative area 2:1
Because the compound has 6 H's and the sum of the relative areas is 3, each absorption is due to twice as many H's.
One signal is due to 4 H's.
The second signal is due to 2 H's (2 x 1).
^1H NMR data:
quintet at 2.2 (2 H's) split by 4 H's
triplet at 3.7 (4 H's) split by 2 H's

21.65

a. The triphenylmethyl radical is highly resonance stabilized, because the radical can be delocalized on each of the benzene rings. As an example using one ring:

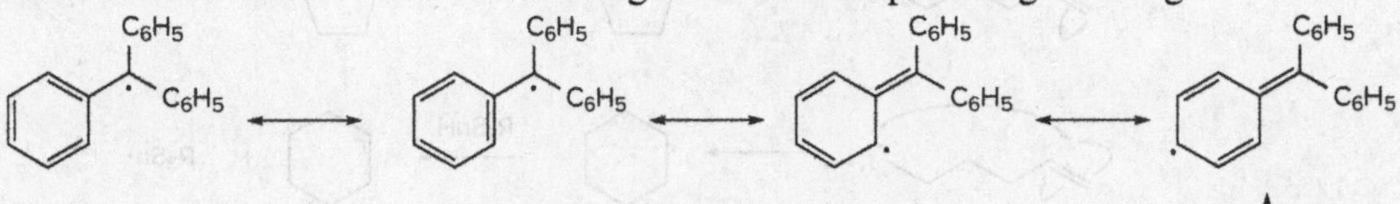

In addition, the radical is very sterically hindered, making it difficult to undergo reactions.

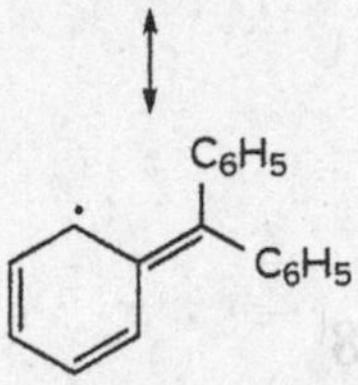

b. First, draw the resonance form of the radical that places the unpaired electron on the C that forms the new C–C bond.

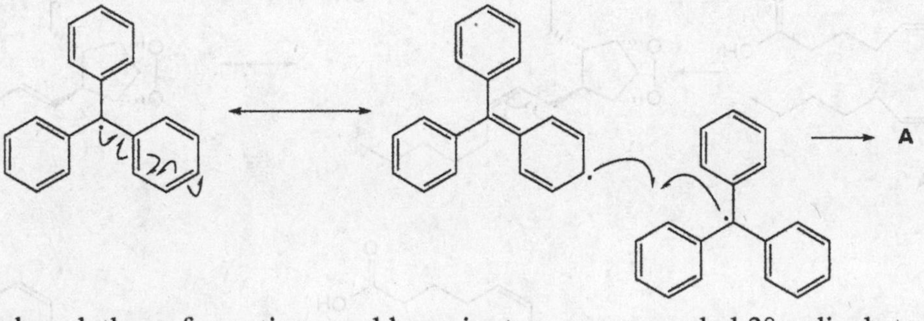

⟶ **A**

c. Hexaphenylethane formation would require two very crowded 3° radicals to combine. The formation of **A** results from a radical on one of the six-membered rings, which is much more accessible for reaction.

d. The ^1H NMR spectrum of hexaphenylethane should show signals only in the aromatic region (7–8 ppm), whereas the ^1H NMR spectrum of **A** will also have signals for the sp^2 hybridized C–H bonds (4.5–6.0 ppm) of the alkenes, as well as the single H on the sp^3 hybridized carbon. The ^{13}C NMR spectrum of hexaphenylethane should consist of lines due to the 4° C's and the aromatic C's. For **A,** the ^{13}C NMR spectrum will also have lines for the sp^3 and sp^2 hybridized C's that are not contained in the aromatic rings.

21.66

$:\ddot{Cl}\cdot$ + Cl_2 $\xrightarrow{h\nu}$ Cl Cl

Cl $:\ddot{Cl}-\ddot{Cl}:$ ⟶ $:\ddot{Cl}$ $\ddot{Cl}:$ + $:\ddot{Cl}\cdot$

21.67

Initiation: R_3SnH + $\cdot Z$ ⟶ $R_3Sn\cdot$ + HZ

Propagation: Br ⟶ + R_3SnBr

$R_3Sn\cdot$

R_3SnH + $R_3Sn\cdot$

R_3SnH + $R_3Sn\cdot$

R_3SnH + $R_3Sn\cdot$

21.68

OH

A

OH ⟷ OH

$\cdot\ddot{O}-\ddot{O}\cdot$

OH

OOH

+ $R\cdot$

$:\ddot{O}-\ddot{O}\cdot$ R–H

Chapter 22 Amines

Chapter Review

General facts

- Amines are organic nitrogen compounds having the general structure RNH_2, R_2NH, or R_3N, with a lone pair of electrons on N (22.1).
- Amines are named using the suffix *-amine* (22.3).
- All amines have polar C–N bonds. Primary (1°) and 2° amines have polar N–H bonds and are capable of intermolecular hydrogen bonding (22.4).
- The lone pair on N makes amines strong organic bases and nucleophiles (22.7).

Summary of spectroscopic absorptions (22.4)

Mass spectra	Molecular ion	Amines with an odd number of N atoms give an odd molecular ion.
IR absorptions	N–H	3300–3500 cm^{-1} (two peaks for RNH_2, one peak for R_2NH)
^{1}H NMR absorptions	NH	0.5–5 ppm (no splitting with adjacent protons)
	CH–N	2.3–3.0 ppm (deshielded Csp^3–H)
^{13}C NMR absorption	C–N	30–50 ppm

Comparing the basicity of amines and other compounds (22.9)

- Alkylamines (RNH_2, R_2NH, and R_3N) are more basic than NH_3 because of the electron-donating R groups (22.9A).
- Alkylamines (RNH_2) are more basic than arylamines ($C_6H_5NH_2$), which have a delocalized lone pair from the N atom (22.9B).
- Arylamines with electron-donor groups are more basic than arylamines with electron-withdrawing groups (22.9B).
- Alkylamines (RNH_2) are more basic than amides ($RCONH_2$), which have a delocalized lone pair from the N atom (22.9C).
- Aromatic heterocycles with a localized electron pair on N are more basic than those with a delocalized lone pair from the N atom (22.9D).
- Alkylamines with a lone pair in an sp^3 hybrid orbital are more basic than those with a lone pair in an sp^2 hybrid orbital (22.9E).

Preparation of amines (22.6)

[1] Direct nucleophilic substitution with NH_3 and amines (22.6A)

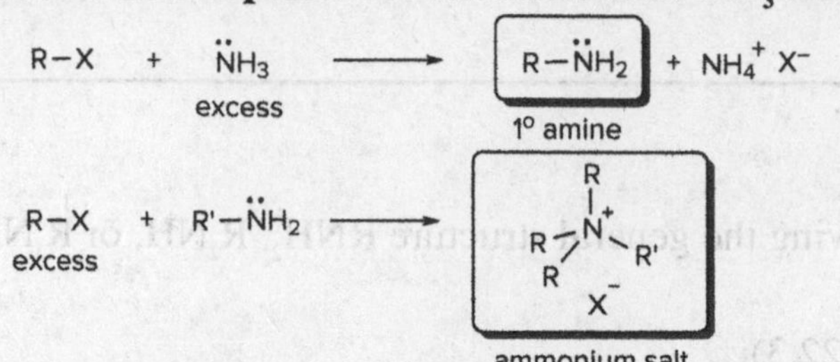

- The mechanism is S_N2.
- The reaction works best for CH_3X or RCH_2X.
- The reaction works best to prepare 1° amines and ammonium salts.

[2] Gabriel synthesis (22.6A)

H–N (phthalimide) → [1] KOH, [2] RX → R–NH_2 (1° amine) + $C_6H_4(CO_2^-)_2$

- The mechanism is S_N2.
- The reaction works best for CH_3X or RCH_2X.
- Only 1° amines can be prepared.

[3] Reduction methods (22.6B)

a. From nitro compounds

R–NO_2 → H_2, Pd-C or Fe, HCl or Sn, HCl → R–NH_2 (1° amine)

b. From nitriles

R–C≡N → [1] $LiAlH_4$, [2] H_2O → RCH_2NH_2 (1° amine)

c. From amides

RCONR'$_2$ (R' = H or alkyl) → [1] $LiAlH_4$, [2] H_2O → RCH_2NR'$_2$ (1°, 2° and 3° amines)

[4] Reductive amination (22.6C)

RCOR' + $R_2''NH$ (R', R" = H or alkyl) → $NaBH_3CN$ → RCH(NR"$_2$)R' (1°, 2°, and 3° amines)

- Reductive amination adds one alkyl group (from an aldehyde or ketone) to a nitrogen nucleophile.
- Primary (1°), 2°, and 3° amines can be prepared.

Reactions of amines

[1] Reaction as a base (22.8)

$$R\text{–}\ddot{N}H_2 + H\text{–}A \rightleftharpoons R\text{–}\overset{+}{N}H_3 + :A^-$$

[2] Nucleophilic addition to aldehydes and ketones (22.10)

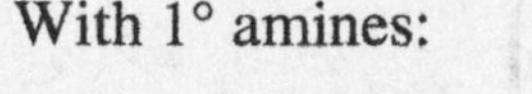

With 1° amines:

R = H or alkyl → ($R'NH_2$) → imine

With 2° amines:

R = H or alkyl → (R'_2NH) → enamine

[3] Nucleophilic substitution with acid chlorides and anhydrides (22.10)

RCOZ + R'_2NH (2 equiv) → $RCONR'_2$

Z = Cl or OCOR
R' = H or alkyl

1°, 2°, and 3° amides

[4] Hofmann elimination (22.11)

[1] CH_3I (excess)
[2] Ag_2O
[3] Δ

alkene

- The less substituted alkene is the major product.

[5] Reaction with nitrous acid (22.12)

With 1° amines:

$R-NH_2$ → ($NaNO_2$, HCl) → $R-\overset{+}{N}\equiv N\text{:}\ Cl^-$

alkyl diazonium salt

With 2° amines:

R_2NH → ($NaNO_2$, HCl) → $R_2N-N=O$

N-nitrosamine

Reactions of diazonium salts

[1] Substitution reactions (22.13)

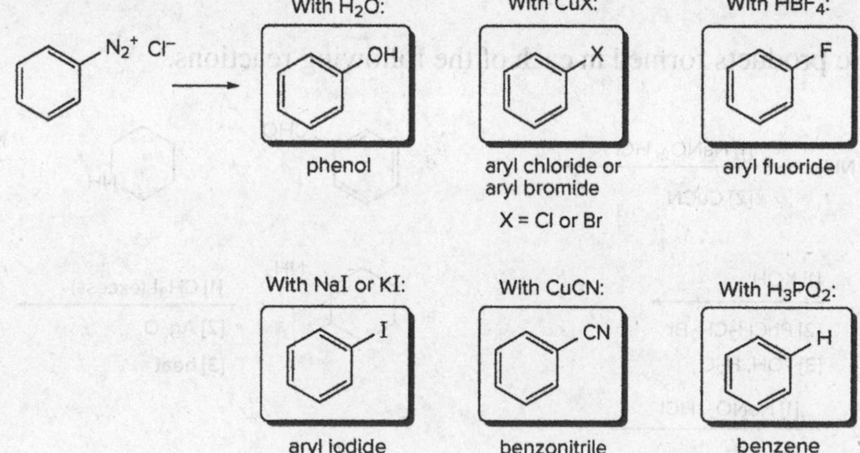

[2] Coupling to form azo compounds (22.14)

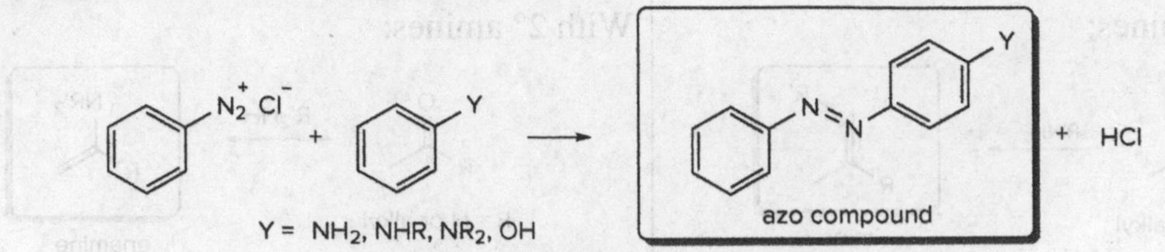

$N_2^+ Cl^-$ + Y → azo compound + HCl

Y = NH_2, NHR, NR_2, OH

(a strong electron-donor group)

Practice Test on Chapter Review

1. Give a systematic name for each of the following compounds.

a. 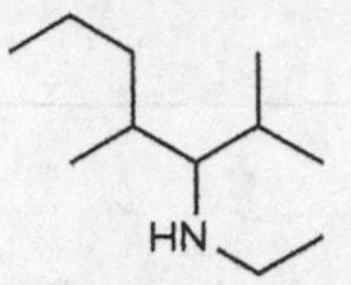 HN

b. 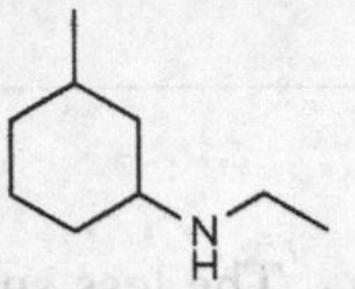 N H

2. (a) Which compound is the weakest base? (b) Which compound is the strongest base?

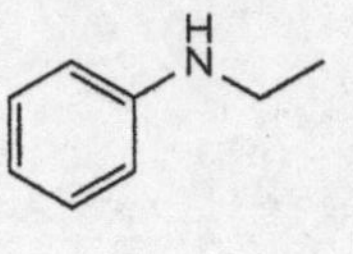

A 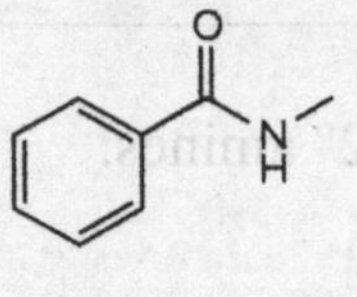 B 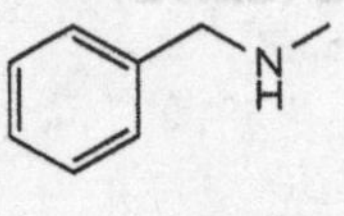 C 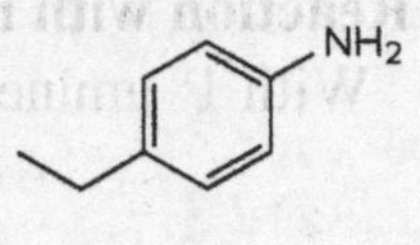 D

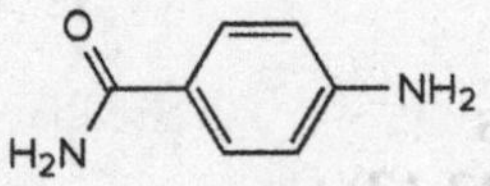

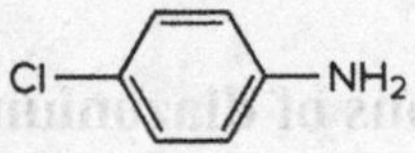

3. (a) Which compound is the weakest base? (b) Which compound is the strongest base?

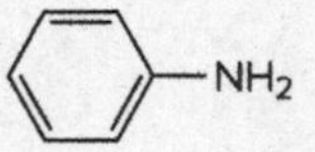

A: NH_2 B: NH_2 C: H_2N, NH_2 D: Cl, NH_2

4. Draw the organic products formed in each of the following reactions.

a. 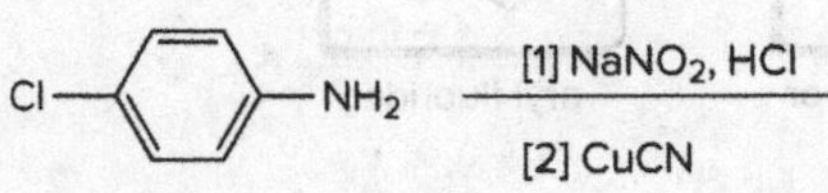 Cl–C_6H_4–NH_2 → [1] $NaNO_2$, HCl [2] CuCN

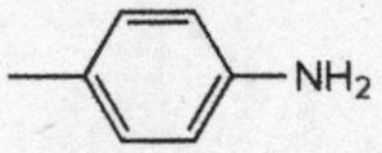

b. 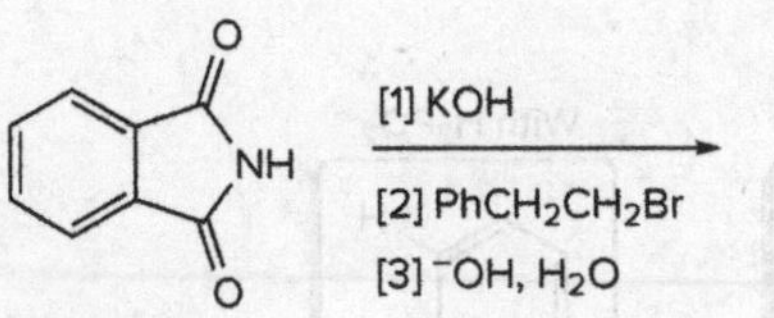 NH → [1] KOH [2] $PhCH_2CH_2Br$ [3] ^-OH, H_2O

c. NH_2, Cl → [1] $NaNO_2$, HCl [2] NaI

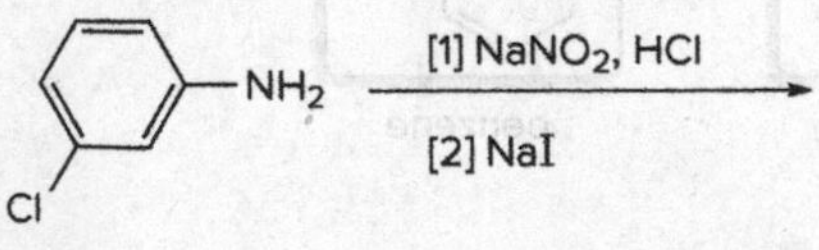

d. 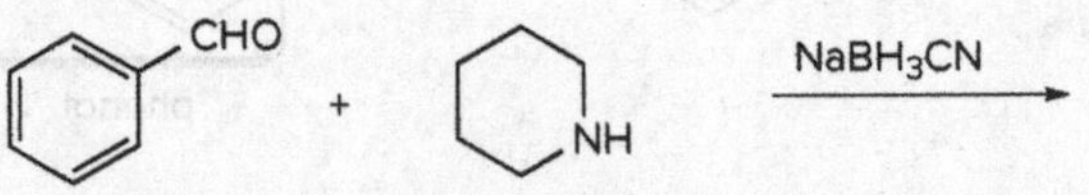 CHO + NH → $NaBH_3CN$

e. 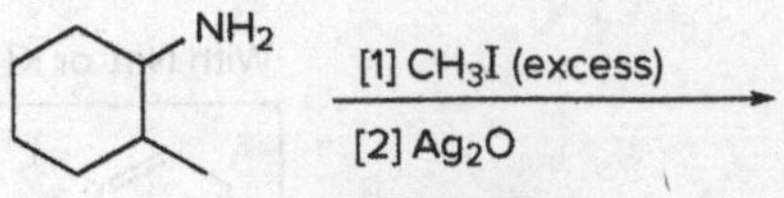 NH_2 → [1] CH_3I (excess) [2] Ag_2O [3] heat

5. Draw the products formed when the given amine is treated with [1] CH_3I (excess); [2] Ag_2O; [3] Δ, and indicate the major product. You need not consider any stereoisomers formed in the reaction.

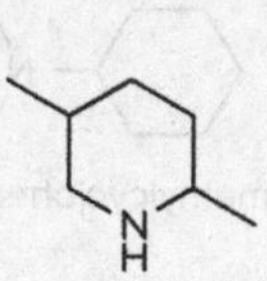

N
H

6. What organic starting materials are needed to synthesize **B** by reductive amination?

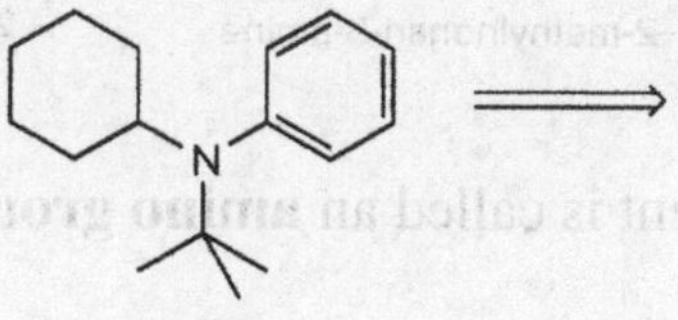

N
B

Answers to Practice Test

1.a. *N*-ethyl-2,4-dimethyl-heptan-3-amine
b. *N*-ethyl-3-methylcyclohexanamine

2.a. **B**
b. **C**

3.a. **C**
b. **B**

4.
a.  Cl—C6H4—CN
b. 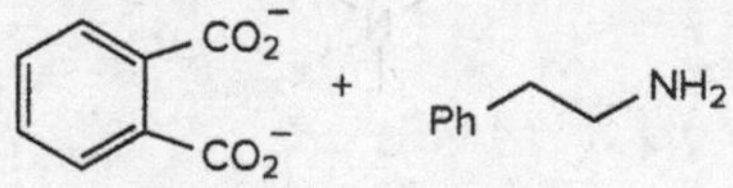 CO_2^- / CO_2^- + Ph NH2
c. 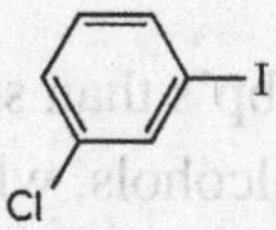 I / Cl
d. 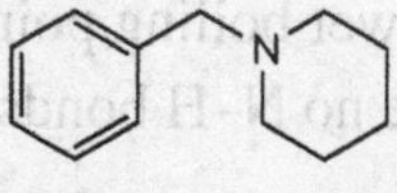 N
e. 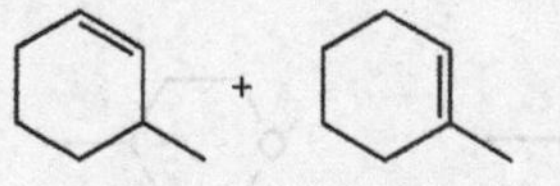 +

5. N + N + N (major)

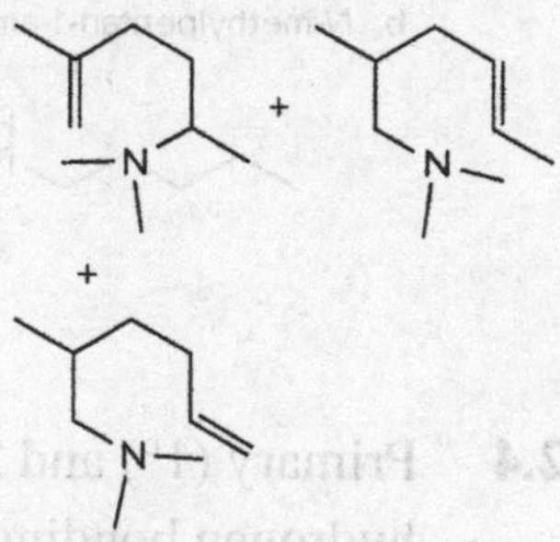

6. O + H–N

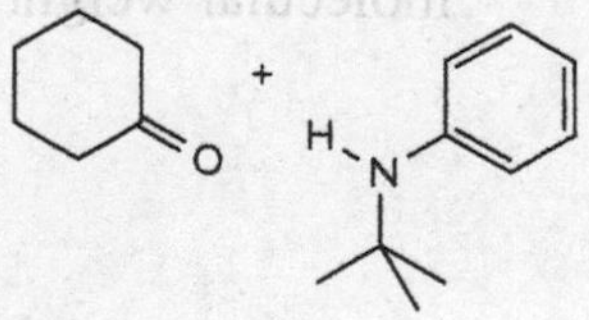

Answers to Problems

22.1 The N atom of an ammonium salt is a stereogenic center when the N is surrounded by four different groups. All stereogenic centers are circled.

a. 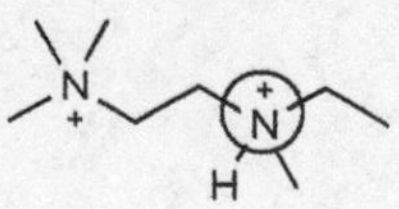 N+ N+ H

b. 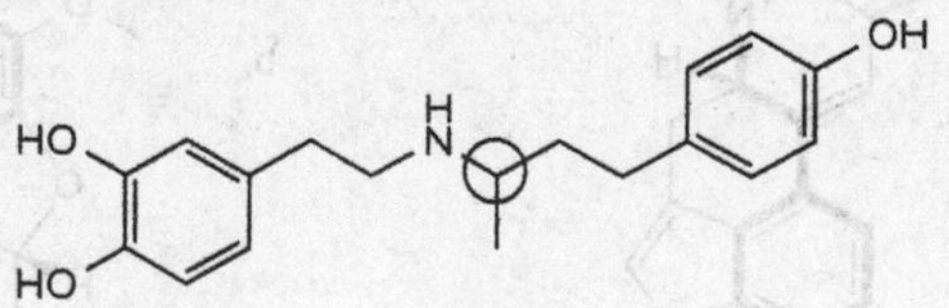 HO HO N H OH

22.2

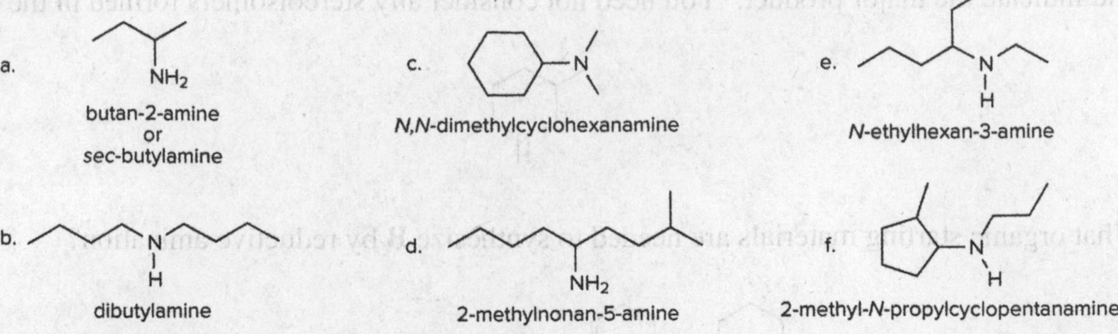

22.3 An $\mathbf{NH_2}$ group named as a substituent is called an **amino group.**

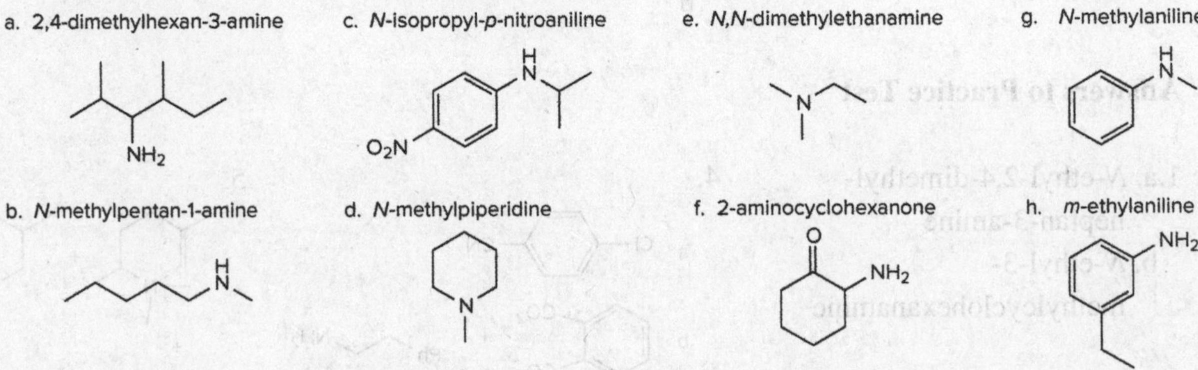

22.4 Primary (1°) and 2° amines have higher bp's than similar compounds (like ethers) incapable of hydrogen bonding, but lower bp's than alcohols, which have stronger intermolecular hydrogen bonds. Tertiary amines (3°) have lower boiling points than 1° and 2° amines of comparable molecular weight because they have no N–H bonds with which to form hydrogen bonds.

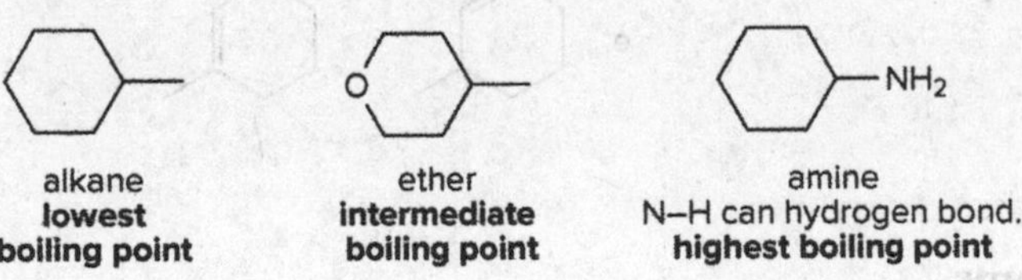

22.5 The atoms of 2-phenylethanamine are in bold.

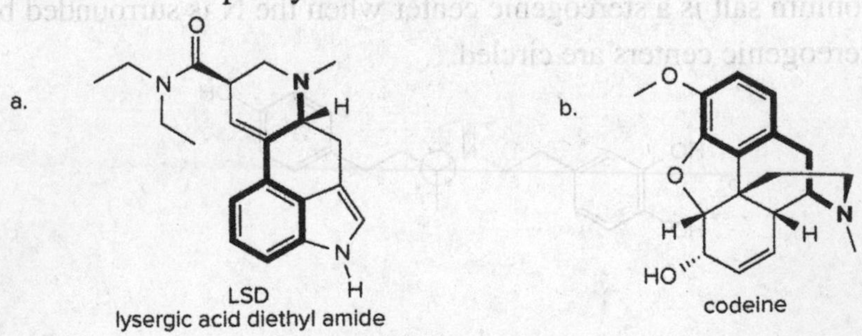

22.6 S_N2 reaction of an alkyl halide with NH_3 or an amine forms an amine or an ammonium salt.

a. 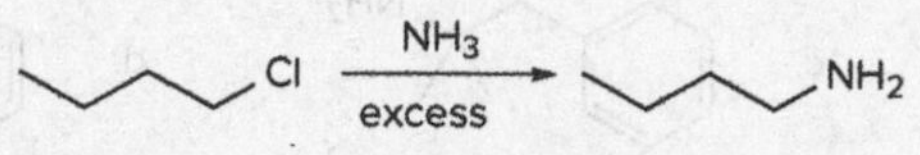 Cl → (NH_3, excess) → NH_2

b. 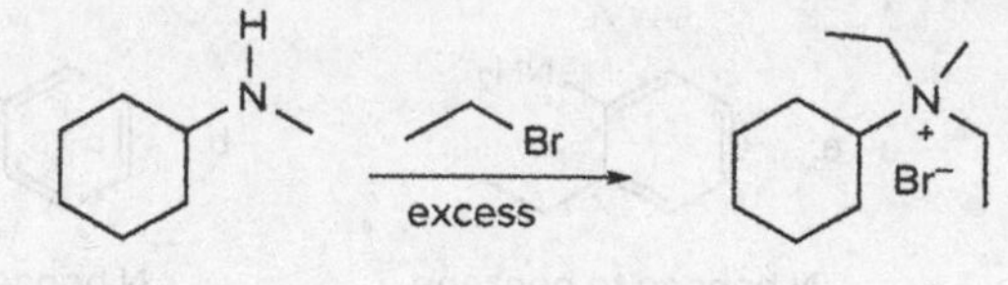 H–N → (Br, excess) → N⁺ Br⁻

22.7 The Gabriel synthesis converts an alkyl halide into a 1° amine by a two-step process: nucleophilic substitution followed by hydrolysis.

a. 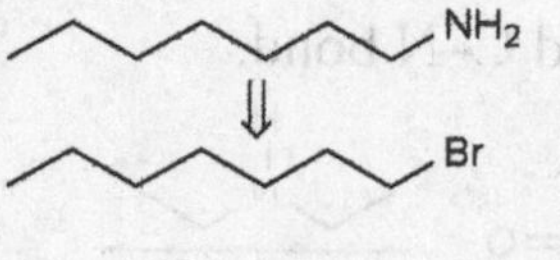 NH_2 ⇓ Br

b. 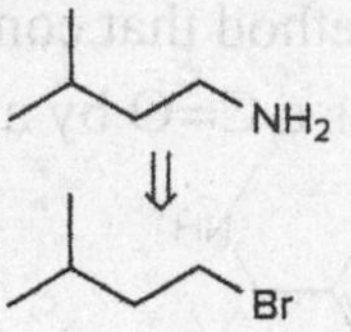 NH_2 ⇓ Br

c. 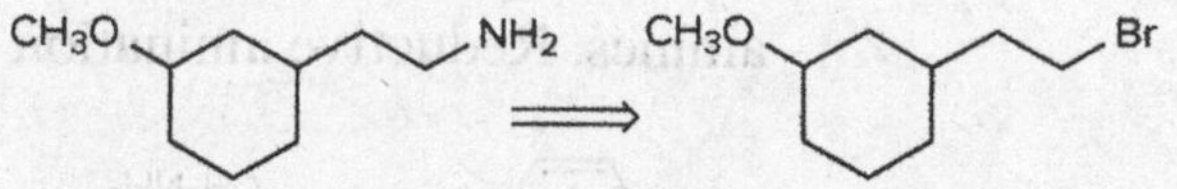 CH_3O … NH_2 ⟹ CH_3O … Br

22.8 The Gabriel synthesis prepares 1° amines from alkyl halides. Because the reaction proceeds by an S_N2 mechanism, the halide must be CH_3 or 1°, and X can't be bonded to an sp^2 hybridized C.

a. 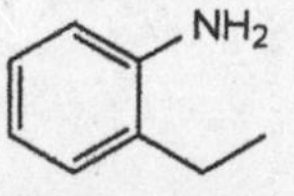 NH_2
aromatic
An S_N2 does not occur on an aryl halide.
cannot be made by Gabriel synthesis

b. 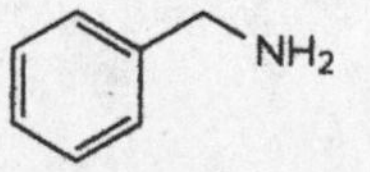 NH_2
can be made by Gabriel synthesis

c. 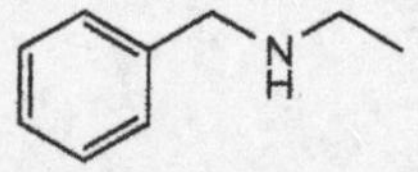 N–H
2° amine
cannot be made by Gabriel synthesis

d. 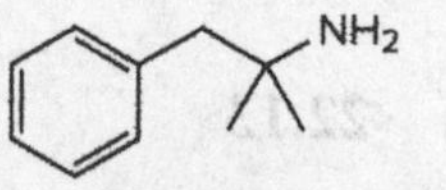 NH_2
N on 3° C
An S_N2 does not occur on a 3° RX.
cannot be made by Gabriel synthesis

22.9 **Nitriles are reduced to 1° amines with $LiAlH_4$. Nitro groups are reduced to 1° amines** using a variety of reducing agents. **Primary (1°), 2°, and 3° amides are reduced to 1°, 2°, and 3° amines** respectively, using $LiAlH_4$.

a. 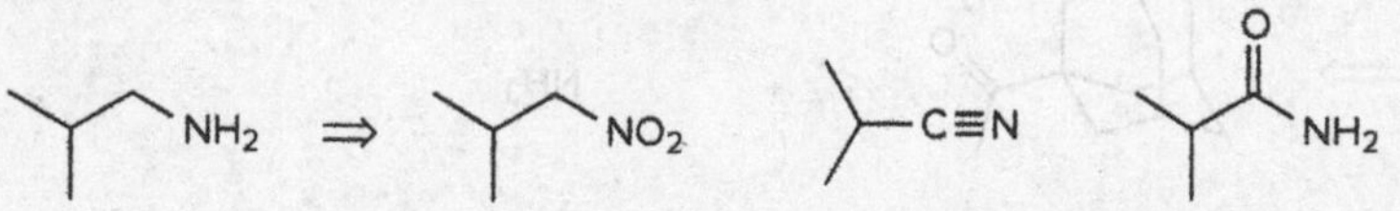 NH_2 ⟹ NO_2 C≡N O NH_2

b. NH_2 ⟹ NO_2 C≡N O NH_2

c. NH_2 ⟹ NO_2 CN O NH_2

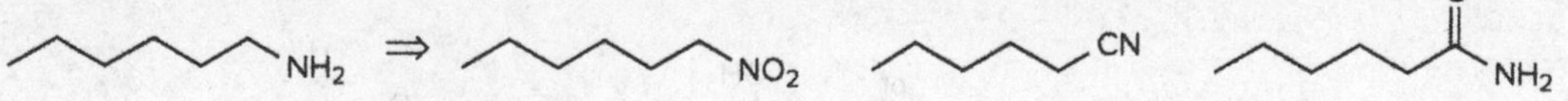

22.10 **Primary (1°), 2°, and 3° amides are reduced to 1°, 2°, and 3° amines** respectively, using $LiAlH_4$.

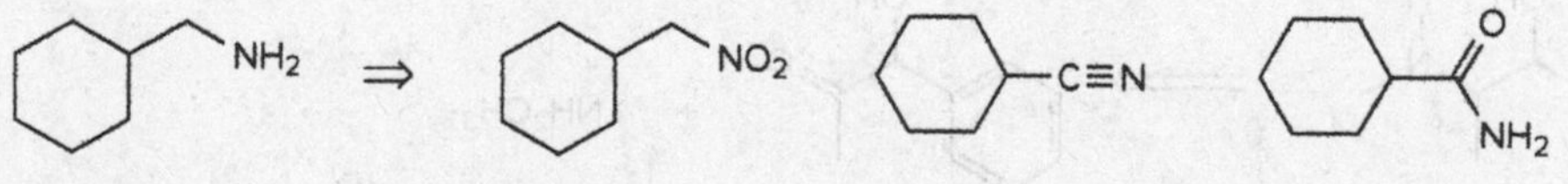

a. 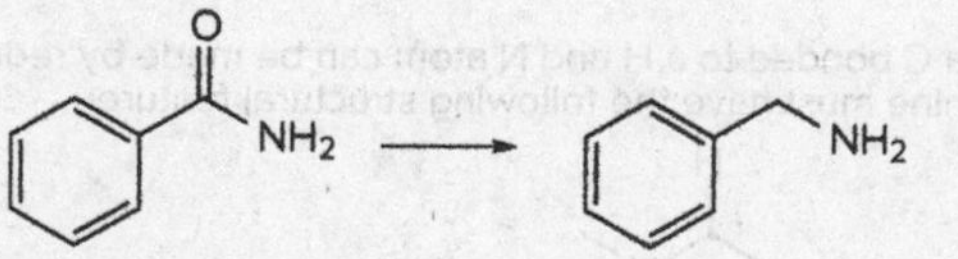 O NH_2 → NH_2

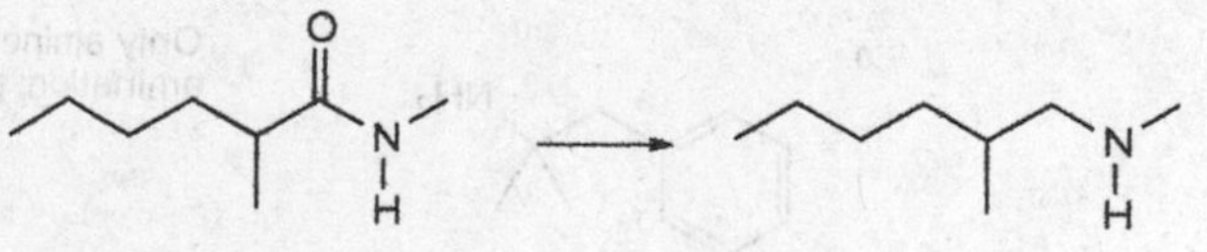

b. 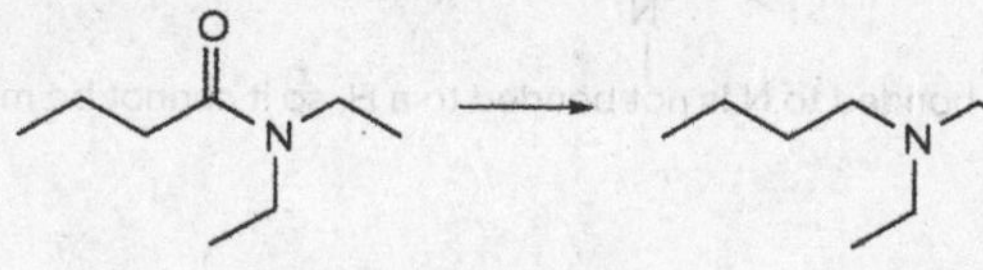 O N → N

c. O N–H → N–H

22.11 Only amines with a CH_2 or CH_3 bonded to the N can be made by reduction of an amide.

a. N bonded to benzene cannot be made by reduction of an amide

b. N bonded to CH_2 can be made by reduction of an amide

c. N bonded to a 3° C cannot be made by reduction of an amide

d. N with 2° C on both sides cannot be made by reduction of an amide

22.12 Reductive amination is a two-step method that converts aldehydes and ketones into 1°, 2°, and 3° amines. Reductive amination replaces a C=O by a C–H and C–N bond.

a. CHO → (CH_3NH_2, $NaBH_3CN$) → NH

b. O → (NH_3, $NaBH_3CN$) → NH_2

c. =O + HN(Et)$_2$ → ($NaBH_3CN$) → N

d. O + NH_2 → ($NaBH_3CN$) → HN

22.13

A + B (HN) → reductive amination → maraviroc

22.14

a. NH_2 ⟹ O + NH_3

b. OH, N–H ⟹ OH, O + NH_2CH_3

or OH, NH_2 + H(C=O)H

22.15

a. phentermine (NH_2)

Only amines that have a C bonded to a H and N atom can be made by reductive amination; that is, an amine must have the following structural feature: H–C–N

In phentermine, the C bonded to N is not bonded to a H, so it cannot be made by reductive amination.

b. systematic name: 2-methyl-1-phenylpropan-2-amine

22.16 The pK_a of many protonated amines is 10–11, so the pK_a of the starting acid must be **less than 10** for equilibrium to favor the products. Amines are thus readily protonated by strong inorganic acids (e.g., HCl and H_2SO_4) and by carboxylic acids.

a. $CH_3CH_2CH_2CH_2NH_2$ + HCl ⇌ $CH_3CH_2CH_2CH_2\overset{+}{N}H_3$ + Cl^-
pK_a = −7; pK_a ≈ 10, weaker acid, **products favored**

b. Benzoic acid (pK_a = 4.2) + $(CH_3)_2NH$ ⇌ $(CH_3)_2\overset{+}{N}H_2$ (pK_a = 10.7, weaker acid, **products favored**) + benzoate

c. Piperidine + H_2O ⇌ protonated piperidine (pK_a ≈ 10) + HO^-
pK_a = 15.7, weaker acid, **reactants favored**

22.17 An amine can be separated from other organic compounds by converting it to a water-soluble ammonium salt by an acid–base reaction. In each case, the extraction procedure would employ the following steps:

- Dissolve the amine and either **X** or **Y** in CH_2Cl_2.
- Add a solution of 10% HCl. The amine will be protonated and dissolve in the aqueous layer, whereas **X** or **Y** will remain in the organic layer as a neutral compound.
- Separate the layers.

a. Cyclohexylamine and methylcyclohexane (**X**) —H–Cl→ cyclohexyl-$\overset{+}{N}H_3$ Cl^- + **X**
- Ammonium salt: **• soluble in H_2O • insoluble in CH_2Cl_2**
- **X**: • insoluble in H_2O • soluble in CH_2Cl_2

b. Tributylamine and dibutyl ether (**Y**) —H–Cl→ tributylammonium Cl^- + **Y**
- Ammonium salt: **• soluble in H_2O • insoluble in CH_2Cl_2**
- **Y**: • insoluble in H_2O • soluble in CH_2Cl_2

22.18 Primary (1°), 2°, and 3° alkylamines are more basic than NH_3 because of the electron-donating inductive effect of the R groups.

a. $(CH_3)_2NH$ and NH_3
2° alkylamine
CH_3 groups are electron donating.
stronger base

b. $CH_3CH_2NH_2$: 1° alkylamine, **stronger base**
$ClCH_2CH_2NH_2$: 1° alkylamine, Cl is electron withdrawing. **weaker base**

22.19 Arylamines are less basic than alkylamines because the electron pair on N is delocalized. Electron-*donor* groups add electron density to the benzene ring making the arylamine *more* basic than aniline. Electron-*withdrawing* groups remove electron density from the benzene ring, making the arylamine *less* basic than aniline.

a.

electron-withdrawing group **least basic**

arylamine **intermediate basicity**

electron-donating group **most basic**

b.

electron-withdrawing group **least basic**

arylamine **intermediate basicity**

alkylamine **most basic**

22.20 Amides are much less basic than amines because the electron pair on N is highly delocalized.

amide **least basic**

arylamine **intermediate basicity**

alkylamine **most basic**

22.21 The higher the pK_a, the weaker the acid, and the stronger the conjugate base.

A electron-donating CH_3 group

B strong electron-withdrawing NO_2 group, strongest acid, lowest pK_a

C alkylammonium ion, weakest acid, highest pK_a

D weak electron-withdrawing Cl

Increasing pK_a: **B** < **D** < **A** < **C**

22.22 HCl protonates the more basic N atom.

a.

delocalized electron pair that is part of an amide

HCl

Cl^-

stronger base
sp^3 hybridized N

b.

HCl

Cl^-

stronger base
sp^3 hybridized N
25% *s*-character

sp^2 hybridized N
33% *s*-character

22.23 Amines attack carbonyl groups to form products of nucleophilic addition or substitution.

a. NH_2 → =N $\underset{H}{N}$ → N

b. NH_2 → $\underset{H}{N}$ $\underset{H}{N}$ → N

c. Cl NH_2 → $\underset{H}{N}$ Cl $\underset{H}{N}$ → N

22.24 [1] Convert the amine (aniline) into an amide (acetanilide).
[2] **Carry out the Friedel–Crafts reaction.**
[3] **Hydrolyze the amide** to generate the free amino group.

a. NH_2 → Cl → NH → Cl, $AlCl_3$ → NH (+ ortho isomer) → H_3O^+ → NH_2

b. NH_2 → Cl → $\underset{H}{N}$ → Cl, $AlCl_3$ → NH (+ para isomer) → H_3O^+ → NH_2

22.25

a. NH_2 → [1] CH_3I (excess) [2] Ag_2O [3] Δ

b. NH_2 → [1] CH_3I (excess) [2] Ag_2O [3] Δ

c. NH_2 → [1] CH_3I (excess) [2] Ag_2O [3] Δ

22.26 In a Hofmann elimination, the base removes a proton from the less substituted, more accessible β carbon atom, because of the bulky leaving group on the nearby α carbon.

a. NH_2 → [1] CH_3I (excess) [2] Ag_2O [3] Δ → (+ *Z* isomer) + major product

22.60

X + Y $\xrightarrow[\text{overall reaction}]{RCO_2H}$ Z =

abbreviated as:

proton transfer

$H-O_2CR$

+ RCO_2^-

+ H_2O

$^-O_2CH_2R$

proton transfer

$H-O_2CR$

+ RCO_2^-

+ H_2O

$^-O_2CH_2R$

Z + RCO_2H

22.61

a. Benzene $\xrightarrow{HNO_3,\ H_2SO_4}$ nitrobenzene (NO_2) $\xrightarrow{Br_2,\ FeBr_3}$ m-bromonitrobenzene (Br, NO_2) $\xrightarrow{H_2,\ Pd\text{-}C}$ m-bromoaniline (Br, NH_2) $\xrightarrow{Cl\text{–}C(=O)CH_3}$ N-(3-bromophenyl)acetamide (Br, N–H, O)

b. Benzene $\xrightarrow{Br_2,\ FeBr_3}$ bromobenzene (Br) $\xrightarrow{HNO_3,\ H_2SO_4}$ o-bromonitrobenzene (NO_2, Br) (+ para isomer) $\xrightarrow{H_2,\ Pd\text{-}C}$ o-bromoaniline (NH_2, Br) $\xrightarrow{[1]\ NaNO_2,\ HCl;\ [2]\ H_2O}$ o-bromophenol (OH, Br) $\xrightarrow{[1]\ NaH;\ [2]\ CH_3CH_2Cl}$ o-bromoethoxybenzene (O–ethyl, Br)

$CH_3CH_2OH \xrightarrow{SOCl_2}$ (CH_3CH_2Cl)

c. Benzene $\xrightarrow{HNO_3,\ H_2SO_4}$ nitrobenzene (NO_2) $\xrightarrow{H_2,\ Pd\text{-}C}$ aniline (NH_2) $\xrightarrow{[1]\ NaNO_2,\ HCl;\ [2]\ H_2O}$ phenol (OH)

$C_6H_5NH_2 \xrightarrow{NaNO_2,\ HCl} C_6H_5N_2^+Cl^-$

Phenol + $C_6H_5N_2^+Cl^-$ → C_6H_5–N=N–C_6H_4–OH (para)

22.62

a. Aniline (NH_2) $\xrightarrow{Cl_2}$ p-chloroaniline (NH_2, Cl) (+ ortho isomer) $\xrightarrow{[1]\ NaNO_2,\ HCl;\ [2]\ NaCN}$ p-chlorobenzonitrile (CN, Cl) $\xrightarrow{H_3O^+}$ p-chlorobenzoic acid (O, OH, Cl) $\xrightarrow{[1]\ SOCl_2;\ [2]\ NH_2CH_3}$ N-methyl-p-chlorobenzamide (O, N–H, Cl)

b. m-Toluidine (NH_2) $\xrightarrow{CH_3COCl}$ acetanilide (N–H, O) $\xrightarrow{Br_2,\ FeBr_3}$ p-bromoacetanilide (Br, N–H, O) (+ ortho isomer) $\xrightarrow{CH_3Cl,\ AlCl_3}$ (Br, CH_3, N–H, O) $\xrightarrow{^-OH,\ H_2O}$ (Br, CH_3, NH_2) $\xrightarrow{[1]\ NaNO_2,\ HCl;\ [2]\ H_3PO_2}$ m-bromotoluene (Br)

c. Aniline (H_2N) $\xrightarrow{[1]\ NaNO_2,\ HCl;\ [2]\ H_2O}$ phenol (HO) $\xrightarrow{CH_3Cl,\ AlCl_3}$ p-cresol (HO, CH_3) (+ ortho isomer) $\xrightarrow{C_6H_5N_2^+Cl^-\ \text{(from a.)}}$ C_6H_5–N=N–(HO, CH_3)aryl

22.63

Starting alkene + mCPBA → epoxide; [1] $LiAlH_4$ [2] H_2O → alcohol (OH); PBr$_3$ → bromide (Br); CH_3NH_2 → MDMA [part (a)]

Alcohol + PCC → ketone; CH_3NH_2, $NaBH_3CN$ → MDMA [part (b)]

22.64

a. Benzene + CH_3Cl, $AlCl_3$ → toluene; HNO_3, H_2SO_4 → p-methyl-NO_2 (+ ortho isomer); H_2, Pd-C → p-methyl-NH_2; [1] $NaNO_2$, HCl [2] NaCN → p-methyl-CN; [1] $LiAlH_4$ [2] H_2O → p-methyl-CH_2NH_2

b. Benzene + CH_3CH_2Cl, $AlCl_3$ → ethylbenzene; HNO_3, H_2SO_4 → p-ethyl-NO_2 (+ ortho isomer); H_2, Pd-C → p-ethyl-NH_2; [1] $NaNO_2$, HCl [2] CuCN → p-ethyl-CN; H_3O^+ → p-ethyl-COOH

c. Benzene + HNO_3, H_2SO_4 → NO_2; H_2, Pd-C → NH_2; CH_3COCl → $NHCOCH_3$; HNO_3, H_2SO_4 → $NHCOCH_3$ / NO_2; Cl_2, $FeCl_3$ (2X) → $NHCOCH_3$, Cl, Cl, NO_2; ^-OH, H_2O → NH_2, Cl, Cl, NO_2; [1] $NaNO_2$, HCl [2] H_3PO_2 → Cl, Cl, NO_2; H_2, Pd-C → Cl, Cl, NH_2; [1] $NaNO_2$, HCl [2] aniline (NH_2) → Cl, Cl, N=N, NH_2 (azo compound)

MDMA

[part (b)]

MDMA

[part (a)]

22.65

a. benzene $\xrightarrow{HNO_3,\ H_2SO_4}$ O_2N–C₆H₅ $\xrightarrow{H_2,\ Pd\text{-}C}$ H_2N–C₆H₅ $\xrightarrow{[1]\ NaNO_2,\ HCl;\ [2]\ H_2O}$ HO–C₆H₅ $\xrightarrow{HNO_3,\ H_2SO_4}$ HO–C₆H₄–NO_2 (+ ortho isomer) $\xrightarrow{H_2,\ Pd\text{-}C}$ HO–C₆H₄–NH_2 → (acetic anhydride) HO–C₆H₄–NHCOCH₃

b. benzene $\xrightarrow{CH_3CH_2COCl,\ AlCl_3}$ $\xrightarrow{Cl_2,\ H_2O,\ HCl}$ Cl $\xrightarrow{NH_2CH_3}$ $\xrightarrow{NaBH_4,\ CH_3OH}$ OH

22.66

a. $(CH_3)_3COH$ $\xrightarrow{HCl}$ $(CH_3)_3CCl$

benzene $\xrightarrow{(CH_3)_3CCl,\ AlCl_3}$ $\xrightarrow{Br_2,\ FeBr_3}$ Br (+ ortho isomer) $\xrightarrow{[1]\ Mg;\ [2]\ CH_2{=}O}$ $\xrightarrow{H_2O}$ OH $\xrightarrow{PCC}$ **A** (H)

CH_3OH $\xrightarrow{PCC}$ $CH_2{=}O$

benzene $\xrightarrow{HNO_3,\ H_2SO_4}$ NO_2 $\xrightarrow{H_2,\ Pd\text{-}C}$ NH_2 $\xrightarrow{[1]\ NaNO_2,\ HCl;\ [2]\ H_2O}$ OH → (acetic acid) → phenyl acetate $\xrightarrow{HNO_3,\ H_2SO_4}$ O_2N–C₆H₄–OAc $\xrightarrow{H_2,\ Pd\text{-}C}$ H_2N–C₆H₄–OAc **B**

ethanol (OH) $\xrightarrow{CrO_3}$ acetic acid (OH)

A (H) + **B** (H_2N) $\xrightarrow{\text{mild } H^+}$ imine (N)

b.

$CH_3OH \xrightarrow{SOCl_2} CH_3Cl$; benzene $\xrightarrow{CH_3Cl,\ AlCl_3}$ toluene $\xrightarrow{HNO_3,\ H_2SO_4}$ p-nitrotoluene (+ ortho) $\xrightarrow{Br_2,\ h\nu}$ p-nitrobenzyl bromide $\xrightarrow{NaOH}$ p-nitrobenzyl alcohol $\xrightarrow{H_2,\ Pd\text{-}C}$ p-aminobenzyl alcohol $\xrightarrow{[1]\ NaNO_2,\ HCl;\ [2]\ NaCN}$ p-cyanobenzyl alcohol $\xrightarrow{[1]\ LiAlH_4;\ [2]\ H_2O}$ p-(aminomethyl)benzyl alcohol

$CH_3OH \xrightarrow{PCC} CH_2{=}O$; $\xrightarrow{NaBH_3CN}$ p-[(methylamino)methyl]benzyl alcohol $\xrightarrow{CrO_3}$ **A**

Benzene $\xrightarrow{HNO_3,\ H_2SO_4}$ nitrobenzene $\xrightarrow{Cl_2,\ FeCl_3}$ m-chloronitrobenzene $\xrightarrow{H_2,\ Pd\text{-}C}$ m-chloroaniline $\xrightarrow{[1]\ NaNO_2,\ HCl;\ [2]\ H_2O}$ m-chlorophenol **B**

A + **B** $\xrightarrow{H^+}$ ester

c. Probably a strong enough activator that the Friedel–Crafts reaction will still occur.

Make two parts:

p-acetamidobenzonitrile + p-iodobenzoyl chloride $\xrightarrow{AlCl_3}$ product

$CH_3CH_2OH \xrightarrow{CrO_3,\ H_2SO_4,\ H_2O} CH_3COOH \xrightarrow{SOCl_2} CH_3COCl$

Benzene $\xrightarrow{[1]\ HNO_3,\ H_2SO_4;\ [2]\ H_2,\ Pd\text{-}C}$ aniline $\xrightarrow{CH_3COCl}$ acetanilide $\xrightarrow{[1]\ HNO_3,\ H_2SO_4;\ [2]\ H_2,\ Pd\text{-}C}$ p-aminoacetanilide (+ ortho isomer) $\xrightarrow{[1]\ NaNO_2,\ HCl;\ [2]\ CuCN}$ p-acetamidobenzonitrile

$CH_3OH \xrightarrow{SOCl_2} CH_3Cl$; benzene $\xrightarrow{CH_3Cl,\ AlCl_3}$ toluene $\xrightarrow{HNO_3,\ H_2SO_4}$ p-nitrotoluene $\xrightarrow{[1]\ H_2,\ Pd\text{-}C;\ [2]\ NaNO_2,\ HCl;\ [3]\ NaI}$ p-iodotoluene $\xrightarrow{[1]\ KMnO_4;\ [2]\ SOCl_2}$ p-iodobenzoyl chloride

d. [Reaction scheme: benzene —HNO_3, H_2SO_4→ nitrobenzene (O_2N) —H_2, Pd-C→ aniline (H_2N) —[1] $NaNO_2$, HCl; [2] H_2O→ phenol (HO) —CH_3Cl, $AlCl_3$ (from CH_3OH, $SOCl_2$)→ p-cresol (HO) (+ ortho isomer); butan-1-ol —CrO_3→ butanoic acid —$SOCl_2$→ butanoyl chloride; p-cresol + butanoyl chloride → p-tolyl butanoate —Br_2, hv→ 4-(bromomethyl)phenyl butanoate (Br) —NH_3→ 4-(aminomethyl)phenyl butanoate (NH_2) —CH_3CHO, $NaBH_3CN$ (from CH_3CH_2OH, PCC)→ 4-[(ethylamino)methyl]phenyl butanoate (N–H)]

22.67 The spectrum of butan-1-amine shows a peak at $m/z = 73$ due to the molecular ion $(C_4H_{11}N)^+$. The base peak at $m/z = 30$ is due to fragmentation to form the resonance-stabilized cation $(CH_2NH_2)^+$.

$$(CH_3CH_2CH_2CH_2NH_2)^{+\bullet} \longrightarrow CH_3CH_2CH_2^{\bullet} + H_2\overset{+}{C}-\ddot{N}H_2 \longleftrightarrow H_2C=\overset{+}{N}H_2$$

This bond is cleaved.

$m/z = 30$

22.68

[Structures: A = N-ethylaniline (NH); B = N-methylbenzylamine (HN—); C = 4-ethylaniline (H_2N—)]

Compound **A**: $C_8H_{11}N$
IR absorption at 3400 cm^{-1} → 2° amine
^{1}H NMR signals at (ppm):
1.3 (triplet, 3 H) CH_3 adjacent to 2 H's
3.2 (quartet, 2 H) CH_2 adjacent to 3 H's
3.6 (singlet, 1 H) amine H
6.8–7.2 (multiplet, 5 H) benzene ring

Compound **B**: $C_8H_{11}N$
IR absorption at 3310 cm^{-1} → 2° amine
^{1}H NMR signals at (ppm):
1.4 (singlet, 1 H) amine H
2.4 (singlet, 3 H) CH_3
3.8 (singlet, 2 H) CH_2
7.3 (multiplet, 5 H) benzene ring

Compound **C**: $C_8H_{11}N$
IR absorption at 3430 and 3350 cm^{-1} → 1° amine
^{1}H NMR signals at (ppm):
1.3 (triplet, 3 H) CH_3 near CH_2
2.5 (quartet, 2 H) CH_2 near CH_3
3.6 (singlet, 2 H) amine H's
6.7 (doublet, 2 H)] para disubstituted
7.0 (doublet, 2 H)] benzene ring

22.69 Guanidine is a strong base because its conjugate acid is stabilized by resonance. This resonance delocalization makes guanidine easily donate its electron pair; thus it's a strong base.

[Scheme: guanidine (H_2N–C(=NH)–NH_2) —HA→ protonated guanidinium ion, shown as resonance structures ($\overset{+}{N}H_2$, H_2N, NH_2), $pK_a = 13.6$]

guanidine

$pK_a = 13.6$

22.70 The compound with the most available electron pair or the compound with the highest electron density on an atom (N in this case) is the strongest base. Pyrrole is the weakest base because its lone pair is delocalized on the five-membered ring to make it aromatic. Both imidazole and thiazole contain sp^2 hybridized N atoms with electron pairs that are localized on N. Imidazole is a stronger base than thiazole, because its second N atom is more basic than thiazole's S atom, so it places more electron density on N by a resonance effect.

imidazole

This N atom is the strongest base.

thiazole

This form contributes less to the hybrid than the equivalent resonance structure in imidazole.

pyrrole **least basic**

thiazole **intermediate basicity**

imidazole **most basic**

22.71

[1] CH_3I (excess) → I^- → [2] Ag_2O [3] Δ → $N(CH_3)_2$ → [1] CH_3I (excess) → I^- $^+N(CH_3)_3$ → [2] Ag_2O [3] Δ → C_8H_{10} **Y**

22.72

One possibility:

a. CH_3COCl, $AlCl_3$ → (+ isomer) → Br_2, CH_3CO_2H → Br → $H_2NC(CH_3)_3$ → NaBH$_4$, CH_3OH → OH → H_3O^+ → HO, HO, OH, albuterol

b. + → HO, N, **A**

HNO_3, H_2SO_4 → O_2N → H_2, Pd-C → H_2N → [1] $NaNO_2$, HCl [2] H_2O → HO → [1] NaH [2] Cl → O → Br_2 → Br (+ ortho isomer) → Mg → CO_2 → H_3O^+ → COOH → H_2SO_4, HNO_3 → O_2N COOH → $SOCl_2$ → O_2N COCl → **A** → O_2N → H_2, Pd-C → H_2N

22.73 $CH_2{=}O$ reacts with the amine to form an intermediate imine, which undergoes an intramolecular Diels–Alder reaction.

X (H_2N, O, Ph) → $CH_2{=}O$, H_2O → $CH_2{=}N$ → **Y** $C_{17}H_{23}NO$ → one step → lupinine + epilupinine

Chapter 23 Amino Acids and Proteins

Chapter Review

Preparation of optically active amino acids

[1] Resolution of enantiomers by forming diastereomers (23.2A)

- Convert a racemic mixture of amino acids into a racemic mixture of *N*-acetyl amino acids [(*S*)- and (*R*)-$CH_3CONHCH(R)COOH$].
- React the enantiomers with a chiral amine to form a mixture of diastereomers.
- Separate the diastereomers.
- Regenerate the amino acids by protonation of the carboxylate salt and hydrolysis of the *N*-acetyl group.

[2] Kinetic resolution using enzymes (23.2B)

(*S*) isomer —Ac_2O→ AcNH…OH —acylase→ (*S*) isomer

(*R*) isomer —Ac_2O→ AcNH…OH —acylase→ recovered amide, **No reaction**

enantiomers — enantiomers — separate

[3] By enantioselective hydrogenation (23.3)

AcNH…OH —H_2, Rh*→ *S* enantiomer —H_2O, ^-OH→ *S* amino acid

Rh* = chiral Rh hydrogenation catalyst

Summary of methods used for peptide sequencing (23.5)

- Complete hydrolysis of all amide bonds in a peptide gives the identity and amount of the individual amino acids.
- Edman degradation identifies the N-terminal amino acid. Repeated Edman degradations can be used to sequence a peptide from the N-terminal end.
- Cleavage with carboxypeptidase identifies the C-terminal amino acid.
- Partial hydrolysis of a peptide forms smaller fragments that can be sequenced. Amino acid sequences common to smaller fragments can be used to determine the sequence of the complete peptide.
- Selective cleavage of a peptide occurs with trypsin and chymotrypsin to identify the location of specific amino acids (Table 23.2).

Adding and removing protecting groups for amino acids (23.6)

[1] Protection of an amino group as a Boc derivative

$H_3\overset{+}{N}$–CH(R)–CO_2^- $\xrightarrow[Et_3N]{(Boc)_2O}$ Boc–NH–CH(R)–COOH

[2] Deprotection of a Boc-protected amino acid

Boc–NH–CH(R)–COOH $\xrightarrow[\text{HCl or HBr}]{CF_3CO_2H \text{ or}}$ $H_3\overset{+}{N}$–CH(R)–CO_2^-

[3] Protection of an amino group as an Fmoc derivative

$H_3\overset{+}{N}$–CH(R)–CO_2^- + Fmoc–Cl $\xrightarrow[H_2O]{Na_2CO_3}$ Fmoc–NH–CH(R)–COOH

[4] Deprotection of an Fmoc-protected amino acid

Fmoc–NH–CH(R)–COOH $\xrightarrow{\text{piperidine}}$ $H_3\overset{+}{N}$–CH(R)–CO_2^-

[5] Protection of a carboxy group as an ester

$H_3\overset{+}{N}$–CH(R)–CO_2^- $\xrightarrow{CH_3OH,\ H^+}$ H_2N–CH(R)–$COOCH_3$

methyl ester

$H_3\overset{+}{N}$–CH(R)–CO_2^- $\xrightarrow[H^+]{PhCH_2OH}$ H_2N–CH(R)–$COOCH_2Ph$

benzyl ester

[6] Deprotection of an ester group

H_2N–CH(R)–$COOCH_3$ $\xrightarrow[H_2O]{^-OH}$ H_2N–CH(R)–CO_2^-

methyl ester

H_2N–CH(R)–$COOCH_2Ph$ $\xrightarrow[\text{or } H_2,\ Pd\text{-}C]{H_2O,\ ^-OH}$ $H_3\overset{+}{N}$–CH(R)–CO_2^-

benzyl ester

Synthesis of dipeptides (23.6)

[1] Amide formation with DCC

Boc–NH (R) OH (O) + H_2N (R) O Ph $\xrightarrow{\text{DCC}}$ Boc–NH (O) NH (R) O Ph

[2] Four steps are needed to synthesize a dipeptide:

a. **Protect** the amino group of one amino acid using a Boc or Fmoc group.
b. **Protect** the carboxy group of the second amino acid using an ester.
c. Form the amide bond with **DCC.**
d. **Remove both protecting groups** in one or two reactions.

Summary of the Merrifield method of peptide synthesis (23.7)

[1] Attach an Fmoc-protected amino acid to a polymer derived from polystyrene.
[2] Remove the Fmoc protecting group.
[3] Form the amide bond with a second Fmoc-protected amino acid using DCC.
[4] Repeat steps [2] and [3].
[5] Remove the protecting group and detach the peptide from the polymer.

Practice Test on Chapter Review

1.a. Which statement is true about the peptide Ala–Gly–Tyr–Phe?
1. The N-terminal amino acid is Ala.
2. The N-terminal amino acid is Phe.
3. The peptide contains four peptide bonds.
4. Statements (1) and (3) are true.
5. Statements (2) and (3) are true.

b. Which of the following peptides is hydrolyzed by trypsin?
1. Glu–Ser–Gly–Arg
2. Arg–Gln–Trp–Asp
3. Glu–Val–Leu–Lys
4. Peptides (1) and (2) are hydrolyzed.
5. Peptides (1), (2), and (3) are all hydrolyzed.

c. In which types of protein structure is hydrogen bonding observed?
1. α-helix
2. β-pleated sheet
3. 3° structure
4. Hydrogen bonding is present in (1) and (2).
5. Hydrogen bonding is present in (1), (2), and (3).

2. Answer the following questions about peptides.

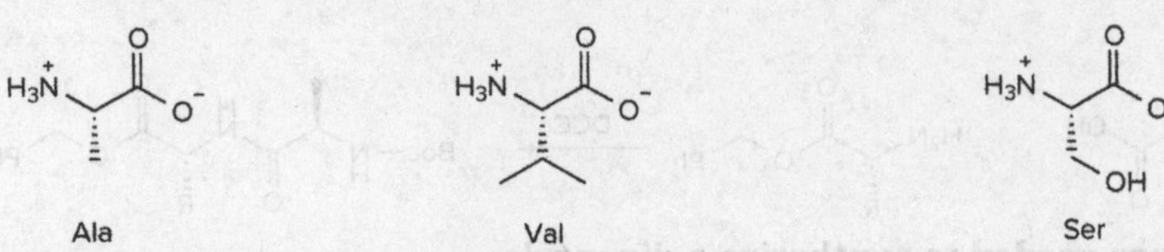

a. Draw the structure of the following tripeptide: Val–Ser–Ala.

b. Give the three-letter abbreviation for the N-terminal amino acid.

c. Give the three-letter abbreviation for the C-terminal amino acid.

3. Answer the following questions about the amino acid leucine (2-amino-4-methylpentanoic acid), which has pK_a's of 2.33 and 9.74 for its ionizable functional groups.

a. Draw a Fischer projection for L-leucine and label the stereogenic center as *R* or *S*.

b. What is the p*I* of leucine?

c. Draw the structure of the predominant form of leucine at its isoelectric point.

d. Draw the structure of the predominant form of leucine at pH 10.

e. Is leucine an acidic, basic, or neutral amino acid?

4. What product is formed when the amino acid phenylalanine is treated with each reagent?

a. $PhCH_2OH, H^+$

b. Ac_2O, pyridine

c. PhCOCl, pyridine

d. $(Boc)_2O$

e. $C_6H_5N{=}C{=}S$

5. Draw the amino acids and peptide fragments formed when the octapeptide Tyr–Gly–Ala–Lys–Val–Ser–Phe–Met is treated with each reagent or enzyme:

a. chymotrypsin

b. trypsin

c. carboxypeptidase

d. $C_6H_5N{=}C{=}S$

Answers to Practice Test

1.a. 1

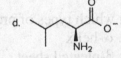

b. 2

c. 5

2.a

$H_3\overset{+}{N}$ — Val — Ser — Ala (structure)

Val Ser Ala

b. Val

c. Ala

3.

a. Fischer projection: COOH; H_2N–**S**–H; $CH_2CH(CH_3)_2$

b. 6.04

c. (structure with $\overset{+}{N}H_3$ and O^-)

d. (structure with NH_2 and O^-)

e. neutral

4.

a.

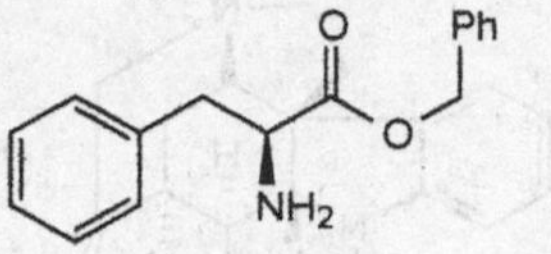

b.

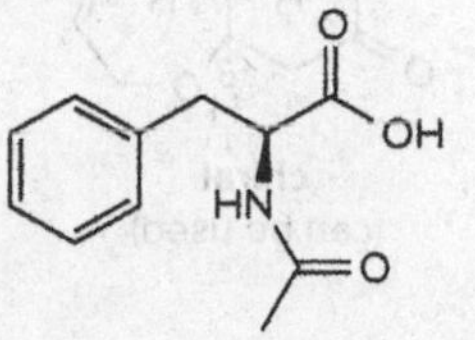

c.

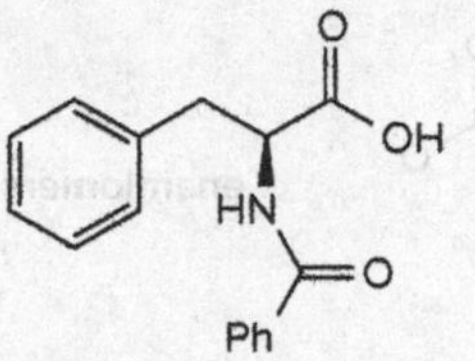

d.

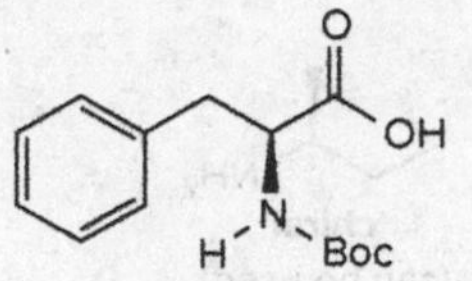

e.

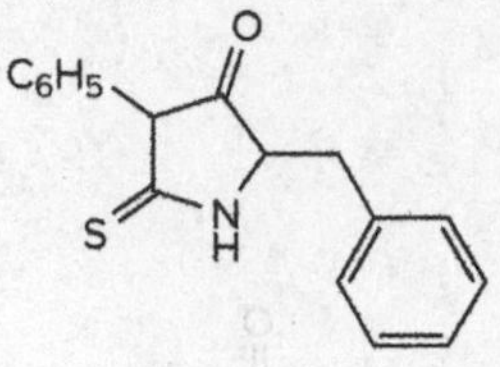

5.a. Tyr, Gly–Ala–Lys–Val–Ser–Phe, Met
b. Tyr–Gly–Ala–Lys, Val–Ser–Phe–Met
c. Tyr–Gly–Ala–Lys–Val–Ser–Phe, Met
d. Tyr, Gly–Ala–Lys–Val–Ser–Phe–Met

Answers to Problems

23.1

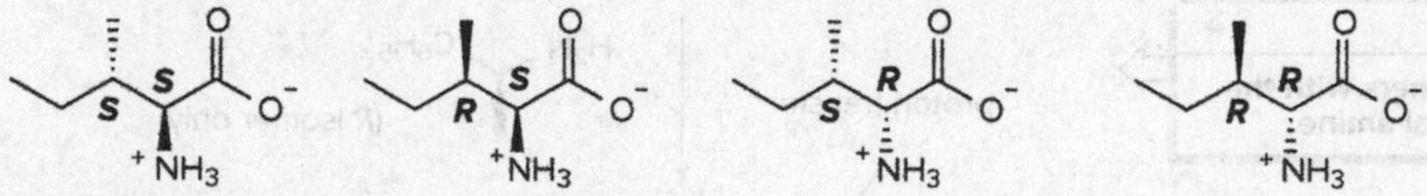

L-isoleucine

23.2

a. 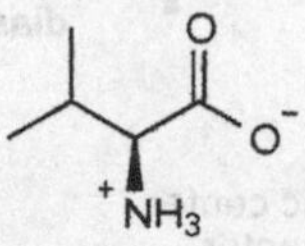 b. 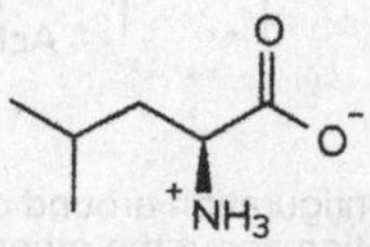 c. 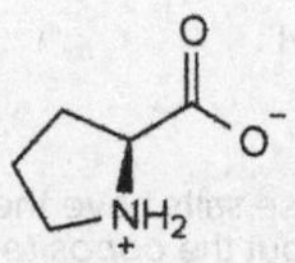 d.

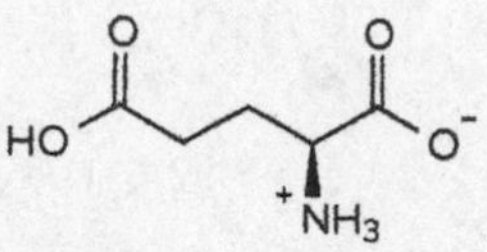

23.3 In an amino acid, the electron-withdrawing carboxy group destabilizes the ammonium ion ($-NH_3^+$), making it more readily donate a proton; that is, it makes it a stronger acid. Also, the electron-withdrawing carboxy group removes electron density from the amino group ($-NH_2$) of the conjugate base, making it a weaker base than a 1° amine, which has no electron-withdrawing group.

23.4

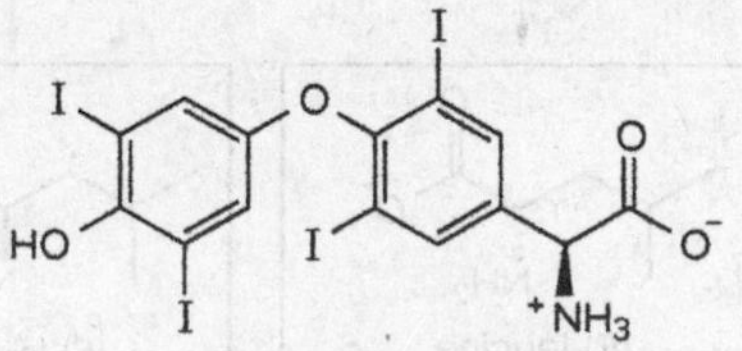

zwitterionic form

23.5 A chiral amine must be used to resolve a racemic mixture of amino acids.

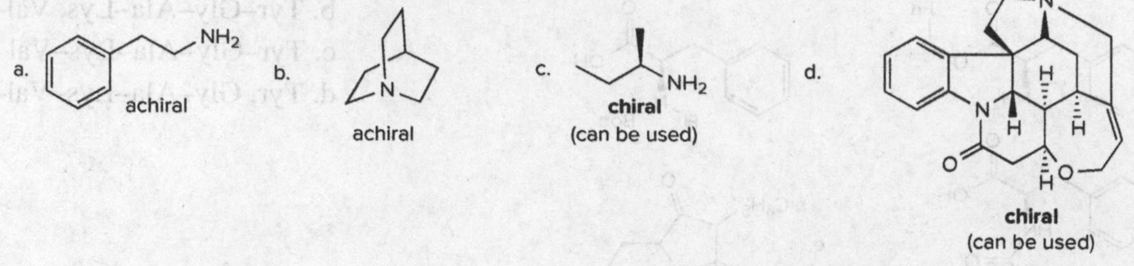
a.
achiral
b.
achiral
c.
chiral
(can be used)
d.
chiral
(can be used)

23.6

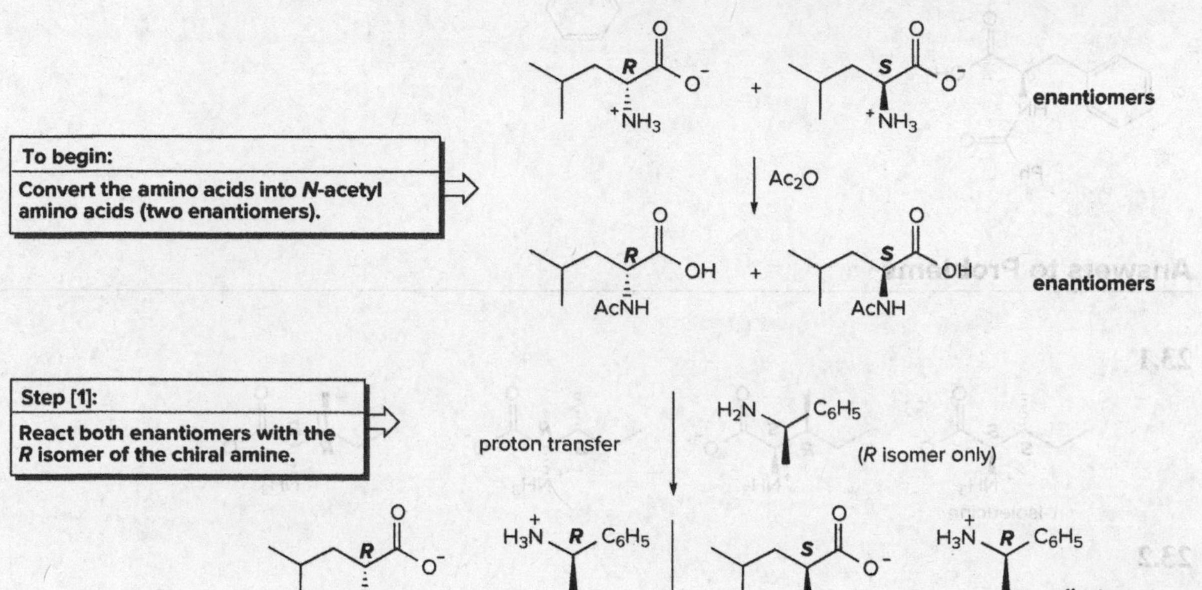
enantiomers
To begin:
Convert the amino acids into N-acetyl amino acids (two enantiomers).
Ac₂O
enantiomers
Step [1]:
React both enantiomers with the R isomer of the chiral amine.
proton transfer
(R isomer only)
diastereomers
These salts have the same configuration around one stereogenic center, but the opposite configuration about the other stereogenic center.

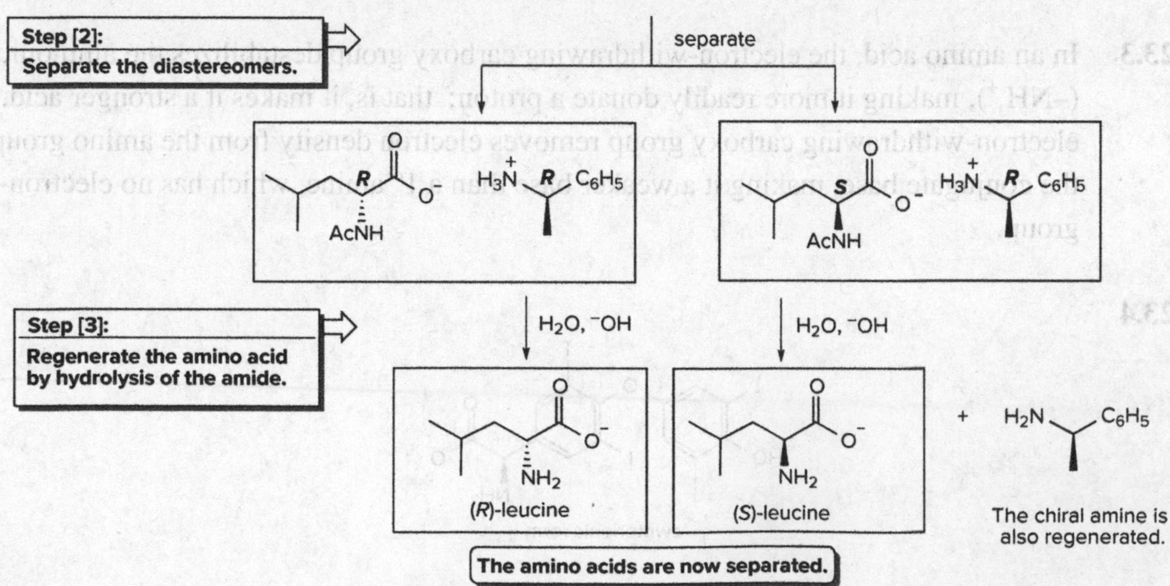
Step [2]:
Separate the diastereomers.
separate
Step [3]:
Regenerate the amino acid by hydrolysis of the amide.
H₂O, ⁻OH
H₂O, ⁻OH
(R)-leucine
(S)-leucine
The chiral amine is also regenerated.
The amino acids are now separated.

23.7

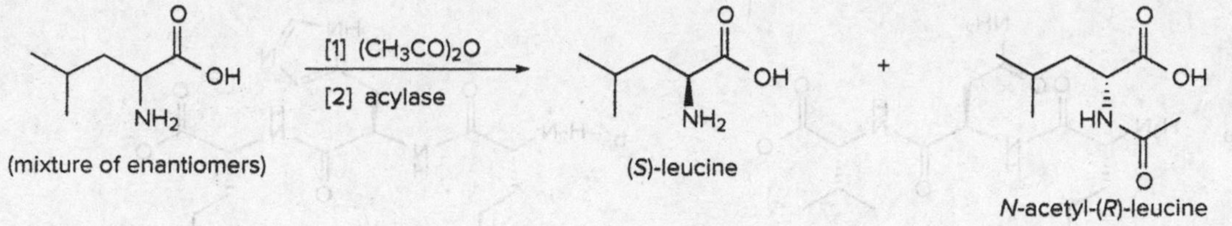

(mixture of enantiomers) — [1] $(CH_3CO)_2O$, [2] acylase → (*S*)-leucine + *N*-acetyl-(*R*)-leucine

23.8

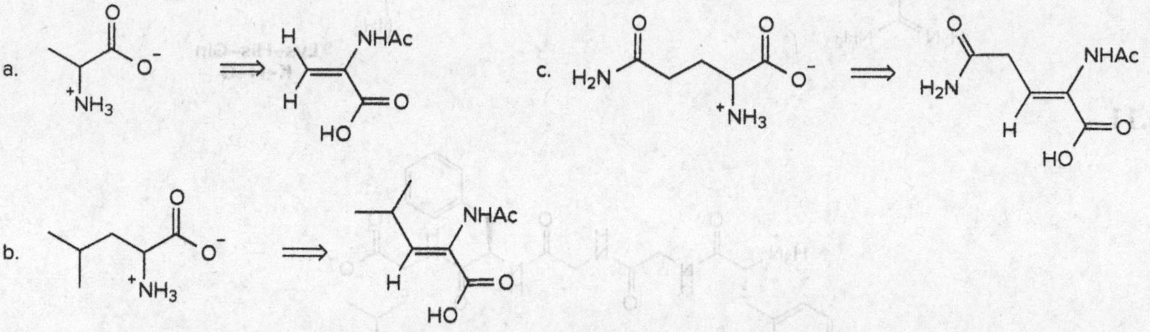

a. ⟹ NHAc

c. ⟹ NHAc

b. ⟹ NHAc

23.9 Draw the peptide by joining adjacent COOH and NH_2 groups in amide bonds.

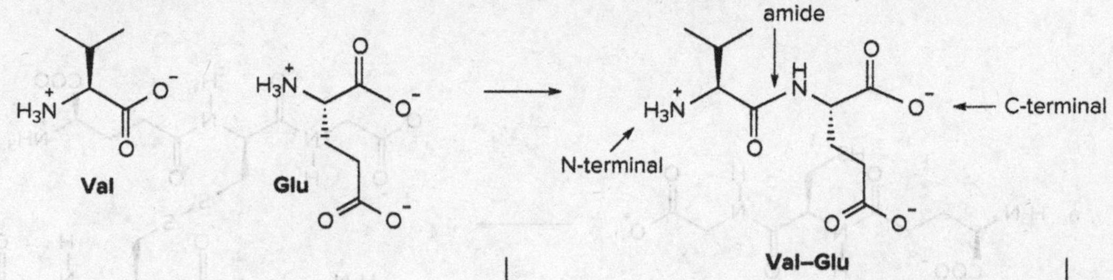

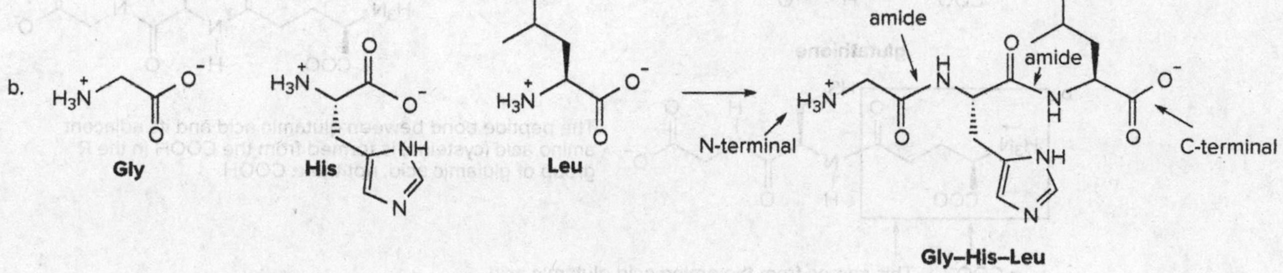

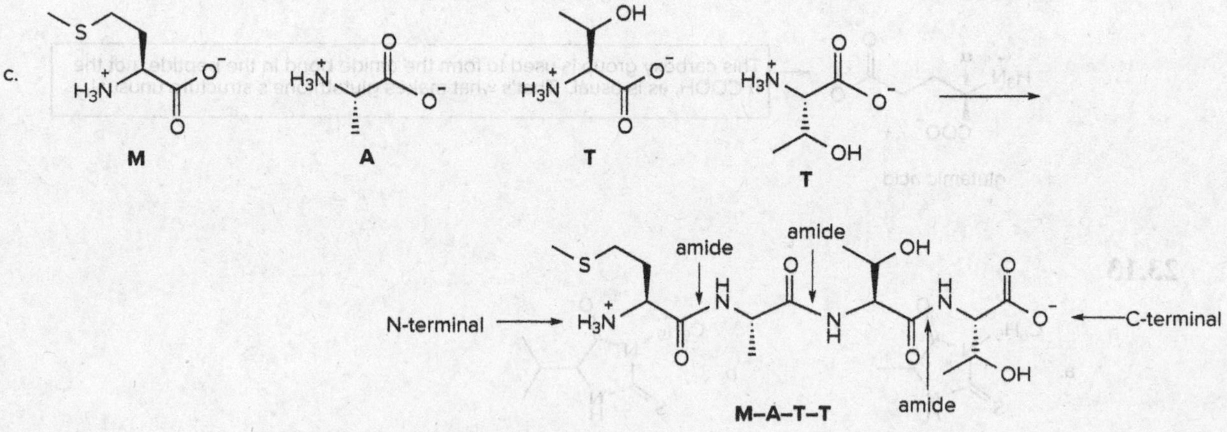

a. Val + Glu → Val–Glu (amide; N-terminal; C-terminal)

b. Gly + His + Leu → Gly–His–Leu (amide; amide; N-terminal; C-terminal)

c. M + A + T + T → M–A–T–T (N-terminal; amide; amide; amide; C-terminal)

23.10

a. 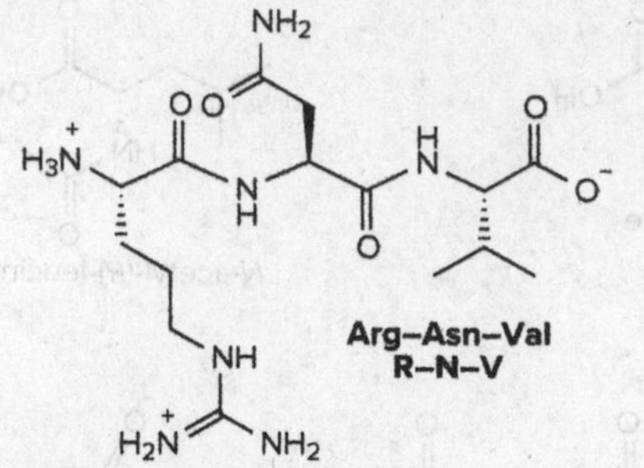 Arg–Asn–Val
R–N–V

b. 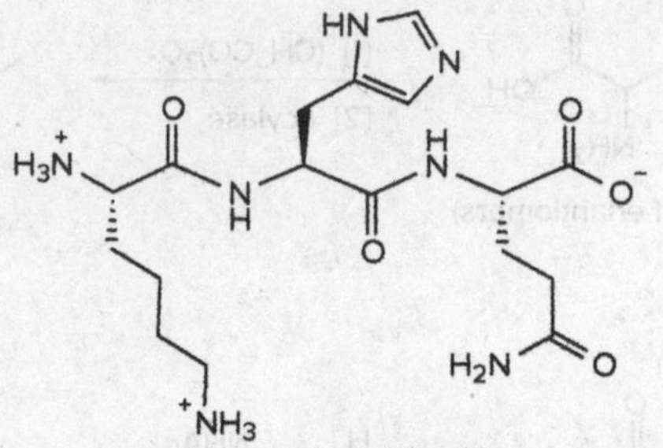 Lys–His–Gln
K–H–Q

23.11

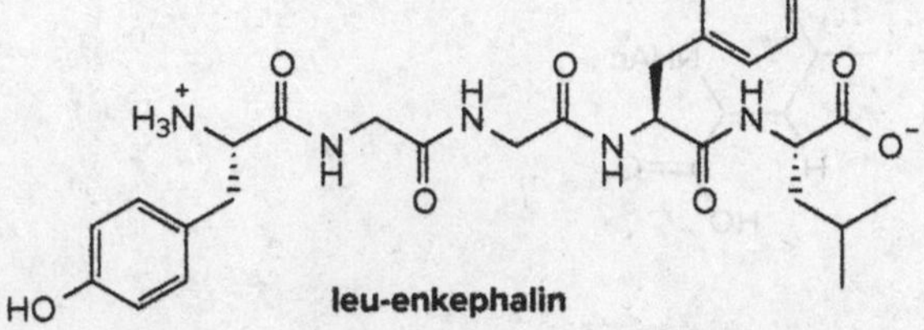

leu-enkephalin

23.12

a. 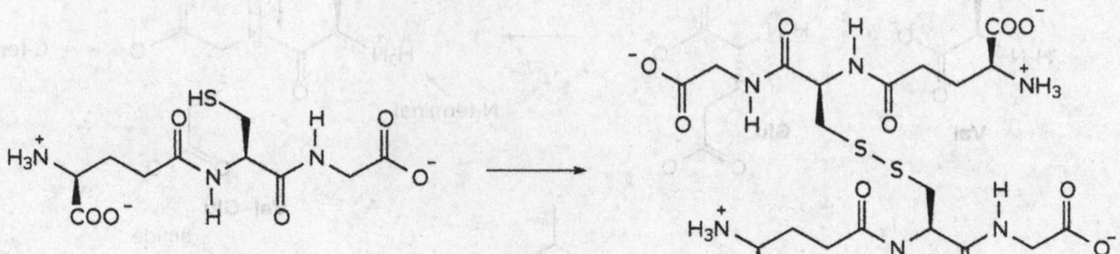 glutathione

b. 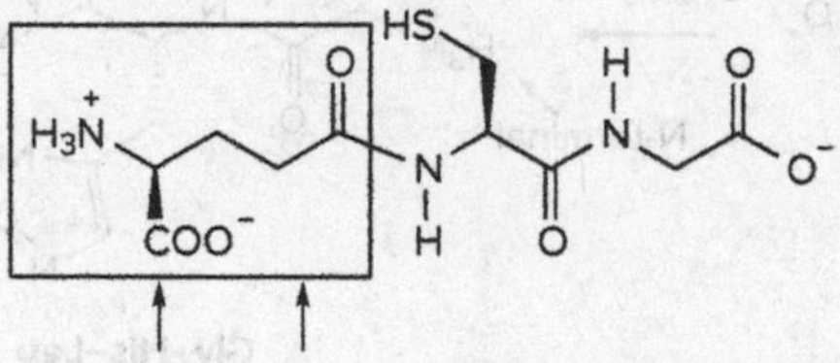 The peptide bond beween glutamic acid and its adjacent amino acid (cysteine) is formed from the COOH in the R group of glutamic acid, not the α COOH.

α COO^- This comes from the amino acid glutamic acid.

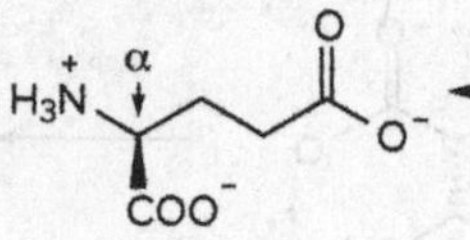

This carboxy group is used to form the amide bond in the peptide, not the α COOH, as is usual. That's what makes glutathione's structure unusual.

glutamic acid

23.13

a. 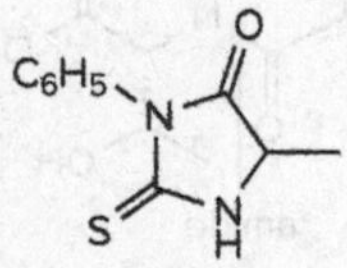 from Ala

b. 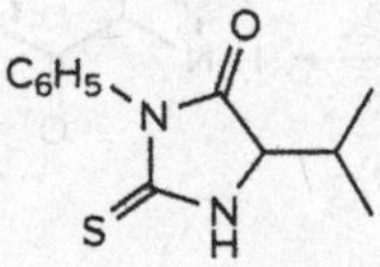 from Val

23.14 Determine the sequence of the octapeptide as in Sample Problem 23.2. Look for overlapping sequences in the fragments.

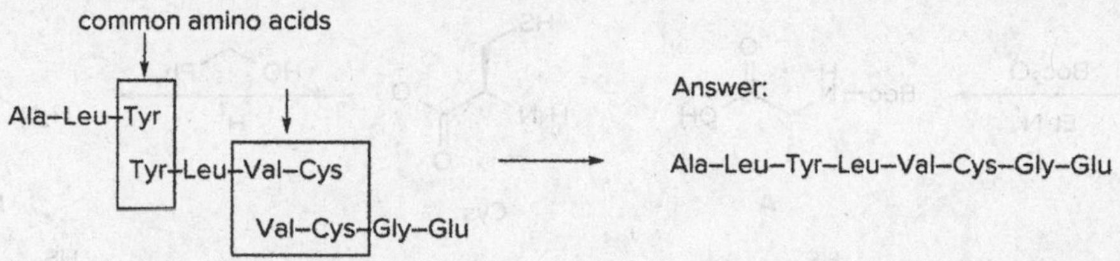

23.15 Trypsin cleaves peptides at amide bonds with a carbonyl group from Arg and Lys. Chymotrypsin cleaves at amide bonds with a carbonyl group from Phe, Tyr, and Trp.

a. [1] Gly–Ala–Phe–Leu–Lys + Ala
[2] Phe–Tyr–Gly–Cys–Arg + Ser
[3] Thr–Pro–Lys + Glu–His–Gly–Phe–Cys–Trp–Val–Val–Phe

b. [1] Gly–Ala–Phe + Leu–Lys–Ala
[2] Phe + Tyr + Gly–Cys–Arg–Ser
[3] Thr–Pro–Lys–Glu–His–Gly–Phe + Cys–Trp + Val–Val–Phe

23.16

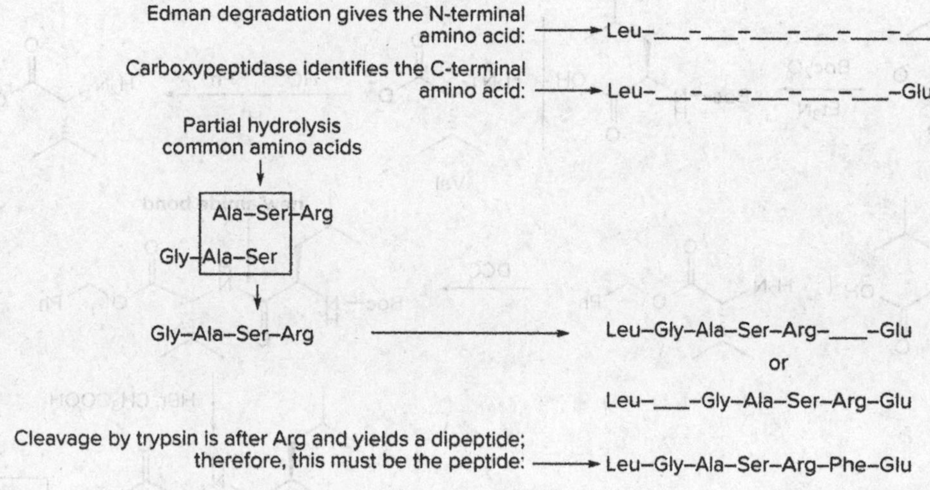

23.17 The dipeptide depicted in the 3-D model has alanine as the N-terminal amino acid and cysteine as the C-terminal amino acid.

Ala $\xrightarrow[Et_3N]{Boc_2O}$ A

Cys $\xrightarrow[H^+]{HO\text{–}CH_2Ph}$ B

A + B $\xrightarrow{DCC}$ Boc-Ala-Cys-OCH$_2$Ph $\xrightarrow{HBr,\ CH_3COOH}$ Ala–Cys

23.18

a.

Leu $\xrightarrow[Et_3N]{Boc_2O}$ Boc-Leu-OH

Val $\xrightarrow[H^+]{HO\text{–}CH_2Ph}$ Val-OCH$_2$Ph

Boc-Leu-OH + H$_2$N-Val-OCH$_2$Ph $\xrightarrow{DCC}$ Boc-Leu-Val-OCH$_2$Ph (new amide bond) $\xrightarrow{HBr,\ CH_3COOH}$ **Leu–Val**

b. $H_3\overset{+}{N}$–Ala–O^- (Ala) $\xrightarrow[Et_3N]{Boc_2O}$ Boc–NH–Ala–OH **A**

$H_3\overset{+}{N}$–Ile–O^- (Ile) $\xrightarrow[H^+]{HO\frown Ph}$ H_2N–Ile–$O\frown Ph$ **B**

A + **B** $\xrightarrow{DCC}$ Boc–NH–Ala–NH–Ile–$O\frown Ph$ (**new amide bond**) $\xrightarrow[Pd\text{-}C]{H_2}$ Boc–NH–Ala–NH–Ile–OH **C**

$H_3\overset{+}{N}$–Gly–O^- (Gly) $\xrightarrow[H^+]{HO\frown Ph}$ H_2N–Gly–$O\frown Ph$

C + H_2N–Gly–$O\frown Ph$ $\xrightarrow{DCC}$ Boc–NH–Ala–NH–Ile–NH–Gly–$O\frown Ph$ (**new amide bond**) $\xrightarrow[CH_3COOH]{HBr}$

Ala–Ile–Gly $H_3\overset{+}{N}$–Ala–NH–Ile–NH–Gly–O^-

23.19

All Fmoc-protected amino acids are made by the following general reaction:

$H_3\overset{+}{N}$–CH(R)–CO_2^- $\xrightarrow[Na_2CO_3,\ H_2O]{Fmoc\text{–}Cl}$ Fmoc–NH–CH(R)–COOH

The steps:

Fmoc-Gly → [1] base, [2] Cl–CH2–POLYMER → Fmoc-Gly–O–CH2–POLYMER → [1] piperidine, [2] DCC, Fmoc-Ile → Fmoc-Ile-Gly–O–CH2–POLYMER → [1] piperidine, [2] DCC, + Fmoc-Leu → Fmoc-Leu-Ile-Gly–O–CH2–POLYMER → [1] piperidine, [2] DCC, Fmoc-Ala → Fmoc-Ala-Leu-Ile-Gly–O–CH2–POLYMER → [1] piperidine, [2] HF → H_3N^+-Ala-Leu-Ile-Gly-O^-

Ala–Leu–Ile–Gly

+ F–CH2–POLYMER

23.20 a. Met–Gly–Leu–Phe–Phe–Gln–Ala has no charged functional groups and no OH groups in the side chains, so it exhibits predominately van der Waals interactions in the side chains.

b. Phe–Asn–Leu–Leu–Met exhibits predominately van der Waals forces for the same reason.

23.21 a. The R group for glycine is a hydrogen. The R groups must be small to allow the β-pleated sheets to stack on top of each other. With large R groups, steric hindrance prevents stacking.

b. Silk fibers are water insoluble because most of the polar functional groups are in the interior of the stacked sheets. The β-pleated sheets are stacked one on top of another, so few polar functional groups are available for hydrogen bonding to water.

23.22 The reaction in part (a) is catalyzed by a decarboxylase, a type of lyase, because CO_2 is lost from the reactant. The reaction in part (b) is catalyzed by an isomerase because a cis isomer is isomerized to a trans isomer.

a. 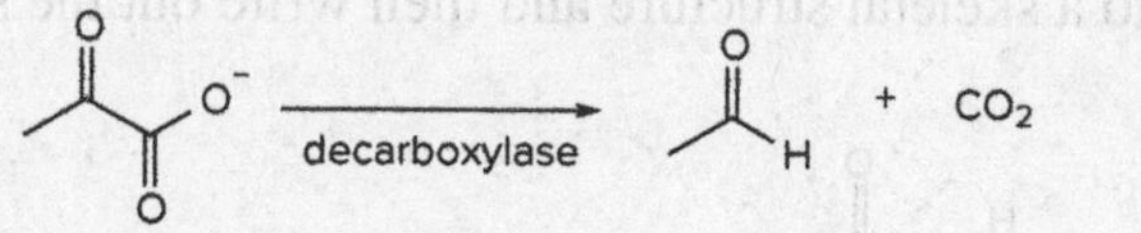 decarboxylase; + CO_2
b. 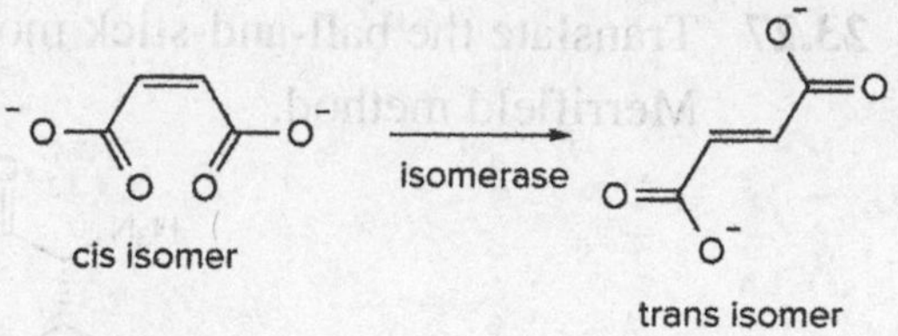 cis isomer; isomerase; trans isomer

23.23 Use Table 23.3 to determine the class of enzyme from its name.
a. chymotrypsin—a protease, a type of hydrolase
b. alcohol dehydrogenase—a dehydrogenase, a type of oxidoreductase
c. phosphofructokinase—a kinase, a type of transferase

23.24 The anionic carboxylate of aspartate forms a strong hydrogen bond with an N–H bond of histidine at the active site, increasing the basicity of the histidine. A similar type of hydrogen-bonding opportunity with the $CONH_2$ group of asparagine is much less effective, so the histidine is less able to remove an O–H proton and enzyme activity is lost.

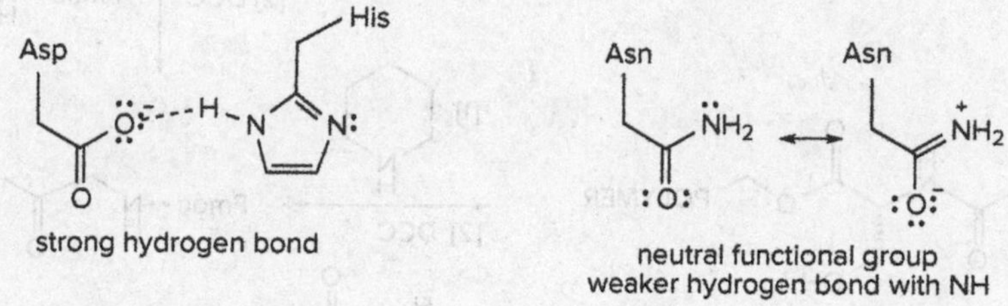

Asp, His — strong hydrogen bond

Asn, Asn — neutral functional group weaker hydrogen bond with NH

23.25

a. N-terminal amino acid: alanine
C-terminal amino acid: serine
b. A–Q–C–S

c. Amide bonds are **bold** (not wedges).

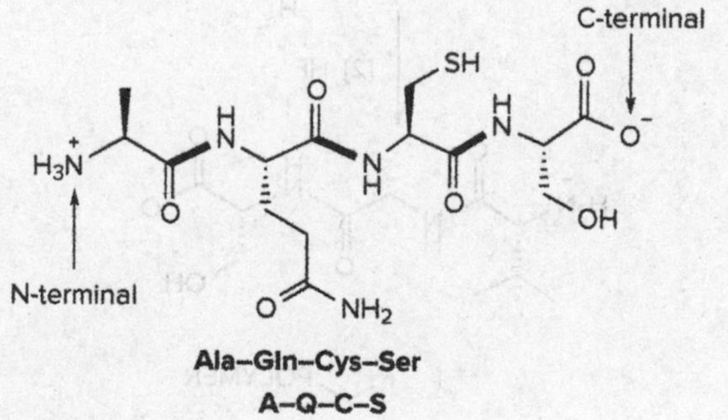

C-terminal; N-terminal

Ala–Gln–Cys–Ser
A–Q–C–S

23.26 The dipeptide is composed of phenylalanine and leucine.

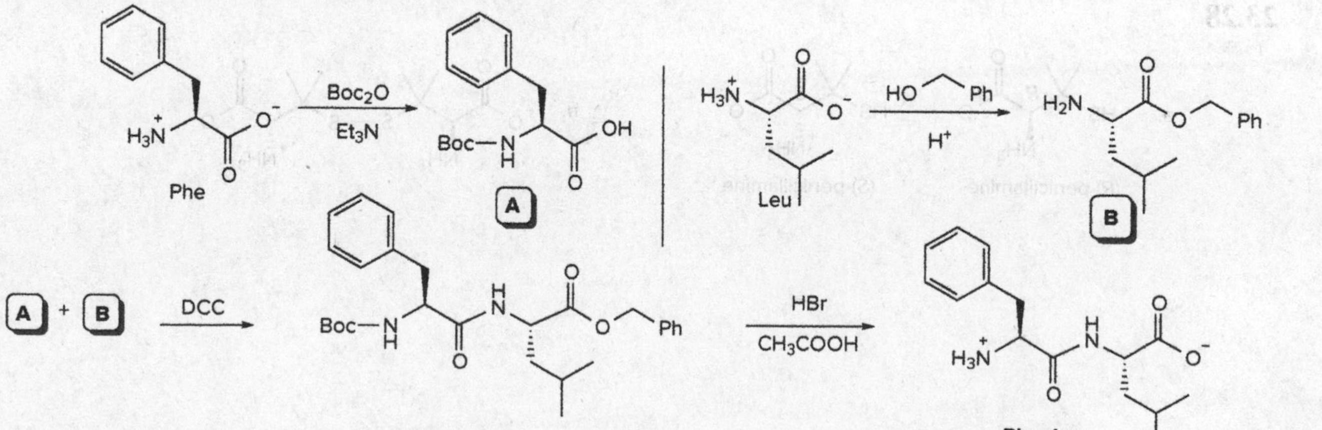

Phe → (Boc_2O, Et_3N) → **A**

Leu → (HO–CH_2Ph, H^+) → **B**

A + **B** → (DCC) → (HBr, CH_3COOH) → **Phe–Leu**

23.27 Translate the ball-and-stick model to a skeletal structure and then write out the steps for the Merrifield method.

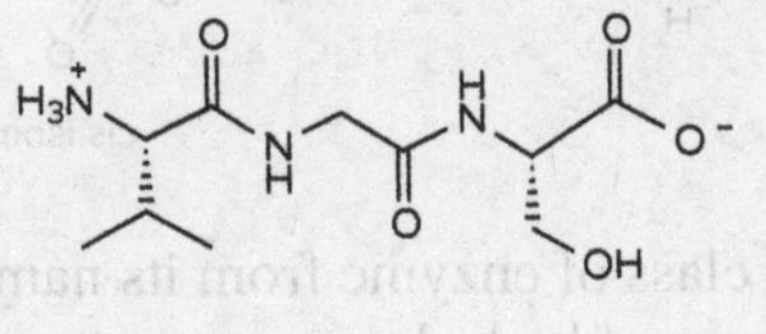

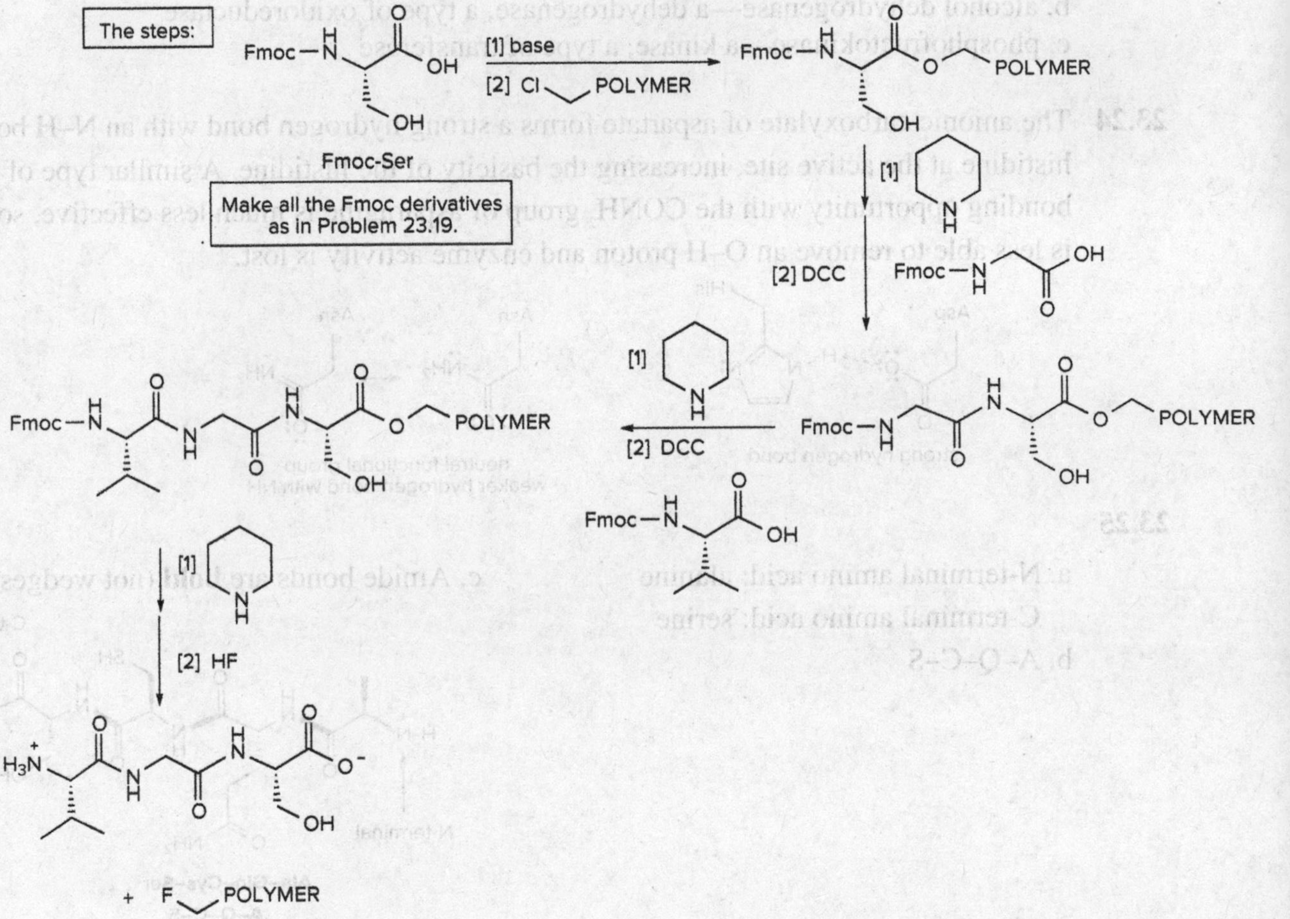

$H_3\overset{+}{N}$ N H O O H N O O^- OH

The steps:

Fmoc—N(H) ... OH, OH — Fmoc-Ser — [1] base [2] Cl–CH₂–POLYMER → Fmoc—N(H) ... O–CH₂–POLYMER, OH

Make all the Fmoc derivatives as in Problem 23.19.

[1] piperidine; [2] DCC, Fmoc—N(H)–CH₂–COOH →

Fmoc—N(H) ... N(H) ... O–POLYMER, OH

[1] piperidine; [2] DCC, Fmoc—N(H)–(Val)–OH →

Fmoc—N(H) ... N(H) ... N(H) ... O–POLYMER, OH

[1] piperidine; [2] HF →

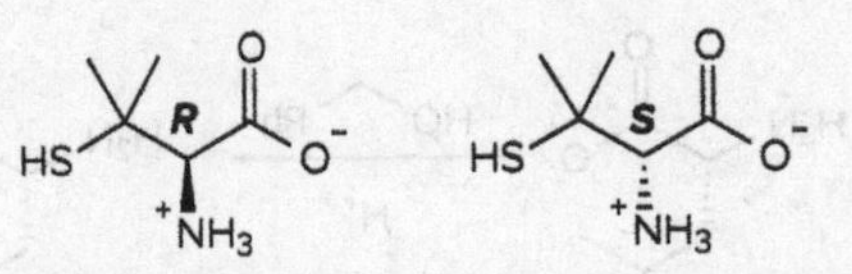

$H_3\overset{+}{N}$ ... N(H) ... N(H) ... O^-, OH

\+ F–CH₂–POLYMER

23.28

a. HS ... *R* ... O^-, $\overset{+}{N}H_3$ (*R*)-penicillamine; HS ... *S* ... O^-, $\overset{+}{N}H_3$ (*S*)-penicillamine

b. ^-O ... S—S ... O^-, $\overset{+}{N}H_3$, $\overset{+}{N}H_3$

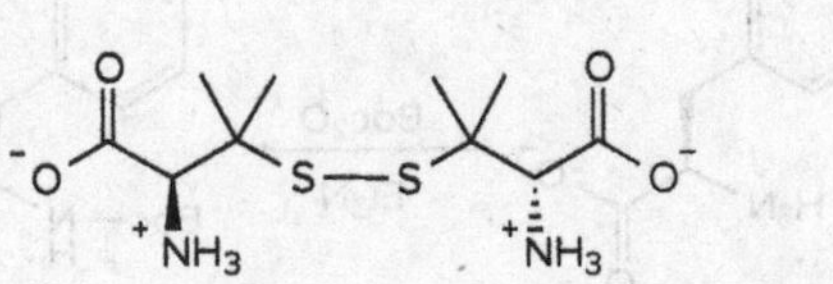

23.29 The electron pair on the N atom not part of a double bond is delocalized on the five-membered ring, making it less basic.

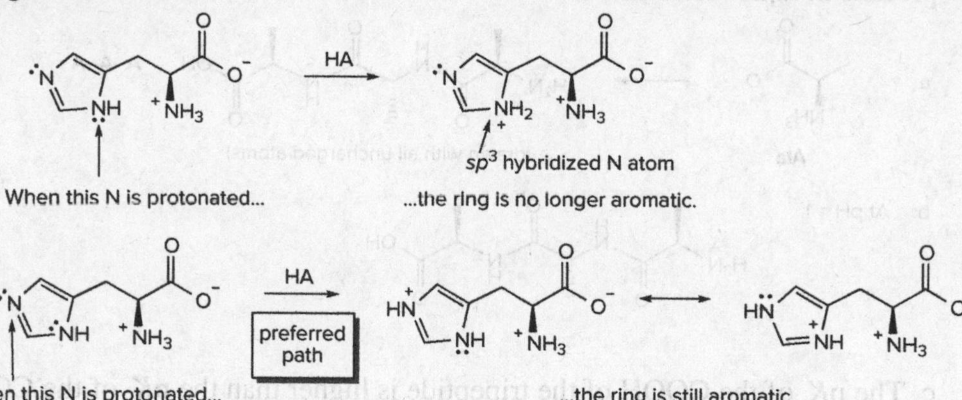

23.30

The ring structure on tryptophan is aromatic because each atom contains a *p* orbital. Protonation of the N atom would disrupt the aromaticity, making this a less favorable reaction.

no *p* orbital on N

This electron pair is delocalized on the bicyclic ring system (giving it 10 π electrons), making it less available for donation, and thus less basic.

23.31 At its isoelectric point, each amino acid is neutral.

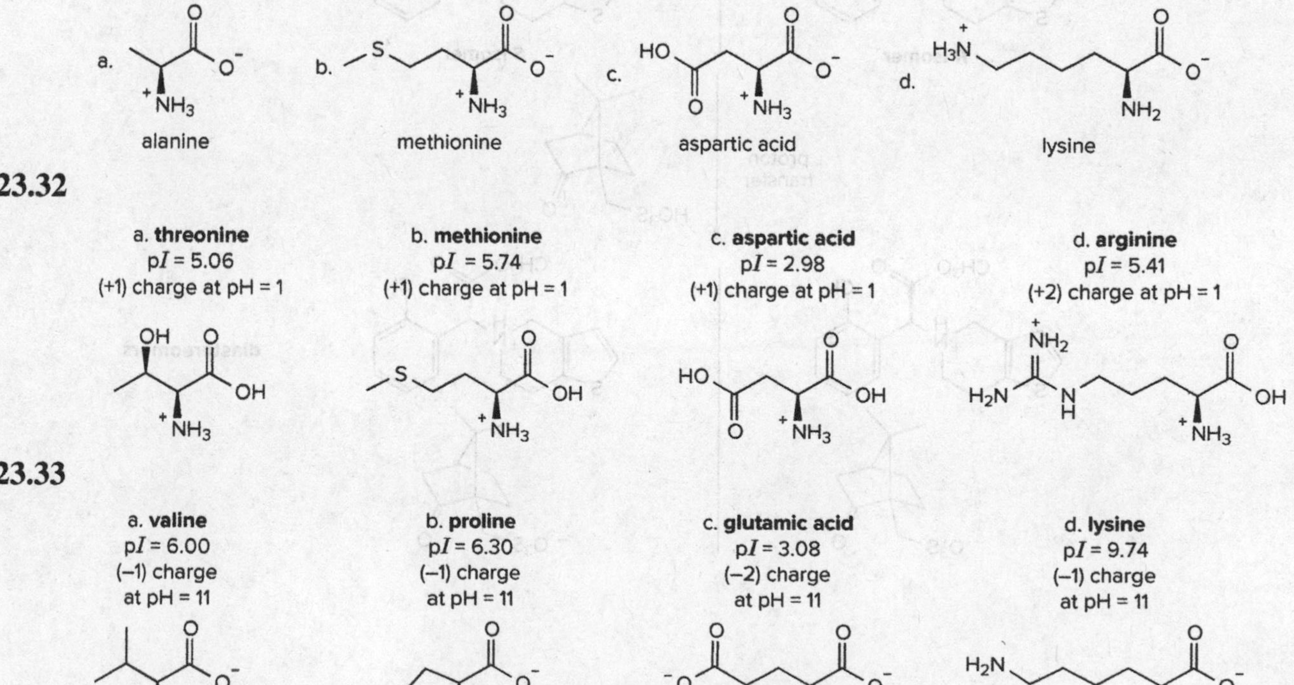

23.32

23.33

23.34 The terminal NH_2 and COOH groups are ionizable functional groups, so they can gain or lose protons in aqueous solution.

a. Ala → A–A–A (drawn with all uncharged atoms)

b. At pH = 1

c. The pK_a of the COOH of the tripeptide is higher than the pK_a of the COOH group of alanine, making it less acidic. This occurs because the COOH group in the tripeptide is farther away from the $-NH_3^+$ group. The positively charged $-NH_3^+$ group stabilizes the negatively charged carboxylate anion of alanine more than the carboxylate anion of the tripeptide because it is so much closer in alanine. The opposite effect is observed with the ionization of the $-NH_3^+$ group. In alanine, the $-NH_3^+$ is closer to the COO^- group, so it is more difficult to lose a proton, resulting in a higher pK_a. In the tripeptide, the $-NH_3^+$ is farther away from the COO^-, so it is less affected by its presence.

23.35

R isomer | *S* isomer — enantiomers

proton transfer

diastereomers

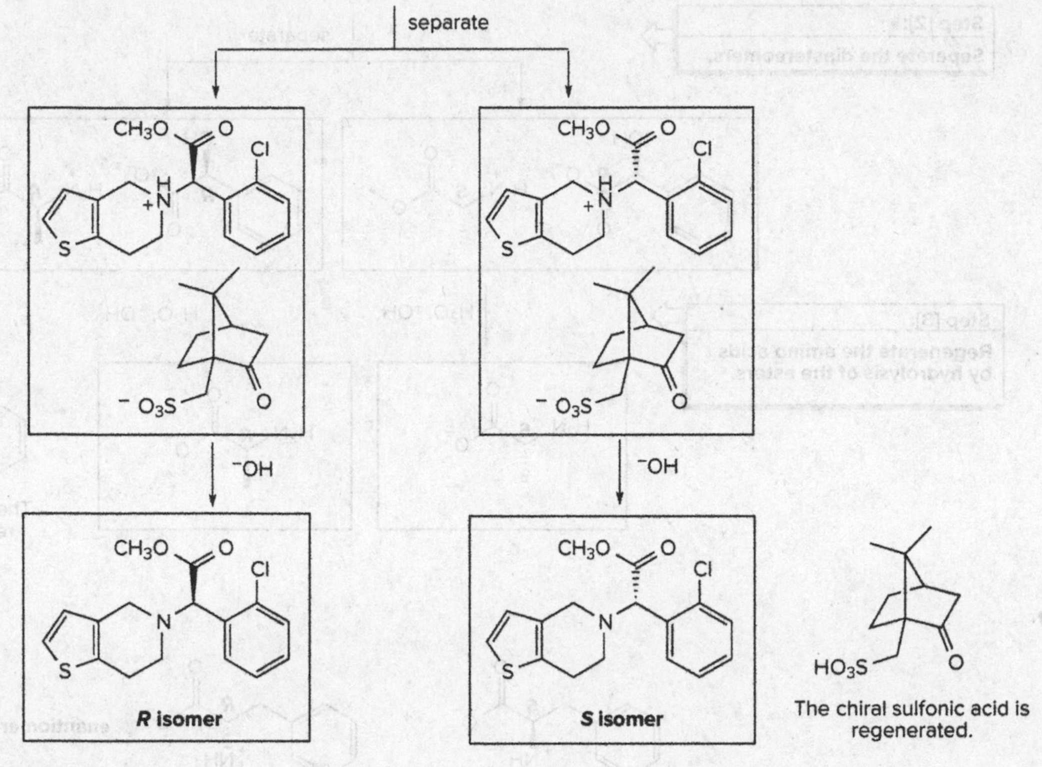

23.36

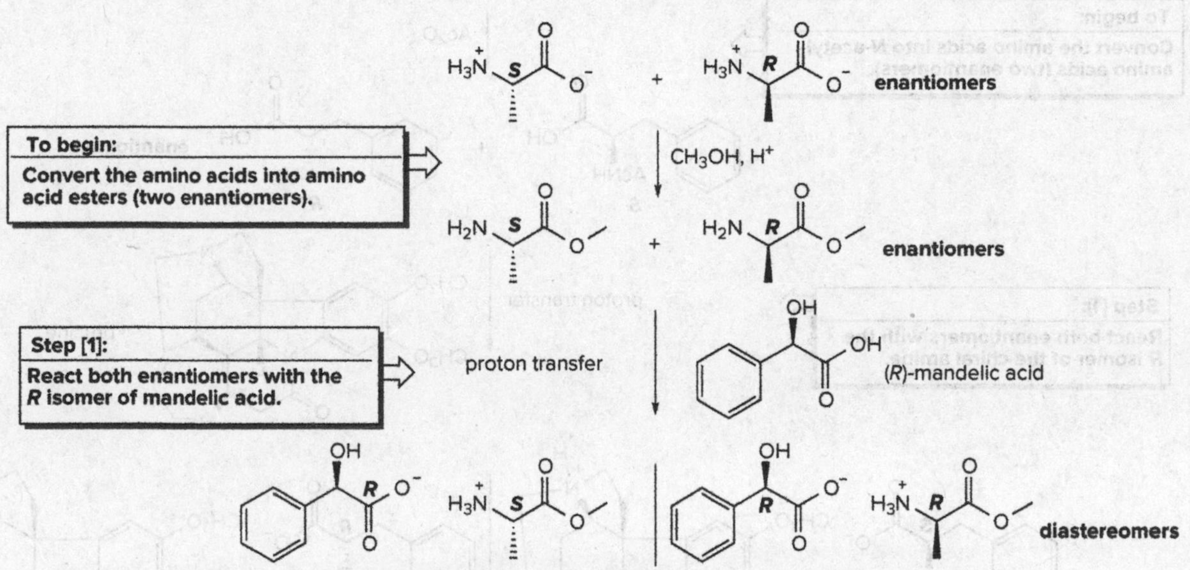

These salts have the *same* configuration around one stereogenic center, but the *opposite* configuration about the other stereogenic center.

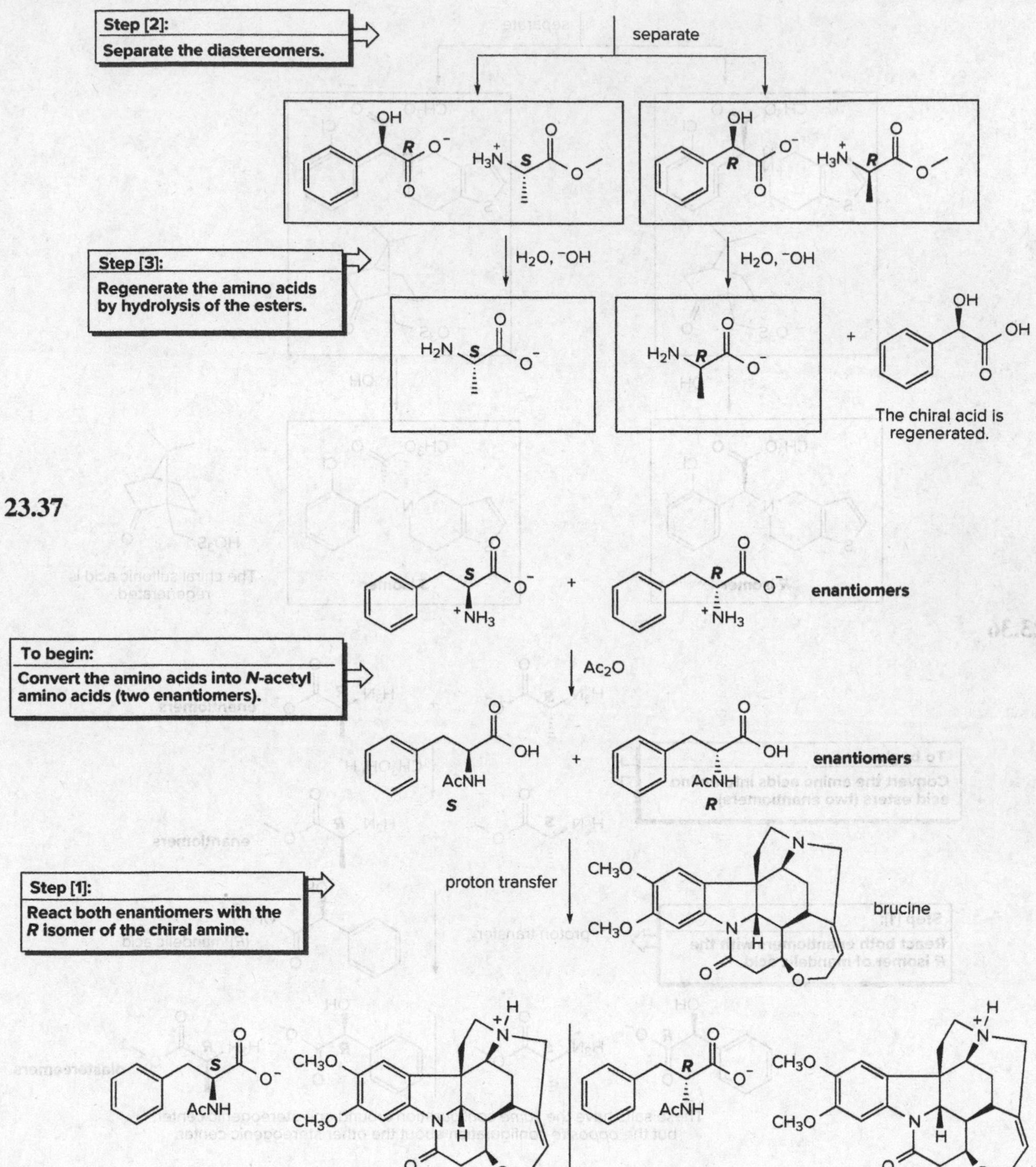
Step [2]:
Separate the diastereomers.
separate
OH
R
O⁻
H3N
S
O
H2O, ⁻OH
Step [3]:
Regenerate the amino acids by hydrolysis of the esters.
H2N
S
R
+
The chiral acid is regenerated.
23.37
NH3
enantiomers
Ac2O
To begin:
Convert the amino acids into N-acetyl amino acids (two enantiomers).
AcNH
proton transfer
Step [1]:
React both enantiomers with the R isomer of the chiral amine.
CH3O
N
H
brucine

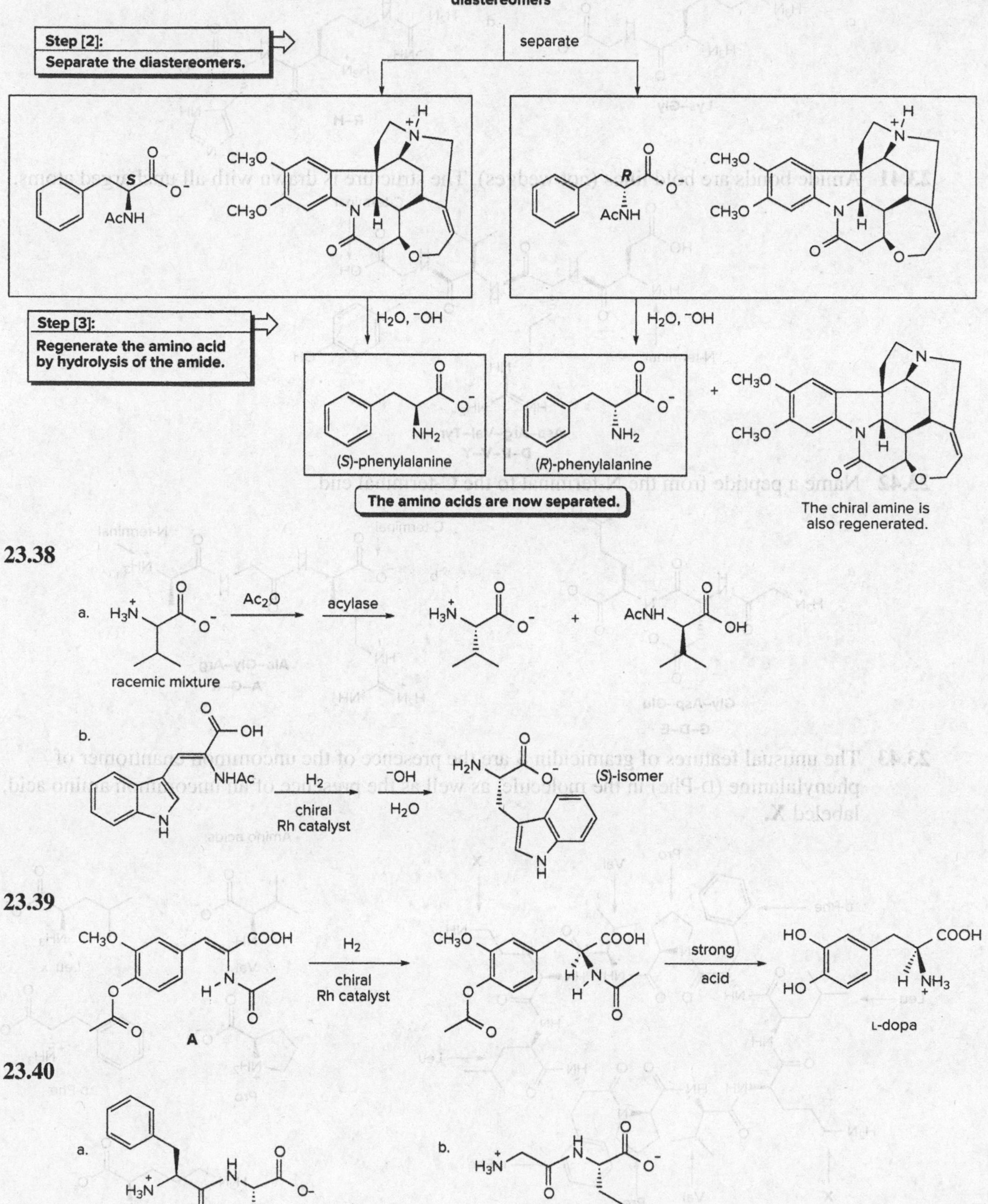
diastereomers
Step [2]:
Separate the diastereomers.
separate
S
AcNH
CH_3O
R
Step [3]:
Regenerate the amino acid by hydrolysis of the amide.
H_2O, ^-OH
(S)-phenylalanine
(R)-phenylalanine
The amino acids are now separated.
The chiral amine is also regenerated.
23.38
a.
racemic mixture
Ac_2O
acylase
AcNH
b.
NHAc
H_2
chiral
Rh catalyst
^-OH
H_2O
(S)-isomer
23.39
COOH
A
strong
acid
L-dopa
23.40
Phe–Ala
Gly–Gln

c.

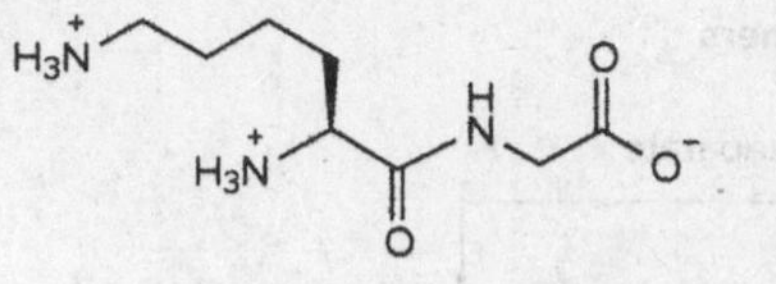

Lys–Gly

d.

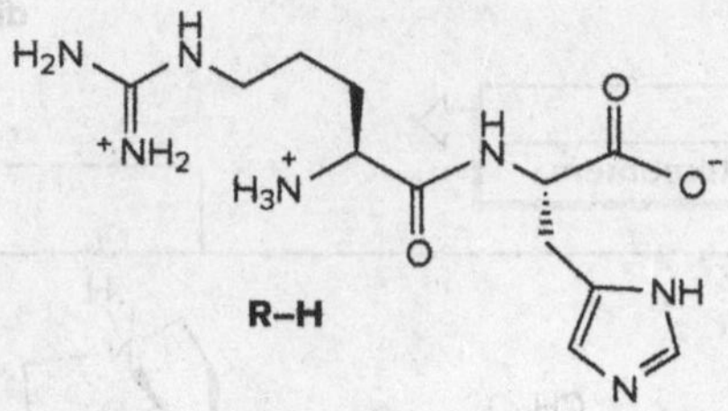

R–H

23.41 Amide bonds are bold lines (not wedges). The structure is drawn with all uncharged atoms.

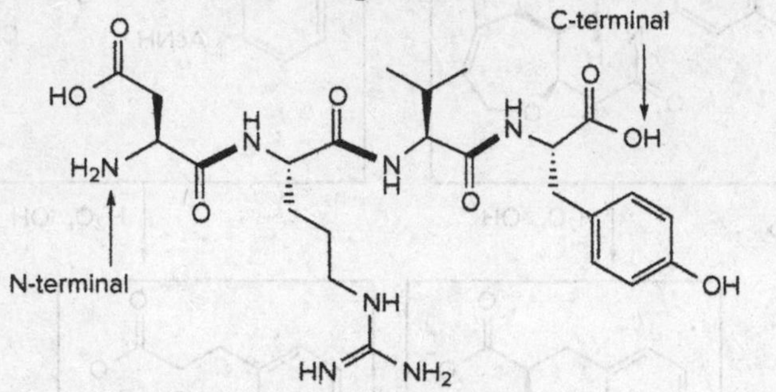

C-terminal

N-terminal

Asp–Arg–Val–Tyr

D–R–V–Y

23.42 Name a peptide from the N-terminal to the C-terminal end.

a.

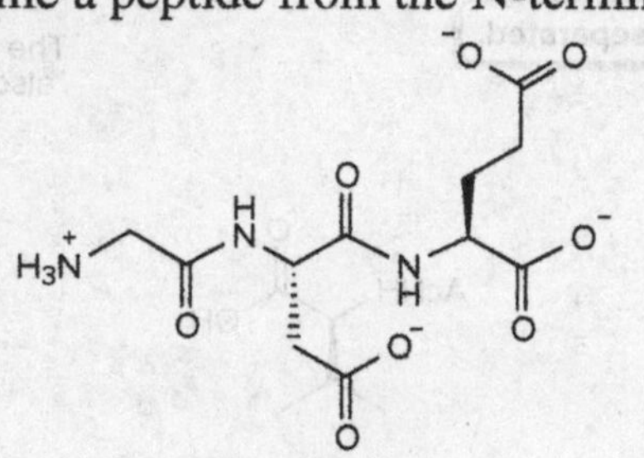

Gly–Asp–Glu

G–D–E

b.

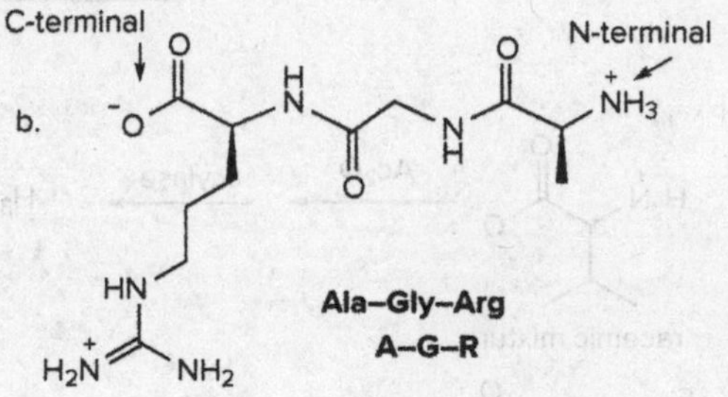

C-terminal

N-terminal

Ala–Gly–Arg

A–G–R

23.43 The unusual features of gramicidin S are the presence of the uncommon enantiomer of phenylalanine (D-Phe) in the molecule, as well as the presence of an uncommon amino acid, labeled **X.**

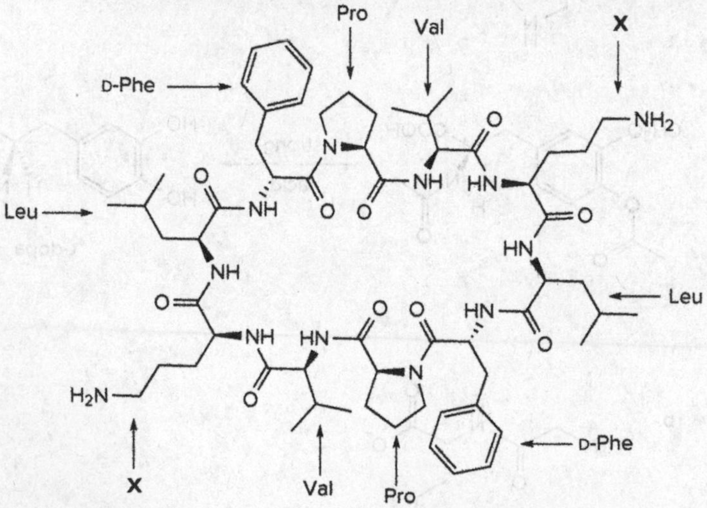

D-Phe, Pro, Val, X, Leu, Leu, X, Val, Pro, D-Phe

gramicidin S

Amino acids:

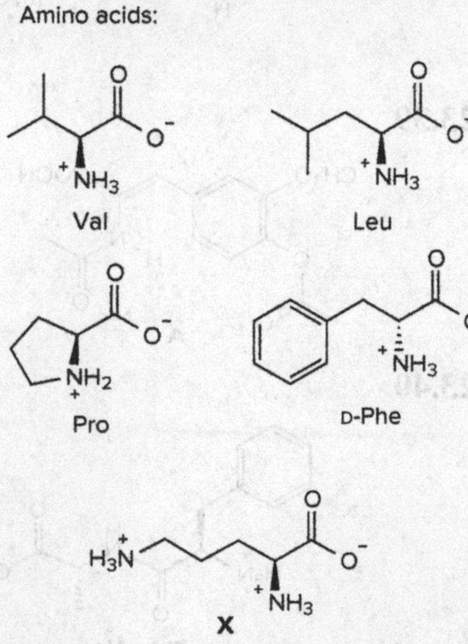

Val

Leu

Pro

D-Phe

X

23.44

Tyr–Gly–Gly–Phe–Leu–Arg–Arg–Ile–Arg–Pro–Lys–Leu–Lys
A

a. **A** $\xrightarrow{\text{chymotrypsin}}$ Tyr + Gly–Gly–Phe + Leu–Arg–Arg–Ile–Arg–Pro–Lys–Leu–Lys

b. **A** $\xrightarrow{\text{trypsin}}$ Tyr–Gly–Gly–Phe–Leu–Arg + Arg + Ile–Arg + Pro–Lys + Leu–Lys

c. **A** $\xrightarrow{\text{carboxypeptidase}}$ Tyr–Gly–Gly–Phe–Leu–Arg–Arg–Ile–Arg–Pro–Lys–Leu + Lys

d. **A** $\xrightarrow{C_6H_5NCS}$ [C_6H_5–N, NH, S, O, OH] + Gly–Gly–Phe–Leu–Arg–Arg–Ile–Arg–Pro–Lys–Leu–Lys

23.45

angiotensin I

Three-letter abbreviation: Asp–Arg–Val–Tyr–Ile–His–Pro–Phe–His–Leu

a. Trypsin cleavage products: Asp–Arg + Val–Tyr–Ile–His–Pro–Phe–His–Leu

b. Chymotrypsin cleavage products: Asp–Arg–Val–Tyr + Ile–His–Pro–Phe + His–Leu

c. ACE cleavage products: Asp–Arg–Val–Tyr–Ile–His–Pro–Phe (angiotensin II) + His–Leu

23.46

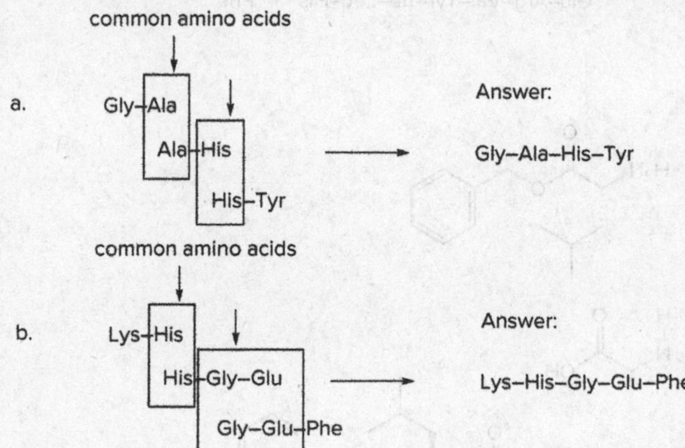

23.47 To sequence a peptide of this length, look for overlapping sequences of the fragments formed when the peptide is treated with enzymes. It is best to start with the cleavage products that have the largest number of amino acids. Then recall that chymotrypsin cleaves amide bonds with a carbonyl group from Phe, Tyr, or Trp whereas trypsin cleaves amide bonds with a carbonyl group from Arg or Lys. The sequence of amino acids in glucagon as well as the cleavage sites is shown.

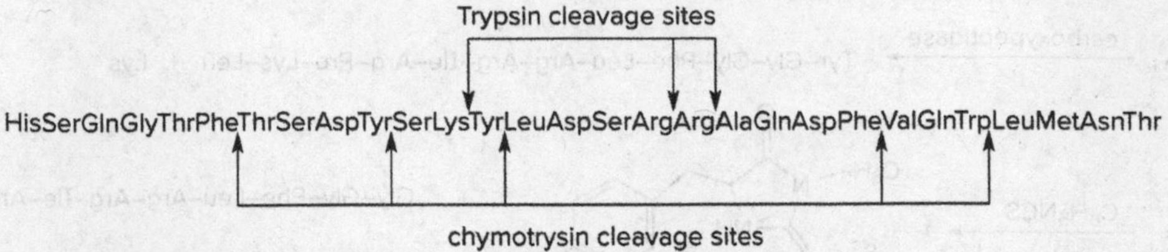

23.48 Gly is the N-terminal amino acid (from Edman degradation), and Leu is the C-terminal amino acid (from treatment with carboxypeptidase). Partial hydrolysis gives the rest of the sequence.

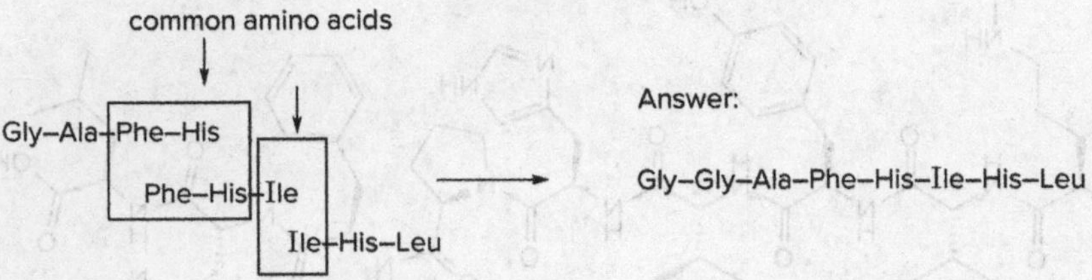

23.49 Edman degradation data give the N-terminal amino acid for the octapeptide and all smaller peptides.

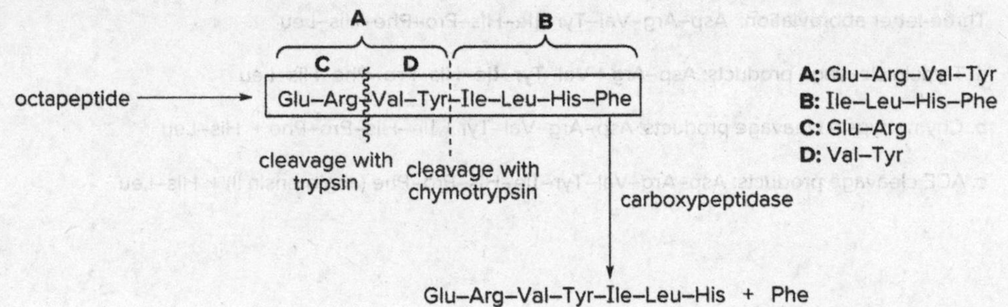

23.50

a. H_3N^+–CH(CH$_2$CH(CH$_3$)$_2$)–COO$^-$ $\xrightarrow[H^+]{Ph\text{–}CH_2\text{–}OH}$ H_2N–CH(CH$_2$CH(CH$_3$)$_2$)–CO–O–CH$_2$Ph

b. H_3N^+–CH$_2$–COO$^-$ $\xrightarrow[Et_3N]{(Boc)_2O}$ Boc–NH–CH$_2$–COOH

c. product in (a) + product in (b) $\xrightarrow{DCC}$ Boc–NH–CH$_2$–CO–NH–CH(CH$_2$CH(CH$_3$)$_2$)–CO–O–CH$_2$Ph

d. 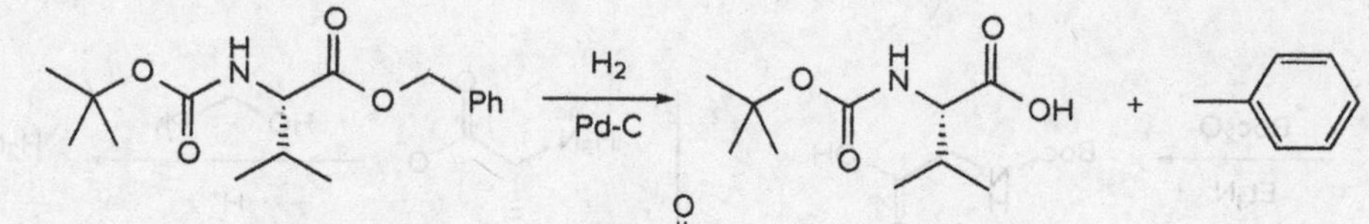 Boc-Val-OCH$_2$Ph $\xrightarrow[\text{Pd-C}]{H_2}$ Boc-Val-OH + toluene

e. 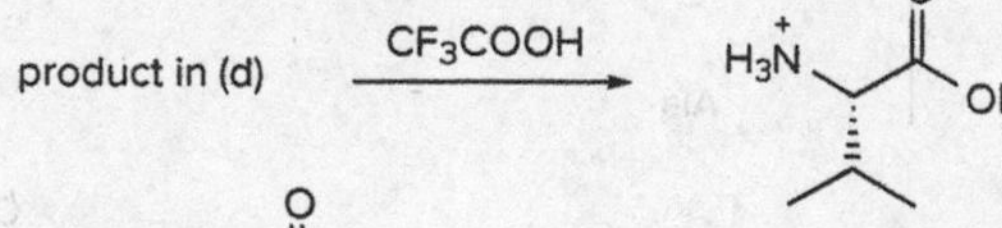 product in (d) $\xrightarrow{CF_3COOH}$ $H_3\overset{+}{N}$-Val-OH

f. 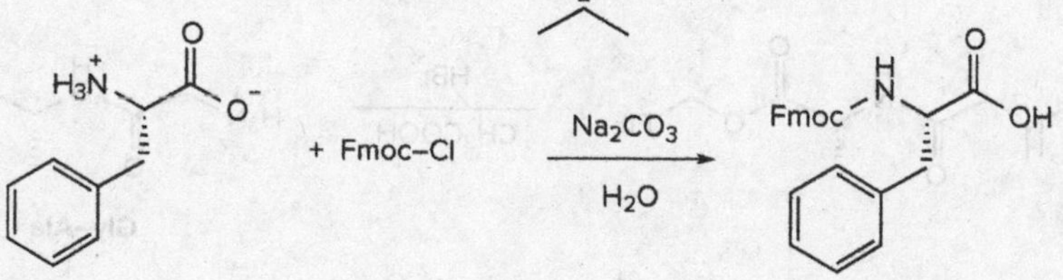 Phe ($H_3\overset{+}{N}$, O^-) + Fmoc–Cl $\xrightarrow[H_2O]{Na_2CO_3}$ Fmoc-NH-Phe-OH

23.51

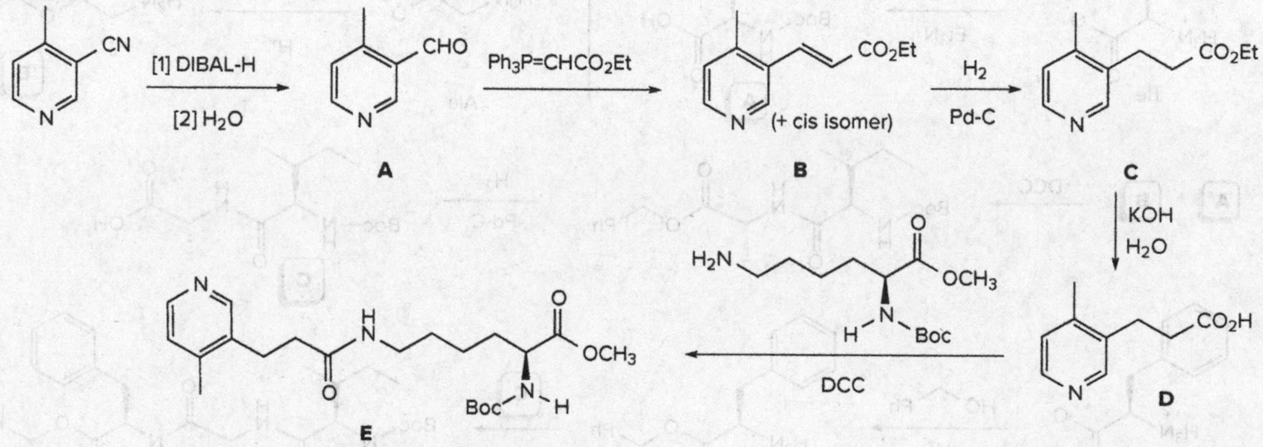

4-methyl-3-cyanopyridine (CN) $\xrightarrow[\text{[2] } H_2O]{\text{[1] DIBAL-H}}$ **A** (CHO) $\xrightarrow{Ph_3P{=}CHCO_2Et}$ **B** (CO_2Et, (+ cis isomer)) $\xrightarrow[\text{Pd-C}]{H_2}$ **C** (CO_2Et) $\xrightarrow[H_2O]{KOH}$ **D** (CO_2H)

D + H_2N-(CH$_2$)$_4$-CH(NH-Boc)-CO-OCH$_3$ $\xrightarrow{DCC}$ **E** (amide; OCH$_3$, Boc-N-H)

23.52

a. 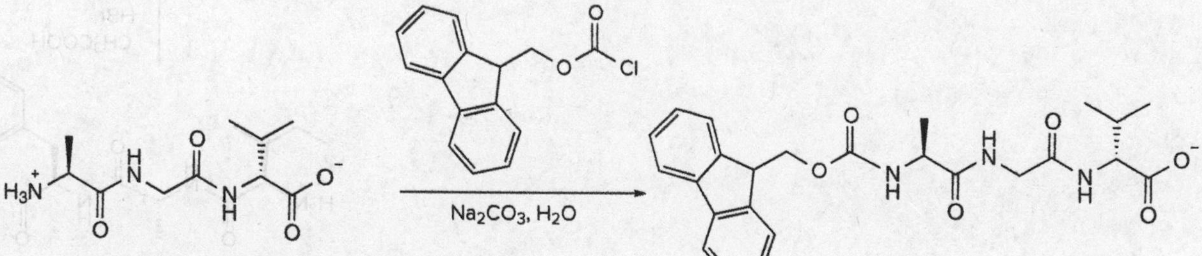 Ala-Gly-Val ($H_3\overset{+}{N}$ … O^-) + Fmoc-Cl $\xrightarrow{Na_2CO_3,\ H_2O}$ Fmoc-NH-Ala-Gly-Val-O^-

b. 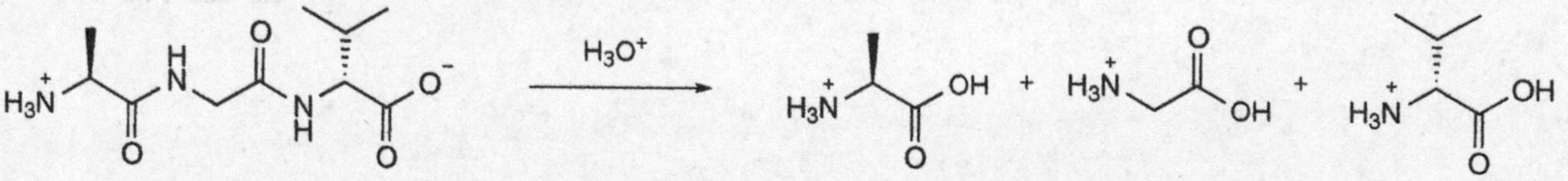 Ala-Gly-Val $\xrightarrow{H_3O^+}$ $H_3\overset{+}{N}$-Ala-OH + $H_3\overset{+}{N}$-Gly-OH + $H_3\overset{+}{N}$-Val-OH

c. 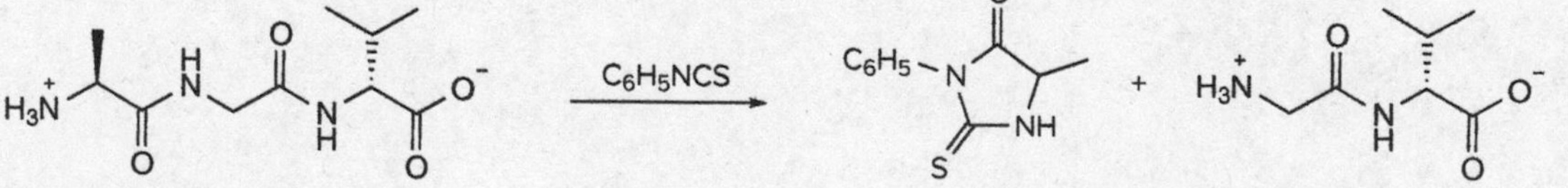 Ala-Gly-Val $\xrightarrow{C_6H_5NCS}$ C_6H_5–N (thiohydantoin, NH, S) + $H_3\overset{+}{N}$-Gly-Val-O^-

d. 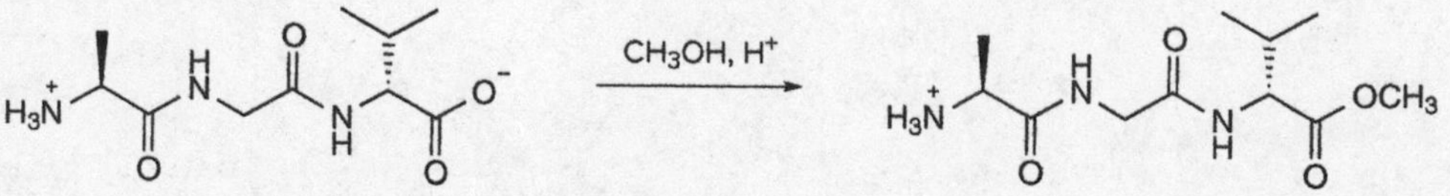 Ala-Gly-Val $\xrightarrow{CH_3OH,\ H^+}$ $H_3\overset{+}{N}$-Ala-Gly-Val-OCH$_3$

23.53

a. $H_3N^+CH_2CO_2^-$ (Gly) —[Boc_2O, Et_3N]→ Boc–NH–CH_2–COOH (**A**)

H_3N^+–CH(CH_3)–CO_2^- (Ala) —[HO–CH_2Ph, H^+]→ H_2N–CH(CH_3)–CO_2CH_2Ph (**B**)

A + **B** —[DCC]→ Boc–Gly–Ala–OCH_2Ph —[HBr, CH_3COOH]→ H_3N^+–Gly–Ala–O^-

Gly–Ala

b. Ile (H_3N^+, O^-) —[Boc_2O, Et_3N]→ Boc–Ile–OH (**A**)

Ala (H_3N^+, O^-) —[HO–CH_2Ph, H^+]→ H_2N–Ala–OCH_2Ph (**B**)

A + **B** —[DCC]→ Boc–Ile–Ala–OCH_2Ph —[H_2, Pd-C]→ Boc–Ile–Ala–OH (**C**)

Phe (H_3N^+, O^-) —[HO–CH_2Ph, H^+]→ H_2N–Phe–OCH_2Ph —[**C**, DCC]→ Boc–Ile–Ala–Phe–OCH_2Ph —[HBr, CH_3COOH]→ H_3N^+–Ile–Ala–Phe–O^-

Ile–Ala–Phe

23.54 Make all the Fmoc derivatives as described in Problem 23.19.

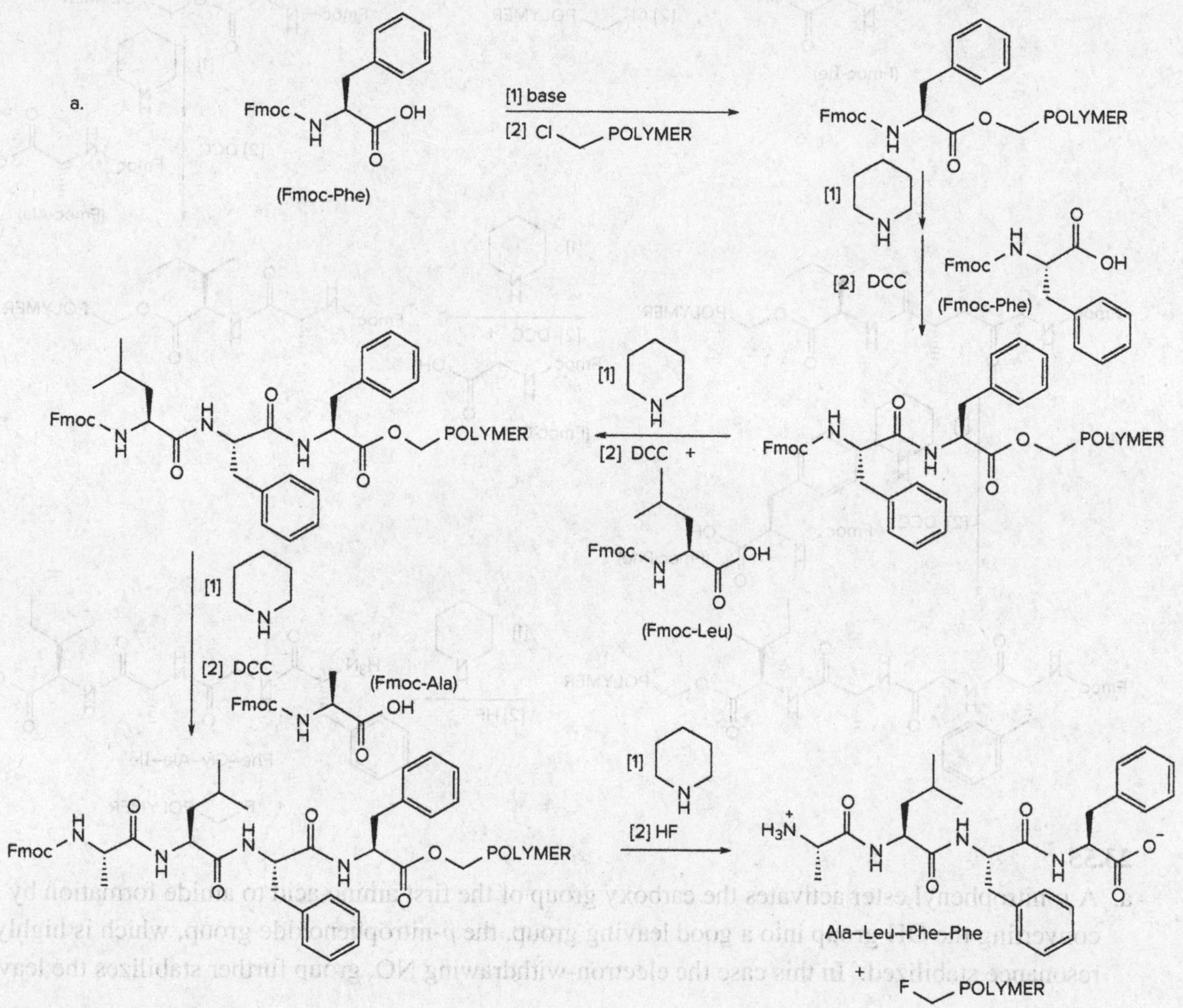

a.

(Fmoc-Phe)

[1] base
[2] Cl–CH2–POLYMER

[1] piperidine
[2] DCC (Fmoc-Phe)

[1] piperidine
[2] DCC + (Fmoc-Leu)

[1] piperidine
[2] DCC (Fmoc-Ala)

[1] piperidine
[2] HF

Ala–Leu–Phe–Phe

+ F–CH2–POLYMER

b.

Fmoc-Ile → [1] base, [2] Cl–CH₂–POLYMER → Fmoc-Ile-O-CH₂-POLYMER → [1] piperidine, [2] DCC + Fmoc-Ala → Fmoc-Ala-Ile-O-CH₂-POLYMER → [1] piperidine, [2] DCC + Fmoc-Gly → Fmoc-Gly-Ala-Ile-O-CH₂-POLYMER → [1] piperidine, [2] DCC + Fmoc-Phe → Fmoc-Phe-Gly-Ala-Ile-O-CH₂-POLYMER → [1] piperidine, [2] HF → Phe–Gly–Ala–Ile + F–CH₂–POLYMER

(Fmoc-Ile) (Fmoc-Ala) (Fmoc-Gly) (Fmoc-Phe)

23.55

a. A *p*-nitrophenyl ester activates the carboxy group of the first amino acid to amide formation by converting the OH group into a good leaving group, the *p*-nitrophenoxide group, which is highly resonance stabilized. In this case the electron-withdrawing NO_2 group further stabilizes the leaving group.

p-nitrophenoxide

The negative charge is delocalized on the O atom of the NO_2 group.

b. The *p*-methoxyphenyl ester contains an electron-donating OCH_3 group, making $CH_3OC_6H_4O^-$ a poorer leaving group than $NO_2C_6H_4O^-$, so this ester does *not* activate the amino acid to amide formation as much.

23.56

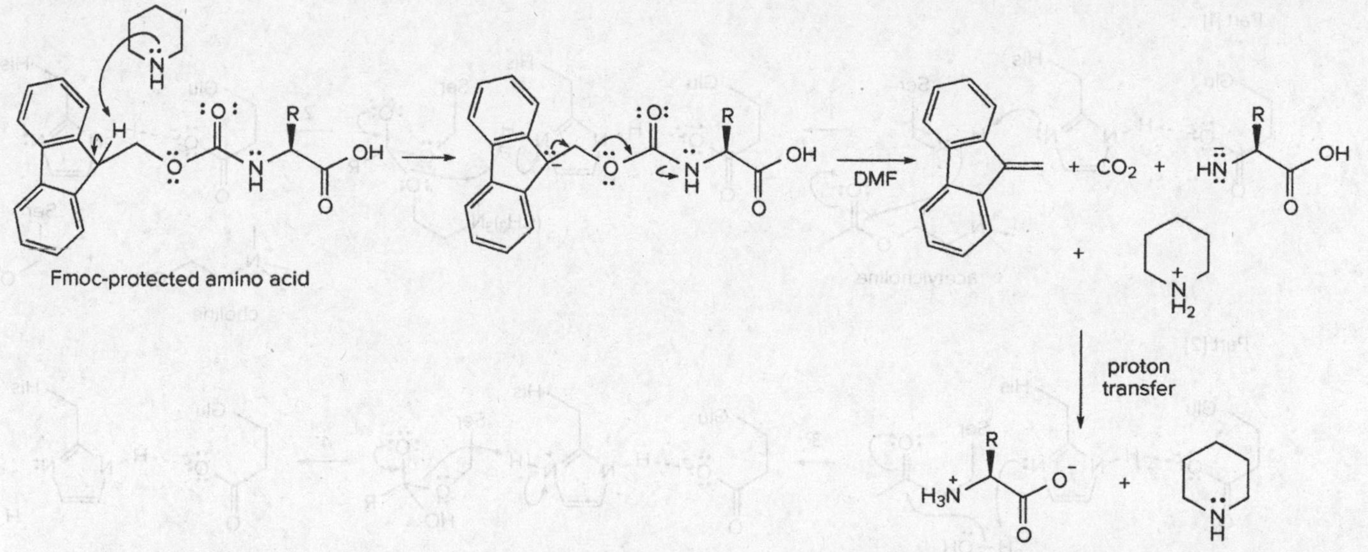

Fmoc-protected amino acid

DMF

$+ CO_2 +$

proton transfer

23.57 Amino acids commonly found in the interior of a globular protein have nonpolar or weakly polar side chains: isoleucine and phenylalanine. Amino acids commonly found on the surface have COOH, NH_2, and other groups that can hydrogen bond to water: aspartic acid, lysine, arginine, and glutamic acid.

23.58

a. VLLFGEDEK — VLLF: hydrocarbon side chains hydrophobic; EDEK: charged side chains hydrophilic

b. RKYSFLGAA — RKYS: OH groups or charged side chains hydrophilic; FLGAA: hydrocarbon side chains hydrophobic

23.59 The proline residues on collagen are hydroxylated to increase hydrogen-bonding interactions.

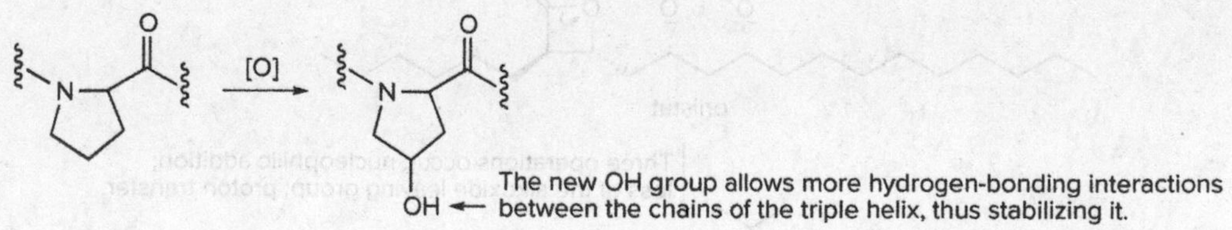

[O]

OH ← The new OH group allows more hydrogen-bonding interactions between the chains of the triple helix, thus stabilizing it.

23.60

a. protease, a type of hydrolase

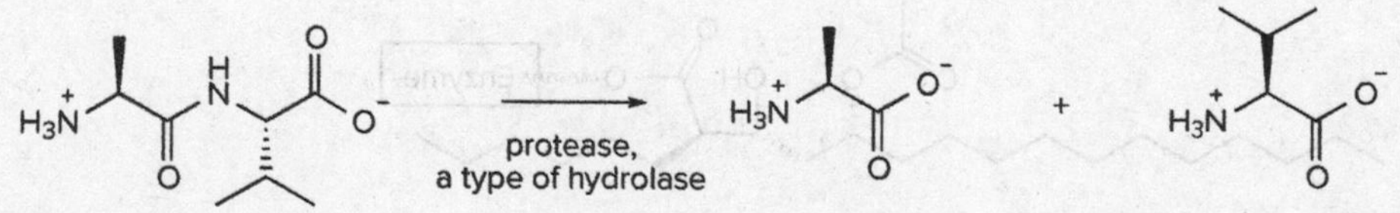

b. isomerase

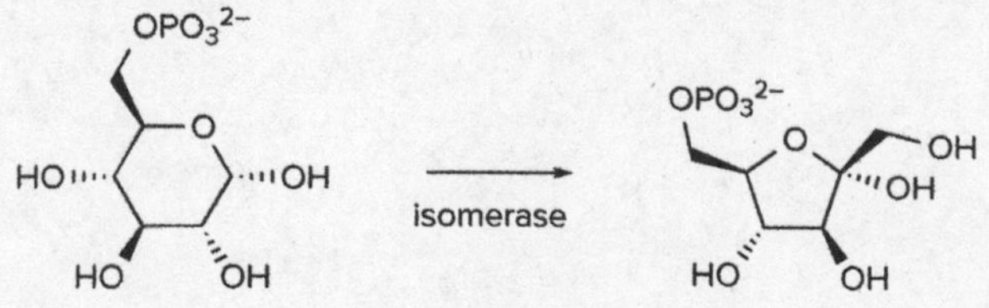

23.61

Part [1]

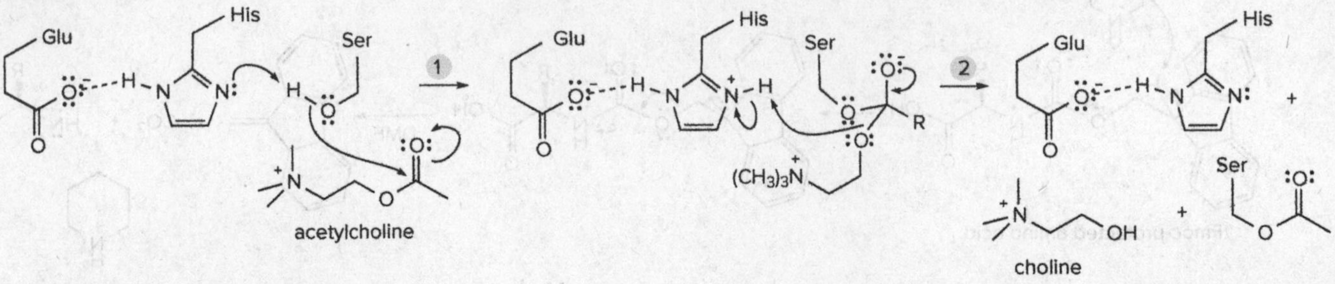

Glu His Ser 1 Glu His Ser R 2 Glu His + Ser + choline

acetylcholine

$(CH_3)_3N$

Part [2]

Glu His Ser 3 Glu His Ser R HO 4 Glu His Ser

HB^+ + acetate 5 :B

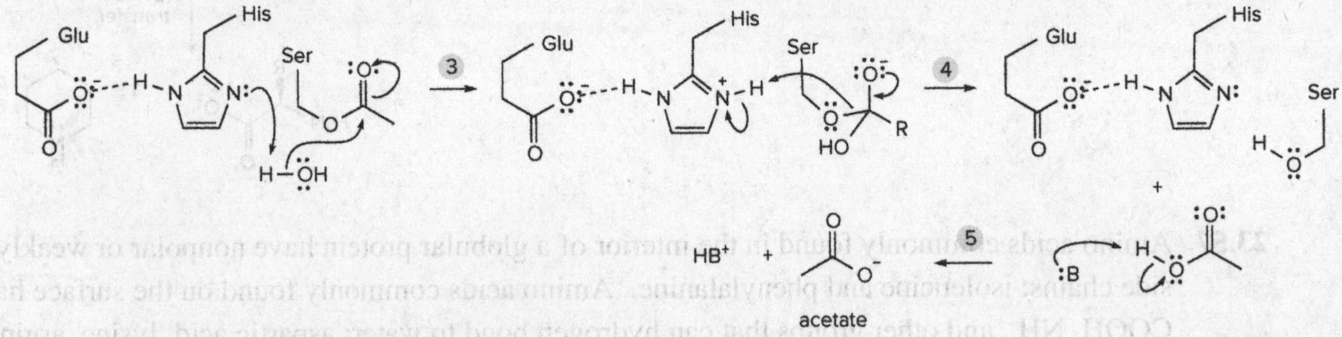

23.62 This reaction is similar to the reaction of penicillin with the glycopeptide transpeptidase enzyme discussed in Section 16.13. Serine has a nucleophilic OH, which can open the strained β-lactone to form a covalently bound, inactive enzyme.

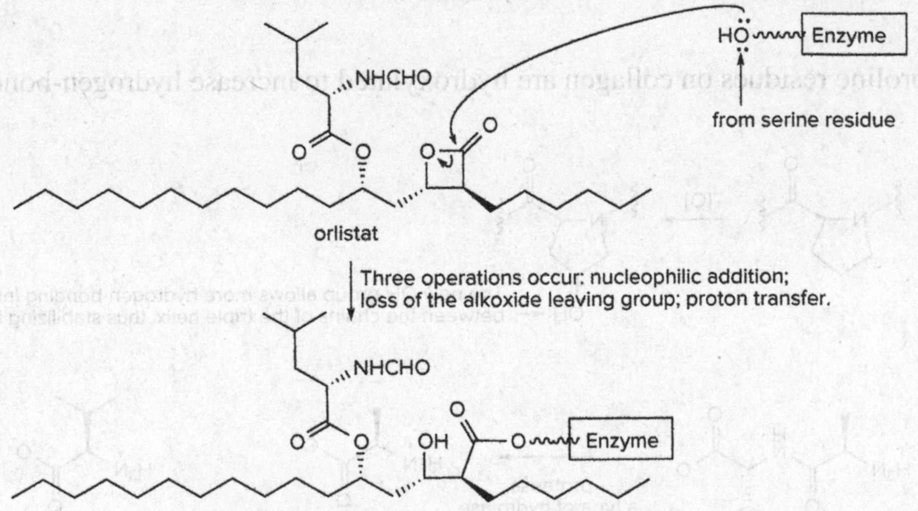

HO Enzyme

from serine residue

NHCHO

orlistat

Three operations occur: nucleophilic addition; loss of the alkoxide leaving group; proton transfer.

NHCHO

OH O Enzyme

23.63

thiazolinone

proton transfer

proton transfer

$H_2\ddot{O}$ +

+ H_3O^+

N-phenylthiohydantoin

Chapter 24 Carbohydrates

Chapter Review

Important terms

- **Aldose** A monosaccharide containing an aldehyde (24.2)
- **Ketose** A monosaccharide containing a ketone (24.2)
- **D-Sugar** A monosaccharide with the O bonded to the stereogenic center farthest from the carbonyl group drawn on the right in the Fischer projection (24.2C)
- **Epimers** Two diastereomers that differ in configuration around one stereogenic center only (24.3)
- **Anomers** Monosaccharides that differ in configuration at only the hemiacetal OH group (24.6)
- **Glycoside** An acetal derived from a monosaccharide hemiacetal (24.7)

Acyclic, Haworth, and 3-D representations for D-glucose (24.6)

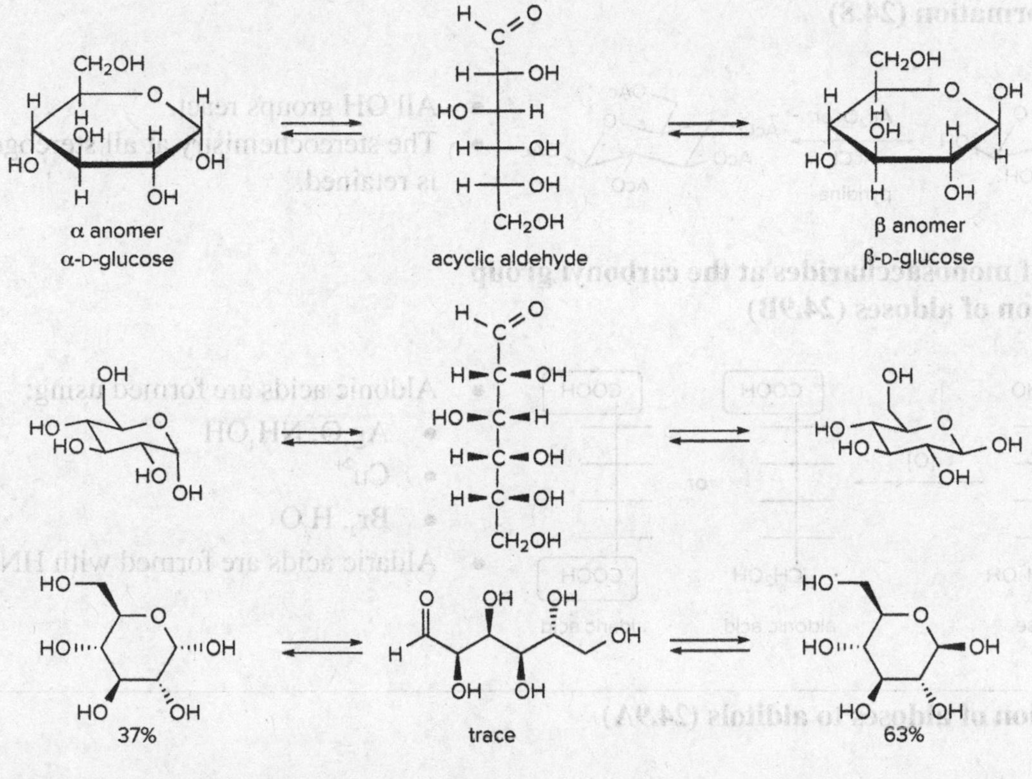

α anomer
α-D-glucose

acyclic aldehyde

β anomer
β-D-glucose

37%

trace

63%

Reactions of monosaccharides involving the hemiacetal

[1] Glycoside formation (24.7A)

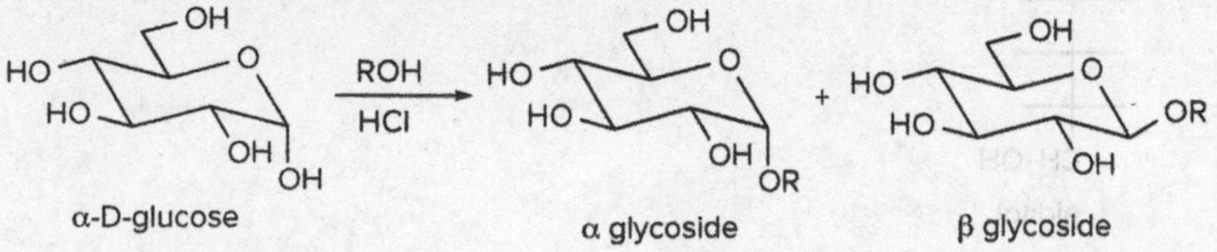

α-D-glucose —ROH / HCl→ α glycoside + β glycoside

- Only the hemiacetal OH reacts.
- A mixture of α and β glycosides forms.

[2] Glycoside hydrolysis (24.7B)

H_3O^+

α anomer + β anomer + ROH

- A mixture of α and β anomers forms.

Reactions of monosaccharides at the OH groups

[1] Ether formation (24.8)

Ag_2O, RX

- All OH groups react.
- The stereochemistry at all stereogenic centers is retained.

[2] Ester formation (24.8)

Ac_2O or AcCl, pyridine

- All OH groups react.
- The stereochemistry at all stereogenic centers is retained.

Reactions of monosaccharides at the carbonyl group

[1] Oxidation of aldoses (24.9B)

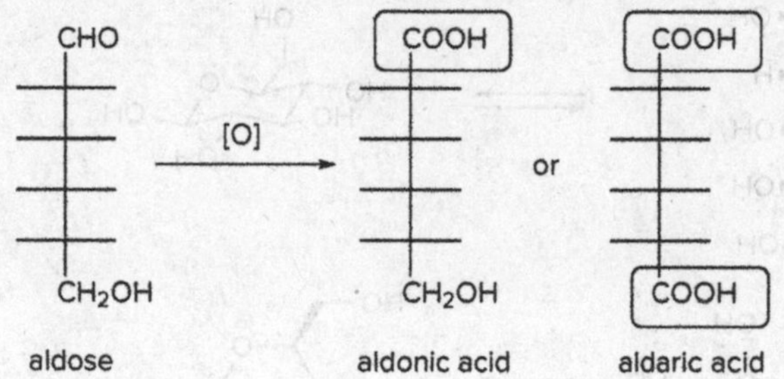

- Aldonic acids are formed using:
  - Ag_2O, NH_4OH
  - Cu^{2+}
  - Br_2, H_2O
- Aldaric acids are formed with HNO_3, H_2O.

[2] Reduction of aldoses to alditols (24.9A)

CHO → ($NaBH_4$, CH_3OH) → CH_2OH

aldose → alditol

[3] Wohl degradation (24.10A)

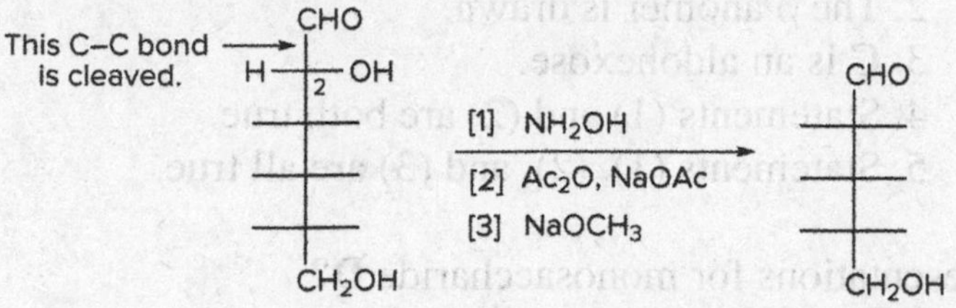

This C–C bond is cleaved.

[1] NH_2OH
[2] Ac_2O, NaOAc
[3] $NaOCH_3$

- The C1–C2 bond is cleaved to shorten an aldose chain by one carbon.
- The stereochemistry at all other stereogenic centers is retained.
- Two epimers at C2 form the same product.

[4] Kiliani–Fischer synthesis (24.10B)

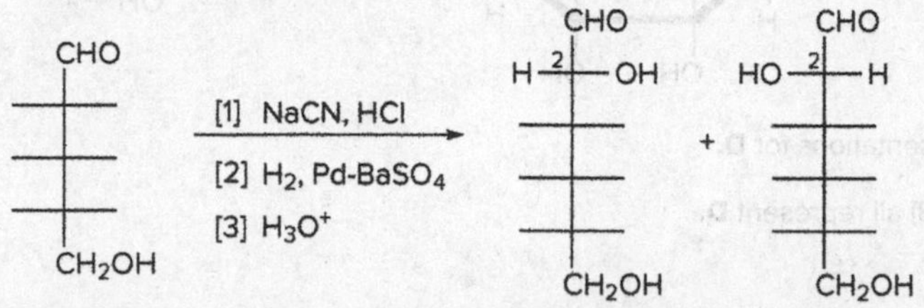

[1] NaCN, HCl
[2] H_2, Pd-$BaSO_4$
[3] H_3O^+

- One carbon is added to the aldehyde end of an aldose.
- Two epimers at C2 are formed.

Other reactions

[1] Hydrolysis of disaccharides (24.11)

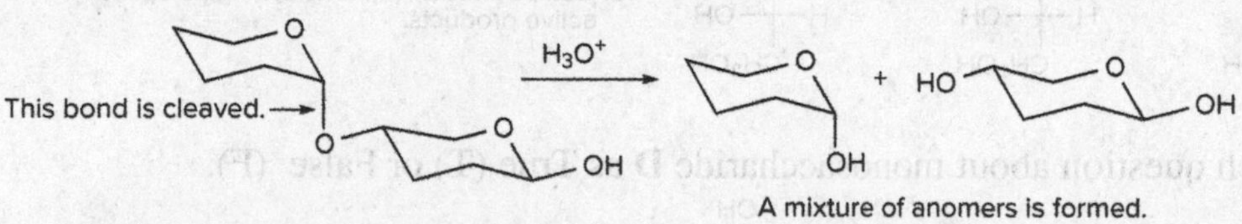

This bond is cleaved.

H_3O^+

A mixture of anomers is formed.

[2] Formation of *N*-glycosides (24.13B)

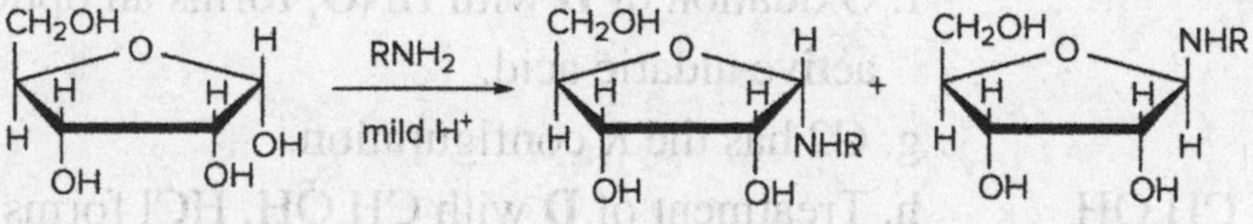

RNH_2
mild H^+

- Two anomers are formed.

Practice Test on Chapter Review

1.a. How are the following two representations related to each other?

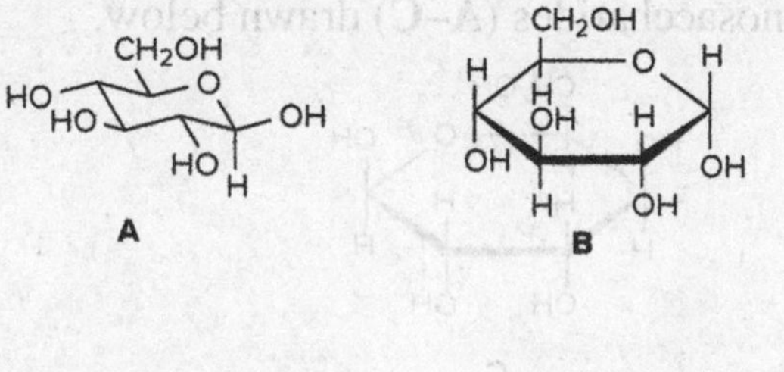

A B

1. **A** and **B** are anomers of each other.
2. **A** and **B** are epimers of each other.
3. **A** and **B** are diastereomers of each other.
4. Statements (1) and (2) are both true.
5. Statements (1), (2), and (3) are all true.

b. Which of the following statements is (are) true about monosaccharide **C**?

1. **C** is a D-sugar.
2. The β anomer is drawn.
3. **C** is an aldohexose.
4. Statements (1) and (2) are both true.
5. Statements (1), (2), and (3) are all true.

c. Which of the following are different representations for monosaccharide **D**?

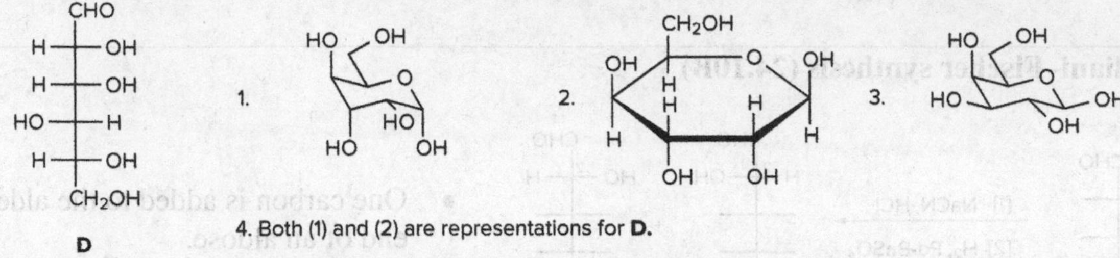

4. Both (1) and (2) are representations for **D**.
5. Compounds (1), (2), and (3) all represent **D**.

d. Which aldoses give an optically active compound upon reaction with $NaBH_4$ in CH_3OH?

4. Both (1) and (2) give an optically active product.
5. Compounds (1), (2), and (3) all give optically active products.

2. Answer each question about monosaccharide **D** as True (T) or False (F).

a. **D** is a D-sugar.

b. **D** is drawn as an α anomer.

c. **D** is an aldohexose.

d. Reduction of **D** with $NaBH_4$ in CH_3OH forms an optically inactive alditol.

e. Oxidation of **D** with Br_2, H_2O forms an optically active aldonic acid.

f. Oxidation of **D** with HNO_3 forms an optically active aldaric acid.

g. C2 has the *R* configuration.

h. Treatment of **D** with CH_3OH, HCl forms two products.

i. Treatment of **D** with Ag_2O, and CH_3I (excess) forms two products.

j. An epimer of **D** at C3 has an axial OH group.

3. Answer the following questions about the three monosaccharides (**A–C**) drawn below.

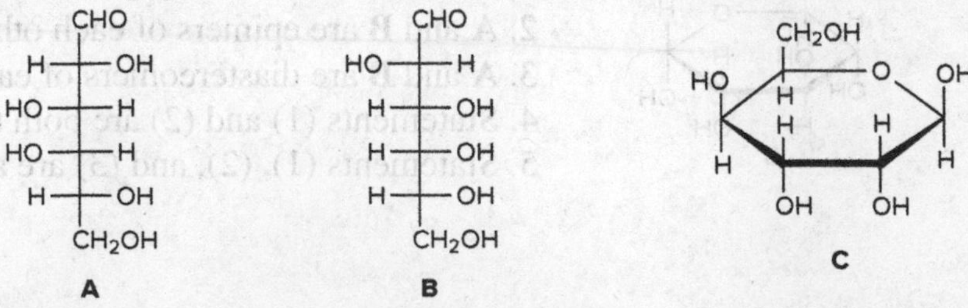

a. Draw the α anomer of **A** in a Haworth projection.
b. Draw the β anomer of **B** in a three-dimensional representation using a chair conformation.
c. Convert **C** into the acyclic form of the monosaccharide using a Fischer projection.
d. Which two aldoses yield **A** in a Wohl degradation?

4. Draw the product of each reaction with the starting material D-xylose.

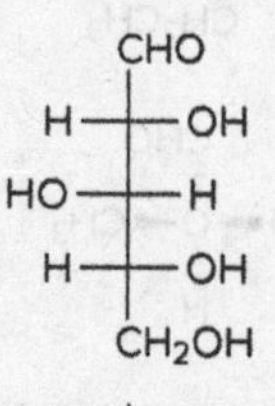

a. CH_3OH, HCl
b. $NaBH_4$, CH_3OH
c. Br_2, H_2O
d. [1] NaCN, HCl; [2] H_2, Pd-$BaSO_4$; [3] H_3O^+
e. Ac_2O, pyridine

Answers to Practice Test

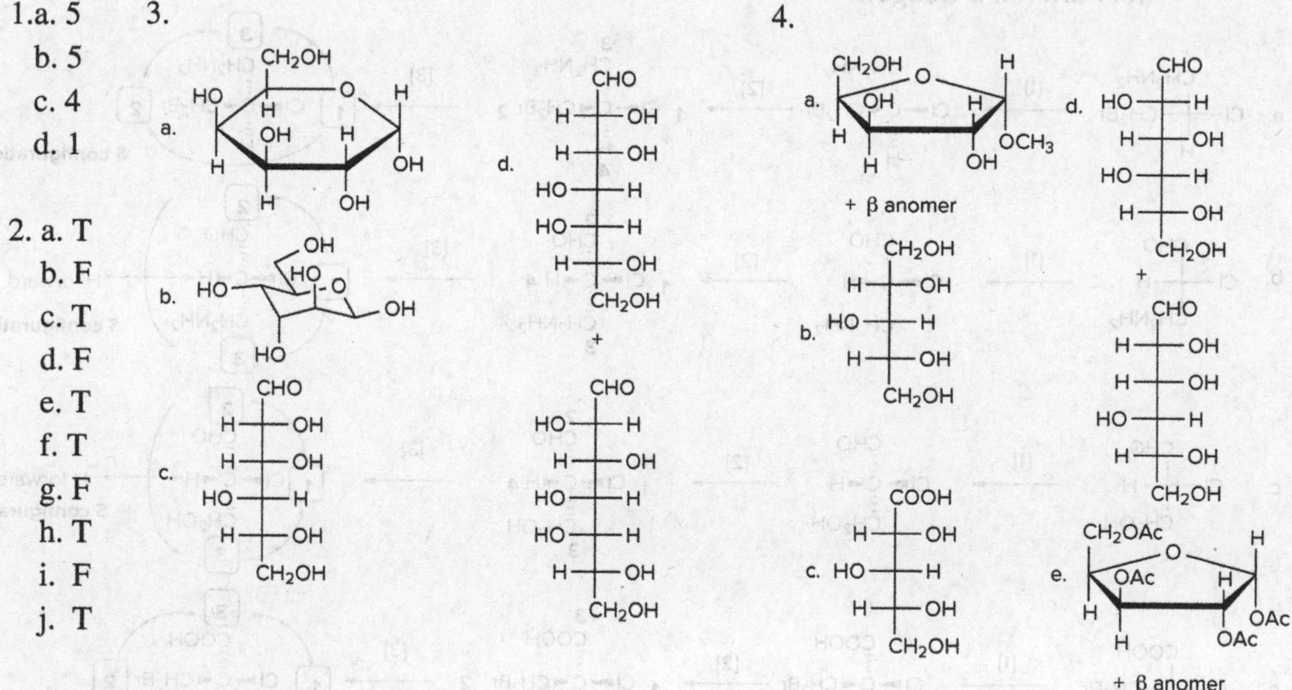

1.a. 5
b. 5
c. 4
d. 1

2. a. T
b. F
c. T
d. F
e. T
f. T
g. F
h. T
i. F
j. T

Answers to Problems

24.1 A *ketose* is a monosaccharide containing a ketone. An *aldose* is a monosaccharide containing an aldehyde. A monosaccharide is called: a *triose* if it has three C's, a *tetrose* if it has four C's, a *pentose* if it has five C's, a *hexose* if it has six C's, and so forth.

a. a ketotetrose

CH_2OH
C=O
H–C–OH
CH_2OH

b. an aldopentose

CHO
H–C–OH
H–C–OH
H–C–OH
CH_2OH

c. an aldotetrose

CHO
H–C–OH
H–C–OH
CH_2OH

24.2 Rotate and re-draw each molecule to place the horizontal bonds in front of the plane and the vertical bonds behind the plane. Then use a cross to represent the stereogenic center in a Fischer projection formula.

a. CH_3–C(–OH)(C(=O)OH)(CH_2CH_2OH) = Fischer projection: HO–C=O top, CH_3 left, OH right, CH_2CH_2OH bottom

c. re-draw: CHO top, H left, CH_2OH right, CH_2CH_3 bottom = H–|–CH_2OH (CHO top, CH_2CH_3 bottom)

b. re-draw: CHO top, HO left, CH_3 right, H bottom = HO–|–CH_3 (CHO top, H bottom)

d. re-draw: CHO top, HO left, CH_3 right, H bottom = HO–|–CH_3 (CHO top, H bottom)

24.3 For each molecule:

[1] Convert the Fischer projection formula to a representation with wedges and dashes.

[2] Assign priorities (Section 5.6).

[3] Determine *R* or *S* in the usual manner. Reverse the answer if priority group [4] is oriented forward (on a wedge).

a. CH_2NH_2 (3), Cl (1), CH_2Br (2), H (4) — **S configuration**

b. CHO (2), Cl (1), H (4), CH_2NH_2 (3) — H forward — **S configuration**

c. CHO (2), Cl (1), H (4), CH_2OH (3) — H forward — **S configuration**

d. COOH (3), Cl (1), CH_2Br (2), H (4) — **S configuration**

24.4

rotate → Convert staggered to eclipsed. → re-draw

CHO
HO–C–H
HO–C–H
HO–C–H
H–C–OH
CH_2OH

=

CHO
HO–|–H
HO–|–H
HO–|–H
H–|–OH
CH_2OH

24.5

CHO — R
H–C–OH — S
HO–C–H — R
H–C–OH — R
H–C–OH
CH_2OH

D-glucose

24.6 A D sugar has the OH group on the stereogenic center farthest from the carbonyl on the right. An L sugar has the OH group on the stereogenic center farthest from the carbonyl on the left.

a.

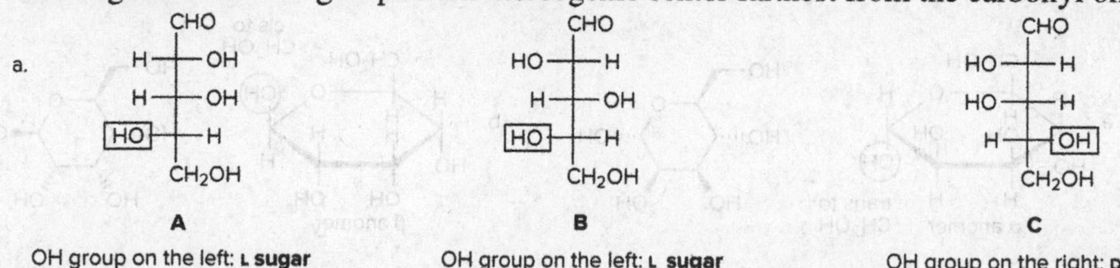

OH group on the left: **L sugar** | OH group on the left: **L sugar** | OH group on the right: **D sugar**

b. **A** and **B** are diastereomers.
A and **C** are enantiomers.
B and **C** are diastereomers.

24.7 There are 32 aldoheptoses; 16 are D sugars.

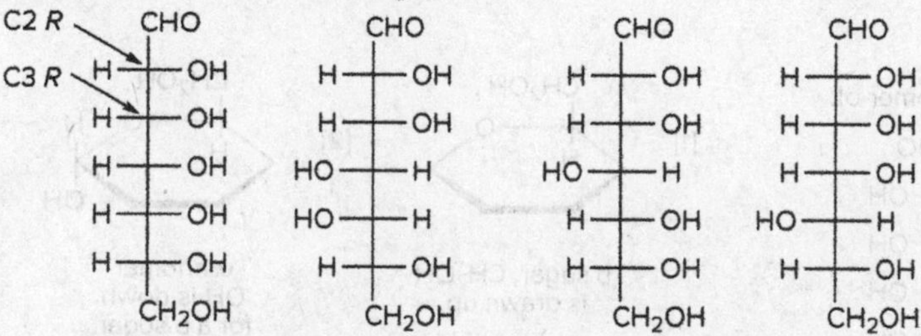

24.8 *Epimers* are two diastereomers that differ in the configuration around only one stereogenic center.

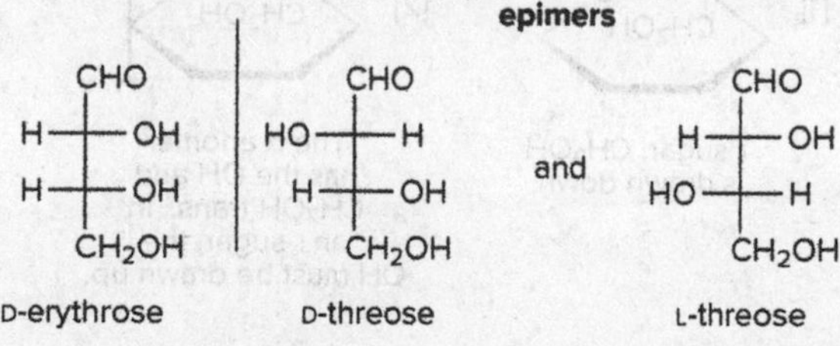

D-erythrose D-threose L-threose

24.9 a. D-allose and L-allose: **enantiomers**

b. D-altrose and D-gulose: **diastereomers** but not epimers

c. D-galactose and D-talose: **epimers**

d. D-mannose and D-fructose: **constitutional isomers**

e. D-fructose and D-sorbose: **diastereomers** but not epimers

f. L-sorbose and L-tagatose: **epimers**

24.10

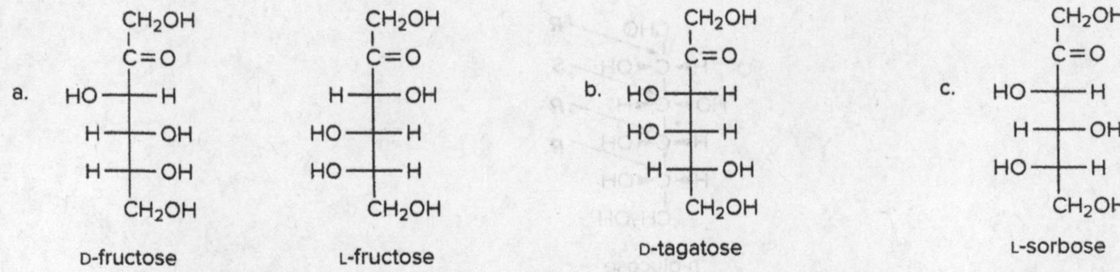
a.
b.
c.
D-fructose
L-fructose
D-tagatose
L-sorbose

enantiomers

24.11

a.

α anomer

trans to CH_2OH

b.

cis to CH_2OH

β anomer

24.12 Step [1]: Place the O atom in the upper right corner of a hexagon, and add the CH_2OH group on the first carbon counterclockwise from the O atom.

Step [2]: Place the anomeric carbon on the first carbon clockwise from the O atom.

Step [3]: Add the substituents at the three remaining stereogenic centers, clockwise around the ring.

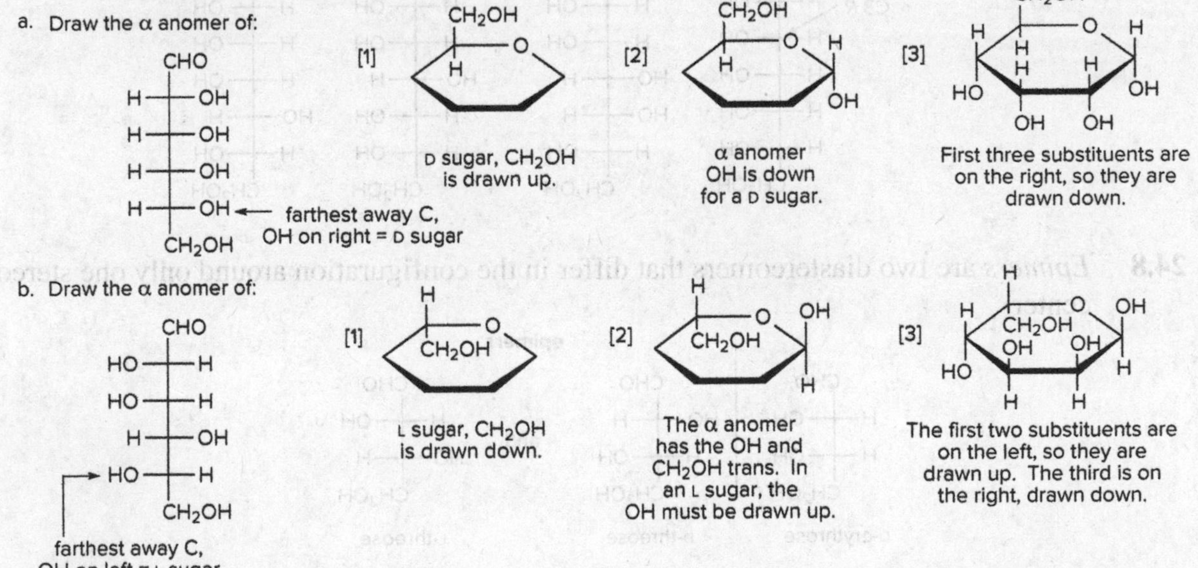
a. Draw the α anomer of:
farthest away C, OH on right = D sugar
D sugar, CH2OH is drawn up.
α anomer OH is down for a D sugar.
First three substituents are on the right, so they are drawn down.
b. Draw the α anomer of:
farthest away C, OH on left = L sugar
L sugar, CH2OH is drawn down.
The α anomer has the OH and CH2OH trans. In an L sugar, the OH must be drawn up.
The first two substituents are on the left, so they are drawn up. The third is on the right, drawn down.

c. Draw the β anomer of:

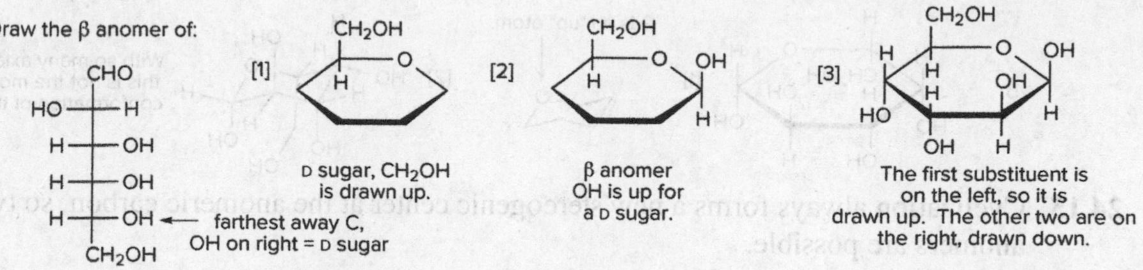

CHO; HO—H; H—OH; H—OH; H—OH ← farthest away C, OH on right = D sugar; CH₂OH

[1] D sugar, CH_2OH is drawn up.

[2] β anomer OH is up for a D sugar.

[3] The first substituent is on the left, so it is drawn up. The other two are on the right, drawn down.

24.13 To convert each Haworth projection into its acyclic form:

[1] Draw the C skeleton with the CHO on the top and the CH_2OH on the bottom.

[2] Draw in the OH group farthest from the C=O.

A CH_2OH group drawn up means a D sugar; a CH_2OH group drawn down means an L sugar.

[3] Add the three other stereogenic centers.

"Up" groups go on the left, whereas "down" groups go on the right.

a.

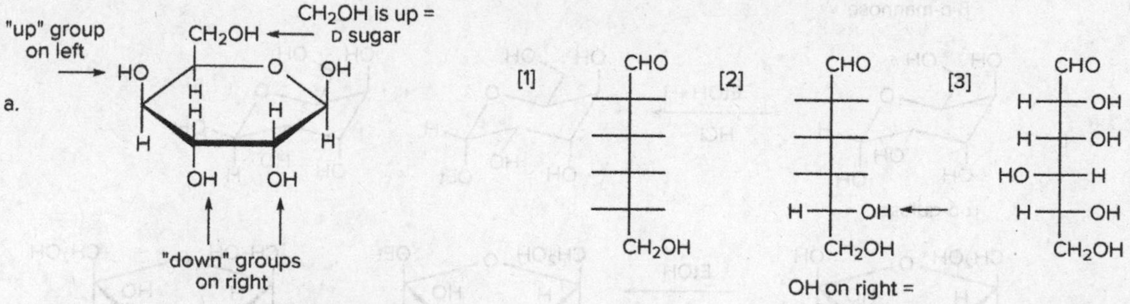

"up" group on left

CH_2OH is up = D sugar

"down" groups on right

[1] [2] [3]

OH on right = D sugar

CHO; H—OH; H—OH; HO—H; H—OH; CH_2OH

b.

CH_2OH is down = L sugar

"down" groups on right

"up" group on left

[1] [2] [3]

OH on left = L sugar

CHO; HO—H; H—OH; H—OH; HO—H; CH_2OH

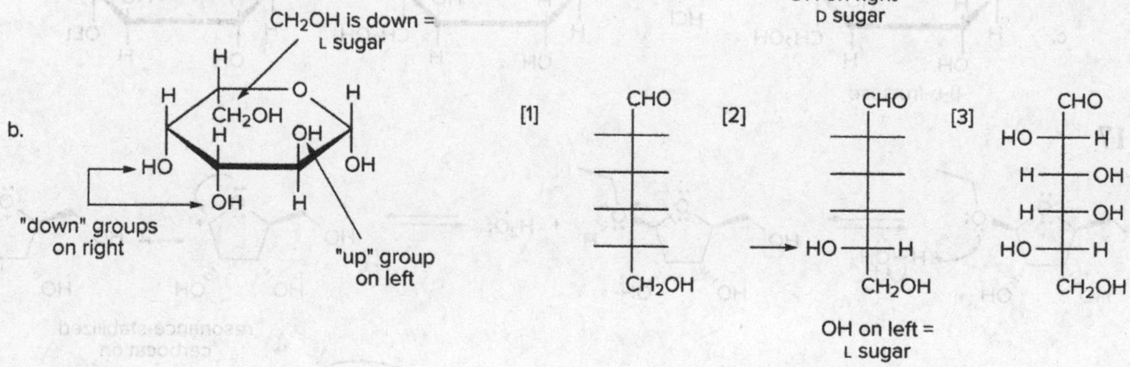

24.14 To convert a Haworth projection into a 3-D representation with a chair cyclohexane:

[1] Draw the pyranose ring as a chair with the O as an "up" atom.

[2] Add the substituents around the ring.

a.

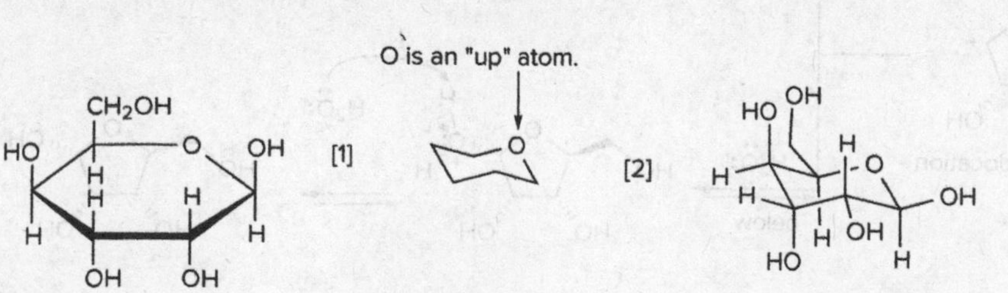

O is an "up" atom.

[1] [2]

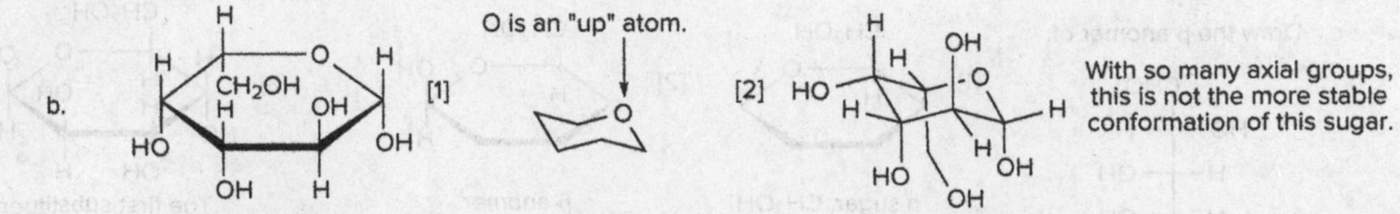

b. [1] O is an "up" atom. [2] With so many axial groups, this is not the more stable conformation of this sugar.

24.15 **Cyclization always forms a new stereogenic center at the anomeric carbon, so two different anomers are possible.**

Two anomers of D-erythrose:

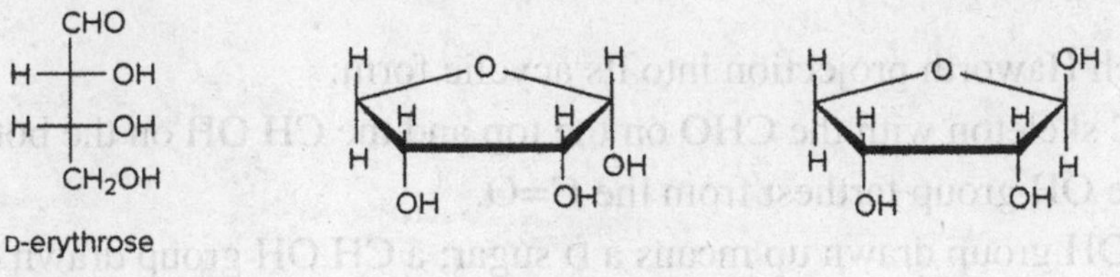

D-erythrose

24.16

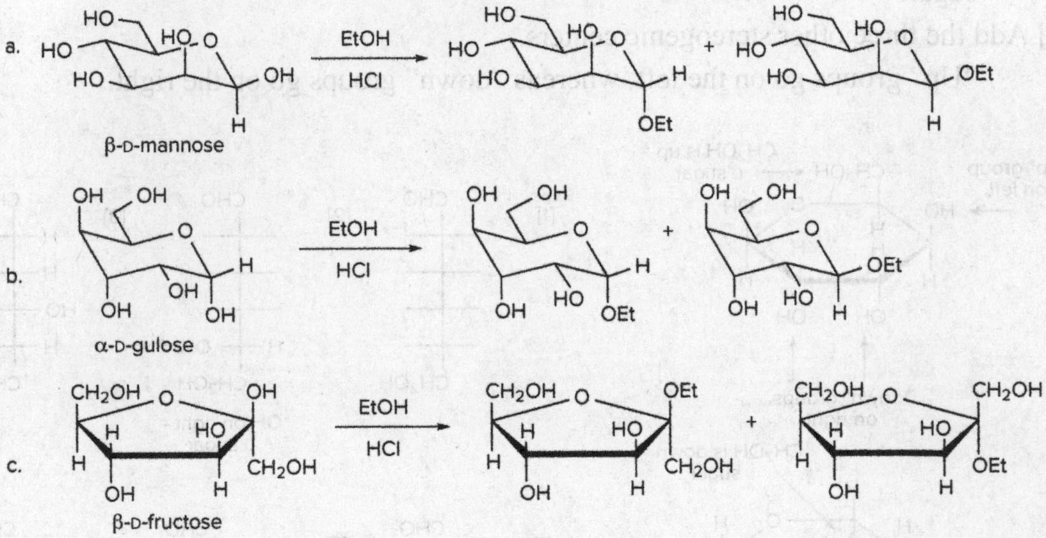

a. β-D-mannose → EtOH, HCl

b. α-D-gulose → EtOH, HCl

c. β-D-fructose → EtOH, HCl

24.17

resonance-stabilized carbocation

planar carbocation

above

below

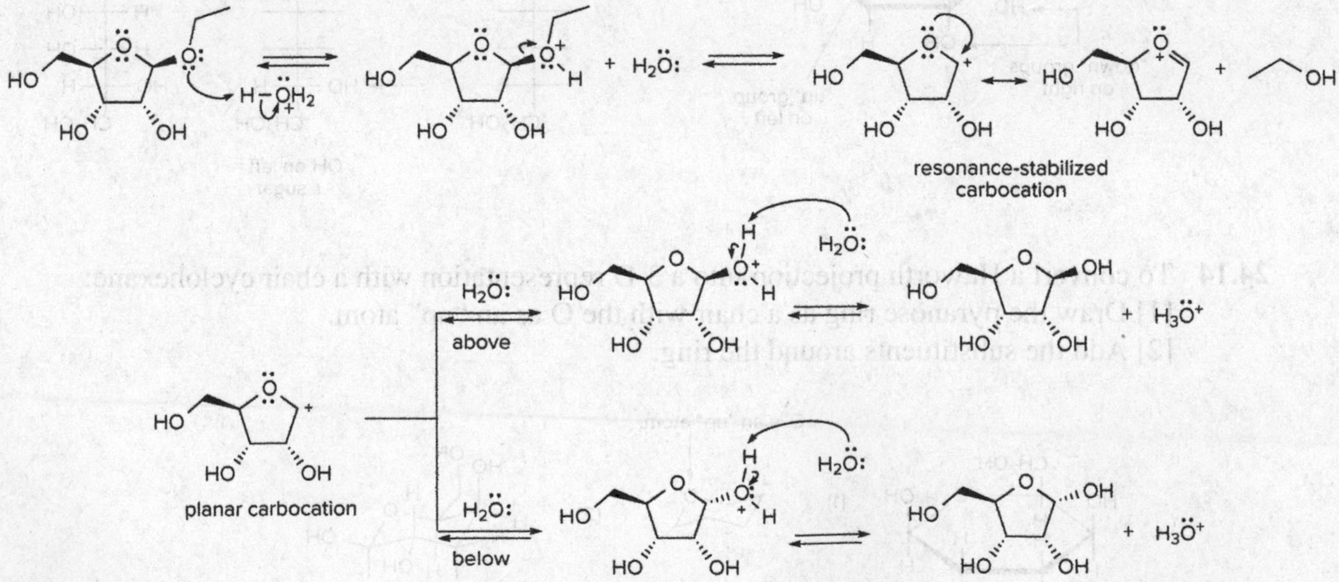

24.18

a. All circled O atoms are part of a glycoside.

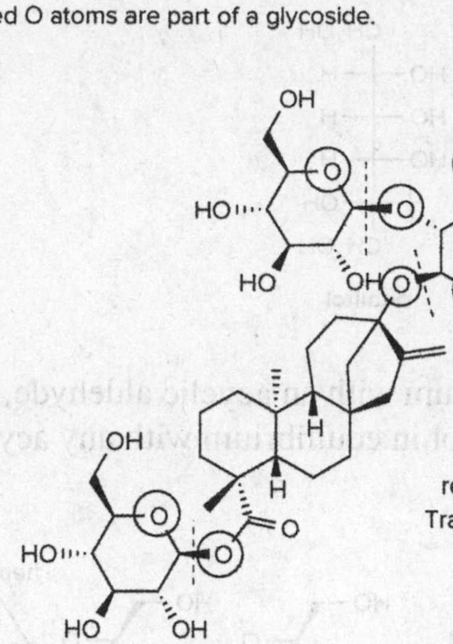

rebaudioside A
Trade name Truvia

b. Hydrolysis of rebaudioside A breaks each bond indicated with a dashed line and forms four molecules of glucose and the aglycon drawn.

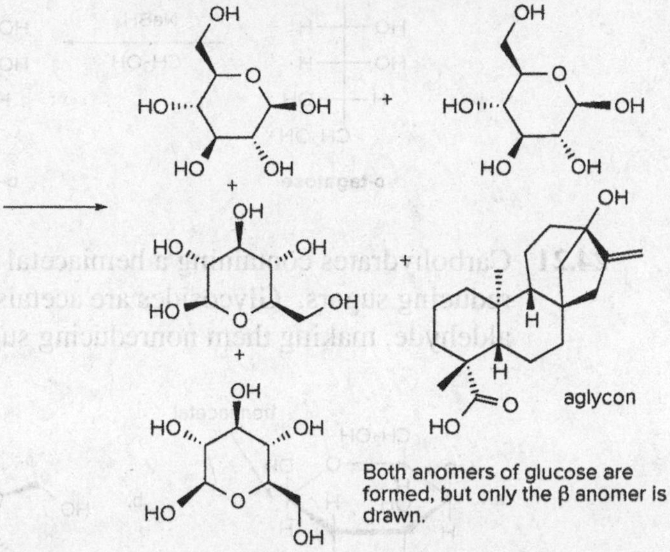

aglycon

Both anomers of glucose are formed, but only the β anomer is drawn.

24.19

a. Ag_2O, CH_3I

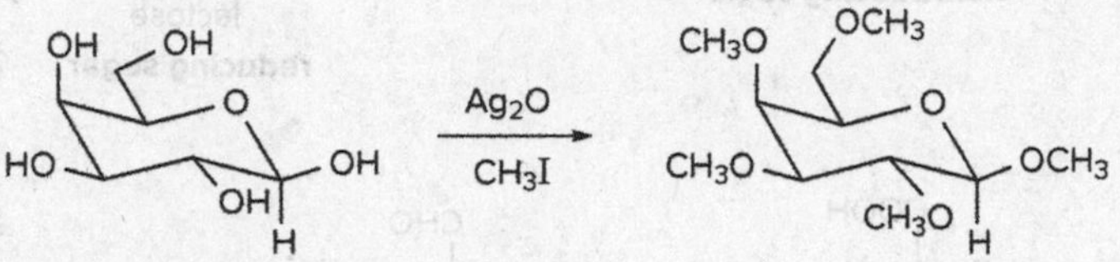

b. NaH, $C_6H_5CH_2Cl$

c. H_3O^+ → + α anomer + $HOCH_2C_6H_5$

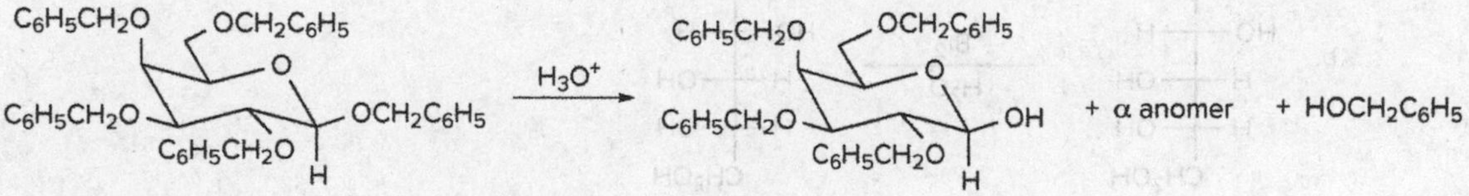

d. Ac_2O, pyridine

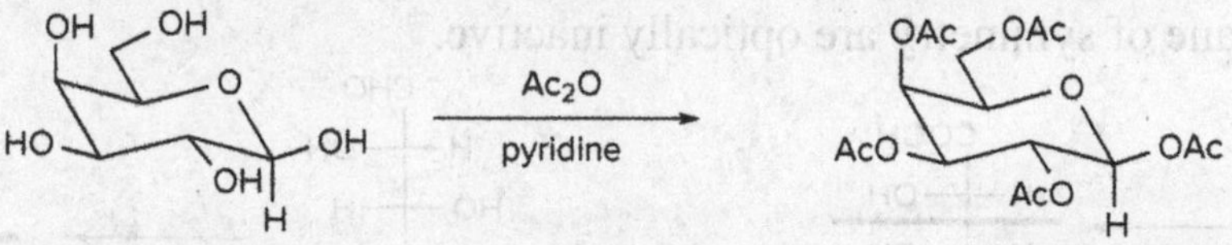

e. + C_6H_5COCl, pyridine

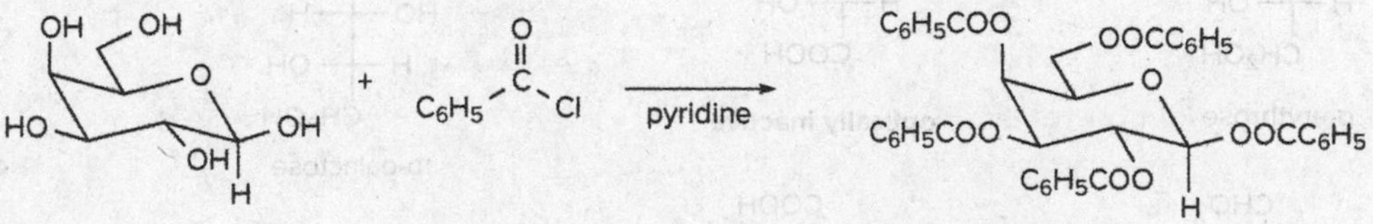

f. product in (c) + C_6H_5COCl → pyridine → + α anomer

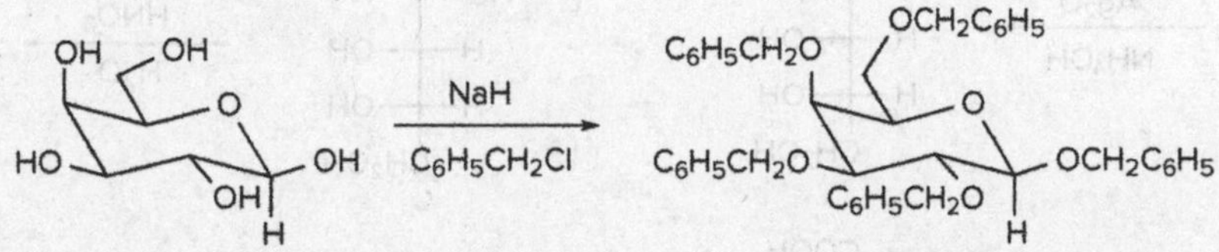

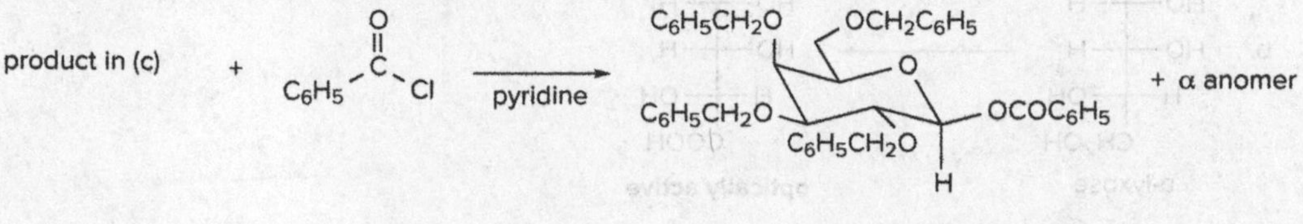

24.20

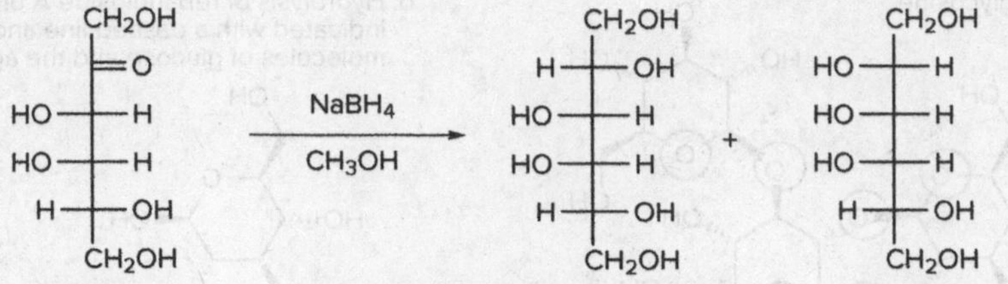

D-tagatose → ($NaBH_4$, CH_3OH) → D-galactitol + D-talitol

24.21 Carbohydrates containing a hemiacetal are in equilibrium with an acyclic aldehyde, making them reducing sugars. Glycosides are acetals, so they are not in equilibrium with any acyclic aldehyde, making them nonreducing sugars.

a. 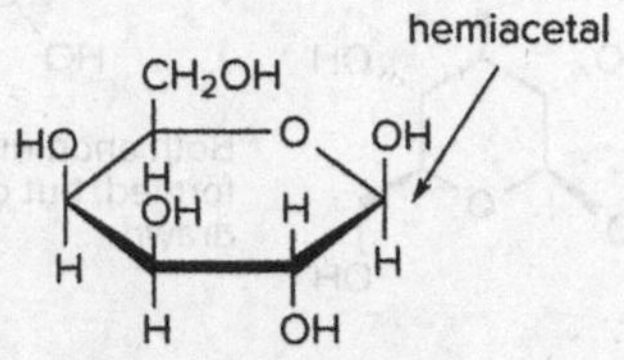 hemiacetal — **reducing sugar**

b. 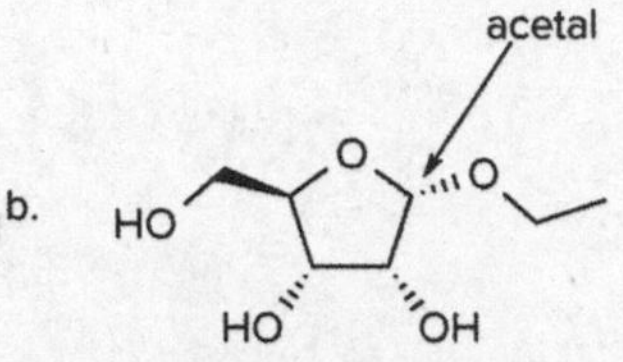 acetal — **nonreducing sugar**

c. 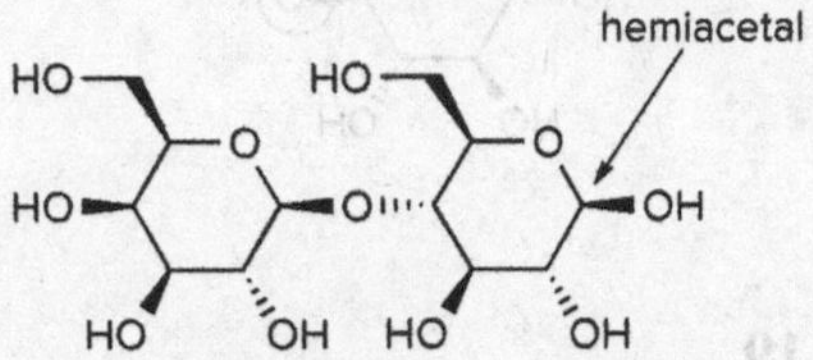 hemiacetal — lactose **reducing sugar**

24.22

a. 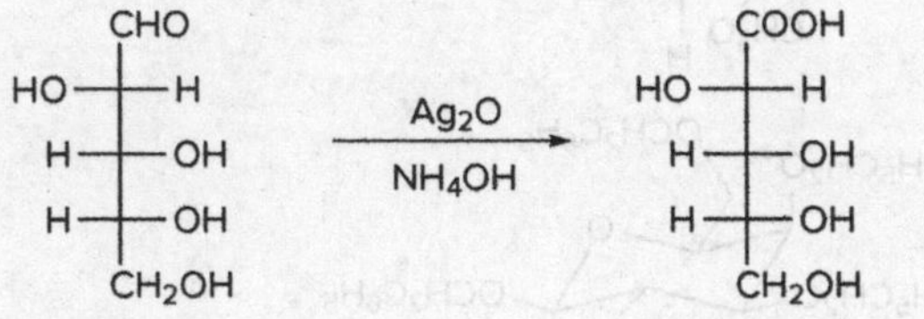 → (Ag_2O, NH_4OH)

c. 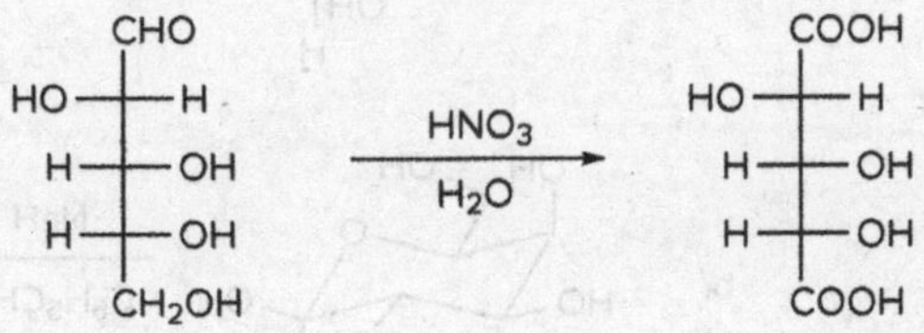 → (HNO_3, H_2O)

b. 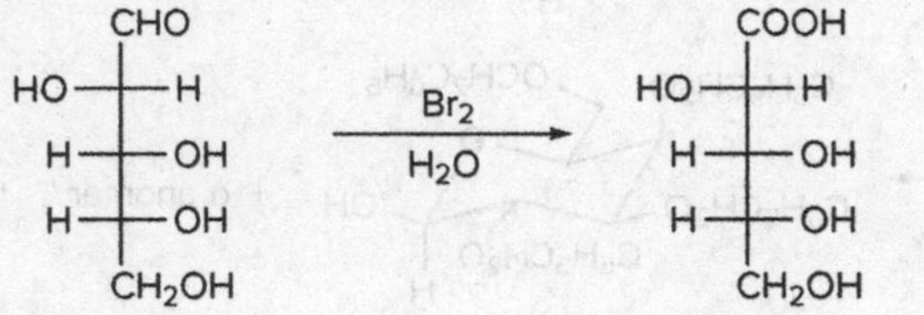 → (Br_2, H_2O)

24.23 Molecules with a plane of symmetry are optically inactive.

a. 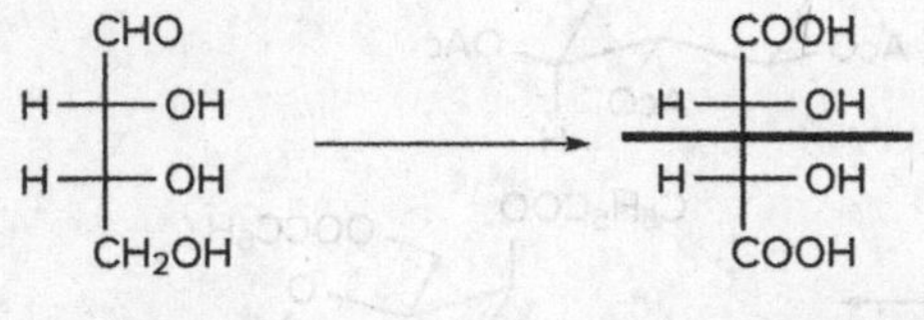 D-erythrose → **optically inactive**

c. 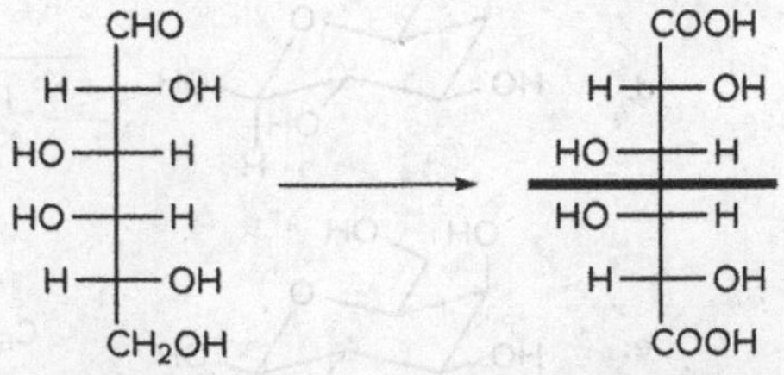 D-galactose → **optically inactive**

b. 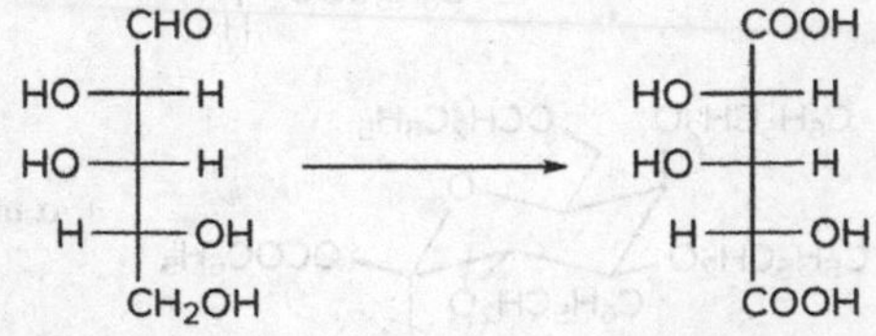 D-lyxose → **optically active**

24.24

D-idose or D-gulose → D-xylose

24.25

a. D-threose →

b. D-ribose →

c. D-galactose →

24.26

Possible optically inactive D-aldaric acids:

A'

plane of symmetry

This OH is on the **right** for a D sugar.

A"

There are two possible structures for the D-aldopentose (A' and A"), and the Wohl degradation determines which structure corresponds to A.

A' —Wohl→ —[O]→ optically inactive

optically active ←[O]— B ←Wohl— A"

This is **A**.

Because this compound has no plane of symmetry, its precursor is **B**, and thus **A"** = **A**.

24.27

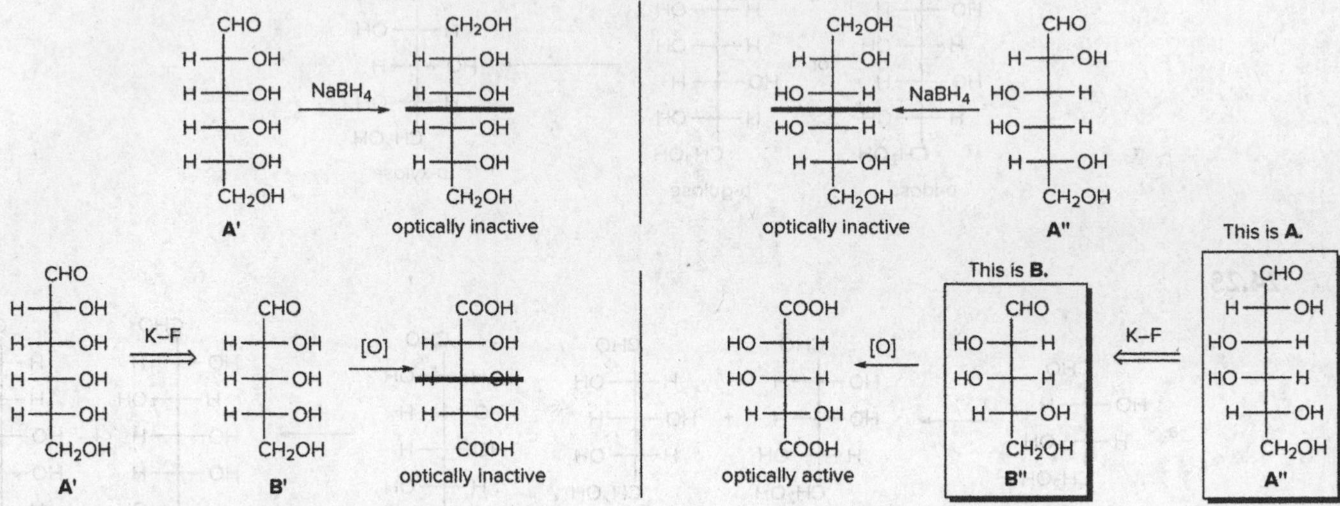

Two D-aldohexoses (**A'** and **A"**) give optically inactive alditols on reduction. **A"** is formed from **B"** by Kiliani–Fischer synthesis. Because **B"** affords an optically active aldaric acid on oxidation, **B"** is **B** and **A"** is **A.** The alternate possibility (**A'**) is formed from an aldopentose **B'** that gives an optically inactive aldaric acid on oxidation.

24.28

H_3O^+

α anomer

α-D-glucose + β-D-glucose

The same products are formed on hydrolysis of the α and β anomers of maltose.

24.29

β glycoside bond

cellobiose

Two anomers are possible here; the β OH is drawn.

24.30

a.

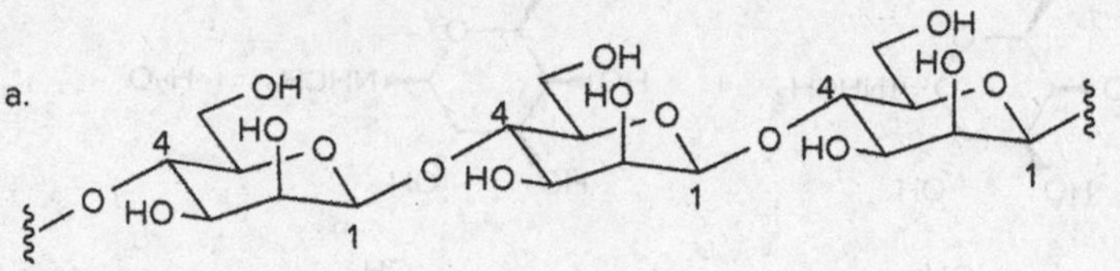

b.

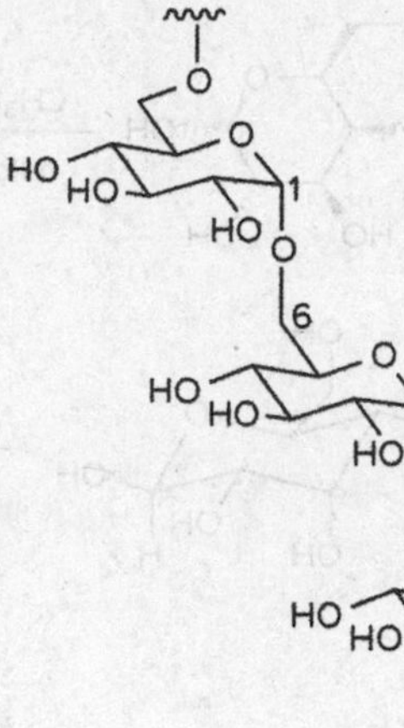

dextran

24.31

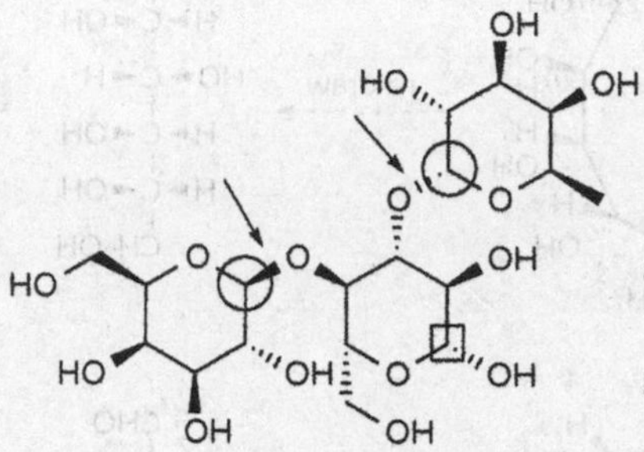

a. 3-Fucosyllactose contains two acetals and one hemiacetal. The acetal C's are circled, and the hemiacetal C is boxed in.

b. H_3O^+ cleaves acetal bonds, labeled with two arrows.

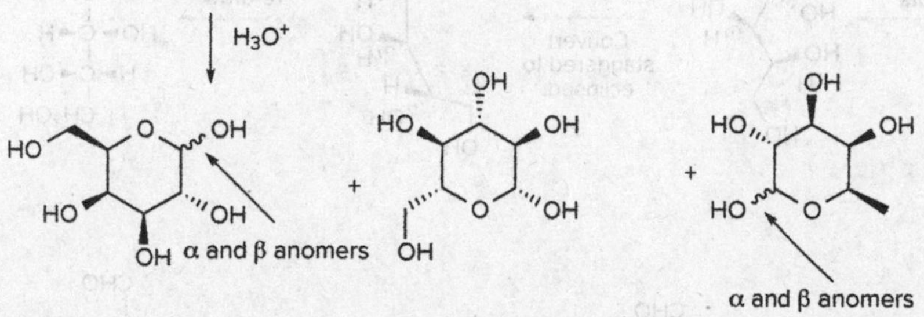

3-fucosyllactose

H_3O^+

α and β anomers

α and β anomers

24.32

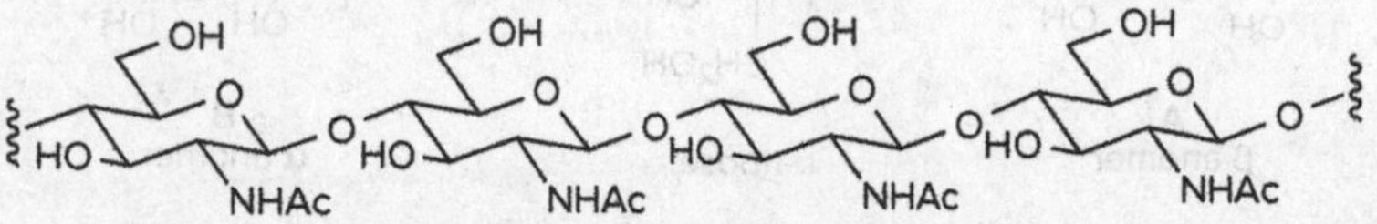

chitin—a polysaccharide composed of NAG units

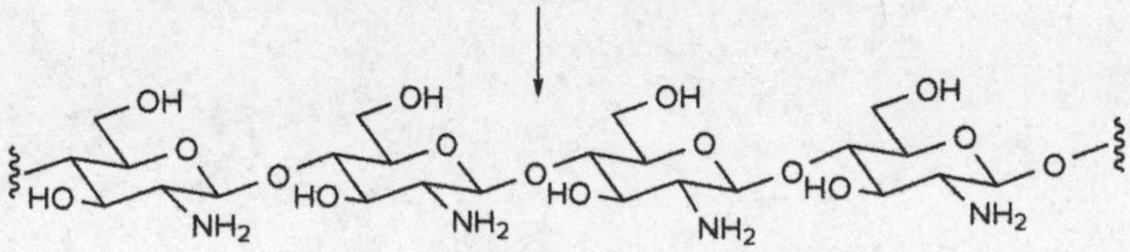

chitosan

24.33

a. HO, HO, HO, OH, OH, O (ring) —CH_3NH_2, mild H^+→ NHCH3 product + NHCH3 product + H_2O

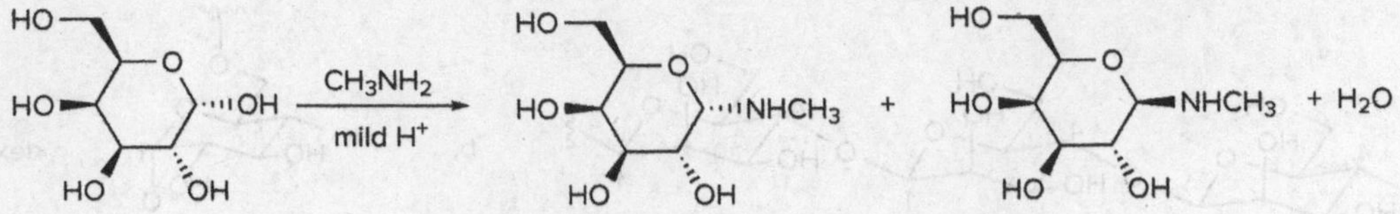

b. —$C_6H_5NH_2$, mild H^+→ NHC_6H_5 product + NHC_6H_5 product + H_2O

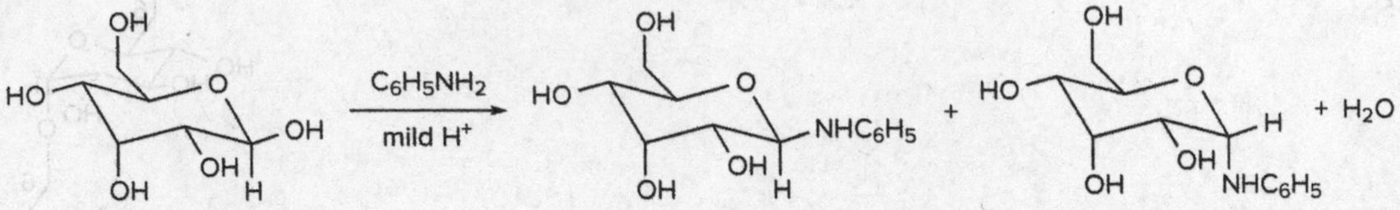

24.34

a. rotate → Convert staggered to eclipsed. → re-draw =

CHO
H—OH
HO—H
H—OH
H—OH
CH_2OH

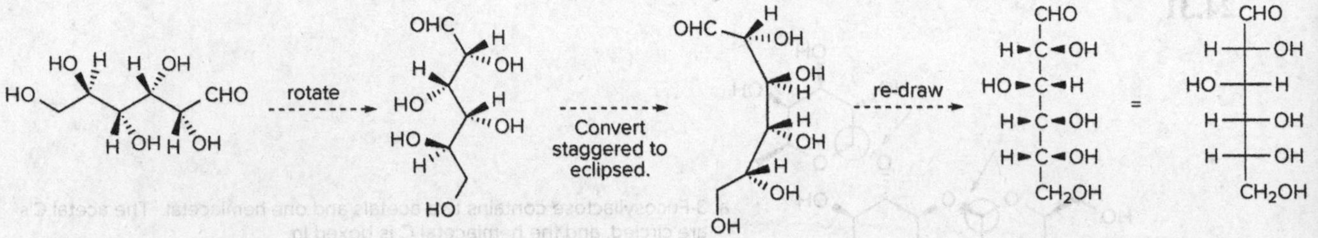

b. rotate → Convert staggered to eclipsed. → re-draw =

CHO
H—OH
HO—H
HO—H
H—OH
CH_2OH

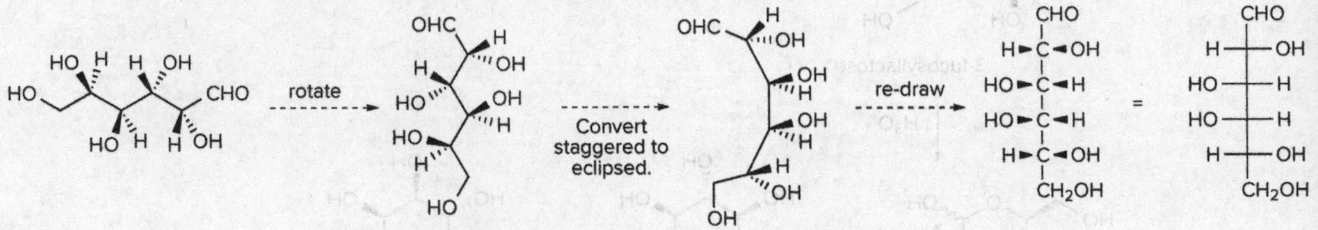

24.35

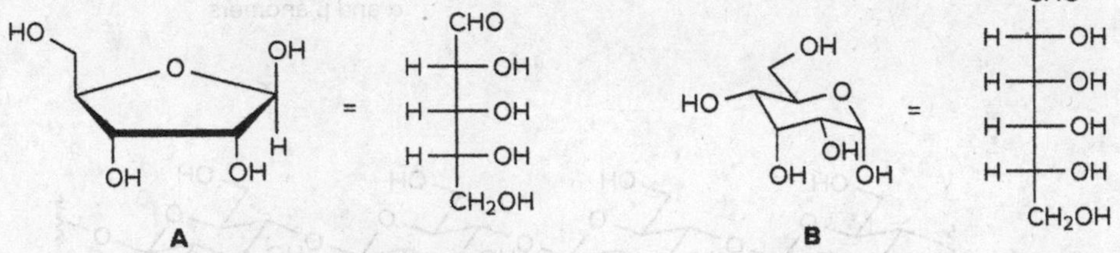

A β anomer = D-ribose

B α anomer = D-allose

24.36 Use the directions from Answer 24.2 to draw each Fischer projection.

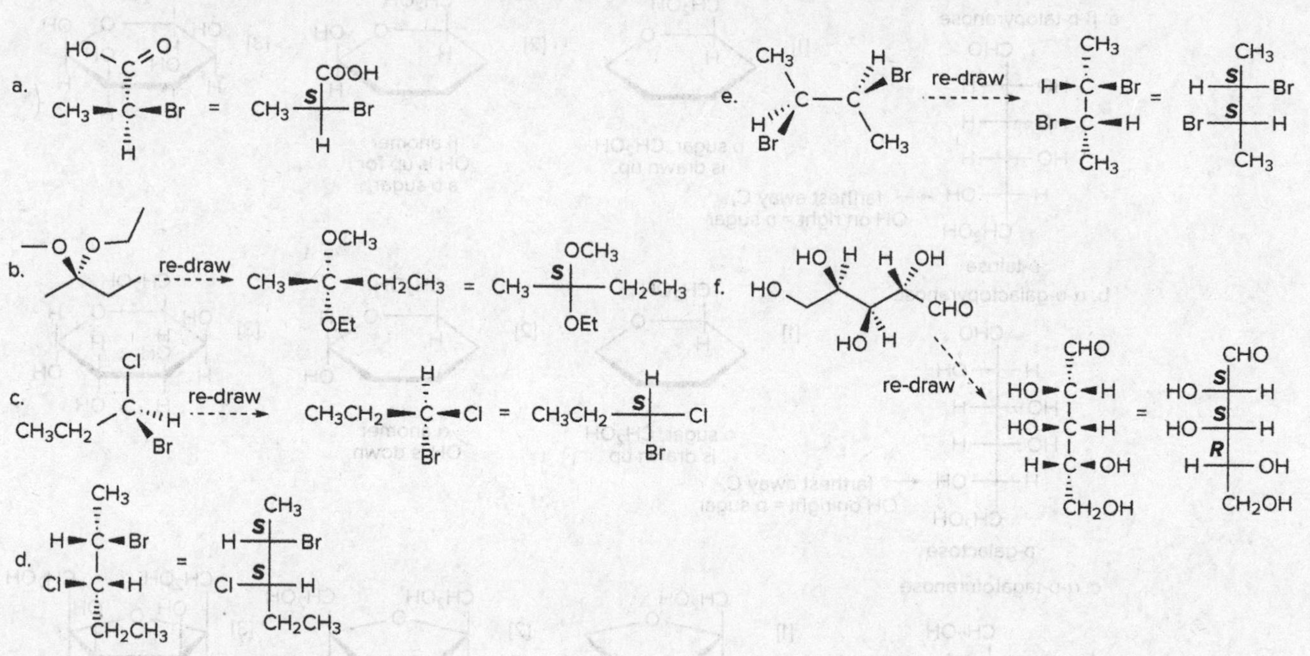

a. CH_3–C(OH=O… COOH)–Br, H = Fischer projection: COOH top, CH_3 left, Br right, H bottom (S)

b. re-draw → CH_3–C(OCH_3)(OEt)–CH_2CH_3 = Fischer projection: OCH_3 top, CH_3 left, CH_2CH_3 right, OEt bottom (S)

c. re-draw → CH_3CH_2–C(H)(Br)–Cl = Fischer projection: H top, CH_3CH_2 left, Cl right, Br bottom (S)

d. Fischer projection: CH_3 top; H–C–Br (S); Cl–C–H (S); CH_2CH_3 bottom

e. re-draw → CH_3 top; H–C–Br (S); Br–C–H (S); CH_3 bottom

f. re-draw → CHO top; HO–C–H (S); HO–C–H (S); H–C–OH (R); CH_2OH bottom

24.37

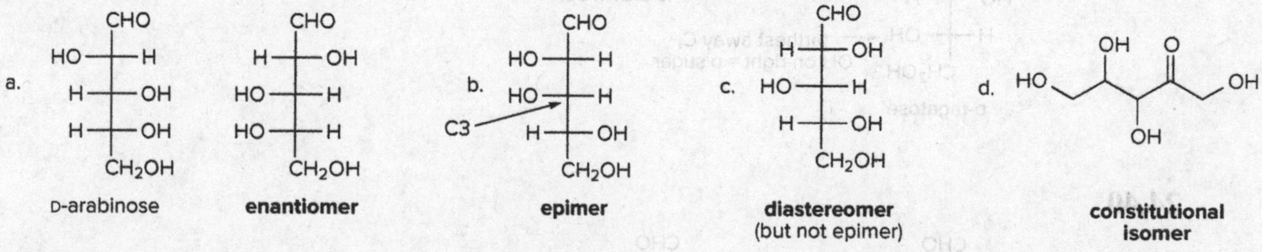

a. D-arabinose: CHO; HO–H; H–OH; H–OH; CH_2OH
enantiomer: CHO; H–OH; HO–H; HO–H; CH_2OH

b. epimer (C3): CHO; HO–H; HO–H; H–OH; CH_2OH

c. diastereomer (but not epimer): CHO; H–OH; HO–H; H–OH; CH_2OH

d. constitutional isomer: $HOCH_2CH(OH)CH(OH)C(=O)CH_2OH$

24.38

A: CHO; HO–H; H–OH; H–OH; H–OH; CH_2OH
B: CHO; HO–H; H–OH; H–OH; HO–H; CH_2OH
C: CHO; H–OH; HO–H; HO–H; H–OH; CH_2OH
D: CH_2OH; C=O; H–OH; H–OH; H–OH; CH_2OH
E (chair form), F (pyranose)

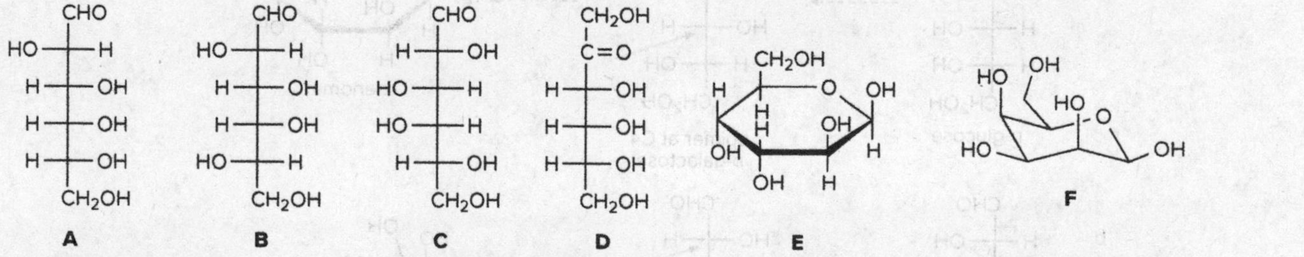

a. **A** and **B** **epimers**
b. **A** and **C** **diastereomers**
c. **B** and **C** **enantiomers**
d. **A** and **D** **constitutional isomers**
e. **E** and **F** **diastereomers**

24.39 Use the directions from Answer 24.12.

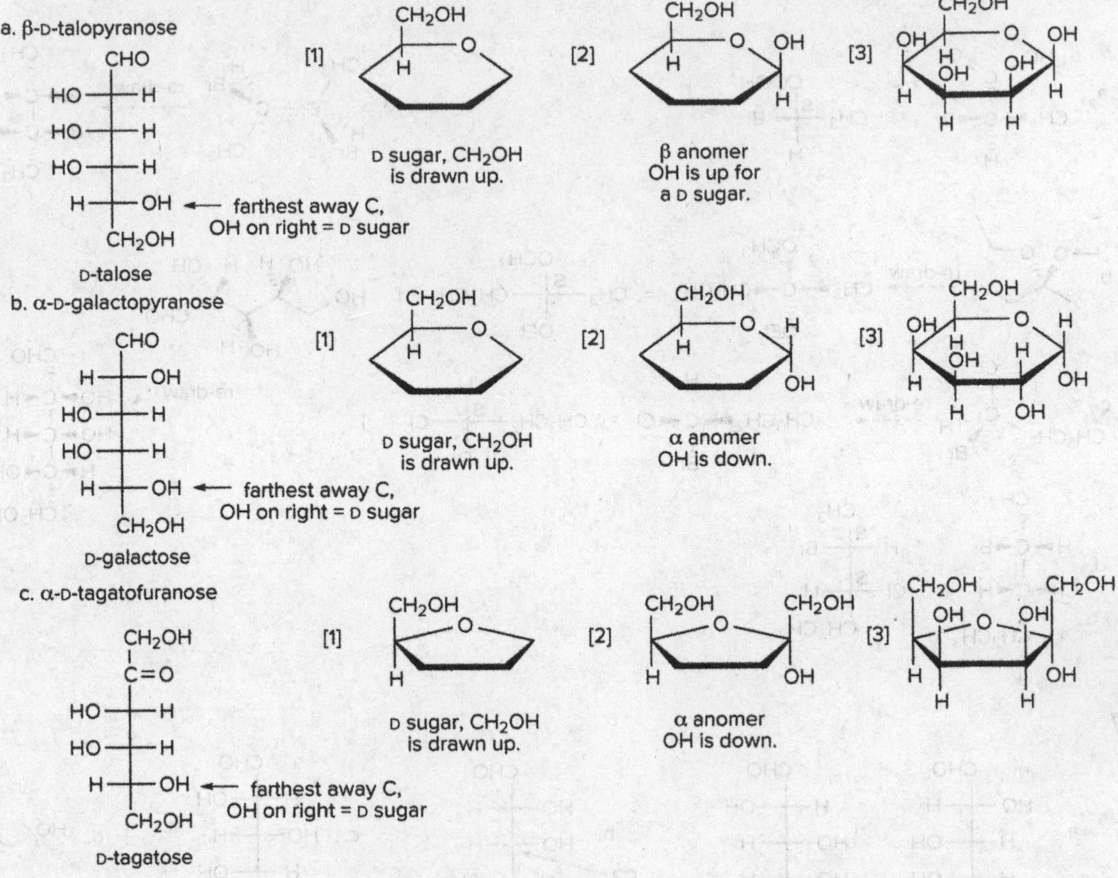

24.40

a.

D-glucose

epimer at C4
D-galactose

α-anomer

b.

D-gulose

epimer at C2
D-idose

β-anomer

24.41

a.

CHO
H—OH
H—OH
HO—H
H—OH ←
CH_2OH

farthest away C,
OH on right = D sugar

OH ← D sugar
α anomer

OH ← D sugar
β anomer

b.

CHO
HO—H
H—OH
HO—H
H—OH ←
CH_2OH

farthest away C,
OH on right = D sugar

OH ← D sugar
α anomer

OH ← D sugar
β anomer

24.42 Use the directions from Answer 24.13.

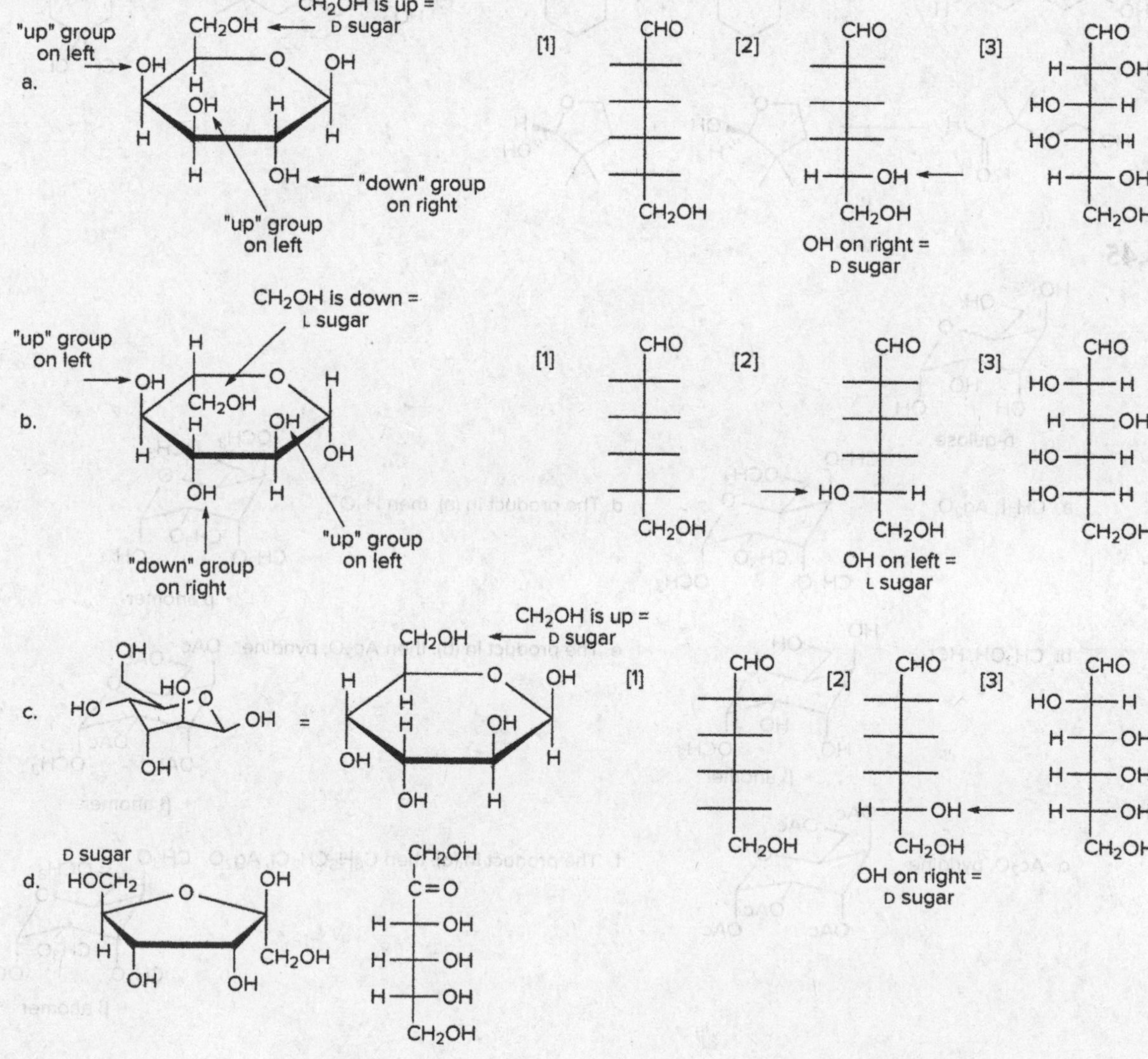

24.43

Two anomers of D-idose, as well as two conformations of each anomer:

α anomer

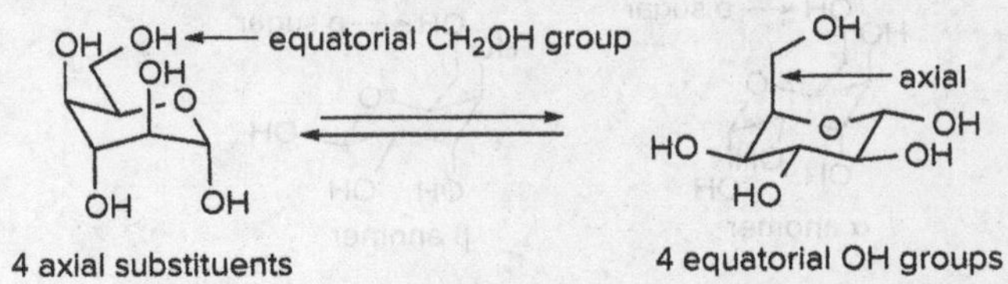

4 axial substituents

4 equatorial OH groups

More stable conformation for the α anomer—the CH_2OH is axial, but all other groups are equatorial.

β anomer

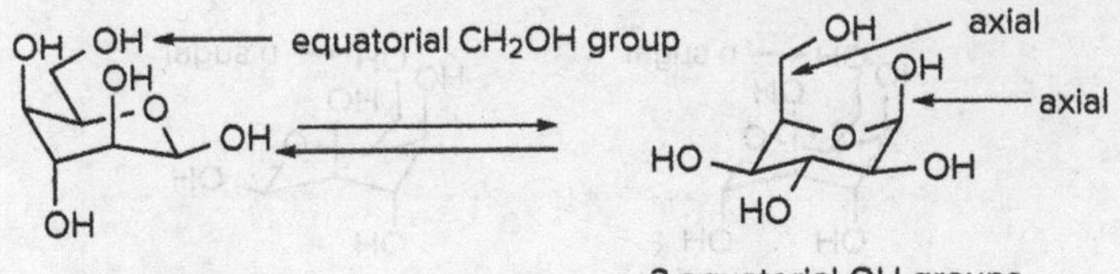

3 axial substituents

3 equatorial OH groups

The more stable conformation for the β anomer—the CH_2OH is axial, as is the anomeric OH, but three other OH groups are equatorial.

24.44

a.

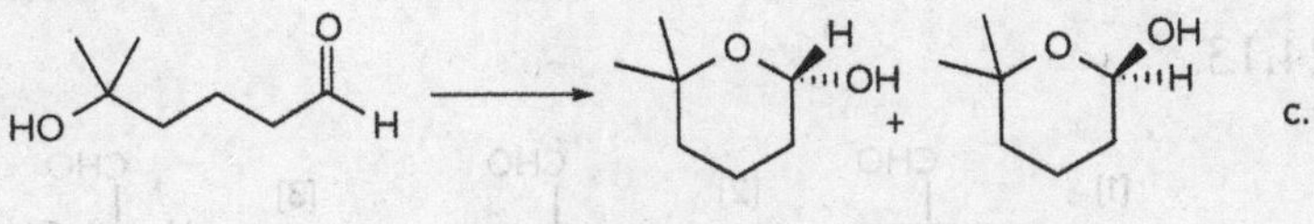

c.

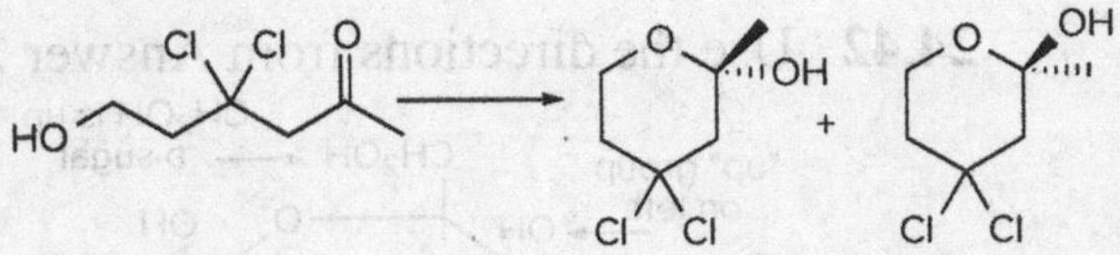

b.

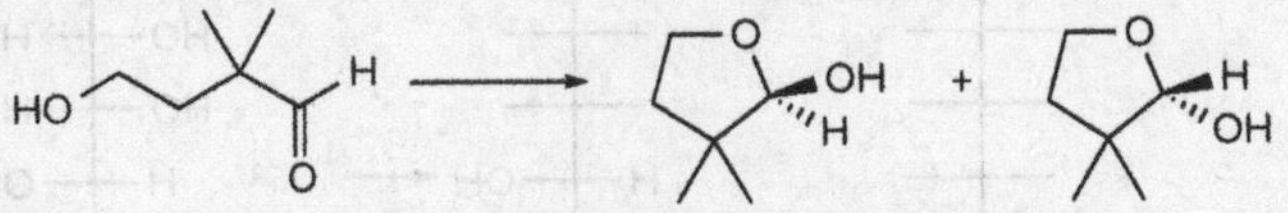

24.45

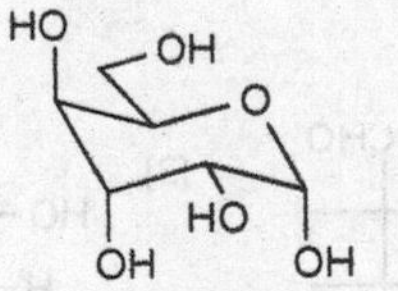

D-gulose

a. CH_3I, Ag_2O

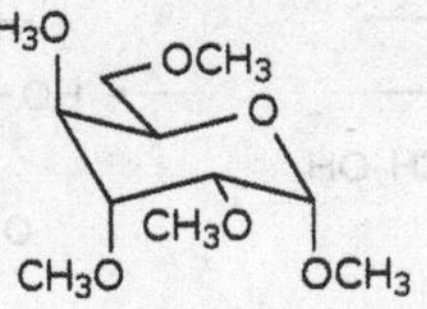

d. The product in (a), then H_3O^+

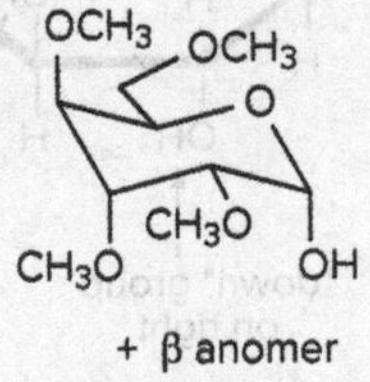

+ β anomer

b. CH_3OH, HCl

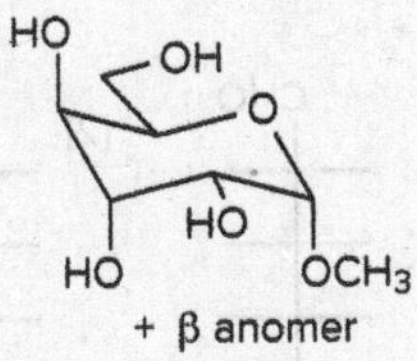

+ β anomer

e. The product in (b), then Ac_2O, pyridine

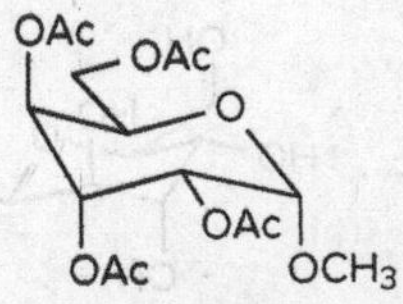

+ β anomer

c. Ac_2O, pyridine

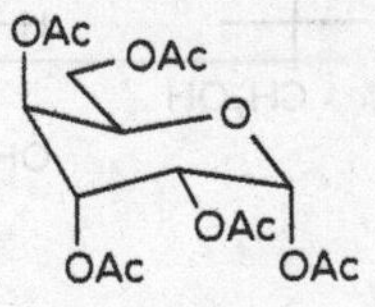

f. The product in (d), then $C_6H_5CH_2Cl$, Ag_2O

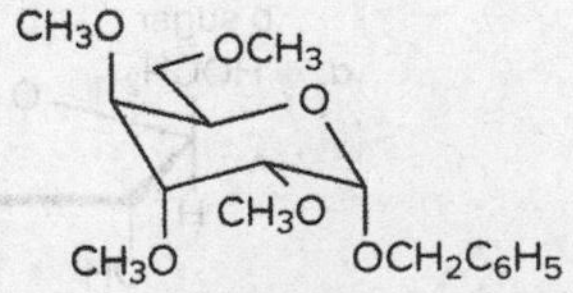

+ β anomer

24.46

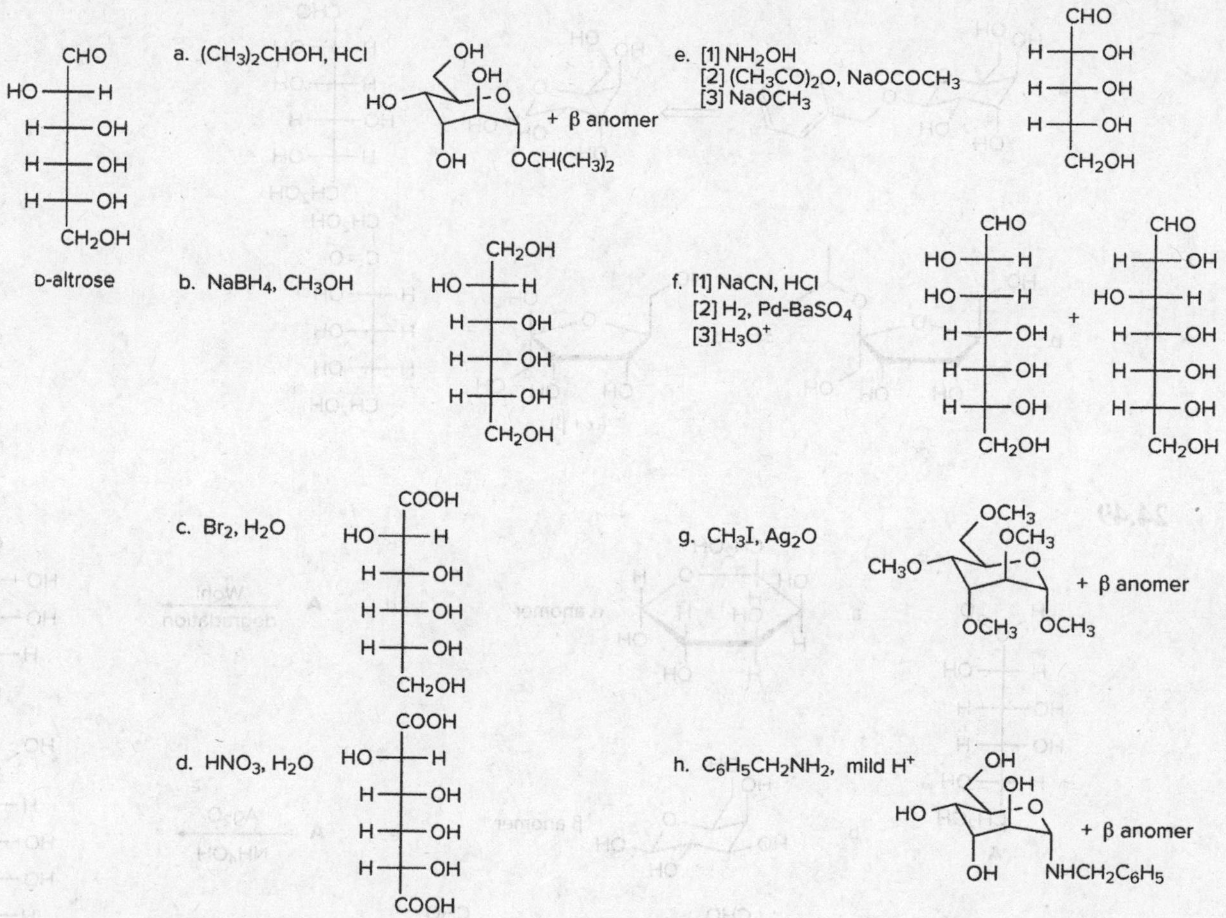

D-altrose

a. $(CH_3)_2CHOH$, HCl → + β anomer

b. $NaBH_4$, CH_3OH

c. Br_2, H_2O

d. HNO_3, H_2O

e. [1] NH_2OH
[2] $(CH_3CO)_2O$, $NaOCOCH_3$
[3] $NaOCH_3$

f. [1] NaCN, HCl
[2] H_2, Pd-$BaSO_4$
[3] H_3O^+

g. CH_3I, Ag_2O → + β anomer

h. $C_6H_5CH_2NH_2$, mild H^+ → + β anomer

24.47

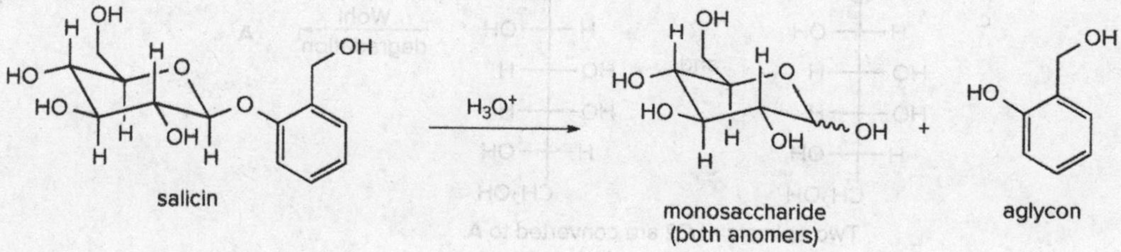

salicin → H_3O^+ → monosaccharide (both anomers) + aglycon

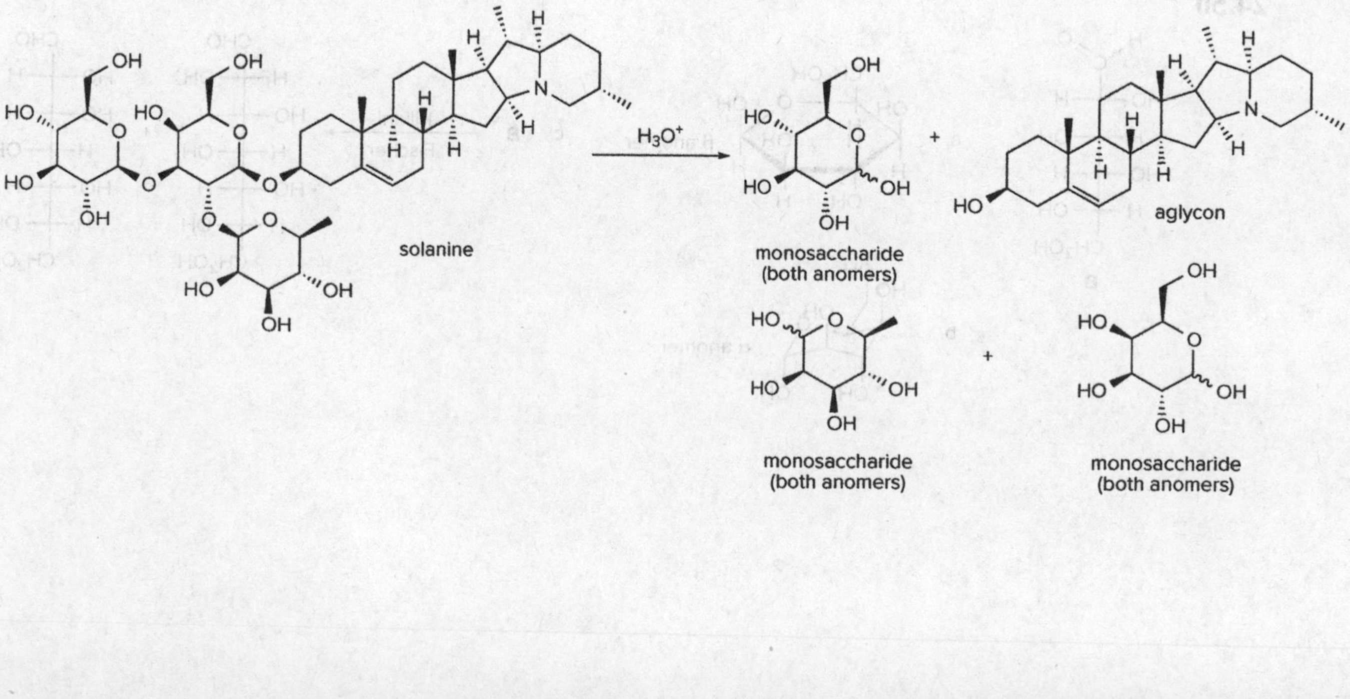

solanine → H_3O^+ → monosaccharide (both anomers) + aglycon + monosaccharide (both anomers) + monosaccharide (both anomers)

24.48

a. ⟹ (β anomer) =

CHO
H—OH
H—OH
HO—H
H—OH
CH_2OH

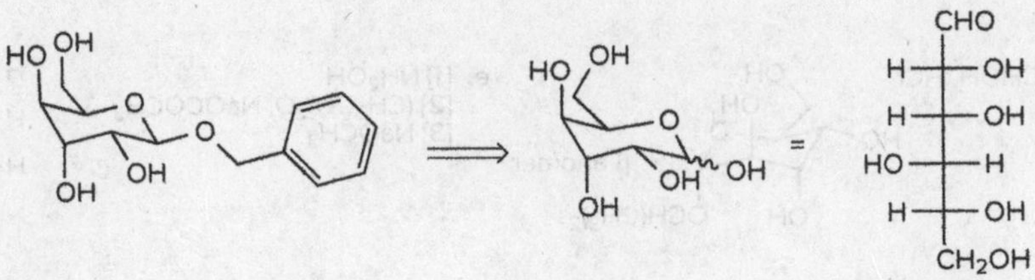

b. 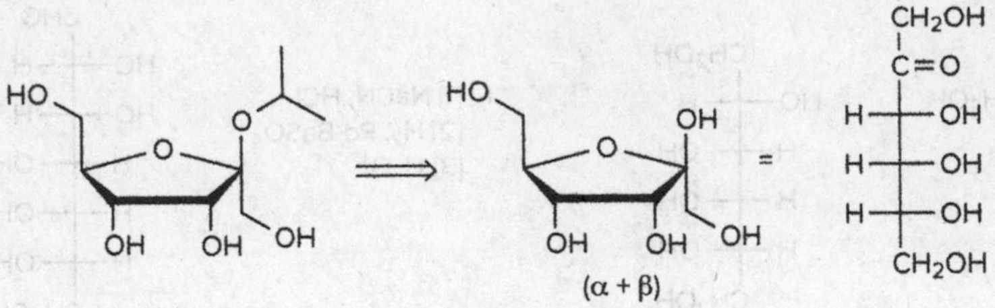 ⟹ (α + β) =

CH_2OH
C=O
H—OH
H—OH
H—OH
CH_2OH

24.49

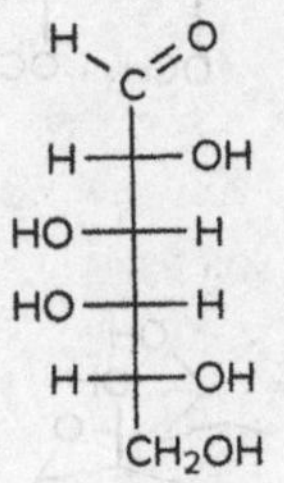

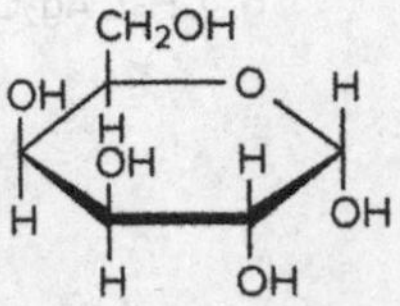

A:

CHO
H—OH
HO—H
HO—H
H—OH
CH_2OH

a. α anomer

b. β anomer

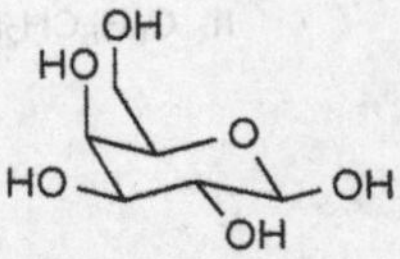

c.

CHO
H—OH
H—OH
HO—H
HO—H
H—OH
CH_2OH

and

CHO
HO—H
H—OH
HO—H
HO—H
H—OH
CH_2OH

Wohl degradation → A

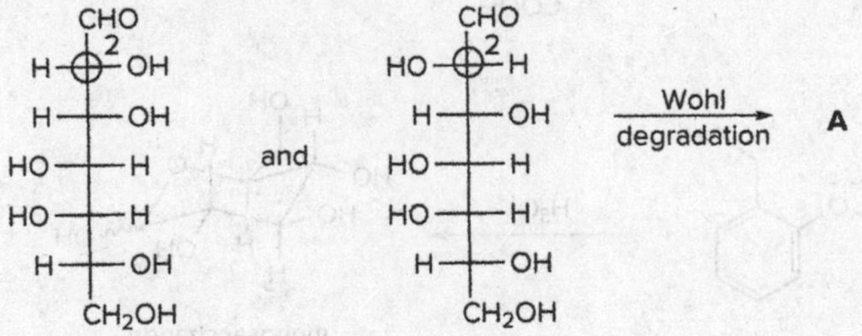

Two epimers at C2 are converted to **A**.

d. A —Wohl degradation→

CHO
HO—H
HO—H
H—OH
CH_2OH

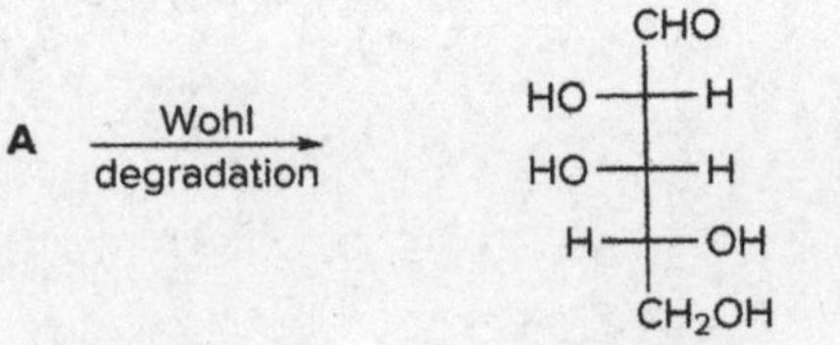

e. A —Ag_2O, NH_4OH→

COOH
H—OH
HO—H
HO—H
H—OH
CH_2OH

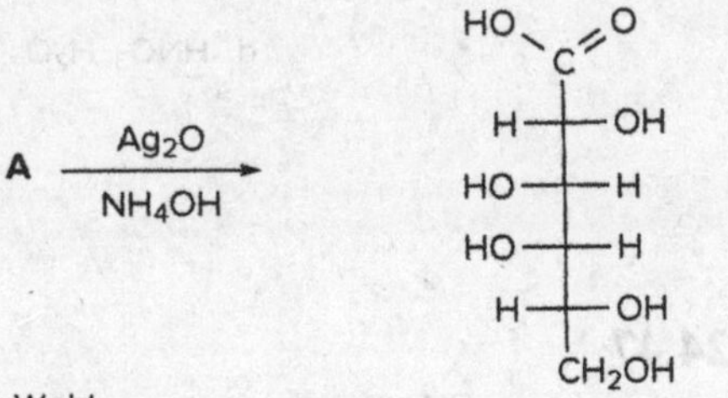

24.50

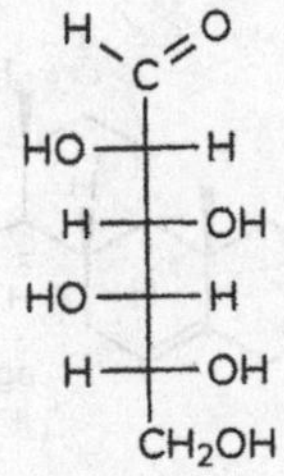

B:

CHO
HO—H
H—OH
HO—H
H—OH
CH_2OH

a. β anomer

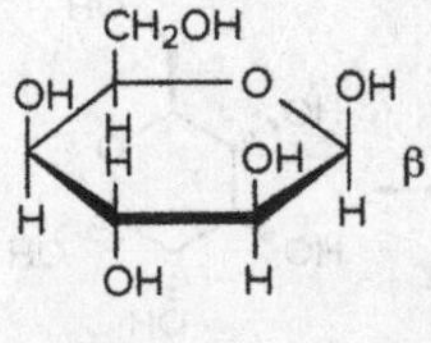

b. α anomer

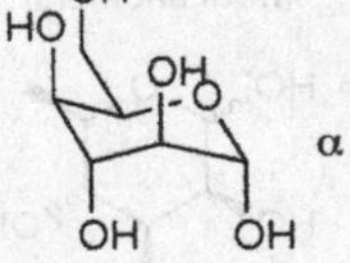

c. B —Kiliani–Fischer→

CHO
H—OH
HO—H
H—OH
HO—H
H—OH
CH_2OH

+

CHO
HO—H
HO—H
H—OH
HO—H
H—OH
CH_2OH

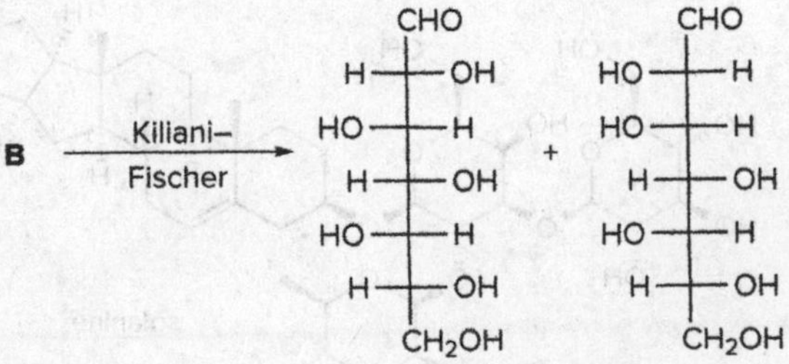

d. 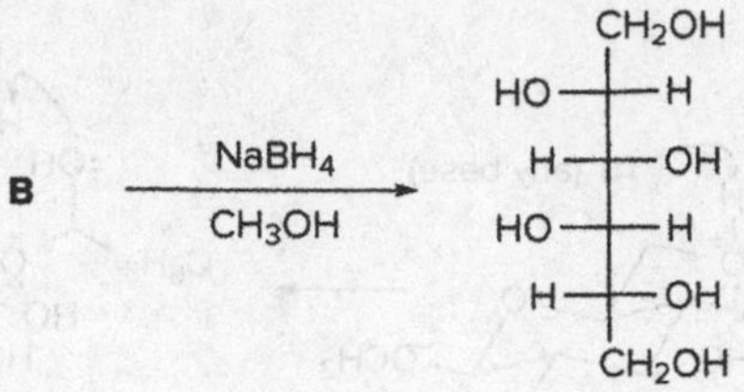 B $\xrightarrow[CH_3OH]{NaBH_4}$ (Fischer projection: CH_2OH; HO–H; H–OH; HO–H; H–OH; CH_2OH)

e. 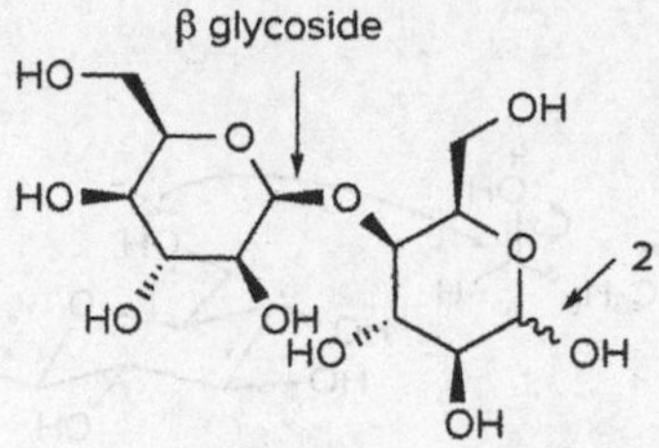 (disaccharide structure) β glycoside; 2 anomers

24.51

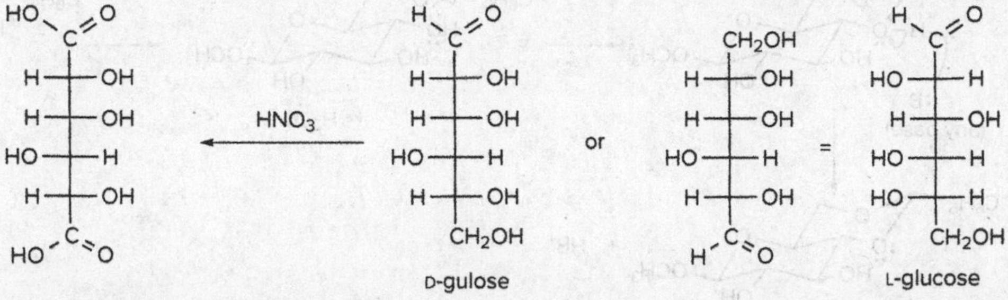

(Fischer projections) HNO_3 oxidation of D-gulose to the aldaric acid; or; = ; D-gulose; L-glucose

24.52

a. 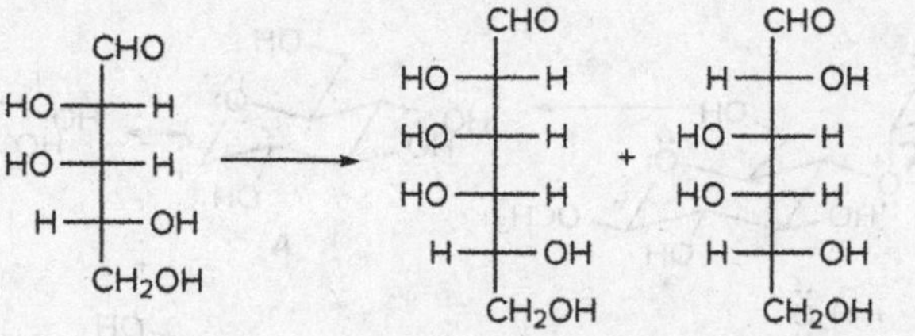 (Fischer projections of aldose and its two chain-extended products)

b. 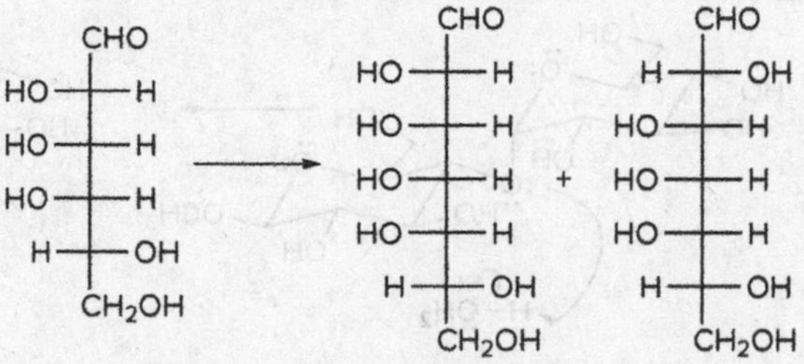 (Fischer projections of aldose and its two chain-extended products)

24.53

a. 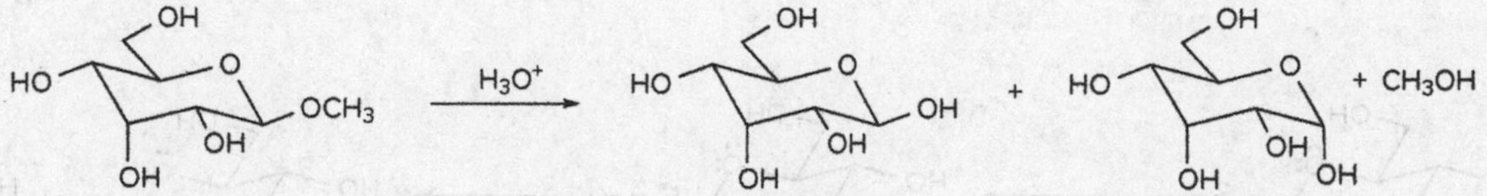 $\xrightarrow{H_3O^+}$ + + CH_3OH

b. 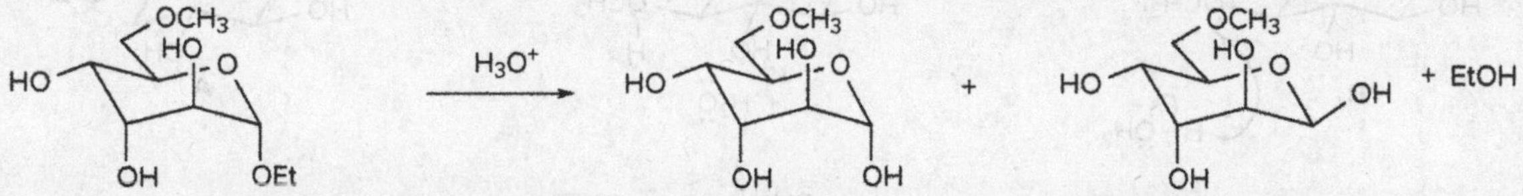 $\xrightarrow{H_3O^+}$ + + EtOH

c. 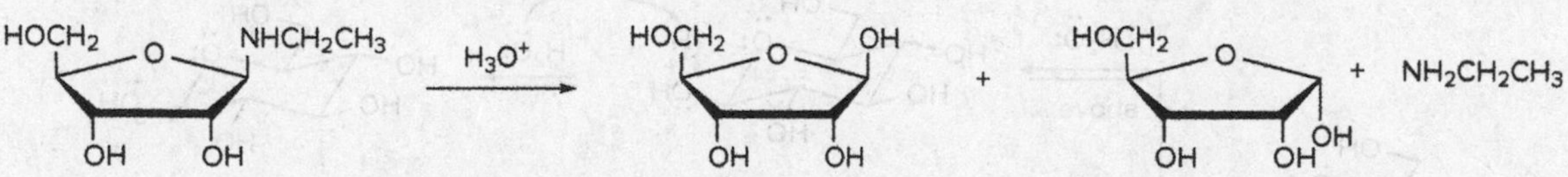 $\xrightarrow{H_3O^+}$ + + $NH_2CH_2CH_3$

24.54

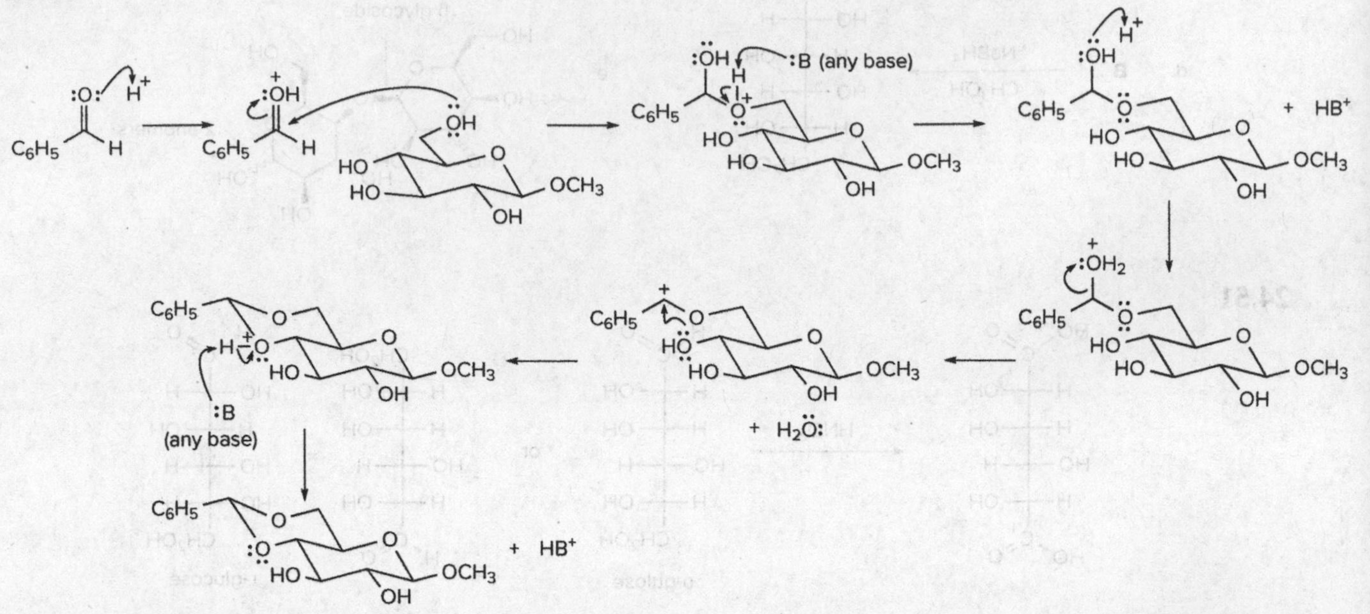

(any base)

(any base)

C_6H_5 ... $+ HB^+$ ... $+ H_2\ddot{O}$: ... $+ HB^+$

24.55

$+ H_2\ddot{O}$:

A + **B**

B → **A** + $CH_3\ddot{O}H$

A

$H_2\ddot{O}$: above

$H_2\ddot{O}$: below

$+ H_3\ddot{O}^+$

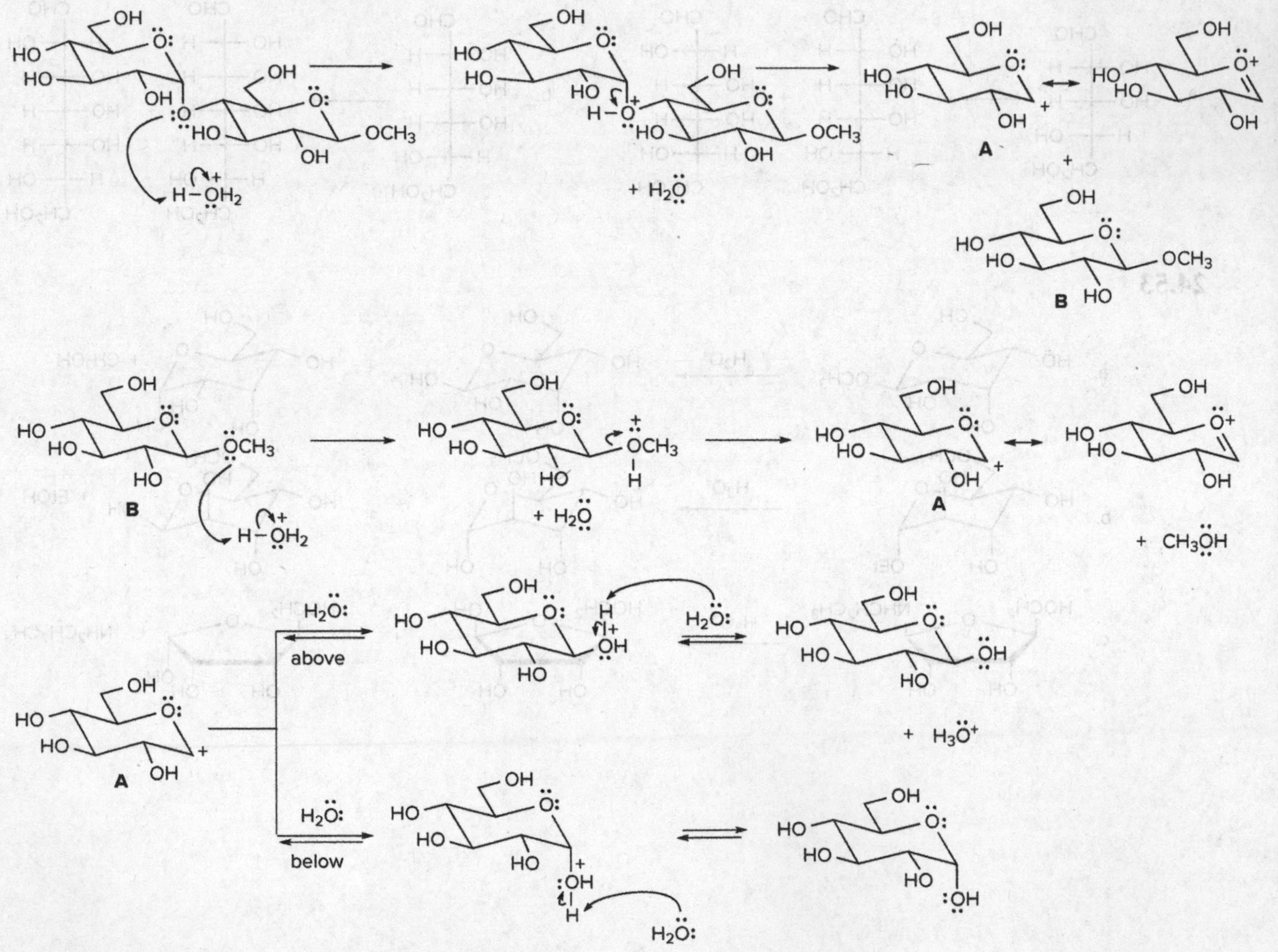

24.56

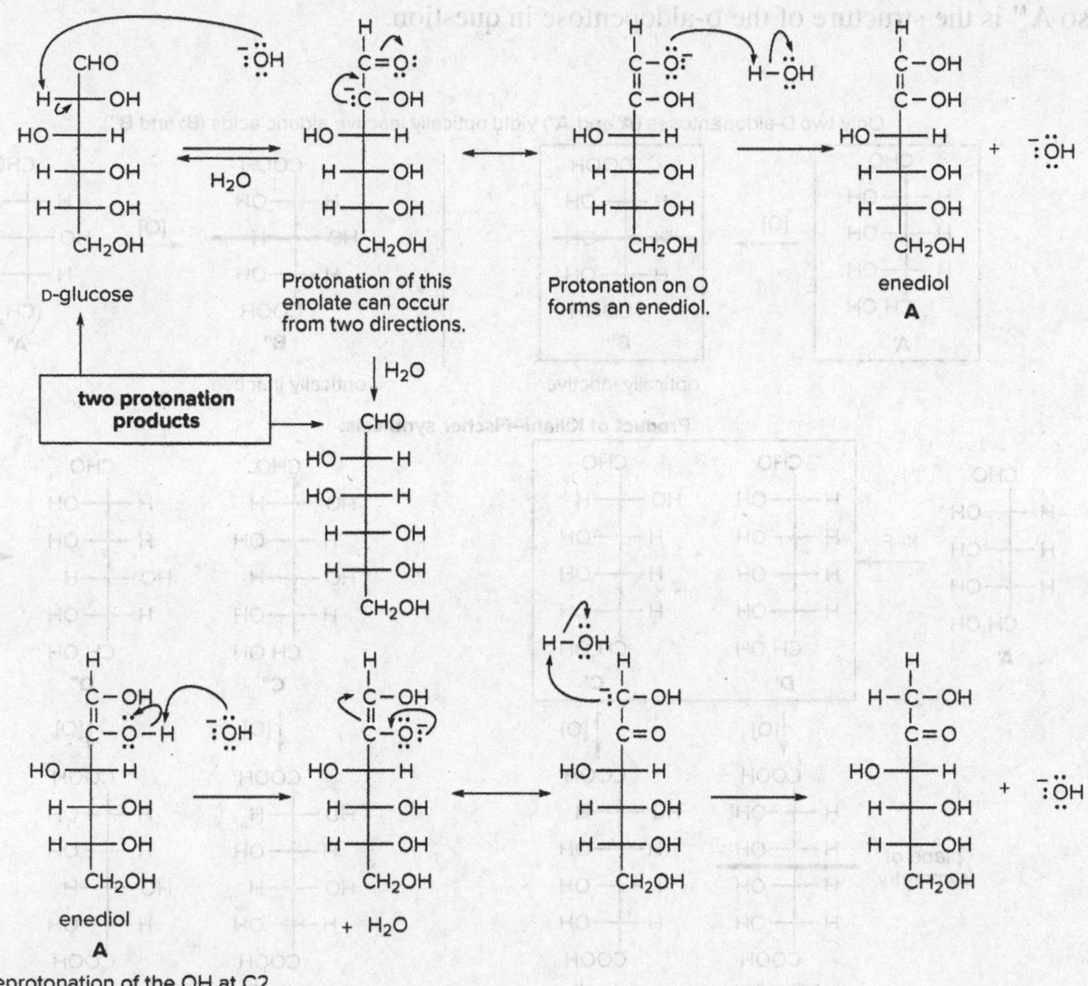
D-glucose
H_2O
Protonation of this enolate can occur from two directions.
Protonation on O forms an enediol.
enediol
A
two protonation products
H_2O
enediol
A
+ H_2O
Deprotonation of the OH at C2 of the enediol forms a new enolate that goes on to form the ketohexose.

24.57

Two D-aldopentoses (**A'** and **A"**) yield optically active aldaric acids when oxidized.

Optically active D-aldaric acids:

A'
[O]
optically active

optically active
[O]
A"
D-lyxose

A'
Wohl
[O]
optically inactive

no plane of symmetry
optically active
[O]
Wohl
A"

Only A'' undergoes Wohl degradation to an aldotetrose that is oxidized to an optically active aldaric acid, so A'' is the structure of the D-aldopentose in question.

24.58

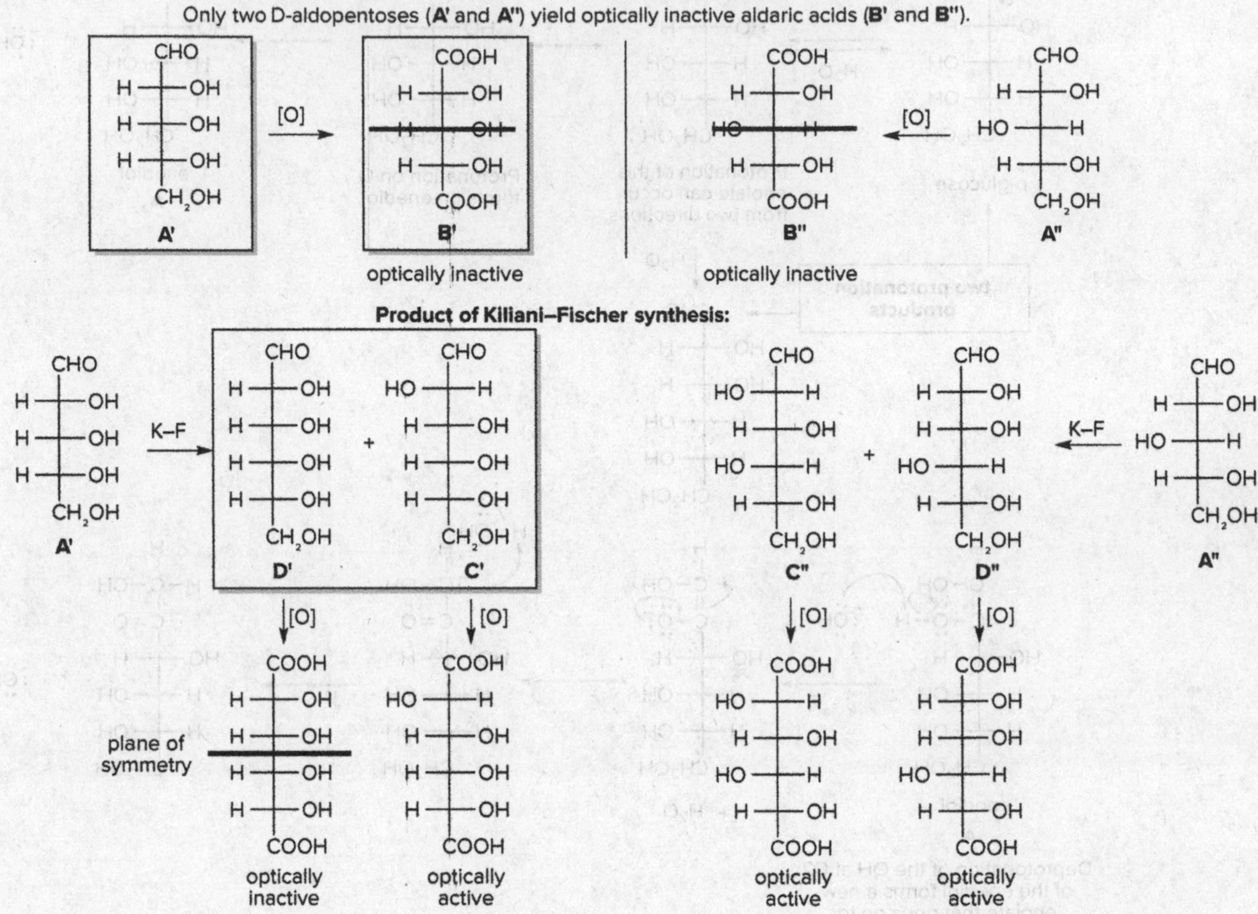

Only A' fits the criteria. Kiliani–Fischer synthesis of A' forms C' and D', which are oxidized to one optically active and one optically inactive aldaric acid. A similar procedure with A'' forms two optically active aldaric acids. Thus, the structures of A–D correspond to the structures of A'–D'.

24.59

Only two D-aldopentoses (**A'** and **A''**) are reduced to optically active alditols.

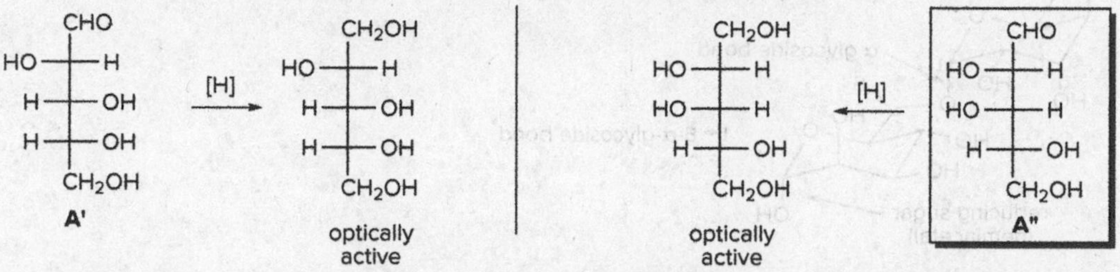

Product of Kiliani–Fischer synthesis:

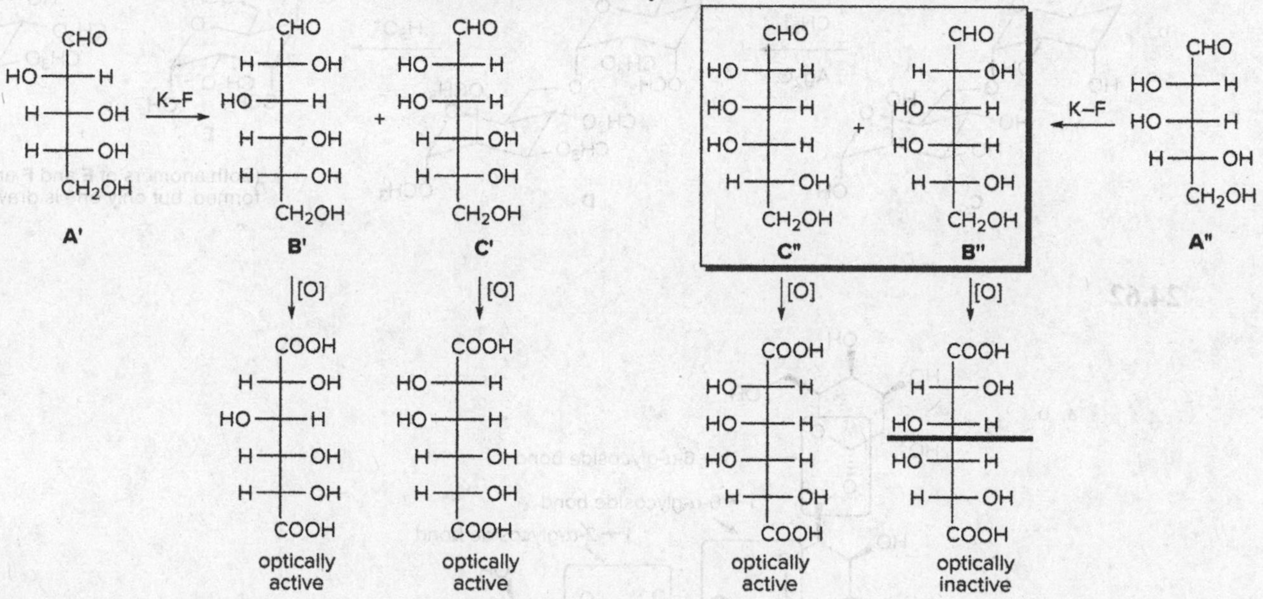

Only **A''** fits the criteria. Kiliani–Fischer synthesis of **A''** forms **B''** and **C''**, which are oxidized to one optically inactive and one optically active diacid. A similar procedure with **A'** forms two optically active diacids. Thus, the structures of **A–C** correspond to **A''–C''**.

24.60 A disaccharide formed from two mannose units in a 1→4-α-glycosidic linkage:

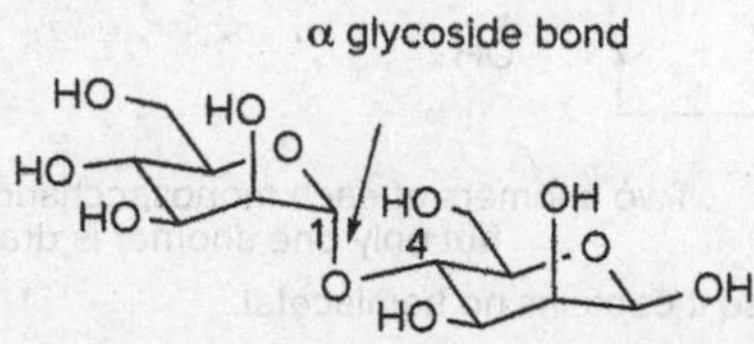

24.61

a.

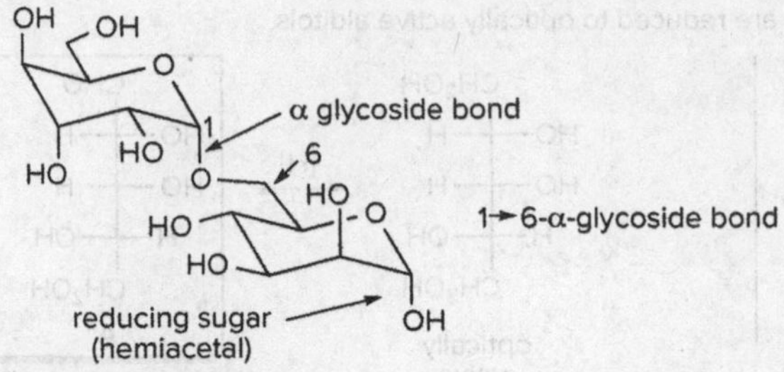

b.

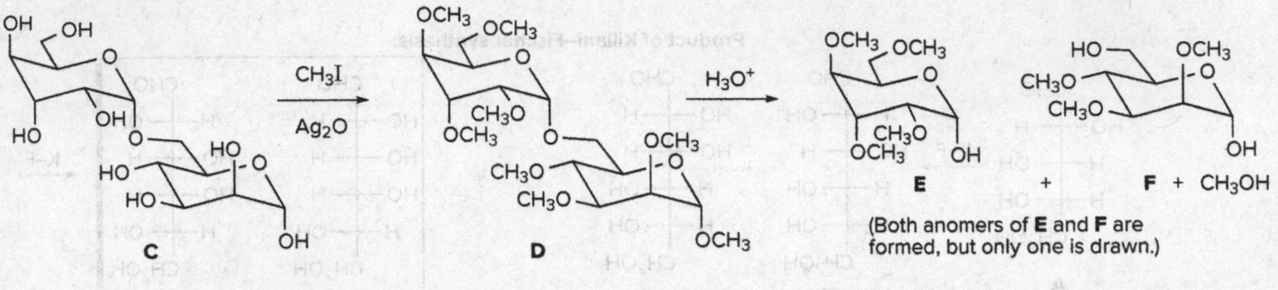

(Both anomers of **E** and **F** are formed, but only one is drawn.)

24.62

a, b.

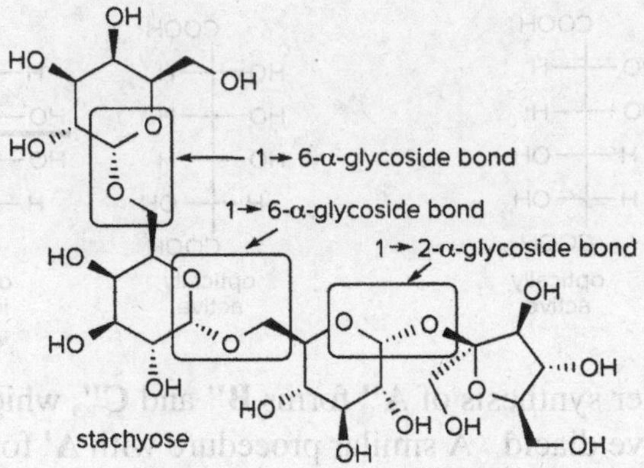

c.

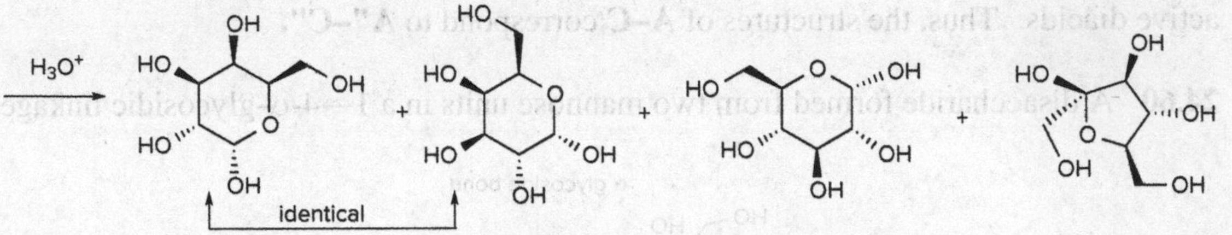

Two anomers of each monosaccharide are formed, but only one anomer is drawn.

d. Stachyose is not a reducing sugar because it contains no hemiacetal.

e. $\xrightarrow[Ag_2O]{CH_3I}$

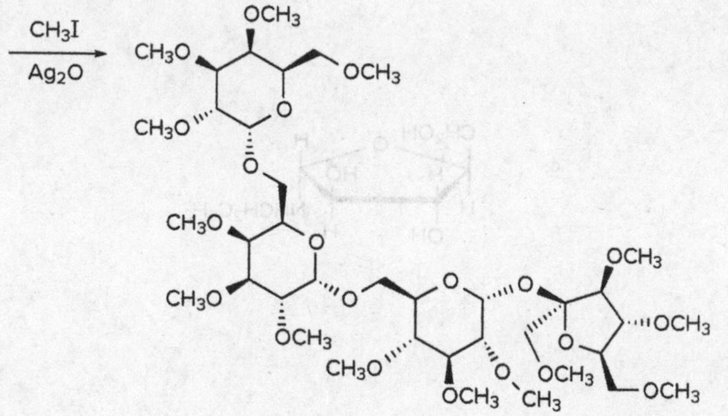

f. product in (e) $\xrightarrow{H_3O^+}$ + + +

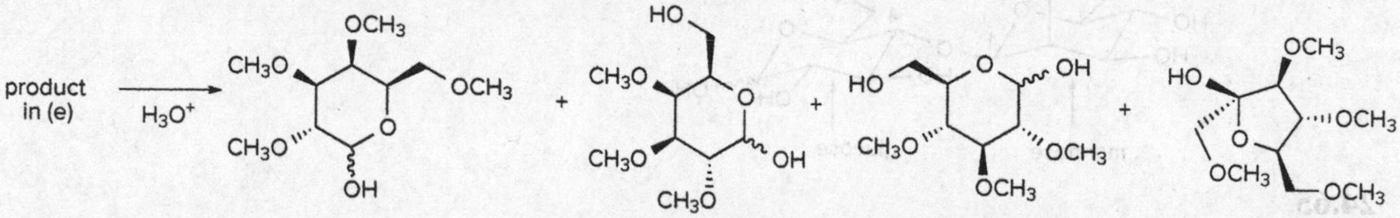

Two anomers of each monosaccharide are formed.

24.63

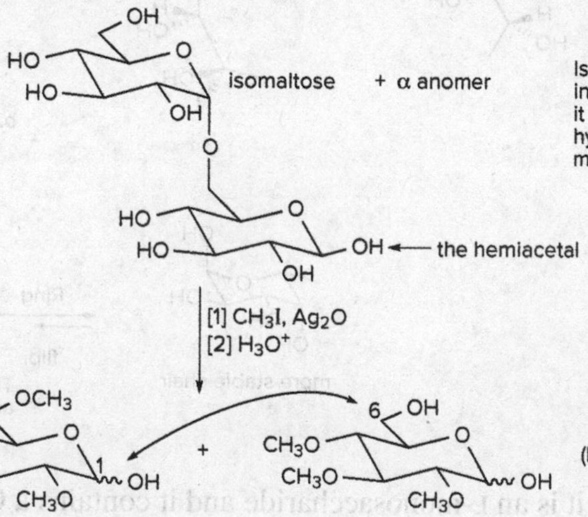

isomaltose + α anomer

Isomaltose must be composed of two glucose units in an α-glycosidic linkage. Because it is a reducing sugar it contains a hemiacetal. The free OH groups in the hydrolysis products show where the two monosaccharides are joined.

the hemiacetal

[1] CH_3I, Ag_2O
[2] H_3O^+

(Both anomers are present.)

24.64

a. HO, HO, H_2N, 1, ← α, O, 6, HO, HO, NH_2, O

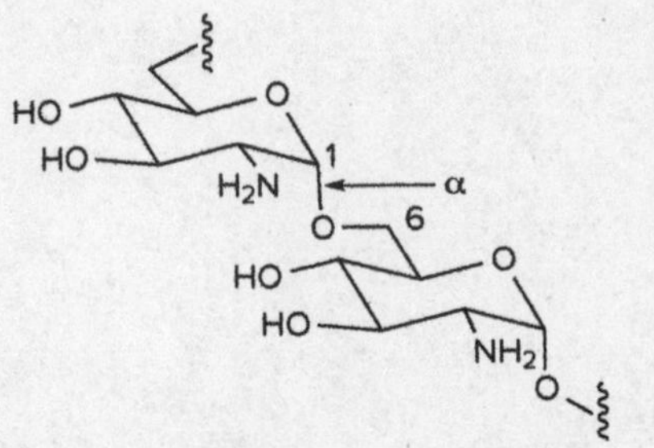

c. CH_2OH, O, H, H, HO, H, $NHCH_2C_6H_5$, H, OH, H

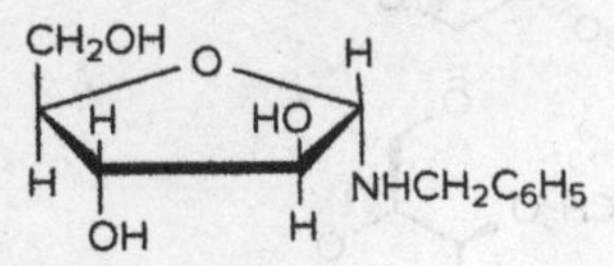

b. OH, OH, O, HO, HO, 1, O, 4, OH, O, HO, OH, OH

mannose glucose

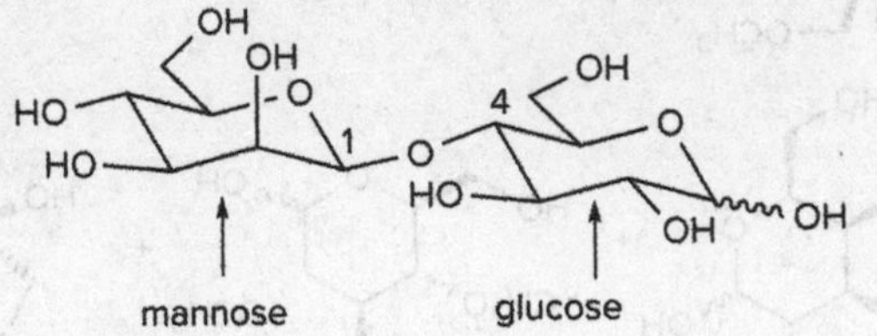

24.65

a. OHC, H OH HO H, H OH HO H — rotate → OHC, OH, H, HO, H, H, OH, H, HO → OHC, OH, H, H, OH, H, OH, OH, H — re-draw →

CHO	
HO	H
H	OH
H	OH
HO	H
CH_3	

b. OH on left in Fischer L-monosaccharide

Ring flip.

more stable chair

The α anomer has the CH_3 on C5 and the anomeric OH trans.

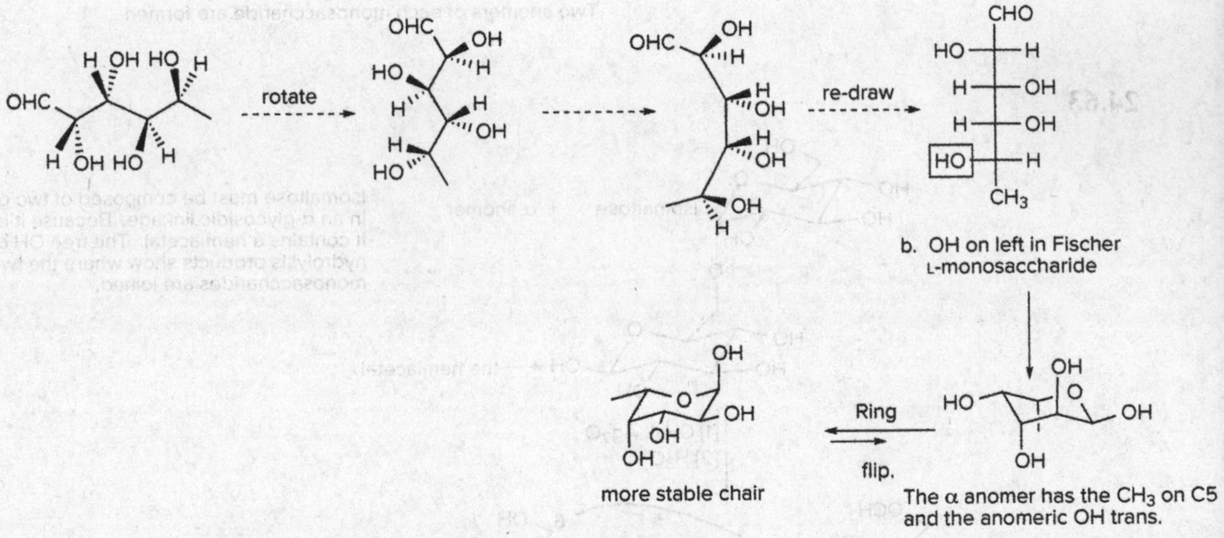

c. Fucose is unusual because it is an L-monosaccharide and it contains a CH_3 group rather than a CH_2OH group on its terminal carbon.

24.66

Ignoring stereochemistry along the way:

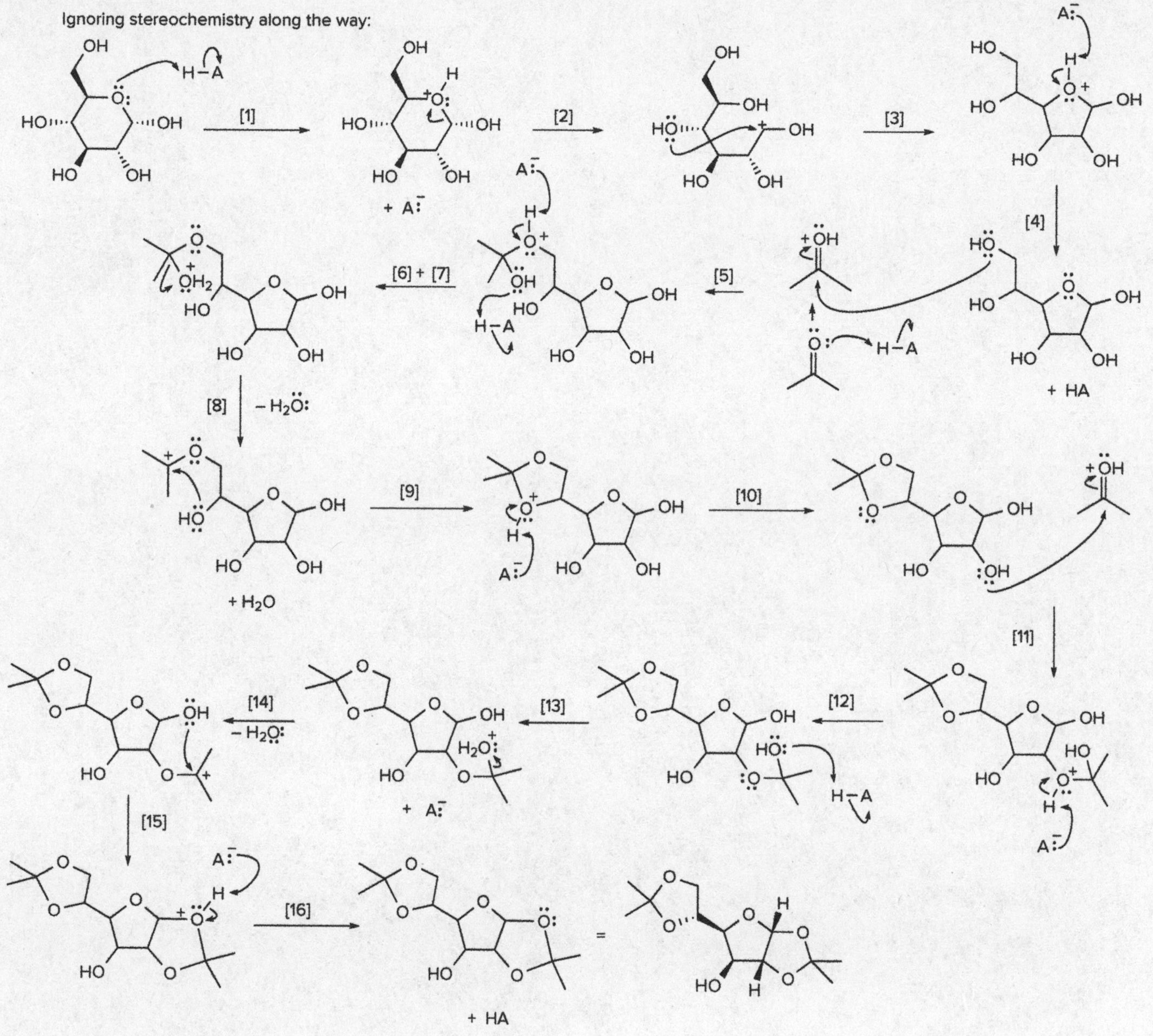

[1] + A:⁻ [2] [3] [4] + HA [5] [6] + [7] [8] − H_2O: + H_2O [9] [10] [11] [12] [13] + A:⁻ [14] − H_2O: [15] [16] + HA =

25.34

a.

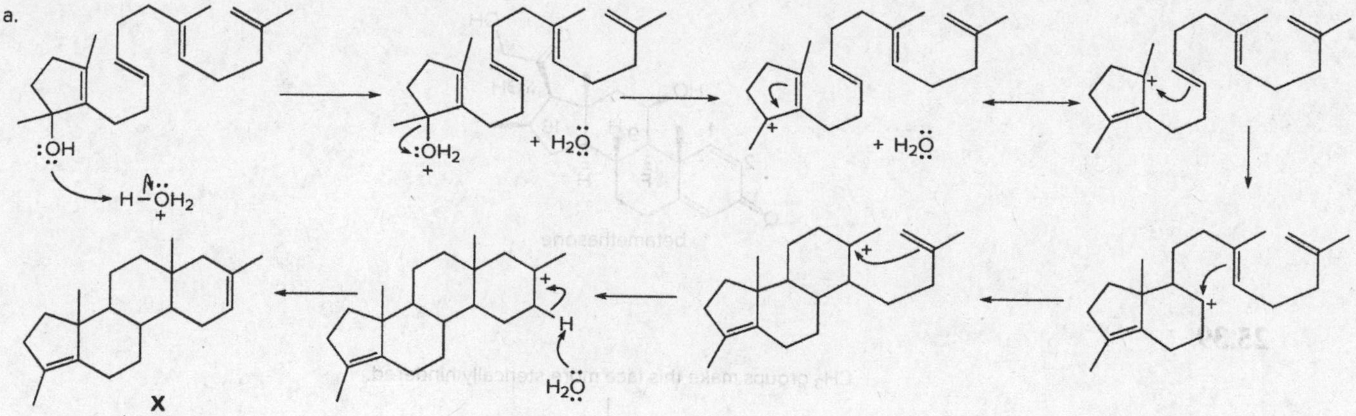

:ÖH
H–ÖH₂ +
:ÖH₂ +
+ H₂Ö:
+ H₂Ö:
H
H₂Ö:
X

b.

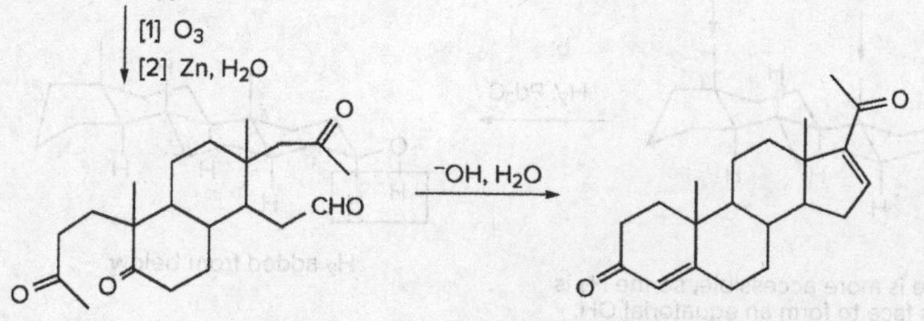

[1] O₃
[2] Zn, H₂O
CHO
⁻OH, H₂O
16,17-dehydroprogesterone

25.35

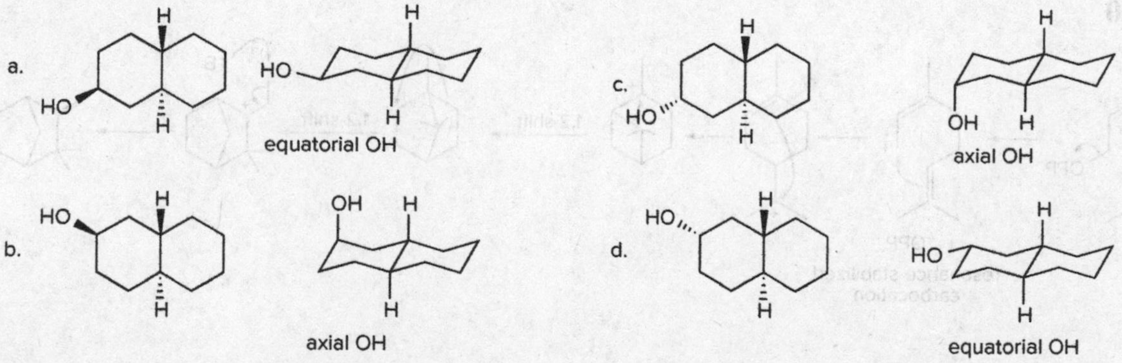

a.
HO
H
H
equatorial OH

b.
HO
H
H
OH
axial OH

c.
HO
H
H
OH
axial OH

d.
HO
H
H
HO
equatorial OH

25.36

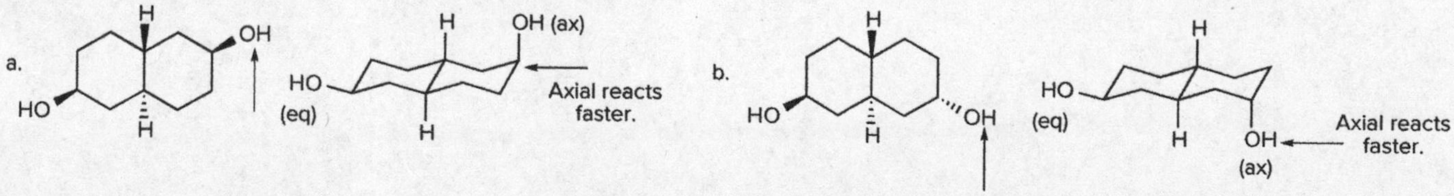

a.
OH
HO
(eq)
OH (ax)
Axial reacts faster.

b.
HO
OH
(eq)
OH (ax)
Axial reacts faster.

25.37

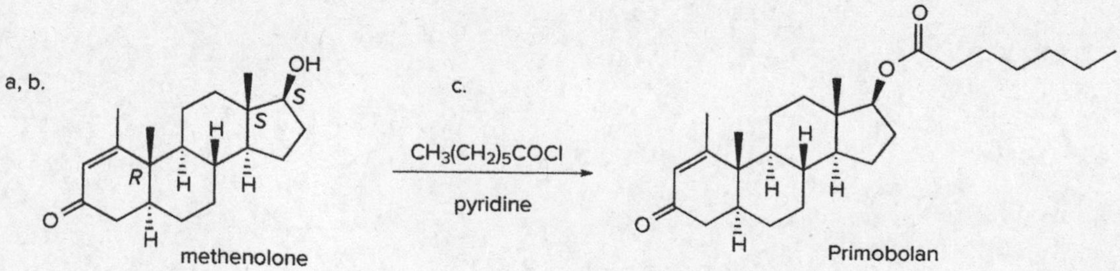

a, b.
OH
R
S
methenolone

c.
$CH_3(CH_2)_5COCl$
pyridine
Primobolan

25.38

betamethasone

25.39

CH_3 groups make this face more sterically hindered.

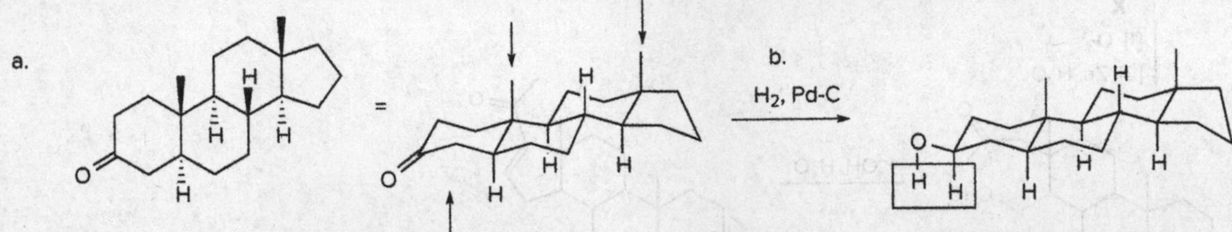

H₂ added from below

The bottom face is more accessible, so the H_2 is added from this face to form an equatorial OH.

25.40

OPP

+ ⁻OPP

resonance-stabilized carbocation

1,2-shift

1,2-shift

:B

25.41 Re-draw the starting material in a conformation that suggests the structure of the product.

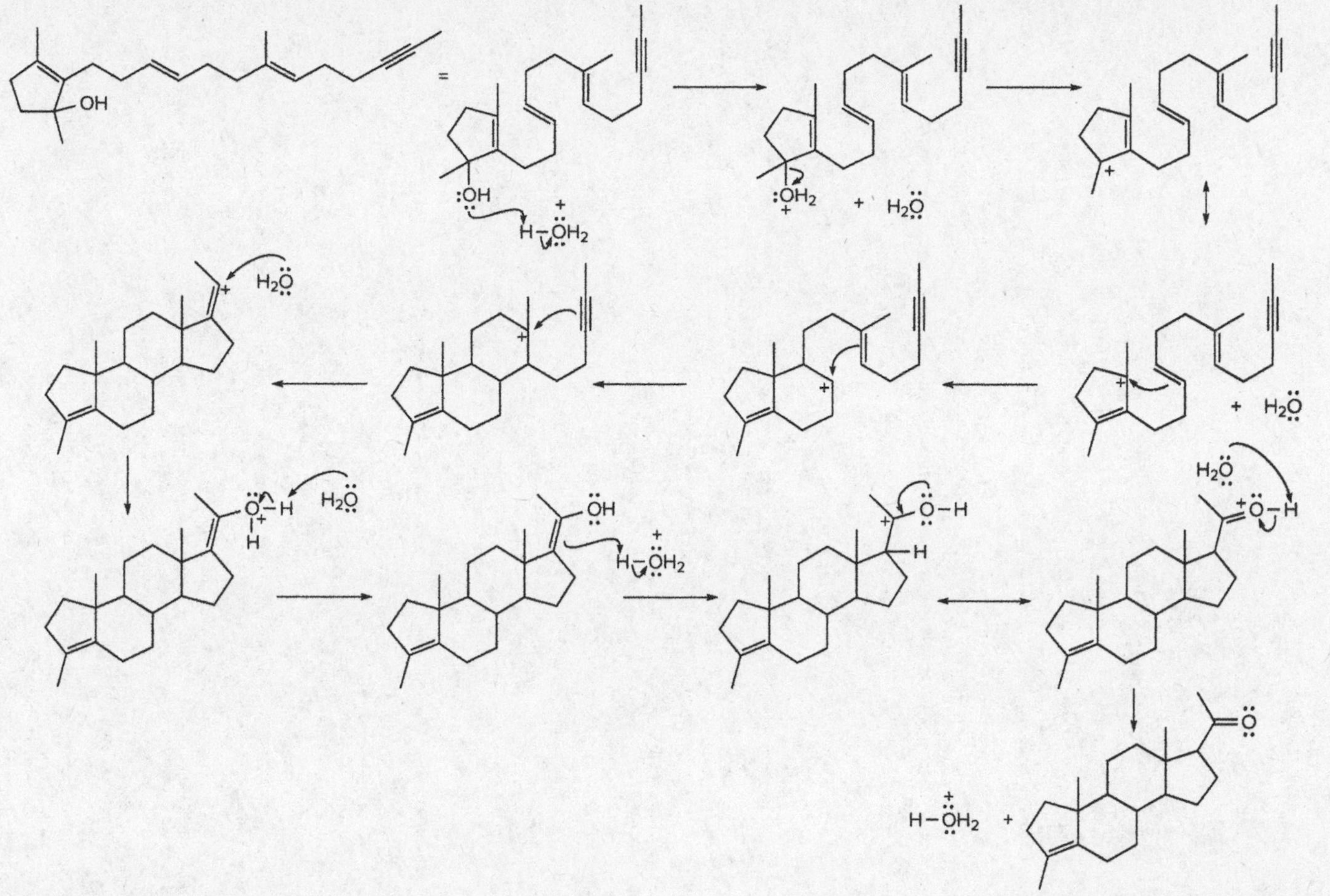

OH =

:OH

$H-\overset{+}{O}H_2$

$:\overset{+}{O}H_2$ + H_2O

+

+ H_2O

H_2O

H_2O

H_2O

OH

$H-\overset{+}{O}H_2$

O–H

H

H_2O

O–H

O

$H-\overset{+}{O}H_2$ +

25.42

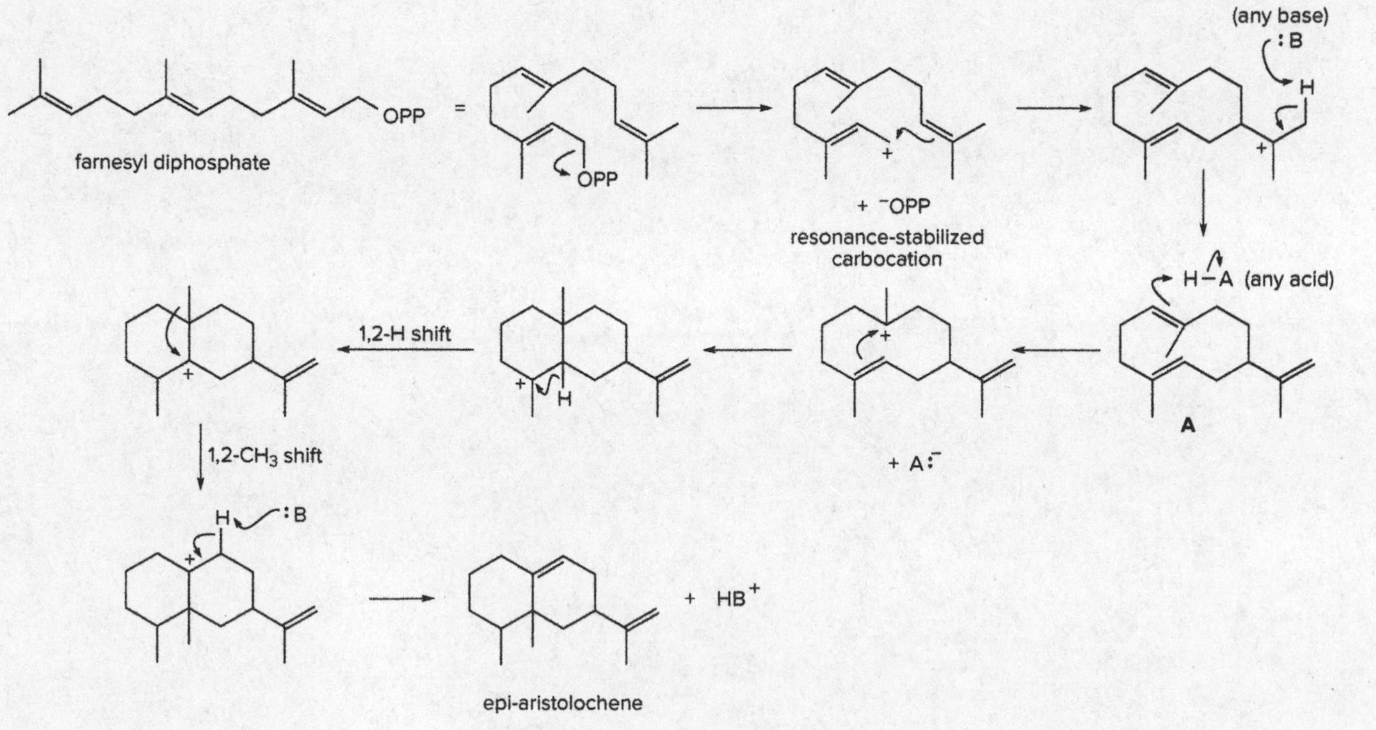

(any base)
:B

H

OPP =

farnesyl diphosphate

OPP

+ ^{-}OPP
resonance-stabilized
carbocation

H–A (any acid)

A

+ A:$^{-}$

1,2-H shift

H

1,2-CH_3 shift

H :B

+ HB^{+}

epi-aristolochene

Chapter 26 Nucleic Acids

Chapter Review

A comparison between DNA and RNA (26.1, 26.3, 26.5)

Deoxyribonucleic acid (DNA)	Ribonucleic acid (RNA)
• composed of deoxyribonucleotides	• composed of ribonucleotides
• double stranded	• single stranded
• monosaccharide—2'-deoxy-D-ribose	• monosaccharide—D-ribose
• bases—C, G, A, and T	• bases—C, G, A, and U

Important terms

Nucleoside	• An *N*-glycoside formed by joining the anomeric carbon of a monosaccharide with a purine or pyrimidine base in a β-glycosidic linkage (26.1)
Nucleotide	• A nucleoside that also contains a phosphate bonded to the OH group on the 5' carbon of the monosaccharide (26.1)
Replication	• The process by which a parent DNA molecule produces two daughter DNA molecules, each of which contains one strand from the parent DNA and one new strand (26.4)
Transcription	• The process which converts DNA to an RNA molecule that contains a base sequence that is complementary to the DNA template from which it is made (26.5)
Translation	• The synthesis of a protein from RNA using the sequence of three-nucleotide codons in RNA that determine the identity of the amino acids in the protein (26.6)

The structure of polynucleotides (26.2)

- Polynucleotides are formed by joining the 3'-OH group of one nucleotide with the 5'-phosphate of a second nucleotide in a **phosphodiester** linkage
- **A polynucleotide is named from the 5' end to the 3' end,** using the one-letter abbreviations for the bases it contains.

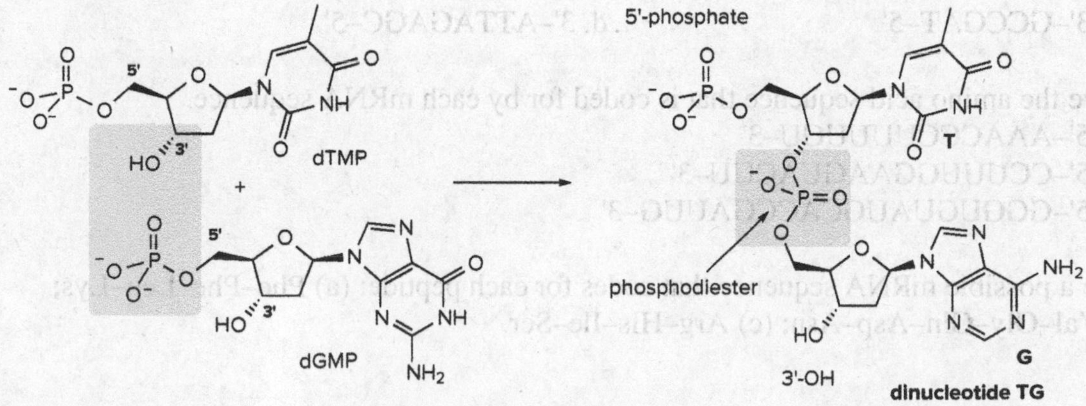

dTMP + dGMP → dinucleotide TG (5'-phosphate, phosphodiester, 3'-OH)

Comparing the sequences of DNA, mRNA, tRNA, and polypeptide

- DNA consists of two strands of nucleotides—a coding strand and a template strand—that have complementary base pairs: A pairs with T and G pairs with C (26.3).
- mRNA is synthesized from the template strand of DNA. The nucleotide sequence of DNA determines the nucleotide sequence of mRNA: G pairs with C, T pairs with A, and A (on DNA) pairs with U (on RNA) (26.5).
- tRNA contains anticodons that are complementary to the codons in mRNA and identify a specific amino acid (26.5).
- The sequence of codons in mRNA (from the 5' end to the 3' end) determines the sequence of amino acids in a polypeptide chain. A polypeptide is synthesized from the N-terminal end to the C-terminal end (26.6).

DNA coding strand:	5' end	ACT	GAG	AAT	GGA	TTA	TCA	3' end
DNA template strand:	3' end	TGA	CTC	TTA	CCT	AAT	AGT	5' end
mRNA:	5' end	ACU	GAG	AAU	GGA	UUA	UCA	3' end
tRNA anticodons:		UGA	CUC	UUA	CCU	AAU	AGU	
Polypeptide:	N-terminal	Thr	Glu	Asn	Gly	Leu	Ser	C-terminal

Practice Test on Chapter Review

1. If 19% of the nucleotides in a sample of DNA contain the base cytosine (C), what are the percentages of bases G, A, and T?

2. What is the sequence of a newly synthesized DNA segment if the template strand has a sequence of 3'–CGCGATTAGATATTGCGC–5'?

3. For each DNA segment: [1] What is the sequence of the mRNA molecule synthesized from each DNA template? [2] What is the sequence of the coding strand of the DNA molecule?
   a. 3'–GGCCTATA–5'
   b. 3'–GCCGAT–5'
   c. 3'–CTACTG–5'
   d. 3'–ATTAGAGC–5'

4. Derive the amino acid sequence that is coded for by each mRNA sequence.
   a. 5'–AAACCCUUUUGU–3'
   b. 5'–CCUUUGGAAGUACUU–3'
   c. 5'–GGGUGUAUGCACCGAUUG–3'

5. Write a possible mRNA sequence that codes for each peptide: (a) Phe–Phe–Leu–Lys; (b) Val–Gly–Gln–Asp–Asn; (c) Arg–His–Ile–Ser.

6. Fill in the base, codon, anticodon, or amino acid needed to complete the following table that relates the sequences of DNA, mRNA, tRNA, and the resulting polypeptide.

DNA coding strand:	5' end		ACA				3' end
DNA template strand:	3' end	CAA				GTC	5' end
mRNA:	5' end			UAC			3' end
tRNA anticodons:					ACA		
Polypeptide:	N-terminal					Lys	C-terminal

Answers to Practice Test

1. 19% G, 31% A, 31% T

2. 5'–GCGCTAATCTATAACGCG–3'

3. a. [1] 5'–CCGGAUAU–3'
   [2] 5'–CCGGATAT–3'
   b. [1] 5'–CGGCUA–3'
   [2] 5'–CGGCTA–3'
   c. [1] 5'–GAUGAC–3'
   [2] 5'–GATGAC–3'
   d. [1] 5'–UAAUCUCG–3'
   [2] 5'–TAATCTCG–3'

4. a. Lys–Pro–Phe–Cys
   b. Pro–Leu–Glu–Val–Leu
   c. Gly–Cys–Met–His–Arg–Leu

5. a. UUCUUCUUAAAA
   b. GUUGGUCAAGAUAAU
   c. AGACAUAUUUCU

6.

DNA coding strand:	5' end	GTT	ACA	TAC	TGT	AAA	CAG	3' end
DNA template strand:	3' end	CAA	TGT	ATG	ACA	TTT	GTC	5' end
mRNA:	5' end	GUU	ACA	UAC	UGU	AAA	CAG	3' end
tRNA anticodons:		CAA	UGU	AUG	ACA	UUU	GUC	
Polypeptide:	N-terminal	Val	Thr	Tyr	Cys	Lys	Gln	C-terminal

Answers to Problems

26.1 The monosaccharide is D-ribose if there is an OH group at C2' and 2'-deoxy-D-ribose if there is no OH group at C2'.

a. HO O N NH O O 2' HO OH uracil D-ribose

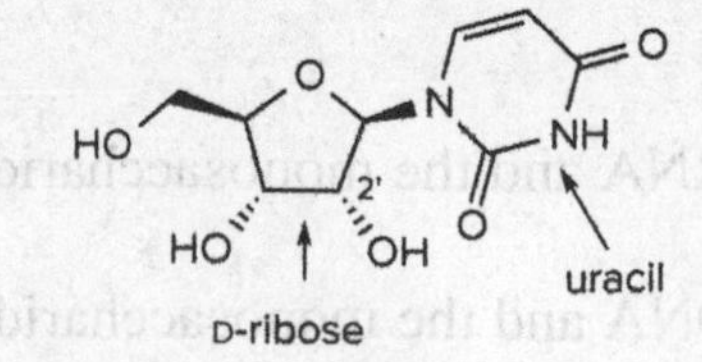

uridine

b. HO O N N O NH N NH₂ guanine 2' HO 2'-deoxy-D-ribose

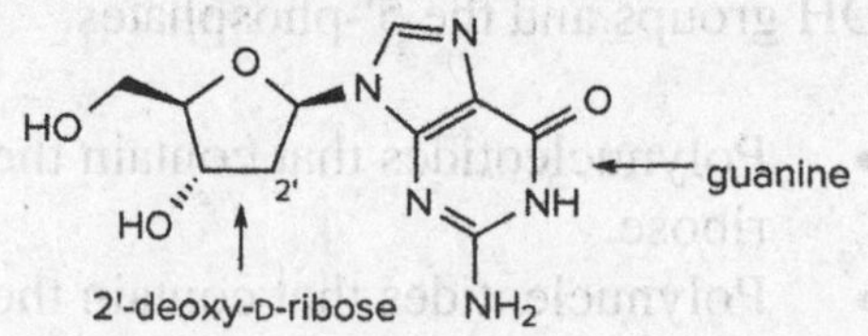

2'-deoxyguanosine

26.2 The monosaccharide in vidarabine has an OH group at C2', but the configuration is different from D-ribose; that is, the OH is drawn *above* the furanose ring, not *below* the ring as in ribose. Figure 24.4 shows that the epimer of D-ribose at C2 is D-arabinose.

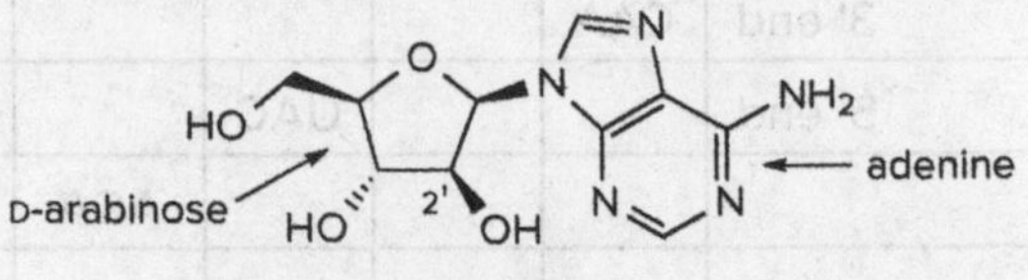

vidarabine

26.3 Convert the three- or four-letter abbreviation to a name.

- An abbreviation that begins with "d" indicates a deoxyribonucleoside and the monosaccharide is 2'-deoxy-D-ribose.
- The first upper-case letter identifies the base (C = cytosine; T = thymine; U = uracil; A = adenine; G = guanine).

a. dTMP = 2'-deoxythymidine 5'-monophosphate

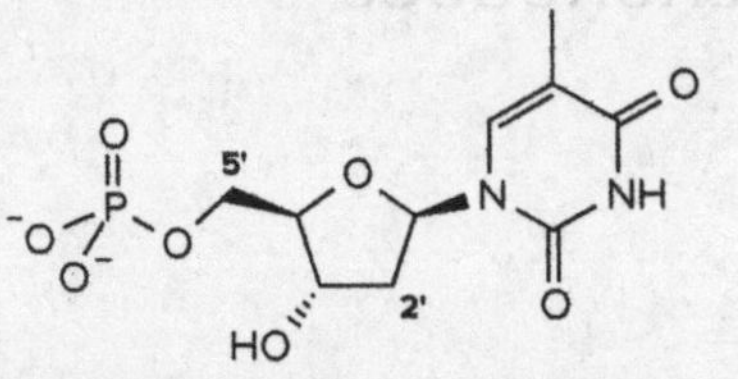

2'-deoxythymidine 5'-monophosphate

b. AMP = adenosine 5'-monophosphate

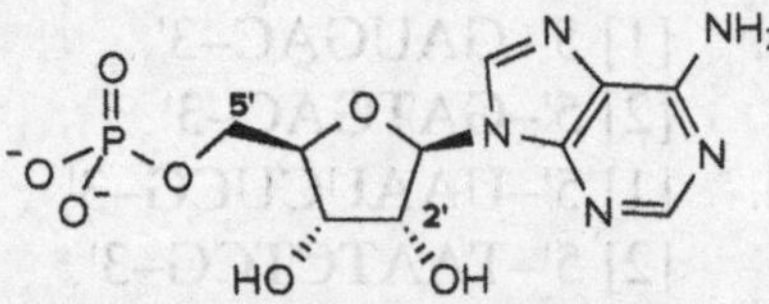

adenosine 5'-monophosphate

26.4 DNA contains (b) 2'-deoxyribose, (c) the base T, and (f) the nucleotide dCMP. RNA contains (a) ribose, (d) the base U, and (e) the nucleotide GMP.

26.5 In the name of a nucleotide, the second uppercase letter indicates the type of phosphate (M = monophosphate; D = diphosphate; T = triphosphate).

a. GTP: guanosine 5'-triphosphate
b. dCDP: 2'-deoxycytidine 5'-diphosphate
c. dTTP: 2'-deoxythymidine 5'-triphosphate
d. UDP: uridine 5'-diphosphate

26.6 To draw a polynucleotide, join nucleotides together with phosphodiester bonds between the 3'-OH groups and the 5'-phosphates.

- Polynucleotides that contain the base U occur in RNA and the monosaccharide is D-ribose.
- Polynucleotides that contain the base T occur in DNA and the monosaccharide is 2'-deoxy-D-ribose.
- In converting the abbreviation of a polynucleotide to a structure, the first nucleotide has a free 5'-phosphate and the last nucleotide has a free 3'-OH group.

a.

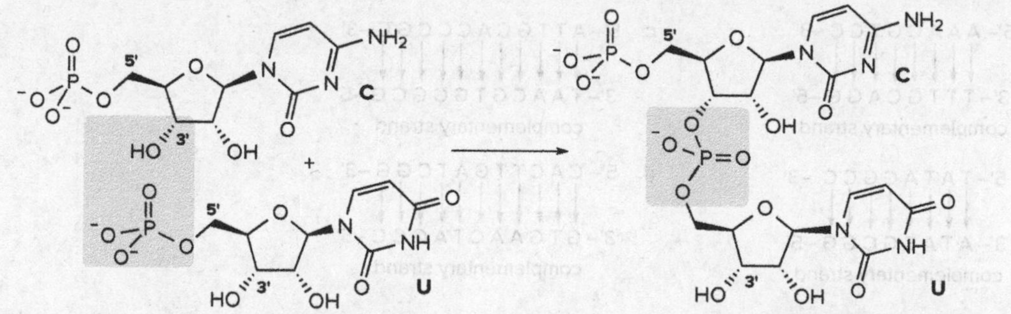

dinucleotide CU

b.

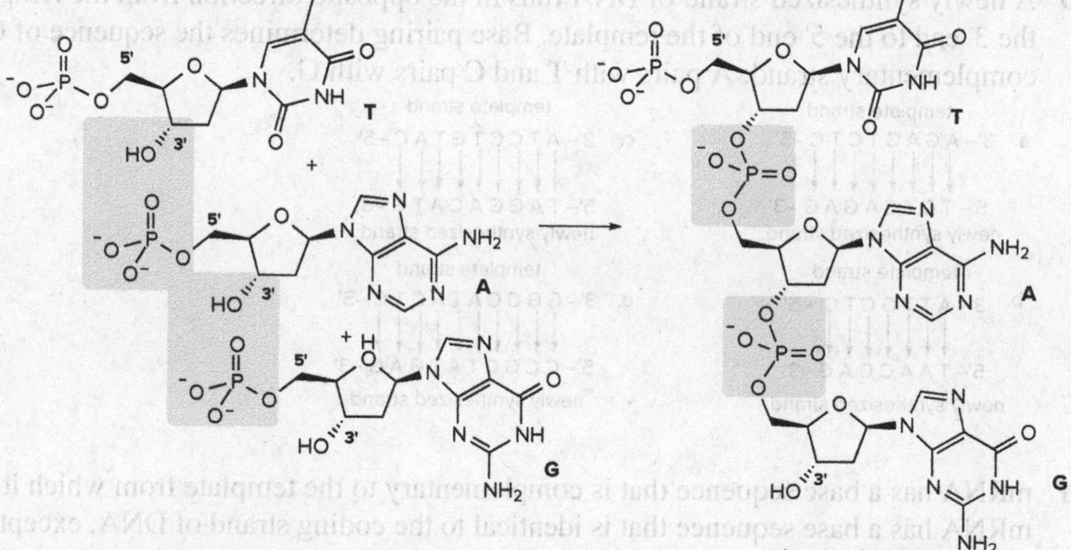

trinucleotide TAG

26.7 For the polynucleotide ATGGCC:

a. Five phosphodiester bonds join the six nucleotides together.

b. The 5' end of the polynucleotide has the base A, a purine.

c. The polynucleotide could be part of a DNA molecule because it contains the base T, a base that occurs only in DNA.

26.8 DNA is arranged with the sugar–phosphate backbone on the outside of the double helix in order to maximize the interaction of the ionic phosphodiesters with the aqueous environment of the cell. The base pairs are stacked one on top of another to maximize the interaction of the uncharged bases with each other.

26.9 The complementary strand of DNA runs in the opposite direction, from the 3' end to the 5' end. Base pairing determines the sequence of the complementary strand: A pairs with T and C pairs with G.

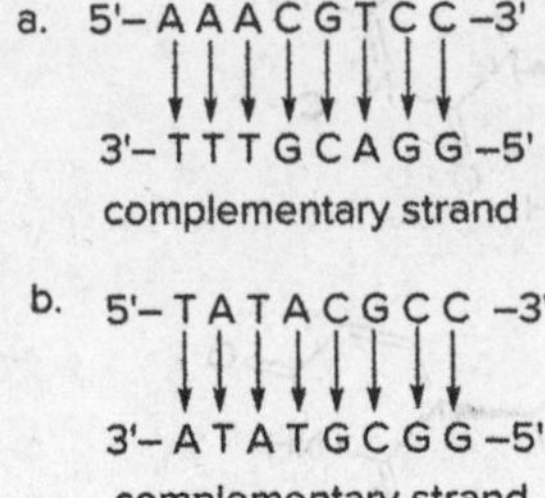

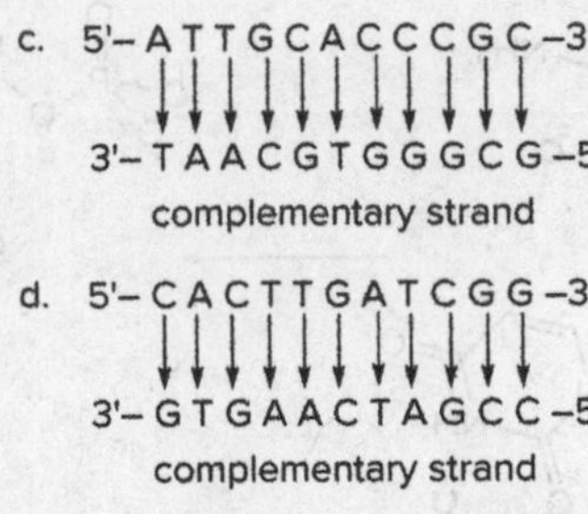

b. 5'– T A T A C G C C –3'

3'– A T A T G C G G –5'

complementary strand

d. 5'– C A C T T G A T C G G –3'

3'– G T G A A C T A G C C –5'

complementary strand

26.10 A newly synthesized strand of DNA runs in the opposite direction from the template strand, from the 3' end to the 5' end of the template. Base pairing determines the sequence of the complementary strand: A pairs with T and C pairs with G.

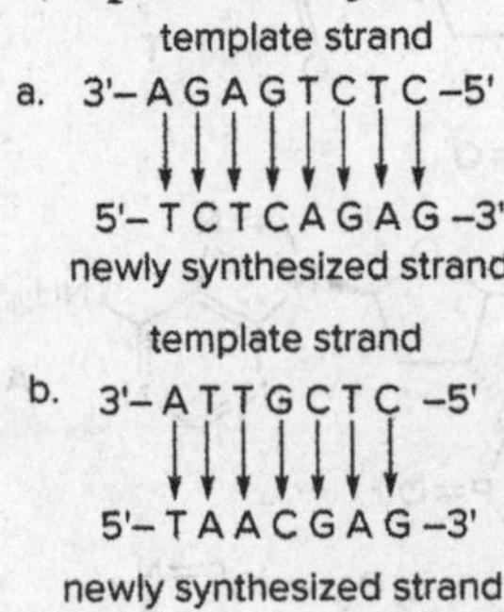

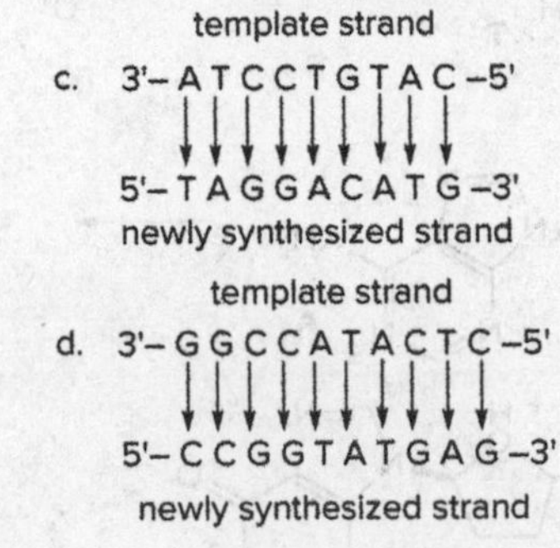

template strand

b. 3'– A T T G C T C –5'

5'– T A A C G A G –3'

newly synthesized strand

template strand

d. 3'– G G C C A T A C T C –5'

5'– C C G G T A T G A G –3'

newly synthesized strand

26.11 mRNA has a base sequence that is complementary to the template from which it is prepared. mRNA has a base sequence that is identical to the coding strand of DNA, except that it contains the base U instead of T.

a. template strand: 3'– T G C C T A A C G –5'
[1] mRNA sequence: 5'– A C G G A U U G C –3'
[2] coding strand: 5'– A C G G A T T G C –3'

b. template strand: 3'– G A C T C C –5'
[1] mRNA sequence: 5'– C U G A G G –3'
[2] coding strand: 5'– C T G A G G –3'

c. template strand: 3'– T T A A C G C G A –5'
[1] mRNA sequence: 5'– A A U U G C G C U –3'
[2] coding strand: 5'– A A T T G C G C T –3'

d. template strand: 3'– C A G T G A C C G T A C –5'
[1] mRNA sequence: 5'– G U C A C U G G C A U G –3'
[2] coding strand: 5'– G T C A C T G G C A T G –3'

26.12 Work backwards from the mRNA sequence to the DNA template strand from which it was prepared. The base G pairs with C, the base U (on mRNA) pairs with A (on DNA), and the base A (on mRNA) pairs with T (on DNA).

a. mRNA sequence: 5'– U G G G G C A U U –3'
template strand: 3'– A C C C C G T A A –5'

b. mRNA sequence: 5'– G U A C C U –3'
template strand: 3'– C A T G G A –5'

c. mRNA sequence: 5'– C C G A C G A U G –3'
template strand: 3'– G G C T G C T A C –5'

d. mRNA sequence: 5'– G U A G U C A C G –3'
template strand: 3'– C A T C A G T G C –5'

26.13 The 5' end of an mRNA molecule codes for the N-terminal amino acid in a protein, and the codon at the 3' end of the mRNA codes for the C-terminal amino acid. (a) The codon at the 5' end of the mRNA segment (5'-CAUAAAACGGAG-3') is CAU, which codes for the N-terminal amino acid His. (b) The codon at the 3' end of the mRNA segment is GAG, which codes for the C-terminal amino acid Glu.

26.14 Looking at the identity of the codons for similar types of amino acids shows that there are some similarities among them.

- Aspartic acid (Asp) and glutamic acid (Glu) are similar in structure in that both acidic amino acids have a side chain containing an ionized carboxy group (CO_2^-). The codons for Asp are GAU and GAC, and the codons for Glu are GAA and GAG. Thus, each codon begins with the same two bases, G and A.

- Leucine (Leu), phenylalanine (Phe), and valine (Val) are neutral amino acids with nonpolar side chains. The codons for these amino acids are as follows: Leu (UUA, UUG, CUU, CUC, CUA, CUG); Phe (UUU, UUC); and Val (GUU, GUC, GUA, GUG). Each of these codons has the pyrimidine base U as the second of the three bases in the codon.

26.15 Use the DNA sequence to determine the transcribed mRNA sequence: C pairs with G, T pairs with A, and A (on DNA) pairs with U (on mRNA). Use the codons in Table 26.2 to determine what amino acids are coded for by a given codon in mRNA.

a. template strand: 3'– TCT CAT CGT AAT GAT TCG –5'
mRNA sequence: 5'– AGA GUA GCA UUA CUA AGC –3'
↓ ↓ ↓ ↓ ↓ ↓
polypeptide: Arg Val Ala Leu Leu Ser

b. template strand: 3'–GCT CCT AAA TAA CAC TTA –5'
mRNA sequence: 5'–CGA GGA UUU AUU GUG AAU –3'
↓ ↓ ↓ ↓ ↓ ↓
polypeptide: Arg Gly Phe Ile Val Asn

26.16 Use the codons in Table 26.2 to determine what amino acids are coded for by a given codon in mRNA. Anticodons in tRNA are complementary to the codons of mRNA.

a. mRNA sequence: 5'– CCA CCG GCA AAC GAA GCA –3'
polypeptide: Pro Pro Ala Asn Glu Ala
tRNA codons: GGU GGC CGU UUG CUU CGU

b. mRNA sequence: 5'– GCA CCA CUA AGA GAC –3'
polypeptide: Ala Pro Leu Arg Asp
tRNA codons: CGU GGU GAU UCU CUG

26.17 Use the information in Answer 26.15.

template strand: 3'– ATG AAA GCC TTC TGT –5'
a. coding strand: 5'– TAC TTT CGG AAG ACA –3'
b. mRNA sequence: 5'– UAC UUU CGG AAG ACA –3'
c. polypeptide: Tyr Phe Arg Lys Thr

26.18 Use the information in Answers 26.11 and 26.15.

DNA coding strand:	5' end	AAC	GTA	TCA	ACT	CAC	ATG	3' end
DNA template strand:	3' end	TTG	CAT	AGT	TGA	GTG	TAC	5' end
mRNA:	5' end	AAC	GUA	UCA	ACU	CAC	AUG	3' end
tRNA anticodons:		UUG	CAU	AGU	UGA	GUG	UAC	
Polypeptide:	N-terminal	Asn	Val	Ser	Thr	His	Met	C-terminal

26.19 Restriction endonucleases cut DNA at specific sequences of bases. HindIII cuts DNA between A and A in the sequence AAGCTT, and HaeIII cuts DNA between G and C in the sequence GGCC.

HindIII cuts here. HindIII cuts here.

5'-CGCGAATTGGCCGTAAGCTTACGTCCTAGGGCTACTCCTCGGCCCAATAAAGCTT-3'

HaeIII cuts here. HaeIII cuts here.

26.20 Lamivudine is formed by joining **A** with cytosine in an *N*-glycoside linkage. Lamivudine can act as an antiviral drug by interfering with reverse transcription. Because lamivudine lacks a 3'-OH group, no additional nucleotide can be added to the growing DNA chain, and this halts viral DNA synthesis.

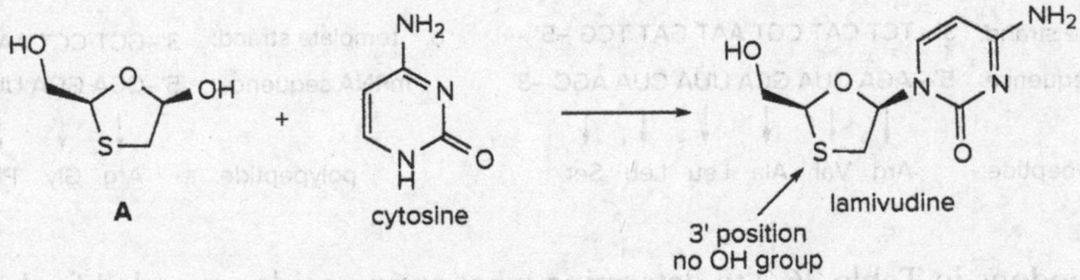

A + cytosine → lamivudine

3' position no OH group

26.21 Translate the ball-and-stick models to skeletal structures. Both **A** and **B** are nucleosides because they are composed of a monosaccharide and a base (but no phosphate). Compounds that contain D-ribose are components of RNA. Compounds that contain 2'-deoxy-D-ribose are components of DNA.

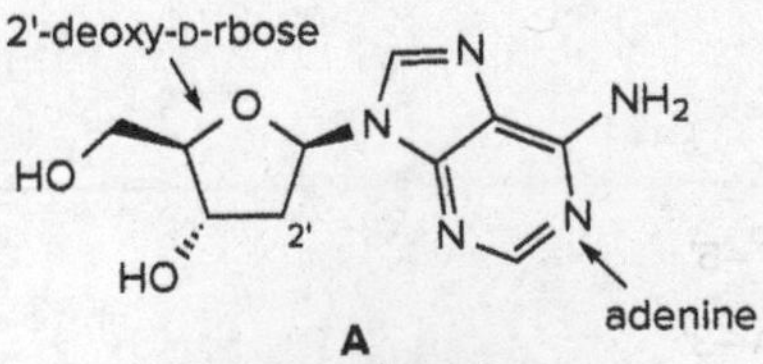

2'-deoxy-D-rbose; adenine

A
2'-deoxyadenosine
component of DNA

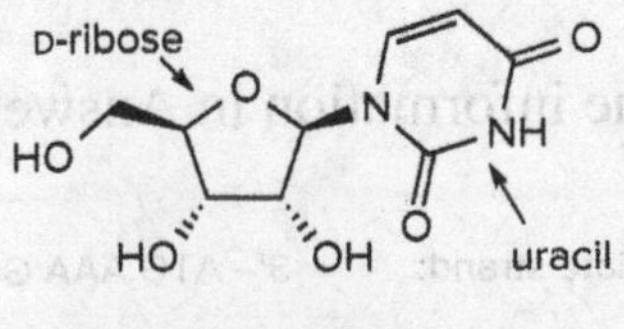

D-ribose; uracil

B
uridine
component of RNA

26.22 Translate the ball-and-stick models to skeletal structures. Compounds with one phosphate group are named as monophosphates and those with two phosphates are named as diphosphates.

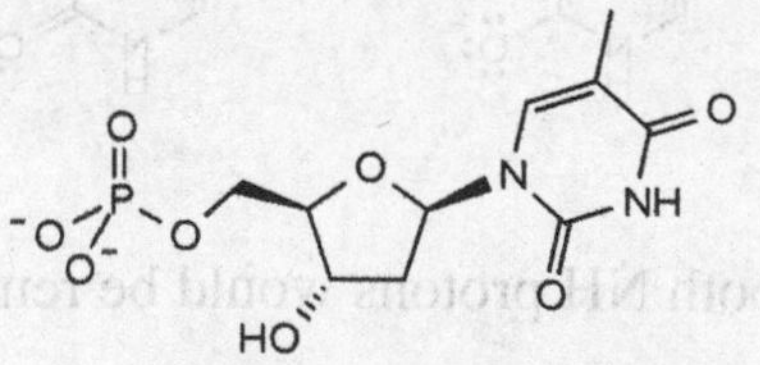

2'-deoxythymidine 5'-monophosphate
dTMP

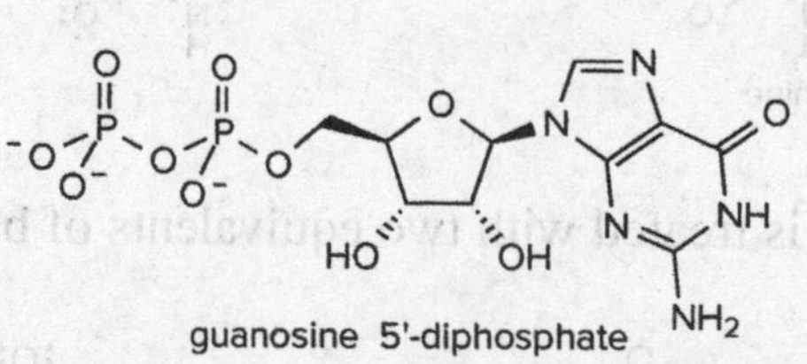

guanosine 5'-diphosphate
GDP

26.23

a. Mono enol tautomers for thymine:

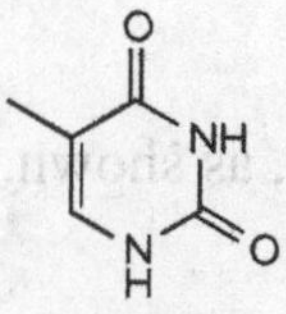

thymine

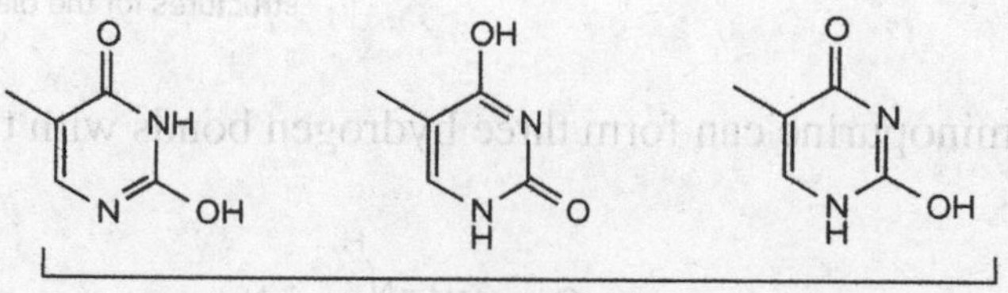

mono enol tautomers

b. Dienol tautomer for uracil:

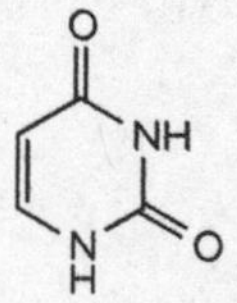

uracil

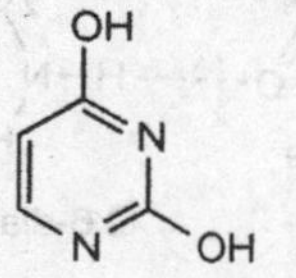

dienol tautomer of uracil

c. No enol tautomers can be drawn for caffeine because caffeine has no H atoms on the N's that are bonded to the carbonyl groups.

d. Caffeine is an aromatic compound because a resonance structure can be drawn with 10 π electrons—eight electrons from four π bonds and two electrons on the labeled N atom.

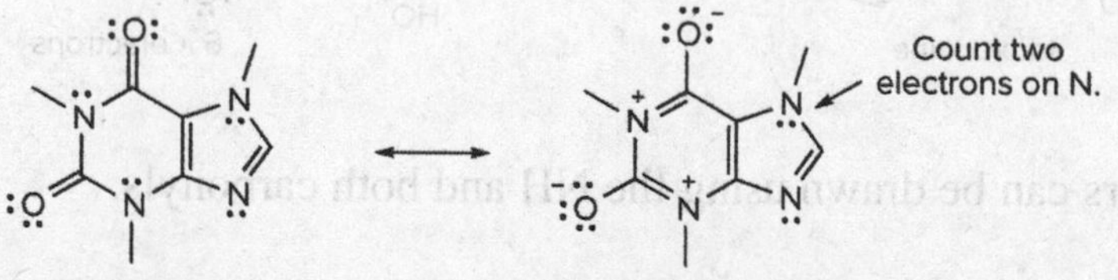

Count two electrons on N.

caffeine

resonance structure showing 10 π electrons

26.24

a. The most acidic proton in thymine is the H atom on the N between the two carbonyls because its conjugate base is the most resonance stabilized.

O, loss of H^+, NH, N, H, thymine → :O:, N:⁻, N, H, O: ↔ :O:, N:, N, H, O:⁻ ↔ :O:⁻, N:, N, H, O:

b. If thymine is treated with two equivalents of base, both NH protons would be removed to form a dianion.

O, loss of H^+, NH, N, H, O, loss of H^+ → :O:, N:⁻, N:⁻, O: ↔ :O:⁻, N:, N:⁻, O:⁻

two of the possible resonance structures for the dianion

26.25 2,6-Diaminopurine can form three hydrogen bonds with thymine, as shown.

O-----H–N(H), N–H-----N, HN, O-----H–N(H), 6, 2, N, NH, N

thymine 2,6-diaminopurine

26.26

a. A second resonance structure can be drawn for the six-membered ring of idoxuridine that gives it three π bonds, thus making it aromatic with six π electrons.

I, HO, O, N:, N, H, :O:, HO, idoxuridine ↔ I, :O:⁻, HO, O, N, NH, +, :O:⁻, HO, 6 π electrons

b. Two enol tautomers can be drawn using the NH and both carbonyls.

I, O, HO, O, N, N, H, O, HO, idoxuridine; I, O, HO, O, N, N, HO, HO; I, OH, HO, O, N, N, O, HO

two enol tautomers

26.27 Alkylation occurs on oxygen because the O atom is nucleophilic, which can be seen by drawing a second resonance structure for guanine which places a negative charge on the oxygen.

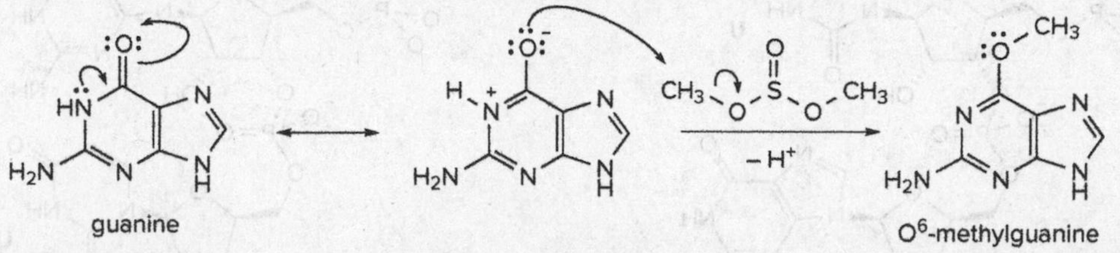

guanine

O^6-methylguanine

O^6-methylguanine can form only two hydrogen bonds with cytosine because it lacks the NH bond used for hydrogen bonding in guanine.

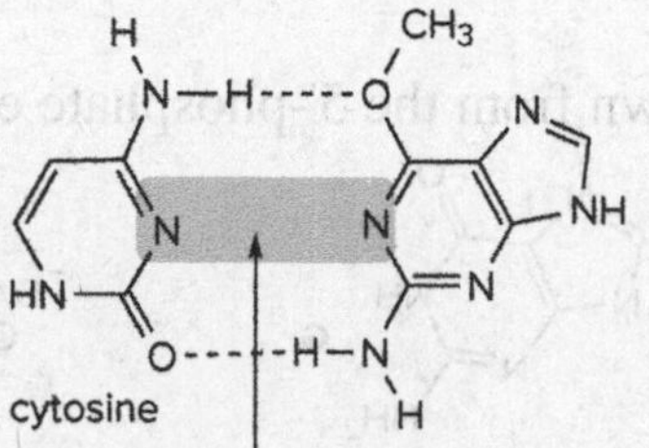

cytosine

no possibility for hydrogen bonding

26.28 The base is thymine and the monosaccharide is D-arabinose.

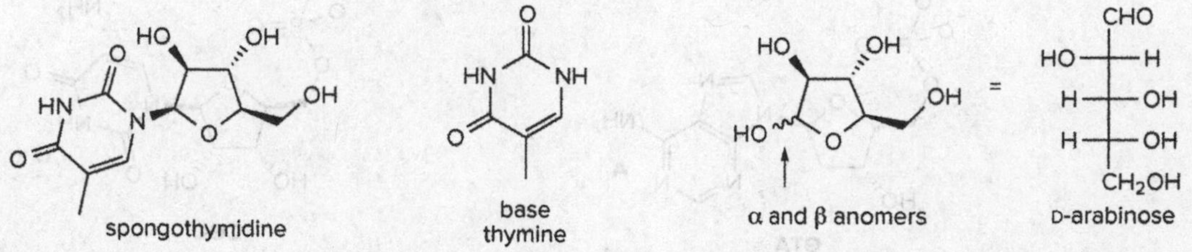

spongothymidine

base thymine

α and β anomers

D-arabinose

26.29

a. Two dinucleotides can be formed from dTMP ad dAMP.

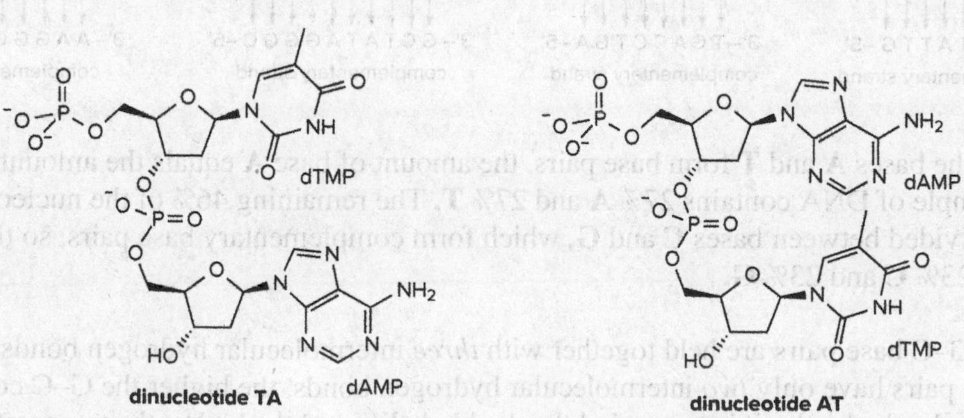

dinucleotide TA

dinucleotide AT

b. Two dinucleotides can be formed from uridine 5'-monophosphate and guanosine 5'-monophosphate.

dinucleotide UG

dinucleotide GU

26.30 The trinucleotides are drawn from the 5'-phosphate end to the 3'-OH end.

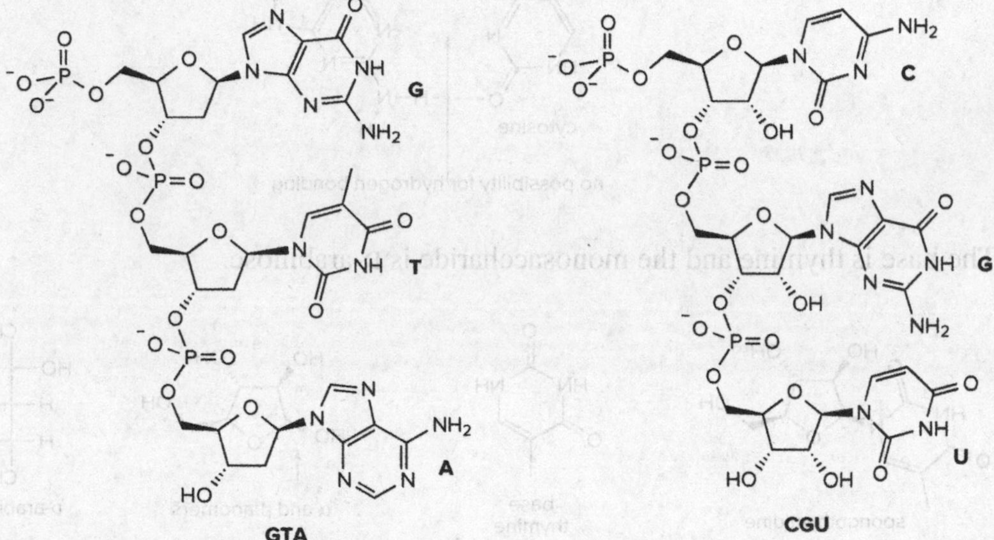

26.31 Use the information in Answer 26.9.

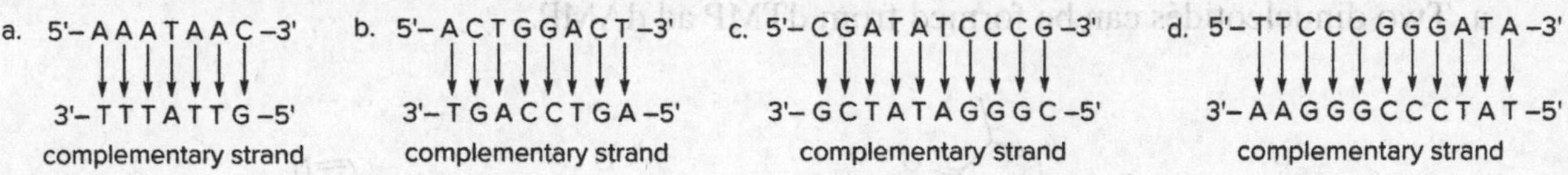

26.32 Because the bases **A** and **T** form base pairs, the amount of base **A** equals the amount of base **T**, so the sample of DNA contains 27% **A** and 27% **T.** The remaining 46% of the nucleotides are equally divided between bases **C** and **G,** which form complementary base pairs, so the sample contains 23% **C** and 23% **G.**

26.33 Because G–C base pairs are held together with *three* intermolecular hydrogen bonds, whereas A–T base pairs have only *two* intermolecular hydrogen bonds, the higher the G–C content of the DNA sample, the harder it is to unwind the double helix, and the higher the temperature required for melting.

26.34 Ribonucleotide codons are written from the 5' end to the 3' end and D-ribose is the monosaccharide.

26.35 Use the information in Answer 26.10.

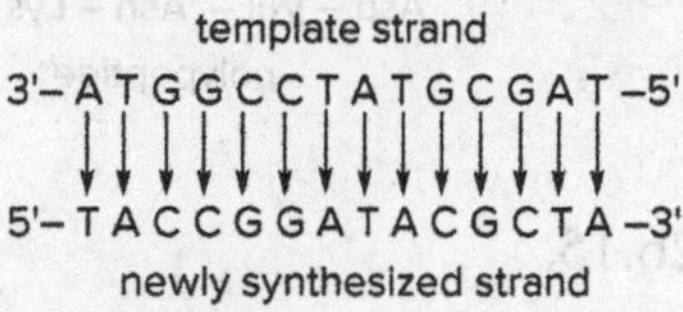

26.36 Use the information in Answer 26.11.

a. template strand: 3'–ATGGCTTA–5'
[1] mRNA sequence: 5'–UACCGAAU–3'
[2] coding strand: 5'–TACCGAAT–3'

b. template strand: 3'–CGGCGCTTA–5'
[1] mRNA sequence: 5'–GCCGCGAAU–3'
[2] coding strand: 5'–GCCGCGAAT–3'

c. template strand: 3'–GGTATACCG–5'
[1] mRNA sequence: 5'–CCAUAUGGC–3'
[2] coding strand: 5'–CCATATGGC–3'

d. template strand: 3'–TAGGCCGTA–5'
[1] mRNA sequence: 5'–AUCCGGCAU–3'
[2] coding strand: 5'–ATCCGGCAT–3'

26.37 When the second base in a codon is one of the pyrimidine bases (C and U), the codon generally codes for amino acids such as Leu, Phe, Ile, Thr, and Val, which contain nonpolar side chains. When the second base in a codon is a purine base (G and A), the codon codes for more polar amino acids, especially those that have charged side chains such as Asp, Glu, and Lys.

26.38 If each codon occurs with equal frequency, the least common amino acids would be those that have the smallest number of codons—methionine and tryptophan—because each of these amino acids has only one codon that codes for it.

26.39 Use the codons in Table 26.2 to determine what amino acids are coded for by a given codon in mRNA.

a. 5'– CCA ACC UGG GUA GAA –3'
Pro – Thr – Trp – Val – Glu

b. 5'– AUG UUU UUA UGG UGG –3'
Met – Phe – Leu – Trp – Trp

c. 5'– GUC GAC GAA CCG CAA –3'
Val – Asp – Glu – Pro – Gln

26.40 Work backwards from the polypeptide to the mRNA sequence using the codons in Table 26.2.

a. Ile – Met – Lys – Ser – Tyr
↓
AUU AUG AAA AGU UAU
possible mRNA sequence

b. Pro – Gln – Glu – Asp – Phe
↓
CCU CAA GAA GAU UUU
possible mRNA sequence

c. Thr – Ser – Asn – Arg
↓
ACU AGU AAU CGU
possible mRNA sequence

26.41 Use the information from Answers 26.12 and 26.39.

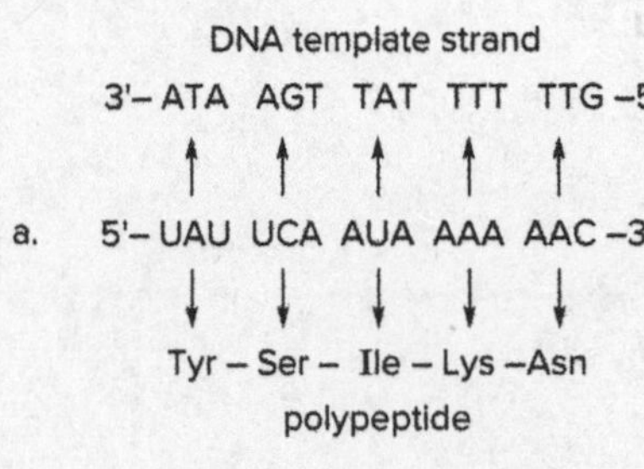

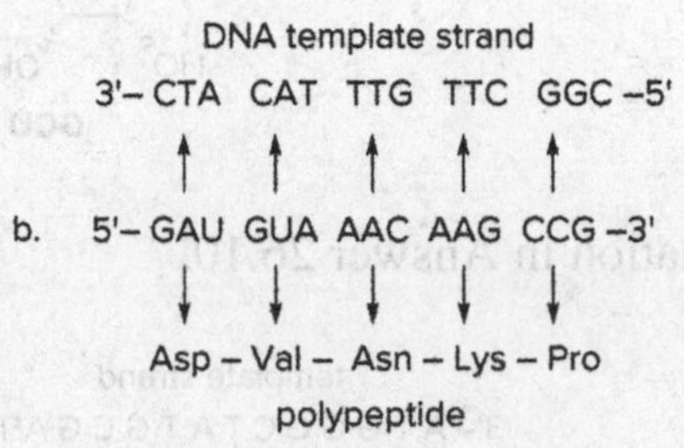

26.42 Use the information from Answer 26.15.

template strand: 3'– AAC GTC CTC ACG ATT –5'.
mRNA sequence: 5'– UUG CAG GAG UGC UAA –3'
↓
polypeptide: Leu – Gln – Glu – Cys STOP

26.43 Work backwards from the polypeptide to a possible mRNA sequence using the codons in Table 26.2. Then work backwards from the mRNA sequence to the DNA template strand.

polypeptide: Tyr – Gly – Gly – Phe – Met
↓
mRNA sequence: 5'– UAU GGU GGU UUU AUG –3'
template strand: 3'– ATA CCA CCA AAA TAC –5'.

26.44 Identify the amino acids in each peptide, and draw the peptide from the N-terminal to the C-terminal end. Work backwards from the peptide to a possible mRNA sequence using the codons in Table 26.2, and then use complementary base pairing to determine the sequence of the DNA template strand.

a.

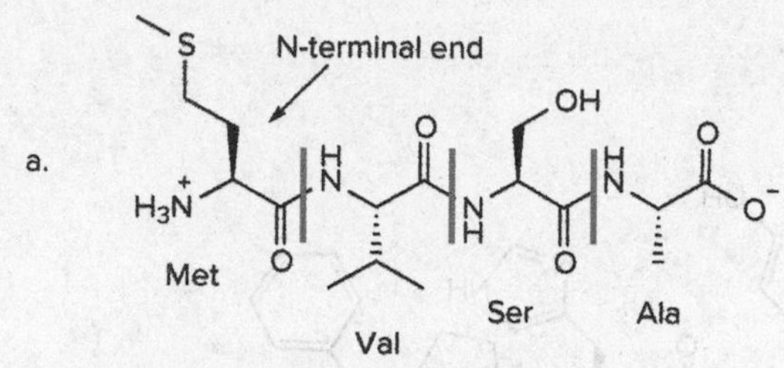

Met, Val, Ser, Ala

N-terminal end

peptide: Met – Val – Ser – Ala

possible mRNA sequence: AUG GUU UCU GCU

DNA template strand: TAC CAA AGA CGA

b.

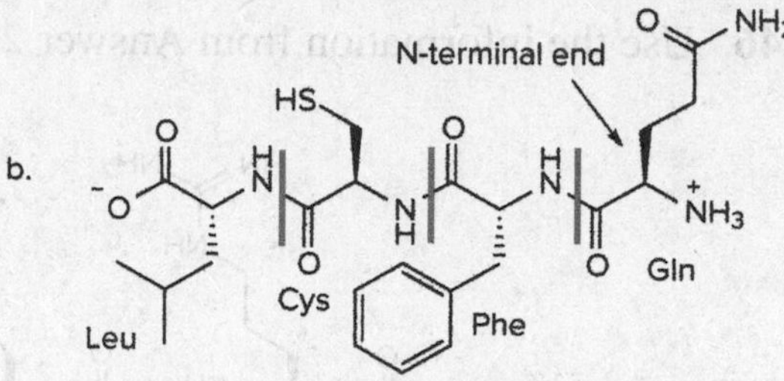

Leu, Cys, Phe, Gln

N-terminal end

peptide: Gln – Phe – Cys – Leu

possible mRNA sequence: CAA UUU UGU UUA

DNA template strand: GTT AAA ACA AAT

26.45 Work backwards from the peptide to a possible mRNA sequence using the codons in Table 26.2, and then use complementary base pairing to determine the sequence of the DNA template strand. Draw the polynucleotide from the 5'-phosphate end to the 3'-OH end. The polynucleotide is composed of six nucleotides.

Met – Trp ⟶ **AUG UGG** ⟶ **TAC ACC**

dipeptide — mRNA sequence — DNA template strand

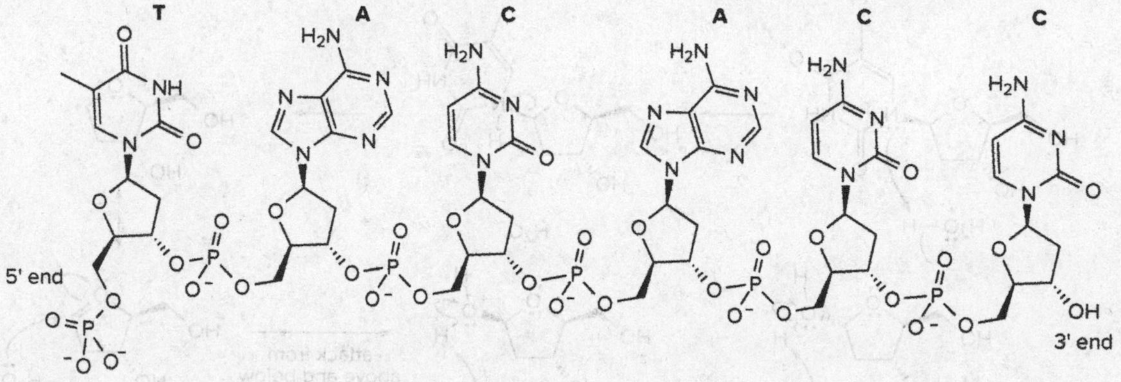

T **A** **C** **A** **C** **C**

5' end

3' end

27.33

a. ATP (or GTP) is directly formed in Steps [7] and [10] of glycolysis and Step [5] of the citric acid cycle.

b. A net of four molecules of ATP are formed directly when glucose is converted to CO_2.

c. A total of 28 molecules of ATP are generated from the reduced coenzymes after Stage [4] of catabolism. As a result, most of the ATP formed during glucose catabolism comes from Stage [4].

27.34 Only one of the carboxy groups in oxalosuccinate is part of a β-keto carboxylate, which readily decarboxylates, as we learned in Chapter 17. Loss of the labeled CO_2 forms a resonance-stabilized enolate, which is protonated to α-ketoglutarate.

β α
decarboxylation
H–A
β-keto carboxylate
oxalosuccinate
resonance-stabilized enolate
α-ketoglutarate

27.35 Use the steps in Answer 27.22 to determine the ATP yield for each carboxylic acid.

For **A**, $CH_3(CH_2)_6CO_2H$: **A** has eight C atoms, so it forms four molecules of acetyl CoA by three cycles of β-oxidation, resulting in 3 NADH and 3 $FADH_2$. **ATP yield:** (–2 ATP) for formation of the acyl CoA; 7.5 ATP from 3 NADH; 4.5 ATP from 3 $FADH_2$; 40 ATP from 4 $CH_3COSCoA$ = **50 ATPs.**

For **B**, $CH_3(CH_2)_{22}CO_2H$: **B** has 24 C atoms, so it forms 12 molecules of acetyl CoA by 11 cycles of β-oxidation, resulting in 11 NADH and 11 $FADH_2$. **ATP yield:** (–2 ATP) for formation of the acyl CoA; 27.5 ATP from 11 NADH; 16.5 ATP from 11 $FADH_2$; 120 ATP from 12 $CH_3COSCoA$ = **162 ATPs.**

27.36 We must take into account the reactions that convert glycerol to dihydroxyacetone phosphate, as well as the catabolism dihydroxyacetone phosphate by glycolysis, conversion to acetyl CoA, and the citric acid cycle.

glycerol —[1] ATP → ADP→ glycerol 3-phosphate —[2] NAD^+ → NADH→ dihydroxyacetone phosphate —[3] 2 ADP → 2 ATP, NAD^+ → NADH→ pyruvate —[4] NAD^+ → NADH→ acetyl CoA —[5]→ CO_2

[1] 1 ATP required for conversion of glycerol to glycerol 3-phosphate –1 ATP

[2] glycerol 3-phosphate → dihydroxyacetone phosphate:

1 NADH → 2.5 ATPs 2.5 ATPs

[3] dihydroxyacetone phosphate → pyruvate:

2 ATPs + 1 NADH → 2.5 ATPs = 4.5 ATPs

[4] pyruvate → acetyl CoA:

1 NADH → 2.5 ATPs 2.5 ATPs

[5] acetyl CoA → CO_2:

1 acetyl CoA × 10 ATP/acetyl CoA = 10 ATPs

Total steps [1]–[5]: **18.5 ATPs**

27.37 Section 27.6C showed that the ATP yield from glucose is 32 ATP/glucose. Problem 27.22 showed that the ATP yield from stearic acid is 120 ATPs. To compare these values, we divide by the number of C's in each compound to give the amount of ATP formed per carbon. Because stearic acid produces more ATP per carbon than glucose (6.7 versus 5.3 ATPs per carbon), this supports the fact that lipids are more effective energy-storing molecules than carbohydrates.

$$\frac{32\text{ ATP}}{6\text{C's in glucose}} = \textbf{5.3 ATPs}\text{ per C in glucose} \qquad \frac{120\text{ ATP}}{18\text{C's in stearic acid}} = \textbf{6.7 ATPs}\text{ per C in stearic acid}$$

27.38 The mechanism consists of two parts: a reverse Claisen reaction that opens the five-membered ring and ester hydrolysis in base.

HO:⁻ CO₂Et HO: :O:⁻ CO₂Et H O OEt HO:⁻ H−OH :OH OEt :O: OEt :O: :OH O H EtO:⁻ H_3O^+ :O: :O: $H-\overset{+}{O}H_2$ $H-\overset{+}{O}H_2$ + EtOH HO OH + 2 H_2O

27.39 Compound **X** is formed by a Claisen reaction between acetyl CoA acting as an electrophile and an enolate formed from oxaloacetate.

B: H O⁻ oxaloacetate :O: SCoA + HB⁺ H−A SCoA O O⁻ X + HSCoA

27.40 The reaction occurs by a Claisen reaction with simultaneous loss of CO_2.

:O: SCoA + :O: SCoA H−A SCoA O SCoA + CO_2 SCoA + HSCoA acetoacetyl CoA

27.41 Conjugate addition of an OH group followed by protonation converts *cis*-aconitate to isocitrate.

cis-aconitate → (2*R*,3*S*)-isocitrate + :A⁻

27.42 To determine the amount of ATP obtained from the given triacylglycerol, we must draw the products formed from hydrolysis and then look at how much ATP is formed from each component.

hydrolysis → glycerol + $C_{10}H_{20}O_2$ + $C_{14}H_{28}O_2$ + $C_{12}H_{24}O_2$

For **glycerol,** 18.5 ATPs produced (Problem 27.36).

For $CH_3(CH_2)_8CO_2H$: $CH_3(CH_2)_8CO_2H$ has 10 C atoms, so it forms five molecules of acetyl CoA by four cycles of β-oxidation, resulting in 4 NADH and 4 $FADH_2$. **ATP yield:** (–2 ATP) for formation of the acyl CoA; 10 ATP from 4 NADH; 6 ATP from 4 $FADH_2$; 50 ATP from 5 $CH_3COSCoA$ = **64 ATPs.**

For $CH_3(CH_2)_{12}CO_2H$: $CH_3(CH_2)_{12}CO_2H$ has 14 C atoms, so it forms seven molecules of acetyl CoA by six cycles of β-oxidation, resulting in 6 NADH and 6 $FADH_2$. **ATP yield:** (–2 ATP) for formation of the acyl CoA; 15 ATP from 6 NADH; 9 ATP from 6 $FADH_2$; 70 ATP from 7 $CH_3COSCoA$ = **92 ATPs.**

For $CH_3(CH_2)_{10}CO_2H$: $CH_3(CH_2)_{10}CO_2H$ has 12 C atoms, so it forms six molecules of acetyl CoA by five cycles of β-oxidation, resulting in 5 NADH and 5 $FADH_2$. **ATP yield:** (–2 ATP) for formation of the acyl CoA; 12.5 ATP from 5 NADH; 7.5 ATP from 5 $FADH_2$; 60 ATP from 6 $CH_3COSCoA$ = **78 ATPs.**

Total: 18.5 + 64 + 92 + 78 = **252.5 ATPs.**

27.43 Like the conversion of glucose 6-phosphate to fructose 6-phosphate, this reaction proceeds through an enediol intermediate.

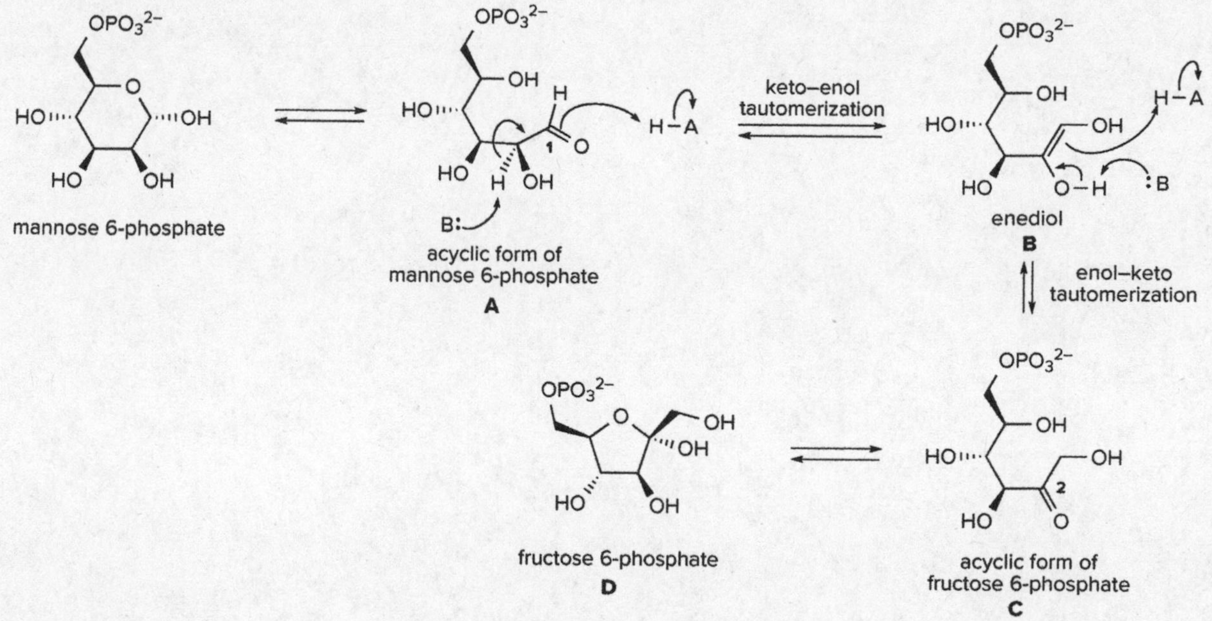

mannose 6-phosphate

acyclic form of mannose 6-phosphate
A

keto–enol tautomerization

enediol
B

enol–keto tautomerization

fructose 6-phosphate
D

acyclic form of fructose 6-phosphate
C

27.44 Isomerization proceeds through an enediol.

fructose

+ H_2O

re-draw

acyclic form of fructose 6-phosphate

enediol

H^+ can bond from the front or the back.

A
H^+ bonds from the back. OH at the labeled C is in front, on a wedge.

B
H^+ bonds from the front. OH at the labeled C is behind, on a dashed wedge.

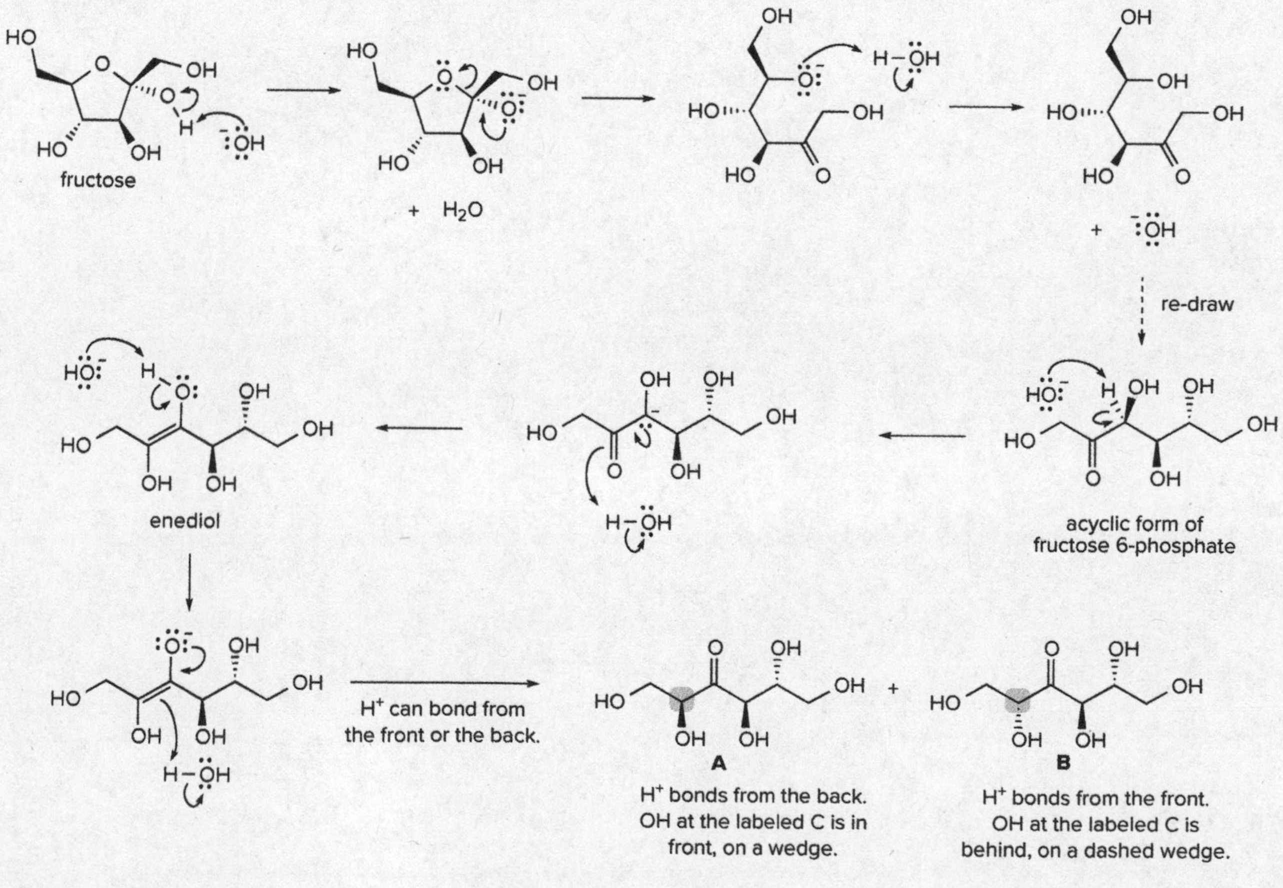

Chapter 28 Carbon–Carbon Bond-Forming Reactions in Organic Synthesis

Chapter Review

Coupling reactions

[1] Coupling reactions of organocuprate reagents (28.1)

$$R'{-}X + R_2CuLi \longrightarrow \boxed{R'{-}R} + RCu + LiX$$

X = Cl, Br, I

- R'X can be CH_3X, RCH_2X, 2° cyclic halides, vinyl halides, and aryl halides.
- X may be Cl, Br, or I.
- With vinyl halides, coupling is stereospecific.

[2] Suzuki reaction (28.2)

$$R'{-}X + R{-}BY_2 \xrightarrow[\text{NaOH}]{Pd(PPh_3)_4} \boxed{R'{-}R} + HO{-}BY_2 + NaX$$

X = Br, I

- R'X is most often a vinyl halide or aryl halide.
- With vinyl halides, coupling is stereospecific.

[3] Heck reaction (28.3)

$$R'{-}X + CH_2{=}CH{-}Z \xrightarrow[\text{P(o-tolyl)}_3,\ Et_3N]{Pd(OAc)_2} \boxed{R'{-}CH{=}CH{-}Z} + Et_3\overset{+}{N}H\ X^-$$

X = Br, I

- R'X is a vinyl halide or aryl halide.
- Z = H, Ph, COOR, or CN
- With vinyl halides, coupling is stereospecific.
- The reaction forms trans alkenes.

Cyclopropane synthesis

[1] Addition of dihalocarbenes to alkenes (28.4)

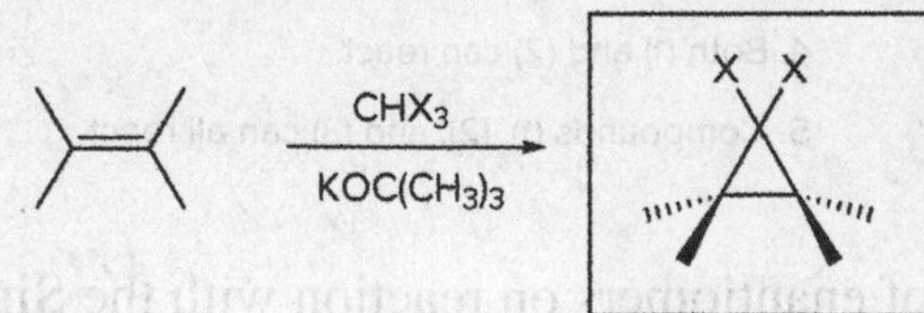

- The reaction occurs with syn addition.
- The position of substituents in the alkene is retained in the cyclopropane.

[2] Simmons–Smith reaction (28.5)

$$R_2C{=}CR_2 \xrightarrow[\text{Zn(Cu)}]{CH_2I_2} \text{cyclopropane (H, H)} + ZnI_2$$

- The reaction occurs with syn addition.
- The position of substituents in the alkene is retained in the cyclopropane.

Metathesis (28.6)

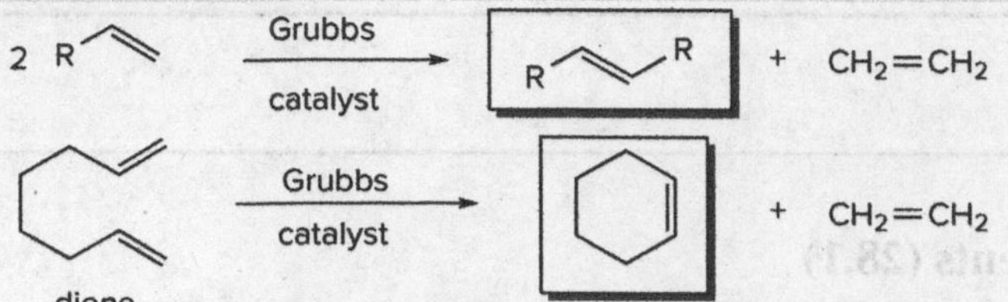

- Metathesis works best when CH_2=CH_2, a gas that escapes from the reaction mixture, is formed as one product.

Practice Test on Chapter Review

1.a. Which functional groups react with lithium dialkyl cuprates?

1. epoxides
2. vinyl halides
3. acid chlorides
4. Compounds (1) and (2) both react with R_2CuLi.
5. Compounds (1), (2), and (3) all react with R_2CuLi.

b. Which of the following statements is (are) true for the Suzuki reaction?

1. Arylboranes can serve as one reactant.
2. The reaction is stereospecific.
3. The reaction occurs between a vinyl or aryl halide and an alkene in the presence of a palladium catalyst.
4. Statements (1) and (2) are both true.
5. Statements (1), (2), and (3) are all true.

c. Which of the following compounds can react with CH_2=CHCN in a Heck reaction?

1. Br 2. I 3. CH_2=CH_2

4. Both (1) and (2) can react.
5. Compounds (1), (2), and (3) can all react.

d. Which of the following compounds yield(s) a pair of enantiomers on reaction with the Simmons–Smith reagent?

1. 2. 3.

4. Compounds (1) and (2) both yield a pair of enantiomers.
5. Compounds (1), (2), and (3) all yield a pair of enantiomers

e. Which of the following compounds can be made by a ring-closing metathesis reaction?

1. 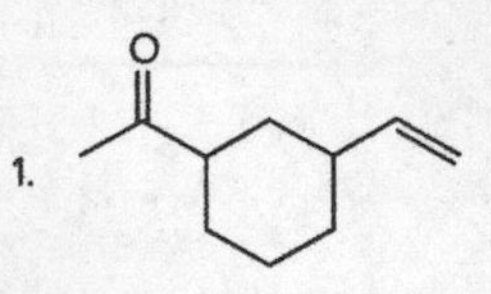

2. 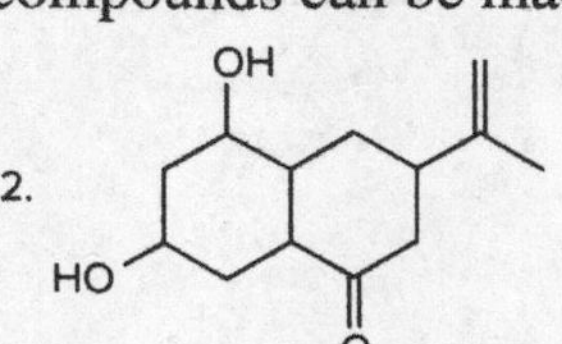

3. 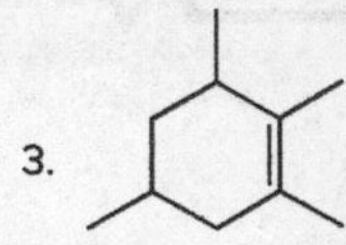

4. Compounds (1) and (2) can both be prepared.
5. Compounds (1), (2), and (3) can all be prepared.

f. Which of the following compounds can be prepared from $CH_3–C≡C–H$ by a Suzuki reaction? You may use other organic compounds or inorganic reagents.

1. 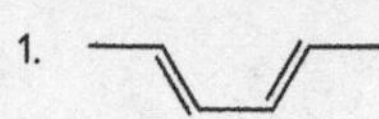 2. 3.

4. Compounds (1) and (2) can both be prepared.
5. Compounds (1), (2), and (3) can all be prepared.

2. Draw the product formed in each reaction. Indicate the stereochemistry around double bonds and stereogenic centers when necessary.

a. [1] catecholborane; [2] I, $Pd(PPh_3)_4$, NaOH

b. Br; [1] 2 Li; [2] 0.5 CuI; [3] Ph, Br

c. Br; [1] 2 Li; [2] $B(OCH_3)_3$; [3] Br, $Pd(PPh_3)_4$, NaOH

d. OCH_3, CH_3O, I + CO_2CH_3; $Pd(OAc)_2$; P(o-tolyl)$_3$; Et_3N

e. $CHBr_3$; $KOC(CH_3)_3$

f. Grubbs cataylst

g. Grubbs cataylst

3. What starting material is needed to synthesize each compound by ring-closure metathesis?

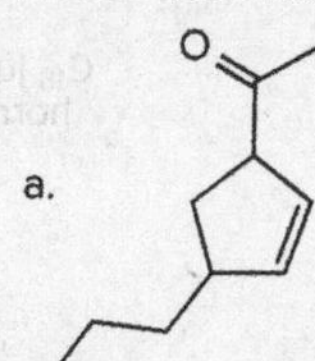

a. b. HO, OH, OH c. O, O

Answers to Practice Test

1.a. 5
b. 4
c. 4
d. 2
e. 3
f. 4

2.
a.
b. Ph
c.
d. OCH_3, CH_3O, CO_2CH_3
e. Br Br
f.
g.

3.
a. O
b. HO, OH, OH
c. O, O

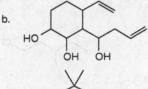

Answers to Problems

28.1 A new C–C bond is formed in each coupling reaction.

a. 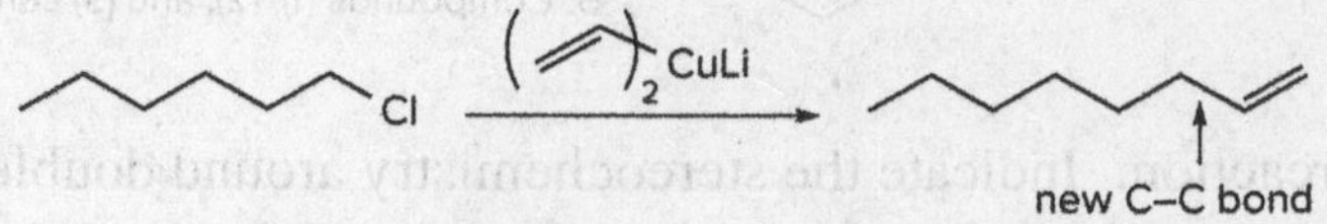 Cl —[(CH_2=CH)$_2$CuLi]→ new C–C bond

b. 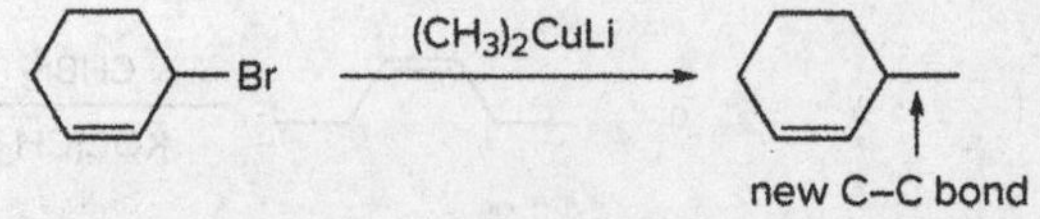 Br —[$(CH_3)_2CuLi$]→ new C–C bond

c. CH_3O, CH_3O, I —[(CH$_3$CH$_2$CH$_2$CH$_2$)$_2$CuLi]→ CH_3O, CH_3O, new C–C bond

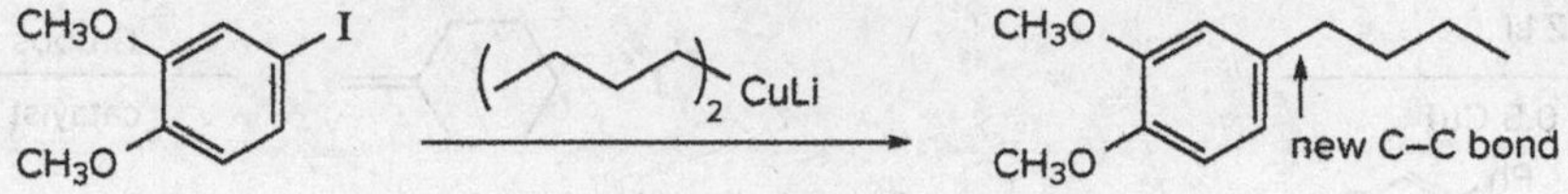

d. 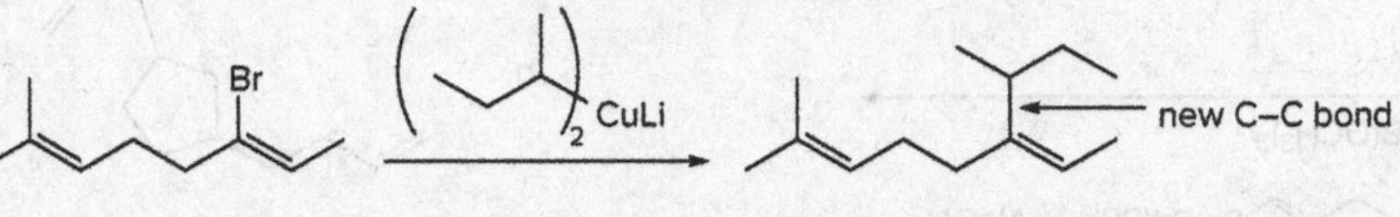 Br —[CuLi, 2]→ new C–C bond

28.2

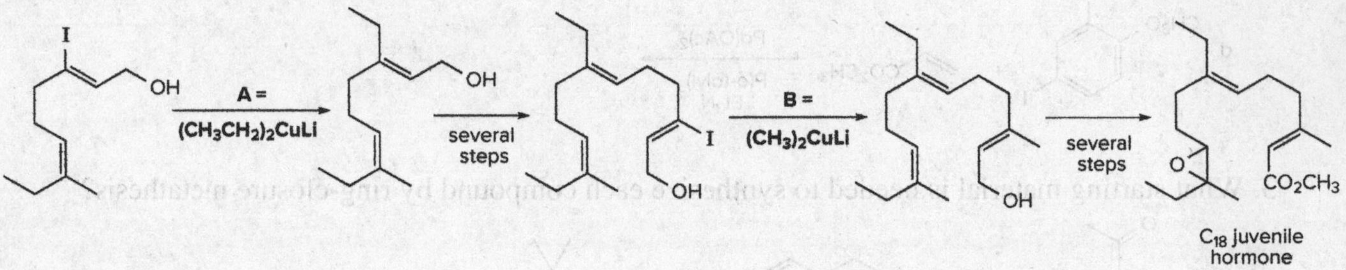

I, OH —[A = $(CH_3CH_2)_2CuLi$]→ OH —[several steps]→ I, OH —[B = $(CH_3)_2CuLi$]→ OH —[several steps]→ O, CO_2CH_3

C_{18} juvenile hormone

28.3

a. 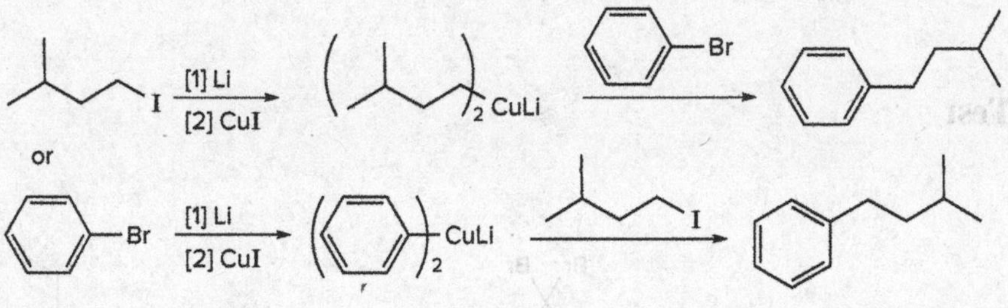 I —[[1] Li, [2] CuI]→ ( )$_2$CuLi —[Br-benzene]→

or

Br —[[1] Li, [2] CuI]→ ( )$_2$CuLi —[I]→

b. 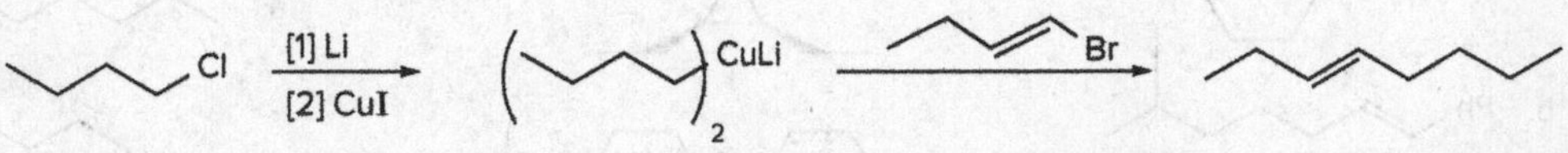 Cl —[[1] Li, [2] CuI]→ ( )$_2$CuLi —[Br]→

c. 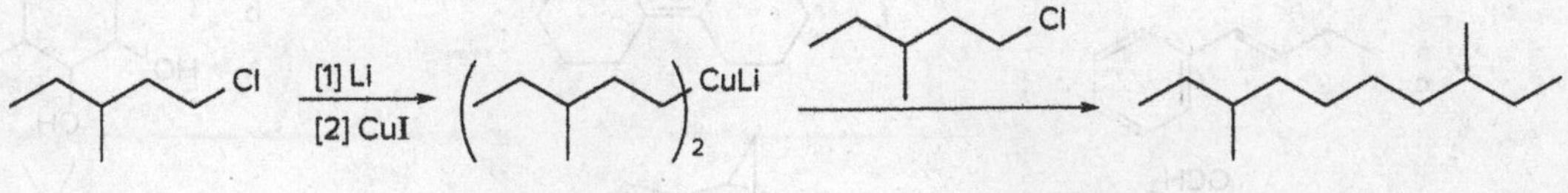 Cl —[[1] Li, [2] CuI]→ ( )$_2$CuLi —[Cl]→

28.4

a. I + (boronate ester) —Pd(PPh$_3$)$_4$, NaOH→

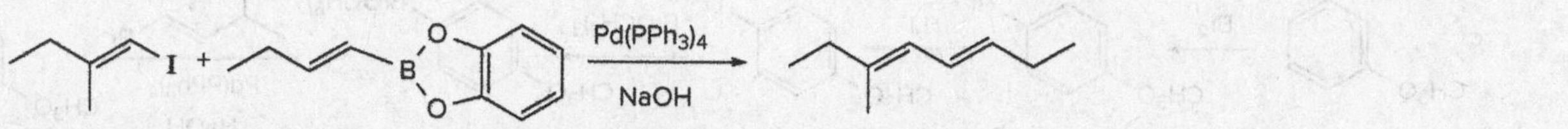

b. B(OCH$_3$)$_2$ + Br —Pd(PPh$_3$)$_4$, NaOH→

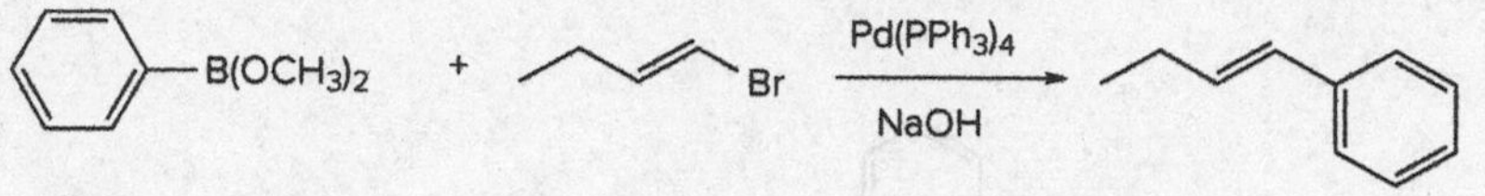

c. Br —[1] Li, [2] B(OCH$_3$)$_3$→ B(OCH$_3$)$_2$ + LiOCH$_3$

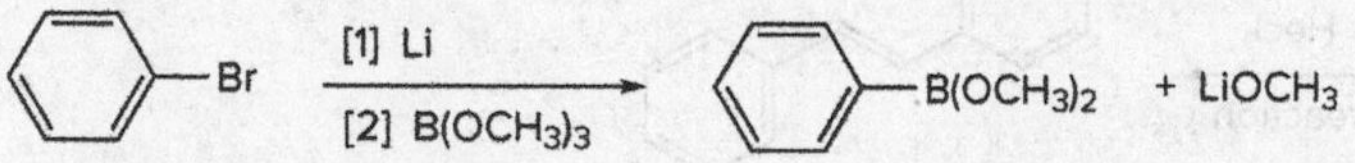

d. + H–B(catechol) →

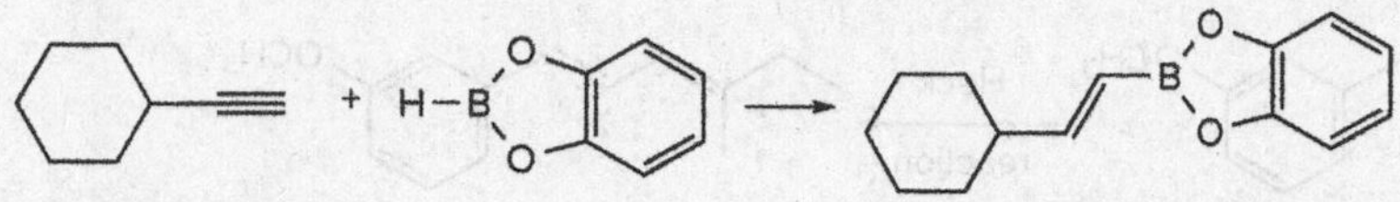

28.5 The Suzuki reaction forms a new carbon–carbon bond between a vinyl halide and an arylborane.

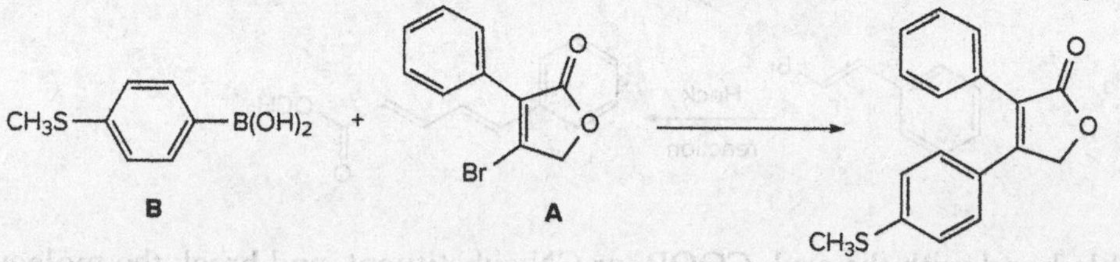

CH_3S–C$_6$H$_4$–B(OH)$_2$ (**B**) + Br (**A**) → CH_3S

28.6

a. [1] catecholborane (B–H); [2] Br, Pd(PPh$_3$)$_4$, NaOH

b. Br —[1] Li (2 equiv), [2] B(OCH$_3$)$_3$→ (CH$_3$O)$_2$B —Br, Pd(PPh$_3$)$_4$, NaOH→

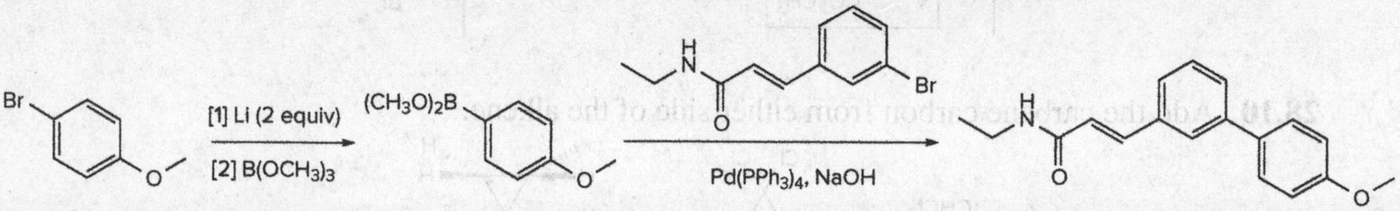

28.7

a. H–B(catechol) → —I, Pd(PPh$_3$)$_4$, NaOH→

b. I —Li→ Li —B(OCH$_3$)$_3$→ B(OCH$_3$)$_2$ + I —Pd(PPh$_3$)$_4$, NaOH→

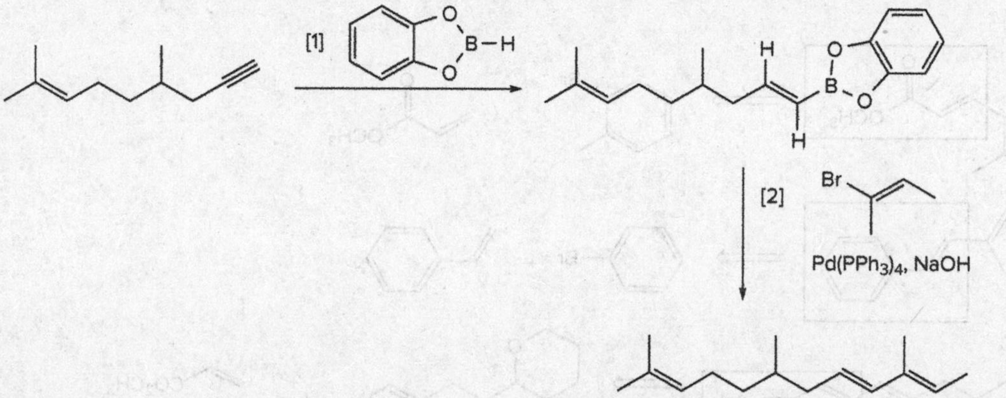

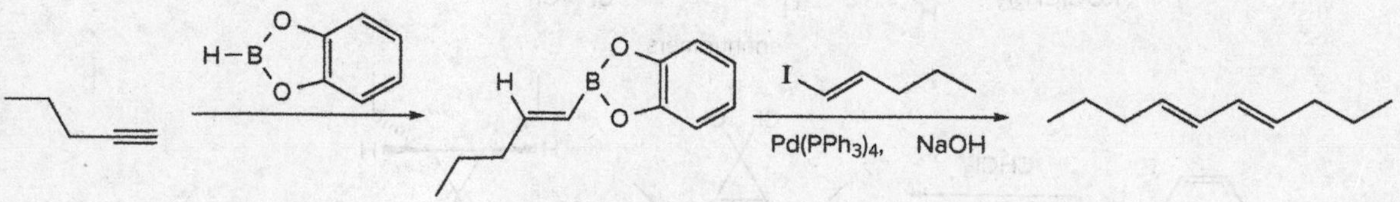

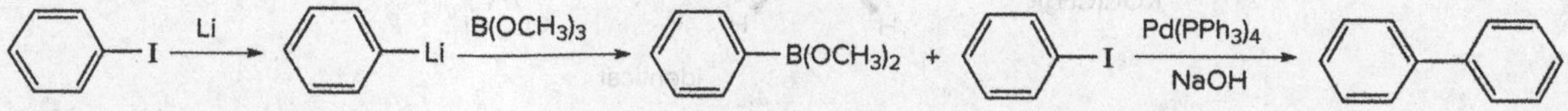

c. CH_3O–C₆H₅ → Br_2 → (+ ortho isomer) → Li → $B(OCH_3)_3$ → Pd(PPh$_3$)$_4$, NaOH

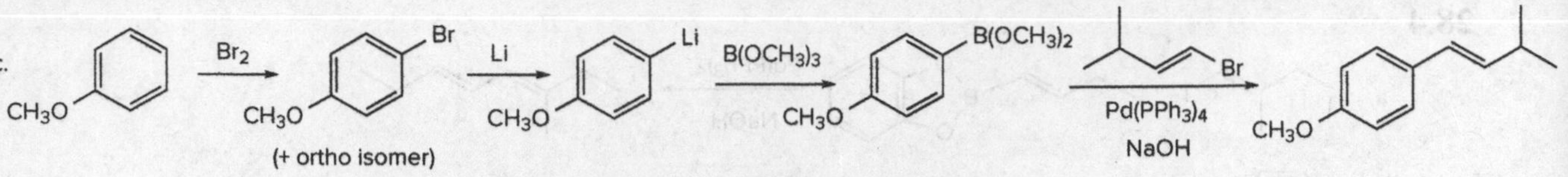

28.8

a. Heck reaction

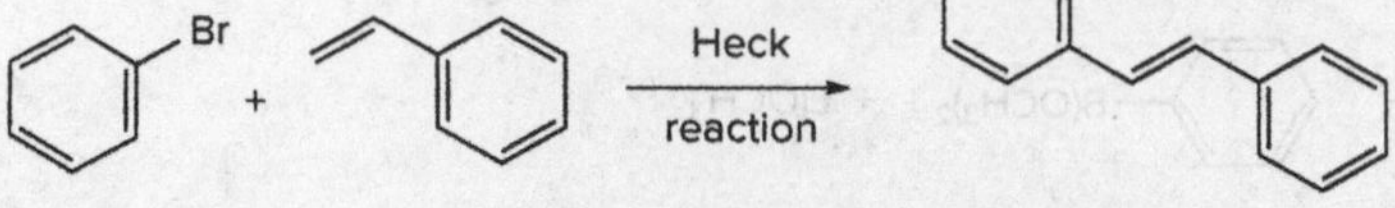

b. Heck reaction

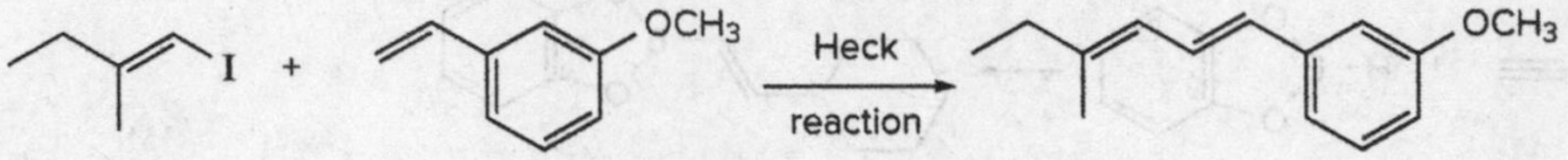

c. Heck reaction

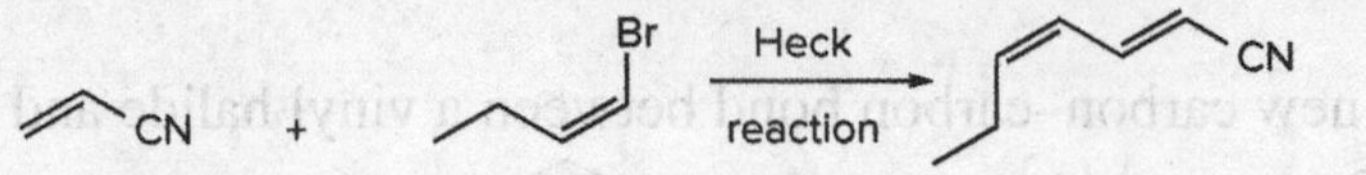

d. Heck reaction

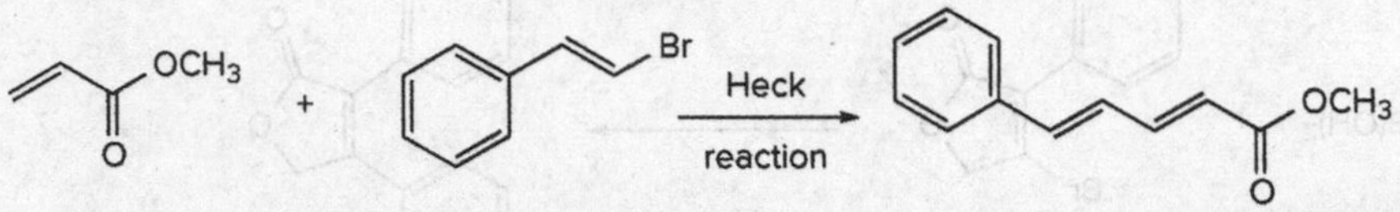

28.9 Locate the double bond with the aryl, COOR, or CN substituent, and break the molecule into two components at the end of the C=C not bonded to one of these substituents.

a.

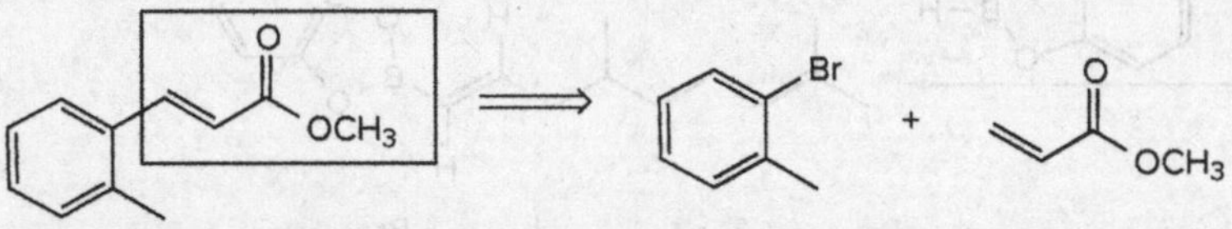

b.

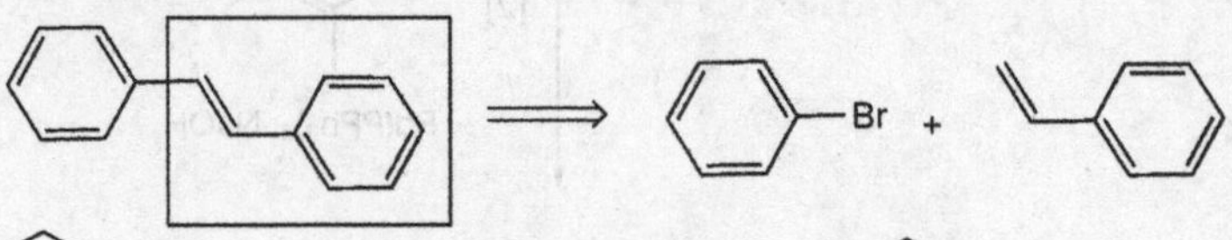

c.

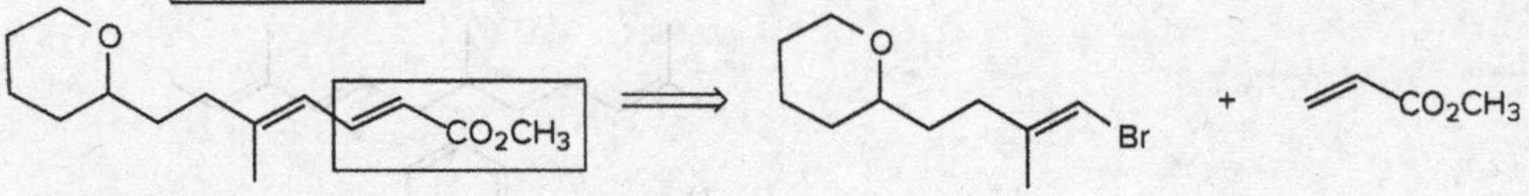

28.10 Add the carbene carbon from either side of the alkene.

a. $CHCl_3$, $KOC(CH_3)_3$ — enantiomers

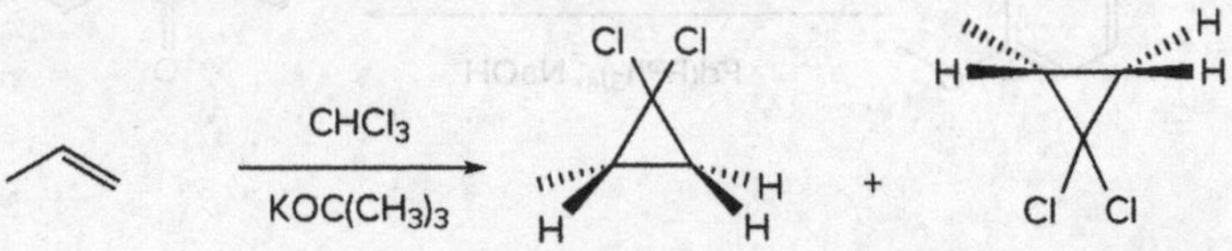

b. $CHCl_3$, $KOC(CH_3)_3$ — identical

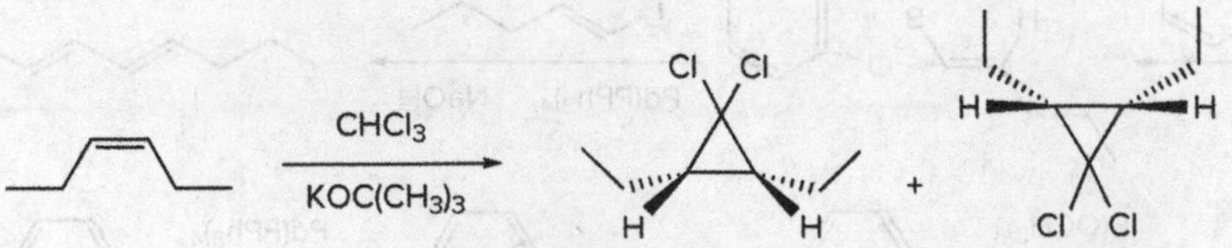

c. 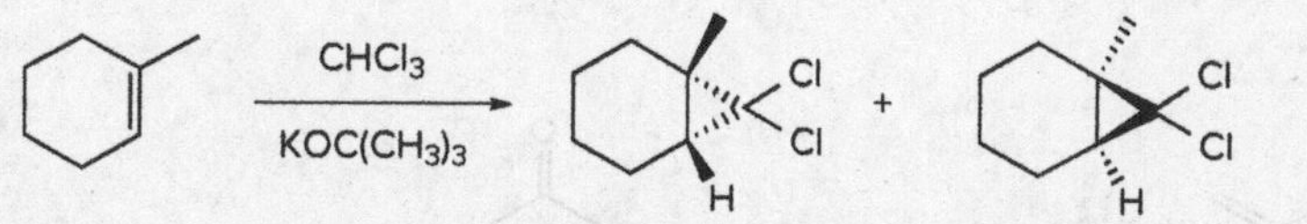 $CHCl_3$, $KOC(CH_3)_3$ → Cl, Cl, H + Cl, Cl, H

enantiomers

28.11

a. 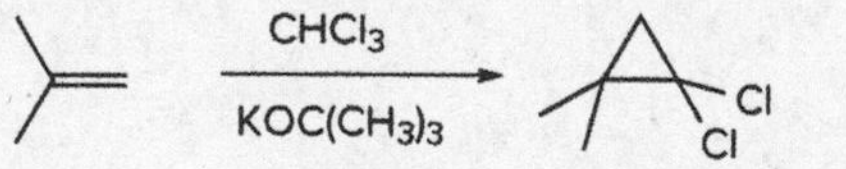 $CHCl_3$, $KOC(CH_3)_3$ → Cl, Cl

b. $CHBr_3$, $KOC(CH_3)_3$ → Br, Br

c. 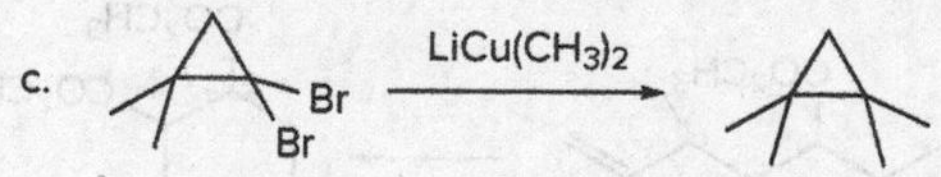 Br, Br (from b.) $LiCu(CH_3)_2$ →

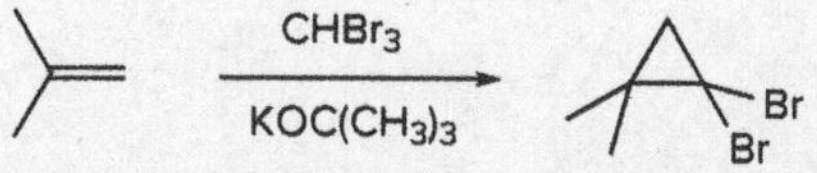

28.12

a. 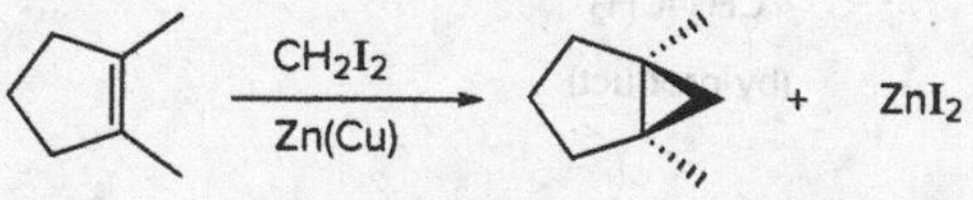 CH_2I_2, Zn(Cu) → + ZnI_2

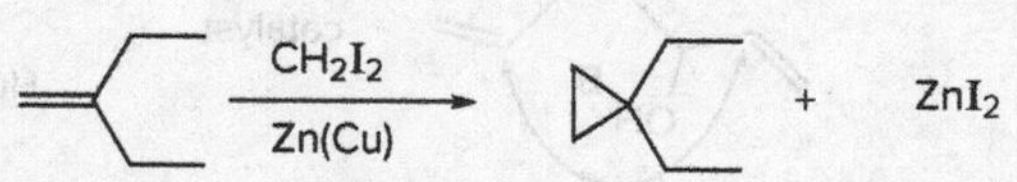

b. 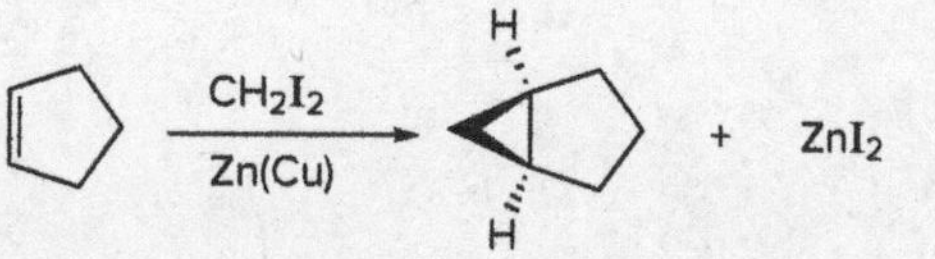 CH_2I_2, Zn(Cu) → H, H + ZnI_2

c. CH_2I_2, Zn(Cu) → + ZnI_2

28.13 The relative position of substituents in the reactant is retained in the product.

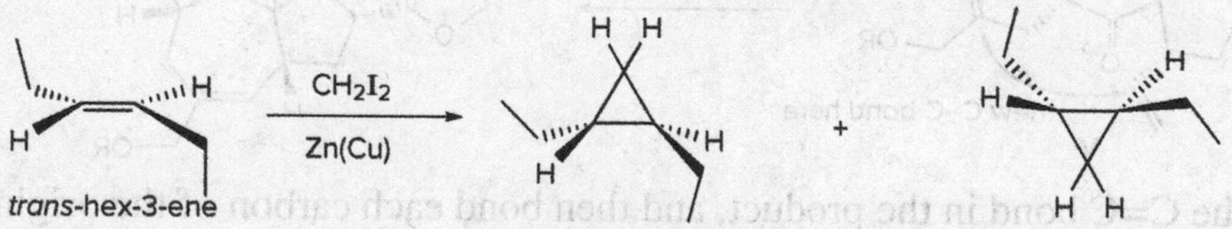

H, H *trans*-hex-3-ene CH_2I_2, Zn(Cu) → H H, H, H + H, H, H H

two enantiomers of *trans*-1,2-diethylcyclopropane

28.14

a. 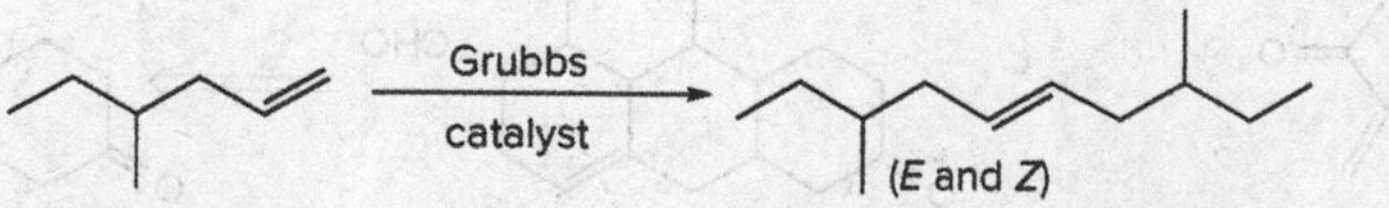 Grubbs catalyst → (*E* and *Z*)

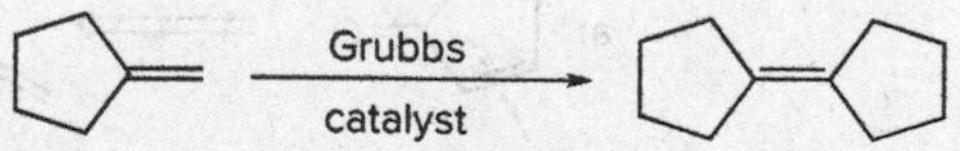

b. 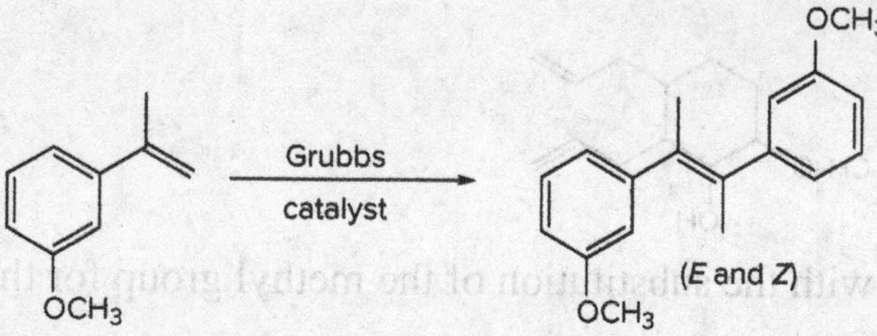 OCH_3 Grubbs catalyst → OCH_3, OCH_3 (*E* and *Z*)

c. Grubbs catalyst →

($CH_2{=}CH_2$ is also formed in each reaction.)

28.15

cis-pent-2-ene → + + +

There are four products formed in this reaction, including stereoisomers, so it is not a practical method to synthesize 1,2-disubstituted alkenes.

28.16

a. 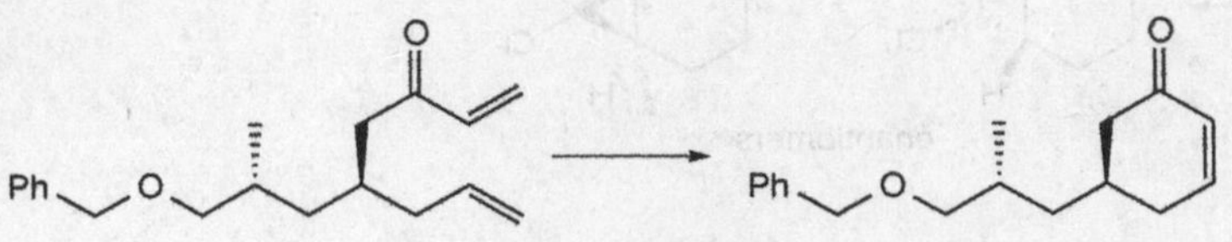 Ph O → Ph O (ring-closing metathesis to cyclohexenone)

b. 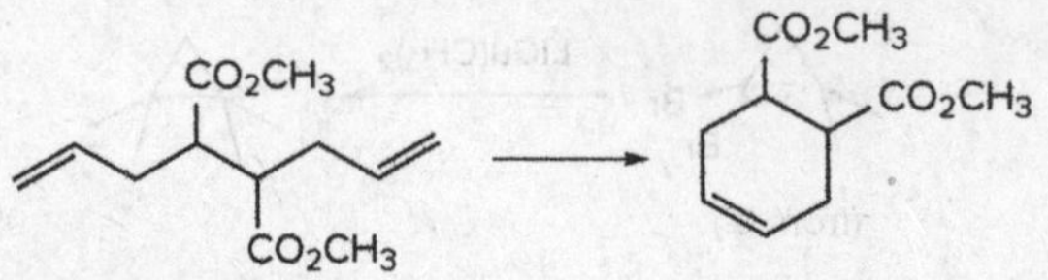 CO_2CH_3, CO_2CH_3 → CO_2CH_3, CO_2CH_3

28.17

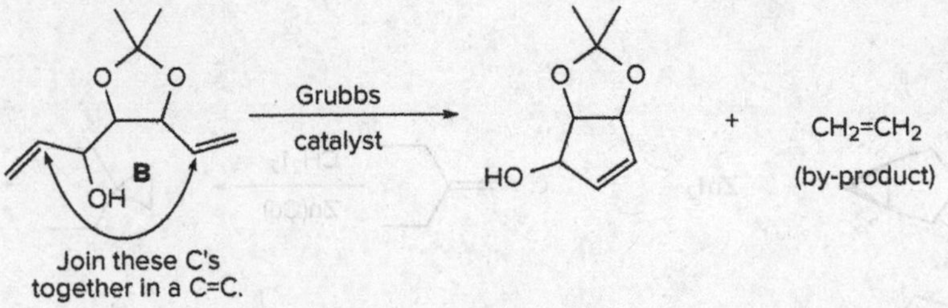

Grubbs catalyst

B OH → HO + $CH_2{=}CH_2$ (by-product)

Join these C's together in a C=C.

28.18

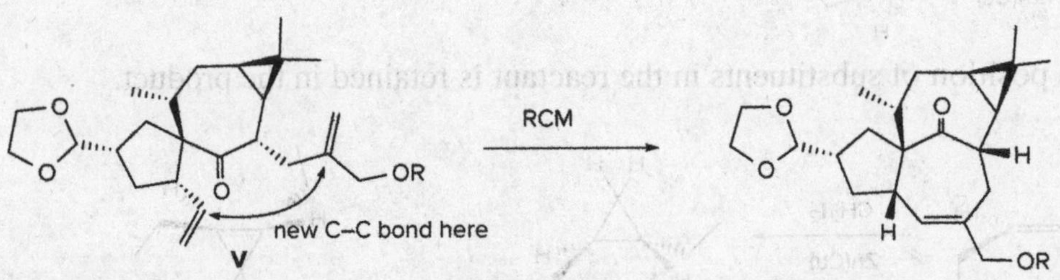

RCM

O, O, OR, V → O, O, H, H, OR

new C–C bond here

28.19 Cleave the C=C bond in the product, and then bond each carbon of the original alkene to a CH_2 group using a double bond.

a. 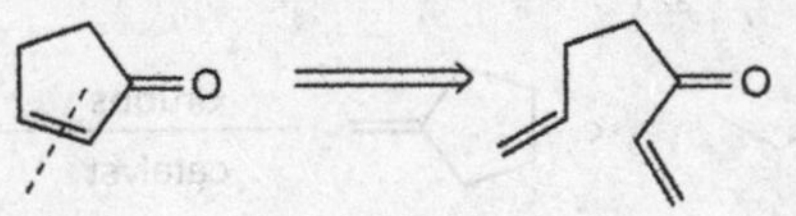 =O ⟹ =O

c. 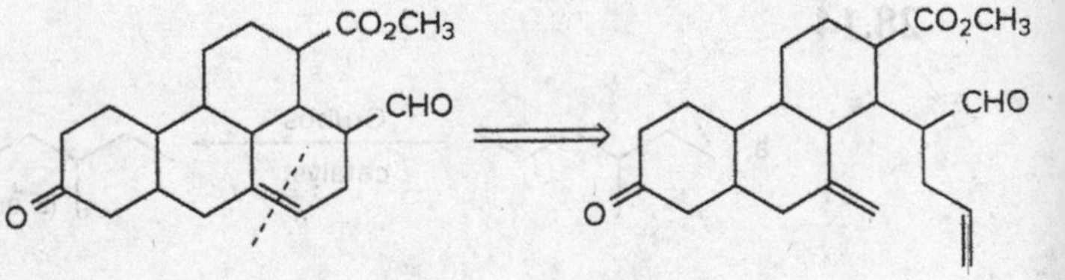 CO_2CH_3, CHO, O ⟹ CO_2CH_3, CHO, O

b. 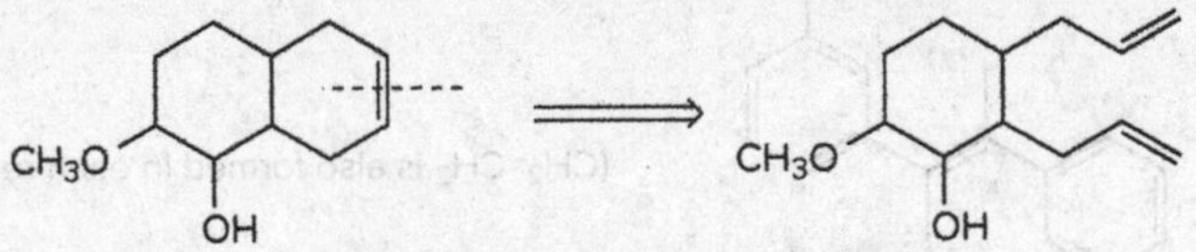 CH_3O, OH ⟹ CH_3O, OH

28.20 Inversion of configuration occurs with the substitution of the methyl group for the tosylate.

a. 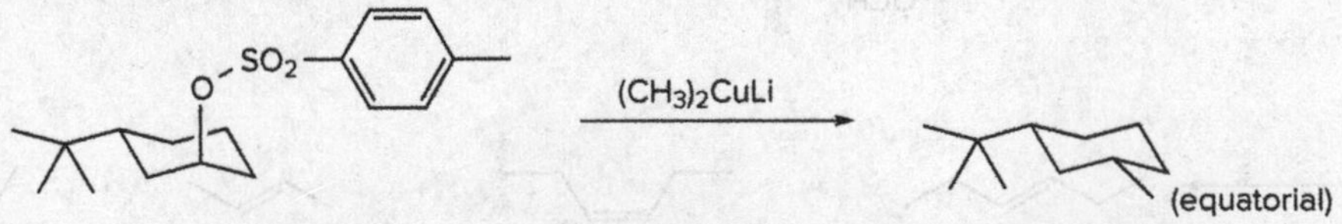 O–SO_2 —$(CH_3)_2CuLi$→ (equatorial)

b. 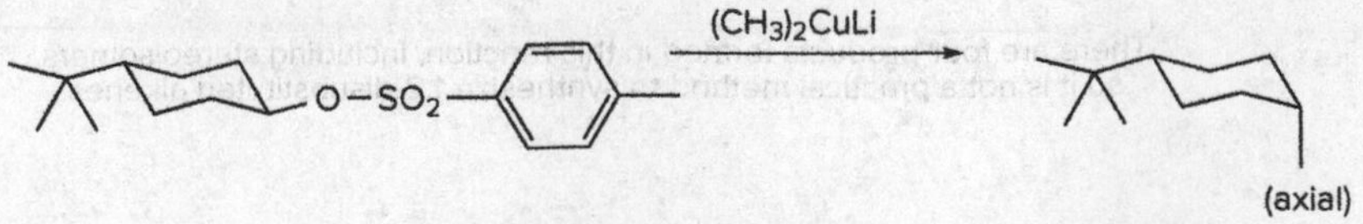 O–SO_2 —$(CH_3)_2CuLi$→ (axial)

28.21

a. 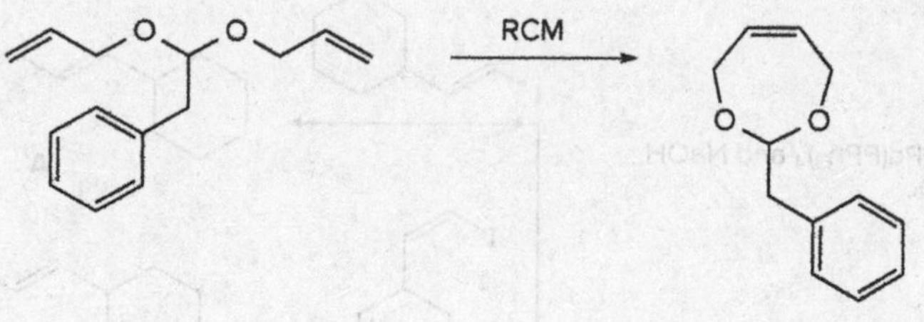 RCM

b. 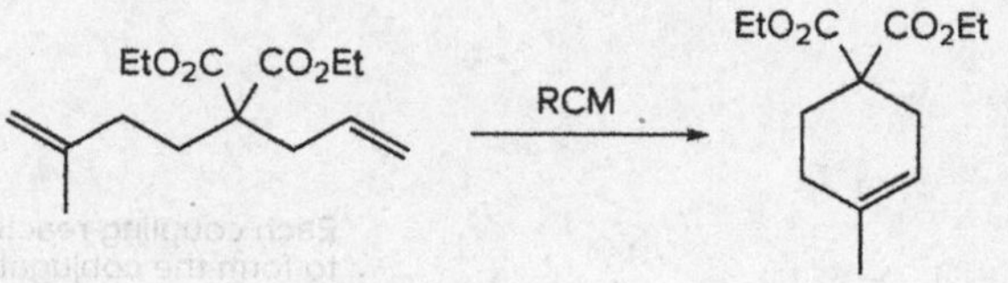 EtO_2C CO_2Et RCM EtO_2C CO_2Et

28.22

a. 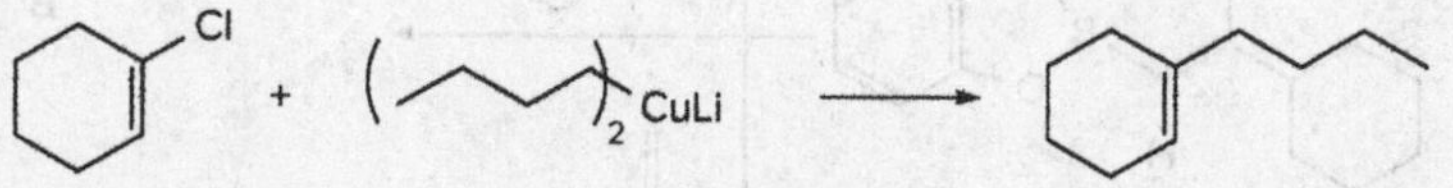 Cl + ( )$_2$CuLi

b. 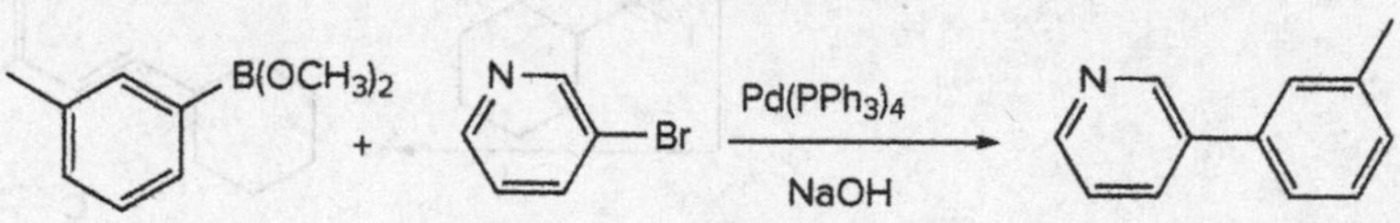 $B(OCH_3)_2$ + N Br $Pd(PPh_3)_4$ NaOH N

c. 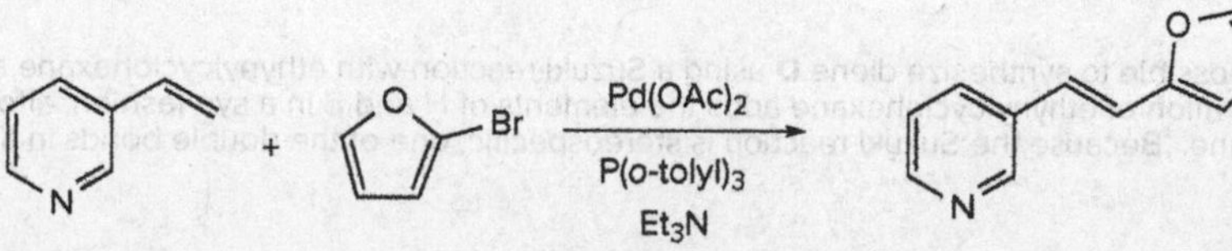 + O Br $Pd(OAc)_2$ P(*o*-tolyl)$_3$ Et_3N O N N

d. 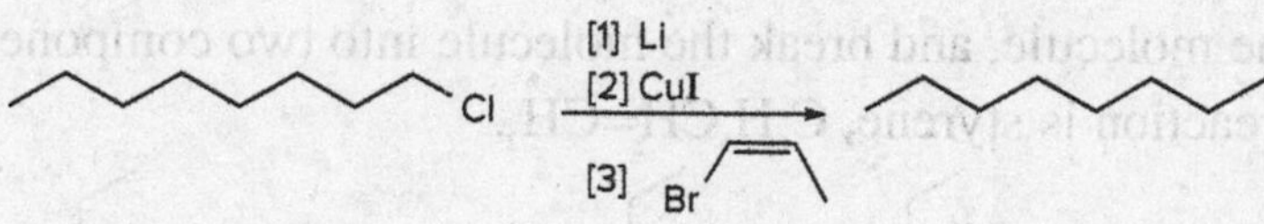 Cl [1] Li [2] CuI [3] Br

e. 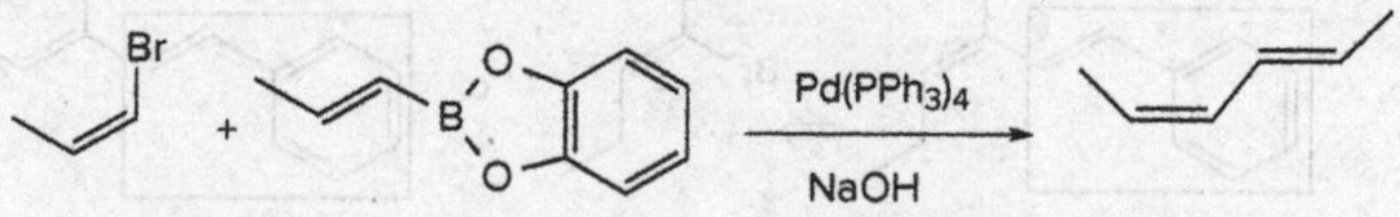 Br + B O O $Pd(PPh_3)_4$ NaOH

f. 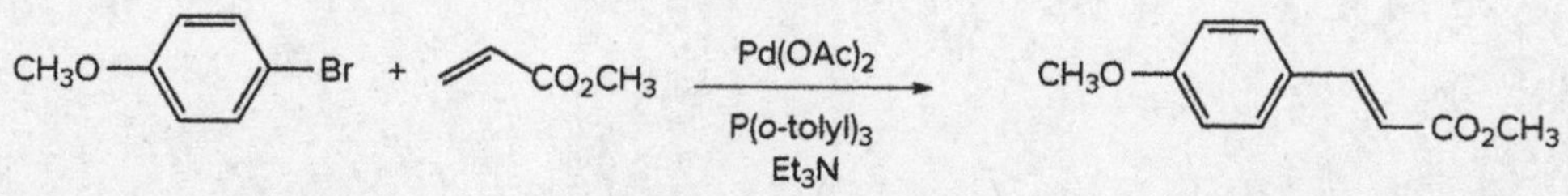 CH_3O Br + CO_2CH_3 $Pd(OAc)_2$ P(*o*-tolyl)$_3$ Et_3N CH_3O CO_2CH_3

g. 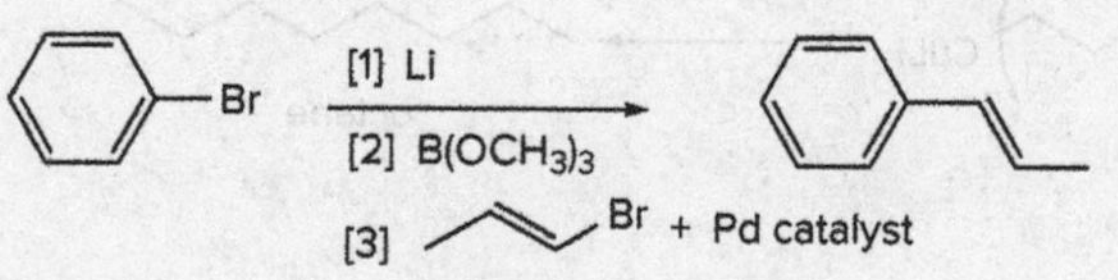 Br [1] Li [2] $B(OCH_3)_3$ [3] Br + Pd catalyst

h. 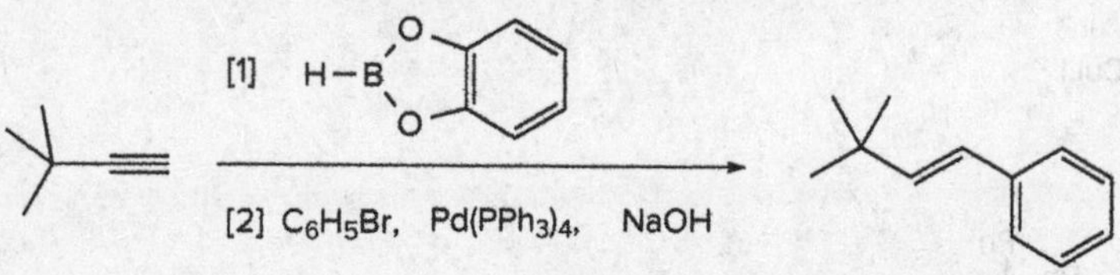 [1] H–B O O [2] C_6H_5Br, $Pd(PPh_3)_4$, NaOH

28.23

a. 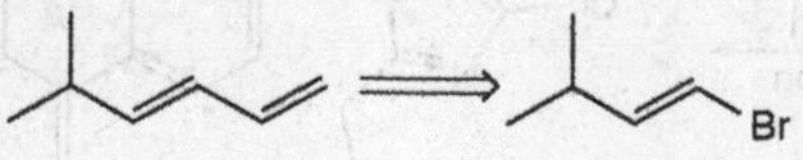 ⟹ Br

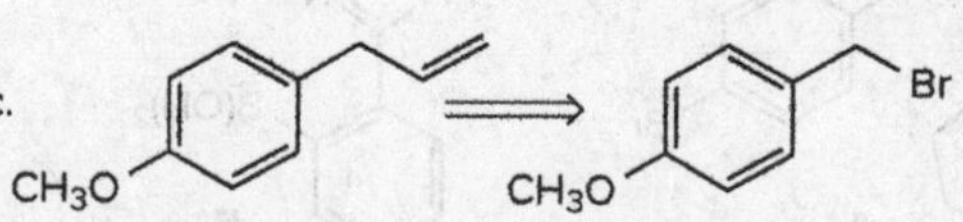

b. 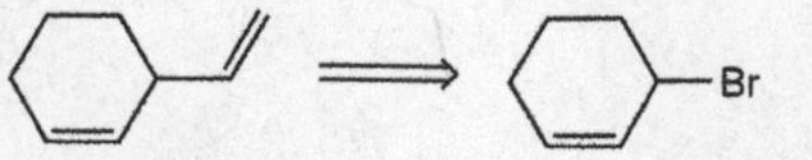 ⟹ Br

c. CH_3O ⟹ CH_3O Br

28.24

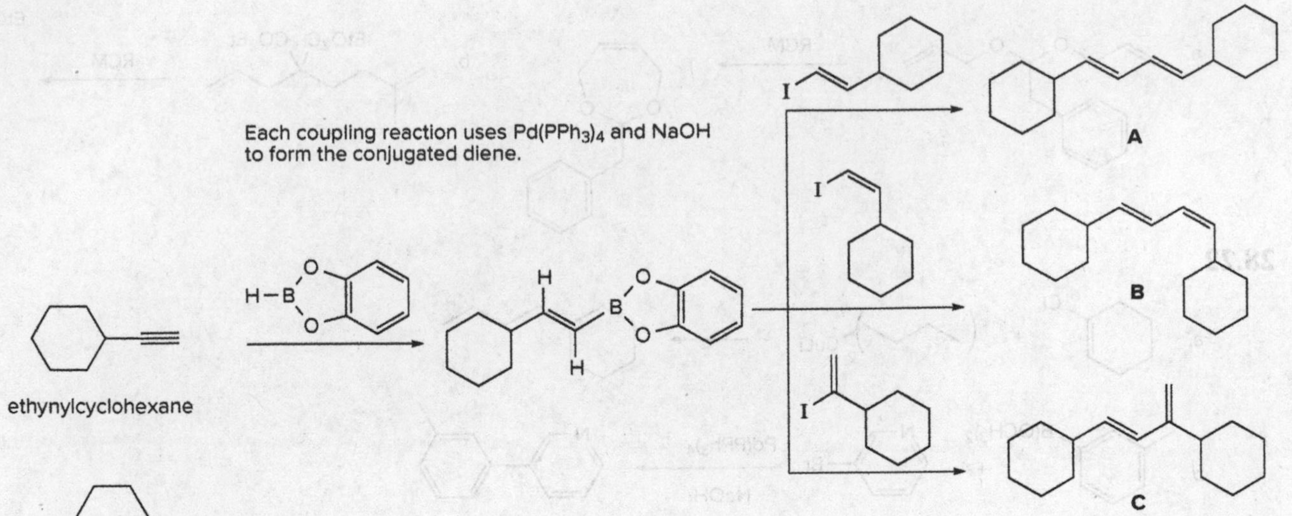

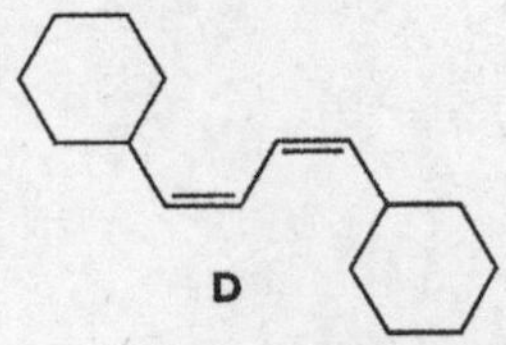

Each coupling reaction uses $Pd(PPh_3)_4$ and NaOH to form the conjugated diene.

ethynylcyclohexane

A

B

C

D

It is not possible to synthesize diene **D** using a Suzuki reaction with ethynylcyclohexane as starting material. Hydroboration of ethynylcyclohexane adds the elements of H and B in a syn fashion, affording a trans vinylborane. Because the Suzuki reaction is stereospecific, one of the double bonds in the product must be trans.

28.25

28.25 Locate the styrene part of the molecule, and break the molecule into two components. The second component in each reaction is styrene, $C_6H_5CH{=}CH_2$.

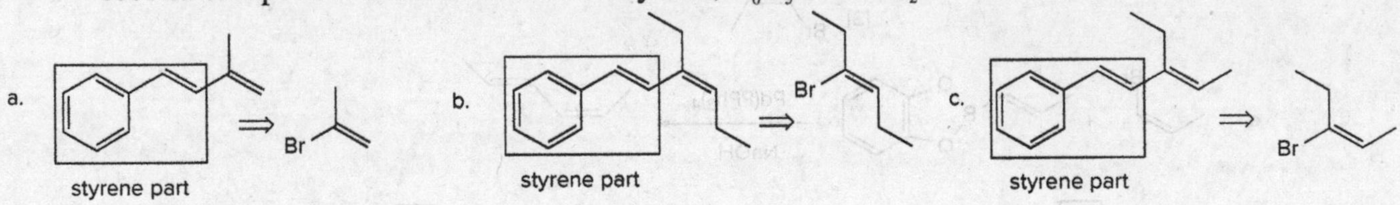

a. styrene part ⟹ Br

b. styrene part ⟹ Br

c. styrene part ⟹ Br

28.26

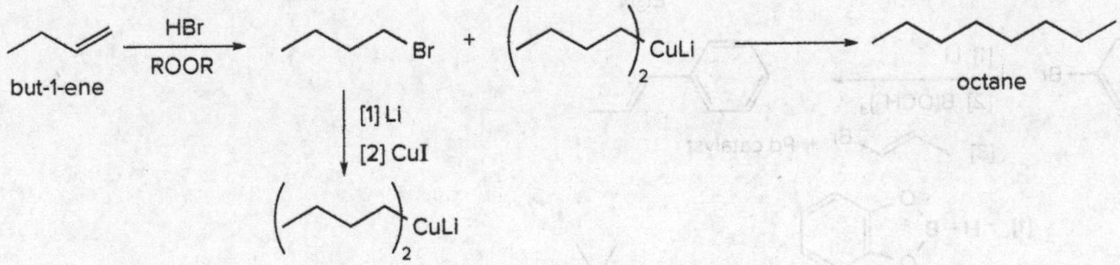

but-1-ene —HBr, ROOR→ Br + $(\ )_2$CuLi → octane

[1] Li

[2] CuI

$(\ )_2$CuLi

28.27

A + B —Suzuki reaction→

HO, Cl, N, Br, Ph, B(OH)$_2$

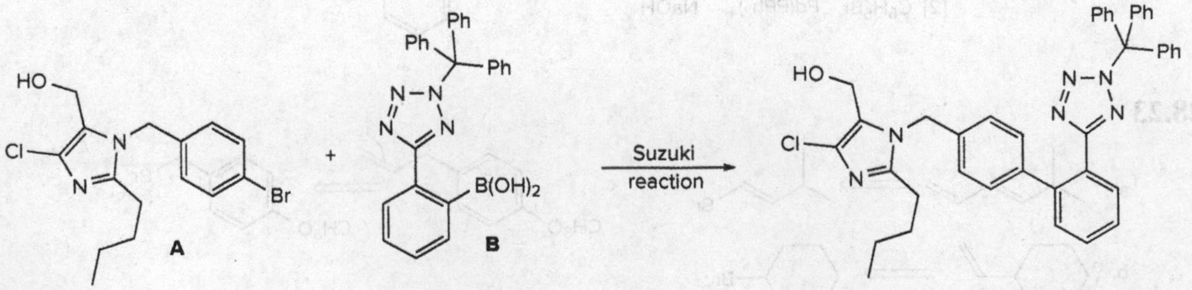

28.28 Join the two labeled C's together. Keep the *E* configuration at the vinyl iodide and form a new *E* C=C during coupling.

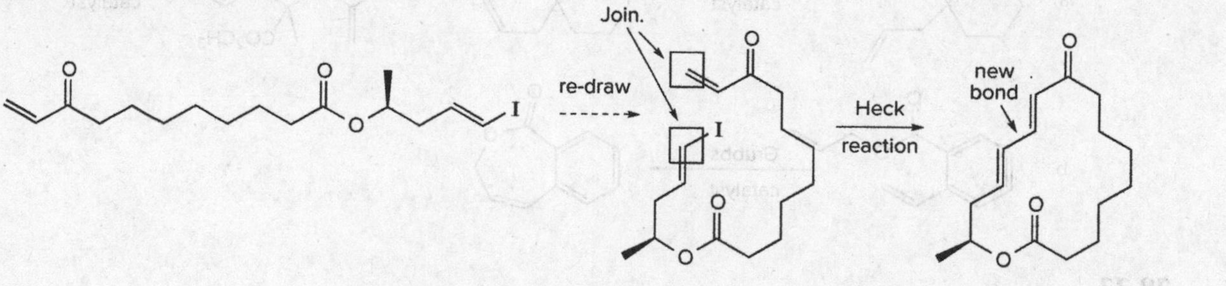

re-draw

Join.

I

Heck reaction

new bond

28.29

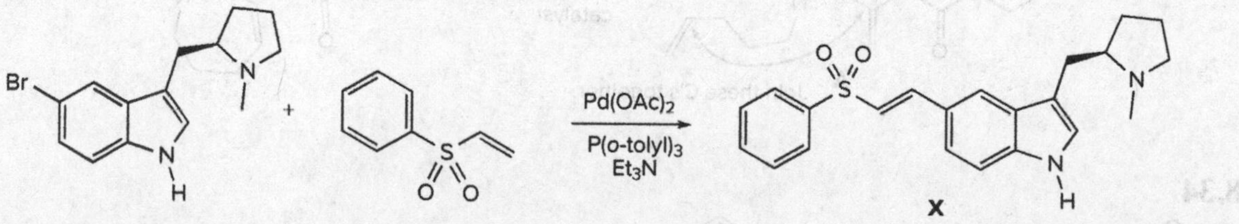

Br

N

H

N

+

$Pd(OAc)_2$

P(o-tolyl)$_3$

Et_3N

X

28.30 Add the carbene carbon from either side of the alkene.

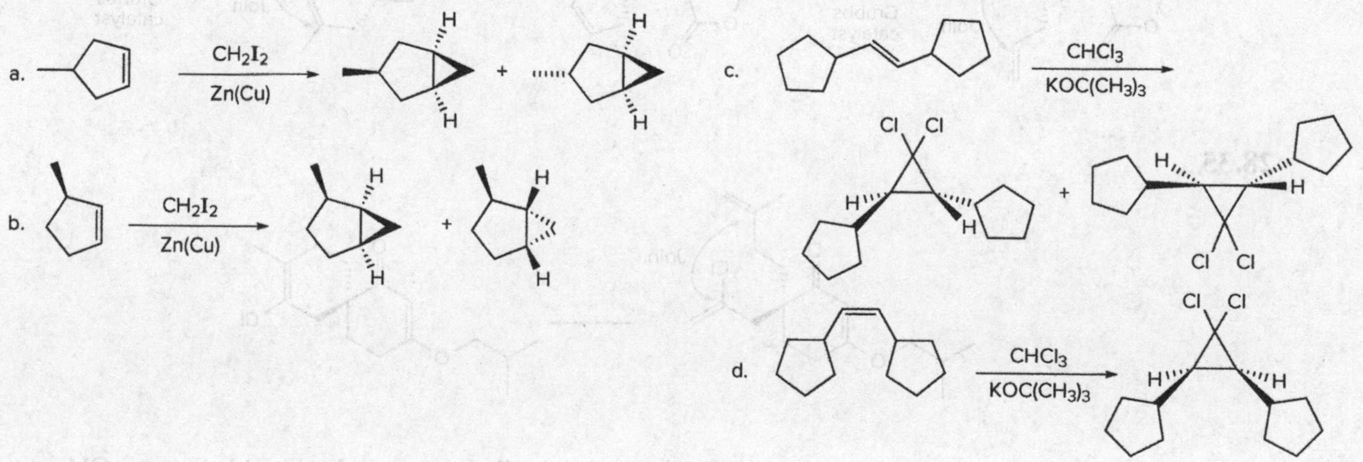

a. CH_2I_2 / Zn(Cu)

b. CH_2I_2 / Zn(Cu)

c. $CHCl_3$ / $KOC(CH_3)_3$

d. $CHCl_3$ / $KOC(CH_3)_3$

28.31 The new three-membered ring has a stereogenic center on the C bonded to the phenyl group, so the phenyl group can be oriented in two different ways to afford two stereoisomers. These products are diastereomers of each other.

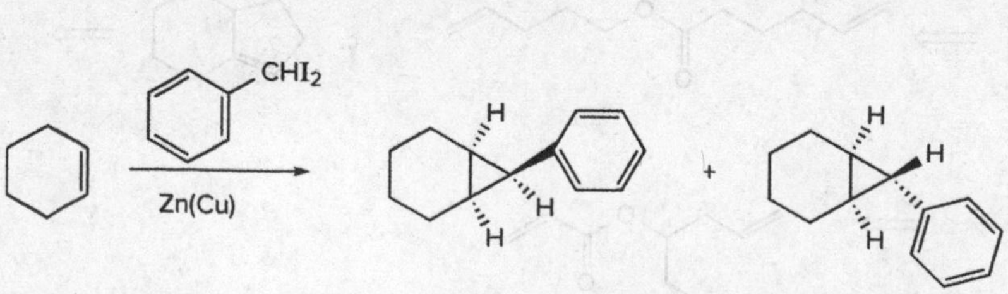

CHI_2

Zn(Cu)

28.32 High-dilution conditions favor intramolecular metathesis.

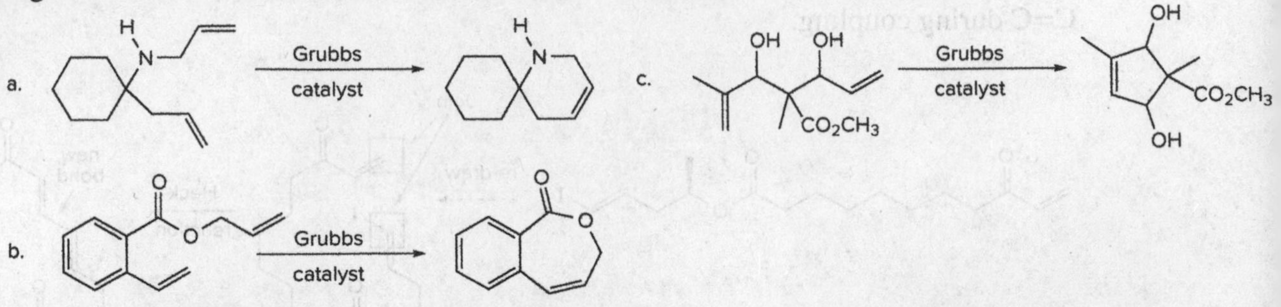

a. Grubbs catalyst

c. OH OH CO$_2$CH$_3$ Grubbs catalyst OH CO$_2$CH$_3$ OH

b. O O Grubbs catalyst O O

28.33

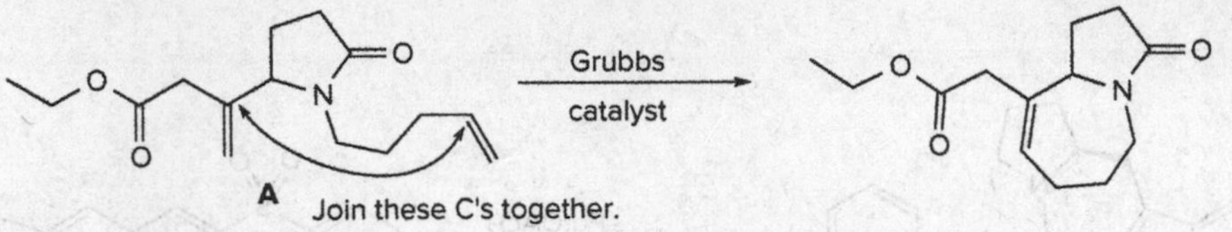

O N O O Grubbs catalyst O O N O

A Join these C's together.

28.34

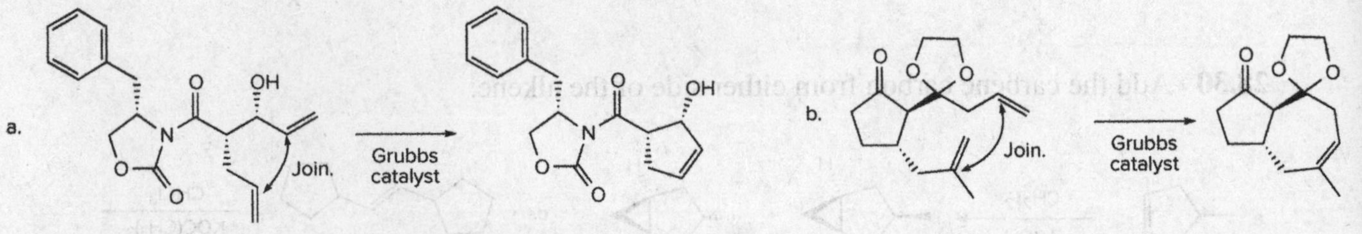

a. O N O O O OH Join. Grubbs catalyst O N O O O OH

b. O O O Join. Grubbs catalyst O O O

28.35

O Cl Join. O O Cl O

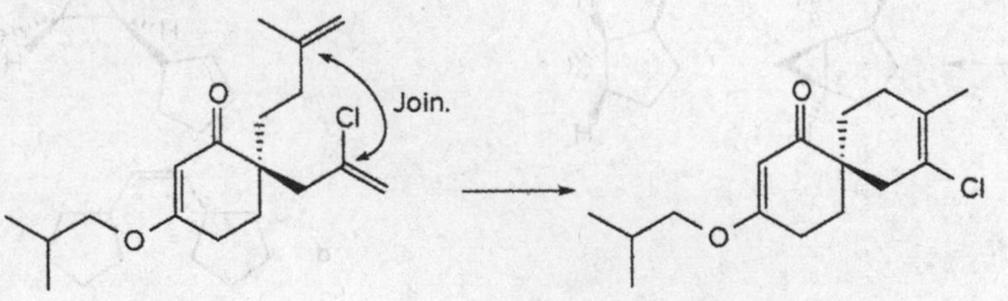

28.36 Retrosynthetically break the double bond in the cyclic compound and add a new =CH$_2$ at each end to find the starting material.

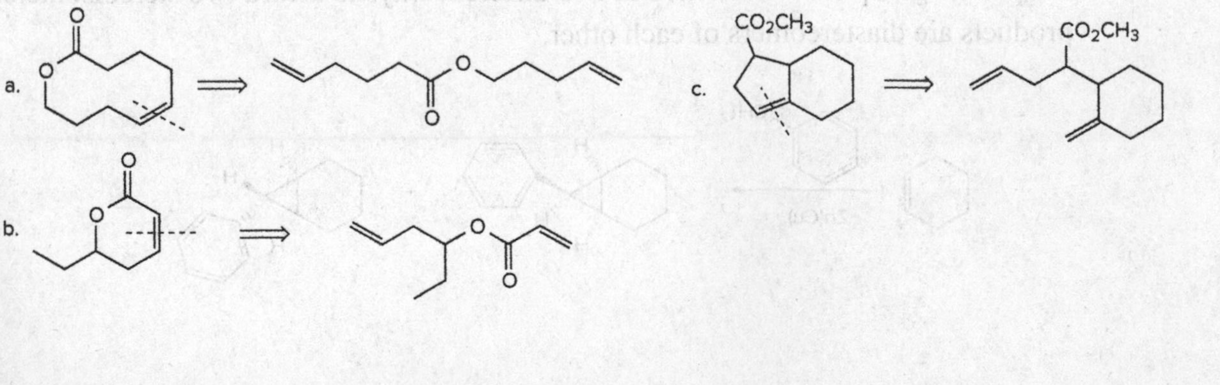

a. O O ⟹ O O

c. CO$_2$CH$_3$ ⟹ CO$_2$CH$_3$

b. O O ⟹ O O

28.37 Alkene metathesis with two different alkenes is synthetically useful only when both alkenes are symmetrically substituted; that is, the two groups on each end of the double bond are identical to the two groups on the other end of the double bond.

a. + → (Z + E) + (Z + E) + (Z + E) +

b. + → This reaction is synthetically useful because it yields only one product.

c. + → + + + + + + + (Z + E) + (Z + E)

28.38 All double bonds can have either the *E* or *Z* configuration.

a.

b.

c.

28.39

a. [1] $CHBr_3$, $KOC(CH_3)_3$ → Br Br; [2] $(CH_3)_2CuLi$ →

b. Br, COOH + CO_2CH_3 → $Pd(OAc)_2$, P(o-tolyl)$_3$, Et_3N → CH_3O_2C, COOH

c. $B(OCH_3)_2$ + CH_3O–Br → $Pd(PPh_3)_4$, NaOH → CH_3O

d. O, B, O + Br → $Pd(PPh_3)_4$, NaOH →

e. O, O → Grubbs catalyst → O, O

f. → CH_2I_2, Zn(Cu) →

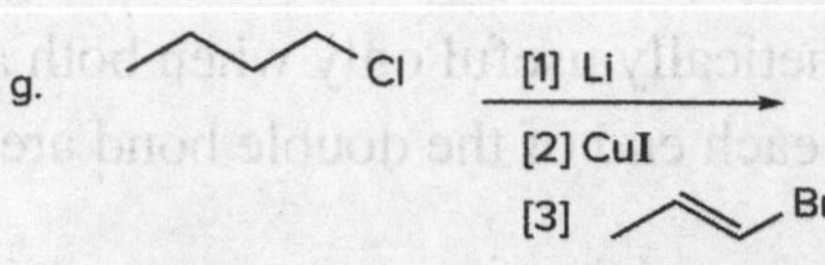

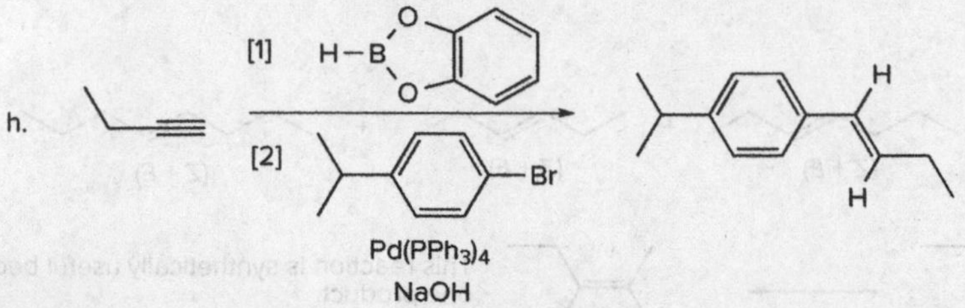

g. [1] Li, [2] CuI, [3] (E)-1-bromopropene

h. [1] catecholborane, [2] 4-isopropylbromobenzene, $Pd(PPh_3)_4$, NaOH

H H

28.40

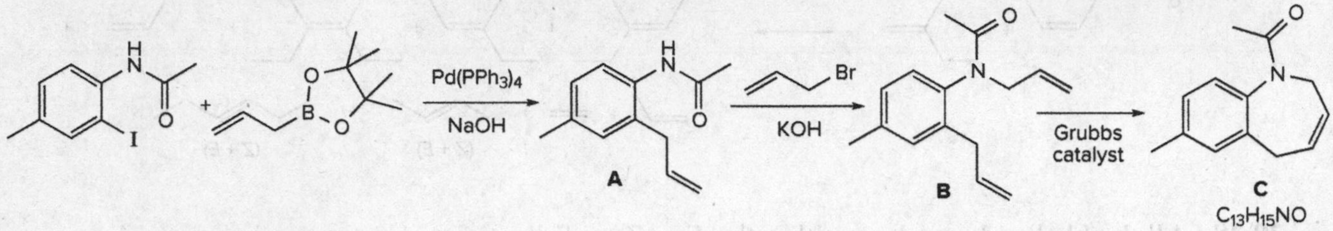

$Pd(PPh_3)_4$ NaOH → A; allyl bromide, KOH → B; Grubbs catalyst → C

A B C $C_{13}H_{15}NO$

28.41

Cl_3C :O: :O: Na^+ Δ → Cl Cl Cl Na^+ + O=C=O → Cl Cl + NaCl

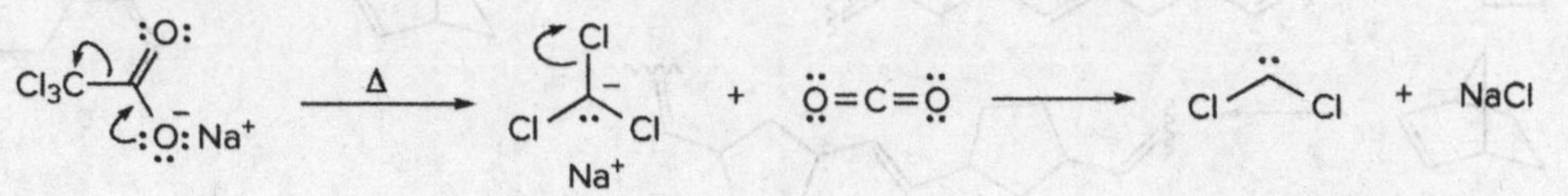

28.42 This reaction follows the Simmons–Smith reaction mechanism illustrated in Mechanism 28.5.

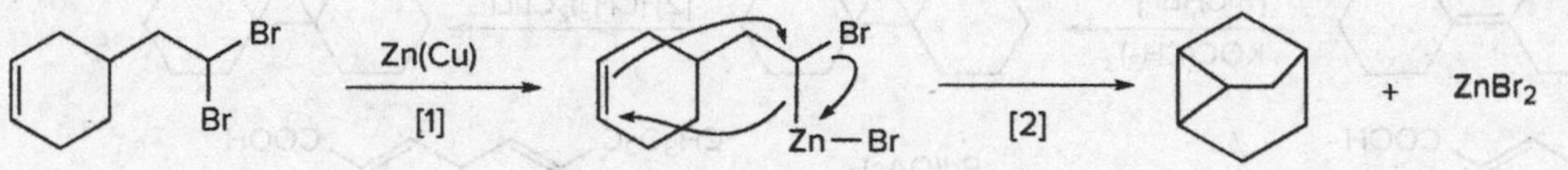

Br Br Zn(Cu) [1] → Br Zn—Br [2] → + $ZnBr_2$

28.43

Join → Grubbs catalyst → A → H^+ → H → + OH → B + OH_2

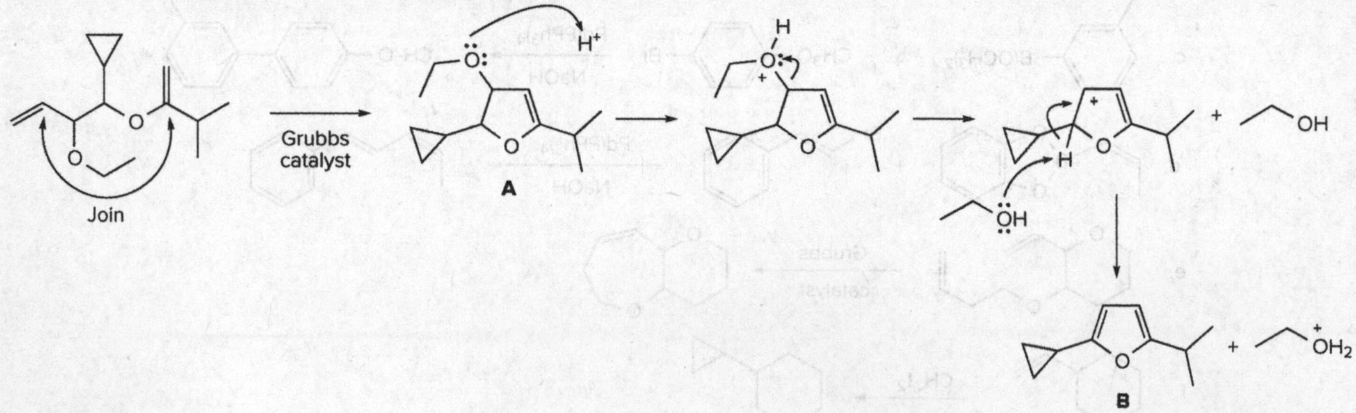

28.44

a sulfur ylide

1,4-addition of the sulfur ylide to the β carbon

a resonance-stabilized enolate

methyl *trans*-chrysanthemate

28.45

A

diazo compound

B

28.46

[1] $Pd(PAr_3)_2$ oxidative addition

[2] syn addition

[3] syn elimination

E

[4] reductive elimination

$Pd(PAr_3)_2$ + HI

The H cis to Pd is eliminated.

This H is trans to Pd.

b. This suggests that the stereochemistry in Step [3] must occur with syn elimination of H and Pd to form **E.** Product **F** cannot form because the only H on the C bonded to the benzene ring is trans to the Pd species, so it cannot be removed if elimination occurs in a syn fashion.

28.47

(*Z*)-2-bromostyrene

$NaNH_2$ (– HBr)

$Pd(PPh_3)_4$ NaOH

A

28.48

HO~~~~OH → ($SOCl_2$) → HO~~~~Cl

TBDMS–Cl, imidazole

$HC{\equiv}CH$ → (NaH) → $HC{\equiv}C^-$ + TBDMSO~~~~Cl

→ alkyne~~~~OTBDMS → (catecholborane, BH) → B~~~~OTBDMS

Bu_4NF → B~~~~OH

28.49

a. CH_3O–C6H4–CH=CH–Ph ⟹ CH_3O–C6H4–Br + styrene

Synthesize these two components, and then use a Heck reaction to synthesize the product.

Benzene → (HNO_3, H_2SO_4) → NO_2 → (H_2, Pd-C) → NH_2 → ($NaNO_2$, HCl) → $N_2^+Cl^-$ → (H_2O) → OH

Phenol → ([1] NaH, [2] CH_3Br) → CH_3O–Ph → (Br_2) → CH_3O–C6H4–Br

CH_3OH → (PBr_3) → CH_3Br

Ethanol (OH) → ($SOCl_2$) → Cl

Benzene → (Cl, $AlCl_3$) → ethylbenzene → (Br_2, $h\nu$) → Br → ($KOC(CH_3)_3$) → styrene

CH_3O–C6H4–Br + styrene → ($Pd(OAc)_2$, $P(o\text{-tolyl})_3$, Et_3N) → CH_3O–C6H4–CH=CH–Ph

b. CH_3O–C6H4–Br (from a.) → (Cl, $AlCl_3$) → Br, CH_3O (ethyl-substituted)

OH → ($SOCl_2$) → Cl

Br-arene (CH_3O) + CO_2Et (ethyl acrylate) → ($Pd(OAc)_2$, $P(o\text{-tolyl})_3$, Et_3N) → CO_2Et, CH_3O → (H_2, Pd-C) → CO_2Et, CH_3O

Benzene → (Br_2, $FeBr_3$) → Br → (Mg) → MgBr

[1] (2 equiv) [2] H_2O → OH, CH_3O (diphenyl tertiary alcohol)

28.50

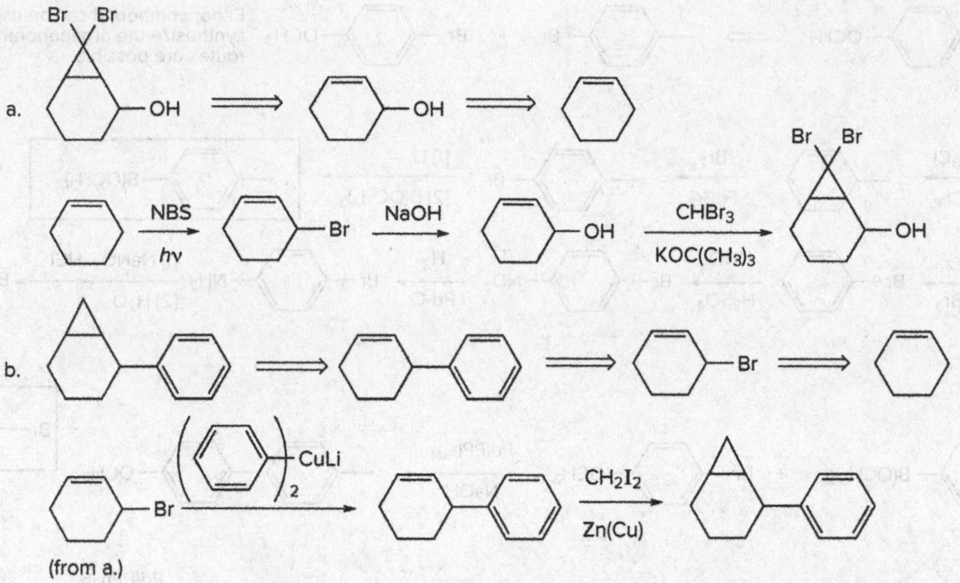

a.

Br Br OH ⟹ OH ⟹

NBS, $h\nu$ → Br; NaOH → OH; $CHBr_3$, $KOC(CH_3)_3$ → Br Br OH

b.

⟹ ⟹ Br ⟹

Br (from a.) → $(C_6H_5)_2CuLi$ → ; CH_2I_2, Zn(Cu) →

28.51

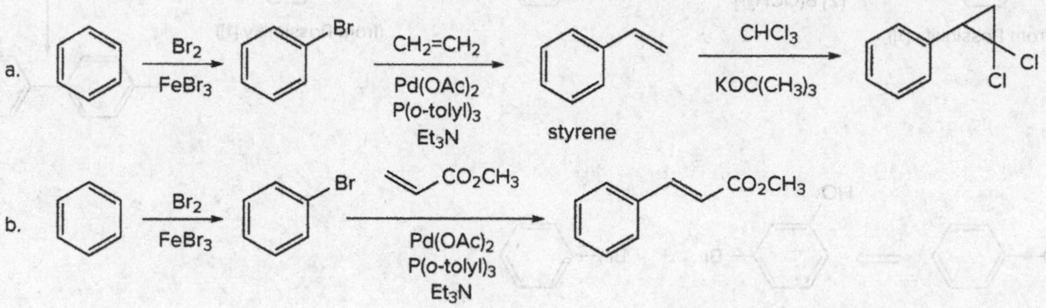

a. Br_2, $FeBr_3$ → Br; $CH_2{=}CH_2$, $Pd(OAc)_2$, P(o-tolyl)$_3$, Et_3N → styrene; $CHCl_3$, $KOC(CH_3)_3$ → Cl Cl

b. Br_2, $FeBr_3$ → Br; $CH_2{=}CHCO_2CH_3$, $Pd(OAc)_2$, P(o-tolyl)$_3$, Et_3N → CO_2CH_3

28.52

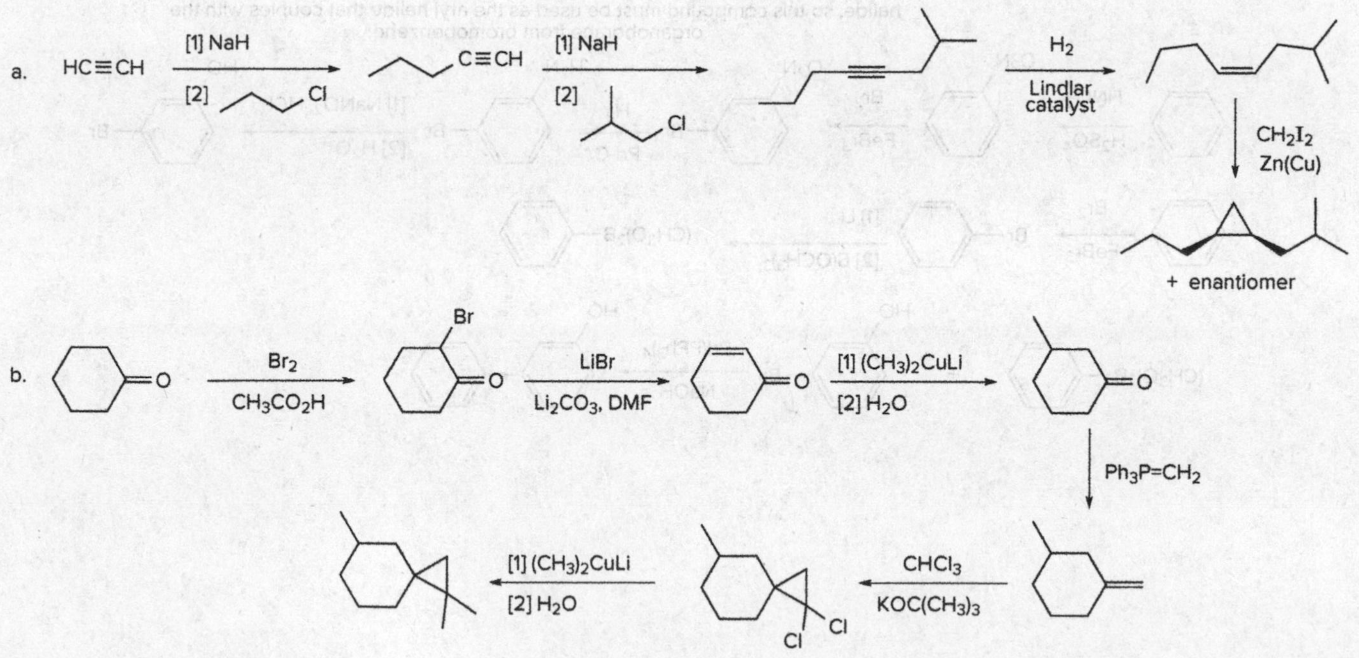

a. $HC{\equiv}CH$ → [1] NaH [2] Cl → $C{\equiv}CH$ → [1] NaH [2] Cl → ; H_2, Lindlar catalyst → ; CH_2I_2, Zn(Cu) → + enantiomer

b. O → Br_2, CH_3CO_2H → Br O; LiBr, Li_2CO_3, DMF → O; [1] $(CH_3)_2CuLi$ [2] H_2O → O; $Ph_3P{=}CH_2$ → ; $CHCl_3$, $KOC(CH_3)_3$ → Cl Cl; [1] $(CH_3)_2CuLi$ [2] H_2O →

28.53

a. $CH_3C_6H_4–C_6H_4OCH_3$ ⟹ $CH_3C_6H_4Br$ + $BrC_6H_4OCH_3$

Either compound can be used to synthesize the organoborane, so two routes are possible.

Possibility [1]:

Benzene —(CH_3Cl, $AlCl_3$)→ toluene —(Br_2, $FeBr_3$)→ Br —([1] Li, [2] $B(OCH_3)_3$)→ $B(OCH_3)_2$

Benzene —(Br_2, $FeBr_3$)→ Br —(HNO_3, H_2SO_4)→ Br, NO_2 —(H_2, Pd-C)→ Br, NH_2 —([1] $NaNO_2$, HCl, [2] H_2O)→ Br, OH —([1] NaH, [2] CH_3I)→ Br, OCH_3

$B(OCH_3)_2$ + Br, OCH_3 —($Pd(PPh_3)_4$, NaOH)→ OCH_3

Possibility [2]:

Br, OCH_3 (from Possibility [1]) —([1] Li, [2] $B(OCH_3)_3$)→ $(CH_3O)_2B$, OCH_3 + Br (from Possibility [1]) —($Pd(PPh_3)_4$, NaOH)→ OCH_3

b. HO-biphenyl ⟹ HO, Br + Br

The acidic OH makes it impossible to prepare an organolithium reagent from this aryl halide, so this compound must be used as the aryl halide that couples with the organoborane from bromobenzene.

Benzene —(HNO_3, H_2SO_4)→ O_2N —(Br_2, $FeBr_3$)→ O_2N, Br —(H_2, Pd-C)→ H_2N, Br —([1] $NaNO_2$, HCl, [2] H_2O)→ HO, Br

Benzene —(Br_2, $FeBr_3$)→ Br —([1] Li, [2] $B(OCH_3)_3$)→ $(CH_3O)_2B$

$(CH_3O)_2B$ + HO, Br —($Pd(PPh_3)_4$, NaOH)→ HO

29.8 Use the rules for electrocyclic reactions found in Answers 29.5 and 29.6.

a. re-draw
3 π bonds
Δ disrotatory
hν conrotatory
+ enantiomer

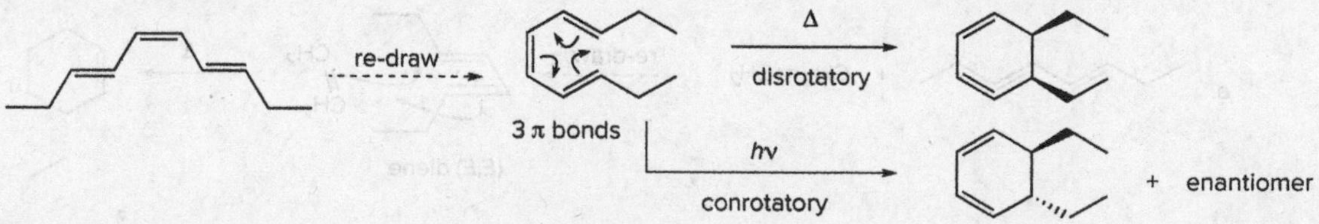

b. re-draw
3 π bonds
Δ disrotatory
+ enantiomer
hν conrotatory

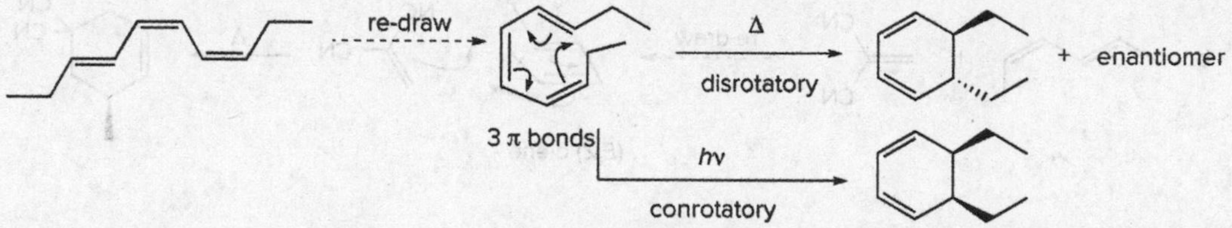

29.9 Use the rules for electrocyclic reactions found in Answer 29.5. A reaction with three π bonds and a disrotatory cyclization is thermal.

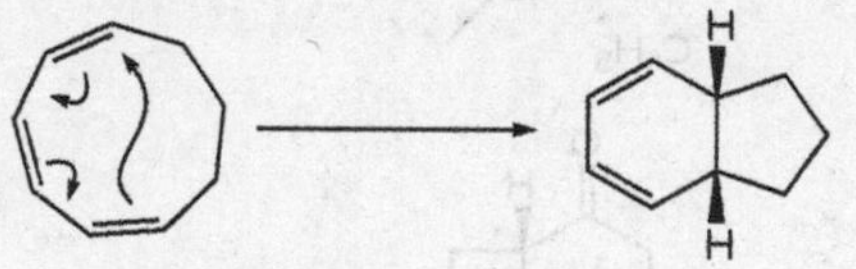

29.10 Count the number of π electrons in each reactant to classify the cycloaddition.

a. 2 π electrons + 2 π electrons → [2 + 2] cycloaddition (one possibility)

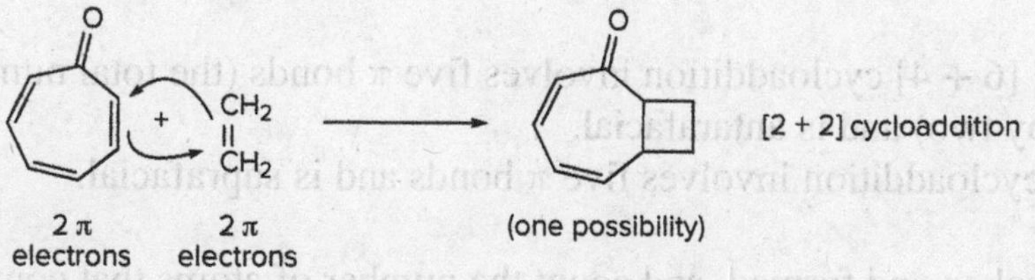

b. 4 π electrons + 2 π electrons → [4 + 2] cycloaddition

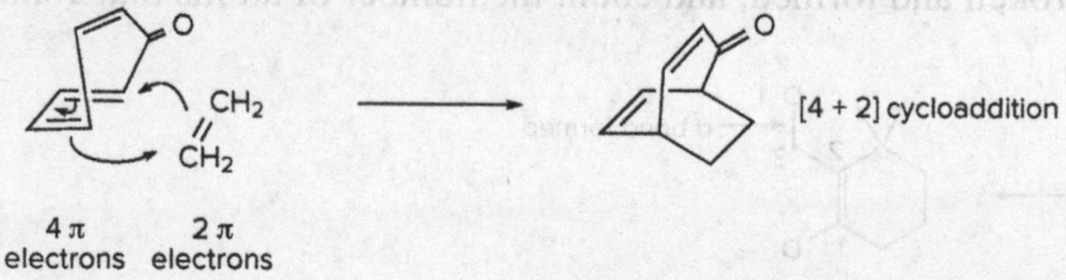

c. 6 π electrons + 2 π electrons → [6 + 2] cycloaddition

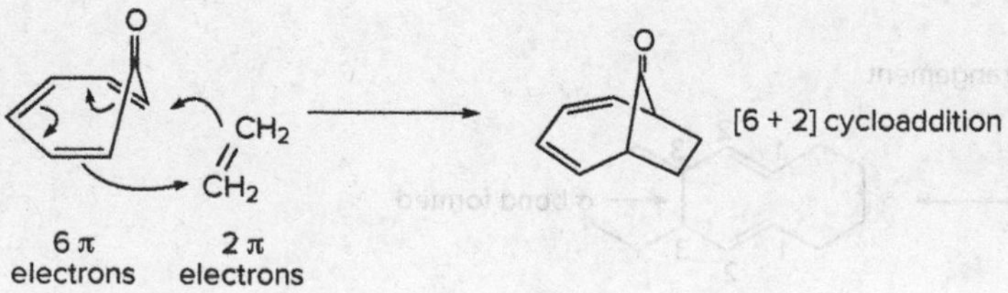

29.11 A thermal [4 + 2] cycloaddition is suprafacial.

a. 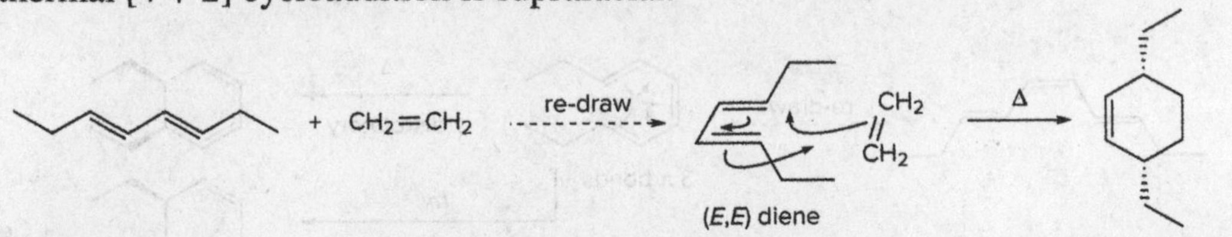 + $CH_2{=}CH_2$ re-draw CH_2 // CH_2 Δ

(*E,E*) diene

b. 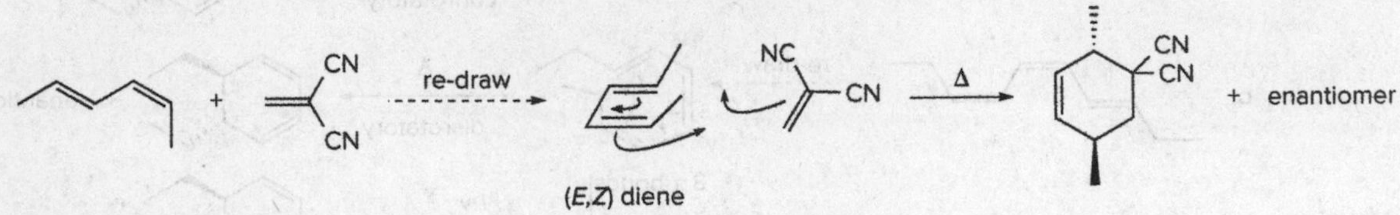 + CN CN re-draw NC CN Δ CN CN + enantiomer

(*E,Z*) diene

29.12 A photochemical [2 + 2] cycloaddition is suprafacial.

a. 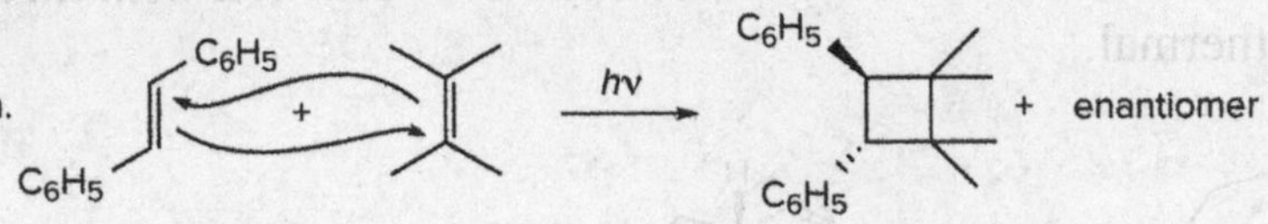 C_6H_5 C_6H_5 + *hν* C_6H_5 C_6H_5 + enantiomer

b. 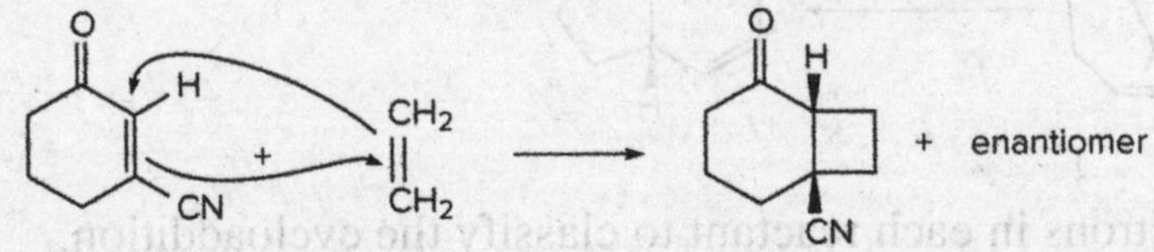 O H + CH_2 || CH_2 CN O H CN + enantiomer

29.13

a. The photochemical [6 + 4] cycloaddition involves five π bonds (the total number of π electrons divided by two) and is antarafacial.

b. A thermal [8 + 2] cycloaddition involves five π bonds and is suprafacial.

29.14 Locate the σ bonds broken and formed, and count the number of atoms that connect them.

a. 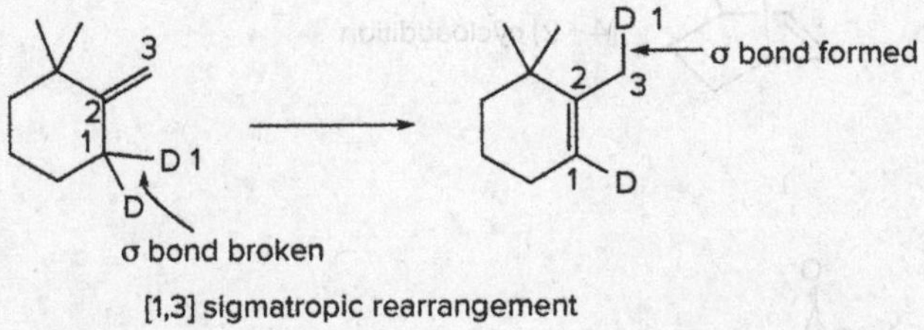 3 2 1 D 1 D σ bond broken D 1 2 3 σ bond formed 1 D

[1,3] sigmatropic rearrangement

b. 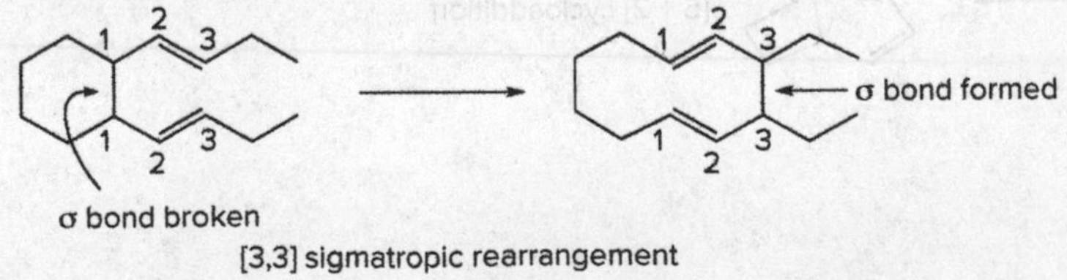 1 2 3 1 2 3 σ bond broken 1 2 3 σ bond formed 1 2 3

[3,3] sigmatropic rearrangement

29.15

a. re-draw [1,7]

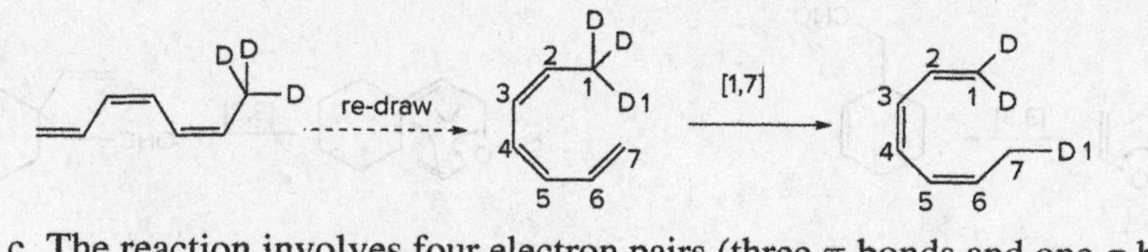

b, c. The reaction involves four electron pairs (three π bonds and one σ bond), so it proceeds by an antarafacial pathway under thermal conditions, and by a suprafacial pathway under photochemical conditions.

29.16 Draw the products of each reaction.

a. re-draw [3,3]

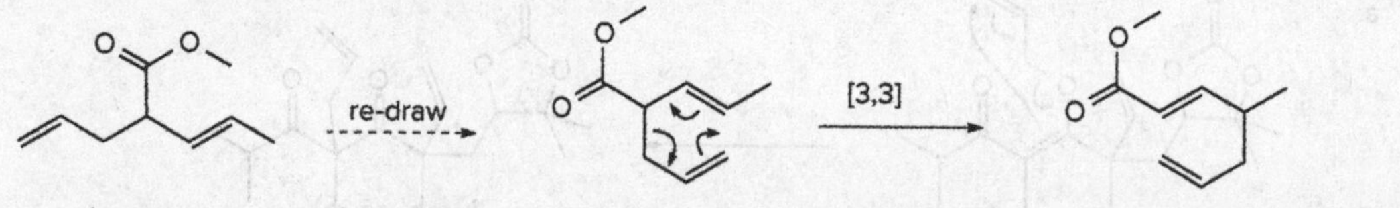

b. [3,3]

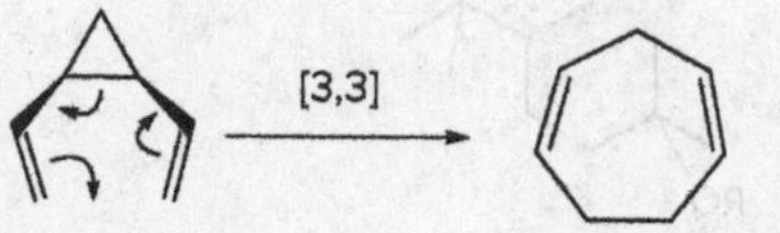

c. [3,3] tautomerize

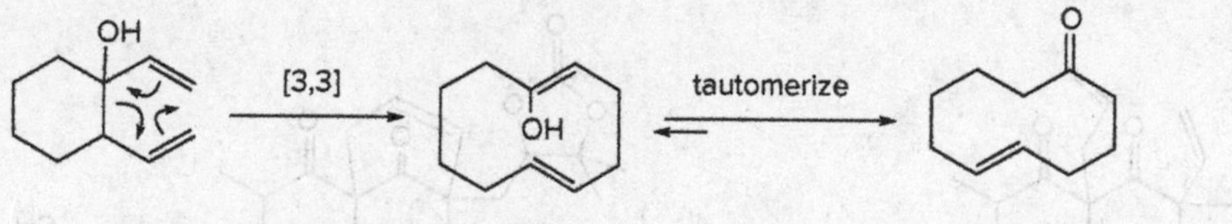

29.17 Draw the product after protonation.

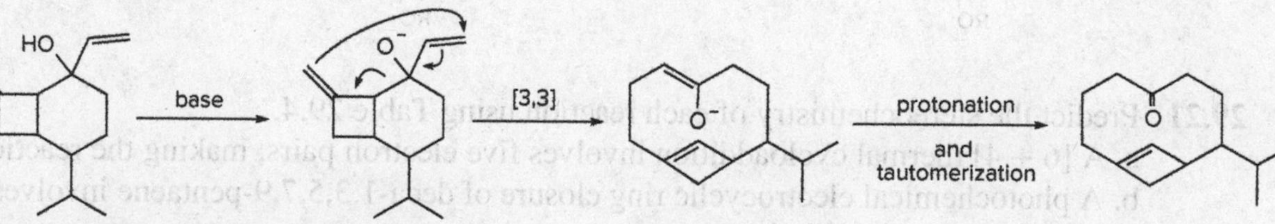

base [3,3] protonation and tautomerization

29.18 Re-draw geranial to put the ends of the 1,5-diene close together. Then draw three curved arrows, beginning at a π bond.

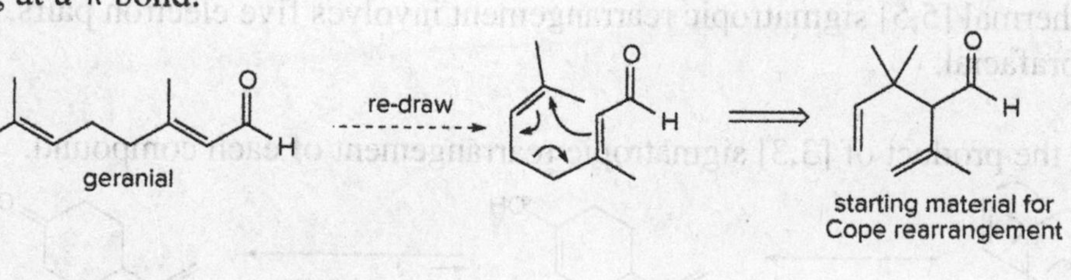

geranial re-draw

starting material for Cope rearrangement

29.19 Draw the product of Claisen rearrangement.

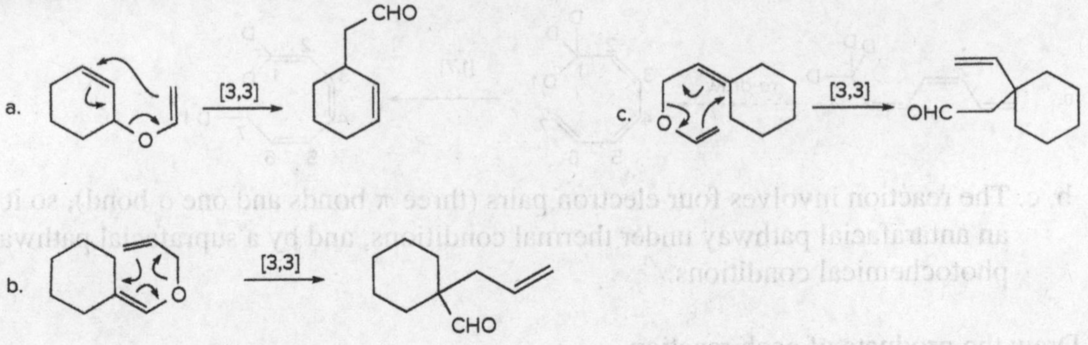

a. [3,3] CHO

c. [3,3] OHC

b. [3,3] CHO

29.20

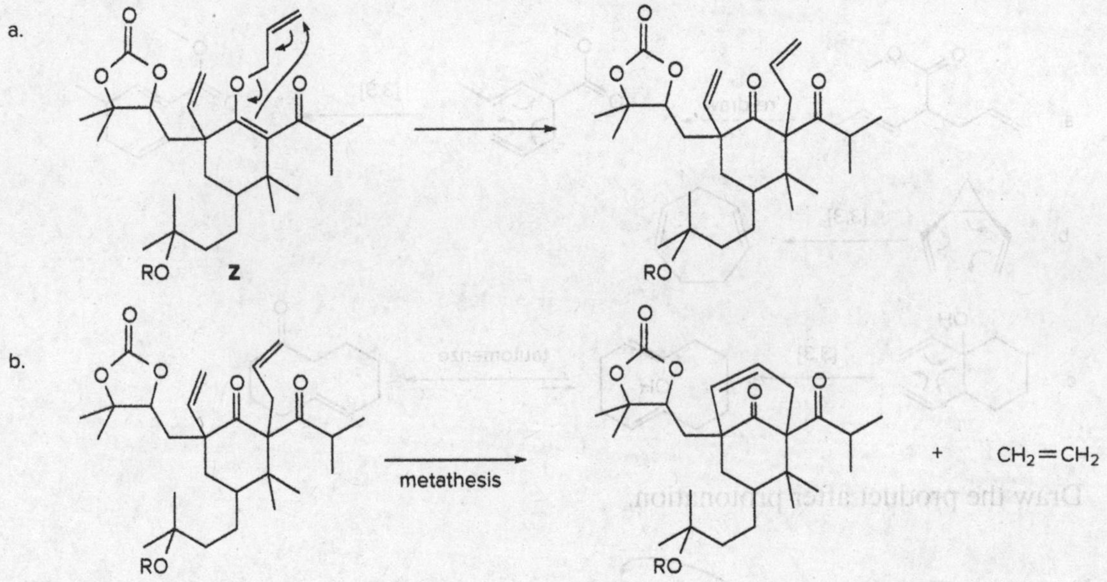

a. RO Z RO

b. metathesis RO RO + $CH_2{=}CH_2$

29.21 Predict the stereochemistry of each reaction using Table 29.4.

a. A [6 + 4] thermal cycloaddition involves five electron pairs, making the reaction suprafacial.

b. A photochemical electrocyclic ring closure of deca-1,3,5,7,9-pentaene involves five electron pairs, making the reaction conrotatory.

c. A [4 + 4] photochemical cycloaddition involves four electron pairs, making the reaction suprafacial.

d. A thermal [5,5] sigmatropic rearrangement involves five electron pairs, making the reaction suprafacial.

29.22 Draw the product of [3,3] sigmatropic rearrangement of each compound.

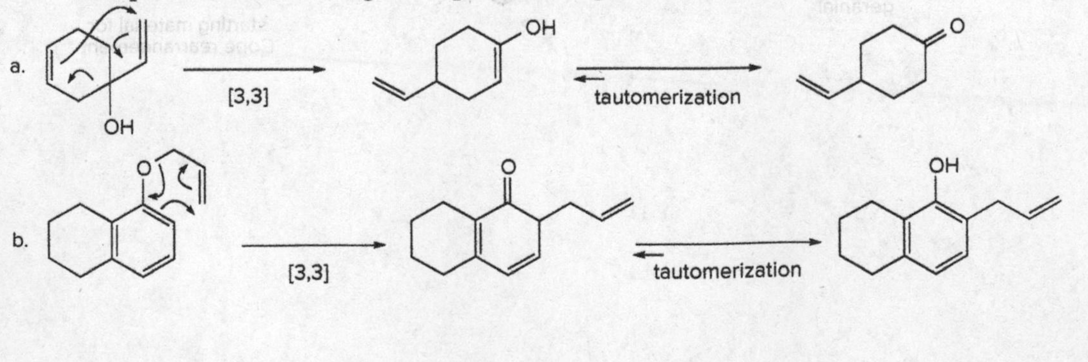

a. OH [3,3] OH tautomerization O

b. O [3,3] O tautomerization OH

29.23 An electrocyclic reaction forms a product with one more or one fewer π bond than the starting material. A cycloaddition forms a ring with two new σ bonds. A sigmatropic rearrangement forms a product with the same number of π bonds, but the π bonds are rearranged. Use Table 29.4 to determine the stereochemistry.

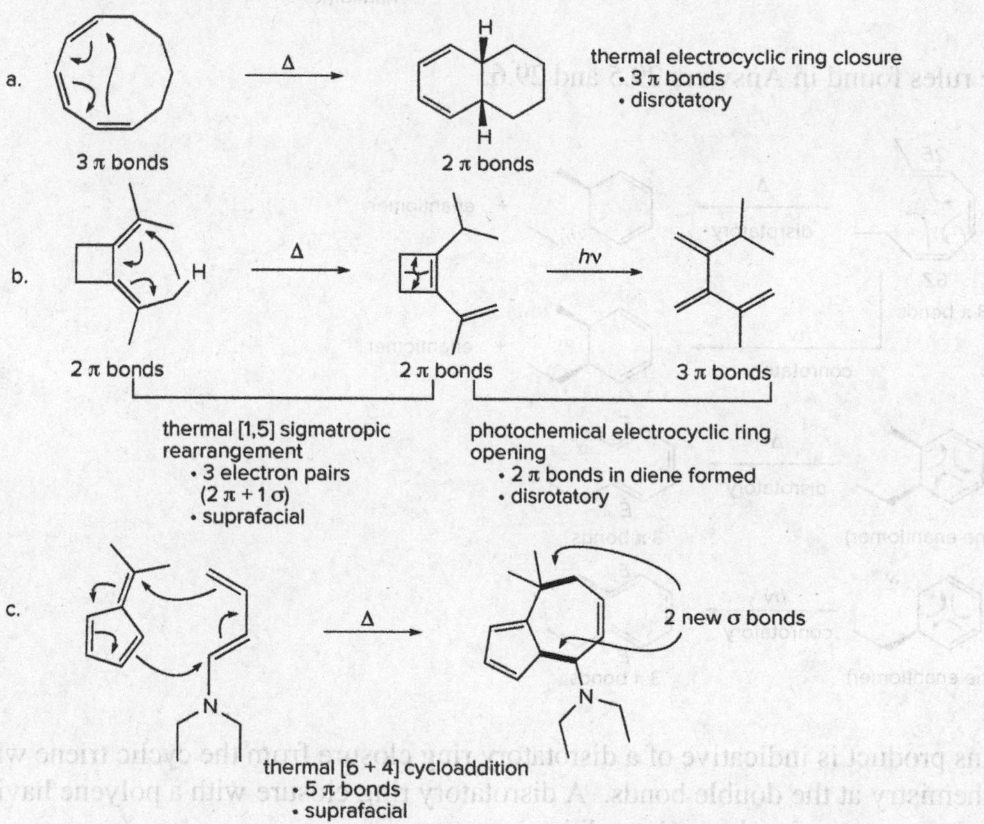

29.24 Use the rules for thermal electrocyclic reactions found in Answer 29.5.

a. Δ, disrotatory — *E*, *Z*, *Z* — 3 π bonds

b. re-draw — Δ, conrotatory — 2 π bonds

29.25 Use the rules for photochemical electrocyclic reactions found in Answer 29.6.

a. hν, conrotatory — *E*, *Z*, *E* — 3 π bonds

Although conrotatory ring opening could also form, at least in theory, an all-(*Z*) triene, steric hindrance during ring opening would cause the terminal CH_3's to crash into one another, making this process unlikely.

b. re-draw — $h\nu$ disrotatory — + enantiomer

29.26 Use the rules found in Answers 29.5 and 29.6.

a, b. 2E, 4Z, 6Z, 3 π bonds — Δ disrotatory — + enantiomer; $h\nu$ conrotatory — + enantiomer

c. (one enantiomer) — Δ disrotatory — E, E, 3 π bonds

d. (one enantiomer) — $h\nu$ conrotatory — E, E, 3 π bonds

29.27 The trans product is indicative of a disrotatory ring closure from the cyclic triene with the given stereochemistry at the double bonds. A disrotatory ring closure with a polyene having three π bonds must occur under thermal conditions.

E, Z — H, H, trans

29.28 Use the rules found in Answers 29.5 and 29.6.

a, c. N, 3 π bonds — Δ disrotatory — H, H, cis

b, c. N, 3 π bonds — $h\nu$ conrotatory — H, H, trans + enantiomer

29.29 Use the rules found in Answers 29.5 and 29.6.

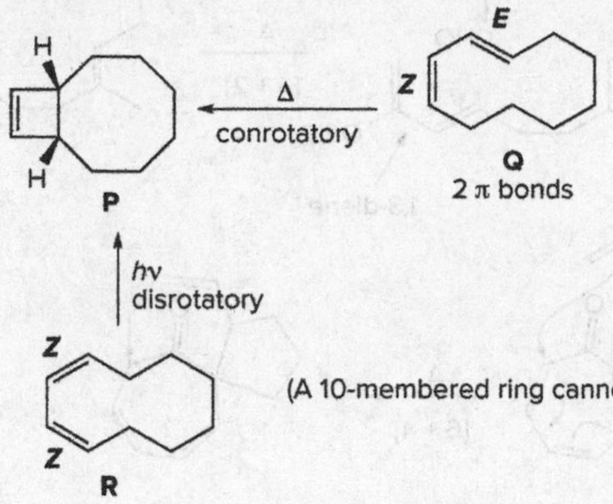

conrotatory

Q

2 π bonds

P

hν
disrotatory

(A 10-membered ring cannot contain two *E* double bonds.)

R

29.30 The reaction involves three π bonds in one reactant and two π bonds in the second reactant, so the reaction is a [6 + 4] cycloaddition. A suprafacial cycloaddition with five π bonds must proceed under thermal conditions.

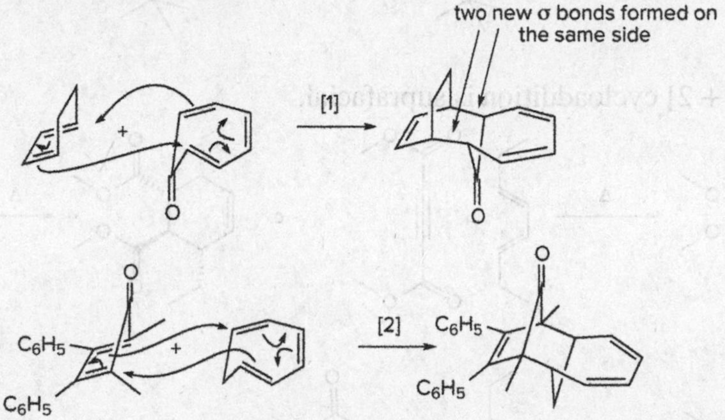

two new σ bonds formed on the same side

[1]

[2]

29.31 The Diels–Alder reaction is a thermal, suprafacial [4 + 2] cycloaddition.

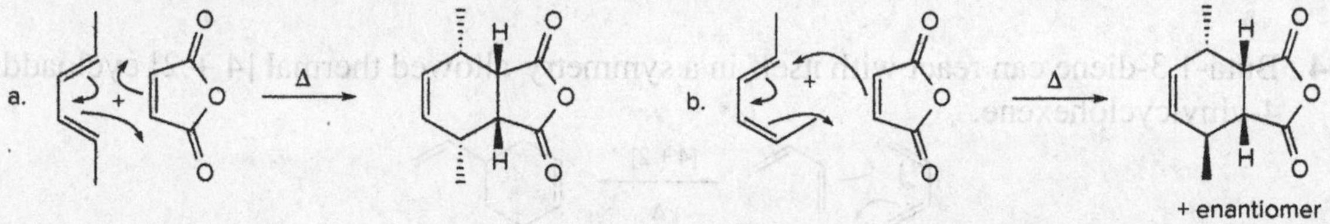

a.

b.

+ enantiomer

29.32

a.

re-draw

hν
[2 + 2]

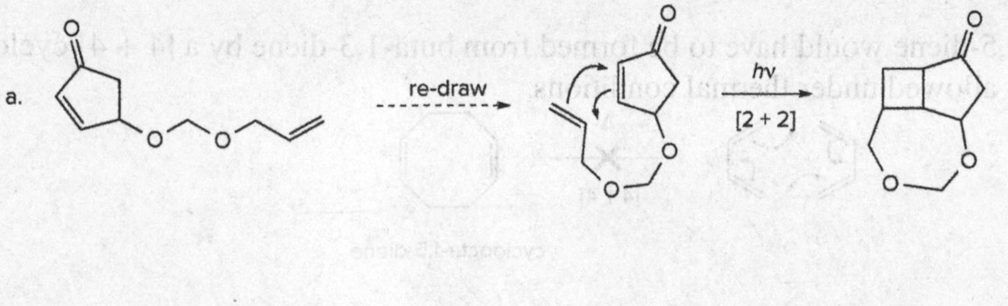

b. re-draw [4 + 2] CHO OSiR$_3$ 1,3-diene

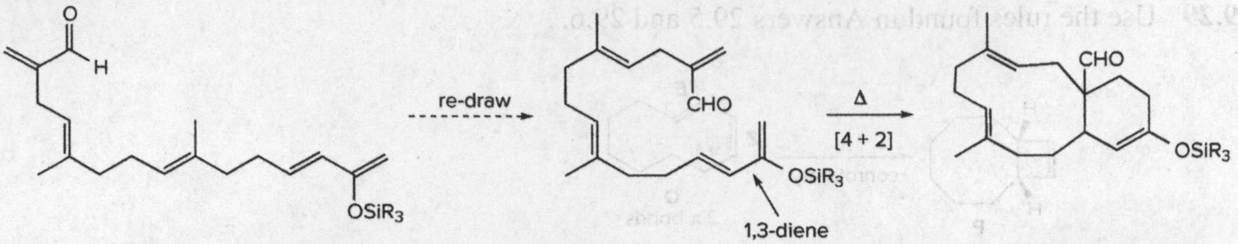

c. re-draw [6 + 4]

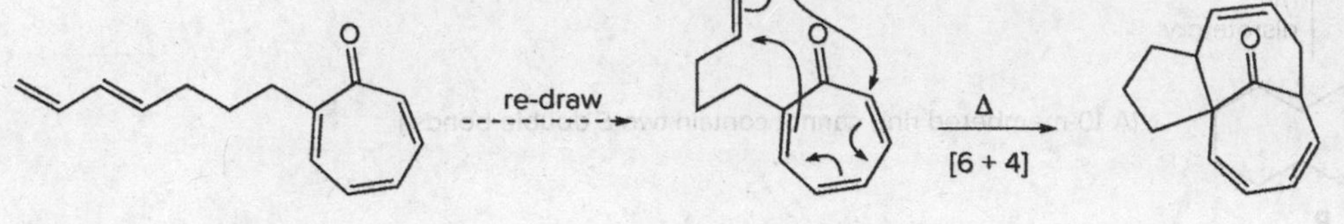

d. re-draw $h\nu$ [2 + 2]

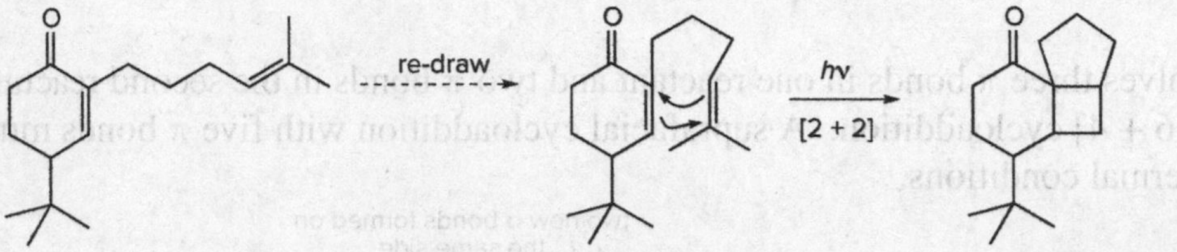

29.33 A thermal [4 + 2] cycloaddition is suprafacial.

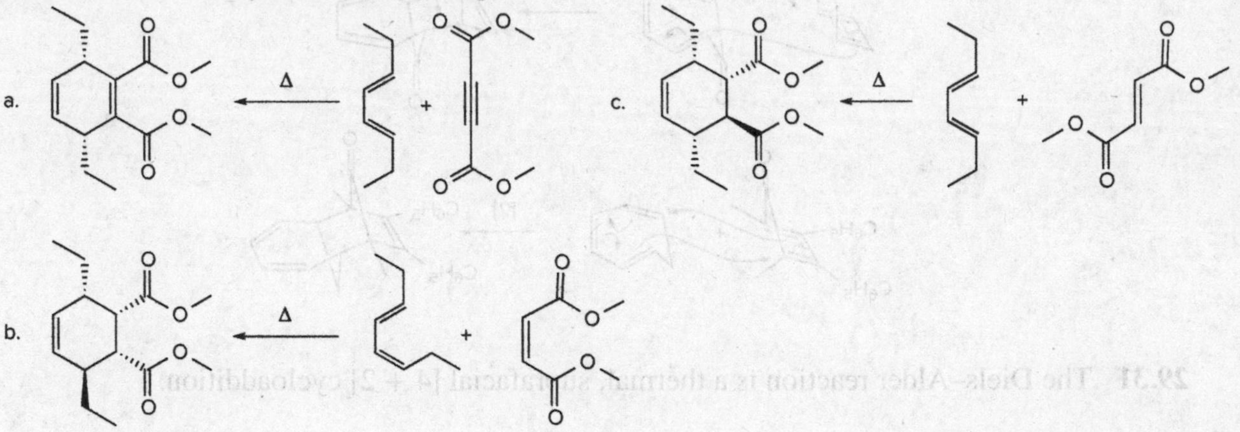

a. b. c.

29.34 Buta-1,3-diene can react with itself in a symmetry-allowed thermal [4 + 2] cycloaddition to form 4-vinylcyclohexene.

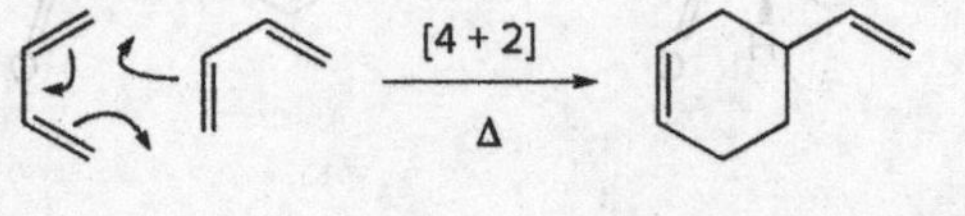

[4 + 2] Δ

4-vinylcyclohexene

Cycloocta-1,5-diene would have to be formed from buta-1,3-diene by a [4 + 4] cycloaddition, which is not allowed under thermal conditions.

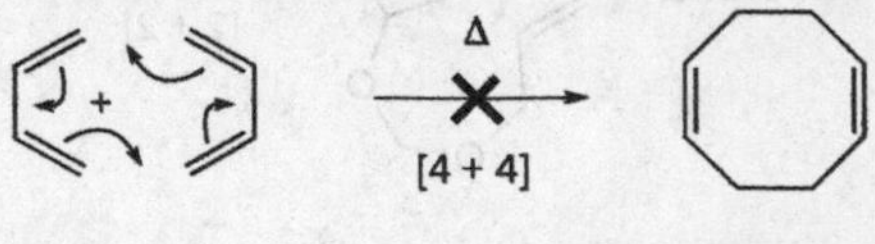

Δ [4 + 4]

cycloocta-1,5-diene

29.35 Re-draw the reactant and product to more clearly show the relative location of the bonds broken and formed.

a. D D re-draw 5 D 4 D 3 2 1 H 1 [1,5] D 4 5 D 3 2 1

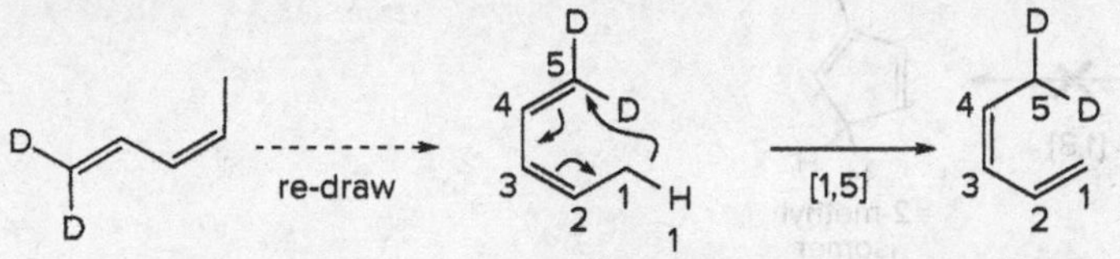

b. O SC_6H_5 re-draw 2 3 1 :S: 2 O: 1 C_6H_5 [2,3] 2 3 1 +S O: 2 1 C_6H_5

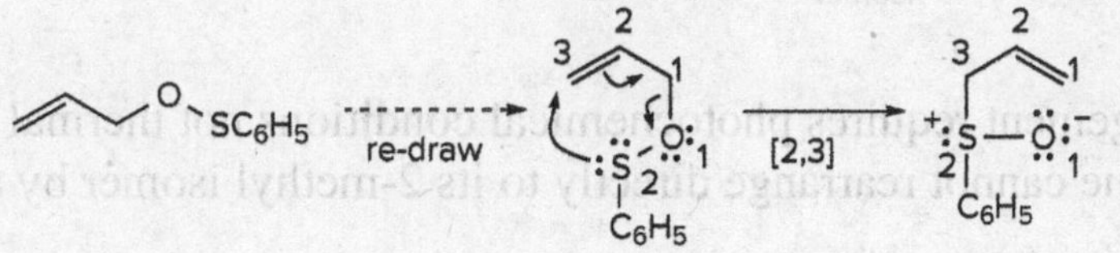

29.36 Draw the products of each reaction.

a. O [3,3] O

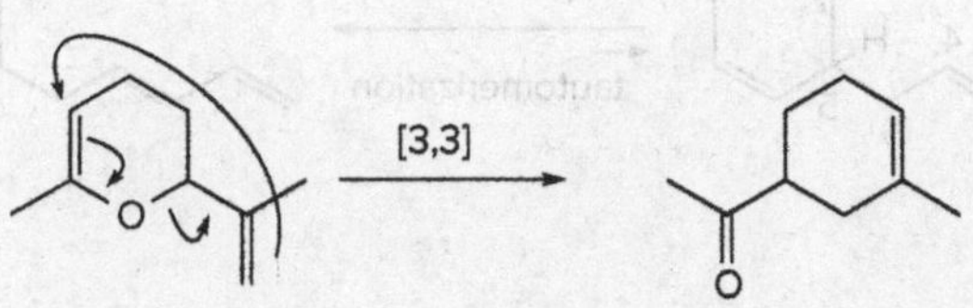

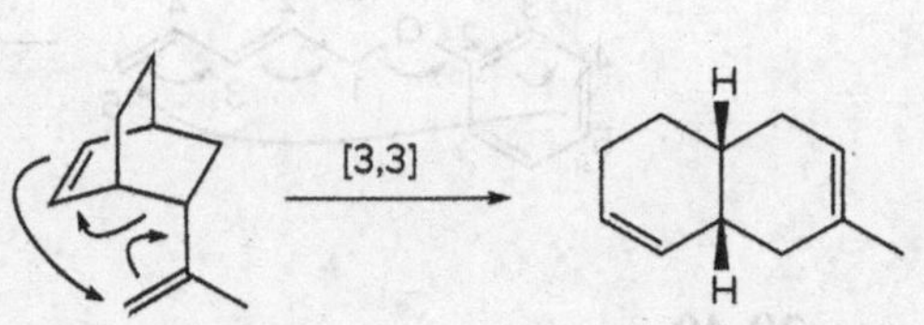

b. OH [3,3] OH tautomerize O

c. [3,3] H H

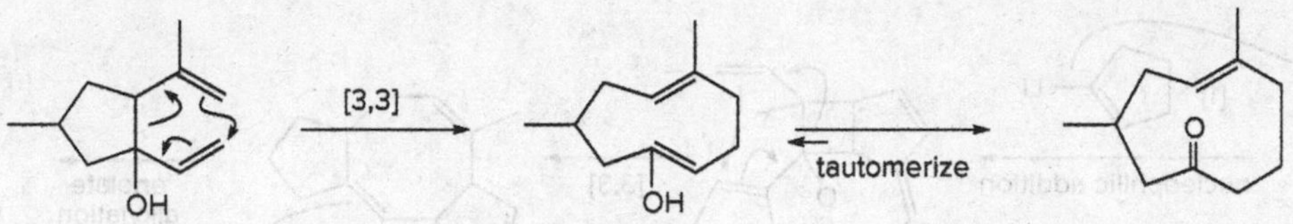

29.37

O O [1] MgBr [2] H_2O HO O C [1] KH, Δ $^-$O O [3,3] O$^-$ O [2] H_2O O O D

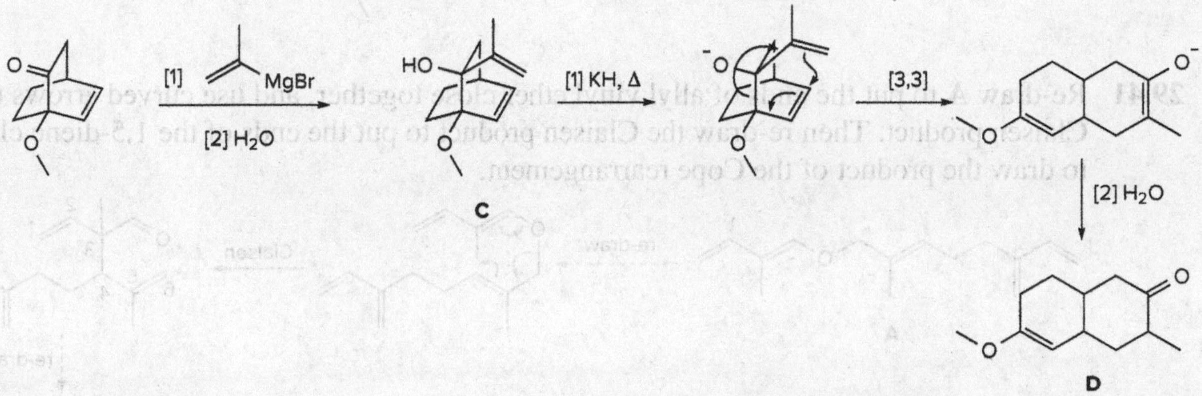

29.38 a. Two [1,5] sigmatropic rearrangements occur.

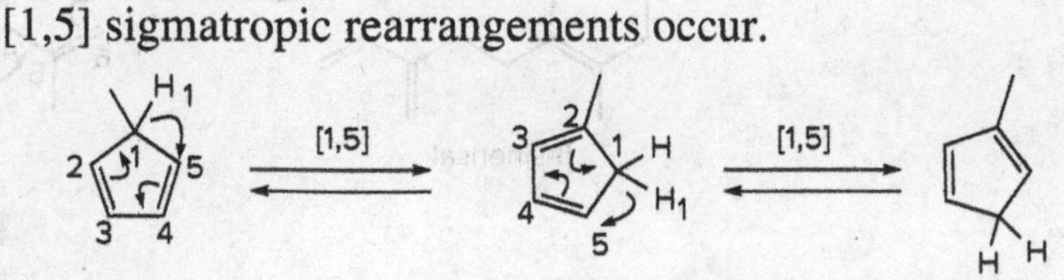

H 1 2 1 5 3 4 [1,5] 3 2 1 H H 1 4 5 [1,5] H H

5-methyl-cyclopenta-1,3-diene 1-methyl-cyclopenta-1,3-diene 2-methyl-cyclopenta-1,3-diene

b.

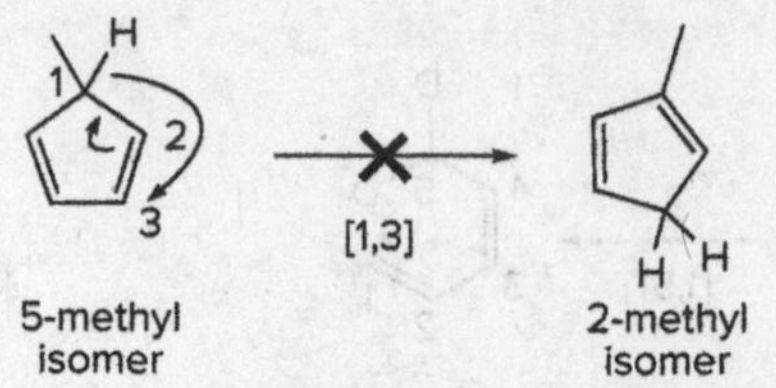

5-methyl isomer —[1,3]✕→ 2-methyl isomer

A [1,3] sigmatropic rearrangement requires photochemical conditions not thermal conditions, so 5-methylcyclopenta-1,3-diene cannot rearrange directly to its 2-methyl isomer by a [1,3] shift.

29.39

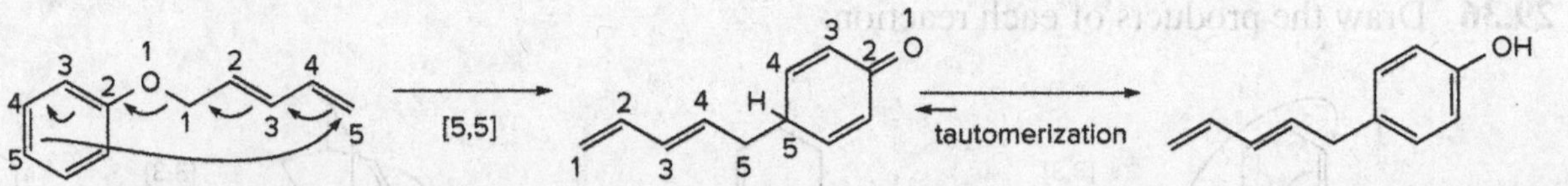

[5,5] — tautomerization

29.40

[1] nucleophilic addition — [3,3] — enolate alkylation — CH_3–I

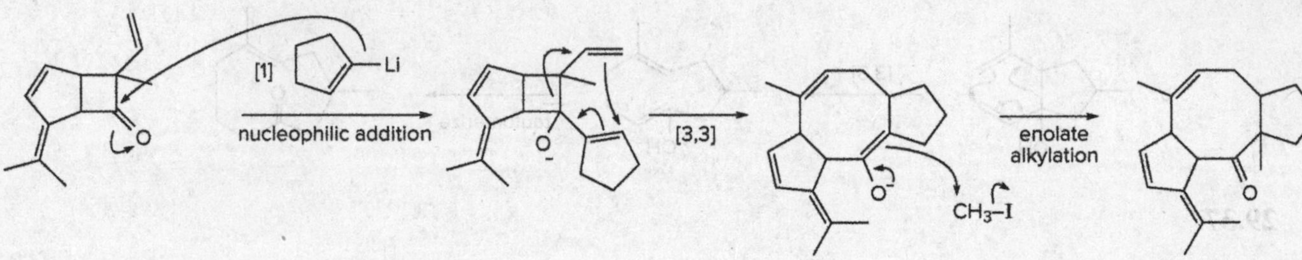

29.41 Re-draw **A** to put the ends of allyl vinyl ether close together, and use curved arrows to draw the Claisen product. Then re-draw the Claisen product to put the ends of the 1,5-diene close together to draw the product of the Cope rearrangement.

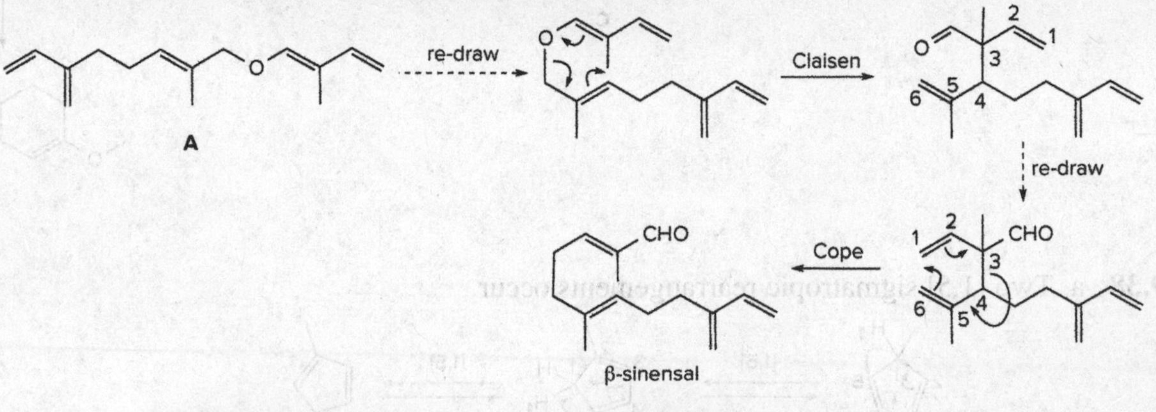

A — re-draw — Claisen — re-draw — Cope — CHO — β-sinensal

29.42 Use the definitions found in Answer 29.1.

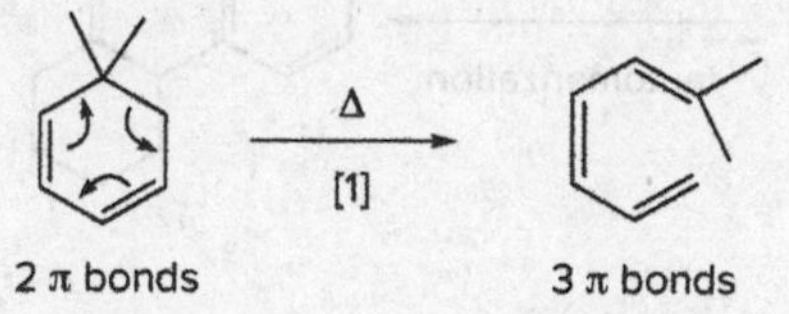

Δ [1] — 2 π bonds → 3 π bonds — an electrocyclic ring opening

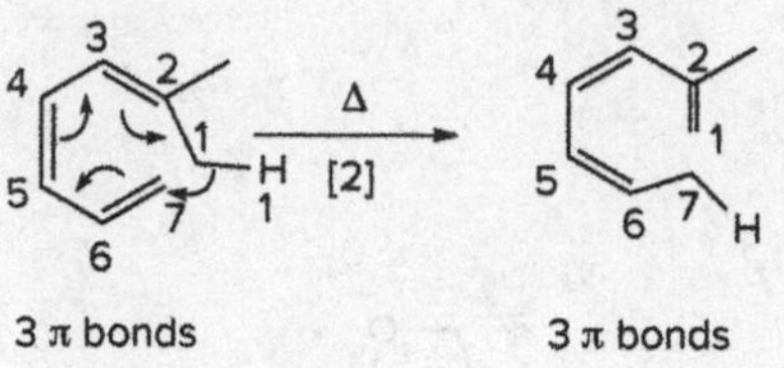

Δ [2] — 3 π bonds → 3 π bonds — a [1,7] sigmatropic rearrangement

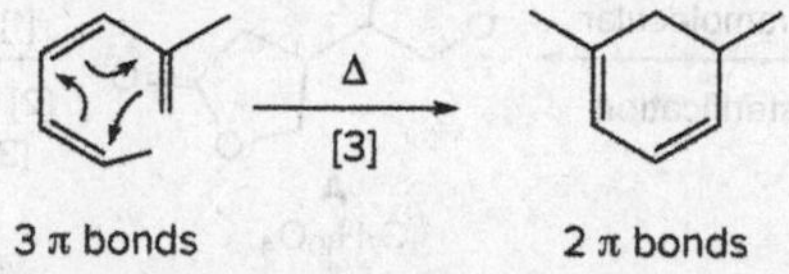

Δ [3] — 3 π bonds → 2 π bonds — an electrocyclic ring closure

29.43

a. 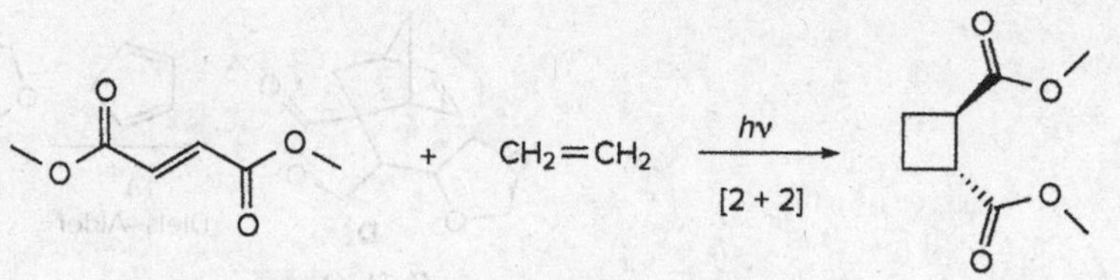 $CH_2{=}CH_2$ — hν, [2 + 2] — + enantiomer

b. 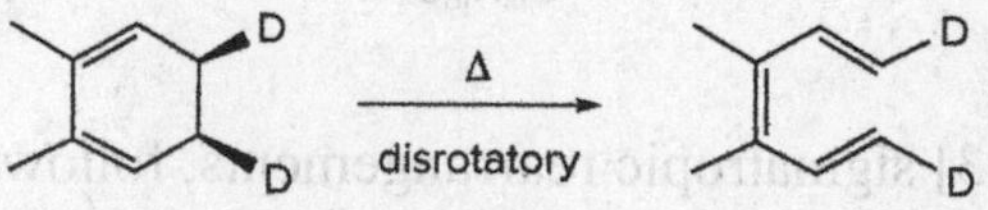 Δ, disrotatory

c. 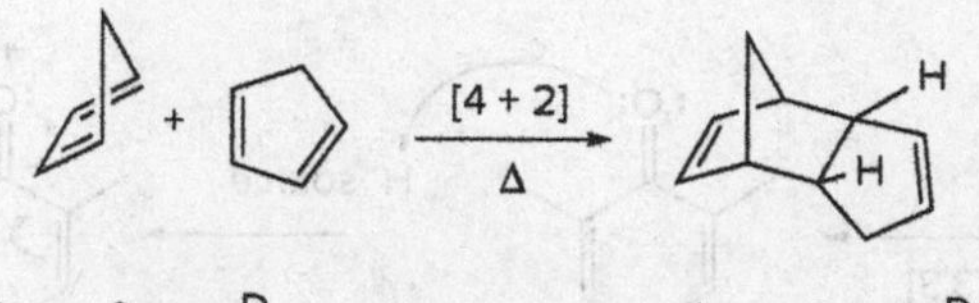 [4 + 2], Δ

d. 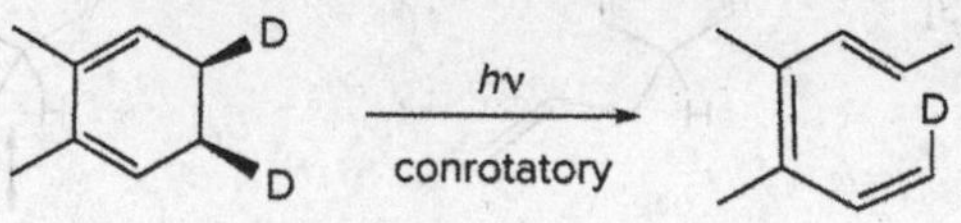 hν, conrotatory

29.44

a. 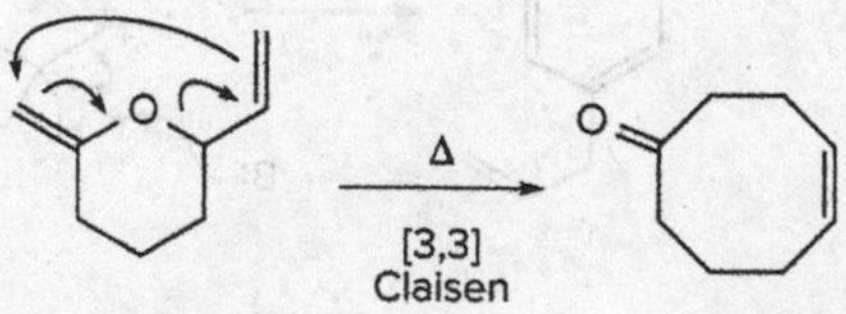 Δ, [3,3] Claisen

b. 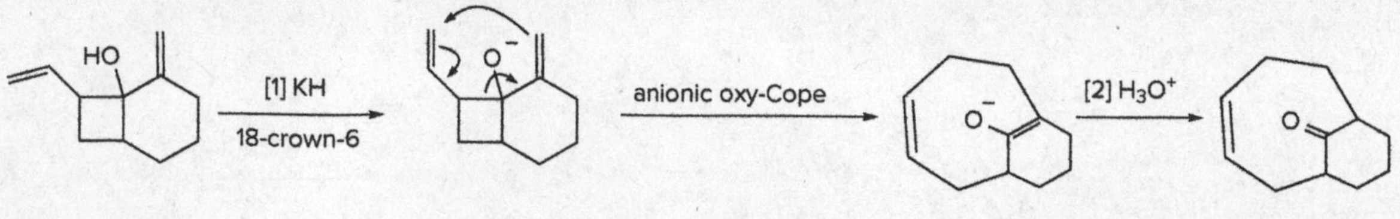 [1] KH, 18-crown-6 — anionic oxy-Cope — [2] H_3O^+

c. 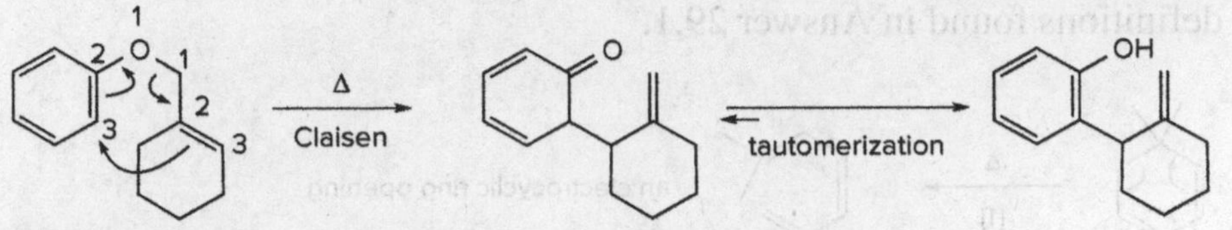

1
2
1
2
3
3
Δ
Claisen
O
tautomerization
OH

d. 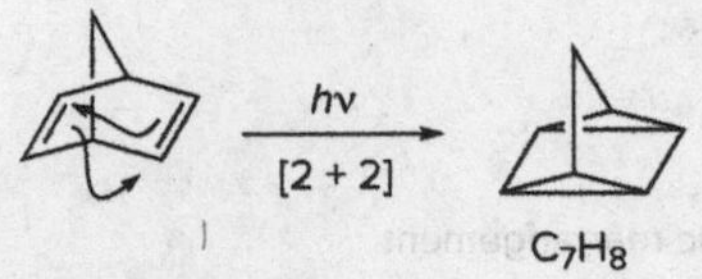

hv
[2 + 2]
C7H8

29.45

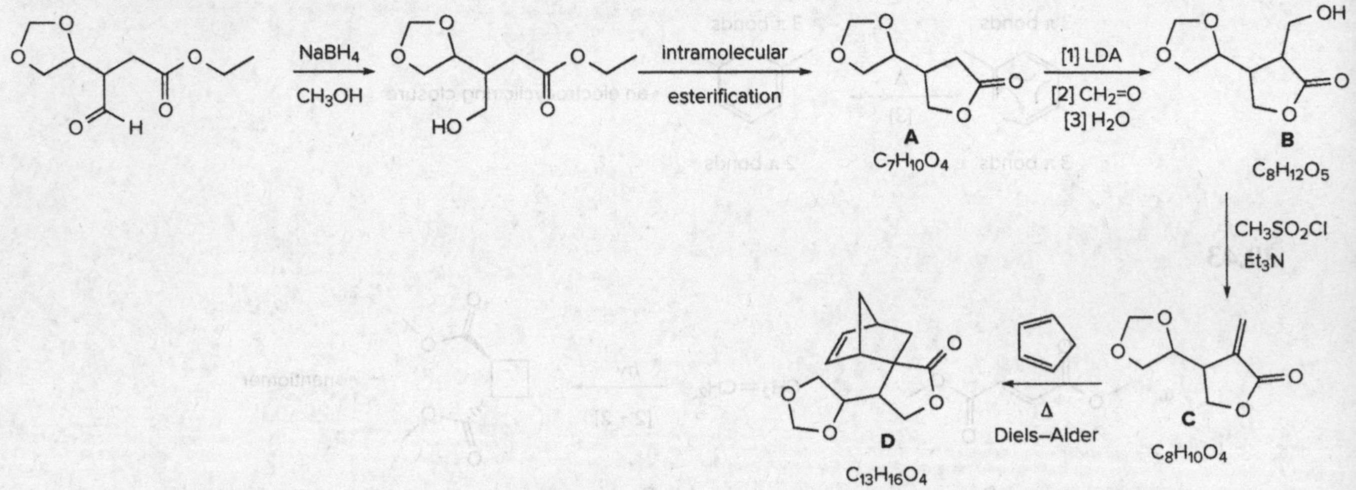

NaBH4
CH3OH
HO
intramolecular
esterification
A
C7H10O4
[1] LDA
[2] CH2=O
[3] H2O
OH
B
C8H12O5
CH3SO2Cl
Et3N
C
C8H10O4
Δ
Diels–Alder
D
C13H16O4

29.46 The mechanism consists of sequential [3,3] sigmatropic rearrangements, followed by tautomerization.

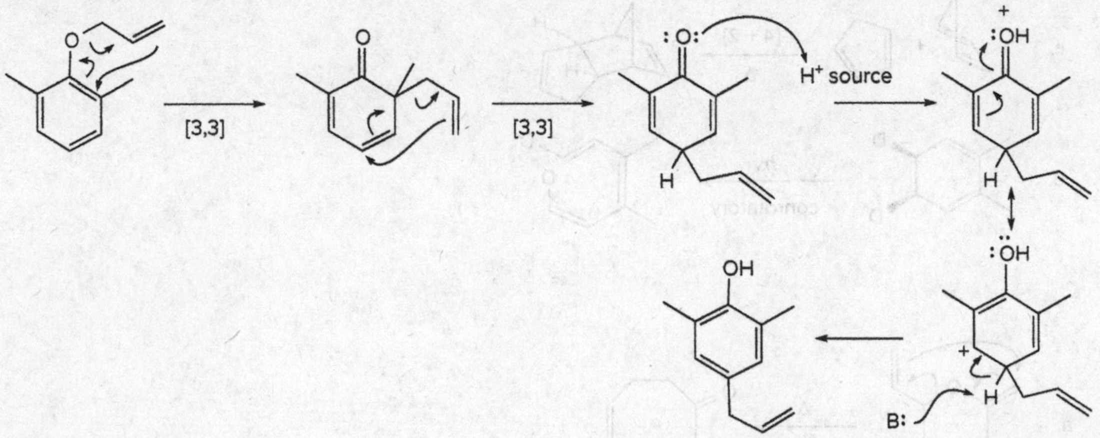

[3,3]
[3,3]
H+ source
:OH
H
B:
OH

29.47

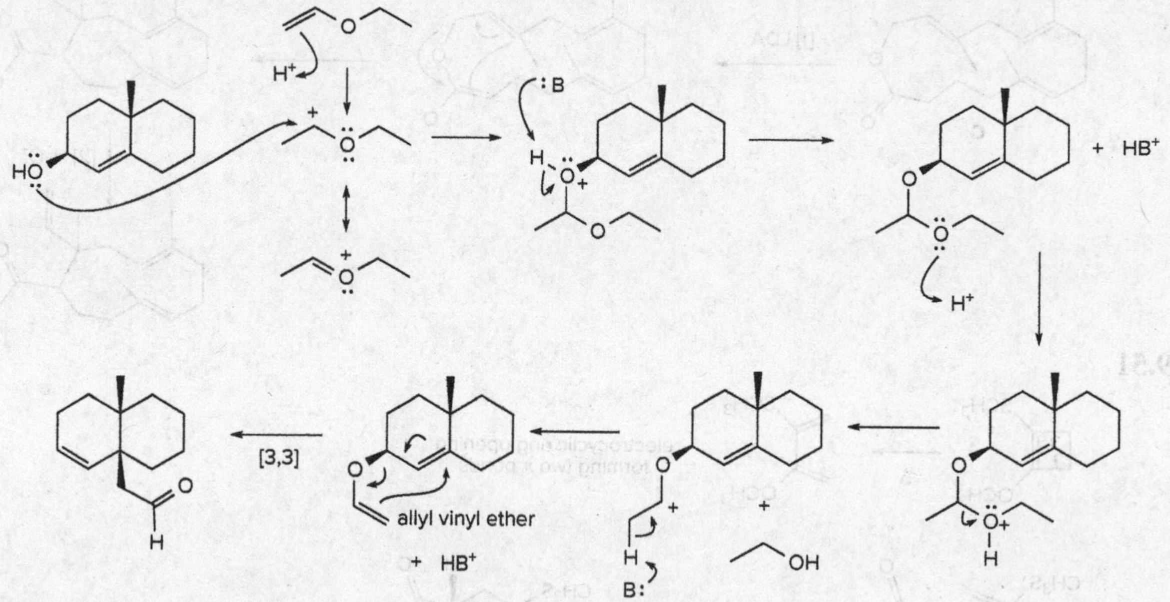

H⁺ : B
HB⁺
H⁺
[3,3]
allyl vinyl ether
+ HB⁺
B:
OH

29.48

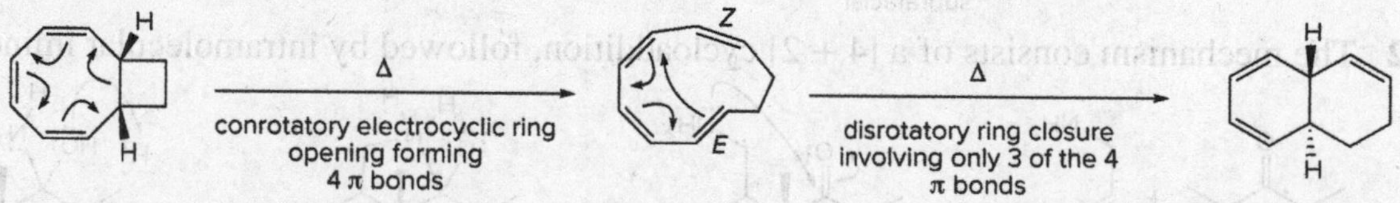

Δ
conrotatory electrocyclic ring opening forming 4 π bonds
Z
E
Δ
disrotatory ring closure involving only 3 of the 4 π bonds

29.49

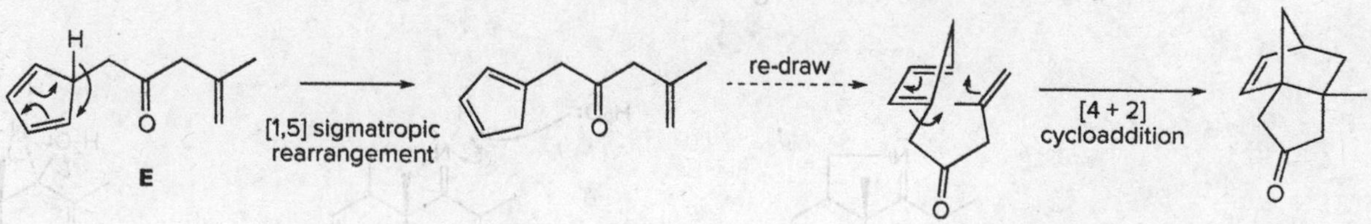

E
[1,5] sigmatropic rearrangement
re-draw
[4 + 2] cycloaddition

29.50

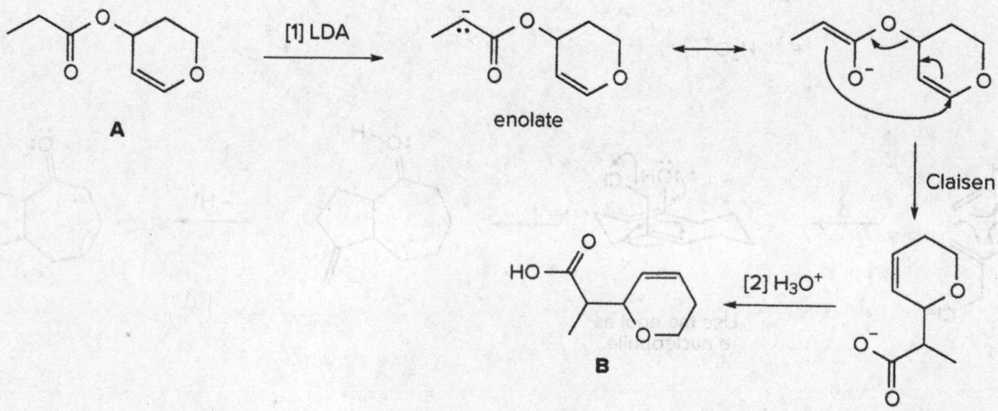

A
[1] LDA
enolate
Claisen
[2] H_3O^+
B

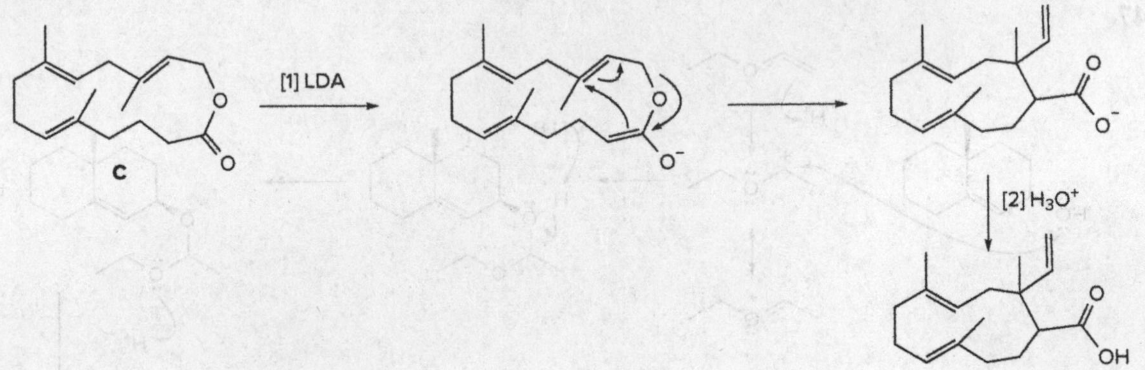

[1] LDA

C

[2] H_3O^+

29.51

SCH_3

OCH_3

Δ

electrocyclic ring opening forming two π bonds

CH_3S

CH_3O

Δ

[4 + 2] cycloaddition suprafacial

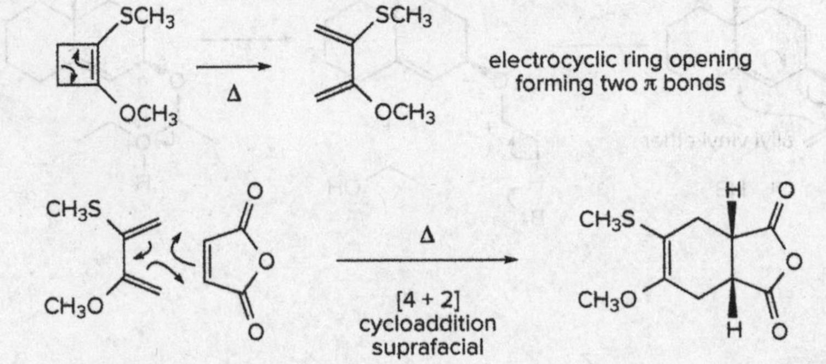

29.52 The mechanism consists of a [4 + 2] cycloaddition, followed by intramolecular imine formation.

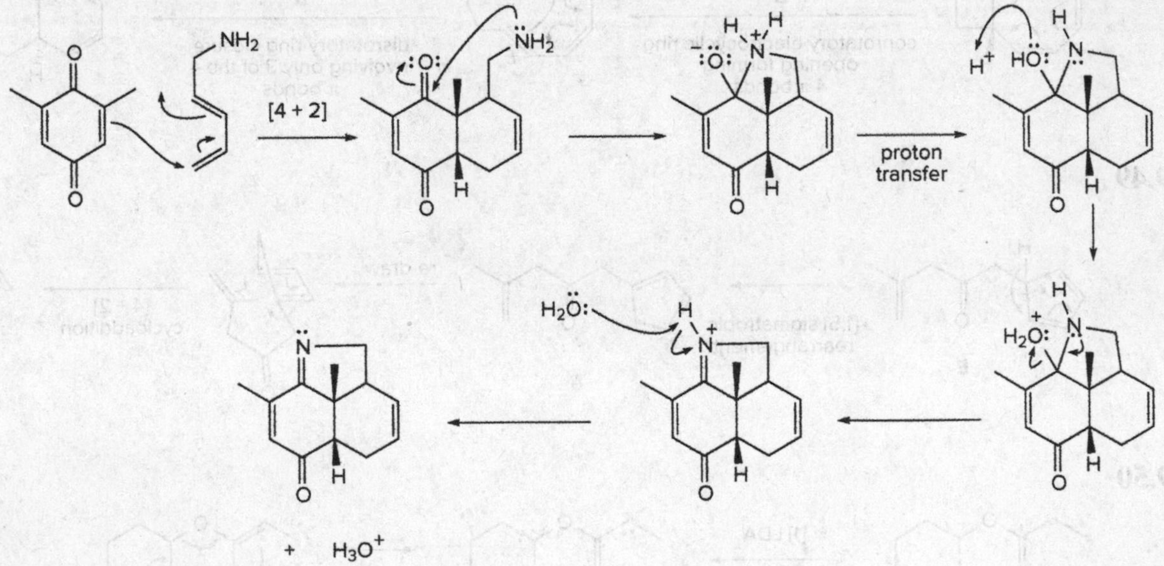

NH_2

[4 + 2]

proton transfer

$H_2\ddot{O}$:

+ H_3O^+

29.53

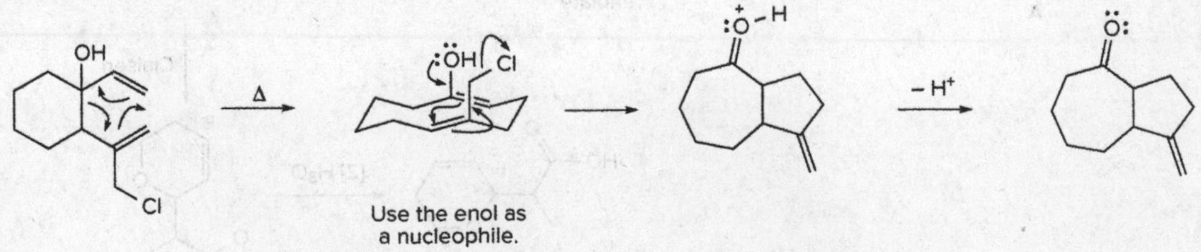

OH

Cl

Δ

Use the enol as a nucleophile.

– H^+

29.54

anionic oxy-Cope

protonation and tautomerization

B

C

D + $^{-}$OEt

29.55

+ HB^+

H_2O

+ CO_2

+ HO^-

29.56 Conrotatory cyclization of **Y** using four π bonds forms **X**. Disrotatory ring closure of **X** can occur in two ways—on the top face or the bottom face of the eight-membered ring to form diastereomers.

C_6H_5

CO_2CH_3

Y

Δ

4 π bonds conrotatory

C_6H_5

CO_2CH_3

X

+ enantiomer

disrotatory

H

C_6H_5

diene

CO_2CH_3

dienophile

Only this stereoisomer has the side chain close to the six-membered ring for Diels–Alder.

=

H H

C_6H_5

CO_2CH_3

endiandric acid E methyl ester

+

H H

C_6H_5

CO_2CH_3

endiandric acid D methyl ester

[4 + 2]

C_6H_5

H H H H H H

CO_2CH_3

methyl ester of endiandric acid A

Endiandric acid A has a free COOH group instead of the CO_2CH_3 group.

Chapter 30 Synthetic Polymers

Chapter Review

Chain-growth polymers—Addition polymers

[1] Chain-growth polymers with alkene starting materials (30.2)

- General reaction:

(alkene with Z) —initiator→ polymer chain with Z substituents

- Mechanism—three possibilities, depending on the identity of Z:

Type	Identity of Z	Initiator	Comments
[1] radical polymerization	Z stabilizes a radical. Z = R, Ph, Cl, etc.	A source of radicals (ROOR)	Termination occurs by radical coupling or disproportionation. Chain branching occurs.
[2] cationic polymerization	Z stabilizes a carbocation. Z = R, Ph, OR, etc.	H–A or a Lewis acid (BF_3 + H_2O)	Termination occurs by loss of a proton.
[3] anionic polymerization	Z stabilizes a carbanion. Z = Ph, COOR, COR, CN, etc.	An organo-lithium reagent (R–Li)	Termination occurs only when an acid or other electrophile is added.

[2] Chain-growth polymers with epoxide starting materials (30.3)

(epoxide with R) —^{-}OH→ polyether chain with R substituents

- The mechanism is S_N2.
- Ring opening occurs at the less substituted carbon of the epoxide.

Examples of step-growth polymers—Condensation polymers (30.6)

Polyamides

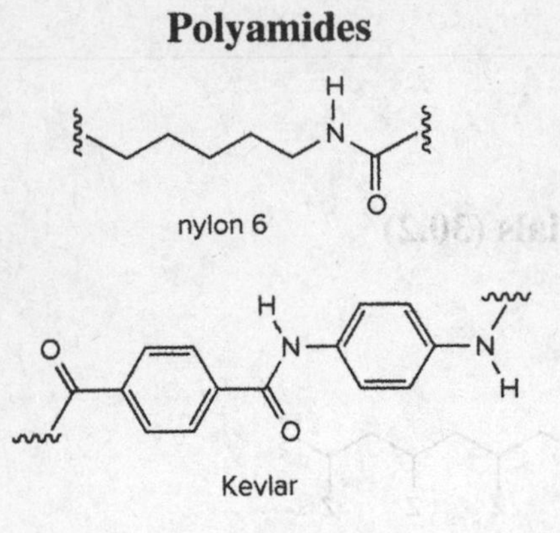

nylon 6

Kevlar

Polyesters

polyethylene terephthalate

copolymer of glycolic and lactic acids

Polyurethanes

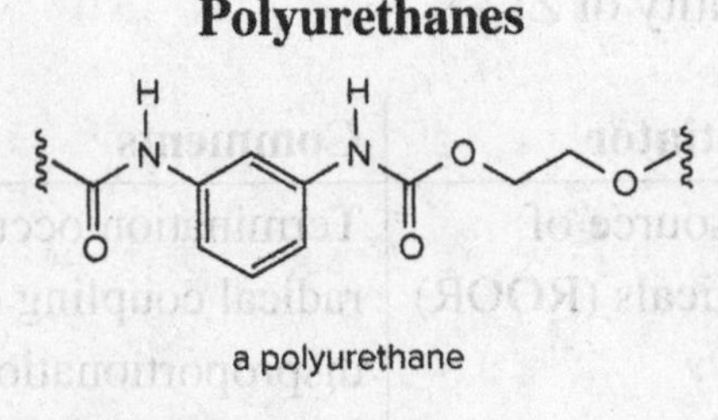

a polyurethane

Polycarbonates

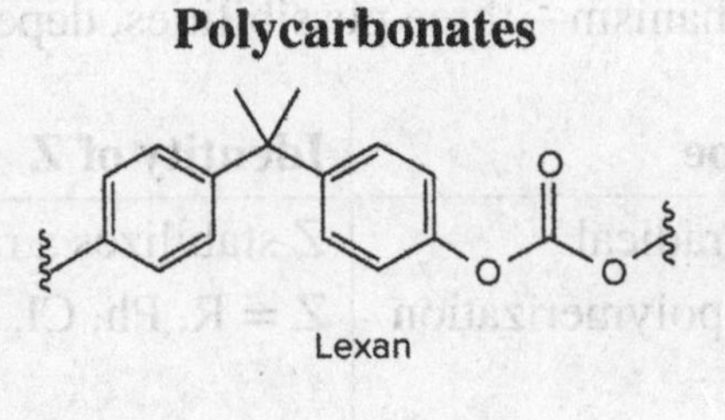

Lexan

Structure and properties

- Polymers prepared from monomers having the general structure CH_2=CHZ can be **isotactic, syndiotactic,** or **atactic,** depending on the identity of Z and the method of preparation (30.4).
- **Ziegler–Natta catalysts** form polymers without significant branching. Polymers can be isotactic, syndiotactic, or atactic, depending on the catalyst. Polymers prepared from 1,3-dienes have the *E* or *Z* configuration, depending on the monomer (30.4, 30.5).
- Most polymers contain ordered crystalline regions and less ordered amorphous regions (30.7). The greater the crystallinity, the harder the polymer.
- **Elastomers** are polymers that stretch and can return to their original shape (30.5).
- **Thermoplastics** are polymers that can be molded, shaped, and cooled such that the new form is preserved (30.7).
- **Thermosetting polymers** are composed of complex networks of covalent bonds, so they cannot be melted to form a liquid phase (30.7).

Practice Test on Chapter Review

1.a. Which of the following statements is (are) true about chain-growth polymers?

1. The reaction mechanism involves initiation, propagation, and termination.
2. The reaction may occur with anionic, cationic, or radical intermediates.
3. Epoxides can serve as monomers.
4. Statements (1) and (2) are both true.
5. Statements (1), (2), and (3) are all true.

b. Which of the following alkenes is likely to undergo anionic polymerization?
  1. $CH_2{=}CHCO_2CH_3$
  2. $CH_2{=}CHOCH_3$
  3. $CH_2{=}CHCH_2CO_2CH_3$
  4. Both (1) and (2) will react.
  5. Compounds (1), (2), and (3) will all react.

c. Which of the following compounds can serve as an initiator in cationic polymerization?
  1. butyllithium
  2. $(CH_3)_3COOC(CH_3)_3$
  3. BF_3
  4. Both (1) and (2) can serve as initiators.
  5. Compounds (1), (2), and (3) can all serve as initiators.

d. Which of the following statements is (are) true about step-growth polymers?
  1. A small molecule such as H_2O or HCl is extruded during synthesis.
  2. Polycarbonates are an example of a step-growth polymer.
  3. Step-growth polymers are also called addition polymers.
  4. Statements (1) and (2) are both true.
  5. Statements (1), (2), and (3) are all true.

2. Label each statement as True (T) or False (F).
  a. A polyester is the most easily recycled polymer.
  b. Natural rubber is a polymer of repeating isoprene units in which all double bonds have the *E* configuration.
  c. A syndiotactic polymer has all Z groups bonded to the polymer chain on the same side.
  d. A polyether can be formed by anionic polymerization of an epoxide.
  e. An epoxy resin is a chain-growth polymer.
  f. A branched polymer is more amorphous, giving it a higher T_m.
  g. Polystyrene is a thermoplastic that can be melted and molded in shapes that are retained when the polymer is cooled.
  h. A polyurethane is a condensation polymer.
  i. Ziegler–Natta catalysts are used to form highly branched chain-growth polymers.
  j. Using a feedstock from a renewable source is one method of green polymer synthesis.

3. What monomer(s) are needed to synthesize each polymer?

a.

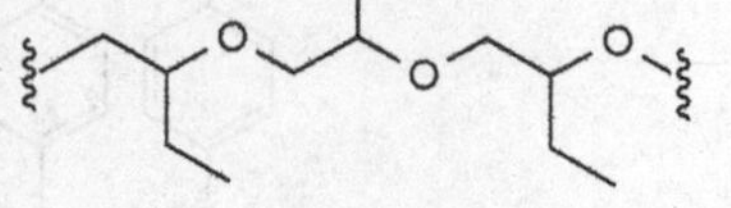

b.

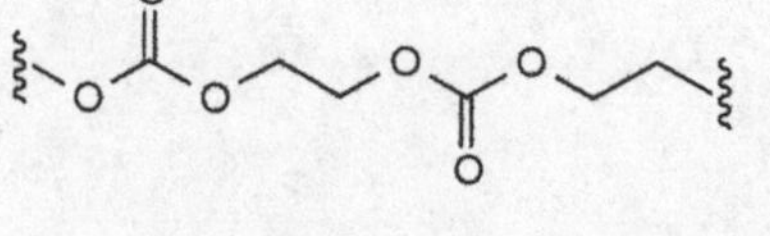

c. 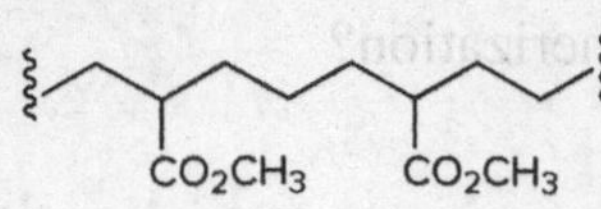 CO_2CH_3 CO_2CH_3

e. 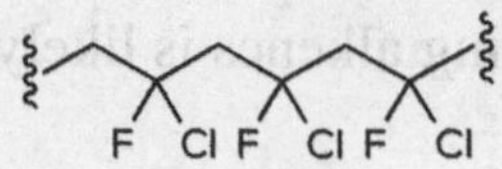 F Cl F Cl F Cl

d.

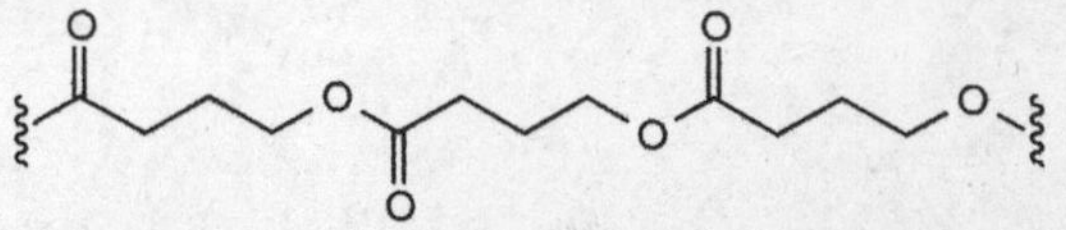

Answers to Practice Test

1.a. 5
b. 1
c. 3
d. 4

2.a. T
b. F
c. F
d. T
e. F
f. F
g. T
h. T
i. F
j. T

3.
a. 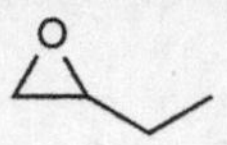
b. 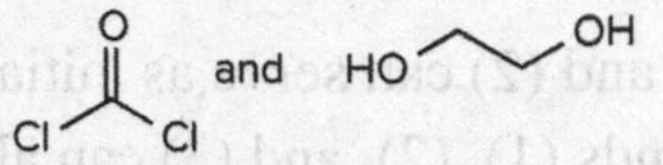 Cl Cl and HO OH
c. 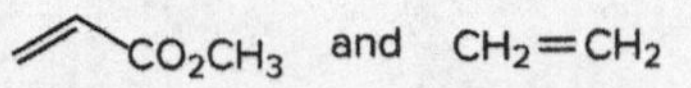 CO_2CH_3 and $CH_2{=}CH_2$
d. 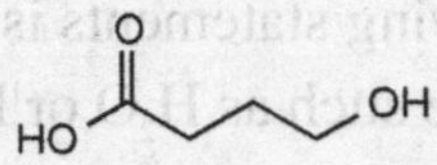 HO OH
e. 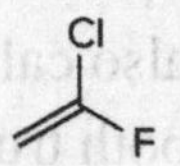 Cl F

Answers to Problems

30.1 Place brackets around the repeating unit that creates the polymer.

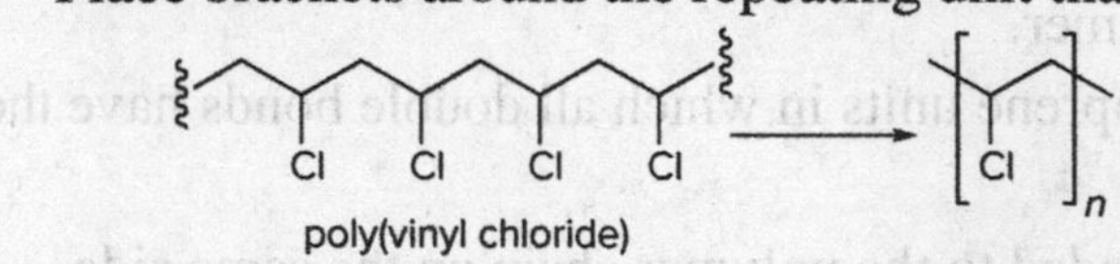

Cl Cl Cl Cl → Cl n

poly(vinyl chloride)

nylon 6,6

H N N H n

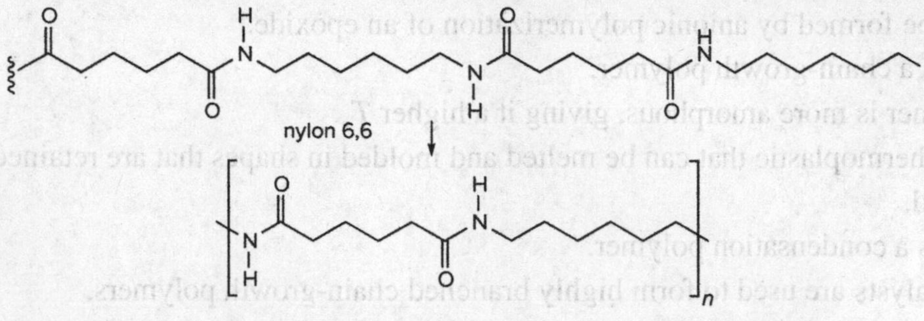

30.2 Draw each polymer formed by chain-growth polymerization.

a. 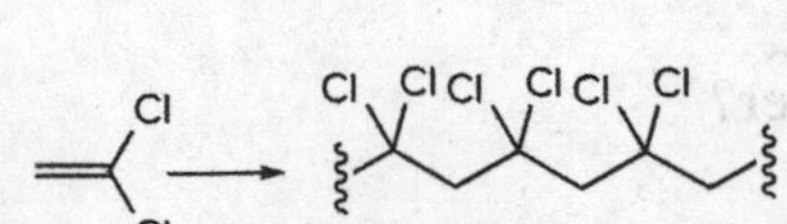 Cl Cl → Cl Cl Cl Cl Cl Cl

b. 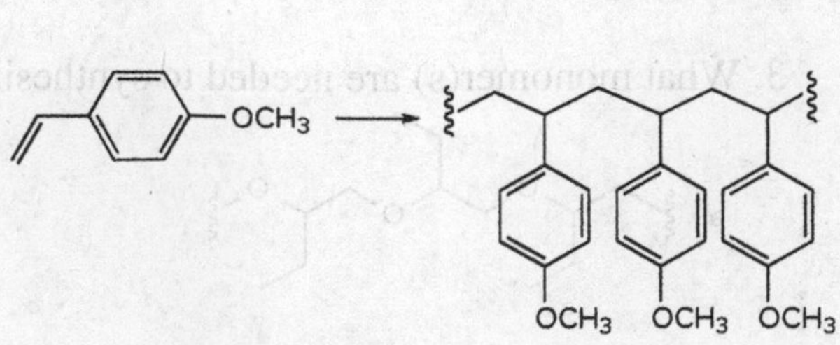 OCH_3 → OCH_3 OCH_3 OCH_3

c. OCH3 OCH3 OCH3 OCH3 d. O O O O O O

30.3 Draw each polymer formed by radical polymerization.

a. b. CN CN CN CN

30.4 Radical polymerization forms a long chain of polystyrene with phenyl groups bonded to every other carbon. To form branches on this polystyrene chain, a radical on a second polymer chain abstracts a H atom. Abstraction of H_a forms a resonance-stabilized radical **A'**. The 2° radical **B'** (without added resonance stabilization) is formed by abstraction of H_b. Abstraction of H_a is favored, therefore, and this radical goes on to form products with 4° C's (**A**).

RO· Ph H_b H_a Ph Ph Ph abstraction of H_b Ph Ph Ph 2° radical no resonance stabilization **B'** Ph 3° C Ph Ph Ph **B** abstraction of H_a Ph Ph Ph resonance-stabilized benzylic radical **A'** Ph 4° C Ph Ph Ph Ph **favored** **A**

Ph Ph

30.5 Cationic polymerization proceeds via a carbocation intermediate. Substrates that form more stable 3° carbocations react more readily in these polymerization reactions than substrates that form less stable 1° carbocations. $CH_2{=}C(CH_3)_2$ will form a more substituted carbocation than $CH_2{=}CH_2$.

3° carbocation | 1° carbocation

30.6 Cationic polymerization occurs with alkene monomers having substituents that can stabilize carbocations, such as alkyl groups and other electron-donor groups. Anionic polymerization occurs with alkene monomers having substituents that can stabilize a negative charge, such as COR, COOR, or CN.

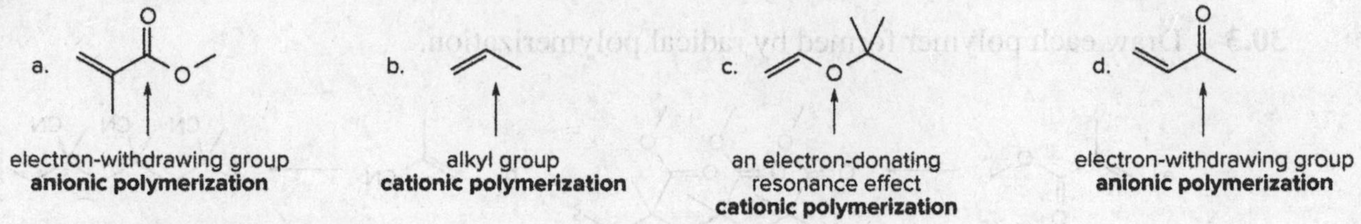

a. electron-withdrawing group **anionic polymerization**

b. alkyl group **cationic polymerization**

c. an electron-donating resonance effect **cationic polymerization**

d. electron-withdrawing group **anionic polymerization**

30.7

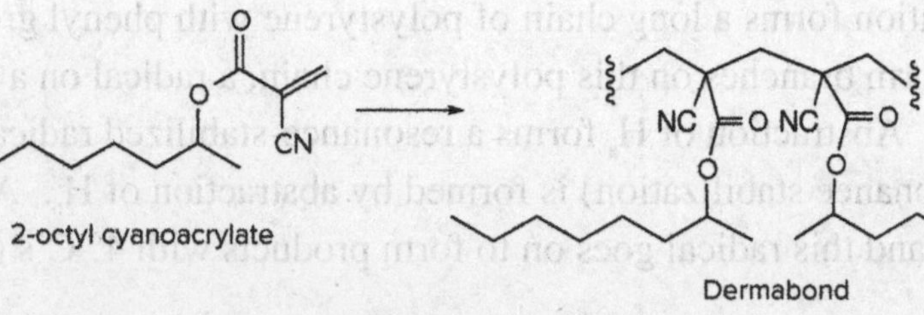

2-octyl cyanoacrylate → Dermabond

30.8 Styrene (CH_2=CHPh) can by polymerized by all three methods of chain-growth polymerization because a benzene ring can stabilize a radical, a carbocation, or a carbanion by resonance delocalization.

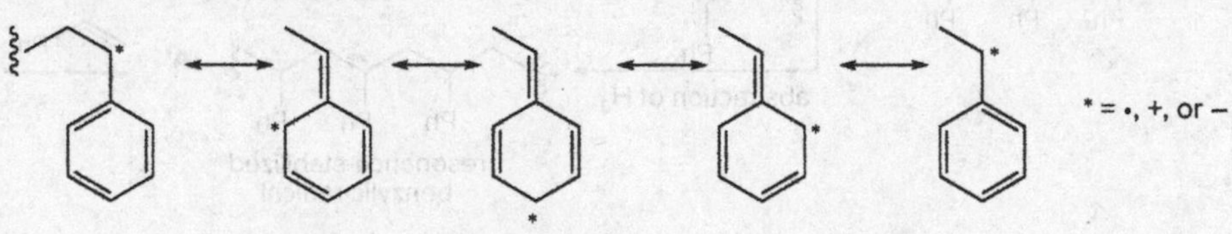

* = •, +, or –

30.9 Draw the copolymers formed in each reaction.

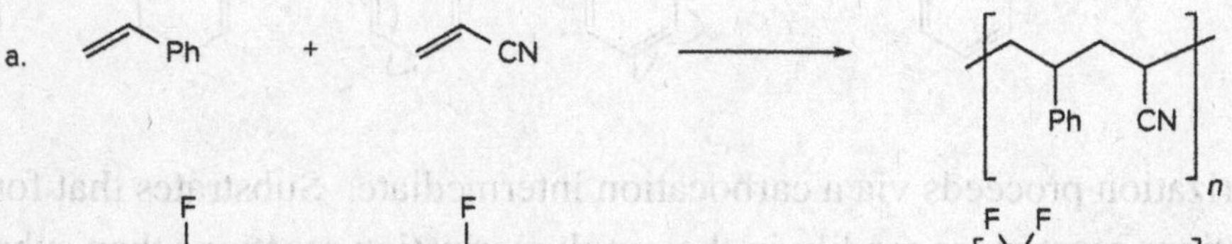

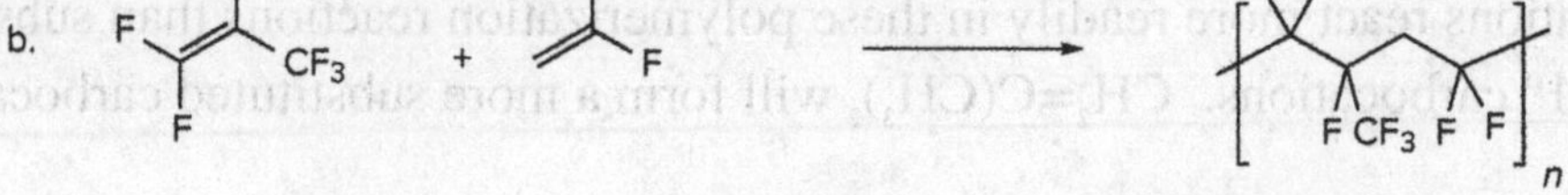

a. Ph + CN →

b. F, F, CF₃, F + F →

30.10

CN ABS Ph

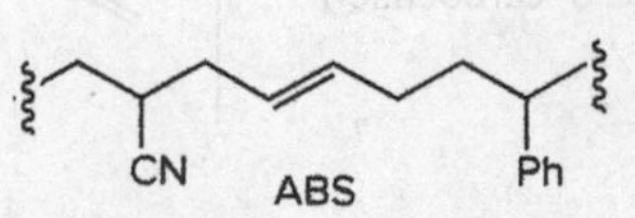

30.11

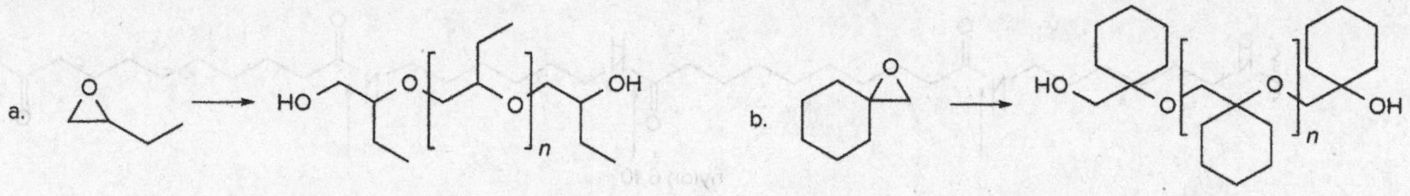

a. HO, OH, n

b. HO, OH, n

30.12

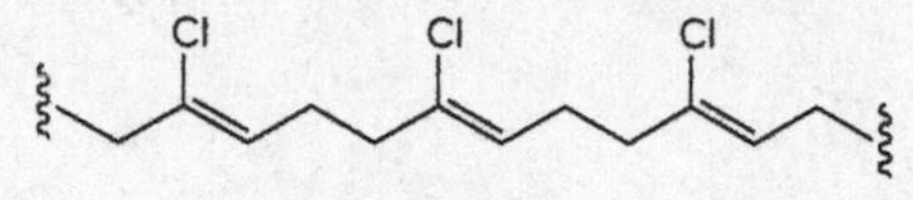

Cl Cl Cl

neoprene

All double bonds have the *Z* configuration.

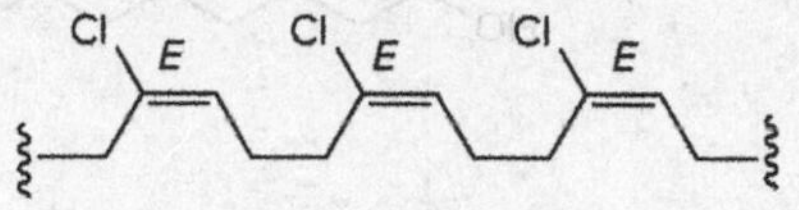

1 Cl, 1 CH_2, CH_2 2, H 2

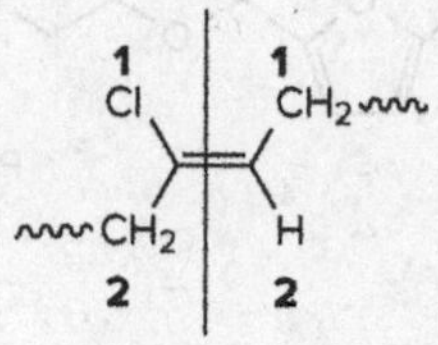

Two higher priority groups (1's) are on the same side of the double bond → *Z* configuration.

Cl *E* Cl *E* Cl *E*

E configuration of each double bond

30.13

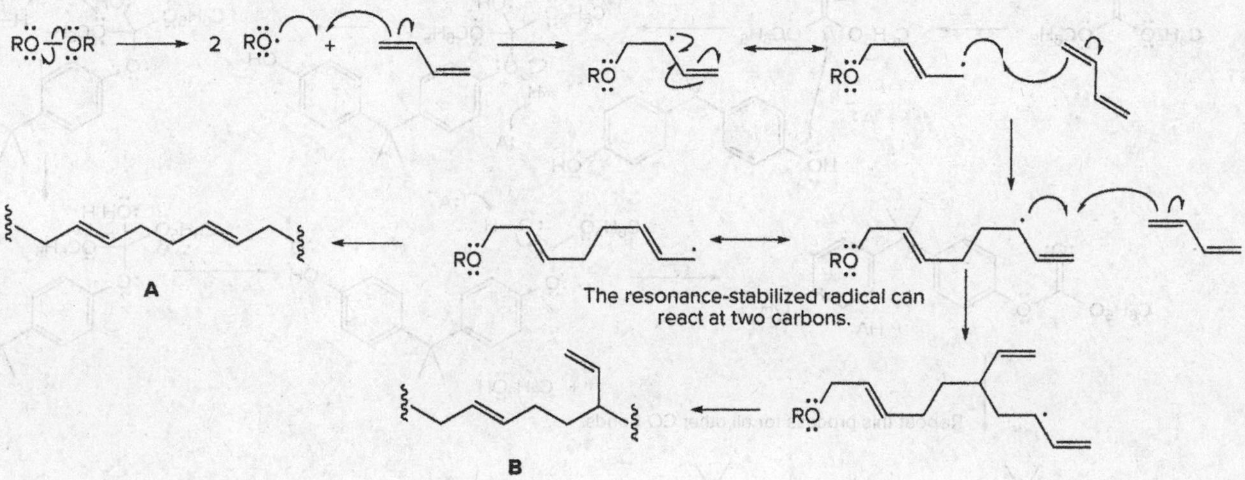

RO–OR → 2 RO· +

RÖ

A

The resonance-stabilized radical can react at two carbons.

B

30.14

a. HO, OH + H_2N, NH_2 → N–H, H–N

b. NH, H_2O → HO, NH_2 → N–H, N–H

c. H_2N, NH_2 + HO, OH → H–N, H–N

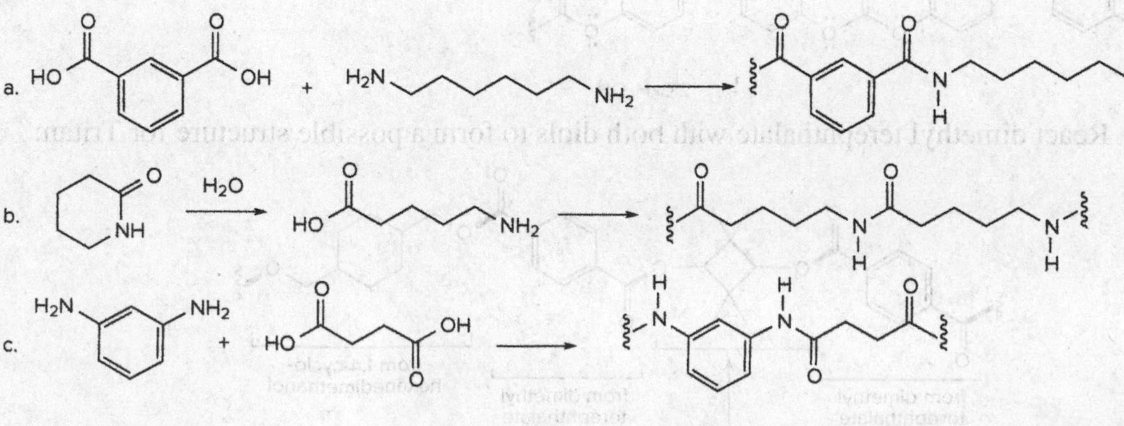

30.15

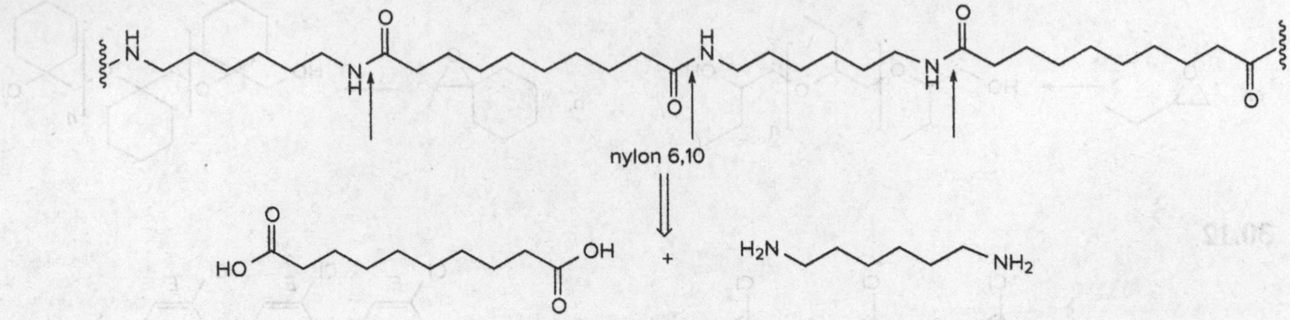

nylon 6,10

30.16

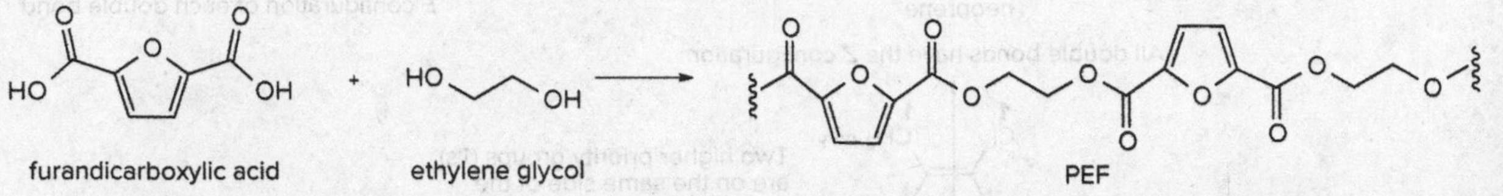

furandicarboxylic acid + ethylene glycol → PEF

30.17

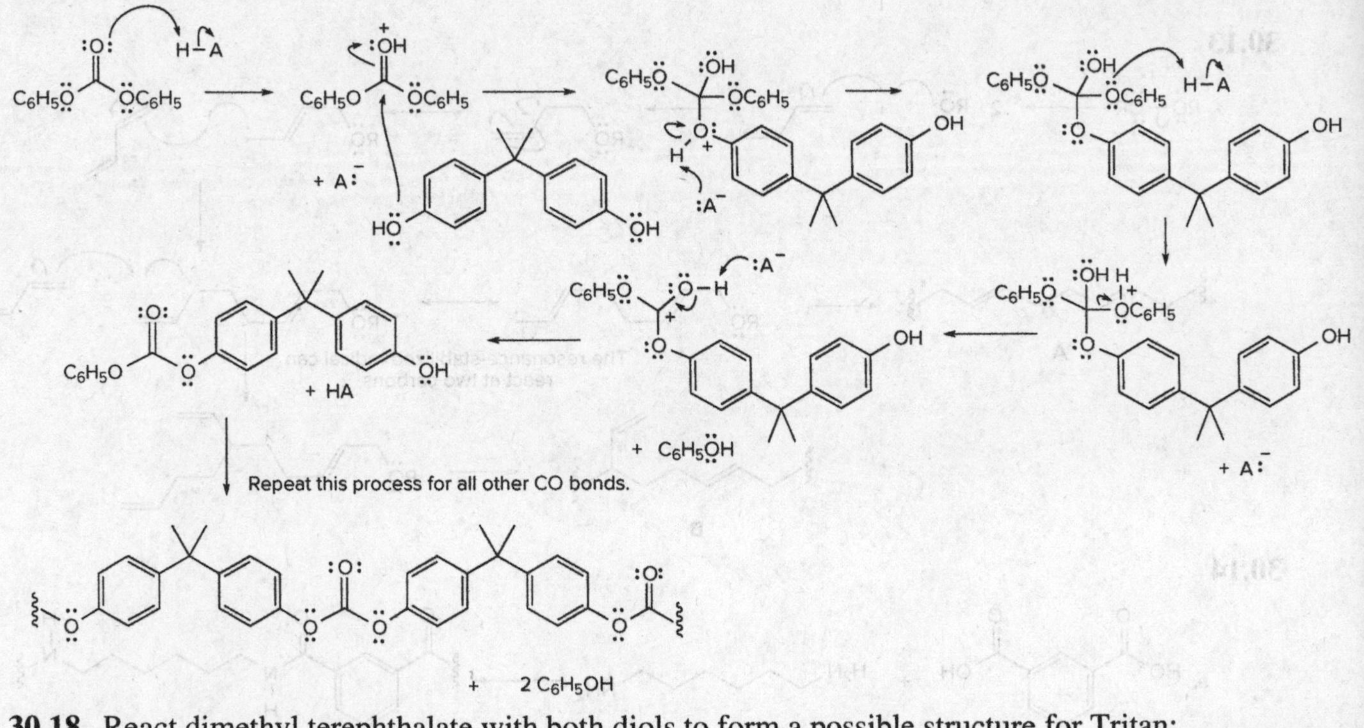

+ A:⁻

+ HA

+ C_6H_5OH

Repeat this process for all other CO bonds.

+ 2 C_6H_5OH

30.18 React dimethyl terephthalate with both diols to form a possible structure for Tritan:

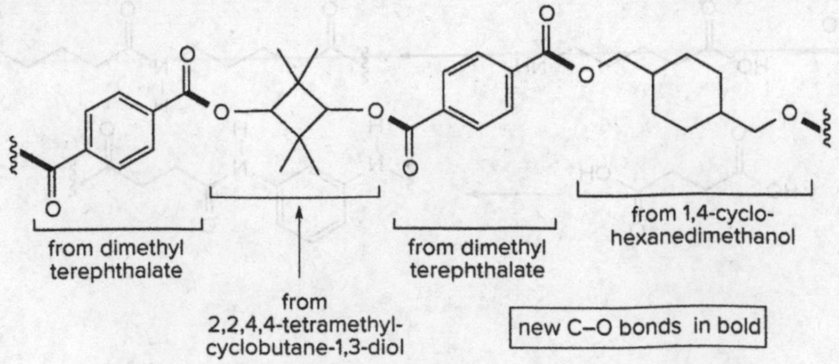

from dimethyl terephthalate

from 2,2,4,4-tetramethyl-cyclobutane-1,3-diol

from dimethyl terephthalate

from 1,4-cyclo-hexanedimethanol

new C–O bonds in bold

30.19

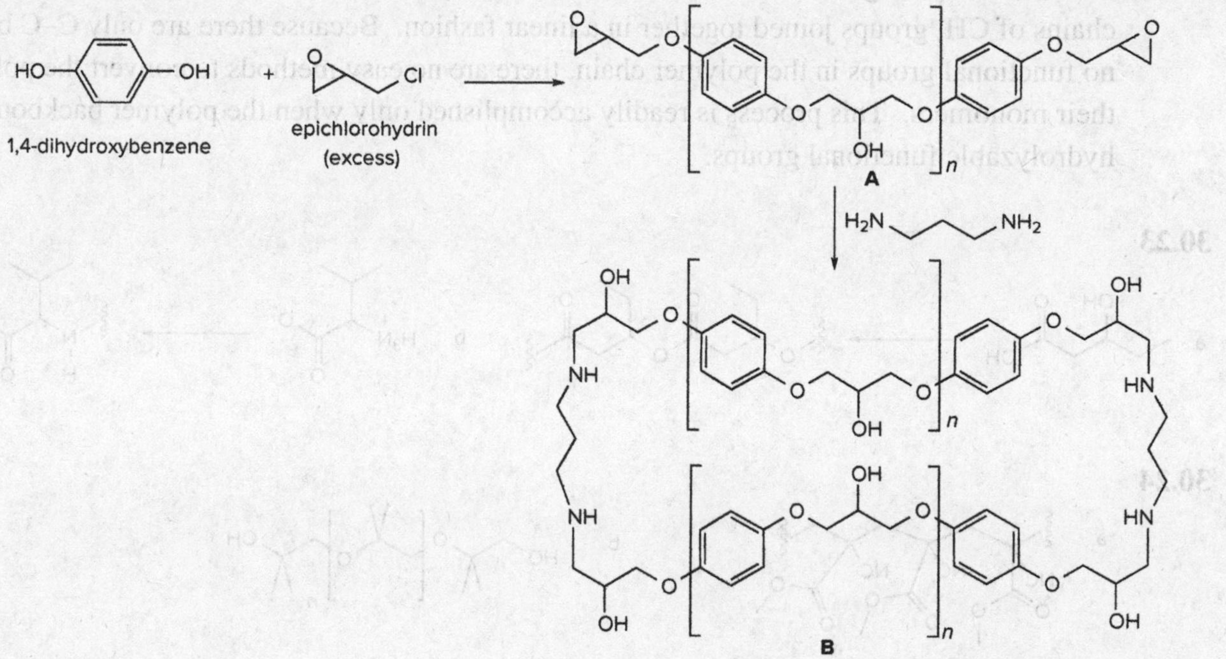

HO–C6H4–OH + epichlorohydrin (excess) → A

1,4-dihydroxybenzene

epichlorohydrin (excess)

A

H_2N–(CH2)3–NH_2

B

30.20

$AlCl_3$

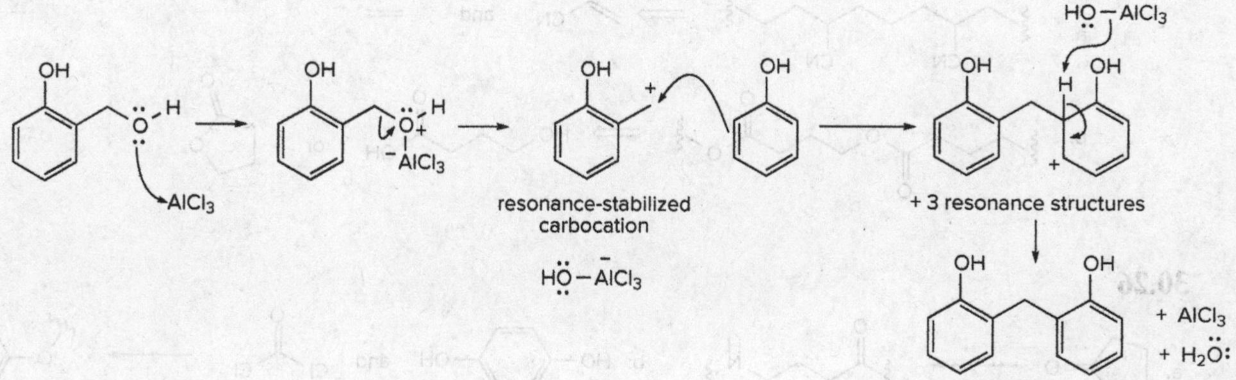

resonance-stabilized carbocation

$H\ddot{O}-AlCl_3$

+ 3 resonance structures

+ $AlCl_3$

+ $H_2\ddot{O}$:

30.21

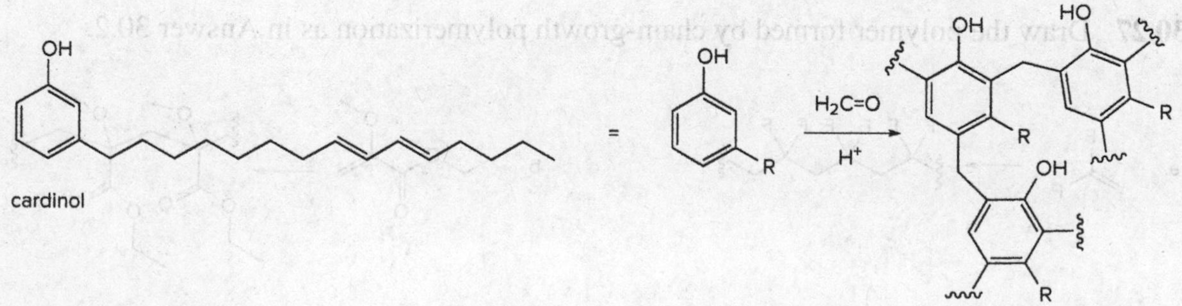

cardinol

=

$H_2C=O$

H^+

30.22 Chemical recycling of HDPE and LDPE is not easily done because these polymers are both long chains of CH_2 groups joined together in a linear fashion. Because there are only C–C bonds and no functional groups in the polymer chain, there are no easy methods to convert the polymers to their monomers. This process is readily accomplished only when the polymer backbone contains hydrolyzable functional groups.

30.23

a. 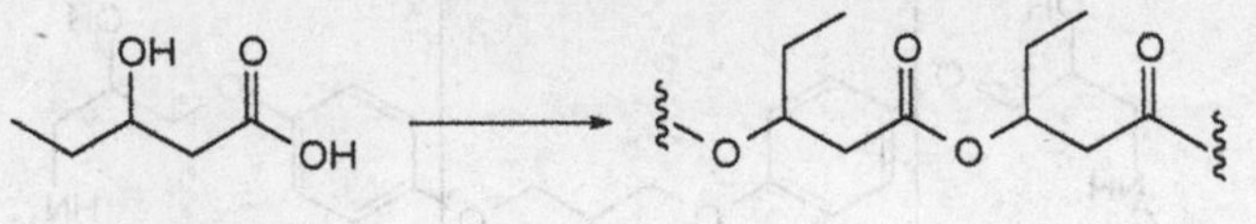 OH O OH → O O O

b. 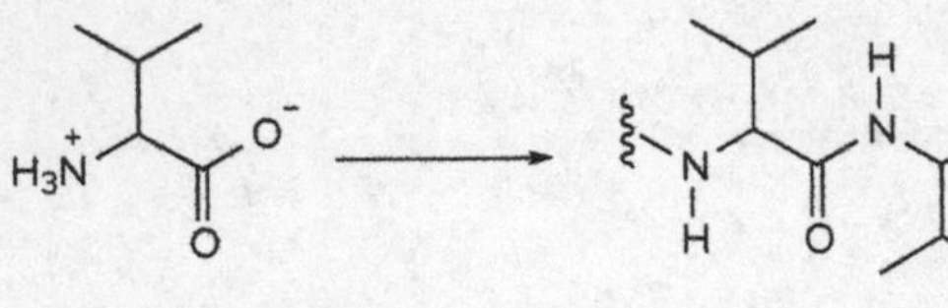 $H_3\overset{+}{N}$ O^- O → N H O H N O

30.24

a. 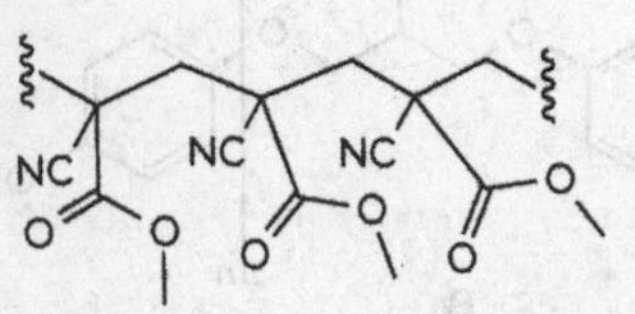 NC NC NC O O O O O O

b. 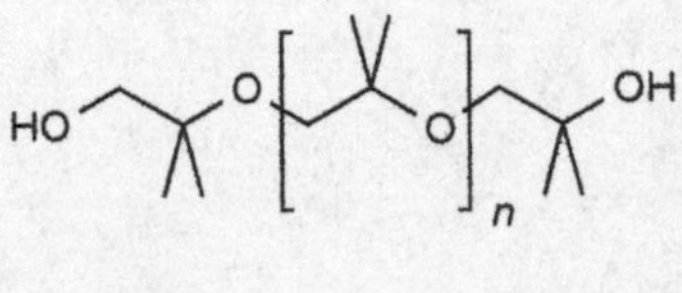 HO O O OH *n*

30.25

a. 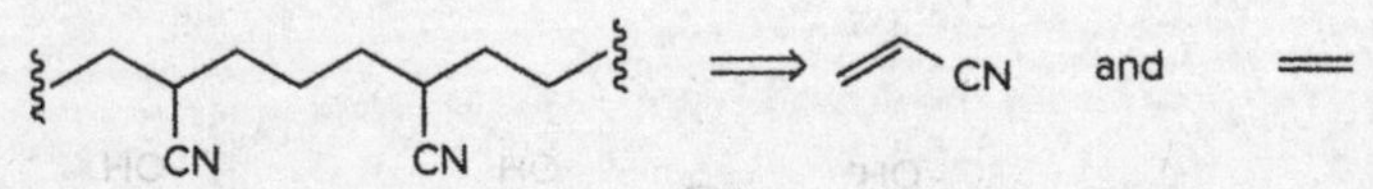 CN CN ⟹ CN and

b. 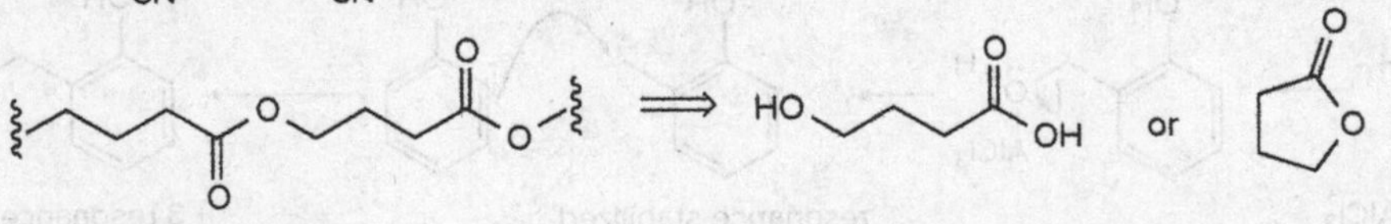 O O O ⟹ HO O OH or O O

30.26

a. 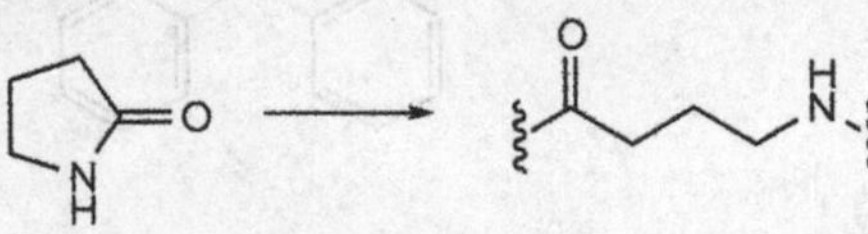 O N H → O H N

b. 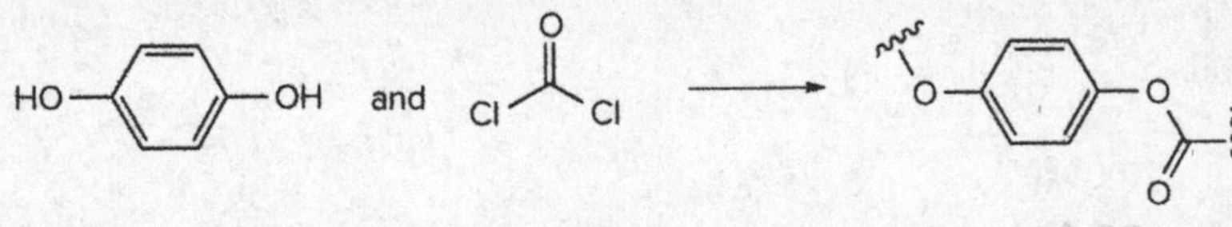 HO OH and O Cl Cl → O O O

30.27 Draw the polymer formed by chain-growth polymerization as in Answer 30.2.

a. 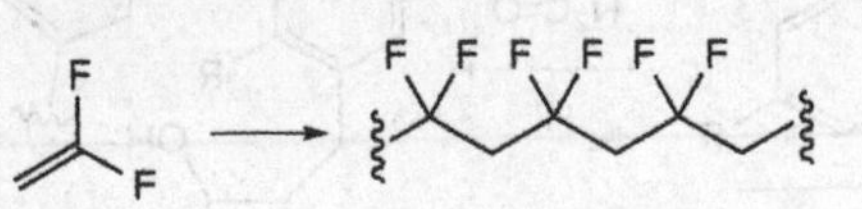 F F → F F F F F F

b. 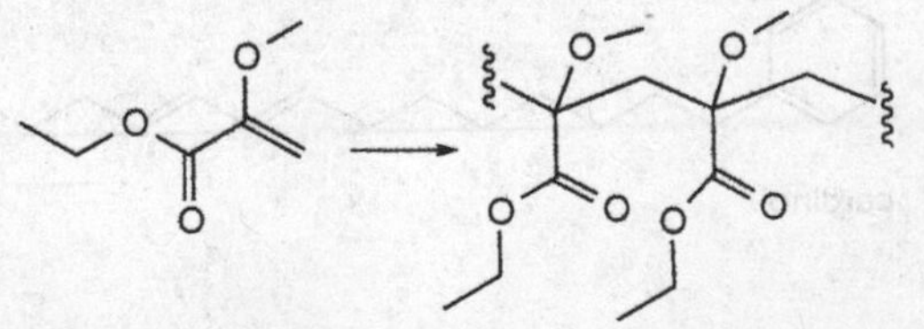 O O O → O O O O O O

c. 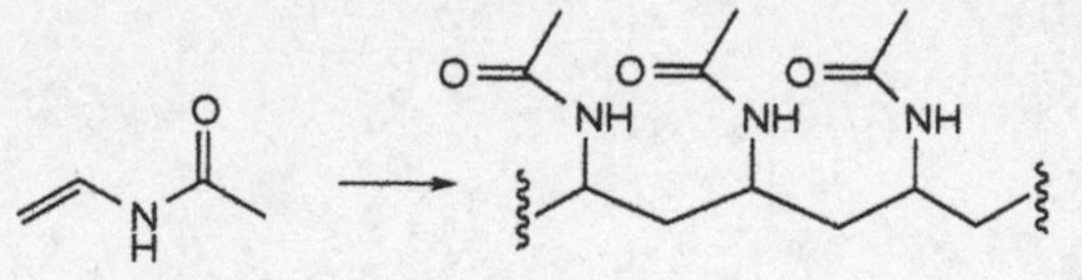 [chemical structure] ⟶ [chemical structure: NH, O, NH, O, NH, O]

d. 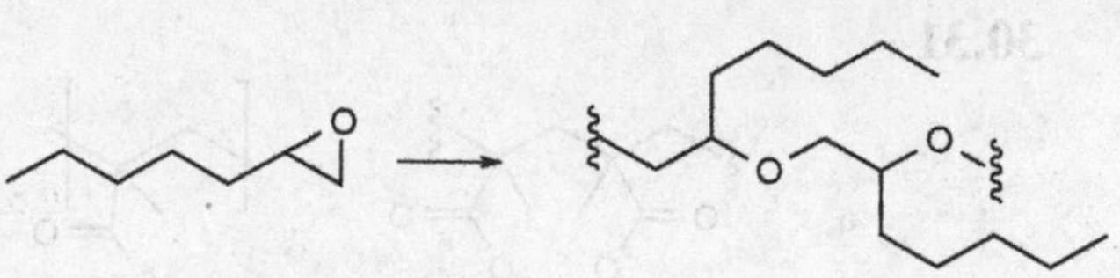 [chemical structure: O] ⟶ [chemical structure: O, O]

30.28 Electron-*donating* groups on a C=C increase the rate of *cationic* polymerization. Electron-*withdrawing* groups on a C=C increase the rate of *anionic* polymerization.

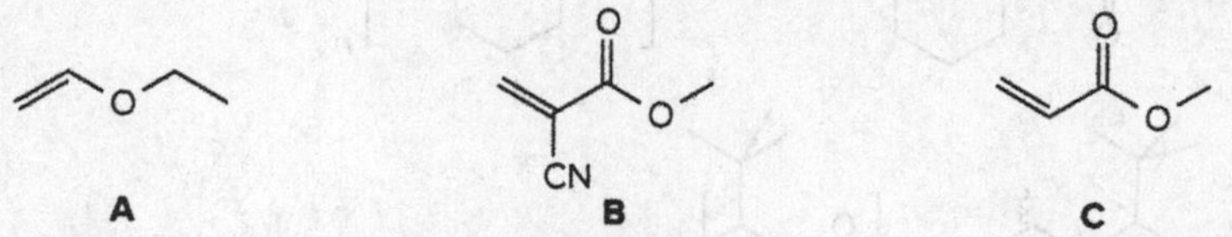

A electron-donating OR group

B two electron-withdrawing groups (CN)

C one electron-withdrawing COOR group

a. Rate in cationic polymerization: **B < C < A**
b. Rate in anionic polymerization: **A < C < B**

30.29 Draw the copolymers.

a. 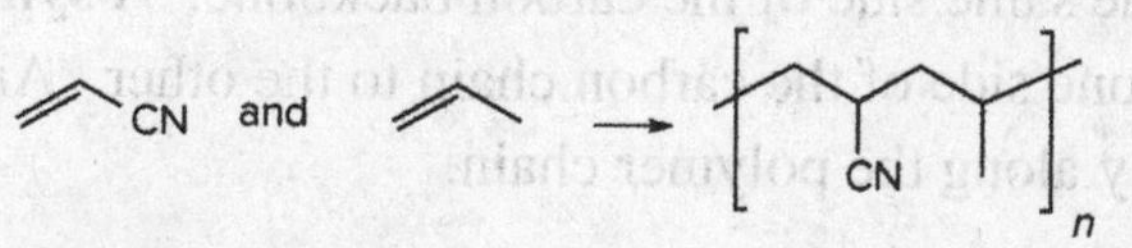 [CN] and [propene] ⟶ [polymer, CN]$_n$

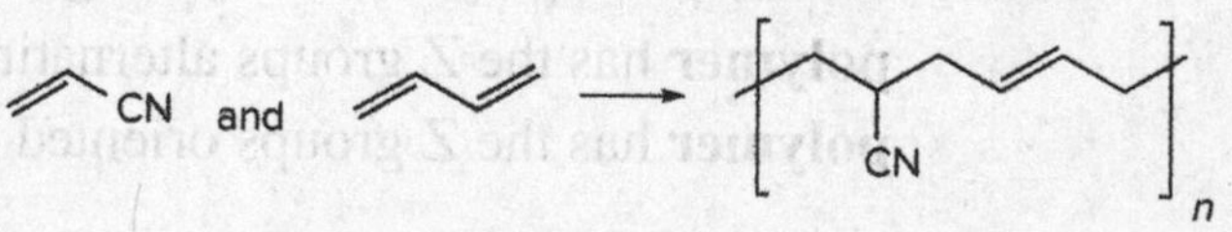

b. 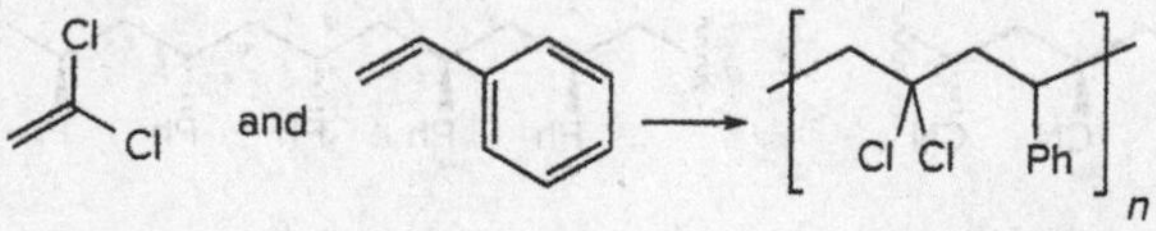 [Cl, Cl] and [styrene] ⟶ [polymer, Cl Cl, Ph]$_n$

c. [CN] and [butadiene] ⟶ [polymer, CN]$_n$

d. 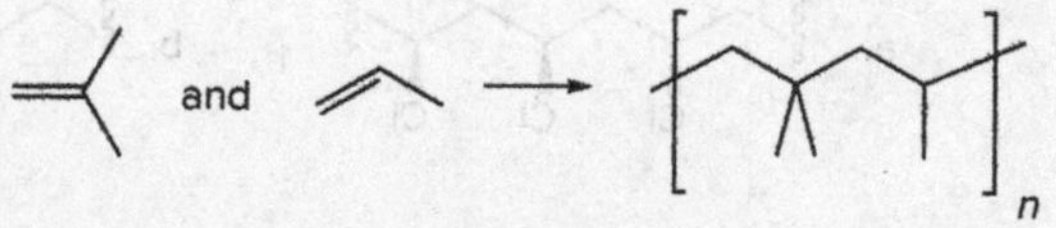 [isobutylene] and [propene] ⟶ [polymer]$_n$

30.30

a. 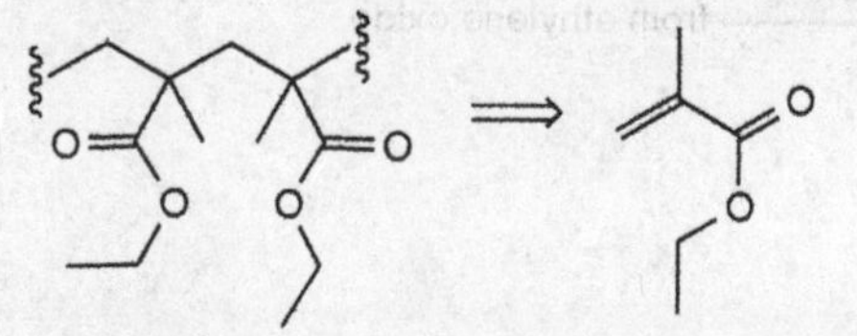 [polymer] ⟹ [monomer]

b. 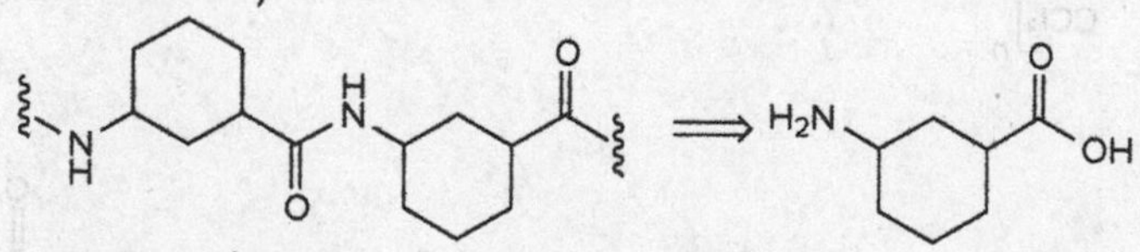 [polymer, NH, HN, O] ⟹ H_2N–[cyclohexane]–COOH (OH)

c. 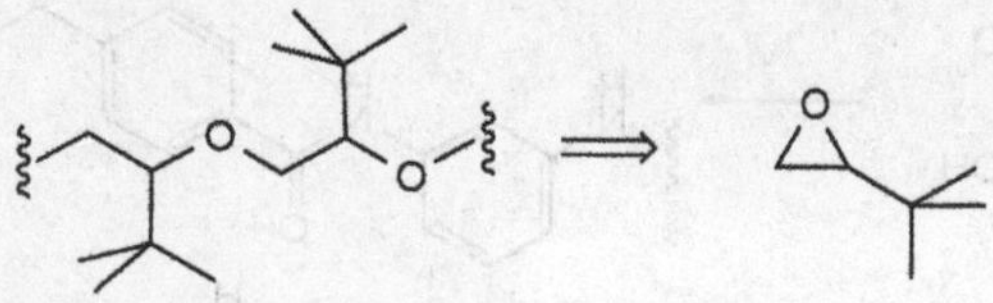 [polymer, O] ⟹ [epoxide]

d. 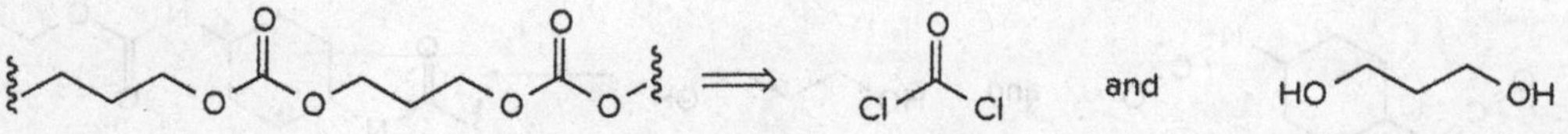 [polymer, O] ⟹ Cl–C(=O)–Cl and HO–(CH$_2$)$_4$–OH

30.31

a.

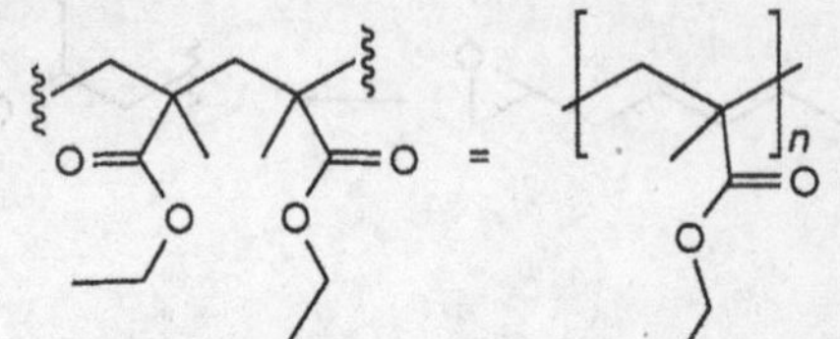

b.

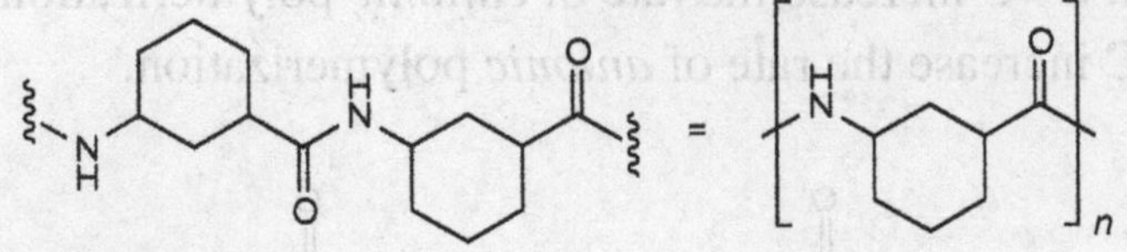

c.

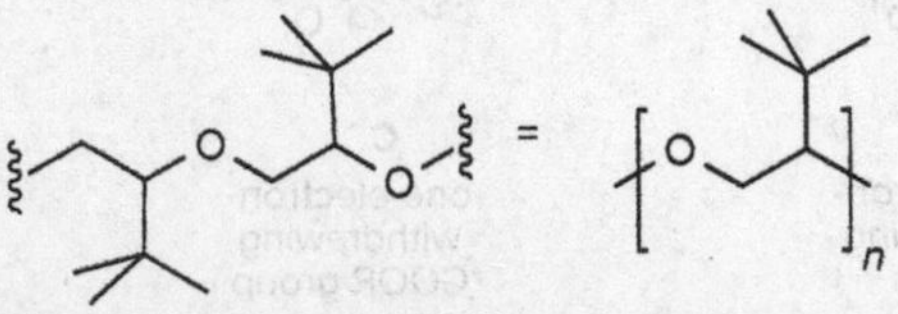

d.

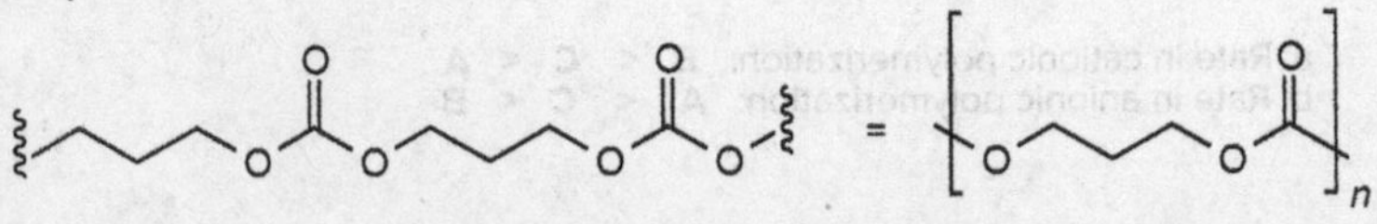

30.32 An **isotactic polymer** has all Z groups on the same side of the carbon backbone. A **syndiotactic polymer** has the Z groups alternating from one side of the carbon chain to the other. An **atactic polymer** has the Z groups oriented randomly along the polymer chain.

a. 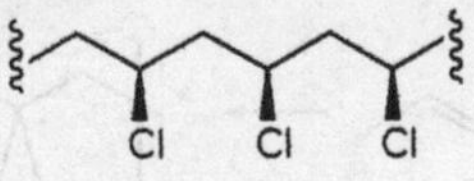 Cl Cl Cl

b. 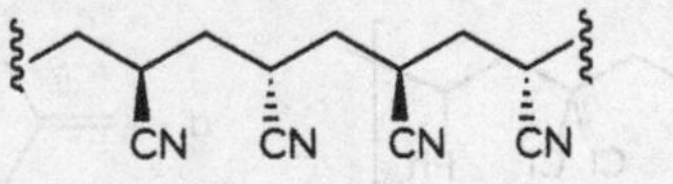 CN CN CN CN

c. 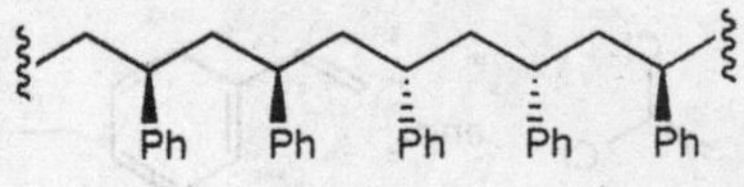 Ph Ph Ph Ph Ph

30.33

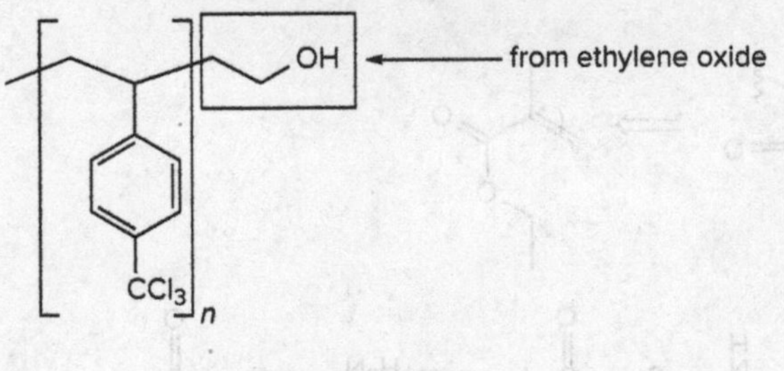

OH ← from ethylene oxide

CCl_3 n

30.34

a. 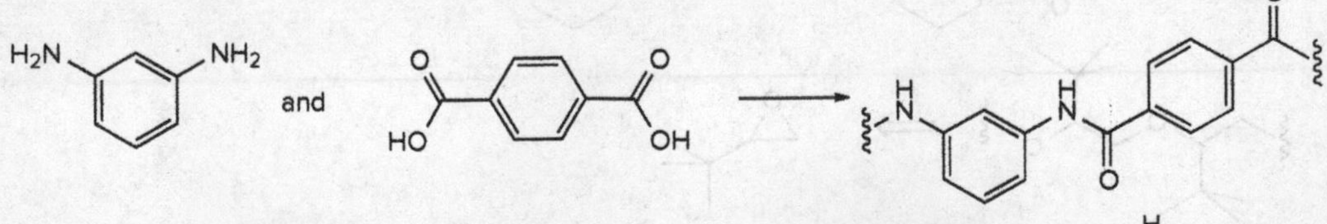 H_2N NH_2 and HO OH →

b. 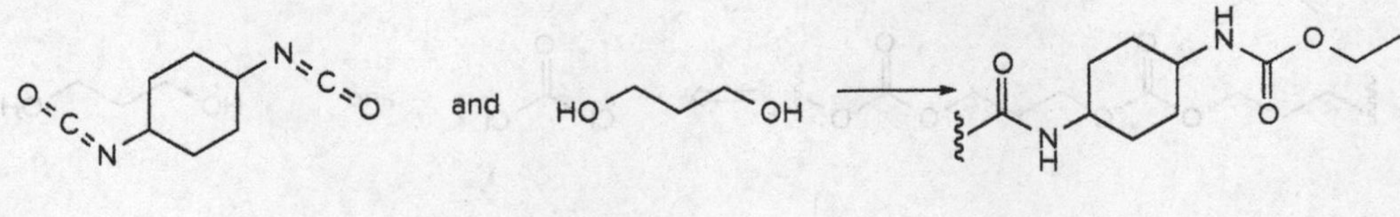 and HO OH →

c. 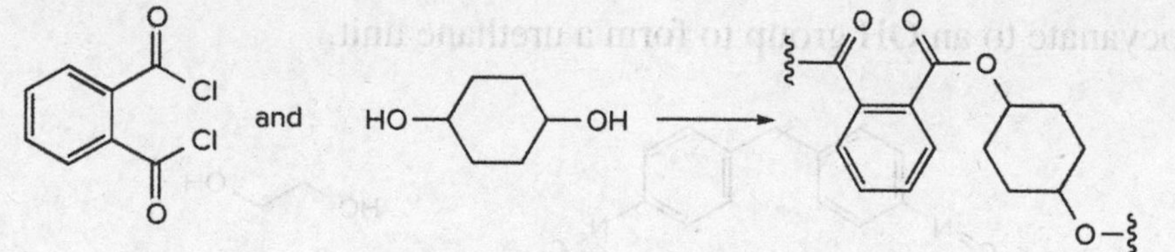 and HO—cyclohexane—OH →

d. 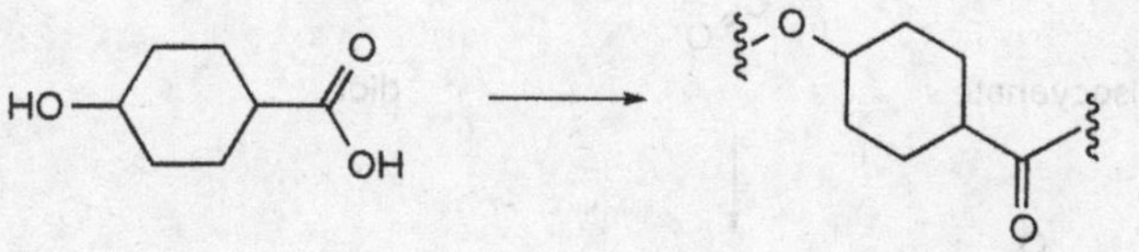 HO—cyclohexane—COOH →

30.35

a. 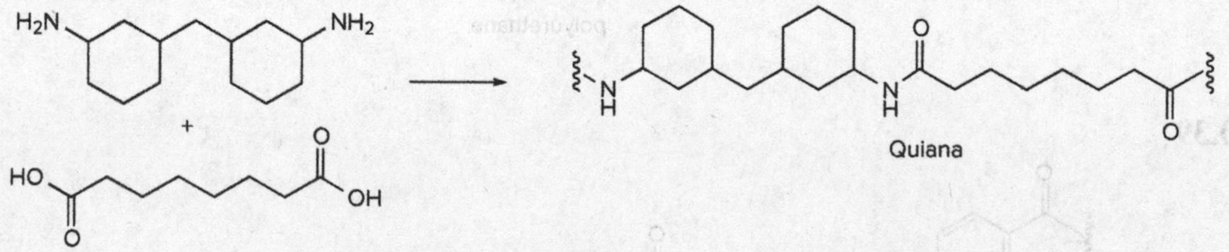 + HO(C=O)…(C=O)OH →

Quiana

b. 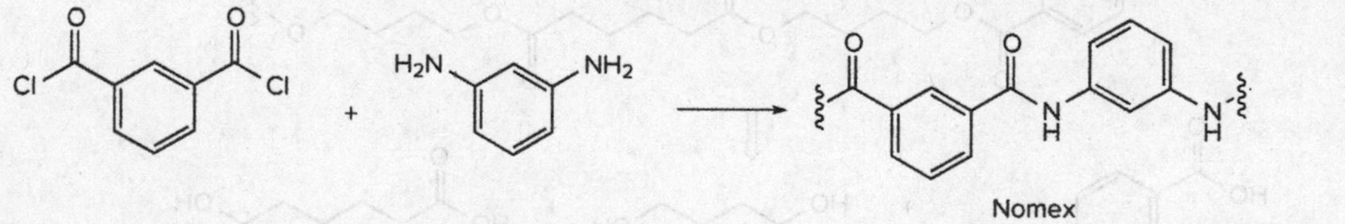 + H₂N…NH₂ →

Nomex

30.36 Draw the structure of a random copolymer formed from twice as many $CH_2{=}CH_2$ monomers as $CH_2{=}CHOCOCH_3$.

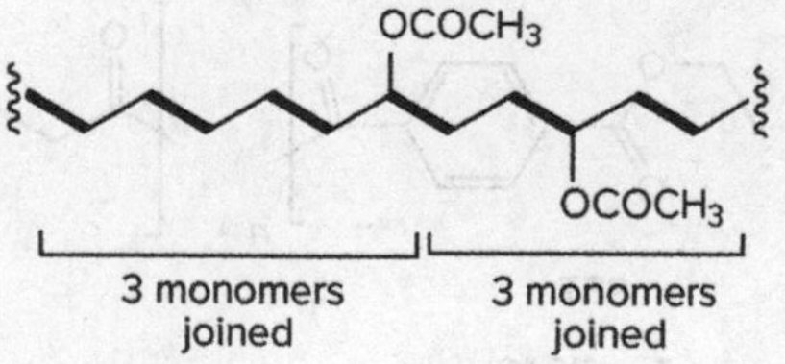

OCOCH₃

OCOCH₃

3 monomers joined

3 monomers joined

30.37

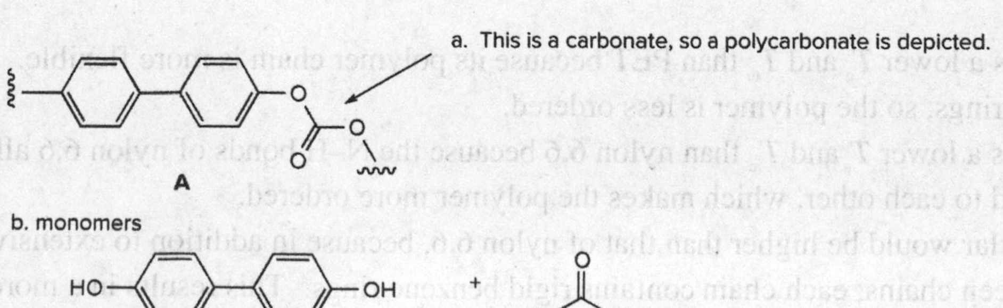

a. This is a carbonate, so a polycarbonate is depicted.

A

b. monomers

HO—C₆H₄—C₆H₄—OH + Cl(C=O)Cl

30.38 Join each isocyanate to an OH group to form a urethane unit.

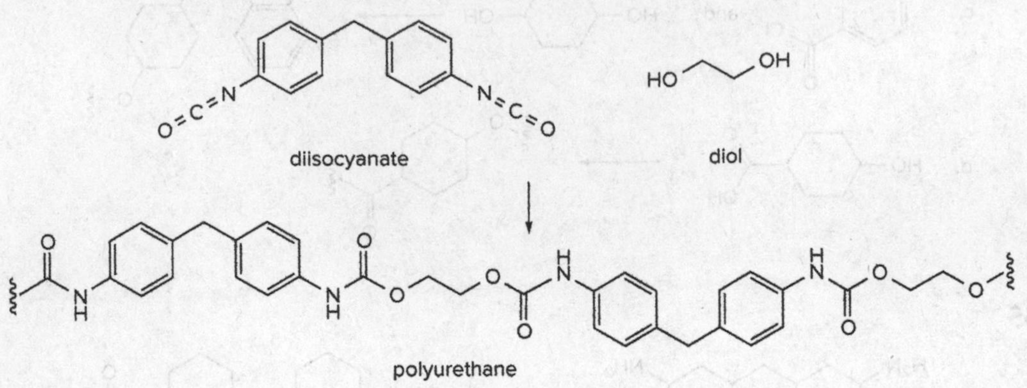

diisocyanate

diol

polyurethane

30.39

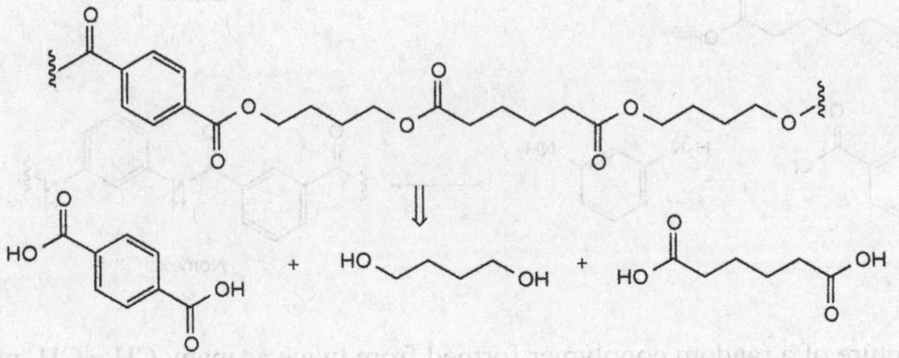

30.40

polyester **A**

T_g = < 0 °C
T_m = 50 °C

PET

T_g = 70 °C
T_m = 265 °C

nylon 6,6

T_g = 53 °C
T_m = 265 °C

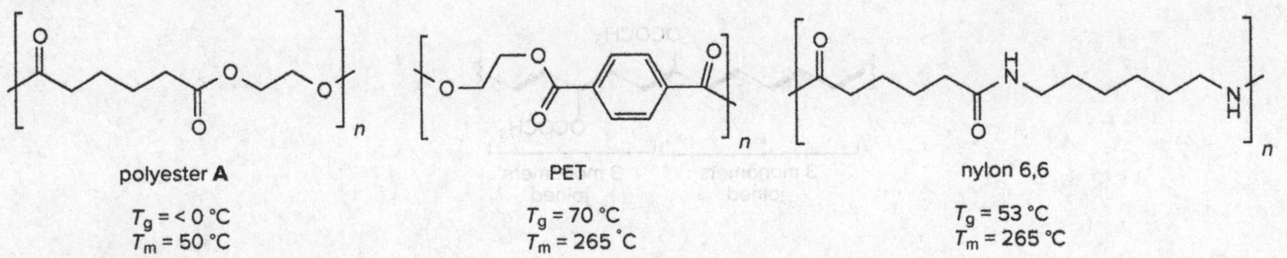

a. Polyester **A** has a lower T_g and T_m than PET because its polymer chain is more flexible. There are no rigid benzene rings, so the polymer is less ordered.

b. Polyester **A** has a lower T_g and T_m than nylon 6,6 because the N–H bonds of nylon 6,6 allow chains to hydrogen bond to each other, which makes the polymer more ordered.

c. The T_m for Kevlar would be higher than that of nylon 6,6, because in addition to extensive hydrogen bonding between chains, each chain contains rigid benzene rings. This results in a more ordered polymer.

30.41

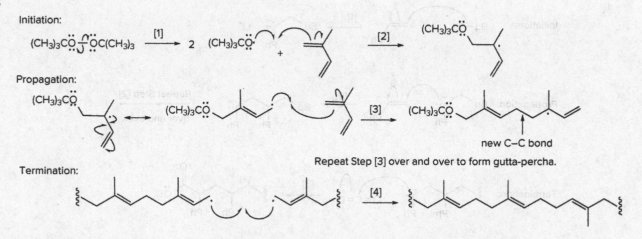

A

dibutyl phthalate

Diester A is often used as a plasticizer in place of dibutyl phthalate because it has a higher molecular weight, giving it a higher boiling point. A should therefore be less volatile than dibutyl phthalate, so it should evaporate from a polymer less readily.

30.42

Initiation:

$(CH_3)_3CO{-}OC(CH_3)_3 \xrightarrow{[1]} 2\ (CH_3)_3CO\cdot$ + isoprene $\xrightarrow{[2]}$ $(CH_3)_3CO$–radical

Propagation:

$(CH_3)_3CO$ … $\xrightarrow{[3]}$ $(CH_3)_3CO$ …

new C–C bond

Repeat Step [3] over and over to form gutta-percha.

Termination:

[4]

30.43

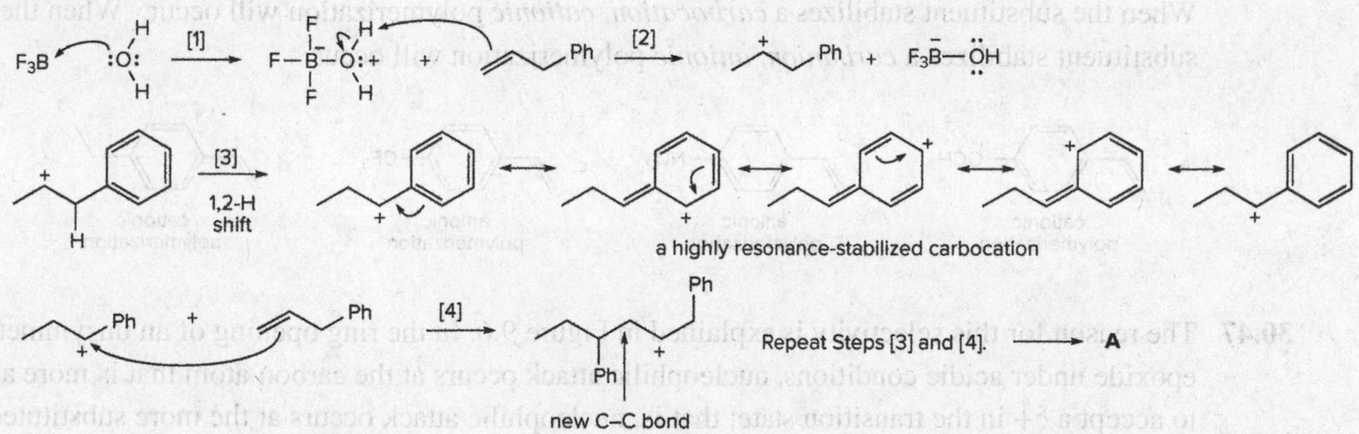

A

major product

B

A is the major product formed due to the 1,2-H shift (Step [3]) that occurs to form a resonance-stabilized carbocation.
B is the product that would form without this shift.

30.44

$F_3\bar{B}-O^+$ (H)(H) + $CH_2{=}CHCN$ → carbocation next to C≡N (δ+)

This carbocation is unstable because it is located next to an electron-withdrawing CN group that bears a δ+ on its C atom. This carbocation is difficult to form, so $CH_2{=}CHCN$ is only slowly polymerized under cationic conditions.

$F_3\bar{B}-O^+$ (H)(H) + allyl cyanide → 2° carbocation (C≡N, H)

This 2° carbocation is more stable because it is not directly bonded to the electron-withdrawing CN group. As a result, it is more readily formed. Thus, cationic polymerization can occur more readily.

30.45

Initiation: Bu—Li + $CH_2{=}CHPh$ —[1]→ Bu–CH_2–$\ddot{C}^-$H(Ph) Li^+

Propagation: Bu–CH_2–$\ddot{C}^-$H(Ph) + $CH_2{=}CHPh$ —[2]→ Bu–CH_2CH(Ph)–CH_2–$\ddot{C}^-$H(Ph) — Repeat Step [2] over and over.

Termination: ~~CH_2CH(Ph)$CH_2\ddot{C}^-$H(Ph) + $\ddot{O}{=}C{=}\ddot{O}$ —[3]→ ~~CH_2CH(Ph)CH_2CH(Ph)–C(=:O:)–$\ddot{O}$:$^-$

30.46 The substituent on styrene determines whether cationic or anionic polymerization is preferred. When the substituent stabilizes a *carbocation, cationic* polymerization will occur. When the substituent stabilizes a *carbanion, anionic* polymerization will occur.

a. p-OCH_3 styrene — cationic polymerization
b. p-NO_2 styrene — anionic polymerization
c. p-CF_3 styrene — anionic polymerization
d. p-ethyl styrene — cationic polymerization

30.47 The reason for this selectivity is explained in Figure 9.6. In the ring opening of an unsymmetrical epoxide under acidic conditions, nucleophilic attack occurs at the carbon atom that is more able to accept a δ+ in the transition state; that is, nucleophilic attack occurs at the more substituted carbon. The transition state having a δ+ on a C with an electron-donating CH_3 group is more stabilized (lower in energy), permitting a faster reaction.

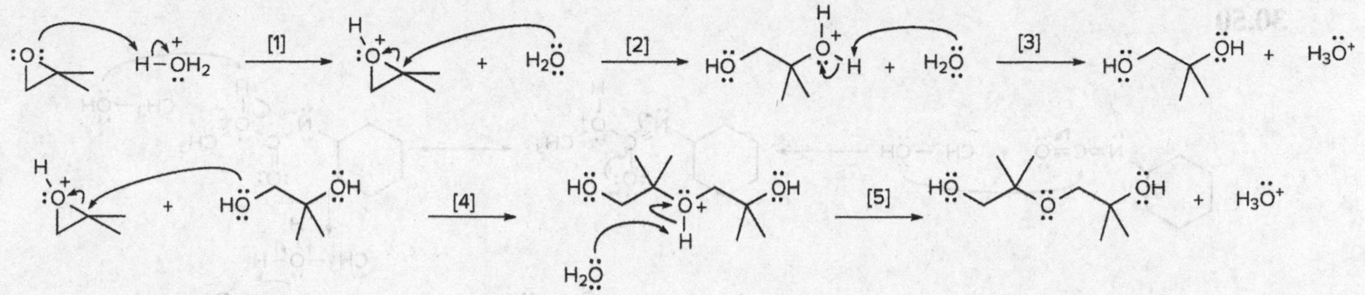

[1] [2] [3]

$H_2\ddot{O}$ + H_3O^+

[4] [5]

Repeat Steps [4] and [5] over and over.

30.48

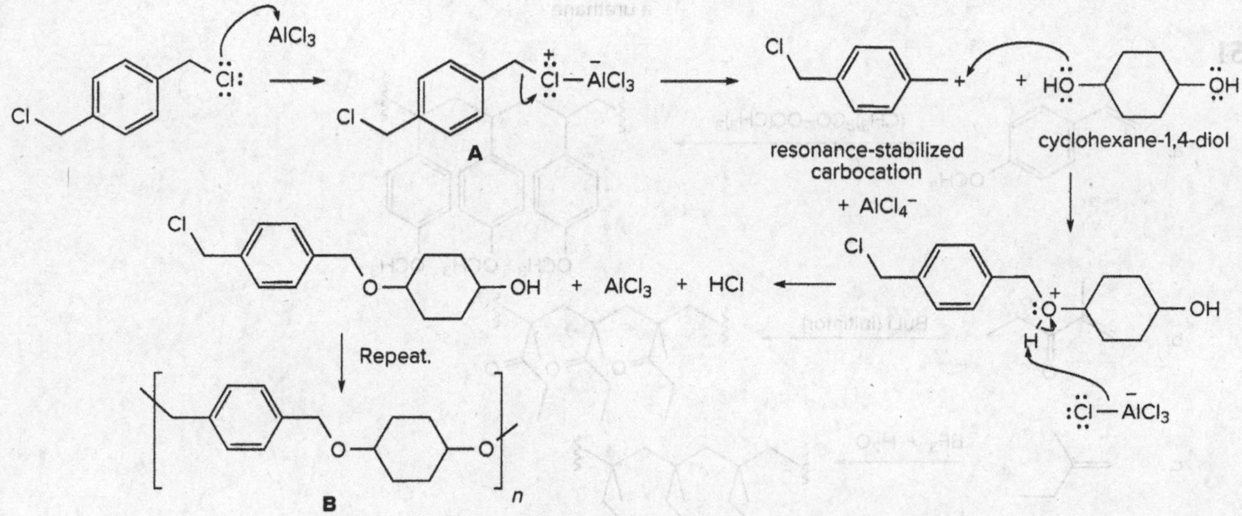

$AlCl_3$

A

resonance-stabilized carbocation

+ $AlCl_4^-$

cyclohexane-1,4-diol

+ $AlCl_3$ + HCl

Repeat.

B

30.49

$H-OSO_3H$

resonance-stabilized cation

HSO_4^-

(+ 3 additional resonance structures)

(+ 5 additional resonance structures)

+ $H_2\ddot{O}$

+ HSO_4^-

(+ 3 additional resonance structures)

bisphenol A

+ H_2SO_4

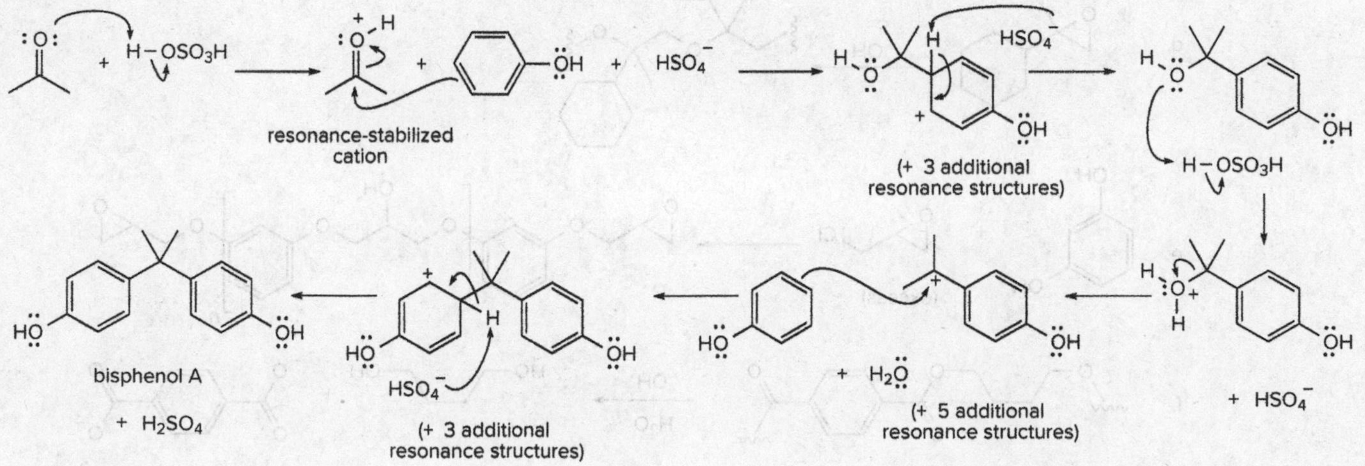

30.50

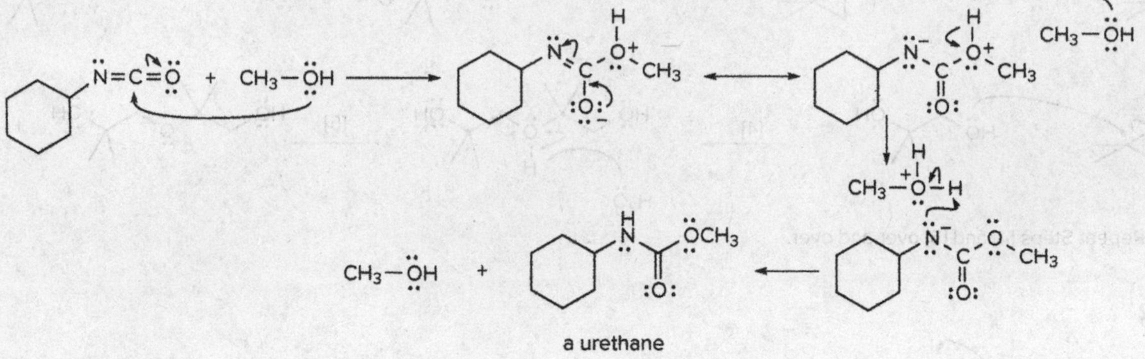

CH$_3$–ÖH
CH$_3$–ÖH
a urethane

30.51

a. (CH$_3$)$_3$CO–OC(CH$_3$)$_3$
OCH$_3$ OCH$_3$ OCH$_3$ OCH$_3$

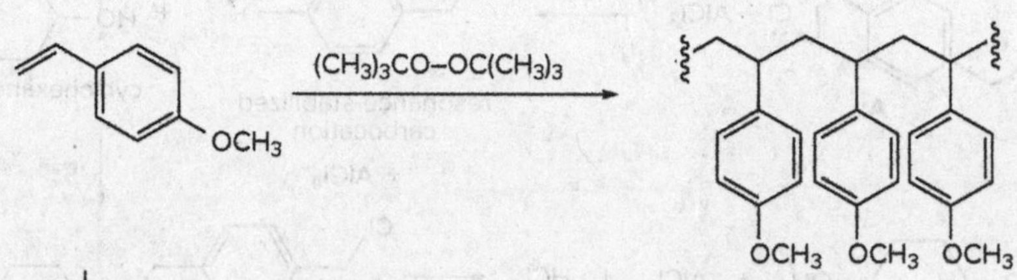

b. BuLi (initiator)

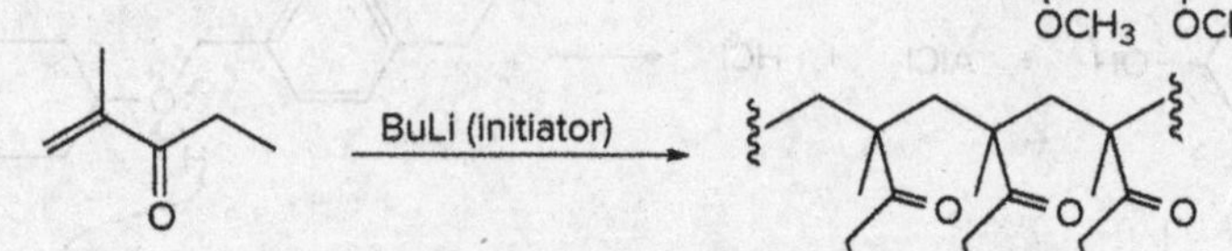

c. BF$_3$ + H$_2$O

d. $^{-}$OH

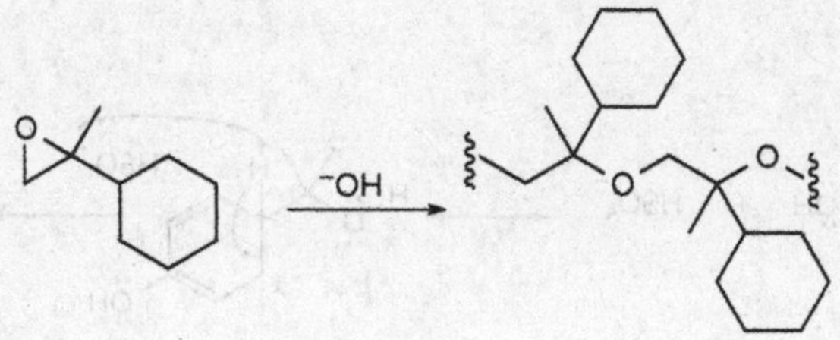

e. (excess)
OH OH OH n

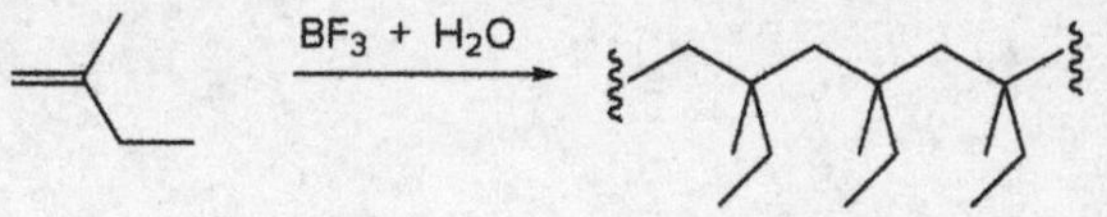

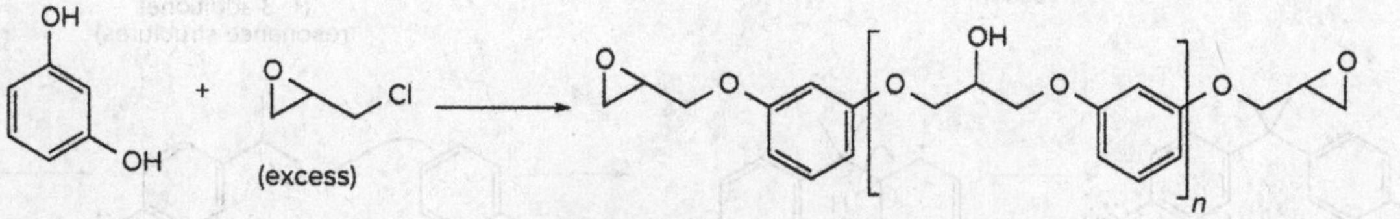

f. $^{-}$OH H$_2$O
HO OH +

g. OCN NCO + HO OH

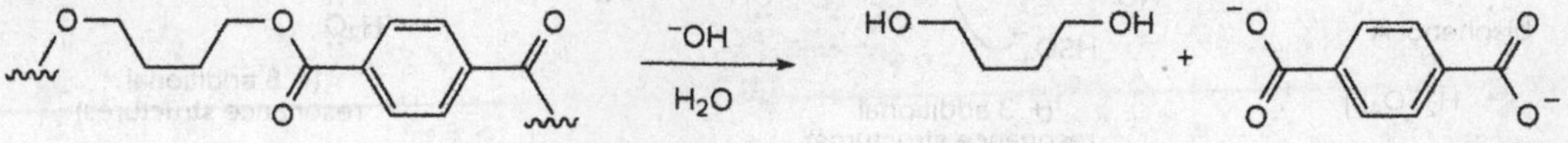

h. HO OH Cl$_2$C=O

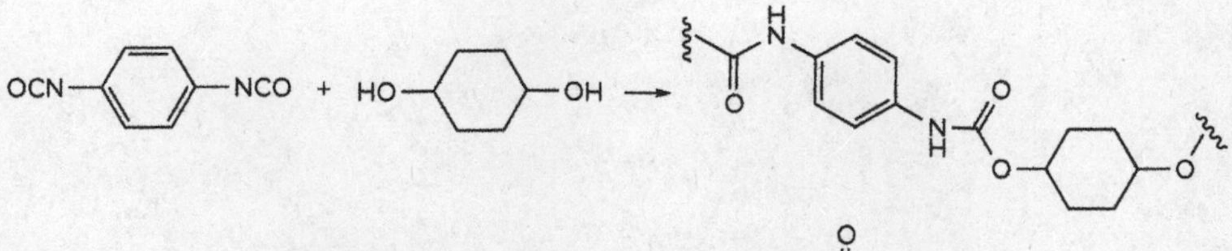

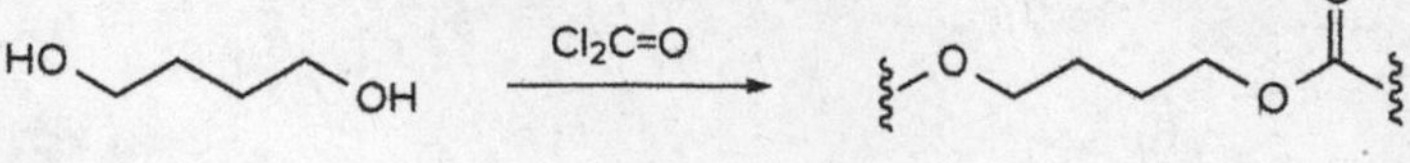

30.52 Polyethylene bottles are resistant to NaOH because they are hydrocarbons with no reactive sites. Polyester shirts and nylon stockings both contain functional groups. Nylon contains amides and polyester contains esters, two functional groups that are susceptible to hydrolysis with aqueous NaOH. Thus, the polymers are converted to their monomer starting materials, creating a hole in the garment.

30.53

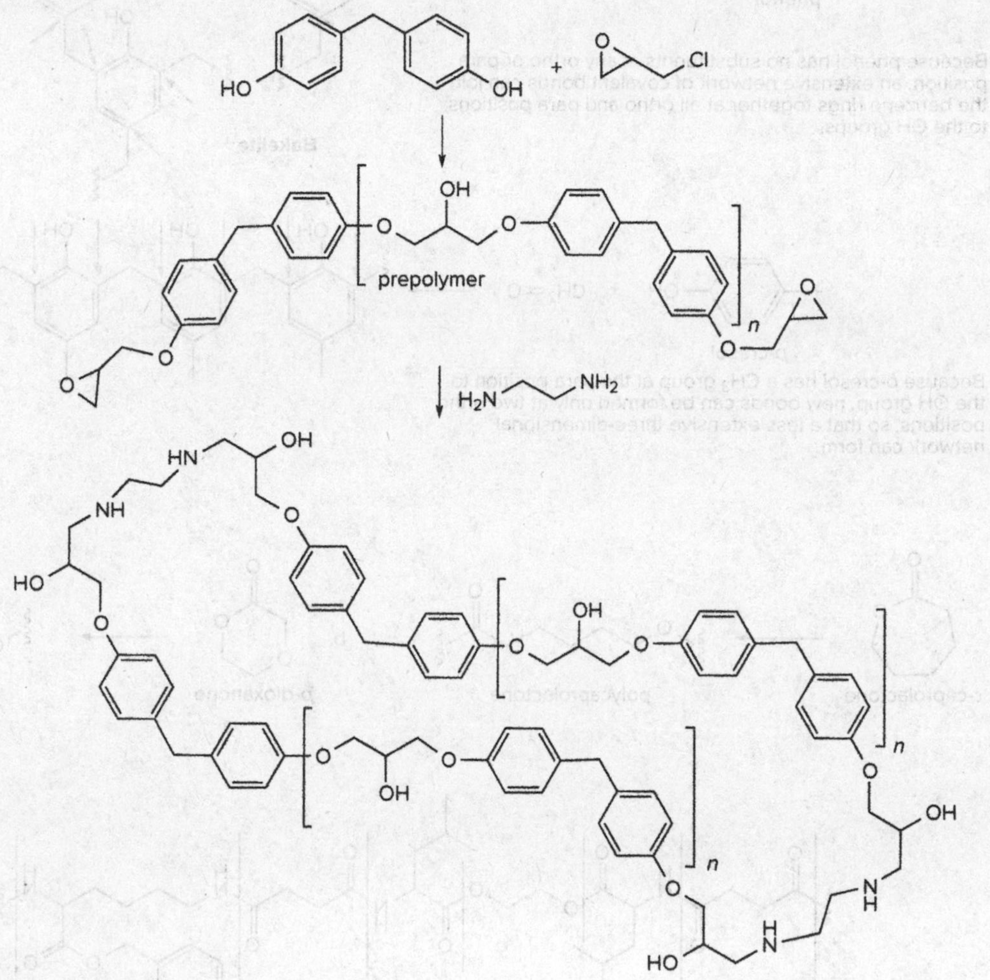

HO OH + O Cl

OH O O

prepolymer

n

H_2N NH_2

HN OH NH HO O O OH O O OH n OH O n O N H HO N H

30.54

CH_3Cl, $AlCl_3$ → CH_3Cl, $AlCl_3$ → $KMnO_4$ →

terephthalic acid

$CH_2{=}CH_2$ [1] OsO_4 [2] $NaHSO_3$, H_2O → HO OH

ethylene glycol

or

mCPBA → H_2O, ^-OH → HO OH

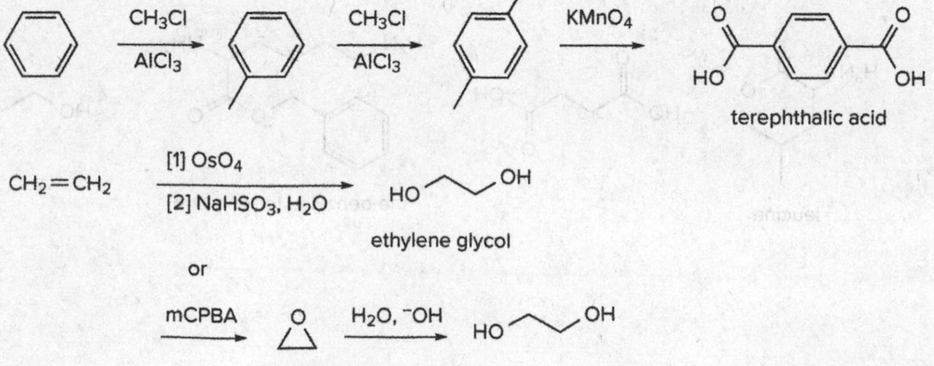

30.55

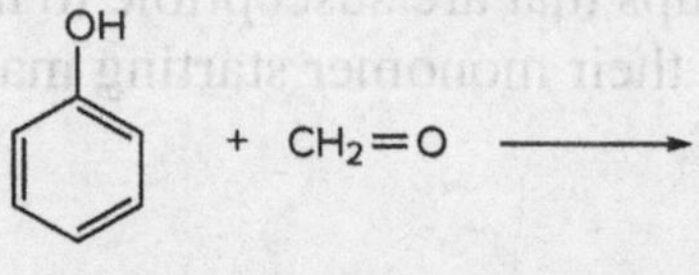

phenol + $CH_2{=}O$ ⟶ Bakelite

Because phenol has no substituents at any ortho or para position, an extensive network of covalent bonds can join the benzene rings together at all ortho and para positions to the OH groups.

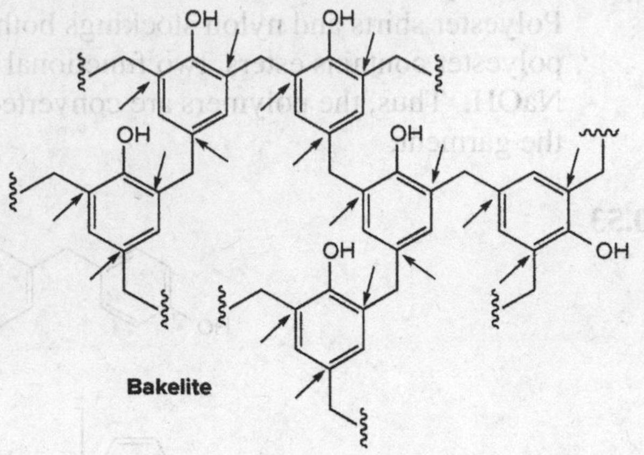

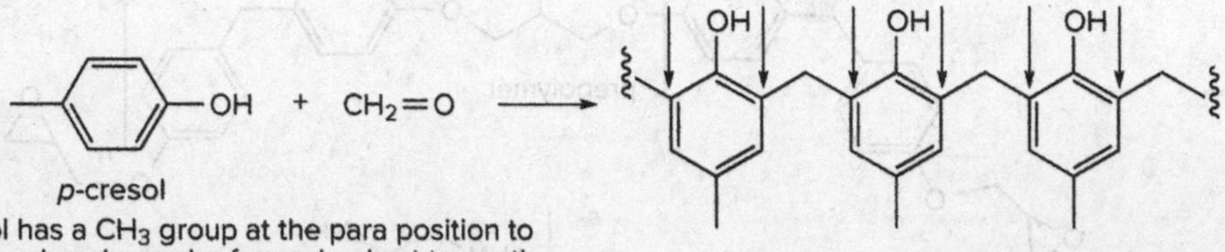

p-cresol + $CH_2{=}O$ ⟶

Because *p*-cresol has a CH_3 group at the para position to the OH group, new bonds can be formed only at two ortho positions, so that a less extensive three-dimensional network can form.

30.56

a. ε-caprolactone ⟶ polycaprolactone

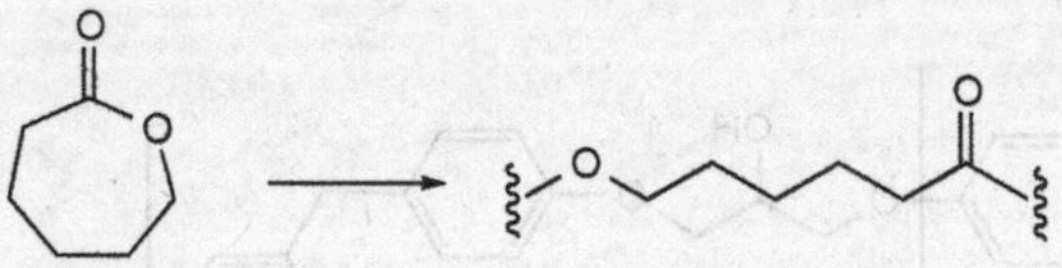

b. *p*-dioxanone ⟶ polydioxanone

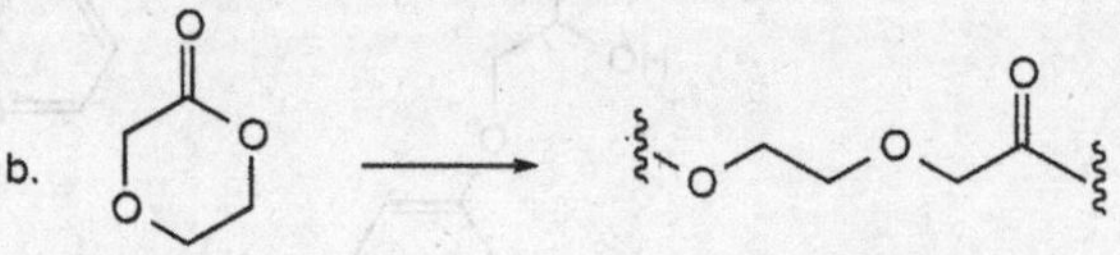

30.57

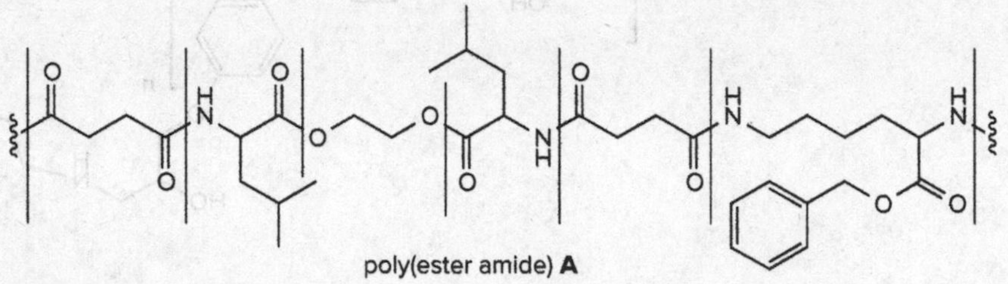

poly(ester amide) **A**

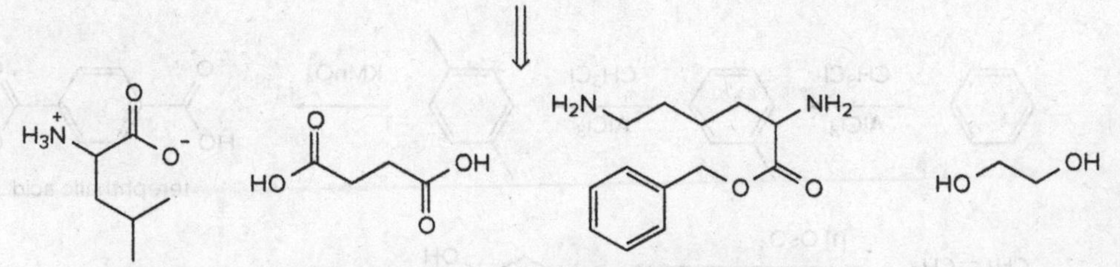

⇓

leucine

the benzyl ester of lysine

30.58

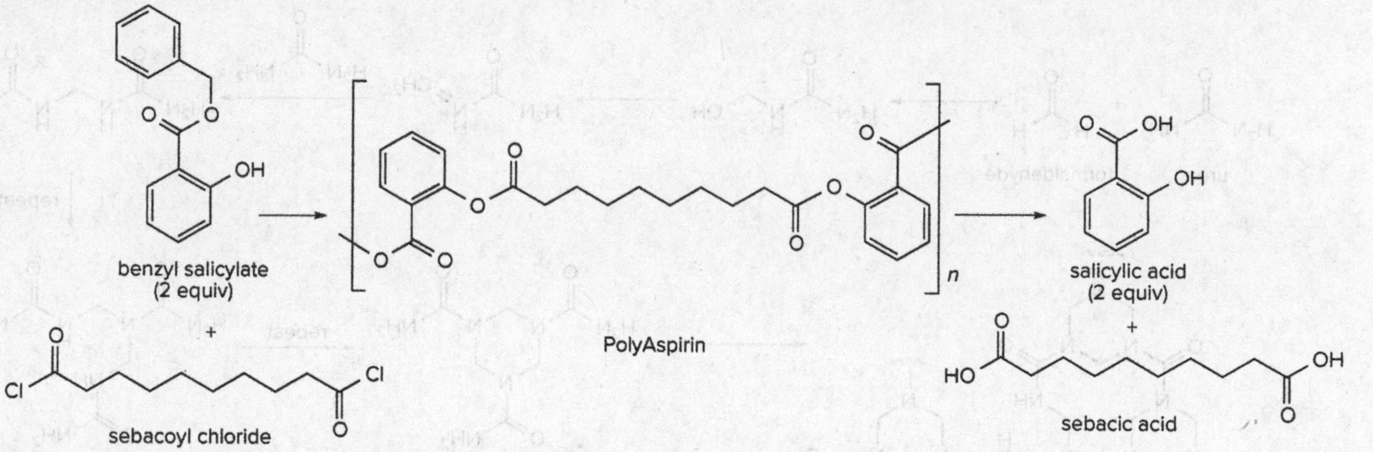

benzyl salicylate
(2 equiv)
+
sebacoyl chloride

PolyAspirin

n

salicylic acid
(2 equiv)
+
sebacic acid

30.59

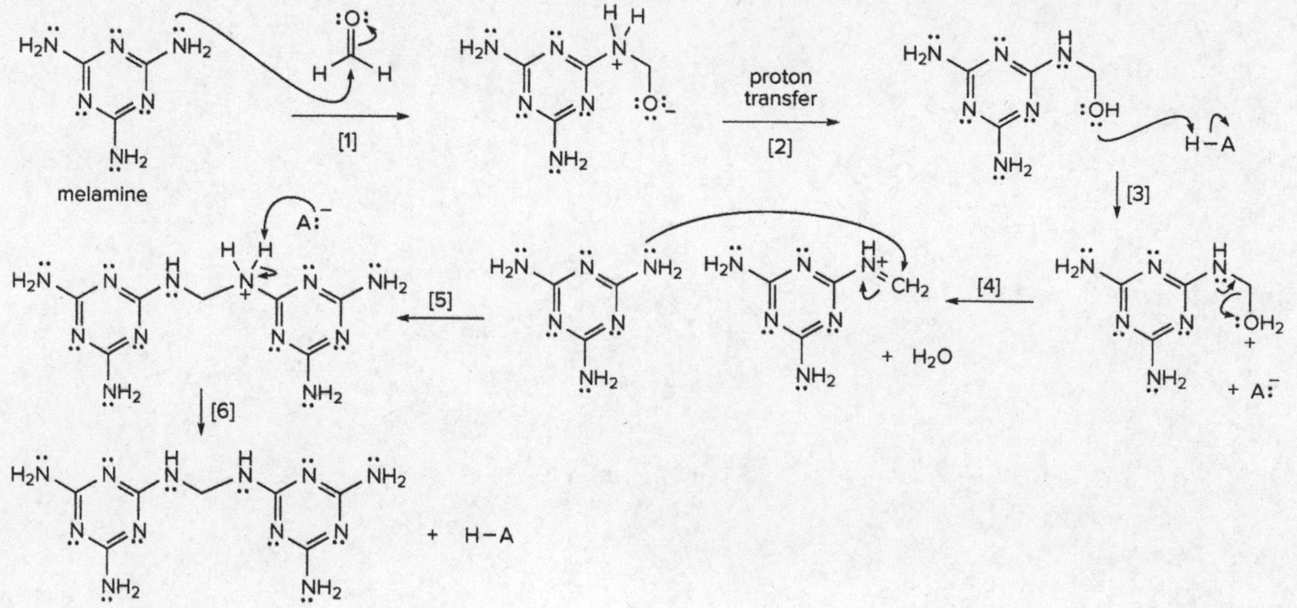

melamine

[1]

proton transfer

[2]

[3]

[4]

+ H_2O

+ A:⁻

[5]

[6]

+ H–A

30.60

a.

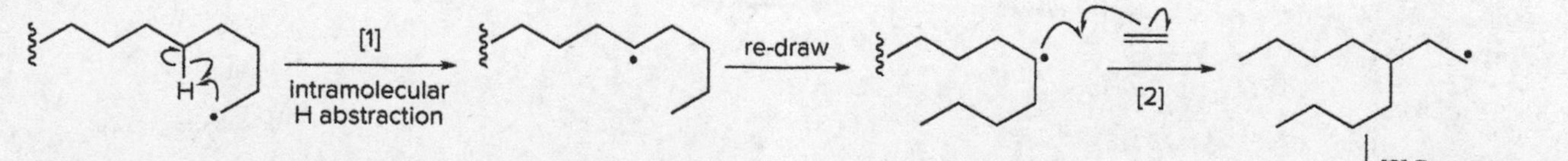

[1]
intramolecular
H abstraction

re-draw

[2]

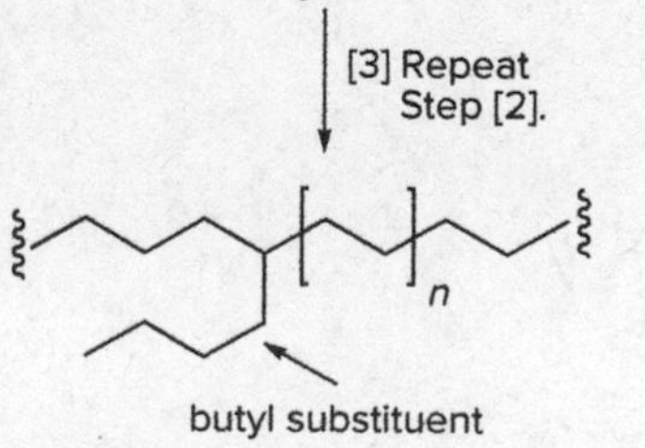

[3] Repeat
Step [2].

n

butyl substituent

b. Abstraction of the H is more facile than abstraction of the other H's because the H atom that is removed is six atoms from the radical. The transition state for this intramolecular reaction is cyclic, and resembles a six-membered ring, the most stable ring size. Other H's are too far away or the transition state would resemble a smaller, less stable ring.

30.61

urea

formaldehyde

repeat

repeat